MINITAB Commands Used Within the Text  W9-CUI-508

C denotes a column, K denotes a constant

9. REGRESSION
  REGRESS   in C using K predictors in C, . . . , C
    Subcommands: COEF into C
                RESIDUALS into C
                PREDICT for K, . . . , K
      or PREDICT for C, . . . , C
    Example: REGRESS C1 2 C2 C3 will regress Y in C1 using 2 predictors in
        C2 and C3.
  BRIEF output at level K
  STEPWISE regression of C using predictors   , . . . , C
    Subcommands: FENTER = K
                FREMOVE = K
                FORCE C, . . . , C
                ENTER C, . . . , C
                REMOVE C, . . . , C

10. ANALYSIS OF VARIANCE
  ONEWAY using data in C, levels in C
  TWOWAY using data in C, levels in C, blocks in C
  (or TWOWAY using data in C, factor A levels in C, factor B levels in C)

11. NONPARAMETRIC STATISTICS
  CHISQUARE using C, . . . , C
  RUNS above and below K for data in C
  MANN-WHITNEY using C and C
    Subcommand: ALTERNATIVE = K
  WTEST using C
    Subcommand: ALTERNATIVE = K
  KRUSKAL-WALLIS test for data in C, levels in C
  RANK the values in C, put ranks into C

12. TIME SERIES
  ACF of C
  DIFFERENCES [of lag K] using C, put into C
  LAG [by K] using C, put into C

13. QUALITY CONTROL
  XBARCHART using data in C, sample size is K
    Subcommands: SIGMA   K
                SLIMITS 1 2 3
                TEST 1 : 8
  RCHART using data in C, sample size is K
    Subcommand: SIGMA = K
  PCHART using data in C, sample size is K
  C CHART using data in C

14. MISCELLANEOUS
  RANDOM K integers, put into C
    Subcommand: INTEGER from K to K
  ERASE C, . . . , C

15. SUBCOMMANDS
  Some MINITAB commands have subcommands to convey additional information. To use a subcommand, include a semicolon at the end of the main command; the subcommands follow (one per line). End the final subcommand with a period.

Introduction to Business Statistics

A COMPUTER INTEGRATED APPROACH

THIRD EDITION

Introduction to Business Statistics

A COMPUTER INTEGRATED APPROACH

THIRD EDITION

Alan H. Kvanli
C. Stephen Guynes
Robert J. Pavur

UNIVERSITY OF NORTH TEXAS

West Publishing Company   ST. PAUL ■ NEW YORK ■ LOS ANGELES ■ SAN FRANCISCO

Copyediting: Pam McMurry
Proofreading: Lynn Reichel
Composition: G&S Typesetters, Inc.
Cover image and design: Lois Stanfield, LightSource Images

Library of Congress Cataloging-in-Publication Data

Kvanli, Alan H.
  Introduction to business statistics : a computer integrated
approach / Alan H. Kvanli, C. Stephen Guynes, Robert J. Pavur.
—3rd ed.
    p. cm.
  Includes index.
  ISBN (invalid) 0-314-89946-6
  1. Commercial statistics—Data processing. 2. Statistics
—Data processing. I. Guynes, C. Stephen, (Carl
Stephen) II. Pavur, Robert J. III. Title.
HF1017.K83 1991
519.5′0285—dc20   91-29920
                              ∞ CIP

To our parents,
  Chet and Donna
  Carl and Lorrain
  Joseph and Shirley
who have always provided love, guidance and
encouragement along life's way.

Contents

Preface

As mentioned in the preface to the first two editions, we feel that a statistics text that *fully* integrates the use of computers with statistics is a necessity in today's marketplace. This edition has retained the "nonintimidating" approach to describing the concepts and applications of statistics while giving students the opportunity to observe (or actually carry out) computer-generated solutions using a mainframe or microcomputer statistical package. The text has also been designed so that those requiring or desiring a more traditional calculator-based approach will find an abundance of exercises and examples that can be solved in this manner.

A new chapter on Quality Control has been added to this edition. We believe an introduction to this material is of vital importance as more and more American businesses are adopting quality improvement procedures to gain a competitive advantage or at the very least, survive in an ever-increasing quality conscious marketplace. This material could be covered in the second semester, following Analysis of Variance, for a two-semester course. However, due to the flexible nature of this material, it can be covered at any point after Chapter 8.

This edition once again includes microcomputer versions (along with mainframe versions) of SAS, SPSS, and MINITAB. A considerable number of revised exercises have been included. The probability chapter (Chapter 4) has been revised to include a discussion of tree diagrams and using a contingency table approach for certain probability problems.

The text is intended to be an undergraduate or M.B.A. introduction to basic statistics. We assume that the student has a good understanding of basic algebra. Reference is made on a few occasions to calculus applications, but no calculus background is required to read the material. The reading level is interesting and easy-to-understand without sacrificing any credibility in the descriptive material. It is a non-mathematical, but not a "black box," approach to teaching the appreciation and application of statistics. We've included a large number of new examples to better guide the student to an understanding of statistical concepts and applications. These examples include more realistic illustrations, many taken from the process improvement (quality control) area.

To the Instructor

This text can be used for either a one- or two-semester introduction to business statistics. Suggested material to be covered in the first semester would be Chapters 1 through 8, in order, which concludes with an introduction to hypothesis testing. Chapters 9 and 10 could be included in a second-semester course, along with those remaining chapters that you feel are particularly relevant and of interest to your students. We encourage you to include the new chapter on Quality Control at any point following the completion of Chapter 8.

The text has intentionally been written in a somewhat conversational style to make it less intimidating to the student. Our intent is for the student to read the text; not just to use it as a source of homework exercises.

The text fully integrates three popular statistical packages: MINITAB, SAS, and SPSS. New to the third edition is the availability of Business MYSTAT, for those instructors not having access to the other three packages. MYSTAT instructions are described in the Instructor's Manual. The featured package throughout all of the chapter examples is MINITAB, since these commands are simple English statements and are illustrative of computer capabilities whether or not you use the MINITAB package in your course instruction. Corresponding SAS and SPSS descriptions are contained at the ends of chapters—a feature unique to this text. We have fully integrated these packages throughout the text, making it possible for you to include computer usage as part of your course without having to spend a great deal of time explaining the mechanics of a particular package. Introductions to each of these three packages and MYSTAT are presented at the end of the text. For instructors who wish to avoid computer usage, the text allows for a calculator-based approach—the exercises do not require a computer package and contain reasonably sized data sets.

Other Features of the Text Include:

a Look Back/Introduction at the start of each chapter to tie the chapter to the relevant material from the preceding chapters. Each chapter closes with a summary section containing the key words (in boldface print) introduced in the chapter.

an abundance of exercises (over 1200) using realistic business situations. Each chapter also includes a case study containing an actual application of the chapter material and requiring an in-depth discussion. Nearly all the case studies have been replaced or updated in the third edition.

additional exercises within each chapter using actual applications in a business setting (and the source of these applications).

a full treatment of the use of p-values to make statistical decisions. These are derived and discussed throughout the entire text.

three continuous distributions (normal, uniform, and exponential), along with three discrete distributions (binomial, hypergeometric, and Poisson).

various sampling procedures, along with corresponding sample estimators and confidence intervals, as separate sections in two of the earlier chapters. With this arrangement, the instructor is able to cover this often-neglected material without having to spend the time to cover an entire chapter.

separate chapters for inference regarding normal parameters (μ, σ) and inference on a binomial parameter (p). Chapters 7, 8, and 9 are strictly devoted to normal inference, both one population (7 and 8) and two populations (9). Binomial inference (one and two populations) is covered in Chapter 10.

an entire chapter devoted to forecasting using time series data (Chapter 17). It includes several exponential smoothing models and discusses the pros and cons of using multiple regression versus time series modeling techniques for such data.

an entire chapter on statistical decision theory. This chapter is placed near the end of the text (Chapter 18) but can be covered at any time, including the first semester, if desired.

a large database (1140 observations) containing data on family income, family size, total indebtedness, monthly utility expenditures, and other variables. This is an end-of-text appendix and is available to adopters on a floppy disk.

a second database containing 1000 observations selected from companies listed in the Moody's Investor Service Industrial Manual (also available on floppy disk).

appendixes that provide an introduction to each of the three statistical packages utilized in the text.

New to the Third Edition Are:

a new chapter on Quality Control with many new examples and exercises throughout the text from this growing area of applied statistics.

a summary of formulas at the end of each chapter.

an earlier discussion of tree diagrams (Chapter 4).

nearly all new case studies.

microcomputer software for time series decomposition (Chapter 16) and forecasting (Chapter 17) that can be run directly from DOS. No additional software or compiler is necessary to run these programs. Earlier versions of this text provided these as FORTRAN subroutines.

The Following Material Is Also Available:

an instructor's manual containing solutions to all exercises, a chapter glossary, and a discussion of MYSTAT commands that pertain to each chapter.

a test bank containing nearly all new true/false questions, completion exercises, and additional application problems.

a student guide written to put students at ease and guide them through applications of the chapter material.

We certainly hope that this text will meet your classroom needs. If you care to offer comments and suggestions, we would like to hear from you. Address any correspondence to Al Kvanli, College of Business Administration, University of North Texas, Denton, Texas 76203.

To the Student

We believe you will find this text to be a readable, easily understood treatment of business statistics. Our intent is to carefully explain the various statistical concepts and strategies without getting bogged down in unnecessary mathematics. We have included many examples within each chapter to allow you to see how each procedure works. At the beginning of each chapter you will find a Look Back/Introduction section which will set up the chapter and tie it in with the previous chapters. At the end of each chapter is a summary containing all of the key definitions and concepts introduced within the chapter, along with a summary of formulas. At the end of the book you will find introductions to the three computer packages integrated into the text: MINITAB, SPSS, and SAS (either mainframe or microcomputer version). For those of you using MYSTAT, there is an introduction to this package at the end of the text.

As the old adage goes, "practice makes perfect," and mastering statistics is no exception. To this end, we have included a large number of exercises to help you along the road to perfection. Also, you will find the solutions to the odd-numbered exercises at the end of the text. A study guide, which contains additional examples along with their solutions, has been prepared. These solutions take you step-by-step through the applications of the various statistical techniques with many blanks where you supply the missing number or word.

Acknowledgments

We are very much indebted to the people who helped in the production and preparation of this text. The editorial advice and assistance of Denise Simon were once again very timely, professional, and of great help. Mélina Brown was always there to keep things on schedule, yet showed tremendous patience when we were on overload. Many thanks to Beverly Kenney who typed the entire solutions manual and the many graduate students at the University of North Texas who supplied valuable input.

A new addition to the "team" is James Pinto of Northern Arizona University. We found his watchful eye to be so useful that we asked him to serve as an accuracy checker (both content and calculations) for exercises in this edition. With his help, we hope this edition comes much closer to containing "zero defects."

We'd like to thank Laura Massey and Joanne Tucker for providing an expanded (and nearly all new) test bank. Their efforts went way beyond what we (and they) expected. A major assistance in this edition came from Stuart Warnock, whose expertise in quality control was a valuable addition.

A heartfelt thanks goes to Jitendra Sarhad who once again has enriched our work by providing us with excellent case studies. His creative efforts and amazing capacity for statistics and the English language are very much appreciated.

Wilke English has again authored a very helpful and entertaining study guide to accompany the text. We feel that his study guide has been (and will continue to be) a big plus for the textbook and we are most appreciative for having his time and talent.

Last but certainly not least we would like to thank the reviewers who had a multitude of excellent suggestions for this edition. The following list contains the names and affiliations of these individuals:

David Auer—Western Washington University

Rick Edgeman—Colorado State University

Jamshid Hosseini—Marquette University

Hossein Kamarei—Indiana University, South Bend

Hassan Pourbabaee—Central State University (Oklahoma)

Susan Simmons—The Citadel, Charleston, SC

Patrick Thompson—University of Florida

Ken Towson—Capilano College

George Wesolowsky—McMaster University

Roy Williams—Memphis State University

Chiou-nah Yeh—Alabama State University

Dale Zimmerman—University of Iowa

Introduction to Business Statistics

A COMPUTER INTEGRATED APPROACH

THIRD EDITION

C H A P T E R

A First Look at Statistics

A FIRST LOOK AT STATISTICS

Many people probably think a statistician is someone who helps figure batting averages during a baseball game broadcast. You might wonder how we can devote an entire textbook to compiling numbers and making simple calculations. Surely it cannot be that complicated!

Statistics is the science comprising rules and procedures for collecting, describing, analyzing, and interpreting numerical data. The applications of statistics are evident everywhere. Hardly a day goes by in which we are not bombarded by such statements as:

> Results show that Crest toothpaste helps prevent tooth decay.

> The chance of a NASA rocket failure is higher than was quoted originally.

> The state court has ruled that the XYZ Company is guilty of age discrimination in its termination procedure.

> The surgeon general has determined that cigarette smoking is dangerous to your health.

Or how about:

> American companies are beginning to place more emphasis on statistical quality control in an attempt to offer better products that can be delivered on time at less cost.

Besides using statistics to inform the public, statisticians help businesses make forecasts for planning and decision making.

The use of statistics began as early as the first century A.D., when governments used a census of land and properties for tax purposes. Census taking was gradually extended to include such local events as births, deaths, and marriages. The *science* of statistics, which uses a sample to predict or estimate some characteristics of a population, began its development during the nineteenth century.

Use of statistical methods has undergone a dramatic change as computers and powerful calculators have entered the research environment. Companies can store and manipulate large collections of data, and once-formidable statistical calculations are reduced to a few keystrokes. Sophisticated computer software allows users merely to specify the type of analysis desired and input the necessary data. This textbook concentrates on three of these statistical software packages: MINITAB (a statistical computer package originally designed at Penn State University specifically for students), SAS (Statistical Analysis System), and SPSS (Statistical Package for the Social Sciences).

Although most statistical functions are performed by professional statisticians, it may be your job to draw a valid conclusion from a statistical report. Occasionally, however, statistics can obscure the truth or give an erroneous impression. Anyone who has ever changed plans due to a 90% chance of rainy weather only to sit home on a sunny day can attest to this fact. You often can avoid a bad decision by recognizing statistical errors and bias in the results that you review.

In addition, you may be asked to perform a statistical analysis. Although you may elect to obtain outside assistance, you will need to know when to consult a statistician and how to tell him or her what you need.

1.1
Uses of Statistics in Business

Modern businesses have more need to predict future operations than did those of the past, when businesses were smaller. Small-business managers often can solve problems simply through personal contact. Managers in large corporations, however, must try to summarize and analyze the various data available to them. They do this by using modern statistical methods.

Areas of business that rely on statistical information and techniques include:

1. *Quality control.* Statistical quality-control procedures assure high product quality and enhance productivity.
2. *Product planning.* Statistical methods are used to analyze economic factors and business trends and to prepare detailed sales budgets, inventory-control systems, and realistic sales quotas.
3. *Forecasting.* Statistics are used to predict sales, productivity, and employment trends.
4. *Yearly reports.* Annual reports for stockholders are based on statistical treatment of the many cost and revenue factors analyzed by the business comptroller.
5. *Personnel management.* Statistical procedures are used in such areas as age- and sex-discrimination lawsuits, performance appraisals, and workforce-size planning.
6. *Market research.* Corporations that develop and market products or services use sophisticated statistical procedures to describe and analyze consumer purchasing behavior.

◢ 1.2
Some Basic Definitions

Statistics has specialized definitions for terms crucial to statistical reasoning. In **descriptive statistics,** you collect data and describe them. If you analyze and interpret the data, you are using **inferential statistics.**

Descriptive statistics are used to describe a large set of data. For example, you can reduce the set of data values to one or more single numbers, such as the average of 150 test scores, or you can construct a graph that represents some feature of the data.

You use inferential statistics to form conclusions about a large group—a population—by collecting a portion of it—a sample. Thus, a **population** is the set of all possible measurements (generally pertaining to a group of people or objects) that is of interest. A **sample** is the portion of the population about which information is gathered.

The analyst decides what the population is. Typically, the population is so large that it would be nearly impossible to obtain information about every item in it. Instead, we obtain information about selected population members and attempt to draw a conclusion about all members. In other words, we attempt to infer something about the population using information about only some of the members of this population.

To make an early prediction of the election results for governor of California, for example, analysts could use a sample of voters leaving the voting booths, as illustrated in Figure 1.1. The population is all the votes cast in the election. To make a valid statistical inference using a sample, it is crucial that the sample **represent** the population; that is, the values in the sample must be representative (typical) of values in the population. One way to make sure the sample is representative is to collect a sample of size *n,* where each set of *n* people has the same chance of being selected for the sample. This is a **simple random sample** (Figure 1.1). It is akin to drawing names out of a hat; each name in the hat has the same chance of being pulled out. Thus, if our population is all votes cast on the day of the gubernatorial election, a sample of votes cast in only one city would not be representative, because we would have no guarantee that these votes would represent the votes of the entire state. A random sample obtained across the entire state would better represent this population.

As another illustration, assume that Calcatron, a producer of electronic calculators, orders 50,000 components from GLC. Calcatron instructs GLC that they will accept the shipment if an outside laboratory that randomly selects 100 components from the batch finds that fewer than 3 are defective. Calcatron relies on inferential statistics; they infer that the population of components is of satisfactory qual-

■ FIGURE 1.1
Population versus
a sample.

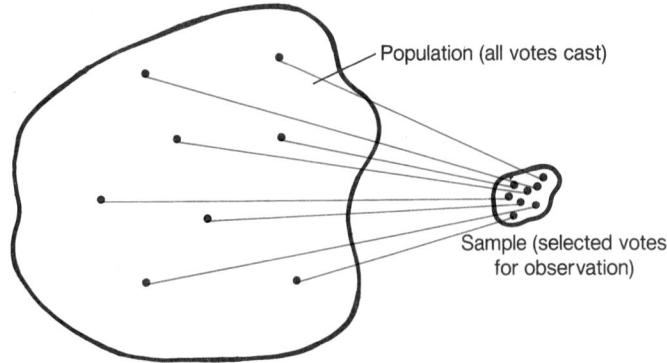

■ FIGURE 1.1
Population versus
a sample.

ity if the sample is satisfactory. Note that it is possible that the sample could contain fewer than 3 defective components even if the population contains, say, 80% defective parts. Whenever we attempt to infer something about a population from a sample, there is always a chance of drawing an incorrect conclusion. The only way of being 100% sure is to list the entire population. Such a sample is called a **census.**

In the Calcatron example, there are two proportions of interest. The first proportion, the proportion of defective components in the *population,* is referred to as a **parameter.** The second proportion, the proportion of defective components in the *sample,* is referred to as a **statistic.** In general, any value describing a population (such as the average or a particular proportion) is a parameter. Parameters typically are unknown and are estimated using the corresponding statistic derived from a statistical sample.

1.3
Discrete and Continuous Numerical Data

Proper use of numerical data can be a great aid in making a critical decision. However, using an improper technique or "bad data" can lead you down the wrong path. Generally, the technique we use to analyze data in statistics depends on the nature of the data. We can distinguish between two types of numerical data.

How do the following two sets of numbers differ?

3, 5, 2, 1, 4, 4, 3, 5, 5, 1, 2, 4

4.31, 11.62, 5.37, 1.55, 3.71, 6.88, 7.23, 9.52, 2.36, 7.42, 6.11, 4.85

The primary difference is that the values in the first data set consist of *counting numbers,* or *integers.* Such data are **discrete.** For example, these data may be the coded responses from 12 people who answered a particular question in a marketing survey where 1 = strongly agree, 2 = agree, 3 = uncertain, 4 = disagree, and 5 = strongly disagree. Note that discrete data may contain a decimal point. Nevertheless, discrete data have *gaps* in their possible values. For example, if you throw a single die twice and record the average of the two throws, the possible values are 1, 1.5, 2, 2.5, 3, 3.5, 4, 4.5, 5, 5.5, and 6. If you repeatedly averaged two throws of the die, you would obtain discrete data.

Examples of discrete data that have integer values are the number of automobiles that arrive at a drive-up window over a 5-minute period, the number of children in your family, and the total of the two numbers appearing on a throw of two dice. Note that although the first two have infinite (theoretically, at least) possible values, the data are discrete. Your family cannot have 2.5 children.

Now consider the second data set. These data might represent the weights of 12 parcels received at a post office. A list of all the possible values of package weights would be long—if our scale were completely accurate, the list would be infinite and any value would be possible. Such data are **continuous**: *any value* over some particular range is possible. There are no gaps in possible values for continuous data. For example, although we may say Sandra is 5.5 feet tall, we mean her height is about 5.5 feet. In fact, it might be 5.50372 feet. Height data are continuous. Or consider the contents of a coffee cup filled by a vending machine. Will the machine release exactly 6 ounces every time? Certainly not. In fact, if you were to observe the machine fill five such cups and measure the contents to the nearest .001 ounce, you might observe values of 6.031, 5.932, 5.871, 6.353, and 5.612 ounces. Here again, any value between, say, 5.5 ounces and 6.5 ounces is possible: these are continuous data. *Data such as weights, heights, age (actual), and time are generally continuous data and will be used in the examples in the chapters to follow.*

It is important to remember that *discrete data* can be the result of observing a *continuous variable.* For example, actual age is a continuous variable, but if a recorded age is the age at the last birthday, the data will be discrete. Very often, measurements on a continuous variable (such as height in inches) result in discrete data. However, discrete data can also be the result of observing a discrete variable, such as the number of traffic tickets a person has received during the past three years.

◢ 1.4
Level of Measurement for Numerical Data

In addition to classifying numerical data as discrete or continuous, we can also classify these data according to their level of measurement. We will discuss them in order of strength, beginning with the weakest. **Nominal data** are really not numerical at all but are merely labels or assigned values. Examples include: sex (1 = male, 2 = female), manufacturer of automobile (1 = General Motors, 2 = Ford, 3 = Chrysler), or color of eyes (1 = blue, 2 = green, 3 = brown). Assigning a numerical code to such data is merely a convenience so that, for example, one can store the information in a computer. Therefore, it makes no sense to perform calculations with such numbers, such as finding their average. What would it mean to claim that "the average eye color is 2.73"? This statement is meaningless. Generally, we are interested in the **proportion** of such data in each category. Consider Calcatron's shipment, in which each component is either defective or not defective. We could assign the code 1 = defective, 0 = not defective. The parameter of interest here is p, where p = proportion of defective components in the population of 50,000 components. If Calcatron believes p is too large, they will not accept the shipment. We will consider what is "too large" in Chapter 10.

Ordinal data can be arranged in order, such as worst to best or F to A (grades on an exam). A classic example of ordinal data is the result of a cross-country race, where ten people compete and 1 = the fastest (the winner), 2 = the runner-up, and so on, with 10 = the slowest. Here, the *order* of the values is important (3 finished before 4) but the *difference* of the values is not. For example, $2 - 1 = 1$ and $10 - 9 = 1$, but this does not imply that 1 and 2 were just as close in the final results as were 9 and 10.

The difference between values of **interval data** *does* have meaning. It is meaningful to add and average such data. The classic example is *temperature,* where it is true that the difference in heat between 60°F and 61°F is the same as that between 80°F and 81°F. Many of the techniques used to analyze data in statistics require data that are at least of this strength.

Ratio data differ from interval data in that there is a definite *zero point.* To decide if your data are interval or ratio, ask yourself whether twice the value is twice the strength. For example, is 100°F twice as hot as 50°F? The answer is no, so these data are interval. Is a 4-acre field twice as large as a 2-acre field? The answer is yes, so these are ratio data. Here the zero point is a field of 0 acres. Typically, data consisting of areas, counts, volumes, and weights are ratio data.

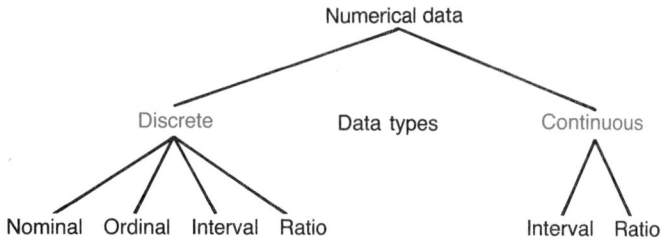

EXAMPLES OF DISCRETE DATA

1. Nominal: Ownership status of resident dweller
   (1 = own, 2 = rent)
2. Ordinal: Level of customer satisfaction
   (1 = very dissatisfied, 2 = somewhat dissatisfied, 3 = somewhat satisfied, 4 = very satisfied)
3. Interval: Person's score on an IQ test
4. Ratio: Number of defective light bulbs in a carton

EXAMPLES OF CONTINUOUS DATA

1. Interval: Actual temperature, °F
2. Ratio: Weight of packaged dog food

The techniques used in statistics generally do not distinguish between interval and ratio data. A summary of the various data classifications is shown in Figure 1.2. Notice that discrete data can result from any of the four levels of measurement, whereas continuous data must be interval or ratio.

▲ Summary   Decision making using statistical procedures continues to grow in popularity, since calculators and computers make it easy to avoid "seat-of-the-pants" decisions by analyzing sample results in a scientific manner. Contemporary applications (such as, should a particular company accept an outside shipment of components based upon a sample of these components?) can be found in a variety of business disciplines.

The science of statistics comprises a set of rules and procedures used to describe numerical data or to make decisions based on these data. The group of measurements that are of interest define the **population.** The portion of a population selected for observation is a **sample.** Most statistical methods assume that a **simple random sample** of size n has been collected, in which each set of n measurements has the same chance of being selected for the sample. A characteristic (such as the average) of the population is referred to as a **parameter,** and the corresponding sample characteristic is a **statistic.** For example, the average age of a sample of 100 people passing the most recent CPA exam is a statistic. The average age of *all* people passing this exam is a parameter. A sample that contains the entire population is a **census.**

Descriptive statistics is concerned only with collecting and describing data. **Inferential statistics** is used when tentative conclusions about a population are drawn on the basis of data contained in a representative sample. The question of whether to accept the shipment of components (the population) based on the sample of 100 components is an example of inferential statistics.

Numerical data are either discrete or continuous. **Discrete data** have limited, specific possible values. **Continuous data** can assume any value over some range. A further classification of data is their level of measurement. At the lowest level,

nominal data are categorical data that are assigned numeric codes. **Ordinal data** are ranked—the order of the data values is meaningful. In **interval data,** both the order of the data and the difference between any two data values have meaning. Finally, **ratio data** have all the properties of interval data and also contain a definite zero point. Most statistical techniques do not distinguish between interval and ratio data but do require the data to have at least an interval level of measurement.

Review Exercises

1.1   The manager of Computer Solutions, Inc., wants to determine the demand for computer consultants in small businesses in Los Angeles. The manager selects 50 small businesses at random in the Los Angeles metropolitan area. Analysis of the questionnaire sent to these small businesses revealed the extent of computerization and the willingness of the companies to utilize consulting services. What is the population? What is the sample?

1.2   The senior management of a fast-food restaurant chain hired a market researcher to obtain information on the eating patterns of families in the northeastern United States. The researcher first randomly selected 36 population centers across this area. At each of the population centers, 100 families were randomly selected to complete a questionnaire. The results of the surveys were used to determine the expansion plans for the restaurant chain. Which group of people did the researcher select for the population? The sample?

1.3   The manager of Easy Fly Airlines took a sample of 200 people who regularly fly on Easy Fly Airlines and collected information on the salaries of these people. The manager then graphed the data to get an idea of the distribution of their incomes. After studying the graph, the manager concluded that most Easy Fly Airlines customers are in the middle income bracket.

a.   Explain how the manager used descriptive statistics.

b.   Did the manager use inferential statistics?

1.4   Explain whether each of the following groups of people or objects represents a population or a sample.

a.   A list of 500 employees of General Motors (*Hint:* Could this be either a sample or a population? Explain.)

b.   Forty students who were randomly stopped and questioned on a university campus

c.   Two hundred people who were selected randomly from the telephone book to receive a marketing questionnaire

d.   The list of all possible choices of 2 cards from a deck of 52 cards

e.   A batch of electronic parts ready for inspection

1.5   Explain whether the following data are continuous or discrete.

a.   The diameters of wire cables in an incoming lot of 10 wire cables

b.   The number of long-distance calls made each month

c.   The length of time for each long-distance call made in a particular month

d.   The number of defective items in a manufacturing process

e.   The ratio of debt to property value for a certain municipality

1.6   Explain why inferential statistics is not needed if data are collected by taking a census.

1.7   What is the lowest level of measurement for a set of data in order to permit the valid calculation of a proportion?

1.8   Your student record contains information about your age, sex, race, current grade point average, and current classification (freshman, sophomore, . . .). State whether each of these data is nominal, ordinal, interval, or ratio.

1.9   Do you think nominal data would usually be continuous data or discrete data?

1.10   Give an example of ordinal data that would not be interval data.

1.11   What is the highest level of measurement for each of the following data sets?

a.   The chain of command for officers in the army

b. The closing prices of the stocks in the Dow Jones industrial average

c. The temperatures in degrees Fahrenheit of several classrooms

d. The social security numbers of 12 randomly selected people

e. A listing of the college graduates and non-college graduates working for a company

f. The length of time required for people arriving at a local hotel to check in

Computer Exercises Using the Database

Exercise 1 -- Appendix I For each of the variables defined in the database of household financial variables, determine if the corresponding data would be classified as discrete or continuous.

Exercise 2 -- Appendix I For each of the variables in the database of household financial variables, what is the highest level of measurement for the corresponding data?

Exercise 3 -- Appendix J Answer exercises 1 and 2 for each variable defined in the database using financial variables on companies.

C H A P T E R

Descriptive Graphs

Chapter 1 introduced you to some of the basic terms used in statistics. One of the key concepts was the idea of acquiring data using a sample from a population. It was also emphasized that the proper use of statistics depends on the nature of the data involved. Are the data discrete or continuous? Are the values nominal, ordinal, interval, or ratio?

Once the data have been gathered, the problem becomes learning whatever we can from them. One method is to describe the data by means of a graph. A graph allows us to discuss intelligently the "shape" of the data.

Everyone has heard the expression that a picture is worth a thousand words (or, more appropriately here, a thousand numbers). This is especially true in statistics, where it may be vital to reduce a large set of numbers to a graph (or picture) that illustrates the structure underlying the data. For example, in a business meeting a quick glance at a graph dem-

onstrates a point much more easily than does a page filled with numbers and words.

Let's illustrate why you may want to describe a data set by using a graph. Suppose that the television department of Q-Mart sells color and black-and-white televisions and home videocassette recorders (VCRs). They decide to take a sample of 50 of their customers over a three-month period. For each sale, they record (1) what the customer purchased, (2) the purchase price, and (3) the number of channels that each customer receives on his or her home set. The results are shown in Table 2.1.

How can you summarize and present these data in a form that is easily understood? There are many graphical methods that can be used, depending on the nature of the data and what you are trying to demonstrate about them. When presenting data graphically, the first step usually is to combine the data values into a frequency distribution.

2.1
Frequency Distributions

We need to reduce a large set of data to a much smaller set of numbers that can be more easily comprehended. If you have recorded the population sizes of 500 randomly selected cities, there is no easy way to examine these 500 numbers visually and learn anything. It would be easier to examine a condensed version of this set of data, such as that presented in Table 2.2.

This type of summary, called a **frequency distribution,** consists of *classes* (such as "10,000 and under 15,000") and *frequencies* (the number of data values within each class). What do you gain using this procedure? You reduce 500 numbers to ten classes and frequencies. You can study the frequency distribution in Table 2.2 and learn a great deal about the shape of this data set. For example, approximately 50% of the cities in your sample have a population between 20,000 and 35,000. Also, only 1% of the cities contain 50,000 people or more.

Frequency Distribution for Continuous Data

A frequency distribution is typically condensed from data having an interval or ratio level of measurement. When you construct a frequency distribution for continuous data, you need to decide how many classes to use (ten in Table 2.2) and the class width (5000 in Table 2.2).

There is no "correct" **number of classes (K)** to use in a frequency distribution. However, you can best condense a set of data by using between 5 and 15 classes. The usual procedure is to choose what you think would be an adequate number of

■ TABLE 2.1
Data from
Q-Mart survey.

ITEM	NO. OF SALES	PURCHASE PRICE/NO. OF CHANNELS						
Color TV	30	460.04	538.13	477.18	475.96	715.93	436.68	643.55
		3	9	6	12	5	8	9
		495.57	515.62	712.26	463.36	676.84	620.24	561.63
		18	6	15	10	8	12	4
		375.94	516.82	434.27	397.95	481.45	517.79	520.24
		16	10	5	13	8	7	12
		488.37	840.57	624.63	419.19	782.57	485.15	812.36
		8	11	20	15	6	9	8
		583.82	388.70					
		5	11					
Black-and-White TV	6	345.88	255.46	295.77	318.91	362.81	405.16	
		7	14	5	10	4	8	
VCR	14	478.03	715.71	450.36	488.34	582.36	657.41	684.71
		9	4	17	8	11	6	19
		631.78	521.48	515.61	540.44	528.57	564.16	745.28
		10	6	9	4	8	13	3

■ TABLE 2.2
Frequency distribution
of the populations of
500 cities.

CLASS NUMBER	SIZE OF CITY	FREQUENCY
1	Under 10,000	4
2	10,000 and under 15,000	51
3	15,000 and under 20,000	77
4	20,000 and under 25,000	105
5	25,000 and under 30,000	84
6	30,000 and under 35,000	60
7	35,000 and under 40,000	45
8	40,000 and under 45,000	38
9	45,000 and under 50,000	31
10	50,000 and over	5
		500

classes and to construct the resulting frequency distribution. A quick look at this distribution will tell you if you have reduced the data too much (not enough classes; K is too small) or not enough (too many classes; K is too large). If you have a very large set of data, you can use a larger number of classes than you would for a smaller data set. Whenever you construct frequency distributions using a computer, select several different values of K and look at the effects of the different choices.

Having chosen a value for K, the next step is to examine

$$\frac{\text{range}}{\text{number of classes}} = \frac{H - L}{K}$$

where H = the highest value in your data and L = the lowest value in your data. Round the result to a value that provides an easy-to-interpret frequency distribution. This is the **class width (CW).** The width of each class should be the same. Later we will discuss possible exceptions to this rule for the first and last classes.

DEFINITION

Class Width (CW) = the value of $\dfrac{H - L}{K}$ rounded (up or down)
to a value that is easy to interpret

Suppose that, for a particular set of data, you have elected to use $K = 10$ classes in your frequency distribution and that $H = 106$ and $L = 10$. Then

$$\frac{H - L}{K} = \frac{106 - 10}{10} = 9.6$$

The desirable class width to use here is CW $= 10$.

Here are some additional examples of rounding to determine the class width:

$\dfrac{H - L}{K}$	ROUNDED VALUE (THE CW)
89.6	100
1.38	1.5
48.2	50
12.4	10

Now let us use the 50 purchase prices in Table 2.1 to construct a frequency distribution of the purchase prices, using six classes. Our first step should be to arrange the data from smallest to largest. This arrangement is called an **ordered array.** Both the original data and the ordered data are **raw data,** since they are not grouped into classes. The ordered purchase prices are listed in Table 2.3. Using the ordered data, $H = 840.57$ and $L = 255.46$. Since $K = 6$, we compute CW:

$$\frac{840.57 - 255.46}{6} = 97.5$$

The best choice for CW is CW $= 100$.

There are two rules to remember in selecting the first class: this class must contain L, your lowest data value, and it should begin with a value that makes the frequency distribution easy to interpret. Because $L = 255.46$, our first class should begin with either 200 or 250—we will use 250. The resulting frequency distribution is shown in Table 2.4.

Perhaps you think that six classes are not enough; that is, this set of data has been condensed too much. One indication of this would be that a large portion of

TABLE 2.3

Fifty purchase prices from Q-Mart survey arranged as an ordered array.

RAW DATA					ORDERED DATA				
460.04	463.36	520.24	345.88	582.36	255.46	434.27	488.34	538.13	657.41
538.13	676.84	488.37	255.46	657.41	295.77	436.68	488.37	540.44	676.84
477.18	620.24	840.57	295.77	684.71	318.91	450.36	495.57	561.63	684.71
475.96	561.63	624.63	318.91	631.78	345.88	460.04	515.61	564.16	712.26
715.93	375.94	419.19	362.81	521.48	362.81	463.36	515.62	582.36	715.71
436.68	516.82	782.57	405.16	515.61	375.94	475.96	516.82	583.82	715.93
643.55	434.27	485.15	478.03	540.44	388.70	477.18	517.79	620.24	745.28
495.57	397.95	812.36	715.71	528.57	397.95	478.03	520.24	624.63	782.57
515.62	481.45	583.82	450.36	564.16	405.16	481.45	521.48	631.78	812.36
712.26	517.79	388.70	488.34	745.28	419.19	485.15	528.57	643.55	840.57

TABLE 2.4

Frequency distribution of purchase prices using six classes.

CLASS NUMBER	CLASS	FREQUENCY
1	250 and under 350	4
2	350 and under 450	8
3	450 and under 550	20
4	550 and under 650	8
5	650 and under 750	7
6	750 and under 850	3
		50

■ TABLE 2.5
Frequency distribution
of purchase prices
using ten classes.

CLASS NUMBER	CLASS	FREQUENCY	RELATIVE FREQUENCY
1	250 and under 310	2	.04
2	310 and under 370	3	.06
3	370 and under 430	5	.10
4	430 and under 490	12	.24
5	490 and under 550	10	.20
6	550 and under 610	4	.08
7	610 and under 670	5	.10
8	670 and under 730	5	.10
9	730 and under 790	2	.04
10	790 and under 850	2	.04
		50	

the data (say, nearly 50%) lies in one class. Table 2.5 summarizes this set of data using $K = 10$ classes. Here, the class width chosen is CW $= 60$ because

$$\frac{H - L}{10} = \frac{840.57 - 255.46}{10} = 58.5$$

As before, the first class begins at 250.

This table also contains each **relative frequency,** where

$$\text{relative frequency} = \frac{\text{frequency}}{\text{total number of values in data set}}$$

For example, in class 2 the relative frequency is .06; this class contains 3 of the 50 values. The advantage of using relative frequencies is that the reader can tell immediately what percentage of the data values lies in each class.

COMMENTS Another alternative for this set of data is to use CW $= 50$, because an increment of 50 produces classes easier to comprehend. This would produce 12 classes, as shown in Table 2.6. We could argue that 12 classes are too many, considering that the data set has only 50 values. Many classes contain only one or two data values.

The highest and lowest values in a class are the **class limits.** For example, in Table 2.4, the lower class limit of class 2 is 350, and the upper class limit is 450. The **class midpoints** are those values in the center of the class.* Each midpoint in a sense "represents" its class. These values often are used in a statistical graph as well as for calculations performed on the information contained within a frequency distribution. The midpoint of class 2 in Table 2.4 is $(350 + 450)/2 = 400$.

Often a set of data contains one or two very small or very large numbers quite unlike the remaining data values. Such values are called **outliers.** It is generally better to include these values in one or two **open-ended classes.** The distribution in Table 2.2 contains two open-ended classes: class 1 (under 10,000) and class 10 (50,000 and over). *You may need an open-ended class if your data set includes one or more outliers or your present frequency distribution has too many empty classes on the low or high end.*

Constructing a Frequency Distribution

1. Gather the sample data.
2. Arrange the data in an ordered array.
3. Select the number of classes to be used.
4. Determine the class width.
5. Determine the class limits for each class; begin by assigning to the first class a lower class limit that will make the frequency distribution easy to interpret.
6. Count the number of data values in each class (the class frequencies).
7. Summarize the class frequencies in a frequency distribution table.

*Class midpoints are often referred to as *class marks.*

■ TABLE 2.6
Frequency distribution
of purchase prices
using CW = 50. This
format is used for
continuous data.

CLASS NUMBER	CLASS	FREQUENCY
1	250 and under 300	2
2	300 and under 350	2
3	350 and under 400	4
4	400 and under 450	4
5	450 and under 500	11
6	500 and under 550	9
7	550 and under 600	4
8	600 and under 650	4
9	650 and under 700	3
10	700 and under 750	4
11	750 and under 800	1
12	800 and under 850	2
		50

■ TABLE 2.7
Frequency distribution
of the number of
channels received. This
format is used for
discrete data.

CLASS NUMBER	CLASS	FREQUENCY	RELATIVE FREQUENCY
1	3–5	10	.20
2	6–8	15	.30
3	9–11	12	.24
4	12–14	6	.12
5	15–17	4	.08
6	18–20	3	.06
		50	1.0

Frequency Distribution for Discrete Data

When your data are discrete, the procedure is almost the same as when they are continuous, except (1) we define the class width CW to be the difference between the lower class limits and not the difference between an upper and lower limit (this will also work for continuous data) and (2) the description of each class is slightly different because we no longer use the "and under" definition of each class. Thus, if CW = 5 and *the data are continuous,* our classes might be 5 and under 10, 10 and under 15, and 15 and under 20. *If the data are discrete,* they might be 5 to 9, 10 to 14, and 15 to 19. Note that for the continuous data, the class midpoints are 7.5, 12.5, and 17.5. For the discrete data, however, the midpoints are 7, 12, and 17.

Using the data in Table 2.1, we can construct a frequency distribution using six classes for the number of channels each customer receives on his or her television set. First we develop an ordered array:

3, 3, 4, 4, 4, 4, 5, 5, 5, 5, 6, 6, 6, 6, 6, 7, 7, 8, 8, 8, 8, 8, 8, 8, 8, 9, 9, 9, 9, 9, 10, 10, 10, 10, 11, 11, 11, 12, 12, 12, 13, 13, 14, 15, 15, 16, 17, 18, 19, 20

So, $H = 20$ and $L = 3$. Since

$$\frac{H - L}{K} = \frac{20 - 3}{6} = 2.83$$

we use CW = 3. The resulting frequency and relative frequency distribution are shown in Table 2.7.

Exercises 2.1 The following are the scores of the students of Oceanspray College on a statistics exam:

69, 47, 82, 73, 99, 97, 55, 18, 100, 85, 77, 80, 94, 79, 66, 81, 81, 88, 94, 70,
62, 58, 43, 21, 85, 68, 50, 43, 91, 85, 60, 45, 88, 95, 46, 59, 75, 80, 74, 71, 70

a. Convert the raw data into an ordered array.

b. If you have to transform the data into a frequency distribution, what value of K would you use? (K = number of classes)

c. Calculate the class width for the frequency distribution.

d. Present the data in the form of a frequency distribution.

e. Calculate the relative frequencies of the scores of the college students.

f. Comment on the shape of the frequency distribution and on the shape of the relative frequency distribution.

2.2 The number of hours per day that the secretary of an accounting firm spends on the telephone is recorded. The following data are the number of hours per day over a 30-day period.

5.21, 2.12, 1.33, 7.10, 4.30, 4.20, 5.20, 2.50, 4.10, 1.50, 3.22, 3.51, 2.45, 2.54, 1.80, 1.70, 1.80, 7.10, 3.20, 2.50, 4.11, 6.20, 6.79, 1.67, 3.30, 2.90, 7.20, 5.00, 2.80, 3.90

a. Are these data discrete or continuous?

b. Construct a frequency and a relative frequency distribution.

c. What are the class midpoints?

2.3 A survey was conducted to find out how long homemakers spend shopping for groceries each week. One hundred homemakers were selected randomly in a telephone survey and asked to state the amount of time they spent in grocery food stores during the past week. The results were:

HOURS SHOPPING	FREQUENCY
0 and under 2	38
2 and under 4	31
4 and under 6	21
6 and under 8	6
8 and under 10	3
10 and over	1

a. Construct a relative frequency distribution.

b. What are the class limits?

c. What are the class midpoints?

d. Before the survey, the researchers believed that most homemakers spent no more than 2 hours per week shopping for groceries. Do the data for these 100 homemakers support that opinion?

2.4 A retail store charges its customers 15% of the value of a check on any check that bounces. In 1 week, 30 checks bounced and the following fees were collected (in dollars);

1.02, .50, 6.00, 7.21, 2.34, 2.51, 10.91, 5.95, 2.59, 4.31, 6.31, 8.30, 1.03, 8.62, 9.71, 5.21, 25.91, 10.91, 7.30, 12.51, 8.51, 6.51, 7.31, 1.19, 11.60, 12.51, 2.24, 5.41, 7.20, 8.51

a. Construct a frequency distribution using class intervals of $0 and under $2, $2 and under $4, and so on.

b. Can the value of $25.91 be considered to be an outlier?

2.5 Harberts Wholesale Plumbing Supply receives special orders for parts not in stock at local retail stores. The following data are the number of special orders received per day over a 30-day period.

2, 4, 10, 1, 12, 15, 5, 6, 9, 18, 14, 13, 19, 22, 18, 6, 4, 10, 20, 7, 15, 10, 6, 2, 7, 17, 23, 11, 16, 4

a. Are the data discrete or continuous?

b. Construct a relative frequency distribution.

2.6 Comment on the "correct" number of classes to be used in a frequency distribution.

2.7   The following are the unemployment rates (percent of available labor) for the year 1984 for selected northern cities in the United States:

6.8, 2.6, 4.6, 5.4, 6.8, 4.3, 17.1, 6.7, 6.2, 6.4, 4.7, 5.1, 2.7, 16.8, 3.4, 2.2, 4.1, 9.8, 10.4, 11.5

a.   Calculate the class width for a frequency distribution for the data.

b.   What do you hope to achieve by tabulating the data in the form of a frequency distribution?

c.   Prepare a frequency distribution table.

2.2
Histograms

After you complete a frequency distribution, your next step will be to construct a "picture" of these data values using a histogram. A **histogram** is a graphical representation of a frequency distribution. It describes the shape of the data. You can use it to answer quickly such questions as, are the data symmetric? and where do most of the data values lie? For the frequency distribution in Table 2.4, the corresponding histogram is illustrated in Figure 2.1. The height of each bar represents the frequency of that particular class.

In a histogram, the bars must be adjoining (no gaps). For discrete data (such as that in Table 2.7), the right edge of each bar is midway between the upper limit of the class contained in the bar and the lower limit of the next class. For example, a histogram of Table 2.7 will contain a bar between 2.5 and 5.5 (with a height of 10), the next bar between 5.5 and 8.5 (height of 15), and so forth. The final bar (height of 3) will extend from 17.5 to 20.5.

Avoid constructing a "squashed" histogram by using the vertical axis wisely. The top of this axis (21 in Figure 2.1) should be a value close to your largest class frequency (20). Notice also that, for this example, you obtain a more concise picture by starting the horizontal axis at 250 rather than at zero and putting a scale break (—⌁—) before the 250 mark.

A histogram can be constructed using the relative frequencies rather than the frequencies. A **relative frequency histogram** of the data in Table 2.4 is shown in Figure 2.2. Notice that the shape of a frequency histogram (Figure 2.1) and a relative frequency histogram (Figure 2.2) are the same. One advantage of using a relative frequency histogram is that the units on the vertical axis are always between zero and one, so the reader can tell at a glance what percentage of the data lies in each class.

Most standard statistical software packages will construct a histogram from your data. Using MINITAB, you can specify the class width and the starting class

■ **FIGURE 2.1**
Frequency histogram for the frequency distribution shown in Table 2.4. Twenty (out of 50) purchases were between $450 and $550, with 18 people spending $550 or more.

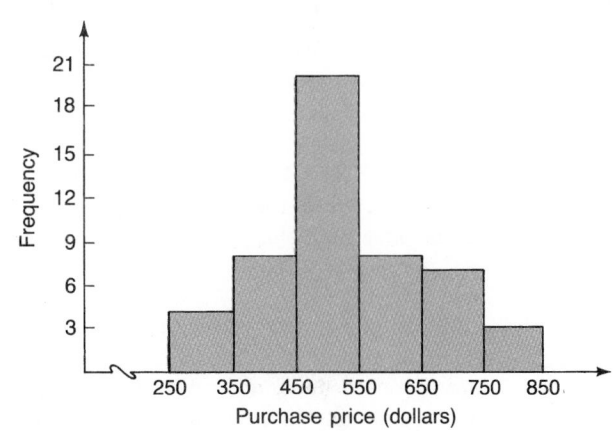

■ **FIGURE** 2.2
Relative frequency
histogram of the
frequency distribution
in Table 2.4. This
histogram shows that
40% of the purchases
were between $450
and $550; 36% of
the purchases were
$550 or more.

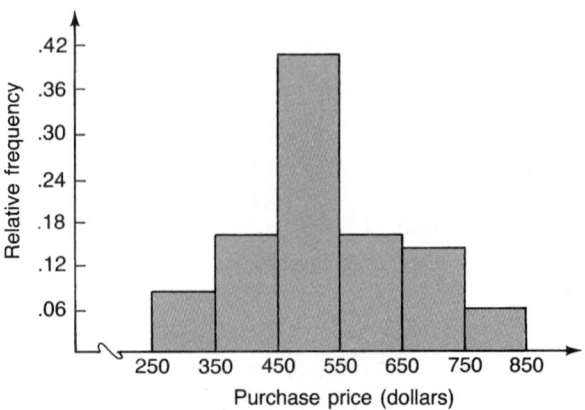

■ **FIGURE** 2.3
Histogram using
MINITAB, where
CW and the first
class midpoint are
not specified.

```
MTB > SET INTO C1
DATA> 460.04 538.13 477.18 475.96 715.93 436.68 643.55 495.57 515.62 712.26
DATA> 463.36 676.84 620.24 561.63 375.94 516.82 434.27 397.95 481.45 517.79
DATA> 520.24 488.37 840.57 624.63 419.19 782.57 485.15 812.36 583.82 388.70
DATA> 345.88 255.46 295.77 318.91 362.81 405.16 478.03 715.71 450.36 488.00
DATA> 582.36 657.41 684.71 631.78 521.48 515.61 540.44 528.57 564.16 745.28
DATA> END
MTB > PRINT C1

C1
   460.04    538.13    477.18    475.96    715.93    436.68    643.55    495.57
   515.62    712.26    463.36    676.84    620.24    561.63    375.94    516.82
   434.27    397.95    481.45    517.79    520.24    488.37    840.57    624.63
   419.19    782.57    485.15    812.36    583.82    388.70    345.88    255.46
   295.77    318.91    362.81    405.16    478.03    715.71    450.36    488.00
   582.36    657.41    684.71    631.78    521.48    515.61    540.44    528.57
   564.16    745.28

MTB > HISTOGRAM OF C1

Histogram of C1   N = 50

Midpoint    Count
     250       1    *
     300       2    **
     350       2    **
     400       5    *****
     450       5    *****
     500      14    **************
     550       5    *****
     600       4    ****
     650       3    ***
     700       5    *****
     750       1    *
     800       2    **
     850       1    *
```

midpoint, or you can let MINITAB select these values. Your output will contain the
frequency distribution as well as a graphical representation in the form of a histogram (using symbols instead of bars). MINITAB will provide each class frequency
next to the corresponding class midpoint (not class limits). Figure 2.3 contains the
necessary MINITAB statements and the resulting output, where the class width and
the midpoint of the first class were not specified. Figure 2.4 specified CW = 100
and the first midpoint to be 300. We can use the output as it appears or use
this information to construct Figure 2.1, which is a graphical representation of
Table 2.4.

When constructing a histogram for discrete data, the boxes should be constructed so that there are no gaps between them. This can best be accomplished by
extending each upper and lower class limit to a value midway between this limit
and the adjoining limit. In this way, the histogram takes on the appearance of a

■ **FIGURE 2.4**
MINITAB histogram
using specified classes.
CW = 100, and the
first midpoint is 300.

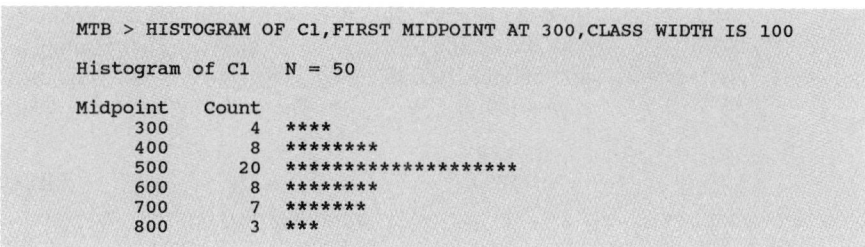

```
MTB > HISTOGRAM OF C1,FIRST MIDPOINT AT 300,CLASS WIDTH IS 100

Histogram of C1    N = 50

Midpoint   Count
     300      4   ****
     400      8   ********
     500     20   ********************
     600      8   ********
     700      7   *******
     800      3   ***
```

■ **FIGURE 2.5**
Histogram for discrete
data using Table 2.7.

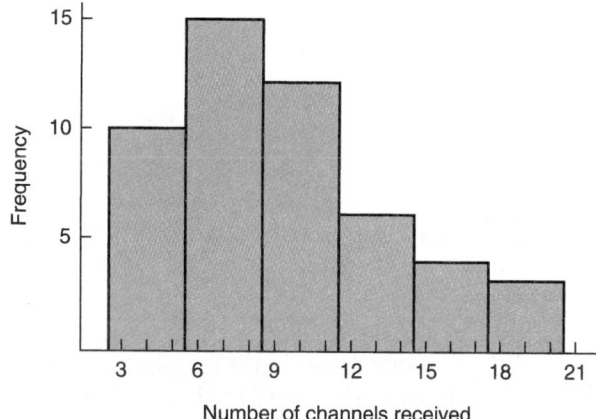

continuous data histogram with no space between successive boxes. The histogram for the discrete frequency distribution in Table 2.7 is shown in Figure 2.5.

In a production process, certain requirements referred to as *specification limits* are often imposed on the product. For example, the inside diameter of a certain machined part must be between 10.1 millimeter and 10.3 millimeter. The value 10.1 is the lower specification (spec) limit and 10.3 is the upper specification (spec) limit. These spec limits can be written as $10.2 \pm .1$ millimeter. Any part with an inside diameter outside these limits is called *nonconforming* and is considered unacceptable.

By gathering a sample and constructing a histogram, production personnel can learn a great deal about the process, in particular, whether the process is capable of meeting these specifications. This situation is illustrated in the following example.

EXAMPLE 2.1

A sample of 100 machine parts having spec limits of $10.2 \pm .1$ millimeters was obtained and contained the following (ordered) values.

10.216	10.221	10.226	10.228	10.230	10.230	10.231	10.231
10.234	10.237	10.239	10.239	10.240	10.240	10.244	10.245
10.249	10.252	10.253	10.254	10.254	10.255	10.255	10.256
10.257	10.258	10.259	10.260	10.262	10.264	10.264	10.266
10.266	10.267	10.267	10.269	10.269	10.270	10.270	10.271
10.271	10.271	10.271	10.271	10.271	10.272	10.274	10.274
10.274	10.276	10.276	10.277	10.277	10.278	10.278	10.279
10.279	10.279	10.280	10.280	10.281	10.281	10.281	10.282
10.283	10.284	10.285	10.286	10.289	10.290	10.290	10.291
10.291	10.291	10.293	10.293	10.293	10.293	10.294	10.296
10.298	10.300	10.300	10.301	10.304	10.309	10.310	10.311
10.311	10.311	10.312	10.312	10.314	10.315	10.315	10.318
10.326	10.328	10.333	10.338				

Here the largest value is $H = 10.338$ and the smallest value is $L = 10.216$. Suppose we decide to use $K = 13$ classes. The resulting class width is determined by first finding $(10.338 - 10.216)/13 = .0094$, which is rounded to $.01$. That is, the class width is $CW = .01$. The resulting frequency distribution is

CLASS NUMBER	CLASS	FREQUENCY	RELATIVE FREQUENCY
1	10.21 and under 10.22	1	.01
2	10.22 and under 10.23	3	.03
3	10.23 and under 10.24	8	.08
4	10.24 and under 10.25	5	.05
5	10.25 and under 10.26	10	.10
6	10.26 and under 10.27	10	.10
7	10.27 and under 10.28	21	.21
8	10.28 and under 10.29	11	.11
9	10.29 and under 10.30	12	.12
10	10.30 and under 10.31	5	.05
11	10.31 and under 10.32	10	.10
12	10.32 and under 10.33	2	.02
13	10.33 and under 10.34	2	.02
		100	1.00

A relative frequency histogram representing this distribution is shown in Figure 2.6. From the histogram it becomes clear at a glance that the process is struggling to meet specifications. The histogram is "shifted too far to the right" with 17% of the parts (the last four classes, excluding the value 10.30) exceeding the upper spec limit. A suggestion for production would be to try shifting the process to the left by whatever means are available, such as making a machine adjustment. With such an adjustment, the process will be more capable of meeting the required specifications.

Histograms are a very simple, yet powerful, tool for analyzing and improving product quality. Additional graphical and measurement techniques will be discussed in Chapter 12, "Quality Control."

■ FIGURE 2.6
Histogram of
100 inside diameters
(Example 2.1).

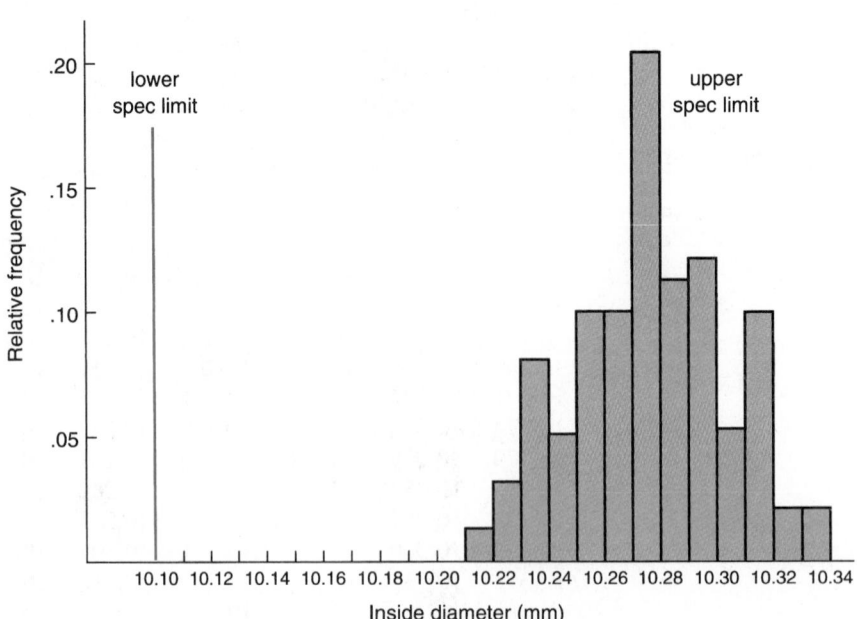

■ *2.3*
Frequency
Polygons

Although a histogram does demonstrate the shape of the data, perhaps the shape can be more clearly illustrated by using a **frequency polygon.** Here, you merely connect the centers of the tops of the histogram bars (located at the class midpoints) with a series of straight lines. The resulting multisided figure is a frequency polygon. Figure 2.7 is an example; once again, the data in Table 2.4 were used.

COMMENTS The polygon can also be constructed from the relative frequency histogram. The shape will not change, but the units on the vertical axis will now represent relative frequencies.

The polygon must begin and end at zero frequency (as in Figure 2.7). To accomplish this, imagine a class at each end of the corresponding histogram that is empty (contains no data values). Begin and end the polygon with the class midpoints of these imaginary classes. Thus, your vertical axis *must* begin at zero. This need not be true for the horizontal axis.

How do you handle an open-ended class? The easiest way is to construct a frequency polygon of the closed classes and place a footnote at each open-ended class location indicating the frequency of that particular class. Figure 2.8 demonstrates this, using the data from Table 2.2.

Frequency polygons are usually better than histograms for comparing the shape of two (or more) different frequency distributions. For example, Figure 2.9 demonstrates at a glance that salaries at Texcom Electronics are higher (for the most part) for management personnel who have a college degree.

Both histograms and frequency polygons represent the actual number of data values in each class. Suppose that your annual salary is one of the values contained in a sample of 250 salaries. One question of interest might be, what fraction of the people in the sample have a salary *less than* mine? Such information can be displayed using a statistical graph called an *ogive*.

■ **FIGURE 2.7**
Frequency polygon using the frequency distribution in Table 2.4. Twenty (out of 50) purchases were between $450 and $550, with 18 people spending $550 or more.

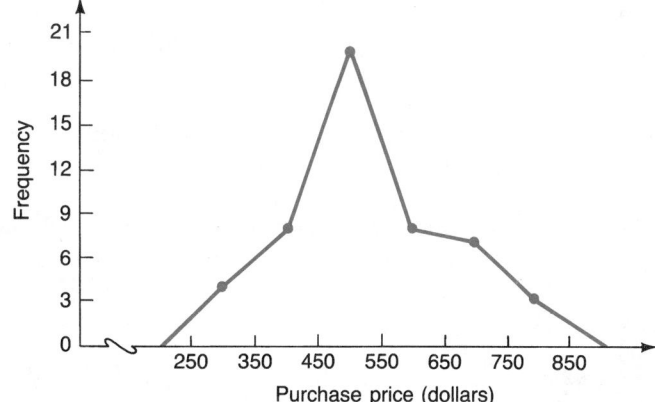

■ **FIGURE 2.8**
Frequency polygon using footnotes to handle open-ended classes. The data are from Table 2.2.

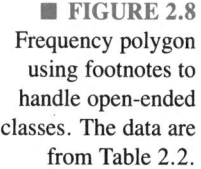
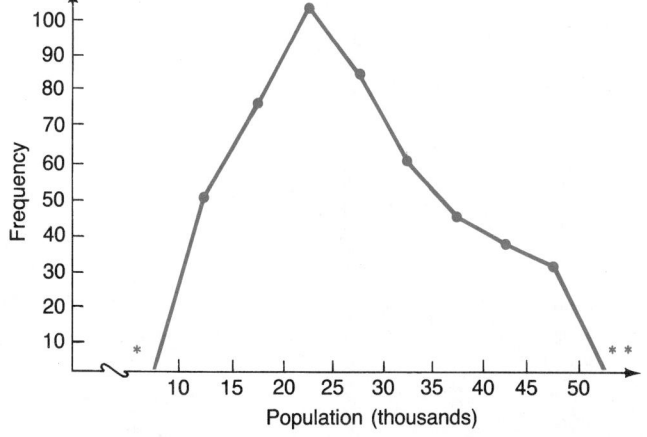

* 4 cities had populations of less than 10,000.
* * 5 cities had populations of 50,000 or greater.

■ **FIGURE 2.9**
Frequency polygon
showing annual salaries
for Texcom Electronics
management personnel.
Higher salaries are
observed in the college
degree sample.

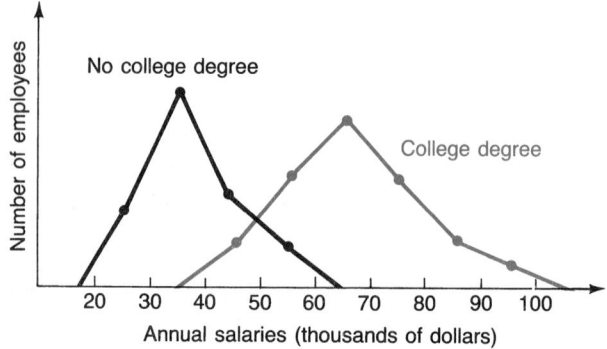

2.4
Cumulative Frequencies (Ogives)

Another method of examining a frequency distribution is to list the number of observations (data values) that are *less than* each of the class limits rather than how many are *in* each of the classes. You are then determining **cumulative frequencies.** Table 2.8 shows the cumulative frequencies for the data in Table 2.4. Notice that you can determine cumulative frequencies (column 4) or cumulative relative frequencies (column 6). The results in Table 2.8 can be summarized more easily in a simple graph called an **ogive** (pronounced oh'-jive). The ogive is useful whenever you want to determine what percentage of your data lies *below* a certain value. Figure 2.10 is constructed by noting that

4 values (4/50 = .08) are less than 350

4 + 8 = 12 values (12/50 = .24) are less than 450

12 + 20 = 32 values (32/50 = .64) are less than 550, and so on.

The ogive allows you to make such statements as "Eighty percent of the purchase prices were less than \$650," and "Fifty percent of the purchase prices were under \$515."

You always begin at the lower limit of the first class (250 here). The cumulative relative frequency at that point is always 0 because the number of data values less

■ **TABLE 2.8**
Purchase prices from
Table 2.3 with
cumulative frequencies
analyzed.

CLASS NUMBER	CLASS	FREQUENCY	CUMULATIVE FREQUENCY	RELATIVE FREQUENCY	CUMULATIVE RELATIVE FREQUENCY
1	250 and under 350	4	4	.08	.08
2	350 and under 450	8	12	.16	.24
3	450 and under 550	20	32	.40	.64
4	550 and under 650	8	40	.16	.80
5	650 and under 750	7	47	.14	.94
6	750 and under 850	3	50	.06	1.00
		50		1.0	

■ **FIGURE 2.10**
Ogive for cumulative
relative frequencies
using data from
Table 2.8. One-half of
the sample spent less
than \$515, and 80% of
the sample spent less
than \$650.

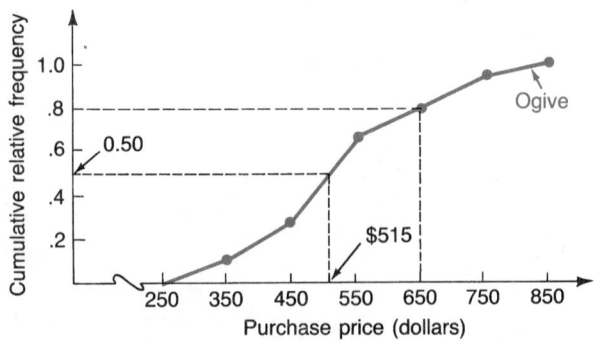

than this number is 0. You always end at the upper limit of the last class (850 here). The cumulative relative frequency at the upper limit is always 1 because all the data values are less than this upper limit. This ogive value would be n = the number of data values (n = 50 here) if you are constructing a frequency ogive rather than a relative frequency ogive. *However, the shape of the ogive is the same for both procedures.*

Exercises

2.8 The following were the daily maximum temperatures in Dallas, Texas, for the month of June (in degrees Fahrenheit):

84, 84, 94, 97, 97, 89, 90, 95, 99, 94, 88, 91, 90, 97, 93, 91, 88, 89, 102, 100, 88, 85, 88, 106, 102, 86, 93, 90, 105, 99

a. Convert the data into an ordered array.

b. Present the data in the form of a frequency distribution, using six classes.

c. Calculate the relative frequencies and the cumulative relative frequencies.

2.9 Construct a frequency histogram for the data in Exercise 2.2.

2.10 Draw a frequency polygon for the data in Exercise 2.3.

2.11 Construct the cumulative frequency distribution for the data in Exercise 2.4. Draw the ogive.

2.12 Does the shape of an ogive change if the cumulative relative frequencies are used instead of the cumulative frequencies?

2.13 The following is the distribution of the population between ages 5 and 39 for a certain town in the year 1988. ("Age" is defined as the age of the person at the person's last birthday.)

AGE	NUMBER
5–9	30,116
10–14	14,633
15–19	29,424
20–24	40,146
25–29	29,424
30–34	44,555
35–39	40,100

a. Construct a frequency histogram.

b. What does the shape of the histogram indicate?

c. If a histogram was constructed from the relative frequency distribution, would the shape of the histogram change? Try it if you are not sure.

2.14 Draw a frequency histogram that indicates the scores of the students on the statistics exam given in Exercise 2.1.

2.15 The price:earnings (P/E) ratio is important to investors selecting a diversified equity mutual fund. The following is a list of P/E ratios for several mutual funds that have performed well during 1990.

INDEX COMPANIES	P/E RATIO	INDEX COMPANIES	P/E RATIO
Mather's Fund	15.6	Cigna Value	20.4
AIM Charter	18.8	Dreyfus Capital Value	22.3
Axe-Houghton Stock	27.2	Righttime Blue Chip	17.4
IDS Growth	25.5	Lindner Dividend	9.7
IDS New Dimensions	21.7	Pax World	18.6
Gateway Index Plus	17.0	SoGen Internal	21.1
Phoenix Balanced Series	21.1	Thomson McKinnon-Growth	19.4
Vanguard World-U.S. Growth	19.1	Hartwell Emerging Growth	37.2
Phoenix Growth Series	21.4		

(*Source:* Adapted from *Money,* October 1990, p. 56)

a.   Construct a frequency distribution.

b.   Construct a frequency histogram for the P/E ratios.

c.   If an investor wished to select mutual funds that had a P/E ratio below 20 from the preceding list, what percentage of the funds could the investor choose?

2.16   A quality-control engineer has been gathering a sample of cylinders that have completed the manufacturing process. The cylinders must be manufactured such that the inside diameter is between 10.6 centimeters and 11 centimeters. These two limits are called the specification limits. Any cylinder with an inside diameter outside of these limits is called nonconforming and is considered a defective cylinder.

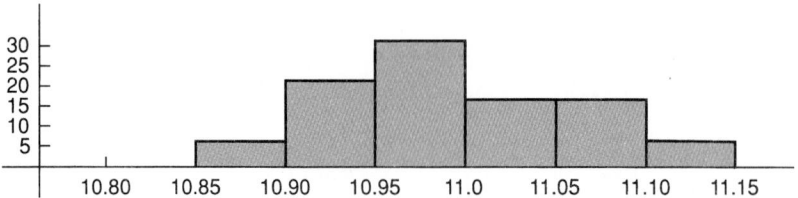

The engineer can make a machine adjustment to shift the process to the right or left. How far and in what direction should the process be shifted to minimize the number of nonconforming cylinders?

2.17   A sample of 50 metal tubes having spec limits of 30.5 ± .2 mm was received from a supplier of machine parts. Construct a histogram of the data. How could a quality control inspector interpret the pattern observed in the histogram?

30.33	30.47	30.34	30.63	30.42	30.53	30.56	30.61	30.36	30.67
30.43	30.37	30.43	30.36	30.62	30.43	30.63	30.38	30.58	30.37
30.58	30.44	30.41	30.63	30.39	30.62	30.37	30.57	30.37	30.61
30.37	30.63	30.57	30.64	30.62	30.38	30.56	30.61	30.36	30.66
30.59	30.36	30.67	30.51	30.69	30.36	30.62	30.41	30.67	30.41

2.18   A county library's records show the following information regarding the number of patrons who used the library during the past 30 days.

100, 87, 44, 53, 17, 34, 88, 67, 31, 40, 98, 77, 55, 41, 73, 62, 88, 28, 70, 51, 82, 44, 32, 50, 33, 49, 59, 67, 79, 84

a.   Construct a cumulative frequency distribution.

b.   Convert the cumulative frequency distribution in part (a) into an ogive graph.

c.   The number of patrons attending the library was less than what value 80% of the time?

2.19   The following is a frequency distribution of the number of daily automobile accidents reported for a month in Newark, New Jersey.

ACCIDENTS PER DAY	FREQUENCY
0–3	12
4–7	10
8–11	7
12–15	1
16–19	1

a.   Construct a cumulative relative frequency distribution for the data.

b.   What percentage of the time do eight or more daily accidents occur?

2.20   The profitability index is widely used by big corporations in making capital investment decisions. It is defined as the ratio of the present value of a project to its cost and should be at least equal to one. The following is a schedule of profitability indices developed by J. Conway, financial analyst of Control Systems:

PROFITABILITY INDEX	PROJECT NAME	PROFITABILITY INDEX	PROJECT NAME
1.70	A	3.50	I
.41	B	1.41	J
2.44	C	2.20	K
2.98	D	5.98	L
4.00	E	6.90	M
1.01	F	1.78	N
5.13	G	6.00	O
2.96	H		

a.   Construct a cumulative relative frequency distribution of the profitability indices.

b.   Construct a frequency histogram.

2.21   Price elasticity of electricity measures the responsiveness of customers to changes in the price of electricity. It can be expressed as the percentage change in quantity demanded (of electricity) over percentage change in the price (of electricity). Because it is always a negative number, it is expressed only in absolute value. The following are the price elasticities (of electricity) for various utility companies in the country:

.40, 71, .33, .08, .14, .24, .38, .27, .44, .35, .22, .05, .39, .18, .22, .70, .52, .31, .21, .36, .15, .38, .41, .23, .55, .61, .52, .35, .48, .62

a.   Construct a frequency polygon.

b.   Construct an ogive curve for the data.

2.22   An econometric model is a statistical model that predicts an econometric measure such as gross national product (GNP), unemployment rate, or inflation rate for a specific time span. The effectiveness of an econometric model is judged by the percentage of wrong predictions made by the model. The following is a frequency distribution for a list of 25 econometric models and the percentage of errors (wrong predictions) created by them:

PERCENTAGE OF WRONG PREDICTIONS	NUMBER OF MODELS
0 and under 5	7
5 and under 10	10
10 and under 15	4
15 and under 20	2
20 and under 25	1
25 and under 30	1
	25

Construct a frequency polygon using this information.

2.23   The following are the test scores of freshmen on the first exam in an economics course at a local university:

62, 67, 74, 48, 100, 93, 49, 57, 77, 63, 82, 10, 78, 88, 99, 44, 51, 80, 71, 39, 58, 76, 89, 94, 70, 41, 66, 82, 18, 73

a.   Construct a relative frequency histogram.

b.   Draw an ogive curve.

c.   How do you interpret the distribution of the test scores?

2.24   David Bannerman, the president of Bannerman Automobile Manufacturing, has gathered the following data on the company's new sports car, the Chariot. The data show the numbers of cars (in hundreds) sold by the 22 top dealers during the past year. Transform the data into an appropriate graph to help David make management decisions in areas such as advertising expenditure and plant expansion. How would you describe the distribution of the data in your report to David Bannerman?

CARS SOLD	DEALERS
0 and under 5	4
5 and under 10	8
10 and under 15	2
15 and under 20	2
20 and under 25	3
25 and under 30	2
30 and under 35	1

2.25 Metro Power manufactures a high-powered copper coil to be used in giant power transformers. Tensile strength (given in thousands of pounds per square inch) is of critical importance in the manufacture of the copper coil. The following data are from a sample of copper coils tested for tensile strength:

$$5, 8, 12, 10, 15, 18, 21, 24, 7, 26, 7, 18, 10, 6, 4, 11, 15, 9, 22, 10$$

a. Construct an ogive.

b. Find an appropriate value, X, in units of pounds per square inch such that more than one-half of the coils sampled have tensile strengths greater than X.

2.26 The following list summarizes the number and value of stocks (in dollars) that make up the investment portfolio of a mutual fund corporation:

VALUE OF STOCKS	NUMBER OF STOCKS
10,000–14,999	7
15,000–19,999	4
20,000–24,999	13
25,000–29,999	6
30,000–34,999	10

a. Draw a frequency histogram.

b. Construct an ogive.

2.27 The leading automotive loan rates from banks in the 24 largest metropolitan areas in the United States are given below

REGION	RATE ON SEPT. 1, 1990	REGION	RATE ON SEPT. 1, 1990
Atlanta	11.50%	New York City	12.50%
Baltimore	11.25%	N. New Jersey	11.25%
Boston	10.75%	Philadelphia	10.75%
Chicago	11.00%	Phoenix	12.50%
Cleveland	11.75%	Pittsburgh	11.00%
Dallas	10.75%	San Diego	10.75%
Denver	9.90%	San Francisco	12.75%
Detroit	11.50%	Seattle	11.00%
Houston	11.00%	St. Louis	10.50%
Los Angeles	11.25%	S. W. Connecticut	12.25%
Miami	11.25%	Tampa	10.50%
Minneapolis	11.75%	Washington, D.C.	11.00%

(*Source:* Adapted from *Money,* October 1990, p. 22)

a. Construct a frequency histogram.

b. Construct a frequency polygon.

c. What is a "typical" automotive loan rate for the 24 largest metropolitan regions?

d. Are there metropolitan regions with exceptionally high automotive loan rates? What can you say about these regions based on the frequency histogram?

◢ 2.5
Bar Charts

Histograms, frequency polygons, and ogives are used for data having an interval or ratio level of measurement. For data having a **nominal** level, we use a bar chart. For situations producing a sample of **ordinal** level data with a reasonable set of possible values (such as 1 = strongly agree, 2 = agree, . . . , 5 = strongly disagree), a bar chart can be used to summarize the sample. A bar chart is similar to a histogram, in that the height of each bar is proportional to the frequency of that class. Such a graph is most helpful when you have many categories to represent.

Consider the data in Table 2.1. If you are interested in the number of sales for each of the three products (color televisions, black-and-white televisions, and VCRs), a bar chart will do a good job of summarizing this information (Figure 2.11). Notice that a gap is inserted between each of the bars in a bar chart. The data here are nominal, so the length of this gap is arbitrary.

Figure 2.12 is an example of a bar chart in which the bars are constructed horizontally rather than vertically. This form enables you to label each category *within* the bar.

EXAMPLE 2.2

The head of Quality Assurance at Microtech (a fictional company) has categorized company costs related to quality improvement into three categories: prevention, appraisal, and failure. These three costs for the past fiscal year were:

Prevention costs: $ 3,600

Appraisal costs: $38,400

Failure costs: $78,000

■ **FIGURE 2.11**
Bar chart for number of sales of each of three items in Table 2.1. Sixty percent of the purchases were color TVs; the smallest number of sales was for black-and-white TVs (12%).

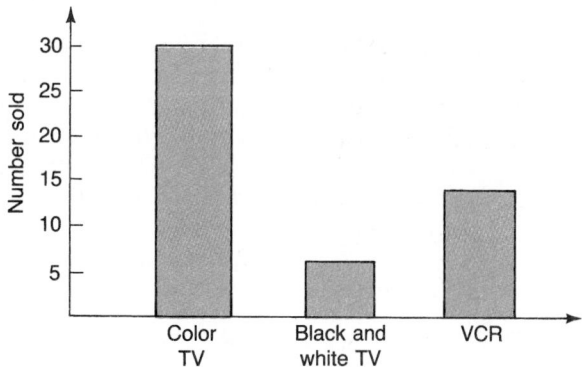

■ **FIGURE 2.12**
Bar chart drawn horizontally; note that it is easy to place labels within the boxes.

Q. If the price of natural gas goes down by 25% in the next few years, would you and your family use more or less?

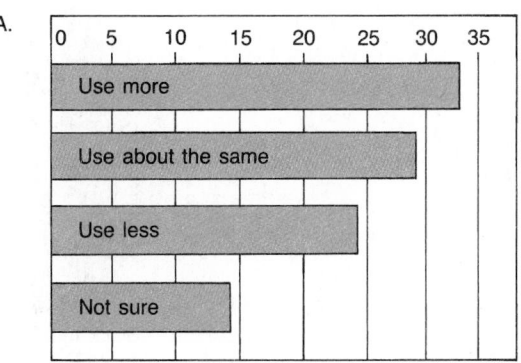

■ **FIGURE 2.13**
Bar chart of quality
costs for Microtech
(Example 2.2).

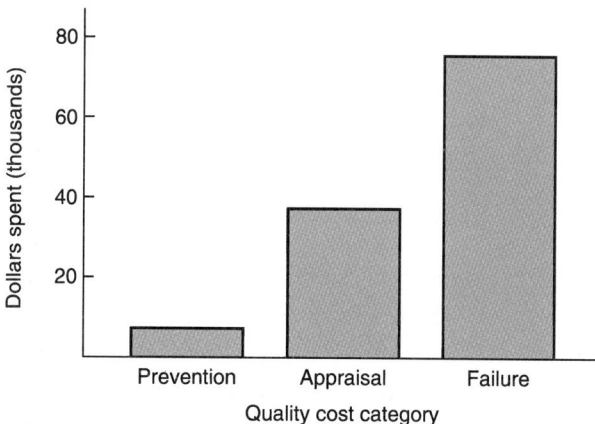

She wants to demonstrate at a glance the small amount spent on prevention measures and the large amount spent on failure costs (largely due to warranty claims but also due to having to rework defective components). Construct a bar chart to illustrate this information.

SOLUTION   The bar chart consists of three boxes (bars), where the height of each box represents the dollar amount for that category. It is shown in Figure 2.13.   ■

◢ 2.6
Pie Charts

A **pie chart** is used to split a particular quantity into its component pieces, typically at some specified point in time or over a specified time span. It is a convenient way of representing percentages or relative frequencies (rather than frequencies). Figure 2.14 shows a pie chart of the 50 sales in Table 2.1. To construct a pie chart, draw a line from the center of the circle to the outer edge. Then construct the various pieces of the pie chart by drawing the corresponding angles. For example, the black-and-white televisions represent 12% of the total number of sales (6 out of 50), so angle A in Figure 2.14 is 12% of 360°, or 43.2°. Angle B is 28% of 360°, or 100.8°, and angle C is 60% of 360°, or 216°.

◢ *Exercises*

2.28   Consumers spend their incomes on a vast array of goods and services. The following figures provide a quick summary of how the average consumer dollar is spent:

■ **FIGURE 2.14**
Pie chart showing
number of sales using
data from Table 2.1.
Sixty percent of the
purchases were for color
TVs; the smallest
number of sales was
for black-and-white
TVs (12%).

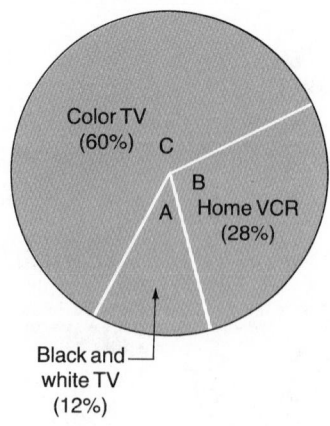

CATEGORY	PERCENT OF INCOME
Medical care	5
Clothing	5
Entertainment	4
Housing	46
Food	17
Transportation	19
Others	4
	100

a. Summarize the information in the form of a pie chart.

b. What area represents the largest piece of the pie? Is it very much larger than the next piece? How much?

2.29 Millions of business firms supply goods and services to us. Usually they are organized as proprietorships, partnerships, or giant corporations. In 1981, there were roughly 14.741 million firms producing goods and services; of these, 11.346 and 1.153 million were proprietorships and partnerships, respectively. The remainder were corporations. Present these data in the form of a pie diagram.

2.30 Use the data on econometric models in Exercise 2.22 to construct a pie chart.

2.31 Reexamine the investment portfolio of the mutual fund corporation in Exercise 2.26. Present the information in the form of a pie chart.

2.32 A survey of the production-inventory systems of Hungarian industrial companies showed that generally well known up-to-date concepts of operation of production-inventory systems have not penetrated Hungarian industry. The data below show the percentage of companies surveyed that have been exposed to various levels of Material Requirements Planning (MRP) and Just In Time (JIT).

EXPOSURES TO MRP/JIT	MRP (%)	JIT (%)
A. Never heard of it	39.7	56.4
B. Using it and benefiting from it	10.3	0.0
C. Using it but not benefiting from it	2.6	19.2
D. Understand it, but feel no necessity of introducing it	20.5	0.0
E. Just starting to introduce it	2.6	1.3
F. Trying to introduce it, but having difficulty doing so	3.8	11.5
G. Considering its introduction	17.9	0.0
H. No response to the question	2.6	6.4

(*Source:* Attila Chikan, "Characterization of Production-Inventory Systems In The Hungarian Industry," *Engineering Cost and Production Economics,* Vol. 18, No. 3 (1990), p. 285–292)

Construct a bar chart for the percentage of exposure to MRP and another bar chart for the percentage exposure to JIT. Comment on the differences in the two charts.

2.33 In a national survey by *Training* magazine, 2,614 respondents answered various questions. The table below shows the respondents' departmental classifications within their companies.

DEPARTMENT	NUMBER OF PERSONS
Training	1510
Personnel/human resources	458
Operations	269
Marketing/sales	125
Customer service	91
Production/manufacturing	91
Data processing	47
Other	23

(*Source:* "Industry Report," *Training,* October 1990, p. 31)

Summarize the data in the form of a bar graph. Can a histogram be used to illustrate the data?

2.34 A supplier of mechanical parts for farm machinery wishes to analyze the types of defects found in various mechanical parts that the buyers have returned. The supplier has compiled a list of the types of defects associated with each mechanical part that has been returned over the past three months.

TYPE OF DEFECT	NUMBER OF MECHANICAL PARTS RETURNED
A. Lubrication problems	60
B. Misaligned component	35
C. Defective component	20
D. Missing component	7
E. Dimensions of part outside of specification limits	5
F. Moving part broken	4
G. Other	5
	136

The supplier used the following Pareto Diagram to analyze the data.

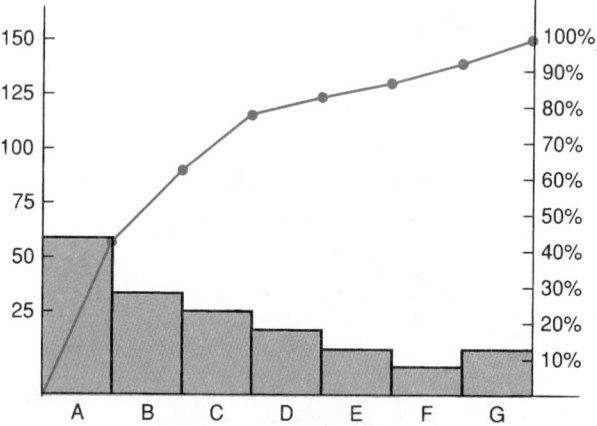

The supplier selected the Pareto Diagram to display the data since this diagram tends to focus on the few problems that are responsible for the majority of the quality-related costs. Note that the type-of-defect category is displayed in descending order of frequency. The cumulative frequency graph is displayed along with the histogram.

a. Draw a pie chart to summarize the data.

b. What information is available through the Pareto Diagram that is not available in the pie chart?

2.35 A senior manager at Four-Mile-Island Utility Company collected various complaints from customers during the past 6 months. The following table shows the types of complaints and the frequency of each.

TYPE OF COMPLAINT	NUMBER OF CUSTOMERS WITH THIS COMPLAINT
A. Customer's check was lost	10
B. Customer's bill was mailed late	80
C. The serviceman was rude to the customer	20
D. The meter was read incorrectly	60
E. The electric utility service was mistakenly disconnected	5
F. Other	25

a. Draw a Pareto Diagram (see Exercise 2.34) to summarize the data.

b. Draw a pie diagram to summarize the data.

c. What course of action does the Pareto Diagram suggest that the senior manager take?

2.36 A successful businessperson receives the following yearly incomes (in dollars) from seven business partnerships.

BUSINESS PARTNERSHIP	YEARLY INCOME
A	23,160
B	30,070
C	32,732
D	35,900
E	37,304
F	43,608
G	60,014
Total	262,788

Express the yearly incomes from each partnership as a percentage of the businessperson's total income and summarize this information using a pie chart.

2.37 The following data indicate the percentage of United States petroleum imports by source for the year 1988:

NATION	PERCENTAGE OF U.S. PETROLEUM IMPORTS
Nigeria	10.5
Saudi Arabia	19.7
Venezuela	14.4
Other OPEC	18.8
Canada	16.5
Mexico	12.4
United Kingdom	5.6
Virgin Islands/Puerto Rico	2.1
	100.00

(*Source: The World Almanac and Book of Facts*, 1990, p. 380)

Construct a pie chart to illustrate these percentages.

2.38 The following table contains the total number of production workers (in thousands) in major U.S. industries in 1987.

INDUSTRY	NUMBER OF WORKERS	INDUSTRY	NUMBER OF WORKERS
Food	1,029.9	Rubber	662.5
Tobacco	32.7	Leather	107.8
Textile	590.3	Stone	400.6
Apparel	904.2	Primary metal	541.8
Lumber	578.6	Fabricated metal	1,085.8
Furniture	409.9	Machinery	1,155.8
Paper	470.4	Electric equipment	1,021.1
Printing	799.8	Transportation equipment	1,213.6
Chemicals	467.3	Instruments	491.8
Petroleum	79.0	Miscellaneous	273.7

(*Source: World Almanac and Book of Facts*, 1990, p. 128)

Present the data in the form of a bar graph. In your opinion, which industries employ an unusually large number of production workers?

2.39   The following are gross average weekly earnings of manufacturing workers in terms of current dollars and in terms of the buying power of 1977 dollars.

	CURRENT DOLLARS	1977 DOLLARS
1970	133.33	208.0
1975	190.79	214.9
1980	288.62	212.0
1985	386.37	220.15
1986	396.01	222.23
1987	406.31	220.10
1988	418.40	217.80
1989, Jan	425.17	216.26

(*Source: The World Almanac and Book of Facts,* 1990, p. 128)

a.   Are the data discrete or continuous?

b.   Present average weekly earnings in terms of current dollars in the form of a bar chart.

c.   Repeat part (b) for 1977 dollars. Compare the two bar charts. What insight do we gain by comparing these two bar charts?

2.40   The amount of time it takes to order a special automotive part from the manufacturer is of great concern to a local automotive dealer. The following data are the average delivery times (in days) from nine different stores for parts that were special-ordered.

STORE	1	2	3	4	5	6	7	8	9
AVERAGE DELIVERY TIMES	2.5	4.5	3.0	6.0	3.0	7.0	4.0	3.5	4.0

a.   Draw a bar chart and describe the shape of the chart.

b.   How does a bar chart differ from a histogram?

◢ 2.7
Deceptive Graphs

You might be tempted to be creative in your graphical displays by using, for example, a three-dimensional figure. Such originality is commendable, but does your graph accurately represent the situation? Consider Figure 2.15, which someone drew in an attempt to demonstrate that there are twice as many men as women in management positions. The artist constructed a box for the category "men" twice as high—but also twice as deep—as that for the category "women." The result is a rectangular solid for men that is, in fact, four times the volume of the one for women. The illustration is misleading—it appears that there are four times as many men as women in management.

When data values correspond to specific time periods—such as monthly sales

■ **FIGURE 2.15**
The illustrator wished to show that there are twice as many men as women in management positions. However, box B is twice the height *and* twice the depth of box A and thus is four times the volume.

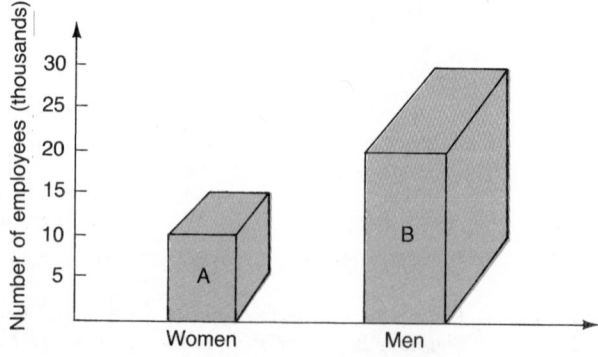

■ FIGURE 2.16
Time-series graph of
the performance of two
mutual funds. The graph
is misleading because
the vertical axis does
not start at zero.

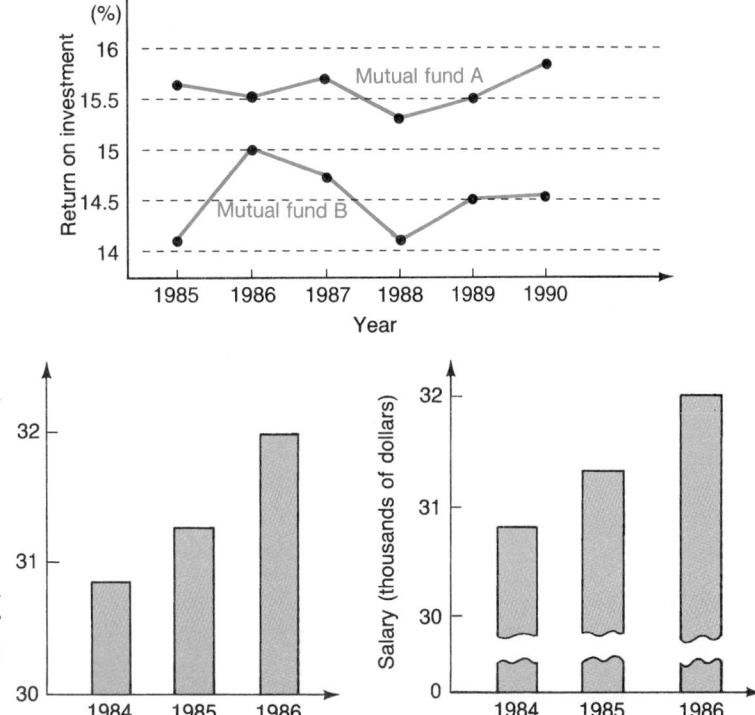

■ FIGURE 2.17
Two misleading bar
charts. The vertical axis
of the left chart does not
begin at zero, and the
bars in the right chart
are chopped without
a corresponding
adjustment in the
vertical axis.

or annual expenditures—the resulting data collection is a **time series.** A time series is represented graphically by using the horizontal axis for the time increments. For example, Figure 2.16 contains a return-on-investment time series for two mutual funds, plotted over a six-year period. A glance at this figure might lead you to believe that mutual fund A is performing nearly twice as well as mutual fund B. A closer look, however, reveals that *the vertical axis does not start at zero;* such a construction can seriously distort the information contained in such a graph. The 1990 return for fund A appears to be roughly twice that for fund B. However, the actual returns are 15.8% for fund A and 14.5% for fund B. Granted, fund A is outperforming fund B, but not nearly as dramatically as Figure 2.16 seems to indicate.

Such examples, and many others, are contained in an entertaining and enlightening book by Darrell Huff entitled *How to Lie with Statistics.** Other deceptive graphs described by Huff include bar charts similar to those in Figure 2.17. Here, you may be tempted to conclude that there is a significant difference in bar heights, either because the vertical axis does not begin at zero (left side) or because the bars are chopped in the middle without a corresponding adjustment of the vertical axis (right side). *As an observer, beware of such trickery. As an illustrator, do not intentionally mislead your reader by disguising the results through the use of a misleading graph.* This practice tends to give statisticians a bad name!

*Darrell Huff, *How to Lie with Statistics* (New York: W. W. Norton, 1954 [and 1982 by Darrell Huff with Irving Geis, illustrator]). More recent discussions are included in *Statistics: Concepts and Controversies* by David S. Moore, 2d ed. (New York: Freeman, 1985) and "How to Display Data Badly" by Howard Wainer, *The American Statistician,* (The American Statistical Association: May 1984).

■ FIGURE 2.18
Frequency polygon
using Harvard Graphics.
See Figure 2.7.

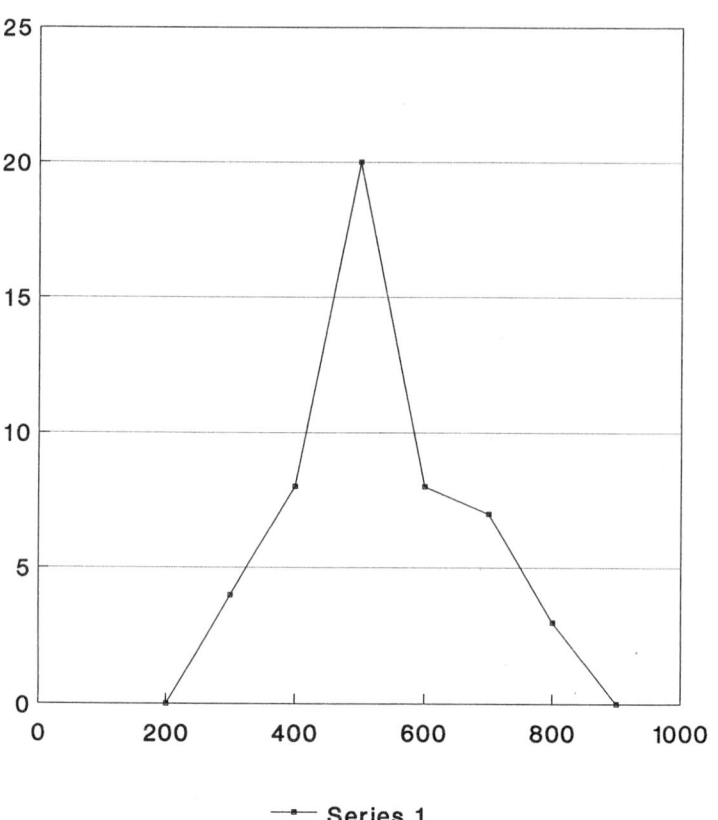

■ FIGURE 2.18
Frequency polygon
using Harvard Graphics.
See Figure 2.7.

 **2.8**

**Computer
Graphics on the
Microcomputer**

Now that you are ready to invest in graph paper, a straight edge, a protractor, and colored pens, you will be happy to learn that there is a much easier method of preparing professional-looking statistical graphs. There are programs available for practically all microcomputers that allow you to construct a variety of multicolored bar charts, pie charts, and so on.

Figures 2.18, 2.19, and 2.20 were constructed using Harvard Graphics, a very versatile, user-friendly graphics package. This particular package provides colored output and three dimensional output (Figures 2.19 and 2.20). Figure 2.18 is another version of Figure 2.7, Figure 2.19 corresponds to Figure 2.13 and Figure 2.20 is the Harvard Graphics version of Figure 2.14.

If you think you will have to create many graphical summaries, try to obtain access to a computer graphics package and its output. No good report is complete without at least one such graph!

Summary

This chapter examined methods of summarizing and presenting a large set of data using a graph. You begin by placing the sample data in order, from smallest to largest (an **ordered array**). The next step is to summarize the data in a **frequency distribution,** which consists of a number of classes (such as "150 and under 250") and corresponding frequencies.

■ **FIGURE 2.19**
Bar chart using Harvard
Graphics (3D option).
See Figure 2.13.

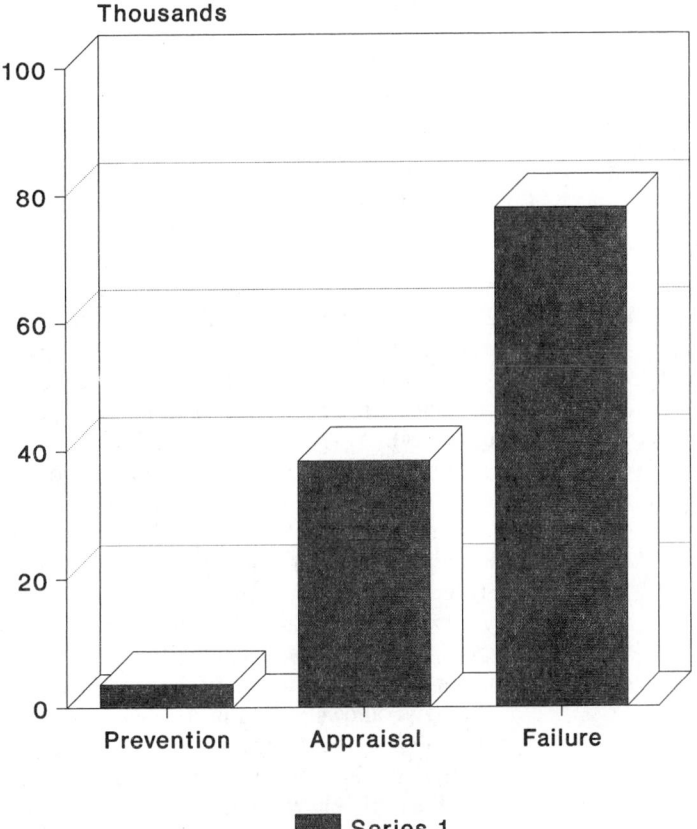

Bar Chart of Quality Costs at Microtech

■ **FIGURE 2.20**
Pie chart using Harvard
Graphics (3D option).
See Figure 2.14.

Type of Appliance Sold

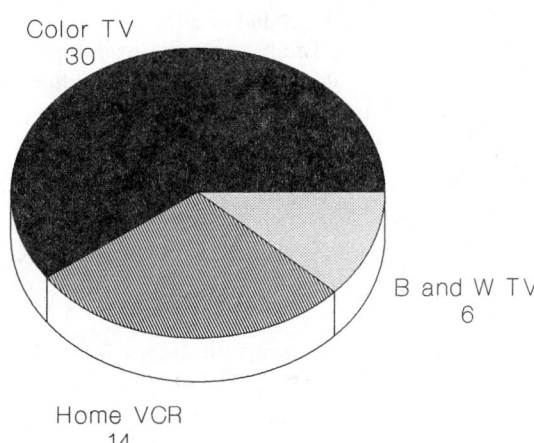

The data summary can then be displayed using an appropriate graph. We discussed four kinds of graphs:

1. A **histogram,** or **frequency polygon,** is a graphical view of a frequency distribution.
2. A **bar chart** summarizes categorical (nominal) or ordinal data.
3. An **ogive** allows you to illustrate "less-than" percentages or frequencies.
4. A **pie chart** presents a percentage breakdown of a particular quantity.

A frequency distribution provides a summary of the data by placing them into groups called **classes.** The number of values in each class is the **class frequency.** For example, there may be ten values in the class "150 and under 250." The numbers 150 and 250 here are the **class limits,** and the difference between consecutive lower class limits is the **class width.** The center of this class [(150 + 250)/2 = 200] is the **class midpoint.** All classes should have the same width, except possibly the first and last class, which may be open-ended if you have a few outliers. For comparisons, the same data can be summarized using **relative frequencies,** which indicate the fraction of data values in each class rather than actual counts (**frequencies**).

A **histogram** is a graphical representation of a frequency distribution and is generally used for data having an interval or ratio level of measurement. When the data are nominal or ordinal, a **bar chart** provides a graphical summary. When constructing a bar chart, gaps are inserted between the bars due to the nature of this data type.

An excellent way to indicate the shape of the data values is to use a **frequency polygon,** which is constructed by replacing the bars in the histogram with straight lines connecting the midpoint of the top of each bar.

An **ogive** allows you to make such statements as 40% of the data values are less than 500. Like a frequency polygon, an ogive consists of many straight lines; in an ogive, the line increases from 0 to 1 on the vertical axis. The final graph we discussed is a **pie chart.** This circular graph can be used to represent percentages (relative frequencies) at some point in time or over a certain time period.

What's Next?

A graph such as a frequency polygon is an excellent method of describing a set of data, but it does have its limitations. For example, we might look at Figure 2.7 and ask, where is the middle (center) of the data? One person might argue that it is "somewhere around 500," whereas someone else might decide that it is some value closer to 550. The point is that we need to define what the word *middle* means and define some method of calculating this value, so that we all get the *same* result. Such a value is called a *numerical measure.*

The next chapter examines a variety of such numerical measures. Rather than reducing a set of data to a graph, we will reduce the data to one or more *numbers* that give us some information about the data.

 *Review Exercises*

2.41 A quality inspector is monitoring the weights of packages of oats. A sample of 24 packages yield the following results.

29.3	30.4	30.7	30.0	29.4	30.9	30.2	30.4	30.1
29.6	30.5	30.0	29.7	29.7	30.8	31.3	30.4	30.2
30.1	30.2	29.8	30.7	30.0	29.8			

a. Construct a frequency distribution using the class intervals of 29.0 and under 29.5, 29.5 and under 30, and so on.

b. Construct a frequency histogram.

c. Draw an ogive.

2.42 An economist has a model to forecast the weekly money supply. The following values represent the difference between the forecasted money supply and the actual money-supply figures over a period of 25 weeks. Units are in hundreds of thousands of dollars.

11.4, 2.5, −50.5, −12.4, −5.1, 4.5, 13.6, 29.8, 51.6, −10.8, −17.8, 30.1, 33.8, 39.6, 44.7, −40.1, −35.6, −37.1, 46.7, 21.6, 18.2, −24.5, −20.5, −15.4, 53.4

a. Construct a frequency distribution.

b. Construct a cumulative relative frequency distribution.

c. Draw the ogive.

2.43 A large real-estate firm has 20 agents. The following data are the yearly salaries of each agent. Units are in thousands of dollars.

13.5, 19.6, 29.8, 43.4, 50.2, 18.7, 7.5, 24.6, 20.3, 27.4, 30.5, 34.6, 12.7, 31.7, 45.8, 41.4, 32.7, 22.6, 27.8, 20.1

a. Construct a frequency distribution.

b. Construct a cumulative relative frequency distribution.

c. Draw the ogive.

2.44 An investor owns several thousand shares of Computer Graphics stock. Because the price of the stock is so volatile, the investor records the closing price of the stock every day to get an idea of the distribution of the price of the stock. The closing prices of this stock for 25 days are:

13.125, 13.5, 12.875, 12.25, 12.375, 13.00, 13.75, 13.375, 14.25, 15.00, 15.25, 15.375, 15.00, 14.75, 15.125, 15.375, 15.75, 16.125, 16.375, 16.50, 16.00, 15.50, 15.75, 16.25, 16.50

a. Construct a frequency distribution.

b. Construct a frequency histogram.

c. Construct a frequency polygon.

2.45 A mutual fund has its assets spread over seven sectors of the economy. The following data are the total value (in millions of dollars) of the stocks in which the fund is invested for each sector.

STOCK	VALUE
Electronics and electrical equipment	2.116
Aerospace and defense	10.375
Food and beverage	4.864
Utilities	2.713
Insurance and finance	6.538
Health care	3.675
Oil and gas	1.532

a. Express the amount invested in each sector of the economy as a percent.

b. Summarize the list in a pie chart.

2.46 The following data represent the scores on a computer-graded multiple-choice test. There are 20 questions, and each question is worth 5 points. Construct a frequency polygon for the grades of the 30 students.

95, 90, 80, 55, 90, 45, 50, 75, 75, 60, 55, 90, 50, 85, 95, 55, 60, 50, 45, 95, 70, 60, 50, 85, 90, 55, 45, 95, 100, 90

2.47 An independent oil firm recently hired ten engineers, five geologists, three accountants, one statistician, four computer scientists, and one chemist. Present these data in the form of a pie chart.

2.48 The numbers of hours that a repair machine is used daily at Pat's Shoe Repair are listed below for 20 days.

3.2, 4.6, 3.1, 3.6, 2.5, 4.3, 2.1, 5.7, 6.1, 4.3, 3.0, 2.5, 1.3, 1.7, 2.4, 5.6, 5.1, 4.2, 1.9, 2.6

a.   Construct a frequency distribution.

b.   Construct a cumulative relative frequency distribution.

c.   What can you say about machine usage (in hours per day)?

2.49   The ages (in years) of the 20 loan officers, 4 vice presidents, and the president of American Bank are

47, 52, 55, 65, 42, 37, 29, 52, 47, 36, 60, 50, 48, 42, 45, 35, 38, 45, 57, 43, 39, 41, 33, 58, 60

a.   Construct a frequency distribution.

b.   Construct a cumulative frequency distribution.

c.   Write a summary statement about the distribution of the ages of the loan officers.

2.50   A psychologist has designed a technique to improve a person's memory. Certain material is given to 30 people to memorize before they learn the technique. Similar material is given to the 30 people after the technique has been taught to them. The difference in the amount of time that it took to memorize the material (before − after) is given in minutes.

5, 10, 15, 11, 13, 20, 14, 5, 23, 18, 17, 4, 1, 5, 29, 18, 15, 21, 24, 16, 2, 15, 19, 30, 24, 21, 14, 18, 26, 10

a.   Construct a frequency distribution.

b.   Construct a frequency histogram.

c.   Construct a frequency polygon.

d.   Take one class interval and write out, in words, exactly what it tells you.

2.51   A manufacturing firm would like to determine the distribution of defective fuses in each package of fuses that it manufactures. Twenty boxes of 50 fuses were randomly selected and the following number of defective fuses were noted for each box:

3, 5, 10, 12, 0, 6, 17, 1, 0, 7, 3, 15, 21, 9, 13, 24, 12, 10, 6, 16

a.   Construct a frequency distribution.

b.   Construct a cumulative relative frequency distribution.

c.   Make several statements about the number of defectives usually found in a box of 50 fuses.

2.52   Custom House Products has seven stores. The following list indicates the total yearly sales for each store (in thousands of dollars):

STORE	SALES	STORE	SALES
1	60.5	5	88.7
2	70.3	6	142.6
3	44.6	7	104.2
4	59.8		

Draw a pie chart to represent the data.

2.53   The following records give the number of workers absent each day for a 30-day period in a steel factory:

10, 5, 2, 13, 17, 3, 16, 5, 7, 10, 3, 19, 22, 14, 11, 6, 9, 18, 23, 14, 7, 8, 20, 17, 13, 7, 24, 2, 6, 15

a.   Construct a frequency distribution for the data.

b.   Assume that the total work force is 100 people. If you were a supervisor, would you inform management of an absentee problem? If so, what would you say?

2.54   In a survey by U.S. News and World Report, 1200 adults were asked how many of their dinners during a typical week were eaten at home and prepared mostly from scratch using fresh ingredients.

NUMBER OF DINNERS PER WEEK	PERCENT OF ADULTS PREPARING THIS NUMBER OF DINNERS FROM SCRATCH
0	5%
1	5%
2	6%
3	10%
4	13%
5	20%
6	15%
7	25%

(*Source:* "A Taste for Real Meals," *U.S. News and World Report,* Jan. 15, 1990, p. 67)

a. Using intervals of less than .5, and .5 and less than 1.5, and so on, construct a relative frequency histogram.

b. Construct an ogive. Make a statement that summarizes the shape of the distribution of the data.

2.55 To improve the service of Excel Hotel, management collected complaints from customers over a one-year time frame. The table below lists the complaints.

TYPE OF COMPLAINT	NUMBER OF CUSTOMERS WITH THIS COMPLAINT
Room not ready on time	40
Room not clean enough	80
Items missing in the room	10
Noise from other rooms	100
Problems with room service	5
Bell boys not helpful enough	30
Problems with final bill	10
Ice machines not working properly	5
Other	20

a. Draw a Pareto Diagram (see Exercise 2.34) to summarize the data.

b. Draw a pie chart to summarize the data.

c. What insight does management gain from the use of the charts in part a and b?

2.56 Delays per 1000 takeoffs and landings at 10 busy U.S. airports are given below. Construct a Pareto chart (see Exercise 2.34). Do a few of the airports account for an unusually large percentage of the delays?

CITY	DELAYS
Atlanta	25
Boston	31
Chicago (O'Hare)	88
Denver	27
Kennedy	76
La Guardia	115
Newark	98
Philadelphia	24
San Francisco	68
St. Louis	27

(*Source:* "Why Plane Travel Isn't Always Fast," *The New York Times,* January 21, 1990, p. 1)

2.57 The Franklin Growth Fund was invested in various sectors of the U.S. economy on September 30, 1990. The following data show the value of the fund's allocation to each sector.

SECTOR OF ECONOMY	VALUE
Aerospace/defense	4,270,625
Automobiles/auto parts	3,805,000
Chemicals/specialty	7,018,750
Communications	10,945,375
Containers	3,628,125
Data processing and services	14,805,000
Diversified manufacturers	7,034,475
Electronics and electrical equipment	11,717,000
Employment and education	1,294,000
Energy	11,308,638
Environmental protection and purification	10,725,901
Health care and cosmetics	34,476,100
Transportation	14,029,375
Other investments, including cash	34,462,421

(*Source:* Adapted from the annual report of the Franklin Custodian Funds, September 1990, pp. 13–15)

a.   Summarize the values of the above investments in a pie chart.

b.   What percentage of the total value of the portfolio does the value of the six largest dollar investment sectors represent?

2.58   The data below show the annual compound growth rate of various countries along the Pacific rim in the agricultural and manufacturing areas over the period 1970–1984.

	COMPOUND GROWTH RATE PER ANNUM	
COUNTRY	**Agriculture**	**Manufacturing**
Australia	+ .5	+ .47
Canada	+ 2.6	+ .94
Hong Kong	− 3.6	+ 4.8
Indonesia	− 3.4	+ 8.4
Japan	+ 3.4	+10.3
South Korea	− 3.8	+ 6.4
Malaysia	+ 6.6	+16.5
Mexico	−14.3	−10.4
Philippines	− 3.6	− .3
Singapore	+ 4.6	−11.6
Thailand	+ 3.1	+ 7.9
United States	+ .34	+ 3.3

(*Source:* Jan Kolm, "Regional and National Consequences of Globalizing Industries of the Pacific Rim," *Technological Forecasting and Social Change* 35 (1989), pp. 63–91).

Draw a bar chart for the compound growth rate of agriculture and a separate bar chart for the compound growth rate of manufacturing. Compare these two graphs.

2.59   The population of 24 metropolitan areas of Canada in 1986 are as follows in units of 1000.

METROPOLITAN AREA	1986	METROPOLITAN AREA	1986
1. Calgary, AB	671	13. Saskatoon, SK	200
2. Chicoutimi-Jonquiere, QC	158	14. Sherbrooke, QC	129
3. Edmonton, AB	785	15. St. Catharines-Niagara, ON	343
4. Halifax, NS	295	16. St. John's, NF	161
5. Hamilton, ON	557	17. Sudbury, ON	148
6. Kitchener, ON	311	18. Thunder Bay, ON	122
7. London, ON	342	19. Toronto, ON	3,427
8. Montreal, QC	2,921	20. Trois-Rivieres, QC	128
9. Oshawa, ON	203	21. Vancouver, BC	1,380
10. Quebec, QC	603	22. Victoria, BC	255
11. Regina, SK	186	23. Windsor, ON	253
12. St. John, NB	121	24. Winnipeg, MB	625

(*Source: Canadian Almanac and Directory,* 1989, p. 75. Toronto: Canadian Almanac and Directory Publishing Company Limited.)

a. Construct a frequency histogram using a statistical package.

b. Interpret the distribution. In your judgment, what percentage of the metropolitan areas are large?

2.60 The average 1988 automobile insurance premiums for the 50 states and Washington, D.C. are given below:

STATE	PREMIUM	STATE	PREMIUM	STATE	PREMIUM
Alabama	278.33	Louisiana	490.50	Ohio	376.82
Alaska	576.25	Maine	435.20	Oklahoma	444.73
Arizona	580.46	Oregon	444.48	Maryland	604.41
Arkansas	613.58	Massachusetts	834.76	Pennsylvania	620.33
California	673.18	Michigan	509.33	Rhode Island	604.28
Colorado	474.46	Minnesota	469.60	South Carolina	526.75
Connecticut	560.27	Mississippi	360.28	South Dakota	324.90
Delaware	581.45	Missouri	473.76	Tennessee	338.46
Florida	462.66	Montana	405.86	Texas	494.66
Georgia	529.75	Nebraska	367.02	Utah	436.10
Hawaii	551.59	Nevada	691.05	Vermont	452.03
Idaho	358.95	New Hampshire	516.16	Virginia	469.54
Illinois	448.00	New Jersey	733.66	Washington	455.25
Indiana	414.42	New Mexico	439.45	Washington, D.C.	606.39
Iowa	292.51	New York	601.84	West Virginia	494.06
Kansas	379.89	North Carolina	445.19	Wisconsin	421.15
Kentucky	431.73	North Dakota	343.85	Wyoming	359.53

(*Source:* "Unprotected on the Road: Uninsured Motorists," *The New York Times,* September 3, 1990, p. 10)

a. Construct a frequency distribution of car insurance premiums for the 50 states and Washington, D.C.

b. Construct a relative frequency distribution of car insurance premiums for the 50 states and Washington, D.C.

c. Approximately what percentage of the states have average car insurance premiums of $500 or less?

2.61 A manufacturing process has produced 25 ball bearings during one shift. The spec limits for the ball bearings are 35.0 ± .2 centimeters. A histogram of the data is given below. This solid bar histogram illustrates another graphics capability of MINITAB. What interpretation can one give to the pattern observed? Would a shift in the process help reduce the number of nonconforming ball bearings?

34.81	34.80	35.29	35.14	34.92
34.86	35.08	34.75	34.99	35.01
34.91	35.29	35.21	34.82	35.20
34.99	35.12	34.91	35.22	34.80
35.03	35.19	35.02	35.10	35.11

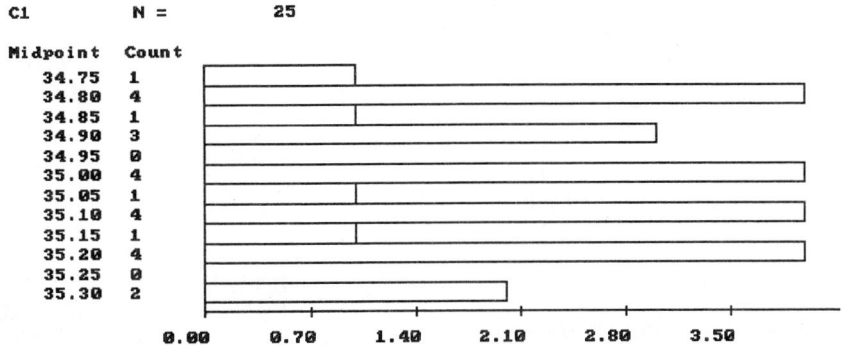

2.62 The vice president of the Association for Manufacturing Excellence wished to determine the use of computerized manufacturing planning and control systems using just-in-time (JIT) manufacturing methods. A survey was sent to 100 factories in the industrialized northeastern part of the United States. Under PRINT C1 in the following MINITAB output, the percentage of parts manufactured with the use of the JIT method at each factory is given. Next, a frequency histogram of the values under PRINT C1 is given; the first class interval midpoint is at 20. What additional insights does one gain about the frequency distribution from the second frequency histogram, in which the first class interval midpoint is at 15?

```
MTB > PRINT C1
C1
   65    78    74    33    81    55    49    77    82    39    71    25    42
   61    50    79    56    71    20    73    88    43    31    73    79    62
   45    73    86    68    28    49    57    81    78    43    75    68    71
   60    23    70    57    42    51    78    62    15    38    28    73    61
   59    33    76    83    19    68    54    58    38    67    73    27    74
   52    63    75    78    63    79    30    47    72    61    73    53    48
   70    58    56    32    74    84    21    92    86    38    58    44    70
   57    68    63    71    61    53    77    25    75

MTB > HISTOGRAM OF C1, FIRST MIDPOINT AT 20, CLASS WIDTH IS 10

Histogram of C1    N = 100

Midpoint    Count
    20.0        5    *****
    30.0       10    **********
    40.0        9    *********
    50.0       11    ***********
    60.0       20    ********************
    70.0       23    ***********************
    80.0       18    ******************
    90.0        4    ****

MTB > HISTOGRAM OF C1, FIRST MIDPOINT AT 15, CLASS WIDTH IS 10

Histogram of C1    N = 100

Midpoint    Count
    15.0        2    **
    25.0        8    ********
    35.0        9    *********
    45.0       10    **********
    55.0       16    ****************
    65.0       16    ****************
    75.0       30    ******************************
    85.0        8    ********
    95.0        1    *
```

Computer Exercises Using the Database

Exercise 1 -- Appendix I   Select 50 observations at random from the database. Using a convenient statistical package, construct a frequency histogram on the variable HPAYRENT (house payment or house/apartment rent). Using this same set of observations, construct separate frequency histograms on variable HPAYRENT for those who own their residence, and for those who rent their residence. Comment on the shapes of the frequency histograms.

Exercise 2 -- Appendix I   Choose at random 30 observations from families living in the

NE sector and then choose another 30 observations at random from families living in the SW sector. Using a convenient statistical computer package, construct frequency histograms on variable INCOME1 (income of principal wage earner) for each group of 30 observations and comment on the frequency distribution of each.

Exercise 3 -- Appendix J   Select 100 observations at random from the database. Construct a frequency distribution and a histogram of the values of the variable ASSETS (current assets).

Exercise 4 -- Appendix J   Repeat Exercise 3 using the variable LIABIL (current liability).

 Case Study

A Look at Some Airline Performance Data

Air traffic has increased dramatically in the wake of airline deregulation, the passenger total was about 465 million in 1990, and is expected to exceed 650

million by 1997. Almost everybody is interested in flights being on time. Many a weary tale of woe could be told about flights delayed, connections missed, or baggage lost.

According to the Federal Aviation Administration (FAA), at least 60% of flight delays are caused by bad weather. Overcrowding of airspace, with flights bunching together at peak hours, contributes to delays. The National Transportation Safety Board (NTSB) has expressed criticism about a shortage of air traffic controllers. The U.S. Department of Transportation (DOT) has looked into allegations that some airlines publish deceptive or unrealistic schedules to gain an advantage in computer reservation systems. Finally, a wave of airline mergers has disrupted organizations and affected the quality of service.

In response to all this, the U.S. Congress is try-ing to get airlines to make public their on-time, baggage handling, and other performance statistics so that consumers can make informed decisions about choosing airlines. Consider the following airline performance data. Table 2.9 shows the percentage of on-time arrivals for selected airlines at the 28 largest airports in the USA. Table 2.10 shows a comparison of passenger reports of mishandled baggage against airlines for April 1991 and April 1990, both as absolute numbers and per 100,000 passengers.

■ TABLE 2.9   Percentage of On-Time Arrivals by Airline at the 28 Largest Airports.

	AMERICAN	CONTI-NENTAL	DELTA	NORTH-WEST	WORLD	UNITED	USAIR	ARRIVALS
Atlanta	89.2	91.2	85.3	85.9	80.8	66.6	83.8	85.1
Boston	79.3	75.6	73.7	72.8	57.8	66.4	79.8	76.4
Charlotte	92.9	NS	84.4	NS	81.7	54.3	86.5	86.1
Chicago	84.0	74.2	73.1	77.0	78.2	69.8	84.3	75.8
Denver	83.1	75.5	70.7	76.7	62.1	70.2	67.5	72.7
Dallas/Fort Worth	90.9	86.4	82.1	82.2	76.1	69.0	74.7	86.4
Detroit	91.6	87.0	82.9	91.0	78.9	71.0	86.6	88.8
Houston	89.0	84.0	82.6	82.5	NS	60.6	87.9	83.8
Kansas City	88.4	83.3	90.5	88.3	75.9	69.7	84.3	83.1
Las Vegas	69.4	70.3	60.5	59.0	47.7	47.8	82.1	69.9
Los Angeles	78.2	53.2	66.0	66.8	44.5	60.5	71.6	65.3
Newark	83.0	77.4	71.5	78.6	68.4	61.7	75.4	75.1
NY-JFK	79.4	NS	60.2	73.3	72.5	73.1	72.2	73.9
NY-La Guardia	86.0	87.3	84.3	86.0	76.2	68.2	86.2	84.5
Memphis	90.4	NS	83.7	87.1	NS	72.1	78.7	86.1
Miami	89.3	84.2	81.0	80.5	75.1	76.0	86.9	84.2
Minneapolis/St. Paul	90.4	86.3	83.1	85.5	77.7	72.9	77.1	84.5
Orlando	90.0	80.0	84.1	80.9	82.6	73.2	85.4	83.0
Philadelphia	87.8	88.6	80.4	86.3	68.4	72.8	81.3	81.6
Phoenix	74.4	63.3	66.1	68.9	57.8	59.4	78.8	67.9
Pittsburgh	85.1	83.4	85.2	87.0	78.7	54.9	85.6	84.9
St. Louis	83.1	72.6	75.8	78.4	77.5	64.7	76.5	77.1
Salt Lake	75.9	80.2	84.2	76.1	59.2	49.3	NS	81.5
San Diego	73.8	50.2	56.4	69.5	49.1	58.0	83.3	65.6
San Francisco	63.8	54.1	51.2	53.2	23.1	55.4	61.4	55.4
Seattle	87.8	74.1	72.8	85.9	75.4	73.0	86.7	79.9
Tampa	89.5	88.4	87.2	81.5	79.6	72.0	84.9	84.2
Washington National	85.4	89.2	81.6	84.0	72.6	65.2	92.5	86.6

NS: Airline does not serve this city

■ TABLE 2.10
Mishandled Baggage
Reports Filed by
Passengers.

Airline	APRIL 1991		APRIL 1990	
	Total	Reports Per 100,000 Passengers	Total	Reports Per 100,000 Passengers
Southwest	7,734	4.03	6,841	3.75
Pan American	3,086	4.46	3,937	5.21
American	26,134	4.55	33,486	5.77
Alaska	2,038	4.58	2,398	6.05
Northwest	14,588	4.61	15,462	5.32
USAir	24,227	4.77	36,438	6.55
Continental	14,508	4.91	15,216	5.46
Trans World	8,290	5.26	10,192	5.48
United	26,799	5.80	23,559	5.40
Delta	37,672	6.20	32,893	6.18
America West	9,717	6.53	7,917	6.29

(*Source*: Air Travel Consumer Report, US Department of Transportation, Office of Consumer Affairs, May, 1991 (Table 2.9) and June, 1991 (Table 2.10))

Case Study Questions

1. From Table 2.9, use only the 28 observations in the last column (on-time arrivals for *all* airlines) to construct a frequency distribution. Next, construct another frequency distribution using all of the values in the first seven data columns. (There are 28 observations for each airline, except where the airline does not serve that city.)

2. Prepare frequency histograms and frequency polygons from each of the above two frequency distributions. Does a comparison of these graphs give you any indication about whether the on-time performance of the second group (selected airlines) is similar to the overall performance of all airlines? Note that the number of data items in the first distribution is smaller. Does that make a difference?

3. Now prepare *relative* frequency histograms and *relative* frequency polygons, and address the same issue raised in Question 2.

4. For a graphical comparison of the on-time arrivals of the seven selected airlines at Chicago versus Los Angeles, which of the following would be appropriate for this purpose: histograms, bar charts or pie charts?

5. From Table 2.10, use the absolute number of passenger reports of mishandled baggage (under the column heading of "Total") to prepare two pie charts, one each for 1990 and 1991, showing the share of reports for each airline. What would the pie charts lead you to believe about the relative rankings of American Airlines and Trans World Airlines with respect to passenger baggage reports?

6. Consider now the columns showing passenger baggage reports per 100,000 passengers. If pie charts were prepared using these figures, what happens to the relative rankings for the above two airlines? Which of the pie charts are deceptive—these or the ones from question 5? Explain where the deception lies.

SPSS

Example
Constructing a
Histogram

You can use SPSS to construct a histogram of the prices of television sets and VCRs purchased from the Q-Mart department store. The SPSS program listing in Figure 2.21 requests a histogram of the data in Table 2.1. As you can see, it is similar to the procedures in the SPSS and SPSS/PC appendices at the end of the text. In this problem the SPSS commands are the same for both the mainframe and PC versions. (Remember to end each command line with a period when using the PC version.)

The TITLE command names the SPSS run.

The DATA LIST command gives each variable a name and describes the data as being in free form.

The BEGIN DATA command indicates to SPSS that the input data immediately follow.

The next ten lines contain the data values, which are the prices of the items in the sample.

The END DATA statement indicates the end of the data entry.

The FREQUENCIES statement specifies the variable from which we wish to produce a histogram, with the HISTOGRAM statement generating the actual graph.

Figure 2.22 shows the SPSS output obtained by executing the listing in Figure 2.21.

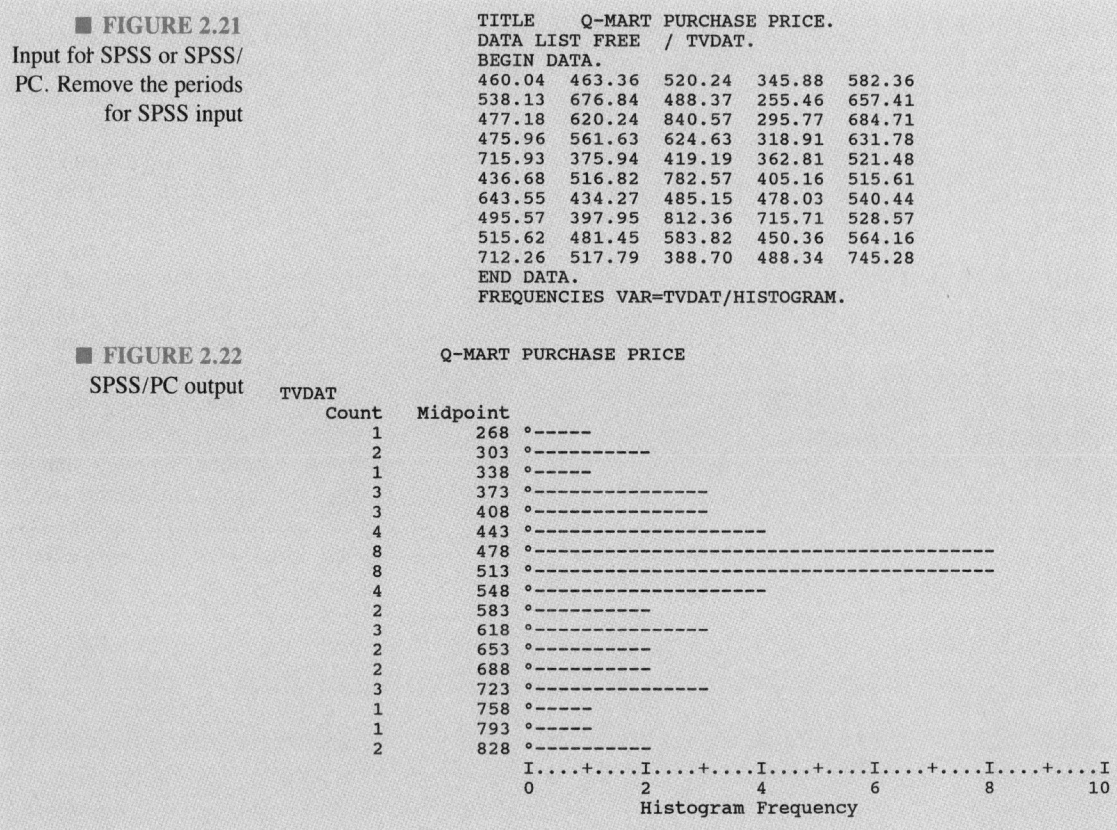

■ **FIGURE 2.21**
Input for SPSS or SPSS/
PC. Remove the periods
for SPSS input

```
TITLE        Q-MART PURCHASE PRICE.
DATA LIST FREE    / TVDAT.
BEGIN DATA.
460.04   463.36   520.24   345.88   582.36
538.13   676.84   488.37   255.46   657.41
477.18   620.24   840.57   295.77   684.71
475.96   561.63   624.63   318.91   631.78
715.93   375.94   419.19   362.81   521.48
436.68   516.82   782.57   405.16   515.61
643.55   434.27   485.15   478.03   540.44
495.57   397.95   812.36   715.71   528.57
515.62   481.45   583.82   450.36   564.16
712.26   517.79   388.70   488.34   745.28
END DATA.
FREQUENCIES VAR=TVDAT/HISTOGRAM.
```

■ **FIGURE 2.22**
SPSS/PC output

```
                    Q-MART PURCHASE PRICE
TVDAT
     Count  Midpoint
         1       268  °------
         2       303  °-----------
         1       338  °------
         3       373  °----------------
         3       408  °----------------
         4       443  °---------------------
         8       478  °------------------------------------------------
         8       513  °------------------------------------------------
         4       548  °---------------------
         2       583  °-----------
         3       618  °----------------
         2       653  °-----------
         2       688  °-----------
         3       723  °----------------
         1       758  °------
         1       793  °------
         2       828  °-----------
              I....+....I....+....I....+....I....+....I....+....I
              0         2         4         6         8        10
                            Histogram Frequency
```

SAS

✔ **Example**
Constructing a
Histogram

You can use SAS to construct a histogram using the prices of television sets and VCRs purchased from the Q-Mart department store. The SAS program listing in Figure 2.23 requests a histogram of the data in Table 2.1. As you can see, it is similar to the procedures in the SAS and SAS/PC appendices at the end of the text. In this problem the SAS commands are the same for both the mainframe and PC versions.

The TITLE command names the SAS run (enclose in single quotes).

The DATA command gives the data a name.

The INPUT command names and gives the correct order for the different fields on the data lines.

The CARDS command indicates to SAS that the input data immediately follow.

The next 50 lines contain the data. The first line, for example, represents the price of the first item in the sample. The remaining lines indicate the prices of the other 49 items.

The PROC CHART command requests a SAS procedure to print a histogram. VBAR PRICE generates a histogram of the variable PRICE. The resulting output contains the class midpoints and frequencies.

Figure 2.24 shows the SAS output obtained by executing the listing in Figure 2.23.

■ **FIGURE 2.23**
Input for SAS
(mainframe or
micro version)

```
TITLE  'Q-MART PURCHASE PRICES';
DATA TVDAT;
INPUT PRICE;
CARDS;
460.04
538.13
477.18
      .
      .
      .
528.57
564.16
745.28
PROC CHART;
 VBAR PRICE;
```

■ **FIGURE 2.24**   SAS/PC output

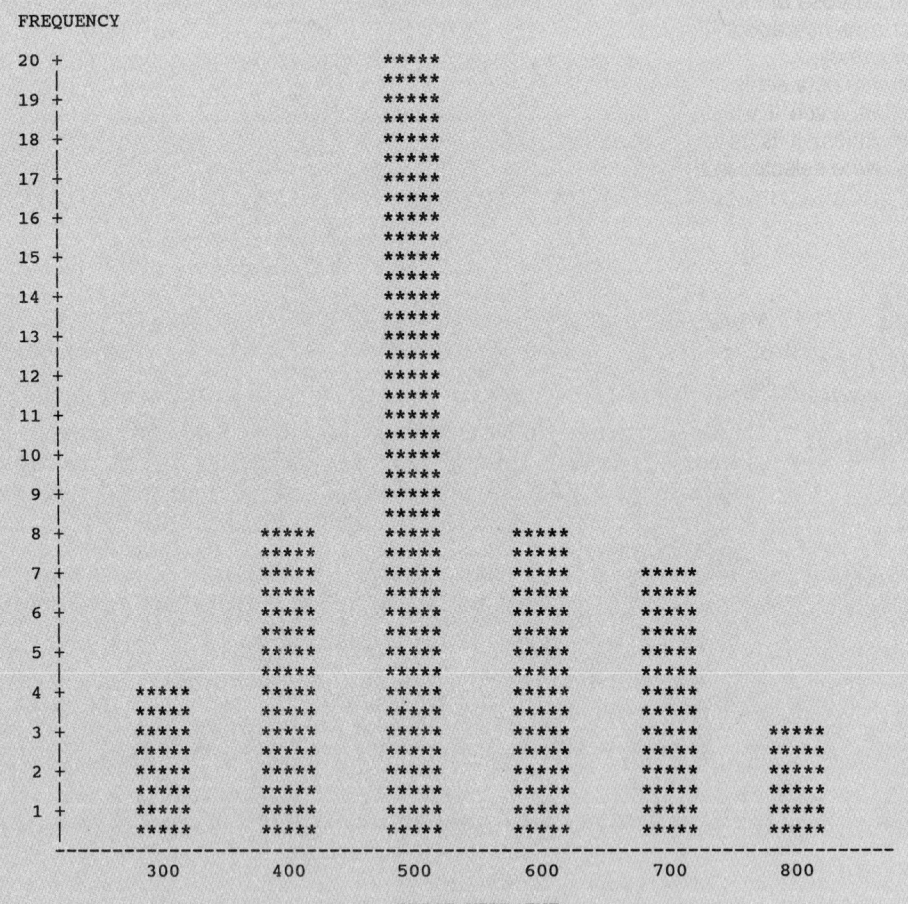

```
                        Q-MART PURCHASE PRICES

                        FREQUENCY OF PRICE

  FREQUENCY
     20 +                          *****
        |                          *****
     19 +                          *****
        |                          *****
     18 +                          *****
        |                          *****
     17 +                          *****
        |                          *****
     16 +                          *****
        |                          *****
     15 +                          *****
        |                          *****
     14 +                          *****
        |                          *****
     13 +                          *****
        |                          *****
     12 +                          *****
        |                          *****
     11 +                          *****
        |                          *****
     10 +                          *****
        |                          *****
      9 +                          *****
        |                          *****
      8 +              *****        *****        *****
        |              *****        *****        *****
      7 +              *****        *****        *****        *****
        |              *****        *****        *****        *****
      6 +              *****        *****        *****        *****
        |              *****        *****        *****        *****
      5 +              *****        *****        *****        *****
        |              *****        *****        *****        *****
      4 +     *****    *****        *****        *****        *****
        |     *****    *****        *****        *****        *****
      3 +     *****    *****        *****        *****        *****        *****
        |     *****    *****        *****        *****        *****        *****
      2 +     *****    *****        *****        *****        *****        *****
        |     *****    *****        *****        *****        *****        *****
      1 +     *****    *****        *****        *****        *****        *****
        |     *****    *****        *****        *****        *****        *****
        ----------------------------------------------------------------------
            300      400         500         600         700         800

                            PRICE MIDPOINT
```

Descriptive Measures

The first two chapters focused on different types of numerical sample data and methods of summarizing and presenting data. A frequency distribution is used to condense data from a sample into groups (called classes). Different types of statistical graphs can be used to illustrate sample data in different ways. The types of graphs we have discussed so far include the histogram, bar chart, ogive, frequency polygon, and pie chart. The purpose of these graphs is to convey information at a glance about the distribution of the values in your sample.

Every sample data set is a small part of a much larger population. Even if we don't always mention the word population, it is always there, since the sample values were selected from this group of inter-est. Every population has properties (called parameters) that describe it. By collecting a set of sample data, we can then estimate these properties by computing statistics and making graphs.

We have seen how to reduce a set of sample data to a graph. It also is helpful to reduce data to one or more numbers (such as an average). Such a number is called a descriptive measure. Because this number is derived from a sample, it also can be called a sample statistic. In this chapter we discuss the commonly used descriptive measures and explain what you can expect to learn from each one. In later chapters we discuss how you can use many of these sample statistics to estimate the corresponding population parameters.

3.1
Various Types of Descriptive Measures

A **descriptive measure** is a *single* number that provides information about a set of sample data. The class of descriptive measures described here consists of four types. Which one you select depends on what you want to measure. These types are:

1. *Measures of central tendency.* These answer the questions, where is the "middle" of my data? and, which data value occurs most often?
2. *Measures of dispersion.* These answer the questions, how spread out are my data values? and, how much do the data values jump around?
3. *Measures of position.* These answer the questions, how does my value (score on an exam, for example) compare with all the others? and, which data value was exceeded by 75% of the data values? by 50%? by 25%?
4. *Measures of shape.* These answer the questions, are my data values symmetric? and, if not symmetric, just how nonsymmetric (skewed) are the data?

3.2
Measures of Central Tendency

The purpose of a **measure of central tendency** is to determine the "center" of your data values or possibly the "most typical" data value. *Some measures of central tendency are the mean, median, midrange, and mode.* We will illustrate each of these measures using as data the number of accidents (monthly) reported over a particular 5-month period:

accident data: 6, 9, 7, 23, 5

The Mean

The **mean** is the most popular measure of central tendency. It is merely the average of the data. The mean is easy to obtain and explain, and it has several mathematical properties that make it more advantageous to use than the other three measures of central tendency.

Business managers often use a mean to represent a set of values. They select one value as typical of the whole set of values, such as average sales, average price, average salary, or average production per hour. In economics, the term *per capita* is a measure of central tendency. The income per capita of a certain district, the number of clothes washers per capita, and the number of televisions per capita are all examples of a mean.

The sample mean, $\bar{x}$ (read "x bar"), is equal to the sum of the data values divided by the number of data values. For the accident data set,

$$\bar{x} = \frac{6 + 9 + 7 + 23 + 5}{5} = 10.0$$

In general, let an arbitrary data set be represented as:

$$x_1, x_2, x_3, \ldots, x_n$$

where n is the number of data values. (In the accident data set, $x_1 = 6$, $x_2 = 9$, $x_3 = 7$, $x_4 = 23$, $x_5 = 5$, and n is 5.) Then,

$$\bar{x} = \frac{x_1 + x_2 + \cdots + x_n}{n} = \frac{\Sigma x}{n} \qquad (3.1)$$

The symbol Σ (sigma) means "the sum of." In this case, the sample mean, $\bar{x}$, is the sum of the x values divided by n.* When dealing with discrete data, the sample mean is very often *not* an integer (such as 10, here) and should *not* be rounded to an integer. For example, remove the last value (5) from the accident data set. The sample mean is now

$$\bar{x} = \frac{6 + 9 + 7 + 23}{4} = 11.25$$

In subsequent chapters, we will be concerned with the mean of the *population*. The symbol for the population mean is μ (mu). For a population consisting of N elements, denoted by

$$x_1, x_2, x_3, \ldots, x_N$$

the population mean is defined to be

$$\mu = \frac{x_1 + x_2 + \cdots + x_N}{N} = \frac{\Sigma x}{N} \qquad (3.2)$$

---

*In another application of this symbol, we square each of the sample values and sum these values. For the accident data, this operation would be written as

$$\Sigma x^2 = 5^2 + 6^2 + 7^2 + 9^2 + 23^2$$
$$= 25 + 36 + 49 + 81 + 529$$
$$= 720$$

For these data, then, $\Sigma x = 50$ and $\Sigma x^2 = 720$.

- Population: $x_1, x_2, \ldots, x_N$
- Population mean $= \mu = \dfrac{x_1 + x_2 + \cdots + x_N}{N} = \dfrac{\Sigma x}{N}$

- Sample values (selected from the population): $x_1, x_2, \ldots, x_n$, where $n \leq N$
- Sample mean $= \bar{x} = \dfrac{x_1 + x_2 + \cdots + x_n}{n} = \dfrac{\Sigma x}{n}$

The Median

The **median** of a set of data is the value in the center of the data values when they are arranged from smallest to largest. Consequently, it is in the center of the ordered array.

Using the accident data set, the median **Md** is found by first constructing an ordered array:

$$5, 6, \mathbf{7}, 9, 23$$

The value that has an equal number of items to the right and the left is the median. Thus, Md = 7.

In general, if n is *odd,* Md is the center data value of the ordered set:

$$\text{Md} = \left(\frac{n + 1}{2}\right)\text{st ordered value}$$

Here, the median is the $(5 + 1)/2 = $ 3rd value in the ordered array. Note that for these data, the *position* of the median is 3, and the *value* of the median is 7. If n is *even,* Md is the average of the two center values of the ordered set. Thus, the median of the array 3, 8, 12, 14 is $(8 + 12)/2 = 10.0$.

In our accident data set, one of the five values (23) is much larger than the remaining values—it is an outlier. Notice that the median (Md = 7) was much less affected by this value than was the mean ($\bar{x} = 10$). *When dealing with data that are likely to contain outliers (for example, personal incomes or prices of residential housing), the median usually is preferred to the mean as a measure of central tendency, since the median provides a more "typical" or "representative" value for these situations.*

Finally, note that newspaper and magazine articles often refer to the mean and the median as an "average" value. Care must be taken not to always interpret this word as representing the sample mean unless this is specified in the discussion.

The Midrange

Although less popular than the mean and median, the **midrange (Mr)** provides an easy-to-grasp measure of central tendency. Notice that it also is severely affected (even more than $\bar{x}$) by the presence of an outlier in the data. In general:

$$\text{Mr} = \frac{(\text{smallest value}) + (\text{largest value})}{2} \tag{3.3}$$

Using the accident data set,

$$\text{Mr} = \frac{5 + 23}{2} = 14.0$$

Compare this to $\bar{x} = 10$ and Md = 7.

The Mode

The **mode (Mo)** of a data set is the value that occurs more than once and the most often. The mode is not always a measure of central tendency; this value need not occur in the "center" of your data. One situation in which the mode is the value of interest is the manufacturing of clothing. The *most common* hat size is what you would like to know, not the *average* hat size. Can you think of other applications where the mode would provide useful information?

Note that there is no mode for our accident data set because all values occur only once. Instead, consider the data set

$$4, 8, 7, 6, 9, 8, 10, 5, 8$$

Mo = 8 (occurs three times).

There may be more than one mode if several numbers occur the same (and the largest) number of times.

EXAMPLE 3.1   A sample of ten was taken to determine the typical completion time (in months) for the construction of a particular model of Brockwood Homes:

$$4.1, 3.2, 2.8, 2.6, 3.7, 3.1, 9.4, 2.5, 3.5, 3.8$$

We find the average completion time as follows:

$$\bar{x} = \frac{4.1 + 3.2 + \cdots + 3.8}{10} = \frac{38.7}{10} = 3.87 \text{ months}$$

Notice that there is an outlier in the data, namely, 9.4 months. To be safe, you should double-check this figure to make sure that it is, in fact, correct, that is, that there was no mistake in recording or transcribing this value. In the presence of one or two outliers, the median generally provides a more reliable measure of central tendency, so we construct an ordered array:

$$2.5, 2.6, 2.8, 3.1, \mathbf{3.2}, \mathbf{3.5}, 3.7, 3.8, 4.1, 9.4$$

Consequently,

$$\text{Md} = \frac{3.2 + 3.5}{2} = 3.35 \text{ months}$$

Also, the midrange is given by

$$\text{Mr} = \frac{2.5 + 9.4}{2} = 5.95 \text{ months}$$

This value is severely affected by the presence of the outlier; the midrange value of nearly 6 months is a poor measure of central tendency for this application.

Finally, no mode exists because there are no repeats in the data values. These results are summarized in the graph in Figure 3.1, a **dot array diagram.** Each data value is represented as a dot on the horizontal line.

■ **FIGURE 3.1**
Dot array diagram of the measure of central tendency for a sample of ten housing construction times. See text for explanation.

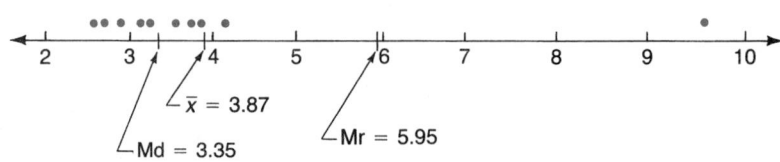

EXAMPLE 3.2

Example 2.1 concerned the inside diameter of 100 machined parts. These 100 (ordered) values (in millimeters) are repeated below. Determine the sample mean, median, midrange, and mode. Also, the machined part is supposed to have an inside diameter of 10.2 millimeters, with specification limits of 10.1 millimeters to 10.3 millimeters. Comment on this process.

10.216	10.221	10.226	10.228	10.230	10.230	10.231	10.231
10.234	10.237	10.239	10.239	10.240	10.240	10.244	10.245
10.249	10.252	10.253	10.254	10.254	10.255	10.255	10.256
10.257	10.258	10.259	10.260	10.262	10.264	10.264	10.266
10.266	10.267	10.267	10.269	10.269	10.270	10.270	10.271
10.271	10.271	10.271	10.271	10.271	10.272	10.274	10.274
10.274	10.276	10.276	10.277	10.277	10.278	10.278	10.279
10.279	10.279	10.280	10.280	10.281	10.281	10.281	10.282
10.283	10.284	10.285	10.286	10.289	10.290	10.290	10.291
10.291	10.291	10.293	10.293	10.293	10.293	10.294	10.296
10.298	10.300	10.300	10.301	10.304	10.309	10.310	10.311
10.311	10.311	10.312	10.312	10.314	10.315	10.315	10.318
10.326	10.328	10.333	10.338				

SOLUTION

The sample mean is found by first finding the total of these 100 values. This is $10.216 + 10.221 + \cdots + 10.338 = 1027.5$. Consequently, the sample mean is

$$\bar{x} = \frac{1027.5}{100} = 10.275$$

The sample median will be the average of the 50th (that is, 100/2) and the 51st value. Since the values are ordered here, these values can be obtained from the preceding array. Both the 50th and 51st values are 10.276, and so the median will be

$$Md = \frac{10.276 + 10.276}{2} = 10.276$$

The sample midrange is the average of the lowest (L) and highest (H) sample values. Here, $L = 10.216$ (the first ordered sample value) and $H = 10.338$ (the last ordered sample value), and so

$$Mr = \frac{10.216 + 10.338}{2} = 10.277$$

The sample mode is that data value occurring the most often. This value is $Mo = 10.271$, which occurs six times in the sample.

This machined part is supposed to have an inside diameter of 10.2 mm. The specification limits were from 10.1 millimeters to 10.3 millimeters, that is, these parts are of acceptable quality only if the diameter is between these two values. All the measures of central tendency here are much larger than 10.2 millimeters (very nearly 10.3 millimeters) and so we come to the same conclusion as in Example 2.1: the process is "off center" and needs adjustment.   ∎

Exercises   **3.1**   The following data were collected.

15, 18, 12, 16, 10, 12

Calculate the mean, median, mode, and midrange for this data set.

3.2   A quality control inspector has selected ten thin sheets of aluminum for inspection. The thickness of these sheets should be 1.5 millimeter, with specification limits of 1.4 and 1.6.

Calculate the mean and median of the data below. Does the process appear to be "off center"?

$$1.5, 1.5, 1.7, 1.7, 1.8, 1.7, 1.4, 1.9, 1.8, 1.6$$

3.3 The appraised values of ten newly built homes in a suburb of Kansas City are given below in units of thousands of dollars:

$$95, 99, 110, 90, 95, 99, 115, 99, 125, 92$$

a. Calculate the mean, median, mode, and midrange of this data set.

b. Now remove the value of 125 and recalculate the statistics in part a using the nine remaining values.

3.4 A manufacturing process involves the manufacture of ball bearings that have a diameter of 20.5 millimeters, with specification limits of 20.4 to 20.6 millimeters. The data below were sampled from the process. Use the various measures of central tendency to determine if the process is "off center."

$$20.4, \ 20.6, \ 20.5, \ 20.3, \ 20.2, \ 20.5, \ 20.4,$$
$$20.2, \ 20.3, \ 20.4, \ 20.1, \ 20.3, \ 20.3, \ 20.2$$

3.5 The distribution of income for full-time women workers in the United States in 1988 was as follows:

EARNINGS	NUMBER OF WOMEN WORKERS (IN THOUSANDS)
$ 2,999 or less	572
$ 3,000 to $ 4,999	350
$ 5,000 to $ 6,999	1,059
$ 7,000 to $ 9,999	2,754
$10,000 to $14,999	7,111
$15,000 to $19,999	6,601
$20,000 to $24,999	4,685
$25,000 and over	6,676

(*Source: Information Please Almanac, 1990: The New Universe of Information*, p. 64.)

Estimate the median. Give an interpretation of the value of the median.

3.6 The vacancy rates for rental housing units across the United States are as follows for the time period 1980–89.

YEAR	NATIONAL VACANCY RATE (PERCENT)	YEAR	NATIONAL VACANCY RATE (PERCENT)
1980	5.4	1985	6.5
1981	5.0	1986	7.3
1982	5.3	1987	7.7
1983	5.7	1988	7.7
1984	5.9	1989	7.1

(*Source: Economic Indicators*, August 1990, p. 19.)

Compute the mean, median, and mode for the vacancy rates for this time period. Does the mode give a "typical" value?

3.7 The following data set is a "symmetrical" set of numbers.

$$35 \quad 38 \quad 40 \quad 45 \quad 50 \quad 55 \quad 60 \quad 62 \quad 65$$

a. Compute the mean and the median for this set of data.

b. Do you think the mean and median will be approximately equal for any nearly "symmetrical" set of numbers?

3.8 Give an appropriate measure of central tendency for the following data on the monthly commissions (in hundreds of dollars) of eight salespersons:

$$0.5, 12.3, 15.9, 16.1, 16.2, 16.3, 16.4, 18.6$$

3.9 Compute the mean, median, and mode for the daily advertising expenses of a car dealer using the following data, which give the expenses in dollars for 20 days. Which measure of central tendency is most appropriate? Why?

38, 60, 20, 130, 55, 150, 47, 35, 86, 95, 31, 46, 112, 130, 55, 42, 130, 35, 60, 130

3.10 New issues of three-month U.S. Treasury securities had the following yields during 1989 and 1990.

PERIOD	YIELD (IN PERCENT)	PERIOD	YIELD (IN PERCENT)
August, 1989	7.91	February, 1990	7.76
September, 1989	7.72	March, 1990	7.87
October, 1989	7.63	April, 1990	7.78
November, 1989	7.65	May, 1990	7.78
December, 1989	7.64	June, 1990	7.74
January, 1990	7.64	July, 1990	7.66

(*Source: Economic Indicators,* August 1990, p. 30.)

Calculate the mean and median of the yields. Interpret the meaning of these two statistics.

3.11 Outstanding debt of third world nations totaled over 1.3 trillion at the end of 1988, according to a survey by the World Bank. The following were the leading debtor nations in 1988.

COUNTRY	DEBT (IN BILLIONS OF DOLLARS)	COUNTRY	DEBT (IN BILLIONS OF DOLLARS)
Brazil	120	Morocco	22
Mexico	107	Chile	21
Argentina	60	Peru	19
Venezuela	35	Colombia	17
Nigeria	31	Cote D'Ivoire	14
Philippines	30	Ecuador	11
Yugoslavia	22		

(*Source: World Almanac and Book of Facts,* 1990, p. 75.)

a. Calculate the mean and median of the debt of the 13 countries.

b. Omit Brazil's debt and recalculate the mean and median of the remaining 12 nations.

c. Which value, the mean or median, is affected more by the omission of Brazil's debt? If Ecuador's debt was omitted, which value, the mean or median, would be affected the most?

◢ 3.3
Measures of Dispersion

A measure of central tendency, such as the mean, is certainly useful. However, the use of any single value to describe a complete distribution fails to reveal important facts.

The more homogeneous a set of data is, the better the mean will represent a "typical" value. **Dispersion** is the tendency of data values to scatter about the mean, $\bar{x}$. If all the data values in a sample are identical, then the mean provides perfect information, and the dispersion is zero. This is rarely the case, however, so we need a measure of this dispersion that will increase as the scatter of the data values about $\bar{x}$ increases.

Knowledge of dispersion can sometimes be used to control the variability of data values in the future. Industrial production operations maintain quality control by observing and measuring the dispersion of the units produced. If there is too much variation in the production process, the causes are determined and corrected using an inspection control procedure.

Some measures of dispersion are the **range, mean absolute deviation, vari-**

ance, standard deviation, and **coefficient of variation.** To illustrate the various dispersion measures, we will use the accident data from the previous section: 6, 9, 7, 23, 5.

The Range

The simplest measure of dispersion is the **range** of the data, which is the numerical difference between the largest value and the smallest value. For the accident data,

$$\text{range} = 23 - 5 = 18$$

The range is a rather crude measure of dispersion, but it is an easy number to calculate and contains valuable information for many situations. Stock reports generally give prices in terms of their ranges, citing the high and low prices of the day. The value of the range is strongly influenced by an outlier in the sample data.

Mean Absolute Deviation (MAD)

The purpose of a measure of dispersion is to determine the variability of data. The more variation there is in the data, the larger this measure should become. Take a look at the accident data illustrated in Figure 3.2. To measure the variation about the sample mean, $\bar{x}$, consider the distance from each data value to $\bar{x}$ (that is, $x - \bar{x}$) and its absolute value:

DATA VALUE (x)	$x - \bar{x}$	$\lvert x - \bar{x} \rvert$
5	-5	5
6	-4	4
7	-3	3
9	-1	1
23	13	13
	$\Sigma (x - \bar{x}) = 0$	$\Sigma \lvert x - \bar{x} \rvert = 26$

As a possible measure, consider the average of the $(x - \bar{x})$ values:

$$\frac{\Sigma (x - \bar{x})}{5} = \frac{0}{5} = 0$$

This value is *always* zero for any set of data because the positive deviations from the sample mean always balance out the negative ones. To overcome this, use the actual distance from each data value to the sample mean, paying no attention to the side of the mean on which it lies, by taking the **absolute value** of each deviation. The **mean absolute deviation (MAD)** is the average of these distances.

$$\text{MAD} = \frac{\Sigma \lvert x - \bar{x} \rvert}{n} \tag{3.4}$$

Using the accident data,

$$\text{MAD} = \frac{5 + 4 + 3 + 1 + 13}{5} = \frac{26}{5} = 5.2$$

that is, the average distance from each point to the mean is 5.2 (accidents).
What is the MAD for these data without the value 23?

$$\bar{x} = \frac{5 + 6 + 7 + 9}{4} = 6.75$$

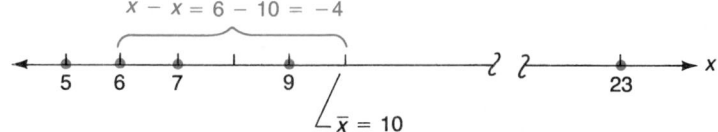

■ FIGURE 3.2
This presentation of the accident data shows their variation.

and so

$$\text{MAD} = \frac{|5 - 6.75| + |6 - 6.75| + |7 - 6.75| + |9 - 6.75|}{4}$$

$$= \frac{1.75 + .75 + .25 + 2.25}{4} = 1.25$$

Now the MAD is much lower, indicating that the smaller data set has much less variation than does the one containing the outlier.

The Variance and Standard Deviation

By far the most widely used measures of dispersion are the **variance** and **standard deviation.** They resemble the MAD in that they are based on the deviations of all the values from the sample mean, $\bar{x}$. The problem encountered earlier in examining the sum of each $(x - \bar{x})$ was that the negative deviations balanced out the positive ones. The MAD handled this situation by taking the absolute value of each deviation. Another possibility is to *square* each of these deviations, thereby removing all the negative signs. Using our example data and recalling that $\bar{x} = 10$, we have

DATA VALUE (x)	$(x - \bar{x})$	$(x - \bar{x})^2$
5	-5	25
6	-4	16
7	-3	9
9	-1	1
23	13	169
	$\Sigma (x - \bar{x}) = 0$	$\Sigma (x - \bar{x})^2 = 220$

So, $\Sigma (x - \bar{x})^2 = 220$.

The obvious thing to do next would be to find the average of these squared deviations:

$$(1/n) \cdot \Sigma (x - \bar{x})^2$$

One use of this particular statistic in subsequent chapters is as an *estimator.* In particular, we will need to estimate the variation within an entire population, using sample data collected from the population. However, a better estimator is obtained by dividing the sum of the squared deviations by $n - 1$ rather than by n. This leads to the **sample variance, s^2.** In general,

Definition

$$s^2 = \frac{\Sigma (x - \bar{x})^2}{n - 1}$$

(3.5)

Using the accident data,

$$s^2 = \frac{220}{5 - 1} = \frac{220}{4} = 55.0$$

The square root of the variance is referred to as the **sample standard deviation, s.** In general,

$$s = \sqrt{\frac{\Sigma (x - \bar{x})^2}{n - 1}} \qquad (3.6)$$

Using the accident data,

$$s = \sqrt{55.0} = 7.416$$

As previously mentioned, the sample variance, s^2, is used to estimate the variance of the entire population. The symbol for the population variance is σ^2 (read as sigma squared). For a population consisting of N elements,

$$x_1, x_2, x_3, \ldots, x_N$$

the population variance is defined to be

$$\sigma^2 = \frac{\Sigma (x - \mu)^2}{N} \qquad (3.7)$$

where μ is the population mean, defined in equation 3.2.

As we saw, the *population* variance can be obtained by dividing the sum of the squared deviations about μ by the population size N. The *sample* variance is calculated by dividing the sum of the squared deviations about $\bar{x}$ by the sample size (n) minus one. Had we chosen to divide by n rather than by $n - 1$, the resulting estimator would (on the average) underestimate σ^2. For this reason, we use $n - 1$ in the denominator of s^2.

- Population: $x_1, x_2, \ldots, x_N$
- Population variance $= \sigma^2 = \dfrac{(x_1 - \mu)^2 + \cdots + (x_N - \mu)^2}{N}$

$$= \frac{\Sigma (x - \mu)^2}{N}$$

- Population standard deviation $= \sigma = \sqrt{\dfrac{\Sigma (x - \mu)^2}{N}}$

- Sample values (selected from the population): $x_1, x_2, \ldots, x_n$, where $n \leq N$
- Sample variance $= s^2 = \dfrac{(x_1 - \bar{x})^2 + \cdots + (x_n - \bar{x})^2}{n - 1}$

$$= \frac{\Sigma (x - \bar{x})^2}{n - 1}$$

- Sample standard deviation $= s = \sqrt{\dfrac{\Sigma (x - \bar{x})^2}{n - 1}}$

Now consider what the units of measurement are for s and s^2. The units of s are the same as the units on the data. If the data are measured in pounds, the units of s are pounds. Consequently, the units of the variance, s^2, would be (pounds)²—a rather difficult unit to grasp at best.

For the accident data,

$$s = 7.416 \text{ accidents}$$

$$s^2 = 55 \text{ (accidents)}^2$$

For this reason, s (rather than s^2) is typically the preferred measure of dispersion.

There is another way to compute the sample variance. Using equation 3.5 to compute the value of s^2 may have appeared easy enough, but the computation was helped in part by the fact that the sample mean, $\bar{x}$, was an integer (10). When $\bar{x}$ is not an integer, it is easier to find s^2 using

---

Computing Formula for s²

$$s^2 = \frac{\Sigma x^2 - (\Sigma x)^2/n}{n - 1} \tag{3.8}$$

As before, the standard deviation is the square root of the variance. To illustrate the use of equation 3.8, consider the accident data:

x	x^2
5	25
6	36
7	49
9	81
23	529
50	720

So, $n = 5$, $\Sigma x = 50$, and $\Sigma x^2 = 720$. Consequently, using equation 3.8:

$$s^2 = \frac{720 - (50)^2/5}{5 - 1}$$

$$= \frac{720 - 500}{4} = 55.0 \quad \text{(as before)}$$

Also

$$s = \sqrt{55.0} = 7.416 \quad \text{(as before)}$$

Finally, you may wish to interpret the magnitude of the value of s or s^2—that is, whether your value of s (or s^2) is large. This is difficult to determine because the values of s and s^2 depend on the magnitude of the data values. In other words, large data values generally lead to large values of s. For example, which of the following two data sets exhibits more variation?

data set 1:   5, 6, 7, 9, 23   (accident reports)

data set 2:   5000, 6000, 7000, 9000, 23,000

As we have already seen, for data set 1, $\bar{x} = 10.0$ and $s = 7.416$. For data set 2, $\bar{x} = 10,000$ and $s = 7416$ (we will discuss this later).

Do these results mean that data set 2 has a great deal more variation, given that its standard deviation is 1000 times that of data set 1? Another look at the values reveals that the large value of s for data set 2 is due to the large values within this set. In fact, considering the size of the numbers within each data set, the *relative* variation within each group of values is the same. So comparing the standard deviations or variances of two data sets is not a good idea unless you know that their mean values $(\bar{x})$ are approximately equal. The next section deals with another statistical measure that will allow you to compare the relative variation within two data sets.

The Coefficient of Variation

Consider again our two data sets, which appear to have the same variation (relative to the size of the data values) yet have vastly different standard deviations:

data set 1:   5, 6, 7, 9, 23 $(\bar{x} = 10, s = 7.416)$

data set 2:   5000, 6000, 7000, 9000, 23,000 $(\bar{x} = 10,000, s = 7416)$

To compare their variation, we need a measure of dispersion that will produce the same value for both of them. The solution here is to measure the standard deviation in terms of the mean; that is, what percentage of $\bar{x}$ is s? This measure of dispersion is the **coefficient of variation, CV.** In general, for samples containing nonnegative values,

$$CV = \frac{s}{\bar{x}} \cdot 100 \qquad\qquad (3.9)$$

For our example data sets:

data set 1: $CV = \dfrac{7.416}{10} \cdot 100 = 74.16$

data set 2: $CV = \dfrac{7,416}{10,000} \cdot 100 = 74.16$

So our conclusion here is that both data sets exhibit the same relative variation; s is 74.16% of the mean for both sets. As a final word here, we must point out that for data sets with *extreme* variation, it is possible to obtain a coefficient of variation larger than 100%.

EXAMPLE 3.3

To review the various measures of dispersion, let's use the data on housing construction time in Example 3.1.

completion time: 4.1, 3.2, 2.8, 2.6, 3.7, 3.1, 9.4, 2.5, 3.5, 3.8 (months)

First, compute the range:

(largest value) − (smallest value) = 9.4 − 2.5 = 6.9 months

To determine the MAD, recall that $\bar{x}$ is 3.87 months:

$$MAD = (1/10)(|4.1 - 3.87| + |3.2 - 3.87| + \cdots + |3.8 - 3.87|)$$
$$= (1/10)(11.52) = 1.152 \text{ months}$$

Now find the variance and the standard deviation:

$$\Sigma x = 4.1 + 3.2 + \cdots + 3.8 = 38.7$$

and

$$\Sigma x^2 = (4.1)^2 + (3.2)^2 + \cdots + (3.8)^2 = 186.25$$

Hence,

$$s^2 = \frac{186.25 - (38.7)^2/10}{10 - 1}$$

$$= \frac{186.25 - 149.77}{9} = 4.05 \text{ (months)}^2$$

and

$$s = \sqrt{4.05} = 2.01 \text{ months}$$

To calculate the coefficient of variation, use the previously obtained values of s and $\bar{x}$, where

$$CV = \frac{2.01}{3.87} \cdot 100 = 51.9$$

The standard deviation is 51.9% of the sample mean.

EXAMPLE 3.4

Using the 100 machine part diameters in Example 3.2, determine the range, mean absolute deviation, variance, standard deviation, and coefficient of variation.

SOLUTION

The range is found by subtracting the largest value (10.338) from the smallest value (10.216); that is, the range is .122 millimeter. The mean absolute deviation (MAD) is calculated by first subtracting the mean (10.275) from each value, taking the absolute value of each difference, and summing the results:

$$|10.216 - 10.275| + |10.221 - 10.275| + \cdots + |10.338 - 10.275|$$

$$= .059 + .054 + \cdots + .063$$

$$= 2.095$$

Consequently, MAD $= 2.095/100 = .02095$ millimeter and so the average distance from the mean is .02095 millimeter.

The variance is derived by first squaring each of the preceding differences and summing the results; that is

$$(10.216 - 10.275)^2 + (10.221 - 10.275)^2 + \cdots + (10.338 - 10.275)^2$$

$$= (-.059)^2 + (-.054)^2 + \cdots + (.063)^2$$

$$= .07049$$

The value of s^2 is found by dividing this result by one less than the sample size, that is

$$s^2 = .07049/99 = .000712 \text{ mm}^2$$

As a result, the sample standard deviation is

$$s = \sqrt{.000712} = .0267 \text{ mm}$$

Finally, the coefficient of variation is

$$CV = \frac{s}{\bar{x}} \cdot 100 = \frac{.0267}{10.275} \cdot 100 = .26$$

and so the sample standard deviation is .26% of the sample mean.

So far, you can reduce a set of sample data to a number that indicates a typical or average value (a measure of central tendency) or one that describes the amount of variation within the data values (a measure of dispersion). The next section examines yet another set of statistics—measures of position.

Exercises   3.12   The following data were collected.

$$2, 1, 5, 9, 8$$

Calculate the:

a. Range
b. Mean absolute deviation
c. Variance

d. Standard deviation
e. Coefficient of variation

3.13   The following are the scores made on an aptitude test by a group of job applicants:

53, 55, 43, 14, 64, 39, 65, 22, 17, 74, 36, 24, 13, 28, 40, 96, 92, 32, 92, 36, 18, 100, 84, 65

Calculate the:

a. Range
b. Mean absolute deviation
c. Variance

d. Standard deviation
e. Coefficient of variation

3.14   The following percentage increases/decreases in daily circulation for the 12-month period ending March 31, 1989, were reported by the following major daily U.S. newspapers.

NEWSPAPER	PERCENT CHANGE IN CIRCULATION
Wall Street Journal	− 4.6
Los Angeles Times	− 1.2
New York Times	3.6
Washington Post	.3
Chicago Tribune	− 4.4
Newsday	4.9
Detroit News	− 1.6
Chicago Sun-Times	−11.3
Boston Globe	2.3
Miami Herald	2.8
Houston Chronicle	3.9

(*Source: The World Almanac and Book of Facts*, 1990, p. 365.)

a.   Calculate the mean percentage change in circulation.

b.   Recalculate the mean percentage change in circulation, but omit the percentage change for the Chicago Sun-Times.

c.   Calculate the standard deviation for the percentage change in circulation.

d.   Recalculate the standard deviation of the percentage change in circulation, but omit the Chicago Sun-Times.

e.   Compare the answers in parts (a) and (b) and in parts (c) and (d) and comment on the differences.

3.15   When making capital-investment decisions, firms frequently consider the dispersion of the estimated future cash flows. Usually, the project with lesser dispersion is preferred to the one with more dispersion. Given the estimated cash flows (after tax) for projects A and B, which one would you prefer? Why?

MONTH	PROJECT A CASH FLOW	PROJECT B CASH FLOW
January	4,000	700
February	7,200	1,100
March	8,800	600
April	2,400	1,300
May	7,400	800
June	4,100	650
July	10,800	710
August	9,100	450
September	2,000	580
October	14,000	640
November	7,700	330
December	3,900	210

3.16 The following values represent the number of years of experience of 12 auditors in a small accounting firm.

$$2.3, 4.1, 7.3, 5.2, 5.8, 3.6, 11.6, 5.0, 4.7, 3.2, 5.5, 4.5$$

a. Calculate the standard deviation.

b. Which one value in the data set, if omitted, would change the value of the standard deviation the most?

c. Which one value in the data set, if omitted, would change the value of the standard deviation the least?

d. Compare your answers in parts (b) and (c) and comment on their values.

3.17 Show that

$$s^2 = \frac{\Sigma(x - \bar{x})^2}{n - 1}$$

is equivalent to

$$s^2 = \frac{\Sigma x^2 - (\Sigma x)^2/n}{n - 1}$$

3.18 If a data set consists of five observations and if the variance of the observations in the data set is zero, what can we say about the value of each of the five observations?

3.19 The game of cricket, which originated in England, is popular in Australia, India, and the West Indies. The following are the runs (strikes) scored by two players, A and B, in various innings:

PLAYER A	PLAYER B
47	66
0	10
14	11
33	22
101	88
68	32
87	40
14	38
22	18
46	41
Total 432	366

a. Who is the better player? On what basis?

b. Who is more consistent? Why?

3.20 The yield on a stock is equal to the dividend per share/market price × 100. It is one of the popular measures used by individual investors when they make investment decisions. The following are the yield figures (in percent) for two stocks, A and B, for the past 10 years:

STOCK A	STOCK B	STOCK A	STOCK B
7	12	18	9
9	13	7	18
14	13	11	12
13	17	12	17
11	11	10	13

Which is a more stable stock in terms of yields? Why?

3.21 Calculate the variance for the daily advertising expense of the car dealer in Exercise 3.9.

3.4
Measures of Position

Suppose that you think you are drastically underpaid compared with other people with similar experience and performance. One way to attack the problem is to obtain the salaries of these other employees and demonstrate that *comparatively* you are way down the list. To evaluate your salary compared with the entire group, you

■ **TABLE 3.1**
Ordered array of
aptitude test scores for
50 applicants ($\bar{x} =$
60.36, $s = 18.61$).

22	44	56	68	78
25	44	57	68	78
28	46	59	69	80
31	48	60	71	82
34	49	61	72	83
35	51	63	72	85
39	53	63	74	88
39	53	63	75	90
40	55	65	75	92
42	55	66	76	96

would use a measure of position. **Measures of position** are indicators of how a particular value fits in with all the other data values. Two commonly used measures of position are (1) a percentile (and quartile), and (2) a Z score.

To illustrate these measures, we suppose that the personnel manager of Texon Industries has administered an aptitude test to 50 applicants. The ordered data are shown in Table 3.1. The mean of the data is $\bar{x} = 60.36$, and the standard deviation is $s = 18.61$. Ms. Jenson received the score of 83. She wishes to measure her performance in relation to all the applicant scores. We will return to this illustration in Example 3.5.

Percentiles

A **percentile** is the most common measure of position. The value of, for example, the 35th percentile is essentially the value that exceeds 35% of all the data values. More precisely, the 35th percentile is that value (say, P_{35}) such that at most 35% of the data values are less than P_{35} and at most 65% of the data values are greater than P_{35}. We will use the Texon Industries applicant data to determine the 35th percentile. Which data value is 35% of the way between the smallest and largest value? Here the number of data values is $n = 50$ and the percentile is $P = 35$. We define the *position* of the 35th percentile as follows:

$$n \cdot \frac{P}{100} = 50 \cdot .35 = 17.5$$

To satisfy the more precise definition of a percentile, whenever $n \cdot P/100$ is *not* a counting number, it should be rounded *up* to the next counting number. So, 17.5 is rounded up to 18, and the 35th percentile is the 18th value *of the ordered values*. Referring to Table 3.1, the 35th percentile is $P_{35} = 53$.

In general, to find the **location** of the Pth percentile, determine $n \cdot P/100$ and use one of the following two location rules.

Location rule 1. If $n \cdot P/100$ *is not* a counting number, round it *up*, and the Pth percentile will be the value in this position of the ordered data.

Location rule 2. If $n \cdot P/100$ *is* a counting number, the Pth percentile is the average of the number in this location (of the ordered data) and the number in the next largest location.

Now we can use the applicant data to determine the 40th percentile. Here $n \cdot P/100 = (50)(.4) = 20$. Then, using the second rule,

$$P_{40} = \text{40th percentile} = \frac{(\text{20th value}) + (\text{21st value})}{2}$$

$$= \frac{55 + 56}{2} = 55.5$$

Notice here that the 40th percentile is *not* one of the data values but is an average of two of them. Now work out the 50th percentile yourself. What measure of central tendency uses the same procedure? From our previous discussion, you should realize that *the 50th percentile is the median.*

EXAMPLE 3.5

Recall that Ms. Jenson received a score of 83. What is her percentile value?

SOLUTION Her value is the 45th largest value (out of a total of 50). An initial guess of her percentile would be:

$$P = \frac{45}{50} \cdot 100 = 90$$

However, due to the percentile rules used here, this guess may be slightly incorrect. Your next step should be to examine this value of P, along with the next two smaller values. The following calculations of $P = 88$, $P = 89$, and $P = 90$ reveal that Ms. Jenson's score is the 89th percentile.

P	$n \cdot P/100$	Pth PERCENTILE
88	$50 \cdot .88 = 44$	$(82 + 83)/2 = 82.5$
89	$50 \cdot .89 = 44.5$	45th value = 83
90	$50 \cdot .90 = 45$	$(83 + 85)/2 = 84$

EXAMPLE 3.6

What is the 50th percentile for the applicant data in Table 3.1?

SOLUTION Here, $n \cdot P/100 = 50 \cdot .5 = 25$. The 50th percentile is an average of the 25th and 26th ordered data values:

$$P_{50} = \text{50th percentile} = \frac{61 + 63}{2} = 62$$

Quartiles

Quartiles are merely particular percentiles that divide the data into quarters, namely:

$$Q_1 = \text{1st quartile} = \text{25th percentile } (P_{25})$$
$$Q_2 = \text{2nd quartile} = \text{50th percentile} = \text{median } (P_{50})$$
$$Q_3 = \text{3rd quartile} = \text{75th percentile } (P_{75})$$

They are used as benchmarks, much like the use of A, B, C, D, and F on examination grades. Using the applicant data in Table 3.1, we can determine the first quartile by first calculating:

$$n \cdot \frac{P}{100} = 50 \cdot .25 = 12.5$$

This result is rounded up to 13, and $Q_1 = $ 13th ordered value = 46.

$$Q_2 = \text{median} = 62$$

from Example 3.6. Finally,

$$n \cdot \frac{P}{100} = 50 \cdot .75 = 37.5$$

This is rounded up to 38, and $Q_3 = $ 38th ordered value = 75.

Another measure commonly used in conjunction with quartiles is the **interquartile range (IQR)**, defined as

$$IQR = Q_3 - Q_1$$

In the applicant data, the interquartile range is

$$IQR = 75 - 46 = 29$$

Consequently, the middle 50% of the data are between 46 and 75.

Strictly speaking, the interquartile range is a measure of dispersion since it can be expected to increase as the data become more "spread out." It is not a commonly used measure of dispersion, although it is certainly easy to compute (much like the range of a sample data set). Its primary disadvantage is that it measures the spread within the middle of the data, not within the entire data set. The interquartile range can be illustrated in a simple graph called a box and whisker plot, discussed in section 3.8.

Z Scores

Another measure of position is a sample **Z score**, which is based on the mean ($\bar{x}$) and standard deviation of the data set. Like a percentile, a Z score determines the relative position of any particular data value x; it is expressed in terms of the number of standard deviations above or below the mean. The Z score of x is defined as

$$Z = \frac{x - \bar{x}}{s} \tag{3.10}$$

Recall from Example 3.5 that Ms. Jenson had a score of 83 on the test. For this data set, $\bar{x} = 60.36$ and $s = 18.61$. Her score of 83 is in the 89th percentile. The corresponding Z score is

$$Z = \frac{83 - 60.36}{18.61} = 1.22$$

This Z score means that Ms. Jenson's score of 83 is 1.22 standard deviations to the *right* of the mean, or above the group's average. Thus, if Z is positive, it indicates how many standard deviations x is to the right of the mean.

A negative value implies that x is to the *left* of the mean. Again referring to Table 3.1, what is the Z score for the individual who obtained a total of 35 on the aptitude examination?

$$Z = \frac{35 - 60.36}{18.61} = -1.36$$

This individual's score is 1.36 standard deviations to the left of the mean, or below the group's average.

EXAMPLE 3.7

Example 3.2 contained data listing 100 measurements for the inside diameter of a certain machined part. The specification limits for this part are 10.1 millimeters (the lower spec limit, or LSL) and 10.3 millimeter (the upper spec limit, or USL). Any part falling outside this range is said to be nonconforming and is not of acceptable quality. Of interest here is the following question: What is the Z score for both the LSL and the USL and which Z score has the smaller absolute value?

■ FIGURE 3.3
MINITAB procedure for describing sample data.

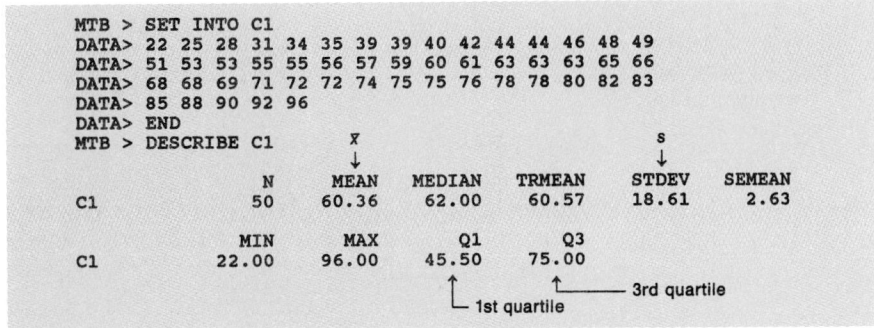

SOLUTION

The Z score for each of these limits is found by subtracting the sample mean and dividing by the standard deviation. In Examples 3.2 and 3.4, the sample mean was found to be $\bar{x} = 10.275$ and the sample standard deviation was $s = .0267$. The Z score for the USL (10.3) is

$$\frac{10.3 - 10.275}{.0267} = .94$$

and the Z score for the LSL is

$$\frac{10.1 - 10.275}{.0267} = -6.55$$

The absolute values of these two Z scores are .94 and 6.55. The minimum of these two absolute values is .94, indicating that the nearer spec limit is the upper spec limit and that the sample mean is .94 standard deviations away from this limit. *A very general rule here is that in order to consistently produce products of acceptable quality, the minimum absolute value should be at least three.* In this case, the product is not capable of meeting these specifications—a result consistent with Examples 3.2 and 3.4. ■

When you derive sample statistics, you have essentially two options: use a calculator or use a computer. Calculators work well for small data sets but involve too much time (and opportunity for error) for moderate or large sample sizes. Practically all statistical computer packages will provide you with the basic sample statistics (mean, median, variance, and so on) in response to only a few commands once the data have been read in. Figure 3.3 contains the MINITAB commands (along with the output) necessary to derive the basic statistics for the data in Table 3.1. The appendices at the end of the chapter demonstrate this procedure using SPSS and SAS.

▟ 3.5
Measures of Shape

A basic question in many applications is whether your data exhibit a **symmetric** pattern. **Measures of shape** determine skewness and kurtosis.

Skewness

The histogram in Figure 3.4 demonstrates a perfectly symmetric distribution. When the data are symmetric, the sample mean, $\bar{x}$, the sample median, Md, and the sample mode, Mo, are the same. As the data tend toward a nonsymmetric distribution, referred to as **skewed,** the mean and median drift apart. The easiest method of determining the degree of skewness present in your sample data is to calculate a

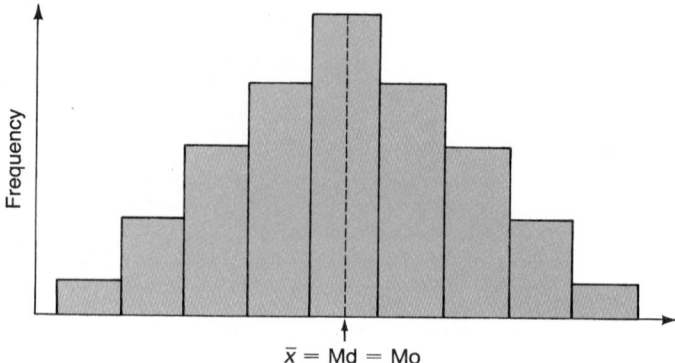

$$\bar{x} = \text{Md} = \text{Mo}$$

measure referred to as the **Pearsonian coefficient of skewness, Sk.** Its value is given by:

$$\text{Sk} = \frac{3(\bar{x} - \text{Md})}{s} \tag{3.11}$$

where s is the standard deviation of the sample data.

The value of Sk ranges from -3 to 3.* If the data are perfectly symmetric (a rare event), Sk = 0, because $\bar{x}$ = Md. For Figure 3.4, Sk is zero. If Sk is positive, the mean is larger than the median, and we say that the data are *skewed right*. This merely means the data exhibit a pattern with a right tail, as illustrated in Figure 3.5. We know the mean is affected by extreme values, so we would expect the mean to move toward the right tail, above the median, resulting in a positive value of Sk. Similarly, if Sk is negative, the data are *skewed left* and the mean is smaller than the median. Figure 3.6 shows a data distribution exhibiting a left tail and negative skew.

Using the aptitude examination scores in Table 3.1, we have $\bar{x}$ = 60.36, s = 18.61, and Md = 62.

$$\text{Sk} = \frac{3(60.36 - 62)}{18.61} = -.26$$

Consequently, a histogram of these data should be just slightly skewed left.

Kurtosis

Sk measures the tendency of a distribution to stretch out in a particular direction. Another measure of shape, referred to as the **kurtosis,** measures the *peakedness* of your distribution. The calculation of this measure is a bit cumbersome, and the kurtosis value is not needed in the remaining text material.[†] Briefly, this value is small if the frequency of observations close to the mean is high and the frequency of observations far from the mean is low.

---

*A recent proof of this statement can be found in Colm Art O'Cinneide, "The Mean is Within One Standard Deviation of Any Median," *The American Statistician* 44, no. 4 (1990), p. 292.

[†]The following texts contain an alternate method of computing sample skewness as well as a procedure for computing the sample kurtosis; L. Ott and D. K. Hildebrand, *Statistical Thinking for Managers,* 2d ed. (Boston: Duxbury Press, 1987); C. L. Olsen, and M. J. Picconi, *Statistics for Business Decision Making* (Glenview, Ill.: Scott, Foresman, 1983).

■ **FIGURE 3.5**
Histogram showing
right (positive) skew.

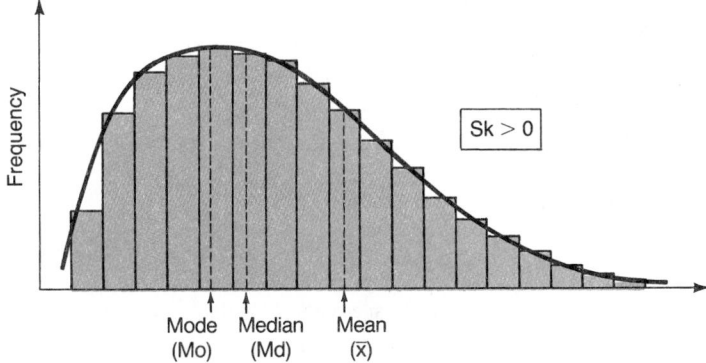

■ **FIGURE 3.6**
Histogram showing left
(negative) skew.

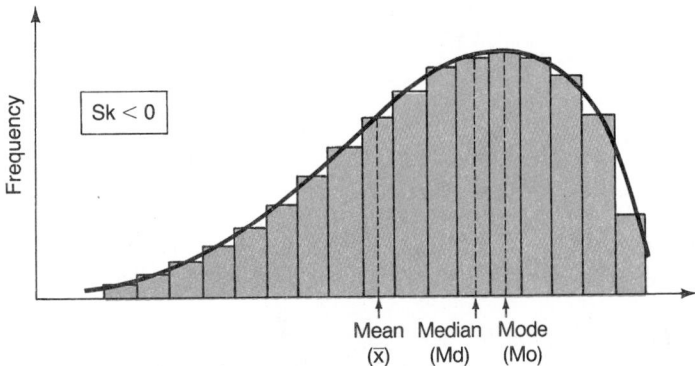

Exercises

3.22 The following table indicates the scores of the students at Hillside College on a statistics examination:

18, 58, 71, 83, 89, 96, 21, 62, 74, 84, 90, 97, 43, 66, 75, 86, 92, 98, 47, 66, 77, 86, 94, 100, 55, 68, 78, 88, 95, 100

a. Calculate the 40th percentile.

b. Calculate the 77th percentile.

c. Interpret the meaning of the numbers calculated for the two percentiles above.

d. Calculate the interquartile range and explain what the value means.

3.23 In Exercise 3.22, how would you evaluate the performance of a student who scored 74 on the exam?

3.24 Using the data in Exercise 3.22, calculate the first and third quartiles. State the result in words.

3.25 Consider the following data:

7, 8, 8, 9, 11, 14, 15, 16, 18, 19, 21, 27, 28, 30, 32, 35

a. Calculate the 90th percentile.

b. Calculate the 58th percentile.

c. Calculate the interquartile range.

3.26 Calculate the 40th percentile and the interquartile range for the following set of numbers.

4, 6, 8, 10, 15, 8, 3, 2

3.27 Assume a particular set of sample data such that

$$\bar{x} = 49$$

$$s = 18$$

Consider one particular value of $x = 63$.

a. Calculate the Z score.

b. Interpret the Z score.

3.28 If the Z score for an observation is 1.50, the standard deviation is 14, and the observation is 32, what is the mean?

3.29 If the mean is 40, $x = 35$, and its Z score is -2, what is the value of the standard deviation?

3.30 If the Z score for an observation is -1.22, $\bar{x} = 83$, and the variance is 84, what is the value of the observation?

3.31 If the Z score is 1.78 and the variance is 64, what is $x - \bar{x}$?

3.32 Assume the following about a set of sample data:

$$s = 14$$
$$\bar{x} = 21$$
$$Md = 18.5$$

a. What do you observe about the pattern of the data?

b. Calculate the coefficient of skewness. What does this value suggest?

3.33 One of the most important indicators of the economic health of a country is the gross national product (GNP). The GNP is the market value of all final goods and services produced by an economy in a given time period. The GNPs (in billions of current dollars) for the years 1980 through 1989 were as follows:

YEAR	GNP	YEAR	GNP
1980	2,732.0	1985	4,014.9
1981	3,052.6	1986	4,231.6
1982	3,166.0	1987	4,515.6
1983	3,405.7	1988	4,873.7
1984	3,772.2	1989	5,200.8

(*Source: Economic Indicators,* August 1990, p. 1.)

a. Calculate the mean, median, variance, and standard deviation of the GNP values.

b. Calculate the coefficient of skewness. Interpret its value.

◣ 3.6
Interpreting $\bar{x}$ and s

Now that you have gone through several pencils determining the sample mean and standard deviation, what can you learn from these values? The type of question that you can answer is, how many of the data values are within two standard deviations of the mean?

Take a look at the aptitude test scores in Table 3.1. Here, $\bar{x} = 60.36$ and $s = 18.61$, and so we obtain

$$\bar{x} - s = 60.36 - 18.61 \qquad \bar{x} + s = 60.36 + 18.61$$
$$= 41.75 \qquad\qquad\qquad = 78.97$$

$$\bar{x} - 2s = 60.36 - 37.22 \qquad \bar{x} + 2s = 60.36 + 37.22$$
$$= 23.14 \qquad\qquad\qquad = 97.58$$

$$\bar{x} - 3s = 60.36 - 55.83 \qquad \bar{x} + 3s = 60.36 + 55.83$$
$$= 4.53 \qquad\qquad\qquad = 116.19$$

Examine these data and observe that (1) 33 out of the 50 values (66%) lie between $\bar{x} - s$ and $\bar{x} + s$; (2) 49 out of the 50 values (98%) lie between $\bar{x} - 2s$ and $\bar{x} + 2s$; and (3) 50 out of the 50 values (100%) lie between $\bar{x} - 3s$ and $\bar{x} + 3s$. Or, put another way: (1) 66% of the data values have a Z score between -1 and 1; (2) 98%

■ **FIGURE 3.7**
A bell-shaped (normal)
population.

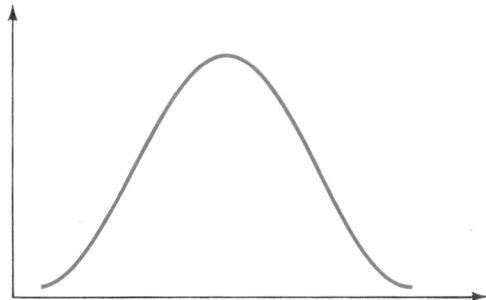

have a Z score between -2 and 2, and (3) 100% have a Z score between -3 and 3.

What can we say in general for any data set? There are two types of statements we can make. One of these, *Chebyshev's inequality,* is usually conservative but makes *no assumption* about the population from which you obtained your data. Following are the components of Chebyshev's inequality.

Chebyshev's Inequality

1. At least 75% of the data values are between $\bar{x}$ and 2s and $\bar{x} + 2s$.
2. At least 89% of the data values are between $\bar{x} - 3s$ and $\bar{x} + 3s$.
3. In general, there are at least $(1 - 1/k^2)$ of your data values between $\bar{x} - ks$ and $\bar{x} + ks$, for $k > 1$.

Note that if $k = 1$, $1 - 1/k^2 = 0$; so Chebyshev's inequality provides no information on the number of data values to expect between $\bar{x} - s$ and $\bar{x} + s$.

The other type of statement is called the **empirical rule.** We make a key assumption here, namely, that the population from which you obtain your sample has a *bell-shaped distribution;* that is, it is symmetric and tapers off smoothly into each tail. Such a population is called a **normal population** and is illustrated in Figure 3.7. Thus, the data set should have a skewness measure, Sk, near zero and a histogram similar to that in Figure 3.4. However, the empirical rule is still quite accurate even if your distribution is not exactly bell-shaped. The empirical rule has three components.

Empirical Rule

Under the assumption of a bell-shaped population:

1. Approximately 68% (large majority) of the data values lie between $\bar{x} - s$ and $\bar{x} + s$.
2. Approximately 95% (19 out of 20) of the data values lie between $\bar{x} - 2s$ and $\bar{x} + 2s$.
3. Approximately 99.7% (nearly all) of the data values lie between $\bar{x} - 3s$ and $\bar{x} + 3s$.

Returning to Table 3.1, we can summarize our previous results along with the information provided by Chebyshev's inequality and the empirical rule. The actual percentages of the sample values in each interval, as well as the percentages specified using each of the two rules, are shown in Table 3.2.

■ TABLE 3.2
Summary of
percentages of sample
values by interval, using
data from Table 3.1.

BETWEEN	ACTUAL PERCENTAGE	CHEBYSHEV INEQUALITY PERCENTAGE	EMPIRICAL RULE PERCENTAGE
$\bar{x} - s$ and $\bar{x} + s$	66% (33 out of 50)	—	≈68%
$\bar{x} - 2s$ and $\bar{x} + 2s$	98% (49 out of 50)	≥75%	≈95%
$\bar{x} - 3s$ and $\bar{x} + 3s$	100% (50 out of 50)	≥89%	≈100%

As you can see, Chebyshev's inequality is very conservative, but it always works. The empirical rule predicted results close to what was observed. This is not surprising, because the skewness measure is only slightly different from zero (Sk = −.26).

EXAMPLE 3.8   In a random sample of 200 automobile insurance claims obtained from Pearson Insurance Company, $\bar{x} = \$615$ and $s = \$135$.

1. What statement can you make using Chebyshev's inequality?
2. If you have reason to believe that the population of all insurance claims is bell-shaped (normal), what does the empirical rule say about these 200 values?

SOLUTION 1   Chebyshev's inequality provides information regarding the number of sample values within a specified number of standard deviations of the mean. For $k = 2$, we have:

$$\bar{x} - 2s = 615 - 2(135) = \$345$$
$$\bar{x} + 2s = 615 + 2(135) = \$885$$

We conclude that at least 75% of the sample values lie between $345 and $885. Because $.75 \cdot 200 = 150$, this implies that at least 150 of the claims are between $345 and $885.

For $k = 3$,

$$\bar{x} - 3s = 615 - 3(135) = \$210$$
$$\bar{x} + 3s = 615 + 3(135) = \$1020$$

and we conclude that at least 8/9 (89%) of the data values are between $210 and $1020. Here, $8/9 \cdot 200 = 177.8$, and so at least 178 of the claims are between $210 and $1020.

SOLUTION 2   If the distribution of automotive claims at Pearson Insurance Company is believed to be bell-shaped, the empirical rule allows us to draw stronger conclusions. In particular, for $k = 1$, we have

$$\bar{x} - s = 615 - 135 = \$480$$
$$\bar{x} + s = 615 + 135 = \$750$$

and we conclude that approximately 68% of the data values ($.68 \cdot 200 = 136$) are between $480 and $750.

For $k = 2$,

$$\bar{x} - 2s = \$345$$
$$\bar{x} + 2s = \$885$$

and we conclude that approximately 95% of the data values ($.95 \cdot 200 = 190$) will lie between $345 and $885.

Exercises

3.34 Use the mean, median, and standard deviation obtained for the data in exercise 3.33. Using Chebyshev's inequality, find the range of GNP that will include 89% of the data.

3.35 Assume that a random sample yielded a sample mean of 60 and a sample standard deviation of 10. Between what two values will at least 75% of the data values lie?

3.36 A random sample of 30 observations from a bell-shaped population yielded a sample mean of 100 and a sample standard deviation of 12. Approximately 99.7% of the data values lie between what two values?

3.37 What do you think will be the value of the coefficient of skewness for a symmetric distribution? Will the mean and median be equal?

3.38 The mean of the wages of a sample of production workers in a company is $18,600, and the standard deviation is $445. Assuming the corresponding population is normally distributed, estimate the range of wages within which about 95% of the wages in the sample are expected to lie.

3.39 A survey was taken to assess the public's risk perceptions of the effects of the Chernobyl nuclear plant accident. Eight items were addressed. Questions were asked regarding degree of concern about health threat, satisfaction with information released about the risk involved, changes in normal daily activities, and other matters related to the eight topics. Respondents used a five-category scale ranging from "not at all" or "very unlikely" to "very great extent" or "very likely." Results of the survey are given below:

ITEM	MEAN	STANDARD DEVIATION
1. Health concern	2.61	1.29
2. Expectation of cancer	1.55	.90
3. Expectation of genetic effects	1.40	.88
4. Satisfaction: releases	3.77	1.05
5. Satisfaction: protective actions	3.16	1.11
6. Media frequency	3.40	.67
7. Change in normal activities	1.44	.84
8. Number of protective actions	.27	.55

(*Source:* "Effects of the Chernobyl Accident on Public Perceptions of Nuclear Accident Risks," *Risk Analysis* 10, no. 3 (1990) pp. 393–399.)

a. Calculate the coefficient of variation for each variable. Do the data sets for each variable exhibit the same relative variation?

b. Use an appropriate rule to calculate ranges within which most of the responses lie for each variable.

3.40 The mean score of students in an accounting class was 78.5, and the standard deviation was 22. Using the empirical rule, estimate the range of scores within which about 95% of the data values are expected to lie.

3.41 If the mean of a sample is 28.2 with a standard deviation of 6.2, what data values would lie within two standard deviations of the mean?

3.42 One hundred people participated in the annual spaghetti eating contest of a small town. The contestants' spaghetti consumption averaged 120 feet with a standard deviation of 14 feet.

a. At least how many contestants ate between 92 and 148 feet of spaghetti?

b. At least how many contestants ate between 78 and 162 feet of spaghetti?

c. Did you have to make any assumptions about the data to answer parts a and b?

3.43 Use the empirical rule to determine if the following data set was generated from a population other than a bell-shaped population.

6, 5, 6, 7, 1, 11, 5, 6, 5, 7

3.44 The average number of visitors to the local museum is 48 per day, and the standard deviation is 22. Using the empirical rule, estimate the range within which about 95% of the data values are expected to lie.

3.45 A shoe salesperson computes that his mean daily sales of shoes is $280 with a standard deviation of $40. Give two bounds within which the salesperson can expect his daily sales to lie at least 75% of the time.

◢ 3.7
Grouped Data

Sometimes we may have to work with data in the form of a frequency distribution, called **grouped data,** when the raw data are not available. This situation can arise when a magazine or newspaper article displays a histogram or frequency distribution but does not include the actual raw data used to construct the histogram. We do not have the data values used to make up this frequency distribution, so we are forced to approximate the sample statistics, in particular the mean, median, and standard deviation.

Approximating the Sample Mean, $\bar{x}$

Assume we obtain the frequency distribution shown in Table 3.3, which contains the ages of 36 individuals who recently passed a CPA examination. The 36 data values are not available, so we cannot add them up. A procedure that works well for estimating $\bar{x}$ is simply to pretend that the 36 data values are equal to their respective class midpoints. Consequently, there are

5 values at $(20 + 30)/2 = 25$
14 values at $(30 + 40)/2 = 35$
$\vdots$
2 values at $(60 + 70)/2 = 65$

We can then estimate the value of $\bar{x}$ ($\cong$ means "is approximately equal to"):

$$\bar{x} \cong \frac{(25 + 25 + 25 + 25 + 25) + \cdots + (65 + 65)}{36}$$

$$= \frac{(5)(25) + (14)(35) + (9)(45) + (6)(55) + (2)(65)}{36}$$

$$= \frac{1480}{36} = 41.1$$

Our estimate of the average age of these 36 individuals is

$$\bar{x} \cong 41.1 \text{ years}$$

In general,

$$\bar{x} \cong \frac{\Sigma f \cdot m}{n} \tag{3.12}$$

where n = sample size, f = frequency of each class, and m = midpoint of each class.

■ TABLE 3.3
Age of 36 individuals who recently passed a CPA examination.

CLASS NUMBER	CLASS (AGE IN YEARS)	FREQUENCY
1	20 and under 30	5
2	30 and under 40	14
3	40 and under 50	9
4	50 and under 60	6
5	60 and under 70	2
		36

Approximating the Sample Standard Deviation, s

Using the same fictitious data set at the various class midpoints, the variance, s^2, can be found in the usual way, using equation 3.8.

$$s^2 = \frac{\Sigma(\text{each data value})^2 - [\Sigma(\text{each data value})]^2/n}{n-1}$$

$$\Sigma(\text{each data value})^2 = \overbrace{(25^2 + 25^2 + \cdots + 25^2)}^{5 \text{ times}}$$

$$+ \overbrace{(35^2 + 35^2 + \cdots + 35^2)}^{14 \text{ times}} + \cdots$$

$$+ (65^2 + 65^2)$$

$$= (5)(25^2) + (14)(35^2) + (9)(45^2) + (6)(55^2) + (2)(65^2)$$

$$= 65,100$$

Also, Σ (each data value) = 1480, as we determined previously when approximating $\bar{x}$.

$$s^2 \cong \frac{65,100 - (1480)^2/36}{35} = \frac{4255.56}{35} = 121.59$$

and

$$s \cong \sqrt{121.59} = 11.03$$

In general,

$$s^2 \cong \frac{\Sigma f \cdot m^2 - (\Sigma f \cdot m)^2/n}{n-1} \tag{3.13}$$

where f, m, and n are as defined in equation 3.12.

The calculations necessary to approximate $\bar{x}$ and s are more easily performed if you construct a table similar to Table 3.4.

Approximating the Sample Median, Md

The sample median can best be approximated by approximating the $(n/2)$th ordered value. Using the previous example, we have

$$\text{Md} \cong \left(\frac{36}{2}\right)\text{th ordered value} = 18\text{th ordered value}$$

Where is this value in the frequency distribution? The first class contains the five smallest values, and the first two classes contain the first 19 ordered values (5 + 14 = 19). So the 18th value is in the second class.

■ TABLE 3.4
Summary of calculations for grouped data.

CLASS NUMBER	CLASS	f	m	f · m	f · m²
1	20 and under 30	5	25	125	3,125
2	30 and under 40	14	35	490	17,150
3	40 and under 50	9	45	405	18,225
4	50 and under 60	6	55	330	18,150
5	60 and under 70	2	65	130	8,450
		36		$\Sigma f \cdot m = 1,480$	$\Sigma f \cdot m^2 = 65,100$

We can better approximate the median by assuming that the values in this class (and all classes) are spread *evenly* between the lower and upper limits. Because the first class contains five values, the median is 13 ($18 - 5$) values into the second class. This class begins at 30, has a width of 10, and has 14 values in it. So we want to go 13 values into a class of width 10 containing 14 values. The resulting estimate of the median is

$$Md \cong 30 + \frac{13}{14}(10) = 39.3$$

In general,

$$Md \cong L + \frac{k}{f} \cdot W \qquad\qquad (3.14)$$

where L = lower limit of the class containing the median (called the **median class**); $k = n/2 -$ (the number of data values preceding the median class); f = frequency of the median class; and W = class width.

In the previous example, $L = 30$, $f = 14$, $W = 10$, and thus

$$k = \frac{36}{2} - 5 = 13$$

If n is odd (say, $n = 25$), then $n/2 = 12.5$, and you need to estimate the 12.5th ordered value, which is halfway between the 12th and the 13th value. The procedure to follow here is exactly the same, except that k will not be a counting number.

Remember that these procedures for approximating the sample statistics are used only when the raw data are not available and your only information is a frequency distribution or corresponding histogram. *If the actual data values are available, these statistics can be determined exactly, and the approximation procedures described in this section should not be used.*

Exercises

3.46 Compute the mean and median for the following grouped data.

CLASS NUMBER	CLASS	FREQUENCY
1	50 and under 60	4
2	60 and under 70	13
3	70 and under 80	11
4	80 and under 90	5
5	90 and under 100	3

3.47 The following information summarizes family income (in dollars) for a specific neighborhood of a major city.

INCOME LEVEL	PERCENT OF FAMILIES
0 and under 10,000	11
10,000 and under 20,000	19
20,000 and under 30,000	30
30,000 and under 40,000	15
40,000 and under 50,000	10
50,000 and under 60,000	15

Find the median family income.

3.48 Advertising expenditures constitute one of the important components of the cost of goods sold. From the following data giving the advertising expenditures (in millions of dollars) of 50 companies, find the median advertising expenditure.

ADVERTISING EXPENDITURE	NUMBER OF COMPANIES
25 and under 35	5
35 and under 45	11
45 and under 55	18
55 and under 65	6
65 and under 75	10
	50

3.49 The number of shareholders of publicly owned issues of common and preferred stocks in the United States has increased over the years. The following table shows the number of shareholders in various age groups in 1985.

AGE	NUMBER OF SHAREHOLDERS (IN THOUSANDS)
Under 21 years	2,260
21–34 years	11,093
35–44 years	10,982
45–54 years	7,899
55–64 years	8,217
65 years and older	6,589

(*Source: The World Almanac and Book of Facts*, 1990, p. 80.)

Estimate the median. Is the median an appropriate summary number for these data? Why or why not?

3.50 Glen's Mufflers and Exhaust has been experiencing a steady growth in sales. The following grouped data show the frequencies of daily sales volume (in hundreds of dollars):

DAILY SALES	NUMBER OF DAYS
.5 and under 1	2
1 and under 1.5	2
1.5 and under 2	3
2 and under 2.5	11
2.5 and under 3	20
3 and under 3.5	12
3.5 and under 4	5
4 and under 4.5	3
4.5 and under 5	2

a. Calculate the mean daily sales.

b. Calculate the standard deviation of the mean daily sales.

3.51 The following is the final distribution of grades in an introductory economics course at a local university:

GRADE	NUMBER OF STUDENTS
90–99	5
80–89	14
70–79	11
60–69	4
50–59	7
	41

a. Calculate the mean score.

b. Calculate the standard deviation.

c. Interpret the value of the mean and standard deviation.

3.52 The industrial engineer of the Bright Light company is interested in examining the average burning hours (life) of the 100-watt bulbs manufactured. Using the following data, determine the mean burning hours of a 100-watt bulb:

BURNING HOURS	NUMBER OF BULBS
0 and under 40	231
40 and under 50	168
50 and under 60	244
60 and under 70	300
70 and under 80	111
80 and under 90	48
90 and under 100	98
	1200

◢ 3.8
Exploratory Data Analysis

Exploratory data analysis (EDA) is a recently developed technique for providing easy-to-construct pictures that summarize and describe a set of sample data. Two popular diagrams that fall under this category are *stem and leaf diagrams* and *box and whisker plots*. Stem and leaf diagrams are like histograms, since both allow you to see the "shape" of the data. Box and whisker plots are graphical representations of the quartile measures of position discussed in Section 3.4.

Stem and Leaf Diagrams

Stem and leaf diagrams were originally developed by John Tukey (pronounced Too′key) of Princeton University. They are extremely useful in summarizing reasonably sized data sets (under 100 values as a general rule) and, unlike histograms, result in no loss of information. By this we mean that it is possible to retrieve the original data set from a stem and leaf diagram, which is not the case when using a histogram—some of the information in the original data is lost when a histogram is constructed. Nevertheless, histograms provide the best alternative when attempting to summarize a large set of sample data.

To illustrate the construction of a stem and leaf diagram, suppose that a study reports the after-tax profits of 12 selected companies. The profits (recorded as cents per dollar of revenue) are as follows:

$$3.4 \quad 4.5 \quad 2.3 \quad 2.7 \quad 3.8 \quad 5.9 \quad 3.4 \quad 4.7 \quad 2.4 \quad 4.1 \quad 3.6 \quad 5.1$$

The stem and leaf diagram for these data is shown in Figure 3.8. Each observation is represented by a **stem** to the left of the vertical line and a **leaf** to the right of the vertical line. For example, the stems and leaves for the first and last observation would be:

Stem	Leaf		Stem	Leaf
3	.4		5	.1

In a stem and leaf diagram, the stems are put *in order* to the left of the vertical line. The leaf for each observation is generally the last digit (or possibly the last two digits) of the data value, with the stem consisting of the remaining first digits. The value 562 could be represented as $5|62$ or as $56|2$ in a stem and leaf diagram, depending upon the range of the sample data. Whether or not the raw data are ordered, the stem and leaf diagram provides at least a partial ordering of the data. If the diagram is rotated counterclockwise, it has the appearance of a histogram and clearly describes the shape of the sample data.

The 50 job applicants' aptitude scores in Table 3.1 are represented by a stem and leaf diagram in Figure 3.9. From this diagram we observe that the minimum

■ **FIGURE 3.8**
Stem and leaf diagram for after-tax profits.

2	.3	.7	.4	
3	.4	.8	.4	.6
4	.5	.7	.1	
5	.9	.1		

■ **FIGURE 3.9**
Stem and leaf diagram
for after-tax profits.

```
2 | 2   5   8
3 | 1   4   5   9   9
4 | 0   2   4   4   6   8   9
5 | 1   3   3   5   5   6   7   9
6 | 0   1   3   3   3   5   6   8   8   9
7 | 1   2   2   4   5   5   6   8   8
8 | 0   2   3   5   8
9 | 0   2   6
```

score is 22, the maximum score is 96, and the largest group of scores is between 60 and 69. Also, the 5 leaves in stem row 3 indicate that 5 people scored at least 30 but less than 40. The 3 leaves in stem row 9 tells us at a glance that 3 people scored 90 or better.

For larger data sets, you may want to consider spreading out the stem column by repeating the stem value two or three times. To illustrate, the same 50 test scores are used to construct another stem and leaf diagram in Figure 3.10, where each stem value is repeated twice. The first stem value contains leaves between 0 and 4, the second stem contains leaves between 5 and 9. A MINITAB version of the stem and leaf diagram using these data is shown in Figure 3.11. Notice that MINITAB used the double stems as in Figure 3.10.

■ **FIGURE 3.10**
Stem and leaf diagram
for aptitude test scores
using repeated stems.

```
2 | 2
2 | 5   8
3 | 1   4
3 | 5   9   9
4 | 0   2   4   4
4 | 6   8   9
5 | 1   3   3
5 | 5   5   6   7   9
6 | 0   1   3   3   3
6 | 5   6   8   8   9
7 | 1   2   2   4
7 | 5   5   6   8   8
8 | 0   2   3
8 | 5   8
9 | 0   2
9 | 6
```

■ **FIGURE 3.11**
Stem and leaf diagram
of aptitude test scores
(Table 3.1) using
MINITAB.

```
MTB > SET INTO C1
DATA> 22 25 28 31 34 35 39 39 40 42
DATA> 44 44 46 48 49 51 53 53 55 55
DATA> 56 57 59 60 61 63 63 63 65 66
DATA> 68 68 69 71 72 72 74 75 75 76
DATA> 78 78 80 82 83 85 88 90 92 96
DATA> END
MTB > STEM AND LEAF USING C1

Stem-and-leaf of C1          N  = 50
Leaf Unit = 1.0

        1     2 2
        3     2 58
        5     3 14
        8     3 599
       12     4 0244
       15     4 689
       18     5 133
       23     5 55679
      (5)     6 01333
       22     6 56889
       17     7 1224
       13     7 55688
        8     8 023
        5     8 58
        3     9 02
        1     9 6
```

```
5 | 21   50   36   72
6 | 71   33   47   62   55
7 | 83   62   11   31   40   21   57
8 | 44   16   35   92
```

In many situations the leaves in your diagram may consist of a *pair* of digits. In Figure 3.12, the tons of chemical produced by Sarhad Industries are illustrated for 20 randomly selected days. Here each stem consists of a single digit and each leaf represents the last two digits of the production amount.

Box and Whisker Plots

A **box and whisker plot** is a graphical representation of a set of sample data that illustrates the lowest data value (L), the first quartile (Q_1), the median (Q_2, Md), the third quartile (Q_3), the interquartile range (IQR), and the highest data value (H).

In Section 3.4, the following values were determined for the aptitude test scores in Table 3.1:

$$L = 22$$
$$Q_1 = 46$$
$$Q_2 = \text{Md} = 62$$
$$Q_3 = 75$$
$$\text{IQR} = 75 - 46 = 29$$
$$H = 96$$

A box and whisker plot of these values is shown in Figure 3.13. The ends of the *box* are located at the first and third quartile with a vertical bar inserted at the median. Consequently, the length of the box is the interquartile range. The dotted lines are the *whiskers* and connect the highest and lowest data values to the ends of the box. Thus approximately 25% of the data values will lie in each whisker and in each portion of the box. If the data are symmetric, the median bar should be located at the center of the box. *Consequently, the bar location indicates the* skewness *of the data; if located in the left half of the box, the data are skewed right, and if located in the right half, the data are skewed left.*

According to Figure 3.13, the data appear to be nearly symmetric (with a slight left skew), which is supported by the stem and leaf diagram in Figure 3.9. A box and whisker plot using MINITAB is shown in Figure 3.14.

In Examples 3.2, 3.4, and 3.7, we examined the 100 inside diameters of a particular machined part and concluded that the process was "off center" and too near the upper spec limit of 10.3 millimeters. A stem and leaf diagram and a box and whisker plot of these data using MINITAB are shown in Figure 3.15. Again it is clear from the stem and leaf diagram that the process is shifted to the right (all the values are to the right of the target of 10.2 millimeters), with the 17 values in the last four leaves (excluding values of 10.300) exceeding the upper spec limit. At a glance we see that 17% of the values exceed the upper spec limit and that none of the values are below the lower spec limit of 10.1 millimeters.

■ FIGURE 3.13

Box and whisker plot
for 50 aptitude test
scores (data in
Table 3.1).

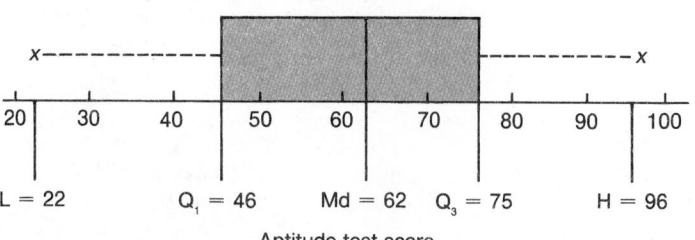

Aptitude test score

■ **FIGURE 3.14**
Box and whisker plot of aptitude test scores (Table 3.1) using MINITAB.

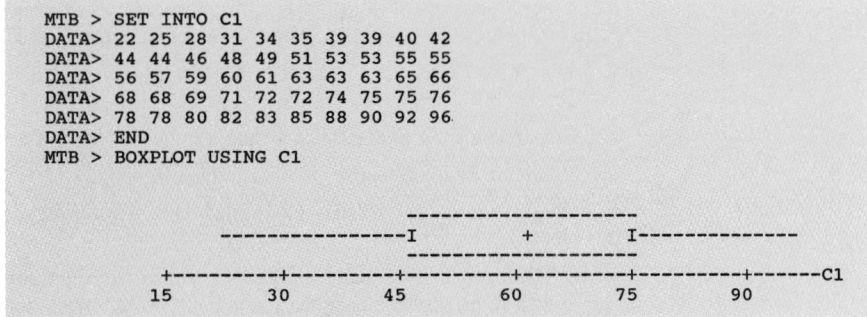

```
MTB > SET INTO C1
DATA> 22 25 28 31 34 35 39 39 40 42
DATA> 44 44 46 48 49 51 53 53 55 55
DATA> 56 57 59 60 61 63 63 63 65 66
DATA> 68 68 69 71 72 72 74 75 75 76
DATA> 78 78 80 82 83 85 88 90 92 96
DATA> END
MTB > BOXPLOT USING C1


                                         ---------------------
                 -----------------I        +        I---------------
                                         ---------------------
            +---------+---------+---------+---------+---------+------C1
            15        30        45        60        75        90
```

■ **FIGURE 3.15**
Stem and leaf diagram and box and whisker plot for 100 machine part diameters, using data from Example 3.2.

```
MTB > STEM AND LEAF USING C1

Stem-and-leaf of C1        N = 100
Leaf Unit = 0.0010


     1 1021 6
     4 1022 168
    12 1023 00114799
    17 1024 00459
    27 1025 2344556789
    37 1026 0244667799
   (21)1027 001111112444667788999
    42 1028 00111234569
    31 1029 001113333468
    19 1030 00149
    14 1031 0111224558
     4 1032 68
     2 1033 38


MTB > BOXPLOT USING C1


                            -----------------
            -----------------I      +      I------------------
                            -----------------
          --------+---------+---------+---------+---------+--------C1
          10.225    10.250    10.275    10.300    10.325
```

In Figure 3.15 the median bar for the box and whisker plot is nearly centered inside the box, indicating that the data are nearly symmetric. This observation is supported by the fact that the two whiskers are nearly the same length. These two plots are very powerful tools to use for examining the shape, location and variability of sample data.

Exercises

3.53 The average monthly expenditures on newspaper advertising by 36 real estate offices in the New Orleans metropolitan area are as follows:

590, 593, 651, 765, 503, 921, 833, 580, 804, 841, 865, 905, 881, 845, 807, 593, 551, 845, 650, 884, 920, 596, 630, 838, 889, 655, 585, 751, 885, 908, 586, 843, 654, 583, 847, 589.

a. List the stem possibilities, first two digits, in order.

b. Form a stem and leaf display by attaching the appropriate leaves to stems.

c. Comment on the shape of the distribution of the data.

3.54 The number of no-shows for four daily commuter flights between Boston's Logan Airport and JFK Airport in New York City is given below for 20 randomly selected business days.

4, 15, 11, 16, 24, 31, 12, 5, 11, 20, 21, 13, 7, 12, 15, 9, 13, 14, 22, 16

a. Construct a stem and leaf display of the data.

b. Comment on the shape of the distribution of the no-shows.

3.55 Forty samples of boxes of electronic resistors were checked for defective resistors. Each box contains 200 resistors. The number of defectives in each box is given below.

11, 13, 8, 12, 9, 12, 11, 37, 13, 48, 14, 19, 12, 14, 15, 13, 19, 14, 12, 17, 12, 16, 15, 33, 18, 11, 13, 15, 26, 56, 22, 14, 17, 18, 12, 7, 15, 12, 14, 23

a. Construct a stem and leaf display and use the first digit of each number as the stem.

b. What number of defectives is most frequently found?

c. Construct a box and whisker plot of this data. What can you say about the data based on this plot?

3.56 The manager of a small restaurant wished to determine how long the average customer had to wait to be served during the lunch hour. At the lunch hour on a particular "typical" day, 20 customers experienced the following waiting times (in minutes):

5.5, 10.3, 7.5, 8.1, 6.8, 11.0, 10.2, 9.0, 7.0, 5.8, 12.5, 7.5, 6.0, 13.7, 5.5, 14.0, 7.0, 6.9, 6.3, 7.4

a. Construct a box and whisker plot of the data.

b. Should the manager feel comfortable in advertising that meals are served in 10 min or less, or else the customer eats for free?

3.57 The yields in percent on the 15 utility stocks included in the Dow Jones utility average are given below for August 6, 1990:

8.5, 9.0, 4.6, 10.1, 7.9, 3.8, 6.4, 9.2, 0.0, 6.9, 4.5, 7.4, 7.9, 8.4, 7.0

(*Source: Wall Street Journal,* August 7, 1990.)

a. Which measure of central tendency do you think is most appropriate for the preceding data set? If the yield of 0.0 for one of the troubled utilities is considered an outlier, what would be the value of this measure of central tendency after eliminating this yield from the data set?

b. Compute the standard deviation and the mean absolute deviation. Compare these two values.

c. Construct a stem and leaf display of the yields.

d. Construct a box and whisker plot of the yields.

e. What do the stem and leaf display in part c and the box and whisker plot in part d tell you about the shape of the distribution?

3.58 The following data reflect the number of consecutive hours that a sample of 20 medical interns worked in a local emergency room:

14, 21, 35, 11, 18, 28, 26, 38, 21, 19, 17, 27, 18, 22, 16, 30, 17, 26, 20, 25

Construct a stem and leaf diagram of these data.

3.59 The scores on a newly devised instrument for measuring life satisfaction could range from a minimum of 100 to a maximum of 800. A sample of 25 scores contained the following values:

512, 587, 563, 542, 556, 577, 515, 520, 532, 526, 560, 584, 530, 593, 542, 518, 570, 588, 534, 542, 557, 538, 562, 548, 525

The researcher noticed that all the scores were between 500 and 600; the lack of variation in these scores is not a good sign. Illustrate this by constructing a stem and leaf diagram of the instrument scores.

3.60 The ages of 40 individuals working at a telephone crisis center were recorded as follows:

33, 40, 35, 24, 31, 32, 45, 23, 45, 34, 34, 54, 25, 36, 38, 44, 62, 31, 56, 47, 30, 42, 43, 52, 48, 56, 64, 40, 61, 39, 26, 34, 37, 43, 47, 32, 45, 27, 35, 29

a. Construct a stem and leaf diagram of the data. What conclusions can you make from this picture?

b. Construct a box and whisker plot of these data. What can you say about the data based on this plot?

■ **FIGURE 3.16**
Data coding.

Subtracting or adding a constant

Actual data		Adjusted data
1005, 1006, 1007, 1009, 1023	—— subtract 1000 ——→	5, 6, 7, 9, 23
$\bar{x} = 1000 + 10 = 1010$ ←——	add 1000 ——	$\bar{x} = 10$
$s = 7.416$ ←——	is the same as ——	$s = 7.416$

Dividing or multiplying by a constant

Actual data		Adjusted data
5000, 6000, 7000, 9000, 23,000	—— divide by 1000 ——→	5, 6, 7, 9, 23
$\bar{x} = 1000 \times 10 = 10,000$ ←——	multiply by 1000 ——	$\bar{x} = 10$
$s = 1000 \times 7.416 = 7,416$ ←——	multiply by 1000 ——	$s = 7.416$

▮ 3.9 Calculating Descriptive Statistics by Coding

When you use a calculator to determine the sample mean or standard deviation, one problem that can occur is that the data values are too large or too small to "fit" into your calculator. To avoid having the calculator self-destruct in your hands, a procedure referred to as **data adjusting,** or **data coding,** allows you to derive these statistics using more reasonable data values. You can then work backward to get the desired statistics. To code, or adjust, the data you subtract (or add) or divide (or multiply) your original data set by a fixed amount. Figure 3.16 demonstrates this procedure using data sets containing several large values. To adjust the data when subtracting (adding) a positive constant to each data value,

actual $\bar{x}$ = adjusted $\bar{x}$ plus (minus) the constant

actual s = adjusted s (no change)

When dividing (multiplying) by a positive constant,

actual $\bar{x}$ = adjusted $\bar{x}$ times (divided by) the constant

actual s = adjusted s times (divided by) the constant

▮ Exercises

3.61 Using the subtraction rule, adjust the following data and calculate the mean and standard deviation.

413, 407, 411, 402, 425, 408, 410, 421

3.62 Using the multiplication rule, adjust the following data and calculate the mean and standard deviation:

0.00119, 0.00101, 0.00121, 0.00108, 0.0010, 0.00114, 0.00117, 0.00104, 0.00123, 0.00124

3.63 Using the division rule, calculate the mean and variance for the following data:

200, 600, 800, 1000, 400, 1200, 10,000, 1400, 1600, 400

3.64 Using the division rule, calculate the mean and standard deviation for the following data:

500, 3000, 600, 1000, 400, 300, 100, 2500, 700, 300

3.65 Using the multiplication rule, adjust the following data and calculate the mean and standard deviation:

0.001182, 0.001104, 0.001270, 0.001251, 0.001407, 0.001553, 0.001177, 0.001333, 0.001489, 0.001505

3.66 Using the subtraction rule, adjust the following data and calculate the mean:

1013, 1007, 1011, 1102, 1025, 1008, 1110, 1021, 1111, 1009

Summary

The purpose of analyzing or describing sample data is to learn more about the population from which it was obtained. Every population has properties that describe it. These properties are referred to as **parameters.** We can estimate these parameters by obtaining a sample and deriving the corresponding sample statistic, which is a particular **descriptive measure.**

This chapter has introduced you to some of the more popular descriptive measures used to describe a set of sample values. **Measures of central tendency** are used to describe a typical value within the sample: the **mean** (the average of the sample data), the **median** (the value in the center of the ordered data), the **mode** (that data value occurring the most often), and the **midrange** (an average of the lowest and highest data values). To measure the variation within a set of sample data, we use **measures of dispersion:** the **range** (difference between the highest and lowest data values), the **mean absolute deviation** (an average of the absolute deviations from the sample mean), the **variance** (sum of the squares of the deviations from the sample mean, divided by $n - 1$), the **standard deviation** (square root of the variance), and the **coefficient of variation** (standard deviation divided by the mean, times 100).

Percentiles and **quartiles** are **measures of position** and indicate the relative position of a particular value. The first quartile (Q_1) and the third quartile (Q_3) are the 25th and 75th percentiles, respectively. The second quartile (Q_2) is the 50th percentile, which is identical to the sample median. The difference between the first and third quartiles is the **interquartile range** (IQR), which is another measure of dispersion, since it measures the spread within the middle 50% of the data values. Another measure of position is the **Z score,** which is derived for a particular observation by subtracting the sample mean and dividing by the sample standard deviation.

Finally, the shape of a data set can be described using various **measures of shape.** Two such measures are the sample **skewness** (the degree of symmetry in the data) and **kurtosis** (the tendency of a distribution to stretch out in a particular direction).

The two most commonly used measures are the sample mean and standard deviation. These two statistics can be used together to describe the sample data by applying **Chebyshev's inequality** or the **empirical rule.** The latter procedure draws a stronger conclusion about the concentration of the data values but assumes that the population of interest is bell-shaped (normal).

We examined how to estimate the sample mean and standard deviation when the only information available is a frequency distribution, or **grouped data. Data coding** can be used to calculate these two measures more easily when you encounter data sets containing extremely large or small values.

Stem and leaf diagrams and **box and whisker plots** are easy-to-construct pictures that illustrate the shape of the sample data. A stem and leaf diagram is similar to a histogram but results in no loss of information. A box and whisker box plot summarizes the lowest and highest data values, along with the three quartiles, in a simple diagram.

Summary of Formulas

Measures of Central Tendency

1. Sample mean

$$\bar{x} = \frac{\Sigma x}{n}$$

2. Population mean

$$\mu = \frac{\Sigma x}{N}$$

3. Midrange

$$Mr = \frac{(\text{smallest value}) + (\text{largest value})}{2}$$

Measures of Dispersion

1. Range

   R = (largest value) − (smallest value)

2. Mean absolute deviation

$$\text{MAD} = \frac{\Sigma|x - \bar{x}|}{n}$$

3. Sample variance

$$s^2 = \frac{\Sigma(x - \bar{x})^2}{n - 1}$$

4. Population variance

$$\sigma^2 = \frac{\Sigma(x - \mu)^2}{N}$$

5. Sample standard deviation

$$s = \sqrt{\frac{\Sigma(x - \bar{x})^2}{n - 1}}$$

6. Population standard deviation

$$\sigma = \sqrt{\frac{\Sigma(x - \mu)^2}{N}}$$

7. Coefficient of variation

$$\text{CV} = \frac{s}{\bar{x}} \cdot 100$$

Measures of Position

1. Z score

$$Z = \frac{x - \bar{x}}{s}$$

Measures of Shape

1. Pearsonian coefficient of skewness

$$\text{Sk} = \frac{3(\bar{x} - \text{Md})}{s}$$

Grouped Data

1. Sample mean

$$\bar{x} \cong \frac{\Sigma f \cdot m}{n}$$

2. Sample variance

$$s^2 \cong \frac{\Sigma f \cdot m^2 - (\Sigma f \cdot m)^2/n}{n - 1}$$

3. Sample median

$$\text{Md} \cong L + \frac{k}{f} \cdot W$$

Review Exercises

3.67 The unemployment rate (U) represents the percent of the civilian labor force that is unemployed. The following are the U.S. unemployment rates for the years 1951 through 1989.

YEAR	U	YEAR	U	YEAR	U
1951	3.3	1964	5.2	1977	7.0
1952	3.0	1965	4.5	1978	6.0
1953	2.9	1966	3.8	1979	5.8
1954	5.6	1967	3.9	1980	7.1
1955	4.4	1968	3.6	1981	7.5
1956	4.1	1969	3.5	1982	9.5
1957	4.3	1970	4.9	1983	9.5
1958	6.8	1971	5.9	1984	7.4
1959	5.5	1972	5.6	1985	7.1
1960	5.5	1973	4.9	1986	6.9
1961	6.7	1974	5.6	1987	6.1
1962	5.6	1975	8.5	1988	5.4
1963	5.6	1976	7.7	1989	5.2

(*Source: Economic Indicators*, August 1990, p. 12 and U.S. Department of Commerce, *Statistical Abstract of the United States*, 1987.)

Calculate the following statistics and explain how these statistics help to describe the distribution of the data:

a. Midrange
b. Mean
c. Median
d. Mode
e. Standard deviation

f. Coefficient of variation
g. First quartile
h. Ninetieth percentile
i. Interquartile range

3.68 The table below gives the projected number of recruits with bachelor's degrees in management information systems that several employers will hire in 1991.

EMPLOYER	NUMBER OF RECRUITS HIRED WITH BACHELOR DEGREES
Arthur Anderson & Company	480
AT&T	286
American Management Systems, Inc.	60
Eli Lilly and Company	65
Aetna Life and Casualty	56
J. C. Penney	50
Pitney Bowes, Inc.	50
UNISYS	50
Hallmark Cards, Inc.	30
State Farm Insurance Companies	30

(*Source:* "The Top One Hundred Employers and The Majors in Demand in 1991," *The Black Collegian,* November/December 1990, p. 148.)

a. Calculate the mean, median, and mode for the given data set. Which measure do you think is most appropriate as a measure of central tendency?

b. For the value of the mean and the value of median calculated in part a, what can you say about the skewness of the data? Calculate the coefficient of skewness and interpret its value.

c. Calculate the standard deviation for the above data set.

d. Find the number of values that are between $\bar{x} - 2s$ and $\bar{x} + 2s$ and also the number of values between $\bar{x} - 3s$ and $\bar{x} + 3s$. Are these numbers consistent with Chebyshev's inequality?

e. Calculate the midrange and the interquartile range. Which is more easily affected by an outlier?

3.69 Answer the questions below for the following data set:

2.0, 1.1, -1.5, .3, 1.9, -2.3, -1.2, .7, $-.5$, 3.1, $-.2$, 1.0, $-.6$, 1.3, $-.9$, $-.1$, .8, -1.1, 2.4, .9, $-.1$, -1.5

a. Do the data appear to be bell-shaped?

b. Calculate the coefficient of skewness.

c. Using the empirical rule, estimate the range of values within which about 68% of the data values are expected to lie.

3.70 The following data show the distribution of the annual incomes (in thousands of dollars) of the households in a neighborhood:

ANNUAL INCOME	NUMBER OF HOUSEHOLDS
0 and under 10	21
10 and under 20	11
20 and under 30	9
30 and under 40	13
40 and under 50	17
50 and under 60	20
60 and under 70	14
70 and under 80	20
80 and under 90	7
90 and under 100	2
	134

a. Calculate the mean annual income.

b. Calculate the variance.

c. Estimate the interval within which at least 75% of the data values are expected to fall.

d. Calculate the coefficient of skewness.

3.71 The mean rate charged by the CPAs in a certain city is about $75 per hour, with a standard deviation of $15. Assuming that the data came from a normal population, estimate the range of rates within which about 95% of the CPAs' charges are expected to lie.

3.72 The Z score is -1.50, the mean is 45, and $x = 15$. What is the value of the variance?

3.73 The mean is 81, the standard deviation is 9, and one particular x value is 45.

a. Calculate the Z score.

b. Interpret the Z score.

3.74 The mean GMAT score of the 65 applicants who were accepted into the MBA program of Xavier Business School was 520 with a standard deviation of 25. About how many applicants scored between 470 and 570 on the GMAT?

3.75 Calculate the mean and standard deviation for these data:

50.2, 53.8, 51.4, 52.2, 50.8, 59.1, 52.8, 57.7, 51.1, 54.3, 55.5, 52.1, 57.6, 55.9, 50.9, 54.7

3.76 Calculate the mean and standard deviation for the following data:

1000, 700, 400, 100, 800, 20,000, 4000, 300, 900, 600, 200, 500, 2000, 700, 2500, 5500

3.77 Ten of the leading U.S. life insurance companies are listed below along with their total sales in millions of dollars for 1988.

LIFE INSURANCE COMPANY	SALES (IN MILLIONS)
Prudential	116,197.0
Metropolitan Life	94,232.0
Equitable Life	50,415.5
Aetna Life	48,884.9
Teachers Insurance & Annuity	38,631.4
New York Life	35,153.8
Connecticut General	31,095.5
Travelers	30,672.2
John Hancock	28,315.2
Northwestern Mutual	25,349.0

(*Source: The World Almanac and Book of Facts,* 1990, p. 86.)

Calculate the following values:

a. Mean
b. Median
c. Mode
d. Variance
e. Standard deviation
f. Mean absolute deviation

g. Range
h. Coefficient of variation
i. Coefficient of skewness
j. First quartile
k. 60th percentile
l. Z score for Aetna

m. Using Chebyshev's inequality, find the values of the sales figures between which at least 75% of the data values will fall.

3.78 In the annual report of Franklin Custodian Funds, Inc., dollar cost averaging is recommended rather than market timing in buying shares of the mutual fund. The following hypothetical example is given in the annual report.

MONTH	SHARE PRICE	MONTHLY AMOUNT INVESTED	SHARES PURCHASED EACH MONTH
January	$ 5.00	$ 500	100.0
February	4.00	500	125.0
March	3.50	500	142.9
April	3.00	500	166.7
May	3.75	500	133.3
June	5.00	500	100.0
TOTAL	$24.25	$3,000	767.9

(*Source: Annual Report* for Franklin Custodian Funds, Inc., September 30, 1990, p. 7.)

a. Calculate the average price of the shares during the six months for which figures are given.

b. Calculate the median price of the shares during these six months.

c. Calculate the investor's average cost for the shares purchased. How does this figure compare to the values given in parts a and b?

d. If the same number of shares were purchased each month, how would the average share cost and average share price compare?

3.79 The insurance agent at Central Insurance Company wished to determine the distribution of the number of minutes that the company's secretary was spending on phone calls about insurance rates. The times for 30 randomly selected calls were analyzed in the MINITAB computer printout given below.

```
MTB > PRINT C1
C1
  17.8    3.8    1.5    6.5   15.1    3.4   13.0    6.5   13.4    3.2    9.8
   9.4    2.8    8.0    7.1    2.4    7.4    5.3    2.1   19.3    5.1    2.2
   7.3    4.3    1.7    6.8    4.1    2.7    7.4    4.8

MTB > DESCRIBE C1

                 N      MEAN    MEDIAN    TRMEAN     STDEV    SEMEAN
C1              30     6.807     5.900     6.304     4.743     0.866

               MIN       MAX        Q1        Q3
C1           1.500    19.300     3.100     8.350

MTB > HISTOGRAM OF C1

Histogram of C1    N = 30

Midpoint     Count
     2         7    *******
     4         6    ******
     6         5    *****
     8         5    *****
    10         2    **
    12         0
    14         2    **
    16         1    *
    18         1    *
    20         1    *
```

a. Compute the coefficient of skewness and comment on the shape of the distribution.

b. What are the range, midrange, and interquartile range? How do these values help describe the distribution?

c. Within what limits would you expect the times of most phone calls to fall?

3.80 A training manager wishes to analyze the test scores of 40 employees who have completed an intensive three-day training course. The test scores are analyzed using the MINITAB printout below.

```
MTB > print c1

C1
   98     87     95     55     42     95     84     44     86     60
   99     78     54     33     76     88     97     65     86     80
   58     93     94     75     32     87     90     65     87     97
   78     98     91     54     87     93     42     84     97     91

MTB > describe c1

                 N      MEAN    MEDIAN    TRMEAN     STDEV    SEMEAN
C1              40     77.38     86.00     78.69     19.84      3.14

               MIN       MAX        Q1        Q3
C1           32.00     99.00     61.25     93.00
```

```
MTB > stem and leaf using c1

Stem-and-leaf of C1          N = 40
Leaf Unit = 1.0

      2      3 23
      2      3
      5      4 224
      5      4
      7      5 44
      9      5 58
     10      6 0
     12      6 55
     12      7
     16      7 5688
     19      8 044
    (7)      8 6677778
     14      9 011334
      8      9 55777889

MTB > boxplot using c1
```

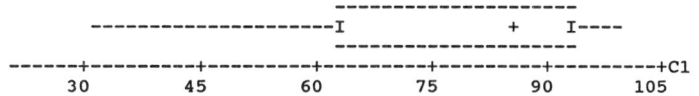

```
                                  ---------------------
            ---------------------I              +    I----
                                  ---------------------
      ------+---------+---------+---------+---------+---------+---------+C1
           30        45        60        75        90       105
```

a. What are the range, midrange, interquartile range, and coefficient of skewness of the data set.

b. The manager would like to include a brief summary of the shape and characteristics of the data in a report to the vice president of the company to help in examining the effectiveness of the training course. Write up an appropriate summary.

Computer Exercises Using the Database

Exercise 1 -- Appendix I Randomly select 100 observations of variable INCOME1 (income of principal wage earner) from the database.

a. Use a convenient statistical computer package to determine the various descriptive measures that describe the distribution of INCOME1.

b. What are the actual proportions of observations that are between ±2.0 and ±3.0 standard devia- tions of the mean of the data set? Are these propor- tions consistent with:

(i) Chebyshev's inequality?

(ii) the empirical rule?

Exercise 2 -- Appendix J Randomly select 100 observations of the variable SALES from the database. Use a convenient statistical computer package to find the mean, median, range, variance, coefficient of variation, and coefficient of skewness for this variable.

 *Case Study*

"I Love the Smell of Falling Stocks in the Morning"

The title of this case study borrows a famous line from the movie *Apocalypse Now*. It is a quote attrib- uted by *Financial Planning* magazine to Peter Su- tro, vice president of the Axe-Houghton Stock Fund, in a December 1990 article. Sutro's comment was a reflection on stock market volatility, which tests the mettle of financial planners managing port- folios of equities. The article in question was a review of the performance of various equity

funds, also known as mutual funds, summarized in Table 3.5.

The tables show the performance of the top 20 performers in each of four types of equity funds: *international* equity funds, *growth* equity funds, *growth and income* equity funds, and *aggressive growth* equity funds. As the names suggest, each type of fund has a dominant investment philosophy. Some funds concentrate on international stock markets, while others focus on aggressive growth stocks.

It should be noted that any equity fund is not necessarily a hundred percent invested in the stock

■ TABLE 3.5   The Top 20 International Equity Funds—1990

AS OF SEPTEMBER 30, 1990	PERCENT RETURN CURRENT YEAR TO DATE	PERCENT RETURN 1989	ASSETS (MIL $)	ANNUAL EXPENSE RATE	MAX LOAD FEE	12 MONTH DIVIDEND YIELD
International Funds						
G.T. Global Health Care	0.4	n/a	114	n/a	4.75	0.2
SoGen International Fund	−2.7	17.2	195	0.94	3.75	3.9
Templeton Foreign Fund	−3.7	30.8	881	0.81	8.50	3.1
Merrill Lynch Global Alloc A	−4.3	n/a	53	1.37	6.50	8.0
GAM International	−4.3	22.1	26	n/a	5.00	0.0
SLB European Portfolio	−4.6	23.8	26	2.00	0.00	1.0
Oppenheimer Global Fund	−4.7	34.7	809	1.79	8.50	0.4
Merrill Lynch Global Alloc B	−4.9	n/a	128	1.40	0.00	7.3
GAM Global Fund	−5.6	24.1	26	n/a	5.00	0.2
Europacific Growth Fund	−6.4	24.1	803	1.08	5.75	2.1
FSP-European Portfolio	−7.0	24.2	90	1.78	0.00	1.3
Fidelity Europe	−7.3	32.3	450	n/a	2.00	1.2
New Perspective Fund	−8.4	25.6	1563	0.63	5.75	2.7
European Emerging Cos Fund	−9.2	26.6	35	2.25	4.50	0.3
MFS Lifetime Global Equity	−9.4	27.5	79	2.05	0.00	0.0
Merrill Lynch Eurofund A	−9.5	25.4	83	1.11	6.50	1.9
Counsellors International Equity	−9.5	n/a	34	1.42	0.00	2.0
Scudder Global	−10.1	37.3	257	1.81	0.00	1.3
Merrill Lynch Eurofund B	−10.1	24.1	529	1.16	0.00	1.4
Fidelity Intl Gr & Inc	−10.2	19.1	36	n/a	1.00	1.3
Growth Funds						
Fidelity Select Energy	6.9	42.8	86	n/a	2.00	1.2
Vanguard Energy	6.4	43.4	102	0.38	0.00	2.4
Putnam Health Sciences	1.8	41.9	348	1.14	5.75	1.3
Rightime Fund	1.7	11.1	149	2.55	0.00	1.1
Rightime Blue Chip Fund	1.0	19.5	123	2.31	4.75	1.6
AIM Charter Fund	0.0	37.9	90	1.37	5.50	3.7
Vanguard SP Port-Health	−0.8	32.9	134	0.39	0.00	2.1
Fidelity Select Food/Agriculture	−1.3	38.9	32	n/a	2.00	0.2
Muirfield Fund	−1.8	13.9	30	1.13	0.00	2.0
Putnam Energy Resources	−2.1	34.8	127	1.40	5.75	3.0
State Farm Growth Fund	−3.3	31.9	445	0.22	0.00	2.5
IDS Managed Retirement	−3.8	34.6	796	0.82	5.00	2.7
Baird Blue Chip Fund	−4.2	27.2	33	1.35	5.75	1.9
Vanguard World-US Growth	−4.3	37.7	347	0.95	0.00	1.3
IDS New Dimensions Fund	−4.4	31.5	916	0.78	5.00	1.9
Sentry Fund	−4.7	24.0	52	0.65	8.00	2.9
Dreyfus Strategic Investment	−4.7	33.0	112	1.79	4.50	1.8
Kemper Blue Chip Fund	−4.7	27.2	34	1.48	4.50	1.0
AIM Weingarten Fund	−5.2	35.9	684	1.27	5.50	0.5
Fidelity Blue Chip Growth	−5.3	36.2	149	n/a	2.00	0.7

market. Indeed, one of the major tasks of the fund managers is to decide when to stay invested, and when to bail out of equities into cash positions. Unlike stockbrokers, who may tend to "churn" their accounts with more frequent trades, financial planners tend to push a philosophy of diversification, or, as they like to put it, asset allocation. They tend to adopt a strategy of holding longer-term conservative portfolios. This is reflected in the data, which shows funds holding out despite poorer performance in the recent short term (1990) as compared to 1989, undoubtedly attributable to declines in the stock market.

Case Study Questions

1. a. Compute the mean and median values for 1989 percentage return, fund assets, and 12-month dividend yield, for each of the four types of funds.

■ TABLE 3.5 (*Continued*)

AS OF SEPTEMBER 30, 1990	PERCENT RETURN CURRENT YEAR TO DATE	PERCENT RETURN 1989	ASSETS (MIL $)	ANNUAL EXPENSE RATE	MAX LOAD FEE	12 MONTH DIVIDEND YIELD
Growth and Income Funds						
Mathers Fund	7.3	10.4	244	1.00	0.00	6.2
Gabelli Convertible Securities	2.4	n/a	75	2.50	4.50	2.8
Gateway Index Plus Fund	1.2	19.3	49	1.25	0.00	2.6
Flex FD-Growth	−0.4	10.2	26	1.38	0.00	2.7
Phoenix Convertible Fund	1.1	20.0	152	1.02	6.90	5.1
Phoenix Growth Fund SER	−1.8	27.3	756	1.04	6.90	3.1
FBL-Growth Common Stock	−1.9	13.7	36	n/a	0.00	5.9
FBA Paramount Fund Inc.	−2.6	21.8	220	n/a	6.50	4.1
Kemper Total Return Fund	−2.8	19.8	914	0.82	5.75	4.6
Parkstone High Inc. Equity	−4.3	27.9	96	n/a	4.50	4.6
SLB Utilities Portfolio	−4.6	20.6	618	1.05	0.00	7.3
Sovereign Investors	−4.7	23.6	80	0.94	5.00	5.0
Dean Witter Strategist	−4.7	22.8	179	0.98	0.00	2.4
Analytic Optioned Equity	−4.9	17.8	108	1.09	0.00	3.9
IDS Stock Fund	−5.3	29.4	1372	0.61	5.00	4.6
Invest Portfolio-Total Ret	−5.6	20.0	572	0.92	0.00	4.5
Transamerica Gr & Inc. Fund	−5.6	22.5	69	0.87	4.75	2.3
Paine Webber Classic Gr & Inc.	−5.9	24.5	65	1.39	4.50	2.3
Putnam Fund for Growth, Inc.	−6.1	20.7	2138	0.72	5.75	5.3
Merrill Lynch Capital Fund A	−6.2	22.8	945	0.60	6.50	6.3
Aggressive Growth Funds						
Fidelity Select Plotech	23.8	44.0	157	n/a	2.00	0.0
Fidelity Select Energy Services	18.8	59.5	80	n/a	2.00	0.0
Fidelity Select Health	7.1	42.5	292	n/a	2.00	0.5
Merrill Lynch Nat Resource B	4.5	26.6	404	0.94	0.00	1.8
Fidelity Select Health Sciences	3.7	59.4	88	1.42	0.00	0.5
AXE-Houghton Stock Fund	−2.6	29.8	70	1.15	0.00	0.8
Franklin Dynatech	−3.3	29.5	45	0.82	4.00	0.8
New England Growth Fund	−3.5	22.2	640	1.21	6.50	1.5
IDS Growth Fund	−3.7	36.4	788	0.68	5.00	1.3
Brandywine Fund	−3.8	32.9	287	1.12	0.00	0.2
Mainstay Capital Appreciation	−4.1	26.1	41	1.54	0.00	0.2
Fidelity Select Medical Del	−4.7	58.0	45	n/a	2.00	0.3
WPG Tudor Fund	−4.9	24.7	203	n/a	0.00	0.6
Fidelity Select Computer	−5.3	6.8	34	n/a	2.00	0.0
Fidelity Contrafund	−6.1	43.2	537	n/a	3.00	1.5
Fidelity Select Electronics	−6.6	15.7	43	n/a	2.00	0.0
Legg Mason Spec Invmt Tr	−6.6	32.1	79	1.24	0.00	1.3
Twentieth Century Ultra	−6.9	37.0	398	n/a	0.00	2.5
Janus Twenty Fund	−7.2	50.8	206	1.32	0.00	0.1
AIM Summit Fund	−7.7	30.7	306	0.82	5.50	2.1

(*Source*: Data tracked by CDA Investment Technologies, reported in "Anxious Fund Investors Are Advised to Stay Calm," *Financial Planning,* December 1990, pp. 39–47.)

b. Which of these, the mean or the median, seems to be a better measure of central tendency for these particular sets of data?

2. Compute the variance and standard deviation for 1989 percentage return, fund assets, and 12-month dividend yield for each of the four types of funds.

3. Which of these four types of funds has the largest relative variation in 1989 percentage return?

4. a. Which values represent +2 and −2 standard deviations from the mean 1989 percentage return for *aggressive growth* equity funds?

b. Assuming a bell-shaped distribution, what percent of individual funds in the aggressive growth category should be within the values found in 4a, and what percent actually are?

5. If it cannot be assumed that we have a bell-shaped distribution, how does that influence your answer for question 4b?

SPSS

▨ Solution
Table 3.1

We can use SPSS for the aptitude test scores in Table 3.1. The SPSS program listing in Figure 3.17 was used to request the calculation of the mean, standard deviation, and other descriptive statistics. The mainframe and micro commands are the same. (Remember to end each command line with a period when using the PC version.)

The TITLE command names the SPSS run.

The DATA LIST command gives each variable a name and describes the data as being in free form. The variable name is APTEST.

The BEGIN DATA command indicates to SPSS that the input data immediately follow.

The next ten lines contain the data values. Each line represents the test score of 5 of the 50 applicants.

The END DATA statement indicates the end of the data entry.

The DESCRIPTIVE statement requests an SPSS procedure to compute simple descriptive statistics for the variable(s) in the applicant data set.

Figure 3.18 shows the SPSS output obtained by executing the listing in Figure 3.17.

■ FIGURE 3.17
Input for SPSS or
SPSS/PC. Remove
the periods for
SPSS input.

```
TITLE     DESCRIPTIVE STATISTICS.
DATA LIST FREE     /  APTEST.
BEGIN DATA.
22    44    56    68    78
25    44    57    68    78
28    46    59    69    80
31    48    60    71    82
34    49    61    72    83
35    51    63    72    85
39    53    63    74    88
39    53    63    75    90
40    55    65    75    92
42    55    66    76    96
END DATA.
DESCRIPTIVE APTEST.
```

■ FIGURE 3.18
SPSS output.

DESCRIPTIVE STATISTICS

Number of Valid Observations (Listwise) =   50.00

Variable	Mean	Std Dev	Minimum	Maximum	N	Label
APTEST	60.36	18.61	22.00	96.00	50	

SAS

▨ Solution
Table 3.1

We can use SAS for the aptitude test scores in Table 3.1. The SAS program listing in Figure 3.19 was used to request the calculation of the mean, standard deviation, and other descriptive statistics. The mainframe and micro commands are the same.

The TITLE command names the SAS run (enclose it in single quotes).

The DATA command gives the data a name.

The INPUT command names and gives the correct order for the different fields on the data lines. The variable name is APTEST.

The CARDS command indicates to SAS that the input data immediately follow.

The next 50 lines contain the data. Each line represents one applicant's aptitude test score.

The PROC MEANS command requests an SAS procedure to print simple descriptive statistics for the variable APTEST.

Figure 3.20 shows the SAS output obtained by executing the listing in Figure 3.19.

■ FIGURE 3.19
Input for SAS
(mainframe or
micro version).

```
TITLE     'DESCRIPTIVE STATISTICS';
DATA APTDAT;
INPUT APTEST;
CARDS;
22
25
28
 .
 .
 .
90
92
96
PROC MEANS;
```

■ FIGURE 3.20
SAS output.

DESCRIPTIVE STATISTICS

Analysis Variable : APTEST

N Obs	N	Minimum	Maximum	Mean	Std Dev
50	50	22.0000000	96.0000000	60.3600000	18.6052001

CHAPTER

Probability Concepts

You use descriptive statistics to summarize or present data that consist of observations that have already occurred. If these data are drawn from a population, then you describe your sample in some way. If you wish to infer something about the population using the smaller sample, you must deal with uncertainty. To measure the chance that something will occur, you use its **probability.** The concepts of probability form the foundation of all decision making in statistics. By using probabilities, you are able to deal with uncertainty because you are able, at least, to measure it.

To illustrate this idea, suppose that a recent report contained the results of a random sample of 100 homes within a large metropolitan city and stated that the average electric bill was $185. However, the electric company claims that the average bill for all of its customers is $110. Is something wrong here? Do you believe that the *population* mean is $110 based on the fact that the *sample* mean is $185? Here we need a probability, in particular the probability of observing a sample mean at least this large (that is, $185 or more), assuming that the electric company is correct in their claim. If we decide that this probability is extremely small, we can infer that the *population* claim is incorrect, based on the *sample* results.

As mentioned in Chapter 1, there is always the possibility of arriving at the wrong decision (maybe the electric company *is* correct) when using sample results to infer something concerning a population. This particular type of question will be addressed in Chapter 8. We begin the journey into probability in this chapter by introducing some basic concepts and discussing various ways of determining probabilities.

4.1
Events and Probability

An activity for which the outcome is uncertain is an **experiment.** An experiment need not involve mixing chemicals in the laboratory; it could be as simple as throwing two dice and observing the total of the faces turned up. At the completion of an experiment, a measurement of some kind is obtained. An **event** consists of one or more possible outcomes of the experiment; it is usually denoted by a capital letter.* The following are examples of experiments and some corresponding events:

1. *Experiment:* Rolling two dice; *events:* A = rolling a total of 7, B = rolling a total greater than 8, C = rolling two 4s.
2. *Experiment:* Taking a CPA exam; *events:* A = pass, B = fail.
3. *Experiment:* Observing the number of arrivals at a drive-up window over a 5-minute period; *events:* A_0 = no arrivals, A_1 = one arrival, A_2 = two arrivals, and so on.

When you estimate a probability, you are estimating the probability *of an event.* For example, when rolling two dice, the probability that you will roll a total of 7 (event A) is the probability that event A occurs. It is written $P(A)$. The probability of any event is always between 0 and 1, inclusive.

Notation
$P(A)$ = probability that event A occurs

---

*The set of all possible outcomes of an experiment is often referred to as the *sample space.*

Classical Definition of Probability

Suppose a particular experiment has n possible outcomes and event A occurs in m of the n outcomes. The *classical definition* of the probability that event A will occur is

$$P(A) = m/n \qquad\qquad (4.1)$$

This definition assumes that all n possible outcomes have the same chance of occurring. Such outcomes (events) are said to be **equally likely,** and each has probability $1/n$ of occurring. If this is not the case, the classical definition does not apply.

Consider the experiment of tossing a nickel and a dime into the air and observing how they fall. Event A is observing one head and one tail. The possible outcomes are (H = head, T = tail):

NICKEL	DIME
H	H
H	T
T	H
T	T

Thus, there are two ($m = 2$) outcomes that constitute event A of the four possible outcomes ($n = 4$). These four outcomes are equally likely, so each occurs with probability 1/4. Consequently,

$$P(A) = 2/4 = .5$$

Relative Frequency Approach

Another method of estimating a probability is referred to as the **relative frequency** approach. It is based on observing the experiment n times and counting the number of times an event (say, A) occurs. If event A occurs m times, your estimate of the probability that A will occur in the future is

$$P(A) = m/n \qquad\qquad (4.2)$$

Suppose that a particular production process has been in operation for 250 days; 220 days have been accident-free. Let A = a randomly chosen day in the future is free of accidents. Using the relative frequency definition, then

$$P(A) = 220/250 = .88$$

Subjective Probability

Another type of probability is **subjective probability.** It is a measure (between 0 and 1) of your belief that a particular event will occur. A value of one indicates that you believe this event will occur with complete certainty.

Examples of situations requiring a subjective probability are:

The probability that the Dow Jones closing index will be below 2500 at some time during the next 6 months.

The probability that your newly introduced product will capture at least 10% of the market.

The probability that an audited voucher will contain an error.

The probability that your recently married cousin, divorced five times already, will once again go down alimony lane.

Although no two people may agree on a particular subjective probability, these probabilities are governed by the same rules of probability, which are developed later in the chapter.

◢ 4.2
Basic Concepts

Datacomp has recently conducted a survey of 200 selected purchasers of their new microcomputer to obtain a sex-and-age profile of their customers. The results obtained are shown in Table 4.1, which is a **contingency**, or **cross-tab table.** Such tables are a popular method of summarizing a group by means of two categories—in this case, age and sex. The numbers within the table represent the frequency, or number of individuals, within each pair of subcategories and so the contingency table allows you to see how these two categories interact.

There are 60 purchasers who are male *and* under 30; 10 purchasers are female *and* over 45. One person from the total group of 200 is to be selected at random to receive a free software package. We can define the following events:

$$M = \text{a male is selected}$$
$$F = \text{a female is selected}$$
$$U = \text{the person selected is under 30}$$
$$B = \text{the person selected is between 30 and 45}$$
$$O = \text{the person selected is over 45}$$

Because there are 200 people, there are 200 possible outcomes to this experiment. All 200 outcomes are equally likely (the person is randomly selected), so the classical definition provides an easy way of determining probabilities.

The probability of any one single event used to define the contingency table is a **marginal probability.** When you use a contingency table, you can obtain the marginal probabilities by merely counting. For example, of the 200 purchasers, 120 are males. So the probability of selecting a male is

$$P(M) = 120/200 = .6$$

Similarly,

$$P(F) = 80/200 = .4$$
$$P(U) = .5$$
$$P(B) = .25$$
$$P(O) = .25$$

Notice that $P(O) = 50/200 = .25$, which implies that (1) if you repeatedly selected a person at random from this group, 25% of the time the person selected would be over 45 years of age and (2) 25% of the people in this group are over 45 years old. So, a probability here is simply a **proportion.**

■ TABLE 4.1
Datacomp survey of microcomputer purchasers.

SEX	AGE (YEARS)			TOTAL
	<30 (U)	30–45 (B)	>45 (O)	
Male (M)	60	20	40	120
Female (F)	40	30	10	80
Total	100	50	50	200

Complement of an Event   The **complement** of an event A is the event that A does *not* occur. This event is denoted by $\bar{A}$. For example, $A =$ it rains tomorrow, $\bar{A} =$ it does not rain tomorrow; or $A =$ stock market rises tomorrow, $\bar{A} =$ stock market does not rise tomorrow.

In our Datacomp survey, $P(M) = .6$ and so

$$P(\bar{M}) = P(F) = .4$$

Notice that $P(M) + P(\bar{M}) = .6 + .4 = 1.0$. In general, for any event A, either A or $\bar{A}$ must occur. Consequently,

$$P(A) + P(\bar{A}) = 1$$

and so

$$P(\bar{A}) = 1 - P(A)$$

Written another way,

$$P(A) = 1 - P(\bar{A})$$

How can we determine what proportion of the purchasers are age 45 or younger?

$$P(\bar{O}) = 1 - P(O) = 1 - .25 = .75$$

Joint Probability

What if we wish to know the probability of selecting a purchaser who is female *and* under age 30? Such a person is selected if events F *and* U occur. This probability is written $P(F \text{ and } U)$ and is referred to as a **joint probability.*** There are 40 purchasers who are female and under 30, so

$$P(F \text{ and } U) = 40/200 = .2$$

What proportion are males between 30 and 45? This is the same as

$$P(M \text{ and } B) = 20/200 = .1$$

because 20 out of 200 satisfy both requirements.

Probability of A or B

In addition to calculating joint probabilities involving two events, we can also determine the probability that *either* of the two events will occur. In our discussion, "either A or B" will refer to the event that A occurred, B occurred, or both occurred. This probability is written as

$$P(A \text{ or } B)$$

for any two events A and B.†

Now we will calculate the probability of selecting someone who is a male *or* under 30 years of age. This is $P(M \text{ or } U)$. How many people qualify? There are 120 males and there are 100 people under 30. Is the answer $(120 + 100)/200 = 1.1$? You should realize that this is not correct because *a probability is never greater than 1*. What is the mistake here? The problem is that the 60 males under age 30

---

*The joint probability of events A and B is often written as $P(A \cap B)$, read as "the probability of A intersect B."

†The probability $P(A \text{ or } B)$ can be written as $P(A \cup B)$, read as "the probability of A union B."

were counted *twice*. How many purchasers are male or under 30? The answer is the 120 males plus the 40 females under age 30. So

$$P(M \text{ or } U) = (120 + 40)/200 = .8$$

What is $P(F \text{ or } B)$? The people in the shaded area in Table 4.1 qualify. So,

$$P(F \text{ or } B) = (80 + 20)/200 = .5$$

Conditional Probability

Suppose that someone has some inside information about who has been selected from the group of 200 purchasers. This person informs you that the selected individual is under 30 years of age; that is, event U occurred. Armed with this information, we can calculate the probability that the selected person is a male. Given that event U occurred, we have immediately narrowed the number of possible outcomes from 200 to the 100 people under age 30. Each of these 100 people is equally likely to be chosen, and 60 of them are male. So the answer is $60/100 = .6$.

Whenever you are given information and are asked to find a probability based on this information, the result is a **conditional probability.** This probability is written as

$$P(A \mid B)$$

where B is the event that you know occurred and A is the uncertain event whose probability you need, given that event B has occurred. The vertical line indicates that the occurrence of event B is given, so the expression is read as the "probability of A given B." In the example, $P(M \mid U) = .6$.

Suppose that you were given *no information* about U and were asked to find the probability that a male is selected. This is a marginal probability. We earlier determined that $P(M) = .6$. For our example, note that

$$P(M) = P(M \mid U) = .6$$

This means that being given the information that the person selected is under 30 has *no effect* on the probability that a male is selected. In other words, whether U happens has no effect on whether M occurs. Such events are said to be independent. *Thus, events* A *and* B *are* **independent** *if the probability of event* A *is unaffected by the occurrence or nonoccurrence of event* B.

There are a number of ways to demonstrate that any two events A and B are independent.

Definition

Events A and B are **independent** if and only if:

1.	$P(A \mid B) = P(A)$	(assuming $P(B) \neq 0$), or	(4.3)
2.	$P(B \mid A) = P(B)$	(assuming $P(A) \neq 0$), or	(4.4)
3.	$P(A \text{ and } B) = P(A) \cdot P(B)$.		(4.5)

You need not demonstrate all three conditions. If one of the equations is true, they are all true; if one is false, they are all false (in which case A and B are not independent). Events that are not independent are **dependent** events.

In our example, are events F and O independent? We previously showed that

$$P(O) = 50/200 = .25$$

Since $P(O \mid F) = 10/80 = .125$, then $P(O) \neq P(O \mid F)$, and these events are dependent. Put another way, if someone informs you that event F (a female) has occurred, this *does* have an effect on whether the person selected is over 45 years of age. If you are told that *F* occurred, the probability that the selected person is over 45 *drops* from .25 to .125. These events do affect each other and so are dependent events.

We could also approach this by showing that $P(F \mid O)$ is not the same as $P(F)$:

$$P(F \mid O) = 10/50 = .2$$
$$P(F) = 80/200 = .4$$

These are not the same values, so events *F* and *O* are not independent.

The final option is to show that $P(F \text{ and } O)$ is not the same as $P(F) \cdot P(O)$. This follows since

$$P(F \text{ and } O) = 10/200 = .05$$
$$P(F) \cdot P(O) = (.4)(.25) = .1$$

In our discussion of joint probabilities, we showed that

$$P(F \text{ and } U) = 40/200 = .2$$

Consequently, events *F* and *U can both occur* because their joint probability is not zero.

How would you calculate $P(F \text{ and } M)$? One cannot be both a male and a female, so $P(F \text{ and } M) = 0$. Because events *M* and *F* cannot both occur, these events are said to be mutually exclusive.

Definition

Events *A* and *B* are **mutually exclusive** if *A* and *B* cannot both occur simultaneously. To demonstrate that two events *A* and *B* are mutually exclusive, you must show that their joint probability is zero: $P(A \text{ and } B) = 0$.

EXAMPLE 4.1

The quality-control department of Lectron has selected ten devices for testing purposes. Which of these outcomes are mutually exclusive?

$$A = \text{exactly one device is defective}$$
$$B = \text{more than two devices are defective}$$
$$C = \text{fewer than four devices are defective}$$

SOLUTION

A and *B* are mutually exclusive events—they cannot both occur.

A and *C* are *not* mutually exclusive—if *A* occurs, so does event *C*.

B and *C* are *not* mutually exclusive—if three devices are defective, both events *B* and *C* will occur.   ∎

By "not mutually exclusive" we do not mean that both of these events *must* occur, only that both *could* occur. *Also, be sure to distinguish between the terms mutually exclusive and independent.* Loosely, mutually exclusive means that they cannot both occur and independent means that one event occurring has no effect on the other. For example, when drawing a single card from a deck of 52 playing cards, the events *K* = drawing a king and H = drawing a heart are *not* mutually exclusive since they can both occur, namely when drawing the king of hearts. However, they *are* independent, since $P(K) = 4/52 = 1/13$, since there are four kings out of 52 cards, and $P(K \mid H) = 1/13$, since there are 13 hearts and one of them is a king. Consequently, knowing that a heart was selected has *no effect* on whether this card was a king, and so these events are independent.

Summary of Probability Definitions

1. **Experiment:** An experiment is any process that yields a measurement (observation).
2. **Outcome:** An outcome is any particular result of an experiment.
3. **Event:** An event consists of one or more possible outcomes of an experiment.
4. **Complement:** The complement of event A is the event that A does not occur. This is written $\bar{A}$.
5. **Mutually exclusive events:** Two events are mutually exclusive if they cannot both occur simultaneously.
6. **Independent events:** Two events are independent if the probability of one event occurring is unaffected by the occurrence or nonoccurrence of the other.
7. **Probability:** A probability is a measure of the likelihood that an event will occur when the experiment is performed.
8. **Marginal probability:** A marginal probability is the probability that any one single event used to define a contingency table will occur.
9. **Joint probability:** The joint probability of events A and B is the probability that both A and B will occur. This is written as P(A and B).
10. **Conditional probability:** The conditional probability of A given B is the probability that event A occurs given that event B occurs. This is written P(A | B).

 *Exercises*

4.1 Which of the following values cannot be a probability? Why?

a. .02 **b.** 0 **c.** 5/4 **d.** 985/1051

4.2 Explain what the following terminologies mean about two events A and B.

a. Probability of A and B

b. Probability of A or B

c. Probability of the complement of A

4.3 If there are 20 sophomores, 10 juniors, and 5 seniors in a classroom, what is the probability of choosing a junior at random? Is this the relative frequency approach to estimating a probability?

4.4 Assume that 20 doctors are chosen at random from the Houston telephone directory. Of these 20, there are 15 who recommend Little's pills and 5 who do not. If a doctor was chosen at random from the city of Houston, estimate the probability that this doctor would recommend Little's pills.

4.5 Let A and B be two mutually exclusive events. Are the complement of A and the event B mutually exclusive?

4.6 Four hundred randomly sampled automobile owners were asked whether they selected the particular make and model of their present car mainly because of its appearance or because of its performance. The results were as follows:

OWNER	APPEARANCE	PERFORMANCE	TOTALS
Male	95	55	150
Female	85	165	250
Both	180	220	400

a. What is the probability that an automobile owner buys a car mainly because of its appearance?

b. What is the probability that an automobile owner buys a car mainly because of its appearance and that the automobile owner is a male?

c. What is the probability that a female automobile owner purchases the car mainly because of its appearance?

4.7   A large sports chain wants to know whether it should concentrate its advertising on the serious athlete or on the "weekend" athlete. The sports store also wants to know which sports are the most popular. The marketing department gathered the following information on 500 randomly selected customers.

ATHLETE	TENNIS	RUNNING	BASKET-BALL	SWIMMING	SOCCER	RACQUET-BALL	TOTAL
Serious	46	17	60	43	59	50	275
Weekend	54	63	20	37	11	40	225

a.   What is the probability that a customer's favorite sport is basketball?

b.   What is the probability that a customer is a weekend athlete?

c.   What is the probability either that a customer is a serious athlete or that a customer's favorite sport is running?

d.   What is the probability that a customer's favorite sport is not swimming?

4.8   The employment center at a university wanted to know the proportion of students who worked and also the proportion of those who lived in the dorm. The following data were collected:

LIVING ARRANGEMENTS	WORK FULL TIME	WORK PART TIME	DO NOT WORK	TOTAL
In dorm	19	22	20	61
Not in dorm	25	9	5	39
				100

a.   What is the probability of selecting a student at random who works either full or part time?

b.   What is the probability that a student who works lives in the dorm?

c.   What is the probability that a student either works full time or else does not live in the dorm?

d.   Is the event that a student lives in the dorm independent of the event that a student works full time? Discuss what your answer means.

4.9   An investment-newsletter writer wanted to know in which investment areas her subscribers were most interested. A questionnaire was sent to 331 randomly selected professional clients, with the following results:

BUSINESS	STOCKS	BONDS	COMMERCIAL PAPER	COMMODITIES	STOCK OPTIONS	TOTAL
Doctors	30	25	15	2	0	72
Lawyers	29	34	12	0	5	80
Bankers	50	35	29	5	10	129
Others	21	14	10	3	2	50
						331

a.   What is the probability that an investment client is neither a doctor nor a lawyer?

b.   What is the probability that an investment client is a banker and that the investment client's main investment interest is in commodities?

c.   If an investment client's main investment interest is commodities, what is the probability that he or she is a banker?

d.   What is the probability that an investment client's main investment interest is not in stock options?

e.   Let A be the event that an investment client is a lawyer. Let B be the event that an investment client's main investment interest is in commodities. Are the events A and B mutually exclusive?

4.10   If events A and B are mutually exclusive, is the occurrence of event A affected by the occurrence of event B? Can one say that if two events are mutually exclusive, they are not independent?

4.11 A large supermarket has 67 employees classified by job and by number of years of schooling. The following contingency table gives the categories:

NUMBER OF YEARS OF SCHOOLING

JOB	≤8	9–10	11–12	13–14	15–16	TOTAL
Stocker	1	5	8	1	0	15
Checker	0	5	6	3	0	14
Meat cutter	1	3	7	1	0	12
Cashier	0	0	4	10	5	19
Manager	0	0	1	3	3	7
						67

a. What is the probability that an employee selected at random has 11 or more years of schooling?

b. What is the probability that an employee is either a manager or has 13 or more years of schooling?

c. What is the probability that a cashier has 15 to 16 years of schooling?

d. Let A be the event that an employee is a meat cutter. Let B be the event that an employee has 15 to 16 years of schooling. Are the events A and B mutually exclusive? Are the events A and B independent?

4.12 A statistics instructor wishes to find out the relationship between the classification of a student and the student's grade in the course. The following is a breakdown of the grades and classification of students in three sections of an introductory statistics course:

GRADE	FRESHMAN	SOPHOMORE	JUNIOR	SENIOR
A	0	7	9	10
B	0	6	8	11
C	1	7	9	12
D	2	4	1	4
F	0	6	2	1
Total	3	30	29	38

a. Suppose that one student is randomly selected. What is the probability that the student is a junior and makes at least a B in the course?

b. What is the probability that a senior does not make an A in the course?

c. What is the probability that the student makes a D or F in the course?

d. Let A be the event that a sophomore is taking the course. Let B be the event that the student makes a C in the course. Are the events A and B independent? Are the events A and B mutually exclusive?

4.13 A local bank has 5276 accounts cross-classified by type of account and average account balance. The summarized results are (in dollars):

ACCOUNT BALANCE	CHECKING ACCOUNT	SAVINGS ACCOUNT	NEW ACCOUNT	MONEY-MARKET ACCOUNT	TOTAL
<500	1020	803	21	90	1934
500–1000	640	774	452	112	1978
>1000	51	659	538	116	1364
				Total	5276

a. What is the probability that an account does not have over $1000 in it and that the account is not a money-market account?

b. What is the probability that a new account's balance is between $500 and $1000?

c. What is the probability that an account has less than $500 in it or that the account is a savings account?

d. Given that an account is not a savings account, what is the probability that the account has $1000 or less in it?

4.14 If the probability that it is going to rain today is .3, what can you say about the probability that it is not going to rain today?

4.15 Give an example of two events that are mutually exclusive. Explain why they are mutually exclusive. Give an example of two events that are independent. Explain why they are independent.

◢ 4.3
Going Beyond the Contingency Table

Our Datacomp survey served as an intuitive introduction to probability definitions. The classical approach was used to derive probabilities by dividing the number of outcomes favorable to an event by the total number of (equally likely) outcomes. Not all probability problems, however, are concerned with randomly selecting an individual from a contingency table.

When dealing with two or more events in general, one approach is to illustrate these events by means of a **Venn diagram.** A Venn diagram representing any two events A and B is shown in Figure 4.1.

In a Venn diagram, the probability of an event occurring is its corresponding area. This may sound complicated, but it really is not. The Venn diagram for $P(A) = .4$ is shown in Figure 4.2. The area of the rectangle is 1; it represents all possible outcomes. The shaded area is the complement of A, namely, $\overline{A}$. Here, $P(\overline{A}) = 1 - P(A) = 1 - .4 = .6$. No effort is made to construct a circle with an area of .4; it is simply labeled .4. The shaded area then represents $\overline{A}$, and the corresponding area must be .6.

Figure 4.3 shows $P(A \text{ and } B)$, and Figure 4.4 shows $P(A \text{ or } B)$.

■ **FIGURE 4.1**
Venn diagram for events A and B. The rectangle represents all possible outcomes of an experiment.

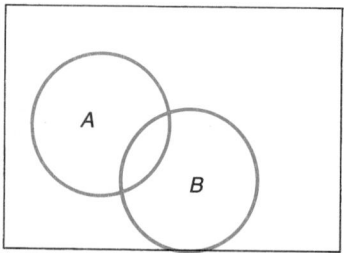

■ **FIGURE 4.2**
Venn diagram for $P(A) = .4$.

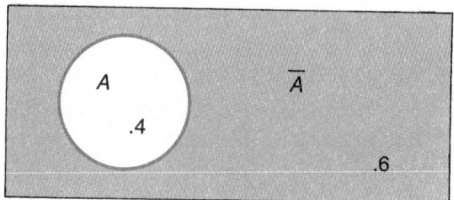

■ **FIGURE 4.3**
$P(A \text{ and } B)$. The points in the shaded area are in A and B.

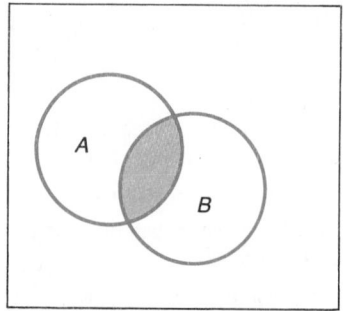

■ FIGURE 4.4
P(*A* or *B*). The points
in the shaded area
are in *A* or *B*.

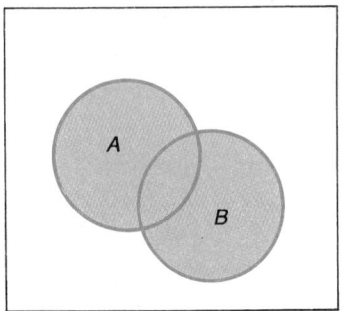

■ FIGURE 4.5
Venn diagram of
mutually exclusive
events. *P*(*A* and *B*)
= 0. *P*(*A* or *B*) =
P(*A*) + *P*(*B*) = .2
+ .25 = .45.

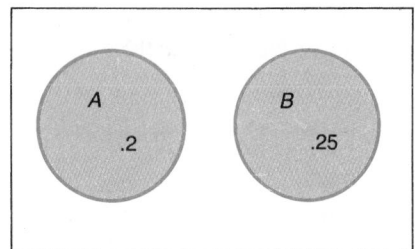

If *A* and *B* are mutually exclusive (they cannot both occur), *P*(*A* and *B*) = 0. For example, an auto dealer has data that indicate that 20% of all new cars ordered contain a red interior and 25% have a blue interior. Only one interior color is allowed. Let *A* be the event that a red interior is selected and *B* be the event that a blue interior is selected. A Venn diagram for this situation is shown in Figure 4.5.

Each person can select only one color, so events *A* and *B* are mutually exclusive, and the resulting circles do not overlap in the Venn diagram. What is the probability that a person selects red *or* blue? This is *P*(*A* or *B*) and is represented by the shaded area in the circles in Figure 4.5. The Venn diagram allows us to see clearly that this shaded area is *P*(*A*) + *P*(*B*) = .2 + .25 = .45. In other words, 45% of the people will purchase either red or blue interiors. We thus have the following rule.

Special Case

If events *A* and *B* are **mutually exclusive,** then

$$P(A \text{ or } B) = P(A) + P(B) \qquad \text{(4.6)}$$

This rule does *not* work when *A* and *B* can both occur, but there is an easy way to devise another solution. Look at the Venn diagram for this situation, shown in Figure 4.6. By adding *P*(*A*) + *P*(*B*), we do not obtain *P*(*A* or *B*) because we have

■ FIGURE 4.6
A Venn diagram
illustrating *P*(*A* or *B*)
and *P*(*A* and *B*).

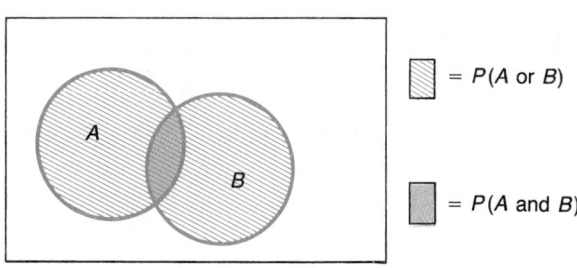

counted $P(A$ and $B)$ *twice*. So we need to subtract $P(A$ and $B)$ to obtain the actual area corresponding to $P(A$ or $B)$. This is the **additive rule of probability.**

Additive Rule
For **any** two events, A and B,
$P(A$ or $B) = P(A) + P(B) - P(A$ and $B)$ **(4.7)**

Notice that if A and B are mutually exclusive, then $P(A$ and $B) = 0$, and we obtain the previous rule; namely, that $P(A$ or $B) = P(A) + P(B)$.

EXAMPLE 4.2

Draw a single card from a deck of 52 playing cards. Let S be the event that the card is a seven and H be the event that the card is a heart. What is $P(S$ or $H)$?

SOLUTION

First, determine $P(S$ and $H)$. $P(S$ and $H)$ is the probability of selecting a seven of hearts from the deck. There is only one such card, so

$$P(S \text{ and } H) = 1/52$$

A Venn diagram for this situation is shown in Figure 4.7. Using the additive rule, the proportion of draws (probability) on which a seven *or* a heart will be selected from the deck is

$$P(S \text{ or } H) = P(S) + P(H) - P(S \text{ and } H)$$
$$= 4/52 + 13/52 - 1/52$$
$$= 16/52$$

Refer back to the Datacomp survey data in Table 4.1. Does the additive rule work here also? It does—this rule works for *any* two events—but it certainly is a hard way to solve this problem. Suppose we want to find the probability (from our previous example) that the person selected is a male or is under age 30. By inspection, we previously found that

$$P(M \text{ or } U) = 160/200 = .8$$

Using the additive rule, we obtain the same result:

$$P(M \text{ or } U) = P(M) + P(U) - P(M \text{ and } U)$$
$$= 120/200 + 100/200 - 60/200$$
$$= 160/200 = .8$$

Conditional Probabilities

Using the Datacomp survey data, we found that the probability the person selected is a male (M), given the information that the person selected is under 30 (U), was $P(M \mid U) = .6$. Our reasoning here was: (1) There are 100 people under 30 years

◼ FIGURE 4.7
$P(S) = 4/52;$
$P(H) = 13/52.$

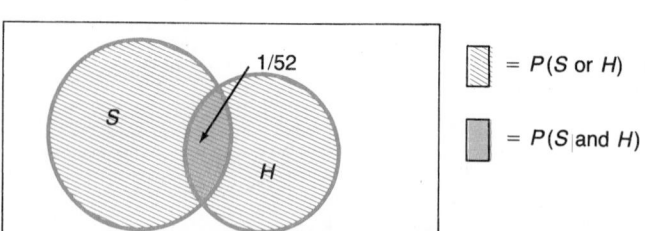

■ **FIGURE 4.8**
A Venn diagram
illustrating a conditional
probability.

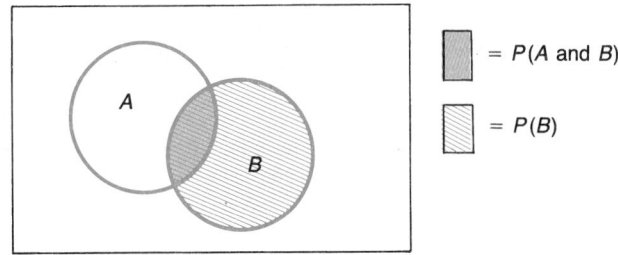

of age, (2) 60 of them are male, (3) each of these 100 people is equally likely to be selected, and so (4) the result is $60/100 = .6$. Notice that

$$P(U) = 100/200 = .5$$

$$P(M \text{ and } U) = 60/200 = .3$$

$$P(M \mid U) = P(M \text{ and } U)/P(U) = .3/.5 = .6$$

This procedure for finding a conditional probability applies to *any* two events. Use the Venn diagram in Figure 4.8 to determine $P(A \mid B)$. Given the information that event B occurred, we are immediately restricted to the lined area (B). What is the probability that a point in B is also in A (that is, event A occurs)? A point is also in A if it lies in the shaded area, and

$$P(A \mid B) = \frac{\text{shaded area}}{\text{striped area}}$$

$$= \frac{P(A \text{ and } B)}{P(B)}$$

This is the rule for conditional probabilities.

Rule for Conditional Probabilities

For any two events, A and B,

$$P(A \mid B) = \frac{P(A \text{ and } B)}{P(B)} \qquad (P(B) \neq 0) \qquad \textbf{(4.8)}$$

and

$$P(B \mid A) = \frac{P(A \text{ and } B)}{P(A)} \qquad (P(A) \neq 0) \qquad \textbf{(4.9)}$$

Independent Events

In the Datacomp example, equations 4.3, 4.4, and 4.5 provided a summary of how to demonstrate that two events are independent. *One need demonstrate only that one of these equations holds to verify independence.* These three methods of proving independence apply to *any two events,* not just to contingency table applications.

To summarize, events A and B are **independent** if any of the following statements can be verified:

$$P(A \mid B) = P(A)$$

$$P(B \mid A) = P(B)$$

$$P(A \text{ and } B) = P(A) \cdot P(B)$$

In many situations, it is unnecessary (or impossible) to prove independence of two events. However, one can often argue convincingly that two events are independent or dependent without resorting to a mathematical proof. Consider these events:

A = Procter and Gamble's new laundry detergent will capture at least 5% of the market next year

and

B = General Motors will introduce a new line of compact automobiles next year

Whether event B happens should have no effect on whether event A occurs. So $P(A \mid B) = P(A)$, and these events are independent. Next, change event A to: Toyota automobile sales will drop next year. Now whether event B occurs could very well have an effect on whether event A occurs. It is not safe to assume that $P(A \mid B) = P(A)$—it seems reasonable that $P(A \mid B)$ is *larger* than $P(A)$. Notice that we have not discussed the values of $P(A)$ and $P(A \mid B)$. The probability values are not necessary to show that the events are dependent. The important thing is that $P(A \mid B) \neq P(A)$, so these events are clearly dependent events.

Joint Probabilities

The rule for conditional probabilities in equations 4.8 and 4.9 can be rewritten as

Multiplicative Rule
For **any** two events A and B,

$$P(A \text{ and } B) = P(A \mid B) \cdot P(B) \qquad (4.10)$$
$$= P(B \mid A) \cdot P(A) \qquad (4.11)$$

This is the **multiplicative rule of probability.** Using equation 4.5, we also have the following rule for two independent events.

Special Case
For any two independent events A and B,

$$P(A \text{ and } B) = P(A) \cdot P(B) \qquad (4.12)$$

You may be wondering how we can use the same equation to define the rule for $P(A \mid B)$ (equation 4.8) and the rule for $P(A \text{ and } B)$ (equation 4.10). This is not a bad question! It appears that we have used the same rule twice to make two different statements—and in fact we have. However, for any application you encounter, either $P(A \mid B)$ or $P(A \text{ and } B)$ must be provided or can be determined without resorting to formulas. We can clarify this using our card-drawing example:

$$S = \text{select a seven}$$
$$H = \text{select a heart}$$

Here $P(S \text{ and } H)$ (the probability of selecting a seven of hearts) is 1/52. No formulas were necessary to determine this, only a little head scratching.

Now, what is $P(S \mid H)$? Using equation 4.8,

$$P(S \mid H) = P(S \text{ and } H)/P(H)$$
$$= (1/52)/(13/52)$$
$$= 1/13$$

Assume that you select a card from a deck, examine it, and then discard it. You then select another card. This procedure is called **sampling without replacement.** Let

$$A = \text{selecting a seven on the first draw}$$
$$B = \text{selecting a seven on the second draw}$$

What is the probability of drawing two sevens [$P(A \text{ and } B)$]? If you selected a seven on the first draw, then, of the 51 cards remaining, three are sevens. So $P(B \mid A) = 3/51$. Again, we used no formulas.

Next, we use the multiplicative rule, equation 4.11:

$$P(A \text{ and } B) = P(B \mid A) \cdot P(A)$$
$$= \left(\frac{3}{51}\right) \cdot \left(\frac{4}{52}\right) \cong .0045$$

Notice that $P(A) = 4/52$ because there are four sevens available on the first draw. So you would expect to draw two sevens from a card deck about 45 times out of 10,000, if you are drawing without replacement.

Now suppose you select a card from a deck but replace it before selecting the second card. This procedure is called **sampling with replacement.** What is $P(B \mid A)$? There are still 52 cards in the deck when you select your second card, and four of these are sevens. So

$$P(B \mid A) = 4/52 = P(B)$$

If event A occurs, the probability of a seven on the second draw is unaffected. This probability is 4/52 *whether or not A* occurs; these events are now independent. For this situation,

$$P(A \text{ and } B) = P(A \mid B) \cdot P(B)$$
$$= P(A) \cdot P(B) \quad \text{(since they are independent)}$$
$$= 4/52 \cdot 4/52 = .0059$$

The probability of getting two sevens is higher when drawing cards with replacement—not a surprising result.

 **4.4**

***Applying
the Concepts***   Using the Formulas

EXAMPLE 4.3   In a particular city, 20% of the people subscribe to the morning newspaper, 30% subscribe to the evening newspaper, and 10% subscribe to both. Determine the probability that an individual from this city subscribes to the morning newspaper, the evening newspaper, or both.

SOLUTION   *The most important step in solving a wordy probability problem is to set up the problem correctly.* Your first step should always be to *define* the events clearly using capital letters. Your initial step should be to define

$$M = \text{person subscribes to the morning newspaper}$$
$$E = \text{person subscribes to the evening newspaper}$$

We do not need to define another event for a person subscribing to both newspapers, as we shall see.

We now have

$$P(M) = .2$$
$$P(E) = .3$$

The probability that a selected individual subscribes to the morning *and* the evening newspaper is given as .10. This is a *joint* probability:

$$P(M \text{ and } E) = .1$$

We want to find the probability of *M* or *E*. Using the additive rule,

$$P(M \text{ or } E) = P(M) + P(E) - P(M \text{ and } E)$$
$$= .2 + .3 - .1$$
$$= .4$$

So 40% of the people in this city subscribe to at least one of the two newspapers.

Suppose we also know that 1/3 of the evening newspaper subscribers are also morning newspaper subscribers. How can you translate this statement into a probability? We can restate the preceding sentence as "Given that a randomly selected individual subscribes to the evening newspaper, the probability that this person also subscribes to the morning newspaper is 1/3." In other words, this is a *conditional* probability:

$$P(M \mid E) = 1/3$$

EXAMPLE 4.4

Referring to the subscription data in Example 4.3, what percentage of the evening subscribers do not subscribe to the morning newspaper?

SOLUTION 1

A Venn diagram for this problem is shown in Figure 4.9. Notice that *M* (the morning subscribers) is made up of two components: (1) those people in *E* (the evening subscribers) and (2) those not in *E*. Since $P(M \text{ and } E) = .1$, the area of *M* that is striped is

$$P(M) - P(M \text{ and } E) = P(M \text{ and } \bar{E})$$
$$= .2 - .1 = .1$$

Similarly, the area of *E* that is striped is

$$P(E) - P(M \text{ and } E) = P(E \text{ and } \bar{M})$$
$$= .3 - .1 = .2$$

Our question could be stated, "Given that a person subscribes to the evening newspaper, what is the probability that this person does not subscribe to the morning newspaper?" This is the *conditional* probability

$$P(\bar{M} \mid E)$$

■ FIGURE 4.9
Venn diagram for
Example 4.4.

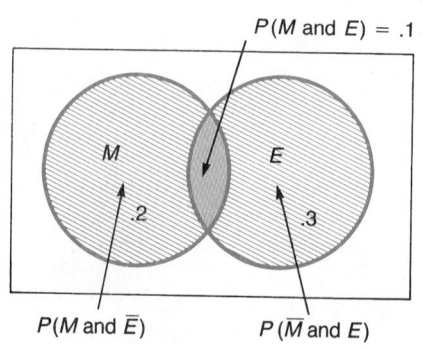

Look at the Venn diagram. You know that E occurred, so the outcome is in the E circle. What is the probability that the outcome is not in M? We know that the total area of E is .3 and that the area that is not in M but is in E is .2. So

$$P(\bar{M} \mid E) = .2/.3 = 2/3$$

Another approach here is to utilize the formulas in the section by noting that given event E has occurred, either event M occurs or it doesn't. Consequently,

$$P(\bar{M} \mid E) = 1 - P(M \mid E)$$
$$= 1 - [P(M \text{ and } E)]/P(E)$$
$$= 1 - (.1/.3) = 2/3 \qquad \blacksquare$$

Using a Contingency Table

SOLUTION 2   Although Example 4.3 made no mention of a random sample, a useful device here is to imagine that a random sample of say, $n = 100$ people is obtained (actually, any sample size (n) could be used). Next, construct a contingency table like the one in Section 4.2, by *assuming that the population percentages given in the problem apply to these 100 people*. In Example 4.3, we would assume that 20% of the sample (20 people) subscribe to the morning newspaper and 30% of the sample (30 people) subscribe to the evening newspaper. So far, the contingency table would be

	M	$\bar{M}$	
E			30
$\bar{E}$			
	20		100

Since the sample size is 100, the totals for $\bar{M}$ and $\bar{E}$ are 80 and 70, respectively, producing the following table

	M	$\bar{M}$	
E			30
$\bar{E}$			70
	20	80	100

The final piece of information is that 10% of the people (10 people) subscribe to *both* newspapers. So, 10 people are in the cell in the upper left corner, corresponding to E *and* M, and the table is now

	M	$\bar{M}$	
E	10		30
$\bar{E}$			70
	20	80	100

By using the row and column totals, the remaining cells can be filled in.*

	M	$\bar{M}$	
E	10	20	30
$\bar{E}$	10	60	70
	20	80	100

Once the table is filled in, probabilities become very easy to derive using the approach used in Section 4.2. To illustrate, the solution to Example 4.3 is

$$P(M \text{ or } E) = \frac{10 + 10 + 20}{100} = \frac{40}{100} = .4$$

---

*This procedure is easier to apply and explain if the sample size (n) is chosen so that all numbers in the contingency table are counting numbers (integers).

The solution to Example 4.4 is

$$P(\bar{M} \mid E) = \frac{20}{30} = 2/3$$

since there are 30 people who subscribe to the evening newspaper (E), 20 of whom do not subscribe to the morning newspaper ($\bar{M}$). ∎

Helpful Hints for Probability Applications

USING THE FORMULAS

1. Define each event using capital letters.
2. Translate each statement into a probability. Does a particular statement tell you $P(A)$? $P(B)$? $P(A$ and $B)$? $P(A$ or $B)$? $P(A \mid B)$? $P(B \mid A)$?
3. Determine the answer by identifying the probability rule that applies and by using a Venn diagram. Using both allows you to check your logic and your arithmetic.

USING A CONTINGENCY TABLE

1. Select *any* sample size, say $n = 100$ or 1000.
2. Using the information given, fill in the row and column totals.
3. Using the final piece of information given, fill in the appropriate cell and complete the contingency table.
4. Determine the answer by dividing the proper value by the appropriate total.

EXAMPLE 4.5

In a certain northeastern state going through financial difficulties, it is believed that 5% of the banks will fail. It is known that the deposits of 90% of the banks in this state are insured by the Federal Depository Insurance Company (FDIC). It is also believed, from past experience, that 3% of the banks protected by FDIC will fail. A bank examiner employed by the federal government would like to know:

1. What is the probability that, for a randomly chosen bank, the bank has deposits protected by FDIC and the bank will fail?
2. What is the probability that, for a randomly chosen bank, the bank has deposits covered by FDIC or the bank will fail?
3. What percentage of the banks that go under have deposits protected by FDIC?

Using the Formulas

SOLUTION 1 The first step is to define appropriate events:

$$A = \text{bank has deposits protected by FDIC}$$
$$B = \text{bank will fail}$$

We now translate each of the statements into a probability. We have the following marginal probabilities:

$$P(A) = .90$$
$$P(B) = .05$$

The last statement in the problem can be written as, "Given that a bank has accounts protected by FDIC, the probability that the bank will fail is .03." So this is a conditional probability, namely,

$$P(B \mid A) = .03$$

What does question 1 ask for? $P(A$ or $B)$? $P(A \mid B)$? $P(A$ and $B)$? The examiner wishes to know the probability that a bank is protected by FDIC *and* will fail.

This is $P(A \text{ and } B)$. Using the multiplicative rule,

$$P(A \text{ and } B) = P(B \mid A) \cdot P(A)$$
$$= (.03)(.90) = .027$$

SOLUTION 2 For question 2, we wish to know the probability that A or B occurs. By the additive rule,

$$P(A \text{ or } B) = P(A) + P(B) - P(A \text{ and } B)$$
$$= .90 + .05 - .027 = .923$$

Thus, 92.3% of the banks are covered by the FDIC, will fail, or both.

SOLUTION 3 Question 3 can be phrased as, "Given that a bank has failed, what is the probability that this bank has deposits protected by FDIC?" This is $P(A \mid B)$.

$$P(A \mid B) = [P(A \text{ and } B)]/P(B) = .027/.05 = .54$$

Therefore, 54% of those banks that fail have deposits protected by FDIC.

Using a Contingency Table

To obtain a table containing all counting numbers, a sample size of $n = 1000$ is used here. Remember that any sample size can be used with this approach. Five percent of the banks failed (.05 × 1000 = 50 banks) and 90% of the banks are insured by FDIC (.90 × 1000 = 900 banks). Filling in the remaining row and column totals, the table is

	FDIC	$\overline{\text{FDIC}}$	
FAIL			50
$\overline{\text{FAIL}}$			950
	900	100	1000

Finally, 3% of the banks *protected by FDIC* failed. There are 900 banks protected by FDIC and so we find .03 × 900 = 27 banks.* So, 27 of the banks are protected by FDIC and failed. This value goes into the upper left cell and the table is

	FDIC	$\overline{\text{FDIC}}$	
FAIL	27		50
$\overline{\text{FAIL}}$			950
	900	100	1000

Filling in the remaining cells, the completed contingency table is

	FDIC	$\overline{\text{FDIC}}$	
FAIL	27	23	50
$\overline{\text{FAIL}}$	873	77	950
	900	100	1000

SOLUTION 1 $$P(\text{FDIC and FAIL}) = 27/1000 = .027$$

SOLUTION 2 $$P(\text{FDIC or FAIL}) = \frac{27 + 873 + 23}{1000} = \frac{923}{1000} = .923$$

SOLUTION 3 $$P(\text{FDIC} \mid \text{FAIL}) = 27/50 = .54$$

since 50 of the banks failed, 27 of which are protected by FDIC.

---

*A sample size of $n = 100$ would have produced a value of 2.7 here. The problem can still be solved using $n = 100$, but is easier to explain using counting numbers in each cell; hence, $n = 1000$ was selected.

Exercises

4.16   If $P(A) = .5$, $P(B) = .3$, and $P(A \mid B) = .4$, what is $P(B \mid A)$?

4.17   Let $P(A) = .7$, $P(B) = .3$, and $P(A \text{ and } B) = .2$. Find the following probabilities:

a. $P(\bar{A})$ d. $P(A \text{ and } \bar{B})$ f. $P(\bar{A} \text{ and } \bar{B})$

b. $P(A \text{ or } B)$ e. $P(A \mid \bar{B})$ g. $P(\overline{A \text{ and } B})$

c. $P(B \mid A)$

4.18   If $P(A) = .4$, $P(B) = .5$, and $P(A \text{ or } B) = .8$, what is $P(A \text{ and } B)$?

4.19   If $P(A \mid B) = .8$ and $P(A \text{ and } B) = .6$, what is the $P(B)$?

4.20   If $P(A) = .5$, $P(B) = .2$, and $P(A \text{ or } B) = .7$, are the events A and B mutually exclusive? Explain.

4.21   If $P(A) = .5$ and $P(B) = .6$, are the events A and B mutually exclusive? Explain.

4.22   If $P(A) = .4$, $P(B) = .3$, and $P(A \text{ and } B) = .12$, what is the probability of A given B? Are the events A and B independent?

4.23   If a penny, a nickel, and a dime are flipped, what is the probability of getting three heads, given that the flip of the penny resulted in a head?

4.24   Suppose one card is randomly picked from a deck of 52 playing cards. Event A is the occurrence of a king. Event B is the occurrence of a spade.

a.   What is the probability of A and B?

b.   What is the probability of A or B?

c.   What is the probability of A given B?

4.25   A manufacturer of widgets historically has produced 80 good widgets out of every 100 widgets. If two widgets are randomly selected off the assembly line, what is the probability that both widgets will be nondefective? What is the probability that two randomly selected widgets will be defective?

4.26   Two quality-control inspectors need to examine boxes of electrical components at random to determine if certain specifications are being met during the manufacturing process. Assume that the first inspector and the second inspector have a 90% and 60% chance, respectively, of noticing a flawed component if the component is truly flawed. Also, assume that the percentage of defective components in each box is 20%. The inspectors inspect only one component at random from each box.

a.   If a box is selected at random by one of the two inspectors, what is the probability that a defective component will be noticed in that box?

b.   If both inspectors inspect the same box, what is the probability that at least one will notice a defective component?

4.27   If the probability that a person orders the morning newspaper is .5 and the probability that a person orders the evening newspaper is .3, and if the probability that a person orders at least one of the two newspapers is .7, then what is the probability that a person orders both the morning and evening newspapers?

4.28   At a certain university, 30% of the students major in mathematics. Of the students majoring in mathematics, 60% are males. Of all the students at the university, 70% are males.

a.   What is the probability that a student selected at random in the university is a male majoring in mathematics?

b.   What is the probability that a student selected at random in the university is a male or is majoring in mathematics?

c.   What proportion of the males are majoring in mathematics?

4.29   At a semiconductor plant, 60% of the workers are skilled and 80% of the workers are full-time. Ninety percent of the skilled workers are full-time.

a.   What is the probability that an employee selected at random is a skilled full-time employee?

b.   What is the probability that an employee selected at random is a skilled worker or a full-time worker?

c.   What percentage of the full-time workers are skilled?

4.30   A supermarket has 40% of its merchandise on sale. Twenty percent of its merchandise consists of nonedible items. Fifty percent of the sale items consists of nonedible items.

a.   What is the probability that an item selected at random in the supermarket is nonedible and on sale?

b.   What is the probability that an item selected at random is either nonedible or on sale?

c.   What proportion of nonedible items are on sale?

4.31   For every person who visits the leasing office of an apartment community near a certain university, there is an 80% chance that the person will lease an apartment if the person is a student and a 50% chance that the person will lease an apartment if the person is not a student. If two people, one of whom is a student and the other is not, enter the office, what are the chances of leasing an apartment to at least one of the two people? What assumption did you have to make here?

4.32   An independent oil-drilling company drills wildcat oil wells. So far, 60% of the wells have been oil-producing and 40% have been dry. A private investor wishes to go into a partnership with the oil company in two wells. Assuming that the outcome of one well does not affect the outcome of the other well, what is the probability that both of the oil wells in the partnership will produce oil? What is the probability that at least one of the two wells will produce oil?

4.33   Use the additive rule for the data in Exercise 4.7(c) to find the probability that a customer is either a serious athlete or that a customer's favorite sport is running.

4.34   A growing proportion of women regard their work as a career rather than "just a job." In a recent survey, working women were asked the question, "Do you think of your work as a career?" The percentage of yes answers for each of several age groups is given below:

AGE	PERCENTAGE OF WOMEN IN THIS CATEGORY ANSWERING YES
18 to 29	37%
30 to 39	52%
40 to 49	48%
50 and older	43%

(*Source:* Adapted from B. Townsend and K. O'Neil, "American Women Get Mad," *American Demographics* (August 1990): 27.)

Assume that a randomly selected working woman is equally likely to be in any one of the four age categories, that is, the probability that she will be in each class is .25.

a.   What is the probability that a working woman selected at random considers her work as career?

b.   What is the probability that a working woman selected at random is 40 years old or more and does not think of her work as a career?

4.35   In 1989, only 6 of the 50 states in the United States had a state gas tax that was less than 10 cents a gallon. If two states, A and B, were sampled at random from the 50 states in the United States without replacement, what is the probability that either state, A or B, has a gas tax less than 10 cents a gallon?

(*Source: The World Almanac and Book of Facts,* 1990, p. 142.)

◢ 4.5
Tree Diagrams

Another useful device for determining probabilities is a **tree diagram.** Notice that in Example 4.5, both marginal probabilities $P(\text{FDIC})$ and $P(\text{Fail})$ were known; that is, the percentage of banks protected by FDIC was known (90%) and the percentage of banks that would fail was known (5%). The contingency table approach illustrated in the previous section works well for this situation.

Suppose instead that one of the marginal probabilities is missing, say, $P(\text{Fail})$, but that we *do* know that

1. 90% of the banks are protected by FDIC (and 10% are not)
2. 3% of the banks protected by FDIC will fail, and
3. 23% of the banks not protected by FDIC will fail.

So, we can write

$$P(\text{FDIC}) = .90$$
$$P(\overline{\text{FDIC}}) = .10$$
$$P(\text{Fail} \mid \text{FDIC}) = .03$$
$$P(\text{Fail} \mid \overline{\text{FDIC}}) = .23$$

This information can be summarized in the following picture, which we refer to as a tree diagram.

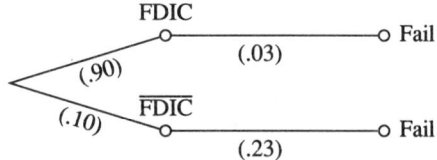

Each number in parentheses is the probability associated with that particular branch. For example, .03 is the probability of a bank failing *given* that it is protected by FDIC, and so this value is placed on the Fail branch corresponding to FDIC.

To find the remaining marginal probability, $P(\text{Fail})$, the following rule is used.

Rule #1 When Using a Tree Diagram

The probability of the event on the right side (say, event *B*) of the tree is equal to the sum of the paths; that is, all probabilities along a path leading to event *B* are multiplied, and then summed over *all* paths leading to *B*.

For the above tree diagram, Rule #1 states that

$$P(\text{Fail}) = (.90)(.03) + (.10)(.23)$$
$$= .027 + .023 = .05$$

Consequently, we conclude that 5% of the banks will fail, the same conclusion we reached in Example 4.5.

Another question of interest might be, given that a bank fails, what is the probability it is protected by FDIC? Or put another way, what percentage of banks that fail are protected by FDIC? This can be written

$$P(\text{FDIC} \mid \text{Fail})$$

Recall that earlier we were given that $P(\text{FDIC}) = .90$; that is, 90% of the banks are protected by FDIC. For this probability we were given no conditions at all; it is referred to as a **prior probability.** Now we are asked to determine the probability of this event having been given some information, namely that the bank failed. The probability $P(\text{FDIC} \mid \text{Fail})$ is called a **posterior probability** and is always a conditional probability; in particular, we are given that the event on the right side of the tree diagram (Fail) did, in fact, occur. Since FDIC lies on the first branch and $\overline{\text{FDIC}}$

■ **FIGURE 4.10**
General form
of a tree diagram.

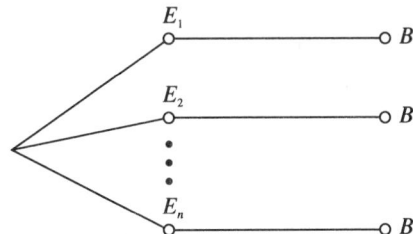

lies on the second branch, we are asked to determine the probability that we got to event Fail along the first path, given that event Fail occurred.

The tree diagram used in this example has two paths. In general, there can be any number of paths, as illustrated in Figure 4.10

To determine a posterior probability, written as $P(E_i \mid B)$ for some i, we use the following rule, usually referred to as Bayes' Rule (named after Thomas Bayes, an English Presbyterian minister and mathematician).

Rule #2 for Tree Diagrams (Bayes' Rule)

The posterior probability for the ith path is

$$P(E_i \mid B) = \frac{i\text{th path}}{\text{sum of paths}}$$

where the "sum of paths" is found using Rule #1.

To illustrate,

$$P(\text{FDIC} \mid \text{Fail}) = \frac{1\text{st path}}{\text{sum of paths}}$$

since FDIC lies on the first path. By "1st path" we mean the product of all probabilities along this path. So,

$$P(\text{FDIC} \mid \text{Fail}) = \frac{(.9)(.03)}{(.9)(.03) + (.1)(.23)}$$

$$= \frac{.027}{.05} = .54$$

Consequently, 54% of the banks that fail are protected by FDIC (the same result we obtained in Example 4.5).

EXAMPLE 4.6   Zetadyne Corporation produces electrical components utilizing three nonoverlapping work shifts. It is known that 50% of the components are produced during shift 1, 20% during shift 2, and 30% during shift 3. A further look at product quality reveals that 6% of the components produced during shift 1 are defective. The corresponding percentage for shift 2 is 8%. Shift 3, the late-night shift, produces a relatively large percentage, 15%, of defective components. Determine

1. What percentage of all components is defective?
2. Given that a defective component is found, what is the probability that it was produced during shift 3?

SOLUTION 1   The tree diagram for this example is

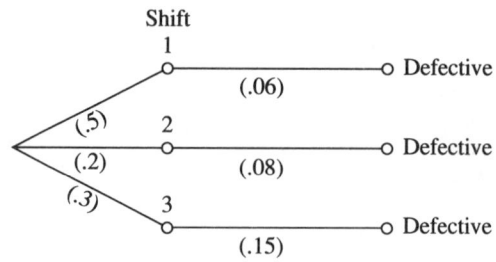

Using Rule #1, we find that

$$P(\text{Defective}) = \text{sum of paths}$$
$$= (.5)(.06) + (.2)(.08) + (.3)(.15)$$
$$= .030 + .016 + .045$$
$$= .091$$

Consequently, 9.1% of the components produced are defective.

SOLUTION 2   We know that $P(\text{shift 3})$ is .3; that is, 30% of the components produced are produced during shift 3. This is a *prior probability.* To determine the *posterior probability,* $P(\text{shift 3} \mid \text{defective})$, we use Rule #2.

$$P(\text{shift 3} \mid \text{defective}) = \frac{\text{3d path}}{\text{sum of paths}}$$
$$= \frac{(.3)(.15)}{.091}$$
$$= \frac{.045}{.091} = .495$$

So, approximately half of the defective components produced are produced during shift 3. This means that once a component is identified as defective, the probability that it came from shift 3 increases from .3 (the prior probability) to .495 (the posterior probability).

The tree diagram illustrated in Figure 4.10 is actually a simplified version of more elaborate decision trees that can be used for more complicated decision problems. A full treatment of decision trees and how they can be used to structure more complex types of decision analyses is provided in Chapter 18.   ■

Exercises   **4.36**   Given the following probabilities, construct a tree diagram and compute $P(E)$.

$$P(A) = .4 \quad P(B) = .6 \quad P(E \mid A) = .10 \quad P(E \mid B) = .05$$

4.37   Given the following probabilities, construct a tree diagram and compute $P(E)$.

$$P(A) = .6 \qquad P(B) = .1 \qquad P(C) = .3$$
$$P(E \mid A) = .03 \qquad P(E \mid B) = .01 \qquad P(E \mid C) = .05$$

4.38   Given the following probabilities, construct a tree diagram and compute $P(C \mid E)$.

$$P(A) = .2 \qquad P(B) = .3 \qquad P(C) = .5$$
$$P(E \mid A) = .15 \qquad P(E \mid B) = .02 \qquad P(E \mid C) = .06$$

4.39   Forty percent of the new employees that a certain firm hires do not have college degrees. Of those employees who do not have college degrees, 30% get promoted to mid-level manager in 3 years. Of those employees who have college degrees, 60% get promoted to mid-level manager in 3 years. What is the probability that a new employee selected at random will move up to mid-level manager in 3 years?

4.40   Seventy percent of all Big Burger chain stores decided to advertise in their local newspapers. Of those chain stores that advertised in their local newspapers, 60% had an increase in sales. Of those chain stores that did not advertise in their local newspapers, 25% had an increase in sales. What is the probability that a randomly selected store with an increase in sales advertised in its local newspaper?

4.41   A large manufacturing company is in the process of training its personnel in quality control procedures. At present, 40% of the assembly lines use control charts, 40% use inspection techniques, and 20% do not use any method for controlling quality. The assembly lines that use control charts have a 1% defective rate. The assembly lines that use inspection techniques have a 5% defective rate. The assembly lines that do not use any quality control techniques have a 12% defective rate. What is the probability that an item produced by this company is defective?

4.42   A software company surveyed office managers to determine the probability that they would buy a new graphics package that includes three-dimensional graphs. Eighty percent of the office managers claimed that they would buy the graphics package. Of those managers who would buy the graphics package, 40% were also interested in upgrading their computer hardware. Of those managers who were not interested in purchasing the graphics package, only 10% were interested in upgrading their computer hardware. What is the probability that an office manager who is interested in upgrading his or her computer hardware is also interested in purchasing the graphics package?

4.43   Materials for a food-processing plant are supplied by four companies. The following table lists the percentage of defective items from each company and the percentage of materials supplied by that company to the food-processing plant.

	PERCENTAGE OF MATERIALS SUPPLIED	PERCENTAGE OF DEFECTIVE MATERIALS
Supplier 1	40	2
Supplier 2	5	10
Supplier 3	20	8
Supplier 4	35	3

a.   Determine the percentage of all materials that are defective.

b.   Given that a material supplied to the plant is defective, what is the probability that it came from supplier 3?

◢ 4.6
Probabilities for More Than Two Events

We illustrate what happens when you encounter more than two events by considering three events, *A*, *B*, and *C*. The following rules can easily be extended to any finite number of events. In the applications of probability in the chapters that follow, we typically will be dealing with multiple events that are either mutually exclusive or independent.

Mutually Exclusive Events

Events *A*, *B*, and *C* are mutually exclusive if no two events can occur simultaneously. A Venn diagram of this situation is shown in Figure 4.11. When dealing with

■ **FIGURE 4.11**
Three mutually exclusive events.

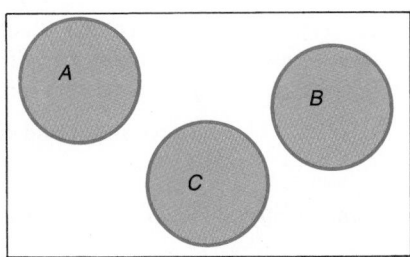

mutually exclusive events, we usually will be interested in the probability that *one* of these events will occur, that is $P(A$ or B or $C)$. We can use a simple rule here:

For mutually exclusive events, A, B, and C,

$$P(A \text{ or } B \text{ or } C) = P(A) + P(B) + P(C) \tag{4.13}$$

Thus, to determine "or" probabilities when the events are mutually exclusive, you add the respective probabilities.

Independent Events

Events A, B, and C are independent if all the following are true:

$$P(A \text{ and } B) = P(A) \cdot P(B)$$
$$P(A \text{ and } C) = P(A) \cdot P(C)$$
$$P(B \text{ and } C) = P(B) \cdot P(C)$$
$$P(A \text{ and } B \text{ and } C) = P(A) \cdot P(B) \cdot P(C)$$

Thus the events are independent if the "and" probability for *any* subset of the events (including the set containing all the events) is equal to the corresponding product of marginal probabilities. When dealing with independent events, the probability of interest usually is that *all* of the events occur, that is, $P(A$ and B and $C)$. Using the fourth condition above, we can make the following statement.

For independent events, A, B, and C,

$$P(A \text{ and } B \text{ and } C) = P(A) \cdot P(B) \cdot P(C) \tag{4.14}$$

Thus, to determine "and" probabilities when the events are independent, multiply the respective probabilities.

EXAMPLE 4.7   Dellex Industries makes memory units for a microcomputer. Dellex customers have agreed to select three units randomly from a very large shipment and test them. If none of the units is defective, the customer will accept the shipment.

Usually, 2% of all Dellex units are defective. Determine the probability that the shipment will be accepted, that is, that all three units tested will be nondefective.

SOLUTION   Let

$$A = \text{first unit is nondefective}$$
$$B = \text{second unit is nondefective}$$
$$C = \text{third unit is nondefective}$$

We know that 2% of the units produced are defective, so 98% of them are not defective; consequently, $P(A) = P(B) = P(C) = .98$.

We want to find $P(A$ and B and $C)$. Are the events independent? There is no need to use fancy formulas here. The answer is yes, simply because the units are selected randomly from the very large shipment. Therefore,

$$P(A \text{ and } B \text{ and } C) = P(A) \cdot P(B) \cdot P(C)$$
$$= .98 \cdot .98 \cdot .98 = .94$$

Consequently, there is a 94% chance that the customer will accept the shipment.

■

Exercises

4.44 Let $P(A) = P(B) = P(C) = .5$ and let $P(A \text{ and } B) = P(A \text{ and } C) = P(B \text{ and } C) = P(A \text{ and } B \text{ and } C) = .25$. Are the events A, B, and C independent events?

4.45 If three events are independent, are the three mutually exclusive? If three events are mutually exclusive, then are the three events independent?

4.46 A. W. Lawrence Securities Corporation is selling corporate bonds that are rated A plus, A, B plus, B, and C. The following is the proportion of bonds for sale in each of these categories.

CATEGORY	PROPORTION (IN PERCENTAGE)
A plus	30
A	10
B plus	20
B	25
C	15

Assume that a novice bond buyer is unaware of the ratings and chooses a bond at random. What is the probability that the investor chooses a bond rated B plus or better?

4.47 In Exercise 4.9, let A be the event that a subscriber is a banker. Let B be the event that a subscriber is a lawyer. Let C be the event that the subscriber is a doctor.

a. Are the events A, B, and C mutually exclusive?

b. What is $P(A \text{ or } B \text{ or } C)$?

c. What is $P(A \text{ and } B \text{ and } C)$?

4.48 In Exercise 4.11, let A be the event that an employee has 8 years or less of schooling. Let B be the event that the employee has 9–10 years of schooling. Let C be the event that an employee has 15–16 years of schooling.

a. Are the events A, B, and C mutually exclusive?

b. What is $P(A \text{ or } B \text{ or } C)$?

c. What is $P(A \text{ and } B \text{ and } C)$?

4.49 Three cards are picked with replacement from a deck of 52 playing cards. What is the probability that the first card will be a queen, that the second will be a spade, and that the third will be a king?

4.50 At a university, 40% of the accounting majors, 20% of the marketing majors, and 15% of the finance majors are from out of state. If a student is selected randomly from each of these three majors, what is the probability that all three are from out of state?

4.51 A retailer receives, on the average, one defective calculator out of every 20. What is the probability that three calculators randomly selected by the retailer are nondefective?

4.7
Counting Rules (Optional)

Counting rules determine the number of possible outcomes that exist for a certain broad range of experiments. They can be extremely useful in determining probabilities. For instance, consider an experiment that has 200 possible outcomes, all of which are equally likely to occur. The probability of any one such outcome is $1/200 = .005$.

The question we wish to answer here is, for a particular experiment, how many possible outcomes are there? No set of rules applies to all situations, but we will consider three very popular counting procedures: (1) filling slots, (2) permutations (a special case of filling slots), and (3) combinations.

Filling Slots

We use **counting rule 1** to fill k different slots. Let

n_1 = the number of ways of filling the first slot
n_2 = the number of ways of filling the second slot *after* the first slot is filled.

n_3 = the number of ways of filling the third slot *after* the first two slots are filled

$\vdots$

n_k = the number of ways of filling the kth slot *after* filling slots 1 through $k - 1$.

The number of ways of filling all the k slots is

$$n_1 \cdot n_2 \cdot n_3 \cdot \;\cdots\; \cdot n_k$$

EXAMPLE 4.8   When ordering a new car, you have a choice of eight interior colors, ten exterior colors, and four roof colors. How many possible color schemes are there?

SOLUTION   There are three slots to fill here, (eight) interior color, (ten) exterior color, and (four) roof color. To answer the question, you simply *multiply* the number of ways of filling each slot. So the answer is $8 \cdot 10 \cdot 4 = 320$ different color schemes.

   The order in which you fill the slots is unimportant. So $n_2 = 10$, regardless of whether you have filled the first slot. For some applications, this is not the case. Consider the following example.   ■

EXAMPLE 4.9   A local PTA group is selecting their officers for the current year. There are 15 individuals in the group, whom we label as $I_1, I_2, \ldots, I_{15}$. They need to select a president, vice president, secretary, and treasurer. How many possible groups of officers are there?

SOLUTION   We have four slots to fill here, president (n_1), vice president (n_2), secretary (n_3), and treasurer (n_4). We know that n_1 is 15. After a president is elected, only 14 people remain, so $n_2 = 14$. By a similar argument, $n_3 = 13$ and $n_4 = 12$. The answer is $15 \cdot 14 \cdot 13 \cdot 12 = 32,760$ different slates of officers.   ■

Permutations

Example 4.9 is a counting situation in which you select people *without replacement*. If a particular person, say I_3, is elected president, then I_3 is not available to fill the remaining slots. Another way of stating the result is that there are 32,760 ways of selecting 4 people out of 15, where the **order of selection** is important. For example,

$$I_2 = \text{president}$$
$$I_6 = \text{vice president}$$
$$I_{12} = \text{secretary}$$
$$I_7 = \text{treasurer}$$

is not the same slate of officers as

$$I_7 = \text{president}$$
$$I_{12} = \text{vice president}$$
$$I_2 = \text{secretary}$$
$$I_6 = \text{treasurer}$$

even though the same four people are involved.

   The number of ways of selecting k objects (or people) from a group of n distinct objects, where the order of selection is important, is referred to as the number of permutations of n objects using k at a time. This is written

$$_nP_k$$

In Example 4.9, $_{15}P_4 = 32,760$. Determining the number of permutations is just a special case of counting rule 1; this is also a slot-filling application.

The symbol $n!$ is read as "n factorial." Its value is determined by multiplying n by all the positive integers smaller than n.

$$n! = (n)(n - 1)(n - 2) \cdots (2)(1) \qquad \textbf{(4.15)}$$

For example,

$$5! = (5)(4)(3)(2)(1) = 120$$
$$1! = 1$$
$$0! = 1 \qquad \text{(by definition)}$$

Notice that $n!$ is the number of ways of filling n slots using n objects. There are n ways of filling the first slot, $(n - 1)$ ways of filling the second slot, $(n - 2)$ ways for the third slot, and so on.

In Example 4.9, the result was obtained by finding $15 \cdot 14 \cdot 13 \cdot 12 = 32,760$. This also can be written as

$$\frac{15 \cdot 14 \cdot 13 \cdot 12 \cdot \cancel{11} \cdot \cancel{10} \cdot 9 \cdot 8 \cdot 7 \cdot 6 \cdot 5 \cdot 4 \cdot 3 \cdot 2 \cdot \cancel{1}}{\cancel{11} \cdot \cancel{10} \cdot 9 \cdot 8 \cdot 7 \cdot 6 \cdot 5 \cdot 4 \cdot 3 \cdot 2 \cdot \cancel{1}} = 32,760$$

This is an application of **counting rule 2:** The number of **permutations** of n objects using k objects at a time is

$$_nP_k = \frac{n!}{(n - k)!} = (n)(n - 1) \cdots (n - k + 1) \qquad \textbf{(4.16)}$$

EXAMPLE 4.10

How many two-digit numbers can you construct using the digits 1, 2, 3, and 4, without repeating any digit?

SOLUTION

The order of selection is certainly important here—the number 42 is not the same as 24. The answer is $_4P_2$, where

$$_4P_2 = \frac{4!}{(4 - 2)!} = \frac{(4)(3)(2!)}{2!} = 12$$

These 12 permutations are

12	21	31	41
13	23	32	42
14	24	34	43

∎

Combinations

Take another look at Example 4.9, where we selected 4 people from a group of 15. This time, however, choose a committee of 4 people from a group of 15 where the order of selection does not matter. Each such committee is one *combination* of the 15 people, using four at a time. For example,

$$I_2 \quad I_6 \quad I_{12} \quad I_7$$

and

$$I_7 \quad I_{12} \quad I_2 \quad I_6$$

are different permutations but are the same combination. These two arrangements are made up of the same individuals; hence they form the same committee or combination.

Clearly, there are not as many combinations (using I_2, I_6, I_{12}, I_7) as there are permutations. The two preceding permutations form the same combination. There are now 24 possible permutations of this combination ($4 \cdot 3 \cdot 2 \cdot 1 = 24$).

Now we wish to determine how many possible committees (combinations) of 4 there are for the group of 15. This is written as

$$_{15}C_4$$

Each combination has 24 permutations, so

$$_{15}C_4 = \frac{_{15}P_4}{24} = \frac{32,760}{24} = 1365$$

There are 1365 possible committee combinations. Notice that 24 is the number of *permutations* of these four numbers (2, 6, 7, and 12); that is, $24 = {_4}P_4 = 4!$

Counting rule 3 is used to count the number of possible combinations. The number of **combinations** of n objects using k at a time is

$$_nC_k = \frac{_nP_k}{k!} = \frac{n!}{k!(n-k)!} \qquad\qquad \textbf{(4.17)}$$

EXAMPLE 4.11

A company must select 5 employees from a department of 40 people to attend a national conference. How many possible delegations are there?

SOLUTION

The order of selection is not a factor here, so this is a combination problem rather than a permutation problem. The answer is

$$_{40}C_5 = \frac{40!}{5!35!}$$

$$= \frac{(40)(39)(38)(37)(36)(\cancel{35!})}{(5!)(\cancel{35!})} = \frac{(\overset{8}{\cancel{40}})(39)(\overset{13}{\cancel{38}})(37)(\overset{9}{\cancel{36}})}{(\cancel{5})(\cancel{4})(\cancel{3})(\cancel{2})(\cancel{1})} = 658,008 \qquad\blacksquare$$

 **Exercises**

4.52 A cafeteria serves four different vegetables, five different main dishes consisting of either fish or meat, and three different desserts. If a customer chooses one serving from each of these three categories, how many different combinations of vegetables, main dishes, and desserts are possible?

4.53 How many different three-digit numbers can be constructed using the digits 3, 5, 7, and 9 if no digit can be repeated?

4.54 How many different ways can you select 4 playing cards from a deck of 52 such that the first is a heart, the second is a diamond, the third is a club, and the fourth is a spade?

4.55 Five offices are available for five recently hired junior executives. How many different ways can the five junior executives be assigned to the offices?

4.56 Seven assistant professors have applied for tenure at a university. However, only two assistant professors can be granted tenure. How many different sets of two assistant professors can be selected to receive tenure?

4.57 A shoe salesperson has ten different western boots to display in her showcase window. She can display only four at one time. How many different sets of four western boots can the salesperson select?

4.58 Six chairs are available for the six typists at a certain firm. How many seating arrangements are possible?

4.59 A committee of ten people needs to select a committee chairperson and committee secretary. How many different ways can the chairperson and secretary be selected from this committee?

4.60 A builder has five different house styles and three lots on which to build. If each lot has a different style of house on it, how many sets of the five house styles are possible on these three lots?

4.61 A firm has 100 laborers, 20 salespersons, and 10 executives. If an employee is chosen from each of these categories, how many different sets of three employees are possible?

4.62 To cut down on overhead, a small business decides to reduce the number of key-punchers employed. How many different sets of five people can be selected from 25 key-punchers if the firm decides to reduce the number of keypunchers by 5?

4.63 A person has six different-colored shirts and ten different-colored trousers. How many color schemes are possible if one shirt and one pair of trousers are chosen?

4.64 A company is trying to encourage women to fill executive positions. For the latest batch of executive trainees, it wishes to fill seven vacancies with 5 women and 2 men. The company has 7 women and 8 men, making a total of 15 finalists for these seven vacancies.

a. In how many ways can the 7 vacancies be filled if sex is disregarded?

b. In how many ways can the 7 vacancies be filled if the company insists on 5 women and 2 men?

c. What is the probability that the company gets the combination it wants if positions are filled randomly?

4.65 Ten employees have the option of selecting the day shift or the night shift. How many different sets of employees can be found such that only two choose the night shift?

4.66 Initial computer screening has given the IRS auditor 12 income tax returns to work on. The auditor intends to select 5 returns randomly from these 12. How many different sets of 5 returns are possible? If John, Mary, Tom, Ahmed, and Lim are among the 12 originally screened by the computer, what is the probability that all five of them will be selected by this auditor?

◢ 4.8
Simple Random Samples (Optional)

In later chapters, practically all the applications that use probabilities derived from sample results to make a decision concerning the population are based on the assumption of a simple random sample or, more simply, a random sample.

We introduced this concept in Chapter 1. A sample of size n, selected from a population of size N, constitutes a *simple random sample* if every possible sample of size n has the same probability of being selected.

In Example 4.11, we determined that there were 658,008 possible delegations when selecting 5 people from a group of 40. If we view this group as the population of interest, then $n = 5$ and $N = 40$. Our concern here is to determine the probability that any specified group of 5 individuals will be the designated delegation if simple random sampling is used.

Notice here that we do not allow any one person to be selected more than once (that is, all five members of the delegation are different people). In effect, we randomly select the first individual from the group of 40, randomly select the second individual from the remaining group of 39, randomly select the third individual from the remaining group of 38, and so on. Consequently, at each step we do *not* replace the people previously selected for the sample when selecting the next individual. Such a sampling procedure is called **sampling without replacement.** This procedure can be simplified in practice by randomly selecting 5 different individuals *at one time* from the group of 40.

When the sample data are obtained one at a time by returning a person or object to the population prior to selection of the next sample value, this procedure is termed **sampling with replacement.** Using this scheme, a particular person (or object) can be selected *more than once* in the sample. These sampling procedures will be discussed further in Chapter 7.

For this illustration there are $_{40}C_5 = 658,008$ possible delegations when selecting without replacement, and each has the same probability of being selected. Therefore, the probability that any one combination of people will be picked is $1/658,008 = .000002$.

In general, when employing a simple random sample of size n, selected without replacement from a population of size N, the total number of possible random samples is

$$_NC_n$$

Also, each of these samples has a probability of being selected equal to

$$\frac{1}{_NC_n}$$

EXAMPLE 4.12   Your task is to obtain a random sample of two individuals selected without replacement from a group of five employees (E_1, E_2, E_3, E_4, and E_5). What is the probability that you select E_2 and E_5 as your sample?

SOLUTION   There are

$$_5C_2 = \frac{5!}{2!3!} = 10$$

possible random samples. They are:

SAMPLE NUMBER	SAMPLE
1	E_1, E_2
2	E_1, E_3
3	E_1, E_4
4	E_1, E_5
5	E_2, E_3
6	E_2, E_4
7	E_2, E_5
8	E_3, E_4
9	E_3, E_5
10	E_4, E_5

Each random sample (including the one containing E_2 and E_5) has a probability of being selected of $1/10 = .1$.   ∎

Obtaining a Random Sample (Without Replacement)

When N is small, we can put the N names in a hat and pick out n of them for our sample. This will constitute a random sample (if you do not have a hat, improvise). When N is moderately large (for example, 10,000), we need a more practical method of selecting n items from this population. A common procedure is to select n **random numbers** between 1 and 10,000 using a table of random numbers or a computer-generated list of n random numbers. A list of random numbers is provided in Table A.13 at the end of the text. To generate a list of random numbers between 1 and 10,000, one procedure you could use is to:

1. Start in any arbitrary position, such as row 5 of column 3.
2. Select a list of random numbers by reading either across or down the table.

■ **FIGURE 4.12**
Generating 100 random numbers using
MINITAB.

```
MTB > RANDOM 100 VALUES INTO C1;
SUBC> UNIFORM FROM 1 TO 10000.
MTB > ROUND C1, PUT INTO C2
MTB > PRINT C2

C2
   9156   4544   7950   3762    182   2669   3586   8231   8881    120   4850
   6145   8037   4632   8888    962    183   2716   9453   1648   5936   1996
   9380   2463   7727   5891   6272   3957   4551   8820   2448   5848    974
   1675   9297   2097   1998   9601     67   8277   4607   5766    701   7511
   8885    657   2017   2017   2039   4777   7038   9663   7872   3987   8268
   3457   2045   5490   6221   7593   9108   8505   3133   1484   5333   6628
   8457   7116   9514   9285    636   9327   5902   7754   9197   9555   4414
   1683    313   8998   4747   3334   6647    838   4674   4206   5647   5787
   3377   2006    608   5907   8313   9157   4674   4156   9388   3477   4518
   4671
```

3. For each five-digit number selected, place a decimal before the final digit and round this value to the nearest counting number; for example, 24127 would become 2412.7, which is then rounded to 2413.

A computer-generated list of random numbers is easiest to use. Figure 4.12 contains the instructions for generating 100 random numbers between 1 and 10,000 using MINITAB. We give you the necessary commands in SAS and SPSS at the end of the chapter.

Using either procedure, let us assume that the resulting set of random numbers is 2413, 6246, and 5418 (for $n = 3$). Then you must (1) code your population from 1 to 10,000 in some manner and (2) select individuals 2413, 6246, and 5418 for your random sample. This topic and extensions of simple random sampling are further discussed in Chapter 7.

MINITAB, SAS, and SPSS all have options that allow you to sample randomly from a stored data set and save the results for further analysis. You can find the necessary commands to carry out this procedure in the appropriate user's manual.

To obtain a representative sample when N is extremely large or unknown requires good judgment. Stopping the first ten people you meet on the street is a very poor way of sampling your population.

You sometimes may be forced to select items for the sample in order to represent the population as accurately as possible, realizing that a poorly gathered sample can easily lead to an incorrect decision that may be important. Accountants often encounter this problem when performing a statistical audit. However, when such a sample is *not* a random sample, it is not correct to use probability theory in your analysis.

⬛ *Exercises*

4.67 An instructor requires each of 35 students in a class to write a term paper comparing nonparametric with parametric statistics. Two students are selected to present their term papers in class. What is the probability that Georgia and Fred (two students in the class) will be picked?

4.68 There are eight identically shaped objects in a jar. Two objects are selected randomly without replacement. If the objects are numbered one through eight, what is the probability that a six and a seven are drawn?

4.69 A firm has 12 skilled technicians, 8 of whom have college degrees. If a group of 4 technicians is chosen randomly to form a production team, what is the probability that none of the 4 will have a college degree? Do the 4 selected at random constitute a simple random sample?

4.70 Explain how to select a simple random sample for a population of 100,000 using the computer-generated random numbers in Table A.13.

4.71 The school-newspaper photographer takes ten different pictures of the homecoming queen at the school's football game. All ten pictures are excellent, so the photographer chooses two at random to place in the school newspaper. What is the probability that the first two pictures taken will be selected?

Summary   This chapter has examined methods of dealing with uncertainty by applying the concept of **probability.** An activity that results in an uncertain outcome is called an **experiment;** the possible outcomes are **events.** Uncertainty is measured in terms of the probabilities of events. To determine the value of a particular probability, we used the classical approach, the relative frequency method, and the subjective probability approach. The **classical** definition for the probability of event A occurring assumes that the experiment has n equally likely outcomes and event A occurs in m of the n outcomes with a resulting probability of occurring equal to m/n. When using the **relative frequency** approach, the experiment is observed n times. Letting m represent the number of times event A occurs out of the n times, then the resulting probability of event A occurring in the future is m/n. A **subjective probability** is a measure of your belief that a particular event will occur, and like all probabilities, ranges from zero to one, inclusive.

When examining more than one event, say A and B, several types of probabilities can be derived. The probability of A and B occurring is a **joint probability** and is written $P(A \text{ and } B)$. The **multiplicative rule** is a method of determining a joint probability. The probability of A or B (or both) occurring is written $P(A \text{ or } B)$ and can be obtained using the **additive rule.** When asked to find a probability given particular information about events, you determine a **conditional probability.** For example, the probability that B occurs given that A has occurred is a conditional probability, written $P(B \mid A)$. A variation of the multiplicative rule provides a method of determining a conditional probability. The probability of a single event, such as P(the person selected is a female) or P(an individual subscribes to the *Wall Street Journal*), is a **marginal probability.**

Two events are said to be **independent** if the occurrence of the one event has no effect on the probability that the other event occurs. Do not confuse "has no effect" with "will never occur simultaneously." If two events can never occur simultaneously, they are **mutually exclusive.** For example, these two events are certainly independent but are not mutually exclusive (since both events could occur): A, the stock market drops more than two points during a particular week, and B, your company's copying machine breaks down during the same week.

An effective method of determining a probability in complicated situations is to use a **Venn diagram.** When you represent the various events visually, you can often obtain a seemingly complex probability easily. Another useful device for structuring a decision problem is a **decision tree.** Using this approach, one or more **prior probabilities** are provided; they state the probability of these events occurring when no additional information is available. If such information is available, these probabilities can be revised, producing **posterior probabilities.** By using **Bayes' Rule,** these posterior probabilities are easily derived, once the decision tree is constructed.

We discussed various counting rules, including **permutations** and **combinations.** These rules are used to count the number of possible outcomes for experiments that select a certain number of people or objects (k) from a large group of n such objects. When determining the corresponding number of permutations (written $_nP_k$), the order of selection is considered. The number of combinations for this situation (written $_nC_k$) ignores the order of selection and counts only the number of groups that can be obtained.

We also discussed the number of random samples that exists when the population size is known, and we examined methods of obtaining such a sample. *In the chapters to follow, any results using a statistical sample assume that the sample is obtained randomly.*

Summary of Formulas

1. Additive rule

$$P(A \text{ or } B) = P(A) + P(B) - P(A \text{ and } B)$$

Special case: If A and B are *mutually exclusive,*

$$P(A \text{ or } B) = P(A) + P(B)$$

2. Multiplicative rule

$$P(A \text{ and } B) = P(A \mid B) \cdot P(B)$$
$$= P(B \mid A) \cdot P(A)$$

Special case: If A and B are *independent,*

$$P(A \text{ and } B) = P(A) \cdot P(B)$$

3. Conditional probability

$$P(A \mid B) = \frac{P(A \text{ and } B)}{P(B)}$$

4. Independence Two events A and B are independent if one of the following can be shown:

$$P(A \mid B) = P(A)$$
$$P(B \mid A) = P(B)$$
$$P(A \text{ and } B) = P(A) \cdot P(B)$$

5. Mutually exclusive Two events A and B are mutually exclusive if $P(A \text{ and } B)$ is zero.

6. Posterior probability (Bayes' Rule) For any event E_i, the posterior probability of event E_i, given that the final event B occurred is

$$P(E_i \mid B) = \frac{i\text{th path}}{\text{sum of paths}}$$

where the sum of paths is obtained by multiplying all probabilities along a path leading to event B and summing the results over all paths leading to event B.

7. Permutations and combinations
The number of permutations of n objects taking k objects at a time is

$$_nP_k = \frac{n!}{(n-k)!}$$

The number of combinations of n objects taking k objects at a time is

$$_nC_k = \frac{n!}{k!(n-k)!}$$

 *Review*

4.72 Assume events A and B are mutually exclusive. Find the following probabilities if $P(A) = 0.4$ and $P(B) = 0.15$:

a. $P(A \text{ or } B)$ c. $P(\bar{A} \text{ or } \bar{B})$ e. $P(A \mid B)$
b. $P(A \text{ or } \bar{B})$ d. $P(\bar{A} \text{ and } B)$ f. $P(A \mid \bar{B})$

4.73 Consider the experiment in which a single die is tossed.

a. What is the probability that an even number occurs?

b. What is the probability that an even number occurs, given that the number is greater than three?

c. What is the probability that an even number occurs or that a number greater than three occurs?

4.74 If you are selecting playing cards at random without replacement from a deck of 52 and you have already drawn a king of spades, queen of spades, ten of spades, and nine of spades, what is the probability of drawing a jack of spades?

4.75 A marketing-research group conducted a survey to find out where people did their holiday shopping. Out of a group of 110 randomly selected shoppers, 70 said that they shopped exclusively at the local mall, 30 said that they shopped exclusively in the downtown area, and 10 said that they shopped both at the local mall and in the downtown area.

a. What is the probability that a customer shops both at the local mall and in the downtown area?

b. What proportion of customers who shop at the local mall also shop in the downtown area?

c. What is the probability that a customer shops downtown but not at the local mall?

4.76 An electronics firm decides to market three different software packages for its personal computers. The marketing analyst gives each of the three packages an 80% chance of success. The outcomes for each of the software packages are independent.

a. What is the probability that all three will be a success?

b. What is the probability that only two of the packages will be a success?

c. What is the probability that none will be successful?

4.77 A payroll record with an error in it is placed in a filing cabinet with six error-free payroll records. Two payroll records are randomly selected, without replacement, by an auditor.

a. What is the probability of drawing the payroll record with the error on the first draw?

b. What is the probability of drawing the payroll record with the error on the second draw?

c. What is the probability of drawing the payroll record with the error on the first or second draw?

d. What is the probability of drawing an error-free payroll record on the first or second draw?

4.78 Each state in the United States was recently categorized with regard to projected likelihood of overall tax increases in the next two years. The following four categories were used:

A: No major change
B: Moderate chance of an increase
C: Strong probability of an increase
D: A sure bet that there will be an increase

The results for the fifty states are illustrated below:

	STATES HAVING A STATEWIDE SALES TAX NO GREATER THAN 4%	STATES HAVING A STATEWIDE SALES TAX GREATER THAN 4%
A	2	1
B	6	14
C	4	9
D	5	9
Total	17	33

(*Source:* Adapted from John Sims, "Is Your State A Haven or Hell?" *Money*, January 1991, pp. 87–91.)

a. What is the probability that a state has a statewide sales tax no greater than 4%?

b. What is the probability that a state has a statewide sales tax no greater than 4% and does not belong to the D category?

c. What is the probability that a state has a statewide sales tax greater than 4% and belongs to the B category?

d. What is the probability that a state has a statewide sales tax greater than 4% or does not belong to either the A or B category?

4.79 For a marketing survey, 200 customers were classified according to their age (in years) and their favorite type of donut.

AGE OF CUSTOMER	GLAZED	CHOCOLATE-COVERED	CREME-FILLED	CAKE	
<21	3	25	10	7	
21–30	5	23	26	10	
31–45	15	12	3	20	
>45	29	5	1	6	
Total	52	65	40	43	(200)

a. What is the probability that a person prefers creme-filled donuts and is age 45 years or less?

b. What is the probability that a person's favorite donut is not glazed or that the person is less than 21 years of age?

c. What is the probability that a person is between 21 and 30 years of age if that person favors chocolate-covered donuts?

d. Are age and favorite donut independent variables?

4.80 The probability that a person buys a car after receiving a sales pitch is .10. After a customer decides to buy a car, the probability that the customer will arrange financing through the dealer is .75. What is the probability that a customer who hears a sales pitch will buy a car and arrange financing through the dealer?

4.81 An instructor has 40 questions from which she will make up a 30-question test. How many different tests can the instructor design?

4.82 A busy executive has to meet with five production managers during the day. The executive needs to decide in which order to see the managers. How many different orderings can the executive choose? What is the probability of the executive's choosing any one ordering, if the choice is random?

4.83 A defective tape recorder is inspected by two service representatives. If one representative has a 50% chance of finding the defect, and the other has a 60% chance, what is the probability that at least one will find the defect if both check the tape recorder independently? What is the probability that neither will spot the defect?

4.84 A student forgot the combination for his bike lock. The combination consists of a sequence of three numbers and each number can range from zero to nine. How many different sequences are possible?

4.85 The worldwide production of automobiles by Japan's automobile industry in 1989 is presented below:

	THOUSANDS OF VEHICLES WORLDWIDE
Toyota	4,448
Nissan	3,009
Honda	1,861
Mitsubishi	1,560
Mazda	1,270
Suzuki	868
Daihatsu	664
Fuji Heavy	563
Isuzu	559

(*Source:* "Not All Japanese Carmakers Are Powerhouses," *Business Week*, February 1990, pp. 46–47.)

a. Are the preceding classifications mutually exclusive?

b. If a Japanese vehicle is selected at random from the total worldwide production of Japanese automobiles, what is the probability that it is neither a Toyota, a Nissan, nor a Honda?

c. Suppose two vehicles, say vehicle A and vehicle B, are each selected at random from the entire worldwide production of Japanese vehicles. What is the probability that both vehicles are Toyotas or that both vehicles are Nissans?

d. For the two vehicles in part c, what is the probability that at least one of the vehicles is a Toyota?

4.86 Forty percent of the students in an economics class major in business and 70% are from St. Louis, Missouri. Also, 20% are neither business majors nor from St. Louis. What is the probability that a student selected at random from the economics class is a business major from St. Louis?

4.87 A record 41.2 million American travelers journeyed abroad in 1988. Approximately 8% traveled to Great Britain and approximately 5% traveled to Germany.

(*Source:* Adapted from *U.S. News and World Report*, February 26, 1990, p. 63.)

a. Suppose that it can be assumed that 90% of the 41.2 million American travelers did not travel to Great Britain or Germany in 1988. What is the probability that an American traveler in 1988 journeyed to Great Britain or did not journey to Germany?

b. Using the same assumption made in part a, what is the probability that an American traveler in 1988 journeyed to either Great Britain or Germany?

c. Using the same assumption made in part a, what is the probability that an American traveler in 1988, who journeyed to Great Britain, did not journey to Germany?

4.88 In 1989, the commercial real estate market had been strong in Montreal and Toronto. Montreal experienced a 10 percent vacancy rate in office markets while Toronto had an 8 percent vacancy rate.

(*Source:* Adapted from *1990 Guide To Industrial and Office Real Estate Markets,* 1990, pp. 194–95.)

a. Assume that if an office was randomly selected from either Montreal or Toronto, there is a 60% chance that the office is in Toronto. Given this assumption, what is the percentage of all vacant offices in the two cities?

b. Using the assumption in part a, what is the probability that an office selected at random from the two cities is from Toronto if the office is known to be vacant?

4.89 The proportion of people in several age categories is given below for the population of the United States in 1989.

AGE CATEGORY	PROPORTION
under 5 years old	7.5
5 to 14 years old	14.1
15 to 64 years old	66.0
65 years old and over	12.4

(*Source: Statistical Abstract of the United States,* 1990, p. 834.)

a. What is the probability that a person selected at random from the population of the United States in 1989 is no older than 15 years?

b. What is the probability that a person selected at random from the population of the United States in 1989 is 65 years old or over, if the person is at least 5 years of age?

c. If two persons are selected, each at random, from the entire population of the United States in 1989, what is the probability that at least one person is 15 to 64 years old?

d. For the two persons in part c, what is the probability that one person is under 5 years old and that the other person is 65 years old or older?

4.90 A computer-generated list of numbers between 0.5 and 6.5 is given in the MINITAB printout. Round these numbers off into integers. Using the relative frequency approach to finding a probability value, find the probability of an odd number. In your judgment, does this value tend to support the assumption that the numbers are truly random?

```
MTB > random 100 values into c1;
SUBC> uniform A=.5 and B=6.5.
MTB > PRINT C1
C1
    3.18409    6.01078    5.34752    6.43991    4.98901    3.62637    1.29673
    4.09175    5.46874    3.59305    3.13126    5.40799    1.99838    1.79702
    0.62798    4.49777    2.22069    5.58658    6.32294    2.36695    5.86879
    5.59827    1.78393    4.99184    3.97935    3.41907    5.38338    4.92242
    1.30241    4.80093    4.11574    2.46772    6.46524    2.15536    3.41979
    5.47431    4.28828    6.03483    2.35320    4.15030    0.78688    6.36033
    1.04070    2.08737    0.92095    5.11850    1.81216    2.52028    1.03504
    1.38061    2.57570    1.96239    3.29816    2.26940    5.67520    5.39965
    0.95661    3.57579    0.97396    5.74543    2.17879    6.34877    5.59579
    1.47414    2.26745    5.43085    4.85600    5.00017    5.02085    1.60677
    0.84645    1.80587    1.73359    4.69857    3.32107    5.13373    3.71653
    0.56630    2.78776    4.46958    4.69761    3.20144    2.18042    0.55235
    1.04325    2.40601    4.75175    3.96839    2.04846    2.05730    3.16237
    3.29612    2.01459    3.82397    1.99612    1.51512    1.39047    3.80817
    6.02175    0.71840
```

4.91 The data given in the following MINITAB printout represent the number of years of managerial experience and the ages of 20 midlevel managers of a department store chain. The plot with the shaded areas can be considered a Venn diagram of the set A of midlevel managers between 32 and 40 years of age, inclusively, and the set B of midlevel managers with greater than 5 years of managerial experience.

a.   What is the probability that a midlevel manager is in set A or set B?

b.   What is the probability that a midlevel manager is in set A and not in set B?

c.   What is the probability that a midlevel manager is not in set A and not in set B?

```
MTB > print c1 c2
ROW    age    exper.

 1      35       5
 2      33       4
 3      30       2
 4      41       6
 5      42       7
 6      45       5
 7      29       2
 8      28       1
 9      38       4
10      40       5
11      43       4
12      38       2
13      35       8
14      34       6
15      38       3
16      40       5
17      35       5
18      45       7
19      40       3
20      38       6
```

```
MTB > plot c2 c1;
SUBC> xincrement = 1;
SUBC> xstart at 25;
SUBC> yincrement = 1;
SUBC> ystart at 0.
* Increment increased to cover range
* Increment increased to cover range
```

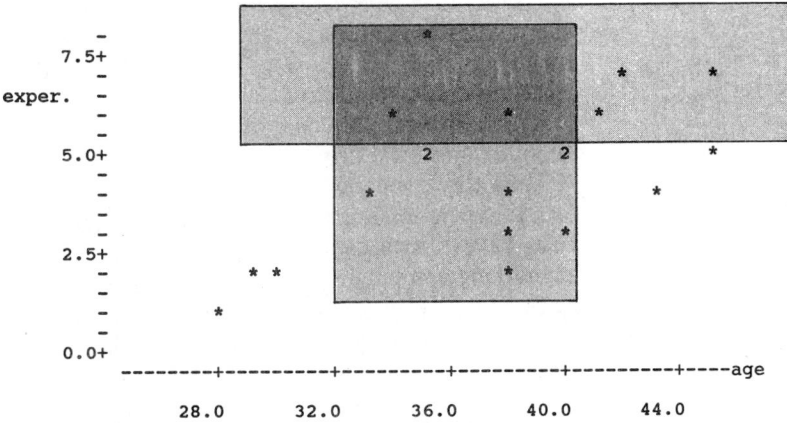

4.92 People are often amazed when they find someone who was born on the same day as themselves. To keep the solution simple, assume in this problem that leap year can be ignored and thus that all years have 365 days. Also, assume that the probability of being born on any day of the year is the same for all days of the year.

a.   If two persons are selected at random, what is the probability that these two persons were born on the same day?

b.   If three persons are selected at random, what is the probability that at least two of the three persons were born on the same day?

c. In general, the probability that at least two of n people have the same birthday is given by the formula

$$\text{Probability} = 1 - \left(\frac{365}{365}\right)\left(\frac{364}{365}\right)\cdots\left(\frac{365 - n + 1}{365}\right)$$

Use this formula to determine the probability that at least two people out of a class of 40 people have the same birthday.

d. The following MINITAB output simulates a group of 40 people with birthdays randomly chosen between 1 and 365. Does the output seem reasonable?

```
MTB > random 40 values into c1;
SUBC> integers 1 to 365.
MTB > name c1 is 'birthday'
MTB > tally c1
```

birthday	COUNT	birthday	COUNT
5	1	226	1
29	1	232	1
43	1	233	1
44	1	243	1
49	1	283	1
52	1	288	1
57	1	307	1
79	1	308	1
94	1	315	1
95	1	316	1
97	1	321	1
118	1	328	1
136	2	329	1
148	1	343	1
149	1	359	1
180	1	363	1
181	1	365	1
182	1	N=	40
195	1		
199	1		
204	1		
225	1		

e. To use simulation to estimate the probability given in part c, use the following commands to simulate twenty sets of 40 birthdays:

```
RANDOM 40 VALUES INTO C1-C20;
INTEGERS 1 TO 365.
TALLY C1-C20;
```

Computer Exercises Using the Database

Exercise 1 -- Appendix I Use the following MINITAB computer printout of random numbers between 1 and 1140, select a random sample of 100 observations from the database. Using the relative frequency approach of finding a probability value, find the probability that a family owns its home.

```
MTB > RANDOM 100 VALUES INTO C1;
SUBC> UNIFORM A=1 AND B=1140.
MTB > ROUND C1, PUT IN C1
MTB > PRINT C1
```

C1									
176	248	130	209	1002	924	289	754	728	862
521	114	491	853	516	602	1050	169	491	905
250	424	422	219	1072	623	261	661	552	578
366	4	326	743	432	284	77	440	224	568
254	825	498	568	319	1031	1138	857	1036	675
1049	1130	1001	879	396	440	181	821	14	465
1035	597	480	667	133	517	699	690	757	1119
848	1088	368	354	845	766	1129	851	283	1034
384	9	1031	13	342	446	994	1081	604	155
1035	579	543	582	821	36	1006	388	581	774

Exercise 2 -- Appendix J Generate 200 random numbers from a uniform distribution and use them to select 200 observations from the database. Using the relative frequency approach,

find the probability that a company has an A bond rating. Also find the probability that a company has a B bond rating.

Case Study

Analysis of Character Frequencies to Compress Files for Data Communication

Two major technologies have been converging at an accelerating pace in the last decade: telecommunications and data processing. So powerful is the convergence that AT&T surrendered its monopoly in telecommunications to enter the field of computers and data processing. As technology marches on, the demarcation between voice communication and data communication is being blurred. Desktop computer workstations, facsimile (FAX) machines, electronic mail facilities, and local area networks (LANs) are very common today.

Salespersons can upload daily sales data to their company's mainframe computer via a modem; reservations clerks can access remote databases of flight bookings; copies of new advertising layouts can be sent to clients via FAX (long-distance photocopying!)—the common element is the use of data communications technology. However, as the volume of data being transferred increases, so does the cost of the transfer.

One way to cut costs is to sacrifice speed and go for cheaper, low-speed channel links. Another approach is to reduce the quantity of data by compressing it. For example, in files containing text, there is a lot of "white space," that is, spaces and blank lines. From the point of view of the computer, a space is just another character, such as A, B, or C. Instead of sending 50 character codes for a space, a special code could be sent, followed by the number 50, to indicate 50 blank spaces. Fifty characters would thus be compressed into 2 or 3 characters. At the receiving end, the compressed characters would be expanded back to their original form.

Held and Marshall (1987) discuss various techniques of data compression. Not all techniques are universally applicable. For example, a method called relative encoding is efficient when applied to telemetry data or facsimile digital scan codes. To look for character patterns in the data to decide which compression technique will work best, Held and Marshall used a computer program called DATANALYSIS on various data files. Some of the results pertaining to one file are given in Tables 4.2–4.3. Our focus, of course, will not be on data

compression techniques, but on the probabilistic aspects of how the characters are distributed in the file.

Table 4.2 lists the frequency of occurrence of individual characters. The total number of characters in the file analyzed was 99,132. Note that SP refers to the space (blank) character. Character codes denoted as SH, EX, SX, ET, EQ, DL, and so on, are so-called control characters and are not generally printable. Table 4.3 gives the frequency analysis for 138 of the most common pair sequences of characters, that is, two characters occurring together in sequence. Counting multiple occurrences of these pairs, a total of 4528 such combinations were found. When answering the questions that follow, consider carefully which table you should use.

Case Study Questions

1. Which is the most frequently occurring individual character? What proportion of the file does it occupy?
2. If *one* character is selected at random from this file, what is the probability that it is:
a. a space (blank)
b. an uppercase letter (*A* to *Z*)
c. a lowercase letter (*a* to *z*)
d. *not* a number (0 to 9)
e. alphanumeric (upper- or lowercase letter or number)
3. If *two* characters, not necessarily sequential pairs, are randomly and independently selected from this file, what is the probability that:
a. both are blank spaces
b. one is an upper- or lowercase letter and the other is a number
c. neither one is a blank space
4. In computing the above probabilities, which view of probability are you assuming: the classical view, the relative frequency approach, or subjective probability?
5. If a pair of characters occurring in sequence is randomly selected from this file, what is the probability that it will be one of the pairs shown in Table 4.3?
(*Hint:* Determine how many *sequential* pair combinations are possible in this file.)

TABLE 4.2
System standard frequency of occurrence—table of characters found in sysout file.

CHAR.	COUNT	%	CHAR.	COUNT	%	CHAR.	COUNT	%
SH	0.	0.	,	1327.	1.34	W	197.	0.20
EX	0.	0.	-	135.	0.14	X	196.	0.20
SX	0.	0.	.	135.	0.14	Y	190.	0.19
ET	0.	0.	/	60.	0.06	Z	62.	0.06
EQ	0.	0.	0	1066.	1.08	[	1.	0.00
AK	0.	0.	1	905.	0.91	/	1.	0.00
BL	0.	0.	2	1427.	1.44	]	1.	0.00
BS	0.	0.	3	405.	0.41		1.	0.00
HT	0.	0.	4	354.	0.36		53.	0.05
LF	0.	0.	5	353.	0.36	@	0.	0.
VT	0.	0.	6	248.	0.25	a	0.	0.
FF	0.	0.	7	281.	0.28	b	0.	0.
CR	0.	0.	8	312.	0.31	c	0.	0.
SO	0.	0.	9	242.	0.24	d	0.	0.
SI	0.	0.	:	21.	0.02	e	0.	0.
DE	0.	0.	;	4.	0.00	f	0.	0.
D1	0.	0.	<	1.	0.00	g	0.	0.
D2	0.	0.	=	152.	0.15	h	0.	0.
D3	0.	0.	>	1.	0.00	i	0.	0.
D4	0.	0.	QM	5.	0.01	j	0.	0.
NK	0.	0.	@	2.	0.00	k	0.	0.
SY	0.	0.	A	713.	0.72	l	0.	0.
EB	0.	0.	B	252.	0.25	m	0.	0.
CN	0.	0.	C	427.	0.43	n	0.	0.
EM	0.	0.	D	290.	0.29	o	0.	0.
SB	0.	0.	E	928.	0.94	p	0.	0.
EC	0.	0.	F	279.	0.28	q	0.	0.
FS	0.	0.	G	192.	0.19	r	0.	0.
GS	0.	0.	H	1095.	1.10	s	0.	0.
RS	0.	0.	I	1223.	1.23	t	0.	0.
US	0.	0.	J	97.	0.10	u	0.	0.
SP	73509.	74.15	K	121.	0.12	v	0.	0.
EP	13.	0.01	L	355.	0.36	w	0.	0.
"	201.	0.20	M	332.	0.33	x	0.	0.
#	2.	0.00	N	644.	0.65	y	0.	0.
$	2.	0.00	O	694.	0.70	z	0.	0.
%	7.	0.01	P	432.	0.44	{	0.	0.
&	105.	0.11	Q	98.	0.10	|	0.	0.
'	1.	0.00	R	813.	0.82	}	0.	0.
(	542.	0.55	S	573.	0.58	~	0.	0.
)	542.	0.55	T	1022.	1.03	DL	0.	0.
*	1055.	1.06	U	511.	0.52	TOTAL	99132.	100.00
+	74.	0.07	V	95.	0.10			

SPSS

Solution
Random Number Generation

You can use SPSS to generate random numbers. The program listing in Figure 4.13 was used to request the generation of 100 random numbers between 1 and 10,000. Mainframe and PC versions are the same, except for the period(.) at the end of each line.

The TITLE command names the SPSS run.

The DATA LIST FREE / A, BEGIN DATA, the 100 values of 1, and the END DATA statements will provide 100 random numbers when the COMPUTE RANNUM = RND(UNIFORM(10000)) command is executed. These 100 values are from a uniform distribution between 1 and 10,000, and are rounded to the nearest integer. If say, 250 random numbers are

TABLE 4.3 Paired character compression analysis.	PAIR/COUNT		PAIR/COUNT		PAIR/COUNT		PAIR/COUNT	
	___ I	156	___ P	22	___ C	72	___ A	21
	E ___	80	RO	20	MA	60	AB	20
	RE	68	LY	19	FO	53	YS	19
	___ D	55	HG	18	D ___	46	HO	18
	ON	50	HU	18	AR	42	___ B	17
	EN	45	HY	17	___ S	37	GO	17
	NT	39	CU	16	IO	33	UE	16
	IT	34	EG	16	TO	29	MI	15
	RA	30	EL	15	IM	26	RR	14
	UB	28	HB	14	NE	24	UR	14
	AC	25	OS	14	PR	23	HF	13
	LI	24	RI	110	NA	21	___ W	85
	TA	22	IN	75	HS	20	___ F	68
	NG	21	AT	66	HC	19	SE	56
	CE	19	IR	53	HI	18	O ___	52
	RD	18	R ___	46	HQ	17	S ___	45
	IP	18	CO	43	HM	17	___ T	39
	HW	17	SI	38	G ___	16	AN	35
	PI	16	PA	33	CS	16	HE	31
	AP	16	IA	30	PE	14	ME	28
	HN	15	EQ	26	HV	14	OU	25
	TS	14	CH	24	HR	13	NU	24
	ES	14	OM	23	PU	96	IL	22
	TE	149	DA	22	OR	70	AI	21
	TI	75	MP	20	ER	59	NO	19
	N ___	66	RS	19	T ___	52	BE	18
	AL	55	IX	18	HA	46	FI	18
	L ___	48	ST	18	IS	40	DI	17
	SU	45	HK	17	TH	37	___ N	16
	LA	39	HD	16	___ R	33	UN	16
	___ E	33	WO	16	WE	29	ED	15
	RM	30	HH	14	H ___	25	TR	14
	LE	27	X ___	14	___ O	24	___ G	14
	GE	24	EP	14	ND	22	HT	13
	CT	23	UT	108				

TOTAL COMBINATIONS FOUND: 4528

Note: The underline ___ represents a space in Table 4.3.

(*Source* for Tables 4.2 and 4.3: Gilbert Held and Thomas R. Marshall. *Data Compression: Techniques and Applications, Hardware and Software Considerations,* 2d ed., New York: John Wiley, 1987. Tables 3.3, 3.8, pp. 128, 137. Reprinted by permission of John Wiley & Sons, Ltd.)

desired, the only change necessary is to replace the 100 values of 1 with 250 such values.

The LIST RANNUM command will output the 100 random numbers.

Figure 4.14 shows the output obtained by executing the program listing in Figure 4.13.

FIGURE 4.13
Input for SPSS or SPSS/PC to generate 100 random numbers between 1 and 10,000. Remove the periods for SPSS input.

```
TITLE     RANDOM NUMBERS.
DATA LIST FREE / A.
BEGIN DATA.
1 1 1 1 1 1 1 1 1 1 1 1 1 1 1 1 1 1 1 1 1 1 1 1 1
1 1 1 1 1 1 1 1 1 1 1 1 1 1 1 1 1 1 1 1 1 1 1 1 1
1 1 1 1 1 1 1 1 1 1 1 1 1 1 1 1 1 1 1 1 1 1 1 1 1
1 1 1 1 1 1 1 1 1 1 1 1 1 1 1 1 1 1 1 1 1 1 1 1 1
END DATA.
COMPUTE RANNUM = RND(UNIFORM(10000)).
LIST RANNUM.
```

■ FIGURE 4.14
SPSS output.

RANDOM NUMBERS

RANNUM	RANNUM	RANNUM	RANNUM	RANNUM
4337.00	1133.00	7426.00	9176.00	3755.00
2572.00	5140.00	2881.00	7731.00	3028.00
1669.00	6690.00	730.00	1796.00	956.00
3944.00	7695.00	7237.00	5624.00	2030.00
4411.00	4830.00	7938.00	5333.00	166.00
530.00	866.00	9270.00	6017.00	9944.00
2765.00	9790.00	8736.00	9103.00	8757.00
7305.00	993.00	7454.00	9168.00	1245.00
9733.00	1802.00	8213.00	540.00	947.00
9669.00	6648.00	5012.00	9203.00	6834.00
1621.00	919.00	8765.00	8761.00	6623.00
5991.00	8868.00	9443.00	5159.00	610.00
9949.00	4152.00	6766.00	7416.00	1032.00
3767.00	9926.00	9213.00	6654.00	8333.00
444.00	1009.00	9363.00	7535.00	856.00
7003.00	6096.00	4917.00	578.00	4850.00
3585.00	2547.00	6280.00	8880.00	7309.00
8158.00	9173.00	8813.00	5710.00	4282.00
5606.00	4994.00	5604.00	8091.00	9366.00
3887.00	4651.00	7975.00	448.00	4818.00

SAS

▮ Solution
Random Number
Generation

You can generate random numbers using SAS. The program listing in Figure 4.15 was used to request the generation of 100 random numbers between 1 and 10,000. The Mainframe and PC commands are the same.

The TITLE command names the SAS run (enclose in single quotes).

The DATA command gives the data a name.

The DO statement sets up a loop that is terminated by the END statement. In this example, we are looping 100 times to compute 100 different random numbers.

The $Y = UNIFORM(0) * 10000$ statement generates a random number from a uniform distribution between 1 and 10,000. The zero in parentheses acts as a seed for the random number generator. This value can be changed to obtain a different set of random numbers.

The $X = ROUND(Y)$ statement rounds the random number to the nearest integer.

The OUTPUT statement stores the X and Y values.

The END statement marks the end of the loop used to generate the random numbers.

The PROC PRINT command prints the (rounded) random numbers as specified by the VAR X statement.

Figure 4.16 shows the output obtained by executing the program listing in Figure 4.15.

■ FIGURE 4.15
Input for SAS
(Mainframe or
micro version)

```
TITLE    'RANDOM NUMBERS';
DATA NEW;
DO I = 1 TO 100;
Y = UNIFORM (0) * 10000;
X = ROUND (Y);
OUTPUT;
END;
PROC PRINT;
  VAR X;
```

■ FIGURE 4.16   RANDOM NUMBERS

SAS output.   OBS   X

OBS	X				
1	2693	34	9295	67	1255
2	2753	35	678	68	6689
3	5864	36	6793	69	4023
4	9707	37	662	70	4100
5	336	38	40	71	59
6	9556	39	5902	72	4257
7	6447	40	7429	73	430
8	1729	41	4820	74	9499
9	2951	42	4346	75	316
10	1413	43	1882	76	9067
11	8581	44	3503	77	2561
12	3118	45	6987	78	8085
13	1926	46	9694	79	8850
14	6617	47	4223	80	1914
15	5246	48	4179	81	3217
16	661	49	1717	82	266
17	41	50	741	83	1551
18	7095	51	5644	84	5468
19	1782	52	2503	85	2572
20	8040	53	5512	86	4222
21	1408	54	7960	87	3118
22	3385	55	9561	88	1672
23	5985	56	2402	89	2657
24	2737	57	1052	90	7475
25	6874	58	2015	91	5430
26	5926	59	1505	92	2927
27	9792	60	4458	93	305
28	4090	61	1783	94	2702
29	6737	62	128	95	8300
30	4587	63	6194	96	8054
31	3633	64	652	97	1842
32	8467	65	6804	98	2826
33	8665	66	3890	99	1558
				100	5059

CHAPTER

5

Discrete Probability Distributions

A LOOK BACK/INTRODUCTION

The early chapters were concerned with describing sample data that had been gathered from a previous experiment, a printed report, or some other source. The data were summarized using one or more numerical measures (for example, a sample mean, variance, or correlation) or using a statistical graph (such as a histogram, bar chart, or scatter diagram).

Chapter 4 introduced you to methods of dealing with uncertainty by using a probability to measure the chance of a particular event occurring. Rules were defined that enabled you to compute various probabilities of interest, such as a conditional or a joint probability. However, so far we have defined only the probability of a certain event happening.

Whenever an experiment results in a numerical outcome, such as the total value of two dice, we can represent the various possible outcomes and their corresponding probabilities much more conveniently by using a **random variable,** the topic of this chapter. Suppose that your company manufactures a product that is sometimes defective and is returned for repair in 10% of the cases. An excellent way of describing the chance that 3 of 20 products will be returned before the warranty runs out is to use the concept of a random variable.

Random variables can be classified into two categories: discrete and continuous. This chapter introduces both but concentrates on the discrete type. Several commonly used discrete random variables will be discussed, as will methods of describing and applying them.

Definition
A **random variable** is a function that assigns a numerical value to each outcome of an experiment.

In Chapter 3 we used various statistics (numerical measures) to describe a set of sample data. For example, the sample mean and standard deviation provide measures of a "typical" value and variation within the sample, respectively. Similarly, we will use a random variable and its corresponding distribution of probabilities to describe a *population*. Just as a sample has a mean and standard deviation, so does the population from which the sample was obtained. We will use the basic concepts from Chapter 4 to derive probabilities related to a random variable.

▰ 5.1 Discrete Random Variables

Random Variables

The probability laws developed in the previous chapter provide a framework for the discussion of random variables. We will still be concerned about the probability of a particular event; often, however, some aspect of the experiment can be easily represented using a random variable. The result of a simple experiment can sometimes be summarized concisely by defining a discrete random variable to describe the possible outcomes.

Flip a coin three times. The possible outcomes for each flip are heads (H) and tails (T). According to counting rule 1 from Chapter 4, there are $2 \cdot 2 \cdot 2 = 8$

possible results. These are TTT, TTH, THT, HTT, HHT, HTH, THH, and HHH. Let

A = event of observing 0 heads in 3 flips (TTT)

B = event of observing 1 head in 3 flips (TTH, THT, HTT)

C = event of observing 2 heads in 3 flips (HHT, HTH, THH)

D = event of observing 3 heads in 3 flips (HHH)

We wish to find $P(A)$, $P(B)$, $P(C)$, and $P(D)$.

Consider one outcome, say, HTH. The coin flips are independent, so we use equation 4.14:

$P(\text{HTH})$ = (probability of H on 1st flip) · (probability of T on 2nd flip)

· (probability of H on 3rd flip)

= $(1/2) \cdot (1/2) \cdot (1/2) = 1/8$

This same argument applies to all eight outcomes. These outcomes are all equally likely, and each occurs with probability 1/8.

Event A occurs only if you observe TTT. It has the probability of occurring one time out of eight:

$$P(A) = 1/8$$

Event B will occur if you observe HTT, TTH, or THT. It would be impossible for HTT and TTH *both* to occur, so $P(\text{HTT and TTH}) = 0$. This is true for any combination of these three outcomes so these three events are all mutually exclusive. Consequently, according to equation 4.13,

$P(B)$ = $P(\text{HTT or TTH or THT})$

= $P(\text{HTT}) + P(\text{TTH}) + P(\text{THT})$

= $1/8 + 1/8 + 1/8 = 3/8$

By a similar argument,

$P(C) = 3/8$ (using HHT, HTH, THH)

$P(D) = 1/8$ (using HHH)

The variable of interest in this example is X, defined as

X = number of heads out of three flips

We defined all the possible outcomes of X by defining the four events A, B, C, and D. This method works but is cumbersome. Consider having to do this for 100 flips of a coin! A more convenient way to represent probabilities is to examine the value of X for each possible outcome.

OUTCOME	VALUE OF X	
TTT	0	1 outcome
THT	1	
TTH	1	3 outcomes
HTT	1	
HHT	2	
HTH	2	3 outcomes
THH	2	
HHH	3	1 outcome

Each outcome has probability 1/8, so the probability that X will be 0 is 1/8, written:

$$P(X = 0) = P(0) = 1/8$$

The probability that X will be 1 is 3/8, written:

$$P(X = 1) = P(1) = 3/8$$

The probability that X will be 2 is 3/8, written:

$$P(X = 2) = P(2) = 3/8$$

The probability that X will be 3 is 1/8, written:

$$P(X = 3) = P(3) = 1/8$$

Notice that

$$P(X = 0) + P(X = 1) + P(X = 2) + P(X = 3)$$
$$= 1/8 + 3/8 + 3/8 + 1/8 = 1$$

because 0, 1, 2, and 3 represent *all the possible values* of X.

The values and probabilities for this random variable can be summarized by listing each value and its probability of occurring.

$$X = \begin{cases} 0 \text{ with probability } 1/8 \\ 1 \text{ with probability } 3/8 \\ 2 \text{ with probability } 3/8 \\ 3 \text{ with probability } 1/8 \end{cases}$$

This list of possible values of X and the corresponding probabilities is a **probability distribution.**

In any such formulation of a problem, the variable X is a **random variable.** Its value is not known in advance, but there is a probability associated with each possible value of X. Whenever you have a random variable of the form

$$X = \begin{cases} x_1 \text{ with probability } p_1 \\ x_2 \text{ with probability } p_2 \\ x_3 \text{ with probability } p_3 \\ \vdots \\ x_n \text{ with probability } p_n \end{cases}$$

where $x_1, \ldots, x_n$ is the set of possible values of X, then X is a **discrete random variable.** In the coin-flipping example, $x_1 = 0$ and $p_1 = 1/8$; $x_2 = 1$ and $p_2 = 3/8$; $x_3 = 2$ and $p_3 = 3/8$, and $x_4 = 3$ and $p_4 = 1/8$.

Other examples of a discrete random variable include:

$X =$ the number of cars that drive up to a bank within a 5-minute period ($X = 0$, 1, 2, 3, . . .).

$X =$ the number of people out of a group of 50 who will suffer a fatal accident within the next 10 years ($X = 0, 1, 2, \ldots, 50$).

$X =$ the number of people out of 200 who make an airline reservation and then fail to show up ($X = 0, 1, 2, \ldots, 200$).

$X =$ the number of calls arriving at a telephone switchboard over a two-minute period ($X = 0, 1, 2, 3, \ldots$).

Notice that, for each example, the discrete random variable is a count *of the number of people, calls, accidents, and so on that can occur.*

EXAMPLE 5.1

You roll two dice, a red die and a blue die. What is a possible random variable X for this situation? What are its possible values and corresponding probabilities? (*Hint:* Roll the dice and observe a particular number. This number is your value of the random variable, X. What observations are possible from the roll of two dice?)

SOLUTION   There are many possibilities here, including

X = total of the two dice

X = average of the two dice

X = the higher of the two numbers that appear (possible values: 1, 2, 3, 4, 5, 6)

X = the number of dice with 3 appearing (possible values: 0, 1, 2)

Suppose that the random variable X equals the total of the two dice. The next step is to determine the possible values of X and the corresponding probabilities. When you roll the two colored dice, there are $6 \cdot 6 = 36$ possible outcomes, using counting rule 1 from Chapter 4.

OUTCOME	RED DIE	BLUE DIE	VALUE OF X
1	1	1	2
2	1	2	3
3	1	3	4
4	1	4	5
5	1	5	6
6	1	6	7
7	2	1	3
8	2	2	4
9	2	3	5
⋮	⋮	⋮	⋮
34	6	4	10
35	6	5	11
36	6	6	12

$P(X = 3) = 2/36$

The 36 outcomes are equally likely because the number appearing on each die (1, 2, 3, 4, 5, or 6) has the same chance of appearing. Notice that we are *not* saying that each value of X is equally likely, as the following discussion will make clear. Each of the above 36 outcomes has probability 1/36 of occurring. If you write down all 36 outcomes and note what can happen to X, your random variable, you will observe:

VALUE OF X	NUMBER OF POSSIBLE OUTCOMES	
2	1	(rolling a 1, 1)
3	2	(rolling a 1, 2, or 2, 1)
4	3	(rolling a 1, 3 or 3, 1 or 2, 2)
5	4	(and so on)
6	5	
7	6	
8	5	
9	4	
10	3	
11	2	
12	1	

Consequently,

$$
X = \begin{cases}
2 \text{ with probability } \dfrac{1}{36} \\[2mm]
3 \text{ with probability } \dfrac{2}{36} \\[2mm]
4 \text{ with probability } \dfrac{3}{36} \\[2mm]
5 \text{ with probability } \dfrac{4}{36} \\[2mm]
6 \text{ with probability } \dfrac{5}{36} \\[2mm]
7 \text{ with probability } \dfrac{6}{36} \\[2mm]
8 \text{ with probability } \dfrac{5}{36} \\[2mm]
9 \text{ with probability } \dfrac{4}{36} \\[2mm]
10 \text{ with probability } \dfrac{3}{36} \\[2mm]
11 \text{ with probability } \dfrac{2}{36} \\[2mm]
12 \text{ with probability } \dfrac{1}{36}
\end{cases}
$$

Total 1.0

Because 2 through 12 represent all possible values of X, the total of all probabilities is equal to 1.

Suppose instead X is defined to be the *average* of the two dice, rather than the total. Now the possible values of X are 1 (with probability 1/36), 1.5 (with probability 2/36), . . . , 5.5 (with probability 2/36), and 6 (with probability 1/36). Notice that X is still a discrete random variable, since there are gaps in the possible values (a value of 4.2 is not possible, for example). However, the possible values of X are *not* all counting numbers. In general, the possible values of a discrete random variable need not be positive integers but generally are since the discrete random variable typically counts the number of occurrences of a particular event. ∎

Continuous Random Variables

The previous section introduced you to the discrete random variable, where the possible values of X can be listed along with corresponding probabilities. Characteristic of this type of random variable is the presence of *gaps* in the list of possible values. For example, when throwing two dice, a total of 8.5 cannot occur.

The other type of random variable is the **continuous random variable,** for which *any* value is possible over some range of values. For a random variable of this type, there are no gaps in the set of possible values. As a simple example, consider two random variables: X is the number of days that it rained in Boston during any particular month and Y is the amount of rainfall during this month. X is a *discrete* random variable, because it counts the number of days, and consequently there are gaps in the possible values (7.4, for example, is not possible). Y, on the

■ **FIGURE 5.1**
Example of a
continuous random
variable. X = height
in feet of a randomly
selected adult male
in the United States.

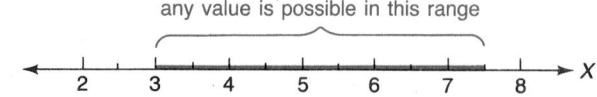

other hand, is a *continuous* random variable because (at least in principle) the amount of rainfall could be any nonnegative value.

Suppose the heights of all adult males in the United States range from 3 feet to 7.5 feet. Your task is to describe these heights using such statements as:

15% of the heights are under 5.5 ft.

88% of the heights are between 5 ft and 6 ft.

We first define the random variable

X = height of a randomly selected adult male in the United States

Figure 5.1 shows the range of X.

We are unable to list all possible values of X, since *any* height is possible over this range. However, we can still discuss probabilities associated with X. For example, the two preceding statements can be described by using the probability statements

$$P(X < 5.5) = .15$$
$$P(X \text{ is between 5 ft and 6 ft}) = P(5 < X < 6) = .88$$

For this situation, X is a continuous random variable. Probabilities for continuous random variables can be found only for *intervals*. (Probabilities of exact values are meaningful only for discrete random variables.) Determining probabilities for a continuous random variable is discussed in Chapter 6.

The discussion in Chapter 1 on discrete and continuous data is directly related to our present topic. *When you observe a discrete random variable, you obtain discrete data. When you observe a continuous random variable (such as 100 heights), you obtain continuous data.*

 *Exercises*

5.1 If a bank is interested in the business received from customers daily, what random variables would be of interest to the bank?

5.2 If a student is taking a multiple-choice test with ten questions and is interested in the final score, what random variable would be of interest to the student?

5.3 Classify the following random variables as discrete or continuous:

a. number of nonconformities per sampling unit

b. the percentage of defective units in a production process

c. total time spent on a project after 3 months

d. closing price of Exxon on the New York Stock Exchange

e. number of deposits made at a local bank in one day

f. the life of an electronic component

5.4 Consider an experiment in which two dice are rolled. Let X be the total of the numbers on the two dice. What is the probability that X is equal to two or four?

5.5 Consider an experiment in which a coin is tossed and a die is rolled. Let X be the number observed from rolling the die. Let Y be the value 1 if a head appears and 0 if a tail appears. List the values that the random variables X and Y can have, along with the corresponding probabilities.

5.6 Consider an experiment in which a coin is tossed four times. List all possible outcomes. Let X be the number of heads in each outcome. What is the value of $P(X = 2)$? of $P(X = 3)$?

5.7 If the random variable X can take on the values of 2, 3, 4, and 5 with equal probability, what is the probability that X is equal to 3? Assume that X cannot take on any other values.

5.8 Consider an experiment in which two dice are rolled. Let X take on the value of 1 if both dice have the same number and of 0 otherwise. What is the probability that X is equal to 1?

◢ 5.2
Representing Probability Distributions for Discrete Random Variables

There are three popular methods of describing the probabilities associated with a discrete random variable X. They are:

List each value of X and its corresponding probability.

Use a histogram to convey the probabilities corresponding to the various values of X.

Use a function that assigns a probability to each value of X.

Remember our coin-flipping example, in which X = number of heads in three flips of a coin. We can list each value and probability:

$$X = \begin{cases} 0 \text{ with probability } 1/8 \\ 1 \text{ with probability } 3/8 \\ 2 \text{ with probability } 3/8 \\ 3 \text{ with probability } 1/8 \end{cases}$$

This works well when there are only a small number of possible values for X; it would not work well for 100 flips of a coin.

Using a histogram also is a convenient way to represent the shape of a discrete distribution having a small number of possible values. For this situation, you construct a histogram in which the height of each bar is the probability of observing that value of X (Figure 5.2). It is easier to determine the shape of the probability distribution by using such a chart. The distribution in Figure 5.2 is clearly symmetric and concentrated in the middle values.

Using a function (that is, an algebraic formula) to assign probabilities is the most convenient method of describing the probability distribution for a discrete random variable. For any given application of such a random variable, however, this function may or may not be known. Later in the chapter we identify certain useful discrete random variables, each of which has a corresponding function that assigns these probabilities.

The function that assigns a probability to each value of X is called a **probability mass function (PMF).** Denoting a particular value of X as x, this function is of the form

■ **FIGURE 5.2**
A histogram representation of a discrete random variable, where X = number of heads in three coin flips.

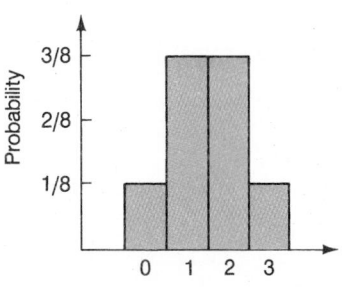

X = number of heads

$$P(X = x) = \text{some expression (usually containing } x)$$
$$\text{that produces the probability of observing } x$$
$$= P(x)$$

Not every function can serve as a PMF. The requirements for a PMF function are:

1. $P(x)$ is between 0 and 1 (inclusively) for each x
2. $\Sigma P(x) = 1$

EXAMPLE 5.2   Consider a random variable X having possible values of 1, 2, or 3. The corresponding probability for each value is:

$$X = \begin{cases} 1 \text{ with probability } 1/6 \\ 2 \text{ with probability } 1/3 \\ 3 \text{ with probability } 1/2 \end{cases}$$

Determine an expression for the PMF.

SOLUTION   Consider the function

$$P(X = x) = P(x) = x/6 \qquad \text{for } x = 1, 2, 3$$

This function provides the probabilities

$$P(X = 1) = P(1) = 1/6 \qquad (\text{OK})$$
$$P(X = 2) = P(2) = 2/6 = 1/3 \qquad (\text{OK})$$
$$P(X = 3) = P(3) = 3/6 = 1/2 \qquad (\text{OK})$$

This function satisfies the requirements for a PMF: each probability is between 0 and 1, and $P(1) + P(2) + P(3) = 1/6 + 1/3 + 1/2 = 1$. Consequently, the function

$$P(x) = x/6 \qquad \text{for } x = 1, 2, 3 \qquad (\text{and zero elsewhere})$$

is the PMF for this discrete random variable. ∎

EXAMPLE 5.3   Consider Example 5.1, where X is the total of two dice. Determine the PMF for this discrete random variable.

SOLUTION   Consider the expression

$$P(x) = \frac{x - 1}{36} \qquad \text{for } x = 2, 3, 4, \ldots, 12 \qquad (\text{and zero elsewhere})$$

If this is the proper PMF, then, for example,

$$P(2) = P(X = 2) = \frac{2 - 1}{36} = 1/36$$

This does appear to be correct, so far. Also,

$$P(5) = P(X = 5) = \frac{5 - 1}{36} = 4/36$$

This also is correct. But now consider

$$P(10) = P(X = 10) = \frac{10 - 1}{36} = 9/36$$

According to our previous solution, we know that $P(10) = 3/36$, not 9/36. So this particular function is not the PMF for this random variable; the PMF must work for *all* values of X.

Consider the expression

$$P(x) = \frac{6 - |x - 7|}{36} \qquad \text{for } x = 2, 3, \ldots, 12 \qquad \text{(and zero elsewhere)}$$

where $| \, |$ represents the absolute value of a number. See if you can demonstrate that this function is a bona fide PMF for this example (it is). Do not worry about where this expression came from, but do verify that it works. The truth of the matter is that often PMFs are derived by trial and error. ∎

Notice that a probability mass function provides a theoretical "model" of the population by describing the chance of observing any particular value of the random variable. *You can view the population as what you would obtain if you observed the corresponding random variable indefinitely.*

Exercises

5.9 Let X be the value observed from rolling a die.

a. What is the probability mass function of X?

b. Construct a histogram in which the height of each bar is the probability of X.

5.10 A salesperson calls on three customers who request to see his product. The salesperson has a 50% probability of selling his product. What is the probability mass function of X, where X is the number of customers (out of three) who buy his product?

5.11 Five copies of the minutes of a business meeting are available for circulation. Suppose two of the five copies have a blurred spot over an important dollar figure. The secretary chooses one of the copies at random with replacement each time an employee requests to read a copy. Let X be the number of times that a blurred copy is chosen for two requests. What is the probability distribution of X?

5.12 Is the following function a probability mass function? Why or why not?

$$P(X = x) = (x - 2)/6 \qquad \text{for } x = 1, 4, 7 \qquad \text{(and zero elsewhere)}$$

5.13 Is the following function a probability mass function? Why or why not?

$$P(X = x) = x^2/10 \qquad \text{for } x = -2, -1, 1, 2 \qquad \text{(and zero elsewhere)}$$

5.14 Let X be equal to the number of heads from tossing a coin twice. Verify that the following is the probability mass function of X.

$$P(X = x) = \frac{1}{2(2 - x)! \, x!} \qquad \text{for } x = 0, 1, 2 \qquad \text{(and zero elsewhere)}$$

5.15 Do you recognize the following probability mass function? What is it?

$$P(X = x) = 1/6 \qquad \text{if } x = 1, 2, 3, 4, 5, 6 \qquad \text{(and zero elsewhere)}$$

5.16 A real estate broker needs to advertise two townhouses, two duplexes, and two single family homes. However, the broker decides to choose at random only one of the six properties for open house on a certain weekend. Let the random variable X take on the value 1 if a townhouse is chosen, 2 if a duplex is chosen, and 3 if a single family home is chosen. Write the probability mass function of X.

5.17 Suppose that in Exercise 5.16, the random variable X is assigned the value of 2 if a townhouse is chosen, 4 if a duplex is chosen, and 6 if a single family home is chosen. Write the probability mass function of X.

5.18 Suppose a probability mass function is defined to be nonzero at three points, $X = 1$, 2, and 3. If $P(X = 1) = .2$ and $P(X = 2) = .3$, what is $P(X = 3)$?

5.19 A quality-control inspector is inspecting incoming lots of materials to check for excessive numbers of defective items. The inspector needs to choose at random one of six incoming lots, and then needs to inspect only that lot. Suppose one lot has zero defective items, two of the lots have three defective items, and the remaining three each have ten defective items. Let the random variable X be equal to 1 if the lot with zero defective items is drawn, 2 if a lot with three defective items is drawn, and 3 if a lot with ten defective items is drawn. Verify that the following function is the probability mass function of X.

$$P(X = x) = x/6 \quad \text{if } x = 1, 2, 3 \quad \text{(and zero elsewhere)}$$

◢ 5.3   Mean of Discrete Random Variables

Mean and Variance of Discrete Random Variables

Chapter 3 introduced you to the mean and variance of a set of sample data consisting of n values. Suppose that these values were obtained by observing a particular random variable n times. The sample mean, $\bar{X}$, represents the *average* value of the sample data. In this section, we determine a similar value, the **mean of a discrete random variable**, written as μ. The value of μ represents the average value of the random variable if you were to observe this variable over an indefinite period of time.

Reconsider our coin-flipping example, where X is the number of heads in three flips of a coin. Suppose you flip the coin three times, record the value of X, flip the coin three times again, record the value of X, and repeat this process ten times. Now you have ten observations of X. Suppose they are

$$2, 1, 1, 0, 2, 3, 2, 1, 1, 3$$

The mean of these data is the *statistic* $\bar{X}$, where

$$\bar{x} = \frac{2 + 1 + 1 + \cdots + 1 + 3}{10}$$

$$= 1.6 \text{ heads}$$

If you observed X *indefinitely,* what would X be on the average?

$$X = \begin{cases} 0 \text{ with probability } 1/8 \\ 1 \text{ with probability } 3/8 \\ 2 \text{ with probability } 3/8 \\ 3 \text{ with probability } 1/8 \end{cases}$$

So 1/8 of the time you should observe the value 0; 3/8 of the time, the value 1; 3/8 of the time, the value 2; and 1/8 of the time, the value 3. In a sense, each probability represents the *relative frequency* for that particular value of X. So the average value of X is

$$(0)(1/8) + (1)(3/8) + (2)(3/8) + (3)(1/8) = 1.5 \text{ heads}$$

Notice that X cannot be 1.5; this is merely the value of X on the average.

Definition

The average value of the discrete random variable X (if observed indefinitely) is the mean of X. The symbol for this parameter is μ.

We found that $\mu = 1.5$ by multiplying each value of X by its corresponding probability and summing the results:

$$\mu = 1.5 = 0 \cdot P(0) + 1 \cdot P(1) + 2 \cdot P(2) + 3 \cdot P(3)$$

This procedure applies to any discrete random variable, and so we define*

$$\mu = \Sigma xP(x) \qquad\qquad\qquad \textbf{(5.1)}$$

EXAMPLE 5.4   A personnel manager in a large production facility is investigating the number of reported on-the-job accidents over a period of one month. We define the random variable

$$X = \text{number of reported accidents per month}$$

Based on past records, she has derived the following probability distribution for X:

$$X = \begin{cases} 0 \text{ with probability } .50 \\ 1 \text{ with probability } .25 \\ 2 \text{ with probability } .10 \\ 3 \text{ with probability } .10 \\ 4 \text{ with probability } \underline{.05} \\ \phantom{4 \text{ with probability }} 1.0 \end{cases}$$

During 50% of the months there were no reported accidents, 25% of the months had one accident, and so on. (Notice that deriving an algebraic expression for the PMF for this distribution would be extremely difficult, if not impossible. This poses no problem, however.)

What is the mean (average value) of X?

SOLUTION   Using equation 5.1,

$$\mu = (0)(.5) + (1)(.25) + (2)(.1) + (3)(.1) + (4)(.05)$$
$$= .95$$

There is .95 (nearly 1) accident reported on the average per month.   ■

Variance of Discrete Random Variables

We previously considered ten observations of the random variable that counted the number of heads in three flips of a coin. These data were 2, 1, 1, 0, 2, 3, 2, 1, 1, 3. We used the notation from Chapter 3 to define the mean of these data, and we obtained $\bar{x} = 1.6$. The variance of these data, using equation 3.8, is $s^2 = .933$. Since s^2 describes a sample, it is a statistic.

Once again, consider observing X indefinitely. For this situation, the average value of X is defined as the mean of X, μ. When we observe X indefinitely, this particular variance is defined to be the variance of the random variable, X, and is written σ^2 (read as "sigma squared").

$$\sigma^2 = \text{variance of the discrete random variable, } X$$

The **variance of a discrete random variable,** X, is a parameter describing the variation of the corresponding population. It is the average (expected) value of $(X - \mu)^2$ if X were observed indefinitely, and it can be obtained by using one of the following expressions, which are mathematically equivalent:[†]

---

*μ is often referred to as the *expected* value of the random variable, X, and is written $\mu = E(X)$.
[†]Using the expectation notation, σ^2 is the expected value of $(X - \mu)^2$ and can be written $\sigma^2 = E(X - \mu)^2$ or $\sigma^2 = E(X^2) - [E(X)]^2$.

$$\sigma^2 = \Sigma(x - \mu)^2 \cdot P(x) \qquad (5.2)$$
$$\sigma^2 = \Sigma x^2 P(x) - \mu^2 \qquad (5.3)$$

Equation 5.3 generally provides an easier method of determining the variance and will be used in all of the examples to follow. For the coin-flipping example,

$$\sigma^2 = \Sigma x^2 P(x) - \mu^2$$
$$= [(0)^2 \cdot (1/8) + (1)^2 \cdot (3/8) + (2)^2 \cdot (3/8) + (3)^2 \cdot (1/8)] - (1.5)^2$$
$$= 3 - 2.25 = .75$$

So our final results would be:

USING THE SAMPLE OF TEN OBSERVATIONS	FOR THE RANDOM VARIABLE, X (INDEFINITE NUMBER OF OBSERVATIONS)
$\bar{x} = 1.6$ $s^2 = .933$	$\mu = 1.5$ $\sigma^2 = .75$
mean variance	mean variance
statistics	parameters

In Chapter 3, the square root of the variance, s, was defined to be the standard deviation of the data. The same definition applies to a random variable. The **standard deviation of a discrete random variable, X**, is denoted σ, where:

$$\sigma = \sqrt{\Sigma(x - \mu)^2 \cdot P(x)} \qquad (5.4)$$
$$\sigma = \sqrt{\Sigma x^2 P(x) - \mu^2} \qquad (5.5)$$

EXAMPLE 5.5

Determine the variance and standard deviation of the random variable concerning on-the-job accidents in Example 5.4.

SOLUTION

A convenient method of determining both the mean and variance of a discrete random variable is to summarize the calculations in tabular form:

x	$P(x)$	$x \cdot P(x)$	$x^2 \cdot P(x)$
0	.5	0	0
1	.25	.25	.25
2	.1	.2	.4
3	.1	.3	.9
4	.05	.2	.8
	1.00	.95	2.35

So,

$$\mu = \Sigma x P(x) = .95 \text{ accident}$$

and

$$\sigma^2 = \Sigma x^2 P(x) - \mu^2 = 2.35 - (.95)^2$$
$$= 1.45$$

Also

$$\sigma = \sqrt{1.45} = 1.20 \text{ accidents}$$

Exercises   **5.20**   To assess the attitudes of students in an introductory statistics course at Iowa State University, students were asked to respond to several statements using a five-category scale from "strongly disagree" to "strongly agree." The following arbitrary coding was used:

$$
\begin{aligned}
\text{Strongly disagree} &= -2 \\
\text{Disagree} &= -1 \\
\text{Neither agree nor disagree} &= 0 \\
\text{Agree} &= +1 \\
\text{Strongly agree} &= +2
\end{aligned}
$$

The responses to two questions are given below:

	KNOWING HOW TO USE A COMPUTER IS IMPORTANT (% OF STUDENTS WITH THIS RESPONSE)	KNOWING HOW TO ANALYZE DATA IS IMPORTANT (% OF STUDENTS WITH THIS RESPONSE)
+2	30.95	18.60
+1	52.38	67.44
0	14.29	13.96
-1	2.38	0.0
-2	0.00	0.0

(*Source:* Adapted from W. Robert Stephenson, "A Study of Student Reaction to the Use of MINITAB in an Introductory Statistics Course," *The American Statistician* 44, no. 3 (1990): 231.)

a.   From observing the data, what do you think the mean and standard deviation is for each of the two statements given above?

b.   Compute the average score for each statement.

c.   Compute the standard deviation of the scores for each statement.

d.   Interpret the meaning of the values given in parts b and c.

5.21   Several students in a finance class subscribe to the *Wall Street Journal*. If two students are chosen at random, the probability of choosing no students who subscribe is .81. The probability of choosing one student who subscribes and one who does not subscribe is .18, and the probability of choosing two students who subscribe is .01. X is a random variable equal to the number of students who subscribe from the two chosen at random. Find the mean value of X and the variance of X.

5.22   Find the mean and variance of a random variable X, which has the following probability mass function:

X	$P(x)$
-2	.12
-1	.3
0	.1
1	.3
2	.18

5.23   Suppose that a coin is flipped three times. Define the random variable X to be equal to twice the number of heads that appear. Determine the mean and variance of X.

5.24   Show that $\Sigma (x - \mu)P(x) = 0$, where the summation is over all outcomes of X, for any discrete random variable.

5.25   Determine the mean and standard deviation of the random variable for which the probability mass function is defined as follows:

$$P(X = x) = (x - 2)/30 \quad \text{if } x = 3, 12, 21 \quad \text{(and zero elsewhere)}$$

5.26   A discrete random variable X has the following probability mass function:

$$P(X = x) = 1/4 \text{ if } x = 1, 2, 3, 4 \text{ (and zero elsewhere)}$$

a.   Draw a histogram representation of a discrete random variable that has the above probability mass function.

b.   From observing the shape of the histogram, what do you think the mean of the random variable X might be? Does a standard deviation of 3.5 seem reasonable for this distribution?

c.   Determine the exact values of the mean and standard deviation of the random variable X and compare your answer to your answers to part b.

5.27   To examine the attitudes of information systems professionals on whistle-blowing, a sample survey was taken of members of the Data Processing Management Association. Their responses to the question, "What do you think about your company's encouragement of employees to blow the whistle on illegal or unethical computer usage?" are given below:

TOO MUCH	ABOUT RIGHT	NOT ENOUGH
39.1%	39.8%	21.1%

(*Source:* J. K. Pierson and J. C. Roderick, "A Study of Attitudes Toward Whistle-blowing in the Computer Environment," *Proceedings of the 1990 Annual Meeting of Decision Sciences Institute* (1990): 1017.)

a.   Let X be a random variable with a value of 1 for "not enough," 2 for "about right," and 3 for "too much." Compute the mean and standard deviation of X.

b.   If the scale is shifted by 1 in part a, that is $X = 2$ for "not enough," $X = 3$ for "about right," and $X = 4$ for "not enough," how do you think the mean and standard deviation of the random variable X will change?

c.   Compute the mean and standard deviation of X in part b and compare them to the answers given in part b.

◢ 5.4
Binomial Random Variables

The random variable X representing the number of heads in three flips of a coin is a special type of discrete random variable, a **binomial** random variable.

We next list the conditions for a binomial random variable in general and as applied to our coin-flipping example:

A BINOMIAL SITUATION	FOR EXAMPLE 5.1
1. Your experiment consists of n repetitions, called **trials.**	1. n = three flips of a coin
2. Each trial has two mutually exclusive possible outcomes, (or can be considered as having two outcomes), referred to as **success** and **failure.**	2. Success = head, failure = tail (this is arbitrary)
3. The n trials are *independent*.	3. The results on one coin flip do not affect the results on another flip.
4. The probability of a success for each trial is denoted p; the value of p remains the same for each trial.	4. p = the probability of flipping a head on a particular trial = 1/2
5. The random variable X is the number of *successes* out of n trials.	5. X = the number of heads out of three flips

You encounter a binomial random variable when a certain experiment is repeated many times (n trials), the trials are independent, and each experiment results in one of two mutually exclusive outcomes. For example, a randomly selected individual is either male or female, is on welfare or is not, will vote Republican or will not, and so on.

The two outcomes for each experiment are labeled as *success* or *failure*. A success need not be considered "good" or "desirable." Instead, it depends on what you are counting at the completion of the n trials. If, for example, the object of the experiment is to determine the probability that 3 people out of 20 randomly selected individuals *are* on welfare, then a success on each of the $n = 20$ trials is the event that the person selected on each trial *is* on welfare.

EXAMPLE 5.6   In Example 4.3, it was noted that 30% of the people in a particular city read the evening newspaper. Select four people at random from this city. Consider how many of these four people read the evening paper. Does this situation satisfy the requirements of a binomial situation? What is your random variable here?

SOLUTION   Refer to conditions 1 through 5 in our list for a binomial situation.

1. There are $n = 4$ trials, where each trial consists of selecting one individual from this city.
2. There are two outcomes for each trial. We are interested in counting the number of people, out of the four selected, who *do* read the evening paper, so define

$$\text{success} = \text{read the evening newspaper}$$

$$\text{failure} = \text{do not read the evening newspaper}$$

3. The trials are independent since the people are selected randomly.
4. $p =$ probability of a success on each trial $= .3$.
5. The random variable here is X, where

$$X = \text{number of successes in } n \text{ trials}$$

$$X = \text{number of people (out of four) who read the evening newspaper}$$

All the requirements are satisfied. Thus, X is a binomial random variable (it is also discrete). ∎

Counting Successes for a Binomial Situation

How many ways are there of getting two heads out of four flips of a coin? There are six: HHTT, HTHT, HTTH, THHT, THTH, and TTHH. How many ways can you select two people from a group of four people, where the order of selection is unimportant (say, you are selecting a two-person committee)? Label the individuals as I_1, I_2, I_3, and I_4. You want to find the number of combinations of four people using two at a time:

$$_4C_2 = \frac{4!}{2!2!} = 6$$

Put these results side by side. The scheme for matching the two results is to select I_1 if H appears on the first flip, select I_2 if H appears on the second flip, and so on.

TWO HEADS OUT OF FOUR FLIPS	TWO PEOPLE FROM A GROUP OF FOUR
HHTT	I_1, I_2
HTHT	I_1, I_3
HTTH	I_1, I_4
THHT	I_2, I_3
THTH	I_2, I_4
TTHH	I_3, I_4

You should see a direct correspondence between the two solutions. Our conclusion is that the number of ways of getting two heads out of four flips of a coin is $_4C_2$. Extending this to any number of flips of a coin, the number of ways of getting k heads out of n flips of a coin is $_nC_k$. Finally, for any binomial situation, the number of ways of getting k successes out of n trials is $_nC_k$. We are thus able to determine the probability mass function (PMF) for any binomial random variable.

Once again, let X equal the number of heads out of three flips. Here X is a binomial random variable, with $p = .5$. Consider any value of X, say, $X = 1$. Then the probability of any one outcome where $X = 1$, such as HTT, is 1/8 and the number of ways of getting one head (success) out of three flips (trials) is $_3C_1 = 3$. Consequently, the probability that X will be 1 is:

$$P(1) = {_3C_1}(1/8) = 3/8$$

The resulting PMF for this situation can be written as

$$P(x) = {_3C_x} \cdot (1/8) \quad \text{for } x = 0, 1, 2, 3 \quad \text{(and zero elsewhere)}$$

Using this function, we obtain the same results as before:

$$P(0) = {_3C_0}(1/8) = 1 \cdot (1/8) = 1/8$$
$$P(1) = {_3C_1}(1/8) = 3 \cdot (1/8) = 3/8$$
$$P(2) = {_3C_2}(1/8) = 3 \cdot (1/8) = 3/8$$
$$P(3) = {_3C_3}(1/8) = 1 \cdot (1/8) = \underline{1/8}$$
$$1$$

EXAMPLE 5.7

In Example 5.6, the binomial random variable X is the number of people (out of four) who read the evening newspaper. Also, there are $n = 4$ trials (people) with $p = .3$ (30% of the people read the evening newspaper). Let S denote a success and F a failure. Then define:

$$S = \text{a person reads the evening newspaper}$$
$$F = \text{a person does not read the evening newspaper}$$

What is the probability that exactly two people (out of four) will read the evening paper?

SOLUTION

This is $P(X = 2)$, or $P(2)$. Consider any one result where $X = 2$, such as SFSF. The probability of this result, using equation 4.14, is (probability of S on first trial) $\cdot$ (probability of F on second trial) $\cdot$ (probability of S on third trial) $\cdot$ (probability of F on fourth trial), which is

$$(.3)(.7)(.3)(.7) = (.3)^2(.7)^2$$

Also note that the probability of *each* result with two S's and two F's ($X = 2$) also is $(.3)^2(.7)^2 = p^2(1 - p)^2$. How many ways can we get two successes out of four trials? This is:

$$_4C_2 = \frac{4!}{2!2!} = 6$$

So, the final result here is

$$P(2) = \text{(number of ways of getting } X = 2)\text{(probability of each one)}$$
$$= {_4C_2}(.3)^2(.7)^2$$
$$= (6)(.09)(.49) = .265$$

So, 26.5% of the time, exactly two people out of four will read the evening newspaper. ∎

We can extend the results of Example 5.7 to obtain the PMF for a binomial random variable:

$$P(x) = {}_nC_x p^x (1 - p)^{n-x} \text{ for } x = 0, 1, 2, \ldots, n \text{ (and zero elsewhere)} \quad \textbf{(5.6)}$$

where n is the number of trials and p is the probability of a success for each trial.

For the newspaper example, $x = 2$, $n = 4$, and $p = .3$. The complete list of probabilities for this example is:

$$X = \begin{cases} 0 \text{ with probability } {}_4C_0(.3)^0(.7)^4 = & .240 \\ 1 \text{ with probability } {}_4C_1(.3)^1(.7)^3 = & .412 \\ 2 \text{ with probability } {}_4C_2(.3)^2(.7)^2 = & .265 \\ 3 \text{ with probability } {}_4C_3(.3)^3(.7)^1 = & .076 \\ 4 \text{ with probability } {}_4C_4(.3)^4(.7)^0 = & \underline{.008} \end{cases}$$

$$1.001$$

Note that the total value may be slightly greater or less than 1.0, due to rounding. A graphical representation of this PMF is shown in Figure 5.3.

Using the Binomial Table

The binomial PMFs have been tabulated in Table A.1 for various values of n and p. The maximum number of trials in this table is $n = 20$. For binomial situations where $n > 20$, one alternative is to use an approximation to a binomial probability. This is considered in the next section and in Chapter 6.

For the evening newspaper illustration in Example 5.6, $n = 4$ and $p = .3$. To find $P(2)$, locate $n = 4$ and $x = 2$. Go across the table to $p = .3$ and you will find the corresponding probability (after inserting the decimal in front of the number). This probability is .265. Similarly, $P(0) = .240$, $P(1) = .412$, $P(3) = .076$, and $P(4) = .008$, as before.

The probability that no more than two people will read the evening paper is written $P(X \leq 2)$, where

$$\begin{aligned} P(X \leq 2) &= P(X = 0) + P(X = 1) + P(X = 2) \\ &= P(0) + P(1) + P(2) \\ &= .240 + .412 + .265 \\ &= .917 \end{aligned}$$

This is a **cumulative probability** and is obtained by summing $P(x)$ over the appropriate values of X.

■ **FIGURE 5.3**
Probability mass function for $n = 4$, $p = .3$.

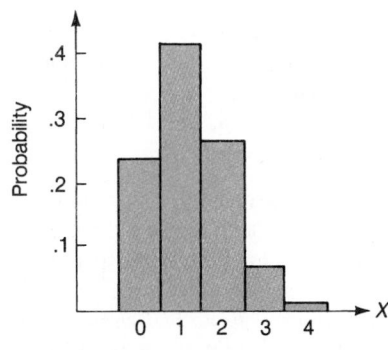

X = number of students

■ **FIGURE 5.4**   Shape of the binomial distribution.

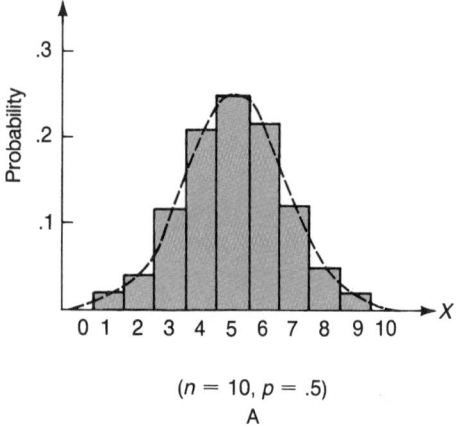

$(n = 10, p = .5)$
A

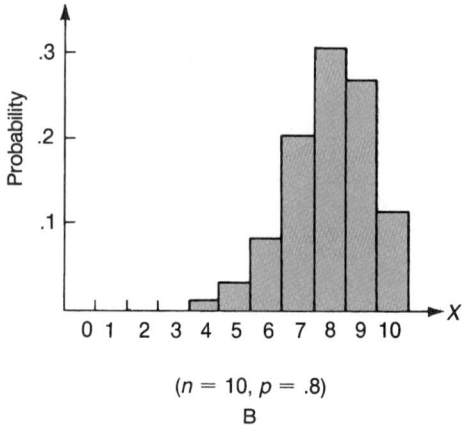

$(n = 10, p = .8)$
B

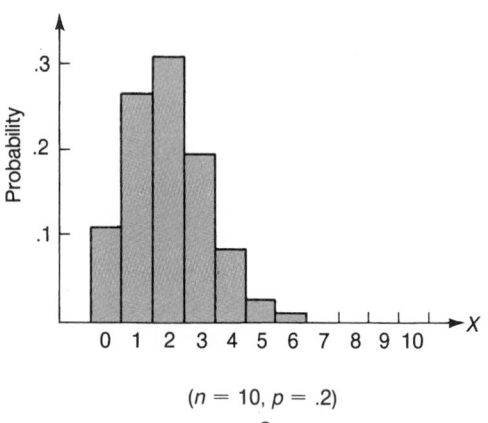

$(n = 10, p = .2)$
C

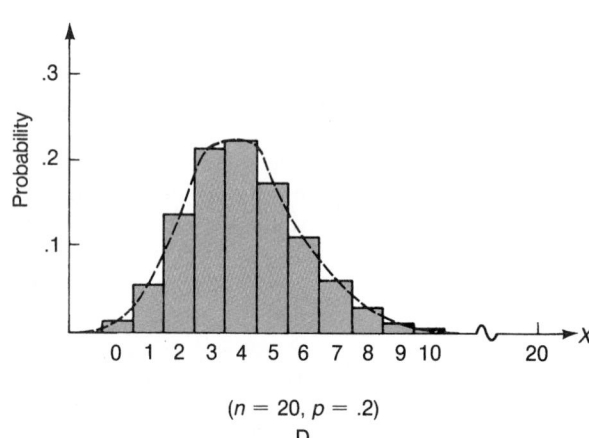

$(n = 20, p = .2)$
D

Shape of the Binomial Distribution

Figure 5.4 contains a graphical representation of four binomial distributions. In particular, notice that:

1. When $p = .5$, the shape is perfectly *symmetrical* and resembles a bell-shaped (normal) curve.
2. When $p = .2$, the distribution is *skewed right*. This skewness increases as p decreases.
3. For $p = .8$, the distribution is *skewed left*. As p approaches 1, the amount of skewness increases.

Compare Figure 5.4c and d. Notice that, in both cases, p is .2; however, the number of trials increased from $n = 10$ (in c) to $n = 20$ (in d). For the larger value of n, the shape of this distribution is nearly bell-shaped, *despite the small value of p. This implies that, regardless of the value of p, the shape of a binomial distribution approaches a bell-shaped distribution as the number of trials (n) increases.* We will use this fact in the next chapter, when we demonstrate an approximation to the binomial distribution using a bell-shaped (normal) curve for large samples.

In summary, the shape of a binomial distribution is:

1. Skewed left for $p > 1/2$ and small n.
2. Skewed right for $p < 1/2$ and small n.
3. Approximately bell-shaped (symmetric) if p is near $1/2$ or if the number of trials is large.

Mean and Variance of Binomial Random Variables

In Example 5.6, we examined the binomial random variable X representing the number of people (out of four) who read the evening newspaper. If you select four people, observe X, select four more people, observe X, and repeat this procedure indefinitely, what will X be on the average? This is the mean of X, where, using equation 5.1,

$$\mu = \Sigma xP(x)$$
$$= (0)(.240) + (1)(.412) + (2)(.265) + (3)(.076) + (4)(.008)$$
$$= 1.2 \text{ people}$$

Also, using equation 5.3, the variance of X is

$$\sigma^2 = \Sigma x^2 P(x) - \mu^2$$
$$= [(0)^2(.240) + (1)^2(.412) + (2)^2(.265)$$
$$+ (3)^2(.076) + (4)^2(.008)] - (1.2)^2$$
$$= 2.28 - 1.44 = .84$$

and so σ = standard deviation of $X = \sqrt{.84} = .92$ people. (Watch the units.)

The good news is that there is a convenient shortcut for finding the mean and variance of a binomial random variable. For this situation, you need not use equations 5.1 and 5.3. Instead, for any binomial random variable,

$$\mu = np \qquad\qquad (5.7)$$
$$\sigma^2 = np(1 - p) \qquad\qquad (5.8)$$

How these expressions were derived is certainly not obvious, but let us verify that they work for Example 5.6. Here $n = 4$ and $p = .3$, so

$$\mu = (4)(.3) = 1.2 \qquad \text{(OK)}$$
$$\sigma^2 = (4)(.3)(.7) = .84 \qquad \text{(OK)}$$

EXAMPLE 5.8   If you repeat Example 5.6 using $n = 50$ people (rather than $n = 4$ people), how many evening newspaper readers will you observe on the average?

SOLUTION   Now, X is the number of people (out of 50) who read the evening paper. Consequently,

$$\mu = np = (50)(.3) = 15$$

So, on the average, X will be 15 people. For this situation, the variance of X is

$$\sigma^2 = np(1 - p) = (50)(.3)(.7) = 10.5$$

Also,

$$\sigma = \sqrt{10.5} = 3.24 \text{ people}$$

∎

EXAMPLE 5.9   Airline overbooking is a common practice. Many people make reservations on several flights due to uncertain plans and then cancel at the last minute or simply fail to show up. Eagle Air is a small commuter airline. Their planes hold only 15 people. Past records indicate that 20% of the people making a reservation do not show up for the flight.

Suppose that Eagle Air decides to book 18 people for each flight.

1. Determine the probability that on any given flight, at least one passenger holding a reservation will not have a seat.
2. What is the probability that there will be one or more empty seats for any one flight?
3. Determine the mean and standard deviation for this random variable.

SOLUTION 1   The binomial random variable for this situation is X = the number of people (out of 18) who book a flight and actually do appear. For this binomial situation, $n = 18$ (18 reservations are made) and $p = 1 - .2 = .8$ (the probability that any one person will show up). At least one passenger will have no place to sit if X is 16 or more. Using Table A.1,

$$P(X \geq 16) = P(X = 16) + P(X = 17) + P(X = 18)$$
$$= .172 + .081 + .018 = .271$$

We see that if the airline follows this policy, 27% of the time one or more passengers will be deprived of a seat—not a good situation.

SOLUTION 2   We want to find the probability that the number of people who actually arrive (X) is 14 or less. Using Table A.1 (where $n = 18$, $p = .8$),

$$P(X \leq 14) = .215 + .151 + \cdots + .003 + .001 = .50$$

(Notice that the four remaining probabilities are nearly zero.) With this booking policy, the airline will have flights with one or more empty seats approximately one-half of the time.

SOLUTION 3   The mean of X is

$$\mu = np = (18)(.8) = 14.4 \text{ people}$$

which implies that the average number of people who book a flight and do appear is 14.4.

The standard deviation of X is

$$\sigma = \sqrt{np(1 - p)} = \sqrt{(18)(.8)(.2)} = 1.70 \text{ people} \qquad \blacksquare$$

EXAMPLE 5.10   It is estimated that one out of ten vouchers examined by the audit staff employed by a branch of the Department of Health and Human Services will contain an error. Define X to be the number of vouchers in error out of 20 randomly selected vouchers.

1. What is the probability that at least three vouchers will contain an error?
2. What is the probability that no more than one contains an error?
3. Determine the mean and standard deviation of X.

SOLUTION 1   The random variable X satisfies the requirements for a binomial random variable with $n = 20$ and $p = .1$. For this situation, a "success" is defined to be that a voucher contains an error. The probability that at least three vouchers will contain an error is the probability that X is *3 or more*, which is

$$P(X \geq 3) = 1 - P(X < 3)$$
$$= 1 - P(X \leq 2)$$
$$= 1 - [P(0) + P(1) + P(2)]$$

$$= 1 - (.122 + .270 + .285)$$
$$= .323$$

Consequently, the probability that at least three vouchers will contain an error is .323.

SOLUTION 2   The chance that no more than one voucher is in error is the probability that X is *1 or less*, which is

$$P(X \leq 1) = P(0) + P(1) = .122 + .270 = .392$$

So, this event will occur with probability .392.

SOLUTION 3   The mean of the random variable X is

$$\mu = np = (20)(.1) = 2 \text{ vouchers}$$

and the standard deviation of X is

$$\sigma = \sqrt{np(1 - p)} = \sqrt{(20)(.1)(.9)} = 1.34 \text{ vouchers}$$

This implies that, on the average, the audit staff will encounter 2 vouchers containing an error (out of 20 randomly selected vouchers).   ∎

EXAMPLE 5.11   One situation that requires the use of a binomial random variable is **lot acceptance sampling,** in which you decide whether to accept or send back a lot (batch) of many electrical components, machine parts, or whatever.

A shipment of 500 calculator chips arrives at Cassidy Electronics. The contract specifies that Cassidy will accept this lot if a sample size of ten from the shipment has no more than one defective chip. What is the probability of accepting the lot if, in fact, 10% of the lot (50 chips) are defective? If 20% are defective?

SOLUTION   This is approximately a binomial situation where:

1. There are $n = 10$ trials.
2. Each trial has two outcomes:

$$\text{success} = \text{chip is defective}$$
$$\text{failure} = \text{chip is not defective}$$

(*Note:* Since the object is to count the number of *defective* chips in the shipment, a success on each trial (chip) will be that the chip is defective. As mentioned earlier, a success need not be a desirable event.)

3. p = probability of a success = .10.*
4. The random variable here is X = number of successes out of n trials = number of defective chips out of ten. Cassidy accepts the lot of chips if X is 0 or 1. The corresponding probability is a cumulative probability:

$$P(\text{accept}) = P(X \leq 1)$$
$$= P(0) + P(1)$$

---

*If the lot size is large (500 here) and the sample size is small (10 here), then the value of p is nearly, although not completely, unaffected by the previous trials. For example, if 10% of the chips are defective, then on the first trial, p is $50/500 = .10$. On the second trial, p is either $50/499 = .1002$ (if the first chip was nondefective) or $49/499 = .098$ (if the first chip was defective). We typically ignore this minor problem in lot sampling from large populations, but this is why at the start of the solution we mentioned that this is "approximately a binomial situation." Situations in which the value of p is severely affected by what occurred on previous trials will be dealt with in the next section, where we discuss the hypergeometric distribution.

Using Table A.1 (for $n = 10$, $p = .10$), you obtain

$$P(0) = .349 \quad \text{and} \quad P(1) = .387$$

The resulting probability of accepting the lot is $.349 + .387 = .736$, so this sampling procedure will result in Cassidy accepting the entire batch of chips 73.6% of the time.

If $p = .20$, then $P(0) = .107$ and $P(1) = .268$, again using Table A.1; Cassidy now accepts the lot with probability $.107 + .268 = .375$. ∎

The concept of lot acceptance sampling was originally presented in Chapter 1 to illustrate the distinction between a population and a sample. It also serves as a brief introduction to the area of inferential statistics, discussed at length in Chapter 7. In Example 5.11, we inferred something about a population (the lot of 500 chips) using a sample (the 10 chips selected for testing). The sample does not include all elements of the population, so there is a risk of making an incorrect decision, such as (1) accepting the lot of chips when in fact it should be rejected or (2) rejecting the lot of chips when in fact it was satisfactory. *Such possibilities for error always exist when a statistical sample is used as a basis for an assertion about a population.*

 *Exercises*

5.28 An investment advisor predicts that five stocks will grow over the next 18 months. From the advisor's records, 40% of the stocks she recommends are profitable.

a. What is the probability that exactly two of the five stocks are profitable?

b. What is the probability that at least three of the five stocks are profitable?

5.29 If four trials are independently conducted and each trial has a 1/3 probability of a success, what is the probability that exactly two of the four trials result in successes?

5.30 A survey reveals that 60% of the eligible people in a certain county vote during a county election.

a. If 20 county residents who are eligible to vote are chosen at random, what is the probability that exactly 12 vote during the next county election?

b. What is the probability that exactly 10 people out of the 20 chosen at random vote during this election?

5.31 The vice president of a business firm has reviewed the records of the firm's personnel and has found that 70% of the employees read the *Wall Street Journal*. If the vice president was to choose 12 employees at random, what is the probability that the number of these employees who read the *Wall Street Journal* is:

a. at least equal to 5

b. between 4 and 10, inclusive

c. no more than 7

5.32 A lawyer estimates that 40% of the cases in which she represented the defendant were won. If the lawyer is presently representing ten defendants in different cases, what is the probability that at least five of the cases will be won? What are you assuming here?

5.33 A market-research firm has discovered that 30% of the people who earn between $25,000 and $50,000 per year have bought a new car within the past two years. In a sample of 12 people earning between $25,000 and $50,000 per year, what is the probability that between four and ten people, inclusive, have bought a new car within the past two years?

5.34 A newsstand owner has calculated that 80% of the midday newspapers are sold. If the owner orders 25 midday newspapers daily, what is the probability that on any day 23 or more of the newspapers will be sold?

5.35 The sales manager of an insurance company knows that the company's best salesperson can sell an insurance policy 60% of the time. If this salesperson were to make 15 calls to sell insurance, what is the probability that at least 10 insurance policies would be sold?

5.36 Let the random variable X represent the number of loans that have gone into default from a sample of eight loans made five years ago. The probability of a loan going into default within five years is equal to .15.

a. What are the mean and standard deviation of the random variable X?

b. What is $P(X = 2)$?

5.37 Let the random variable X represent the number of correct responses on a multiple-choice test that has 15 questions. Each question has five multiple-choice answers.

a. What is the probability that the random variable X is greater than 8 if the person taking the test randomly guesses?

b. What is the mean value of X if the person randomly guesses?

c. What is the standard deviation of X if the person randomly guesses?

d. Estimate the probability that X will fall within the limits $\mu \pm 2\sigma$.

5.38 The *Professional Technician* recommends stocks each month. If 40% of the stocks recommended advance at least 20%, what is the probability that of the five stocks most recently recommended, at least three will advance at least 20%?

5.39 The manager of a retail store knows that 10% of all checks written are "hot" checks. Of the next 25 checks written at the retail store, what is the probability that no more than 3 checks are hot?

5.40 In addressing the human side of quality, F. W. Smith, chief executive officer of Federal Express Corporation, encourages promotions from within his corporation. Activated through weekly job postings, a thorough search for qualified candidates occurs inside his corporation before looking outside. Approximately 75% of all positions are filled by Federal Express's employees through this process.

(*Source:* F. W. Smith, "Our Human Side of Quality," *Quality Progress* (Oct. 1990): 19.)

a. For a random sample of eight recently filled positions in the Federal Express Corporation, how many would you expect to have been filled by Federal Express employees?

b. What is the probability that all eight positions in part a were filled by Federal Express employees? Would this outcome be considered a rare event?

c. Within what limits would you expect the number of positions filled by Federal Express employees to fall at least 75% of the time, if repeated samples of size eight were taken? (*Hint:* use Chebyshev's inequality.)

5.41 In view of the generational profile of chief executive officers of corporations, military service constitutes an important socialization attribute. Approximately 78% of the chief executive officers have served in the military.

(*Source:* "The American Corporate Elite: A Profile," *Business Horizon*, (May/June 1990): 61.)

a. For a random sample of ten chief executive officers, let X represent the number who have prior military service. What do you think the probability would be that X is greater than 8? Compute it.

b. Compute and interpret the mean and standard deviation of X in part a.

5.5
The Hyper-geometric Distribution

Another type of discrete random variable that fits many sampling situations is the **hypergeometric random variable.** It bears a strong resemblance to the binomial random variable since the experiment once again consists of n trials, with each trial having two possible outcomes (success or failure).

The conditions for a hypergeometric random variable are:

1. Population size $= N$. In this population, k members are S (successes) and $N - k$ are F (failures).
2. Sample size $= n$ trials, obtained *without replacement*.
3. $X =$ the number of successes out of n trials (a hypergeometric random variable).

The main distinction between a hypergeometric and a binomial situation is that the trials in the former *are not independent*. As a result, the probability of a success on each trial is affected by the results of the previous trials. This situation occurs when sampling *without replacement* from a *finite* population.

The situation surrounding a hypergeometric random variable is similar to the binomial situation in that you count "successes" in both cases. However, for the hypergeometric situation, you have a *finite* population (of size N) and you know the number of successes (k) and failures ($N - k$) that make up this population. For example, you might select a random sample of $n = 8$ from a group of $N = 30$ unionized workers, of which $k = 20$ are in favor of a strike and $N - k = 10$ are not. For this situation, the hypergeometric random variable is $X =$ the number of workers (of the 8) who favor the strike.

We can repeat Example 5.11 using 50 chips (instead of 500), 10 of which are selected for testing. Suppose that 10% of these chips (5 chips) are defective. As before, define

$$S = \text{success} = \text{chip is defective}$$

$$F = \text{failure} = \text{chip is not defective}$$

In Example 5.11, we used $p = P(S) = .10$ for each trial. Here, out of the 50 chips, 5 are defective. So

$$P(S \text{ on first trial}) = 5/50 = .10$$

The conditional probability of S on the second trial is:

$$5/49 = .102 \text{ if first chip was not defective}$$

$$4/49 = .082 \text{ if first chip was defective}$$

The probability of a success on the second trial is affected by what occurred on the first trial; this is a hypergeometric situation.

The PMF for the hypergeometric random variable is:

$$P(x) = \frac{{}_kC_x \cdot {}_{N-k}C_{n-x}}{{}_NC_n} \tag{5.9}$$

for $x = a, a + 1, a + 2, \ldots, b$, where a is the maximum of 0 and $n + k - N$ and b is the minimum of k and n. The value of $P(x)$ is zero for all other values of X. Also, n is the sample size and N is the population size, k of which are successes.

EXAMPLE 5.12   Determine the probability of observing exactly one defective chip out of a sample of size ten.

SOLUTION   Imagine two containers (the population). One contains 5 S's and the other has 45 F's. The sample consists of 10 chips, randomly selected from these two containers. If x chips are selected from the success container, then $10 - x$ chips are selected from the failure container. For this situation, $N = 50$, $k = 5$ and $n = 10$. The possible values for X are from $a =$ maximum of 0 and -35 (0) to $b =$ minimum of 5 and 10 (5). The probability of obtaining one S and nine F's in your sample is

$$P(X = 1) = P(1) = \frac{{}_5C_1 \cdot {}_{45}C_9}{{}_{50}C_{10}}$$

As you will quickly see, the term ${}_NC_n$ gets very large—in fact, it becomes too large for many calculators. The only practical way to evaluate a hypergeometric probability, short of relying on a computer, is to cancel as many terms as possible in the expression.

The final result here is $P(1) = .431$; 43% of the time, you will obtain exactly 1 defective chip in your sample of size 10.   ∎

EXAMPLE 5.13

A local group of 30 unionized workers contains 20 people who are in favor of a strike and 10 who are not. Determine the probability that a random sample of 8 workers contains 5 individuals who favor the strike and 3 who are opposed.

SOLUTION

This situation fits the requirements for a hypergeometric random variable, where X is the number of workers (out of 8) who favor a strike, $n = 8$, $N = 30$, and $k = 20$. Consequently,

$$P(X = 5) = P(5) = \frac{{}_{20}C_5 \cdot {}_{10}C_3}{{}_{30}C_8}$$

$$= \frac{\dfrac{20!}{5!15!} \cdot \dfrac{10!}{3!7!}}{\dfrac{30!}{8!22!}}$$

$$= \frac{(15,504)(120)}{5,852,925} = .318$$

Approximately 32% of the time, in a sample of size 8 from this group, 5 people would favor a strike. ∎

Mean and Variance of a Hypergeometric Random Variable

As we did with the binomial random variable, we could use the definition of the mean and variance of a discrete random variable contained in equations 5.1 and 5.3. For example,

$$\mu = \Sigma x P(x)$$

where $P(x)$ is the PMF given in equation 5.9.

As in the binomial situation, simpler expressions exist for both the mean and the variance of the hypergeometric random variable. These are:

$$\mu = \Sigma x P(x) = \frac{nk}{N} \tag{5.10}$$

and

$$\sigma^2 = \Sigma x^2 P(x) - \mu^2$$
$$= \frac{k(N - k)n(N - n)}{N^2(N - 1)} = \left[n\left(\frac{k}{N}\right)\left(1 - \frac{k}{N}\right) \right]\left(\frac{N - n}{N - 1}\right) \tag{5.11}$$

For Example 5.12, $N = 50$, $k = 5$, $n = 10$. Consequently,

$$\mu = \frac{(10)(5)}{50} = 1 \text{ chip}$$

$$\sigma^2 = \frac{(5)(45)(10)(40)}{(50)^2(49)} = .735$$

and so

$$\sigma = \sqrt{.735} = .857 \text{ chip}$$

This means that if we observed this process of sampling 10 chips out of a batch of 50 indefinitely, we would obtain one $(= \mu)$ defective chip on the average. Also, $\sigma = .857$ (or $\sigma^2 = .735$) is our measure of the variation in the observations of this random variable if we observe it over an indefinite period.

Using the Binomial to Approximate the Hypergeometric

Whenever $n/N < .05$, the binomial distribution will provide a good approximation to the hypergeometric distribution. Here, define

$$p = \frac{\text{number of successes in population}}{\text{size of population}} = \frac{k}{N}$$

Then X is the number of successes in the sample. X is approximately a binomial random variable with n trials and probability of success p. Briefly, the binomial approximation works well if your sample size is *less than 5%* of your population size; that is, $n/N < .05$. This was the case in Example 5.11, which was, in fact, a hypergeometric application but was treated as an approximate binomial situation since $n/N = 10/500 = .02$.

What probability would you obtain had you treated Example 5.12 as a binomial situation, where $p = k/N = 5/50 = .10$? Here you have a binomial situation with $n = 10$ and $p = .10$. Using equation 5.6,

$$P(1) = {}_{10}C_1(.1)^1(.9)^9 = (10)(.1)(.387) = .387$$

The same result is obtained using Table A.1.

For this example, .431 is the *exact* probability using the hypergeometric distribution and .387 is the *approximate* probability using the binomial distribution. We did not obtain a very good approximation here. The problem is that the population size is $N = 50$ and the sample size is $n = 10$, which is 20% of the population size.

◢ 5.6
The Poisson Distribution (Optional)

The Poisson distribution, named after the French mathematician Simeon Poisson, is useful for counting the number of times a particular event occurs over a specified period of time. It also can be used for counting the number of times an event (such as a manufacturing defect) occurs over a specified area (such as a square yard of sheet metal) or in a specified volume. **We will restrict our discussion to counting over time, although any unit of measurement is permissible.**

The random variable X for this situation is the number of occurrences of a particular event over a specified period of time. The possible values are 0, 1, 2, 3, ... For X to be a **Poisson random variable** over a given interval of time, the occurrences of this event need to occur *randomly,* as summarized by the following three conditions:

1. The number of occurrences in one interval of time is unaffected by (statistically independent of) the number of occurrences in any other nonoverlapping time interval. For example, what took place between 3:00 and 3:20 P.M. is unaffected by what took place between 9:00 and 10:00 A.M.
2. The expected (or average) number of occurrences over any time period is proportional to the size of this time interval. For example, we would expect half as many occurrences between 3:00 and 3:30 P.M. as between 3:00 and 4:00 P.M.

    This condition also implies that the probability of an occurrence must be constant over any intervals of the same length. A situation in which this is usually *not* true is at a restaurant from 12:00 noon to 12:10 P.M. and 2:00 to 2:10 P.M. Due to the differences in traffic flow for these two intervals, we would not expect the arrivals between, say, 11:30 A.M. and 2:30 P.M. to satisfy the requirements of a Poisson situation.
3. Events cannot occur exactly at the same time. More precisely, there is a unit of

time sufficiently small (such as one second) that no more than one occurrence of an event is possible during this time.

Four situations that usually meet these conditions are:

The number of arrivals at a local bank over a five-minute interval.

The number of telephone calls arriving at a switchboard over a one-minute interval.

The number of daily accidents reported along a 20-mile stretch of an intercity toll road.

The number of trucks in a fleet that break down over a one-month period.

For each situation, the (discrete) random variable X is the number of occurrences over the time period T. If all the assumptions are satisfied, then X is a Poisson random variable. Define μ to be the expected (or average) number of occurrences over this period of time.* For any application, the value of μ must be specified or estimated in some manner. The Poisson PMF for X follows.

Poisson Probability Mass Function

X = number of occurrences over time period T.　　　(5.12)

$$P(x) = \frac{\mu^x e^{-\mu}}{x!} \quad \text{for } x = 0, 1, 2, 3, \ldots$$

where μ = expected number of occurrences over T.

Equation 5.12 contains the number e, which is an interesting and useful number in mathematics and statistics. To get an idea how this number is derived, consider the following sequence:

$$(1 + 1/2)^2 = 2.25$$
$$(1 + 1/3)^3 = 2.37$$
$$(1 + 1/4)^4 = 2.44$$
$$(1 + 1/5)^5 = 2.49$$
$$\vdots$$
$$(1 + 1/100)^{100} = 2.705$$
$$\vdots$$
$$(1 + 1/1000)^{1000} = 2.717$$
$$\vdots$$

This sequence of numbers is approaching e. The actual value of e is

$$e = 2.71828\ldots$$

One interesting application of the number e occurs when calculating compound interest. For example, if you invest $100 at 12% compounded annually, then at the end of the year you will have $112. However, if your interest is compounded not monthly, not daily, but continuously, the amount in your account will be $(100)(e^{.12}) = (100)(1.1275) = \$112.75$. The difference in these amounts is not as large as you might expect! We will use e again in Chapter 6.

---

*The symbol λ (lambda) often is used to denote this parameter.

Mean and Variance of a Poisson Random Variable

Once again, we could use the definition of the mean and variance of a discrete random variable in equations 5.1 and 5.3. However, this is not necessary. It is fairly easy to show, using equation 5.12, that

$$\text{mean of } X = \Sigma x P(x)$$
$$= \mu$$

This is hardly a surprising result, given how μ was originally defined. Also,

$$\text{variance of } X = \sigma^2$$
$$= \Sigma x^2 P(x) - \mu^2$$
$$= \mu$$

So, *both the mean and the variance of the Poisson random variable X are equal to μ.* Recall that the Poisson random variable is the number of occurrences of a particular event (such as a traffic accident) over a given time period (such as an hour). If the time period is doubled to two hours, then the mean of the "new" Poisson random variable is twice the original mean; if the time period is halved to 30 minutes, the corresponding mean is halved, and so on. This is illustrated in the nxt two examples.

Applications of a Poisson Random Variable

EXAMPLE 5.14   Handy Home Center specializes in building materials for home improvements. They recently constructed an information booth in the center of the store. Define X to be the number of customers who arrive at the booth over a five-minute period. Assume that the conditions for a Poisson situation are satisfied with

$$\mu = 4 \text{ customers over a five-minute period}$$

A graph of the Poisson probabilities for $\mu = 4$ is contained in Figure 5.5.

1. What is the probability that over any five-minute interval, exactly four people arrive at the information booth?
2. What is the probability that more than one person will arrive?
3. What is the probability that exactly six people arrive over a ten-minute period?

SOLUTION 1   First, this probability is not 1 because $\mu = 4$ is the *average* number of arrivals over this time period. The actual number of arrivals over some five-minute period may

■ **FIGURE 5.5**
Poisson probabilities for $\mu = 4$.

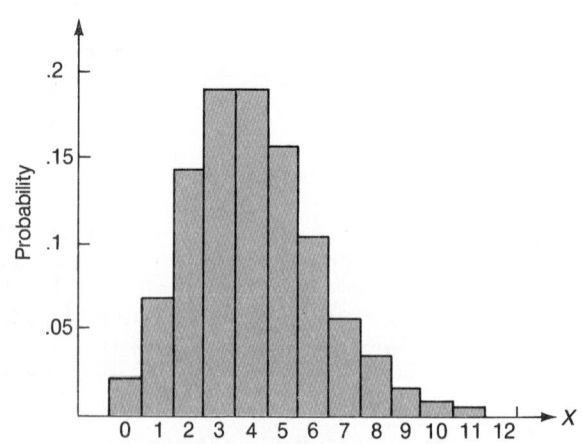

be fewer than four, more than four, or exactly four. The fraction of time that you observe exactly four people is, using Table A.3,

$$P(4) = \frac{4^4 e^{-4}}{4!} = .1954$$

If you stand in the booth for many five-minute periods, 19.5% of the time you will observe four people arrive.

SOLUTION 2   This is $P(X > 1) = P(X \geq 2)$. We could try

$$P(X \geq 2) = P(X = 2) + P(X = 3) + \cdots$$
$$= P(2) + P(3) + \cdots$$
$$= .1465 + .1954 + \cdots$$

There is an infinite number of terms here, however, so this is *not* the way to find this probability. A much better way is to use the fact that these probabilities sum to 1. Consequently,

$$P(X \geq 2) = 1 - P(X < 2)$$
$$= 1 - P(X \leq 1)$$
$$= 1 - [P(0) + P(1)]$$
$$= 1 - \left[\frac{4^0 e^{-4}}{0!} + \frac{4^1 e^{-4}}{1!} \right]$$
$$= 1 - [.0183 + .0733] = .9084$$

SOLUTION 3   For this time interval,

μ = expected (average) number of people over a ten-minute time period

$\mu = 8$ (we expect four people over a five-minute period)

Therefore, the probability of observing six people over a ten-minute period is

$$\frac{8^6 e^{-8}}{6!} = .1221$$

using Table A.3.   ■

The Poisson distribution is widely used in the area of quality control for describing the number of nonconformities observed in a sampling unit. A *nonconformity* is defined as a failure to conform to a particular specification, such as "no scratches on a strip of sheet metal" or "no leaks in an automobile radiator." If a sampling unit is a square yard of sheet metal, then the number of nonconformities might be the number of observed scratches. Or if the sampling unit is a radiator, the number of nonconformities is the number of observed leaks. If the occurrences of a nonconformity are relatively rare (compared to the number that could occur if everything went wrong), then the Poisson distribution typically works well to describe the random variable X = number of nonconformities per sampling unit, as illustrated in the following example.

EXAMPLE 5.15   A certain process produces 100-foot long sheets of vinyl composed of a simulated wood grain top layer and a black bottom layer. A blemish (nonconformity) occurs when the black layer shows through or the wood grain pattern is not distinct. A 10-foot sample is obtained by trimming the end of a roll, at which point the number of blemishes is observed and recorded. It is believed that the number of blemishes per sample follows a Poisson distribution with an average of two blemishes per 10-foot sample. Determine the probability that:

1. There will be no blemishes observed in a 10-foot sample.
2. There will be more than eight blemishes observed if a 30-foot sample is used.

SOLUTION 1 The Poisson variable X for this situation is the number of observed blemishes. The average number of observed blemishes in a 10-foot sample is two, so

$$P(X = 0) = \frac{2^0 e^{-2}}{0!} = .1353$$

using Table A.3. This result means that 13.5% of the 10-foot samples will contain no blemishes.

SOLUTION 2 The average number of blemishes in a 30-foot sample is six, given that the average is two for a 10-foot sample. Therefore, using Table A.3 with $\mu = 6$,

$$P(X > 8) = 1 - P(X \le 8)$$

$$= 1 - \left[\frac{6^0 e^{-6}}{0!} + \frac{6^1 e^{-6}}{1!} + \cdots + \frac{6^7 e^{-6}}{7!} + \frac{6^8 e^{-6}}{8!} \right]$$

$$= 1 - (.0025 + .0149 + \cdots + .1377 + .1033) = .153$$

We can expect more than eight blemishes in a 30-foot sample about 15% of the time. ∎

Poisson Approximation to the Binomial

There will be many times when you are in a binomial situation but n is too large to be tabulated. In such situations, you can use a computer, but if that is not convenient, there are methods of *approximating* these probabilities without sacrificing much accuracy. One method is to pretend that your binomial random variable, X, is a Poisson random variable having the same mean. The corresponding Poisson probability may be much simpler to derive and will serve as an excellent approximation to the binomial probability.

In certain situations, a good approximation to a binomial probability is obtained using the Poisson distribution. Ideally, this approximation should be used when n is large ($n > 20$) and p is small ($p < .05$). However, for most situations, this approximation will be reasonably accurate if $n > 20$ and $np \le 7$. An illustration using $n = 20$ and $p = .10$ is shown in Figure 5.6. The binomial probabilities are from Table A.1 and the Poisson probabilities are from Table A.3.

EXAMPLE 5.16 In Example 5.11, Cassidy Electronics received a batch (lot) of 500 calculator chips, 10 of which they sampled. Suppose instead that they receive a batch of 2500 chips and test 100 of them. They will accept the lot if the sample contains no more than one defective chip. If we assume that 5% of the chips are defective, what is the probability that they will accept the lot?

SOLUTION We can treat this as a binomial situation (rather than the more complicated hypergeometric situation) because

$$\frac{n}{N} = \frac{100}{2500} = .04 < .05$$

This is approximately a binomial situation with $n = 100$ trials and $p = .05$. The binomial random variable X here is the number of defective chips out of 100. So

$$P(\text{accept}) = P(X \le 1) = P(0) + P(1)$$

Using the PMF in equation 5.6:

$$P(\text{accept}) = {}_{100}C_0 \cdot (.05)^0 (.95)^{100} + {}_{100}C_1 \cdot (.05)^1 (.95)^{99}$$

■ **FIGURE 5.6** Poisson distributions provide a good approximation of binomial probabilities where $n > 20$ and $np \leq 7$. Here, $n = 20$, $p = .10$.

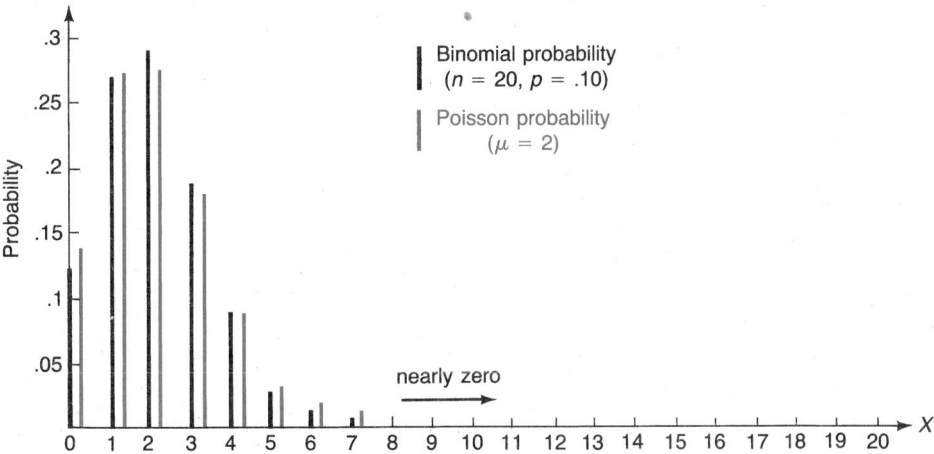

Table A.1 does not contain values of n larger than 20, so you need another means of determining these probabilities. A computer or a calculator will allow you to determine these exactly. The exact answer here is $P(\text{accept}) = .037$.

The other alternative is to pretend that X is a Poisson random variable with the same mean as the actual binomial random variable; that is,

$$\mu = np = (100)(.05) = 5$$

The approximation should work quite well because $n > 20$, p is .05, and np is ≤ 7. Using Table A.3, with $\mu = 5$,

$$P(0) = .0067$$

and

$$P(1) = .0337$$

Therefore,

$$P(X \leq 1) = .0067 + .0337 = .0404$$

Using the Poisson approximation,

$$P(\text{accept}) \cong .0404$$

which is quite close to the binomial value of .037. ■

Exercises

5.42 Ten people apply for jobs as bookkeepers. Six of the applicants have college degrees and the remainder do not. If four of the applicants are randomly selected for the jobs, what is the probability that exactly three have college degrees?

5.43 Six vegetables are available at a cafeteria. Four vegetables are green and the other two are not green. If three different vegetables are ordered at random, what is the probability that at least two of the vegetables are green?

5.44 A population consists of eight round and seven square objects. Let the random variable X be equal to the number of round objects selected randomly without replacement from a sample of nine objects.

a. Find $P(2 \le X \le 5)$.   c. Find the mean of the random variable X.

b. Find $P(X > 4)$.   d. Find the standard deviation of the random variable X.

5.45 A batch of 350 resistors is to be shipped if a random sample of 15 resistors has 2 or fewer defective resistors. If it is known that there are 50 defective resistors in the batch, what is the probability that 2 or fewer of the sample of 15 resistors will be defective?

5.46 The Good Olde Boys used-car lot has 20 cars for sale. It is known that 8 of the cars get over 28 miles per gallon on the highway and 12 do not. Let X be the random variable equal to the number of cars out of the next five cars sold that get over 28 miles per gallon. Assume that each car is equally likely to sell.

a. What is $P(X \le 2)$?

b. What is $P(1 \le X \le 3)$?

c. Find the mean and variance of X.

5.47 In a sample of ten men, it is found that six are physically fit. If four men are randomly selected from this sample of ten, what is the probability that no more than three are physically fit?

5.48 A box contains eight golf balls. Four of these balls are not perfectly round. If three balls are randomly selected without replacement from the eight golf balls, what is the probability that at least one is not perfectly round?

5.49 The 102d Congress of the United States has a total of 100 senators. The number of Democrats is 56 and the number of Republicans is 44. A sample of 6 senators is randomly chosen. Let the random variable X be the number of Democrats in the sample.

(*Source: The World Almanac and Book of Facts*, 1991, p. 82)

a. What is the probability that X is greater than 2?

b. What would be required to have the binomial distribution provide a good approximation to the answer in part a.

c. Use the binomial approximation in part a to determine how good the approximation is.

5.50 A textbook copy editor is reviewing a manuscript for grammatical errors. Let the random variable X represent the number of grammatical errors made in a particular chapter. Assume that the conditions of a Poisson distribution are satisfied and that there are an average of ten grammatical errors per chapter.

a. What is the probability that X is less than seven?

b. What is the mean value of X?

c. What is the standard deviation of X?

5.51 Let the random variable X be binomially distributed with $n = 60$ and $p = 0.05$. Use the Poisson distribution to approximate the probability that X is greater than or equal to three.

5.52 A manufacturer of carpet produces 100-foot long rolls of plush carpet. A nonconformity occurs when the pattern is not consistent. A 20-foot sample from each roll is inspected. It is believed that the number of nonconformities per sample follows a Poisson distribution with an average of one nonconformity per 20-foot sample.

a. What is the probability that at least two nonconformities will appear in a 20-foot sample?

b. What is the probability that three or more nonconformities will appear in a 40-foot sample?

5.53 The auto parts department of an automotive dealership sends out an average of eight special orders daily. The number of special orders is assumed to follow a Poisson distribution.

a. What is the probability that for any day, the number of special orders sent out will be more than four?

b. What is the standard deviation of the number of special orders sent out daily?

5.54 A survey indicates that 10% of the people who earn less than $20,000 per year are homeowners.

a. If a sample of 40 people who earn less than $20,000 per year is randomly selected, what is the probability that more than four people are homeowners?

b. What is the probability that exactly four people from the sample of 40 people who earn less than $20,000 per year are homeowners?

5.55 A certain manufacturer sells a machine that has numerous moving parts. A quality-control inspector counts the number of moving parts that are misaligned as the number of nonconformities for a particular machine. It is believed that the number of nonconformities per machine follows a Poisson distribution, with an average of three nonconformities per machine.

a. Determine the probability that the quality-control inspector finds no more than one non-conformity on a particular machine selected at random.

b. What is the standard deviation of the number of nonconformities per machine.

5.56 Many managers are implementing statistical process control to reduce both common and special causes of accidents on the job. The common causes are those that are the fault of management, and special causes are those that can be traced back to an individual worker or machine. Experts estimate that 85% of the problems in the system are common causes and 15% are special causes. Assume that a certain manager took a random sample of 30 accidents on the job.

(*Source:* T. A. Smith, "Why You Should Put Your Safety Program under Statistical Control," *Professional Safety* (April 1989): 31.)

a. Approximately what is the probability that this manager finds fewer than four accidents that can be traced back to special causes?

b. What outcomes from this sample could the manager consider to be rare events?

 *Summary*

When an experiment results in a numerical outcome, a convenient way of representing the possible values and corresponding probabilities is to use a random variable. A **random variable** takes on a numerical value for each outcome of an experiment. If the possible values of this variable can be listed along with the probability for each value, this variable is said to be a **discrete random variable.** Conversely, if any value of this variable can occur over a specific range, then it is a **continuous random variable.** This chapter concentrated on the discrete type, whereas Chapter 6 discusses the continuous random variable.

For a discrete random variable, the set of possible values and corresponding probabilities is a **probability distribution.** There are several ways of representing such a distribution, including a list of each value and its probability, a histogram, or an expression called a **probability mass function (PMF),** which is a numerical function that assigns a probability to each value of the random variable.

If you could observe a random variable indefinitely, you would obtain the corresponding *population*. A *sample* then consists of a finite number of random variable observations. In Chapter 3, we introduced ways of describing a set of sample data using various statistics, including the sample mean and variance. Similarly, we can describe a random variable using its mean and variance. Since they describe the population, they are parameters. The **mean** of a discrete random variable, μ, is the average value of this variable if observed over an indefinite period. The mean is found by summing the product of each value and its probability of occurring. The **variance** of a discrete random variable, σ^2, is a measure of the variation for this variable. The **standard deviation,** σ, also measures this variation and is the square root of the variance.

Three popular discrete random variables used in a business setting are the binomial, hypergeometric, and Poisson random variables. A **binomial** random variable counts the number of successes out of n independent trials. When the trials are dependent and the population size (N) as well as the number of successes in the population (k) are known, the **hypergeometric** random variable can be used to describe the number of successes out of the n trials. The **Poisson** random variable is used for situations in which you observe the number of occurrences of a particular event over a specified period of time or space. The Poisson distribution is used in

■ **FIGURE 5.7**
Summary of how the
three most common
types of discrete random
variables can be used to
approximate values for
one another.

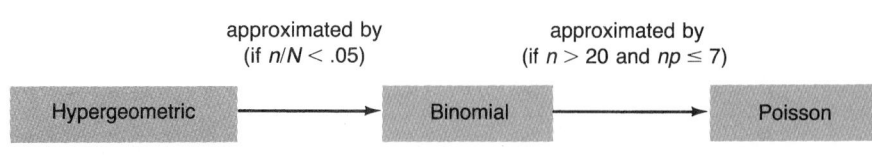

the area of quality control to describe the number of nonconformities observed in a sampling unit, such as the number of surface flaws in a square yard of finished sheet metal.

A table of binomial probabilities is contained at the back of the text in Table A.1. Tables A.2 and A.3 can be used for Poisson probabilities. For these three distributions, shortcut formulas exist for deriving the mean and variance of the random variable. Often, the probabilities for one of these discrete distributions are difficult to calculate due to the magnitude of the numbers involved. In many situations, you can use one discrete distribution to *approximate* the probability for another. Figure 5.7 summarizes how this is done.

Summary of Discrete Random Variables

Any Discrete Random Variable, X

1. Mean $= \mu = \Sigma x P(x)$
2. Variance $= \sigma^2 = \Sigma(x - \mu)^2 P(x) = \Sigma x^2 P(x) - \mu^2$
3. Standard deviation $= \sigma = \sqrt{\Sigma(x - \mu)^2 P(x)} = \sqrt{\Sigma x^2 P(x) - \mu^2}$

Binomial Distribution

1. $X =$ the number of successes out of n independent trials. Each trial results in a success (with probability p) or a failure (with probability $1 - p$).
2. PMF is $P(X = x) = P(x) = {}_nC_x p^x (1 - p)^{n-x}$ for $x = 0, 1, \ldots, n$
3. Mean $= \mu = np$
4. Variance $= \sigma^2 = np(1 - p)$ and standard deviation $= \sigma = \sqrt{np(1 - p)}$
5. Probabilities for the binomial random variable are provided in Table A.1.

Hypergeometric Distribution

1. $X =$ the number of successes in a sample of size n when selecting from a population of size N containing k successes and $N - k$ failures
2. PMF is $P(x) = ({}_kC_x \cdot {}_{N-k}C_{n-x}/{}_NC_n)$ for $x = a, a + 1, \ldots, b$, where $a =$ maximum $\{0, n + k - N\}$ and $b =$ minimum $\{k, n\}$
3. Mean $= \mu = n(k/N)$
4. Variance $= \sigma^2 = [n(k/N)(1 - k/N)][(N - n)/(N - 1)]$

Poisson Distribution

1. $X =$ the number of occurrences of a particular event over a certain unit of time, length, area, or volume
2. PMF is $P(x) = (\mu^x e^{-\mu})/(x!)$ for $x = 0, 1, 2, \ldots$
3. Mean $= \mu$
4. Variance $= \mu$
5. Probabilities for the Poisson random variable are provided in Table A.3.

 *Review Exercises*

5.57 Is the following function a probability mass function? Why or why not?

$$P(X = x) = x^3/153 \qquad \text{for } x = 1, 3, 5 \qquad \text{(and zero elsewhere)}$$

5.58 Assume that a fair die has one blue face, two white faces, and three black faces. Define the random variable X as follows:

$$X = \begin{cases} 1 & \text{if blue} \\ 2 & \text{if white} \\ 3 & \text{if black} \end{cases}$$

Find the probability mass function of X. Construct a histogram in which the height of each bar is the probability of X.

5.59 Find the mean and variance of the following random variable X with probability mass function $P(x)$. Can a value of X be negative as shown in the table?

X	P(x)
−3	.2
0	.1
3	.2
5	.3
10	.2

5.60 A bakery knows that historically the number of cakes sold daily has the following probability distribution:

X: NUMBER OF CAKES DAILY	P(x)
0	.40
1	.30
2	.15
3	.10
4	.05

a. Find the probability that at least two cakes are sold daily.

b. Find the mean and standard deviation of the number of cakes sold daily.

5.61 For a binomially distributed random variable X with 12 trials and with the probability of a success equal to .3, find the following.

a. $P(X = 7)$.   **c.** $P(X > 5)$.
b. $P(4 < X \le 6)$.   **d.** $P(X < 2)$.

5.62 Let the variable X be equal to −1 if stock XYZ declines, 0 if stock XYZ remains unchanged, and 1 if stock XYZ increases in price. If $P(X = x)$ is equal to $(x + 2)/6$, what are the mean and standard deviation of X?

5.63 A manager has ten research projects to assign to either engineer 1 or engineer 2. If each research project is randomly assigned to either one of the two engineers, what is the probability that engineer 1 will be assigned no more than five research projects?

5.64 An average of five books per week are returned to a bookstore. Assume that the number of returned books is Poisson distributed.

a. What is the probability that less than four books will be returned in one week?

b. What is the standard deviation of the distribution of the number of books returned in one week?

5.65 The supervisor of the employees who solder resistors on certain electrical components would like to know what the average number of absentees is daily and also what the standard deviation is of the daily employee absentee rate. Find these two values from the following probability mass function, which was constructed from historical data of the company:

X: NUMBER OF DAILY ABSENTEES	P(x)
0	.50
1	.23
2	.12
3	.10
4	.02
5	.02
6	.01

5.66 Ten employees are being reviewed for promotion. Four of the employees are females. If each employee is equally likely to get promoted, what is the probability that two females and three males will be promoted, if a total of five promotions are given?

5.67 There are 90 drill bits in a box at a machine shop. Fifty of the drill bits are 3/8-inch diameter, and 40 are 7/16-inch diameter. If four drill bits are selected at random, what is the probability that two drill bits of 3/8-inch diameter and two drill bits of 7/16-inch diameter will be chosen?

5.68 A population consists of 15 employees, 6 of whom have less than two years experience. Let X be equal to the number of employees with less than two years experience from a sample of eight employees randomly drawn from this population.

a. Find $P(X = 3)$.   c. Find the average value of X.
b. Find $P(X \leq 2)$.   d. Find the standard deviation of X.

5.69 Blair's Moving Company loads an average of three boxes of damaged merchandise daily. What is the probability that exactly three boxes of damaged merchandise are shipped daily? What is the standard deviation of the number of boxes of damaged merchandise that are shipped daily? Assume a Poisson distribution.

5.70 A person has written seven songs, three of which are ballads. If the songwriter chooses two at random, what is the probability of the following?

a. Exactly one is a ballad.

b. None are ballads.

c. Both are ballads.

5.71 What Poisson expression would you use to approximately evaluate the following expression?

$$_{100}C_3 (.02)^3 (.98)^{97}$$

5.72 The Conference Board, a business-research firm in New York, learned from a national survey of 435 companies that 32% of the companies used a lump-sum payment to reward employees for performance. This lump-sum payment was a one-time reward, never added to base salary.

(Source: "The Great Earnings Gamble," *U.S. News and World Report* (September 17, 1990): 65–68.)

a. If 10 companies were selected randomly from the 435 companies in the survey, could the number of companies in this sample who used a lump-sum payment reward be considered a binomial random variable? Why?

b. Would it be unreasonable to expect that out of these 10 selected companies, at least 7 use a lump-sum payment to reward individuals for performance? Why?

5.73 In a recent Harris Poll, 99% of those surveyed rated safety in automobiles as "very important" or "somewhat important" in deciding which car to buy. Surveys such as this one are prompting Ford and General Motors to supply air bags in their cars by the mid-1990s.

(Source: "Can Detroit Hold Its Lead in Safety?" *Business Week* (November 1990): 127–130.)

a. In a sample of size 20 of automobile drivers across the nation, what is the probability that all 20 consider safety in automobiles "very important" or "somewhat important" in deciding which car to buy?

b. For the sample in part a, what is the probability that no more than 17 people consider safety in automobiles "very important" or "somewhat important" in deciding which car to buy?

5.74 The United States Bureau of Labor Statistics projects that African Americans and Hispanics will account for 22% of the U.S. population in the year 2000.

(Source: "Weathering A Weak Economy," *Black Enterprise* (January 1991): 46–51.)

a. If a sample of 10 persons is selected in the year 2000, what is the probability that at least two persons are African American or Hispanic?

b. What would be considered a rare event in the sample of 10 persons in part a?

5.75 Assume that a procedure for vaccinating personal computers against viruses is properly implemented 95% of the time. What is the probability that of 36 vaccinated personal computers, exactly 34 of them are properly vaccinated?

5.76 In a recent survey by Cowen & Co., 829 respondents indicated they were planning to purchase an IBM AS/400 computer over the next two years. Out of the 829 respondents, 473 of them (57%) were not replacing any existing systems, but 24 (about 2.9%) of the 829 were replacing computers built by IBM's competitors.

(*Source:* "IBM's Proprietary Paradox," *Datamation* (July 1990): 41–42.)

a. If a sample of size 30 is randomly selected from the 839 respondents, what is the probability that at least one respondent intended to replace a computer built by IBM's competitors?

b. To find the probability that at least half of the respondents in the sample in part a were not replacing any existing systems, should one expect the Poisson approximation to give a good estimate?

5.77 The following MINITAB computer printout displays the results of randomly generating 500 observations from a population with a binomial distribution with $n = 17$ and $p = .50$.

a. Find the relative frequency at each possible value of a binomial random variable with $n = 17$ and $p = .50$. How do these compare to the actual probabilities given in the binomial table in the appendix?

b. Calculate the mean of the generated observations by using the formula $\Sigma f_i x_i / 500$ where f_i is the frequency. Compare this value to the population mean.

c. Calculate the variance of the generated observations by using the formula $[\Sigma f_i x_i^2 - (\Sigma f_i x_i)^2/500)]/(500 - 1)$. Compare this value to the population variance.

```
MTB > random 500 c1;
SUBC> binomial n=17 and p=.50.
MTB > tally c1

     C1    COUNT
      3       6
      4      11
      5      25
      6      43
      7      72
      8      93
      9      87
     10      69
     11      56
     12      26
     13       8
     14       4
     N=     500
```

5.78 The following MINITAB computer printout displays the results of randomly generating 500 observations from a population with a Poisson distribution and a mean of 3.

a. Find the relative frequencies at each possible value in the population. Compare these relative frequencies to the probabilities in the Poisson table in the appendix.

b. Answer part b from Exercise 5.77 with respect to the randomly generated data given.

c. Answer part c from Exercise 5.77 with respect to the data given.

```
MTB > random 500 c1;
SUBC> poisson mean = 3.
MTB > tally c1

     C1    COUNT
      0      25
      1      83
      2     100
      3     106
      4      91
      5      57
      6      17
      7      11
      8       8
      9       2
     N=     500
```

Computer Exercises Using the Database

Exercise 1 -- Appendix I Generate 100 random numbers and select a sample of 100 observations from the database. Consider the variable FAMLSIZE (family size). Let p represent the proportion of observations in your set of 100 observations in which the family size is no greater than 2. If you randomly select 10 observations (with replacement) from the 100 possible, what is $P[X \leq 5]$, where X is the number of observations (out of 10) in which the family size is no greater than 2? What type of random variable is X?

Exercise 2 -- Appendix I Repeat Exercise 1, where 10 observations are selected without replacement.

Exercise 3 -- Appendix I Referring to Exercise 2, if you were to obtain samples of size 10 indefinitely, what would X be on the average? What is the standard deviation of this random variable?

Exercise 4 -- Appendix I Estimate the proportion of homeowners in a randomly selected set of 100 observations. From this set of 100 observations, select with replacement a random set of 10 observations. Can this be considered a binomial experiment with $n = 10$ and p equal to the proportion of homeowners in the set of 100 observations? Estimate the probability that the number of homeowners in the set of 10 observations is greater than or equal to 5.

Exercise 5 -- Appendix I Randomly select 200 observations from the database. From this set, estimate the proportion of observations in which the location of residence is in one of the northern sectors. From this set of observations, randomly select without replacement 8 observations. Find the probability that, in this sample of size 8, the residences of 4 or fewer observations are in the northern sectors. Can the binomial approximation be used? Why?

Exercise 6 -- Appendix J Select 200 observations at random from the database. Let p be the proportion of companies with a positive net income. If you were to randomly select (with replacement) 15 observations from the 200 possible, what is the probability of selecting at least 9 companies with a positive net income?

Exercise 7 -- Appendix J For the 20 observations in Exercise 6, let p be equal to the proportion of companies in which the number of employees exceeds 10,000. If you were randomly to select 15 observations (with replacement) from the 200 possible, what is the probability of selecting at least 7 companies in which the number of employees exceeds 10,000?

Case Study

Could Aviation Fuel Be Like Fast-Moving Consumer Goods?

Purchasing behavior for aviation fuel is an industrial business-to-business marketing activity, whereas buyer behavior in fast-moving consumer goods (FMCG) markets is an individual consumer activity. On the face of it, the suggestion that these two markets could display similar characteristics might seem surprising. Yet, Mark Uncles and Andrew Ehrenberg, writing in the *International Journal of Research in Marketing,* demonstrate that similarities do exist.

According to them, oil companies and airlines generally enter into contracts of one to three years, awarded on the basis of sealed bids, for the supply of aviation fuel at individual airports. They examined data pertaining to various oil companies, such as Shell, Total, Mobil, Esso, and Chevron, and 249 airlines, such as Air France, British Airways, KLM, and Swissair, at 16 major European international airports that averaged almost 100,000 international commercial aircraft movements annually. The data were provided by Shell, so airports where Shell had no presence were excluded.

The pattern of purchasing behavior that emerged was found to resemble previously established patterns in FMCG markets ranging from breakfast cereals to detergents to motor oil. The FMCG markets have been found to follow a pattern predicted by a mathematical model known as the Dirichlet model. This model is somewhat complicated; in any case, it is not our primary focus here. Suffice it to say that the Dirichlet model describes purchase incidence and brand choice by assuming a mixture of probability distributions such as the multivariate beta, binominal, Poisson, gamma, and negative binominal distributions. For the aviation fuel contract study, the authors used a less restrictive "empirical Dirichlet" model. The purchase incidence is given in Table 5.1 and is interpreted in the following manner: 19% of airlines had $n = 1$ contract at the 16 airports, 8% had $n = 2$ contracts, and so on.

The Dirichlet model was used to generate predicted behavior for the aviation fuel market. Table

■ TABLE 5.1	Number of contracts	1	2	3	4	5	6	7	8	9	10
The observed frequency	% of airlines	19	8	8	6	6	3	8	7	3	2
distribution of the	Number of contracts	11	12	13	14	15	16	17	18	19	20
number of contracts	% of airlines	3	4	4	2	5	4	2	1	1	0
per airline.	Number of contracts	21	22	23	24	25	26	27	28	29	30 +
	% of airlines	0.8	0	0.8	0	0.4	0	0.8	0	0.4	0.4

(*Source:* Mark D. Uncles and Andrew S. C. Ehrenberg, "Industrial Buying Behavior: Aviation Fuel Contracts," *International Journal of Research in Marketing* 7 (1990): 57.

5.2 shows how the theoretical predictions of the model, designated by T, compare with the observed distribution, designated by O.

Case Study Questions

1. Do the data in Table 5.1 constitute a discrete probability distribution? If X = number of contracts, should the random variable $X = 0$ appear in Table 5.1?

2. Compute the mean, variance, and standard deviation for the distribution in Table 5.1.

3. Based on Table 5.1, what is the probability that an airline randomly selected from the 249 included in this study had:

a. more than 10 contracts at the 16 airports?

b. between 4 and 8 contracts, inclusive, at the 16 airports?

c. no contracts at the 16 airports?

4. Compare the observed and theoretical distributions in Table 5.2. How closely do the observations seem to follow the pattern predicted by the Dirichlet model?

5. Compute the mean and standard deviation for Shell, Mobil, and Chevron in Table 5.2.

6. (*Postscript:* After completing Chapter 12 on the chi-square test for a distributional form, return to this case study and revisit question 4.)

■ TABLE 5.2		NUMBER OF CONTRACTS (% CUSTOMERS)								
The distribution of contracts (by customers).	**SUPPLIERS**	**1**	**2**	**3**	**4**	**5**	**6**	**7**	**8**	**9 +**
	Shell $O(\%)$	30	22	12	10	6	3	4	4	8
	$T(\%)$	32	18	13	10	7	5	4	3	8
	"Others" $O(\%)$	30	16	19	7	10	5	2	2	9
	$T(\%)$	38	20	13	9	6	4	3	2	5
	BP $O(\%)$	42	24	14	7	5	2	2	1	3
	$T(\%)$	46	21	12	7	5	3	2	1	3
	Total $O(\%)$	36	33	11	7	4	1	6	0	1
	$T(\%)$	52	21	11	6	4	2	1	1	2
	Mobil $O(\%)$	47	19	19	6	3	1	0	1	3
	$T(\%)$	52	21	11	6	4	2	1	1	2
	Esso $O(\%)$	51	18	17	8	4	0	0	0	1
	$T(\%)$	53	21	10	6	3	2	1	1	2
	Chevron $O(\%)$	54	29	8	8	0	0	0	0	0
	$T(\%)$	58	20	9	5	3	2	1	1	1
	Average $O(\%)$	42	23	14	8	5	2	2	1	3
	$T(\%)$	47	20	11	7	4	3	2	1	3

(*Source:* Mark D. Uncles and Andrew S. C. Ehrenberg, "Industrial Buying Behavior: Aviation Fuel Contracts," *International Journal of Research in Marketing* 7 (1990): 57.

C H A P T E R

Continuous Probability Distributions

A LOOK BACK/INTRODUCTION

After we discussed the use of descriptive statistics, we introduced you to the area of uncertainty by using probability concepts and random variables. Random variables offer you a convenient method of describing the various outcomes of an experiment and their corresponding probabilities.

When each value of the random variable as well as its probability of occurring can be listed, the random variable is discrete. The other type of random variable, a continuous random variable, can assume any value over a particular range. Continuous random variables include such variables as X = height, X = weight, and X = time. For such variables it is impossible to list all values of X, yet you can still make probability statements regarding X if you can make certain assumptions about the type of population.

In statistics, making decisions from sample information is called **statistical inference.** In subsequent chapters, we will develop a formal set of rules to offer you as a guide in making statistical decisions. Making such a decision typically involves one or more assumptions about the population from which the sample was obtained. One such assumption, widely used in statistics, is that the data came from a normal population, which means that you are dealing with a normal random variable.

The concept of a continuous random variable was introduced in Chapter 5. What distinguishes a discrete random variable from one that is continuous is the presence of *gaps* in the possible values for a discrete random variable. To illustrate, X = total of two dice is a discrete random variable; there are many gaps over the range of possible values, and a value of 10.4, for example, is not possible. One can list the possible values of a discrete variable, along with the probability that each value will occur.

Determining probabilities for a continuous random variable is quite different. For such a variable, any value over a specific range is possible. Therefore we are unable to list all the possible values of this variable. Probability statements for a continuous random variable (such as X) are not concerned with specific values of X (such as the probability that X will equal 50) but rather deal with probabilities over a range of values, such as the probability that X is *between* 40 and 50, *greater than* 65, or *less than* 20, for example.

Such probabilities can be determined by first making an assumption regarding the nature of the population involved. We assume that the population can be described by a curve having a particular shape—such as normal, uniform, or exponential. Once this curve is specified, a probability can be determined by finding the corresponding area under this curve. As an illustration, Figure 6.1 shows a particular curve (called the normal curve) for which the probability of observing a value of X between 20 and 60 is the area under this curve between these two values. *The entire range of probability is covered using such a curve, since, for any continuous random variable, the total area under the curve is equal to 1.*

The following sections examine the normal, uniform, and exponential distributions, since these are the most widely encountered random variables in practice. The graphs and descriptive statistics discussed in the previous chapters can help determine if one of these random variables might be appropriate for a particular situation. If a histogram of the sample data appears nearly flat, the population might be represented by a uniform random variable. If the histogram is symmetric with decreasing tails at each end, a normal random variable may be in order. If the sample histogram steadily decreases from left to right, the population of all possible values perhaps can be described using an exponential random variable. In the first two cases, the mean and median should be nearly equal (the population is symmetric), providing a skewness measure near zero. For the exponential case, the median should be less than the mean with a corresponding positive measure of skewness.

177

■ FIGURE 6.1
Finding a probability
for a continuous
random variable.

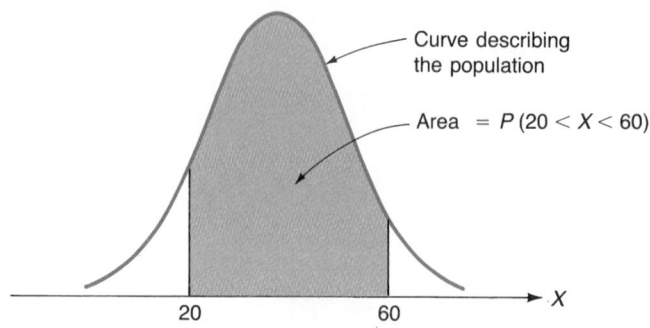

Curve describing
the population

Area $= P(20 < X < 60)$

6.1
Normal Random Variables

The normal distribution is the most important of all the continuous distributions. You will find that this distribution plays a key role in the application of many statistical techniques. When attempting to make an assertion about a population by using sample information, a major assumption often is that the population has a normal distribution.

When discussing measurements such as height, weight, thickness, or time, the resulting population of all measurements often can be assumed to have a probability distribution that is normal.

A histogram constructed from a large *sample* of such measurements can help determine whether this assumption is realistic. Assume, for example, that data were collected on the length of life of 200 Everglo light bulbs. Let X represent the length of life (in hours) of an Everglo bulb. One thing we are interested in is the *shape* of the distribution of the 200 lifetimes. Where are they centered? Are they symmetric? The easiest way to approach such questions is to construct a histogram of the 200 values, as illustrated in Figure 6.2. This histogram indicates that the data are nearly symmetric and are centered at approximately 400.

The curve in Figure 6.2 is said to be a **normal curve** because of its shape. A normal curve is characterized by a **symmetric, bell-shaped appearance,** with tails that "die out" rather quickly. We use such curves to represent the **assumed population** of all possible values. This example contained 200 values observed in a *sample*. Consequently:

1. A histogram represents the shape of the sample data.
2. A smooth curve represents the assumed shape (distribution) of the population.

If all possible values of a variable X follow an assumed normal curve, then X is said to be a **normal random variable,** and the population is **normally distributed.**

When you assume that a particular population follows a normal distribution,

■ FIGURE 6.2
Histogram of 200
Everglo light bulb
lifetimes (in hours).
The curve represents
all possible values
(population). The
histogram represents the
sample (200 values).

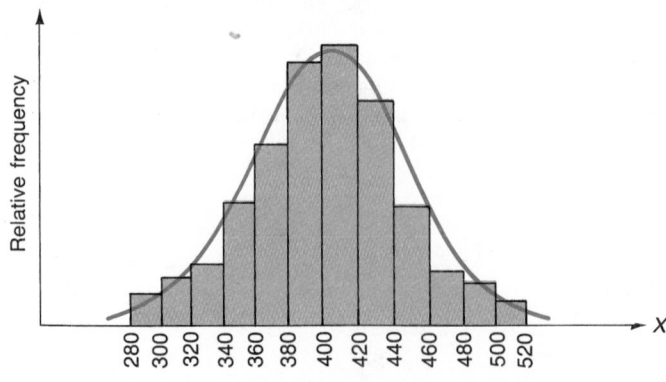

you assume that X, an observation randomly obtained from this population, is a normal random variable. Based on the histogram in Figure 6.2, it appears to be a reasonable assumption that the smooth curve describing the population of *all* Everglo bulbs can be approximated using a normal curve centered at 400 hours. Therefore, we will assume that X is a normal random variable, centered at 400 hours.

There are two numbers used to describe a normal curve (distribution); they tell where the curve is centered and how wide it is. The **center** of a normal curve is called the *mean* and is represented by the symbol μ (mu). The **width** of a normal curve can be described using the *standard deviation,* which is represented by the symbol σ (sigma).

These descriptions are illustrated in Figure 6.3, which shows the normal curve representing the lifetime of Everglo light bulbs. Another way of stating this situation is: X is a normal random variable with $\mu = 400$ hours and $\sigma = 50$ hours. Notice that the units of μ and σ are the same as the units of the data (hours).

In Figure 6.3, there is a point P on the normal curve. Above this point P, the curve resembles a bowl that is upside down, and below P the curve is "right side up." In calculus, this point is referred to as an **inflection point.** The distance between vertical lines through μ and P is the value of σ.

Because μ and σ represent the location and spread of the normal distribution, they are called **parameters.** The parameters are used to define the distribution completely. The values of μ and σ of a normal population are all you need to distinguish it from all other normal populations that have the same bell shape but different location and/or variability. The values of the parameters must be specified in order to make probability statements regarding X. As a result, there are infinitely many normal curves (populations), one for each pair of values of μ and σ.

In Chapter 5, we discussed the mean of (say) ten observations of the random variable X, written as $\bar{X}$. If you were to observe X indefinitely, then you could obtain the mean of the population, μ. The same concept applies to continuous random variables; for the Everglo example, $\bar{X}$ represents the mean of the 200 bulbs (the sample) and μ is the mean of all Everglo bulbs (the population).

MEAN		STANDARD DEVIATION	
Sample	Population	Sample	Population
$\bar{X}$	μ	s	σ
the average of the sample	the average of the population	the standard deviation of the sample	the standard deviation of the population

In our Everglo example, the average lifetime of all bulbs is *assumed* to be $\mu = 400$ hours. The standard deviation of the population, σ, just like s, is a measure of **variability.** The larger σ is, the more variation (jumping around) we would see

■ FIGURE 6.3
Distribution of the lifetime of Everglo bulbs showing the mean ($\mu = 400$), the standard deviation ($\sigma = 50$), and the inflection point (P).

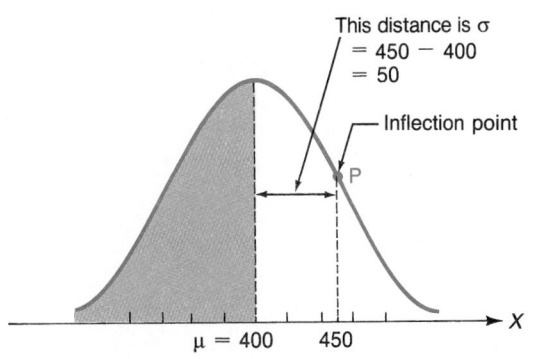

This distance is σ
= 450 − 400
= 50

Inflection point

P

$\mu = 400$ 450

X

■ **FIGURE 6.4**
Two normal curves
with unequal means
and equal standard
deviations.

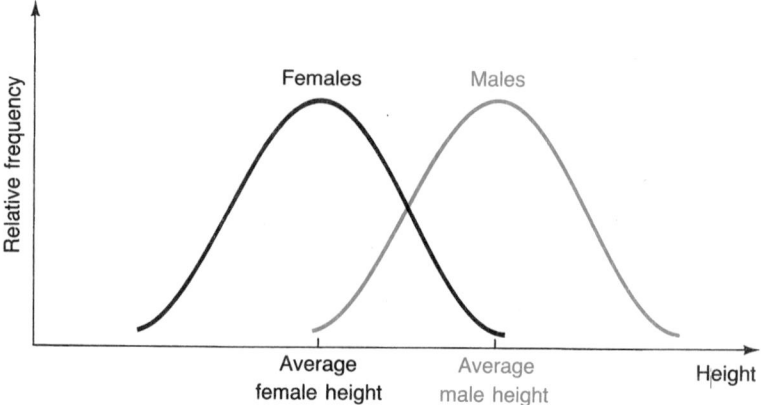

if X were observed indefinitely. For both the sample and the population, the square of the standard deviation is referred to as the variance. It is another measure of the variability of X. The **variance** of a random variable, X, is represented by σ^2.

Consider whether the sample average ($\overline{X}$) of the 200 values in our example is the *same* as μ. It is not. Do not confuse the average lifetime of all light bulbs (μ) with the average lifetime of just 200 bulbs ($\overline{X}$). This is an important distinction in statistics. However, if our assumed normal distribution (with $\mu = 400$ and $\sigma = 50$) is correct, then $\overline{X}$ most often will be "close to" μ. We examine this again in Chapter 7.

The curve in Figure 6.3 is an illustration of a normal random variable with a mean of 400 hours and a standard deviation of 50 hours. We can compare normal curves that may differ in mean, standard deviation, or both. The normal curves in Figure 6.4 indicate that, on the average, males are taller than females. The mean of the male curve is to the right of the mean of the female curve. The male heights "jump around" about as much as female heights. In other words, there is about the same amount of *variation* in male and female heights because the standard deviation of each curve is the same, that is, each curve is equally wide.

In Figure 6.5, the two normal curves represent the ages of the employees at two large companies. It appears that:

1. The average age of employees for the two companies is the same.
2. The ages in Company B have more variability. This simply means that there are more old people and more young people in Company B than in Company A.

■ **FIGURE 6.5**
Two normal curves
with equal means
and unequal standard
deviations.

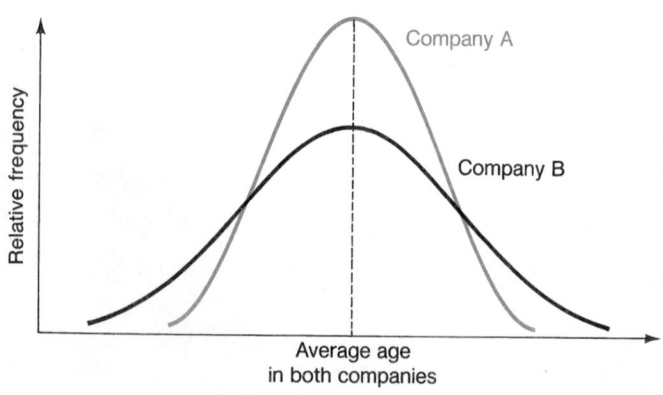

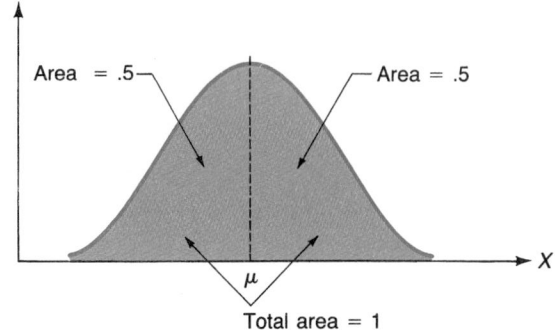

■ **FIGURE 6.7**
Normal curve for
Everglo light bulbs
showing $P(X < 360)$.
The shaded area is the
percentage of time that
X will be less than 360.
(X = lifetime of
Everglo bulb.)

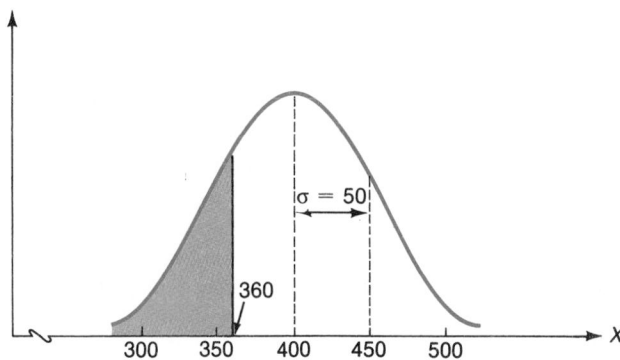

◢ 6.2

Determining a Probability for a Normal Random Variable

So you have assumed that the lifetime of an Everglo light bulb is a normal random variable with $\mu = 400$ and $\sigma = 50$. Now what? This brings us back to the subject of probability. Before we describe probabilities for a normal random variable, consider one important property of *any* normal curve (or of any curve representing a continuous random variable, for that matter), namely, that the total area under the curve is 1 (see Figure 6.6). When we described the normal curve as bell-shaped, we also determined that it was symmetrical. If the halves are identical, then the probability above the mean (μ) is equal to .5 and is the same as the probability below the mean. Thus, in Figure 6.3 the shaded area is equal to the nonshaded area under the curve.

Returning to the Everglo bulb example, what percentage of the time will the burnout time, X, be less than 360? This probability is written as

$$P(X < 360)$$

We discuss how to determine this area (a simple procedure) later in the chapter, but for now, just remember that when dealing with a normal random variable, a **probability** is represented by an **area** under the corresponding normal curve. The value of $P(X < 360)$ is illustrated in Figure 6.7. It appears that roughly 20% of the total area has been shaded, so we can conclude that (1) roughly 20% of the Everglo bulbs will burn out in less than 360 hours, and (2) the probability that X is less than 360 is approximately .2.

◢ 6.3

Finding Areas under a Normal Curve

Areas under the Standard Normal Curve

We begin our discussion by finding the area under a special normal curve—namely, one that is centered at 0 ($\mu = 0$) and has a standard deviation of 1 ($\sigma = 1$). This random variable is typically represented by the letter Z and is referred to as the **standard normal random variable.** As Figure 6.8 demonstrates, Z is as likely to

■ **FIGURE 6.8**
Standard normal curve.

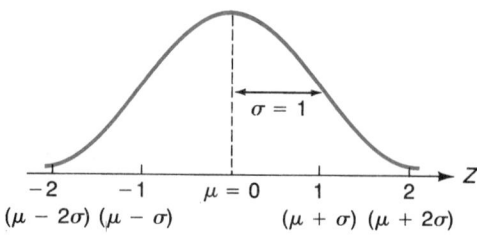

■ **FIGURE 6.9**
Shaded area = .4474,
from Table A.4.

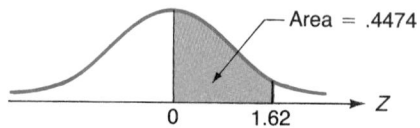

be negative as positive; that is, $P(Z \leq 0) = P(Z \geq 0) = .5$. Although you probably never will observe a random variable like Z in practice, it is a useful normal random variable. In fact, an area under *any* normal curve (as in Figure 6.7) can be determined by finding the corresponding area under the standard normal curve.

To derive the area under the standard normal curve requires the use of integral calculus. Unfortunately, the integral of the function describing the standard normal curve does not have a simple (closed form) expression. By using excellent approximations of this integral, however, we can tabulate these areas—see Table A.4 and Figure 6.9.

For example, suppose that we want to determine the probability that a standard normal random variable will be between 0 and 1.62. This is written as

$$P(0 < Z < 1.62)$$

The value of this probability is obtained from Table A.4, which contains the area under the curve between the mean of zero and the particular value of Z. The far left column of Table A.4 identifies the first decimal place for Z, and you read across the table to obtain the second decimal place.

In our example, we find the intersection between 1.6 on the left and .02 on the top, because $Z = 1.62$. Look at Table A.4; the value .4474 is the *area* between 0 and 1.62. In other words, the probability that Z will lie between 0 and 1.62 is .4474.

You can begin to see why it is a good idea to sketch the curve and shade in the area when dealing with normal random variables. It gives you a clear picture of what the question is asking and cuts down on mistakes.

EXAMPLE 6.1

What is the probability that Z will be greater than 1.62?

SOLUTION

We wish to find $P(Z > 1.62)$. Examine Figure 6.10. The area under the right half of the Z curve is .5, so, using our value from Table A.4, the desired area here is

$$.5 - .4474 = .0526$$

So the probability that Z will exceed 1.62 is .05.   ■

What if we wish to know the probability that Z is equal to a particular value, such as $P(Z = 1.62)$? There is no area under the curve corresponding to $Z = 1.62$, so

$$P(Z = 1.62) = 0$$

In fact,

$$P(Z = \text{any value}) = 0$$

■ **FIGURE 6.10**
The shaded area
represents the
probability that Z will
be greater than 1.62
$[P(Z > 1.62)]$.

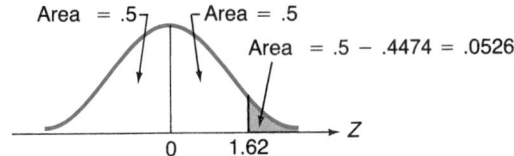

One nice thing about this fact is that $P(Z \geq 1.62)$ is the *same* as $P(Z > 1.62)$ (that is, .0526). So putting the equal sign on the inequality ($\geq$ or $\leq$) has *no* effect on the resulting probability.

By looking at the Z curve in Figure 6.11, you can see that

$$P(Z < 1.62) = .5 + .4474 = .9474$$

As before, this also is $P(Z \leq 1.62)$.

Figure 6.12 shows $P(1.0 < Z < 2.0)$ (areas from Table A.4). We see that

$$P(1.0 < Z < 2.0) = P(0 < Z < 2.0) - P(0 < Z < 1.0)$$
$$= .4772 - .3413$$
$$= .1359$$

By subtracting the two areas, we find that the probability that Z will lie between 1.0 and 2.0 is .1359.

We use Figure 6.13 and Table A.4 to determine $P(-1.25 < Z < 1.15)$:

$$P(-1.25 < Z < 1.15) = P(-1.25 < Z < 0) + P(0 < Z < 1.15)$$
$$= A_1 + A_2$$

Using the symmetry of the Z curve and Figure 6.14, the area of A_1 is the same as $P(0 < Z < 1.25)$ and thus is .3944. The area of A_2, from Table A.4, is .3749. So we add A_1 and A_2:

$$.3944 + .3749 = .7693$$

■ **FIGURE 6.11**
Area under the Z curve
for $P(Z < 1.62)$.

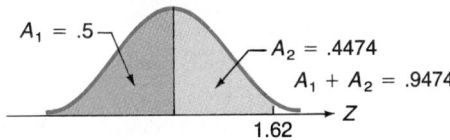

■ **FIGURE 6.12**
Area under the Z curve
for $P(1.0 < Z < 2.0)$.

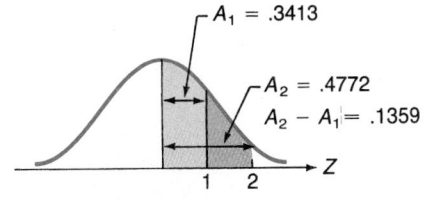

■ **FIGURE 6.13**
Area under the Z curve
for $P(-1.25 < Z < 1.15)$.

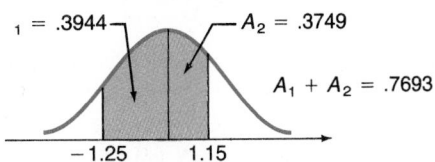

■ FIGURE 6.14
Z curve for
$P(0 < Z < 1.25) =$
$P(-1.25 < Z < 0)$.

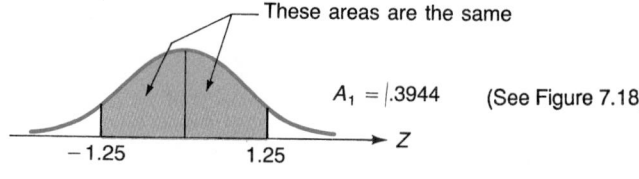

These areas are the same

$A_1 = |.3944$ (See Figure 7.18)

-1.25   1.25   Z

■ FIGURE 6.15
Area under the Z curve
for $P(Z < -1.45)$.

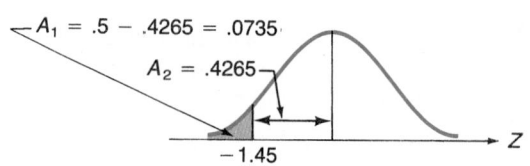

$A_1 = .5 - .4265 = .0735$

$A_2 = .4265$

-1.45   Z

Finally, we can determine $P(Z < -1.45)$ using Figure 6.15. This can be written as $P(Z < 0) - P(-1.45 < Z < 0)$. Using the discussion from Figure 6.14, the area between 0 and -1.45 is .4265 (from Table A.4). As a result, Z will be less than (or equal to) -1.45 approximately 7.35% of the time.

Exercises

6.1   Explain how the parameters μ and σ determine the graph of a normal distribution.

6.2   Find the area under the standard normal curve for the following Z values. Sketch the corresponding area.

a. $Z \le 0$ c. $Z \ge 1.0$
b. $Z \le 1.0$ d. $Z \le -1.0$

6.3   Find the area under the standard normal curve bounded by the following Z values. Sketch the corresponding area.

a. $Z = 0$ to 1.0 c. $Z = -1.0$ to 1.0
b. $Z = 1.0$ to 1.5 d. $Z = -2.5$ to -1.5

6.4   Find the following probabilities. Sketch the corresponding area.

a. $P(Z \le 1.75)$ c. $P(-1.0 \le Z \le 2.5)$
b. $P(Z \ge 1.96)$ d. $P(-.5 \le Z \le .5)$

6.5   Find the probability that an observation taken from a standard normal population will be

a. between -3 and 1.6 c. between $.76$ and 1.96
b. less than -2.1 d. between -1.65 and 1.65

6.6   Find the value of z for the following probability statements and sketch the corresponding area.

a. $P(Z \le z) = .95$ c. $P(Z \ge z) = .025$
b. $P(Z \le z) = .10$ d. $P(Z \ge z) = .55$

6.7   Find the value of z for the following probability statements and sketch the corresponding area.

a. $P(-1.8 \le Z \le z) = .6$ c. $P(1.0 \le Z \le z) = .1$
b. $P(0 \le Z \le z) = .25$ d. $P(-2.8 \le Z \le z) = .05$

6.8   Find the two Z values such that

a.   The area bounded by them is equal to the middle 40% of the standard normal distribution.

b.   The area bounded by them is equal to the middle 80% of the standard normal distribution.

6.9   Find the Z values such that the area under the standard normal curve between the Z value and $Z = 1.0$ is equal to .10. Find both Z values that make this possible.

6.10 The output from a monitor that measures the amperage of an electronic circuit follows a normal distribution with mean 0 and variance 1. What proportion of the data would be outside the interval from -2 to 2?

Areas under Any Normal Curve

Take another look at the histogram of the 200 Everglo light bulb lifetimes in Figure 6.2. A normal curve with $\mu = 400$ hours and $\sigma = 50$ hours was used to describe the population of *all* Everglo lifetimes. So, X = Everglo lifetime is a normal random variable with $\mu = 400$ and $\sigma = 50$.

What happens to the shape of the data if we take each of the 200 lifetimes in this example and subtract 400 (that is, subtract μ)? As you can see in Figure 6.16, the histogram (and corresponding normal curve) is merely "shifted" to the left by 400. It resembles the normal curve for X, except the "new" mean is 0. The random variable defined by $Y = X - 400$:

1. is a normal random variable
2. has a mean equal to zero
3. has a standard deviation equal to that of X, that is, 50

Figure 6.17 shows what happens to the shape of 200 Y values if each of them is *divided* by 50 (that is, by σ). Notice the horizontal axis in the histogram and the corresponding normal curve. The resulting normal curve resembles a normal curve with a mean of 0 and a standard deviation equal to 1.

Thus, if X is a normal random variable with mean 400 and standard deviation 50, then the random variable defined by

$$Z = \frac{X - 400}{50}$$

■ **FIGURE 6.16**
Histogram obtained by subtracting $\mu = 400$ (compare with Figure 6.2).

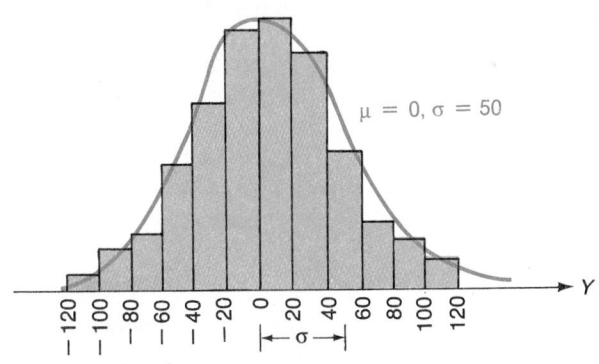

■ **FIGURE 6.17**
Histogram obtained by subtracting μ and dividing by σ (compare with Figures 6.2 and 6.16).

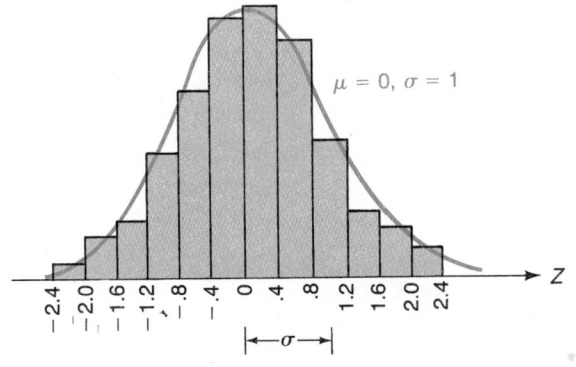

1. is a normal random variable
2. has a mean equal to zero
3. has a standard deviation equal to 1

In general, for *any normal* random variable X,

$$Z = \frac{X - \mu}{\sigma}$$

is a **standard normal random variable.** This procedure of subtracting μ and dividing by σ is referred to as **standardizing** the normal random variable X. *It allows us to determine probabilities for any normal random variable by first standardizing it and then using Table A.4. So the standard normal distribution turns out to be much more important than you might have expected!*

EXAMPLE 6.2   The normal curve in Figure 6.7 represented the lifetime of all Everglo bulbs, with $\mu = 400$ hours and $\sigma = 50$ hours. What percentage of the bulbs will burn out in less than (or equal to) 360 hours? Or, put another way, what is the probability that any particular bulb will last less than 360 hours?

SOLUTION   This probability is written as

$$P(X < 360)$$

This random variable is continuous, so $P(X < 360) = P(X \le 360)$. To determine the probability, you need to standardize this variable: *

$$P(X < 360) = P\left(\frac{X - 400}{50} < \frac{360 - 400}{50}\right)$$
$$= P(Z < -.8)$$

where $Z = (X - 400)/50$ (Figure 6.18).

Earlier, by examining Figure 6.7, we estimated this area to be roughly 20%. The actual area, from Figure 6.18, is .2119; that is, it is 21.19% of the total area. The conclusion here is that

$$P(X < 360) = .2119$$

and so 21% of all Everglo bulbs will have a lifetime of less than 360 hours.   ∎

Interpreting Z

What does a Z value of $-.8$ imply in Example 6.2? It simply means that 360 is .8 standard deviations to the left (Z is negative) of the mean. So,

$$\mu - .8(\sigma) = 400 - .8(50) = 360$$

Recall that a Z score was defined in exactly the same way in Chapter 3 using a sample mean ($\bar{X}$) and standard deviation (s). In this chapter, we use the population mean (μ) and standard deviation (σ). In general:

1. A *positive* value of Z designates how many standard deviations (σ) X is to the *right* of the mean (μ).

---

*Since 400 is subtracted from *both* sides of the inequality and *both* sides are divided by 50, the events described by the original inequality $P(X < 360)$ and the standardized inequality $P(Z < -.8)$ are the same.

■ **FIGURE 6.18**
Compare the areas for
(a) the X and (b) the Z
normal curves to find
$P(X < 360)$.

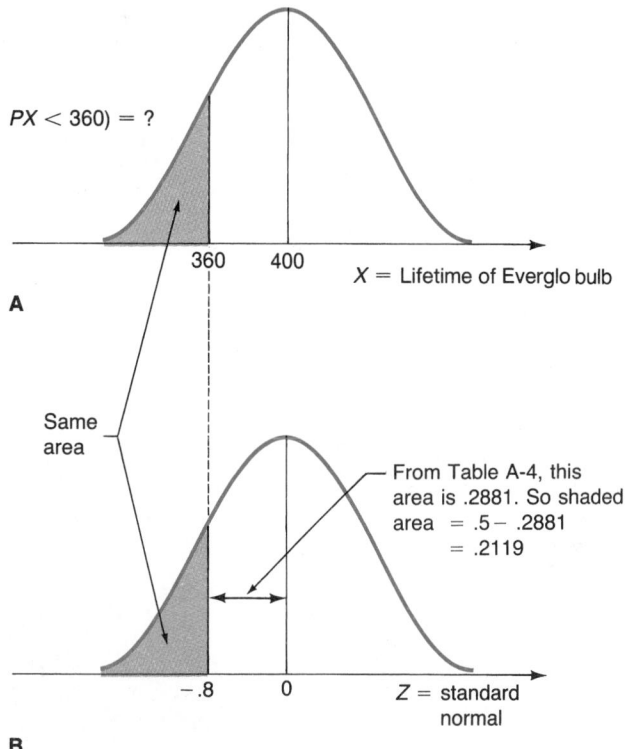

$PX < 360) = ?$

360   400

X = Lifetime of Everglo bulb

A

Same
area

From Table A-4, this
area is .2881. So shaded
area = .5 − .2881
= .2119

− .8   0 Z = standard
normal

B

2. A *negative* value of Z designates how many standard deviations X is to the *left* of the mean.

EXAMPLE 6.3   The inside diameter of a manufactured machine bearing is believed to follow a normal distribution centered at 2.52 in with a standard deviation of .022 in. For the bearing to be of acceptable quality it must be between 2.45 in (the lower specification limit) and 2.55 in (the upper specification limit). A bearing within this range is said to be *conforming* and a bearing with an inside diameter smaller than 2.45 in or larger than 2.55 in is *nonconforming*. What proportion of the bearings are nonconforming?

SOLUTION   The probability that a bearing is conforming can be written as

$$P(2.45 < X < 2.55)$$

Using the standardizing procedure,

$$P(2.45 < X < 2.55) = P\left(\frac{2.45 - 2.52}{.022} < \frac{X - 2.52}{.022} < \frac{2.55 - 2.52}{.022}\right)$$

$$= P(-3.18 < Z < 1.36)$$

where Z once again represents the *standardized* normal random variable, which for this example is defined by

$$Z = \frac{X - 2.52}{.022}$$

Refer to Table A.4 and Figure 6.19 and note that the shaded areas in Figures 6.19a and b are equal:

$$.4993 + .4131 = .9124$$

■

■ FIGURE 6.19
(a) The probability that
X is between 2.45 and
2.55 (b) The probability
that Z is between
−3.18 and 1.36.

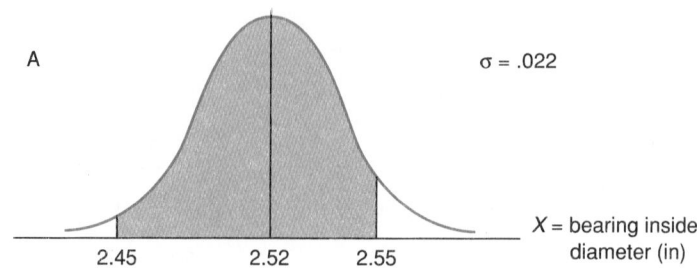

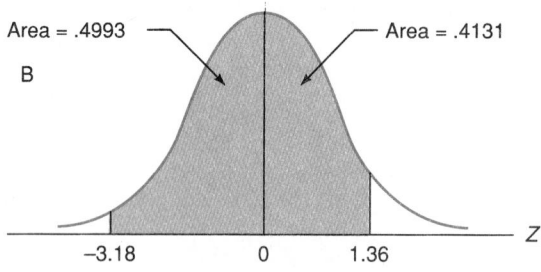

Thus 91.24% of the bearings are conforming, and so 8.76% are nonconforming. Also note that:

1. The upper specification limit is 1.36 standard deviations to the right of the mean: $Z = 1.36$ and $2.55 = 2.52 + (1.36)(.022)$
2. The lower specification limit is 3.18 standard deviations to the left of the mean: $Z = -3.18$ and $2.45 = 2.52 - (3.18)(.022)$
3. $P(X = 2.45)$ and $P(X = 2.55)$ both are zero, so $P(2.45 < X < 2.55) = P(2.45 \leq X \leq 2.55) = .9124$

EXAMPLE 6.4

Actuarial scientists in an insurance company formulate insurance policies that will be both profitable and marketable. For a particular policy, the lifetimes of the policyholders follow a normal distribution with $\mu = 66.2$ years and $\sigma = 4.4$ years. One of the options with this policy is to receive a payment following the 65th birthday and a payment every 5 years thereafter.

1. What percentage of policyholders will receive at least one payment using this option?
2. What percentage will receive two or more payments?
3. What percentage will receive exactly two payments?

SOLUTION 1

The normal curve for the policyholder lifetimes is shown in Figure 6.20. To receive at least one payment, the policyholder must live beyond 65 years of age. So we need to determine (see Figure 6.21):

$$P(X > 65) = P[(X - 66.2)/4.4 > (65 - 66.2)/4.4]$$
$$= P(Z > -.27) = .1064 + .5$$
$$= .6064$$

So nearly 61% of the policyholders will receive at least one payment.

■ **FIGURE 6.20**
The normal curve for
policyholder lifetimes.
X = age at death (in
years).

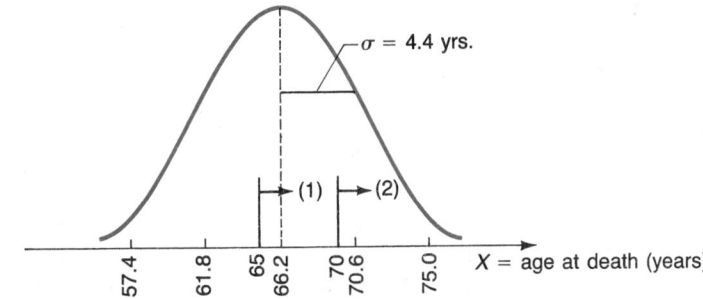

■ **FIGURE 6.21**
Z curve for
$P(Z > -.27)$.

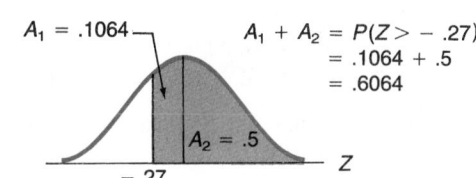

■ **FIGURE 6.22**
Z curve for $P(Z > .86)$.

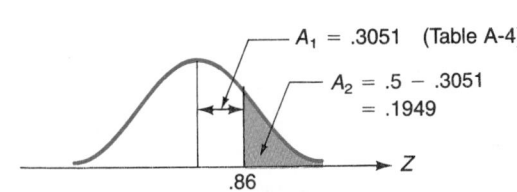

■ **FIGURE 6.23**
Z curve for
$P(.86 < Z < 2.00)$.

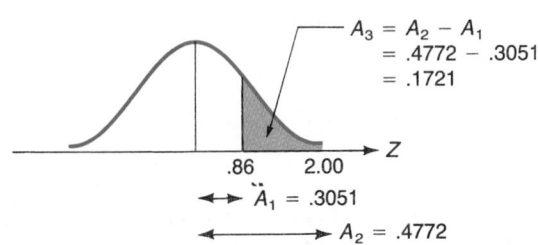

SOLUTION 2 Because the policyholder receives a payment every 5 years, he or she will receive two or more payments provided he or she lives to be older than 70 years of age. Thus the probability of two or more payments is determined by (see Figure 6.22):

$$P(X > 70) = P[(X - 66.2)/4.4 > (70 - 66.2)/4.4]$$
$$= P(Z > .86) = .5 - .3051$$
$$= .1949$$

Thus, 19.5% of the policyholders will survive long enough to collect two payments.

SOLUTION 3 To receive exactly two payments, the policyholder must live longer than 70 years and less than 75 years. This probability is

$$P(70 < X < 75)$$

Using the same standardization procedure (see Figure 6.23):

$$P(70 < X < 75) = P[(70 - 66.2)/4.4 < (X - 66.2)/4.4 < (75 - 66.2)/4.4]$$
$$= P(.86 < Z < 2.00)$$
$$= .4772 - .3051 = .1721$$

So 17.21% of the policyholders will receive exactly two payments. ■

6.4
Applications Where the Area Under a Normal Curve Is Provided

Another twist to dealing with normal random variables is a situation where you are given the area under the normal curve and asked to determine the corresponding value of the variable. This is a common application of a normal random variable. For example, the manufacturer of a product may want to determine a warranty period during which the product will be replaced if it becomes defective, so that at most 5% of the items are returned during this period. Or, in a grocery store on any given day, the demand for a freshly made food item may or may not exceed the supply. The owner may want to determine how much to supply each day, such that the demand (a normal random variable) will exceed this value 10% of the time (in other words, the customers will be disappointed no more than 10% of the time).

EXAMPLE 6.5

Referring to Example 6.2, after how many hours will 80% of the Everglo bulbs burn out? Recall that $\mu = 400$ and $\sigma = 50$.

SOLUTION

The first step here is to sketch this curve (Figure 6.24a) and estimate the value of X (say X_0) so that

$$P(X < X_0) = .8$$

Because .8 is larger than .5, X_0 must lie to the *right* of 400.

Next, find the point on a standard normal (Z) curve such that the area to the left is also .8 (Figure 6.24b). Using Table A.4, the area between 0 and .84 is .2995. This means that

$$P(Z < .84) = .5 + .2995$$
$$= .7995$$
$$= .8 \quad \text{(approximately)}$$

By standardizing X, we conclude that

$$\frac{X_0 - 400}{50} = .84$$
$$X_0 - 400 = 42$$
$$X_0 = 400 + 42 = 442$$

So 80% of the Everglo bulbs will burn out within 442 hours.   ■

■ FIGURE 6.24
(a) $P(X < X_0) = .8$.
(b) $P(Z < .84) = .8$.

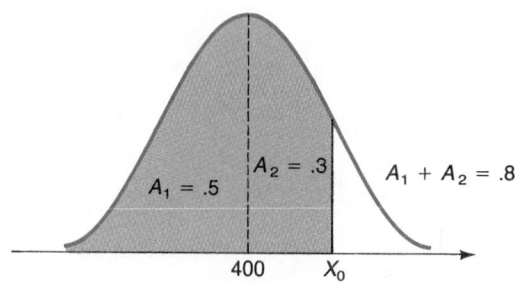

A

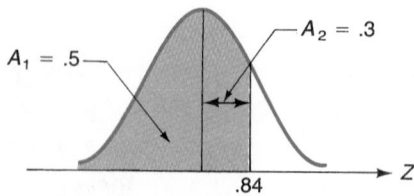

B

EXAMPLE 6.6

A bakery shop sells loaves of freshly made French bread. Any unsold loaves at the end of the day are either discarded or sold elsewhere at a loss. The demand for this bread has followed a normal distribution with $\mu = 35$ loaves and $\sigma = 8$ loaves.

How many loaves should the bakery make each day so that they can meet the demand 90% of the time?

SOLUTION

The normal random variable X here is the demand for French bread (measured in loaves) (Figure 6.25a). To meet the demand 90% of the time, the bakery must determine an amount, say X_0 loaves, such that:

$$P(X \leq X_0) = .90$$

Proceeding as before, examine a Z curve having an area of .90 to the *left* of X_0 (Figure 6.25b). Using Table A.4,

$$P(0 \leq Z \leq 1.28) = .4 \qquad \text{(more accurately, .3997)}$$

which means that

$$P(Z \leq 1.28) = .4 + .5 = .9$$

So

$$\frac{X_0 - 35}{8} = 1.28$$

and

$$X_0 = 35 + (1.28)(8) = 45.24$$

To be conservative, round this value up to 46 loaves. By stocking 46 loaves each day, the bakery will meet the demand for this product 90% of the time. ∎

6.5
Another Look at the Empirical Rule

In Chapter 3, the empirical rule specified that when sampling from a bell-shaped distribution (which means a normal distribution):

1. Approximately 68% of the data values should lie between $\bar{X} - s$ and $\bar{X} + s$.
2. Approximately 95% of them should lie between $\bar{X} - 2s$ and $\bar{X} + 2s$.
3. Approximately 99.7% of them should lie between $\bar{X} - 3s$ and $\bar{X} + 3s$.

Nothing was said at that time about the origin of these numbers. They actually came directly from Table A.4. To see this, consider Figure 6.26, in which

$$P(-1 < Z < 1) = .68$$

■ **FIGURE 6.25**
(a) $P(X \leq X_0) = .90$.
(b) $P(Z \leq 1.28) = .90$.

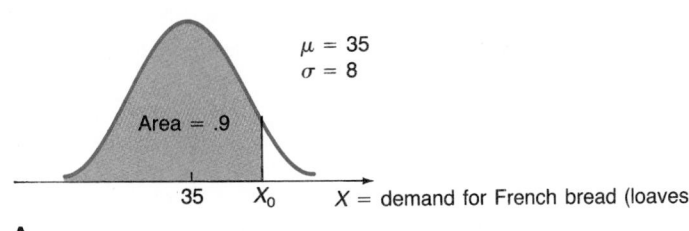

A

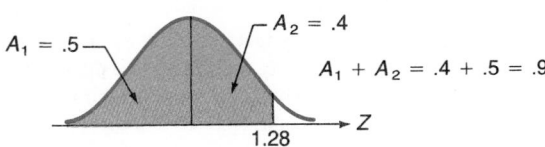

B

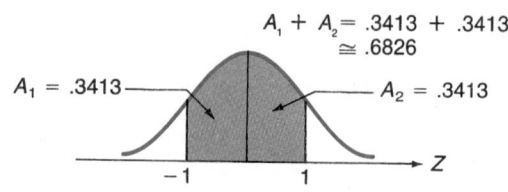

$A_1 + A_2 = .3413 + .3413$
$\cong .6826$

$A_1 = .3413$ ———

——— $A_2 = .3413$

−1   1

This implies that, for any normal random variable X,

$$P[-1 < (X - \mu)/\sigma < 1] = .68$$

That is,

$$P[(\mu - \sigma) < X < (\mu + \sigma)] = .68$$

As a result, for a set of data from a normal population where $\bar{X}$ is the sample mean and s is the sample standard deviation, we would expect approximately 68% of the data to lie between $\bar{X} - s$ and $\bar{X} + s$.

Similarly, $P(-2 < Z < 2) = .4772 + .4772 = .9544$, so you can expect (approximately) 95% of the data points from a normal (bell-shaped) population to lie between $\bar{X} - 2s$ and $\bar{X} + 2s$.

Finally, $P(-3 < Z < 3) = .4987 + .4987 = .9974$, which leads to the third conclusion of the empirical rule.

 *Exercises*

6.11 Let the random variable X be normally distributed with mean 5 and variance 4. Find the following probabilities.

a. $P(X \geq 5.7)$ c. $P(2.8 \leq X \leq 5.1)$
b. $P(X \leq 3.4)$ d. $P(5.7 \leq X \leq 6.8)$

6.12 Find the value of x if the random variable X is normally distributed with mean 10 and variance 9.

a. $P(X \leq x) = .51$ c. $P(10 \leq X \leq x) = .05$
b. $P(X \geq x) = .805$ d. $P(8 \leq X \leq x) = .13$

6.13 High-Tech, Inc. produces an electronic component, GX-7, that has an average life span of 4500 hours. The life span is normally distributed with a standard deviation of 500 hours. The company is considering a 3800 hours warranty on GX-7. If this warranty policy is adopted, what proportion of GX-7 components should High-Tech expect to replace under warranty?

6.14 The estimated miles-per-gallon (on the highway) ratings of a class of trucks are normally distributed with a mean of 12.8 and a standard deviation of 3.2. What is the probability that one of these trucks selected at random would get

a. between 13 and 15 miles-per-gallon?   b. between 10 and 12 miles-per-gallon?

6.15 The yearly cost of dental claims for the employees of D. S. Inc. is normally distributed with a mean of $75 and a standard deviation of $30. At least what yearly cost would be expected for 40 percent of the employees?

6.16 The diameter of $\frac{1}{2}$-inch bolts produced by a workshop is normally distributed with a mean of .5 inch and a standard deviation of .04 inch. What is the probability that a bolt selected at random will fit in a hole whose diameter is between .475 and .525 inch?

6.17 The inside diameter of a manufactured cylinder is believed to follow a normal distribution centered at 5.00 inches with a standard deviation of .03 inches. The diameter of the cylinder must be within 4.95 inches (the lower specification limit) and 5.07 inches (the upper specification limit). If the diameter is outside these two specification limits, then the cylinder is nonconforming. What proportion of the manufactured cylinders are nonconforming?

6.18 The thickness of a manufactured sheet of metal is believed to follow a normal distribution with a mean of .30 inches and a standard deviation of .02 inches. The lower specification limit is 2.5 standard deviations to the left of the mean. The upper specification limit is 3.0 standard deviations to the right of the mean. Nonconforming sheets of metal have thicknesses outside of these two specification limits. What proportion of the sheets of metal are nonconforming?

6.19 Career success can depend on personal characteristics that contribute to effective job performance. A recent survey of students in business schools revealed adaptive and innovative orientations of students in different majors. The Kirtow Adaption-Innovation Inventory (KAI) was used to measure these orientations. Accounting majors scored more to the adaptive end of the KAI scale than other business majors. For the accounting majors, the mean score was 86.8 with a standard deviation of 14.6. Assume that a normal distribution can be used to approximate the distribution of the KAI scores.

(*Source:* W. L. King and R. J. Masters, "Differences Between the Work Orientations of College Accounting Majors and Those Who Are Most Successful in Accounting," *The Journal of Applied Business Research* 6, no. 3 (1990): 8.)

a. Find the probability that a randomly selected accounting student has a KAI score between 80 and 90.

b. If a randomly selected accounting student has a KAI score of 60, would this be considered unusual?

6.20 The vice president of Offshore Oil and Gas, a consulting firm, notices that the average length of time that a consultant spends on the telephone with a client at any one time is 40 minutes with a standard deviation of 18 minutes. Assuming that the length of such conversations is normally distributed, what percent of the consultant's phone calls would take longer than 50 minutes?

6.21 As part of an experiment conducted for a graduate class in organizational behavior, the time taken to complete an assembly task was measured for two groups of workers. For the first group, the mean time was 10 minutes with a standard deviation of 1.5 minutes. For the second group, the mean time was 11.5 minutes with a standard deviation of 2.0 minutes. Assume that the times for both groups follow normal distributions. A worker from each group is randomly selected. What is the probability that the assembly time for this worker is less than 9 minutes, if the worker is from

a. the first group?   **b.** the second group?

6.22 Find the value of k such that $P(\mu \leq X \leq \mu + k\sigma) = .251$, for a random variable X having a normal distribution with mean μ and standard deviation σ.

6.23 If X is a normally distributed random variable with standard deviation of 10, find the mean μ given that $P(X \leq .35) = .182$.

6.24 If X is normally distributed with a mean of 100, find the standard deviation given that $P(X \geq 110) = .123$.

6.25 If X is a normally distributed random variable with $P(X \geq 2) = .1$ and $P(X \leq 1) = .3$, find both the mean and standard deviation.

◢ 6.6
Normal Approximation to the Binomial

The binomial random variable was introduced in Chapter 5. It is a discrete random variable used to count the number of successes in a binomial situation.

Characteristics of a Binomial Situation
1. You have n independent identical trials.
2. Each trial is a success (with probability p) or a failure (with probability $1 - p$).
3. The binomial random variable X is the number of successes out of n trials.
4. The mean of X is $\mu = np$, and the standard deviation of X is $\sigma = \sqrt{np(1 - p)}$.

Examples included:

X = the number of heads (successes) out of three flips (trials) of a coin

X = the number of people who read the evening newspaper (successes) out of a sample of 50 people (trials)

X = the number of defectives (successes) out of a sample of ten electrical components (trials)

Table A.1 contains values of n (the number of trials) only up to $n = 20$. In Chapter 5, we used the Poisson approximation to determine binomial probabilities for values of $n > 20$. In other words, we pretend that X is a Poisson random variable *having the same mean* as the actual binomial random variable. This is a good approximation, provided n is large (>20), p is small, and $np \leq 7$.

We can also use the **normal approximation** to the binomial random variable. Here you pretend that X is a normal random variable *having the same mean and standard deviation* as the actual binomial random variable. This approximation works well when p is near .5 and in general offers a good estimate when both $np > 5$ and $n(1 - p) > 5$.

Approximations to the Binomial

- Poisson approximation: Use when $n > 20$ and $np \leq 7$.
- Normal approximation: Use when $np > 5$ and $n(1 - p) > 5$.

Consider 12 flips of a coin. We want to determine (1) the probability of observing no more than 4 heads and (2) the probability of observing more than 5 heads. First, notice that a normal approximation is not necessary here. This is a binomial situation with $n = 12$ and $p = .5$, and Table A.1 does contain probabilities for this set of values. We chose this illustration to compare the actual binomial probability to the approximated probability using the normal distribution. Look at Figure 6.27, which demonstrates how we estimate binomial probabilities using a normal curve.

To solve question 1, let X = the number of heads in 12 flips, so X is a binomial random variable. We want to determine $P(X \leq 4)$. We can obtain an exact solution using Table A.1:

$$P(X \leq 4) = P(0) + P(1) + P(2) + P(3) + P(4)$$
$$= 0 + .003 + .016 + .054 + .121$$
$$= .194$$

In Figure 6.27, this value is the sum of the areas of the boxes corresponding to $X = 0, 1, 2, 3,$ and 4. Note that the width of each box is 1, and so the height of the box (the probability) is the same as the area of the box. As a result, the total area of the boxes is 1.

■ **FIGURE 6.27**
Approximating binomial probabilities using a normal curve.

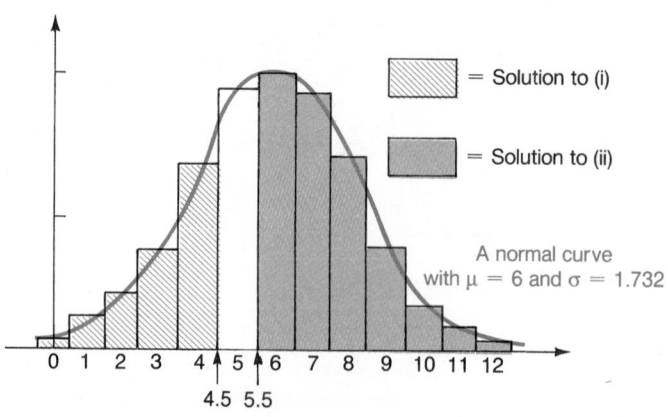

= Solution to (i)

= Solution to (ii)

A normal curve with $\mu = 6$ and $\sigma = 1.732$

We can also obtain an approximate solution. For this binomial random variable,

$$\mu = np = (12)(.5) = 6$$

and

$$\sigma = \sqrt{np(1 - p)}$$
$$= \sqrt{3} = 1.732$$

To obtain an approximation, treat X as a normal random variable with $\mu = 6$ and $\sigma = 1.732$, illustrated in Figure 6.27. Note that both the total area of the boxes and the total area under the normal curve are 1. The area under the normal curve that approximates $P(X \leq 4)$ is the area to the left of 4.5. So we obtain a better approximation here if we find the area under the normal curve to the left of 4.5, not 4.0. This .5 adjustment is referred to as an **adjustment for continuity.** This adjustment is necessary whenever you approximate a *discrete* random variable (such as a binomial random variable) using a *continuous* distribution (such as the normal distribution). Remember that the discrete distribution has gaps, whereas the continuous does not, so we must assign a portion of the space (probability) between 4 and 5 when we use a continuous distribution to approximate a discrete one. Using Table A.4,

Binomial   **Normal**

($n = 12, p = .5$)   ($\mu = 6, \sigma = 1.732$)

$P(X \leq 4) \cong P(X \leq 4.5)$

$$= P\left[Z \leq \frac{4.5 - 6}{1.732}\right]$$

$$= P(Z \leq -.87) = .1992$$

Notice that the approximate solution of .1922 is very close to the actual probability of .194. This is helped in part by the fact that $p = .5$ for this situation, which means that the binomial distribution is perfectly symmetric. *As the value of* p *moves away from .5, larger values of n are necessary to achieve an approximation this good.*

Now consider question 2, the probability of observing more than 5 heads in 12 flips, or $P(X > 5) = P(X \geq 6)$. Using Table A.1, we can obtain an *exact* solution:

$$P(X \geq 6) = P(6) + P(7) + \cdots + P(11) + P(12)$$
$$= .226 + .193 + \cdots + .003 + 0 = .613$$

We can also obtain an approximate solution. Using Figure 6.27, the area under the normal curve that corresponds to the lined area representing the exact solution is the area to the right of 5.5. So, using Table A.4:

Binomial   **Normal**

($n = 12, p = .5$)   ($\mu = 6, \sigma = 1.732$)

$P(X \geq 6) \cong P(X \geq 5.5)$

$$= P\left(Z \geq \frac{5.5 - 6}{1.732}\right)$$

$$= P(Z \geq -.29) = .6141$$

Again, we obtain a very good approximation, helped by the fact that we are using a perfectly symmetrical binomial distribution.

How to Adjust for Continuity

If X is a binomial random variable with n trials and probability of success $= p$, then:

1. $P(X \leq b) \cong P\left(Z \leq \dfrac{b + .5 - \mu}{\sigma}\right)$

2. $P(X \geq a) \cong P\left(Z \geq \dfrac{a - .5 - \mu}{\sigma}\right)$

3. $P(a \leq X \leq b) \cong P\left(\dfrac{a - .5 - \mu}{\sigma} \leq Z \leq \dfrac{b + .5 - \mu}{\sigma}\right)$

where

$$\mu = np, \qquad \sigma = \sqrt{np(1 - p)}$$

and Z is a standard normal random variable

4. Be sure to convert a $<$ probability to a $\leq$, and convert a $>$ probability to a $\geq$ before switching to the normal approximation.

EXAMPLE 6.7

In Chapter 5, we discussed a binomial situation (approximated by the Poisson) in which we had a sample of 100 chips to be tested. Each chip was either defective (a success) or not defective (a failure). Therefore, X was the number of defective chips (out of 100) and was a binomial random variable. We assumed that $p = .05$, which resulted in a very good Poisson approximation because n was large and p was small. Suppose, instead, that 10% of these chips are defective; that is, $p = .10$. Now, $np = 10$ and, because this is greater than 7, the Poisson distribution cannot be expected to provide a good approximation. However, we can obtain a good normal approximation here because $np = 10$ and $n(1 - p) = 90$, both of which are >5.

For this situation, what is the probability that you observe one or fewer defective chips in a sample of 100, in which case the lot of chips is accepted?

SOLUTION X is a binomial random variable with

$$\mu = np = (100)(.10) = 10$$

and

$$\sigma = \sqrt{np(1 - p)} = \sqrt{9} = 3$$

Therefore, using Table A.4:

Binomial	Normal
$(n = 100, p = .10)$	$(\mu = 10, \sigma = 3)$

$$P(X \leq 1) \cong P(X \leq 1.5)$$

$$= P\left[Z \leq \frac{1.5 - 10}{3}\right]$$

$$= P(Z \leq -2.83) = .0023$$

Consequently, there is a very small chance of accepting the lot, since you can expect to accept it only 23 times out of 10,000 using this procedure. ∎

EXAMPLE 6.8

In Chapter 5, we discussed a binomial situation in which Eagle Air was intentionally overbooking their flights. On a particular flight from Dallas to El Paso, they use a much larger aircraft that holds 200 people. As in our previous example, 20% of the people who make reservations do not show up. If Eagle Air accepts 235 reservations, what is the probability that at least one passenger will end up without a seat on this flight?

SOLUTION The binomial random variable X here is the number of people (out of 235) who show up for the flight. For this situation, $n = 235$, and $p = .8$ represents the probability that any one passenger *will* show up. The mean of this random variable is

$$\mu = (235)(.8) = 188$$

and the standard deviation is

$$\sigma = \sqrt{(235)(.8)(.2)} = 6.13$$

At least one person holding a reservation will be deprived of a seat if $X \geq 201$ because the plane holds only 200 people. Once again, we use the normal approximation (Table A.4) to obtain the following probability:

Binomial   **Normal**

$(n = 235, p = .8)$ $(\mu = 188, \sigma = 6.13)$

$P(X \geq 201) \cong P(X \geq 200.5)$

$$= P\left[Z \geq \frac{200.5 - 188}{6.13}\right]$$

$$= P(Z \geq 2.04)$$

$$= .5 - .4793$$

$$= .0207$$

So on approximately 2 flights out of 100, at least one person will be unable to secure a seat. ∎

Exercises **6.26** A random variable X has a binomial distribution with the probability of a success, p, equal to .25.

a. Would it be appropriate to use the normal approximation to the binomial if $n = 30$? if $n = 15$?

b. With $n = 40$, use the normal approximation to find $P(2 \leq X \leq 10)$.

c. What is the smallest value that n can be and still have the normal distribution be appropriate for approximating the binomial distribution?

6.27 Let the random variable X indicate the number of female students chosen (with replacement) in a sample of 15 from a student body with 40% female students.

a. Using the binomial table, find the probability that X is greater than 4 and less than 9.

b. Use the normal approximation to answer part a.

c. Compare the answers in parts a and b.

6.28 Thirty percent of the computer programmers who are hired to work for Techtronics do not have work experience in programming. If a random sample of 35 computer programmers is selected, what is the probability that fewer than 20 have had experience in computer programming before being hired by Techtronics?

6.29 A travel agency promotes vacation packages by phoning households at random in the evening hours. Historically, only 65% of heads of households are at home when the agency phones. If 30 households are phoned on a given evening, what is the probability that the agency will find between 15 and 25 households, inclusive, with the head of the household at home?

6.30 A recent survey by the Gallup Organization examined employees' attitudes toward quality improvement activities. In the survey, 34% of the employees responded that at their companies it is very important to have management trust employees to make good decisions about quality.

(*Source:* J. Ryan, "Quality: A Job With Many Vacancies," *Quality Progress* (November 1990): 23.)

a. Suppose 50 employees were selected at random. What is the probability that between 10 and 20 employees, inclusive, responded that it is important for their companies' management to trust employees to make decisions about quality?

b. If 50 employees were selected at random, what is the probability that fewer than 12 employees responded that it is important for their companies' management to trust employees to make decisions about quality?

6.31 Fifty-eight percent of the cars sold at Lance Holey's used-car lot required financing. If 30 car buyers at this lot are randomly selected, what is the probability that between 15 and 25 buyers (inclusive) financed their car?

6.32 Trying to tailor benefit plans to fit the needs of employees, many companies have been using flexible plans in which employees can select more or less health coverage. For example, it may be possible to take less health insurance in exchange for more vacation days or child care assistance. According to a recent survey of the 5000 largest employers in the United States, 47% offer or plan to offer flexible plans.

(*Source:* "Employee Benefits for a Changing Work Force," *Business Week* (November 1990): 31–40.)

a. In a random sample of 40 employers from the largest employers in the United States, what is the probability that between 15 and 25 employers offer or plan to offer flexible plans?

b. Suppose that in the sample of 40 employers in part a it was found that either fewer than 10 or more than 30 employers offer or plan to offer flexible plans. Would this make one question whether the 47% figure given in the survey was accurate? Why?

6.33 If a pair of fair dice is rolled 70 times, what is the probability that a pair of snake eyes (a one on each die) will appear between five and ten times, inclusive? Is the normal approximation appropriate here? What other approximation should work well?

6.7 Other Continuous Distributions (Optional)

The normal distribution is one example of a continuous distribution. A normal random variable X is a continuous random variable; that is, over some specific range, *any* value of X is possible. We used X to represent the lifetime of an Everglo bulb to illustrate a continuous random variable because any value between 280 hours and 520 hours (see Figure 6.2) is possible. In fact, any value less than 280 or more than 520 is also possible, although not likely to occur.

In the Everglo example, a normal distribution seemed appropriate because the histogram of 200 sample bulbs in Figure 6.2 revealed a concentration of burnout times in the "middle" and not nearly as many burnout times around 300 or 500. These features give the normal curve its "mound" in the center and "tails" on each end.

There are many continuous distributions that do not resemble a normal curve in appearance. For example, consider the following two situations, in which a random variable, X, ranges from 1 to 10.

Situation 1   The chance that X is between 1.0 and 1.5

$$= \text{the chance that } X \text{ is between 1.5 and 2.0}$$
$$= \text{the chance that } X \text{ is between 2.0 and 2.5}$$
$$\vdots$$
$$= \text{the chance that } X \text{ is between 9.0 and 9.5}$$
$$= \text{the chance that } X \text{ is between 9.5 and 10.0}$$

Situation 2   The larger X is, the less likely it is to occur. Thus, the chance that X is between 1.0 and 1.5

> the chance that X is between 1.5 and 2.0
> the chance that X is between 2.0 and 2.5

$\vdots$

> the chance that X is between 9.0 and 9.5
> the chance that X is between 9.5 and 10.0

These two cases can be represented by two other popular continuous distributions. Situation 1 can be represented by a uniform random variable, whereas situation 2 could be described using an exponential random variable.

Although there are other random variables that apply to these two situations, the uniform and exponential distributions most often fit the applications encountered in business.

The Uniform Distribution

Consider spinning the minute hand on a clock face. Define a random variable X to be the stopping point of the minute hand. It seems reasonable to assume that, for example, the probability that X is between two and four is *twice* the probability of observing a value of X between eight and nine. In other words, the probability that X is in any particular interval is *proportional* to the width of that interval.

A random variable of this nature is a **uniform random variable.** The values of such a variable are evenly distributed over some interval because the random variable occurs *randomly* over this interval. Unlike the normal random variable, values of the uniform random variable do not tend to be concentrated about the mean.

Assume that the manager of Dixie Beverage Service is concerned about the amount of soda that is released by the dispensing machine that the company is now using. He is considering the purchase of a new machine that electronically controls the cutoff time and is supposed to be very accurate. The present machine cuts off mechanically, and he suspects that the device shuts off the fluid flow *randomly* anywhere between 6 and 8 ounces. To test the present system, a sample of 150 cups is taken from the machine, and the amount of soda released into each cup is recorded. The relative frequency histogram made from these 150 observations is shown in Figure 6.28.

Would you be tempted to describe the population of *all* cup contents using a normal curve? We hope not, because there is no evidence of a declining number of observations in the tails. As a word of warning here, we often have a tendency to think of all continuous random variables as being normally distributed. As this application demonstrates, this is certainly not the case. Instead, this distribution is a flat or uniform distribution. The random variable $X = $ cup contents is a uniform random variable. The corresponding smooth curve describing the population is shown in Figure 6.29.

■ **FIGURE 6.28**
Relative frequency histogram of a sample of 150 cups of soda.

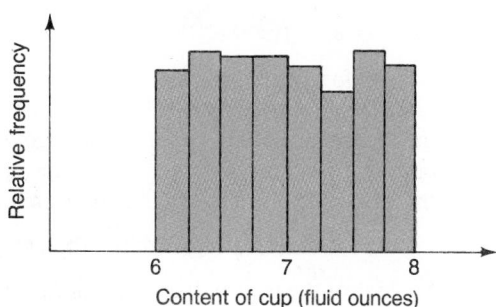

Content of cup (fluid ounces)

■ FIGURE 6.29
Uniform distribution for
X = soda content
(compare with
Figure 6.28).

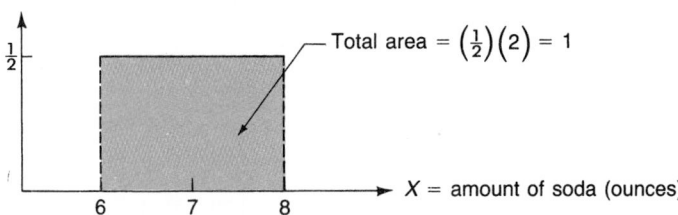

■ FIGURE 6.30
Total area for a uniform
distribution.

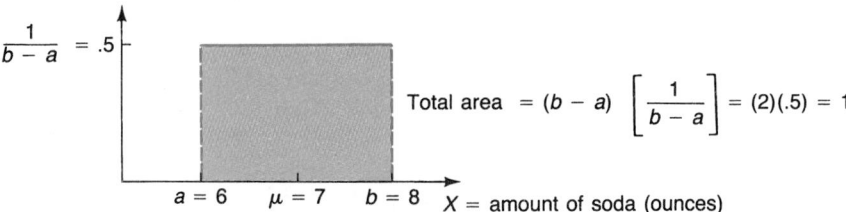

Notice that the total area here is given by a rectangle, and, as is true of all continuous random variables, this total area must be 1. The area of a rectangle is given by (width) · (height). By making the height of this curve (a straight line, actually) equal to .5, the total area is

$$(8 - 6)(.5) = 1.0$$

In general, the curve defining the probability distribution for a uniform random variable is shown in Figure 6.30. The total area is

$$(b - a) \left[\frac{1}{b - a} \right] = 1.0$$

Mean and Standard Deviation

Refer to Figure 6.30. The mean (μ) of X is the value midway between a and b, namely,

$$\mu = \frac{a + b}{2}$$

The standard deviation (σ) of X is, as before, a measure of how much variation there would be in X if you were to observe it indefinitely. Unlike the standard deviation of a normal distribution, σ for a uniform distribution is hard to represent graphically as a particular distance on the probability curve. Its value, however, is given by

$$\sigma = \frac{b - a}{\sqrt{12}}$$

Determining Probabilities

As it is for all continuous random variables, a probability based on a uniform random variable is determined by finding an area under a curve. Suppose, for example, the manager of Dixie Beverage Service would like to know what percentage of the cups will contain more than 7.5 ounces, using the present machines. In Figure 6.31, the shaded area is a rectangle, so its area is easy to find:

$$\text{area} = (\text{width}) \cdot (\text{height}) = (8 - 7.5) \cdot .5 = .25$$

So 25% of the cups will contain more than 7.5 ounces.

■ FIGURE 6.31
The probability that X
exceeds 7.5. The shaded
area represents the
percentage of cups
containing more than
7.5 ounces.

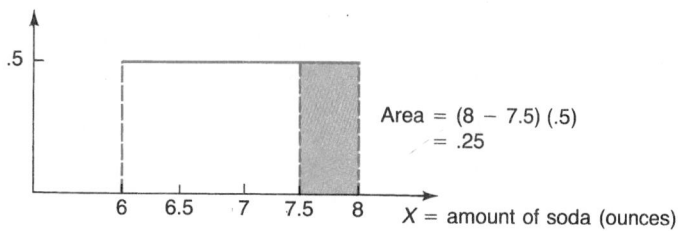

■ FIGURE 6.32
The probability that X is
between 6.5 and 7.5.

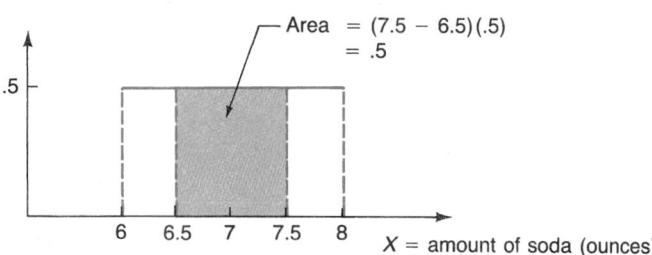

EXAMPLE 6.9

What is the probability that a cup will contain between 6.5 and 7.5 ounces? What is the average content?

SOLUTION

The first result is the same as the percentage of cups containing between 6.5 and 7.5 ounces. Based on Figure 6.32, we conclude that

$$P(6.5 < X < 7.5) = .5$$

The average cup content (mean of X) is

$$\mu = \frac{6 + 8}{2} = 7 \text{ ounces}$$

The standard deviation of X is

$$\sigma = \frac{8 - 6}{\sqrt{12}} = 0.58 \text{ ounce}$$

Notice that, as with the normal random variable, the probability that X is equal to any particular value is zero. So,

$$P(X = 6.5) = P(X = 7.5) = 0$$

As a result,

$$P(6.5 \le X \le 7.5) = P(6.5 < X < 7.5) = .5 \qquad ■$$

Simulation is an area of statistics that relies heavily on the uniform distribution. In fact, the uniform distribution is the underlying mechanism for this often complex procedure. So, although not as many "real-world" populations have uniform distributions as have normal ones, the uniform distribution is extremely important in the application of statistics.

The Exponential Distribution

The final continuous distribution we will discuss is the **exponential distribution.** Similar to the uniform random variable, the exponential random variable is used in a variety of applications in statistics. One application is observing the time between arrivals at, for example, a drive-up bank window. Another situation that often fits the exponential distribution is observing the lifetime of certain components in a machine.

Chapter 5 discussed the Poisson random variable, which often is used to describe the *number* of arrivals over a specified time period. If the random variable *Y*, representing the number of arrivals over time period *T*, follows a Poisson distribution, then *X*, representing the *time between* successive arrivals, will be an **exponential random variable.** The exponential random variable has many applications when describing any situation in which people or objects have to wait in line. Such a line is called a **queue.** People, machines, or telephone calls may wait in a queue.

The Exponential Random Variable   The shape of the exponential distribution is represented by a curve that steadily decreases as the value of the random variable, *X*, increases. Thus, the larger *X* is, the probability of observing a value of *X* at least this large decreases exponentially. This type of curve is illustrated in Figure 6.33.

Determining Probabilities   Determining areas for exponential random variables is not as simple as for uniform ones, but it is easier than for normal random variables because exponential probabilities can be derived on a calculator. Table A.2 also can be used to determine the probability for an exponential random variable.

As Figure 6.34 illustrates, for an exponential random variable, *X*, the probability that *X* exceeds or is equal to a specific value, X_0, is

$$P(X \geq X_0) = e^{-A \cdot X_0}$$

The parameter *A* is related to the Poisson random variable we used when discussing arrivals. In fact, the Poisson distribution for arrivals per unit time and the exponential distribution for time *between* arrivals provide two alternative ways of describing the same thing. For example, if the number of arrivals per unit time follows a Poisson distribution with an average of $A = 6$ per hour, then an alternate way of describing this situation is to say that the time between arrivals is exponentially distributed with mean time between arrivals equal to $1/A = 1/6$ hour (10 minutes).

In general, $1/A$ is the average (mean) value of the exponential random variable, *X*. It is also equal to the standard deviation of *X*. So,

$$\mu = 1/A$$
$$\sigma = 1/A$$

In applications using this distribution, the value of *A* either will be given or can be estimated in some way.

■ **FIGURE 6.33**
Curve showing the
distribution of an
exponential random
variable.

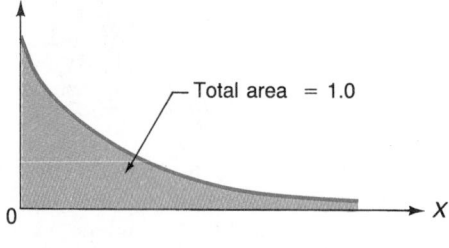

■ **FIGURE 6.34**
Curve used for
determining a
probability for an
exponential random
variable.

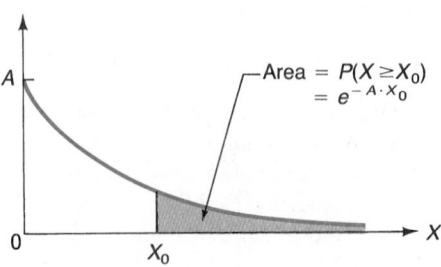

EXAMPLE 6.10

The owner of the Downtown Haircut Emporium believes the best way to run his barbershop is to rely on walk-in customers and not schedule appointments. From past experience, the arrival of customers follows a Poisson distribution with an average arrival rate of $A = 4$ customers per hour.

1. If the owner just witnessed the arrival of a customer, what is the probability that there will be a new arrival within 30 minutes?
2. If X represents the time between successive arrivals, what are the mean and standard deviation of X?

SOLUTION 1

To determine this probability, we must first convert 30 minutes to .5 hour, since the arrival rate is 4 **per hour.** The desired probability then is $P(X \leq .5)$. Referring to Figure 6.35, the probability that X *exceeds* .5 is

$$P(X > .5) = P(X \geq .5)$$
$$= e^{-(4)(.5)}$$
$$= e^{-2}$$
$$= .135$$

Consequently, $P(X \leq .5) = 1 - .135 = .865$, and so 86.5% of the time, the time between successive arrivals will not exceed 30 minutes.

SOLUTION 2

Both the mean and standard deviation of X (the time between successive arrivals) are $1/A = 1/4$ hour (15 minutes). ∎

EXAMPLE 6.11

The exponential distribution is widely used in the area of **reliability engineering** to describe the time to failure of a component or system. The parameter μ is called the *mean time to failure* and $A = 1/\mu$ is the *failure rate of the system.* Suppose that an automobile battery has a useful life described by the exponential distribution with a mean of 1000 days.

1. What is the probability that a battery will fail before its expected lifetime of 1000 days?
2. If the battery has a 12-month (365-day) warranty, what fraction of the batteries fail during the warranty period?

SOLUTION 1

The battery lifetime (X) follows an exponential distribution with $\mu = 1000$ and $A = 1/1000 = .001$. Referring to Figure 6.36, we wish to find the probability that X is less than 1000:

$$P(X < 1000) = 1 - e^{-(.001)(1000)} = 1 - e^{-1} = .632$$

Consequently, there is a 63% chance that the battery will fail prior to its mean lifetime of 1000 days. This value is larger than 50% since this distribution is not symmetric and is positively skewed.

■ **FIGURE 6.35**
Curve showing the probability that X exceeds $.5[P(X > .5)]$.

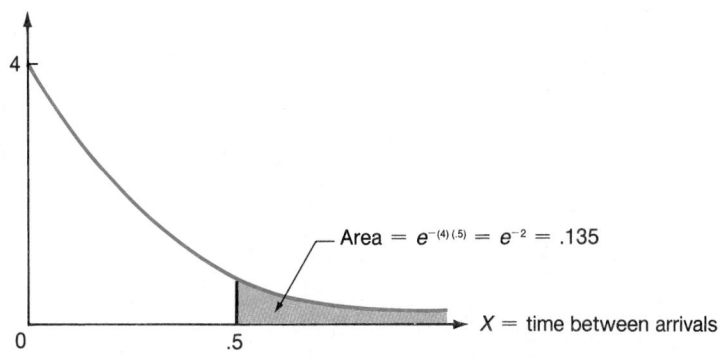

■ **FIGURE 6.36**
Curve showing the
probability that X is less
than 1000 days.

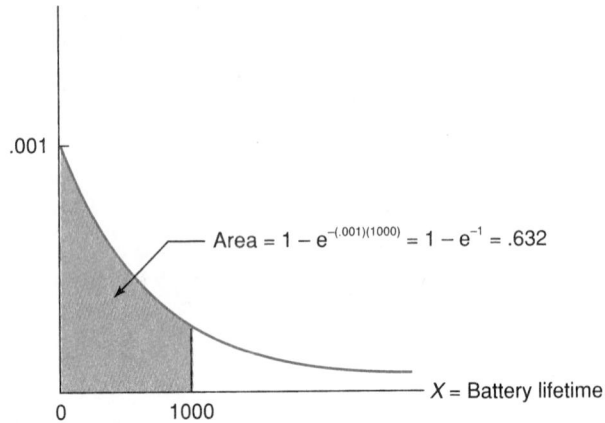

SOLUTION 2   According to Figure 6.37, the probability that the battery fails before the warranty expires is

$$P(X \leq 365) = 1 - e^{-(.001)(365)} = 1 - e^{-.365} = .306$$

The manufacturer will be forced to replace 30.6% of the batteries during the one-year warranty period. This large percentage is encountered despite the fact that the average lifetime of the battery is nearly three years because of the positive skewness and the heavy concentration of probability on the "low end" of the distribution.

■

Exercises   **6.34** A random variable X has a uniform distribution between the values 0 and 4.

a. What is the mean of X?

b. What is the standard deviation of X?

c. What is the height of the probability distribution of X?

d. What is the probability that X is greater than 1.23?

6.35 The errors from a forecasting technique appear to be uniformly distributed between − 3 and 3.

a. Find the probability that the errors deviate by no more than 1.5 from the mean.

b. Find the value x such that 60% of the errors occur between $− x$ and x.

6.36 The temperature of a warming tray is uniformly distributed between the values of 100°F and 104°F.

■ **FIGURE 6.37**
Curve showing the
probability that X is less
than 365 days.

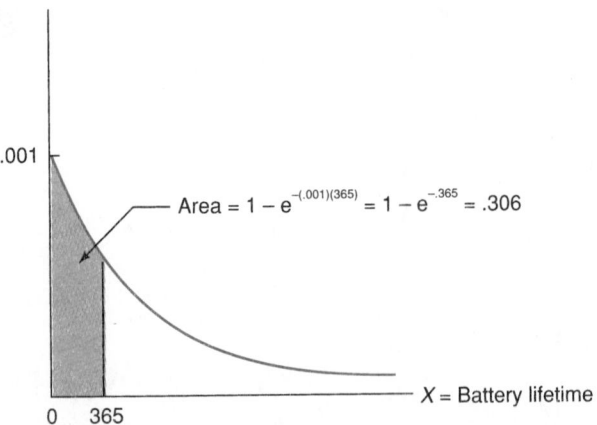

a. What percent of the time is the warming tray temperature less than 101.5°?

b. What is the mean temperature of the warming tray?

c. What is the standard deviation of the temperature of the warming tray?

6.37 The rate at which a swimming pool is filled is uniformly distributed between 20 and 26.3 gallons per minute.

a. What is the probability that the filling rate at any one time is between 21.3 and 24.6 gallons per minute?

b. What is the mean rate at which the swimming pool is filled?

c. What is the standard deviation of the rate at which the swimming pool is filled?

6.38 A random variable X has an exponential distribution with mean 5.

a. What is the standard deviation of X?

b. What is the probability that X is greater than 5?

c. What is the probability that X is between 3 and 6?

6.39 The time intervals (in operating hours) between successive failures of air-conditioning equipment in certain aircraft is believed to follow an exponential distribution. Assume that the mean time between failures is 300 hours. What is the probability that air conditioning equipment in a particular aircraft will fail after 200 operating hours?

6.40 If the amount of time that a customer spends in Ricky's Hide-Away restaurant follows an exponential distribution and if the average time spent by a customer is 0.75 hours, what is the probability that a customer will spend more than an hour in the restaurant? What is the standard deviation of the amount of time spent by a customer in the restaurant?

6.41 Yellow Rose taxi company estimates that it makes an average of $415 in profits per day. Assuming that the daily profit follows an exponential distribution, what is the probability that on a given day at least $500 in profits will be made?

6.42 The president of Bright-Light Candles estimates that the average burning time of their "medium-K" candles is 40 hours. Assuming that burning time follows an exponential distribution, calculate the probability that a given medium-K candle will burn for at least 50 hours?

6.43 If the amount of time ships spend at the Philadelphia dockyard follows an exponential distribution and if the average ship spends 3.1 days there, what is the probability that a given ship spends no more than 1.5 days at the dockyard?

6.44 The Mylapore County fire department has determined that the amount of time per month spent fighting fires follows an exponential distribution. If the average firefighting time per month is 10.4 hours, what is the probability that in a given month no more than 15 hours will be spent fighting fires?

6.45 Use a convenient computer package to generate randomly 50 uniformly distributed observations between 2.0 and 4.0. Find the mean and standard deviation of this sample of 50 observations. How closely do the sample mean and sample standard deviation agree with the true population mean and standard deviation? Repeat this procedure with 100 uniformly distributed observations. In MINITAB, the format is as follows:

```
MTB  >  RANDOM   50 observations, put in C1;
SUBC >  UNIFORM with continuous uniform on 2.0 to 4.0 .
MTB  >  Describe C1
```

 *Summary*   A random variable that can assume any value over a specific range is a **continuous random variable.** Many business applications have continuous probability distributions that can be approximated using a normal, uniform, or exponential random variable. Each of these distributions has a unique curve that can be used to determine probabilities by finding the corresponding area under this curve. The **normal** distribution is characterized by a bell-shaped curve with values concentrated about the mean. The **uniform** distribution (curve) is flat; values of this random variable

are evenly distributed over a specified range. The **exponential** distribution has a shape that steadily decreases as the value of the random variable increases. Table A.2 (or a good calculator) can be used to derive probabilities for the exponential distribution.

We discussed examples illustrating the shape of each distribution. The exact curve for a particular random variable is specified using one or two *parameters* that describe the corresponding population. As in the case of a discrete random variable, the population consists of what you would obtain if the random variable was observed indefinitely. The resulting average value and standard deviation represent the **mean** and **standard deviation** of the random variable and corresponding population.

There are infinitely many normal random variables, one for each mean (μ) and positive standard deviation (σ). If $\mu = 0$ and $\sigma = 1$, this normal random variable is the **standard normal** random variable, Z. Consequently, there is only *one* normal random variable of this type. Table A.4 gives the probabilities (areas) under the standard normal curve. You can also use this table to determine a probability for any normal random variable if you first **standardize** the variable by defining $Z = (X - \mu)/\sigma$. For this situation, Z represents the number of standard deviations that X is to the right (Z is positive) or left (Z is negative) of the mean.

The normal distribution can be used to approximate binomial probabilities for a large number of trials, n. Because the normal distribution is continuous and the binomial is discrete, the approximation can be significantly improved by **adjusting for continuity** before applying the normal approximation.

Summary of Formulas

1. Standardizing a normal random variable (X)

$$Z = \frac{X - \mu}{\sigma}$$

2. Approximating a binomial random variable (X) using a standard normal random variable (Z)

$$P(X \leq b) = P\left(Z \leq \frac{b + .5 - \mu}{\sigma}\right)$$

$$P(X \geq a) = P\left(Z \geq \frac{a - .5 - \mu}{\sigma}\right)$$

$$P(a \leq X \leq b) =$$
$$P\left(\frac{a - .5 - \mu}{\sigma} \leq Z \leq \frac{b + .5 - \mu}{\sigma}\right)$$

where $\mu = np$ and $\sigma = \sqrt{np(1 - p)}$

3. Uniform Distribution

$$\mu = \frac{a + b}{2} \quad \text{and} \quad \sigma = \frac{b - a}{\sqrt{12}}$$

$$P(X \leq X_0) = \frac{X_0 - a}{b - a} \quad \text{for } a \leq X_0 \leq b$$

4. Exponential Distribution

$$P(X \leq X_0) = 1 - e^{-AX_0} \quad \text{for } X_0 \geq 0$$
$$\text{where } A = 1/\mu$$

Review Exercises

6.46 Determine each of the following for a standard normal curve. Sketch the corresponding area.

a. $P(0 < Z < 1.5)$ c. $P(Z < -1.88)$
b. $P(Z > -3)$ d. $P(-2.5 < Z < 2.5)$

6.47 Calculate and sketch the area under the standard normal curve between the following Z values.

a. 2.2 and 3.25   b. -1.5 and 1.5   c. $-.75$ and 0

6.48   A commodities broker has a record of being correct 30% of the time in transactions which the broker solicits. From a random sample of 35 different recommendations to clients, what is the probability that less than 11 of the recommendations by the broker are profitable?

6.49   A quality-control engineer of a high tech company noted that 4% of all parts provided by the company's main supplier were of inferior quality (nonconforming items). Out of a sample of 300 parts provided by the company's main supplier, what is the probability that more than 10 parts are of inferior quality?

6.50   The mean length of certain gauges manufactured by a firm is 20 inches with a standard deviation of .44 inch. A random sample of 100 gauges was taken. Assuming that the length of these gauges is approximately normally distributed, what percentage of these gauges measured less than 20 inches in length?

6.51   Let X be a normally distributed random variable. Find the values of X that bound the middle 40% of the distribution of X if the mean is 100 and the variance is 16.

6.52   Scores on the English screening exam for international students are distributed normally with a mean of 68 and a standard deviation of 11. Calculate

a.   the percentage of scores between 70 and 80

b.   the percentage of scores that are less than 60

6.53   The examination committee of the Institute of Chartered Accountants passes only 20% of those who take the examination. If the scores follow a normal distribution with an average of 72 and a standard deviation of 18, what is the passing score?

6.54   The shelf life of cookies made by a small bakery is considered to be exponentially distributed with a mean equal to 3 days. What percentage of the boxes of cookies placed on the shelf today would still be considered marketable after 2.75 days?

6.55   The time that a certain drug has an effect on a normal human being is considered to be exponentially distributed when a standard dose is taken. If the average length of time that the drug has an effect is 30 hours, what is the probability that any given normal person will be affected by the drug for at least 32 hours? What is the standard deviation for the length of time that the drug affects a person?

6.56   Accidents such as the 1987 Amtrak-Conrail disaster in Maryland, in which 16 people died, have heightened employers' awareness of their legal liabilities for personal and property losses caused by workers under the influence of drugs and alcohol. Federal experts estimate that 10% to 23% of all American workers use drugs on the job. Suppose that a certain firm has over 100 manufacturing plants across the country. Let X be the percentage of employees who use drugs on the job at a particular plant. Assume that past data indicate that X has a normal distribution with mean 16% and standard deviation 4%.

(*Source:* Robert J. Aalberts, "Drug Testing—Walking a Legal Tightrope" *Business* (January–March 1988): 52–56.)

a.   What is the probability that X is greater than 10%?

b.   What is the probability that X is less than 5%, which may be considered an acceptable figure with regard to risk taken by the employer?

6.57   Clearvision Company manufactures picture tubes for color television sets and claims that the life spans of their tubes are exponentially distributed with a mean of 1800 hours. What percentage of the picture tubes will last no more than 1600 hours?

6.58   The amount of time each day that the copying machine is used at a certain business is approximately exponentially distributed with a mean of 3.5 hours. What is the probability that the copying machine will be used at least 2 hours a day?

6.59   The diameter of a special aluminum pipe made by Everything Aluminum Inc. is normally distributed with a mean of 3.00 centimeters and a standard deviation of .1 centimeter. Calculate the proportion of pipes whose diameters are more than 3.15 centimeters.

6.60   The Defense Contract Audit Agency (DCAA), which audits defense contractors, has a backlog of unaudited contracts in any given year. Suppose that the dollar amounts of unaudited contracts (in billions of dollars per annum) is normally distributed, with a mean of $74.5 billion and a standard deviation of $11.6 billion. What is the probability that in any given year, the backlog is less than $50 billion?

6.61   A manufacturer of heating elements for water heaters ships boxes that contain 100 elements. A quality-control inspector randomly selects a box in each shipment and accepts the shipment if there are 5 or fewer defective heating elements in the box. Assuming that the manufacturer has had a rate of 6% defective items, what is the probability that a shipment of heating elements will pass the inspection?

6.62 A recent poll of the chief financial officers (CFOs) at 100 major financial institutions in the United States (reported in the "Best Brokers Survey" of *Financial World,* January 1988) revealed that 53% of the CFOs favored abolishing the Glass-Steagall law, which separates commercial banking from investment banking.

a. If 20 CFOs were chosen at random from any of the major financial institutions in the United States, what is the approximate probability that fewer than 18 CFOs will favor abolishing the Glass-Steagall law? Assume that 53% is a close approximation of the population proportion.

b. If in part a, 20 CFOs were chosen at random and more than 15 CFOs favored abolishing the Glass-Steagall law, would this be considered an unusual sample? Why?

6.63 A paint sprayer coats a metal surface with a layer of paint between 0.5 and 1.5 millimeters thick. The thickness of the coat of paint is approximately uniformly distributed.

a. What are the mean and standard deviation of the thickness of the coat of paint on the metal surface?

b. What is the probability that paint from this sprayer on any given metal surface will be between 1.0 and 1.3 mm thick?

6.64 The rate at which a sack of soybeans is filled varies uniformly from 50 pounds per hour to 65 pounds per hour. What percent of the time is the rate greater than 55 pounds per hour?

6.65 If X is a uniform random variable that represents the percentage of time each day that a machine does not work, what is the probability that X is greater than the mean percentage of time that the machine does not work?

6.66 If the random variable X has a uniform distribution between -10 and 10, find the value of x such that $P[X \geq x] = .25.$

6.67 The marketing division of Goodlife Tires determined the average (mean) life of tires to be 30,000 miles with a standard deviation of 5,000 miles. Given that tire life is a normally distributed random variable, find the following:

a. the probability that tires last between 25,000 and 35,000 miles.

b. the probability that tires last between 28,000 and 33,000 miles

c. the probability that tires last less than 28,000 miles

d. the probability that tires last more than 35,000 miles

6.68 The random variable X is normally distributed with mean μ and variance σ^2. Find k if $P(\mu - k\sigma \leq X \leq \mu + k\sigma) = .67.$

6.69 If the random variable X is normally distributed with mean 25, find the variance if $P(X \geq 29) = .27.$

6.70 The random variable X is normally distributed such that $P(X \leq 10) = .12$ and $P(X \geq 15) = .4$. Find the mean and variance of the random variable X.

6.71 In Canada, Quebec is a key part of the chemical industry. Uncertainty about the relationship of Quebec with the rest of the nation presents tough questions about the interprovincial chemical trade between Quebec and the other Canadian provinces. Twenty-four percent of Canada's chemical shipments begin in Quebec. Suppose that petrochemical investors were examining a sample of 30 chemical shipments from Canada, what is the probability that 10 or more shipments are from Quebec?

(*Source:* "Chemical Firms Wait Out Canada's Constitutional Crisis Over Quebec," *Chemical and Engineering News* 68 (July 23, 1990): 13.)

6.72 The mechanics at Quick Brown Fox can tune up a car in an average of 30 minutes with a standard deviation of 5 minutes. If a car arrives for a tune-up 25 minutes before closing, what is the probability that the car will be serviced by closing, assuming that the time it takes for a tune-up is normally distributed.

6.73 According to a recent *Wall Street Journal*/NBC News Poll, 71% of the people who have been paying attention to the Senate Ethics Committee hearing on the Keating Five say that the violations that the Keating Five are accused of are typical of all members of Congress. Suppose that in a sample of 30 people who have been paying attention to the hearings, it was found that fewer than half agree that the violations the Keating Five are accused of are typical of all members of Congress. Would this event be considered unusual? Find the probability of this event happening.

(*Source:* "Keating Five Hearings Have Lawmakers Caught between Public Duty and Loyalty to Senate Club," *Wall Street Journal* (December 24, 1990): 28.)

6.74 There are certain situations where both $np > 5$ and $n(1 - p) > 5$ do not hold but the normal approximation may be a reasonable approximation. The following MINITAB computer printout shows the distribution of a binomial random variable X with $n = 20$ and $p = .20$. Examine the graph and describe how well a bell-shaped curve could fit this distribution. Calculate the probability that $P[X \leq 3]$ and $P[X \geq 5]$ by using both the normal approximation and the binomial table. Would you say that the normal approximation is reasonably close?

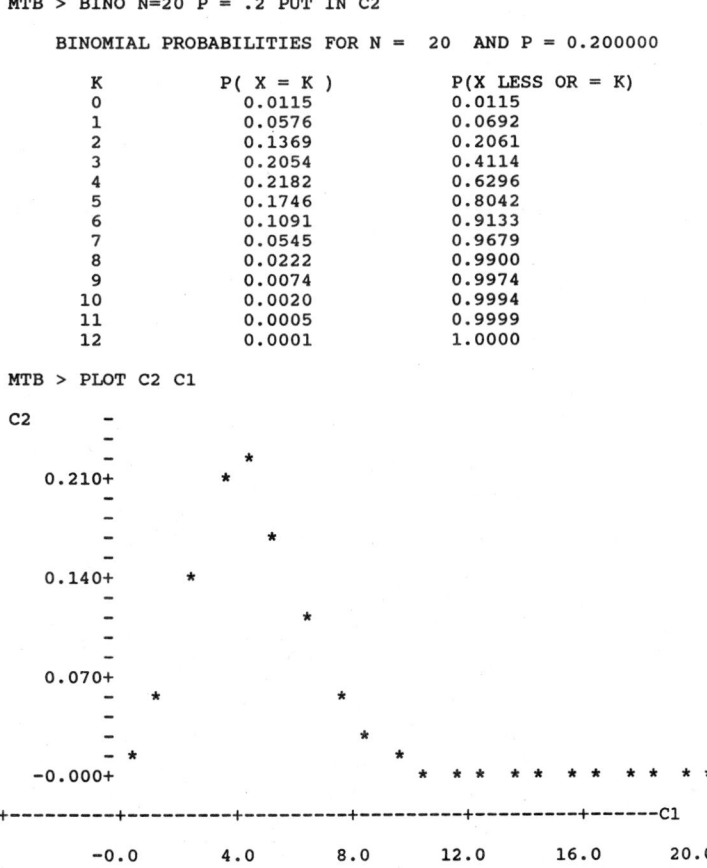

```
MTB > BINO N=20 P = .2 PUT IN C2

    BINOMIAL PROBABILITIES FOR N =  20  AND P = 0.200000

         K            P( X = K )         P(X LESS OR = K)
         0              0.0115              0.0115
         1              0.0576              0.0692
         2              0.1369              0.2061
         3              0.2054              0.4114
         4              0.2182              0.6296
         5              0.1746              0.8042
         6              0.1091              0.9133
         7              0.0545              0.9679
         8              0.0222              0.9900
         9              0.0074              0.9974
        10              0.0020              0.9994
        11              0.0005              0.9999
        12              0.0001              1.0000

MTB > PLOT C2 C1

 C2      -
         -
         -                  *
   0.210+                 *
         -
         -                   *
         -
   0.140+            *
         -                     *
         -
         -
   0.070+
         -     *             *
         -
         -                      *
         -  *               *
  -0.000+                      *  *  *   *  *  *  *   *  *  *
         +---------+---------+---------+---------+---------+------C1
           -0.0       4.0       8.0       12.0      16.0      20.0
```

6.75 Data were collected in a life test study on the life of deep-groove ball bearings. The life of each ball bearing is recorded in millions of revolutions to failure. A MINITAB computer printout displays a histogram of the data. It is believed that the life of each ball bearing follows an exponential distribution. A MINITAB computer printout also displays a graph of an exponential distribution with mean 10. The solid curve illustrates another graphics capability of MINITAB. Does it appear that the exponential distribution is a good approximation of the distribution of the life of the ball bearings?

```
MTB > name c1 'life'
MTB > print c1

life
   6.8942    3.0973    3.4907   17.8592    0.4579    9.0166    3.0514
  20.6228    1.1049    0.7869    6.1572    6.3199    8.1731   15.9860
  13.0033   28.9843    1.1798   24.1269   16.2584    5.2212   18.4759
   3.5372    2.7533    0.8878    9.6401    4.0061    3.0251    9.9500
  15.3466    0.6072    4.5263    7.0539    3.0187    8.4473    3.4397
   4.8021   11.0286    7.1305   13.1512    5.8844

MTB > histogram c1;
SUBC> increment = 4;
SUBC> start = 2.

Histogram of life    N = 40

Midpoint    Count
    2.00      14   **************
    6.00      10   **********
   10.00       6   ******
   14.00       4   ****
   18.00       3   ***
   22.00       1   *
   26.00       1   *
   30.00       1   *

MTB > set c2
DATA> 0:30/.25
DATA> end
MTB > let c3 = (1/10)*exp(-c2/10)
MTB > gplot;
SUBC> line c3 c2.
```

Computer Exercises Using the Database

Exercise 1 -- Appendix I Select 100 observations at random from the database and use a convenient statistical computer package to estimate the mean and standard deviation of the variable HPAYRENT (house payments or apartment/house rents). Find the percentage of the observations between $\bar{x} \pm s$, $\bar{x} \pm 2s$, and $\bar{x} \pm 3s$. Comment on whether these percentages support the conclusion that the data come from a normally distributed population.

Exercise 2 -- Appendix I Select 150 observations at random from the database and, with reference to the variable OWNORENT, calculate the proportion of those observations that indi-cate the house is owned rather than rented. If a random sample of 20 observations were chosen from this set of 150 observations with replacement, what is the probability that more than half of the homes in the sample of 20 are owned by their occupants? Is the normal approximation appropriate for this situation?

Exercise 3 -- Appendix J Select 100 observations at random from the database on the variable EMPLOYEES (number of employees). Use a convenient statistical computer package to construct a histogram. What type of distribution does the histogram approximate? Normal? Uniform? Exponential? None of these?

Exercise 4 -- Appendix J Repeat Exercise 3 using the variable SALES.

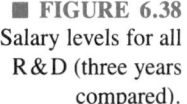

Case Study

The Case of the Bulging, Shifting and Flattening Bell Curve

R&D magazine conducts an annual survey of salaries of professionals in the research and development field. In most recent years, the frequency histogram of salary levels has produced a fairly nice bell curve with a distinct peak and almost-matching halves. Each year, as general salary levels were increased to keep pace with inflation, the bell shape shifted a little to the right, toward the higher income side, but essentially retained its shape.

This trend has been developing over the past couple of years and became especially noticeable after the 1988 *R&D* opinion poll of 1,900 respondents. The 1990 survey of 3,800 readers of the magazine showed the salary range continuing to expand, with the upper end climbing while the lower end remained essentially stationary.

The salary range of $40,000 to $44,999 represented the modal class in previous years, but the size of this group has shrunk from 16.5% of the total respondents in 1986 to 14.5% in 1988. In 1990, with 60.8% of the respondents reporting annual salaries of $45,000 or more, the modal class was $50,000 to $54,999. About 13.2% of respondents belonged in this category. Expressed graphically, the height of the plotted peak and the peak's shoulders also are lower, as if someone were pressing down on the hump from above.

Do you think this would cause the curve to flatten out? Well, yes and no. A greater proportion of those who were "pushed out" from under the peak are now to be found along a flattened right-hand slope, representing the higher income portion of the curve. In 1986, 60.8% of R&D workers fell into the salary range of $35,000 to $59,999. In 1988, the figure for this group declined to 58.7%. Over the same period, those in the group receiving a salary above $59,999 increased from 8.7% to 14.4%.

Consider the histogram in Figure 6.38. The old cliché about the rich getting richer and the poor getting poorer seems to be borne out. Salary is a continuous variable, but since the chart shows three years side by side, gaps have been left between the bars for the sake of readability. The shape seems to follow a normal distribution, but it is clearly becoming skewed, as evidenced by a 4.7% rightward shift of the 1989 median salary of $47,417 to $49,632 for 1990. The *R&D* article did not provide the mean and standard deviation. This is typical of much data published in magazines and newspapers, so we have to learn how to come up with a good estimate ourselves. Remember, statistics is not only the science of numbers but also the art of approximation.

Case Study Questions

1. Which class contains the median salary for 1990? Would the midpoint of this class be a good approximation of the mean? Could the upper class limit be a better estimate?

2. Estimate the percentage in each class from the above discussion and from Figure 6.38. Use these results to estimate the standard deviation. Your answer may differ from someone else's, depending on how you argue your case.

3. Assume a mean of $50,500 and a standard deviation of $14,000 for the year 1990.

a. What should be the proportion of individuals in the R&D field earning $80,000 or more? Does this agree with Figure 6.38?

b. What is the probability that an R&D professional selected at random earns between $30,000 and

■ **FIGURE 6.38**
Salary levels for all R&D (three years compared).

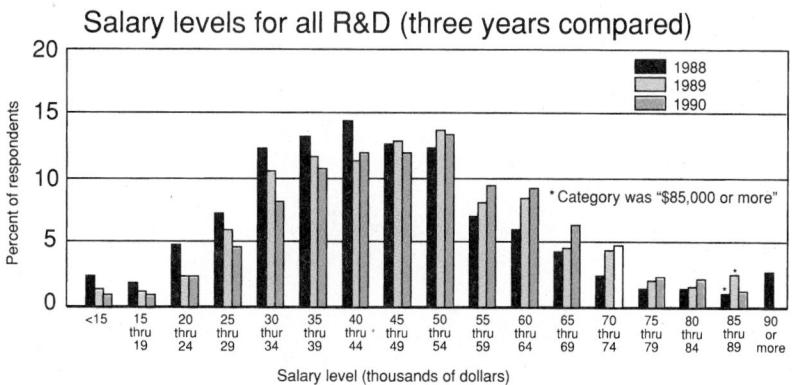

$49,999 annually? Again, how well does this agree with Figure 6.38?

4. Use the concept of skewness (see Chapter 3) to explain the differences between the theoretical proportions indicated by the ideal normal curve and those indicated by Figure 6.38.

5. Consider the quotation, "The R&D salary bell is going flat. This is not to say that it makes a sour note. In fact, there are many people who find it especially attractive this way." What do you think the author means by this statement?

7

Statistical Inference and Sampling

The previous three chapters laid the foundation for using statistical methods in decision making. Any such decision will have uncertainty associated with it, but we can attempt to measure this uncertain outcome using a probability. Random variables (both discrete and continuous) allow you to represent certain outcomes of an experiment and their corresponding probabilities conveniently. If the experiment involves a particular discrete situation (such as a binomial random variable), you can easily determine the probability of certain events or determine the mean (average) value of the related distribution.

If the random variable of interest is continuous, you can make probability statements after assuming the probability distribution involved (such as normal, exponential, uniform, or others we have not discussed). Both discrete and continuous random variables come into play in all areas of decision making.

They allow us to make decisions concerning a large population using the information contained in a much smaller sample.

Such decision making lies in the area of **statistical inference,** which this chapter introduces by demonstrating how to estimate something about a population (such as its average value, μ) by using the corresponding value from a sample (such as the sample average, $\bar{X}$). Recall that μ (belonging to the population) is a parameter and $\bar{X}$ (belonging to the sample) is a statistic. When dealing with a normal population, for example, what does one do if the population mean, μ, is unknown? So far in the text, this value has been specified for you. In this chapter, we discuss methods of estimating population parameters using sample statistics, along with several methods of gathering sample data.

7.1

Random Sampling and the Distribution of the Sample Mean

In Chapter 3, you learned how to calculate the mean of a sample, $\bar{X}$. This sample is drawn from a population having a particular distribution, such as normal, exponential, or uniform. If you were to obtain another sample (you probably will not, as most decisions are made from just one sample), would you get the same value of $\bar{X}$? Assuming that the new sample was made up of different individuals than was the first sample, then almost certainly the two $\bar{X}$'s would not be the same. So, $\bar{X}$ itself is a random variable. We will demonstrate that if a sample is large enough, $\bar{X}$ is very nearly *normally* distributed regardless of the shape of the sampled population. That is, if you were to obtain many large samples, calculate the resulting $\bar{X}$'s, and then make a histogram of these $\bar{X}$'s, *this histogram would always approximately resemble a bell-shaped (normal) curve.*

Simple Random Samples

In Chapter 4, the concept of a simple random sample was introduced. The mechanics of obtaining a random sample range from drawing names out of a hat to using a computer to generate lists of random numbers. For extremely large populations, one is often forced to select individuals (elements) from the population in a *nearly* random manner.

The underlying assumption behind a random sample of size n is that any sample of size n has the same chance (probability) of being selected. To be completely assured of obtaining a random sample from a *finite* population, you should number the members of the population from 1 to N (the population size) and, using a set of *n* random numbers, select the corresponding sample of *n* population elements for your sample.

This procedure was described in Chapter 4 and is often used in practice, particularly when you have a sampling situation that needs to be legally defensible, as is the case in many statistical audits. However, for situations in which the population is extremely large, this strategy may be impractical, and instead you can use a sampling plan that is nearly random. Several other sampling procedures are discussed in the last section of this chapter.

The main point of all this lengthy discussion is that practically all the procedures presented in subsequent chapters relating to decision making and estimation assume that you are using a random sample. In the chapters that follow, the word *sample* will mean *simple random sample*.

Estimation

The idea behind statistical inference has two components:

1. The *population* consists of everyone of interest. By "everyone" we mean all people, machine parts, daily sales, or whatever else you are interested in measuring or observing. The mean value (for example, average height, average income) of everyone in this population is μ and generally is not known.
2. The *sample* is randomly drawn from this population. Elements of the sample thus are part of the population—but certainly not all of it. The exception to this is a *census,* a sample that consists of the entire population.

The sample values should be selected randomly, one at a time, from the entire population. Figure 7.1 emphasizes our central point—namely, an unknown population **parameter** (such as μ = the mean value for the entire population) can be **estimated** using the corresponding sample **statistic** (such as $\bar{X}$ = the mean of your sample).

It makes sense, doesn't it? It would be most desirable to know the average value for everyone in the population, but in practice this is nearly always impossible. It may take too much time or money, we may not be able to obtain values for them all even if we want to, or the process of measuring the individual items may destroy

■ **FIGURE 7.1**
The sample mean, $\bar{X}$, is used to estimate the population mean, μ. In general, sample statistics are used to estimate population parameters.

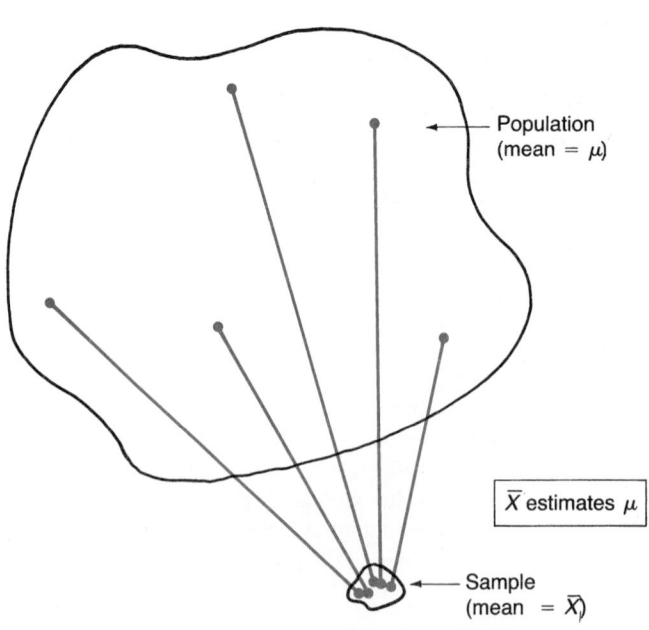

them (such as measuring the lifetime of a light bulb). In many instances, estimating the population value using a sample estimate is the best we can do.

EXAMPLE 7.1

In Example 6.3, the inside diameter of a manufactured machine bearing was believed to follow a normal distribution with a mean of $\mu = 2.52$ inches and a standard deviation of $\sigma = .022$ inches. There is no way of *knowing* that μ is 2.52 inches unless all such bearings are examined. Assume that

$$X = \text{inside diameter of a bearing}$$

is a normal random variable, but do not assume anything about the mean and standard deviation. Ignoring the standard deviation, estimating μ involves obtaining a random sample of bearings and recording their inside diameters. Suppose you obtain a random sample of size $n = 10$, with the following results (in inches):

2.511, 2.540, 2.505, 2.527, 2.488, 2.579, 2.526, 2.524, 2.503, 2.533

What is the estimate of μ, based on these values?

SOLUTION

The sample mean is $\bar{x} = 2.524$ inches. Thus, based on these ten sample values, our best estimate of μ is $\bar{x} = 2.524$ inches.* ∎

Distribution of $\bar{X}$

Referring to Example 7.1, the value of $\bar{X}$ would almost certainly change if you were to obtain another sample. The question of interest here is, if we *were* to obtain many values of $\bar{X}$. how would they behave? If we observed values of $\bar{X}$ indefinitely, where would they center; that is, what is the **mean** of the distribution for the random variable, $\bar{X}$? Is the variation of the $\bar{X}$ values more, less, or the same as the variation of individual observations? This variation is measured by the **standard deviation** of the distribution for $\bar{X}$.

In Example 6.2, it was assumed that the average lifetime of an Everglo light bulb was $\mu = 400$ hours, with a population standard deviation of $\sigma = 50$ hours. This result does not imply that if you obtain a random sample of these bulbs, the resulting sample mean, $\bar{X}$, always will be 400. Rather, a little head scratching should convince you that $\bar{X}$ will not be exactly 400, but $\bar{X}$ should be *approximately* 400.

Twenty samples of 10 bulbs each and the calculated $\bar{X}$ for each sample are shown in Table 7.1. We will assume for now that the population parameters are $\mu = 400$ hours and $\sigma = 50$ hours (Figure 7.2).

The 20 values of $\bar{X}$ are:

384.0, 399.5, 397.3, 399.4, 414.4, 393.2, 398.8, 403.8, 402.8, 404.6, 401.9, 385.6, 408.2, 411.0, 415.6, 408.5, 382.1, 391.3, 428.6, 423.1

They are not each 400, but they are all close to 400. Using a calculator or computer, you would also find that (1) the average (mean) of these 20 values is 402.88 (this is close to $\mu = 400$) and (2) the standard deviation of these 20 values is 12.78 (this is *much smaller* than $\sigma = 50$).

The $\bar{X}$ values appear to be centered at $\mu = 400$ hours but have *much less variation* than the individual observations in each of the samples. A histogram of these 20 values generated by MINITAB is contained in Figure 7.3. Based on the

---

*The notation $\hat{\mu}$ is commonly used (in place of $\bar{x}$) to denote an *estimate* of μ. For this example, the estimate of μ is $\hat{\mu} = 2.524$ inches.

■ TABLE 7.1
Twenty samples of
10 Everglo bulbs.

SAMPLE 1	SAMPLE 2	SAMPLE 3	SAMPLE 4	SAMPLE 5
308	431	416	373	354
419	448	361	451	385
389	380	389	329	449
432	371	497	460	419
362	387	400	481	483
302	410	489	350	396
440	400	406	431	317
430	426	333	356	457
375	381	307	410	404
383	361	375	353	480
$\bar{x} = 384.0$	399.5	397.3	399.4	414.4
$s = 49.30$	28.54	60.51	53.99	54.25

SAMPLE 6	SAMPLE 7	SAMPLE 8	SAMPLE 9	SAMPLE 10
404	372	449	403	354
390	404	389	350	446
390	493	397	565	343
454	344	428	354	458
386	396	374	358	404
385	441	502	412	468
384	373	365	441	416
351	438	402	340	340
392	360	416	359	409
396	367	316	446	408
$\bar{x} = 393.2$	398.8	403.8	402.8	404.6
$s = 25.45$	46.10	50.32	68.93	46.28

SAMPLE 11	SAMPLE 12	SAMPLE 13	SAMPLE 14	SAMPLE 15
329	429	461	448	457
473	286	399	386	432
336	382	416	375	425
356	380	378	488	391
385	423	359	447	429
365	388	408	429	448
419	329	393	377	416
448	438	374	380	429
459	423	440	372	414
449	378	454	408	315
$\bar{x} = 401.9$	385.6	408.2	411.0	415.6
$s = 54.12$	47.91	34.60	40.12	39.73

SAMPLE 16	SAMPLE 17	SAMPLE 18	SAMPLE 19	SAMPLE 20
491	439	331	418	428
353	336	427	422	368
375	425	445	341	445
536	419	420	485	429
447	346	401	442	475
415	408	389	470	437
322	392	363	404	475
350	409	439	370	458
453	313	352	539	308
343	334	346	435	408
$\bar{x} = 408.5$	382.1	391.3	432.6	423.1
$s = 71.46$	45.28	41.35	56.78	51.48

■ **FIGURE 7.2**
Assumed distribution
of Everglo bulbs.

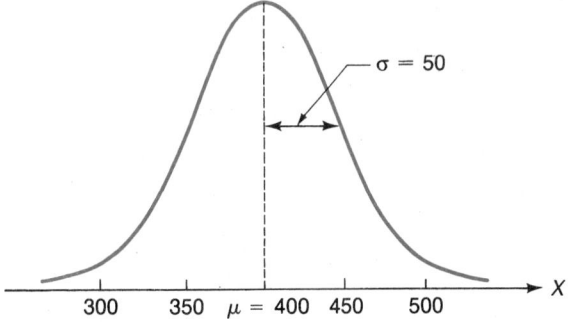

■ **FIGURE 7.3**
Histogram of 20 sample
means generated by
MINITAB. Compare
with Figure 7.2.

```
MTB > SET INTO C1
DATA> 384.0 399.5 397.3 399.4 414.4 393.2 398.8
DATA> 403.8 402.8 404.6 401.9 385.6 408.2 411.0
DATA> 415.6 408.5 382.1 391.3 432.6 423.1
DATA> END
MTB > HISTOGRAM OF C1

Histogram of C1   N = 20

Midpoint   Count
    380      2    **
    390      3    ***
    400      8    ********
    410      4    ****
    420      2    **
    430      1    *
```

shape of this histogram, it seems reasonable to assume that the values of $\bar{X}$ follow a normal distribution, but one that is much *narrower* than the population of individual lifetimes in Figure 7.2.

◢ 7.2
The Central Limit Theorem

Our last example illustrates a useful result, the *Central Limit Theorem* (CLT).

Central Limit Theorem

When using a random sample of size n from a population with mean μ and standard deviation σ, the resulting sample mean, $\bar{X}$, has a normal distribution with mean μ and standard deviation $\sigma/\sqrt{n}$. This is true for any sample size, n, if the underlying population is normally distributed, and it is approximately true for large sample sizes (generally $n > 30$) obtained from any population.

In other words, the distribution of all possible $\bar{X}$ values has an exact or approximate normal distribution with mean μ and standard deviation $\sigma/\sqrt{n}$.

COMMENTS

1. The second part of the Central Limit Theorem is an extremely strong result; it says that you can assume that $\bar{X}$ follows an approximately normal distribution *regardless* of the shape of the population from which the sample was obtained if the sample size (n) is large. For example, if you repeatedly sampled from a population with an exponential distribution, the resulting $\bar{X}$'s would follow a *normal* (not an exponential) curve.
2. For any sample size n, the mean of $\bar{X}$ is μ and the standard deviation is $\sigma/\sqrt{n}$. However, for $\bar{X}$ to be approximately normally distributed, a large value of n is necessary.

In Table 7.1, 20 samples of size ten were obtained and the corresponding values of $\bar{X}$ were determined. Suppose samples of size ten were obtained *indefinitely* and we wished to describe the shape of the resulting $\bar{X}$'s. According to the Central Limit Theorem, $\bar{X}$ will be a normal random variable. We are assuming that the individual lifetimes follow a normal curve (see Figure 7.2), so this will be true for any sample size—in particular, $n = 10$. So the resulting $\bar{X}$'s will describe a normal curve similar to the curve in Figure 7.3.

Where is the curve centered? According to the Central Limit Theorem, the mean of this normal random variable is the *same* as that in Figure 7.2; that is, it is the mean of the population from which you are sampling. This value is $\mu = 400$, and so, on the average, the value of $\bar{X}$ is $\mu_{\bar{x}} = 400$ hours. Notice that the average of the 20 values of $\bar{X}$ that we did observe was 402.88. This value will get closer to, or **tend toward,** 400 as we take more samples of size ten.

What is the standard deviation of the normal curve for $\bar{X}$? As we noted earlier, the 20 values of $\bar{X}$ jump around (vary) much less than do the individual observations in each of the samples. Consequently, the standard deviation of the $\bar{X}$ normal curve will be much less than that of the population curve (describing individual lifetimes) in Figure 7.2. In fact, according to the Central Limit Theorem, this will be

$$\sigma_{\bar{x}} = \frac{\sigma}{\sqrt{n}} \qquad\qquad (7.1)$$

where σ is the standard deviation of the population ($\sigma = 50$ in Figure 7.2). Consequently,

$$\sigma_{\bar{x}} = \frac{50}{\sqrt{10}} = 15.81$$

Recall that the standard deviation of the 20 observed $\bar{X}$ values was 12.78. This value will tend toward 15.81 if we take more samples of size ten. These results are summarized in Figure 7.4, where $\mu_{\bar{x}} = 400$ and $\sigma_{\bar{x}} = 15.81$.

■ **FIGURE 7.4**
Normal curves
for population and
sample mean.

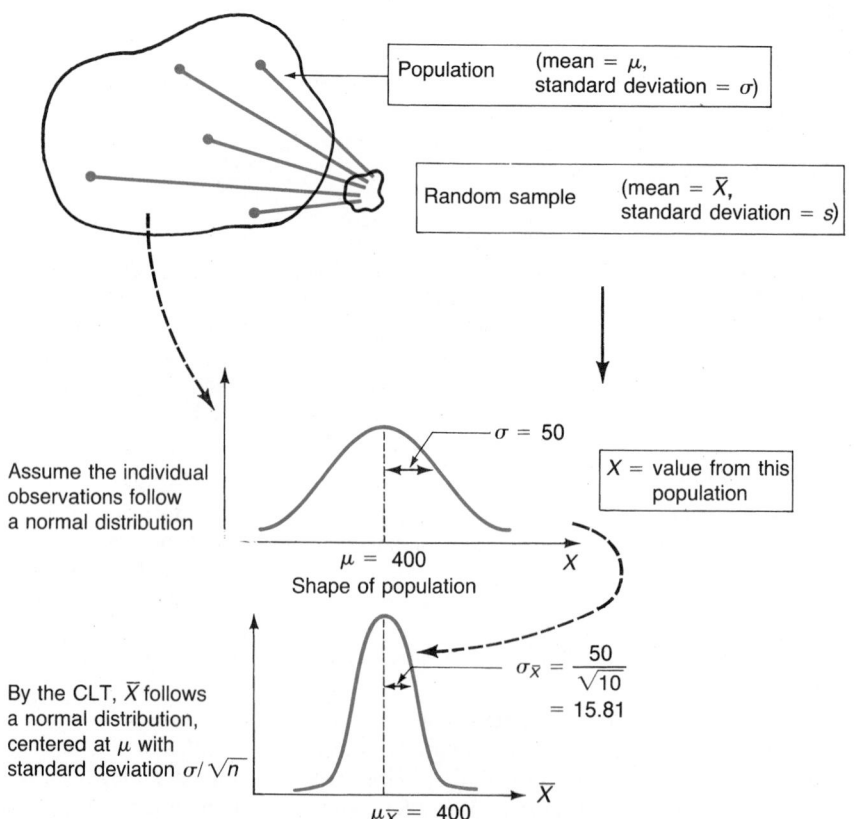

Basically, the Central Limit Theorem says that the normal curve (distribution) for $\bar{X}$ is centered at the same value as the population distribution but has a much smaller standard deviation. Notice that as the sample size, n, increases, $\sigma/\sqrt{n}$ decreases, and so the spread relative to the mean of the $\bar{X}$ curve (that is, the variation in the $\bar{X}$ values) decreases. If we repeatedly obtained samples of size 100 (rather than 10), the corresponding $\bar{X}$ values would lie even closer to $\mu_{\bar{x}} = 400$ because now $\sigma_{\bar{x}}$ would equal $50/\sqrt{100} = 5$ (see Figure 7.5).

For the 20 values of $\bar{X}$ in Table 7.1, it was assumed that the population mean was *known* to be $\mu = 400$, so each of the $\bar{X}$ values estimates μ with a certain amount of error. The more variation in the $\bar{X}$ values, the more error we encounter using $\bar{X}$ as an estimate of μ. Consequently, the standard deviation of $\bar{X}$ also serves as a measure of the error that will be encountered using a sample mean to estimate a population mean. The standard deviation of the $\bar{X}$ distribution is often referred to as the **standard error** of $\bar{X}$.

> Standard error of $\bar{X}$ = standard deviation of the probability
> distribution for $\bar{X}$
>
> $$= \frac{\sigma}{\sqrt{n}}$$

The previous discussion has described the probability distribution of the sample mean, $\bar{X}$. This distribution is referred to as the *sampling distribution* of $\bar{X}$.

■ **FIGURE 7.5**
Normal curves for the sample mean ($n = 10$, 20, 50, 100).

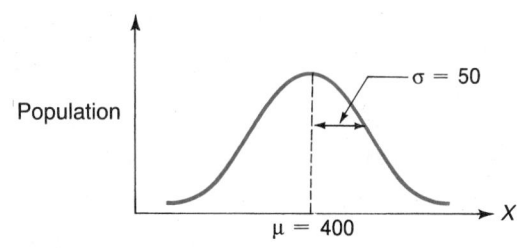

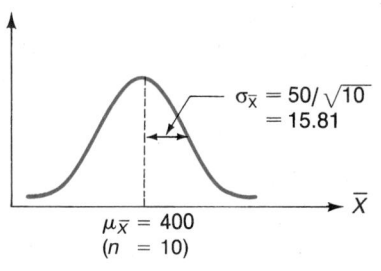

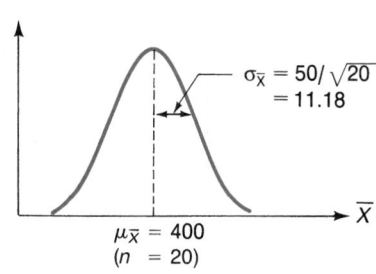

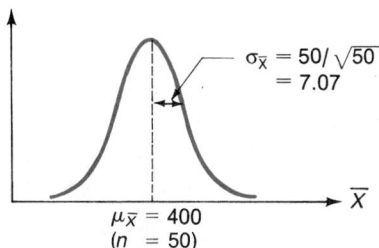

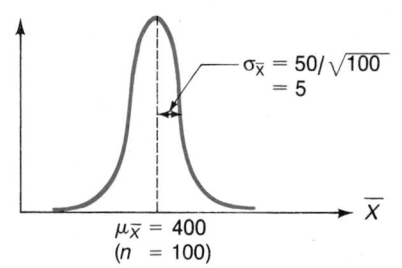

Definition

The probability distribution of a sample statistic is its **sampling distribution**.

Thus, according to the Central Limit Theorem, the sampling distribution of $\bar{X}$ is approximately normal, centered at μ, with a standard deviation (standard error) of $\sigma/\sqrt{n}$, regardless of the shape of the sampled population for large samples.

EXAMPLE 7.2   Electricalc has determined that the assembly time for a particular electrical component is normally distributed with a mean of 20 minutes and a standard deviation of 3 minutes.

1. What is the probability that an employee in the assembly division takes longer than 22 minutes to assemble one of these components?
2. What is the probability that the average assembly time for 15 such employees exceeds 22 minutes?
3. What is the probability that the average assembly time for 15 employees is between 19 and 21 minutes?

SOLUTION 1   The random variable X here is the assembly time for a component. It is assumed to be a normal random variable, with $\mu = 20$ minutes and $\sigma = 3$ minutes (Figure 7.6). We wish to determine $P(X > 22)$. Standardizing this variable and using Table A.4, we obtain

$$P(X > 22) = P\left[\frac{X - 20}{3} > \frac{22 - 20}{3}\right]$$
$$= P(Z > .67)$$
$$= .5 - .2486 = .2514$$

Therefore, a randomly chosen employee will require longer than 22 minutes to assemble the component with probability .25.

SOLUTION 2   Figure 7.6 does *not* apply to this question because we are concerned with the *average* time for 15 employees, not an individual employee. Using the Central Limit Theorem, we know that the curve describing $\bar{X}$ (an average of 15 employees) is normal with

$$\text{mean} = \mu_{\bar{x}} = \mu = 20 \text{ minutes}$$
$$\text{standard deviation (standard error)} = \sigma_{\bar{x}} = \sigma/\sqrt{n}$$
$$= 3/\sqrt{15} = .77 \text{ minutes}$$

(See Figure 7.7.)

■ FIGURE 7.6
Assembly time for the population of electrical components. See Example 7.2.

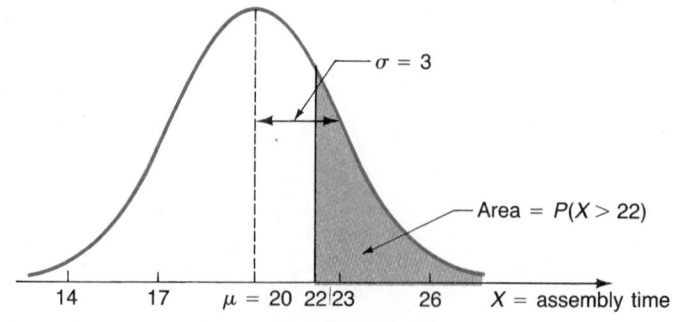

Area = $P(X > 22)$

14   17 $\mu = 20$ 22 23   26 $X =$ assembly time

■ **FIGURE 7.7**
Curve for $\bar{X}$ = average of 15 employees' assembly times. Shaded area shows $P(\bar{X} > 22)$.

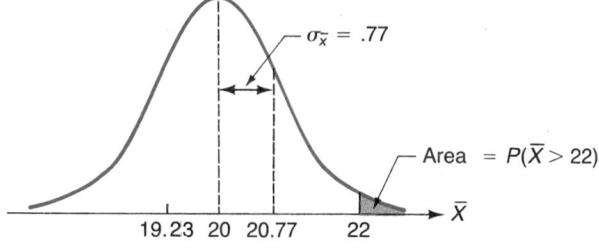

$\sigma_{\bar{x}} = .77$

Area $= P(\bar{X} > 22)$

19.23 20 20.77   22 $\bar{X}$

■ **FIGURE 7.8**
Curve for average assembly time of 15 employees. Shaded area shows $P(19 < \bar{X} < 21)$.

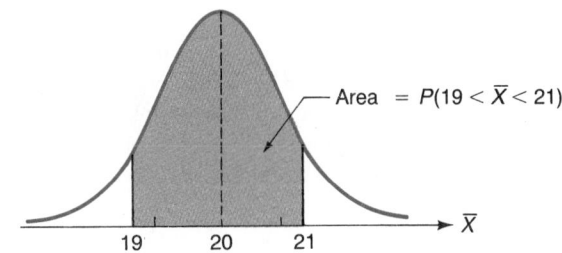

Area $= P(19 < \bar{X} < 21)$

19   20   21 $\bar{X}$

The procedure is the same as in solution 1, except now the standard deviation of this curve is .77 rather than 3:

$$P(\bar{X} > 22) = P\left[\frac{\bar{X} - 20}{.77} > \frac{22 - 20}{.77}\right]$$

$$= P(Z > 2.60)$$

$$= .5 - .4953 = .0047$$

So an average assembly time for a sample of 15 employees will be more than 22 minutes with less than 1% probability; that is, it is very unlikely that an average of 15 assembly times will exceed 22 minutes.

SOLUTION 3   The curve for this solution is shown in Figure 7.8. We wish to find $P(19 < \bar{X} < 21)$.

$$P(19 < \bar{X} < 21) = P\left[\frac{19 - 20}{.77} < \frac{\bar{X} - 20}{.77} < \frac{21 - 20}{.77}\right]$$

$$= P(-1.30 < Z < 1.30)$$

$$= .4032 + .4032 = .8064$$

Thus, a sample of 15 employees will produce an average assembly time between 19 and 21 minutes with probability about .81.   ■

In Example 7.2, it was assumed that the individual assembly times followed a normal distribution. *However, remember that the strength of the Central Limit Theorem is that this assumption is not necessary for large samples.* We can answer questions 2 and 3 for *any* population whose mean is 20 minutes and standard deviation is 3 minutes, provided we take a *large* sample ($n > 30$). In this case, the normal distribution of $\bar{X}$ may not be exact, but it provides a very good approximation.

EXAMPLE 7.3   The price-earnings (P/E) ratio of a stock is usually considered by analysts who put together financial portfolios. Suppose a population of all P/E ratios has a mean of 10.5 and a standard deviation of 4.5.

1. What is the probability that a sample of 40 stocks will have an average P/E ratio less than nine?
2. What assumptions about the population of all P/E ratios are necessary in your answer to question 1?

SOLUTION 1   By the Central Limit Theorem, $\bar{X}$ is approximately a normal random variable with mean $= \mu = 10.5$ and standard deviation $= \sigma/\sqrt{n} = 4.5/\sqrt{40} = .71$. So

$$Z = \frac{\bar{X} - 10.5}{.71}$$

is approximately a standard normal random variable, and consequently

$$P(\bar{X} < 9) = P\left[\frac{\bar{X} - 10.5}{.71} < \frac{9 - 10.5}{.71}\right]$$

$$= P(Z < -2.11) = .0174$$

SOLUTION 2   No assumptions regarding the shape of the P/E ratio population are necessary. This population may be normal or it may not be—it simply does not matter because we are using a fairly large sample ($n = 40$). The distribution of $\bar{X}$ is approximately normal, regardless of the shape of the population of all P/E ratios. Our only assumptions in solution 1 were that $\mu = 10.5$ and $\sigma = 4.5$.   ■

Exercises

7.1   Let $\bar{X}$ be the average of a sample of size 18 from a normally distributed population with mean 37 and variance 16. Find the following probabilities.

a. $P(\bar{X} \leq 35)$ c. $P(34 \leq \bar{X} \leq 36.5)$
b. $P(\bar{X} \geq 38)$ d. $P(36 \leq \bar{X} \leq 38)$

7.2   The manager of Homer and Gordon Realty finds that their four realtors have sold 0, 1, 3, and 4 homes, respectively, in the past month.

a. List the number of homes sold by two realtors selected randomly with replacement for all possible samples of size two.

b. Calculate the sample mean for each sample of size two. Construct the probability distribution for the sample mean.

c. Draw a histogram showing the distribution of the sample mean ($\bar{X}$).

7.3   A southwestern bank issues traveler's checks in denominations of $10, $20, $50, $100, and $500. All five amounts have occurred with equal probability.

a. List all possible samples of three from these five denominations. (Denominations may not be repeated.)

b. Calculate the sample mean for each sample of size three.

c. Construct the probability distribution of the sample mean.

d. Draw a histogram showing the distribution of the sample mean ($\bar{X}$).

7.4   Five machines produce electronic components. The number of components produced per hour is normally distributed with a mean of 25 and a standard deviation of 4.

a. What percentage of the time does a machine produce more than 27 components per hour?

b. What percentage of the time is the average rate of output of the five machines more than 27 components per hour?

7.5   A quality engineer knows from past data that in a shipment lot, the number of wire cables that have tensile strengths below the lower specification limit follows a Poisson distribution with a mean of 50. For a sample of 45 shipment lots, what is the probability that the sample mean for the number of wire cables with tensile strengths below the lower specification limit exceeds 53?

7.6   The average length of actual running time (excluding advertisements) for television feature films is 1 hour and 40 minutes, with a standard deviation of 15 minutes. If a sample of 49 TV feature films is taken at random, what is the probability that the average running time for this group is 1 hour and 45 minutes or more?

7.7   Investment experts have pointed out that small stocks have outperformed large stocks during the 12 months following recessions since the early 1950s. This observation gives hope to investors who bought stocks in 1990. The NASDAQ composite index of smaller stocks was down about 18% in 1990. Assume that the standard deviation of the returns of the stocks in the NASDAQ composite index was 20.0%. What is the probability that the mean return of 45 randomly selected stocks from the NASDAQ composite index was larger than −12.0%?

(*Source:* Adapted from "Big Investors Say The Time Is Nearing For The Davids to Outperform Goliaths," *Wall Street Journal* (December 24, 1990): 13.)

In a survey of members of the American Society for Quality Control (ASQC), a response from 1 to 5 (with 5 representing "strongly disagree" and 1 representing "strongly agree") was given to the statement "I think that it is possible to teach industry-relevant skills for the quality assurance function in an academic course." The mean response was 1.62 with a standard deviation of .84. Assume that this mean and standard deviation are representative of the population mean and standard deviation for the entire 7,500 members of the ASQC. Find the probability that in a sample of 50 ASQC members the sample mean response to the above statement is less than 1.75.

(*Source:* "QA Education in the Business School: Practitioner's Views," *Quality Progress* (May 1990): 87.)

Applying the Central Limit Theorem to Normal Populations

The Central Limit Theorem tells us that $\bar{X}$ tends toward a normal distribution as the sample size increases. If you are dealing with a population that has an assumed normal distribution (as in Example 7.2), then $\bar{X}$ is normal regardless of the sample size. However, as the sample size increases, the variability of $\bar{X}$ decreases, as is illustrated in Figure 7.5.

This means that for large sample sizes, if you were to get many samples and corresponding values of $\bar{X}$, these values of $\bar{X}$ would be more concentrated around the middle, with very few extremely large or extremely small values.

Look at Figure 7.5, which illustrates the assumed normal distribution of all Everglo bulbs. We know that (using Table A.4) 95% of a normal curve is contained within 1.96 standard deviations of the mean. For a sample size of $n = 10$ from a normal population with $\mu = 400$ and $\sigma = 50$, $\sigma_{\bar{x}} = 15.81$. Now,

$$\mu_{\bar{X}} - 1.96\sigma_{\bar{x}} = 400 - 1.96(15.81) = 369.0$$

and

$$\mu_{\bar{X}} + 1.96\sigma_{\bar{x}} = 400 + 1.96(15.81) = 431.0$$

Thus, if we repeatedly obtain samples of size ten, 95% of the resulting $\bar{X}$ values will lie between 369.0 and 431.0.

This result and the corresponding results using $n = 20$, 50, and 100 are contained in Table 7.2, which reemphasizes that for larger samples, you are much more likely to get a value of $\bar{X}$ that is close to $\mu = 400$. In practice, you typically do not know the value of μ. However, by using a larger sample size, you are more apt to obtain an $\bar{X}$ that is a good estimate of the unknown μ.

■ TABLE 7.2

Sampling from a normal population with $\mu = 400$ and $\sigma = 50$; 95% of the time, the value of $\bar{X}$ will be between $\mu_{\bar{x}} - 1.96\sigma_{\bar{x}}$ and $\mu_{\bar{x}} + 1.96\sigma_{\bar{x}}$. Refer to Figure 7.5 for the values of $\sigma_{\bar{x}}$.

SAMPLE SIZE	$\sigma_{\bar{x}}$	$\mu_{\bar{x}} - 1.96\sigma_{\bar{x}}$	$\mu_{\bar{x}} + 1.96\sigma_{\bar{x}}$	CONCLUSION
$n = 10$	15.81	369.0	431.0	95% of the time, the value of $\bar{X}$ will be between 369.0 and 431.0
$n = 20$	11.18	378.1	421.9	95% of the time, the value of $\bar{X}$ will be between 378.1 and 421.9
$n = 50$	7.07	386.1	413.9	95% of the time, the value of $\bar{X}$ will be between 386.1 and 413.9
$n = 100$	5	390.2	409.8	95% of the time, the value of $\bar{X}$ will be between 390.2 and 409.8

Applying the Central Limit Theorem to Nonnormal Populations

The real strength of the Central Limit Theorem is that $\bar{X}$ will tend toward a normal random variable regardless of the shape of your population. You need a large sample ($n > 30$) to obtain a nearly normal distribution for $\bar{X}$. The Central Limit Theorem also holds when sampling from a discrete population.

Figures 7.9, 7.10, and 7.11 illustrate the distribution of $\bar{X}$ for three nonnormal populations. Notice that the uniform population (Figure 7.9) is at least symmetric about the mean, so the distribution of the sample mean, $\bar{X}$, tends toward a normal distribution for much smaller sample sizes. The U-shaped distribution (Figure 7.11) is another continuous distribution. It is characterized by many small and large values, with few values in the middle. This distribution is symmetric about the mean, but its shape is opposite to that of a normal distribution. Here, $\bar{X}$ requires a large sample ($n > 30$) to attain a normal distribution.

Sampling from a Finite Population

In the previous discussion, we assumed that the population was large enough that the sample was extremely small by comparison. We will now consider whether our results, including the Central Limit Theorem, apply when the exact size of the population is known and the sample is a large portion of the population.

Sampling with Replacement

When you return each element of the sample to the population before taking the next sample element, you are sampling with replacement. This sampling procedure is not common; people generally obtain their sample all at once, making it impossible to sample with replacement. When sampling with replacement, it is possible to obtain the same element more than once. For example, the same person could be chosen all three times in a sample of size $n = 3$.

When sampling with replacement, the Central Limit Theorem applies exactly as before, without any adjustments necessary.

■ FIGURE 7.9
Distribution of $\bar{X}$ for a uniform population.

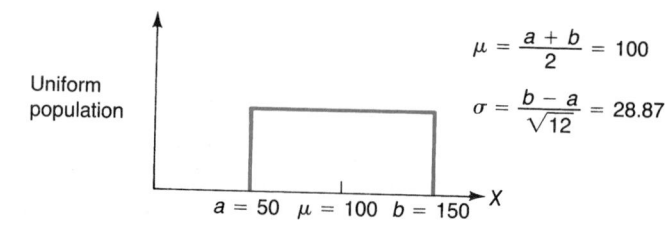

$$\mu = \frac{a + b}{2} = 100$$

$$\sigma = \frac{b - a}{\sqrt{12}} = 28.87$$

$$a = 50 \quad \mu = 100 \quad b = 150$$

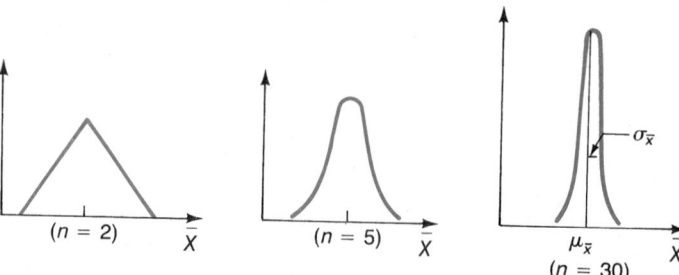

$(n = 2)$ $(n = 5)$ $(n = 30)$

By the CLT, $\mu_{\bar{x}} = \mu = 100$

$$\sigma_{\bar{x}} = \frac{\sigma}{\sqrt{n}} = \frac{28.87}{\sqrt{30}}$$

$$= 5.27$$

■ **FIGURE 7.10**
Distribution of $\bar{X}$ for an
exponential population.

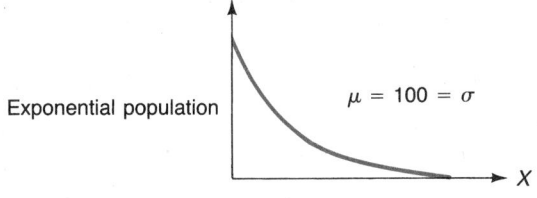

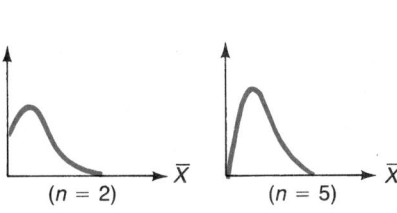

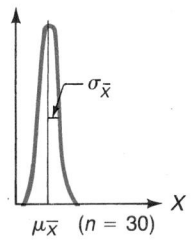

■ **FIGURE 7.10**
Distribution of $\bar{X}$ for an
exponential population.

By the CLT, $\mu_{\bar{X}} = \mu = 100$

$$\sigma_{\bar{X}} = \frac{\sigma}{\sqrt{n}} = \frac{100}{\sqrt{30}}$$

$$= 18.26$$

■ **FIGURE 7.11**
Distribution of $\bar{X}$ for a
U-shaped population.

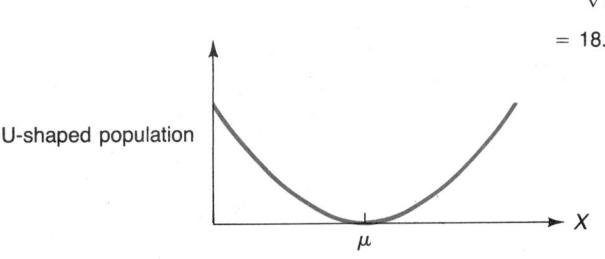

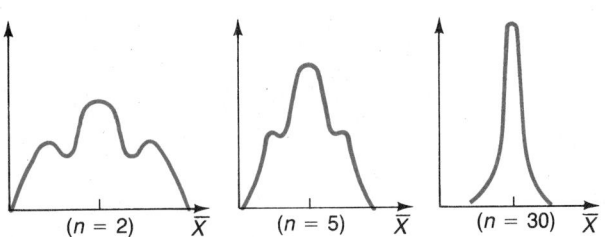

Central Limit Theorem: Sampling with Replacement from a Finite Population

When sampling with replacement from a finite population with mean μ and standard deviation σ, the sample mean $\bar{X}$ tends toward a normal distribution with

$$\text{mean} = \mu_{\bar{x}} = \mu \qquad (7.2)$$

$$\text{standard deviation (standard error)} = \sigma_{\bar{x}} = \frac{\sigma}{\sqrt{n}}$$

where n = sample size.

Sampling without Replacement

We first encountered the problem of sampling without replacement from a finite population in Chapter 5, where the hypergeometric distribution considered the population size (N) and the binomial distribution did not. It is easy to show that, for this situation,

$$\begin{bmatrix} \text{variance of hypergeometric} \\ \text{random variable} \end{bmatrix} = \begin{bmatrix} \text{variance of corresponding} \\ \text{binomial random variable} \end{bmatrix} \cdot \begin{bmatrix} \dfrac{N - n}{N - 1} \end{bmatrix}$$

because

$$\frac{k(N - k)n(N - n)}{N^2(N - 1)} = n\frac{k}{N}\left[1 - \frac{k}{N}\right] \cdot \left[\frac{N - n}{N - 1}\right]$$

$$= np(1 - p) \cdot \left[\frac{N - n}{N - 1}\right]$$

where $p = k/N$. Here, $(N - n)/(N - 1)$ is called the **finite population correction (fpc) factor.** When the sample size, n, is very small compared with the population size, N, the fpc factor is nearly 1 and can be ignored. In fact, as discussed in Chapter 5, the binomial distribution serves as a good approximation to the hypergeometric whenever $n/N < .05$. The same result applies to sampling situations as well. We can express this as a rule: The fpc can be ignored whenever $n/N < .05$.

We can also use the Central Limit Theorem in this situation.

Central Limit Theorem: Sampling Without Replacement from a Large Finite Population

When sampling without replacement from a large, finite population (of size N), with mean μ and standard deviation σ, the sample mean $\bar{X}$ tends toward a normal distribution with

$$\text{mean} = \mu_{\bar{x}} = \mu$$

$$\text{standard deviation (standard error)} = \sigma_{\bar{x}} = \frac{\sigma}{\sqrt{n}} \cdot \sqrt{\frac{N - n}{N - 1}} \quad \textbf{(7.3)}$$

where n = sample size.

The fpc recognizes that our estimate is better for a finite population than for an infinite population and shrinks the size of the standard error. In fact, when the sample size (n) and the population size (N) are the same, the fpc is 0, making the standard error equal to 0. This value is correct, since repeated samples of size $n = N$ would produce the same sample, namely the entire population, and consequently, the same value of $\bar{X}$. Since there is no variation in the $\bar{X}$ values, the standard deviation of $\bar{X}$ (standard error) *should* be zero. *As a final note, remember that although the effect of the fpc is negligible when $n/N < .05$, it can (and some will argue "should") be used to derive a more accurate standard error.*

EXAMPLE 7.4

A group of women managers at Compumart are considering filing a sex-discrimination suit. A recent report stated that the average annual income of all employees in middle management positions at Compumart is $48,000 and the standard deviation is $8500. A random sample of 45 women taken from a population of 350 female middle managers at Compumart had an average income of $\bar{x} =$

■ **FIGURE 7.12**
Distribution of sample
mean of annual salaries
(assuming $\mu =$
$48,000, $\sigma =$ $8500).
The shaded area
represents the solution
to Example 7.4,
$P(\bar{X} \leq 43,900)$.

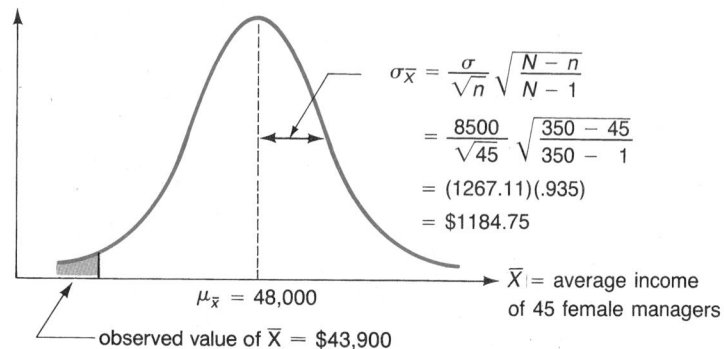

$$\sigma_{\bar{X}} = \frac{\sigma}{\sqrt{n}} \sqrt{\frac{N-n}{N-1}}$$

$$= \frac{8500}{\sqrt{45}} \sqrt{\frac{350-45}{350-1}}$$

$$= (1267.11)(.935)$$

$$= \$1184.75$$

$\mu_{\bar{x}} = 48,000$

$\bar{X}$ = average income of 45 female managers

observed value of $\bar{X} = \$43,900$

$43,900. If the population of all female incomes at this level is assumed to have the same mean ($48,000) and standard deviation ($8500) as the distribution of incomes for all employees, what is the probability of observing a value of $\bar{X}$ this low?

SOLUTION Because we have a large sample, we can assume (using the Central Limit Theorem) that the curve describing $\bar{X}$ is normal, as shown in Figure 7.12. Here, $n = 45$ and $N = 350$. We need to find $P(\bar{X} \leq 43,900)$. Standardizing and using Table A.4, we find that

$$P(\bar{X} \leq 43,900) = P\left[Z \leq \frac{43,900 - 48,000}{1184.75}\right]$$

$$= P(Z \leq -3.46) = .0003$$

So, if the female population has an average salary of $48,000 (and standard deviation of $8500), then the chance of obtaining an $\bar{X}$ as low as $43,900 is extremely small. If we assume that the standard deviation is correct, then, based strictly on this set of data, our conclusion would be that the average salary for women at this level is not $48,000 but is less than $48,000.   ■

With the type of question asked in Example 7.4, there is always the chance that we will reach an incorrect decision using the sample data; there is always the chance of error due to sampling. This possible error will be a concern whenever you test a hypothesis. For now, remember that when dealing with sample data, statistics never *prove* anything. They do, however, *support* or *fail to support* a claim (such as $\mu < \$48,000$).

🖊 *Exercises*   **7.9**   From a finite population of size 300 that is approximately normally distributed with a mean of 50 and a standard deviation of 10, what is the probability that a random sample of size 30 without replacement will yield a sample mean larger than 55? What is the probability that a random sample of size 30 with replacement will yield a sample mean larger than 55?

7.10   Advanced Machinery manufactured 1584 diagnostic machines. The machines can pinpoint electrical problems in a certain type of machinery in 5 minutes on the average with a standard deviation of 2 minutes. If a random sample of 300 diagnostic machines are selected without replacement, what is the probability that the sample mean of the time it takes to pinpoint the electrical problems is greater than 6 minutes?

7.11   The electric bill for 250 households in a small midwestern town was found to have a mean of $120 with a standard deviation of $25 for the month of November. If 10 households are selected at random from the 250 households, what is the probability that the sample mean will be between $110 and $130? What are you assuming about the population?

7.12   General Appliances has 70 microwave ovens that need repair. The mean cost of repair for the 70 microwaves is $80. The standard deviation of the cost is $35. The cost can be considered to be approximately normally distributed.

STATISTICAL INFERENCE AND SAMPLING

a. If a sample of 10 of the 70 microwaves is selected without replacement, what is the probability that the mean cost for the sample is greater than $100?

b. If a sample of 10 of the 70 microwaves is selected with replacement, what is the probability that the mean cost for the sample is greater than $100?

7.13 The mean daily time spent on the telephone by the 60 personnel managers of Retail Products is 1.25 hours; the standard deviation is .62 hours. Assuming that the time spent on the telephone is approximately normally distributed, what is the probability that the mean daily time spent on the telephone by 10 different personnel managers selected at random is greater than 1.5 hours?

7.14 As the finite population size gets large for a fixed sample size, explain how the finite population correction factor is affected.

7.15 National Distributing employs 500 salespersons in 200 territories throughout the United States. The average yearly commission earned by a salesperson is $47,000 and the standard deviation of the yearly commission is $8540. If a random sample of 60 salespersons is selected, what is the probability that the sample mean of their yearly commission is less than $45,000?

7.16 A chemical company produces 275 barrels of a certain chemical under a federal contract. The mean amount of caustic material in the barrels is 35 fluid ounces, and the standard deviation of the amount of caustic material in the barrels is 10 fluid ounces. Twenty barrels are selected at random for inspection. What is the probability that the sample mean is between 33 and 37 fluid ounces? What are you assuming about the distribution of the amount of caustic material?

7.17 Legislators are allowed certain postal privileges to communicate with their constituency. Suppose that for 435 members of Congress, the mean annual postage use is $1630 with a standard deviation of $170. If a sample of 40 members of Congress is obtained, what is the probability that the average annual postal charge for this group of 40 is $1600 or more? If a sample of 20 members of Congress is taken, what is the probability that the average annual charge for this group of 20 is $1600 or more? What must you assume to answer the latter question?

7.18 The weights of sheets of steel are approximately normally distributed with a mean of 20 pounds and a standard deviation of 1 pound. A shipment of 200 sheets of metal arrive at a manufacturing plant. Would it be considered unusual for a quality inspector at the manufacturing plant to find that a random sample of 20 sheets from the 200 sheets had a mean weight of 19.25 pounds?

Confidence Intervals for the Mean of a Normal Population (σ Known)

Return to the situation where we have obtained a sample from a normal population with unknown mean, μ. We first consider a case in which we know σ, the standard deviation of the normal random variable (Figure 7.13). (The situation where both μ and σ are unknown is dealt with in the next section.)

We know that to estimate μ, the average of the entire population, we obtain a sample from this population and calculate $\bar{X}$, the average of the sample. The sample mean, $\bar{X}$, is the estimate of μ and is also called a **point estimate** because it consists of a single number.

In Example 7.2, it was assumed that the assembly time for a particular electrical component followed a normal distribution, with $\mu = 20$ minutes and $\sigma = 3$ minutes. What if μ is not known for *all* workers? A random sample of 25 workers' assembly times was obtained with the following results (in minutes):

22.8, 29.3, 27.2, 30.2, 24.0, 23.2, 22.9, 30.3, 27.1, 31.2, 27.0, 32.0, 28.6, 24.1, 28.9, 26.8, 26.6, 23.4, 25.1, 26.6, 25.7, 28.1, 31.5, 24.8, 25.2

Based on these data,

$$\text{estimate of } \mu = \text{sample mean, } \bar{X}$$
$$= \frac{22.8 + 29.3 + \cdots + 25.2}{25} = 26.9 \text{ minutes}$$

■ **FIGURE 7.13**
An example where the
standard deviation σ is
known, but the mean μ
is unknown.

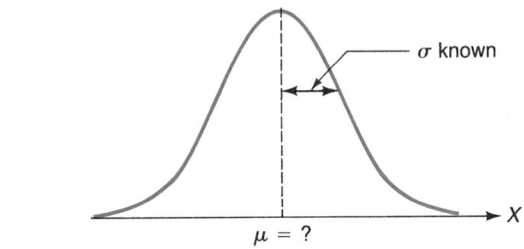

■ **FIGURE 7.14**
Distribution of $\bar{X}$ if
$\mu = 20$ minutes.

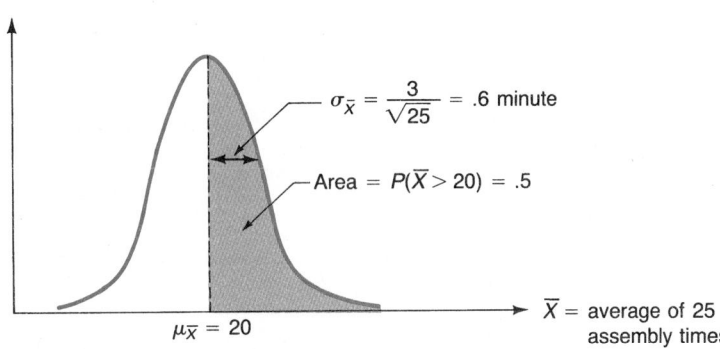

Is this large value of $\bar{X}$ ($= 26.9$) due to random chance? We know that 50% of the samples drawn will have $\bar{X}$ larger than 20, even if $\mu = 20$ (Figure 7.14). Or is this value large because μ is a value larger than 20? In other words, does this value of $\bar{X}$ provide just cause for concluding that μ is larger than 20? We tackle this type of question in Chapter 8.

How accurate is a derived estimate of the population mean, μ? The accuracy depends, for one thing, on the sample size. We can measure the precision of this estimate by constructing a **confidence interval.** By providing the confidence interval, one can make such statements as "I am 95% confident that the average assembly time, μ, is between 25.7 minutes and 28.1 minutes." For this illustration, (25.7, 28.1) is called a 95% confidence interval for μ. The following discussion demonstrates how to construct such a confidence interval.

Using the Central Limit Theorem, we know that $\bar{X}$ is approximately a normal random variable with*

$$\mu_{\bar{X}} = \mu$$

$$\sigma_{\bar{X}} = \frac{\sigma}{\sqrt{n}}$$

where μ and σ represent the mean and standard deviation of the population. To standardize $\bar{X}$, you subtract the mean (μ) of $\bar{X}$ and divide by the standard deviation ($\sigma/\sqrt{n}$) of $\bar{X}$. Consequently,

$$Z = \frac{\bar{X} - \mu}{\sigma/\sqrt{n}}$$

is a standard normal random variable. Consider the following statement and refer to Figure 7.15:

$$P(-1.96 \le Z \le 1.96) = .95$$

---

*This discussion ignores the finite population correction (fpc) factor defined in the previous section. For the case of sampling without replacement, where the population size (N) is known, see the discussion of simple random sampling in Section 7.6.

■ **FIGURE 7.15**
$P(-1.96 \leq Z \leq 1.96)$
$= .95.$

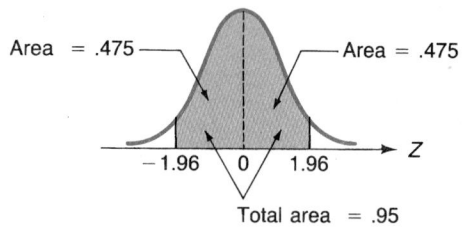

so

$$P\left(-1.96 \leq \frac{\bar{X} - \mu}{\sigma/\sqrt{n}} \leq 1.96\right) = .95$$

After some algebra and rearrangement of terms, we get

$$P\left(\bar{X} - 1.96\frac{\sigma}{\sqrt{n}} \leq \mu \leq \bar{X} + 1.96\frac{\sigma}{\sqrt{n}}\right) = .95$$

How does the last statement apply to a *particular* sample mean, $\bar{x}$? Consider the interval

$$\left(\bar{x} - 1.96\frac{\sigma}{\sqrt{n}}, \ \bar{x} + 1.96\frac{\sigma}{\sqrt{n}}\right) \qquad\qquad (7.4)$$

Using the values from our assembly-time example, we have $\bar{x} = 26.9$, $\sigma = 3$, and $n = 25$. The resulting 95% confidence interval is

$$\left(26.9 - 1.96 \cdot \frac{3}{\sqrt{25}}, \ 26.9 + 1.96 \cdot \frac{3}{\sqrt{25}}\right)$$

or

$$(25.72, 28.08)$$

Since μ is unknown, we do not know whether μ lies between 25.72 and 28.08. However, if you were to obtain random samples repeatedly, calculate $\bar{x}$, and determine the intervals defined by formula 7.4, then 95% of these intervals would contain μ and 5% would not. For this reason, formula 7.4 is called a **95% confidence interval** for μ. Using our assembly-time illustration, we are 95% confident that the average assembly time, μ, lies between 25.72 and 28.08.

Notation Let Z_a denote the value of Z such that the area *to the right* of this value is equal to a. How can we determine $Z_{.025}$, $Z_{.05}$, and $Z_{.1}$ (Figure 7.16)? Using Table A.4, $Z_{.025} = 1.96$, $Z_{.05} = 1.645$, and $Z_{.1} = 1.28$.

When defining a confidence interval for μ, we can define a 99% confidence interval, a 95% confidence interval, a 90% confidence interval, or whatever. The specific percentage represents the **confidence level**. The *higher* the confidence level, the *wider* the confidence interval. The confidence level is written as $(1 - \alpha) \cdot 100\%$, where $\alpha = .01$ for a 99% confidence interval, $\alpha = .05$ for a 95% confidence interval, and so on. Thus, a $(1 - \alpha) \cdot 100\%$ confidence interval for the mean of a normal population, μ, is

$$\left[\bar{x} - Z_{\alpha/2}\left(\frac{\sigma}{\sqrt{n}}\right), \ \bar{x} + Z_{\alpha/2}\left(\frac{\sigma}{\sqrt{n}}\right)\right] \qquad\qquad (7.5)$$

■ FIGURE 7.16
$1.28 = Z_1$, $1.645 = Z_{.05}$, and $1.96 = Z_{.025}$.

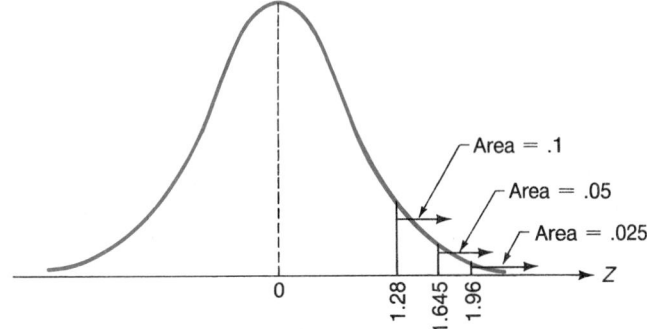

According to the Central Limit Theorem, formula 7.5 provides an approximate confidence interval for the mean of any population, provided the sample size, n, is large (n > 30).

EXAMPLE 7.5   Determine a 90% and a 99% confidence interval for the average assembly time of all workers using the 25 observations given on page 230.

SOLUTION   The sample mean here was $\bar{x} = 26.9$. The population standard deviation is assumed to be 3 minutes. The resulting 90% confidence interval for the population mean μ is

$$26.9 - Z_{.05}\left(\frac{3}{\sqrt{25}}\right) \quad \text{to} \quad 26.9 + Z_{.05}\left(\frac{3}{\sqrt{25}}\right)$$

$$= 26.9 - 1.645\left(\frac{3}{\sqrt{25}}\right) \quad \text{to} \quad 26.9 + 1.645\left(\frac{3}{\sqrt{25}}\right)$$

$$= 26.9 - .99 \quad \text{to} \quad 26.9 + .99$$

$$= 25.91 \text{ minutes} \quad \text{to} \quad 27.89 \text{ minutes}$$

The 99% confidence interval for μ is

$$26.9 - Z_{.005}\left(\frac{3}{\sqrt{25}}\right) \quad \text{to} \quad 26.9 + Z_{.005}\left(\frac{3}{\sqrt{25}}\right)$$

$$= 26.9 - 2.575\left(\frac{3}{\sqrt{25}}\right) \quad \text{to} \quad 26.9 + 2.575\left(\frac{3}{\sqrt{25}}\right)$$

$$= 26.9 - 1.54 \quad \text{to} \quad 26.9 + 1.54$$

$$= 25.36 \text{ minutes} \quad \text{to} \quad 28.44 \text{ minutes}$$

Consequently, we are 90% confident that the mean assembly time for all workers is between 25.91 and 27.89 minutes. We are also 99% confident that this parameter is between 25.36 and 28.44 minutes, based on the results of this sample. Notice that the width of the interval increases as the confidence level increases when using the same sample data.   ■

Discussing a Confidence Interval

The narrower your confidence interval, the better, for the same level of confidence. Suppose Electricalc spent $50,000 investigating the average time necessary to assemble their electrical components. Part of this study included obtaining a confidence interval for the average assembly time, μ. Which statement would they prefer to see?

1. I am 95% confident that the average assembly time is between 2 minutes and 50 minutes.

2. I am 95% confident that the average assembly time is between 25 minutes and 27 minutes.

The information contained in the first statement is practically worthless, and that's $50,000 down the drain. The second statement contains useful information; μ is narrowed down to a much smaller range.

Given the second statement, can you tell what the corresponding value of $\bar{X}$ was that produced this confidence interval? For any confidence interval for μ, $\bar{X}$ (the estimate of μ) is always *in the center*. So $\bar{X}$ must have been 26 minutes.

For the 90% confidence interval in Example 7.5, the following conclusions are valid:

1. I am 90% confident that the average assembly time for the population (μ) lies between 25.91 and 27.89 minutes.
2. If I repeatedly obtained samples of size 25, then 90% of the resulting confidence intervals would contain μ and 10% would not. (Question from the audience: Does this confidence interval [25.91, 27.89] contain μ? Your response: I don't know. All I can say is that this procedure leads to an interval containing μ 90% of the time.)
3. I am 90% confident that my estimate of μ (namely, $\bar{x} = 26.9$) is within .99 minute of the actual value of μ.

Here .99 is equal to $1.645 \cdot (\sigma/\sqrt{n})$. This quantity is referred to as the **maximum error, E.**

$$E = \text{maximum error} = Z_{\alpha/2}\left(\frac{\sigma}{\sqrt{n}}\right) \qquad \textbf{(7.6)}$$

Be careful! The following statement is *not* correct: The probability that μ lies between 25.91 and 27.89 is .90. What is the probability that the number 27 lies in this confidence interval? How about 24? The answer to the first question is 1, and to the second, 0, because 27 lies in the confidence interval and 24 does not. So what is the probability that μ lies in the confidence interval? Remember that μ is a fixed number; we just do not know what its value is. It is *not* a random variable, unlike its estimator, $\bar{X}$. As a result, this probability is either 0 or 1, not .90. Therefore, remember that once you have inserted your sample results into formula 7.5 to obtain your confidence interval, the word *probability* can no longer be used to describe the resulting confidence interval.

EXAMPLE 7.6

Refer to the 20 samples of Everglo bulbs in Table 7.1. Using sample 1, what is the resulting 95% confidence interval for the population mean, μ? Assume that σ is 50 hours.

SOLUTION

Here, $n = 10$ and $\bar{x} = 384.0$. The confidence level is 95%, so $Z_{\alpha/2} = Z_{.025} = 1.96$ (from Table A.4). Therefore, the resulting 95% confidence interval for μ is

$$384.0 - 1.96\left(\frac{50}{\sqrt{10}}\right) \quad \text{to} \quad 384.0 + 1.96\left(\frac{50}{\sqrt{10}}\right)$$

$$= 384.0 - 31.0 \quad \text{to} \quad 384.0 + 31.0$$

$$= 353.0 \quad \text{to} \quad 415.0$$

So we are 95% confident that μ lies between 353 and 415 hours. Also, we are 95% confident that our estimate of μ ($\bar{x} = 384.0$) is within 31.0 hours of the actual value. A confidence interval constructed using MINITAB is shown in Figure 7.17.

∎

■ FIGURE 7.17
MINITAB solution
to Example 7.6.

```
MTB > SET INTO C1
DATA> 308 419 389 432 362 302 440 430 375 383
MTB > END
MTB > ZINTERVAL USING 95%, SIGMA = 50, DATA IN C1

THE ASSUMED SIGMA =50.0

              N      MEAN    STDEV   SE MEAN   95.0 PERCENT C.I.
C1           10      384.0   49.3      15.8  (   353.0,    415.0)
```

■ FIGURE 7.18
Confidence intervals
constructed using 20
samples in Table 7.1.

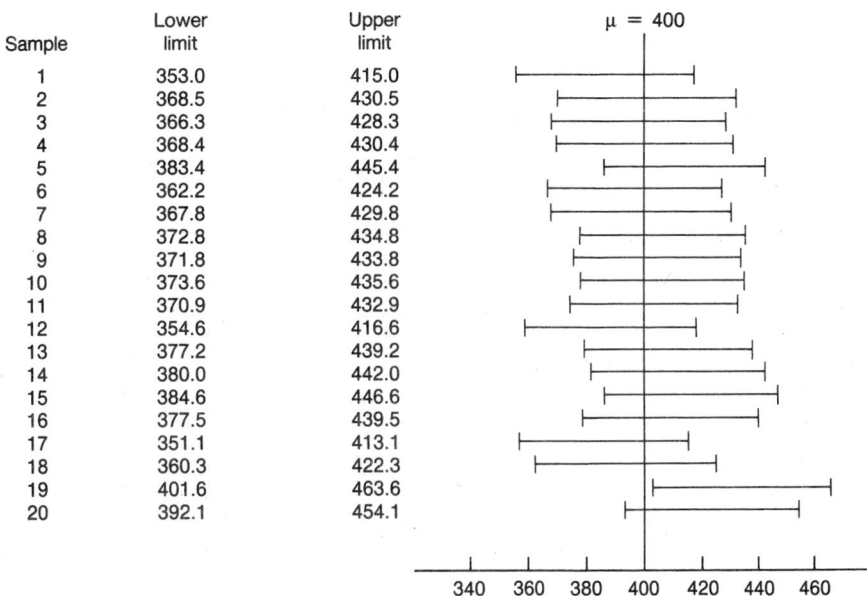

Sample	Lower limit	Upper limit
1	353.0	415.0
2	368.5	430.5
3	366.3	428.3
4	368.4	430.4
5	383.4	445.4
6	362.2	424.2
7	367.8	429.8
8	372.8	434.8
9	371.8	433.8
10	373.6	435.6
11	370.9	432.9
12	354.6	416.6
13	377.2	439.2
14	380.0	442.0
15	384.6	446.6
16	377.5	439.5
17	351.1	413.1
18	360.3	422.3
19	401.6	463.6
20	392.1	454.1

To illustrate the nature of a confidence interval, take a closer look at the 20 samples in Table 7.1. By repeating the procedure in Example 7.6 for the remaining 19 samples, Figure 7.18 can be constructed. If this procedure of obtaining samples of size 10 were repeated indefinitely, we would expect 95% of the resulting confidence intervals to contain the population mean (known to be $\mu = 400$ here), and 5% would not. For these 20 samples, in fact, we observe that all but 1 (sample 19) resulted in a confidence interval containing $\mu = 400$.

 *Exercises*

7.19 A random sample of 125 observations is obtained from a normally distributed population with a standard deviation of 5. Given that the sample mean is 20.6, construct a 90% confidence interval for the mean of the population.

7.20 The following data are the values of a random sample from a normally distributed population. Assume that the population variance is 10.2. Construct a 99% confidence interval for the mean of the population.

50.6, 52.3, 48.6, 45.3, 51.8, 50.8, 46.7, 56.1, 47.7, 49.3, 44.9, 57.0, 50.7, 42.6, 49.8, 46.1, 48.7, 51.8, 54.3, 48.4, 50.5

7.21 A random sample of size 60 from a normally distributed population yields a mean of 100. The standard deviation of the population is 30. Construct a 95% confidence interval for the mean of the population.

7.22 The monthly advertising expenditure of Discount Hardware Store is normally distributed with a standard deviation of $100. If a sample of 10 randomly selected months yields a mean advertising expenditure of $380 monthly, what is a 90% confidence interval for the mean of the store's monthly advertising expenditure?

7.23 In analyzing the operating cost for a huge fleet of delivery trucks, a manager takes a sample of 25 cars and calculates the sample mean and sample standard deviation. Then he finds a 95% confidence interval for the mean cost to be $253 to $320. He reasons that this interval contains the mean operating cost for the fleet of delivery trucks since the sample mean is contained in the interval. Do you agree? How would you interpret this confidence interval?

7.24 The perfectionist owner of Kwik Kar Kare has reduced an oil-change job to a science and wants to keep it that way. The owner constantly monitors the performance of the staff. This week, 15 oil-change jobs were sampled with a sample mean of 9.8 minutes per job. Experience has shown that the times follow a normal distribution, and the standard deviation of the population is known to be 1.2 minutes. Based on this week's sample, construct a 90% confidence interval for the population mean (average time for an oil-change job).

7.25 A manufacturer of ten-speed racing bicycles believes that the average weight of the bicycle is normally distributed with a mean of 22 pounds and a standard deviation of 1.5 pounds. A random sample of 30 bicycles is selected. If the mean from this sample is 22.8, what is a 96% confidence interval for the mean weight of the bicycle?

7.26 A quality-control engineer is concerned about the breaking strength of a metal wire manufactured to stringent specifications. A sample of size 25 is randomly obtained, and the breaking strengths are recorded. The breaking strength of the wire is considered to be normally distributed with a standard deviation of 3. Find a 95% confidence interval for the mean breaking strength of the wire.

   26, 27, 18, 23, 24, 20, 21, 24, 19, 27, 25, 20, 24, 21, 26, 19, 21, 20, 25, 20, 23, 25, 21, 20, 21

7.27 An investor would like to bid on a tract of forest land and then clear the land selectively to market the timber. To arrive at an estimate of the total weight of the lumber, a random sample of 50 trees is selected and their diameters are measured. The sample yields a mean diameter of 13.2 inches. Find a 90% confidence interval for the mean diameter of the trees if the diameter of the trees on the tract of land is considered to be normally distributed with a variance of 4.3 inches squared.

7.28 As the sample size increases, would a confidence interval given by equation 7.5 get smaller or larger? For a given random sample, would the confidence interval given by equation 7.5 for a 90% confidence interval be larger or smaller than that for an 80% confidence interval?

7.29 A medical researcher would like to obtain a 99% confidence interval for the mean length of time that a particular sedative is effective. Thirty subjects are randomly selected. The mean length of time that the sedative was effective is found to be 8.3 hours for the sample. Find the 99% confidence interval, assuming that the length of time that the sedative is effective is considered to be approximately normally distributed with a standard deviation of .93 hours.

7.30 A safety council is interested in the age at which a person first obtains his or her driver's license. If this age is considered to be normally distributed with a standard deviation of 2.5 years, what is an 80% confidence interval for the mean age, given that a random sample of 20 new drivers yields a mean age of 19.3 years?

7.4 Confidence Intervals for the Mean of a Normal Population (σ Unknown)

If σ is unknown, it is impossible to determine a confidence interval for μ using formula 7.5 because we are unable to evaluate the standard error $\sigma/\sqrt{n}$. Let us take another look at how we estimate the parameters of a normal population.

When a population mean is unknown, we can estimate it using the sample mean. The logical thing to do if σ is unknown is to replace it by its estimate, the standard deviation of the sample, s. But consider what happens when

$$\frac{\bar{X} - \mu}{\sigma/\sqrt{n}}$$

is replaced by

$$\frac{\bar{X} - \mu}{s/\sqrt{n}}$$

This is no longer a standard normal random variable, Z. However, it does follow another identifiable distribution, the **t distribution.** Its complete name is *Student's t distribution,* named after W. S. Gosset, a statistician in a Guinness brewery who used the pen name Student. The distribution of

$$\frac{\bar{X} - \mu}{s/\sqrt{n}}$$

will follow a t distribution, *provided* the population from which you are obtaining the sample is normally distributed.

The t distribution is similar in appearance to the standard normal (Z) distribution in that it is symmetric about zero. Unlike the Z distribution, however, its shape depends on the sample size, n. Consequently, when you use the t distribution, you must take into account the sample size. This is accomplished by using **degrees of freedom.** For this application using the t distribution,

$$\text{degrees of freedom} = \text{df} = n - 1$$

The value of $\text{df} = n - 1$ can be explained by observing that for a given value of $\bar{X}$, only $n - 1$ of the sample values are free to vary. For example, in a sample of size $n = 3$, if $\bar{x} = 5.0$, $x_1 = 2$, and $x_2 = 7$, then x_3 must be 6 because this is the only value providing a sample mean equal to 5.0.

Two t distributions are illustrated in Figure 7.19. Notice that the t distributions are symmetrically distributed about zero but have wider tails than does the standard normal, Z. Observe that as n increases, the t distribution tends toward the standard normal, Z. In fact, for $n > 30$, there is little difference between these two distributions. Areas under a t curve are provided in Table A.5 for various df. So, for large samples ($n > 30$), it does not matter whether σ is known (Z distribution, Table A.4) or σ is unknown (t distribution, Table A.5) because the t and Z curves are practically the same. For this reason, the t distribution often is referred to as the **small-sample distribution** for $\bar{X}$. The Z table can be used as an approximation even if σ is unknown, provided n is larger than 30. *Remember, however, that a more accurate confidence interval is always obtained using the* t *table when the sample standard deviation (s) is used in the construction of this interval.*

Using the t distribution, then, a $(1 - \alpha) \cdot 100\%$ confidence interval for μ is

$$\bar{x} - t_{\alpha/2,n-1}\left(\frac{s}{\sqrt{n}}\right) \quad \text{to} \quad \bar{x} + t_{\alpha/2,n-1}\left(\frac{s}{\sqrt{n}}\right) \qquad (7.7)$$

where $t_{\alpha/2,n-1}$ denotes the t value from Table A.5 using a t curve with $n - 1$ df and a right-tail area of $\alpha/2$.

Do you remember our sample of 25 assembly times that produced a point estimate for μ having a value of $\bar{x} = 26.9$ minutes? This estimate was used in Example 7.5, where it was assumed that the population standard deviation was $\sigma = 3$, in

■ FIGURE 7.19
The t distribution.

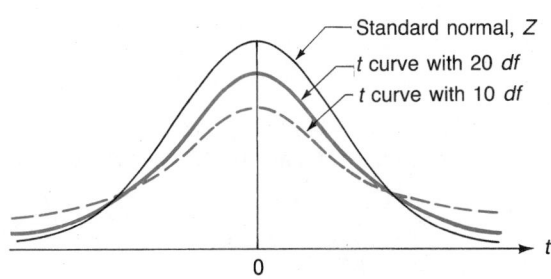

constructing a confidence interval for μ. Furthermore, the assembly times were assumed to follow a *normal* distribution.

Suppose that we do not know σ, either. Then the point estimate of the population standard deviation is

$$s = \sqrt{\frac{(22.8^2 + 29.3^2 + \cdots + 25.2^2) - (22.8 + 29.3 + \cdots + 25.2)^2/25}{24}}$$

$$= \sqrt{\frac{18{,}285.14 - (672.6)^2/25}{24}}$$

$$= \sqrt{7.896} = 2.81 \text{ minutes}$$

Using Table A.5 to find a 90% confidence interval for μ, you first determine that

$$t_{\alpha/2, n-1} = t_{.05, 24} = 1.711$$

The resulting 90% confidence interval is

$$26.9 - 1.711 \left(\frac{2.81}{\sqrt{25}}\right) \quad \text{to} \quad 26.9 + 1.711 \left(\frac{2.81}{\sqrt{25}}\right)$$

$$= 26.9 - .96 \quad \text{to} \quad 26.9 + .96$$

$$= 25.94 \quad \text{to} \quad 27.86$$

Using these data, we are 90% confident that the estimate for the mean of this normal population ($\bar{x} = 26.9$) is within .96 minute of the actual value. Comparing this result with Example 7.5, we notice little difference in the two 90% confidence intervals. Their agreement is due mostly to the fact that the estimate of σ ($s = 2.81$) is very close to the assumed value of $\sigma = 3$.

EXAMPLE 7.7

The output voltage of power supplies manufactured by Clark Products is believed to follow a normal distribution. Of primary concern to the company is the average output voltage of a particular power supply unit, believed to be 10 volts. Eighteen observations taken at random from this unit are shown below:

10.85, 11.40, 10.81, 10.24, 10.23, 9.49, 9.89, 10.11, 10.57,
11.21, 10.10, 11.22, 10.31, 11.24, 9.51, 10.52, 9.92, 8.33

What is the 95% confidence interval for the average output voltage for this power supply unit?

SOLUTION

Your point estimate of σ is $s = .767$ volt. Also, your point estimate of μ is $\bar{x} = 10.331$ volts. A 95% confidence interval for the average output voltage (μ) is

$$10.331 - t_{.025, 17} \left(\frac{.767}{\sqrt{18}}\right) \quad \text{to} \quad 10.331 + t_{.025, 17} \left(\frac{.767}{\sqrt{18}}\right)$$

$$= 10.331 - 2.11 \left(\frac{.767}{\sqrt{18}}\right) \quad \text{to} \quad 10.331 + 2.11 \left(\frac{.767}{\sqrt{18}}\right)$$

$$= 10.331 - .381 \quad \text{to} \quad 10.331 + .381$$

$$= 9.950 \quad \text{to} \quad 10.712$$

We are 95% confident that the average output voltage of this power supply unit is between 9.950 and 10.712 volts. Notice here that the maximum error is

$$E = 2.11 \left(\frac{.767}{\sqrt{18}}\right) = .381 \text{ volt}$$

■ FIGURE 7.20
MINITAB solution
to Example 7.7.

```
MTB > SET INTO C1

DATA> 10.85 11.40 10.81 10.24 10.23 9.49 9.89 10.11 10.57
DATA> 11.21 10.10 11.22 10.31 11.24 9.51 10.52 9.92 8.33
DATA> END
MTB > TINTERVAL WITH 95% CONFIDENCE USING C1

                 N       MEAN    STDEV   SE MEAN    95.0 PERCENT C.I.
     C1          18     10.331   0.7667   0.181    ( 9.949, 10.712)
```

which implies that we are 95% confident that $\bar{X}$ is within .381 volt of the actual average voltage. The MINITAB solution for this example is shown in Figure 7.20.

■

Exercises

7.31 Find the t values for the following α levels and degrees of freedom.

a. $t_{.10,29}$ c. $t_{.05,18}$ e. $t_{.95,25}$
b. $t_{.025,13}$ d. $t_{.90,20}$ f. $t_{.10,40}$

7.32 For a sample of size 21 used to test a hypothesis about a sample mean, what is the t value from a t distribution such that the following are true?

a. Ninety percent of the area under the t distribution is to the right of the t value.

b. Ten percent of the area under the t distribution is to the right of the t value.

c. Five percent of the area under the t distribution is to the left of the t value.

7.33 A random sample of size 15 is selected from a normally distributed population. The sample mean is 30 and the sample variance is 16. Find a 95% confidence interval for the population mean.

7.34 The total value per month of back orders at Harlington Industries is considered to be normally distributed. A random sample of back orders for 12 randomly selected months yields a mean of $115,320 with a standard deviation of $35,000. Construct a 90% confidence interval for the mean value of back orders per month at Harlington Industries.

7.35 The mean monthly expenditure on gasoline per household in Middletown is determined by selecting a random sample of 36 households. The sample mean is $68, with a sample standard deviation of $17.

a. What is a 95% confidence interval for the mean monthly expenditure on gasoline per household in Middletown?

b. What is a 90% confidence interval for the mean monthly expenditure on gasoline per household in Middletown?

7.36 Second Federal Savings and Loan would like to estimate the mean number of years in which 30-year mortgages are paid off. Eighteen paid-off 30-year mortgages are randomly selected and the numbers of years in which the loans were paid in full are

19.6, 20.8, 29.6, 6.3, 3.1, 10.6, 30.0, 21.7, 10.5, 26.3, 10.7, 6.1, 7.3, 12.6, 9.8, 27.4, 20.1, 10.8

Assuming that the number of years in which 30-year mortgages are paid off is normally distributed, construct an 80% confidence interval for the mean number of years in which the mortgages are paid off.

7.37 An apartment-finder service would like to estimate the average cost of a one-bedroom apartment in Kansas City. A random sample of 41 apartment complexes yielded a mean of $310 with a standard deviation of $29. Construct a 90% confidence interval for the mean cost of one-bedroom apartments in Kansas City.

7.38 The degree to which a product can be visualized and provide a clear, concrete image prior to purchase is a working definition of the term *tangibility*. This term is used by marketing strategists in efforts to market new products. The results of a survey on the tangibility score that people give to life insurance, a used car, and carpet are given below, with a low score indicating high tangibility.

	MEAN	STANDARD DEVIATION
Life Insurance	4.2	1.2
Used Car	3.5	1.0
Carpet	2.3	.7

(*Source:* G. McDougall and D. Snetsinger, "The Intangibility of Services: Measurements and Competitive Perspectives," *The Journal of Services Marketing* 4, no. 4 (Fall 1990): 27.)

Construct the following confidence intervals using the preceding data and assuming that the data are approximately normally distributed:

a. The 90% confidence interval for the mean tangibility score of life insurance. Assume a sample of size 28.

b. The 95% confidence interval for the mean tangibility score of a used car. Assume a sample of size 28.

c. The 99% confidence interval for the mean tangibility score of a carpet. Assume a sample of size 28.

d. Work part a assuming a sample of size 55 and compare the confidence intervals.

7.39 An investment advisor believes that the return on interest-sensitive stocks is approximately normally distributed. A sample of 24 interest-sensitive stocks was selected and their yearly return (including dividends and capital appreciation) was as follows (in percentages):

11.1, 12.5, 13.6, 9.1, 8.7, 10.6, 12.5, 15.6, 13.8, 8.0, 10.9, 7.6, 5.2, 1.2, 12.8, 16.7, 13.9, 10.1, 9.6, 10.8, 11.6, 12.3, 12.9, 11.6

Find a 90% confidence interval for the mean yearly return on interest-sensitive stocks.

7.40 The president of Secure Savings and Loan Association would like to estimate the average salaries of vice presidents of savings and loan associations. After selecting a random sample of 41 vice presidents, the following statistics on annual salaries were calculated:

$$\bar{x} = 52{,}100$$
$$s = 10{,}350$$

Construct a 95% confidence interval for the mean annual salaries of vice presidents.

7.41 A quality-control engineer conducted a test of the tensile strength of 20 aluminum wires. The coded data represent the tensile strength of 20 aluminum wires selected at random. Find a 90% confidence interval for the mean value of the coded data. What are you assuming about the distribution of the tensile strengths?

105, 113, 95, 90, 112, 93, 106, 80, 95, 90, 88, 101, 93, 91, 86, 107, 103, 93, 84, 87

7.42 MAXIX has been selling automatic toll-collection systems in the United States at various prices depending on the competition. It is believed that the price at which the systems are sold can be approximated by a normal distribution. A sample of 15 systems was sold at the following values (in dollars):

16,500, 13,200, 14,560, 12,320, 13,640, 12,980, 13,350, 12,130, 11,980, 13,590, 15,670, 16,350, 11,860, 13,400, 13,860

Find a 90% confidence interval for the mean price at which the automatic toll-collection systems are sold.

7.5 Selecting the Necessary Sample Size

Sample Size for Known σ

How large a sample do you need? This is often difficult to determine, although a carefully chosen *large* sample generally provides a better representation of the population than does a smaller sample. Acquiring large samples can be costly and time-consuming; why obtain a sample of size $n = 1000$ if a sample size of $n =$

500 will provide sufficient accuracy for estimating a population mean? This section will show you how to determine what sample size is necessary when the maximum error, E, is specified in advance.

In Example 7.6, we assumed that the lifetime of Everglo bulbs is normally distributed with standard deviation $\sigma = 50$ hours but unknown mean μ. Based on the results of sample 1 from Table 7.1, we concluded that we were 95% confident that the estimate of $\mu(\bar{x} = 384.0)$ was within 31.0 hours of the actual value of μ for $n = 10$. How large a sample is necessary if we want our point estimator $(\bar{x})$ to be within 10 hours of the actual value of μ, with 95% confidence? The value 10 here is the maximum error, E, defined in equation 7.6. We would like the estimate of μ (that is, $\bar{x}$) to be within 10 of the actual value, so

$$E = 10 = Z_{\alpha/2} \left(\frac{50}{\sqrt{n}} \right)$$

Because the confidence level is 95%, $Z_{\alpha/2} = Z_{.025} = 1.96$. Consequently,

$$10 = (1.96) \frac{50}{\sqrt{n}}$$

$$\sqrt{n} = \frac{(1.96)(50)}{10} = 9.8$$

Squaring both sides of this statement produces

$$n = (9.8)^2 = 96.04$$

To be a bit conservative, this number should always be rounded *up*. So, a sample size of $n = 97$ will produce a confidence interval with $E \leq 10$ hours. Your point estimate of μ, $\bar{x}$, will then be within 10 hours of the actual value, with 95% confidence.

This sequence of steps can be summarized by the following expression:

$$n = \left[\frac{Z_{\alpha/2} \cdot \sigma}{E} \right]^2 \qquad \textbf{(7.8)}$$

Sample Size for Unknown σ

Equation 7.8 works if σ is known but does not apply to situations where both μ and σ are unknown. There are two approaches to the latter situation.

A Preliminary Sampling

If you have already obtained a small sample, you have an estimate of σ, namely, the sample standard deviation, s. Replacing σ by s in equation 7.8 gives you the desired sample size, n. Assuming that the resulting value of n is greater than 30, the $Z_{\alpha/2}$ notation in equation 7.8 is still valid because the actual t distribution here will be closely approximated by the standard normal distribution.

When you do obtain the confidence interval using the larger sample, the resulting maximum error, E, may not be exactly what you originally specified because the new sample standard deviation will not be the same as that belonging to the smaller original sample.

EXAMPLE 7.8 In Example 7.7, Clark Products obtained 18 observations of the output voltages for a particular power supply unit. For these data,

$$\bar{x} = 10.331 \text{ volts}$$
$$s = .767 \text{ volt}$$

How large a sample would they need for $\bar{X}$ to be within .2 volt of the actual average output voltage with 95% confidence?

SOLUTION   Based on the results of the original sample, $s = .767$, so

$$n = \left[\frac{Z_{\alpha/2} \cdot s}{E}\right]^2 = \left[\frac{(1.96)(.767)}{.2}\right]^2 = 56.5$$

We round this number up to 57. Consequently, they would need 57 observations to make a statement with this much precision, that is, within .2 volt. Of course, they already have 18 observations that can be included in the larger sample.

Obtaining a Rough Approximation of σ

We know from the empirical rule and Table A.4 that 95.4% of the population will lie between $\mu - 2\sigma$ and $\mu + 2\sigma$. Because $(\mu + 2\sigma) - (\mu - 2\sigma) = 4\sigma$, this is a span of four standard deviations. One method of obtaining an estimate of σ is to ask a person who is familiar with the data to be collected these questions:

1. What do you think will be the highest value in the sample (H)?
2. What will be the lowest value (L)?

The approximation of σ is then obtained by assuming that $\mu + 2\sigma = H$ and $\mu - 2\sigma = L$, so

$$H - L = (\mu + 2\sigma) - (\mu - 2\sigma) = 4\sigma$$

Consequently,

$$\sigma \cong \frac{H - L}{4} \qquad\qquad (7.9)$$

We can use this estimate of σ in equation 7.8 to determine the necessary sample size, n.

EXAMPLE 7.9   The manager of quality assurance for a division that produces hair dryers is interested in the average number of switches that can be tested by the division's employees. Assuming that the number of switches that are tested each hour by an employee follows a normal distribution (centered at μ), the manager wants to estimate μ with 90% confidence. Also, this estimate must be within one unit (switch) of μ. The manager estimates that H is 45 switches and L is 25 switches. How large a sample will be necessary?

SOLUTION   Based on $H = 45$ switches and $L = 25$ switches,

$$\sigma \cong \frac{45 - 25}{4} = 5 \text{ switches}$$

The sample size necessary to obtain a maximum error of $E = 1$ is

$$n = \left[\frac{(1.645)(5)}{1}\right]^2 = 67.7$$

Thus, a sample size of 68 should produce a value of E close to one switch. The value will not be exactly one because the sample standard deviation, s, probably will not be exactly 5. Estimating σ in this manner, however, produces a value that is "in the neighborhood" of σ.   ■

Exercises

7.43 A 99% confidence interval is to be constructed such that $\bar{X}$ is within 1.5 units of the mean of a normal population. Assuming that the population variance is 30, what sample size would be necessary to achieve this maximum error?

7.44 To be 95% confident that $\bar{X}$ is within .65 of the actual mean of a normal population with a standard deviation of 2.5, what sample size would be necessary?

7.45 The Chamber of Commerce of Tampa, Florida, would like to estimate the mean amount of money spent by a tourist to within $100 with 95% confidence. If the amount of money spent by tourists is considered to be normally distributed with a standard deviation of $200, what sample size would be necessary for the Chamber of Commerce to meet their objective in estimating this mean amount?

7.46 Security Savings and Loan Association's manager would like to estimate the mean deposit by a customer into a savings account to within $500. If the deposits into savings accounts are considered to be normally distributed with a standard deviation of $1250, what sample size would be necessary to be 90% confident?

7.47 If a sample size of 70 was necessary to estimate the mean of a normal population to within 1.2 with 90% confidence, what is the approximate value of the standard deviation of the population?

7.48 The marketing agency for computer software of Personal Micro Systems would like to estimate with 95% confidence the mean time that it takes for a beginner to learn to use a standard software package. Past data indicate that the learning time can be approximated by a normal distribution with a standard deviation of 20 minutes. How large a sample size should the marketing agency choose if the mean time to learn to use the software package is to be estimated within 8 minutes with 90% confidence?

7.49 Past data indicate that the distribution of the daily price/earnings ratio of National Health and Medical Services can be approximated by a normal distribution with a variance of 17. How large a sample size would be necessary to estimate the mean price/earnings ratio of National Health and Medical Services to within 2 units with 98% confidence?

7.50 A chemist at International Chemical would like to measure the adhesiveness of a new wood glue. From past experiments, a measure used to indicate adhesiveness has ranged from 7.3 to 11.1 units. To be 98% confident, how large a sample would be necessary to estimate the mean adhesiveness to within .5 units?

7.51 An investor would like to obtain an idea of the profitability of a soft-drink vending machine by taking a random sample of several days and recording the daily number of soft drinks sold. A preliminary sample shows that the highest number of drinks sold daily is 106 and the least sold in a day is 36. What sample size would be necessary to estimate the mean number of soft drinks sold to within 6 soft drinks with 99% confidence?

7.52 Federal law requires companies producing wastewater that includes toxic chemicals to treat and dilute these chemicals before the wastewater reaches municipal sewage treatment plants. Violations are punishable by fines up to $2000 per day in Dallas. If one can assume that the most significant violator in the Dallas metropolitan area paid $10,158 in 1987, find the sample size necessary to estimate the mean amount in fines per company to within $800 with 95% confidence.

(*Source:* "Dallas Businesses Dumped Array of Toxic Pollutants," *Dallas Morning News* (March 7, 1988): 6A.)

7.53 An economist would like to estimate the rise in personal income for a particular quarter. A preliminary study shows that the rise is between 1% and 6%. What sample size would be necessary to obtain a 95% confidence interval to estimate the mean rise in personal income for the quarter to within .1%?

◢ 7.6
Other Sampling Procedures (Optional)

To discuss methods of sampling other than simple random sampling, we need to define several terms. These definitions also apply to simple random sampling.

1. **Population.** As before, population refers to the collection of people or objects about which we are trying to learn something. It may be as large as the set of all voting adults in the United States or as small or smaller than the set of all top-level managers in a particular company. In this section, we will assume that we are sampling from a *finite* population.
2. **Sampling unit.** A sampling unit is a collection of elements or an individual element selected from the population. Elements within one sampling unit must not overlap with the elements in other sampling units.
3. **Cluster.** A cluster is a sampling unit that is a group of elements from the population, such as all adults in a particular city block.
4. **Sampling frame.** A sampling frame is a list of population elements from which the sample is to be selected. Ideally, the sampling frame should be identical to the population. In many situations, however, this is impossible, in which case the frame must be *representative* of the population.
5. **Strata.** Strata are nonoverlapping subpopulations. For example, the population of all cigarette smokers can be split into two strata—men and women. You can then use **stratified sampling,** in which your total sample consists of a sample selected from each individual stratum.
6. **Sampling design.** A sampling design specifies the manner in which the sampling units are to be selected for your sample. Examples include simple random sampling, systematic sampling, stratified sampling, and cluster sampling.

Simple Random Sampling

The results obtained when using a simple random sample were presented earlier and are summarized here for the usual case of sampling without replacement, where every sample of n elements (from a population of size N) has an equal chance of being selected.

According to the Central Limit Theorem, for large samples the distribution of the sample mean, $\bar{X}$, is approximately normal, without making any assumptions concerning the shape of the population being sampled. The resulting confidence interval for the population mean, μ, is an *approximate* confidence interval for this parameter. If you assume that the population has a normal distribution with mean μ, the confidence interval is exact.

Simple Random Sampling

- Population mean: μ
- Estimator:

$$\bar{X} = \frac{\Sigma x}{n} \tag{7.10}$$

- Variance of the estimator:

$$\sigma_{\bar{x}}^2 = \frac{\sigma^2}{n} \cdot \frac{N-n}{N-1} \tag{7.11}$$

The fpc is $(N-n)/(N-1)$ and can be ignored if $n/N < .05$.
- Approximate confidence interval: $\bar{X} \pm Z_{\alpha/2}\sigma_{\bar{x}}$

Systematic Sampling

For large populations, obtaining a random sample can be quite cumbersome. Perhaps you have just informed a group of bank tellers that you need a random sample of their customers over the next few days. For them to select people randomly would be nearly impossible. A much easier scheme would be to have them select, say, every tenth customer to be included in the sample. This is systematic sampling.

Other situations where systematic sampling is advantageous include:

1. The population consists of N records on a magnetic tape or disk. The sample of n is obtained by sampling every kth record, where k is an integer approximately equal to N/n. For example, if there are $N = 9435$ records and you need a sample of size $n = 100$, selecting every $9435/100 \cong 94$th record would result in a systematic sample. Typically, a random starting point (record) is determined, and then every kth record is selected for your sample.
2. The population consists of a collection of files stored consecutively by date of birth. A quick (although not necessarily reliable) method of obtaining a "nearly random" sample is to select every kth file for your sample. (What could cause the sample selected from such a list *not* to be random?)

There are many situations in which it is dangerous to use systematic sampling. If there are obvious patterns contained in the sample frame listing, your sample may be far from random. If elements are stored according to days, for example, your sample could consist of data that all belong to Tuesday. If the data are cyclic, your sample might consist of all the peaks or all the valleys of the population. *Basically, systematic sampling works best when the order of your population is fairly random with respect to the measurement of interest.*

Despite its dangers, systematic sampling can provide an easy method of obtaining a representative sample. If the order of your population is in fact random (no cycles, no obvious patterns of any kind), a systematic sample can be analyzed as though it were a simple random sample.

Stratified Sampling

Suppose that you own a chain of four tire stores in four different cities and you are interested in the average amount due on delinquent accounts. The populations of these four cities differ considerably, ranging from a small store in an east Texas town to a large store in downtown Houston. To obtain a random sample, you could combine the delinquent accounts from all four stores into one large population and obtain your random sample from this group of accounts. On the other hand, because of the different sizes, locations, and credit policies of the stores, you might want to sample the stores individually. You could obtain the largest sample from the Houston store and smaller samples from the smaller stores. This is proportional stratified sampling.

Stratified sampling is used when the population can be physically or geographically separated into two or more groups (strata), where the variation within the strata is less than the variation within the entire population. The cost of obtaining the stratified sample may be less than that of collecting a random sample of the same size, especially if the sampling units are determined geographically.

The advantages of stratified sampling are:

1. By stratifying, we can obtain more information from the sample because data are more homogeneous within each stratum; consequently, confidence intervals are narrower than those obtained through random sampling.
2. We do obtain a cross section of the entire population.

3. We do obtain an estimate of the mean within each stratum as well as an estimate of μ for the entire population. We use the following notation:

$$n_i = \text{sample size in stratum } i$$

$$N_i = \text{number of elements in stratum } i$$

$$N = \text{total population size } = \Sigma N_i$$

$$n = \text{total sample size } = \Sigma n_i$$

$$\overline{X}_i = \text{sample mean in stratum } i$$

$$s_i = \text{sample standard deviation in stratum } i$$

Stratified Sampling

- Population mean: μ
- Estimator:

$$\overline{X}_s = \frac{\Sigma N_i \overline{X}_i}{N} \qquad (7.12)$$

- Variance of the estimator:

$$\sigma_{\overline{X}_s}^2 = \frac{\Sigma N_i^2 \left(\dfrac{N_i - n_i}{N_i - 1}\right) \dfrac{s_i^2}{n_i}}{N^2} \qquad (7.13)$$

- Approximate confidence interval: $\overline{X}_s \pm Z_{\alpha/2}\sigma_{\overline{X}_s}$

One method often used to determine the strata sample sizes, n_i, is to select each sample size proportional to stratum size.* Consequently,

$$n_i = n\left(\frac{N_i}{N}\right)$$

In this way, you obtain larger samples from the larger strata.

Because you desire an estimator with *small* variance (that is, one that will not drastically vary from one data set to the next), you should attempt to create strata such that the individual variances, s_i^2, are as small as possible.

Assume you would like to obtain a sample of size 20 from the chain of four tire stores. You want to use a stratified sample with proportional sample sizes because each of the stores has a different volume of customers, credit policy, and credit ceiling. Here, N_i is the number of delinquent accounts at each store; $N_1 = 72$, $N_2 = 39$, $N_3 = 25$, and $N_4 = 44$. So

$$N = 72 + 39 + 25 + 44 = 180 \quad \text{delinquent accounts}$$

Your sample sizes are:

---

*More precision can often be obtained by considering the variation within each stratum and obtaining larger samples from strata with more variation. In particular, set $n_i = n[N_i s_i / (\Sigma N_i s_i)]$ where s_i is the sample standard deviation in the ith stratum. This method of determining the sample sizes is called *Neyman allocation* and reduces to proportional sampling if the strata standard deviations are equal or ignored.

$$n_1 = 20 \left(\frac{72}{180} \right) \qquad n_3 = 20 \left(\frac{25}{180} \right)$$

$$\cong 8 \qquad\qquad\qquad \cong 3$$

$$n_2 = 20 \left(\frac{39}{180} \right) \qquad n_4 = 20 \left(\frac{44}{180} \right)$$

$$\cong 4 \qquad\qquad\qquad \cong 5$$

The randomly selected accounts are analyzed to find the dollar amounts due on delinquent accounts. The sample results are:

	STORE 1	STORE 2	STORE 3	STORE 4
	$150	$ 82	$186	$321
	175	106	162	285
	216	98	174	306
	205	110		356
	182			332
	240			
	195			
	213			
N_i	72	39	25	44
n_i	8	4	3	5
$\bar{X}_i$	$197.00	$ 99.00	$174.00	$320.00
s_i	$ 27.91	$ 12.38	$ 12.00	$ 26.75

To estimate μ from these data,

$$\bar{X}_s = \frac{(72)(197) + (39)(99) + (25)(174) + (44)(320)}{180}$$

$$= \$202.64$$

Also

$$\sigma^2_{\bar{X}_s} = \left[(72)^2 \left(\frac{72 - 8}{72 - 1} \right) \frac{(27.91)^2}{8} + (39)^2 \left(\frac{39 - 4}{39 - 1} \right) \frac{(12.38)^2}{4} \right.$$

$$\left. + (25)^2 \left(\frac{25 - 3}{25 - 1} \right) \frac{(12.00)^2}{3} + (44)^2 \left(\frac{44 - 5}{44 - 1} \right) \frac{(26.75)^2}{5} \right] \div (180)^2$$

$$= 787,475.21 \div 32,400 = 24.305$$

Consequently,

$$\sigma_{\bar{X}_s} = \sqrt{24.305} = \$4.93$$

The corresponding approximate 95% confidence interval for the average overdue amount, μ, is

$$\$202.64 - (1.96)(4.93) \quad \text{to} \quad \$202.64 + (1.96)(4.93) = \$192.98 \quad \text{to} \quad \$212.30$$

So we are 95% confident that the average delinquent amount for the four stores is between $192.98 and $212.30.

Cluster Sampling

We can sample clusters (groups) within the population rather than collecting individual elements one at a time. For example, to determine the opinions of the members of a particular labor union, you might interview everyone attending several of the local meetings. Of course, the danger here is that possibly (1) the people attending the local meetings that were sampled (clusters) do not represent the population of all voting members, and (2) the people attending the local meetings do not provide an adequate representation of the local members. As a general rule, it is advisable to select many small clusters rather than a few large clusters to obtain a more accurate representation of your population.

Cluster sampling is preferred to (and less costly than) random and stratified sampling when:

1. The only sampling frame that can be constructed consists of clusters (for example, all people in a particular household, city block, or zip code).
2. The population is extremely spread out, or it is impossible to obtain data on all the individual members.

When using cluster sampling, you should *randomly* select a set of clusters (once they have been clearly defined) for sampling. You can then include all individuals within each cluster selected for the sample (**single-stage cluster sampling**) or randomly select individuals from the sampled clusters to be included in the sample (**two-stage cluster sampling**).

We use the following notation:

M = total number of clusters in the population
m = number of clusters randomly selected for the sample
n_i = number of elements in sample cluster i
$\bar{n}$ = average cluster size of the sampled clusters ($\bar{n} = \Sigma n_i/m$)
N = total population size (N = total of all M cluster sizes that make up the population)
$\bar{N}$ = average cluster size for the population ($\bar{N} = N/M$)
T_i = total of all observations within cluster i (required for the sampled clusters only)

Cluster Sampling (Single Stage)

- Population mean: μ
- Estimator:

$$\bar{X}_c = \frac{\Sigma T_i}{\Sigma n_i} \tag{7.14}$$

- Variance of estimator:

$$\sigma^2_{\bar{X}_c} = \left(\frac{M-m}{mM\bar{N}^2}\right) \frac{\Sigma(T_i - \bar{X}_c n_i)^2}{m-1} \tag{7.15}$$

If $\bar{N}$ is unknown, it can be replaced by its estimate, $\bar{n}$.
- Approximate confidence interval: $\bar{X}_c \pm Z_{\alpha/2}\sigma_{\bar{X}_c}$

As marketing director for a cable-television company in a large city, you are trying to decide whether to begin a major advertising campaign to reach tenants in local high-rise apartment buildings. Your staff disagree about whether this is a good idea. One group of your employees feels that people living in high-rise apartments are always on the go and are not likely to spend much time watching television—cable or network. The others tend to believe that such tenants have no grass to mow or leaves to rake and so have a great deal of time to spend watching television.

Rather than drawing a sample from all high-rise tenants, you construct a sampling frame consisting of all 18 $(= M)$ high-rise apartment complexes. From these, you randomly select a sample of $m = 4$ complexes (clusters). Each tenant in these four complexes is then asked how many hours per week he or she watches television. You obtain the following results:

	COMPLEX 1	COMPLEX 2	COMPLEX 3	COMPLEX 4
Number of units (n_i)	260	220	310	274
Total number of hours per cluster (complex)	2475	2750	3160	4110

$N =$ the total number of units in the 18 high-rise complexes (population) $= 4590$

$$\bar{N} = \frac{4590}{18} = 255$$

You begin by noting

$$\Sigma (T_i - \bar{X}_c n_i)^2 = \Sigma T_i^2 - 2\bar{X}_c \Sigma T_i n_i + \bar{X}_c^2 \Sigma n_i^2 \qquad (7.16)$$

Using the sample data,

$$\Sigma T_i = 2475 + \cdots + 4110 = 12,495$$
$$\Sigma n_i = 260 + \cdots + 274 = 1064$$
$$\Sigma T_i^2 = (2475)^2 + \cdots + (4110)^2 = 40,565,825$$
$$\Sigma T_i n_i = (2475)(260) + \cdots + (4110)(274) = 3,354,240$$
$$\Sigma n_i^2 = (260)^2 + \cdots + (274)^2 = 287,176$$

As a result,

$$\bar{X}_c = \frac{\Sigma T_i}{\Sigma n_i} = \frac{12,495}{1064} = 11.743$$

Also, using equation 7.16,

$$\frac{\Sigma (T_i - \bar{X}_c n_i)^2}{m - 1} = [40,565,825 - (2)(11.743)(3,354,240)$$
$$+ (11.743)^2 (287,176)] \div 3$$
$$= 463,051.49$$

Consequently,

$$\sigma_{\bar{X}_c}^2 = \frac{18 - 4}{(4)(18)(255)^2} \cdot 463,051.49 = 1.385$$

and so

$$\sigma_{\bar{X}_c} = \sqrt{1.385} = 1.177$$

The resulting approximate 95% confidence interval for the average number of television hours (for all 18 complexes) is

$$11.743 - 1.96(1.177) \quad \text{to} \quad 11.743 + 1.96(1.177)$$
$$= 9.44 \text{ hours} \quad \text{to} \quad 14.05 \text{ hours}$$

Therefore, we are 95% confident that μ lies between 9.44 and 14.05 hours and that we have estimated μ to within 2.3 hours. This example has illustrated single-stage cluster sampling, where everyone in the selected clusters (apartment complexes) was used in the sample. Another look at this example indicates that it might be more practical to use a two-stage cluster procedure, where each of the four sample clusters is also sampled to obtain the final sample.

Exercises

7.54 A real estate agent would like to estimate the average price of a home in the suburbs of a major metropolitan city. The agent decides to use stratified random sampling. The population of homes is stratified into the five major suburbs. The results of the stratified sample yield the following statistics. Construct a 95% confidence interval for the mean price of a home (in units of thousands of dollars).

STRATUM	N_i: NUMBER OF HOUSES	n_i: SAMPLE SIZE	$\bar{X}_i$ (IN THOUSANDS)	s_i^2
Suburb 1	150	22	101.2	64.2
Suburb 2	220	33	80.7	24.3
Suburb 3	140	21	61.4	20.8
Suburb 4	70	11	139.6	53.5
Suburb 5	90	13	76.8	30.1
	670	100		

7.55 An advertising firm would like to estimate the amount of money spent per month on advertising by certain retail stores in an industrial sector of northeastern New Jersey. Three sizes of retail stores were chosen—small, medium, and large. The random sample for each stratum yielded the following values. Construct a 90% confidence interval for the mean monthly advertising expenditure of retail stores.

STRATUM	N_i: NUMBER OF STORES	n_i: SAMPLE SIZE	MONTHLY ADVERTISING EXPENDITURE (IN THOUSANDS)
Large	40	8	2.1 1.6 1.8 1.2 0.7 2.6 0.9 0.8
Medium	112	22	0.5 0.7 0.9 1.1 0.6 1.4 1.7 0.4
			0.8 0.7 0.9 0.7 0.9 1.3 1.1 1.2
			0.4 0.3 0.8 0.6 0.9 1.1
Small	80	15	0.3 0.4 0.3 0.6 0.5 0.4 0.3 0.4
	232	45	0.1 0.3 0.2 0.4 0.5 0.2 0.7

7.56 Basic Microcomputers would like to market its version of the professional computer. To price the professional computer and its peripheral equipment properly, a survey is taken among the lower-middle, upper-middle, and high income groups to find out what a businessperson would be willing to pay. The survey was restricted to a certain city in an industrial area. A stratified sample among these three groups yielded the following statistics. Construct a 90% confidence interval for the mean price that a professional businessperson would be willing to pay, in units of thousands of dollars.

INCOME LEVEL	N_i	n_i	$\bar{X}_i$	s_i^2
Lower-middle	8,641	56	2.3	1.6
Upper-middle	14,683	95	4.6	1.9
High	7,457	49	4.8	1.4
	30,781	200		

7.57 A market-research firm would like to estimate the average number of hours that a householder spends shopping each week. Four neighborhoods were selected from a total of 24 neighborhoods for sampling purposes. Find a 90% confidence interval for the mean number of hours that a householder spends shopping each week from the following data (units are in hours per week):

1	2	3	4
2.3	5.4	1.6	5.6
1.1	4.2	0.9	4.1
4.3	3.6	4.6	2.3
0.5	7.2	5.4	7.3
3.7	8.4	5.6	6.1
4.6	11.8	4.6	4.7
10.1	2.1	7.1	5.8
6.3	1.5	3.2	4.8
7.8	8.1	4.5	5.3
8.4	4.1	3.1	1.9
7.9	3.4	2.6	8.4
10.6			

7.58 In what situation is the use of systematic sampling appropriate? Explain how a systematic sample would be taken from a file of students listed by social security number.

7.59 The administration of Digital Systems would like to obtain an estimate of the amount of time workers spend on physical exercise. Five departments out of 20 in Digital Systems were selected for sampling purposes. Find a 95% confidence interval for the mean time that an employee spends on physical fitness per week given the following data (units are in hours per week):

1	2	3	4	5
1.1	2.3	3.5	7.9	0.1
0.2	0.4	4.6	1.3	0.0
2.3	0.3	1.5	2.5	7.6
4.6	1.0	0.7	5.7	5.1
0.1	4.6	3.6	7.8	4.0
0.0	8.3	9.5	10.3	3.0
2.6	7.1	0.8	0.6	6.5
6.8	0.2			
1.1	2.7			

Summary

This chapter introduced you to **statistical inference,** an extremely important area of statistics. Inference procedures were used to estimate a certain unknown *parameter* (such as the mean, μ, or the standard deviation, σ) of a population by using the corresponding sample *statistic* (such as the sample mean, $\bar{X}$, or the sample standard deviation, s).

The **Central Limit Theorem** states that for large samples, the sample mean $\bar{X}$ always follows an approximate normal distribution. If, in addition, you assume that the population is normally distributed, then $\bar{X}$ will follow an exact normal distribution. The strength of the Central Limit Theorem is that no assumptions need be made concerning the shape of the population, provided the sample is large ($n > 30$). The Central Limit Theorem allows you to make probability statements concerning $\bar{X}$, such as $P(\bar{X} < 150)$. When sampling without replacement from a finite population, the standard deviation of the normal distribution for $\bar{X}$ (the **standard error**) is obtained by including a **finite population correction factor (fpc),** which adjusts the standard error by including the effect of the known population size, N.

The probability distribution of a sample statistic is its **sampling distribution.** Consequently, according to the Central Limit Theorem, the sampling distribution of $\bar{X}$ is approximately normal, regardless of the shape of the sampled population.

The sample mean, $\bar{X}$, provides a **point estimate** of μ because it estimates this parameter using a single number. A **confidence interval** for μ measures the precision of the point estimate. If the population standard deviation σ is known, the standard normal table (Table A.4) is used to derive the confidence interval. If σ is unknown, it can be replaced by its estimate—the sample standard deviation, s. This

■ FIGURE 7.21
The correct table to use
for constructing a
confidence interval for a
population mean.

Note: In all cases, if the
population size (N) is
known, the fpc can be used
to derive a more precise
standard error. This
adjustment is negligible
if $n/N < .05$.

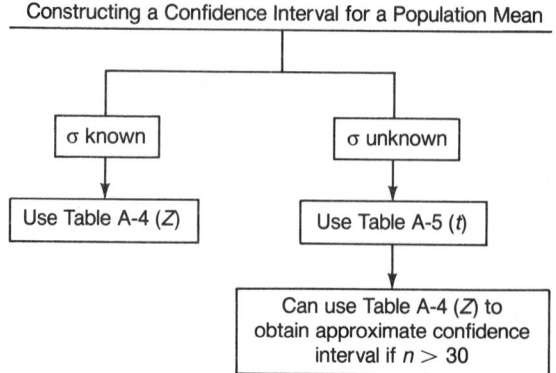

provided an introduction to the **t distribution**. The corresponding confidence interval for μ is constructed using the t table (Table A.5) and assumes that the sampled population is normally distributed (that is, that μ is the mean of a normal population). For sample sizes (n) greater than 30, the standard normal table can be used to construct an approximate confidence interval for the population mean when σ is unknown. A summary of this procedure is contained in Figure 7.21.

For many applications, the precision of the point estimate, $\bar{X}$, is specified using the **maximum error, E.** When constructing a confidence interval, E is the amount that is added to and subtracted from the point estimate to obtain the endpoints of the desired interval. The sample size n necessary to achieve a desired accuracy can be obtained using a specified value of E. The population standard deviation can be estimated from a preliminary sample, or a rough approximation procedure can be used.

Simple random sampling is employed when every sample of n elements has an equal chance of being selected. Nonrandom sampling procedures can often be used to obtain a more precise estimate of the population mean, μ, providing much narrower confidence intervals for this parameter. These sampling techniques include systematic, stratified, and cluster sampling. **Systematic** sampling selects a random starting point and then selects every kth value for some counting number $k > 0$. This procedure assumes that the population is stored sequentially in some manner, such as in a computer file. **Stratified** sampling is used whenever the population can be physically or geographically separated into two or more groups (strata) where the variation within the strata is less than the variation within the entire population. **Cluster** sampling involves sampling groups of people (clusters) within the population, rather than selecting individual elements one at a time.

Summary of Formulas

1. Standard error of $\bar{X}$ (simple random sample).

$$\frac{\sigma}{\sqrt{n}} \quad (N \text{ unknown or infinite})$$

$$\frac{\sigma}{\sqrt{n}} \sqrt{\frac{N-n}{N-1}} \quad (N \text{ known})$$

where $\sqrt{\dfrac{N-n}{N-1}}$ is the finite population correction (fpc)

2. Confidence interval for a population mean (simple random sample, σ known)

$$\bar{X} \pm Z_{\alpha/2} \frac{\sigma}{\sqrt{n}}$$

Note: Include the fpc if N is known.

3. Confidence interval for a population mean (simple random sample, σ unknown)

$$\bar{X} \pm t_{\alpha/2, n-1} \frac{s}{\sqrt{n}}$$

Note: Include the fpc if N is known.

4. Necessary sample size (simple random sample)

$$n = \left[\frac{Z_{\alpha/2}\sigma}{E}\right]^2$$

where E is the specified maximum error

5. Stratified sampling

$\bar{X}_s$ = estimator of population mean

$$= \frac{\Sigma N_i \bar{X}_i}{N}$$

where $\bar{X}_i$ is the ith stratum mean and N_i is the size of the ith stratum in the population

$\sigma^2_{\bar{x}_S}$ = variance of $\bar{X}_s$

$$= \frac{\Sigma N_i^2 \left(\dfrac{N_i - n_i}{N_i - 1}\right) \dfrac{s_i^2}{n_i}}{N^2}$$

where s_i^2 = the sample variance within the ith stratum. Approximate confidence interval:
$\bar{X}_s \pm Z_{\alpha/2}\sigma_{\bar{x}_S}$

6. Cluster sampling.

$\bar{X}_c$ = estimator of the population mean

$$= \frac{\Sigma T_i}{\Sigma n_i}$$

$\sigma^2_{\bar{x}_C}$ = variance of $\bar{X}_c$

$$= \left(\frac{M - m}{mM\bar{N}^2}\right) \frac{\Sigma(T_i - \bar{X}_c n_i)^2}{m - 1}$$

where T_i is the total of the observations in the ith cluster. Approximate confidence interval:
$\bar{X}_c \pm Z_{\alpha/2}\sigma_{\bar{x}_C}$

 **Review Exercises**

7.60 Medical Products Consolidated wishes to estimate the yearly maintenance costs of the mechanical ventilators that hospitals buy from them. Fifteen hospitals were randomly chosen from a total of 45. From the data, construct a 90% confidence interval for the mean yearly maintenance costs (in thousands of dollars) of the ventilators.

HOSPITAL	NUMBER OF VENTILATORS	MAINTENANCE COST	HOSPITAL	NUMBER OF VENTILATORS	MAINTENANCE COST
1	2	1.2	9	5	2.3
2	5	2.3	10	4	1.9
3	2	0.6	11	5	2.8
4	7	4.1	12	7	3.8
5	6	3.0	13	6	3.4
6	5	2.0	14	4	1.8
7	4	1.3	15	5	2.7
8	2	0.9			

7.61 An accounting firm has a large pool of secretaries. It is assumed that the time it takes a secretary to type a certain legal document is normally distributed with a mean time of 30 minutes and a standard deviation of 4 minutes.

a. What is the probability that a secretary will spend less than 27 minutes typing the legal document?

b. In a randomly selected sample of six secretaries, what is the probability that the average time that it takes them to type the legal document is less than 27 minutes?

7.62 At a manufacturing firm approximately 300 similar machines are used in a production process for manufacturing certain steel products. It is discovered that the number of times per day that a machine needs to be readjusted because it went out of control has a Poisson distribution with mean 1.

a. Describe the distribution of the average number of times a daily readjustment is necessary per machine.

b. Find the probability that the average number of readjustments per machine is less than 1.1 on a certain day.

7.63 A quality assurance inspector believes that the weight of a machine bearing produced by a certain manufacturing process approximately follows a uniform distribution. A sample of 50 machine bearings yielded a sample mean of .211 pounds with a standard deviation of

.05. Find a 90% confidence interval for the mean weight of the machine bearings produced by this manufacturing process.

7.64 A local merchant would like to estimate the mean amount of money that a family spends at the state fair to within $15. If the amount spent by a family is considered to be normally distributed with a standard deviation of $27, what sample size would be necessary to be 90% confident?

7.65 A random variable is found to range from a high of 50 to a low of 25 from past data. The distribution of the random variable can be approximated by a normal distribution. To estimate the mean of the random variable to within 2.1, what sample size would be necessary in selecting a random sample to achieve a 99% confidence level?

7.66 A wholesale furniture store has 160 dining tables that have a mean weight of 47.3 pounds with a standard deviation of 9.8 pounds. If 15 tables are randomly selected, what is the probability that the average weight of the 15 tables will be between 41 and 56 pounds? What are you assuming about the distribution of the table weights?

7.67 The research and development department of a large oil company employs 253 engineers who have an average of 6.2 years of practical experience with a standard deviation of 2.1 years.

a. If a sample of 35 engineers is selected randomly without replacement, what is the probability that the average number of years of experience of the sample will be greater than 6.8 years?

b. If a sample of 35 engineers is selected randomly with replacement, what is the probability that the average number of years of experience will be greater than 6.8 years?

7.68 A confidence interval for the mean of a normally distributed population is found to range from 70.1 to 80.2. What is the level of confidence for the confidence interval if the sample size is 36 and the population standard deviation is 13.2?

7.69 The tensile strength of a high-powered copper coil used in giant power transformers is believed to follow a normal distribution. A sample of 14 high-powered copper coils yields the following tensile strength (in units of thousands of pounds per square inch). Construct a 90% confidence interval for the mean tensile strength of copper coils.

6.1, 2.6, 3.5, 4.3, 3.1, 5.2, 3.6, 3.5, 5.4, 4.2, 3.2, 2.8, 4.0, 3.7

7.70 The mean number of defectives found in a box of electrical resistors is 10 with a population standard deviation of 4. If 11 boxes are selected at random, what is the probability that the average number of defectives will be between 9 and 12? What are you assuming about the probability distribution here?

7.71 The United States Department of Labor has been encouraging companies to offer more investment options for the retirement savings plans of their employees. The average employer in Fidelity programs now offers five investment options, up from an average of 2.5 three years ago. Suppose that the population standard deviation of the number of investment options offered by employers who use Fidelity programs is now 2.1. What is the probability that the average number of investment options offered by 36 randomly selected employers currently in Fidelity programs exceeds 6?

(*Source:* "Employee Benefits For A Changing Work Force," *Business Week* (November 5, 1990): 31–40.)

7.72 A personnel administrator for Teltronix would like to estimate the amount of term life insurance that an employee carries. Three strata are used for finding a stratified random sample of all employees. From the data, construct a 95% confidence interval for the mean amount of term life insurance that an employee carries (given in units of thousands of dollars).

STRATUM	N_i TOTAL NUMBER IN COMPANY	n_i	$\bar{X}_i$	s_i
Employees paid by the hour	350	67	28.5	5.7
Engineers and technicians	112	22	80.6	10.3
Management	57	11	125.2	13.6
	519	100		

7.73   A machine at a manufacturing plant fills sacks with 10 pounds of oats. Each case contains ten sacks of oats. The quality-control engineer would like to find a 99% confidence interval for the mean weight per sack of oats. Using cases as clusters, construct the confidence interval if eight cases are chosen at random. Assume that there is a total of 50 cases from which to choose.

CASES	SACKS PER CASE	WEIGHT OF CASE	CASES	SACKS PER CASE	WEIGHT OF CASE
1	10	96.7	5	10	110.8
2	10	99.8	6	10	104.6
3	10	103.5	7	10	93.5
4	10	92.7	8	10	112.3

7.74   In the state of Michigan in 1988, the average number of years that a homeowner stayed in his home was 13.7 years. Suppose that the population standard deviation of this number was 7.0 years. What is the probability that the average number of years that 36 randomly selected homeowners in Michigan in 1988 stayed in their homes is less than 15 years?

(*Source: U.S. News and World Report* (February 5, 1990): 74.)

7.75   A binomial random variable can be thought of as a sum of 1s and 0s, where 1 represents a success on a trial and 0 represents a failure on a trial. Dividing the binomial random variable by n gives an estimate of the proportion of successes. This estimate of the proportion is equivalent to a sample mean of 1s and 0s. Show that the Central Limit Theorem can be applied to the estimate of the proportion of successes in a binomial experiment.

7.76   As described in Exercise 7.75, the estimate of the proportion of successes in a binomial experiment can be treated the same as $\bar{X}$. In the following MINITAB computer printout, 1000 random observations from a binomial distribution with $n = 35$ and $p = .40$ are generated. A histogram is printed to represent the distribution of the estimates of the proportion, that is, the distribution of the values of the binomial random variables divided by $n = 35$. Does the distribution of the estimate of the proportion appear to comply with the empirical rule? Approximately what percentage of the observations are within one standard deviation of the true proportion, two standard deviations of the true proportion, and three standard deviations of the true proportion, respectively? Assume that the population standard deviation of the estimate of the proportion is $\sqrt{p(1 - p)/n}$.

```
MTB > random 1000 c1;
SUBC> binomial n = 35 p = .40.
MTB > let c2 = c1/35
MTB > erase c1
MTB > histogram c2

Histogram of C2    N = 1000
Each * represents 5 obs.

Midpoint   Count
   0.10       1    *
   0.15       2    *
   0.20      10    **
   0.25      51    **********
   0.30     123    *************************
   0.35     246    **************************************************
   0.40     138    ****************************
   0.45     223    *********************************************
   0.50     144    *****************************
   0.55      52    **********
   0.60       5    *
   0.65       4    *
   0.70       1    *
```

7.77   The following MINITAB computer printout shows the results of simulating 50 times a sample of 13 observations from a uniform distribution over the interval 0 to 20. According to the Central Limit Theorem, the distribution of the sample mean should be approximately normal for large sample sizes. However, the distribution of the sample mean from a population with a uniform distribution approaches a normal distribution for fairly small sample sizes. Comment on the shape of the histogram of the sample mean. Repeat this simulation experiment with sample sizes of $n = 18$ and 25. Comment on the shapes of the histogram of the sample mean.

```
MTB > let k2 = 1
MTB > store
MTB > random 13 observations, put into c1;
MTB > uniform with continuous uniform on 0 to 20.
MTB > mean data in c1, put in k1
MTB > let c2(k2) = k1
MTB > let k2 = k2 + 1
MTB > end
MTB > execute 50 times
MTB > random 13 observations, put into c1;
SUBC> uniform with continuous uniform on 0 to 20.
    .
    .
    .
MTB > histogram of c2

Histogram of C2    N = 50

Midpoint   Count
       6       1   *
       7       2   **
       8       6   ******
       9       7   *******
      10      13   *************
      11      10   **********
      12       8   ********
      13       1   *
      14       1   *
      15       1   *
```

Computer Exercises Using the Database

Exercise 1 -- Appendix I Select the first 500 observations of the database as the population of interest. Estimate the mean of HPAYRENT (house payment or house/apartment rent) for this population by taking a simple random sample of size 40. Also estimate the mean by taking a stratified sample proportional to the size of the two strata. Let stratum 1 be the group of observations in which a secondary wage earner (variable INCOME2) has a positive income, and let stratum 2 be the group of observations in which there is no secondary wage earner. Compare the confidence intervals on mean house payment/rent (HPAYRENT) for both the simple random sample and the stratified random sample.

Exercise 2 -- Appendix I Select a simple random sample of 32 observations from the database. Calculate a 95% confidence interval on the mean income of the principal wage earner (variable INCOME1). Select another simple random sample of size 60 from the database in Appendix H. Calculate a 95% confidence interval on the mean income of the principal wage earner. Comment on the widths of these two confidence intervals.

Exercise 3 -- Appendix J Generate 20 random samples of size 10 using the data for the variable ASSETS (current assets). For each sample determine the sample mean. Construct a histogram of these 20 sample means. Also find the mean and the variance of these 20 values. Repeat this procedure for samples of size 25 and 40. Does the Central Limit Theorem appear to be operating correctly here? Discuss.

Case Study

The Case of the Abundant Zeros

In many audit applications, a sample is taken from a population of vouchers (financial transactions), and each voucher is checked to see if it is in error. For example, suppose that a medical doctor charges Medicare $5000 to perform a particular operation and this payment is selected during an audit of Medicare payments. The maximum amount Medicare charge allowed for this operation is stated to be

$4000, resulting in a $1000 error. On the other hand, if the payment was for $4000 *or less,* there is no error and a value of 0 (indicating an error of 0) is recorded. (Consequently, audit samples can contain a great many zero values, often between 50% and 95% of the entire sample.) If the entire payment is disallowed because it represents an operation that Medicare doesn't cover (such as most cosmetic surgery), there is a $5000 error.

The latter situation was encountered in the audit of payments to foster home facilities in the late

1980s. Two hundred monthly payments were audited from a population of $N = 44,712$ payments that covered one calendar year. An error was encountered if the payment was ruled ineligible for a variety of reasons including (1) a foster home was not approved for program participation at the time of payment or (2) children were not in foster care. Of the 200 payments sampled, 88 were in error and these payments were disallowed.

One of the actions resulting from the audit was an effort to seek monetary recovery based on the sample results. The steps involved in determining this recovery amount are outlined in the set of questions to follow. A summary of the sample results (obtained from the report and related working papers) is shown below.

NUMBER OF ZEROS	NUMBER OF ERRORS
112	88
	$\Sigma x = 8724.80$
	$\Sigma x^2 = 933,455$
	$\bar{x} = \$99.15$
	$s = \$28.05$

Case Study Questions

1. Consider the combined sample of 200 values. What is the overall sample mean $(\bar{X}_T)$? $\bar{X}_T$ is the estimate of the population mean, μ_T. Note that the population contains both zero values (no error) and the nonzero error amounts.

2. For the combined sample of 200 values, what is the overall sample standard deviation (s_T)? (*Hint:* In the computational formula for s^2 in Chapter 3 (formula 3.8), Σx and Σx^2 are the same as for the 88 errors, since the zero values don't change these summations. However, the value of n is now 200, rather than 88. The value of $s = \$28.05$ for the error val-

ues is obtained using $n = 88$ in formula 3.8.)

3. Determine a 90% confidence interval for μ_T assuming the Central Limit Theorem applies (the usual procedure for large samples in this chapter).

4. To obtain a 90% confidence interval for the population *total*, you simply multiply both ends of the confidence interval for the mean in question 3 by N, the number of elements in the population. Obtain this interval.

5. The particular government agency conducting the audit (Office of Inspector General, Office of Audit) does *not* seek a recovery amount equal to the point estimate located in the center of the interval in question 4; that is, $N\bar{X}_T$. Rather, it offers some leniency by seeking recovery of an amount equal to the *lower limit* of the 90% confidence interval derived in question 4. What is this recovery amount?

6. The report states, "We conclude that there is a 95% probability that for this 12-month period, at least $1,675,039 in Federal funds were inappropriately reimbursed to the State agency." Ignoring rounding, do you agree with this statement? Comment.

7. Previous simulation studies related to this type of problem have shown that under similar conditions, what we intend to be a 90% confidence interval may have an entirely different confidence level. That is, the resulting confidence interval may, in fact, be an 81% (say) confidence interval due to the "strangeness" of the population, which consists of, in a sense, apples (zero values) and oranges (nonzero values). What does the statement about 90% versus 81% mean, and what are its implications?

(*Source:* Report released by the Office of Inspector General, Office of Audit, "Review of Foster Care Maintenance Payments Claimed Under Title IV-E of the Social Security Act," April 1989.)

Hypothesis Testing for the Mean and Variance of a Population

We have seen that statistical inference is used to estimate a population parameter using a sample statistic. For the rest of this book, the mean (μ) and standard deviation (σ) of the parent population will be unknown and will have to be estimated from the sample. Do not forget that even though you have estimated μ or σ, these values still are unknown and will forever remain unknown.

As a measure of how reliable your point estimate of the population mean μ really is, you can determine a confidence interval for this parameter. For a given confidence level, the narrower your resulting confidence interval is, the more faith you can have in the ability of your sample mean, $\bar{X}$, to provide an accurate estimate of the population mean. Also, when the Central Limit Theorem is applicable, you need not worry about the shape of the parent population (normal, exponential, and so on) when making probability statements regarding $\bar{X}$, provided you have a large sample (generally, $n > 30$). When you do have a large sample, the distribution of $\bar{X}$ closely approxi-

mates the normal. You can thus construct confidence intervals for population means without worrying about the nature of the parent population, simply because it doesn't matter.

Next, we turn to the situation in which someone makes a claim regarding the value of the population mean, μ. For example, when dealing with the lifetime of Everglo light bulbs in Chapter 6, we assumed that the population average of *all* bulbs was $\mu = 400$ hours. Where did this value come from? Suppose that Everglo advertisements claim that the average lifetime of the bulbs is 400 hours. By testing a sample of bulbs, can we prove this statement? The answer is an emphatic no; the only way to know the value of μ exactly is to obtain data for *all* Everglo bulbs; that is, obtain the entire population.

The sample, however, may allow us to reject the claim that μ is 400 hours, but since the sample is only a portion of the population, this conclusion may be incorrect. Such is the nature of hypothesis testing.

8.1
Hypothesis Testing on the Mean of a Population: Large Sample

A newspaper article claims that the average height of adult males in the United States is not the same as it was 50 years ago; it claims the average height is now 5.9 feet (approximately 5'11''). Your firm manufactures clothing, so the value of this population mean is of vital interest to you. To investigate the article's claim, you randomly select 75 males and measure their heights.* Your results for $n = 75$ are $\bar{x} = 5.76$ feet and $s = .48$ feet.

Let μ represent the population average (mean) of all U.S. male heights. We do have a point estimate of μ; $\bar{x} = 5.76$ feet is an estimate of μ. Keep in mind that the actual value of μ is unknown (although it *does exist*) and will remain that way. What we can do is estimate μ using the sample data. This situation can be summarized by considering the following pair of hypotheses:

Null hypothesis:
$H_0: \mu = 5.9$
Alternative hypothesis:
$H_a: \mu \neq 5.9$

---

*The size of this sample is unrealistically small (yet large, statistically).

H_0 asserts that the value of μ that has been claimed to be correct is in fact correct. H_a asserts that μ is some value other than 5.9 feet. The alternative hypothesis typically contains the conclusion that the researcher is attempting to demonstrate using the sample data. In our height example, if you do not believe that the average height is 5.9 feet and you expect the data to demonstrate that μ has some other value, H_a is $\mu \neq 5.9$.

Definition

1. Null hypothesis (H_0). A statement (equality or inequality) concerning a population parameter; the researcher wishes to discredit this statement.
2. Alternative hypothesis (H_a). A statement in contradiction to the null hypothesis; the researcher wishes to support this statement.

The task of all hypothesis testing is to **reject H_0** or **fail to reject H_0**. Notice that we do not say "reject H_0 or accept H_0." This is an important distinction.

In our study of male heights, the (point) estimate of μ is $\bar{x} = 5.76$ feet. Should we reject H_0, given that it claims that μ is 5.9? First, we need not worry about the shape of the underlying population of male heights because, by the Central Limit Theorem, $\bar{X}$ is approximately normally distributed for large samples, regardless of the shape of this population. So, $\bar{X}$ is approximately a normal (and thus continuous) random variable. What is the probability that *any* continuous random variable is equal to a certain value? In particular, what is the probability that $\bar{X}$ is exactly equal to 5.9 feet? The answer to both questions is zero. Thus we see that we cannot reject H_0 simply because $\bar{X}$ is not equal to 5.9 feet. What we do is to allow H_0 to stand, provided $\bar{X}$ is "close to" 5.9 feet, and reject H_0 otherwise. To define what "close" means, we need to take an in-depth look at what happens when you test hypotheses.

Type I and Type II Errors

Because the sample does not consist of the entire population, there always is the possibility of drawing an incorrect conclusion when inferring the value of a population parameter using a sample statistic. When testing hypotheses, there are two types of possible errors:

Type I error   A Type I error occurs if you rejected H_0 when in fact it is true. For example, a Type I error would occur if you were to reject the claim (hypothesis) that the population mean is 5.9 feet when in fact it really is true.

Type II error   A Type II error occurs if you fail to reject H_0 when in fact H_0 is not true. For example, a Type II error occurs if you fail to reject the hypothesis that the population mean is 5.9 feet when in fact the mean is *not* 5.9 feet.

	ACTUAL SITUATION	
CONCLUSION	H_0 True	H_0 False
FAIL TO REJECT H_0	Correct decision	Type II error
REJECT H_0	Type I error	Correct decision

For any test of hypothesis, define

$$\alpha = \text{the probability of rejecting } H_0 \text{ when } H_0 \text{ is true}$$

$$= P(\text{Type I error})$$

β = the probability of failing to reject H_0 when H_0 is false

= P(Type II error)

For any test of hypothesis, you would like to have control over n (the sample size), α (the probability of a Type I error), and β (the probability of a Type II error). However, in reality, you can control only two of these: n and α, n and β, or α and β. *In other words, for a* fixed *sample size, you cannot control both α and β.*

Suppose you decide to set $\alpha = .02$. Then the procedure you use to test H_0 versus H_a will reject H_0 when it is true with a probability of .02. You may wonder why we do not set $\alpha = 0$, so that we would never have a Type I error. The thought of never rejecting a correct H_0 sounds appealing, but the bad news is that β (the probability of a Type II error) is then equal to 1; that is, you will *always* fail to reject H_0 when it is false. If we set $\alpha = 0$, then the resulting test of H_0 versus H_a will automatically fail to reject H_0: $\mu = 5.9$ whenever μ is, in fact, any value other than 5.9 feet. If, for example, μ is 7.5 feet (hardly the case, but interesting), we would still fail to reject H_0—not a good situation at all. We therefore need a value of α that offers a better compromise between the two types of error probabilities. (Note that for the situation where $\alpha = 0$ and $\beta = 1$, $\alpha + \beta = 1$. As later examples will demonstrate, this is *not true in general*.)

The value of α you select depends on the relative importance of the two types of error. For example, consider the following hypotheses and decide if the Type I error or the Type II error is the more serious.

You have just been examined by a physician using a sophisticated medical device, where the hypotheses under consideration are:

H_0: you do not have a particular serious disease
H_a: you do have the disease

$\alpha = P$(rejecting H_0 when it is true)

= P(device indicates that you have the disease when you do not have it)

$\beta = P$(fail to reject H_0 when in fact it is false)

= P(device indicates that you do not have the disease when you do have it)

For this situation, the Type I error (measured by α) is not nearly as serious as the Type II error (measured by β). Provided the treatment for the disease does you no serious harm if you are well, the Type I error is not serious. But the Type II error means you fail to receive the treatment even though you are ill.

We never set β in advance, only α. This will allow us to carry out a test of H_0 versus H_a. *The smaller α is, the larger β is. Consequently, if you want β to be small, you choose a large value of α.* For most situations, the range of acceptable α values is .01 to .1.

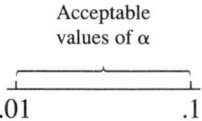

Acceptable
values of α

.01   .1

For the medical-device problem, you could choose a value of α near .1 or possibly larger, due to the seriousness of a Type II error. On the other hand, if you are more worried about Type I errors for a particular test (such as rejecting an expensive manufactured part that really is good), a small value of α is in order. What if there is no basic difference in the effect of these two errors? If there is no significant difference between the effects of a Type I error versus a Type II error, researchers often choose $\alpha = .05$.

Performing a Statistical Test

The claim that the average adult male height is 5.9 feet resulted in the following pair of hypotheses:

$$H_0: \mu = 5.9$$
$$H_a: \mu \neq 5.9$$

We decide to use a test that carries a 5% risk of rejecting H_0 when it is correct; that is, $\alpha = .05$. In hypothesis testing, α is referred to as the **significance level** of your test. Using $n = 75$, $\bar{x} = 5.76$ ft, and $s = .48$ ft, we wish to carry out the resulting statistical test of H_0 versus H_a. We decided to let H_0 stand (not reject it) if $\bar{X}$ was "close to" 5.9 feet. In other words, we will reject H_0 if $\bar{X}$ is "too far away" from 5.9 feet. We write this as follows:

$$\text{reject } H_0 \text{ if } |\bar{X} - 5.9| \text{ is "too large"}$$

or, by standardizing $\bar{X}$, we can

$$\text{reject } H_0 \text{ if } \left|\frac{\bar{X} - 5.9}{s/\sqrt{n}}\right| \text{ is "too large"}$$

We rewrite the last statement as

$$\text{reject } H_0 \text{ if } \left|\frac{\bar{X} - 5.9}{s/\sqrt{n}}\right| > k, \text{ for some } k$$

What is the value of k? Here is where the value of α has an effect. If H_0 is true and the sample size is large, then using the Central Limit Theorem, $\bar{X}$ is approximately a normal random variable with

$$\text{mean} = \mu = 5.9 \quad \text{and} \quad \text{standard deviation} \cong \frac{s}{\sqrt{n}}$$

So, if H_0 is true, $(\bar{X} - 5.9)/(s/\sqrt{n})$ is approximately a standard normal random variable, Z, for large samples.* In this case, we reject H_0 if $|Z| > k$, for some k. Suppose $\alpha = .05$. Then,

$$.05 = \alpha = P(\text{rejecting } H_0 \text{ when it is true})$$
$$= P\left(\left|\frac{\bar{X} - 5.9}{s/\sqrt{n}}\right| > k, \text{ when } \mu = 5.9\right)$$
$$= P(|Z| > k)$$

To find the value of k that satisfies this statement, consider Figure 8.1. When $|Z| > k$, either $Z > k$ or $Z < -k$, as illustrated. Since $P(|Z| > k) = .05$, the total shaded area is .05, with .025 in each tail due to the symmetry of this curve. Consequently, the area between 0 and k is .475, and, using Table A.4, $k = 1.96$. So our test of H_0 versus H_a is

$$\text{reject } H_0 \text{ if } \left|\frac{\bar{X} - 5.9}{s/\sqrt{n}}\right| > 1.96$$

---

*In Chapter 7, we mentioned that $(\bar{X} - 5.9)/(s/\sqrt{n})$ actually follows a t distribution when sampling from a *normal* population, but for large sample sizes, it can be approximated well using the standard normal (Z) distribution. This section deals with large samples from *any* population, so we will use the Z notation to represent this random variable.

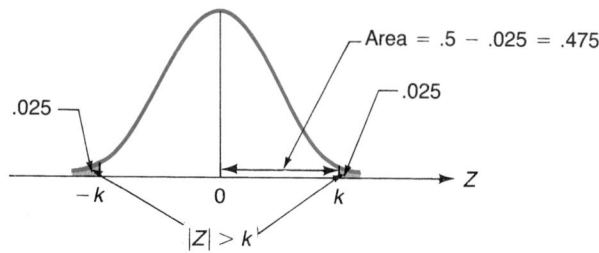

and fail to reject H_0 otherwise. So,

$$\text{reject } H_0 \text{ if } \frac{\bar{X} - 5.9}{s/\sqrt{n}} > 1.96$$

or

$$\text{reject } H_0 \text{ if } \frac{\bar{X} - 5.9}{s/\sqrt{n}} < -1.96$$

This test will reject H_0 when it is true 5% of the time. This means that there is a 5% risk of making a Type I error.

Using the sample data, we obtained $n = 75$, with $\bar{x} = 5.76$ feet and $s = .48$ feet. Is $\bar{x} = 5.76$ feet far enough away from 5.9 feet for us to reject H_0? This was not at all obvious at first glance; it may have seemed that this value of $\bar{X}$ is "close enough to" 5.9 for us not to reject H_0. Such is not the case, however, because

$$Z = \frac{\bar{X} - 5.9}{s/\sqrt{n}} = \frac{5.76 - 5.9}{.48/\sqrt{75}} = -2.53 = Z^*$$

where Z^* is the **computed value** of Z. Because $-2.53 < -1.96$, we reject H_0. We thus conclude that based on the sample results and a value of $\alpha = .05$, the average population male height (μ) is not equal to 5.9 feet.

Another way of phrasing this result is to say that if H_0 is true (that is, if $\mu = 5.9$ feet), the value of $\bar{X}$ obtained from the sample (5.76 feet) is 2.53 standard deviations to the left of the mean using the normal curve for $\bar{X}$ (Figure 8.2). Because a value of $\bar{X}$ this far away from the mean is very unlikely (that is, with probability less than $\alpha = .05$), our conclusion is that H_0 is not true, and so we reject it.

When testing $\mu =$ (some value) versus $\mu \neq$ (some value), the null hypothesis, H_0, always contains the $=$, and the alternative hypothesis, H_a, always contains the $\neq$. In our example, this resulted in splitting the significance level, α, in half and including one-half in each tail of the test statistic, Z. Consequently, a test of H_0: $\mu =$ (some value) versus H_a: $\mu \neq$ (some value) is referred to as a **two-tailed test.**

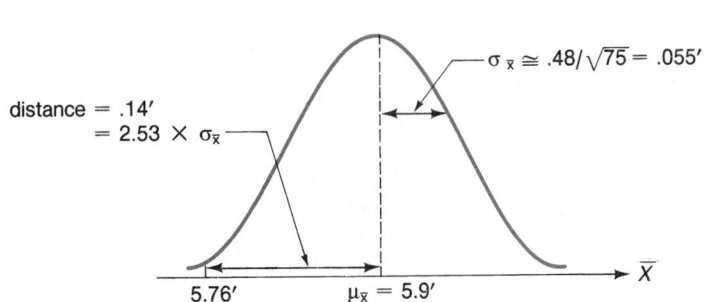

EXAMPLE 8.1

Using the data from our example of male heights, what would be the conclusion using a significance level α of .01?

SOLUTION

The only thing that we need to change from our previous solution is the value of k. Now,

$$P(|Z| > k) = \alpha = .01$$

as shown in Figure 8.3. Using Table A.4, $k = 2.575$, and the test is (see Figure 8.4):

reject H_0 if $Z > 2.575$ or $Z < -2.575$

What is the value of $(\bar{X} - 5.9)/(s/\sqrt{n})$? Our data values have not changed, so the value of this expression is the same: $Z^* = -2.53$.

The region defined by values of Z to the right of 2.575 and to the left of -2.575 in Figure 8.4 is the **rejection region.** The value of k (2.575) defining this region is the **critical value.** Z^* fails to fall in this region, so we fail to reject H_0. In other words, for $\alpha = .01$, the value of $\bar{X}$ is "close enough" to 5.9 to let H_0 stand; there is insufficient evidence to conclude that μ is different from 5.9 feet. ∎

Clearly then, the choice of the significance level, α, is a delicate matter. It is important to remember that a value of α must be selected prior to obtaining the sample and should reflect the impact of a Type I versus a Type II error.

Accepting H_0 or Failing to Reject

It may appear that there is no difference between "accepting" and "failing to reject" a null hypothesis, but there *is* a difference between these two statements. When you test a hypothesis, H_0 is *presumed innocent* until it is demonstrated to be guilty. In Example 8.1, using $\alpha = .01$ we failed to reject H_0. Now, how certain are we that μ is *exactly* 5.9 feet? After all, our estimate of μ is 5.76 feet. Clearly, we do not believe that μ is precisely 5.9 feet. There simply was not enough evidence to *reject* the claim that $\mu = 5.9$ feet.

■ **FIGURE 8.3**
The shaded area is
$\alpha = .01$.

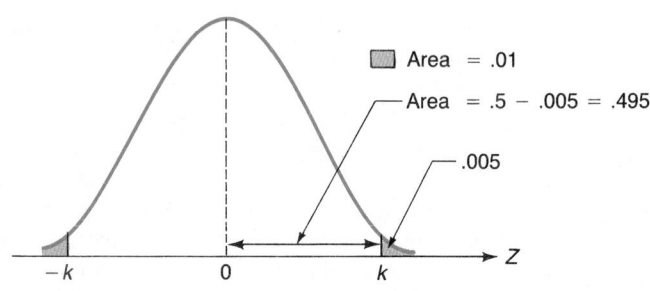

Area = .01
Area = .5 − .005 = .495
.005

■ **FIGURE 8.4**
We reject H_0 if Z^*
falls within either tail—
the rejection region
for $\alpha = .01$.

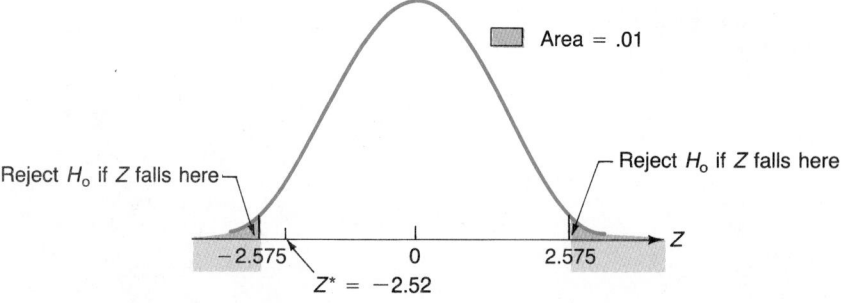

Area = .01
Reject H_0 if Z falls here
Reject H_0 if Z falls here
$Z^* = -2.52$

For any hypothesis-testing application, the only hypothesis that can be *accepted* is the alternative hypothesis, H_a. Either there is sufficient evidence to *support* H_a (we reject H_0) or there is not (we fail to reject H_0). The focus of our attention is whether there is sufficient evidence within the sample data to conclude that H_a is correct. By failing to reject H_0, we are simply saying that the data do not allow us to support the claim made in H_a (such as $\mu \neq 5.9$ feet) and not that we accept the statement made in H_0 (such as $\mu = 5.9$ feet).

The Five-Step Procedure for Hypothesis Testing

The discussion up to this point has concentrated on hypothesis testing on the unknown mean of a particular population. We want to emphasize that the shape of the parent population is not important, provided you have a large sample. In other words, the population may be a normal (bell-shaped) one or it may not—it simply does not matter for large samples. Once the level of significance (α) has been determined, the steps carried out when attempting to reject or failing to reject a claim regarding the population mean μ are:

Step 1. *Set up the null hypothesis, H_0, and the alternative hypothesis, H_a.* If the purpose of the hypothesis test is to test whether the population mean is equal to a particular value (say, μ_0), the "equal hypothesis" always is stated in H_0 and the "unequal hypothesis" always is stated in H_a.

Step 2. *Define the test statistic.* The test statistic will be evaluated, using the sample data, to determine if the data are compatible with the null hypothesis. For tests regarding the mean of a population using a large sample, the test statistic is approximately a standard normal random variable given by the equation

$$Z = \frac{\bar{X} - \mu_0}{s/\sqrt{n}} \qquad (8.1)$$

where μ_0 is the value of μ specified in H_0.

Step 3. *Define a rejection region,* having determined a value for α, the significance level. In this region the value of the test statistic will result in rejecting H_0.

Step 4. *Calculate the value of the test statistic, and carry out the test.* State your decision: to reject H_0 or to fail to reject H_0.

Step 5. *Give a conclusion* in the terms of the original problem or question. This statement should be free of statistical jargon and should merely summarize the results of the analysis.

Steps 1 through 5 apply to all tests of hypothesis in this and subsequent chapters. The form of the test statistic and rejection region change for different applications, but the sequence of steps always is the same.

EXAMPLE 8.2

Remember that Everglo light bulbs are advertised as lasting 400 hours on the average. As manager of the quality assurance department, you need to examine this claim closely. If the average lifetime is, in fact, less than 400 hours, you can expect at least a half-dozen government watchdog agencies knocking on your door. If the light bulbs last longer than the 400 hours (on the average) claimed, you want to revise your advertising accordingly. To check this claim, you have tested the lifetimes of 100 bulbs, each under the same circumstances (power load, room temperature, and so on). The results of this sample are $n = 100$, $\bar{x} = 411$ hours, and $s = 5$ hours. What conclusion would you reach using a significance level of .1?

■ FIGURE 8.5
See Example 8.2; the
rejection region is
$|Z| > 1.645$.

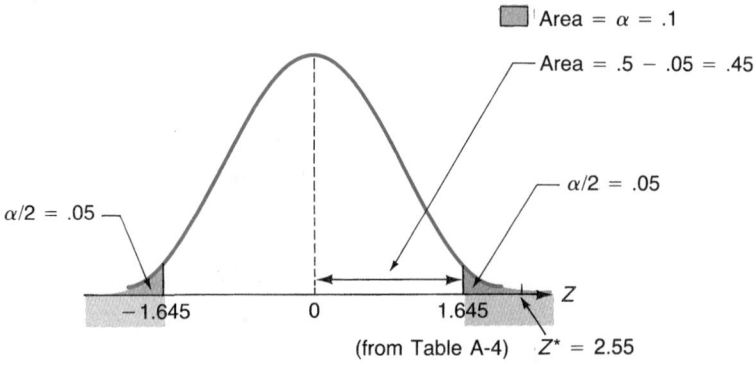

Area = α = .1

Area = .5 − .05 = .45

$\alpha/2$ = .05

$\alpha/2$ = .05

−1.645   0   1.645 Z

(from Table A-4) $Z^* = 2.55$

SOLUTION   **Step 1.** *Define the hypotheses.* We will test H_0: $\mu = 400$ versus H_a: $\mu \neq 400$.

Step 2. *Define the test statistic.* The proper test statistic for this problem is

$$Z = \frac{\bar{X} - 400}{s/\sqrt{n}}$$

Step 3. *Define the rejection region.* The steps for finding the rejection region are shown in Figure 8.5. We conclude:

reject H_0 if $Z > 1.645$ or $Z < -1.645$

Step 4. *Calculate the value of the test statistic and carry out the test.* The computed value of Z is

$$Z^* = \frac{411 - 400}{42.5/\sqrt{100}} = \frac{11}{4.25} = 2.59$$

Since $2.59 > 1.645$, our decision is to reject H_0. In Figure 8.5, Z^* falls in the rejection region.

Step 5. *State a conclusion.* Based on the sample data, there is sufficient evidence to conclude that the average lifetime of Everglo bulbs is not 400 hours. ■

COMMENTS   In Example 8.2, $\bar{X}$ was "far enough away from" 400 for us to reject the claim that the average lifetime is 400 hours (H_0). However, remember that you cannot decide what is "far enough away from" without also considering the value of the standard deviation ($s = 42.5$ hours in Example 8.2). This is why the value of s (or σ, if it is known) is a vital part of the test statistic. Essentially, when the sample contains much variation (s is large), the sample mean ($\bar{X}$) is a less reliable estimator of the population mean and it is more difficult for $\bar{X}$ to be significantly different than the hypothesized value (400, in the previous value).

Examine the test statistic in Example 8.2. Observe that for *small s*, it is "easier" to reject H_0. As s becomes smaller, the absolute value of the test statistic, Z, becomes larger, and the test statistic is more likely to be in the rejection region for a given value of α.

Confidence Intervals and Hypothesis Testing

What is the relationship, if any, between a 95% confidence interval and performing a *two-tailed* test using $\alpha = .05$? There is a very simple relationship here: When testing H_0: $\mu = \mu_0$ versus H_a: $\mu \neq \mu_0$ using the five-step procedure and a significance level, α, H_0 will be rejected if and only if μ_0 lies outside the $(1 - \alpha) \cdot 100\%$ confidence interval for μ.

The five-step procedure and the confidence interval procedure always lead to the same result. In fact, you can think of a confidence interval as that set of values of μ_0 that would not be rejected by a *two-tailed* test of hypothesis.

In our example involving heights of U.S. males, a sample of 75 heights produced $\bar{x} = 5.76$ feet and $s = .48$ feet. The resulting 95% confidence interval for μ is

$$\bar{X} - k\left[\frac{s}{\sqrt{n}}\right] \quad \text{to} \quad \bar{X} + k\left[\frac{s}{\sqrt{n}}\right]$$

What is the value of k? The population standard deviation (σ) is unknown, so we need to use the t table (Table A.5). We do have a large sample, however, so the t value will be closely approximated by the corresponding Z value (Table A.4). Keep in mind that when dealing with large samples, it really does not matter if σ is known or replaced by s. In either case, the standard normal table (Table A.4) gives us the probability points we need.

The value of k that provides a 95% confidence interval here is the *same* value of k that provides a two-tailed area under the Z curve equal to $1 - .95 = .05$. In other words, we use the same k value that we used in a two-tailed test of H_0 versus H_a—namely, $k = 1.96$. So the 95% confidence interval for μ is

$$\bar{X} - 1.96\left(\frac{s}{\sqrt{n}}\right) \quad \text{to} \quad \bar{X} + 1.96\left(\frac{s}{\sqrt{n}}\right)$$

$$= 5.76 - 1.96\left(\frac{.48}{\sqrt{75}}\right) \quad \text{to} \quad 5.76 + 1.96\left(\frac{.48}{\sqrt{75}}\right)$$

$$= 5.76 - .11 \quad \text{to} \quad 5.76 + .11$$

$$= 5.65 \quad \text{to} \quad 5.87$$

The value of μ we are investigating here is $\mu = 5.9$ feet, and the corresponding hypotheses are $H_0: \mu = 5.9$ and $H_a: \mu \neq 5.9$. For $\alpha = .05$, our result using the two-tailed test was to reject H_0. Using the confidence interval procedure, we obtain the same result because 5.9 does not lie in the 95% confidence interval.

Thus, if you already have computed a confidence interval for μ, you can tell at a glance whether to reject H_0 for a two-tailed test, provided the significance level, α, for the hypothesis test and the confidence level, $(1 - \alpha) \cdot 100\%$, match up.

EXAMPLE 8.3

Repeat the heights of U.S. males example using a 99% confidence interval. Is the result the same as in Example 8.1, where we failed to reject $H_0: \mu = 5.9$ using $\alpha = .01$?

SOLUTION

Using $\alpha = .01$, we failed to reject H_0 because the absolute value of the test statistic did not exceed the critical value of $k = 2.575$. The corresponding 99% confidence interval for μ is

$$\bar{X} - 2.575\left(\frac{s}{\sqrt{n}}\right) \quad \text{to} \quad \bar{X} + 2.575\left(\frac{s}{\sqrt{n}}\right)$$

$$= 5.76 - 2.575\left(\frac{.48}{\sqrt{75}}\right) \quad \text{to} \quad 5.76 + 2.575\left(\frac{.48}{\sqrt{75}}\right)$$

$$= 5.76 - .143 \quad \text{to} \quad 5.76 + .143$$

$$= 5.617 \quad \text{to} \quad 5.903$$

Because 5.9 does (barely) lie in this confidence interval, our decision is to fail to reject H_0—the same conclusion reached in Example 8.1.   ∎

The Power of a Statistical Test

Up to this point, the probability of a Type II error, β, has remained a phantom—we know it is there, but we don't know what it is. One thing we can say is that a *wide* confidence interval for μ means that the corresponding two-tailed test of H_0 versus H_a has a *large* chance of failing to reject a false H_0; that is, β is large. Now,

$$\beta = P(\text{fail to reject } H_0 \text{ when } H_0 \text{ is false})$$

which means that

$$1 - \beta = P(\text{rejecting } H_0 \text{ when } H_0 \text{ is false})$$

The value of $1 - \beta$ is referred to as the **power** of the test. Since we like β to be small, we prefer the power of the test to be large. Notice that $1 - \beta$ represents the probability of making a *correct* decision in the event that H_0 is false, because in this case we *should* reject it. The more powerful your test is, the better.

Determining the power of your test (hence, β) is not difficult. We will illustrate this procedure for the previous two-tailed test of H_0: $\mu = \mu_0$ versus H_a: $\mu \neq \mu_0$, for some μ_0. We will first consider the case where σ is known and then discuss the situation where σ is unknown.

Power of the Test: σ Known   In Example 8.2 we looked at the data on Everglo light bulbs, where the hypotheses were H_0: $\mu = 400$ hours and H_a: $\mu \neq 400$ hours. Assume that the actual population standard deviation is known to be $\sigma = 50$ hours. For this situation, our test statistic is (using a sample size of $n = 100$):

$$Z = \frac{\bar{X} - 400}{\sigma/\sqrt{n}} = \frac{\bar{X} - 400}{50/\sqrt{100}}$$

$$= \frac{\bar{X} - 400}{5}$$

Proceeding as in Example 8.2, using $\alpha = .10$, we reject H_0 if $Z > 1.645$ or $Z < -1.645$; that is, $|Z| > 1.645$. So, reject H_0 if $(\bar{X} - 400)/5 > 1.645$ (same as $\bar{X} > 400 + (1.645)(5) = 408.225$) or if $(\bar{X} - 400)/5 < -1.645$ (same as $\bar{X} < 400 - (1.645)(5) = 391.775$). This way of representing the rejection region is illustrated in Figure 8.6, using the shaded area under curve A. The power of this test is

$$1 - \beta = P(\text{rejecting } H_0 \text{ if } H_0 \text{ is false})$$

$$= P(\text{rejecting } H_0 \text{ if } \mu \neq 400)$$

What is the power of this test if μ is not 400 but is 403? What you have here is a value of $1 - \beta$ for *each* value of $\mu \neq 400$.

Recall that we reject H_0 if $\bar{X} > 408.225$ or $\bar{X} < 391.775$. The probability of this occurring if $\mu = 403$ is illustrated as the lined area under curve B in Figure 8.6. Now, if $\mu = 403$ and $\sigma = 50$ (assumed), then

$$Z = \frac{\bar{X} - 403}{50/\sqrt{n}} = \frac{\bar{X} - 403}{5}$$

is a standard normal random variable. So, in Figure 8.6, the striped area to the right of 408.225 is

$$P(\bar{X} > 408.225) = P\left[\frac{\bar{X} - 403}{5} > \frac{408.225 - 403}{5}\right]$$

$$= P\left[Z > \frac{5.225}{5}\right]$$

$$= P(Z > 1.04)$$
$$= .5 - .3508$$
$$= .1492$$

Also, the striped area to the left of 391.775 is

$$P(\bar{X} < 391.775) = P\left[\frac{\bar{X} - 403}{5} < \frac{391.775 - 403}{5}\right]$$
$$= P(Z < -2.24)$$
$$= .5 - .4875$$
$$= .0125$$

Adding these two areas, we find that, if $\mu = 403$, the power of the test of H_0: $\mu = 400$ versus H_a: $\mu \neq 400$ is

$$1 - \beta = .1492 + .0125 = .1617$$

This means that if $\mu = 403$, the probability of making a Type II error (not rejecting H_0) is $\beta = 1 - .1617 = .8383$ (rather high).

This procedure is summarized in the following box. Notice that in the previous discussion, $Z_{\alpha/2} = Z_{.05} = 1.645$, $z_1 = 1.645 - (403 - 400)\sqrt{100}/50 = 1.04$, and $z_2 = -1.645 - (403 - 400)\sqrt{100}/50 = -2.24$.

Power of Test for H_0: $\mu = \mu_0$ versus H_a: $\mu \neq \mu_0$

1. Determine

$$z_1 = Z_{\alpha/2} - \frac{(\mu - \mu_0)\sqrt{n}}{\sigma}$$

and

$$z_2 = -Z_{\alpha/2} - \frac{(\mu - \mu_0)\sqrt{n}}{\sigma}$$

where $Z_{\alpha/2}$ is the value of Z from Table A.4 having a right-tailed area of $\alpha/2$ and μ is the specific value of the population mean (403 in Figure 8.6).

2. Power of test $= P(Z > z_1) + P(Z < z_2)$.

■ FIGURE 8.6
The shaded area is the probability of rejecting H_0 if $\mu = 400$ (that is, $\alpha = .10$), and the striped area is the probability of rejecting H_0 if $\mu = 403$ (that is, the power of the test $1 - \beta$ when $\mu = 403$).

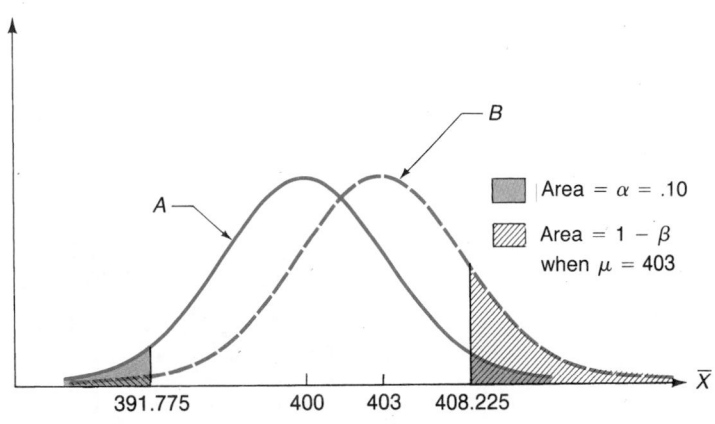

■ FIGURE 8.7
Power curve for
H_0: $\mu = 400$ versus
H_a: $\mu \neq 400$.

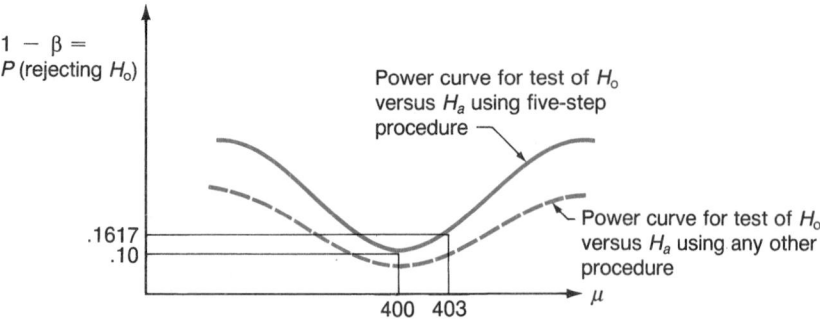

The power of your test increases (β decreases) as μ moves away from 400, as illustrated in Figure 8.7. Using the five-step procedure, which uses the test statistic $Z = (\bar{X} - 400)/(\sigma/\sqrt{n})$, the resulting power curve is the solid line curve in Figure 8.7. It is symmetric, and its lowest point is located at $\mu = 400$. For this value of μ, H_0 is actually true, so that a Type II error was not committed. Nevertheless, the value on the power curve corresponding to $\mu = 400$ is always

$$P(\text{rejecting } H_0 \text{ if } \mu = 400) = \alpha = .10 \text{ (for this example)}$$

The *steeper* your power curve is, the better. You are more apt to reject H_0 as μ moves away from 400—certainly a nice property. If we assume that the sampled population is normally distributed, Figure 8.7 illustrates that the power curve using the five-step procedure lies above (is steeper than) the power curve for any other testing procedure. To illustrate briefly another testing procedure, rather than basing the test statistic on the sample mean $\bar{X}$, we could derive a test statistic using the sample *median*. The resulting power curve for this procedure would lie *below* the one using $\bar{X}$, indicating that the test using the sample median is less powerful and thus inferior. So, in this sense, the five-step procedure defines the best (most powerful) test of H_0: $\mu = \mu_0$ versus H_a: $\mu \neq \mu_0$.

Power of the Test: σ Unknown   When σ is unknown, we are forced to *approximate* the power of the test by replacing σ with the sample estimate, s. We are dealing with large samples, so we can use Table A.4 (the Z table).

In our discussion of the power of our test for Everglo bulb lifetimes, we treated the population standard deviation, σ, as known. If we make no assumptions about this parameter, we need to approximate the power of the test for $\mu = 403$. We assume $s = 42.5$ hours (as before) and use $\alpha = .10$.

We now reject H_0 if

$$\bar{X} > 400 + 1.645\left[\frac{s}{\sqrt{n}}\right] \quad \text{or} \quad \bar{X} < 400 - 1.645\left[\frac{s}{\sqrt{n}}\right]$$

So, H_0 is rejected, provided

$$\bar{X} > 400 + 1.645\left[\frac{42.5}{\sqrt{100}}\right] = 406.99$$

or

$$\bar{X} < 400 - 1.645\left[\frac{42.5}{\sqrt{100}}\right] = 393.01$$

The resulting power of the test for $\mu = 403$ is approximately equal to

$$P(\bar{X} > 406.99 \text{ if } \mu = 403) + P(\bar{X} < 393.01 \text{ if } \mu = 403)$$
$$= P\left[Z > \frac{406.99 - 403}{42.5/\sqrt{100}}\right] + P\left[Z < \frac{393.01 - 403}{42.5/\sqrt{100}}\right]$$

$$= P(Z > .94) + P(Z < -2.35)$$
$$= (.5 - .3264) + (.5 - .4906) = .1736 + .0094$$
$$= .183$$

So, there is an 18% chance of rejecting the null hypothesis if the population mean is, in fact, equal to 403.

Exercises

8.1 The vice president of Metropolitan Bank must decide whether to grant a large loan to an independent energy-exploration company. Consider the null hypothesis: the energy-exploration company will pay back the entire loan.

a. Describe the four possible outcomes from deciding either to fail to reject the null hypothesis or to reject the null hypothesis.

b. Which of the two errors, Type I or Type II, is more serious?

c. If the energy-exploration company does not qualify for the loan, does this "prove" that the energy-exploration company would not pay back the entire loan?

8.2 State what type of error can be made in the following situations:

a. The conclusion is to reject the null hypothesis.

b. The conclusion is to fail to reject the null hypothesis.

c. The calculated value of the test statistic does not fall in the rejection region.

8.3 Explain why the following statements are true or false.

a. The probability of the Type I error and the probability of the Type II error always add to 1.

b. Increasing the value of α increases the value of β.

c. A large value for the power at a specified value of the alternative hypothesis indicates a small value for the probability of a Type II error, given the specified value stated in the alternative hypothesis.

d. The smaller the specified value of α is, the larger the rejection region.

8.4 Hallman Industrial is interested in testing the null hypothesis that a particular applicant is qualified for the position of marketing strategist.

a. Explain what the Type I and Type II errors are for this situation.

b. Which of the two errors in part a is more serious?

8.5 The mean of a normally distributed population is believed to be equal to 50.1. A sample of 36 observations is taken and the sample mean is found to be 53.2. The alternative hypothesis is that the population mean is not equal to 50.1. Complete the hypothesis test, assuming that the population standard deviation is equal to 4. Use a .05 significance level.

8.6 The average life expectancy of males in a developing nation was believed to be 62.5 years. However, the belief was based on data that might be considerably out-of-date. A random survey of 250 deaths in that country revealed an average life span of 64.2 years with a standard deviation of 8.8 years. Does the sample evidence indicate that the mean life expectancy differs from 62.5 years? Use a significance level of .05.

8.7 The weights of fish in a certain pond that is regularly stocked are considered to be normally distributed with a mean of 3.1 pounds and a standard deviation of 1.1 pounds. A random sample of size 30 is selected from the pond and the sample mean is found to be 2.4 pounds. Is there sufficient evidence to indicate that the mean weight of the fish differs from 3.1 pounds? Use a 10% significance level.

8.8 A crime reporter was told that, on the average, 3000 burglaries per month occurred in his city. The reporter examined past data, which was used to compute a 95% confidence interval for the number of burglaries per month. The confidence interval was from 2176 to 2784. At a 5% level of significance, do these data tend to support the alternative hypothesis, $H_a: \mu \neq 3000$?

8.9 A 95% confidence interval for the mean time that it takes a city bus to complete its route is 2.2 hours to 2.6 hours. The time that it takes the bus to complete its route is normally distributed. Is there sufficient evidence to indicate that the mean time to complete the route is different from 2.0 hours? Use a 5% significance level.

8.10 The life span of an electronic chip used in a high-powered microcomputer is estimated to be 625.35 hours from a random sample of 40 chips. The life of an electronic chip is considered to be normally distributed with a population variance of 400 hours.

a. Find a 90% confidence interval for the mean life of the electronic chips.

b. Is the true mean life of the electronic chips different from 633 hours? Use a 10% significance level.

8.11 The manufacturer of a special-purpose industrial pipe is interested in testing the hypothesis that the mean diameter of the pipes is 12.75 inches. A sample of 100 pipes was randomly selected and the diameters were measured. The sample mean was found to be 12.73 inches and the sample standard deviation was found to be .01.

a. Find a 99% confidence interval for the mean diameter of the pipes.

b. Is there evidence that the mean diameter of the pipes is different from 12.75 inches? Use a 1% significance level.

8.12 The hypotheses for a situation are

$$H_0: \mu = 20$$
$$H_a: \mu \neq 20$$

If the population of interest is normally distributed, what is the power of the test for the mean if μ is actually equal to 22? Assume that a sample of size 49 is used and the sample standard deviation is 4.2. Use a significance level of .05.

8.13 Find the power of the test for the mean for the following situations if the true population mean is 30 and the population variance is 25. Use a 10% significance level.

a. $H_0: \mu = 26$, $H_a: \mu \neq 26$, $n = 20$.

b. $H_0: \mu = 36$, $H_a: \mu \neq 36$, $n = 25$.

c. $H_0: \mu = 33$, $H_a: \mu \neq 33$, $n = 25$.

8.14 An electro-optical firm currently uses a laser component in producing sophisticated graphic designs. The time it takes to produce a certain design with the current laser component is 70 seconds, with a standard deviation of 8 seconds. A new laser component is bought by the firm because it is believed that the time it takes this laser to produce the same design is not equal to 70 seconds; the new component also has a standard deviation of 8 seconds. The research-and-development department is interested in constructing the power curve for testing the claim that the time it takes to produce the same design by the new laser component is not equal to 70 seconds. Graph the power function for a sample of size 25 and a significance level of .05.

8.15 Fermet's Soup is interested in knowing how much the average homemaker spends on soup and ingredients to make soup per month. The company's marketing analyst takes a sample of 100 homemakers from a certain city and finds the standard deviation of the amount spent monthly on soup to be $1.50. What would be the power of the test for the hypothesis that the monthly expenditure on soup is equal to $8 if the true monthly expenditure on soup was $10? Assume a significance level of .05.

8.16 Explain why the sample mean, rather than the sample median, is used as a basis for testing the hypothetical mean of a normally distributed population.

◢ 8.2
One-Tailed Test for the Mean of a Population: Large Sample

There are many situations in which you are interested in demonstrating that the mean of a population is *larger* or *smaller* than some specified value. For example, as a member of a consumer-advocate group, you may be attempting to demonstrate that the average weight of a bag of sugar for a particular brand is not 10 pounds (as specified on the bag) but is in fact less than 10 pounds. Because the situation that you (the researcher) are attempting to demonstrate goes into the alternative hypothesis, the resulting hypotheses would be $H_0: \mu \geq 10$ and $H_a: \mu < 10$. *Remember that we said it is standard practice always to put the equal sign in the null hypothesis.* In the testing procedure only the **boundary value** is important, and so the hypotheses may be written as

$$H_0: \mu = 10$$

$$H_a: \mu < 10$$

In this way, we can identify the distribution of $\bar{X}$ when H_0 is true—namely, $\bar{X}$ is a normal random variable centered at 10 with standard deviation $s/\sqrt{n}$ (or $\sigma/\sqrt{n}$ if σ is known). Because the focus of our attention is on H_a (can we support it or not?), which of the two ways you use to write H_0 is not an important issue. The procedure for testing H_0 versus H_a is the same regardless of how you state H_0.

The resulting test is referred to as a **one-tailed test,** and it uses the same five-step procedure as the two-tailed test. The only change we make is to modify the rejection region: all the error is in a single tail.

EXAMPLE 8.4

A foreign car manufacturer advertises that its newest model, the Bullet, rarely stops at gas stations. In fact, they claim its EPA rating for highway driving is at least 32.5 mpg. However, the results of a recent independent study determined the mpg for 50 identical models of the Bullet, with these results: $n = 50$, $\bar{x} = 30.4$ mpg, and $s = 5.3$ mpg. This report failed to offer any conclusion, and you have been asked to interpret these results by someone who has always felt that the 32.5 figure is too high. What would be your conclusion using a significance level of $\alpha = .05$?

SOLUTION

Step 1. *An important point to be made here is that H_0 and H_a (as well as α) must be defined* before *you observe any data.* In other words, *do not let the data dictate your hypotheses;* this approach would introduce a serious bias into your final outcome. For this application, we want to demonstrate that the population mean, μ, is less than 32.5 mpg, and so this goes into H_a. The appropriate hypotheses then are $H_0: \mu \geq 32.5$ and $H_a: \mu < 32.5$.

Step 2. The test statistic for a one-tailed test is the same as that for a two-tailed test, namely,

$$Z = \frac{\bar{X} - \mu_0}{s/\sqrt{n}} \quad \text{or, if } \sigma \text{ is known,} \quad \frac{\bar{X} - \mu_0}{\sigma/\sqrt{n}}$$

$$= \frac{\bar{X} - 32.5}{s/\sqrt{n}}$$

Step 3. What happens to Z when H_a is true? Here we would expect $\bar{X}$ to be <32.5 (because μ is), so the value of Z should be negative. Consequently, our procedure will be to reject H_0 if Z lies "too far to the left" of 0; that is,

$$\text{reject } H_0 \text{ if } Z = \frac{\bar{X} - 32.5}{s/\sqrt{n}} < k \text{ for some } k < 0$$

Since $\alpha = .05$, we will choose a value of k (the critical value) such that the resulting test will reject H_0 (shoot down the mpg claim) when it is true, with a 5% risk of an incorrect decision. This amounts to defining a rejection region in the *left tail* of the Z curve, the shaded area in Figure 8.8. Using Table A.4, we see that the critical value is $k = -1.645$, and the resulting test of H_0 versus H_a is

$$\text{reject } H_0 \text{ if } Z = \frac{\bar{X} - 32.5}{s/\sqrt{n}} < -1.645$$

■ FIGURE 8.8
The one-tailed rejection region is $Z < -1.645$.
We reject H_0 if $Z = (\bar{X} - 32.5)/(s/\sqrt{n})$ < -1.645.

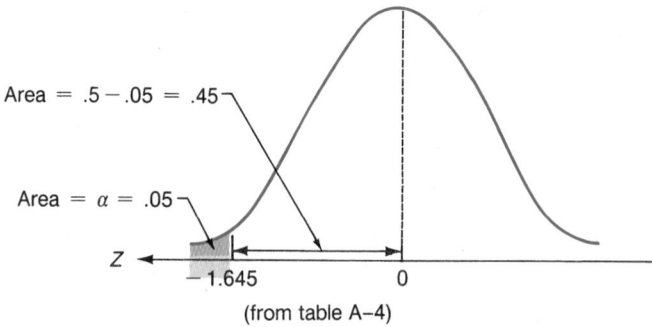

Area $= .5 - .05 = .45$

Area $= \alpha = .05$

-1.645

0

(from table A–4)

Step 4. Using the sample results, the value of the test statistic is

$$Z^* = \frac{30.4 - 32.5}{5.3/\sqrt{50}} = -2.80$$

Because $-2.80 < -1.645$, the decision is to reject H_0.

Step 5. The results of this study support the claim that the average mileage for the Bullet is *less than* 32.5 mpg. This result would provide just cause for claiming false advertising by the auto manufacturer.   ■

One-Tailed Test or Two-Tailed Test?

The decision to use a one-tailed test or a two-tailed test depends on what you are attempting to demonstrate. For example, when the quality-control department of a manufacturing facility receives a shipment from one of its vendors and wants to determine if the product meets minimal specifications, a one-tailed test is appropriate. If the product does not meet specifications, it will be rejected. This type of problem was first encountered in Chapter 5, where we examined lot acceptance sampling. Here, the product is *not* checked to see whether it *exceeds* specifications because any product that exceeds specifications is acceptable.

On the other hand, the vendors who supply the products would generally run two-tailed tests to determine two things. First, they must know if the product meets the minimal specifications of their customers before they ship it. Second, they must determine whether the product greatly exceeds specifications because this can be very costly in production (making a product that uses too much raw material costs them extra money).

The testing of electric fuses is a classic example of a two-tailed test. A fuse must break when it reaches the prescribed temperature or a fire will result. However, the fuse must not break before it reaches the prescribed temperature or it will shut off the electricity when there is no need to do so. Therefore, the quality-control procedures for testing fuses must be two-tailed.

EXAMPLE 8.5   The mean consumption of electricity for the month of June at the Southern States Power Company (SSPC) historically has been 918 kilowatt-hours per residential customer. As part of its request for a rate increase, SSPC is arguing that the power consumption for June of the current year is substantially higher. To demonstrate this, they hired an independent consulting firm to examine a random sample of customer accounts. The results of the sample were $n = 60$ customers, $\bar{x} = 952.36$ kilowatt-hours, and $s = 173.92$ kilowatt-hours. Can you conclude that the average consumption for all users during June of this year (denoted by μ) is larger than 918? Use $\alpha = .01$.

■ **FIGURE 8.9**
One-tailed rejection
region; reject H_0
if $Z > 2.33$.

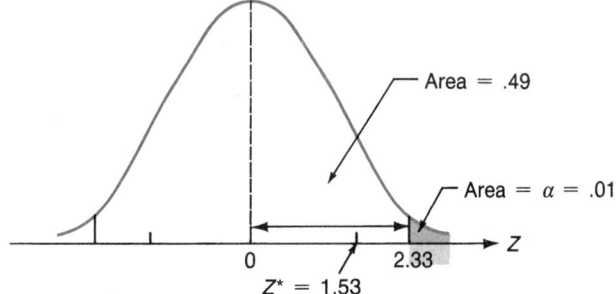

SOLUTION
Step 1. The hypotheses here are $H_0: \mu \leq 918$ and $H_a: \mu > 918$.
Step 2. The correct test statistic is

$$Z = \frac{\bar{X} - 918}{s/\sqrt{n}}$$

Step 3. For this situation, what happens to Z if H_a is true? The value of $\bar{X}$ should then be *larger* than 918 (on the average), resulting in a positive value of Z. So we

$$\text{reject } H_0 \text{ if } Z = \frac{\bar{X} - 918}{s/\sqrt{n}} > k \text{ for some } k > 0$$

Examine the standard normal curve in Figure 8.9, where the area corresponding to α is the shaded part of the *right tail;* using Table A.4, the critical value is $k = 2.33$. The test of H_0 versus H_a will be

$$\text{reject } H_0 \text{ if } Z > 2.33$$

Step 4. The value of your test statistic is

$$Z^* = \frac{952.36 - 918}{173.92/\sqrt{60}} = 1.53$$

Because $1.53 < 2.33$, the decision is to fail to reject H_0.
Step 5. Using this value of α, there is insufficient evidence to support the power company's claim that the power consumption for June has increased. ■

COMMENTS This result is very much tied to the value of α. Using $\alpha = .10$ in Example 8.5, we would obtain the *opposite* conclusion—which you may find somewhat disturbing. You often hear the expression that "statistics lie." This is not true—statistics are merely mistreated, either intentionally or accidentally. One can often obtain the desired conclusion by choosing the value of α that produces the desired conclusion. We therefore reemphasize that you must choose α by weighing the seriousness of a Type I versus a Type II error *before* seeing the data. A partial remedy for this dilemma is discussed in Section 8.3.

Large Samples Taken from a Finite Population

For applications in which we take a large sample from a finite population, we make a slight adjustment to the standard error of $\bar{X}$ by including the finite population correction (fpc) factor.

For the finite population case, the standard error (standard deviation) of $\bar{X}$ is not $s/\sqrt{n}$ but instead

$$\text{standard error of } \bar{X} = s_{\bar{x}} = \frac{s}{\sqrt{n}} \sqrt{\frac{N - n}{N - 1}} \qquad (8.2)$$

Once again we can use the results of the Central Limit Theorem: the test statistic is an *approximate* standard normal random variable, given by

$$Z = \frac{\bar{X} - \mu_0}{s_{\bar{X}}}$$

As a result, the five-step procedure can be carried out exactly as before.

EXAMPLE 8.6

In Example 7.4 we considered a sample of 45 incomes from a group of female managers at Compumart. The women wished to demonstrate that the average income of the population of 350 female middle managers was less than $48,000. For this illustration, it was assumed that the population standard deviation was known. Because this assumption is *not necessary* and perhaps incorrect, a safer procedure would be to use the sample estimate, s. The results of the sample were $n = 45$, $\bar{x} = \$43,900$, and $s = \$7140$.

What would be your conclusion from these results, using a significance level of $\alpha = .05$?

SOLUTION

Step 1. The hypotheses are $H_0: \mu \geq 48,000$ and $H_a: \mu < 48,000$, where $\mu =$ the average annual income for all females in middle-management positions at Compumart.

Step 2. The corresponding test statistic here is

$$Z = \frac{\bar{X} - 48,000}{s_{\bar{X}}}$$

where

$$s_{\bar{X}} = \frac{s}{\sqrt{n}} \sqrt{\frac{N - n}{N - 1}}$$

Step 3. Using Figure 8.8, the rejection region is:

$$\text{reject } H_0 \text{ if } Z < -1.645$$

Step 4. Here,

$$s_{\bar{X}} = \frac{7140}{\sqrt{45}} \sqrt{\frac{350 - 45}{350 - 1}} = 995.01$$

so our computed test statistic is

$$Z^* = \frac{43,900 - 48,000}{995.01} = -4.12$$

Since $-4.12 < -1.645$, we (strongly) reject H_0 in favor of H_a.

Step 5. The sample results strongly support the assertion that the female middle managers are underpaid. We reached the same conclusion in Example 7.4, where we based this decision on the extremely small probability of observing a value of $\bar{X}$ this small if μ was in fact $48,000. ∎

COMMENTS

1. As mentioned in Chapter 7, the fpc factor of $(N - n)/(N - 1)$ can be ignored whenever your sample size is less than 5% of the population size—that is, when $n/N < .05$. Such is also the case when using the fpc in hypothesis testing. In the preceding example, $n/N = 45/350 = .129$. Consequently, ignoring the fpc would have produced a much smaller (and less accurate) value of Z^*.

2. To calculate the power of a one-sided test, refer to the box on page 267. We modify this procedure for a one-sided test by determining $z_1 = Z_\alpha - (\mu - \mu_0)\sqrt{n}/\sigma$ for $H_a: \mu > \mu_0$ or $z_2 = -Z_\alpha - (\mu - \mu_0)\sqrt{n}/\sigma$ for $H_a: \mu < \mu_0$. The resulting power is $P(Z > z_1)$ for $H_a: \mu > \mu_0$ or $P(Z < z_2)$ for $H_a: \mu < \mu_0$.

Large-Sample Tests on a Population Mean

TWO-TAILED TEST

$$H_0: \mu = \mu_0$$

$$H_a: \mu \neq \mu_0$$

reject H_0 if $|Z^*| > Z_{\alpha/2}$

where Z^* is the computed value of $Z = (\bar{X} - \mu_0)/s_{\bar{x}}$ ($Z_{\alpha/2} = 1.96$ for $\alpha = .05$)

ONE-TAILED TEST

$H_0: \mu \leq \mu_0$	$H_0: \mu \geq \mu_0$
$H_a: \mu > \mu_0$	$H_a: \mu < \mu_0$
reject H_0 if $Z > Z_\alpha$	reject H_0 if $Z < -Z_\alpha$
($Z_\alpha = 1.645$ for $\alpha = .05$)	($-Z_\alpha = -1.645$ for $\alpha = .05$)

For a finite population with $n/N > .05$,

$$s_{\bar{x}} = \frac{s}{\sqrt{n}} \sqrt{\frac{N - n}{N - 1}}$$

Otherwise,

$$s_{\bar{x}} = \frac{s}{\sqrt{n}}$$

 *Exercises*

8.17   Find the rejection region of the Z-statistic in a hypothesis test of the population mean for the following situations:

a.   It is believed that the mean monthly advertising expenditure for a company was greater than $2000. A significance level of .05 is used.

b.   It is believed that the average length of sick time taken by an employee of firm XYZ differs from 5.2 days per year. A significance level of .10 is used.

c.   It is believed that the mean age of an applicant applying for a particular job is less than 25 years. A significance level of .01 is used.

8.18   A sample of size 20 is drawn from a finite population of size 225. The finite population can be approximated by a normal distribution. The sample of size 20 yields a sample mean of 75.8. The population variance is equal to 16. Is there sufficient evidence to indicate that the mean of the population is different from 82.5? Use a 10% significance level.

8.19   Carry out the hypothesis test for the mean of the normally distributed population given the following information:

$$H_0: \mu \geq 4.5 \qquad \bar{x} = 3.9$$
$$H_a: \mu < 4.5 \qquad \sigma = 1.12$$
$$n = 30 \qquad \alpha = .07$$

8.20 Bobby Marks is seriously considering investing in the grocery business in the southeastern United States. He believes that the industry's average return on sales (ROS) is less than 5%. A random sample of 46 such businesses in various sectors of the southeastern United States revealed that

$$\bar{x} = 4.6\% \qquad s = 1.2\%$$

Test Marks' belief concerning the ROS of the grocery business in the southeastern United States, using a 5% significance level.

8.21 A production process is working normally if the average weight of a manufactured steel bar is at least 1.3 pounds. A sample of 50 steel bars yields a mean of 1.26 pounds with a standard deviation of .10 pounds. Is there evidence to indicate that the production process needs adjusting? Use a .05 significance level.

8.22 An auditing firm would like to test the belief that the average customer of a small town's utility service pays the utility bill in less than 15 days after receipt of the bill. The town has only 12,352 customers. A sample of size 1325 yields a sample mean of 14.6 days until a customer paid the utility bill. The sample standard deviation is 6 days. Do the data support the belief? Use a 5% significance level.

8.23 Two hundred fifty applicants apply for the same position at an assembly plant. A random sample of 25 applicants is reviewed carefully. The average experience of the 25 applicants is 3.4 years. Can it be concluded that the mean experience of the 250 applicants is greater than 2.5 years at a significance level of .05? Assume that the population standard deviation of the experience of the 250 applicants is 1.3 years.

8.24 The quality-control engineer of a battery-manufacturing firm has been asked to verify the marketing department's claim that the mean life of the multipurpose battery made by the firm is greater than 47 hours. The quality-control engineer takes a random sample of 80 batteries and finds the sample mean to be 47.5 hours with a sample standard deviation of 1.6 hours. Do the data support the marketing department's claim? Use a .05 significance level.

8.25 According to a survey in *American Demographics*, of about 40 leisure hours a week, Americans spend an average of 15 hours a week watching TV. Americans' insatiable appetite for television has steadily increased over the years. Suppose that a marketing research firm believes that full-time working males between the ages of 30 and 40 watch television less than 15 hours a week. Assume that this marketing research firm samples 100 males in this category and finds a sample mean of 14.1 hours and a standard deviation of 6 hours. Is there sufficient evidence to conclude that full-time working males between the ages of 30 and 40 watch television less than 15 hours a week? Use a .05 significance level.

(*Source:* Adapted from "The Leisure Pie," *American Demographics* (November 1990): 39.)

8.26 The International Women's Forum (IWF) was founded in 1982 to give prominent women leaders in various professions around the world a way to share their knowledge with each other. A 1989 survey of the women in IWF revealed that the average yearly income for these women was $140,573. Suppose that an independent polling agency questioned this figure and that a sample of 80 women from IWF yielded a mean of $136,687 with a standard deviation of $26,382.

(*Source:* Adapted from J. Roseac, "Ways Women Lead," *Harvard Business Review* 68 (November–December 1990): 119.)

a. Is there sufficient evidence to conclude that the average yearly income for women in IWF is less than $140,573? Use a 1% significance level.

b. What assumption is necessary about the population size for the statistical test in part a to be valid?

■ **FIGURE 8.10**
Rejection regions for
$\alpha = .01, .05$.

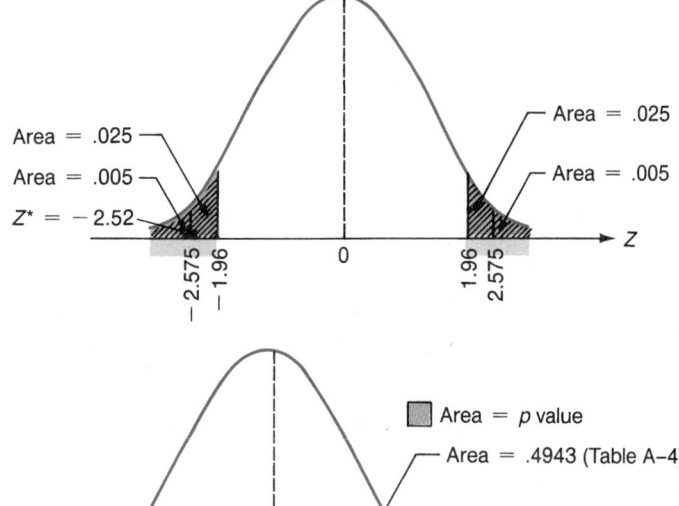

■ **FIGURE 8.11**
p-value is determined by
replacing the area
corresponding to α (see
Figure 8.10) by the area
corresponding to Z^*.
Here $Z^* = -2.53$,
and the p-value $=$
$2 \cdot .0057 = .0114$
(total shaded area).

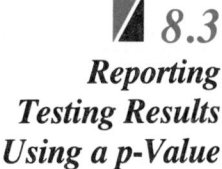

8.3

**Reporting
Testing Results
Using a p-Value**

In Example 8.1, we noted that for one value of α we rejected H_0, and for another (seemingly reasonable) value of α we failed to reject H_0. Is there a way of summarizing the results of a test of hypothesis that allows you to determine whether these results are barely significant (or insignificant) or overwhelmingly significant (or insignificant)? Did we barely reject H_0, or did H_0 go down in flames?

A convenient way to summarize your results is to use a p-value, often called the *observed* α or *observed significance* level.

> The **p-value** is the value of α at which the hypothesis test procedure changes conclusions based on a given set of data. It is the largest value of α for which you will fail to reject H_0.

Consequently, the p-value is the point at which the five-step procedure leads us to switch from rejecting H_0 to failing to reject H_0 for a given set of data.

Determining the *p*-Value

The p-value for *any* test is determined by replacing the area corresponding to α by the area corresponding to the *computed* value of the test statistic. In our discussion and Example 8.1, using $\alpha = .05$ you reject H_0 and using $\alpha = .01$ you fail to reject H_0. We know that the p-value here is between .01 and .05. For this example, the computed value of the test statistic was $Z^* = -2.53$, where the hypotheses are $H_0: \mu = 5.9$ feet and $H_a: \mu \neq 5.9$ feet. The Z curve for this situation is shown in Figure 8.10.

For which value of α does the testing procedure change the conclusions here? In Figure 8.10, if you were using a predetermined significance level α, you would split α in half and put $\alpha/2$ into each tail. So the total tail area represents α. Using Figure 8.11, we reverse this procedure by finding the *total* tail area corresponding to a two-tailed test with $Z^* = -2.53$; we add the area to the left of -2.53 (.0057) to that to the right of 2.53 (also .0057). This total area is .0114, which is the *p*-value for this application. Thus, if you choose a value of $\alpha > .0114$ (such as .05), you will reject H_0. If you choose a value of $\alpha < .0114$ (such as .01), you will fail to reject H_0.

Procedure for Finding the *p*-Value

1. For $H_a: \mu \neq \mu_0$

    $$p = 2 \cdot (\text{area outside of } Z^*)$$

    Reason: When using a significance level α, the value of α represents a *two*-tailed area.

2. For $H_a: \mu > \mu_0$

    $$p = \text{area to the right of } Z^*$$

    Reason: When using a significance level α, the value of α represents a *right*-tailed area.

3. For $H_a: \mu < \mu_0$

    $$p = \text{area to the left of } Z^*$$

    Reason: When using a significance level α, the value of α represents a *left*-tailed area.

EXAMPLE 8.7   What is the *p*-value for Example 8.5?

SOLUTION   The results of the sample were $n = 60$, $\bar{x} = 952.36$ kilowatt-hours, and $s = 173.92$ kilowatt-hours. The corresponding value of the test statistic was

$$Z^* = \frac{952.36 - 918}{173.92/\sqrt{60}} = 1.53$$

The alternative hypothesis is $H_a: \mu > 918$, so the *p*-value will be the area to the *right* of the computed value, 1.53, as illustrated in Figure 8.12. Notice that the inequality in H_a determines the *direction* of the tail area to be found. The *p*-value here is .063, which is consistent with the results of Example 8.5, where we concluded that for $\alpha = .01$, you fail to reject H_0 and for $\alpha = .10$, you reject H_0. That is, the *p*-value is between .01 and .10.   ∎

Most statistical computer packages will provide you with the computed *p*-value when testing the mean of a population. The MINITAB solution to Example 8.5 is provided in Figure 8.13. This procedure assumes that the population standard deviation (σ) is unknown, and so it uses the command *TTEST* (as in *t*-test). The *p*-value in Figure 8.13 is slightly different than the value obtained in Example 8.7, since MINITAB uses the *t* distribution to obtain this value. We discuss this point further in Section 8.4, but for now remember that the *t* random variable is closely approximated by the standard normal, *Z*, when using a large sample.

Interpreting the *p*-Value

We will consider two ways of using the *p*-value to arrive at a conclusion. The first is the **classical approach** that we have used up to this point: We choose a value for

■ **FIGURE 8.12**
p-value for $Z^* = 1.53$.

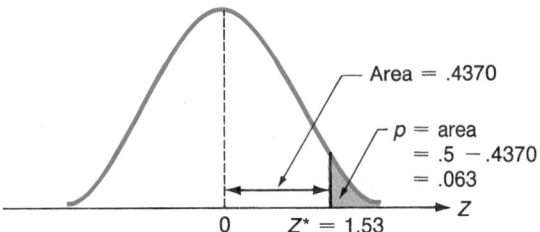

■ **FIGURE 8.13**
MINITAB solution for
Example 8.5.

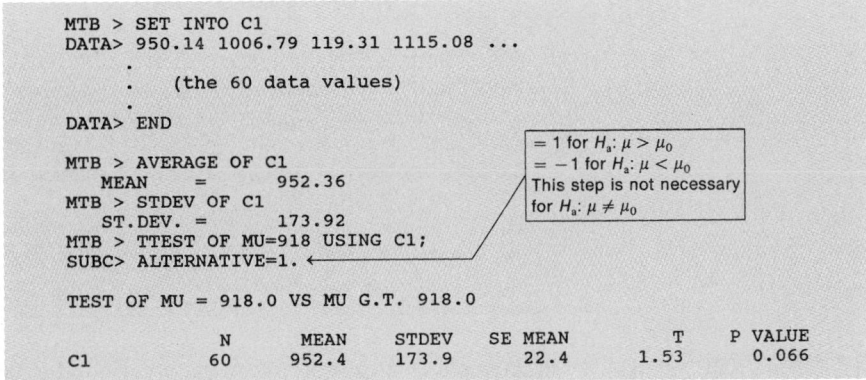

α and base our decision on this value. When using a p-value in this manner, the procedure is:

$$\text{reject } H_0 \text{ if } p\text{-value} < \alpha$$

$$\text{fail to reject } H_0 \text{ if } p\text{-value} \geq \alpha$$

The second approach is a **general rule of thumb** that applies to most applications of hypothesis testing on μ. We previously stated that typical values of α range from .01 to .10, implying that for most applications we will not see values of α smaller than .01 or larger than .1. With this in mind, the following rule can be defined:

$$\text{reject } H_0 \text{ if the } p\text{-value is small } (p < .01)$$

$$\text{fail to reject } H_0 \text{ if the } p\text{-value is large } (p > .1)$$

Consequently, if $.01 \leq p\text{-value} \leq .1$, the data are *inconclusive*.

The advantage of this approach is that you avoid having to choose a value of α; the disadvantage is that you may arrive at an inconclusive result.

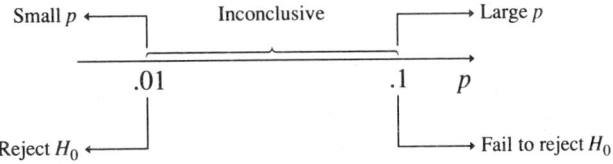

Now for a brief disclaimer: this rule does not apply to all situations. If a Type I error would be extremely serious and you prefer a very small value of α using the classical approach, then you can lower the .01 limit. Similarly, you might raise the .1 limit if the Type II error is extremely critical and you prefer a large value for α. However, this rule gives a working procedure for most applications in business.

What can you conclude if the p-value is $p = .0001$? This value is extremely small compared with *any* reasonable value of α. So we would strongly reject H_0. Consequently, if you are making an investment decision based on these results, for example, you can breathe a little easier. This data set supports H_a overwhelmingly. On the other hand, if $p = .65$, this value is large compared with any reasonable value of α. Without question, we would fail to reject H_0.

There is yet one other interpretation of the *p*-value, summarized in the following box.

Another Interpretation of the *p*-Value

1. For a two-tailed test where H_a: $\mu \neq \mu_0$, the *p*-value is the probability that the value of the test statistic, Z^*, will be at least as large (in absolute value) as the observed Z^*, if μ is in fact equal to μ_0.
2. For a one-tailed test where H_a: $\mu > \mu_0$, the *p*-value is the probability that the value of the test statistic, Z^*, will be at least as large as the observed Z^*, if μ is in fact equal to μ_0.
3. For a one-tailed test where H_a: $\mu < \mu_0$, the *p*-value is the probability that the value of the test statistic, Z^*, will be at least as small as the observed Z^*, if μ is in fact equal to μ_0.

In Example 8.7, we determined the *p*-value to be .063; the computed value of the test statistic was $Z^* = 1.53$; the hypotheses were H_0: $\mu \leq 918$ and H_a: $\mu > 918$. So the probability of observing a value of Z^* as large as 1.53 (that is, $Z^* \geq 1.53$) if μ is 918 is $p = .063$.

Based on this description of the *p*-value, if p is small, conclude that H_0 is not true and reject it. We obtain precisely the same result using the classical and rule-of-thumb options of the *p*-value. *Small values of p favor H_a, and large values favor H_0.*

EXAMPLE 8.8

In Example 8.6 we performed a one-tailed test of H_0: $\mu \geq 48{,}000$ and H_a: $\mu < 48{,}000$. The sample results were $n = 45$, $\bar{x} = \$43{,}900$, and $s = \$7140$. The calculated value of the test statistic was

$$Z^* = \frac{43{,}900 - 48{,}000}{s_{\bar{x}}}$$

where

$$s_{\bar{x}} = \frac{s}{\sqrt{n}} \sqrt{\frac{N - n}{N - 1}}$$

$$= \frac{7140}{\sqrt{45}} \sqrt{\frac{350 - 45}{350 - 1}} = 995.01$$

so $Z^* = -4.12$.

1. What is your conclusion based on the corresponding *p*-value, using $\alpha = .05$?
2. Without specifying a value of α, what would be your conclusion based on the calculated *p*-value?
3. Interpret the *p*-value for this application.

SOLUTION 1

The *p*-value is illustrated in Figure 8.14. We are unable to determine the *p*-value exactly using Table A.4; however, this area is roughly the same as the area to the left of -4.0 under the Z curve—namely, $.5 - .49997 = .00003$. So $p \cong .00003$.

■ FIGURE 8.14
Illustration of the *p*-value for Example 8.8.

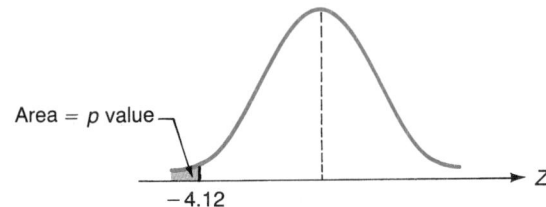

Area = *p* value

-4.12

Z

Because p is less than $\alpha = .05$, we reject H_0. Our conclusion is the same as that of Example 8.6 (which also used $\alpha = .05$), where we concluded that the female managers were underpaid.

SOLUTION 2   We use the general rule of thumb for interpreting the *p*-value. Since $p \cong .00003$, it is extremely small, so we strongly reject H_0 (same conclusion as Solution 1).

SOLUTION 3   We can make the following statements:

1. The significance level at which the conclusion indicated by the testing procedure changes is $\alpha = .00003$.
2. The largest significance level for which you fail to reject the null hypothesis is $\alpha = .00003$.
3. The probability of observing a value of the test statistic as small as the one obtained (≤ -4.12) is .00003 if, in fact, the population mean is $48,000.   ■

Practical Versus Statistical Significance

Researchers often calculate what appears to be a conclusive result without considering the practical significance of their findings. For example, consider a situation similar to the one described in Example 8.4; this time, a sample of 1000 Bullets, tested under normal highway conditions, results in a sample average of $\bar{x} = 32.32$ mpg, with a standard deviation of $s = 2.15$ mpg. Advertising for this car claims that the mpg under test conditions is at least 32.5 mpg. Is there sufficient evidence to reject this claim?

The hypotheses are $H_0: \mu \geq 32.5$ and $H_a: \mu < 32.5$. The value of the test statistic is

$$Z^* = \frac{\bar{X} - 32.5}{s/\sqrt{n}} = \frac{32.32 - 32.5}{2.15/\sqrt{1000}} = -2.65$$

The *p*-value here is the area to the left of -2.65 under the Z curve, as illustrated in Figure 8.15. This value (from Table A.4) is .004. Based on this small *p*-value, we reject H_0 and conclude (as we did in Example 8.4) that the mpg for these cars under normal highway conditions is less than 32.5. Statistically speaking, this is correct, and the data do provide sufficient evidence to support the statement that their mpg claim is overstated. As a consumer, however, how concerned would you

■ FIGURE 8.15
p-value for
$Z^* = -2.65$.

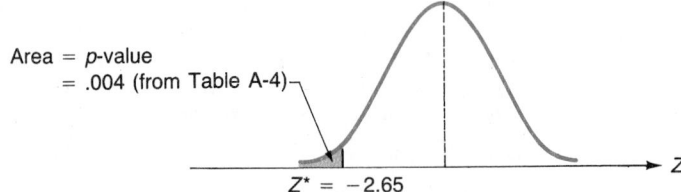

Area = *p*-value
= .004 (from Table A-4)

$Z^* = -2.65$

Z

be that the sample average ($\bar{x} = 32.32$) is (only) .18 mpg under the advertised level? In other words, in a practical sense, how misleading is the Bullet advertising?

What we have seen is that $\bar{X}$ is far enough away from 32.5 (in a statistical sense) to conclude that μ is less than 32.5 mpg. However, perhaps in the eyes of a consumer about to invest \$15,000 in a new car, this value of $\bar{X}$ is really "close enough" to 32.5.

Moral: It is possible for a statistically significant result to be of no particular practical significance, depending on the context of the analysis.

Exercises

8.27 State whether you would reject or fail to reject the null hypothesis in each of the following cases.

a. $p = .12, \alpha = .05.$ c. $p = .001, \alpha = .01.$
b. $p = .03, \alpha = .05.$ d. $p = .01, \alpha = .001.$

8.28 Using the rule-of-thumb option (not selecting a value of α) in the interpretation of the p-value, state whether the test statistic would be statistically significant in the following situations.

a. $p = .57.$ c. $p = .12.$
b. $p = .008.$ d. $p = .04.$

8.29 Explain the difference between "significance" in a statistical sense and "significance" in a practical sense.

8.30 Find p-values for the following situations with calculated test statistics given by Z^*.

a. $H_0: \mu = 30, H_a: \mu \neq 30, Z^* = 2.38$
b. $H_0: \mu \leq 20, H_a: \mu > 20, Z^* = 1.645$
c. $H_0: \mu \geq 15, H_a: \mu < 15, Z^* = -2.54$
d. $H_0: \mu = 50, H_a: \mu \neq 50, Z^* = -1.85$

8.31 Test the belief that the mean of a normally distributed population exceeds 20, assuming that a sample of size 60 yields the following statistics:

$$\bar{x} = 20.4 \qquad s = 3.0$$

Use the p-value criteria.

8.32 The producer of Take-a-Bite, a snack food, claims that each package weighs 175 grams. A representative of a consumer advocate group selected a random sample of 70 packages. From this sample, the mean and standard deviation were found to be 172 grams and 8 grams, respectively.

a. Find the p-value for testing the claim that the mean weight of Take-a-Bite is less than 175 grams.

b. Interpret the p-value in part a.

8.33 A marketing-research analyst is interested in examining the statement made by the makers that brand A cigarettes contain less than 3 milligrams of tar. The marketing-research analyst randomly selected 60 cigarettes and found the mean amount of tar to be 2.75 milligrams with a standard deviation of 1.5 milligrams. Do the data support the claim? Find the p-value.

8.34 The Association of Independent Commercial Producers enlisted the Television Bureau of Advertising in 1987 to get estimates on the average cost of a 30-second TV spot. From surveying 60 production houses, the Television Bureau of Advertising found that on the average, it would cost approximately \$50,000 to shoot a 30-second TV spot. Assume that a group of advertisers wished to verify this claim and that a separate random sample of 55 production houses produced a sample mean of \$57,386 with a standard deviation of \$10,112.

(*Source:* "Spot Discrepancies," *Sales and Marketing Management* (January 1988): 27.)

a. Does the sample evidence indicate that the mean cost to shoot a 30-second TV spot is not \$50,000? Interpret the p-value for the test.

b. From the *p*-value given in part a, would you expect a 99% confidence interval for the mean cost to contain $50,000? Find a 99% confidence interval for the mean cost of shooting a 30-second TV spot.

8.35 Find the *p*-value for the test conducted in Exercise 8.24.

8.36 A recruiter from a large recruiting firm wishes to determine if the mean starting salaries for students with MBAs and no experience is greater than $34,000 in a certain metropolitan area. From a random sample of 50 starting salaries for MBAs without experience, the mean and standard deviation were found to be $34,715 and $2960. Do the data support the belief that the mean starting salary of MBAs without experience is greater than $34,000? Use the *p*-value criteria.

◢ 8.4
Hypothesis Testing on the Mean of a Normal Population: Small Sample

Our approach to hypothesis testing with small samples when the standard deviation, σ is unknown uses the same technique we used for dealing with confidence intervals on the mean of a population: we switch from the standard normal distribution, Z, to the t distribution. However, we need to examine the distribution of the population when the sample is small—the population distribution determines the procedure that we use. In this section, we have reason to believe that the population has a normal distribution. When it does not, we use a nonparametric procedure, which is discussed in Chapter 19.

Certain variations from a normal population *are* permissible with the small-sample test. If a test of hypothesis is still reliable when slight departures from the assumptions are encountered, the test is said to be **robust.** If you believe the parent population to be reasonably symmetric, the level of your confidence interval and Type I error (α) will be quite accurate, even if the population has heavy tails (unlike the normal distribution), as shown in Figure 8.16a. However, when using small samples, the small-sample test is *not* robust for populations that are heavily skewed (see Figure 8.16b). A nonparametric procedure offers a much better solution for this situation. For larger sizes, a histogram of your data often can detect whether a population is heavily skewed in one direction.

To reemphasize, the discussion in this section assumes a normal population. In other words, if X is an observation from this population, then $\bar{X}$ is a normal random variable with unknown mean μ. Also, we assume that σ is unknown. (If σ is known, the resulting test statistic is $Z = (\bar{X} - \mu_0)/(\sigma/\sqrt{n})$, and the five-step procedure of Section 8.1 allows you to do hypothesis tesing on μ.)

The only distinction between using a small and a large sample is the form of the test statistic. Using the discussion from Chapter 7, if we define the test statistic as

$$t = \frac{\bar{X} - \mu_0}{s/\sqrt{n}} \qquad (8.3)$$

we now have a t distribution with $n - 1$ degrees of freedom (df). The procedure to use for testing $H_0: \mu = \mu_0$ and $H_a: \mu \neq \mu_0$ is the same five-step procedure,

■ FIGURE 8.16
(a) Small-sample test is valid. (b) Small-sample test is not valid.

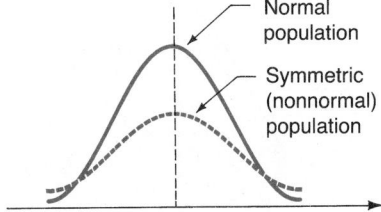

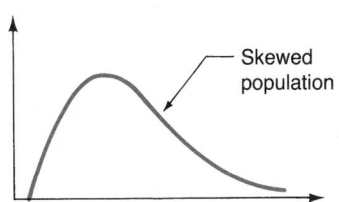

except that the rejection region is defined using the t table (Table A.5) rather than the Z table (Table A.4). This procedure also applies to a one-tailed test. Because we are looking at small samples (typically $n < 30$), we can ignore the finite population correction factor.

EXAMPLE 8.9   You may recall from Example 7.7 that Clark Products manufactures a power supply with an output voltage that is believed to be normally distributed with a mean of 10 volts. During the design stage, the quality-engineering staff recorded 18 observations of the output voltage of a particular power supply unit. They decide to use a significance level of .05 since the implications of making a Type I error (rejecting a correct H_0) and a Type II error (failing to reject an incorrect H_0) appear to be the same. Is there evidence to indicate that the average output voltage is not 10 volts?

SOLUTION   **Step 1.** When a question is phrased "Is there evidence to indicate that . . . ," what follows is the *alternative hypothesis*. For this application then, the alternative hypothesis is that the mean is unequal to 10 volts and the resulting hypotheses are $H_0: \mu = 10$ and $H_a: \mu \neq 10$.

Step 2. The test statistic here is

$$t = \frac{\bar{X} - 10}{s/\sqrt{n}}$$

Step 3. Using a significance level of .05 and Figure 8.17, the corresponding two-tailed procedure is to

reject H_0 if $|t| > t_{.025,17} = 2.11$

because df $= n - 1 = 17$.

Step 4. For these data, $n = 18$, $\bar{x} = 10.331$ volts and $s = .767$ volt. The value of the test statistic is

$$t^* = \frac{10.331 - 10}{.767/\sqrt{18}} = 1.83$$

Because $1.83 < 2.11$, we fail to reject H_0.

Step 5. There is insufficient evidence to indicate that the average output voltage is different from 10 volts.   ■

What is the p-value in Example 8.9, and what can we conclude based on this value? We run into a slight snag when dealing with the t distribution because we are not able to determine precisely the p-value. You can see this in Figure 8.18, using Table A.5 (17 df). The p-value is twice the area to the right of $t^* = 1.83$. The best we can do here is to say that p is *between* (2)(.025) and (2)(.05), that is, between .05 and .10. (*Note:* A reliable computer package or sophisticated calculator will provide the exact p-value. Using MINITAB, this value is $p = .0847$.)

Using the classical approach and $\alpha = .05$ we *can* say that p is greater than

■ FIGURE 8.17
t distribution; the rejection region is the lightly shaded area to the right of 2.11 and to the left of -2.11, for Example 8.9.

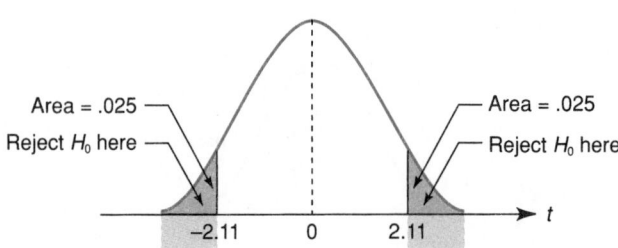

■ **FIGURE 8.18**
t curve with 17 df. The
p-value is twice the area
to the right of
$t^* = 1.83$, so we can
say only that it is
between .05 and .10.

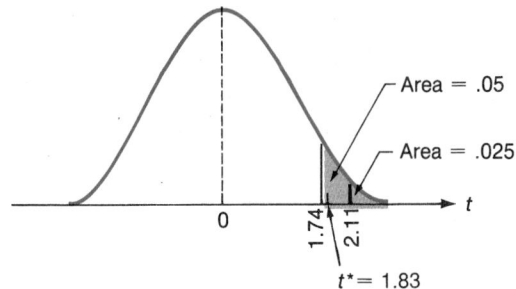

Area = .05

Area = .025

0

1.74

2.11

$t^* = 1.83$

.05, despite not knowing *p* exactly. Consequently, we fail to reject H_0. *This procedure* always *produces the same result as the five-step procedure.*

Suppose we choose not to select a significance level (α) but prefer to base our conclusion strictly on the calculated *p*-value. We use the rule of thumb and decide whether *p* is small ($<.01$), large ($>.1$), or in between. Despite not having an exact value of *p*, we can say that this *p*-value falls in the inconclusive range. These data values do not provide us with any strong conclusion. One approach available to Clark Products is to obtain some additional data.

EXAMPLE 8.10   An auditing firm was hired to determine if a particular defense plant was overstating the value of their inventory items. It was decided that 15 items would be randomly selected. For each item, the recorded amount, the audited (exact) amount, and the difference between these two amounts (recorded − audited) were determined. Of particular interest was whether it could be demonstrated that the average difference exceeds $25, in which case the defense plant would be subject to a loss of contract and financial penalties. The following 15 differences were obtained (in dollars):

17, 35, 31, 22, 50, 42, 56, 23, 27, 38, 20, 25, 43, 45, 21

So $n = 15$, $\bar{x} = \$33.00$, and $s = \$12.15$. Set up the appropriate hypotheses and test them using a significance level of $\alpha = .05$. The population of differences is believed to be normally distributed.

SOLUTION   **Step 1.** The hypotheses are $H_0: \mu \leq 25$ and $H_a: \mu > 25$, where μ is the average difference between the recorded and audited amounts for *all* the inventory items.

Steps 2, 3.

$$\text{reject } H_0 \text{ if } t = \frac{\bar{X} - 25}{s/\sqrt{n}} > t_{.05,14} = 1.761,$$

where the df $= n - 1 = 14$.

Step 4. The calculated *t* is

$$t^* = \frac{33 - 25}{12.15/\sqrt{15}} = 2.55$$

Because 2.55 exceeds the tabulated value of 1.761, we reject H_0. Also, the *p*-value (using Table A.5 and 14 df) is the area to the right of 2.55. It is between .01 and .025, so it is less than $\alpha = .05$, and so (as before) we reject H_0.

Step 5. These data indicate that the defense plant is overstating the value of their inventory items by more than $25.   ■

A MINITAB solution for this example is contained in Figure 8.19. Note that the calculated (exact) *p*-value is .012.

■ FIGURE 8.19
MINITAB solution for
Example 8.10.

```
MTB > SET INTO C1
DATA> 17 35 31 22 50 42 6 23 27 38 20 25 43 45 21
DATA> END
MTB > TTEST OF MU=25 USING C1;
SUBC> ALTERNATIVE=1.

TEST OF MU = 25.000 VS MU G.T. 25.000

              N       MEAN     STDEV    SE MEAN        T      P VALUE
C1           15     33.000    12.148      3.137     2.55        0.012
```

Small-Sample Tests on a Normal Population Mean

TWO-TAILED TEST

$$H_0: \mu = \mu_0$$

$$H_a: \mu \neq \mu_0$$

reject H_0 if $|t^*| > t_{\alpha/2, n-1}$

where n = sample size and t^* is the computed value of

$$t = \frac{\bar{X} - \mu_0}{s/\sqrt{n}}$$

ONE-TAILED TEST

$$H_0: \mu \leq \mu_0 \qquad\qquad H_0: \mu \geq \mu_0$$

$$H_a: \mu > \mu_0 \qquad\qquad H_a: \mu < \mu_0$$

reject H_0 if $t^* > t_{\alpha, n-1}$ reject H_0 if $t^* < -t_{\alpha, n-1}$

Exercises

8.37 Find the rejection region of the t-test used to test the following situations for a normally distributed population:

a. Twenty observations are randomly selected to test the claim that mean yearly maintenance expense on a certain type of lawn mower is less than $28 per year. A significance level of .05 is used.

b. Twenty-five observations are randomly selected to test the claim that managers of convenience stores have an annual income of more than $30,000. A significance level of .10 is used.

c. Fifteen observations are randomly selected to test the claim that the tensile strength of steel rods is different from the tensile strength specified by the firm ordering the steel rods. A significance level of .05 is used.

8.38 Find the p-value for the following situations with calculated test statistics given by t^*.

a. $H_0: \mu = 40$, $H_a: \mu \neq 40$, $t^* = 2.30$, $n = 12$.

b. $H_0: \mu \leq 13.6$, $H_a: \mu > 13.6$, $t^* = 2.73$, $n = 19$.

c. $H_0: \mu \geq 100.80$, $H_a: \mu < 100.80$, $t^* = 1.25$, $n = 20$.

d. $H_0: \mu = 35.6$, $H_a: \mu \neq 35.6$, $t^* = 1.57$, $n = 11$.

8.39 Carry out the hypothesis test for the mean of a normally distributed population given the following information:

$$H_0: \mu \leq 1.6$$

$$H_a: \mu > 1.6$$

$$n = 15 \qquad \bar{x} = 1.8 \qquad s^2 = 1.7 \qquad \alpha = .10$$

8.40 The following sample of seven scores was randomly selected from a finite population of 300 scores that are approximately normally distributed:

$$3, 8, 3, 1, 5, 3, 6$$

Is there sufficient reason to believe that the mean of the population is less than 6? Use a significance level of .05.

8.41 Five measurements are randomly selected from a normal population. Find a 95% confidence interval for the population mean.

$$72, 56, 81, 45, 88$$

8.42 The senior executive of a publishing firm would like to train employees to read faster than 1000 words per minute. A random sample of 21 employees underwent a special speed-reading course. This sample yielded a mean of 1018 words per minute with a standard deviation of 30 words per minute. Do the data support the belief that the speed-reading course will enable the employees to read more than 1000 words per minute at a significance level of .05? Assume that the reading speeds of persons who have taken the course are normally distributed.

8.43 It is believed that the mean aptitude test score for engineers graduating from Safire University is greater than 180. Assume that the scores are normally distributed. A random sample of 26 engineers yielded a mean score of 186 with a standard deviation of 10.2? Do these data support the belief? Use the *p*-value.

8.44 In an effort to control cost, a quality-control inspector is interested in whether the mean number of ounces of sauce dispensed by bottle-filling machines differs from 16 ounces. From the bottling process, the inspector collects the following measurements.

$$16.3, 16.2, 15.8, 15.4, 16.0$$
$$15.6, 15.5, 16.1, 15.9, 16.1$$

Test at a .05 significance level that the bottle-filling machines need adjusting.

8.45 A delivery service is considering delivering Swandorf's ice cream if the average order in a suburban area is greater than 1.5 gallons of ice cream. A random sample of 23 household orders yields a mean of 1.7 gallons with a standard deviation of .5 gallons. Test at the .05 significance level that the mean household order is greater than 1.5 gallons.

8.46 In 1990, the average number of days that homes in Houston stayed on the market before selling was 105. A real estate broker in Houston believes that in a certain subdivision of Houston the average number of days homes stayed on the market before selling was less than 105. Fifteen homes that sold in 1990 were randomly selected from this subdivision and the number of days that each home stayed on the market is given below.

$$100, 110, 83, 82, 115, 75, 120, 91,$$
$$95, 88, 119, 73, 103, 96, 83$$

(*Source:* "Heating Up—Or Cooling Down?" *Money* (January 1991): 32)

a. Do the data provide sufficient evidence to conclude that the mean number of days homes stayed on the market in 1990 in this subdivision is less than 105? Use a .01 significance level.

b. Find the *p*-value and interpret this value.

8.5
Inference for the Variance and Standard Deviation of a Normal Population (Optional)

Our discussion in Chapters 7 and 8 has been concerned with the mean of a particular random variable or population. In other words, we are trying to decide or estimate what is occurring *on the average.* Suppose someone involved with a production process that manufactures 2-inch bolts has just been informed that, without a doubt, these bolts are 2 inches long, on the average. Is there anything else this person might like to know about the production process? Suppose that half of the bolts produced are 1 inch long and the other half are 3 inches. The report was accurate—on the average, they *are* 2 inches long.* However, such a production process certainly will not satisfy the customers, and this company soon will be out of the bolt business.

What was missing in the report was the amount of *variation* in this production process. If the variation was zero, every bolt would be exactly 2 inches long—an ideal situation. In practice, there always will be a certain amount of variation in any

---

*A statistician often is described as someone who thinks that if half of you is in an oven and the other half is in a deep freeze, on the average you are very comfortable.

mechanical or production process. So we are concerned about not only the mean length μ of the population of bolts but also the variance σ^2 or standard deviation σ of the lengths of these bolts. If the variance is *too large*, the process is not operating correctly and needs adjustment. *Consequently, a key element of statistical quality control (Chapter 12) is the act of monitoring and attempting to reduce process variation. A process that is "in control" is one that is consistent and contains only random variation.*

The variance of a population also is of vital interest to someone making investment decisions. Here the *risk* of a venture (or portfolio) often is measured by the variance of the return paid by the venture in the past. Often, financial analysts prefer a financial package with a relatively small average return (based on past history) that appears to be low risk on the basis of only small fluctuations in its past performance.

In the inference procedures for a population variance (and standard deviation) to follow, we will assume that the population of interest is normally distributed. Unlike the *t*-test, the hypothesis testing procedures and confidence intervals for the variance are very sensitive to departures from the normal population—notably, heavy tails in the distribution or heavy skewness will have a large effect. In other words, the following tests of hypothesis are less robust than are those we discussed earlier.

Confidence Interval for the Variance and Standard Deviation

The point estimate of a population variance is the obvious one—namely, the sample variance, which was discussed in Chapter 7, where we used the variance, s^2, of a sample to estimate the variance, σ^2, of the much larger population.[+]

When constructing a confidence interval for μ using a small sample, we used the *t* distribution. Such a distribution is referred to as a **derived distribution** because it was derived to describe the behavior of a particular test statistic. This type of distribution is not used to describe a population, as is the normal distribution in many applications. For example, you will *not* hear a statement such as, "Assume that these data follow a *t* distribution"—normal, exponential, uniform, maybe, but not a *t* distribution. The *t* random variable merely offers us a method of testing and constructing confidence intervals for the mean of a *normal* population when the standard deviation is unknown and is replaced by its estimate.

Another such continuous derived distribution, the **chi-square** (pronounced ky) distribution, written as χ^2, allows us to determine confidence intervals and perform tests of hypothesis on the variance and standard deviation of a normal population. The shape of this distribution is illustrated in Figure 8.20. Notice that unlike the Z and *t* curves, the χ^2 distribution is not symmetric and is definitely skewed right.

For chi-square, as for all continuous distributions, a probability corresponds to an area under a curve. Also, the shape of the chi-square curve, like that of its cousin the *t* distribution, depends on the sample size *n*. As before, this will be specified by the corresponding degrees of freedom (df).

When using the χ^2 distribution to construct a confidence interval or perform a test of hypothesis on a population variance or standard deviation, the degrees of freedom are given by

$$df = n - 1$$

Let $\chi^2_{a,df}$ be the χ^2 value whose area to the right is a, using the proper df.

EXAMPLE 8.11   Using a chi-square curve with 12 df, determine $P(\chi^2 > 18.5494)$ and $P(\chi^2 < 6.30380)$.

[+]The notation $\hat{\sigma}^2$ is often used to represent an estimate of σ^2. Consequently, $s^2 = \hat{\sigma}^2$.

■ **FIGURE 8.20**
Shape of a chi-square distribution.

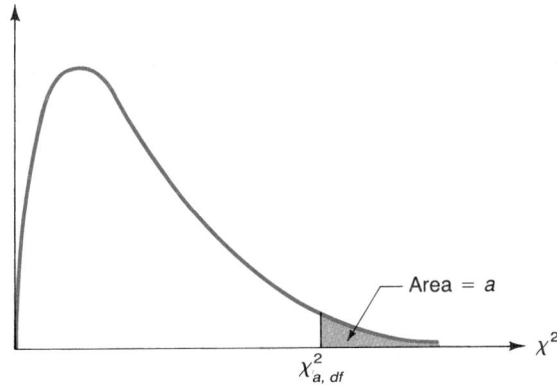

■ **FIGURE 8.21**
χ^2 curve with 12 df. The shaded area represents $P(\chi^2 > 18.5494)$.

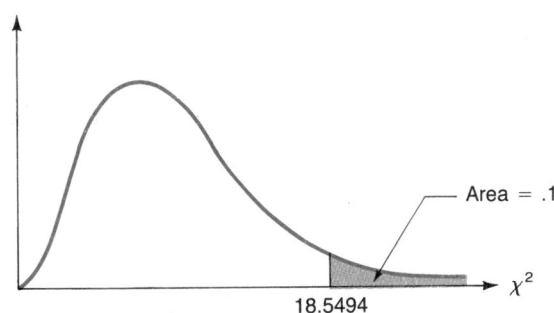

SOLUTION Tabulated values for the χ^2 distribution are contained in Table A.6. This table contains *right-tailed* areas (probabilities). Based on this table (see Figure 8.21),

$$P(\chi^2 > 18.5494) = .1$$

This can be written as

$$\chi^2_{.1,12} = 18.5494$$

For $\chi^2 = 6.30380$, Table A.6 informs us that the area to the right of 6.30380 is .900. Because the total area is 1, the area to the left of 6.30380 is $1 - .900 = .1$, and so $P(\chi^2 < 6.30380) = .1$. As a result, we can say that

$$P(6.30380 \leq \chi^2 \leq 18.5494) = 1 - .1 - .1 = .8$$

That is, 80% of the time a χ^2 value (with 12 df) will be between 6.30380 and 18.5494. ∎

EXAMPLE 8.12 Using Example 8.11, determine a and b that satisfy

$$P(a < \chi^2 < b) = .95, \quad \text{with df} = 12$$

Choose a and b so that an equal area occurs in each tail.

SOLUTION Figure 8.22 shows the areas for a and b. Using Table A.6,

$$a = \text{the } \chi^2 \text{ value whose left-tailed area is .025}$$
$$= \text{the } \chi^2 \text{ value whose area to the right is .975}$$
$$= 4.40$$

and

$$b = \text{the } \chi^2 \text{ value whose right-tailed area is .025}$$
$$= 23.3$$

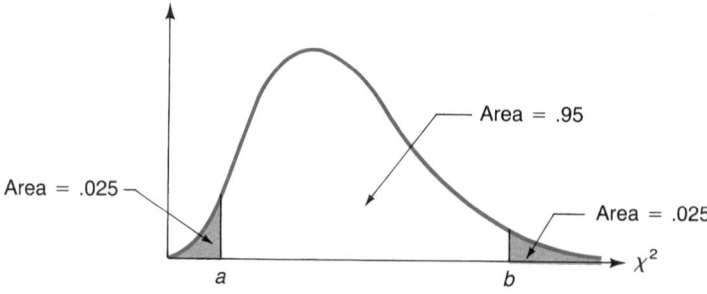

To derive a confidence interval for σ^2, we need to examine the sampling distribution of s^2. If we repeatedly obtained a random sample from a normal population with mean μ and variance σ^2, calculated the sample variance s^2, and made a histogram of these s^2 values, what would be the shape of this histogram? It can be shown that the shape will depend on the sample size n and the value of σ^2 but *not* on the value of the population mean μ. In fact, the values of n and σ^2, along with the random variable s^2, can be combined to define a chi-square random variable, given by

$$\chi^2 = \frac{(n - 1)s^2}{\sigma^2} \qquad (8.4)$$

having a chi-square distribution with $n - 1$ df. Therefore, the sampling distribution for s^2 can be defined using the chi-square distribution in equation 8.4.

For example, a sample size of $n = 13$ results in 12 df. From Example 8.12, it follows that

$$P(4.40 < \chi^2 < 23.3) = .95$$

So,

$$P\left[4.40 < \frac{12s^2}{\sigma^2} < 23.3\right] = .95 \qquad \text{using equation 8.4}$$

or

$$P\left[\frac{12s^2}{23.3} < \sigma^2 < \frac{12s^2}{4.40}\right] = .95$$

As in all confidence interval constructions, the parameter (σ^2) is bounded between two limits defined by a random variable (s^2). This means that a 95% confidence interval for σ^2 is

$$\frac{12s^2}{23.3} \quad \text{to} \quad \frac{12s^2}{4.40}$$

In general, the following procedure can be used to construct a confidence interval for σ^2 or σ. A $(1 - \alpha) \cdot 100\%$ confidence interval for σ^2 is

$$\frac{(n - 1)s^2}{\chi^2_{\alpha/2, n-1}} \quad \text{to} \quad \frac{(n - 1)s^2}{\chi^2_{1-\alpha/2, n-1}} \qquad (8.5)$$

The corresponding confidence interval for σ is

$$\sqrt{\frac{(n-1)s^2}{\chi^2_{\alpha/2,n-1}}} \quad \text{to} \quad \sqrt{\frac{(n-1)s^2}{\chi^2_{1-\alpha/2,n-1}}} \tag{8.6}$$

EXAMPLE 8.13

Vitamix Dog Chow comes in 10-, 25-, and 50-pound bags. The owners are concerned about the variation in the weight of the 50-pound bags because they have recently acquired a new mechanical packaging device. A random sample of the weights of 15 bags (in pounds) was obtained, with the following results:

51.2, 47.5, 50.8, 51.5, 49.5, 51.1, 51.3, 50.7, 46.7, 49.2, 52.1, 48.3, 51.6, 49.2, 51.5

For these data, $\bar{x} = 50.15$ pounds and $s = 1.65$ pounds. Determine a 90% confidence interval for σ^2 and for σ. The bag weights are believed to come from a normal population.

SOLUTION

The corresponding 90% confidence interval for σ^2 is

$$\frac{(15-1)(1.65)^2}{\chi^2_{.05,14}} \quad \text{to} \quad \frac{(15-1)(1.65)^2}{\chi^2_{.95,14}} = \frac{(14)(1.65)^2}{23.7} \quad \text{to} \quad \frac{(14)(1.65)^2}{6.57}$$

$$= 1.61 \quad \text{to} \quad 5.80$$

The 90% confidence interval for σ would be

$$\sqrt{1.61} \quad \text{to} \quad \sqrt{5.80}$$

that is, 1.27 pounds to 2.41 pounds. ■

Hypothesis Testing for the Variance and Standard Deviation

For many applications, we are concerned that the standard deviation or variance of our population may be exceeding some specified value. If this claim is supported, then, for example, we may wish to shut down a production process and make adjustments that will reduce this excessive variation. As you could with the tests of hypothesis examined so far, you can (although this is not the usual case) perform a two-tailed test where either too much variation or too little variation is the topic of concern.

Hypothesis Testing on σ^2

TWO-TAILED TEST

$$H_0: \sigma^2 = \sigma_0^2$$
$$H_a: \sigma^2 \neq \sigma_0^2$$

test statistic: $\chi^2 = \dfrac{(n-1)s^2}{\sigma_0^2}$

reject H_0 if $\chi^{2^*} > \chi^2_{\alpha/2,n-1}$ or if $\chi^{2^*} < \chi^2_{1-\alpha/2,n-1}$

ONE-TAILED TEST

$H_0: \sigma^2 \leq \sigma_0^2$	$H_0: \sigma^2 \geq \sigma_0^2$
$H_a: \sigma^2 > \sigma_0^2$	$H_a: \sigma^2 < \sigma_0^2$
reject H_0 if $\chi^{2^*} > \chi^2_{\alpha,n-1}$	reject H_0 if $\chi^{2^*} < \chi^2_{1-\alpha,n-1}$

EXAMPLE 8.14

Example 8.13 was concerned with the variation of the actual weight of a (supposedly) 50-pound bag of Vitamix Dog Chow. Based on earlier production tests, management is convinced that the average weight of all bags being produced is, in fact, 50 pounds. However, the production supervisor has been informed that at least 95% of the bags produced *must* be within 1 pound of the specified weight (50 pounds). Using a significance level of $\alpha = .1$, what can we conclude? Assume a normal distribution for the bag weights.

What is the supervisor being told about σ? Remember that for a normal population, 95% of the observations will lie within two standard deviations of the mean (empirical rule, Chapter 3). So, if two standard deviations are equivalent to 1 pound, then the supervisor is being told that σ must be no more than .5 pound. Is there any evidence to conclude that this is not the case—that is, that σ is larger than .5 pound? Let's investigate.

SOLUTION

Step 1. The appropriate hypotheses are $H_0: \sigma \le .5$ and $H_a: \sigma > .5$ (production is not meeting required standards).

(Note that these hypotheses are precisely the same as $H_0: \sigma^2 \le .25$ and $H_a: \sigma^2 > .25$. Whether you write H_0 and H_a in terms of σ or σ^2 does not matter; the testing procedure is the same in either case.)

Step 2. The test statistic is

$$\chi^2 = \frac{(15-1)s^2}{(.5)^2} = \frac{14s^2}{.25}$$

which has a chi-square distribution with 14 df.

Step 3. Using $\alpha = .1$ and Table A.6, the rejection region for this test is

$$\text{reject } H_0 \text{ if } \chi^2 > 21.1$$

Step 4. The computed value using the sample data is

$$\chi^{2*} = \frac{(15-1)(1.65)^2}{(.5)^2} = 152.5$$

Since $152.5 > 21.1$, we reject H_0. This is hardly a surprising result; the point estimate of σ is $s = 1.65$, quite a bit larger than .5.

Step 5. We conclude rather convincingly that σ is larger than .5 pound. The bagging procedure has far too much variation in the weight of the bags produced. ∎

Note that the p-value for the test of hypothesis in Example 8.14 is the area to the right of 152.5 under the χ^2 curve with 14 df (illustrated in Figure 8.23). All we are able to determine about this value using Table A.6 is that it is much smaller than .005 (the smallest tabulated value). Using this information, we arrive at the same decision—namely, reject H_0—because (1) using the classical approach, p is less than $\alpha = .10$, or (2) the p-value is extremely small ($<.01$) by the general rule of thumb described in Section 8.3.

■ **FIGURE 8.23**
Illustration for the p-value for Example 8.14.

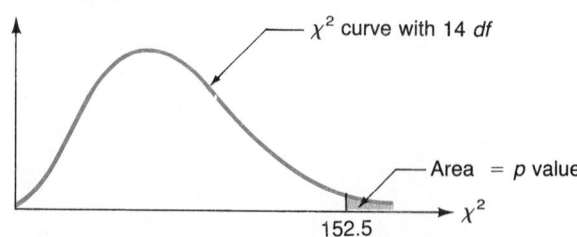

Exercises

8.47 From the tabulated values for the chi-square distribution, find the following values and indicate graphically where the values fall with respect to other values of the chi-square distribution.

a. $\chi^2_{.10,10}$

b. $\chi^2_{.025,30}$

c. $\chi^2_{.95,15}$

d. $\chi^2_{.01,26}$

8.48 A sample of size 25 from a normally distributed population yields a sample standard deviation of 12.8. At the 10% significance level, determine if there is sufficient evidence to indicate that the population standard deviation is greater than 11.3.

8.49 A sample of size 15 from a normally distributed population yields the sample statistic

$$\Sigma(x - \bar{x})^2 = 180.3$$

a. Construct a 90% confidence interval for the population variance.

b. Construct a 90% confidence interval for the population standard deviation.

c. Do the data indicate that the population variance differs from 10? Use a 10% significance level.

8.50 A sample of size five from a normally distributed population yields the following sample statistics:

$$\Sigma x^2 = 135 \qquad \Sigma x = 23$$

a. Construct a 95% confidence interval for the population variance.

b. Construct a 95% confidence interval for the population standard deviation.

c. Is there sufficient evidence to indicate that the population standard deviation differs from 2.8? Use a 5% significance level.

8.51 The production manager of Crystal-Clear Picture Tubes believes that the life of the company's picture tubes is 25,000 hours. However, to maintain the company's reputation for quality, the manager would like to keep the standard deviation of the life span of the picture tubes below 1000 hours. A sample of 24 picture tubes was randomly selected and the sample standard deviation was found to be 928 hours. Do the data indicate that the population standard deviation is less than 1000 hours? Use a 10% significance level. What assumption must be made about the distribution of the life span of the picture tubes?

8.52 For a random variable χ^2, which has chi-square distribution with 18 df, determine the values of a and b such that $P(a \le \chi^2 \le b) = .90$ and such that the areas in each tail are equal, that is, $P(0 \le \chi^2 < a) = P(b < \chi^2)$.

8.53 A production manager in charge of manufacturing plastic discs must maintain a standard deviation of less than 2 millimeters for the diameter of the disc. A sample of 26 plastic discs randomly selected reveals a standard deviation of 1.85 millimeters. Assuming that the diameters of the disc are normally distributed, do the data indicate that the standard deviation of the disc is less than 2 millimeters? Use the p-value criteria.

8.54 The salaries for mathematics teachers in secondary schools in Connecticut are believed to be normally distributed with a variance greater than $3000. Test this belief using the following sample statistics:

$$\Sigma(x - \bar{x})^2 = 45,130$$

$$n = 14$$

where X represents a math teacher's salary. Use a 1% significance level.

8.55 With the widespread use of computer terminals in American society, numerous studies have investigated the interaction between human beings and machines. One study looked at (among other things) the degree of flicker and fuzziness that a human operator can tolerate for characters on the video screen. A scale of measure for flicker and fuzziness was devised by the researchers, who concluded that a variance of .25 was the optimum level. A manufacturer of video terminals claimed that his machines have precisely this level of variance (i.e., .25). A sample of 40 terminals from this manufacturer had a variance of .29. At a 1% significance level, do the data indicate that the true population variance may differ from .25?

8.56 A real estate agent believes that the standard deviation of the prices of homes in Bloomington Hill Estates is less than \$6000. A random sample of 25 homes in Bloomington Hill Estates yields a sample standard deviation of \$5030. Do the data support the agent's belief? Use the p-value criteria.

◢ Summary

In Chapter 7 you were introduced to the topic of **statistical inference** by discussing the concept of estimating a population parameter (such as μ or σ) by using a corresponding sample estimate. The reliability of using the sample mean to estimate μ was measured using a confidence interval. This chapter presented the other side of statistical inference—**hypothesis testing** regarding these two population parameters, along with a method of deriving a confidence interval for the population standard deviation or variance.

For testing against a hypothetical value of the population mean (μ), we introduced a procedure that used the standard normal (Z) distribution for large samples ($n > 30$) and the t distribution for small samples. For small samples, the hypotheses are concerned with the mean of a normal population. However, the Central Limit Theorem allows us to discuss the mean of any continuous population when we have a large sample.

The two hypotheses under investigation are the **null hypothesis, H_0,** and the **alternative hypothesis, H_a.** Typically, a claim that one is attempting to demonstrate goes into the alternative hypothesis.

Since any test of hypothesis uses a sample to infer something about a population, errors can result. Two specific errors are of great concern when you use the hypothesis-testing procedure. A **Type I error** occurs in the event you reject a null hypothesis when in fact it is true; a **Type II error** occurs when you fail to reject a null hypothesis when in fact it is not true.

The probability of a Type I error is the **significance level** of the test and is written as α. The probability of a Type II error is β; large values of β are associated with small values of α and vice versa. To define a test of hypothesis, you **select a value of α** that considers the cost of rejecting a correct H_0 and failing to reject an incorrect H_0. Typical values of α range from .01 to .1.

The **power** of a statistical test is defined as $1 - \beta$ and is equal to the probability of rejecting H_0 when it is in fact false. The value of β (and so $1 - \beta$) depends on the actual value of the parameter under investigation, and so the power of the test can be obtained for each possible value of this parameter. The resulting set of power values defines a **power curve** for this test of hypothesis.

A five-step procedure was defined for any test of hypothesis:

Step 1. Set up H_0 and H_a.

Step 2. Define the **test statistic,** which is evaluated using the sample data.

Step 3. Define a **rejection region,** using the value of α, by selecting a **critical value** from the appropriate table.

Step 4. Calculate the value of the test statistic from the sample data and carry out the test. This will result in rejecting H_0 or failing to reject H_0.

Step 5. Give a conclusion in the language of the problem.

A test such as H_0: $\mu = 50$ versus H_a: $\mu \neq 50$ is called a **two-tailed test** because we reject H_0 whenever the sample estimate of μ ($\bar{X}$) is either too large (test statistic is in the right tail) or too small (test statistic is in the left tail). Similarly, a test on the population variance (or standard deviation) such as H_0: $\sigma^2 = .2$ versus H_a: $\sigma^2 \neq .2$ also is a two-tailed test.

H_0: $\mu \leq 50$ versus H_a: $\mu > 50$ or H_0: $\mu \geq 50$ versus H_a: $\mu < 50$

or

H_0: $\sigma^2 \leq .2$ versus H_a: $\sigma^2 > .2$ or H_0: $\sigma^2 \geq .2$ versus H_a: $\sigma^2 < .2$

are all examples of **one-tailed tests** of hypothesis, since the rejection region lies in either the left tail *or* the right tail.

The tests on a population variance introduced the **chi-square distribution, χ^2.** This distribution was used to construct confidence intervals for σ^2 and σ as well as to define a distribution for the test statistic when performing a test of hypothesis on the variance or standard deviation.

Finally, we discussed why you should always include a *p-value* in the results of any hypothesis test. This value measures the strength of your point estimate (such as $\bar{X}$ or s^2). When using a predetermined significance level, α, you reject H_0 whenever the *p*-value is less than α and fail to reject H_0 otherwise. Another option you can use is not to select the somewhat arbitrary value of α but simply to reject H_0 whenever the *p*-value is "small" (say, $<.01$), fail to reject H_0 if it is "large" (say, $>.1$), or decide that the data are inconclusive if the *p*-value lies between these two values. You can also use the *p*-value to measure the enthusiasm (*p*-value very small) with which you reject H_0 or the authority (*p*-value quite large) with which you fail to reject H_0.

Summary of Formulas

1. Standard error of $\bar{X}$

$$s_{\bar{x}} = \begin{cases} \dfrac{s}{\sqrt{n}} \sqrt{\dfrac{N-n}{N-1}} & \text{if population size, } N, \text{ is known} \\[2ex] \dfrac{s}{\sqrt{n}} & \text{if } N \text{ is unknown or infinite} \end{cases}$$

2. Test statistic for hypothesis testing on a population mean

$$\frac{\bar{X} - \mu_0}{s_{\bar{x}}}$$

This statistic has an approximately standard normal distribution (Z) for *large samples* and a t distribution with $n - 1$ df for *small samples*.

3. Power of a test for $H_0: \mu = \mu_0$ versus $H_a: \mu \neq \mu_0$ (large sample)

$$\text{Power} = P(Z > z_1) + P(Z < z_2)$$

where Z is the standard normal random variable,

$$z_1 = Z_{\alpha/2} - \frac{(\mu - \mu_0)\sqrt{n}}{\sigma},$$

$$z_2 = -Z_{\alpha/2} - \frac{(\mu - \mu_0)\sqrt{n}}{\sigma}$$

and μ is the specified value of the population mean.

4. Confidence interval for a population variance (σ^2)

$$\frac{(n-1)s^2}{\chi^2_{\alpha/2,n-1}} \quad \text{to} \quad \frac{(n-1)s^2}{\chi^2_{1-\alpha/2,n-1}}$$

5. Confidence interval for a population standard deviation (σ)

$$\sqrt{\frac{(n-1)s^2}{\chi^2_{\alpha/2,n-1}}} \quad \text{to} \quad \sqrt{\frac{(n-1)s^2}{\chi^2_{1-\alpha/2,n-1}}}$$

6. Test statistic for hypothesis testing on a population variance or standard deviation

$$\chi^2 = \frac{(n-1)s^2}{\sigma_0^2}$$

where σ_0^2 is the hypothesized variance.

Review Exercises

8.57   Explain how changes in the α level affect the following.

a.   The rejection region.

b.   The Type II error.

8.58   The manager of Jack-Be-Nimble candle company would like to claim that a certain type of their candles burns more than 14 hours. To test this claim, the manager randomly selects 50 candles and finds that the sample mean is equal to 14.75 hours with a standard deviation of 1.8 hours.

a.   What are the null and alternative hypotheses?

b.   Which error would you consider to be more serious, Type I or Type II?

c.   At a significance level of .05, what is your conclusion?

8.59   Given the following statistics from a normally distributed population, is there sufficient evidence to support the claim that the mean of the population differs from 235.6?

$$n = 21$$
$$\bar{x} = 234.1$$
$$\sigma = 2.3$$

Use the p-value to draw your conclusion.

8.60   The manager of the Train Depot Restaurant believes that the average time customers wait before being served is 10 minutes. To test the belief, the manager selects 50 customers at random and records that the average waiting time is 11.9 minutes with a standard deviation of 1.4 minutes.

a.   Find a 95% confidence interval for the mean waiting time of a customer.

b.   Do the data indicate that the mean waiting time differs from 10 minutes, at a 5% significance level?

8.61   Calculate the power of the test for the mean of a normally distributed population with known population variance for the following situations, assuming that the true population mean is 10 and the known population standard deviation is 3.1. Use a significance level of .05.

a. $H_0: \mu = 11, H_a: \mu \neq 11, n = 14$.

b. $H_0: \mu = 9.5, H_a: \mu \neq 9.5, n = 25$.

c. $H_0: \mu = 8, H_a: \mu \neq 8, n = 40$.

8.62   The federal government provides an enormous amount of funding to the states for research and development (R&D), according to a federal official interviewed by a journalist. These funding amounts are believed to follow a normal distribution with a standard deviation of \$45 million. The official said that each state, on the average, gets \$150 million or more per annum. The journalist thought this figure might be too high and felt it was less than \$150 million. Ten states were selected at random. The average amount of federal funding for R&D received per state was \$120 million. Does this provide enough evidence to reject the federal official's claim, using a 5% significance level? Determine the p-value for your test. (*Hint:* Use $N = 50$ in the finite population correction factor.)

8.63   A manufacturer of drugs and medical products claims that a new anti-inflammatory drug will be effective for 4 hours after the drug is administered in the prescribed dosage. A random sample of 50 volunteers demonstrated that the average effective time is 3.70 hours with a sample standard deviation of .606 hours. Use the p-value criteria to determine if there is sufficient evidence to support the hypothesis that the mean effective time of the drug differs from 4 hours.

8.64   Indicate what the p-values are for the following situations, in which the mean of a normally distributed population is being tested.

a. $H_0: \mu = 31.6, H_a: \mu \neq 31.6$ (population variance is known), $Z^* = 2.16$.

b. $H_0: \mu = 4.07, H_a: \mu \neq 4.07$ (population variance is known), $Z^* = -1.35$.

c. $H_0: \mu = 87.6, H_a: \mu \neq 87.6$ (population variance is unknown), $t^* = 2.51, n = 15$.

d. $H_0: \mu = 195.3, H_a: \mu \neq 195.3$ (population variance is unknown), $t^* = -1.71$, $n = 25$.

8.65   Using the following information, perform the hypothesis test for the mean of a normally distributed population:

$$H_0: \mu \geq 7.19$$
$$H_a: \mu < 7.19$$
$$\bar{x} = 6.21$$
$$s^2 = .26$$
$$n = 23$$
$$\alpha = .10$$

8.66 The vice president of academic affairs at a small private college believes that the average full-time student who lives off campus spends about $300 per month for housing. A random sample of 200 full-time students living off campus spent an average of $305 per month with a standard deviation of $70 a month.

a. Find the p-value to determine whether there is sufficient evidence to indicate that a full-time student spends more than $300 per month on housing.

b. Would you reject the null hypothesis for the test in question a if $\alpha = .01$? if $\alpha = .05$? if $\alpha = .10$?

8.67 A marketing analyst is looking at the feasibility of opening a new movie theater in a small town. The town currently has only two movie theaters. The movie theater would be a practical investment if the average family in the town spends at least 14 hours at the movies each year. A random sample of 80 households yielded a sample mean of 14.5 hours per year with a standard deviation of 1.4.

a. Find the 95% confidence interval for the mean time that a family spends per year at the movies.

b. Is there sufficient evidence to indicate that the mean time that a family spends at the movies is greater than 14 hours per year? Use a .05 significance level.

8.68 From a finite population of size 425, a random sample of size 20 is drawn. The finite population can be approximated by a normal distribution. From the sample, it is found that

$$\bar{x} = 43.7 \qquad s^2 = 6.7$$

Do the data support the statement that the mean of the population is greater than 40? Use a 1% significance level.

8.69 A quality-control engineer is interested in the average time of work stoppages in a production process. From a random sample of 15 different work stoppages, the sample mean was 33 minutes and the standard deviation was 11.6 minutes. Find a 90% confidence interval for the mean time of the work stoppages for the production process.

8.70 The owners of a shopping center are contemplating increasing the parking space in front of the shopping center. The owners would like to demonstrate that the average driver parks for more than .75 hours. The length of time parked is considered to be normally distributed. A random sample of 45 parked cars is observed; the average time parked was .80 hours with a standard deviation of .12. Do the data support the idea that the average driver parks for more than .75 hours? Use a 10% significance level.

8.71 There are 420 persons attending a conference on the Strategic Defense Initiative (SDI). A reporter randomly selected 50 persons from those attending, to determine the average income of the participants. From this sample, a mean of 38.6 (thousand dollars) and a standard deviation of 6.7 (thousand dollars) were obtained.

a. Construct the 95% confidence interval for the mean income of the conference participants.

b. Do the data indicate that the average income is less than 40 (thousand dollars)? Use a 5% significance level.

c. Determine the p-value.

8.72 From a normally distributed population, a random sample of size 22 yields the following statistic:

$$\Sigma(x - \bar{x})^2 = 1.67$$

a. Find a 95% confidence interval for the population variance.

b. Find a 95% confidence interval for the population standard deviation.

c. Test the null hypothesis that the population variance is equal to .07. Use a two-tailed test and a .05 significance level.

8.73 Using a significance level of .05, perform the hypothesis test for the standard deviation of a normally distributed population, given the following information:

$$H_0: \sigma \geq 20.6$$
$$H_a: \sigma < 20.6$$

$$\Sigma(x - \bar{x})^2 = 6100$$
$$n = 18$$

8.74   Eastern State Bank currently operates five drive-in teller windows. Management is concerned about the variability of the time spent waiting by a customer using the windows. A sample of 24 customers was taken, and the sample standard deviation was found to be 4.7 minutes. Management would like to keep the standard deviation below 4 minutes and may consider adding another drive-in teller window.

a.   Test the null hypothesis that the standard deviation of a customer's waiting time is less than or equal to 4 minutes. Use a 10% significance level.

b.   What assumption should be made about the distribution of the waiting time of customers who use the drive-in teller windows?

8.75   An investment counselor would like to know how much variability there is in the yield of money market funds. The yields of these funds can be considered to be approximately normally distributed for the time frame of interest. A sample of 21 money market funds yields a sample standard deviation of .7%. At the .05 significance level, is there sufficient evidence to indicate that the standard deviation of the yields of money market funds is greater than .6%?

8.76   A large university had recently converted to a computerized registration system for enrollment. After the first semester, administrators found that the average time spent registering per student was quite satisfactory, yet there still continued to be substantial complaints and dissatisfaction among students. Further study indicated that although the *average* time might seem satisfactory, there might be too much *variation* in the registration times. It was decided to study the situation during the next semester's registration period. If the standard deviation was greater than 20 minutes, six additional computer terminals would be installed; otherwise, two new computer terminals would be installed. From a random sample, the following data were obtained:

$$\Sigma(x - \bar{x})^2 = 6900, \qquad n = 18$$

Assume the population is normally distributed.

a.   Is there sufficient evidence to indicate that the population standard deviation exceeds 20 minutes? Use a 5% significance level.

b.   What is the decision indicated by the test: install six new terminals or two new terminals?

c.   State the *p*-value for the test.

8.77   Employees at Scranton Steel were asked to fill out a questionnaire to determine how they viewed the changes taking place in their company. A score of 3 meant no improvement had taken place. A score above 3 meant that the company has improved, with a score of 5 representing the highest possible improvement. A score below 3, with zero being the lowest, represented the opinion that the company had gotten worse. The sample mean and standard deviation were 3.30 and .65, respectively.

(*Source:* Adapted from M. Bees, R. Eisentet, B. Spector, "Why Change Programs Don't Produce Change," *Harvard Business Review* (November–December 1990): 158.)

a.   Do the data provide sufficient evidence that the company's changes were viewed as improvements? Assume that the sample was taken from 100 randomly selected employees. Use a 1% significance level.

b.   What assumptions are necessary for the statistical test in part a to be valid?

8.78   Economists at the Federal Reserve Bank of Chicago have compiled an index to show that the Midwest did benefit from the surge in exports in 1987. Using factory hours worked and electricity usage as a gauge of output, the economists found that factories in Illinois, Indiana, Iowa, Michigan, and Wisconsin boosted production by 4.2% in 1987. Assume that an economist was interested in whether manufacturers in the Midsouth shared in the same production increase. A sample of 50 manufacturers in the Midsouth was selected and their changes in production were recorded. The results can be summarized as follows:

$$\bar{x} = 3.8\%$$
$$s^2 = 0.81$$

(*Source:* "Why Mid-Western Manufacturers Outpace The Nation," *Business Week* (April 11, 1988): 27.)

a. Is there sufficient evidence to claim that the mean production increase in the Midsouth is less than 4.2% at the .01 significance level?

b. Suppose the standard deviation of the production increase for manufacturers in the Midwest is believed to be equal to 1. Does the sample evidence indicate that the standard deviation for the production increase in the Midsouth differs significantly from 1? Use a .05 significance level.

8.79 Expert systems are special-purpose computer programs that help to solve problems in the same manner as human experts. For 51 different situations involving various production planning problems, the amount of time spent by a manager interacting with a particular expert system on one of these production planning problems is given below in units of hours. The following MINITAB computer printout displays a 95% confidence interval.

```
1.2  1.4  0.9  1.8  0.7  1.1  1.5  0.6  0.8  1.2  1.3
1.7  0.7  0.9  1.2  1.5  0.7  1.0  1.3  1.1  1.4  1.8
1.2  1.1  0.4  0.3  1.8  0.7  1.9  0.9  1.0  0.9  0.7
1.5  0.8  1.1  1.7  1.9  1.2  0.7  0.6  1.0  1.3  1.4
0.7  1.9  1.1  1.2  0.7  1.1  1.2
```

MTB > describe C1

	N	MEAN	MEDIAN	TRMEAN	STDEV	SEMEAN
C1	51	1.1333	1.1000	1.1289	0.4068	0.0570

	MIN	MAX	Q1	Q3
C1	0.3000	1.9000	0.8000	1.4000

MTB > zint with 95 percent, sample stdev = .4068, data in c1

THE ASSUMED SIGMA = 0.407

	N	MEAN	STDEV	SEMEAN	95.0 PERCENT C.I.
C1	51	1.1333	0.4068	0.0570	(1.0215, 1.2451)

a. At the .05 significance level, can one conclude from the computer printout that the data support the belief that the mean time spent by the manager with the expert system is not equal to one hour?

b. If a significance level of .10 is used, can the question in part a be answered using the same computer printout? Why?

c. What justifies the use of the confidence interval with the normal distribution instead of the t distribution in part a?

8.80 The manager at A & A Plumbing and Air-Conditioning Company is concerned about the time that it takes a serviceperson to repair a defective switch in a high-efficiency heat pump found in local residences. The manager believes that the average repair time is greater than 30 minutes. Past data shown in the histogram below give the following times in minutes for a serviceperson to service a defective switch.

MTB > describe c1

	N	MEAN	MEDIAN	TRMEAN	STDEV	SEMEAN
C1	25	31.348	31.200	31.352	4.278	0.856

	MIN	MAX	Q1	Q3
C1	19.800	42.800	29.150	33.350

MTB > histogram of c1

Histogram of C1   N = 25

```
Midpoint   Count
      20      1   *
      22      0
      24      0
      26      1   *
      28      4   ****
      30      6   ******
      32      6   ******
      34      4   ****
      36      1   *
      38      1   *
      40      0
      42      1   *
```

```
MTB > ttest mu = 30, data in c1

TEST OF MU = 30.000 VS MU N.E. 30.000

              N       MEAN    STDEV    SE MEAN          T    P VALUE
C1           25     31.348    4.278      0.856       1.58       0.13

MTB > ttest mu = 30, data in c1;
SUBC> alternate = 1.

TEST OF MU = 30.000 VS MU G.T. 30.000

              N       MEAN    STDEV    SE MEAN          T    P VALUE
C1           25     31.348    4.278      0.856       1.58      0.064

MTB > tinterval with 90 percent confidence, data in c1

              N       MEAN    STDEV   SE MEAN    90.0 PERCENT C.I.
C1           25     31.348    4.278     0.856   ( 29.884,  32.812)
```

a. Do the assumptions needed to perform a *t*-test appear to hold?

b. Do the data support the manager's claim? At what significance level? (Note that the MINITAB printout with "Alternate = 1" indicates a right-tailed test. "Alternate = 0" is the default value if no value is assigned to the alternate.)

c. Should the same conclusion hold in testing the manager's belief for a one-tailed test as for a two-tailed test? What is the conclusion from the two-tailed test? Does the confidence interval support this conclusion at a significance level of 10%?

8.81 To illustrate the difference between the *t* distribution and the normal distribution, the MINITAB computer printout below shows the output of 1000 observations randomly generated from a population with a normal distribution with mean 0 and variance 1 and also from a population with a *t* distribution with degrees of freedom equal to 7. Comment on the differences between the two histograms.

```
MTB > random 1000 observations, put into c1;
SUBC> normal mean = 0, sigma = 1.
MTB > histogram c1

Histogram of C1   N = 1000
Each * represents 5 obs.

Midpoint    Count
    -3.5        1   *
    -3.0        3   *
    -2.5        5   *
    -2.0       28   ******
    -1.5       65   *************
    -1.0      112   **********************
    -0.5      191   **************************************
     0.0      191   **************************************
     0.5      177   ***********************************
     1.0      111   **********************
     1.5       72   ***************
     2.0       31   *******
     2.5        8   **
     3.0        2   *
     3.5        2   *
     4.0        1   *

MTB > random 1000 observations, put into c1;
SUBC> t degrees of freedom = 7.
MTB > histogram of c1

Histogram of C1   N = 1000
Each * represents 10 obs.

Midpoint    Count
      -4        7   *
      -3       18   **
      -2       69   *******
      -1      212   *********************
       0      379   **************************************
       1      225   ***********************
       2       67   *******
       3       18   **
       4        1   *
       5        3   *
       6        0
       7        1   *
```

8.82 Use a convenient statistical computer package to generate 24 observations from a normal population with mean 0 and standard deviation 4, and then find a 90% confidence interval of the mean using the t distribution. Execute this program 200 times. Find the number of times that the confidence intervals contained the true mean. Should the confidence interval always contain the true mean? What does the 90% confidence level indicate? The needed program using MINITAB commands is given below. Simply change the command "execute 1" to "execute 200."

```
MTB > store
MTB > random 24 observations, put into c1;
MTB > normal with mean = 0 , sigma = 4.
MTB > tinterval with confidence level = 90,data in c1
MTB > end
MTB > execute 1
```

Computer Exercises Using the Database

Exercise 1 -- Appendix I Randomly select 100 observations from the database. Use a convenient statistical computer package to determine whether the sample evidence indicates that the mean of the variable TOTLDEBT (total indebtedness) exceeds $10,000 at the .05 significance level. Also determine whether the sample evidence indicates that the standard deviation of the variable TOTLDEBT exceeds $3000 at the .05 significance level.

Exercise 2 -- Appendix I Randomly select from the database 30 observations in which the location of residence is in the NE sector, and also randomly select another 30 observations in which the location of residence is in the NW sector. Find separate 90% confidence intervals on the mean of the variable INCOME1 (income of principal wage earner) from each of these two sets of data. Comment on the difference in the confidence intervals.

Exercise 3 -- Appendix J Randomly select from the database 30 observations from companies with a bond rating of A and 30 observations from companies with a bond rating of C. Find separate 95% confidence intervals on the mean of sales minus cost of sales for each of these two random samples. Compare and comment on the differences.

Case Study

Centering and Variation—A First Look at Quality Control

Inherent in any manufacturing process is a certain amount of shift and variation. For a process to be "in control," the process must be centered approximately on target and not display unusual variation. For example, bags of lawn fertilizer advertised as weighing 25 lbs would have a target of 25 lbs, and ideally all bags would weigh exactly 25 lbs. In practice, however, the production process might be producing bags with an average weight of 24.5 lbs (a shift of .5 lb from target) and can be expected to have a certain amount of natural variation since no two bags weigh exactly the same amount and no production process is perfectly consistent. This topic will be developed further in Chapter 12.

A sample of 45 observations was obtained from the fluid seals division of George Angus Ltd. These observations were obtained by taking nine samples of five observations each, but for this discussion, we will treat the data as a single sample of 45 observations.* Shaft seals are produced by an injection-molding process. Grease is applied to the surface of the seal of the product that will be in contact with the shaft. The amount of grease applied is important and provides initial lubrication for a new engine prior to the oil circulating. Too little grease will cause the seal to run dry and be damaged. An excess of grease will vulcanize the seal and cause malfunction. To assess the efficiency of the grease deposition process, a number of weighted blanks are passed through the machine and reweighed on exit. The gain in weight in grams is recorded and tested against a desired mean of .09 and acceptable standard deviation of .011. The sample of 45 observations (weight gain, in grams) is contained in Table 1.

---

*Chapter 12 will explain how to analyze this sample as nine separate samples of five observations each.

■ Table 1

.094	.072	.072	.079	.104	.074	.078	.110	.076
.067	.064	.096	.085	.088	.089	.097	.066	.084
.089	.105	.102	.072	.079	.064	.089	.085	.098
.075	.096	.084	.108	.099	.088	.099	.096	.106
.108	.110	.085	.082	.098	.090	.115	.123	.076

(*Source:* M. A. A. Cox, "Control Charts on Spreadsheets: A Tutorial Guide or a Poor Man's Persuader," *Quality Engineering* 1, no. 2 (1988–89): 135.

Case Study Questions

1. Determine the sample mean and standard deviation.
2. Using a significance level of .05, is the process centered at .09 grams? Determine the corresponding *p*-value.
3. Using a significance level of .05, is there evidence to indicate that the process standard deviation is larger than .011? Determine the corresponding *p*-value.
4. Determine the probability of a Type II error (β)

for the statistical test in question 2 if the process (population) mean is in fact:

a. .08   b. .082   c. .084   d. .086
e. .088   f. .09   g. .092   h. .094
i. .096   j. .098   k. .10

5. Rather than constructing a power curve for this test, construct a curve with the true mean on the horizontal axis (as in a power curve) but use the Type II error probability (β) on the vertical axis. The resulting curve is called an **operating characteristic (OC)** curve and will be explained in more detail in Chapter 12.
6. Will the OC curve for a test on the mean always be symmetric? Is it necessary to include both halves of the OC curve in question 5?
7. Can you devise a way to construct the OC curve in question 5 that would apply to any normally distributed process, that is, for any mean and standard deviation?

SPSS

▨ Solution
Example 8.10

Example 8.10 was concerned with the computation of means and a *t* statistic to examine average differences between the recorded and audited values of inventory items. You can use SPSS to solve this problem by borrowing a technique that actually belongs in Chapter 9 (and will be used again in that chapter). The program listing in Figure 8.24 was used to request the calculation of the *t* statistic and the corresponding *p*-value. Note that SPSS automatically assumes a two-tailed test; this was a one-tailed test, so the calculated *p*-value needs to be divided by 2. In this problem the SPSS commands are the same for both the mainframe and PC versions (remember to end each command line with a period when using the PC version).

The TITLE command names the SPSS run.

The DATA LIST command gives each variable a name and describes the data as being in free form. The variable AMT (abbreviated form of "amount") is defined as the value of the mean contained in the hypotheses. Actually, AMT is a constant but is treated as a variable in the analysis.

The BEGIN DATA command indicates to SPSS that the input data immediately follow.

The next 15 lines contain the data values. Each line contains the difference (DIFFER) between the stated value and the audited value of one inventory item and the constant 25 (AMT).

The END DATA statement indicates the end of the data.

The T-TEST PAIRS = DIFFER AMT command computes the *t* statistic using the variable DIFFER and the corresponding *p*-value.

Figure 8.25 shows the SPSS output obtained by executing the listing in Figure 8.24.

■ FIGURE 8.24
Input for SPSS or
SPSS/PC. Remove
the periods for SPSS
input.

```
TITLE   DEFENSE.
DATA LIST FREE / DIFFER , AMT.
BEGIN DATA.
17 25
35 25
31 25
22 25
50 25
42 25
56 25
23 25
27 25
38 25
20 25
25 25
43 25
45 25
21 25
END DATA.
T-TEST PAIRS = DIFFER AMT.
```

■ FIGURE 8.25   SPSS output.

```
Paired samples t-test:  DIFFER
                        AMT

Variable    Number                Standard   Standard
            of Cases   Mean       Deviation   Error        for a one-tailed test the
                                                            actual p-value = .023/2 = .012
DIFFER        15       33.0000    12.148     3.137
AMT           15       25.0000     .000       .000

(Difference) Standard   Standard  |   2-Tail   |   t      Degrees of   2-Tail
  Mean       Deviation    Error   | Corr. Prob.|  Value    Freedom     Prob.

  8.0000     12.148      3.137    | 99.00099.000|  2.55      14         .023
-----------------------------------------------------------------------------
```

value of *t*-statistic   *p*-value

SAS

▨ Solution
Example 8.10

Example 8.10 was concerned with the computation of means and a *t* statistic to examine average differences between the recorded and audited values of inventory items. You can use SAS to solve this problem. The program listing in Figure 8.26 was used to request the calculation of the *t* statistic and the resulting *p*-value. Note that SAS automatically assumes a two-tailed test; this was a one-tailed test, so the calculated *p*-value needs to be divided by two. In this problem the SAS commands are the same for both the mainframe and PC versions.

The TITLE command names the SAS run (enclose it in single quotes).

The DATA command gives the data a name.

The INPUT command gives the variable a name.

The HO1 = DIFFER-25.0. statement is used to compute a new variable, HO1, which is the difference between the variable DIFFER and the constant 25.0.

The CARDS command indicates to SAS that the input data immediately follow.

The next 15 lines are the data values, representing the differences between the stated value and the audited value of the inventory items along with the constant 25.

The PROC MEANS command requests an SAS procedure to print the number of observations, the mean, the t statistic, and the p-value.

The VAR statement specifies that the variable HO1 is the variable to be used in computing the statistics.

The TITLE statement specifies the heading for the printout.

Figure 8.27 shows the SAS output obtained by executing the listing in Figure 8.26.

■ **FIGURE 8.26**
Input for SAS
(mainframe or
micro version).

```
TITLE     'DEFENSE';
DATA DIFFDATA;
INPUT DIFFER;
 HO1=DIFFER-25.0;
CARDS;
17.0
35.0
31.0
22.0
50.0
42.0
56.0
23.0
27.0
38.0
20.0
25.0
43.0
45.0
21.0
PROC PRINT;
PROC MEANS N MEAN T PRT;
 VAR HO1;
 TITLE 'ONE-TAILED TEST';
```

■ **FIGURE 8.27**
SAS output.

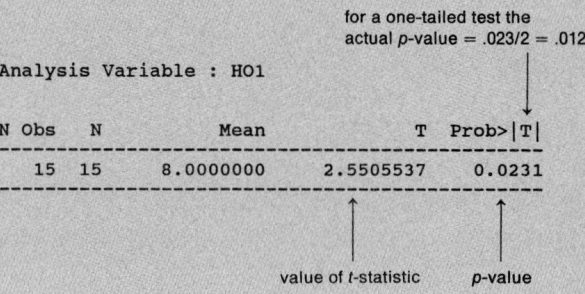

```
                            ONE-TAILED TEST

                                       for a one-tailed test the
                                       actual p-value = .023/2 = .012

        Analysis Variable : HO1
                                                        │
                                                        ▼
        N Obs   N          Mean           T   Prob>|T|
        -------------------------------------------------
         15     15      8.0000000    2.5505537    0.0231
        -------------------------------------------------

                                         ▲            ▲
                                         │            │
                              value of t-statistic  p-value
```

CHAPTER

Inference Procedures for Two Populations

We have learned to describe and summarize data from a single population using a statistic (such as the sample mean, $\bar{X}$) or a graph (such as a histogram). Chapters 7 and 8 introduced you to statistical inference, where we (1) attempted to estimate a parameter (such as the mean, μ) from this population by using the corresponding sample statistic and (2) arrived at a conclusion about this parameter (such as $\mu > 5.9$ ft) by performing a test of hypothesis. The concept behind hypothesis testing was described, and we paid special attention to the errors (Type I and Type II) that can occur when we use a sample to infer something about a population.

Next we learn how to compare two populations. Questions of interest here include:

1. Are the values in population 1 larger, on the average, than those in population 2? (For example, are men taller, on the average, than women?)

2. Do the values in population 1 exhibit more variation than those in population 2? (For example, do male heights vary more than female heights?)

The two populations under observation may or may not be normally distributed. When we compare two population means using large samples, once again using the Central Limit Theorem, the type of distribution simply does not matter. For small samples, we need to examine the distributions of the populations so that we can use the proper procedure to construct confidence intervals and perform tests of hypothesis.

This chapter discusses two different sampling situations. In the first, random samples from two populations are obtained *independently* of each other; in the second, corresponding data values from the two samples are matched up, or paired. Paired samples are *dependent*.

9.1
Independent Versus Dependent Samples

When making comparisons between the means of two populations, we need to pay particular attention to how we intend to collect sample data. For example, how would you determine if tire brand A lasts longer than brand B? You might decide to put one of each brand on the rear wheels of ten cars and measure the tires' wear. Or you might randomly select ten brand A and ten brand B tires, attach them to a machine that wears them down for a certain time, and then measure the resulting tire wear. If you use the first procedure (putting both brands of tire on the same car), you obtain *dependent* samples; in the latter situation, you obtain *independent* samples.

Consider another situation. Suppose you are interested in male heights as compared with female heights. You obtain a sample of $n_1 = 50$ male heights and $n_2 = 50$ female heights. You obtain these data:

OBSERVATION	MALE HEIGHTS	OBSERVATION	FEMALE HEIGHTS
1	5.92 ft	1	5.36 ft
2	6.13 ft	2	5.64 ft
3	5.78 ft	3	5.44 ft
⋮	⋮	⋮	⋮
50	5.81 ft	50	5.52 ft

Is there any need to match up 5.92 with 5.36, 6.13 with 5.64, 5.78 with 5.44, and so on? The male heights were randomly selected and the female heights were obtained independently, so there is no reason to match up the first male height with the first female height, the second male height with the second female height, and so on. Nothing relates male 1 with female 1 other than the accident of their being selected first—these are **independent samples.**

What if you wish to know whether husbands are taller than their wives? To collect data, you select 50 married couples. Suppose you obtain the 100 observations from the previous male and female height example. Now, is there a reason to compare the first male height with the first female height, the second with the second, and so on? The answer is a definite yes, since each pair of heights belongs to a married couple. The resulting two samples are **dependent, or paired, samples.**

In summary,

1. If there is a definite reason for pairing (matching) corresponding data values, the two samples are **dependent** samples.
2. If the two samples were obtained independently and there is no reason for pairing the data values, the resulting samples are **independent** samples.

Why does this distinction matter? *If you are trying to decide whether male heights are, on the average, greater than female heights, the procedure that you use for testing this depends on whether the samples are obtained independently.*

Applications of dependent samples in a business setting include data from the following situations.

1. Comparisons of *before versus after.* Sample 1: person's weight before a diet plan is begun. Sample 2: person's weight 6 months after starting the diet. Why do we pair the data? We pair them because each pair of observations belongs to the same person.
2. Comparisons of people with *matching characteristics.* Sample 1: salary for a male employee at Company ABC. Sample 2: salary for a female employee at Company ABC, where the woman's education and job experience are equal to the man's. Why do we pair the data? We pair them because the two paired employees are identical in their job qualifications.
3. Comparisons of observations *matched by location.* Sample 1: sales of brand A tires for a group of *n* stores. Sample 2: sales of brand B tires for the same group of stores. Why do we pair the data? We pair them because both observations were obtained from the same store. Your data consist of sales (weekly, monthly, and so on) from a sample of stores selling these two brands.
4. Comparisons of observations *matched by time.* Sample 1: sales of restaurant A during a particular week. Sample 2: sales of restaurant B during this week. Why do we pair the data? We pair them because each pair of observations corresponds to the same week of the year.

 *Exercises*

9.1   For each of the following claims, determine whether the paired samples or the independent samples procedure would be appropriate.

a.   The mean of the scores attained by students before a tutorial session is less than the mean of the scores attained by the same students after the tutorial session.

b.   There is a difference between the mean grade point averages of females and males in the MBA program.

c.   The average wage for Japanese auto workers is less than that for European auto workers.

9.2   Two private colleges decided to compare the mean SAT scores of their incoming freshmen. One college gathered 98 scores and the other took a sample of 52 scores. Do the two sets of data represent dependent or independent samples?

9.3   A medical institution is examining the effectiveness of a newly developed drug. The drug was administered to 18 patients whose health condition before and after taking the drug was recorded. Is this a case of dependent or independent samples?

9.4   The advertising division of a chemical company would like to see how two different dishwashing detergents are rated by homemakers. Homemakers are chosen at random. They assign a value from zero to ten to each product. They assign a value of zero to the product if they think the detergent is worthless and ten to the detergent if they believe it is the best on market. How can dependent samples be chosen? How can independent samples be chosen?

9.5   The career placement center at Safire University conducts a survey of beginning salaries for MBAs with no on-the-job experience. Ten pairs of men and women are chosen randomly such that each pair of one man and one woman has nearly identical qualifications. Can the sample of observations from men be independent of the sample of observations from women?

9.6   A retail store would like to compare sales from two different arrangements of displaying its merchandise. Sales are recorded for a 30-day period with one arrangement and then sales are recorded for another 30-day period for the alternative arrangement. Can the data for each of the two 30-day periods be independent or dependent?

9.7   Fifty people were randomly selected to rate particular brands of soft drink on a scale from one to ten, with ten being the highest rating. If 25 people rated brand A and the other 25 people rated brand B, would the samples from these two groups be independent or dependent?

9.8   In Exercise 9.7 suppose the 50 people were each asked to rate both brand A and brand B. Would the sample of 50 observations of brand A be independent of the sample of 50 observations of brand B?

◢ 9.2
Comparing Two Means Using Two Large Independent Samples

When comparing the means of two independent samples from different populations, we can use Figure 9.1 to help visualize the situation. The two populations are shown to be normally distributed, but, because we will be using large samples from these populations, this is *not* a necessary assumption. For these populations,

$$\mu_1 = \text{mean of population 1}$$
$$\mu_2 = \text{mean of population 2}$$
$$\sigma_1 = \text{standard deviation of population 1}$$
$$\sigma_2 = \text{standard deviation of population 2}$$

For example, if we wished to compare U.S. adult male and female heights:

$$\mu_1 = \text{average of all female heights}$$
$$\mu_2 = \text{average of all male heights}$$
$$\sigma_1 = \text{standard deviation of all female heights}$$
$$\sigma_2 = \text{standard deviation of all male heights}$$

■ **FIGURE 9.1**
Example of two populations.
Is $\mu_1 = \mu_2$?

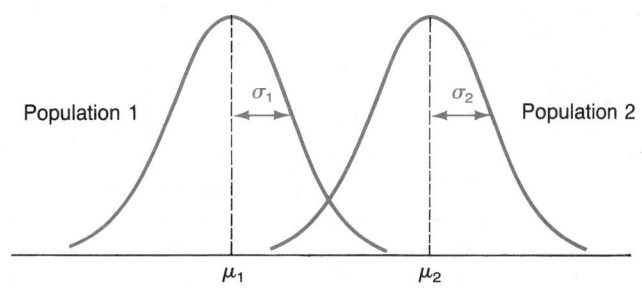

The point estimates discussed in earlier chapters apply here as well—we simply have two of everything because we are dealing with two populations.

The procedure we follow is to obtain a random sample of size n_1 from population 1 and then obtain another sample of size n_2, completely independent of the first sample, from population 2. So, $\bar{X}_1$ is our best (point) estimate of μ_1. Likewise, $\bar{X}_2$ estimates μ_2. The sample standard deviations (s_1 and s_2) provide the best estimates of the population standard deviations (σ_1 and σ_2).

Constructing a Confidence Interval for $\mu_1 - \mu_2$

Ace Delivery Service operates a fleet of delivery vans in the Houston area. They prefer to have all their drivers charge their gasoline using the same brand of credit card. Presently, they all use a Texgas credit card. Ace management has decided that perhaps Quik-Chek, a chain of convenience stores that also sells gasoline but does not accept credit cards, is worth investigating. A random sample of gas prices at 35 Texgas stations and 40 Quik-Chek stores in the Houston area is obtained. The cost of 1 gallon of regular gasoline is recorded; the data are summarized:

Sample 1 (Texgas)	Sample 2 (Quik-Chek)
$n_1 = 35$	$n_2 = 40$
$\bar{x}_1 = \$1.48$	$\bar{x}_2 = \$1.39$
$s_1 = \$ .12$	$s_2 = \$ .10$

Let μ_1 be the average price of regular gasoline at *all* Texgas stations in the Houston area, and let μ_2 be the average price of regular gasoline at all Quik-Chek stores in the Houston area.

When dealing with these two populations, the parameter of interest is $\mu_1 - \mu_2$, rather than the individual values of μ_1 and μ_2. Here, $\mu_1 - \mu_2$ represents the difference between the average gasoline prices at the Texgas stations and Quik-Chek stores. If we conclude that $\mu_1 - \mu_2 > 0$, then $\mu_1 > \mu_2$. In this case, the gasoline *is* more expensive at the Texgas stations.

The point estimator of $\mu_1 - \mu_2$ is the obvious one: $\bar{X}_1 - \bar{X}_2$. For our data, the (point) estimate of $\mu_1 - \mu_2$ is $\bar{x}_1 - \bar{x}_2 = 1.48 - 1.39 = .09$. How much more expensive is the gasoline from all of the Texgas stations, on the average? We do not know because this is $\mu_1 - \mu_2$, but we *do* have an estimate of this value—namely, 9¢.

What kind of random variable is $\bar{X}_1 - \bar{X}_2$? First, because the samples are moderately large, we know by using the Central Limit Theorem that $\bar{X}_1$ is approximately a normal random variable with mean μ_1 and variance σ_1^2/n_1 and that $\bar{X}_2$ is approximately a normal random variable with mean μ_2 and variance σ_2^2/n_2. Because these are two independent samples, it follows that $\bar{X}_1 - \bar{X}_2$ is also approximately a normal random variable with mean $\mu_1 - \mu_2$ and variance $(\sigma_1^2/n_1) + (\sigma_2^2/n_2)$. Note that the variance of $\bar{X}_1 - \bar{X}_2$ is obtained by *adding* the variances for $\bar{X}_1$ and $\bar{X}_2$.

By standardizing this normal distribution, we obtain an approximate standard normal random variable defined by

$$Z = \frac{(\bar{X}_1 - \bar{X}_2) - (\mu_1 - \mu_2)}{\sqrt{\dfrac{\sigma_1^2}{n_1} + \dfrac{\sigma_2^2}{n_2}}} \qquad (9.1)$$

We do not need normal populations. The results of equation 9.1 are approximately valid *regardless* of the shape of the two populations, provided both samples are large (from the Central Limit Theorem). We pointed out that the two populations illustrated in Figure 9.1 need not follow a normal distribution. In fact, they can have any shape, such as exponential, uniform, or possibly a discrete distribution of some sort.

We can now derive a confidence interval for $\mu_1 - \mu_2$. By using Table A.4, we know that for the standard normal random variable Z,

$$P(-1.96 < Z < 1.96) = .95$$

Using equation 9.1 (and rearranging the inequalities), we can make the following statement about a random interval prior to obtaining the sample data:

$$P\left[(\bar{X}_1 - \bar{X}_2) - 1.96\sqrt{\frac{\sigma_1^2}{n_1} + \frac{\sigma_2^2}{n_2}} < \mu_1 - \mu_2 < (\bar{X}_1 - \bar{X}_2) + 1.96\sqrt{\frac{\sigma_1^2}{n_1} + \frac{\sigma_2^2}{n_2}}\right] = .95$$

This produces the following $(1 - \alpha) \cdot 100\%$ confidence interval for $\mu_1 - \mu_2$ (large samples, where σ_1 and σ_2 are *known*):

$$(\bar{X}_1 - \bar{X}_2) - Z_{\alpha/2}\sqrt{\frac{\sigma_1^2}{n_1} + \frac{\sigma_2^2}{n_2}} \quad \text{to} \quad (\bar{X}_1 - \bar{X}_2) + Z_{\alpha/2}\sqrt{\frac{\sigma_1^2}{n_1} + \frac{\sigma_2^2}{n_2}} \quad \textbf{(9.2)}$$

If σ_1 and σ_2 are *unknown*, we have:

$$(\bar{X}_1 - \bar{X}_2) - Z_{\alpha/2}\sqrt{\frac{s_1^2}{n_1} + \frac{s_2^2}{n_2}} \quad \text{to} \quad (\bar{X}_1 - \bar{X}_2) + Z_{\alpha/2}\sqrt{\frac{s_1^2}{n_1} + \frac{s_2^2}{n_2}} \quad \textbf{(9.3)}$$

Notice that this interval is very similar to the confidence interval for a single population mean using a large sample, namely,

$$(\text{point estimate}) \pm Z_{\alpha/2} \cdot (\text{standard deviation of the point estimator})$$

To construct the confidence interval if σ_1 and σ_2 are unknown (the usual case), you simply substitute the sample estimates in their place *provided* you have large samples ($n_1 > 30$ and $n_2 > 30$). Consequently, the confidence interval in equation 9.2 is exact (σ_1, σ_2 known) and the confidence interval in equation 9.3 is approximate (σ_1, σ_2 unknown).

EXAMPLE 9.1   Using the data from the two gas-price samples, construct a 90% confidence interval for $\mu_1 - \mu_2$.

SOLUTION   To begin with, the estimate of μ_1 is $\bar{x}_1 = \$1.48$, and the estimate of μ_2 is $\bar{x}_2 = \$1.39$. We are constructing a 90% confidence interval, so (using Table A.4) we find that $Z_{.05} = 1.645$ (Figure 9.2). The resulting 90% confidence interval for $\mu_1 - \mu_2$ is

$$(\bar{X}_1 - \bar{X}_2) - 1.645\sqrt{\frac{s_1^2}{n_1} + \frac{s_2^2}{n_2}} \quad \text{to} \quad (\bar{X}_1 - \bar{X}_2) + 1.645\sqrt{\frac{s_1^2}{n_1} + \frac{s_2^2}{n_2}}$$

$$= (1.48 - 1.39) - 1.645\sqrt{\frac{(.12)^2}{35} + \frac{(.10)^2}{40}} \quad \text{to} \quad (1.48 - 1.39) + 1.645\sqrt{\frac{(.12)^2}{35} + \frac{(.10)^2}{40}}$$

$$= .09 - (1.645)(.0257) \quad \text{to} \quad .09 + (1.645)(.0257)$$

$$= .09 - .042 \quad \text{to} \quad .09 + .042$$

$$= .048 \quad \text{to} \quad .132$$

■

■ **FIGURE 9.2**
Finding the pair of Z
values containing 90%
of the area under the
curve. The values are
-1.645 and 1.645.

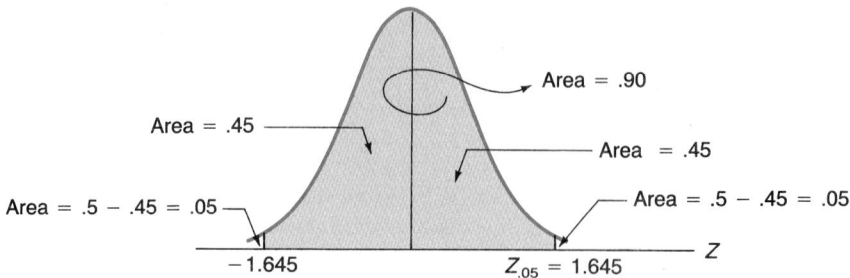

We can summarize this result in several ways:

1. We are 90% confident that $\mu_1 - \mu_2$ lies between .048 and .132.
2. We are 90% confident that the average price of Texgas regular gasoline is between 4.8¢ and 13.2¢ higher than the regular gasoline at Quik-Chek.
3. We are 90% confident that our estimate of $\mu_1 - \mu_2$ ($\bar{X}_1 - \bar{X}_2 = .09$) is within 4.2¢ of the actual value.

The confidence intervals defined in equations 9.2 and 9.3 will contain $\mu_1 - \mu_2$ 90% of the time. In other words, if you repeatedly obtained independent samples and repeated the procedure in Example 9.1, 90% of the corresponding confidence intervals would contain the unknown value of $\mu_1 - \mu_2$, and 10% of them would not.

Sample Sizes

The amount that you add to and subtract from your point estimate to obtain the confidence interval is the **maximum error, E.** For Example 9.1, this value is $E = .042$ (4.2¢). If you think that E is too large and you would like it to be smaller, one recourse is to *obtain larger samples* from your two populations. To determine how large a sample you need, one procedure is to select equal sample sizes. Consider the illustration in Example 9.1, and suppose that you want large enough sample means so that the difference in sample means is within 2¢ (rather than 4.2¢ in Example 9.1) of the difference in population means, with 90% confidence. So, $E = .02$. By insisting on equal sample sizes, where $n_1 = n_2 = n$ (say), then

$$.02 = 1.645 \sqrt{\frac{(.12)^2}{n} + \frac{(.10)^2}{n}}$$

After some algebraic manipulation, we have

$$n = \frac{(1.645)^2[(.12)^2 + (.10)^2]}{(.02)^2} \cong 166 \qquad \text{(by rounding up)}$$

In general, this value is

$$n = \frac{Z_{\alpha/2}^2(s_1^2 + s_2^2)}{E^2} \qquad\qquad \textbf{(9.4)}$$

In this illustration, the total sample size is $n_1 + n_2 = 166 + 166 = 332$. A better way to proceed here is to find the values of n_1 and n_2 that **minimize the total sample size.** The values of n_1 and n_2 that accomplish this are

$$n_1 = \frac{Z_{\alpha/2}^2 s_1 (s_1 + s_2)}{E^2} \tag{9.5}$$

$$n_2 = \frac{Z_{\alpha/2}^2 s_2 (s_1 + s_2)}{E^2} \tag{9.6}$$

For this illustration, $Z_{\alpha/2} = Z_{.05} = 1.645$, $s_1 = .12$, $s_2 = .10$, and $E = .02$. Consequently,

$$n_1 = \frac{(1.645)^2 (.12)(.22)}{(.02)^2} \cong 179$$

$$n_2 = \frac{(1.645)^2 (.10)(.22)}{(.02)^2} \cong 149$$

and the total sample size is $179 + 149 = 328$.

A derivation of this result is contained in Appendix B. Keep in mind that when you use these values of n_1 and n_2, the resulting value of E may not be exactly what you previously specified—because the values of s_1 and s_2 in the new samples will change. If no prior estimates of σ_1 and σ_2 are available, each can be roughly estimated using the high/low procedure discussed in Chapter 8.

Using equations 9.5 and 9.6, observe that if $s_1 = s_2$, your total sample size $(n_1 + n_2)$ will be the smallest when $n_1 = n_2$. If $s_1 > s_2$, you will select $n_1 > n_2$, and if $s_1 < s_2$, you will select $n_1 < n_2$. Finally, note that the ratio of the sample sizes (n_1/n_2) is the same as the ratio of the estimated standard deviations (s_1/s_2).

Hypothesis Testing for μ_1 and μ_2 (Large Samples)

Are men on the average taller than women? How do you answer such a question? We know that we can start by getting a sample of male heights and independently obtaining a sample of female heights. Figure 9.3 shows two such samples.

We proceed as before and put the claim that we are trying to demonstrate into the *alternative* hypothesis. The resulting hypotheses are

$$H_0: \mu_1 \geq \mu_2 \quad \text{(men are not taller, on the average)}$$

$$H_a: \mu_1 < \mu_2 \quad \text{(men are taller, on the average)}$$

We have estimators of μ_1 and μ_2, namely, $\bar{X}_1$ and $\bar{X}_2$. A sensible thing to do would be to reject H_0 if $\bar{X}_2$ is "significantly larger" than $\bar{X}_1$. In this case, the obvious conclusion is that μ_2 (the average of all male heights in your population) is larger than μ_1 (for female heights).

To define "significantly larger," we need to know what chance we are willing to take of rejecting H_0 when in fact it is true. This chance is α (the significance level) and, as before, it is determined prior to seeing any data. Typical values range

■ **FIGURE 9.3**
Hypothesis testing for two populations. Sample 1: size, n_1, mean, $\bar{X}_1$, and standard deviation, s_1. Sample 2: size, n_2, mean, $\bar{X}_2$, and standard deviation, s_2. Is $\mu_2 > \mu_1$?

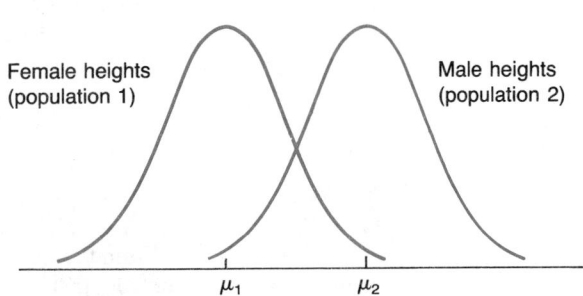

from .01 to .1, with $\alpha = .05$ generally providing a good trade-off between Type I and Type II errors. The test statistic here is the same as the one used to derive a confidence interval for $\mu_1 - \mu_2$. We are dealing with large samples ($n_1 > 30$ and $n_2 > 30$), so the test statistic is approximately a standard normal random variable, defined by

$$Z = \frac{\bar{X}_1 - \bar{X}_2}{\sqrt{\dfrac{s_1^2}{n_1} + \dfrac{s_2^2}{n_2}}} \tag{9.7}$$

EXAMPLE 9.2   The Ace Delivery people suspected that the gasoline at the Quik-Chek stores was less expensive than that at Texgas before they obtained any data. (*Note:* This is important! Do not let the data dictate your hypotheses for you. If you do, you introduce a serious bias into your testing procedure, and the "true" significance level may no longer be the predetermined α.) Here, μ_1 represents the average price at all of the Texgas stations and μ_2 is the average price at the Quik-Chek stores in the area. Is $\mu_2 < \mu_1$? Or, put another way, is $\mu_1 > \mu_2$? Use a significance level of .05.

SOLUTION   **Step 1.** *Define the hypotheses.* The question is whether the data support the claim that $\mu_1 > \mu_2$, so we put this statement in the alternative hypothesis.

$H_0: \mu_1 \leq \mu_2$ (Texgas is less expensive or the same.)

$H_a: \mu_1 > \mu_2$ (Quik-Chek is less expensive.)

As in Chapter 8, the equal sign goes into H_0 for a one-tailed test. In other words, the case where $\mu_1 = \mu_2$ is contained in the null hypothesis.

Step 2. *Define the test statistic.* This is the statistic that you evaluate using the sample data. Its value will either support the alternative hypothesis or it will not. The test statistic for this situation is given by equation 9.7:

$$Z = \frac{\bar{X}_1 - \bar{X}_2}{\sqrt{\dfrac{s_1^2}{n_1} + \dfrac{s_2^2}{n_2}}}$$

Step 3. *Define the rejection region.* In Figure 9.4, where should the null hypothesis H_0 be rejected? We simply ask, what happens to Z when H_a is true? In this case ($\mu_1 > \mu_2$), we *should* see $\bar{X}_1 > \bar{X}_2$. In other words, Z will be positive. So we reject H_0 if Z is "too large," that is,

reject H_0 if $Z > k$ for some $k > 0$

Using $\alpha = .05$, we use Table A.4 to find the corresponding value of Z (that is, k). In Figure 9.4, $k = 1.645$. This is the same value and rejection region we obtained in Chapter 8 when using Z for a one-tailed test in the right tail. The test is

reject H_0 if $Z > 1.645$

Step 4. *Evaluate the test statistic and carry out the test.* The data collected showed $n_1 = 35$, $\bar{x}_1 = 1.48$, $s_1 = .12$ (from the Texgas sample) and $n_2 = 40$, $\bar{x}_2 = 1.39$, $s_2 = .10$ (from the Quik-Chek sample). Based on these sample results, can we conclude that $\bar{x}_1 = 1.48$ is significantly *larger* than

■ **FIGURE 9.4**
Z curve showing
rejection region for
Example 9.2.

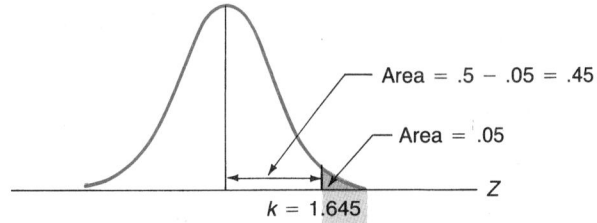

$\bar{x}_2 = 1.39$? If we can, the decision will be to reject H_0. The following value of the test statistic will answer our question.

$$Z = \frac{\bar{X}_1 - \bar{X}_2}{\sqrt{\dfrac{s_1^2}{n_1} + \dfrac{s_2^2}{n_2}}} = \frac{1.48 - 1.39}{\sqrt{\dfrac{(.12)^2}{35} + \dfrac{(.10)^2}{40}}}$$

$$= \frac{.09}{.0257} = 3.50 = Z^*$$

Because $3.50 > 1.645$, we reject H_0; $\bar{x}_1$ *is* significantly larger than $\bar{x}_2$. Therefore, we claim that $\mu_1 > \mu_2$.

Step 5. *State a conclusion.* We conclude that the Quik-Chek stores *do* charge less for gasoline (on the average) than do the Texgas stations. If the locations of these stores are equally convenient to Ace Delivery Service, buying gas from Quik-Chek appears to be a money-saving alternative. ■

Using the corresponding *p*-value for the data in Example 9.2, what would you conclude using the classical approach (with $\alpha = .05$)? For this example, the *p*-value will be the area under the Z curve (Z is our test statistic) to the right (we reject H_0 in the right tail for this example) of the calculated test statistic, $Z^* = 3.50$. In general,

$$p = \text{p-value} = \begin{cases} \text{area to the right of } Z^* \text{ for } H_a: \mu_1 > \mu_2 \\ \text{area to the left of } Z^* \text{ for } H_a: \mu_1 < \mu_2 \qquad \textbf{(9.8)} \\ 2 \cdot (\text{tail area of } Z^*) \text{ for } H_a: \mu_1 \neq \mu_2 \end{cases}$$

These three alternative hypotheses are your choices for this situation. Once again, $H_0: \mu_1 = \mu_2$ versus $H_a: \mu_1 \neq \mu_2$ is a two-tailed test, and the first two alternative hypotheses represent one-tailed tests.

Returning to our example, we can see from Figure 9.5 that the resulting *p*-value is $p = .0002$ (very small). Using the classical approach, because $p <$ the

■ **FIGURE 9.5**
Z curve showing
p-value for $Z^* = 3.50$.

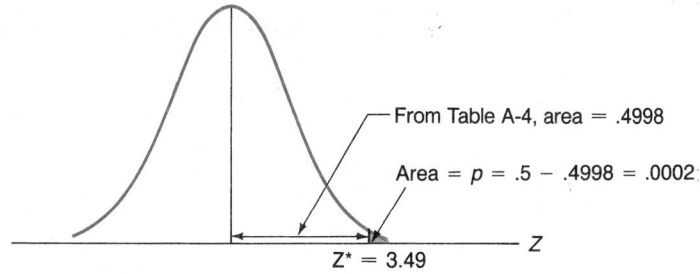

significance level of .05, we reject H_0—the same conclusion as before. In fact, this procedure *always* leads to the same conclusion as the five-step solution, as we saw in Chapter 8.

If we elect not to select a significance level α and instead use only the p-value to make a decision, we proceed as before:

Reject H_0 if p is small ($p < .01$).

Fail to reject H_0 if p is large ($p > .1$).

Data are inconclusive if p is neither small nor large ($.01 \le p \le .1$).

For this example, $p = .0002$ is clearly small, and so we again reject H_0. The Quik-Chek gasoline definitely appears to be less expensive than the Texgas gasoline. As was pointed out in the previous chapter, you often encounter a result that is *statistically* significant but not significant in a *practical sense*. To illustrate, suppose that the p-value of .0002 was the result of two very large samples and that the difference in gasoline price for the two samples was $\bar{x}_1 - \bar{x}_2 = .008$. You might not view this difference (less than 1¢) as being worth the inconvenience of having to pay cash for all gasoline purchases.

COMMENT   There may well be situations where the severity of the Type I error requires a significance level smaller than .01 on the low end, or the impact of a Type II error dictates a significance level larger than .1 on the upper end. This rule is thus only a general yardstick that applies to most, but certainly not all, business applications.

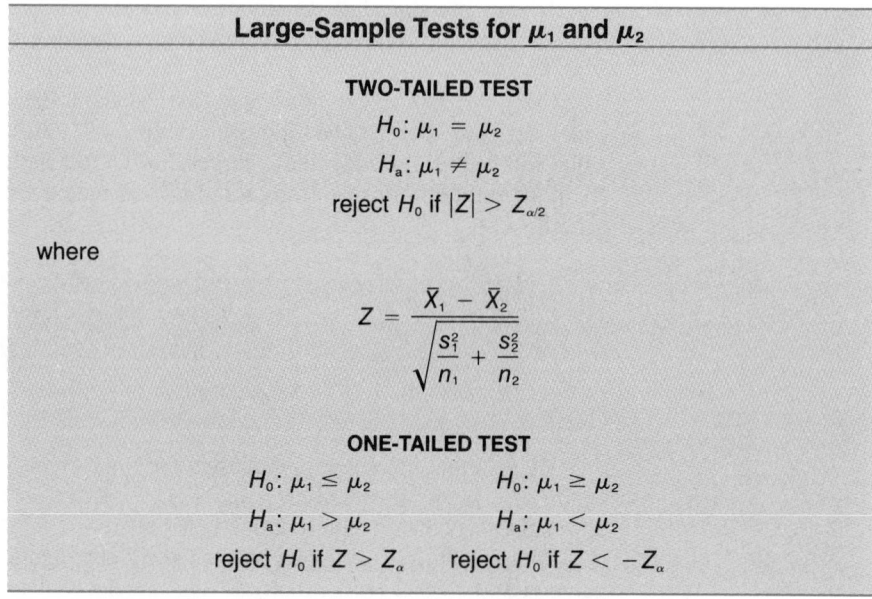

Large-Sample Tests for μ_1 and μ_2

TWO-TAILED TEST

$H_0: \mu_1 = \mu_2$

$H_a: \mu_1 \ne \mu_2$

reject H_0 if $|Z| > Z_{\alpha/2}$

where

$$Z = \frac{\bar{X}_1 - \bar{X}_2}{\sqrt{\dfrac{S_1^2}{n_1} + \dfrac{S_2^2}{n_2}}}$$

ONE-TAILED TEST

$H_0: \mu_1 \le \mu_2$	$H_0: \mu_1 \ge \mu_2$
$H_a: \mu_1 > \mu_2$	$H_a: \mu_1 < \mu_2$
reject H_0 if $Z > Z_\alpha$	reject H_0 if $Z < -Z_\alpha$

Two-Sample Procedure for any Specified Value of $\mu_1 - \mu_2$

The two-tailed hypotheses for large sample tests for μ_1 and μ_2 can be written as

$$H_0: \mu_1 - \mu_2 = 0$$

$$H_a: \mu_1 - \mu_2 \ne 0$$

The right-sided one-tailed hypotheses are

$$H_0: \mu_1 - \mu_2 \le 0$$

$$H_a: \mu_1 - \mu_2 > 0$$

The left-tailed hypotheses can be written in a similar manner. The point is that H_0 (so far) claims that $\mu_1 - \mu_2$ is equal to 0 or lies to one side of 0 (the one-tailed tests).

Suppose the claim is that $\mu_1 - \mu_2$ is more than ten. To demonstrate that this is true, we must make our alternative hypothesis H_a: $\mu_1 - \mu_2 > 10$; the corresponding null hypothesis is H_0: $\mu_1 - \mu_2 \leq 10$.

In general, to test that $\mu_1 - \mu_2 =$ (some specified value, say D_0), the five-step procedure still applies, but the test statistic is now

$$Z = \frac{(\bar{X}_1 - \bar{X}_2) - D_0}{\sqrt{\dfrac{s_1^2}{n_1} + \dfrac{s_2^2}{n_2}}} \qquad (9.9)$$

Equation 9.9 applies to both one-tailed and two-tailed tests. It can be used to compare two means directly (for example, H_0: $\mu_1 = \mu_2$ versus H_a: $\mu_1 \neq \mu_2$) by setting $D_0 = 0$, as in Example 9.2.

EXAMPLE 9.3

In Example 9.2, we decided that Ace Delivery Service would save money if they purchased their gasoline from Quik-Chek because that store's average gasoline price appeared to be less than that of the Texgas stations. Because Quik-Chek does not accept credit cards, the owner of Ace is willing to purchase their gasoline only if their average price is more than 6¢ per gallon less than Texgas's. Do the data indicate that it is? (α is still .05.)

SOLUTION

The question now is whether the data support the claim that the difference between the two means (Texgas and Quik-Chek) is larger than 6¢. So the hypotheses are H_0: $\mu_1 - \mu_2 \leq .06$ and H_a: $\mu_1 - \mu_2 > .06$, where μ_1 is Texgas's mean and μ_2 is Quik-Chek's mean.

The test statistic is

$$Z = \frac{(\bar{X}_1 - \bar{X}_2) - .06}{\sqrt{\dfrac{s_1^2}{n_1} + \dfrac{s_2^2}{n_2}}}$$

The computed value of Z is

$$Z^* = \frac{(1.48 - 1.39) - .06}{\sqrt{\dfrac{(.12)^2}{35} + \dfrac{(.10)^2}{40}}} = \frac{.03}{.0257} = 1.17$$

The testing procedure is exactly as it was previously—reject H_0 if $Z^* > 1.645$. Because $1.17 < 1.645$, we fail to reject H_0. The difference between the two sample means (9¢) was *not* significantly larger than the hypothesized value of 6¢.

These data provide insufficient evidence to conclude that Quik-Chek is more than 6¢ less expensive (on the average) than Texgas. If Ace's owner thinks that not using credit cards would be too much trouble for a savings of less than 6¢ per gallon, Ace should use the Texgas gasoline.

Exercises

9.9 The following data were collected from two independent, normally distributed populations, each having a population standard deviation of 1.5. Find a 90% confidence interval for the difference in the means of the two populations.

Sample 1: 11   15   14   11   12
Sample 2: 18   13   17   13   15

9.10   Alan Nukumi, a designer for high-quality stereo systems, is interested in constructing a 95% confidence interval for the difference between the signal/noise ratios for the two models he developed. Data were obtained from each model as follows. Model A: $\bar{x} = 54$, $n = 45$, $s = 8$. Model B: $\bar{x} = 62$, $n = 60$, $s = 17$. Find the 95% confidence interval.

9.11   Fort Worth and Dallas are two large cities that, despite their geographical closeness, have somewhat different economies. To find out the difference in the amount of unemployment, a statistician randomly interviewed 120 unemployed workers from Fort Worth and 150 unemployed workers from Dallas and asked them how many weeks they had been out of work over the past 52 weeks. The data are as follows. Total number of weeks: Fort Worth, 2031; Dallas, 3713. Standard deviation: Fort Worth, 1.3; Dallas, 2.1. Find a 90% confidence interval for the mean difference in the number of weeks unemployed for the work forces of Fort Worth and Dallas.

9.12   First National Bank and City National Bank are competing for customers who would like to open IRAs (individual retirement accounts). Thirty-two weeks are randomly selected for First National Bank and another 32 weeks are randomly selected for City National. The total amount deposited into IRAs is noted for each week. A summary of data (deposits in thousands of dollars) from the survey is as follows. First National: $\bar{x} = 4.1$, $s = 1.2$. City National: $\bar{x} = 3.5$, $s = 0.9$. Use a 98% confidence interval to estimate the difference in the mean weekly deposits into IRAs for each bank.

9.13   Two discount stores in a popular shopping mall have their merchandise laid out differently. Both stores claim that the arrangement of goods in their store makes the customer buy more on impulse. A survey of 100 customers from each store is taken. Each customer is asked how much money he or she spent on merchandise he or she did not originally intend to buy before walking into the store. The results are as follows. Discount store 1, $\bar{x} = \$15.50$, $s = \$3.20$. Discount store 2: $\bar{x} = \$19.40$, $s = \$4.80$. Find a 90% confidence interval for the difference in the mean amount of cash spent per customer on impulse buying for the two different stores. Is layout affecting impulse buying? How do you know?

9.14   If a 99% confidence interval is found in Exercise 9.13 instead of a 90% confidence interval, would the former confidence interval be larger or smaller than the latter? Explain.

9.15   A personnel director wants to know if the mean length of employment in years with the company is about the same for assembly and clerical workers. A sample of 35 employees was randomly drawn from each of these two groups of workers. Conduct a test of hypothesis to determine whether there is a significant difference in the mean length of employment with the company for the two groups. Use a significance level of .03.

$$\text{Assembly workers: } n = 35, \bar{x} = 4.1, s^2 = 30.2.$$
$$\text{Clerical workers: } n = 35, \bar{x} = 3.2, s^2 = 28.1.$$

9.16   Construct the 90% confidence interval for the difference in the mean length of employment for two groups of workers in Exercise 9.15.

9.17   Determine the value of the test statistic and the p-value that would result from the hypothesis test in each of the following cases:

a. $H_0: \mu_1 - \mu_2 \leq 100$, $H_a: \mu_1 - \mu_2 > 100$, $n_1 = 31$, $n_2 = 34$, $\bar{x}_1 = 190$, $\bar{x}_2 = 80$, $s_1 = 25.1$, $s_2 = 20$.

b. $H_0: \mu_1 - \mu_2 = 0$, $H_a: \mu_1 - \mu_2 \neq 0$, $n_1 = 40$, $n_2 = 64$, $\bar{x}_1 = 4.5$, $\bar{x}_2 = 5.8$, $\sigma_1 = 19.1$, $\sigma_2 = 49.3$.

c. $H_0: \mu_1 - \mu_2 \geq 406$, $H_a: \mu_1 - \mu_2 < 406$, $n_1 = 100$, $n_2 = 100$, $\bar{x}_1 = 1050$, $\bar{x}_2 = 650$, $s_1 = 900$, $s_2 = 330$.

9.18   The computer center managers in Becker Industries would like to know if there is a difference in the weekly computer time (in seconds) used by the employees in the financial-planning department and that used by employees in the legal-aid department. Conduct a test of hypothesis to determine whether there is a significant difference in computer usage for the two departments. Let the significance level be .05. Use the sample statistics that follow, in which 40 weeks were randomly selected for each department.

STATISTIC	n	$\bar{x}$	$\Sigma(x - \bar{x})^2$
Financial Planning	40	2503	2180.4
Legal	40	2510	2291.6

9.19 The average annual pay in 1989 was $21,128 in the state of Rhode Island and $25,233 in the state of Massachusetts. This is a difference of $4105. Suppose that a statistician believes that the difference is much less for employees in the manufacturing industry and takes an independent random sample of employees in the manufacturing industry in each state. The results are as follows:

STATE	$\bar{x}$	s	n
Rhode Island	21,900	3,700	150
Massachusetts	24,800	3,100	190

At the .05 significance level, do the data support the statistician's belief that for employees in the manufacturing industry, the mean annual salary in Rhode Island differs from the mean annual salary in Massachusetts by less than $4105? Report the p-value and interpret it.

(*Source:* Adapted from *The World Almanac and Book of Facts* (1991): 137.)

9.20 The systems manager of Ace Manufacturing is about to purchase a microcomputer and has narrowed her choices to the Alpha and Gemini models. She is seriously concerned about the cost of maintenance of these two models. After interviewing 40 experts on the Alpha model and a different 40 experts on the Gemini model, she obtained the following information on the cost of maintenance. Average annual cost of maintenance: Alpha, $46.50; Gemini, $37.20. Standard deviation: Alpha, 4.20; Gemini, 6.10. Do the data indicate a significant difference in the cost of maintenance for the two models? Use the p-value to justify your answer.

9.21 The financial analyst of Hogan Securities believes that there is no difference in the annual average returns for steel industry stocks and mineral industry stocks. Using the following information, test the hypothesis that there is no significant difference in the average returns for these two types of stocks. Steel industry stocks: $\bar{x} = 9\%$, $n = 33$, $s = 2.4\%$. Mineral industry stocks: $\bar{x} = 11\%$, $n = 41$, $s = 4\%$. Use a 10% significance level.

9.22 From an initial study of a sample of the length of time that it takes for a package to be delivered by two different express mail companies, it was found that the standard deviation of times to send a package is 1.5 days for company A and 2.3 days for company B. Let E be the maximum error of estimating the mean difference in the times with a 90% confidence interval. If E is taken to be .6, find the values of n_1 and n_2 that minimize the total sample size.

9.23 An education analyst is studying the performance of high-school seniors on the SAT examination. He is specifically testing whether there is any difference between the mean SAT scores of seniors who attended public schools and those who attended private schools. He believes that private schools should score more than 50 points higher than public schools. Using the following information, what would be your conclusion at a significance level of .05? Do you agree with the analyst?

Seniors in public schools: $\bar{x} = 590$, $s = 67$, $n = 40$

Seniors in private schools: $\bar{x} = 680$, $s = 110$, $n = 55$

9.24 The department of management is interested in comparing the utilization of two student laboratories. A crude measure of utilization was used, namely, time between sign-in and sign-out, in hours. Attendance records for the two labs from the previous semester were obtained, and 40 students were randomly selected for each lab. The results are summarized as follows:

LAB	n	$\bar{x}$	s
1	40	1.5	0.9
2	40	1.3	1.1

a. Construct a 90% confidence interval for the true population difference $(\mu_1 - \mu_2)$ in the utilization of the two labs.

b. If a 99% confidence interval is wanted but the maximum error E has to remain the same as obtained in part a, what are the optimum samples sizes n_1 and n_2?

c. Conduct a hypothesis test to determine whether there is any difference in the utilization rate of the two labs. Use $\alpha = .05$.

d. Indicate the p-value for the test.

9.3 Comparing Two Normal Population Means Using Two Small Independent Samples

When dealing with *small* samples from two populations, we need to consider the assumed distribution of the populations because the Central Limit Theorem no longer applies. This section is concerned with comparing two population means when two small independent random samples are used. It differs from the previous section in two respects:

1. We are dealing with *small samples*.
2. We have reason to believe that the two populations of interest are *normal* populations. In Figures 9.1 and 9.3, where we had large sample sizes, this assumption was not necessary. When you use small samples from two populations, one or both of which appear to be *not* normally distributed, a nonparametric procedure is the proper method for analyzing such data. This procedure is discussed in Chapter 19.

In Chapter 8, we showed that when going from large samples to small samples from normal populations, the confidence interval and hypothesis-testing procedures both remained exactly the same, except that we used the *t* distribution rather than the *Z* distribution to describe the test statistic. We will use the same approach for small samples from two populations.

Confidence Interval for $\mu_1 - \mu_2$ (Small Independent Samples)

When using large samples from two populations to compare μ_1 and μ_2, we used the *Z*-statistic defined by

$$Z = \frac{\bar{X}_1 - \bar{X}_2}{\sqrt{\dfrac{s_1^2}{n_1} + \dfrac{s_2^2}{n_2}}}$$

When using small samples ($n_1 < 30$ or $n_2 < 30$), this statistic no longer approximates the standard normal. To make matters more complicated, it is not a *t* random variable either. However, this expression is *approximately* a *t* random variable if a somewhat complicated expression is used to derive the degrees of freedom (df). So we define

$$t' = \frac{\bar{X}_1 - \bar{X}_2}{\sqrt{\dfrac{s_1^2}{n_1} + \dfrac{s_2^2}{n_2}}} \qquad (9.10)$$

This statistic approximately follows a *t* distribution with df given by

$$\text{df for } t' = \frac{\left[\dfrac{s_1^2}{n_1} + \dfrac{s_2^2}{n_2}\right]^2}{\dfrac{\left(\dfrac{s_1^2}{n_1}\right)^2}{n_1 - 1} + \dfrac{\left(\dfrac{s_2^2}{n_2}\right)^2}{n_2 - 1}} \qquad (9.11)$$

Admittedly, equation 9.11 is a bit messy, but a good calculator or computer package makes this calculation relatively painless. To be on the conservative side, if df as calculated is not an integer $(1, 2, 3, \ldots)$, it should be rounded *down* to the next integer. As a check of your calculations, the df should be between A and B, where A is the smaller of $(n_1 - 1)$ and $(n_2 - 1)$ and B is $(n_1 - 1) + (n_2 - 1)$.

When finding the df, you can scale *both* s_1 and s_2 any way you wish, provided you scale them both the same way. By scaling, we mean that you can use s_1 and s_2 as is, or you can move the decimal point to the right or left. The resulting df will be the same *regardless* of the scaling used. However, when you evaluate the test statistic, t', or later perform a test of hypothesis, you must return to the *original* values of s_1 and s_2.

To derive an approximate confidence interval for $\mu_1 - \mu_2$, we use the same logic as in the previous (large samples) procedure. Thus, a $(1 - \alpha) \cdot 100\%$ confidence interval for $\mu_1 - \mu_2$ (small samples) is:

$$(\bar{X}_1 - \bar{X}_2) - t_{\alpha/2, \mathrm{df}} \sqrt{\frac{s_1^2}{n_1} + \frac{s_2^2}{n_2}} \quad \text{to} \quad (\bar{X}_1 - \bar{X}_2) + t_{\alpha/2, \mathrm{df}} \sqrt{\frac{s_1^2}{n_1} + \frac{s_2^2}{n_2}} \quad \textbf{(9.12)}$$

where df is specified in equation 9.11. If df is not an integer, round this value *down* to the next integer.

EXAMPLE 9.4

Checkers Cab Company is trying to decide which brand of tires to use for the coming year. Based on current price and prior experience, they have narrowed their choice to two brands, Beltex and Roadmaster. A recent study examined the durability of these tires by using a machine with a metallic device that wore down the tires. The time it took (in hours) for the tire to blow out was recorded.

Because the test for each tire took a great deal of time and the tire itself was ruined by the test, small samples (15 of each brand) were used. Notice that these are *independent* samples; there is no reason to match up the first Beltex tire with the first Roadmaster tire in the sample, the second Beltex with the second Roadmaster, and so on. (As discussed in Section 9.1, they would be dependent samples if the tires were tested by putting one of each brand on the rear wheels of 15 different cars.)

The blowout times (hours) were as follows:

BELTEX	ROADMASTER	BELTEX	ROADMASTER
3.82	4.16	2.84	3.65
3.11	3.92	3.26	3.82
4.21	3.94	3.74	4.55
2.64	4.22	3.04	3.82
4.16	4.15	2.56	3.85
3.91	3.62	2.58	3.62
2.44	4.11	3.15	4.88
4.52	3.45		

Construct a 90% confidence interval for $\mu_1 - \mu_2$, letting μ_1 be the average blowout time for *all* Beltex tires and μ_2 be the average blowout time for *all* Roadmaster tires.

SOLUTION

Here is a summary of the data from these two samples.

Sample 1 (Beltex)	Sample 2 (Roadmaster)
$n_1 = 15$	$n_2 = 15$
$\bar{x}_1 = 3.33$ hours	$\bar{x}_2 = 3.98$ hours
$s_1 = .68$ hours	$s_2 = .38$ hours

Your next step is to get a t-value from Table A.5. To do this, you first must calculate the correct df using equation 9.11:

$$df = \frac{\left[\dfrac{(.68)^2}{15} + \dfrac{(.38)^2}{15}\right]^2}{\dfrac{\left(\dfrac{(.68)^2}{15}\right)^2}{14} + \dfrac{\left(\dfrac{(.38)^2}{15}\right)^2}{14}}$$

$$= \frac{(.0404)^2}{.0000679 + .00000662} = 21.9$$

Rounding down, we use df = 21. Using Table A.5:

$$t_{.10/2,21} = t_{.05,21} = 1.721$$

The resulting 90% confidence interval for $\mu_1 - \mu_2$ is

$$(\bar{X}_1 - \bar{X}_2) - t_{.05,21}\sqrt{\frac{s_1^2}{n_1} + \frac{s_2^2}{n_2}} \quad \text{to} \quad (\bar{X}_1 - \bar{X}_2) + t_{.05,21}\sqrt{\frac{s_1^2}{n_1} + \frac{s_2^2}{n_2}}$$

$$= (3.33 - 3.98) - 1.721\sqrt{\frac{(.68)^2}{15} + \frac{(.38)^2}{15}} \quad \text{to} \quad (3.33 - 3.98) + 1.721\sqrt{\frac{(.68)^2}{15} + \frac{(.38)^2}{15}}$$

$$= -.65 - .35 \quad \text{to} \quad -.65 + .35$$

$$= -1.00 \text{ hr} \quad \text{to} \quad -.30 \text{ hr}$$

So we are 90% confident that the average blowout time for the Beltex tires is between 18 minutes (.3 hours) and 1 hour *less* than the average for the Roadmaster tires. Based on these results, Roadmaster appears to be the better (longer-wearing) tire. ∎

Hypothesis Testing for μ_1 and μ_2 (Small, Independent Samples)

The five-step procedure for testing hypotheses concerning μ_1 and μ_2 with large samples also applies to the small-sample situation. The only difference is that Table A.5 is used (rather than Table A.4) to define the rejection region.

EXAMPLE 9.5

In Example 9.4 a confidence interval was constructed for the difference in average blowout times for Beltex and Roadmaster tires. Can we conclude that these average blowout times are in fact not the same? Use a significance level of .10.

SOLUTION   **Step 1.** We are testing for a difference between the two means (not that Roadmaster is longer-wearing than Beltex or vice versa). The corresponding appropriate hypotheses are $H_0: \mu_1 = \mu_2$ and $H_a: \mu_1 \neq \mu_2$.

Step 2. The test statistic is

$$t' = \frac{\bar{X}_1 - \bar{X}_2}{\sqrt{\dfrac{s_1^2}{n_1} + \dfrac{s_2^2}{n_2}}}$$

which approximately follows a t distribution with df given by equation 9.11.

Step 3. You next need the df in order to determine your rejection region. In Example 9.4 we found that df = 21. Because $H_a: \mu_1 \neq \mu_2$, we will reject H_0

if t' is too large ($\bar{X}_1$ is significantly *larger* than $\bar{X}_2$) or if t' is too small ($\bar{X}_1$ is significantly *smaller* than $\bar{X}_2$). As in previous two-tailed tests using the Z or t statistic, H_0 is rejected if the absolute value of t exceeds the value from the table corresponding to $\alpha/2$. Using Table A.5, the rejection region for this situation will be

$$\text{reject } H_0 \text{ if } |t'| > t_{\alpha/2, df} = t_{.05, 21} = 1.721$$

Step 4. The value of the test statistic is

$$t'* = \frac{3.33 - 3.98}{\sqrt{\dfrac{(.68)^2}{15} + \dfrac{(.38)^2}{15}}} = \frac{-.65}{.20} = -3.25$$

Because $|t'*| = 3.25 > 1.721$, we reject H_0. Consequently, the difference between the sample means ($-.65$) *is* significantly large (in absolute value), which leads to a rejection of the null hypothesis.

Step 5. There *is* a significant difference in the average blowout times for the two brands. ∎

COMMENTS The hypotheses in Example 9.4 could be written as $H_0: \mu_1 - \mu_2 = 0$ and $H_a: \mu_1 - \mu_2 \neq 0$. Having already determined a 90% confidence interval for $\mu_1 - \mu_2$, a much simpler way to perform this two-tailed test (using $\alpha = .10$) would be to reject H_0 if 0 does not lie in the confidence interval for $\mu_1 - \mu_2$ and fail to reject H_0 otherwise. The confidence interval according to Example 9.4 is $(-1.00, -.30)$, which does not contain zero, and so we reject H_0 (as before).

This alternative method of testing H_0 versus H_a holds only for a two-tailed test in which the significance level of the test, α, and the confidence level [$(1 - \alpha) \cdot 100\%$] of the confidence interval "match up." For example, a significance level of $\alpha = .05$ would correspond to a 95% confidence interval, a value of $\alpha = .10$ would correspond to a 90% confidence interval, and so on.

A MINITAB solution to Example 9.5 is provided in Figure 9.6. The calculated p-value is $p = .0038$. Based on this extremely small value, we again reject H_0.

Notice that the procedure in this section for testing μ_1 versus μ_2 and constructing confidence intervals for $\mu_1 - \mu_2$ made no mention as to whether the population variances (or standard deviations) were equal or not. In fact, we can say that this procedure did not assume that $\sigma_1 = \sigma_2$; it also did *not* assume that $\sigma_1 \neq \sigma_2$. Next, we will examine a special case where we have reason to believe that the standard deviations *are* equal. For this situation, we will define another t test to detect any difference between the population means.

■ FIGURE 9.6
MINITAB solution to
Example 9.5.

```
MTB > SET INTO C1
DATA> 3.82 3.11 4.21 2.64 4.16 3.91 2.44 4.52 2.84
DATA> 3.26 3.74 3.04 2.56 2.58 3.15
DATA> END
MTB > SET INTO C2
DATA> 4.16 3.92 3.94 4.22 4.15 3.62 4.11 3.45 3.65
DATA> 3.82 4.55 3.82 3.85 3.62 4.88
DATA> END
MTB > NAME C1 = 'BELTEX' C2 = 'ROADMAST'
MTB > TWOSAMPLE TEST WITH 90% CONFIDENCE USING C1 AND C2

TWOSAMPLE T FOR BELTEX VS ROADMAST
                N      MEAN     STDEV    SE MEAN
BELTEX         15      3.332    0.679    0.18
ROADMAST       15      3.984    0.377    0.097

90 PCT CI FOR MU BELTEX - MU ROADMAST: (-1.00, -0.307)

TTEST MU BELTEX = MU ROADMAST (VS NE): T= -3.25   P=0.0038   DF=  21
```

No subcommands are necessary for a two-tailed test

t' p-value

Special Case of Equal Variances

There are some situations in which we are willing to assume that the population variances (σ_1^2 and σ_2^2) are equal. This situation is common in many long-running production processes for which, based on past experience, you are convinced that the variation within population 1 is the same as the variation within population 2.

Another situation in which we may assume σ_1 and σ_2 arises when we obtain two *additional* samples from the two populations, which we use strictly to determine if the population standard deviations are equal. If there is not sufficient evidence to indicate that $\sigma_1 \neq \sigma_2$, then there is no harm in assuming that $\sigma_1 = \sigma_2$. A procedure for comparing the population standard deviations is discussed in Section 9.4.

Why make the assumption that $\sigma_1 = \sigma_2$? Remember, we are still interested in the means, μ_1 and μ_2. As before, we would like to obtain a confidence interval for $\mu_1 - \mu_2$ and to perform a test of hypothesis. If, in fact, σ_1 *is* equal to σ_2, we can construct a slightly stronger test of μ_1 versus μ_2. By stronger, we mean that we are *more likely* to reject H_0 when it is actually false. This test is said to be more **powerful.**

For this case, because we believe that $\sigma_1^2 = \sigma_2^2 = \sigma^2$ (say), it makes sense to combine—or **pool**—our estimate of σ_1^2 (s_1^2) with the estimate of σ_2^2 (s_2^2) into one estimate of this common variance (σ^2). The resulting estimate of σ^2 is called the **pooled sample variance** and is written s_p^2. This estimate is merely a *weighted average* of s_1^2 and s_2^2, defined by

$$s_p^2 = \frac{(n_1 - 1)s_1^2 + (n_2 - 1)s_2^2}{n_1 + n_2 - 2} \tag{9.13}$$

Notice that s_p^2 gives more weight to the sample variance from the larger sample. Also, if the sample sizes are the same, then s_p^2 is simply the average of s_1^2 and s_2^2.

Constructing Confidence Intervals for $\mu_1 - \mu_2$

To construct the confidence interval, we make two changes in the previous procedure. First, t' is replaced by

$$t = \frac{\bar{X}_1 - \bar{X}_2}{\sqrt{\dfrac{s_p^2}{n_1} + \dfrac{s_p^2}{n_2}}} \tag{9.14}$$

$$= \frac{\bar{X}_1 - \bar{X}_2}{s_p \sqrt{\dfrac{1}{n_1} + \dfrac{1}{n_2}}} \tag{9.15}$$

Here (unlike the previous test statistic), t exactly follows a t distribution (assuming the two populations follow normal distributions).

Second, the df for t are much easier to derive:

$$df = n_1 + n_2 - 2$$

So you avoid the difficult df calculation in equation 9.11, but you need to derive the pooled variance, s_p^2, using the individual sample variances, s_1^2 and s_2^2.

As a check, your resulting pooled value for s_p^2 should be between s_1^2 and s_2^2, since it is a weighted average of these two values.

Hypothesis Testing for μ_1 and μ_2

In hypothesis testing for $\mu_1 - \mu_2$, the previous procedure applies, except that t' is replaced by t and the df used in Table A.5 is df $= n_1 + n_2 - 2$ rather than the df value from equation 9.11.

In Examples 9.4 and 9.5, we examined the blowout times for two brands of tires as measured by a machine performing a stress test of the sampled tires. Assume we have determined from previous tests that the *variation* of the blowout times is not affected by the tire brand. Assuming that σ_1^2 (Beltex) $= \sigma_2^2$ (Roadmaster), how can we construct a 90% confidence interval for $\mu_1 - \mu_2$ and determine whether there is a difference in the mean blowout times?

<table>
<tr><td>**Sample 1 (Beltex)**</td><td>**Sample 2 (Roadmaster)**</td></tr>
<tr><td>$n_1 = 15$</td><td>$n_2 = 15$</td></tr>
<tr><td>$\bar{x}_1 = 3.33$ hr</td><td>$\bar{x}_2 = 3.98$ hr</td></tr>
<tr><td>$s_1 = .68$ hr</td><td>$s_2 = .38$ hr</td></tr>
</table>

Our first step is to pool the sample variances:

$$s_p^2 = \frac{(15 - 1)(.68)^2 + (15 - 1)(.38)^2}{15 + 15 - 2} = \frac{(14)(.4624) + (14)(.1444)}{28}$$

$$= \frac{8.495}{28} = .303$$

$$s_p = \sqrt{.303} = .55 \text{ hr}$$

Is .303 between .1444 and .4624? Yes. Consequently, $s_p^2 = .303$ is our estimate of the common variance (σ^2) of the two tire populations. To find the 90% confidence interval for $\mu_1 - \mu_2$, we use

$$(\bar{X}_1 - \bar{X}_2) - t_{\alpha/2,\text{df}}\sqrt{\frac{s_p^2}{n_1} + \frac{s_p^2}{n_2}} \quad \text{to} \quad (\bar{X}_1 - \bar{X}_2) + t_{\alpha/2,\text{df}}\sqrt{\frac{s_p^2}{n_1} + \frac{s_p^2}{n_2}} \quad (9.16)$$

where df $= n_1 + n_2 - 2$ and $\alpha = .10$.

Because $n_1 + n_2 - 2 = 28$, we find (from Table A.5) that $t_{.05,28} = 1.701$. Next,

$$\sqrt{\frac{s_p^2}{n_1} + \frac{s_p^2}{n_2}} = s_p\sqrt{\frac{1}{n_1} + \frac{1}{n_2}} = .55\sqrt{\frac{1}{15} + \frac{1}{15}} = .20$$

The resulting confidence interval is

$$(3.33 - 3.98) - (1.701)(.20) \quad \text{to} \quad (3.33 - 3.98) + (1.701)(.20)$$

$$= -.65 - .34 \quad \text{to} \quad -.65 + .34$$

$$= -.99 \quad \text{to} \quad -.31$$

Comparing this result to the confidence interval in Example 9.4, you see little difference in the two confidence intervals, although the interval using the pooled variance is a bit narrower. Oftentimes these intervals can differ considerably, depending on the relative sizes of n_1 and n_2 as well as the relative values of s_1^2 and s_2^2.

Now we wish to test $H_0: \mu_1 = \mu_2$ versus $H_a: \mu_1 \neq \mu_2$. For this particular example, we can, as noted earlier, reject H_0 (using $\alpha = .10$) because zero does not lie in the previously derived confidence interval for $\mu_1 - \mu_2$. In the five-step procedure, there are only two changes we need to make when using the pooled sample variances. First, when defining our rejection region, we use $n_1 + n_2 - 2 = 28$ df. From Table A.5, the test procedure is to

$$\text{reject } H_0 \text{ if } |t| > t_{\alpha/2,\text{df}}$$

where $t_{.05,28} = 1.701$.

Second, the value of our test statistic is now

$$t = \frac{\bar{X}_1 - \bar{X}_2}{s_p \sqrt{\dfrac{1}{n_1} + \dfrac{1}{n_2}}} \tag{9.17}$$

Here,

$$t = \frac{3.33 - 3.98}{.55\sqrt{\dfrac{1}{15} + \dfrac{1}{15}}} = \frac{-.65}{.20} = -3.25$$

Because $|-3.25| = 3.25 > 1.701$, we reject H_0; once again the two sample means are significantly different. We conclude that there is a difference in the population mean blowout times for the two brands of tires.

A MINITAB solution for this example is provided in Figure 9.7. As in Example 9.5 (Figure 9.6), we obtain a very small p-value when pooling the sample variances. For this particular example, we observe little difference in the two solutions.

To Pool or Not to Pool? You might think, based on the previous examples, that it really does not matter whether you assume $\sigma_1 = \sigma_2$ or not. The two confidence intervals were nearly the same and the tests of hypothesis results were very close, differing only in their df for the test statistic. However, this is not always the case. *Unless you have strong evidence that the variances are the same,* we suggest you not pool the sample variances *and use the test statistic defined in equation 9.10.* If you assume that $\sigma_1 = \sigma_2$ and use the t test statistic in equation 9.17 but in fact $\sigma_1 \neq \sigma_2$, your results will be unreliable. This test is quite sensitive to this particular assumption. Also, if σ_1 and σ_2 *are* the same, we would expect s_1 and s_2 to be nearly the same. If, in addition, $n_1 = n_2$ (or nearly so), then the computed values of t' and t will be practically identical (including the df). What this means is that you have little to gain by pooling the variances (and using t) but a great deal to lose if your assumption is incorrect.

We will show you in the next section how to use two samples to test the hypothesis that $\sigma_1 = \sigma_2$. With those results in hand, one possible procedure to use when testing the *means* would be: (1) if you reject H_0: $\sigma_1 = \sigma_2$, then use t'

■ FIGURE 9.7
MINITAB pooled variances solution using data from Example 9.4.

```
MTB > TWOSAMPLE TEST WITH 90% CONFIDENCE USING C1 AND C2;
SUBC> POOLED.
                                          This subcommand
TWOSAMPLE T FOR BELTEX VS ROADMAST        is necessary when
           N      MEAN    STDEV   SE MEAN you assume σ₁ = σ₂
BELTEX    15     3.332    0.679    0.18
ROADMAST  15     3.984    0.377    0.097

90 PCT CI FOR MU BELTEX - MU ROADMAST: (-0.99, -0.311)

TTEST MU BELTEX = MU ROADMAST (VS NE): T= -3.25  P=0.0030  DF=  28

POOLED STDEV =       0.549
                                       t            p-value
```

to test $H_0: \mu_1 = \mu_2$, and (2) if you fail to reject $H_0: \sigma_1 = \sigma_2$, then use t to test $H_0: \mu_1 = \mu_2$.

At first glance this may appear to be statistically sound, but it has some problems. The main one is that these two tests use the same data, and so the tests are not performed independently of one another. Also, your actual significance level may not be the α that you had previously chosen before you saw any data. This *can* be a valid procedure if you obtain separate samples—one to test the σ values and the other to test the μ values. Again, however, caution is in order, since the test of $H_0: \sigma_1 = \sigma_2$ is very sensitive to the assumption of normal populations, and so using small samples to carry out this test can be unreliable. Consequently, if there is reason to believe that the standard deviations might not be equal, a safe procedure is to proceed as if they weren't; that is, use the t' statistic to test the means.

The next section provides a procedure for testing the standard deviations from two normal populations using independent samples. By comparing the standard deviations using separate data from the two populations, one can decide whether the pooling procedure should be used when using additional data to test μ_1 versus μ_2. If you reject $H_0: \sigma_1 = \sigma_2$, then the t'-statistic in equation 9.10 is the proper test statistic to use on a test for the means because it does *not* assume that the population standard deviations are equal. On the other hand, if you fail to reject H_0, then the t statistic in equation 9.17, which *does* assume that $\sigma_1 = \sigma_2$, is the recommended test statistic for testing μ_1 versus μ_2.

Small-Sample Tests for μ_1 and μ_2

TWO-TAILED TEST

$$H_0: \mu_1 - \mu_2 = D_0$$
$$H_a: \mu_1 - \mu_2 \neq D_0$$
$$(D_0 = 0 \text{ for } H_0: \mu_1 = \mu_2)$$
$$\text{reject } H_0 \text{ if } |T| > t_{\alpha/2, df}$$

where, not assuming $\sigma_1 = \sigma_2$:

$$T = t' = \frac{(\bar{X}_1 - \bar{X}_2) - D_0}{\sqrt{\dfrac{s_1^2}{n_1} + \dfrac{s_2^2}{n_2}}}$$

$$df = \frac{\left[\dfrac{s_1^2}{n_1} + \dfrac{s_2^2}{n_2}\right]^2}{\dfrac{\left(\dfrac{s_1^2}{n_1}\right)^2}{n_1 - 1} + \dfrac{\left(\dfrac{s_2^2}{n_2}\right)^2}{n_2 - 1}}$$

Or, assuming $\sigma_1 = \sigma_2$:

$$T = t = \frac{(\bar{X}_1 - \bar{X}_2) - D_0}{s_p \sqrt{\dfrac{1}{n_1} + \dfrac{1}{n_2}}}$$

$$df = n_1 + n_2 - 2$$

where

$$s_p = \sqrt{\frac{(n_1 - 1)s_1^2 + (n_2 - 1)s_2^2}{n_1 + n_2 - 2}}$$

ONE-TAILED TEST
$H_0: \mu_1 - \mu_2 \leq D_0$	$H_0: \mu_1 - \mu_2 \geq D_0$
$H_a: \mu_1 - \mu_2 > D_0$	$H_a: \mu_1 - \mu_2 < D_0$
$(D_0 = 0$ for $H_0: \mu_1 \leq \mu_2)$	$(D_0 = 0$ for $H_0: \mu_1 \geq \mu_2)$
reject H_0 if $T > t_{\alpha, df}$	reject H_0 if $T < -t_{\alpha, df}$

Exercises

9.25 Achieving a high score on the LSAT examination is a prerequisite to acceptance to law school. Scores on the LSAT are considered to be normally distributed. Two law schools decided to compare the mean scores on the LSAT for students enrolled in their schools. Is there sufficient evidence to indicate that the average scores differ between the two schools? Law school 1: $\bar{x} = 680$, $s = 84$, $n = 15$. Law school 2: $\bar{x} = 634$, $s = 92$, $n = 21$. Use a 1% significance level. Assume that the population variances are equal for law school 1 and law school 2.

9.26 Construct a 95% confidence interval for $\mu_1 - \mu_2$ in Exercise 9.25. Assume that the population variances are equal for the two law schools.

9.27 The president of a personnel agency is interested in examining the annual mean salary differences between vice presidents of banks and vice presidents of savings and loan institutions. A random sample of eight of each kind of vice president was selected. Their annual salaries (in dollars) were as follows:

n	BANKS	SAVINGS AND LOAN INSTITUTIONS	n	BANKS	SAVINGS AND LOAN INSTITUTIONS
1	84,320	73,420	5	48,940	88,670
2	67,440	49,580	6	56,790	59,640
3	98,590	58,750	7	77,610	65,590
4	111,780	101,400	8	62,000	74,810

Conduct a test of hypothesis to determine if there is a significant difference in the average salary for the two vice president groups. The salaries for both groups are considered to be approximately normally distributed. Use a significance level of .05. Do not assume that the population variances are equal.

9.28 Construct a 90% confidence interval for the difference in the means of the salaries for vice presidents in the banking industry and for vice presidents of savings and loan institutions for Exercise 9.27. Do not assume that the population variances are equal.

9.29 Using the data in Exercise 9.27, test the same hypothesis, but assume that the population variances *are* equal.

9.30 The production supervisor of Dow Plast is conducting a test of the tensile strengths of two types of copper coils. The relevant data are as follows. Coil A: $\bar{x} = 118$, $s = 17$, $n = 9$. Coil B: $\bar{x} = 143$, $s = 24$, $n = 16$. The tensile strengths for the two types of copper coils are approximately normally distributed. Based on the p-value, do the data support the conclusion that the mean tensile strengths of the two coils are different at a significance level of 7%? Do not assume that the population variances are equal.

9.31 Construct a 99% confidence interval for $\mu_A - \mu_B$ in Exercise 9.30. Do not assume that the population variances are equal.

9.32 Using a pooled estimate of the variance, perform the test of hypothesis in Exercise 9.30. Compare the two answers.

9.33 The rapid advances of information and decision support technologies have recently made possible a satisfactory method for improving group decisions. This combination of information and decision support technologies is called *group decision support systems* (GDSS). An experiment was recently conducted to determine whether the decision time will be less in GDSS groups than in non-GDSS groups. The following data were collected.

GROUP TYPE	MEAN	STD. DEV.	SAMPLE SIZE
GDSS	103.25	25.98	8
non-GDSS	110.50	15.82	8

Construct a 90% confidence interval for the difference in the population means of GDSS groups and non-GDSS groups. Assume unequal population variances.

(*Source:* Adapted from H. Hwang and M. K. Raja, "The Effect of GDSS In Large Face-To-Face Group Decision-Making," *Proceedings of Twenty-first Southwest Region Decision Sciences Institute* (March 1990): 65.)

9.34 A machine operator is interested in whether there is a significant difference in the time machine 1 and machine 2 take to produce a particular item of output. The time that it takes to output an item is normally distributed. Ten items produced by machine 1 and then another 10 items produced by machine 2 were recorded. The resulting times in minutes were:

MACHINE 1		MACHINE 2	
40.3	39.7	43.7	41.6
35.6	40.2	42.1	42.3
42.7	38.2	41.8	40.9
41.9	39.6	42.8	43.8
38.6	40.3	40.2	42.7

Without assuming equal population variances, perform the test of hypothesis to determine whether there is a significant difference in the mean time for machine 1 and machine 2 to produce 1 item of output. Use a significance level of .05.

9.35 Construct a 95% confidence interval for $\mu_1 - \mu_2$ for Exercise 9.34.

9.36 Using the pooled estimate of the variance, perform the test of hypothesis in Exercise 9.34. Calculate the p-value.

9.4
Comparing the Variances of Two Normal Populations Using Independent Samples

Once again we concentrate on independent samples from two normal populations, only this time we focus our attention on the *variation* of these populations rather than on their averages (see Figure 9.8). When estimating and testing σ_1 versus σ_2, we will not be concerned about μ_1 and μ_2. They may be equal, or they may not—it simply does not matter for this test procedure.

In business applications, you may want to compare the variation of two different production processes or compare the risk involved with two proposed investment portfolios. As mentioned previously, when testing for population *means* using small independent samples, you must pay attention to the population standard deviations (variances). Based on your belief that σ_1 does or does not equal σ_2, you select your corresponding test statistic for testing the means, μ_1 and μ_2. As a reminder, it is *not* a safe procedure to use the *same data set* to test both $\sigma_1 = \sigma_2$ and $\mu_1 = \mu_2$. A proper procedure would be to test σ_1 and σ_2 using one set of samples (as outlined in this section) and to obtain another set of samples *independently* of the first to test the means.

In the previous section, when trying to decide if $\mu_1 = \mu_2$, we examined the *difference* between the point estimators, $\overline{X}_1 - \overline{X}_2$. If $\overline{X}_1 - \overline{X}_2$ was large enough (in absolute value), we rejected H_0: $\mu_1 = \mu_2$. When looking at the variances, we use the **ratio of the sample variances**, s_1^2 and s_2^2, to derive a test of hypothesis and construct confidence intervals. We do this because the distribution of $s_1^2 - s_2^2$ is

■ **FIGURE 9.8**
Comparing two standard deviations. Is $\sigma_1 = \sigma_2$?

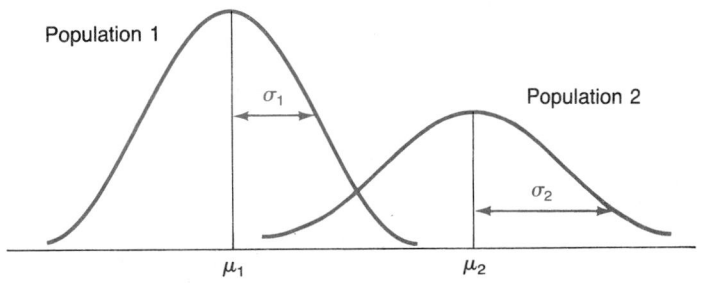

Comparing two standard deviations. Is $\sigma_1 = \sigma_2$?

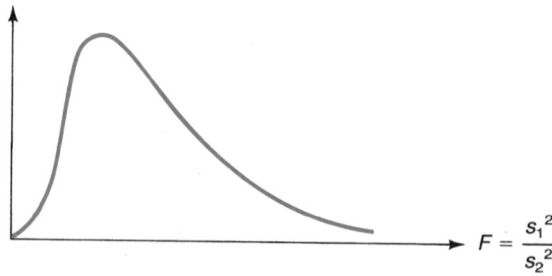

difficult to describe mathematically, but s_1^2/s_2^2 does have a recognizable distribution when in fact σ_1^2 and σ_2^2 are equal. So we define

$$F = \frac{s_1^2}{s_2^2} \qquad (9.18)$$

If you were to obtain sets of two samples repeatedly, calculate s_1^2/s_2^2 for each set, and make a histogram of these ratios, the shape of this histogram would resemble the curve in Figure 9.9, the **F distribution.** Its shape resembles the chi-square curve—it is nonsymmetric, skewed right (right-tailed), and the corresponding random variable is never negative. There are many F curves, depending on the sample sizes, n_1 and n_2. The shape of the F curve becomes more symmetric as the sample sizes, n_1 and n_2, increase. As later chapters will demonstrate, the F distribution has a large variety of applications in statistics. Right-tail areas for this random variable have been tabulated in Table A.7. As a final note here, *the F-statistic in equation 9.18 is highly sensitive to the assumption of normal populations. For larger data sets, it is recommended that you examine the shape of the sample data when using this particular F-statistic.*

When using the t and χ^2 statistics, we needed a way to specify the sample size(s) because the shapes of these curves change as the sample size changes. The same applies to the F distribution. There are two samples here, one from each population, and we need to specify *both* sample sizes. As before, we use the degrees of freedom (df) to accomplish this:

$$\nu_1 = \text{df for numerator} = n_1 - 1$$
$$\nu_2 = \text{df for denominator} = n_2 - 1$$

So, the F-statistic shown in Figure 9.9 follows an F distribution with ν_1 and ν_2 df provided $\sigma_1^2 = \sigma_2^2$ ($\sigma_1 = \sigma_2$). What happens to F when $\sigma_1 \neq \sigma_2$? Suppose that $\sigma_1 > \sigma_2$? Then we would expect s_1 (the estimate of σ_1) to be larger than s_2 (the estimate of σ_2); we should see

$$s_1^2 > s_2^2$$

or

$$F = \frac{s_1^2}{s_2^2} > 1$$

Similarly, if $\sigma_1 < \sigma_2$, then we expect an F-value < 1. We will use this reasoning to define a test of hypothesis for σ_1 versus σ_2.

Hypothesis Testing for $\sigma_1 = \sigma_2$

Is $\sigma_1 = \sigma_2$? We use the usual five-step procedure for testing a hypothesis concerning the two variances. Your choice of hypotheses is (as usual) a two-tailed test or a one-tailed test. For the two-tailed test the hypotheses are H_0: $\sigma_1 = \sigma_2$ ($\sigma_1^2 = \sigma_2^2$)

■ FIGURE 9.10
Unequal population
variances.

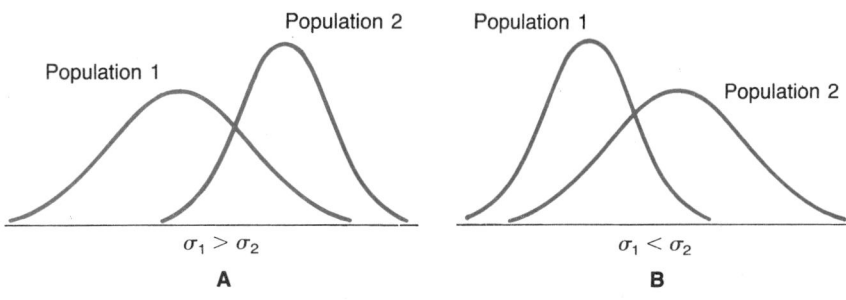

$\sigma_1 > \sigma_2$

A

$\sigma_1 < \sigma_2$

B

■ FIGURE 9.11
F curve with 10 and 12
df for probability that *F*
exceeds 2.75 (2.75 is
from Table A.7b).

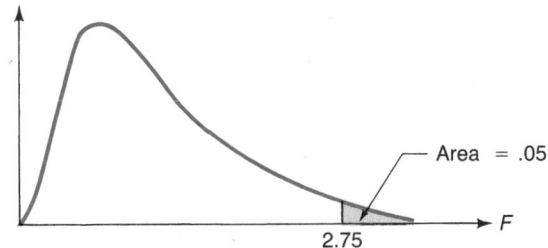

Area = .05

F

2.75

and $H_a: \sigma_1 \neq \sigma_2$ $(\sigma_1^2 \neq \sigma_2^2)$. For the one-tailed test the hypotheses are $H_0: \sigma_1 \leq \sigma_2$ and $H_a: \sigma_1 > \sigma_2$ (Figure 9.10a) or $H_0: \sigma_1 \geq \sigma_2$ and $H_a: \sigma_1 < \sigma_2$ (Figure 9.10b).

Notice that the hypotheses can be written in terms of the standard deviations $(\sigma_1$ and $\sigma_2)$ or the variances $(\sigma_1^2$ and $\sigma_2^2)$; if $\sigma_1 > \sigma_2$, then $\sigma_1^2 > \sigma_2^2$.

Right-tail areas under an *F* curve are provided in Table A.7. Notice that we have a table for areas of .1 (Table A.7(a)), .05 (Table A.7(b)), .025 (Table A.7(c)), and .01 (Table A.7(d)). These are the most commonly used values. For each table, the df for the numerator (ν_1) run across the top, and the df for the denominator (ν_2) run down the left margin.

Suppose we want to know which *F*-value has a right-tail area of .05, using 10 and 12 df. Let the *F*-value whose right-tail area is *a*, where the df are ν_1 and ν_2, be

$$F_{a, \nu_1, \nu_2}$$

For example, $F_{.05, 10, 12} = 2.75$ (Figure 9.11).

Notice that Table A.7 contains *right-tail* areas only. Later, we will show you how to find left-tail areas. We can, however, define each of our tests of hypothesis as a right-tailed test by simply and arbitrarily putting the larger sample variance in the numerator for a two-tailed test. Then *F* always will be greater than or equal to 1. This procedure is summarized in the accompanying box.

Hypothesis Tests for σ_1 and σ_2

TWO-TAILED TEST

$$H_0: \sigma_1 = \sigma_2$$

$$H_a: \sigma_1 \neq \sigma_2$$

$$F = \frac{\text{larger of } s_1^2 \text{ and } s_2^2}{\text{smaller of } s_1^2 \text{ and } s_2^2}$$

reject H_0 if $F > F_{\alpha/2, \nu_1, \nu_2}$

where

$$\nu_1 = \begin{cases} n_1 - 1 & \text{if } s_1^2 \geq s_2^2 \\ n_2 - 1 & \text{if } s_1^2 < s_2^2 \end{cases}$$

and

$$V_2 = \begin{cases} n_2 - 1 & \text{if } s_1^2 \geq s_2^2 \\ n_1 - 1 & \text{if } s_1^2 < s_2^2 \end{cases}$$

ONE-TAILED TEST

$H_0: \sigma_1 \leq \sigma_2$ $H_0: \sigma_1 \geq \sigma_2$

$H_a: \sigma_1 > \sigma_2$ $H_a: \sigma_1 < \sigma_2$

$$F = \frac{s_1^2}{s_2^2}$$ $$F = \frac{s_2^2}{s_1^2}$$

reject H_0 if $F > F_{\alpha, v_1, v_2}$ reject H_0 if $F > F_{\alpha, v_2, v_1}$

where   (be careful about order) where

$$v_1 = n_1 - 1$$ $$v_2 = n_2 - 1$$

and   and

$$v_2 = n_2 - 1$$ $$v_1 = n_1 - 1$$

EXAMPLE 9.6

The management of Case Automotive Products is considering the purchase of some new equipment that will fill 1-quart containers with a recently introduced radiator additive. They have narrowed their choice of brand of filling machine to brand 1 and brand 2. Although brand 1 is considerably less expensive than brand 2, they suspect that the contents delivered by the brand 1 machine will have more variation than would be obtained using brand 2. In other words, brand 1 is more apt to slightly (or severely) overfill or underfill containers. (The Case people realize that they must use a container slightly larger than 1 quart in any event, to allow for heat expansion and overfill of their product.)

The Case production department was able to obtain data on the performance of both brands for a sample of 25 containers using brand 1 and 20 containers using brand 2. Using their summary information, can you confirm Case's suspicions? Use $\alpha = .05$. All mean and standard deviation measurements are in fluid ounces.

Brand 1	Brand 2
$n_1 = 25$	$n_2 = 20$
$\bar{x}_1 = 31.8$	$\bar{x}_2 = 32.1$
$s_1 = 1.21$	$s_2 = .72$

SOLUTION **Step 1.** The purpose of the test is to determine if one standard deviation (or variance) is *larger* than the other; this calls for a one-tailed test. The suspicion is that σ_1 is larger than σ_2, so this statement is put in the alternative hypothesis. The resulting hypotheses are

$$H_0: \sigma_1 \leq \sigma_2 \qquad H_a: \sigma_1 > \sigma_2$$

Step 2. The appropriate test statistic is

$$F = \frac{s_1^2}{s_2^2}$$

Step 3. Because the df are $v_1 = 25 - 1 = 24$ and $v_2 = 20 - 1 = 19$, we find $F_{.05,24,19} = 2.11$. The test of H_0 versus H_a will be to

reject H_0 if $F > 2.11$

Step 4. The computed F-value is

$$F^* = \frac{(1.21)^2}{(.72)^2} = 2.82$$

Because $2.82 > 2.11$, we reject H_0.

Step 5. On the basis of these data and this significance level, Case is correct in its belief that the variation in the containers filled by brand 1 exceeds that of the containers filled by brand 2. ■

EXAMPLE 9.7 Using the blowout times data from Example 9.4, examine the variances *only*. Can you conclude that there is a difference in the two population variances, using a significance level of .05?

SOLUTION From these data, we determined that $s_1 = .68$ hour and $s_2 = .38$ hour, with $n_1 = n_2 = 15$.

Step 1. We are trying to detect a *difference* in the two variances (not whether one exceeds the other); a two-tailed test should be used. We define

$$H_0: \sigma_1 = \sigma_2 \text{ (or } \sigma_1^2 = \sigma_2^2)$$

$$H_a: \sigma_1 \neq \sigma_2 \text{ (or } \sigma_1^2 \neq \sigma_2^2)$$

Step 2. The test statistic here is

$$F = \frac{\text{larger of } s_1^2 \text{ and } s_2^2}{\text{smaller of } s_1^2 \text{ and } s_2^2}$$

Step 3. We need $F_{.025,14,14}$ from Table A.7(c). Unfortunately, it is not there; this table contains only selected df. In this situation, we pick the nearest df, which, for this example, is $F_{.025,15,14} = 2.95$. So our test of H_0 versus H_a is to

$$\text{reject } H_0 \text{ if } F > 2.95$$

Step 4. Since $s_1^2 = (.68)^2$ is larger than $s_2^2 = (.38)^2$, the computed value of the test statistic is

$$F^* = \frac{(.68)^2}{(.38)^2} = 3.20 > 2.95$$

Therefore, reject H_0 in favor of H_a.

Step 5. There *is* sufficient evidence to conclude that the two variances are unequal. If additional data are obtained to test the population *means* (such as $H_0: \mu_1 \geq \mu_2$ versus $H_a: \mu_1 < \mu_2$) using small samples, the correct procedure would be to use the t' statistic described earlier, which does not assume that σ_1 and σ_2 are equal, provided both populations are believed to be normally distributed. ■

Confidence Interval for σ_1^2/σ_2^2

Consider an F curve with v_1 and v_2 df. To construct a 95% confidence interval for σ_1^2/σ_2^2, you first need to find the values of F_L and F_U, where (Figure 9.12)

$$F_L \text{ has a left-tail area } = .025$$

$$F_U \text{ has a right-tail area } = .025$$

F_U can be found directly from Table A.7(c). It is F_L that poses a problem, however, because Table A.7 contains only right-tail areas and the F distribution is *not sym-*

■ FIGURE 9.12
F curve with v_1 and v_2
df showing F values
used for a 95%
confidence interval.

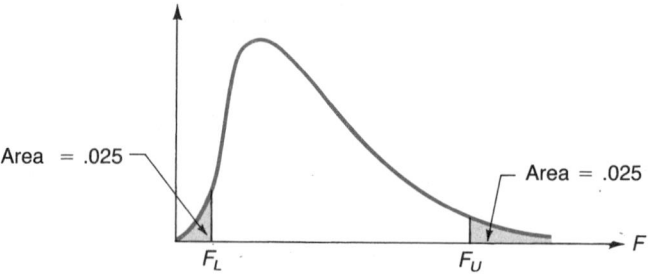

metric. However, we can use the following rule to determine F_L, since for this illustration, F_L can be written as $F_{.975, v_1, v_2}$.

$$F_{1-\alpha/2, v_1, v_2} = \frac{1}{F_{\alpha/2, v_2, v_1}} \qquad (9.19)$$

Consequently,

$$F_L = \frac{1}{F_{.025, v_2, v_1}}$$

Notice that we *switched* the df used when finding F_U because F_U can be written

$$F_U = F_{.025, v_1, v_2}$$

The confidence interval for σ_1^2/σ_2^2 is then

$$\frac{s_1^2/s_2^2}{F_U} \quad \text{to} \quad \frac{s_1^2/s_2^2}{F_L}$$

In general, we have a $(1 - \alpha) \cdot 100\%$ confidence interval for σ_1^2/σ_2^2 (independent samples):

$$\frac{s_1^2/s_2^2}{F_U} \quad \text{to} \quad \frac{s_1^2/s_2^2}{F_L} \qquad (9.20)$$

where

$$F_U = F_{\alpha/2, v_1, v_2}$$
$$F_L = 1/F_{\alpha/2, v_2, v_1}$$
$$v_1 = n_1 - 1$$
$$v_2 = n_2 - 1$$

EXAMPLE 9.8

Using the Case Automotive Products data in Example 9.6, determine a 95% confidence interval for σ_1^2/σ_2^2.

SOLUTION

Here, $n_1 = 25$, $s_1 = 1.21$, and $n_2 = 20$, $s_2 = .72$. So we need

$$F_U = F_{.025, 24, 19} = 2.45$$
$$F_L = 1/F_{.025, 19, 24}$$
$$\cong 1/2.33 \qquad (\text{using } F_{.025, 20, 24})$$
$$= .43$$

The 95% confidence interval for σ_1^2/σ_2^2 is

$$\frac{(1.21)^2/(.72)^2}{2.45} \quad \text{to} \quad \frac{(1.21)^2/(.72)^2}{.43} = 1.15 \quad \text{to} \quad 6.57$$

As a result, we are 95% confident that σ_1^2/σ_2^2 is between 1.15 and 6.57. This means that we are 95% confident that σ_1^2 is between 1.15 and 6.57 *times as large* as σ_2^2.

■

EXAMPLE 9.9   For the Case Automotive Products data in Example 9.6, determine a 95% confidence interval for σ_1/σ_2. Use the results of Example 9.8.

SOLUTION   This is obtained simply by finding the *square root* of each endpoint of the confidence interval for σ_1^2/σ_2^2. Your 95% confidence interval for σ_1/σ_2 will be

$$\sqrt{1.15} \quad \text{to} \quad \sqrt{6.57} = 1.07 \quad \text{to} \quad 2.56 \text{ (fluid ounces)}$$ ■

 *Exercises*

9.37   In evaluating capital-investment projects, the variability of the cash flows of the returns is carefully assessed. The higher the variability, the higher the risk associated with that project. Boone Enterprises is currently evaluating two projects. The mean expected net cash flow for the next 11 years for project 1 is $134,000, compared with $166,000 for project 2 for the next 8 years. The standard deviations of the net cash flows are $28,000 and $37,000 for project 1 and project 2, respectively. Assume that cash flows are normally distributed. Do these sample standard deviations present sufficient evidence to indicate that project 1 and project 2 are not equally risky? Use a significance level of 5%.

9.38   A computer program generates data that are approximately normal. Two independent samples are generated:

SAMPLE 1		SAMPLE 2	
17	26	24	29
25	34	32	34
30	19	29	26
28	25	25	30

Using a significance level of .05, would you reject the null hypothesis that the variance of the population from which sample 1 was taken is less than or equal to the variance of the population from which sample 2 was taken?

9.39   Construct a 95% confidence interval for the ratio of the population variances in Exercise 9.38.

9.40   The Water Pollution Prevention Council (WPPC) had recommended that the discharge of industrial waste and effluents into rivers in the district should be done at a slow and steady rate of 100 pound/hour (i.e., 2400 pound/day). Industrial plants tended to concentrate their effluent discharge activity in the night shift. The WPPC found that although companies might technically achieve an average discharge rate of 2400 pound/day, the rivers could not cope with the erratic rate of discharge. The effluents needed to be released throughout the day, rather than all at night. It was recommended that the true variance of the discharge rate should not exceed 600. The following results were obtained from samples of 21 observations each.

Factory A: variance 585

Factory B: variance 618

At a 5% significance level, is there sufficient evidence to conclude that the variance for factory A is less than that for factory B?

9.41   The transportation of frozen food by trucks requires that the temperature be maintained within a narrow range. If the temperature is kept too low, extra fuel is consumed and unnecessary costs are incurred. If the temperature is too high, health standards would be violated and there is danger of food spoilage and bacterial contamination. Two models of refrigeration units were being compared for their variation from a given temperature setting.

A special sensor attached to the units took readings every hour. The variance of temperatures for model 1 was $1.44°^2$. The variance for model 2 was $2.15°^2$C. The sample consisted of 16 readings for each model.

a.   Construct the 95% confidence interval for the ratio of the two variances.

b.   At a 5% significance level, is the variation of temperature in model 2 significantly greater than that for model 1?

9.42   The statistical quality-control department of a company that manufactures wall clocks is studying the variability of two types of wall clocks that have been recently developed. Using the following information, test the hypothesis that $H_0: \sigma_1 = \sigma_2$, using a significance level of .05. Assume that the samples are taken from populations that are approximately normally distributed. Clock 1: $n = 25$, $s = 1.8$. Clock 2: $n = 21$, $s = 1.39$.

9.43   Construct a 95% confidence interval for σ_1^2/σ_2^2 using the data in Exercise 9.42.

9.44   The following is a summary of the mean annual return ($\bar{X}$) and variance (s^2) of the annual return for common stocks in three different industries. Computer industry: $n = 16$, $\bar{X} = 14.3\%$, $s^2 = 5.6$. Steel industry: $n = 9$, $\bar{X} = 8.5\%$, $s^2 = 11.2$. Oil-and-gas industry: $n = 13$, $\bar{X} = 11.8\%$, $s^2 = 16.4$. Using these data, can we conclude that computer stocks are less risky than oil and gas stocks? Use a significance level of .05. Assume that the mean annual returns for these industries are approximately normally distributed.

9.45   Using the data in Exercise 9.44, test the hypothesis that the steel industry's stocks are just as risky as the computer industry's stocks. Use a significance level of .05.

9.46   The manager of a vending-machine company decided to buy one of two types of dispenser to put in her vending machines. Both dispensers claim to dispense, on the average, 6 ounces of fluid in a plastic cup. The amount dispensed is approximately normally distributed. However, to test this claim, the manager would first like to know whether the variability in the amount of fluid dispensed is the same for both dispensers. Using a significance level of .05, is the manager justified in using a pooled estimate of the population variance, if 16 replications on each dispenser give the following results? Dispenser 1: $s = 1.5$, $n = 16$. Dispenser 2: $s = 3.4$, $n = 16$. If so, derive the pooled estimate.

◢ 9.5 Comparing the Means of Two Normal Populations Using Paired Samples

The final section of this chapter examines the situation in which the two samples are *not* obtained independently. All discussion up to this point has assumed that the two samples *are* independent. By not independent, we mean that the corresponding elements from the two samples are *paired*. Perhaps each pair of observations corresponds to the same city, the same week, the same married couple, or even the same person. Our discussion focuses on comparing the two population means for the situation in which two *dependent* samples are obtained from the two populations.

When attempting to estimate or test for the difference between two population means, your first question always should be, is there any natural reason to pair the first observation from sample 1 with the first observation from sample 2, the second with the second, and so on? If there is no reason to pair these data and the samples were obtained independently, the previous methods for finding confidence intervals and testing μ_1 versus μ_2 apply. If the data were gathered such that pairing the values is necessary, then it is *extremely* important that you recognize this and treat the data in a different manner. We can still determine confidence intervals and perform a test of hypothesis, but the procedure is different.

As an illustration, Metalloy manufactures metal hinges. The hardness of these hinges is tested by pressing a rod with a pointed tip into the hinge with a specified force and measuring the depth of the depression caused by the tip. Two tips are available for the hardness tester, and it is suspected that tip #1 produces higher hardness readings, on the average. To control for variation in the hardness of the hinges, it was decided to use paired samples in which *both* tips were used to test the hardness of the *same* metal hinge. The following coded data were obtained using 12 randomly selected hinges. The letter d represents the *difference* of each pair of hardness values (Tip #1 − Tip #2).

HINGE	1	2	3	4	5	6	7	8	9	10	11	12
Tip #1	39	32	42	49	45	47	45	48	38	48	41	47
Tip #2	35	34	38	48	47	43	41	47	35	46	37	44
d	4	-2	4	1	-2	4	4	1	3	2	4	3
d^2	16	4	16	1	4	16	16	1	9	4	16	9

$$\Sigma d = 4 - 2 + 4 + \cdots + 3 = 26$$
$$\Sigma d^2 = 16 + 4 + 16 + \cdots + 9 = 112$$

Each pair of values was obtained from the same metal hinge, so these data values clearly need to be paired—they are dependent samples. It seems reasonable to examine the difference of the two values for each hinge, so these differences (d), along with the d^2 values, are also shown. We have thus reduced the problem from two sets of values to a single set. The parameter of interest here is the **difference** of the population means, μ_d. Put another way, μ_d is the **mean of the population differences.**

Since we have a single set of sample values (the 12 differences) and a single parameter (μ_d), the results of Sections 7.4 and 8.4 can be used to construct a confidence interval and perform a test of hypothesis, *provided we have reason to believe that the population differences are normally distributed.* As a result, we need not worry about large versus small samples, because we will use the t distribution for our confidence intervals and tests of hypothesis, regardless of the sample sizes. Of course, if the number of differences is large (generally, > 30), this distribution is closely approximated by the standard normal distribution.

If you have reason to suspect that the population of differences is *not* normally distributed, then one alternative is to use a nonparametric procedure, in particular, the Wilcoxon signed rank test (discussed in Chapter 19).

Confidence Interval for μ_d Using Paired Samples

The statistic used to derive a confidence interval for μ_d and perform a test of hypothesis using *dependent* samples is

$$t_D = \frac{\bar{X}_1 - \bar{X}_2}{s_d/\sqrt{n}} = \frac{\bar{d}}{s_d/\sqrt{n}} \qquad (9.21)$$

where

n = the number of pairs of observations

s_d = the standard deviation of the n differences

$$= \sqrt{\frac{\Sigma d^2 - (\Sigma d)^2/n}{n - 1}}$$

df for $t_D = n - 1$

This is a *t random variable* with $n - 1$ df. Notice that the numerator of t_D is the same as before, namely, $\bar{X}_1 - \bar{X}_2$, which is also represented by $\bar{d} = \Sigma d/n$, the mean of the differences. The mean of the differences $\bar{d}$ always is equal to $\bar{X}_1 - \bar{X}_2$ (this can help you in checking your arithmetic when computing the d's).

Based on the discussion in Section 7.4, we obtain a $(1 - \alpha) \cdot 100\%$ confidence interval for μ_d:

$$\bar{d} - t_{\alpha/2, n-1} \frac{s_d}{\sqrt{n}} \quad \text{to} \quad \bar{d} + t_{\alpha/2, n-1} \frac{s_d}{\sqrt{n}} \qquad (9.22)$$

EXAMPLE 9.10   Using the hardness data, derive a 95% confidence interval for μ_d, where

$$\mu_d = \text{average difference in hardness}$$

SOLUTION   We have

$$\bar{d} = \frac{\Sigma d}{n} = \frac{26}{12} = 2.167$$

Notice that

$$\bar{x}_1 = \frac{39 + 32 + \cdots + 47}{12} = 43.417$$

and

$$\bar{x}_2 = \frac{35 + 34 + \cdots + 44}{12} = 41.25$$

so $\bar{d} = \bar{x}_1 - \bar{x}_2 = 2.167$. It checks! Also,

$$s_d = \sqrt{\frac{\Sigma d^2 - (\Sigma d)^2/n}{n-1}} = \sqrt{\frac{112 - (26)^2/12}{11}}$$

$$= \sqrt{\frac{55.667}{11}} = \sqrt{5.061} = 2.250$$

The resulting 95% confidence interval for μ_d is

$$\bar{d} - t_{.025,11}\frac{s_d}{\sqrt{n}} \quad \text{to} \quad \bar{d} + t_{.025,11}\frac{s_d}{\sqrt{n}}$$

$$= 2.167 - 2.201\frac{2.250}{\sqrt{12}} \quad \text{to} \quad 2.167 + 2.201\frac{2.250}{\sqrt{12}}$$

$$= 2.167 - 1.430 \quad \text{to} \quad 2.167 + 1.430$$

$$= .737 \quad \text{to} \quad 3.597$$

Based on these data, we are 95% confident that the hardness reading using tip #1 is between .737 and 3.597 *more* than the tip #2 reading.   ∎

Hypothesis Testing Using Paired Samples

The test statistic for testing the means is the same as that in Section 8.4, except that we use the sample differences.

$$t_D = \frac{\bar{d} - D_0}{s_d/\sqrt{n}} \tag{9.23}$$

where D_0 is the hypothesized value of μ_d. When testing $H_0: \mu_d = D_0$ versus $H_a: \mu_d \neq D_0$, reject H_0 if $|t_D| > t_{\alpha/2,n-1}$. Here, $t_{\alpha/2,n-1}$ is obtained from Table A.5 using $n - 1$ df. One-tailed tests are performed in a similar manner by placing α in either the right tail ($H_a: \mu_d > D_0$) or in the left tail ($H_a: \mu_d < D_0$). A summary is provided in the box on paired sample tests for μ_d and D_0 on page 339.

EXAMPLE 9.11   Consider the previous hardness data collected by Metalloy. Can you confirm the suspicion that the average difference in hardness readings (tip #1 − tip #2) is positive? Use a significance level of $\alpha = .05$.

SOLUTION   **Step 1.** We are attempting to demonstrate that the average difference in hardness readings is positive; this claim goes into the alternative hypothesis. The resulting hypotheses are

$$H_0: \mu_d \le 0$$

$$H_a: \mu_d > 0$$

Step 2. We are dealing with paired data, so the correct test statistic is

$$t_D = \frac{\bar{d}}{s_d/\sqrt{n}}$$

Step 3. What happens to t_D when H_a is true? If $\mu_d > 0$, then we would expect $\bar{d}$ to be *positive*. So, the test procedure is to

reject H_0 if $t_D > k$, for some $k > 0$

What is k? As before, this depends on α, and in the usual manner, we have

reject H_0 if $t_D > t_{\alpha, n-1}$

where $t_{\alpha, n-1}$ is obtained from Table A.5. For this situation, $t_{.05, 11} = 1.796$, and so we

reject H_0 if $t_D > 1.796$

Step 4. Using the sample data,

$$t_D^* = \frac{2.167}{2.250/\sqrt{12}} = 3.34$$

Because $3.34 > 1.796$, we reject H_0.

Step 5. The average hardness reading using tip #1 is higher than that using tip #2. ∎

A MINITAB solution to Example 9.11 is provided in Figure 9.13. Notice that the differences are first derived, and then a standard t test (described in Section 8.4) is used to test that the mean difference μ_d is ≤ 0 versus the alternative $H_a: \mu_d > 0$. The resulting p-value is $p = .0033$, which, using $\alpha = .05$, again results in rejecting H_0 because $p < \alpha$.

What happens if you fail to pair these observations and perform a regular two-sample t test, as we did in Section 9.3 for small *independent* samples? The results

■ **FIGURE 9.13**
MINITAB solution to Examples 9.10 and 9.11. This is the correct way to analyze these data.

```
MTB > SET INTO C1
DATA> 39 32 42 49 45 47 45 48 38 48 41 47
DATA> END
MTB > SET INTO C2
DATA> 35 34 38 48 47 43 41 47 35 46 37 44
DATA> END
MTB > SUBTRACT C2 FROM C1, PUT INTO C3
MTB > NAME C1 = 'TIP1' C2 = 'TIP2' C3 = 'DIFF'
MTB > TINTERVAL WITH 95% CONFIDENCE USING C3

              N      MEAN    STDEV   SE MEAN    95.0 PERCENT C.I.
DIFF         12      2.167   2.250   0.649    (  0.737,   3.596)

MTB > TTEST OF MU = 0 USING C3;
SUBC> ALTERNATIVE = 1. ←        = 1 for H_a: μ_d > D_0
                                = -1 for H_a: μ_d < D_0
                                (not necessary for H_a: μ_d ≠ D_0)

TEST OF MU = 0.000 VS MU G.T. 0.000

              N      MEAN    STDEV   SE MEAN      T     P VALUE
DIFF         12      2.167   2.250   0.649      3.34   0.0033
                       |       |                  |
                       d̄      s_d                t_D*
```

```
MTB > TWOSAMPLE TEST USING C1 AND C2;
SUBC> ALTERNATIVE = 1.

TWOSAMPLE T FOR TIP1 VS TIP2
          N       MEAN      STDEV    SE MEAN
TIP1     12      43.42       5.14       1.5
TIP2     12      41.25       5.26       1.5

95 PCT CI FOR MU TIP1 - MU TIP2: (-2.3, 6.6)

TTEST MU TIP1 = MU TIP2 (VS GT): T= 1.02   P=0.16   DF=   21
                                             |
                                           p-value
```

are summarized in Figure 9.14, where we observe an interesting result. The t value (using the test statistic from equation 9.10) now is 1.02, with a corresponding p-value of $p = .16$. This means that, using this test, we now *fail to reject* H_0. We are unable to demonstrate a difference between the average hardness readings, which, according to Figure 9.13, is *not* a correct conclusion. Figure 9.14 shows convincingly that failing to pair the observations when you should can cause you to obtain an incorrect result. More importantly, there is nothing to warn you that this has occurred.

EXAMPLE 9.12   The market research staff at Allied Foods is considering two different packaging designs for an instant breakfast cereal that Allied is about to introduce. The first type of container under consideration is a rectangular box, whereas the second container type has a cylindrical shape.

The staff decides to conduct a pilot study by placing the product in both containers and locating the two types at opposite ends of the breakfast cereal section in ten different supermarkets. All the containers are placed at eye level to remove any effect due to the height of the display. The main question under consideration is whether there is any difference in the sales of the two types of container. From the following data, can you conclude that there is a difference in sales for the rectangular and cylindrical containers? Use $\alpha = .05$ to define your test.

SUPERMARKET	1	2	3	4	5	6	7	8	9	10
Rectangular	194	152	160	172	118	110	137	126	176	145
Cylindrical	184	161	153	184	105	123	155	111	156	129

SOLUTION   The data were gathered by collecting a pair of observations from each supermarket, so this is a clear-cut case of dependent sampling. Your next step should be to determine the paired differences. Define d to be the rectangular box sales minus the cylindrical box sales.

SUPERMARKET	1	2	3	4	5	6	7	8	9	10	TOTAL
d	10	-9	7	-12	13	-13	-18	15	20	16	29
d^2	100	81	49	144	169	169	324	225	400	256	1917

Step 1. We are attempting to detect a difference in the two means: a two-tailed test is in order. Let

μ_d = average difference in sales for the two container types.

The correct hypotheses are

$$H_0: \mu_d = 0$$
$$H_a: \mu_d \neq 0$$

Steps 2, 3. Using the t_D test statistic, the test will be to

$$\text{reject } H_0 \text{ if } |t_D| > t_{\alpha/2, n-1}$$

where $t_{\alpha/2, n-1} = t_{.025, 9} = 2.262$.

Step 4. Using the sample data,

$$\bar{d} = \frac{\Sigma d}{n} = \frac{29}{10} = 2.9$$

$$s_d = \sqrt{\frac{\Sigma d^2 - (\Sigma d)^2/n}{n-1}}$$

$$= \sqrt{\frac{1917 - (29)^2/10}{9}}$$

$$= \sqrt{\frac{1832.9}{9}} = 14.271$$

From these values we obtain

$$t_D^* = \frac{\bar{d}}{s_d/\sqrt{n}} = \frac{2.9}{14.271/\sqrt{10}} = .643$$

Because $.643 < 2.262$, we fail to reject H_0.

Step 5. Based on these data, there is *insufficient evidence* to conclude that the container type has an effect on sales. ∎

Paired Sample Tests for μ_1 and μ_2

TWO-TAILED TEST

$$H_0: \mu_d = D_0$$

$$H_a: \mu_d \neq D_0$$

$$\text{reject } H_0 \text{ if } |t_D^*| > t_{\alpha/2, n-1}$$

where

1. Each difference, d, is (sample 1 value − sample 2 value)

2. $$t_D = \frac{\bar{d} - D_0}{s_d/\sqrt{n}}$$

3. $$\bar{d} = \bar{X}_1 - \bar{X}_2 = \frac{\Sigma d}{n}$$

4. $$s_d = \sqrt{\frac{\Sigma d^2 - (\Sigma d)^2/n}{n-1}}$$

5. $$\text{df for } t_D = n - 1$$

ONE-TAILED TEST

$H_0: \mu_d \leq D_0$	$H_0: \mu_d \geq D_0$
$H_a: \mu_d > D_0$	$H_a: \mu_d < D_0$
$\text{reject } H_0 \text{ if } t_D^* > t_{\alpha, n-1}$	$\text{reject } H_0 \text{ if } t_D^* < -t_{\alpha, n-1}$

Exercises

9.47 A hospital is experimenting with the effectiveness of a newly developed drug that controls blood pressure. The blood-pressure level is measured using a sphygmomanometer before and after administration of the drug to a sample of hypertensive patients with a history of elevated blood pressure. The question is whether there is a measured decrease in systolic blood pressure (in mm Hg) after administration of the drug. The difference in blood pressure before and after administration of the drug is believed to be approximately normally distributed.

PATIENT	BEFORE DRUG	AFTER DRUG	PATIENT	BEFORE DRUG	AFTER DRUG
1	110	94	6	82	85
2	88	81	7	96	77
3	84	82	8	97	89
4	94	88	9	134	110
5	108	97			

a. Using a significance level of .10, can you conclude that the blood-pressure level is lower after the drug is administered?

b. Should you use an independent or dependent sample t statistic to analyze this experiment?

9.48 Construct a 99% confidence interval for μ_d using the data in Exercise 9.47 ($d =$ before − after).

9.49 Suppose that in Exercise 9.47, the manufacturer of the drug claims that the new drug is effective in reducing the average blood-pressure scores by more than 8 mm Hg. Would you support that statement at a significance level of .01?

9.50 The controller of a fast-food chain is interested in determining whether there is any difference in the weekly sales of restaurant 1 and restaurant 2. The weekly sales are approximately normally distributed. The sales, in dollars, for seven randomly selected weeks are:

WEEK	RESTAURANT 1	RESTAURANT 2
1	4100	3800
2	1800	4600
3	2200	5100
4	3400	3050
5	3100	2800
6	1100	1950
7	2200	3400

a. Should this problem be analyzed using an independent or dependent sample t statistic?

b. Using a significance level of .01, is there evidence to support the conclusion that there is a significant difference in the weekly sales of the two restaurants?

9.51 An expert in carpet cleaning rated the performance of two stain-removing products on ten different stains. Let $\bar{d} = \bar{X}_1 - \bar{X}_2$. The following statistics were calculated from the data:

$$\bar{d} = 3.2, \ \Sigma d^2 = 1899, \ \Sigma d = 28$$

Calculate the p-value for testing that there is sufficient evidence to conclude that there is a difference in the performance of the stain removers. Interpret the p-value. What assumptions about the distribution of the data are necessary for this statistical test to be valid?

9.52 Calculate a 90% confidence interval for the mean difference in the scores given to the two stain removers in Exercise 9.51.

9.53 Smart Look, an exercise program developed by Joni Beauty consultants, is claimed to be effective in reducing the weight of a typical overweight woman by more than 17 pounds. In order to exmaine the validity of this hypothesis, the program was tried on a group of middle-aged women, and their weights (in pounds) were recorded before and after completion of the exercise program. Assume that the difference in weights after completion of the program is approximately normally distributed.

WOMAN	BEFORE	AFTER	WOMAN	BEFORE	AFTER
1	140	115	5	175	165
2	160	130	6	145	125
3	110	100	7	115	101
4	132	109	8	122	105

a.   Using a significance level of .10, what would be your conclusion?

b.   Why did you select the particular test statistic you used to analyze this problem?

9.54   Using the data from Exercise 9.53, test the hypothesis that H_0: $\mu_d = 0$ versus H_a: $\mu_d \neq 0$ at a .01 significance level ($d = $ before $-$ after).

9.55   Ten commonly bought automotive parts that are available at both a dealership and at a local automotive parts shop are randomly selected to determine if the mean price at the dealership is significantly higher than the mean price at the local shop. Prices are as follows:

AUTO PART	PRICE AT LOCAL SHOP (IN DOLLARS)	PRICE AT DEALERSHIP (IN DOLLARS)	AUTO PART	PRICE AT LOCAL SHOP (IN DOLLARS)	PRICE AT DEALERSHIP (IN DOLLARS)
1	31	38	6	45	43
2	25	44	7	21	26
3	36	47	8	69	64
4	51	56	9	58	73
5	42	53	10	34	33

a.   At the .01 significance level, do the data support the belief that the mean price at the dealership is higher than the mean price at the local automotive parts shop?

b.   What assumptions are necessary to ensure the validity of the test procedure?

9.56   To test that white officers receive higher peer ratings than black officers in the United States Army, 76 randomly selected officers were asked to evaluate their peers. Data were collected only on black and white officers who had just completed a six-week advanced training camp. The data yielded the following statistics:

Mean rating received by whites	3.94
Mean rating received by blacks	3.46
Standard deviation of differences	.73
Number of pairs	76

Is there sufficient evidence to conclude that the white army officers receive higher peer ratings than black army officers? Use the p-value to justify your conclusion.

(*Source:* Adapted from J. Lordan, "Race Effects In Peer Ratings of U.S. Army ROTC Cadets," *The Journal of Applied Business Research* 6 (no. 3): 75.)

Summary   This chapter has presented an introduction to **statistical inference for two populations.** We examined tests of hypothesis and confidence intervals for the means and variances (for example, whether they are equal) of two populations, using both independent and dependent samples.

When we used large **independent samples** to test the population means, we defined a test statistic having approximately a standard normal distribution, and we also used this test statistic to define a confidence interval for $\mu_1 - \mu_2$. For small independent samples ($n_1 < 30$ or $n_2 < 30$), hypothesis testing on μ_1 versus μ_2 is concerned with means from two normal populations. For this situation, although we are concerned with the means, we must pay special attention to whether we also have reason to believe that the population standard deviations (σ_1 and σ_2) are equal.

If we do not assume that the σ values are equal, we use a test statistic for μ_1 versus μ_2 having an *approximate t* distribution. This statistic also results in an approximate confidence interval for $\mu_1 - \mu_2$. If we assume that the σ values are equal, we use a procedure that pools the sample variances and results in a test statistic having an *exact t* distribution. We also derived confidence intervals for $\mu_1 - \mu_2$ for this situation.

To determine whether two population variances (or standard deviations) are the same, we introduced the **F distribution.** This distribution is nonsymmetric (right skew) and assumes that two independent samples were obtained from normal populations. Probabilities (areas under the curve) for the F random variable are contained in Table A.7. Using this distribution, we can perform two-tailed tests (such as $H_a: \sigma_1 \neq \sigma_2$) or one-tailed tests (such as $H_a: \sigma_1 > \sigma_2$) on the two standard deviations. We also use it to construct a confidence interval for σ_1^2/σ_2^2 or σ_1/σ_2.

When two samples are obtained such that corresponding observations are paired (matched), the resulting samples are **dependent** or **paired.** When using two such samples, we defined a t-statistic to test the mean of the population differences, μ_d, and to construct a confidence interval for μ_d. We need not be concerned about whether the population standard deviations are equal for this situation because the test statistic uses the differences between the paired observations, a new variable.

Summary of Formulas

1. Large independent samples
Confidence interval for $\mu_1 - \mu_2$ (σ_1, σ_2 known):

$$(\bar{X}_1 - \bar{X}_2) \pm Z_{\alpha/2} \sqrt{\frac{\sigma_1^2}{n_1} + \frac{\sigma_2^2}{n_2}}$$

Confidence interval for $\mu_1 - \mu_2$ (σ_1, σ_2 unknown):

$$(\bar{X}_1 - \bar{X}_2) \pm Z_{\alpha/2} \sqrt{\frac{S_1^2}{n_1} + \frac{S_2^2}{n_2}}$$

Hypothesis testing for μ_1 and μ_2: Test statistic is

$$Z = \frac{(\bar{X}_1 - \bar{X}_2) - D_0}{\sqrt{\frac{S_1^2}{n_1} + \frac{S_2^2}{n_2}}}$$

where D_0 is the hypothesized value of $\mu_1 - \mu_2$

2. Sample sizes (minimizing $n = n_1 + n_2$)

$$n_1 = \frac{Z_{\alpha/2}^2 S_1(S_1 + S_2)}{E^2}$$

$$n_2 = \frac{Z_{\alpha/2}^2 S_2(S_1 + S_2)}{E^2}$$

3. Small independent samples
Confidence interval for $\mu_1 - \mu_2$ (not assuming $\sigma_1 = \sigma_2$):

$$(\bar{X}_1 - \bar{X}_2) \pm t_{\alpha/2,df} \sqrt{\frac{S_1^2}{n_1} + \frac{S_2^2}{n_2}}$$

where

$$df = \frac{\left[\frac{S_1^2}{n_1} + \frac{S_2^2}{n_2}\right]^2}{\frac{\left(\frac{S_1^2}{n_1}\right)^2}{n_1 - 1} + \frac{\left(\frac{S_2^2}{n_2}\right)^2}{n_2 - 1}}$$

Confidence interval for $\mu_1 - \mu_2$ (assuming $\sigma_1 = \sigma_2$):

$$(\bar{X}_1 - \bar{X}_2) \pm t_{\alpha/2,df} \sqrt{\frac{S_p^2}{n_1} + \frac{S_p^2}{n_2}}$$

where

$$df = n_1 + n_2 - 2$$

and

$$S_p^2 = \frac{(n_1 - 1)S_1^2 + (n_2 - 1)S_2^2}{n_1 + n_2 - 2}$$

Hypothesis testing for μ_1 and μ_2 (not assuming $\sigma_1 = \sigma_2$):
Test statistic is

$$t' = \frac{\bar{X}_1 - \bar{X}_2}{\sqrt{\frac{S_1^2}{n_1} + \frac{S_2^2}{n_2}}}$$

where

$$df = \frac{\left[\dfrac{s_1^2}{n_1} + \dfrac{s_2^2}{n_2}\right]^2}{\dfrac{\left(\dfrac{s_1^2}{n_1}\right)^2}{n_1 - 1} + \dfrac{\left(\dfrac{s_2^2}{n_2}\right)^2}{n_2 - 1}}$$

Hypothesis testing for μ_1 and μ_2 (assuming $\sigma_1 = \sigma_2$):
Test statistic is

$$t = \frac{\bar{X}_1 - \bar{X}_2}{s_p \sqrt{\dfrac{1}{n_1} + \dfrac{1}{n_2}}}$$

where

$$df = n_1 + n_2 - 2$$

and

$$s_p = \sqrt{\frac{(n_1 - 1)s_1^2 + (n_2 - 1)s_2^2}{n_1 + n_2 - 2}}$$

4. Comparing variances (or standard deviations):
Confidence interval for σ_1^2/σ_2^2:

$$\frac{s_1^2/s_2^2}{F_U} \text{ to } \frac{s_1^2/s_2^2}{F_L}$$

where $F_U = F_{\alpha/2, v_1, v_2}$ and $F_L = 1/F_{\alpha/2, v_2, v_1}$ and $v_1 = n_1 - 1$, $v_2 = n_2 - 1$

Hypothesis testing for σ_1^2/σ_2^2:
Test statistic is

$$F = \begin{cases} \dfrac{\text{larger of } s_1^2 \text{ and } s_2^2}{\text{smaller of } s_1^2 \text{ and } s_2^2} & (H_a: \sigma_1 \neq \sigma_2) \\[3mm] \dfrac{s_1^2}{s_2^2} & (H_a: \sigma_1 > \sigma_2) \\[3mm] \dfrac{s_2^2}{s_1^2} & (H_a: \sigma_1 < \sigma_2) \end{cases}$$

5. Dependent (paired) samples
Confidence interval for μ_d:

$$\bar{d} \pm t_{\alpha/2, n-1} \frac{s_d}{\sqrt{n}}$$

where

n = number of paired observations
$\bar{d}$ = average of n differences
s_d = standard deviation of n differences

Hypothesis testing for μ_d: Test statistic is

$$t_D = \frac{\bar{d} - D_0}{s_d/\sqrt{n}} \quad (df = n - 1)$$

where D_0 = hypothesized value of μ_d.

Review Exercises

9.57 To evaluate the expected life of two types of tires, a car manufacturer decided to use a randomly selected set of 20 similar cars for testing the mean difference in the amount of wear (in thousandths of an inch) for the two brands of tires after 10,000 miles. The manufacturer placed two tires of the first brand and two tires of the second brand on each car. Will the resulting samples be independent samples or dependent samples? Discuss.

9.58 A sandwich shop wishes to test the effectiveness of its coupons. The manager believes that the business brought in by the responses to the coupon in the *Highland Village Daily* is equal to the business brought in by the responses to the coupon placed in the *Green Sheet*. The amount spent by each customer using a coupon is recorded (in dollars) and can be considered to be normally distributed. Test the manager's belief with a significance level of .01. *Highland Village Daily: n* = 32, $\bar{x}$ = 9.50, *s* = 26.3. *Green Sheet: n* = 39, $\bar{x}$ = 11.80, *s* = 29.4.

9.59 Dairy Castle wanted to boost the sales of their "Country Baskets." They thought that it might be helpful to hang posters that picture the item. They recorded the number of Country Baskets sold during lunchtime for one week at their various stores. They repeated the sampling for another week when the poster advertising was used. Assume that weekly sales are normally distributed. Is there sufficient evidence to say that hanging the posters improved sales of the Country Baskets? Use a .05 significance level.

STORE	BEFORE	AFTER	STORE	BEFORE	AFTER
1218	215	240	1270	201	220
1224	180	220	1282	207	215
1236	150	190	1292	195	219
1252	180	175	1304	180	195

9.60   Denver Hydro-Mulch Company helps lawns grow by spraying a prepared mixture on each lawn. A chemical company sales representative would like to convince Denver Hydro-Mulch that his company has a better fertilizer mixture. He has agreed to give the company enough fertilizer mixture to spray on eight randomly selected lawns. An additional set of eight randomly selected lawns are sprayed with the fertilizer mix that the company currently is using. At the end of four weeks, the eight lawns prepared with the new mixture had an average growth of 32 cm and a standard deviation of 7.8 cm. The eight lawns sprayed with the fertilizer mixture that the company is currently using had an average growth of 25 cm and a standard deviation of 6 cm. The growth of the grass at the end of the four-week period is considered to be normally distributed. Test the claim that the new fertilizer mixture is superior to the current one. Use a 10% significance level. Do not assume that the population variances are equal.

9.61   The suggested 1988 retail prices on a Ford Escort and on a Chevrolet Cavalier are $6895 and $7395, respectively, according to *U.S. News and World Report* (April 1988, p. 83). A manager at a Ford automotive dealership believes that the mean advertised price of the Chevrolet Cavalier exceeds the mean advertised price of the Ford Escort at automotive dealerships in the Chicago metropolitan area. Assume that a random sample of seven Ford dealerships yields a mean of $6291 with a sample standard deviation of $300. Also assume that a random sample of seven Chevrolet dealerships yields a mean of $6981 with a sample standard deviation of $380. At the .10 level of significance, is there sufficient evidence to conclude that the advertised price of the Chevrolet Cavalier exceeds the advertised price of the Ford Escort by $500? Assume equal population variances, and that the data come from populations that are normally distributed.

9.62   A used-car lot manager wished to know how successful car owners were in receiving their asking price when selling their car. The manager selected a random sample of ten car owners who sold a car that was no older than five years. Below are listed both the asking price and the selling price.

CAR OWNER	SELLING PRICE	ASKING PRICE
1	$10,600	$10,950
2	6,450	6,900
3	5,500	5,500
4	12,890	12,200
5	6,700	6,700
6	11,350	11,700
7	8,500	8,600
8	7,540	8,100
9	9,100	9,200
10	12,700	12,900

a.   Is there sufficient evidence to indicate a significant difference in the selling price and the asking price? Use a .10 significance level.

b.   What assumption needs to be made about the distribution of the data for the statistical test in part a to be valid?

9.63   Researchers at Stanton University have said that while older workers were often more productive than their younger counterparts, supervisors tended to rate the older workers lower. Suppose we have the following data after sampling at random the ratings of eight older workers and eight younger workers.

RATINGS

Younger Workers	82	91	50	82	75	89	90	76
Older Workers	77	80	65	80	70	83	79	75

Do the data support the contention the mean rating for younger workers is higher than the mean rating for older workers? Assume the data came from normally distributed populations with equal population variances. Use a significance level of .05.

9.64 Two independent samples of students were asked to give a score on a Likert-type scale to the relative importance of social responsibility in the output quality of a corporation. In the first sample, 161 students were randomly selected and were to assume the role of an investor in a certain company. In the second sample, 61 students were randomly selected and were to assume the role as manager of the company. The following statistics resulted from the data collected:

	MEAN	STANDARD DEVIATION
Sample 1	5.33	.80
Sample 2	5.60	.67

Is there sufficient evidence to conclude that there is a significant difference in the relative importance of social responsibility in the output quality of a corporation when the role of manager is assumed and when the role of investor is assumed? Use a .05 significance level.

(*Source:* Adapted from K. Kraft, "The Relative Importance of Social Responsibility In Determining Organizational Effectiveness: A Student Point of View," *Proceedings of the Annual Meeting of the Decision Sciences Institute,* November 1990, pp. 1422–1424.)

9.65 A new packaging method that is proposed has an average output yield of finished units approximately the same as the existing packaging method. This new packaging method will be adopted if the variability in the number of finished units is less, thus providing greater process control. At a .05 significance level, is there sufficient evidence to conclude that the variance of the number of finished units is less for the new packaging method.

	EXISTING PACKAGING METHOD	NEW PACKAGING METHOD
Days sampled	9	9
s^2	1190	465

9.66 Determine which of the following sets of hypotheses are equivalent.

a. $H_0: \sigma_1^2/\sigma_2^2 \leq 1$ and $H_a: \sigma_1^2/\sigma_2^2 > 1$.

b. $H_0: \sigma_2^2/\sigma_1^2 \geq 1$ and $H_a: \sigma_2^2/\sigma_1^2 < 1$.

c. $H_0: \sigma_2^2 \geq \sigma_1^2$ and $H_a: \sigma_2^2 < \sigma_1^2$.

d. $H_0: \sigma_1 \leq \sigma_2$ and $H_a: \sigma_1 > \sigma_2$.

9.67 A study is designed to determine the effect of an office-training course on typing productivity. Ten typists are randomly selected and are asked to type 15 pages of equally difficult text before and after completing the training course. Their productivity is measured by the total number of errors made.

TYPIST	BEFORE	AFTER	TYPIST	BEFORE	AFTER
1	30	27	6	33	31
2	19	14	7	28	22
3	36	31	8	30	25
4	42	37	9	27	30
5	35	29	10	34	33

Assume that the total number of errors can be approximated by a normal distribution. Test the claim that taking the office-training course leads to a reduction in the average number of errors made by a typist. Use a significance level of .05.

9.68 Suppose that a sample of size 16 is chosen from population 1 and a sample of size 26 is drawn from population 2. Assume that both populations are normally distributed. If a 90% confidence interval for the ratio of the variance of population 1 to the variance of population 2 is .367 to 1.753, what is the point estimate of the ratio of the two population variances?

9.69 The average yearly expenditure for homeowner's insurance is $465 for homes in the New England states and $400 for homes in the mid-Atlantic states. A member of the National Insurance Consumers Organization believes that the difference is much larger for certain cities in these two regions. Assume that the following statistics were the result of a sample of 16 homes from Philadelphia and 16 homes from Boston.

	PHILADELPHIA	BOSTON
Average yearly expenditure for homeowner's insurance	396	487
Standard deviation	25	37

a. Perform a test for equal variances at the ten percent significance level.

b. Using the results of part a, is there sufficient evidence to conclude that the difference between the average yearly expenditure for homeowner's insurance in Philadelphia and Boston exceeds $65? Use a ten percent significance level.

(*Source:* Adapted from "What You Really Spend on Insurance?" *Journal of American Insurance* 60 (1990): 13.)

9.70 The sales of two Stop-N-Go convenience stores are compared for 12 randomly selected weeks. The MINITAB computer printout gives a printout of store 1 in C1 and store 2 in C2. The units are in thousands of dollars

```
MTB > print c1
C1
  16.8    17.3    18.2    17.0    16.2    16.8    15.7    18.3    17.5
  15.0    16.3    17.3

MTB > print c2
C2
  17.2    17.1    18.5    17.9    15.7    17.7    15.9    19.7    18.2
  15.1    16.2    17.7

MTB > let c3 = c2 - c1
MTB > print c3
C3
   0.40000   -0.20000    0.30000    0.90000   -0.50000    0.90000
   0.20000    1.40000    0.70000    0.10000   -0.10000    0.40000

MTB > ttest of mu = 0 using c3;
SUBC> alternative = 0.

TEST OF MU = 0.000 VS MU N.E. 0.000

              N      MEAN     STDEV    SE MEAN      T     P VALUE
C3           12     0.375     0.534      0.154     2.43     0.033
```

a. Is there sufficient evidence to conclude that the two convenience stores' sales differ significantly? Use the p-value to justify your conclusion.

b. Construct a 99% confidence interval on the difference in the mean sales for the two stores. Interpret the confidence interval.

c. What assumptions are necessary for the test procedure in part a to be valid?

9.71 Twelve stock market analysts randomly selected from the Marty Sinch brokerage firm and ten stock market analysts randomly selected from the E. P. Sutton brokerage firm were asked to forecast the percentage change in Standard and Poor's 500 index in one year. The MINITAB computer printout gives several confidence intervals comparing the means of the two groups. The values of the random sample from the Marty Sinch brokerage firm are under C1 and the values of the random sample from the E. P. Sutton brokerage firm are in C2. The figures are in percentages.

```
MTB > PRINT C1
SINCH
    3.4    10.5    20.6    18.5    11.3    12.6    14.7    10.9    9.6
   15.7    25.6     9.5

MTB > PRINT C2
SUTTON
   16.5    11.8    17.3    18.8    11.6     6.1    12.6    22.8    20.6
   14.8

MTB > TWOSAMPLE TEST WITH 99% CONFIDENCE USING C1 AND C2;
SUBC> ALTERNATIVE = 0;
SUBC> POOL.
```

```
TWOSAMPLE T FOR SINCH VS SUTTON
            N      MEAN     STDEV    SE MEAN
SINCH      12     13.57      5.90       1.7
SUTTON     10     15.29      4.95       1.6

99 PCT CI FOR MU SINCH - MU SUTTON: (-8.4, 5.0)
TTEST MU SINCH = MU SUTTON (VS NE): T=-0.73 P=0.47 DF=20.0

MTB > TWOSAMPLE TEST WITH 95% CONFIDENCE USING C1 AND C2;
SUBC> ALTERNATIVE = 0;
SUBC> POOL.

TWOSAMPLE T FOR SINCH VS SUTTON
            N      MEAN     STDEV    SE MEAN
SINCH      12     13.57      5.90       1.7
SUTTON     10     15.29      4.95       1.6

95 PCT CI FOR MU SINCH - MU SUTTON: (-6.6, 3.2)
TTEST MU SINCH = MU SUTTON (VS NE): T=-0.73 P=0.47 DF=20.0
MTB > TWOSAMPLE TEST WITH 90% CONFIDENCE USING C1 AND C2;
SUBC> ALTERNATIVE = 0;
SUBC> POOL.

TWOSAMPLE T FOR SINCH VS SUTTON
            N      MEAN     STDEV    SE MEAN
SINCH      12     13.57      5.90       1.7
SUTTON     10     15.29      4.95       1.6

90 PCT CI FOR MU SINCH - MU SUTTON: (-5.8, 2.3)
TTEST MU SINCH = MU SUTTON (VS NE): T=-0.73 P=0.47 DF=20.0
```

a. Interpret the three confidence intervals. Explain why the width of the confidence intervals changes for each of the confidence levels?

b. What are the necessary assumptions for the confidence intervals to be valid?

9.72 A manager at a manufacturing plant was interested in whether the job proficiency of workers with exactly three years of experience was significantly different from the job proficiency of workers with exactly five years of experience. A random sample of 12 workers is selected from each of these two groups. A job proficiency test is administered to each worker and a score obtained. A MINITAB printout gives an analysis of the data. The printout shows the results of pooling and not pooling in testing that $H_0: \mu_1 - \mu_2 = 0$ versus $H_a: \mu_1 - \mu_2 \neq 0$.

```
MTB > name c1 '3 years'
MTB > name c2 '5 years'
MTB > print c1
3 years
    92    64    79    51    95    83    76    87    58    86
    68    88

MTB > print c2
5 years
    85    86    89    82    91    83    86    83    87    85
    90    84

MTB > twosample test with 95% confidence using c1 and c2;
SUBC> alternative = 0.

TWOSAMPLE T FOR 3 years VS 5 years
             N      MEAN     STDEV    SE MEAN
3 years     12      77.3     14.1       4.1
5 years     12     85.92     2.87      0.83

95 PCT CI FOR MU 3 years - MU 5 years: (-17.8, 0.46)
TTEST MU 3 years = MU 5 years (VS NE): T=-2.09 P=0.061 DF=11.9

MTB > twosample test with 95% confidence using c1 and c2;
SUBC> alternative = 0;
SUBC> pool.

TWOSAMPLE T FOR 3 years VS 5 years
             N      MEAN     STDEV    SE MEAN
3 years     12      77.3     14.1       4.1
5 years     12     85.92     2.87      0.83

95 PCT CI FOR MU 3 years - MU 5 years: (-17.3, -0.07)
TTEST MU 3 years = MU 5 years (VS NE): T=-2.09 P=0.048 DF=22.0
```

a. What conclusion would one make for the test procedure with pooling and without pooling at the .05 significance level?

b. Intuitively, does it appear that the standard deviations of the two groups differ significantly? At the .05 significance level, do the data provide sufficient evidence to conclude that the population standard deviation for the workers with three years experience differs from the population standard deviation for the workers with five years experience? What conclusion would you draw from the MINITAB printout?

Computer Exercises Using the Database

Exercise 1 -- Appendix I Choose at random ten observations from the database in which the family owns their home and ten observations in which the family rents their home. (Refer to the variable OWNORENT.) Do the data support the conclusion that the home payment for homeowners is larger than the home payment for renters? Use a .05 significance level. What assumptions are necessary to ensure that the test procedure is valid? Do not assume equal population variances.

Exercise 2 -- Appendix I Choose at random ten observations from the database from a family of size 2 and ten observations from a family of size 4. Do the data support the conclusion that the monthly utility expenditure (variable UTILITY) is larger for a family of size 4? Use a .05 significance level. Do not assume equal population variances.

Exercise 3 -- Appendix J From the database, choose a random sample of 12 companies with an A bond rating and another random sample of 12 companies with a C bond rating. Do the data support the conclusion that the net income of companies with a C bond rating is less than the net income of companies with an A rating? Use a .05 significance level. Do not assume equal population variances.

Exercise 4 -- Appendix J From the database, choose a random sample of 15 companies with a B bond rating and another random sample of 15 companies with a C bond rating. Do the data support the conclusion that the variances of the current assets of the companies with B bond ratings and C bond ratings differ significantly at the .05 level?

Case Study

Requests for Proposals, and Some Responses—With Enthusiasm

Most government agencies (and sometimes private sector bodies) initiate contracts with requests for proposals (RFPs). The RFP is a technical document providing detailed specifications of certain goods or services that are desired by the buyer and inviting potential suppliers to submit bids. A bid is submitted in the form of a technical proposal that aims to establish the bidder's technical capability to provide the required goods or services and his or her superiority over other suppliers in this respect.

By its very nature, a technical proposal tends to be scientific, objective, nonemotional and dry. Nonetheless, it is also designed to be a persuasive document in order to "make a sale." One might expect to find elements of subjective rhetoric in such a document. Beck and Wegner conducted a study that examined one particular subjective element in technical proposals, the idea of the "enthusiasm" shown by the bidding firm.

Enthusiasm, of course, is a multidimensional variable. The authors of the study mention lexicon, syntax, metaphor, punctuation, and other style variables as some of the dimensions of enthusiasm. They selected lexicon, that is, vocabulary or use of words, as the variable to be analyzed. The first step was to establish a lexicon of enthusiasm and construct a way of measuring enthusiasm. Details of the methods are given in the referenced article. The authors ended up with a list of words representing a preliminary lexicon of enthusiasm, along with a measure called the enthusiasm index (EI). The EI indicates the level of enthusiasm as measured by multiple occurrences of words from the lexicon in samples of 1000, 2000, or 3000 words of text taken from technical proposals.

The enthusiasm index was then checked for validity by comparing ten "statements of work" from government RFPs with ten motivational sales texts. The former are specification documents used by bidders to prepare their proposals and should contain almost no occurrences of the enthusiasm lexicon; the latter should obviously be more enthusiastic. The results of this phase of the study are given in Table 9.1, where Lit-1, Lit-2, etc., refer to motivational sales texts gleaned from such sources as *Becoming A Superstar Seller* by Sheehan and O'Toole, and *Enthusiasm Makes The Difference* by Norman Vincent Peale.

In the next phase of the study, 15 technical proposals were analyzed. All these proposals were winning proposals that had been submitted to federal, state, or local agencies. The proposals varied in length and were grouped into three categories: over 100 pages, 50–100 pages, and less than 50 pages. Proposals not already in word processing files were scanned to capture the text in ASCII files. Finally, all texts were put on diskettes in Microsoft Word format and run through the spell-checker for accuracy. From the larger proposals, three files each of 1000 words were sampled in three- to five-page increments; from the midsize proposals, two files; and from the small proposals, a single file of 1000 words. The length of each sample was determined using the statistical function software of Grammatik II 1.31 to ensure uniform counting. Each file was electronically searched for occurrences of words listed in the previously established lexicon of enthusiasm. Words not in a proper enthusiastic context were excluded. The results are summarized in Tables 9.2 and 9.3.

Case Study Questions

1. a. Conduct a t-test on the two groups in Table 9.1, assuming equal population variances, at the .05 level of significance.
b. Since we already know *a priori* that sales texts are likely to be more enthusiastic than statements of work in RFPs, what is the purpose of such a hypothesis test?
c. If we failed to reject the null hypothesis, how does that affect the rest of the experiment?
2. Table 9.2 shows the results for the 15 proposals that were analyzed. Is there a significant difference in EI means for the 10 RFPs in Table 9.1 and these 15 proposals in Table 9.2? Report the p-value.
3. In Table 9.3, the 15 proposals have been classified into two groups, social science/humanities proposals and science/engineering proposals. Do you find the social science/humanities proposals to be significantly more enthusiastic? Report the p-value.
4. Does the above study establish that winning proposals have a higher EI score, on the average?

■ **TABLE 9.1**
Summary of Computer Search Results for RFPs and Sales Texts.

TEXT FILE	NO. OF OCCURRENCES	ENTHUSIASM INDEX
RFP-1	2	2.0
RFP-2	0	0
RFP-3	0	0
RFP-4	1	1.0
RFP-5	1	1.0
RFP-6	3	3.0
RFP-7	1	1.0
RFP-8	1	1.0
RFP-9	2	2.0
RFP-10	0	0
Mean ($\bar{X}_1$)		1.1
Standard Deviation (s_1)		0.99
Number of Samples (n_1)		10

TEXT FILE	NO. OF OCCURRENCES	ENTHUSIASM INDEX
Lit-1	3	3.0
Lit-2	0	0
Lit-3	14	14.0
Lit-4	3	3.0
Lit-5	17	17.0
Lit-6	13	13.0
Lit-7	21	21.0
Lit-8	17	17.0
Lit-9	8	8.0
Lit-10	14	14.0
Mean ($\bar{X}_2$)		11.0
Standard Deviation (s_2)		7.09
Number of Samples (n_2)		10

(*Source:* Charles E. Beck and Keith A. Wegner, "Enthusiasm in Technical Proposals: Verifying a Method of Lexical Analysts," *IEEE Transactions On Professional Communication* 33 (no. 3, September 1990: 118.

■ **TABLE 9.2**
Summary of Computer Search Results for Technical Proposals Examined.

NO.	TYPE OF PROPOSAL	NO. OF WORDS	NO. OF WORD OCCURRENCES	EI VALUES
1.	Advertising	1000	3	3.0
2.	Arts (Film)	3000	8	2.7
3.	Social Research	3000	7	2.3
4.	Social Research	3000	6	2.0
5.	Chemical Research	3000	6	2.0
6.	Social Research	2000	4	2.0
7.	Social Research	3000	5	1.7
8.	Environ. Consulting	3000	4	1.3
9.	Environ. Consulting	2000	2	1.0
10.	Program Audit	1000	1	1.0
11.	Social Research	2000	2	1.0
12.	Engineering	3000	2	0.7
13.	Data Review	3000	2	0.7
14.	Program Audit	2000	1	0.5
15.	Chemical Research	1000	0	0

Mean ($\bar{X}_3$)   1.46
Standard Deviation (s_3)   0.863
Number of Samples (n_3)   15

(*Source:* Charles E. Beck and Keith A. Wegner, "Enthusiasm in Technical Proposals: Verifying a Method of Lexical Analysis," *IEEE Transactions On Professional Communication* 33 (no. 3, September 1990): 118.

■ **TABLE 9.3**
Summary of Computer Search Results by Technical Proposal Type.

SOCIAL SCIENCE/HUMANITIES PROPOSALS

No.	Type of Proposal	No. of Words	EI Values
1.	Advertising	1000	3.0
2.	Arts (Film)	3000	2.7
3.	Social Research	3000	2.3
4.	Social Research	3000	2.0
6.	Social Research	2000	2.0
7.	Social Research	3000	1.7
11.	Social Research	2000	1.0

Mean ($\bar{X}_4$)   2.10
Standard Deviation (s_4)   0.658
Number of Samples (n_4)   7

SCIENCE/ENGINEERING PROPOSALS

No.	Type of Proposal	No. of Words	EI Values
5.	Chemical Research	3000	2.0
8.	Environ. Consulting	3000	1.3
9.	Environ. Consulting	2000	1.0
10.	Program Audit	1000	1.0
12.	Engineering	3000	0.7
13.	Data Review	3000	0.7
14.	Program Audit	2000	0.5
15.	Chemical Research	1000	0

Mean ($\bar{X}_5$)   0.90
Standard Deviation (s_5)   0.590
Number of Samples (n_5)   8

(*Source:* Charles E. Beck and Keith A. Wegner, "Enthusiasm in Technical Proposals: Verifying a Method of Lexical Analysis," *IEEE Transactions On Professional Communication* 33 (no. 3, September 1990): 118.

5. The authors recommend, as topics for additional research, a paired comparison between winning and losing proposals submitted for the same RFPs; examining the cover letters of proposals in addition to the actual proposal texts; and an elaboration of the measurement of "enthusiasm" to include other dimensions such as lexicon modifiers (e.g., *interested* versus *very interested*), bolding, underlining, punctuation, and syntactical construction. Comment on each of these, and how they illustrate strengths and weaknesses of the study discussed above.

SPSS

✓ Solution
Example 9.4

Example 9.4 was concerned with a *t* test for two independent samples to compare two population means. The problem was to determine if the average blowout times using the test apparatus were different for Beltex and Roadmaster tires ($H_0: \mu_1 = \mu_2$). The SPSS program listing in Figure 9.15 was used to test the variances and means for two independent samples. For the test on the means, the value of the test statistic and the *p*-value are provided both for the case where the population variances are assumed equal and for the case where it is not assumed that the variances are equal.

In this problem the SPSS commands are the same for both the mainframe and PC versions (remember to end each command line with a period when using the PC version).

The TITLE command names the SPSS run.

The DATA LIST command gives each variable a name and describes the data as being in free form.

The BEGIN DATA command indicates to SPSS that the input data immediately follow.

The next 30 lines contain the data values, with each line representing the brand (Roadmaster or Beltex) and the wear factor. The first line, for example, represents brand B (Beltex) and a wear factor of 3.82.

The END DATA statement indicates the end of the data.

The T-TEST command compares two sample means. The GROUPS and VARIABLES subcommands divide the cases into two groups for a comparison of sample means.

Figure 9.16 shows the SPSS output obtained by executing the listing in Figure 9.15.

■ FIGURE 9.15
Input for SPSS or SPSS/PC. Remove the periods for SPSS input.

```
TITLE    BELTEX-ROADMASTER TIRES.
DATA LIST FREE/WEAR BRAND.
BEGIN DATA.
3.82 0
3.11 0
4.21 0
2.64 0
4.16 0
3.91 0
2.44 0
4.52 0
2.84 0
3.26 0
3.74 0
3.04 0
2.56 0
2.58 0
3.15 0
4.16 1
3.92 1
3.94 1
4.22 1
4.15 1
3.62 1
4.11 1
3.45 1
3.65 1
3.82 1
4.55 1
3.82 1
3.85 1
3.62 1
4.88 1
END DATA.
T-TEST GROUPS=BRAND(0,1) / VARIABLES=WEAR.
```

■ **FIGURE 9.16** SPSS output.

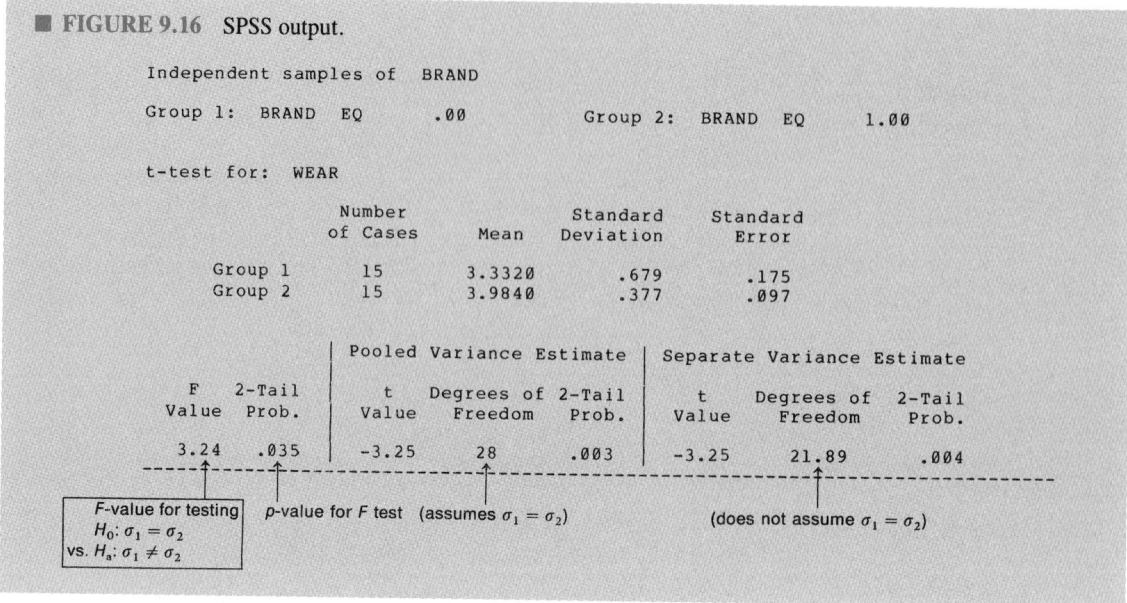

        Independent samples of BRAND

        Group 1: BRAND EQ   .00   Group 2: BRAND EQ   1.00

        t-test for: WEAR

                              Number   Standard   Standard
                              of Cases   Mean   Deviation   Error

                Group 1   15   3.3320   .679   .175
                Group 2   15   3.9840   .377   .097

                         | Pooled Variance Estimate | Separate Variance Estimate

            F   2-Tail | t   Degrees of 2-Tail | t   Degrees of 2-Tail
          Value Prob. | Value   Freedom   Prob. | Value   Freedom   Prob.

          3.24   .035 | -3.25   28   .003 | -3.25   21.89   .004

    ┌─────────────────┐
    │ F-value for testing │   p-value for F test (assumes $\sigma_1 = \sigma_2$)   (does not assume $\sigma_1 = \sigma_2$)
    │ H_0: $\sigma_1 = \sigma_2$ │
    │ vs. H_a: $\sigma_1 \neq \sigma_2$ │
    └─────────────────┘

SPSS

▰ **Solution**
Example 9.11

Example 9.11 was concerned with the computation of the t statistic for the means of two populations using paired (dependent) samples. The problem was to determine whether the average difference in hardness readings (tip #1 − tip #2) is positive (H_a: $\mu_d > 0$).

The SPSS program listing in Figure 9.17 was used to request a mean, t score and p-value. Note that SPSS assumes a two-tailed test. This was a one-tailed test, so the calculated p-value must be divided by two.

In this problem the SPSS commands are the same for both the mainframe and PC versions (remember to end each command line with a period when using the PC version).

The TITLE command names the SPSS run.

The DATA LIST command gives each variable a name and describes the data as being in free form.

The BEGIN DATA command indicates to SPSS that the input data immediately follow.

The next 12 lines contain the data values, which represent the hardness readings using tip #1 and tip #2, respectively. The first line, for example, implies that the hardness reading was 39 using tip #1 and 35 using tip #2.

The END DATA statement indicates the end of the data.

The T-TEST command compares two sample means. The PAIRS subcommand names the variables being compared.

Figure 9.18 shows the SPSS output obtained by executing the listing in Figure 9.17.

■ FIGURE 9.17
Input for SPSS or
SPSS/PC. Remove
the periods for
SPSS input.

```
TITLE   HARDNESS READINGS.
DATA LIST FREE/TIP1 TIP2.
BEGIN DATA.
39 35
32 34
42 38
49 48
45 47
47 43
45 41
48 47
38 35
48 46
41 37
47 44
END DATA.
T-TEST PAIRS=TIP1 TIP2.
```

■ FIGURE 9.18   SPSS output.

HARDNESS READINGS

Paired samples t-test: TIP1
                         TIP2

Variable	Number of Cases	Mean	Standard Deviation	Standard Error
TIP1	12	43.4167	5.143	1.485
TIP2	12	41.2500	5.259	1.518

(Difference) Mean	Standard Deviation	Standard Error	° ° ° °	Corr.	2-Tail Prob.	° ° ° °	t Value	Degrees of Freedom	2-Tail Prob.
2.1667	2.250	.649	°	.907	.000	°	3.34	11	.007

t_D

For a one-tailed test, the actual p-value = .007/2 = .0035

SAS

■ Solution
Example 9.4

Example 9.4 was concerned with a *t*-test for two independent samples to compare two population means. The problem was to determine if the average blowout times using the test apparatus were different for Beltex and Roadmaster tires ($H_0: \mu_1 = \mu_2$). The SAS program listing in Figure 9.19 was used to request the variances and the means for two independent samples. For the test on the means, the value of the test statistic and the *p*-value are provided both for the case where the population variances are assumed equal and for the case where it is not assumed that the variances are equal.

In this problem the SAS commands are the same for both the mainframe and PC versions.

The TITLE command names the SAS run (enclose it in single quotes).

The DATA command gives the data a name.

The INPUT command names and gives the correct order for the different fields on the data lines. The $ implies the BRAND is a character data.

The CARDS command indicates to SAS that the input data immediately follow.

The next 30 lines contain the data values, with each line representing the brand (Roadmaster or Beltex) and the wear factor. The first line, for example, represents brand B (Beltex) and a wear factor of 3.82.

The PROC TTEST command compares the means of two groups of observations. The subcommand CLASS identifies BRAND as the variable to be classified in this example. The subcommand TITLE provides a report heading for the output.

Figure 9.20 shows the SAS output obtained by executing the listing in Figure 9.19.

■ **FIGURE 9.19**
Input for SAS
(Mainframe or
micro version).

```
TITLE     'BELTEX-ROADMASTER TIRES';
DATA BLOWOUT;
INPUT BRAND $ WEAR;
CARDS;
B 3.82
B 3.11
B 4.21
B 2.64
B 4.16
B 3.91
B 2.44
B 4.52
B 2.84
B 3.26
B 3.74
B 3.04
B 2.56
B 2.58
B 3.15
R 4.16
R 3.92
R 3.94
R 4.22
R 4.15
R 3.62
R 4.11
R 3.45
R 3.65
R 3.82
R 4.55
R 3.82
R 3.85
R 3.62
R 4.88
PROC TTEST;
 CLASS BRAND;
 TITLE 'INDEPENDENT SAMPLES TTEST';
```

■ **FIGURE 9.20** SAS output.

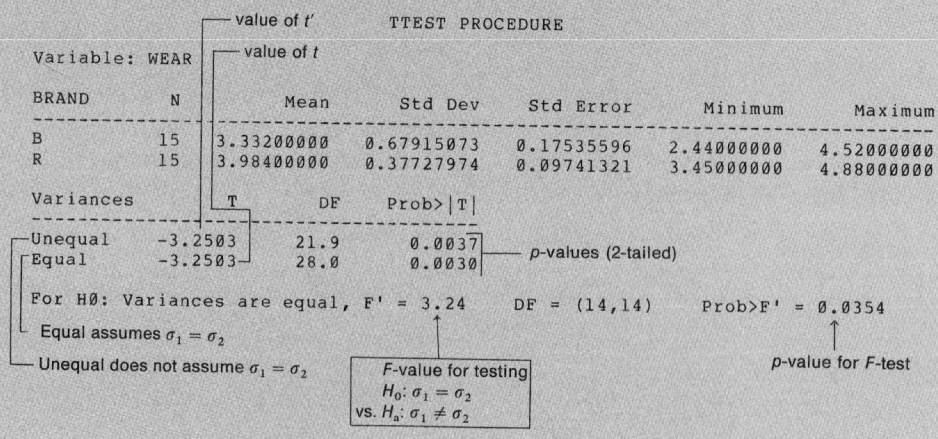

INDEPENDENT SAMPLES TTEST

TTEST PROCEDURE

BRAND	N	Mean	Std Dev	Std Error	Minimum	Maximum
B	15	3.33200000	0.67915073	0.17535596	2.44000000	4.52000000
R	15	3.98400000	0.37727974	0.09741321	3.45000000	4.88000000

Variances	T	DF	Prob>|T|
Unequal	-3.2503	21.9	0.0037
Equal	-3.2503	28.0	0.0030

For H0: Variances are equal, F' = 3.24   DF = (14,14)   Prob>F' = 0.0354

SAS

**Solution
Example 9.11**

Example 9.11 was concerned with the computation of the *t* statistic for the means of two populations using point (dependent) samples. The problem was to determine whether the average difference in hardness readings (tip #1 − tip #2) is positive ($H_a: \mu_d > 0$).

The SAS program listing in Figure 9.21 was used to request a mean, *t* score and *p*-value. Note that SAS assumes a two-tailed test. This was a one-tailed test, so the calculated *p*-value must be divided by two.

In this problem the SAS commands are the same for both the mainframe and PC versions.

The TITLE command names the SAS run (enclose it in single quotes).

The DATA command gives the data a name.

The INPUT command names and gives the correct order for the different fields on the data cards.

The DIFF = TIP1 − TIP2 statement is used to compute a new variable, DIFF, which is the difference between the value of TIP1 and the value of TIP2.

The CARDS command indicates to SAS that the input data immediately follow.

The next 12 lines contain the data values. Each line represents the hardness readings using tip #1 and tip #2, respectively. The first line, for example, implies that the hardness reading using tip #1 was 39 and was 35 using tip #2.

The PROC MEANS command requests an SAS procedure to print simple descriptive statistics for the variable in the following subcommand VAR DIFF. The TITLE subcommand names the output.

Figure 9.22 shows the SAS output obtained by executing the listing in Figure 9.21.

FIGURE 9.21
Input for SAS (Mainframe or micro version).

```
TITLE    'HARDNESS READINGS';
DATA HARDNESS;
INPUT TIP1 TIP2;
DIFF=TIP1-TIP2;
CARDS;
39 35
32 34
42 38
49 48
45 47
47 43
45 41
48 47
38 35
48 46
41 37
47 44
PROC MEANS N MEAN T PRT;
 VAR DIFF;
 TITLE 'TWO DEPENDENT SAMPLES';
```

■ **FIGURE 9.22**
SAS output.

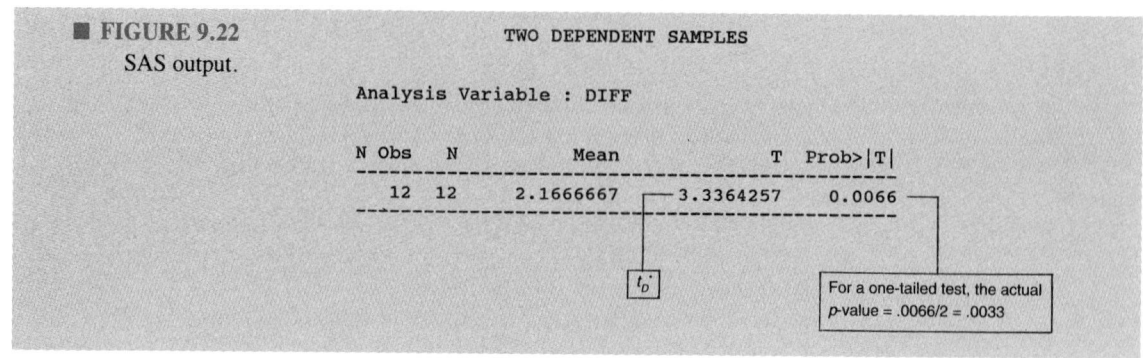

TWO DEPENDENT SAMPLES

Analysis Variable : DIFF

N Obs	N	Mean	T	Prob>|T|
12	12	2.1666667	3.3364257	0.0066

t_D

For a one-tailed test, the actual
p-value = .0066/2 = .0033

CHAPTER

10 *Estimation and Testing for Population Proportions*

By now you should be comfortable with the concepts of estimation and hypothesis testing. If, for example, you have reason to believe that the population is normally distributed, you can then estimate the necessary population parameters (such as the mean and standard deviation) using the corresponding sample statistics. You should be well aware that there always is the risk of arriving at an incorrect conclusion when using sample information to infer something about an entire population. We can use the Central Limit Theorem, to relax the assumptions regarding normality when making inferences about population means if large samples are used.

Chapters 7, 8, and 9 concentrated primarily on normal populations. We provided you with confidence intervals for the mean and the variance of a single normal population. We examined how to check a statement regarding one of these parameters (such as $\mu < 100$ or $\sigma > .5$) using a test of hypothesis. Then this concept was extended to comparing the means or variances of two normal populations.

Now we return to the *binomial* situation, in which we are interested in the *proportion* of your population that has a certain attribute. Attributes can include

personal attributes, such as willingness to buy a product or being in favor of a proposed labor contract. We can also examine proportions as they relate to a particular physical attribute, such as the proportion of defective components in a batch.

We are interested in a single parameter, referred to as **p,** which is the **proportion** of the population having this attribute. For example, suppose that a recent report claims that only 10% of all registered voters in a certain area are in favor of forced busing for school children ($p = .10$). Or suppose it has been reported that a lower proportion of families with children favor busing than do those without children. How can we estimate the actual proportions here and test these claims?

Examining a population proportion plays a *vital* role in describing and monitoring the quality of a production process. Here, the parameter of interest is the proportion (p) of nonconforming units being produced by the process. Based on the results of a statistical sample, we can decide if the parameter p is too large (process is out of control) or acceptable (process is in control). This topic will be discussed further in Chapter 12.

10.1

Estimation and Confidence Intervals for a Population Proportion

A test for a population *proportion* is a binomial situation. Using the definitions from Chapter 5, each member of your population is either a *success* or a *failure*. These words can be misleading; it is necessary only that each person (or object) in your population either have a certain attribute (a success) or not have it (a failure). So we define p to be the proportion of successes in the population—that is, the proportion that have a certain attribute.

Do not confuse the notation p = a population proportion with the previously used notation for a p-value. They do not mean the same thing. We hope that the context will make it clear which of the two p's is being described.

In Chapter 5 we assumed that p is known. For any binomial situation, perhaps p is known, or more likely it was estimated in some way. This chapter examines how you can estimate p by using a sample from the population. Also, we can support (or fail to support) claims concerning the value of p. The final section in this chapter compares two samples from two separate populations.

Point Estimate for a Population Proportion

Suppose that the management of Cassidy Electronics, a manufacturer of calculators and microcomputers, is considering offering a dental plan to their employees. Because the monthly premium will be deducted from employee paychecks, perhaps not all employees will wish to join the plan. The insurance company is interested in the proportion of employees who will want to join. A random sample of 200 employees was interviewed. Of these, 137 said they would purchase the dental insurance if it were offered. What can you say about the proportion (p) of all employees who wish to join?

We view this problem as a binomial situation and define a success as a person who will sign up for the dental insurance and a failure as a person who will not sign up for the dental insurance. Consequently, p is the proportion of successes in the population (proportion of all employees who favor the dental insurance). Remember that p, like μ and σ previously, will remain *unknown* forever. To *estimate p*, we obtain a random sample and observe the proportion of successes in our sample. We use $\hat{p}$ (read "p hat") to denote the estimate of p, which is the proportion of successes in the sample. Here, $\hat{p}$ = proportion of employees in the sample who will sign up for the dental insurance, so $\hat{p} = 137/200 = .685$.

In general,

$$
\begin{aligned}
\hat{p} &= \text{estimate of } p \\
&= \text{proportion of sample having a specified attribute} \qquad \textbf{(10.1)} \\
&= \frac{x}{n}
\end{aligned}
$$

where n = sample size and x = the number of sample observations having this attribute.

The symbol $\hat{}$ is used to denote an *estimate*. Distinguish between $\hat{p}$ obtained from sample information and p, the population proportion being estimated by $\hat{p}$. This is the same type of difference that we previously recognized between a sample mean, $\bar{X}$ (often referred to as $\hat{\mu}$), and a population mean, μ.

Confidence Intervals for a Population Proportion (Using a Small Sample)

The calculations involved in determining a confidence interval for p using a small sample are fairly complex. To make them easier, we have listed 90% and 95% confidence intervals for sample sizes of $n = 5, 6, \ldots, 20$ in Table A.8. For sample sizes other than these, you can (1) use the large sample confidence interval (described next) or (2) extend Table A.8 by consulting your local statistician. Or you can use a computer subroutine to derive additional values for this table. An explanation of the method used to generate these confidence limits is included at the end of the table.

Using Table A.8 is much like using Table A.1, the table of binomial probabilities. Let n = sample size and x = the observed number of successes in your sample. Based on these values, the confidence interval (p_L, p_U) can be obtained directly from the table.

EXAMPLE 10.1   A private company is considering the purchase of 200 Beagle microcomputers to monitor seismic activity. These computers will be placed in outdoor stations where

they must be able to operate in extremely cold weather. If the computers will operate in temperatures as low as $-10°F$, the company will purchase them. Beagle,
anxious to demonstrate the reliability of their system, has agreed to subject 15
computers to a "cold test." Let p = proportion of *all* Beagle computers that will
function at $-10°F$.

Of the 15 sample computers, three of them stopped operating at or above
$-10°F$. What can you say about p? Construct a 95% confidence interval for p.

SOLUTION   Let a success be that a computer *survives* the cold test (still functions at $-10°F$).
We observe 12 successes out of 15 in the sample. So,

$$\hat{p} = \frac{12}{15} = .8$$

Using Table A.8 for $n = 15$, $x = 12$, and $\alpha = .05$, we find $p_L = .519$ and $p_U = .957$. The corresponding 95% confidence interval for p is

$$p_L \quad \text{to} \quad p_U = .519 \quad \text{to} \quad .957$$

So we are 95% confident that the actual (population) percentage of Beagle
computers that can function at $-10°F$ is between 51.9% and 95.7%.   ■

COMMENTS   One of the purposes of this section (omitted in many textbooks) is to demonstrate that
confidence intervals for a population proportion (p) are typically very wide when using a small sample. The
moral of this section is: *To obtain a useful (narrow) confidence interval for p, obtain a large sample.*

Confidence Intervals for a Population Proportion
(Using a Large Sample)

When dealing with large samples, the Central Limit Theorem once again provides
us with a reliable method of determining approximate confidence intervals for a
population proportion. For each element in your sample, assign a value of 1 if this
observation is a success (has the attribute) or 0 if this observation is a failure (does
not have the attribute). Using the dental-plan example to illustrate, for *each* person
in the sample we assign 1 if this person wants the dental insurance and 0 if this
person does not want the dental insurance. So what is $\hat{p}$? We can write it as

$$\hat{p} = \frac{\overbrace{1 + 1 + \cdots + 1}^{137 \text{ times}} + \overbrace{0 + 0 + \cdots + 0}^{63 \text{ times}}}{200} = \frac{137}{200} = .685$$

In this sense, then, $\hat{p}$ is a **sample average:** it is an average of 0s and 1s. *As a result,
we can apply the Central Limit Theorem to $\hat{p}$ and conclude that $\hat{p}$ is (approximately)
a normal random variable for large samples.* This works reasonably well provided
np and $n(1 - p)$ are both greater than 5. So the distribution of $\hat{p}$ [large sample;
$np > 5$ and $n(1 - p) > 5$] can be summarized: $\hat{p}$ is (approximately) a normal
random variable with

$$\text{mean} = p$$

$$\text{standard deviation (standard error)} = \sqrt{\frac{p(1 - p)}{n}}$$

By standardizing this result, we have

$$Z = \frac{\hat{p} - p}{\sqrt{\dfrac{p(1 - p)}{n}}} \qquad (10.2)$$

which is approximately a standard normal random variable. This variable allows us to use Table A.4 to construct a confidence interval for p. This confidence interval is obtained in the identical manner used to construct previous confidence intervals with the standard normal distribution, namely,

$$(\text{point estimate}) \pm Z_{\alpha/2} \cdot (\text{standard deviation of point estimator}) \qquad (10.3)$$

Thus, a $(1 - \alpha) \cdot 100\%$ confidence interval for p (large sample; np and $n(1 - p) > 5$) is

$$\hat{p} - Z_{\alpha/2} \sqrt{\frac{\hat{p}(1 - \hat{p})}{n}} \quad \text{to} \quad \hat{p} + Z_{\alpha/2} \sqrt{\frac{\hat{p}(1 - \hat{p})}{n}} \qquad (10.4)$$

where $\hat{p}$ = sample proportion. Notice that we used $\hat{p}$ and $1 - \hat{p}$ under the square root in equation 10.4 rather than p and $1 - p$. This is necessary because p is *unknown* and must be replaced by its estimate, $\hat{p}$. As we observed in previous chapters, replacing an unknown parameter by its estimate works well provided our sample is large enough. For this situation, both np and $n(1 - p)$ should be greater than 5.

The expression $\sqrt{\hat{p}(1 - \hat{p})/n}$ is the *estimated* standard error (standard deviation) of $\hat{p}$ and can be written

$$s_{\hat{p}} = \sqrt{\frac{\hat{p}(1 - \hat{p})}{n}} \qquad (10.5)$$

The mean of the random variable $\hat{p}$ is the (unknown) value of p. In other words, the average value of $\hat{p}$ is the parameter it is estimating. Such an estimator is said to be **unbiased.** If we obtained random samples indefinitely, the resulting $\hat{p}$'s—on the average—will equal p. This is a desirable property for a sample estimator to have. We have actually discussed two other unbiased estimators previously; $\bar{X}$ is an unbiased estimator of a population mean (μ) and s^2 is an unbiased estimator of a population variance (σ^2).

EXAMPLE 10.2   Using the data regarding employees' desire to join the dental plan, what is a 90% confidence interval for the proportion of all employees who would participate in the dental insurance program?

SOLUTION   Using Table A.4, $Z_{\alpha/2} = Z_{.05} = 1.645$. Also, $\hat{p} = 137/200 = .685$. So the 90% confidence interval for p is

$$.685 - 1.645 \sqrt{\frac{(.685)(.315)}{200}} \quad \text{to} \quad .685 + 1.645 \sqrt{\frac{(.685)(.315)}{200}}$$

$$= .685 - .054 \quad \text{to} \quad .685 + .054$$

$$= .631 \quad \text{to} \quad .739$$

Based on the sample data, we are 90% confident that the percentage of employees who would purchase the dental insurance is between 63.1% and 73.9%. ∎

EXAMPLE 10.3

Remember that in lot acceptance sampling, we either accept or reject a batch (lot) of components, parts, or assembled products based on tests using a random sample drawn from the lot.

Suppose we draw a sample of size 150 from a lot of calculators. We test each of the sampled calculators and find 13 defectives. Determine a 95% confidence interval for the proportion of defectives in the entire batch.

SOLUTION

Let p = proportion of defective calculators in the batch. Based on the sample of 150 calculators, we have

$$\hat{p} = \frac{13}{150} = .0867$$

Because $Z_{.025} = 1.96$, the 95% confidence interval for p is

$$.0867 - 1.96 \sqrt{\frac{(.0867)(.9133)}{150}} \quad \text{to} \quad .0867 + 1.96 \sqrt{\frac{(.0867)(.9133)}{150}}$$

$$= .0867 - .045 \quad \text{to} \quad .0867 + .045$$

$$= .042 \quad \text{to} \quad .132$$

Consequently, we are 95% confident that our estimate $\hat{p} = .0867$ is within .045 of the actual value of p. In other words, this sample estimates the actual percentage of defective calculators to within 4.5%, with 95% confidence. ∎

Choosing the Sample Size (One Population)

Suppose that you want your point estimate, $\hat{p}$, to be within a certain amount of the actual proportion, p. In Example 10.3 the *maximum error*, E, was $E = .045$, that is, 4.5%. What if the buyer's specifications necessitate that we estimate the parameter p to within 2% with 95% confidence? Now,

$$E = 1.96 \sqrt{\frac{p(1 - p)}{n}} \qquad \qquad \textbf{(10.6)}$$

We have an earlier estimate of p ($\hat{p} = .0867$) using the sample of size 150; this value can be used in equation 10.6. The purpose is to extend this sample in order to obtain this specific maximum error, E. The specified value of E is .02, so

$$E = .02 = 1.96 \sqrt{\frac{(.0867)(.9133)}{n}}$$

Therefore,

$$\sqrt{\frac{(.0867)(.9133)}{n}} = \frac{.02}{1.96}$$

Squaring both sides and rearranging leads to

$$n = \frac{(1.96)^2(.0867)(.9133)}{(.02)^2} = 760.5$$

■ **FIGURE 10.1**
Curve of values of
$\hat{p}(1 - \hat{p})$.

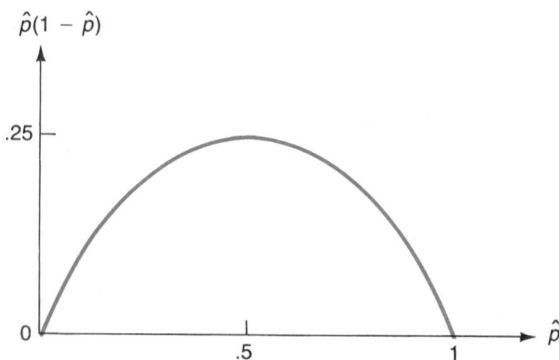

Rounding up (*always*), we come to the conclusion that a sample of size $n = 761$ calculators will be necessary to estimate p to within 2%.

In general, the following equation provides the necessary sample size to estimate p with a specified maximum error, E, and confidence level $(1 - \alpha) \cdot 100\%$:

$$n = \frac{Z_{\alpha/2}^2 \hat{p}(1 - \hat{p})}{E^2} \qquad \textbf{(10.7)}$$

In this illustration, we used an estimate of p from a prior sample to determine the necessary sample size using equation 10.7. If the sample of size n based on this equation is our first and only sample, then we *have no estimate of p*. There is a conservative procedure we can follow here that will guarantee the accuracy (E) that we require. Look at the curve of different values of $\hat{p}(1 - \hat{p})$ in Figure 10.1. Consider these values:

$\hat{p}$	$\hat{p}(1 - \hat{p})$
.2	.16
.4	.24
.5	.25
.7	.21
.9	.09

Note that the largest value of $\hat{p}(1 - \hat{p})$ is .25.

If we make $\hat{p}(1 - \hat{p})$ in equation 10.7 *as large as possible*, we will obtain a value of n that will result in a maximum error that is sure to be less than the specified value. So we can formulate this rule: If no prior estimate of p is available, a conservative procedure to determine the necessary sample size from equation 10.7 is to use $\hat{p} = .5$.

EXAMPLE 10.4   Suppose that the insurance company underwriting the dental plan wishes to obtain a single sample that will estimate, to within 2% with 90% confidence, the proportion (p) of employees who would purchase the dental insurance. They have *no* prior knowledge of this proportion. Their intent is to obtain a large enough sample the first time so that they can estimate the population proportion with this much accuracy. How large a sample is required?

SOLUTION   We have no prior knowledge of p, so we use $\hat{p} = .5$ in equation 10.7 to obtain a sample size of

$$n = \frac{(1.645)^2(.5)(.5)}{(.02)^2} = 1691.3$$

To obtain an estimate of p with a maximum error of $E = .02$, we will need a sample size of $n = 1692$ employees. With a sample of this size, we can safely say that the point estimate, $\hat{p}$, will be within 2% of the actual value of p, with 90% confidence (however, this is a very large sample). ∎

Exercises

10.1 Mary Pharmaceuticals is interested in estimating the proportion (p) of its employees who would accept a substantial increase in benefits instead of an annual raise in salary for a particular year. A random sample of 20 employees was obtained, and 12 welcomed the idea. How can you estimate the proportion in the population who would welcome this plan? Construct a 90% confidence interval for p. Use Table A.8.

10.2 In Exercise 10.1, let the maximum error of the estimating proportion be 4%. Estimate the necessary sample size for a 95% confidence interval. Use the estimate of p from Exercise 10.1 for the value of p.

10.3 In Exercise 10.2, assume that you have no prior knowledge of p. Estimate the sample size that is required to estimate the proportion (p) of employees who would accept the plan with a 95% confidence level, using the same maximum error of estimate for the proportion.

10.4 A company wishes to estimate the proportion, p, of its employees who went on sick leave during the past six months. A random sample of 18 employees was taken; of them, 10 went on sick leave. Construct a 90% confidence interval for p. Use Table A.8.

10.5 An investment firm surveyed 18 randomly selected economists. The survey found that 13 of them felt that the economy of the United States would not slip into a recession for at least a year. Use Table A.8 to construct 90% and 95% confidence intervals for the percentage of economists who believe that a recession will not occur in the United States for at least a year.

10.6 A math workshop will be offered only if the student demand is sufficiently high. What is the required sample size necessary to estimate with 90% confidence the proportion of students who would register for the workshop if we specify a value of .03 for the maximum error, E?

10.7 In Exercise 10.6, if a previous study indicated that the proportion of students who would register for the workshop was .68, estimate the necessary sample size for a maximum error, E, of 3%.

10.8 Winthrop Boat Lines is exploring the possibility of offering a ferry service between the cities of Patna and Madura, provided there is sufficient demand to make it feasible. The firm randomly interviewed 210 commuters from the two cities, and 146 of them indicated they would patronize the ferry service instead of the present bus service. Estimate the population proportion p of commuters from the two cities who would prefer the ferry service. Construct a 95% confidence interval for p.

10.9 In Exercise 10.8, if the maximum error is $E = .01$, estimate the necessary sample size at the 95% confidence level. Use the value obtained in Exercise 10.8 for the estimate of p.

10.10 Suppose a small sample ($n = 12$) was drawn from a lot of electric bulbs. Each bulb was tested; four defectives were found. Determine the 90% confidence interval for the proportion of defectives in the entire batch.

10.11 A manufacturer of microcomputers purchases electronic chips from a supplier that claims its chips are defective only 5% of the time. Determine the sample size that would be required to estimate the true proportion of defective chips if we wanted our estimate, $\hat{p}$, to be within 1.25% of the true proportion, with 99% confidence.

10.12 In a recent survey of 300 randomly chosen subscribers of *Money* magazine, it was found that 75% of those surveyed believed that opportunity for job advancement was better today than it was a generation ago. Suppose that the editor of a local magazine published in Chicago wished to estimate the proportion of workers who could answer yes to the question, "Is opportunity for job advancement in the Chicago metropolitan area better today than it was a generation ago?" Assume that the editor wished to estimate this proportion to within .06 with 95% confidence. Using the estimate from the survey by *Money* magazine as the initial estimate of the proportion, find the sample size necessary to estimate the proportion of workers who would answer yes to this question with this much accuracy.

(*Source:* "How Are We Doing?" *Money Extra* (1990): 18–19.)

10.13   Blackburry Candies is considering withdrawing its product Nutty Bar from the market if Nutty Bar has not captured at least 5% of the candy bar market. A random sample of 115 candy-bar buyers was taken; 4 bought the Nutty Bar. Find a 95% confidence interval for the proportion, p, of the population of candy-bar buyers who choose Nutty Bar.

10.14   Using the data in Exercise 10.13, what is the required sample size necessary to estimate with 95% confidence, and to be within .03, the proportion of candy-bar buyers who would choose the candy bar? Assume that we have no prior knowledge of p.

10.15   A quality-control inspector randomly records a certain measurement of the characteristics of brass rivets that are taken from a conveyor belt during a production process. Rivets that have measurements that fall outside of the range 23.5 to 24.5 are considered to be nonconforming (defective). A sample of 50 measurements are taken from brass rivets coming off the conveyor belt, and the measurements are given below. Construct a 90% confidence interval on the true proportion of nonconforming brass rivets from the production process.

```
23.7  23.6  23.9  24.1  23.7  23.5  24.0  24.7  24.2  23.6
23.2  23.9  24.5  23.6  23.8  24.2  24.1  24.0  23.7  24.1
23.6  23.7  23.7  24.6  24.0  24.2  24.3  23.5  24.1  23.9
24.1  24.3  23.8  24.1  23.6  24.0  24.4  23.6  24.1  24.4
23.4  24.0  23.3  24.0  24.1  23.5  23.3  23.8  23.7  23.5
```

◢ 10.2
Hypothesis Testing for a Population Proportion

How can you statistically reject a statement such as, at least 60% of all heavy smokers will contract a serious lung or heart ailment before age 65? Perhaps someone merely took a wild guess at the value of 60%, and it is your job to gather evidence that will either shoot down this claim or let it stand if there is insufficient evidence to conclude that this percentage actually is less than 60%. We set up hypotheses and test them much as we did before, only now we are concerned about a population proportion, p, rather than the mean or standard deviation of a particular population.

Hypothesis Testing Using a Small Sample

Because confidence intervals can be used to perform a test of hypothesis, we will use Table A.8 to conduct such a test. Table A.8 contains sample sizes of $n = 5$ to 20 and $\alpha = .05$ and .10. If the sample size exceeds 20 and np and $n(1 - p)$ are both greater than 5, the large-sample approximation will provide an accurate test. For sample sizes contained in Table A.8, use the procedure outlined in the accompanying box.

Hypothesis Testing (Small Sample; n Is Between 5 and 20)

TWO-TAILED TEST

$$H_0: p = p_0$$
$$H_a: p \neq p_0$$

1. Obtain the $(1 - \alpha) \cdot 100\%$ confidence interval from Table A.8; that is, (p_L, p_U), using $x =$ the observed number of successes.
2. Reject H_0 if p_0 does not lie between p_L and p_U.
3. Fail to reject H_0 if $p_L \leq p_0 \leq p_U$.

ONE-TAILED TEST

$H_0: p \leq p_0$	$H_0: p \geq p_0$
$H_a: p > p_0$	$H_a: p < p_0$

1. Obtain the $(1 - 2\alpha) \cdot 100\%$ confidence interval from Table A.8; that is, (p_L, p_U), using $x =$ the observed number of successes.
2. Reject H_0 if $p_0 < p_L$.
3. Fail to reject H_0 if $p_0 \geq p_L$.

1. Obtain the $(1 - 2\alpha) \cdot 100\%$ confidence interval from Table A.8; that is, (p_L, p_U), using $x =$ the observed number of successes.
2. Reject H_0 if $p_0 > p_U$.
3. Fail to reject H_0 if $p_0 \leq p_U$.

Notice that for a one-tailed test, we *double* α when finding the confidence interval for p from Table A.8. For example, if $\alpha = .05$, then $2\alpha = .10$, and so we retrieve a 90% confidence interval from the table. As a result, this particular binomial table can be used only when $\alpha = .025$ or $.05$ for a one-tailed test.

EXAMPLE 10.5   In Example 10.1, suppose that the company interested in the Beagle microcomputers will purchase them if Beagle's claim that the proportion, p, of all Beagle computers that can survive these cold temperatures is greater than .75 (75%) can be shown to be true. Do the data support this claim using $\alpha = .05$?

SOLUTION   The claim under investigation goes into the alternative hypothesis. The appropriate hypotheses are

$$H_0: p \leq .75 \quad \text{and} \quad H_a: p > .75$$

We observed, in the sample of 15 computers, $x = 12$ successes (computers that survived). Because $\alpha = .05$, we double this $(2\alpha = .10)$ and refer to Table A.8 for a 90% confidence interval for p when $n = 15$, $x = 12$. This confidence interval is:

$$(p_L, p_U) = (.560, .943)$$

We will reject H_0 provided $p_0 = .75$ lies to the left of p_L. Because .75 is greater than $p_L = .560$, we fail to reject H_0.

Based on the evidence gathered from this sample, we cannot demonstrate that p is greater than the required 75%. Notice that we are not *accepting* H_0—we simply *fail to reject* it. This means that the point estimate $\hat{p} = 12/15 = .8$ is not enough larger than .75 to justify the claim made in H_a. The fact that $\hat{p}$ exceeds .75 may be due to the sampling error that is possible when using a sample statistic ($\hat{p}$) to infer something about a population parameter (p). ∎

COMMENTS   In the same sense that confidence intervals for a proportion, p, are typically very wide, it is usually very difficult to obtain a significant result (reject H_0) when using a small sample. The moral mentioned in the confidence interval section applies here as well: *To obtain a reasonably powerful test on a population proportion, use a large sample.*

Hypothesis Testing Using a Large Sample

The standard five-step procedure is used for testing H_0 versus H_a when attempting to support a claim regarding a binomial parameter, p, using a large sample. The approximate standard normal random variable given by equation 10.2 is used as a test statistic for this situation.

The rejection region for this test is defined by determining the distribution of the test statistic, given that H_0 is true. This means that the unknown value of p in equation 10.2 is replaced by the value of p specified in H_0 (say, p_0). For a one-tailed test, the boundary value of p in H_0 is used. This procedure is summarized in the following box.

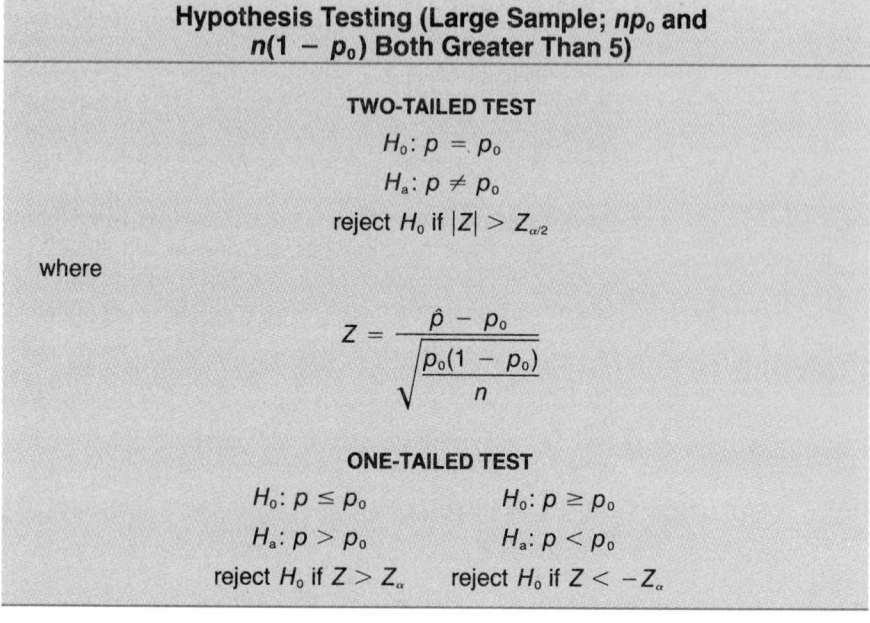

Hypothesis Testing (Large Sample; np_0 and $n(1 - p_0)$ Both Greater Than 5)

TWO-TAILED TEST

$$H_0: p = p_0$$
$$H_a: p \neq p_0$$

reject H_0 if $|Z| > Z_{\alpha/2}$

where

$$Z = \frac{\hat{p} - p_0}{\sqrt{\dfrac{p_0(1 - p_0)}{n}}}$$

ONE-TAILED TEST

$H_0: p \leq p_0$	$H_0: p \geq p_0$
$H_a: p > p_0$	$H_a: p < p_0$
reject H_0 if $Z > Z_\alpha$	reject H_0 if $Z < -Z_\alpha$

Notice that the form of the test statistic is that used in many of the previous large sample test statistics, namely,

$$Z = \frac{\text{(point estimate)} - \text{(hypothesized value)}}{\text{(standard deviation of point estimator)}} \qquad (10.8)$$

EXAMPLE 10.6

In Example 10.2, we estimated the proportion of employees at Cassidy Electronics who would sign up for the dental insurance. The insurance company is not willing to offer such a plan unless more than 60% of the employees will participate. Using the sample of 200 employees, can you conclude that this percentage is greater than the required 60%? Use a significance level of $\alpha = .10$.

SOLUTION   **Step 1.** Your hypotheses should be

$$H_0: p \leq .6$$
$$H_a: p > .6$$

Step 2. Since $np_0 = (200)(.6) = 120$ and $n(1 - p_0) = (200)(.4) = 80$ are both greater than 5, the large-sample test statistic can be used, namely,

$$Z = \frac{\hat{p} - p_0}{\sqrt{\dfrac{p_0(1 - p_0)}{n}}} = \frac{\hat{p} - .6}{\sqrt{\dfrac{(.6)(.4)}{200}}}$$

Step 3. The testing procedure, using $\alpha = .10$, will be to

reject H_0 if $Z > Z_{.10} = 1.28$

■ **FIGURE 10.2**
Z curve showing
p-value for
Example 10.6.

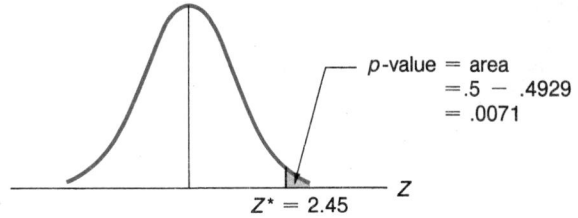

p-value = area
=.5 − .4929
= .0071

Z

$Z^* = 2.45$

Step 4. Using the sample data, $\hat{p} = 137/200 = .685$, so

$$Z^* = \frac{.685 - .6}{\sqrt{\dfrac{(.6)(.4)}{200}}} = \frac{.085}{.0346} = 2.45$$

Because $2.45 > 1.28$, we reject H_0 in favor of H_a.

Step 5. This sample indicates that the population proportion of employees who would participate in the dental insurance plan *is* greater than 60%.   ■

In Example 10.6, the computed test statistic was $Z^* = 2.45$. Figure 10.2 shows the Z curve and the calculated p-value, which is .0071. Using the classical approach, because $.0071 < \alpha = .10$, we reject H_0. If we choose to base our conclusion strictly on the p-value (without choosing a significance level, α), this value would be classified as *small*—it is less than .01. This means that using the classical approach, we would reject H_0 for any α greater than or equal to .01. Consequently, we once again reject H_0.

EXAMPLE 10.7   In Example 10.3, we estimated the proportion of calculators that were defective in a batch (lot). The company has determined that a good target for this defective percentage is 4%. The sample of 150 had 13 defectives. Can we conclude that the actual proportion of defective calculators is different from 4%? Use $\alpha = .05$.

SOLUTION   **Step 1.** We wish to see if p is *different* from 4%, so we should use a two-tailed test with hypotheses

$$H_0: p = .04$$

$$H_a: p \neq .04$$

Step 2. Here $np_0 = (150)(.04) = 6$ and $n(1 - p_0) = (150)(.96) = 144$. Both are > 5, so the appropriate test statistic is

$$Z = \frac{\hat{p} - p_0}{\sqrt{\dfrac{p_0(1 - p_0)}{n}}} = \frac{\hat{p} - .04}{\sqrt{\dfrac{(.04)(.96)}{150}}}$$

Step 3. With $\alpha = .05$, the test procedure of H_0 versus H_a will be to

reject H_0 if $|Z| > 1.96$

Step 4. Using $\hat{p} = 13/150 = .0867$,

$$Z^* = \frac{.0867 - .04}{\sqrt{\dfrac{(.04)(.96)}{150}}} = \frac{.0467}{.016} = 2.92$$

Because $2.92 > 1.96$, we reject H_0.

■ **FIGURE 10.3**
Z curve showing
p-value (twice the shaded
area) for Example 10.7.

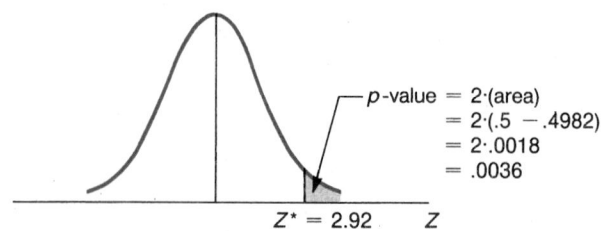

p-value = 2·(area)
= 2·(.5 − .4982)
= 2·.0018
= .0036

Z* = 2.92   Z

Step 5. The company is *not* meeting their target percentage of defectives. As a reminder, because $\alpha = .05$, 5% of the time this particular test will reject H_0 when in fact it is true. ■

In Example 10.7, $Z^* = 2.92$. What is the p-value? This is a two-tailed test, so we need to *double* the right-tail area, as illustrated in Figure 10.3. So $p = 2 \cdot .0018 = .0036$. Thus, using either the classical procedure (comparing the p-value to $\alpha = .05$) or basing our decision strictly on the p-value, we reject H_0 because of this extremely small p-value.

Exercises

10.16   Using the data in Exercise 10.13, test the hypothesis that Nutty Bar captures less than 5% of the market share using a .05 level of significance.

10.17   Calculate the p-value for Exercise 10.16. Based on the p-value, would you reject the null hypothesis at the .05 level?

10.18   An official for a computer firm was told by an independent source that 20% of the employees of the computer firm perceived that there was sex discrimination in the salary structure of the company. A quick random survey by the official of 18 employees found that 3 of the employees thought there was sex discrimination in the salary structure of the company. Is there sufficient evidence in the official's survey to indicate that the figure of 20% given by the independent source is in error? Conduct a hypothesis test using a significance level of .10 and using Table A.8.

10.19   Using the data in Exercise 10.10, test the hypothesis that the defective rate is more than 10%, using a .05 level of significance. Use Table A.8.

10.20   There are about 87 million households in the United States. In 1970, about 71% of the households were occupied by married couples. In the last 15 years, the trend toward living alone has accelerated. One market researcher believes that at present the proportion of households occupied by married couples is 58%. A random sample of 20 households reveals that 12 are occupied by married couples. At a significance level of 10%, are you in a position to contradict the market researcher and say that this person is probably wrong?

10.21   In order for $np > 5$ and $n(1 - p) > 5$, how large must n be if $p = .03$?

10.22   A random sample of 40 coffee drinkers was asked to taste-test a new coffee brand. The responses are listed below with 1 representing "like the brand," 2 representing "indifferent to the brand," and 3 representing "do not like the brand." Do the data support the conclusion that more than half of the coffee drinkers like the new coffee brand? Use a .10 significance level.

1	3	1	3	3	1	2	1	1	1
1	1	1	2	1	1	3	1	2	1
1	2	1	1	1	2	1	1	3	2
1	1	2	1	3	1	3	1	2	1

10.23   In a survey of 600 adults who earn over $100,000 a year, 36 of them said that they feel that it is a necessity for them to fly first class when they travel. Do the data support the belief that more than 5 percent of those adults with an annual income exceeding $100,000 find that it is a necessity to fly first class when traveling? Use a 10% significance level for the test.

(Source: U.S. News and World Report (May 28, 1990): 75.)

10.24 A manager at National Insurance believes that out of the total number of automobile-accident claims settled in a particular month, there are more claims related to speeding by the driver than there are claims that are not related to speeding. From a random sample of 75 claims, 40 were found to be associated with speeding. Test the manager's belief. Use a significance level of .05.

10.25 Calculate the *p*-value for Exercise 10.24 and interpret it.

10.26 An instructor believes that of the students who take a certain course, there are more students who have not taken the prerequisites for the course than students who have taken the prerequisites. The instructor randomly selected 70 students and found that only 30 students had taken the prerequisites for the course. Do these data support the instructor's belief? Use a significance level of .10.

10.27 Based on the *p*-value for Exercise 10.26, would you reject the null hypothesis at the .01 level?

10.28 One of the top regularly scheduled network programs has been "60 Minutes." According to a 1989 Nielsen Report, 22.2 percent of TV households watch this program. Suppose that another independent media-rating agency believed that this percentage was too low. Assume that this media-rating agency sampled 200 households, finding that 52 of the households watch "60 Minutes." Is there sufficient evidence from this agency to conclude that the true proportion of households watching "60 Minutes" is greater than 22.2 percent? Use a 10% significance level.

(Source: Adapted from The 1990 Information Please Almanac (1990): 743.)

10.29 KNNN, a television news channel, claimed that more than 65% of its subscribers had an annual income of $40,000 or more. A random sample of 160 subscribers was interviewed; 71% of them had incomes of $40,000 or more. Does this information support KNNN's claim? Use a significance level of .05.

10.30 Calculate and interpret the *p*-value for Exercise 10.29.

▌ 10.3
Comparing Two Population Proportions (Large Independent Samples)

Consider the following questions:

Is the divorce rate higher in California than it is in New York?

Is there any difference in the proportion of cars manufactured by Henry Motor Company requiring an engine overhaul before 100,000 miles and the proportion of General Auto (GA) automobiles requiring one?

Is there a higher rate of lung cancer among cigarette smokers than there is among nonsmokers?

These questions are concerned with proportions from *two* populations. Our method of estimating these proportions will be exactly as it was for one population. We simply have two of everything—two populations, two samples, two estimates, and so on. In this section, it is assumed that the two samples are obtained *independently*.

For example, consider the question concerning the proportion of cars requiring an engine overhaul. Population 1 is all Henry cars, with p_1 = proportion of Henry cars requiring an engine overhaul before 100,000 miles, n_1 = Henry sample size, and x_1 = number of Henry cars requiring an overhaul before 100,000 miles. Population 2 is all GA cars, with p_2 = proportion of GA cars requiring an engine overhaul before 100,000 miles, n_2 = GA sample size, and x_2 = number of GA cars requiring an overhaul before 100,000 miles.

Define a success to be that a car requires an overhaul before 100,000 miles.

(Keep in mind that "success" is merely a label for the trait you are interested in. It need not be a desirable trait.) Our unbiased point estimator of p_1 will be as before:

$$\hat{p} = \frac{\text{observed number of successes in the sample}}{\text{sample size}}$$

$$= \frac{\text{number of cars in the Henry sample requiring an overhaul}}{n_1}$$

That is, the unbiased point estimator of p_1 is

$$\hat{p}_1 = \frac{x_1}{n_1} \tag{10.9}$$

Similarly, the unbiased point estimator of p_2, obtained from the second sample, is

$$\hat{p}_2 = \frac{x_2}{n_2} \tag{10.10}$$

For the two-population case, the parameter of interest will be the *difference* between the two population proportions, $p_1 - p_2$. The next section discusses a method of estimating $p_1 - p_2$ by using a point estimate along with a corresponding confidence interval.

Confidence Interval for $p_1 - p_2$ (Large Independent Samples)

The logical estimator of $p_1 - p_2$ is $\hat{p}_1 - \hat{p}_2$, the difference between the sample estimators. What kind of random variable is $\hat{p}_1 - \hat{p}_2$? We are dealing with large independent samples (where $n_1\hat{p}_1$, $n_1(1 - \hat{p}_1)$, $n_2\hat{p}_2$, and $n_2(1 - \hat{p}_2)$ are each larger than five) so it follows that $\hat{p}_1 - \hat{p}_2$ is (approximately) a normal random variable with

$$\text{mean} = p_1 - p_2$$

and

$$\text{standard deviation} = \sqrt{\frac{p_1(1 - p_1)}{n_1} + \frac{p_2(1 - p_2)}{n_2}}$$

In a previous section, we observed that $\hat{p}_1$ is a sample mean, where the sample consists of observations that are either a 1 (a particular event occurred) or a 0 (this event did not occur). Because the two samples are obtained independently, the results extend to this situation, leading to the approximate normal distribution for $\hat{p}_1 - \hat{p}_2$. Since $\hat{p}_1$ and $\hat{p}_2$ are unbiased estimators of p_1 and p_2, respectively, the mean of the estimator $\hat{p}_1 - \hat{p}_2$ is $p_1 - p_2$; that is, $\hat{p}_1 - \hat{p}_2$ is an unbiased estimator of $p_1 - p_2$. Notice that the variance of $\hat{p}_1 - \hat{p}_2$ is obtained by *adding* the variance of $\hat{p}_1$, or $p_1(1 - p_1)/n_1$, to the variance of $\hat{p}_2$, or $p_2(1 - p_2)/n_2$.

To evaluate the confidence interval, we are forced to approximate the confidence limits by replacing p_1 by $\hat{p}_1$ and p_2 by $\hat{p}_2$ under the square root sign. This approximation works well provided both sample sizes are large. So we derive a $(1 - \alpha) \cdot 100\%$ confidence interval for $p_1 - p_2$ (large, independent samples; $n_1\hat{p}_1$, $n_1(1 - \hat{p}_1)$, $n_2\hat{p}_2$, and $n_2(1 - \hat{p}_2)$ are each greater than 5):

$$(\hat{p}_1 - \hat{p}_2) - Z_{\alpha/2} \sqrt{\frac{\hat{p}_1(1 - \hat{p}_1)}{n_1} + \frac{\hat{p}_2(1 - \hat{p}_2)}{n_2}}$$

$$\text{to} \quad (\hat{p}_1 - \hat{p}_2) + Z_{\alpha/2} \sqrt{\frac{\hat{p}_1(1 - \hat{p}_1)}{n_1} + \frac{\hat{p}_2(1 - \hat{p}_2)}{n_2}} \quad \textbf{(10.11)}$$

where $\hat{p}_1 = x_1/n_1$ and $\hat{p}_2 = x_2/n_2$ are the sample proportions. Observe that the construction of this confidence interval was the "usual" procedure employing Table A.4 and described in equation 10.3.

EXAMPLE 10.8

Of a random sample of 100 cars manufactured by Henry (population 1), 28 needed an engine overhaul before reaching 100,000 miles. A second sample, obtained independently of the first, consisted of 150 cars produced by GA (population 2); 48 of them required an engine overhaul before 100,000 miles. Both sets of cars were subjected to the same weather conditions, maintenance program, and driving conditions. Construct a 99% confidence interval for $p_1 - p_2$.

SOLUTION

We have $\hat{p}_1 = 28/100 = .28$ and $\hat{p}_2 = 48/150 = .32$. Also, $Z_{\alpha/2} = Z_{.005} = 2.575$, using Table A.4. The resulting confidence interval for $p_1 - p_2$ is

$$(.28 - .32) - 2.575 \sqrt{\frac{(.28)(.72)}{100} + \frac{(.32)(.68)}{150}}$$

$$\text{to} \quad (.28 - .32) + 2.575 \sqrt{\frac{(.28)(.72)}{100} + \frac{(.32)(.68)}{150}}$$

$$= -.04 - .15 \quad \text{to} \quad -.04 + .15$$

$$= -.19 \quad \text{to} \quad .11$$

This confidence interval leaves us unable to conclude that either manufacturer produces a better engine. We are 99% confident that the percentage of Henry engines requiring an overhaul before 100,000 miles is between 19% *lower* and 11% *higher* than for the GA engines. ∎

EXAMPLE 10.9

The Redican Corporation manufactures 1-quart metal cans to hold canned vegetable juice. A can is *nonconforming* if it is out of round or has a leak in the side weld. The cans are produced during two shifts, shift 1 (the day shift) and shift 2 (the night shift). The quality supervisor suspects that the proportion of nonconforming cans produced during the day shift (p_1) is *lower* than that for the night shift (p_2), since the day shift has better qualified workers. To investigate this, random samples of 500 cans were obtained from each shift. The results were as follows. Of the $n_1 = 500$ cans from the day shift, $x_1 = 70$ were nonconforming, and of the $n_2 = 500$ cans from the night shift, $x_2 = 110$ were nonconforming. Determine a 95% confidence interval for $p_1 - p_2$.

SOLUTION

The proportion estimates are

$$\hat{p}_1 = \frac{70}{500} = .14$$

$$\hat{p}_2 = \frac{110}{500} = .22$$

The 95% confidence interval for $p_1 - p_2$ is

$$(.14 - .22) - 1.96 \sqrt{\frac{(.14)(.86)}{500} + \frac{(.22)(.78)}{500}}$$

$$\text{to}\quad (.14 - .22) + 1.96 \sqrt{\frac{(.14)(.86)}{500} + \frac{(.22)(.78)}{500}}$$

$$= -.08 - .047 \quad \text{to} \quad -.08 + .047$$

$$= -.127 \quad \text{to} \quad -.033$$

So we are 95% confident that (1) our estimate of the difference in proportions (shift 1 minus shift 2), namely $\hat{p}_1 - \hat{p}_2 = -.08$, is within 4.7% of the actual value, and (2) the proportion of nonconforming cans during shift 1 is between 3.3% and 12.7% *lower* than for shift 2.   ∎

Choosing the Sample Sizes (Two Populations)

In Chapter 9, we discussed how to select samples from two populations when the desired accuracy of the point estimate of the difference between two population means is specified—this is the maximum error, E. If E is 10 pounds, for instance, then what sample sizes (n_1 and n_2) are necessary for the point estimate of $\mu_1 - \mu_2$ (namely, $\bar{X}_1 - \bar{X}_2$) to be within 10 pounds of the actual value, with 95% (or whatever) confidence? Using the results contained in Appendix B at the end of the text, values of n_1 and n_2 were provided in Chapter 9 to minimize the total sample size, $n_1 + n_2$, for this specific value of E.

We encounter a similar situation when dealing with two population proportions, p_1 and p_2. If a maximum error of $E = .10$, for instance, is specified, then the question of interest is, what sample sizes (n_1 and n_2) are necessary for the point estimate of $p_1 - p_2$ (namely, $\hat{p}_1 - \hat{p}_2$) to be within .10 of the actual value, with 95% (or whatever) confidence?

The maximum error, E, always is the amount that you *add to* and *subtract from* the point estimate when determining a confidence interval. When dealing with two proportions, E is

$$E = Z_{\alpha/2} \sqrt{\frac{p_1(1 - p_1)}{n_1} + \frac{p_2(1 - p_2)}{n_2}} \qquad \text{(10.12)}$$

To evaluate this expression, you will need estimates of p_1 and p_2. You have two options. If you have previously obtained small samples from these two populations, you can use the resulting sample estimates $\hat{p}_1$ and $\hat{p}_2$. The purpose then will be to extend these samples to obtain better accuracy in the point estimate, $\hat{p}_1 - \hat{p}_2$. If no information regarding p_1 and p_2 is available, then you can use the conservative approach discussed in Section 10.1 by letting $\hat{p}_1 = \hat{p}_2 = .5$.

By applying the results of Appendix B to this situation, the sample sizes n_1 and n_2 that minimize the total sample size $n_1 + n_2$ are given by

$$n_1 = \frac{Z_{\alpha/2}^2 (A + B)}{E^2} \qquad \text{(10.13)}$$

$$n_2 = \frac{Z_{\alpha/2}^2 (C + B)}{E^2} \qquad \text{(10.14)}$$

where

$$A = p_1(1 - p_1)$$
$$B = \sqrt{p_1 p_2 (1 - p_1)(1 - p_2)}$$
$$C = p_2(1 - p_2)$$

To determine A, B, and C, estimates of p_1 and p_2 should be substituted for p_1 and p_2 by using one of the two options described.

EXAMPLE 10.10

Using the situation described in Example 10.9, determine what sample sizes are necessary for the estimate of the difference between the two proportions to be within .03 of the actual difference, with 99% confidence, if (1) the results from Example 10.9 are available, and (2) no sample information is available.

SOLUTION 1

The specified maximum error is $E = .03$. Sample data have been collected regarding these proportions, so we use the corresponding estimates to determine the sample sizes necessary to obtain this degree of accuracy. Using Table A.4, $Z_{\alpha/2} = Z_{.005} = 2.575$. Here, $\hat{p}_1 = .14$ and $\hat{p}_2 = .22$. Consequently,

$$A = \hat{p}_1(1 - \hat{p}_1)$$
$$= .1204$$
$$B = \sqrt{\hat{p}_1 \hat{p}_2 (1 - \hat{p}_1)(1 - \hat{p}_2)}$$
$$= .1437$$
$$C = \hat{p}_2(1 - \hat{p}_2)$$
$$= .1716$$

To obtain the *smallest possible* total sample size for the required accuracy, the two sample sizes should be

$$n_1 = \frac{(2.575)^2(.1204 + .1437)}{(.03)^2} \cong 1946$$

(remember—always round up) and

$$n_2 = \frac{(2.575)^2(.1716 + .1437)}{(.03)^2} \cong 2323$$

providing a total sample size of $n_1 + n_2 = 4269$ cans.

SOLUTION 2

If no prior estimates of p_1 and p_2 are available, using $\hat{p}_1 = \hat{p}_2 = .5$ will result in sample sizes n_1 and n_2 that will provide a maximum error *no larger than* the specified value of $E = .03$. Here, $A = (.5)(.5) = .25$. Similarly, $B = C = .25$, so

$$n_1 = n_2 = \frac{(2.575)^2(.25 + .25)}{(.03)^2} \cong 3684$$

Consequently, a total sample size of $n_1 + n_2 = 7368$ cans will be necessary for $\hat{p}_1 - \hat{p}_2$ to be within .03 of the actual value of $p_1 - p_2$, with 99% confidence. ∎

Hypothesis Testing for p_1 and p_2 (Large, Independent Samples)

Suppose that a recent report stated that, based on a sample of 500 people, 35% of all cigarette smokers had at some time in their lives developed a particular fatal disease. On the other hand, 25% of the nonsmokers in the sample acquired the disease. Can we conclude from this sample that, because $\hat{p}_1 = .35 > \hat{p}_2 = .25$, the proportion ($p_1$) of all smokers who will acquire the disease exceeds the proportion (p_2) for nonsmokers? In other words, is $\hat{p}_1$ *significantly* larger than $\hat{p}_2$? After all, even if $p_1 = p_2$, there is a 50–50 chance that $\hat{p}_1$ will be larger than $\hat{p}_2$, because for large samples, the distribution of $\hat{p}_1 - \hat{p}_2$ is approximately a bell-shaped (normal) curve centered at $p_1 - p_2$, which, if $p_1 = p_2$, would be zero.

Are the results of the sample significant, or are they due simply to the sampling error that is always possible when estimating from a sample? Your alternative hypothesis can be that two proportions are *different* (a two-tailed test) or that one *exceeds* the other (a one-tailed test). As before, we will assume that the two random samples are obtained *independently*. The possible hypotheses are these:

For a two tailed test,

$$H_0: p_1 = p_2$$
$$H_a: p_1 \neq p_2$$

and for a one-tailed test,

$$H_0: p_1 \leq p_2$$
$$H_a: p_1 > p_2$$

or

$$H_0: p_1 \geq p_2$$
$$H_a: p_1 < p_2$$

One possible test statistic to use here would be the standard normal (Z) statistic that was used to derive a confidence interval for $p_1 - p_2$, namely,

$$Z = \frac{\hat{p}_1 - \hat{p}_2}{\sqrt{\dfrac{\hat{p}_1(1 - \hat{p}_1)}{n_1} + \dfrac{\hat{p}_2(1 - \hat{p}_2)}{n_2}}} \qquad (10.15)$$

In previous tests of hypothesis, we always examined the distribution of the test statistic when H_0 was *true*. For a one-tailed test, we assumed the boundary condition of H_0, which in this case is $p_1 = p_2$. Because of this, whenever we obtained a value of the test statistic in one of the tails, our decision was to reject H_0 because this value would be very unusual if H_0 were true. This reasoning was used for test statistics that followed a Z, t, χ^2, or F distribution.

We use the same approach here. If $p_1 = p_2 = p$ (for example), we can improve the test statistic in equation 10.15. For this situation, p is the proportion of successes in the combined population. Our best estimate of p is the proportion of successes in the *combined sample*. So define

$$\bar{p} = \frac{x_1 + x_2}{n_1 + n_2}$$

Thus, assuming $p_1 = p_2$, $\hat{p}_1 - \hat{p}_2$ is approximately a normal random variable with

$$\text{mean} = p_1 - p_2 = 0$$

and

$$\text{standard deviation} = \sqrt{\frac{p_1(1 - p_1)}{n_1} + \frac{p_2(1 - p_2)}{n_2}}$$

$$\cong \sqrt{\frac{\bar{p}(1 - \bar{p})}{n_1} + \frac{\bar{p}(1 - \bar{p})}{n_2}}$$

The resulting test statistic for p_1 versus p_2 (large, independent samples; $n_1\hat{p}_1$, $n_1(1 - \hat{p}_1)$, $n_2\hat{p}_2$, and $n_2(1 - \hat{p}_2)$ are each greater than 5) is

$$Z = \frac{\hat{p}_1 - \hat{p}_2}{\sqrt{\dfrac{\bar{p}(1 - \bar{p})}{n_1} + \dfrac{\bar{p}(1 - \bar{p})}{n_2}}} \tag{10.16}$$

where

$$\hat{p}_1 = x_1/n_1$$

$$\hat{p}_2 = x_2/n_2$$

$$\bar{p} = \frac{x_1 + x_2}{n_1 + n_2}$$

Observe that the form of this test statistic is the same as for the single-population case described in equation 10.8. The test procedure is the standard routine when using the Z distribution. For a two-tailed test,

$$H_0: p_1 = p_2$$
$$H_a: p_1 \neq p_2$$
$$\text{reject } H_0 \text{ if } |Z| > Z_{\alpha/2}$$

where Z is defined in equation 10.16. For a one-tailed test,

$$H_0: p_1 \leq p_2$$
$$H_a: p_1 > p_2$$
$$\text{reject } H_0 \text{ if } Z > Z_{\alpha}$$

or

$$H_0: p_1 \geq p_2$$
$$H_a: p_1 < p_2$$
$$\text{reject } H_0 \text{ if } Z < -Z_{\alpha}$$

EXAMPLE 10.11 Use the engine overhaul data from Example 10.8 and determine whether there is any difference between the proportion of Henry cars and the proportion of GA cars that required an engine overhaul before 100,000 miles. Let $\alpha = .01$.

SOLUTION The five-step procedure is the correct one. The confidence interval derived in Example 10.8 would produce the same result as the five-step procedure *if* the test statistic were the one defined in equation 10.15. *The correct procedure here is to use the Z-statistic in equation 10.16 as your test statistic.*

Step 1. Since we are looking for a difference between p_1 and p_2, define

$$H_0: p_1 = p_2$$
$$H_a: p_1 \neq p_2$$

Step 2. The test statistic is

$$Z = \frac{\hat{p}_1 - \hat{p}_2}{\sqrt{\dfrac{\bar{p}(1 - \bar{p})}{n_1} + \dfrac{\bar{p}(1 - \bar{p})}{n_2}}}$$

■ FIGURE 10.4
Z curve showing
p-value (twice the
shaded area) for
Example 10.11.

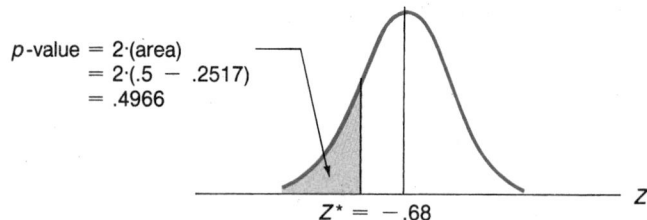

$$p\text{-value} = 2\cdot(\text{area})$$
$$= 2\cdot(.5 - .2517)$$
$$= .4966$$

$$Z^* = -.68$$

Step 3. Using $\alpha = .01$, then $Z_{\alpha/2} = Z_{.005} = 2.575$. The test procedure will be to

$$\text{reject } H_0 \text{ if } |Z| > 2.575$$

Step 4. Since $n_1 = 100$, $x_1 = 28$, and $n_2 = 150$, $x_2 = 48$, then

$$\bar{p} = \frac{x_1 + x_2}{n_1 + n_2} = \frac{76}{250} = .304$$

Therefore, our estimate of the proportion of cars needing an overhaul in the combined population (if $p_1 = p_2$) is $\bar{p} = .304$ (30.4%). Also, $\hat{p}_1 = 28/100 = .28$, and $\hat{p}_2 = 48/150 = .32$. The value of the test statistic is

$$Z^* = \frac{.28 - .32}{\sqrt{\dfrac{(.304)(.696)}{100} + \dfrac{(.304)(.696)}{150}}} = \frac{-.04}{.059} = -.68$$

Because $|Z^*| = .68 < 2.575$, we fail to reject H_0.

Step 5. There is *insufficient evidence* to conclude that a difference exists between the Henry and GA cars as far as engine durability is concerned. ■

The Z curve and calculated p-value for Example 10.11 are shown in Figure 10.4. The p-value is twice the shaded area (this was a two-tailed test) and is .4966, which is extremely large. Using the classical approach, because $.4966 > \alpha = .01$, we fail to reject H_0—there is insufficient evidence to indicate a difference in engine durability. As a reminder, this reasoning *always* leads to the same conclusion as the five-step procedure. Because .4966 exceeds *any* reasonable value of α, we fail to reject H_0 quite strongly for this application.

EXAMPLE 10.12   In Example 10.9, we examined the proportions of nonconforming metal cans produced by Redican during the day shift (p_1) and the night shift (p_2). Based on these data, can you conclude that the proportion of nonconforming cans during the day shift is lower than during the night shift?

SOLUTION   **Step 1.** We wish to know whether the data warrant the conclusion that p_1 is *smaller* than p_2. Placing this in the alternative hypothesis leads to

$$H_0: p_1 \geq p_2$$
$$H_a: p_1 < p_2$$

Steps 2 and 3. Using the test statistic in equation 10.16, the resulting one-tailed test procedure would be to

$$\text{reject } H_0 \text{ if } Z < -Z_{.05} = -1.645$$

Step 4. We have

$$\hat{p}_1 = \frac{70}{500} = .14$$

FIGURE 10.5
Z curve showing the
calculated *p*-value for
Example 10.12.

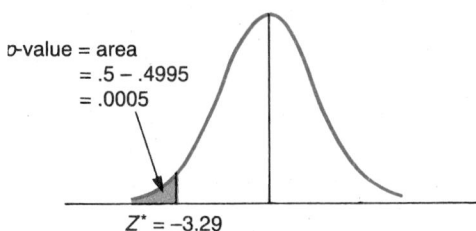

p-value = area
= .5 − .4995
= .0005

$Z^* = -3.29$

and

$$\hat{p}_2 = \frac{110}{500} = .22$$

Also,

$$\bar{p} = \frac{(70 + 110)}{(500 + 500)} = \frac{180}{1000} = .18$$

Consequently,

$$Z^* = \frac{.14 - .22}{\sqrt{\dfrac{(.18)(.82)}{500} + \dfrac{(.18)(.82)}{500}}} = \frac{-.08}{.0243} = -3.29$$

Because $-3.29 < -1.645$, we reject H_0.

Step 5. There *is* evidence that the proportion of nonconforming cans is smaller during the day shift. ∎

The Z curve and calculated *p*-value for Example 10.12 (a one-tailed test) are shown in Figure 10.5. The *p*-value is .0005. This is definitely a very small *p*-value, and (as before) it leads to rejecting H_0 using a significance level of .05. Based on this *p*-value alone, we arrive at the same conclusion—namely, that the proportion of nonconforming cans is lower during the day shift.

Exercises

10.31 A manufacturer of storm windows sampled 250 new (less than five years of age) homes and found that 142 of them had storm windows. Another sample of size 320 of older (at least five years of age) homes was taken; 150 of them had storm windows. The manufacturer believes that the proportion of new homes that have storm windows is larger than the proportion of older homes that have storm windows. Do the sample statistics support the manufacturer's claim at the .05 significance level?

10.32 Using the data in Exercise 10.31, construct a 90% confidence interval for the difference between the proportion of new homes with storm windows and the proportion of older homes with storm windows.

10.33 How does the value of the test statistic given in equation 10.16 change if $\hat{p}_1$, the proportion of successes for sample 1, is replaced by the proportion of failures for sample 1 and if $\hat{p}_2$, the proportion of successes for sample 2, is replaced by the proportion of failures for sample 2?

10.34 Corporations in America have a long tradition of making charitable contributions to support the fine arts. A 1987 Lou Harris poll showed that although Americans seem to want more arts (such as music, ballet, and so on), they seem to be enjoying them less. In 1984, 35% of those surveyed said they attended either an opera or a musical theater, whereas in 1987, the percentage had fallen to 27%. Find a 99% confidence interval on the difference in the proportion of people who attended either an opera or a musical theater in 1984 and in

1987. Assume that 2000 observations were used in each of the surveys in 1984 and 1987. If you were a corporate donor supporting the arts for many years with charitable contributions, would you say that your donations provided support during a time of declining audiences?

(*Source:* "So Much To Do, So Little Time," *Dallas Morning News* (April 17, 1988): 12C.)

10.35 The owner of two hotels in Atlanta is interested in the proportion of "no-shows" on Friday night. The manager believes that the proportion of no-shows does not differ significantly between the two hotels. Assume that preliminary estimates of the proportion of no-shows are 15% at hotel A and 19% at hotel B. How many random Friday nights would the owner need to select from each hotel to estimate the difference in the proportions of no-shows to within .20 with 95% confidence, minimizing the total sample size?

10.36 A random sample of 125 manufacturing firms showed that 64% of them spent more than 75% of their total revenue on salaries and wages. A random sample of 100 wholesale firms showed that 57% of them spent more than 75% of their total revenue on salaries and wages. Let $\alpha = .05$ and test $H_0: p_1 \leq p_2$ and $H_a: p_1 > p_2$, where p_1 and p_2 are the proportions of manufacturing and wholesale firms, respectively, that spent more than 75% of their total revenue on salaries and wages.

10.37 A quality engineer wishes to determine if there is a difference in the number of nonconforming bottle-top seals produced by two different assembly line processes. A bottle-top seal is nonconforming if the seal is not airtight. To investigate this, the engineer samples 300 bottle-top seals from each assembly line process. The samples reveal that one process produced 8 nonconforming items and the other produced 12 nonconforming items. Based on these data, can the engineer conclude that there is a significant difference in the number of nonconforming bottle-top seals produced by the two assembly line processes? Use the *p*-value to support your conclusion.

10.38 A financial analyst compared the performances of individual stocks with the performance of the industry average (in terms of rate of return). The industry average is the average of all stocks that belong to the same industry and are listed on the New York Stock Exchange. The analyst believed that the proportion of stocks that perform better than the industry average is the same for both the oil and the steel industry. In a random sample of 37 oil-industry stocks, 17 performed better than the industry average. Similarly, in a random sample of 30 steel-industry stocks, 11 did better than the industry average. Using a .05 significance level, test the validity of the financial analyst's belief.

10.39 Two manufacturers supply rebuilt motors to an air-conditioning repair company. A preliminary study showed that the proportion of defective motors from one manufacturer was 10% and from the other manufacturer was 16%. Determine what sample sizes are necessary for the estimate of the difference between the proportions of defective motors from the two manufacturers to be within .15 with 95% confidence, minimizing the total sample size.

Summary

You will often encounter a situation in which you are concerned with a population **proportion** rather than the mean or variance. For example, the parameter of interest might be the proportion (p) of executives earning more than $100,000 annually, rather than the average salary (μ) or the standard deviation (σ) of the salaries. The usual procedure of estimating a population parameter using the sample estimator, $\hat{p}$, allows us to derive a point estimate and construct a confidence interval for p. When the sample is small, Table A.8 provides an exact confidence interval for p. For large samples, the Central Limit Theorem can be applied to determine an approximate confidence interval, provided that *both* np and $n(1 - p)$ are greater than 5.

When the desired accuracy of the point estimator, $\hat{p}$, is specified in advance, you can determine the sample size necessary to obtain this degree of accuracy for a certain confidence level. To derive this sample size, an estimate of p is necessary. You can calculate this value using a previous sample estimate or, if no information is available, using a conservative procedure and making $\hat{p} = .5$.

When you investigate a statement concerning a population proportion, you can use a statistical test of hypothesis. For small samples, the confidence interval from Table A.8 provides an exact procedure for either a one- or two-tailed test. For tests of a hypothesis when a large sample is used, a test statistic having an approximate standard normal distribution can be used.

To compare **two population proportions** (p_1 and p_2), two *independent* random samples are obtained, one from each population. Procedures for large, independent samples generally provide an accurate confidence interval or test of hypothesis whenever $n_1 \hat{p}_1$, $n_1(1 - \hat{p}_1)$, $n_2 \hat{p}_2$, and $n_2(1 - \hat{p}_2)$ each exceed 5. Using a standard normal approximation, we can construct a confidence interval for $p_1 - p_2$. If the accuracy of this estimate is specified, the sample sizes necessary to obtain this level of accuracy as well as to minimize the total sample size $n_1 + n_2$ can be obtained.

Two population proportions can be compared by using two large, independent samples to evaluate a test statistic having an approximate standard normal distribution. We examined procedures for a one-tailed test (for example, $H_a: p_1 > p_2$) or a two-tailed test ($H_a: p_1 \neq p_2$). The rejection regions for these tests are defined using the areas from Table A.4.

Summary of Formulas

1. Single population

point estimate of population proportion (p):

$$\hat{p} = \frac{x}{n}$$

where

x = number of sample items having the selected attribute

and

n = sample size

confidence interval for p (large sample):

$$\hat{p} \pm Z_{\alpha/2} \sqrt{\frac{\hat{p}(1 - \hat{p})}{n}}$$

sample size necessary to obtain maximum error (E) with $(1 - \alpha) \cdot 100\%$ confidence:

$$n = \frac{Z_{\alpha/2}^2 \hat{p}(1 - \hat{p})}{E^2}$$

test statistic for hypothesis testing on p (large sample):

$$Z = \frac{\hat{p} - p_0}{\sqrt{\dfrac{p_0(1 - p_0)}{n}}}$$

where p_0 is the hypothesized value of p.

2. Two populations (large, independent samples)

confidence interval for $p_1 - p_2$:

$$(\hat{p}_1 - \hat{p}_2) \pm Z_{\alpha/2} \sqrt{\frac{\hat{p}_1(1 - \hat{p}_1)}{n_1} + \frac{\hat{p}_2(1 - \hat{p}_2)}{n_2}}$$

samples sizes necessary to obtain maximum error (E) with $(1 - \alpha) \cdot 100\%$ confidence (total sample size minimized):

$$n_1 = \frac{Z_{\alpha/2}^2(A + B)}{E^2}$$

$$n_2 = \frac{Z_{\alpha/2}^2(C + B)}{E^2}$$

where

$A = p_1(1 - p_1)$
$B = \sqrt{p_1 p_2(1 - p_1)(1 - p_2)}$
$C = p_2(1 - p_2)$

test statistic for hypothesis testing on p_1 and p_2:

$$Z = \frac{\hat{p}_1 - \hat{p}_2}{\sqrt{\dfrac{\bar{p}(1 - \bar{p})}{n_1} + \dfrac{\bar{p}(1 - \bar{p})}{n_2}}}$$

where

$$\bar{p} = \frac{x_1 + x_2}{n_1 + n_2}$$

*Review
Exercises*

10.40   Ten of the 17 employees who took an in-house speed-reading course can show that the course has substantially increased their efficiency on the job.

a.   Are there significantly more employees who have benefited from the course than have not at the .05 significance level? Use Table A.8.

b.   Find a 95% confidence interval for the proportion of employees who have benefited from the course. Use Table A.8.

10.41   An advertising agent for Computerized Telephone Systems claims that the proportion of installed telephone systems that have maintenance problems during the first three years is less than 10%. A random sample of 19 computerized telephone systems that were installed within the last three years was taken, and one of the telephone systems was found to have needed repairs.

a.   Test the advertising agent's claim at the .05 significance level. Use Table A.8.

b.   Find a 90% confidence interval for the true proportion of installed telephone systems that have maintenance problems.

10.42   Fifteen male customers were asked which of two electric shavers, brand 1 or brand 2, they preferred. Nine of them preferred brand 1.

a.   At the .05 level of significance, can it be concluded that brand 1 was preferred to brand 2 by male shoppers? Use Table A.8.

b.   Find a 95% confidence interval for the proportion of male shoppers who preferred brand 1 over brand 2. Use Table A.8.

c.   Assume you have no prior knowledge of p (the proportion of males who preferred brand 1 over brand 2). Estimate the sample size that is required to estimate p, with 90% confidence, assuming a maximum error of .08.

10.43   A small sample ($n = 15$) is drawn from a lot of dry-cell batteries. Each of the batteries was tested; seven were defective. Determine a 95% confidence interval for the proportion of defectives in the entire batch. Use Table A.8.

10.44   William's Packaging is interested in estimating the proportion of its employees who would attend an alcohol-awareness program. In a random sample of 70 employees, 39 said that they would attend the program. Calculate the estimate of the proportion of all employees who would attend the program. Find a 90% confidence interval for this proportion.

10.45   Using the data from Exercise 10.44, if the maximum error of the estimate for the proportion is 4%, estimate the sample size needed for a 95% confidence level. Use the value of $\hat{p}$ from Exercise 10.44.

10.46   On some issues that come to a vote, the United States Senate takes a roll-call vote. However, when the final version of the savings and loan bailout bill was brought to the floor for a vote, only 21 senators showed up to vote. Many senators had left the Capitol to get an early start on their month-long summer recess. Because of incidents such as this one, 94% of the respondents in a recent national survey favored forcing Congress to use roll-call votes on important legislation. Suppose that a politician wishes to estimate the percentage of voters in his state who favor forcing Congress to use roll-call votes on important legislation.

(*Source:* W. Updegrove, "America's Best (and Worst) Lawmakers," *Money* (October 1990): 124–126.)

a.   What sample size is necessary for the politician to estimate the proportion to within .03 with 95% confidence? Use the proportion figure in the national survey as an estimate of the proportion.

b.   Find the sample size in part a, assuming that the proportion figure in the national survey is not available.

10.47   *People's Choice*, a monthly magazine, claimed that more than 40% of its subscribers had an annual income of $50,000 or more. In a random sample of 62 subscribers, 30 had incomes of $50,000 or more. Does this information substantiate the magazine's claim? Use a significance level of .10.

10.48   Calculate and interpret the p-value for Exercise 10.47.

10.49 A statistician reported to a car insurance company a confidence interval for the proportion of convertible cars that had been involved in major accidents during the past year. The 95% confidence interval for p was reported to be the interval from .10 to .36.

a. What is the statistician's estimate of p?

b. What is the maximum error of estimate (E) of the proportion for this confidence interval?

c. Approximately what sample size did the statistician use?

10.50 A confidence interval is reported for the proportion of male YLU students who belong to the $\alpha\sigma\mu$ fraternity. The confidence interval is based on a sample of 100 students and is given to be .22 to .44. What level of confidence was used in obtaining this interval?

10.51 Must a confidence interval for a proportion contain the true proportion of the population? Explain what the "level of confidence" means for a confidence interval.

10.52 A market-research firm believed that the proportion of households with more than four family members in county 1 was greater than the probability of households with more than four family members in county 2. The firm gathered random samples of size 180 and 155 from counties 1 and 2, respectively. The number of households with more than four members were 74 from county 1 and 61 from county 2. From these data, can we conclude that the proportion of households with more than four members is higher in county 1 than in county 2? Use a significance level of .01.

10.53 Calculate the p-value for Exercise 10.52. Using the p-value, would you reject the null hypothesis at the .05 level?

10.54 Construct a 95% confidence interval for the difference of the two proportions in Exercise 10.52.

10.55 A market-research firm is interested in testing the hypothesis that the proportion of students who own a car is the same for the local state university campus and a local private college. They interviewed 240 students from the state university and 270 from the private college. The number of students who did not own a car was 78 at the state university and 82 at the private college. Using a .02 significance level, test the hypothesis.

10.56 For Exercise 10.55, construct a 95% confidence interval for the difference of the true proportions of students who own cars at the two campuses.

10.57 Two machines are used in a production process to cut metal circles from thin sheets of steel. If the circumference of the steel circles are larger than 9.01 inches or smaller than 9.00 inches, the steel circles are considered to be *nonconforming*. The quality supervisor suspected that the number of nonconforming steel circles produced by the older machine was larger than that for the newer machine. The supervisor randomly selected 400 steel circles from each machine. The data revealed that of the 400 steel circles from the older machine, 86 were nonconforming, and of the 400 steel circles from the newer machine, 56 were nonconforming.

a. Determine a 95% confidence interval for the difference in the proportion of nonconforming steel circles produced by the machines.

b. How much larger would the sample size need to be for the difference between the two proportions in part a to be within .05 of the actual difference with 95% confidence?

10.58 ABC's Monday Night Football had a rating of 19.7 and a share of 32 in 1985. "Rating" is the percentage of homes with television sets that are tuned to a particular program out of all the homes that have television sets in the United States. "Share" is the percentage of sets actually in use that are tuned to a particular program. Although the actual process of determining these figures is fairly sophisticated, let us assume, for the sake of simplicity, that the "share" figure of 32 is computed from a sample of 1000 homes.

a. What is the 95% confidence interval for ABC's 1985 share for Monday Night Football?

b. What is the maximum error of the estimate in part a?

10.59 A credit union randomly selected 110 savings-account customers and found that 85 of them also had checking accounts with the union. Construct a 95% confidence interval for the true proportion of savings-account customers who also have checking accounts.

10.60 A labor-union leader stated that at least 40% of the members of the union had a college degree. A random sample of 440 members revealed that only 170 had a college degree. Test this hypothesis if H_0 is H_0: the proportion of union members with a college degree is greater than or equal to 40%. Use a .05 level of significance.

10.61 A recent survey was conducted to measure how women's attitudes were changing since women have been contributing more to the family income. In this survey, 52 percent of the married women cited "how much my mate helps around the house" as the second biggest cause of resentment. Suppose that there were 300 respondents in this survey. Find a 90% confidence interval on the true proportion of married women who resented how much their mate helped around the house.

(*Source:* B. Townsend and K. O'Neil, "Women Get Mad," *American Demographics* (August 1990): 28.)

10.62 The manager at a manufacturing firm is interested in the proportion of the quality problems that originate with the worker rather than with the system. In a random sample of 200 quality problems, it was found that 90 of the quality problems originated with the workers. From the following MINITAB computer printout, is there sufficient evidence for the manager to conclude that less than 50% of the quality problems originate with the worker? At what significance level? (Note that 90 ones and 110 zeros are placed in C1, and sigma is taken to be the square root of p times $1 - p$, where p is the hypothesized proportion.)

```
MTB > set c1
DATA> 90(1) 110(0)
DATA> end
MTB > let k1 = sqrt (.5 * .5)
MTB > ztest, mu = .5, alternative hypothesis = -1, sigma = k1, data in c1

TEST OF MU = 0.5000 VS MU L.T. 0.5000
THE ASSUMED SIGMA =0.500

              N      MEAN    STDEV   SE MEAN      Z    P VALUE
C1          200    0.4500   0.4987   0.0354    -1.41    0.079
```

10.63 An industrial psychologist conducted a survey of 150 randomly selected middle-level managers 40 years or older and 150 randomly selected middle-level managers younger than 40. The industrial psychologist recorded that 81 of the managers 40 years of age or older did not think they would achieve a top-level management position in their firms. In the group of younger managers, 70 of the managers did not think they would achieve a top-level management position in their firm.

A MINITAB computer printout performs the statistical test procedure to determine if the proportions of the two groups are significantly different. What conclusion can be drawn from the analysis? Using the statistics in the computer printout, construct a 95% confidence interval on the difference of the proportions of the two groups. (Note that 81 ones and 69 zeros are placed in C1 and 70 ones and 80 zeros are placed in C2. Note that sigma is taken to be equal to $\sqrt{[2\bar{p}(1 - \bar{p})]}$. Using the MINITAB commands to perform the Z-test in this fashion requires that n_1 and n_2 be equal.)

```
MTB > set c1
DATA> 81(1) 69(0)
DATA> end
MTB > set c2
DATA> 70(1) 80(0)
DATA> end
MTB > let c3 = c1 - c2
MTB > let k2 = (81 + 70) / (150 + 150)
MTB > let k1 = sqrt(2*k2*(1-k2))
MTB > ztest, mu = 0, alternative hypothesis = 0, sigma = k1, data in c3

TEST OF MU = 0.0000 VS MU N.E. 0.0000
THE ASSUMED SIGMA =0.707

              N      MEAN    STDEV   SE MEAN      Z    P VALUE
C3          150    0.0733   0.2616   0.0577     1.27    0.20
```

Computer Exercises
Using the Database

Exercise 1 -- Appendix I Randomly select 100 observations from the database. Find a 95% confidence interval on the proportion of households that own their homes. (Refer to the variable OWNORENT.)

Exercise 2 -- Appendix I Randomly select 100 observations from the database. Estimate the proportion of observations from the NE sector in which the households own their homes. (Refer to the variables LOCATION and OWNORENT.) Also estimate the proportion of observations from the NW sector in which the households own their homes. Find a 95% confidence interval on the difference of the proportions of house owners for the two sectors.

Exercise 3 -- Appendix J Randomly select 100 observations from the database. Find a 95% confidence interval on the proportion of companies with a positive net income. (Refer to the variable NETINC.)

Exercise 4 -- Appendix J Randomly select 100 observations from the database. Estimate the proportion of observations from companies with an A bond rating that have a positive net income. (Refer to the variables BONDRATE and NETINC.) Also estimate the proportion of observations from companies with a B bond rating that have a positive net income. Find a 95% confidence interval on the difference of the proportions of companies with positive net income between those with A bond ratings and those with B bond ratings.

 *Case Study*

Can Private David Do Business With Government Goliath?

The answer to the above question is of course, but not without some pain. Also, unlike the biblical Goliath, the federal government bureaucracy is virtually indestructible. Any private sector David wishing to play on the federal field must play by Goliath's rules. These rules are daunting, to say the least, and represent a maze of regulations, paperwork, preparation of tenders/proposals, preaward surveys and various other potential obstacles. Large and small firms alike face these difficulties in dealing with the federal government, but small businesses may experience them quite acutely.

Various legislative acts have been enacted to promote participation by small businesses in the federal marketplace. The Small Business Subcontract Act (1979) required federal contractors to subcontract a portion of work to small businesses. The Small Business Administration Office of Advocacy (1979) provides an advocate within the government for small businesses to use for problem resolution. The Certificate of Competency Program (1982) provided a method for preauditing firms for procurement acceptability. The Prompt Payment Act (1982) mandated time deadlines for federal contracting agency bill payments. The Small Business Innovation Research Act (1982) mandated an increasing percentage of research and development procurement go to small businesses. The Commerce Business Daily Act (1983) mandated that all federal procurement be advertised prior to contracting. The Small Business and Federal Procurement Enhance-

ment Act (1984) and the Defense Procurement Reform Act (1984) introduced more competition into federal procurement markets and the Department of Defense procurements, respectively.

Quite a list! The very fact that such abundant legislation is necessary speaks for itself. Indeed, some of the very legislation aimed at helping small businesses may itself add another layer of bureaucracy to the whole process. To understand the problem better, the U.S. Small Business Administration commissioned a study to determine perceived barriers to small business participation in the federal marketplace.

Through a direct mail questionnaire with telephone follow-up interviews for clarification, this study addressed two broad areas. The first area of focus was problems and barriers, framed in the question: Do small and large businesses differ in perceived barriers, complaints, and unnecessary costs in federal procurement versus private sector work? The second area of focus was benefits and profitability: Do small and large businesses differ in terms of perceived relative cost and profit advantages for doing federal work versus private sector work?

The businesses that were surveyed had conducted business with both the federal government and the private sector. Of 570 firms initially contacted, 206 responded to the survey. Forty percent of all small firms and 26% of all large firms responded. There is too much data for all the results to be reproduced here. However, one table should provide the flavor of the study.

The respondents were asked to identify their biggest complaints in doing business with the fed-

	FEDERAL		PRIVATE	
Type of Complaint	Small Firm % (n = 158)	Large Firm % (n = 48)	Small Firm % (n = 158)	Large Firm % (n = 48)
Too much paperwork	32	40	2	2
Too much time between bid & work startup	9	8	4	4
Specifications for work to be done unclear	7	10	11	13
Delayed reimbursement for funds	19	4	30	21
Overall costs of procuring the work	8	19	16	17
Other	20	15	29	17
No answer	5	4	8	8
Total	100	100	100	100

■ TABLE 10.1 Biggest Complaint Reported in Doing Business with the Federal Government and Private Sector.

(*Source:* Jeffrey A. Cantor, "The Small Business And The Federal Marketplace: A Study Of Barriers To Successful Access," *Business Journal,* Fall 1989, pp. 42–50.)

eral government and with private firms. Table 10.1 summarizes their responses.

Case Study Questions

1. The complaint that there is too much paperwork involved in dealing with the federal government is readily supported by the dramatic difference in responses on this issue, regardless of size of firm. However, what about the complaint that "specifications for work to be done are unclear"? Conduct a hypothesis test at .05 significance to determine if, on the issue of unclear specifications, there is any difference between the perceptions reported by small firms dealing with the federal government and large firms dealing with the federal government? What is the p-value?

2. Construct a 90% confidence interval for the difference between the proportion of large firms complaining about too much paperwork and the proportion of small firms complaining about too much paperwork, in their dealings with the federal government.

3. Comment on any other similarities or differences between small and large firms in the above sample.

Analysis of Variance

A LOOK BACK/INTRODUCTION

In Chapter 9, we considered a question of the type, do men have the same heights as women? By this we mean, is the *average* height of males equal to the average height of females? We were interested in the means of two populations and performed a test of hypothesis, using, for example, $H_0: \mu_M = \mu_F$ and $H_a: \mu_M \neq \mu_F$. Such tests work well when dealing with two populations, but how can we compare the means of *more than two* populations? For example, we might wish to examine the average sales of salespeople trained in five different training programs to see whether they are the same. Our hypotheses become

$$H_0: \mu_1 = \mu_2 = \mu_3 = \mu_4 = \mu_5$$

$$H_a: \text{not all } \mu\text{'s are equal}$$

We test such a hypothesis by first collecting five samples, one from each of the training programs (populations). We will see that to compare these five means one pair at a time is *not* the correct approach:

this procedure results in ten different pairwise tests, and what was intended to be a testing procedure with, say, a .05 significance level results in a much higher significance level. In other words, the overall significance level, α, is *larger* than the predetermined value. The correct procedure for this situation is to examine the *variation* of the sales values, both (1) within each of the samples (examining the variability of each sample alone) and (2) among the five samples (for example, are the values in sample 1 larger or smaller, on the average, than the values in the other samples?).

In Chapter 9, we saw that when trying to decide if $\bar{X}_1$ is "significantly different" from $\bar{X}_2$, a key part of the answer rested on the values of s_1 and s_2, the variation *within* the two samples. Both s_1 and s_2 affect the width of the confidence interval for $\mu_1 - \mu_2$. Consequently, we infer something about the *means* of several populations by utilizing the *variation* of the resulting samples. Hence the term *analysis of variance*—our next topic.

◢ 11.1 Comparing Two Means: Another Look

We begin with an example. The manufacturer of a small battery-powered tape recorder decides to include four alkaline batteries with their product. Two battery suppliers are being considered; each has its own brand (brand 1 and brand 2). The supervising inspector of incoming quality wants to know if the average lifetimes of the two brands are the same. Based on past experience, she believes that the battery lifetimes follow a *normal* distribution. A simple experiment is conducted: each of 10 batteries (5 of each brand) is connected to a test device that places a small drain on the battery power and records the battery lifetime. The following results (in hours) are obtained:

BRAND 1	BRAND 2
43	30
48	26
38	37
41	31
51	34

Let μ_1 be the average lifetime (if observed indefinitely) for brand 1 and μ_2 be the average for brand 2. We wish to determine whether the data allow us to conclude that $\mu_1 \neq \mu_2$, using $\alpha = .10$.

We examined the same type of question in Chapter 9; we are dealing with two small independent samples. In Chapter 9 we advised against assuming that σ_1 was equal to σ_2. As a result, we generally used a t-test that did *not* pool the sample variances. However, when examining more than two normal populations (the main concern of this chapter), the following testing procedure for detecting a difference in the population means requires that the populations have the *same* distribution if, in fact, the population means are equal. Consequently, it can be used only when we are willing to assume that the *population variances are equal* (or approximately equal). The analysis of variance procedure is *not* extremely sensitive to departures from this assumption, especially if equal-sized samples are obtained from each population. A procedure for verifying this assumption (similar to the F-test used to compare two variances in Chapter 9) is discussed in this chapter.

As a result, we will assume that we have reason to believe that the variation of the brand 1 lifetimes is the same as for brand 2, that is, $\sigma_1 = \sigma_2$. Using the approach discussed in Chapter 9, we first find

$$s_p^2 = \text{pooled variance} = \frac{(n_1 - 1)s_1^2 + (n_2 - 1)s_2^2}{n_1 + n_2 - 2}$$

where n_1, n_2 = sample sizes for brand 1, brand 2 and s_1^2, s_2^2 = sample variances for brand 1, brand 2. Using the sample data,

BRAND 1	BRAND 2
$n_1 = 5$	$n_2 = 5$
$\bar{x}_1 = 44.2$ hrs	$\bar{x}_2 = 31.6$ hrs
$s_1 = 5.263$ hrs	$s_2 = 4.159$ hrs

Consequently,

$$s_p^2 = \frac{(4)(5.263)^2 + (4)(4.159)^2}{8}$$

$$= \frac{180.0}{8}$$

$$= 22.5$$

and so

$$s_p = \sqrt{22.5} = 4.74 \text{ hrs}$$

The appropriate hypotheses are $H_0: \mu_1 = \mu_2$ and $H_a: \mu_1 \neq \mu_2$. The resulting test statistic is

$$t = \frac{\bar{X}_1 - \bar{X}_2}{s_p \sqrt{\frac{1}{n_1} + \frac{1}{n_2}}}$$

$$= \frac{44.2 - 31.6}{4.74 \sqrt{\frac{1}{5} + \frac{1}{5}}} = \frac{12.6}{2.998}$$

$$= 4.20$$

That is, $t^* = 4.20$.

We are dealing with a two-tailed test using a t-statistic with $(n_1 - 1) + (n_2 - 1) = 4 + 4 = 8$ df, so the test procedure is to

$$\text{reject } H_0 \text{ if } |t^*| > t_{\alpha/2, df} = t_{.05, 8} = 1.86$$

Comparing $t^* = 4.20$ to 1.86, we reject H_0 and conclude that the mean lifetimes for the two brands are not the same. Looking at the sample data, we can say that $\bar{x}_1 = 44.2$ is significantly different from $\bar{x}_2 = 31.6$.

The Analysis of Variance Approach

We need to introduce two new terms. The previous example examined the effect of one **factor** (brand), consisting of two **levels** (brand 1 and brand 2). If you want to extend this to four brands (say, brands 1, 2, 3, and 4), then you still have *one* factor but you now have *four* levels.

The purpose of **analysis of variance (ANOVA)** is to determine whether this factor has *a significant effect* on the variable being measured (battery lifetime, in our example). If, for instance, the brand factor *is* significant, the mean lifetimes for the different brands will not be equal. Consequently, testing for equal means among the various brands is the same as attempting to answer the question, is there a significant effect on lifetime due to this factor?

This section examines the effect of a single factor on the variable being measured, **one-factor ANOVA.** Extensions of this technique include ANOVA procedures that determine the effect of two or more factors operating simultaneously. These factors may be *qualitative* (such as brand in the previous illustration) or *quantitative* (such as several levels of advertising expenditure).

All ten values in the battery lifetimes example are different, and we observe a variation in these values. We will look at two *sources of variation:* (1) variation *within* the samples (levels) and (2) variation *between* the samples.

Within-Sample Variation

When you obtain a sample, you usually obtain different values for each observation. The five sample values for brand 1 vary about the mean $\bar{x}_1 = 44.2$ hours, as measured by $s_1 = 5.26$ hours. Likewise, the five values in the second sample also exhibit some variation ($s_2 = 4.16$) about $\bar{x}_2 = 31.6$ hours. These are the **within-sample variations.** They are used when estimating the common population variance, say σ^2. This procedure provides an accurate estimate of σ^2, whether or not the sample means are equal.

Between-Sample Variation

When you compare the two samples, you observe that the values for brand 1 are *larger,* on the average, than those for brand 2. This is summarized in the sample means, where $\bar{x}_1 = 44.2$ appears to be considerably larger than $\bar{x}_2 \doteq 31.6$. So there is a variation in the ten values due to the *brand;* that is, due to the factor. This is **between-sample variation.** In general, if this variation is large, we expect considerable variation among the sample means. The between-sample variation is also used in another estimate of the common variance, σ^2, *provided the population means are equal. In other words, if the means are equal, the between-sample and within-sample estimates of σ^2 should be nearly the same.* As we will see later in this section, we can derive a test of hypothesis procedure for determining whether the means are equal by comparing these two estimates.

Measuring Variation

When using the ANOVA approach, we measure these two sources of variation by calculating various **sums of squares, SS.** We determine

SS(factor), which measures between-sample variation (also called SS(between))

SS(error), which measures within-sample variation (also called SS(within))

SS(total) = SS(between) + SS(within) = SS(factor) + SS(error)

Each of the first two sums of squares will have corresponding degrees of freedom, df, which are determined from the number of terms that make up this particular SS. The df for our example are given by

$$\text{df for factor} = (\text{number of levels}) - 1$$
$$= (\text{number of brands}) - 1$$
$$= 2 - 1 = 1$$
$$\text{df for error} = (n_1 - 1) + (n_2 - 1)$$
$$= n_1 + n_2 - 2 = 5 + 5 - 2 = 8$$

We will show how to determine these sums of squares and how we combine them and their df into another test statistic for testing $H_0: \mu_1 = \mu_2$ against $H_a: \mu_1 \neq \mu_2$. The beauty of this approach is that it extends nicely to the situation in which you wish to compare more than two means using a *single* test.

Determining SS(factor)   SS(factor) is the sum of squares that determines whether the values in one sample are larger or smaller on the average than the values in the second sample:

$$SS(\text{factor}) = n_1(\bar{x}_1 - \bar{x})^2 + n_2(\bar{x}_2 - \bar{x})^2 \qquad \textbf{(11.1)}$$

where $\bar{x}_1$, $\bar{x}_2$ are the two sample means and

$$\bar{x} = \frac{\Sigma(\text{all data values})}{n} = \frac{n_1\bar{x}_1 + n_2\bar{x}_2}{n_1 + n_2}$$

and $n = n_1 + n_2 = $ total sample size.

There is another method of determining this sum of squares that is much easier using a calculator:

$$SS(\text{factor}) = \left[\frac{T_1^2}{n_1} + \frac{T_2^2}{n_2}\right] - \frac{T^2}{n} \qquad \textbf{(11.2)}$$

where $T_1 = $ total of the sample 1 observations, $T_2 = $ total of the sample 2 observations, and $T = $ grand total $= T_1 + T_2$.

Determining SS(total)   SS(total) is a measure of the variation in all $n = n_1 + n_2$ data values. You obtain its value as though you were finding the *variance* of these n values, except that you do not divide by $n - 1$:

$$SS(\text{total}) = \Sigma(x - \bar{x})^2 \qquad \textbf{(11.3)}$$

or (after some algebra similar to that used in Chapter 3),

$$SS(\text{total}) = \Sigma x^2 - \frac{(\Sigma x)^2}{n} = \Sigma x^2 - \frac{T^2}{n} \qquad \textbf{(11.4)}$$

Determining SS(error) SS(error) is the measure of the variation *within* each of the samples. Its value simply is the *numerator of the pooled variance*, s_p^2, obtained using the previous *t*-test. Thus,

$$SS(error) = \underbrace{\Sigma(x - \bar{x}_1)^2}_{\text{first sample}} + \underbrace{\Sigma(x - \bar{x}_2)^2}_{\text{second sample}} \tag{11.5}$$

and therefore,

$$SS(error) = \Sigma x^2 - \left[\frac{T_1^2}{n_1} + \frac{T_2^2}{n_2}\right] \tag{11.6}$$

Given that

$$SS(total) = SS(factor) + SS(error)$$

a much easier way to find this value is

$$SS(error) = SS(total) - SS(factor) \tag{11.7}$$

Let us return to the battery lifetimes example. To find the SS(factor) here, we first determine

$$T_1 = 43 + 48 + 38 + 41 + 51 = 221$$
$$T_2 = 30 + 26 + 37 + 31 + 34 = 158$$
$$T = T_1 + T_2 = 221 + 158 = 379$$

So, using equation 11.2,

$$SS(factor) = \frac{221^2}{5} + \frac{158^2}{5} - \frac{379^2}{10}$$
$$= 14{,}761 - 14{,}364.1$$
$$= 396.9$$

To find SS(total), the only new term we need to evaluate is

$$\Sigma x^2 = \text{sum of each data value squared}$$
$$= 43^2 + 48^2 + \cdots + 31^2 + 34^2$$
$$= 14{,}941$$

So, using equation 11.4 (the value 14,364.1 was obtained in SS(factor)),

$$SS(total) = \Sigma x^2 - \frac{T^2}{n}$$
$$= 14{,}941 - 14{,}364.1$$
$$= 576.9$$

Finally, we find SS(error) by subtraction:

$$SS(error) = SS(total) - SS(factor)$$
$$= 576.9 - 396.9$$
$$= 180.0$$

ANOVA Test for H_0: $\mu_1 = \mu_2$ Versus H_a: $\mu_1 \neq \mu_2$

To begin with, the procedure we are about to define is valid for a *two-tailed test only*. In other words, the alternative hypothesis must be that the two means differ, not that one is larger than the other (a one-tailed test). (When examining more than two means, the alternative hypothesis will be that *at least* two of the means are unequal and H_0 will be that all the means are equal.) The next step when using the ANOVA procedure is to determine something resembling an "average" sum of squares, referred to as a **mean square.** We compute a mean square for only SS(factor) and SS(error), not for SS(total).

$$MS(\text{factor}) = \frac{SS(\text{factor})}{df \text{ for factor}} = SS(\text{factor})/1 \qquad (11.8)$$

Note that the df for this term always is (number of levels) $-$ 1. In this section, we are dealing with two levels (populations), and so here df is 1.

$$MS(\text{error}) = \frac{SS(\text{error})}{df \text{ for error}} = \frac{SS(\text{error})}{(n_1 + n_2 - 2)} \qquad (11.9)$$

We denote the common variance of the two normal populations as σ^2. So, $\sigma^2 = \sigma_1^2 = \sigma_2^2$. If the null hypothesis (H_0: the means are equal) is true, then, because the populations have identical means and variances, this implies that under H_0 we are dealing with a *single population*. The ANOVA procedure is based on a comparison between two separate estimates of the variance, σ^2. The first estimate is derived using the variation among the sample means (only two in the previous example). The other estimate is determined using the variation *within* each of the samples.

The ANOVA procedure is based on a comparison of these two estimates of σ^2 because they should be approximately equal *provided H_0 is true*. We have derived these two estimates:

MS(factor) = estimate of σ^2 based on the variation among the sample means

MS(error) = estimate of σ^2 based on the variation within each of the samples

Our new test statistic for testing H_0: $\mu_1 = \mu_2$ versus H_a: $\mu_1 \neq \mu_2$ is the *ratio* of these two estimates:

$$F = \frac{\left(\begin{array}{l}\text{estimated population variance based on the variation} \\ \text{among the sample means}\end{array}\right)}{\left(\begin{array}{l}\text{estimated population variance based on the variation} \\ \text{within each of the samples}\end{array}\right)} = \frac{MS(\text{factor})}{MS(\text{error})} \qquad (11.10)$$

This test statistic follows an F distribution, which was first introduced in Chapter 9 as a ratio of two variance estimates. The degrees of freedom (df) for the F statistic in equation 11.10 are the df for factor and the df for error; that is, in our present example, the df for F are 1 and $(n_1 + n_2 - 2)$. *Because the F statistic is based on a comparison of two variance estimates, this technique is called analysis of variance.*

■ **FIGURE 11.1**
Shape of the F
distribution shown by F
curve with 1 and
$n_1 + n_2 - 2$ df.

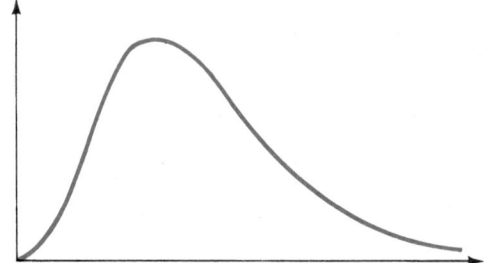

This is our second encounter with the F distribution. In Chapter 9, we used this distribution to compare two population variances (σ_1^2 and σ_2^2). The shape of this distribution is illustrated in Figure 11.1 and is tabulated in Table A.7. Remember that the shape of the F curve is affected by both the df for the numerator (1 here) and the df for the denominator ($n_1 + n_2 - 2$ here).

Defining the Rejection Region

What happens to the F statistic when H_a is true, that is, when $\mu_1 \neq \mu_2$? In this case, we would expect $\bar{X}_1$ and $\bar{X}_2$ to be "far apart." As a result, the estimate of the variance σ^2 using the *between-sample* variation (measured by MS(factor)) will be *larger* than the estimate of σ^2 based on the *within-sample* variation (measured by MS(error)). This implies that we should reject H_0 in favor of H_a whenever the ratio of these two estimates is large—in which case the computed F-value is in the right tail. Consequently, the test procedure will be to

$$\text{reject } H_0 \text{ if } F^* > F_{\alpha, v_1, v_2}$$

where v_1 = df for numerator = (number of levels) − 1 = 1, v_2 = df for denominator = $n_1 + n_2 - 2$, and F_{α, v_1, v_2} is obtained from Table A.7 with a right-tail area = α.

EXAMPLE 11.1

Using the data from the battery lifetimes example and the previously calculated sums of squares, test $H_0: \mu_1 = \mu_2$ versus $H_a: \mu_1 \neq \mu_2$, where μ_1 = average lifetime for brand 1, if observed indefinitely, and μ_2 = average for brand 2. Use a significance level of $\alpha = .10$.

SOLUTION

Step 1. The hypotheses are as defined—$H_0: \mu_1 = \mu_2$ and $H_a: \mu_1 \neq \mu_2$.
Step 2. The test statistic is

$$F = \frac{\text{MS(factor)}}{\text{MS(error)}}$$

Step 3. The rejection region (using Table A.7(a)) is

$$\text{reject } H_0 \text{ if } F > F_{.10, 1, 8} = 3.46$$

Step 4. From the previous calculations, SS(factor) = 396.9 and SS(error) = 180. So,

$$\text{MS(factor)} = \text{SS(factor)}/1 = 396.9/1 = 396.9$$

and

$$\text{MS(error)} = \frac{\text{SS(error)}}{(n_1 + n_2 - 2)} = 180.0/8 = 22.5$$

The resulting value of the test statistic is

$$F^* = \frac{396.9}{22.5} = 17.64$$

Because $17.64 > 3.46$, we reject H_0.

Step 5. These data indicate that the mean lifetimes for brand 1 and brand 2 are *not* the same. ∎

COMMENTS   Compare our first treatment of the battery lifetimes problem with Example 11.1. Both solutions led to the same conclusion, namely, that the two average lifetimes are not the same. In fact, the two solutions *always* lead to the same conclusion when comparing *two* means. Furthermore, the p-values for the two solutions *are the same*, as illustrated in Figure 11.2. The values were obtained using a computer program (available in many statistical packages) that provides an exact p-value for a t or F statistic, given the computed value and corresponding degrees of freedom.

The computed value of the F statistic is equal to the square of the computed value of the t statistic because $17.64 = (4.20)^2$. This is true whenever you have an F statistic or table value with 1 df in the numerator. So,

$$F^* = (t^*)^2$$

Furthermore, the table values satisfy the same relationship, namely,

$$F_{.10,1,8} = 3.46 = (1.86)^2 = [t_{.05,8}]^2$$

We see that the two tests are *identical;* they produce the same conclusion and p-value. Furthermore, the computed value and the table value for the F statistic are the squares of the corresponding values using the t-statistic. This comparison applies *only* when the F statistic has 1 df in the numerator—that is, when there are two factor levels (as in this illustration). As mentioned previously, the advantage of the ANOVA approach is that it extends very easily to the situation of comparing means for more than two populations (covered in the next section).

The ANOVA Table

Rather than carrying out the five-step procedure using the F statistic, an easier method is to use an **ANOVA table** of the various sums of squares. The format of this table is as follows:*

SOURCE	df	SS	MS	F
Factor	1	SS(factor)	$MS(factor) = \dfrac{SS(factor)}{1}$	$\dfrac{MS(factor)}{MS(error)}$
Error	$n - 2$	SS(error)	$MS(error) = \dfrac{SS(error)}{n - 2}$	
Total	$n - 1$	SS(total)		

To fill in this table, you compute the necessary sums of squares along with the mean squares and insert them. Notice that $n = n_1 + n_2 =$ total sample size and that column 3 (MS) = column 2 (SS) divided by column 1 (df).

The ANOVA table for Example 11.1 follows.

SOURCE	df	SS	MS	F
Factor	1	396.9	396.9	17.64
Error	8	180.0	22.5	
Total	9	576.9		

---

*The headings under the "Source" column will vary, depending on the computer package. SS(factor) often is labeled "between groups" (SAS) or "among groups"; SS(error) often is labeled "within groups" (SAS), "residual" (SPSS), or "error" (MINITAB).

■ **FIGURE 11.2**
p-values for the solution to the battery lifetimes example. (a) Solution using pooled variance *t* test. (b) Solution using ANOVA (see Example 11.1).

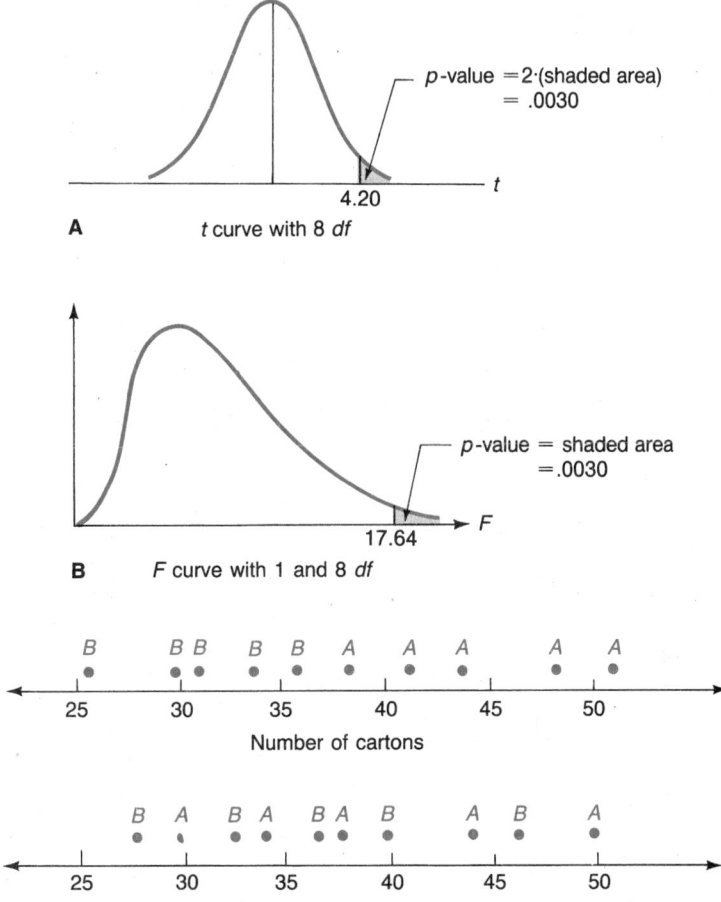

p-value $= 2 \cdot$(shaded area)
$= .0030$

4.20
A   *t* curve with 8 *df*

p-value $=$ shaded area
$= .0030$

17.64
B   *F* curve with 1 and 8 *df*

■ **FIGURE 11.3**
Dot-array diagram of replicates in Example 11.1.

   B   B B   B   B   A   A   A   A   A

25   30   35   40   45   50
Number of cartons

■ **FIGURE 11.4**
Dot-array diagram where between-sample and within-sample variations are nearly the same. The *F* statistic would not lie in the rejection region.

   B   A   B   A   B A   B   A   B   A

25   30   35   40   45   50
Number of cartons

Summary of the ANOVA Approach for One-Factor Tests

In Example 11.1 we concluded that a difference existed between the two *means* because the variation *between* the two samples (measured by MS(factor)) was much greater than the variation *within* the samples (measured by MS(error)). Thus, the ratio of these values was very large and F^* fell in the rejection region. Consequently, we rejected $H_0: \mu_1 = \mu_2$. What this means in the language of ANOVA is that there *is* a significant effect on battery lifetime due to the brand factor.

To carry out the *F* test, we first randomly obtain observations, called **replicates,** from each population. Example 11.1 used five replicates from each of the two battery brand populations. It is *not necessary* to obtain the same number of replicates from each population.

Figure 11.3 is a dot-array diagram of the data in Example 11.1, where the symbol A represents a value from brand 1 and B represents brand 2. You do not need to be an expert statistician to observe that a clear difference exists between the lifetimes of the two brands of batteries. The variation within the As alone and the Bs alone is the within-sample variation. Because the distances from the A values to the B values are much larger than the distances among the A values alone, the between-sample variation is quite large, as we have already observed.

Suppose instead that your dot-array diagram looks like Figure 11.4. Now the two sources of variation appear to be nearly the same and there is no obvious difference between the two brands. The resulting F statistic here would not lie within the rejection region, and we would not be able to demonstrate, using the ANOVA approach, a difference between the two mean lifetimes.

▚ 11.2
One-Factor ANOVA Comparing More Than Two Means

In the previous section, we examined a single *factor* with two *levels*. Our concern was whether there was any difference between the two levels of this factor. We performed a test of hypothesis on the means of two populations. Because we were dealing with the effect of a single factor, this was a one-factor (or one-way) ANOVA.

In general, one-factor ANOVA techniques can be used to study the effect of any single factor on performance, sales, and the like. This factor can consist of any number of levels—say, k levels. To determine if the levels of this factor affect our measured observations, we examine the hypotheses

$$H_0: \mu_1 = \mu_2 = \cdots = \mu_k$$

$$H_a: \text{not all } \mu\text{'s are equal}$$

Suppose we are interested in the average lifetimes of not two but five brands of batteries. Is there any difference in these five mean lifetimes? To answer this question, we test

$$H_0: \mu_1 = \mu_2 = \mu_3 = \mu_4 = \mu_5$$

$$H_a: \text{not all } \mu\text{'s are equal}$$

We have a single factor (brand) consisting of five levels (brand 1, brand 2, . . . , brand 5). One possibility is to examine these samples one pair at a time using the t statistic discussed in the previous section. This appears to be a safe way to proceed here, although there are $_5C_2 = 10$ such pairs of tests to perform this way. The main problem with performing many tests of this nature is determining the probability of making an incorrect decision. In particular, what value does α have, where α is the probability of rejecting H_0: all μ's are equal, when in fact it is true? You set α in advance but, after performing ten of these pairwise tests ($\mu_1 = \mu_2$, $\mu_1 = \mu_3$, . . .), for instance, what is your *overall* probability of concluding that at least one pair of means are not equal when they actually are? This is a difficult question. The overall probability is not the significance level, α, with which you started for just one pair. *So we need an approach that will test for the equality of these five means using a single test. This is what the ANOVA approach does.*

Assumptions Behind the ANOVA Analysis

When using the ANOVA procedure, there are three key assumptions that must be satisfied. They are basically the same assumptions that were necessary when testing two means using small independent samples and the pooled variance approach. These assumptions are:

1. The replicates are obtained *independently* and *randomly* from each of the populations. The value of one observation has no effect on any other replicates within the same sample or within the other samples.
2. The observations (replicates) from each population follow (approximately) a *normal* distribution.
3. The normal populations all have a *common variance, σ^2.* We expect the values in each sample to vary about the same amount. The ANOVA procedure will be

much less sensitive to violations of this assumption when we obtain samples of equal size from each population.

Deriving the Sum of Squares

When examining k populations, for example, the data will be configured somewhat like this:

	Level 1	Level 2	$\cdots$	Level k
	$\vdots$	$\vdots$		$\vdots$
	n_1 replicates	n_2 replicates	$\cdots$	n_k replicates
	$\vdots$	$\vdots$		$\vdots$
Totals	T_1	T_2	$\cdots$	T_k

This resembles the data from Example 11.1, where $k = 2$ and $n_1 = n_2 = 5$ replicates. To derive the sum of squares for this situation, we extend the results in equations 11.2, 11.4, and 11.6 to

$$\text{SS(factor)} = \left[\frac{T_1^2}{n_1} + \frac{T_2^2}{n_2} + \cdots + \frac{T_k^2}{n_k} \right] - \frac{T^2}{n} \qquad (11.11)$$

$$\text{SS(total)} = \Sigma x^2 - \frac{T^2}{n} \qquad (11.12)$$

$$\text{SS(error)} = \Sigma x^2 - \left[\frac{T_1^2}{n_1} + \frac{T_2^2}{n_2} + \cdots + \frac{T_k^2}{n_k} \right] \qquad (11.13)$$

$$= \text{SS(total)} - \text{SS(factor)} \qquad (11.14)$$

Here, n = the total number of observations = $n_1 + n_2 + \cdots + n_k$, and $T = \Sigma x$ = the sum of all n observations = $T_1 + T_2 \cdots + T_k$. Also, to find Σx^2, you square each of the n observations and sum the results.

The ANOVA Table

The good news is that the format of the ANOVA table is the same regardless of the number of populations (levels), k. The only change from the two-population case is that

$$\text{df for factor} = k - 1$$

$$\text{df for error} = n - k$$

As before, the total df are $n - 1$. The resulting ANOVA table follows.

SOURCE	df	SS	MS	F
Factor	$k - 1$	SS(factor)	$\text{MS(factor)} = \dfrac{\text{SS(factor)}}{k - 1}$	$\dfrac{\text{MS(factor)}}{\text{MS(error)}}$
Error	$n - k$	SS(error)	$\text{MS(error)} = \dfrac{\text{SS(error)}}{n - k}$	
Total	$n - 1$	SS(total)		

Note that

$$MS(factor) = \frac{SS(factor)}{df\ for\ factor}$$

$$= \frac{SS(factor)}{k-1} \qquad (11.15)$$

$$MS(error) = \frac{SS(error)}{df\ for\ error}$$

$$= \frac{SS(error)}{n-k} \qquad (11.16)$$

The test statistic for testing $H_0: \mu_1 = \mu_2 = \cdots = \mu_k$ versus H_a: not all μ's are equal is

$$F = \frac{MS(factor)}{MS(error)}$$

which has an F distribution with $k-1$ and $n-k$ df.

As in the two-sample case, the procedure is to reject H_0 when the variation among the sample means (measured by MS(factor)) is *large* compared to the variation within the samples (measured by MS(error)). Consequently, the test will be to reject H_0 whenever F lies in the *right-tailed* rejection region defined by the significance level, α.

EXAMPLE 11.2

The manufacturer of the small battery-powered tape recorder (discussed at the start of this chapter) also manufactures battery powered AM/FM radio sets that include the required batteries. This unit uses a smaller battery, and four suppliers (brands) of this battery are being considered. Past experience has indicated that the battery lifetimes are normally distributed. The supervising inspector of incoming quality again wants to know whether there is any difference among the average lifetimes of the four battery brands. Twenty-four batteries (6 of each brand) are placed on a test device that slowly drains the battery power and records the battery lifetime. The following data (in hours) were obtained:

	BRAND 1	BRAND 2	BRAND 3	BRAND 4
	41	32	35	33
	35	37	30	27
	48	46	24	36
	40	53	26	35
	45	41	28	27
	52	43	31	25
Total (T)	261	252	174	183
Average ($\bar{X}$)	43.5	42.0	29.0	30.5
Variance (s^2)	37.1	52.8	15.2	22.3

The four sample averages are $\bar{x}_1 = 43.5$, $\bar{x}_2 = 42.0$, $\bar{x}_3 = 29.0$, and $\bar{x}_4 = 30.5$. Brands 1 and 2 appear to be outlasting brands 3 and 4. In other words, it appears that there is a significant *between-group variation*. But do these sample means provide sufficient evidence to reject $H_0: \mu_1 = \mu_2 = \mu_3 = \mu_4$, where each μ_i represents the average of *all* lifetimes for brand i? Use the ANOVA procedure to answer this question with $\alpha = .05$.

SOLUTION   The assumptions behind this analysis are (1) the samples were obtained randomly and independently from each of the four populations and (2) the battery lifetimes for each brand follow a *normal* distribution with a *common variance*, say, σ^2.

$$\text{SS(factor)} = \left[\frac{T_1^2}{n_1} + \frac{T_2^2}{n_2} + \frac{T_3^2}{n_3} + \frac{T_4^2}{n_4}\right] - \frac{T^2}{n}$$

So $n = n_1 + n_2 + n_3 + n_4 = 24$, and

$$T = \Sigma x = T_1 + T_2 + T_3 + T_4$$
$$= 261 + 252 + 174 + 183$$
$$= 870$$

Therefore,

$$\text{SS(factor)} = \frac{261^2}{6} + \frac{252^2}{6} + \frac{174^2}{6} + \frac{183^2}{6} - \frac{870^2}{24}$$
$$= 32{,}565 - 31{,}537.5$$
$$= 1027.5$$

$$\text{SS(total)} = \Sigma x^2 - \frac{T^2}{n}$$

$$= [41^2 + 35^2 + \cdots + 27^2 + 25^2] - \frac{870^2}{24}$$

$$= 33{,}202 - 31{,}537.5$$
$$= 1664.5$$

$$\text{SS(error)} = \text{SS(total)} - \text{SS(factor)}$$
$$= 1664.5 - 1027.5$$
$$= 637$$

The ANOVA table for this analysis follows.

SOURCE	df	SS	MS	F
Factor	$k - 1 = 3$	1027.5	$1027.5/3 = 342.5$	$342.5/31.85 = 10.75$
Error	$n - k = 20$	637	$637/20 = 31.85$	
Total	23	1664.5		

The computed F value using the ANOVA table is $F^* = 10.75$. Since $\alpha = .05$, we use Table A.7 to find that $F_{.05,3,20} = 3.10$. Comparing these two values, $F^* = 10.75 > 3.10$, so we reject H_0.

We conclude that the average lifetimes for the four brands are not the same. This confirms our earlier suspicion based on the variation among the four sample means. Our results indicate that the brand factor *does* have a significant effect on battery lifetime.   ■

The Assumptions Behind ANOVA and a Test for Equal Variances

Using *independent random samples* is of extreme importance with the ANOVA procedure. The F test used for comparing the population means in the ANOVA table is very sensitive to departures from this assumption, so the safest way to guard against incorrect conclusions is to use random sampling techniques. In many situations, however, such as when using the same set of people for before-and-after experiments, this may be difficult or impossible. One solution to this problem is to modify your study design, for example, by using a randomized block design, discussed later in this chapter.

Lack of **normality** within the populations is not a critical matter provided the departure is not too extreme. The F test used to test the means is not severely affected by populations that are somewhat nonnormal in nature. One way of making the ANOVA procedure even less sensitive to this assumption is to use *large samples*.

If the *variances* of the population *are not equal*, the F test used in the ANOVA procedure for testing the means is only slightly affected, provided the *sample sizes are equal* (or nearly so). However, for this case, there is a very simple test of hypothesis for verifying this assumption.

In Chapter 9 an F test was defined for determining whether two normal population variances (or standard deviations) are equal. A similar test is used when you are comparing more than two normal population variances, provided the sample sizes are equal.* Here the hypotheses are

$H_0: \sigma_1^2 = \sigma_2^2 = \cdots = \sigma_k^2$

H_a: at least two variances are unequal (the k variances are not the same)

We warned you in Chapter 9 about the dangers of using the same data to test both the variances *and* the means. This warning also applies to tests of more than two populations. A better procedure is to use a different data set for testing H_0: the variances are equal. This requires a much larger data set than is necessary if you use the same data for both tests. The test for equal variances is the Hartley test; the test statistic is defined to be

$$H = \frac{\text{maximum } s^2}{\text{minimum } s^2} \qquad (11.17)$$

which is simply the ratio of the largest sample variance divided by the smallest of these k variances.

If H_0 is false, the test statistic will be "large," so the testing procedure is to reject H_0 if the computed value of H lies in the right tail. The rejection region for a 5% level of significance can be obtained from Table A.14. This region depends on the number (k) of populations or levels and the number of observations in *each* sample.

Suppose we use the battery lifetime data from Example 11.2 only for testing the hypotheses

$H_0: \sigma_1^2 = \sigma_2^2 = \sigma_3^2 = \sigma_4^2$

H_a: at least two variances are unequal

Using $\alpha = .05$ and Table A.14, because $k = 4$ and there are six observations in each sample, the test is to

reject H_0 if $H > 13.7$

*When the sample sizes are unequal, a computationally more difficult test for equal variances can be performed, derived by M. S. Bartlett. For details, see J. Neter, W. Wasserman, and M. Kutner, *Applied Linear Statistical Models*, 3d ed. (Homewood, Ill.: Richard D. Irwin, 1990): 614–18.

Using the data summary in Example 11.2, the minimum s^2 is 15.2 and the maximum s^2 is 52.8. Consequently,

$$H = \frac{52.8}{15.2} = 3.47$$

which is less than 13.7, and so the conclusion is that we have no reason to suspect unequal variances for this situation.

If other data are available for testing the means, the assumption of equal variances behind the ANOVA procedure appears to be safe.

Confidence Intervals in One-Factor ANOVA

When we deal with normal populations, as we do here, we can supply:

1. A point estimate of each mean, μ_i; for example, an estimate of μ_2 is $\bar{X}_2$.
2. A point estimate of each mean difference, $\mu_i - \mu_j$; for example, an estimate of $\mu_1 - \mu_3$ is $\bar{X}_1 - \bar{X}_3$.

When using the ANOVA procedure, the populations are believed to have a common variance, say σ^2. To estimate this variance, we use an estimate of σ^2 that does not depend on whether the population means are equal—the *within*-sample variation, measured by MS(error). The point estimate of σ^2 is

$$s_p^2 = \text{pooled variance}$$

$$= \text{MS(error)}$$

where MS(error) is defined in equation 11.16.

In previous chapters, we always supplied a confidence interval along with a point estimate to provide a measure of how reliable this estimate really is. The narrower the confidence interval, the more faith you have in your point estimate. A $(1 - \alpha) \cdot 100\%$ confidence interval for μ_i is

$$\bar{X}_i - t_{\alpha/2, n-k} s_p \sqrt{\frac{1}{n_i}} \quad \text{to} \quad \bar{X}_i + t_{\alpha/2, n-k} s_p \sqrt{\frac{1}{n_i}} \qquad \textbf{(11.18)}$$

where

$$k = \text{number of populations (levels)}$$

$$n_i = \text{number of replicates in the } i\text{th sample}$$

$$n = \text{total number of observations}$$

$$s_p = \sqrt{\text{MS(error)}}$$

$t_{\alpha/2, \text{df}}$ is the value from Table A.5 with df = df for error = $n - k$, and right-tail area = $\alpha/2$.

A $(1 - \alpha) \cdot 100\%$ confidence interval for $\mu_i - \mu_j$ is

$$(\bar{X}_i - \bar{X}_j) - t_{\alpha/2, n-k} s_p \sqrt{\frac{1}{n_i} + \frac{1}{n_j}} \quad \text{to}$$

$$\qquad\qquad (\bar{X}_i - \bar{X}_j) + t_{\alpha/2, n-k} s_p \sqrt{\frac{1}{n_i} + \frac{1}{n_j}} \qquad \textbf{(11.19)}$$

EXAMPLE 11.3

Using the battery lifetime data from Example 11.2, construct a 95% confidence interval for the average lifetime of brand 1. Also determine a 95% confidence interval for the difference between the average lifetimes of brands 1 and 3.

SOLUTION

First, your point estimate of μ_1 is $\bar{x}_1 = 43.5$. Using the ANOVA table from Example 11.2,

$$s_p^2 = MS(error) = 31.85$$

and so

$$s_p = \sqrt{31.85} = 5.64$$

Because $n = 24$ and $k = 4$, the resulting 95% confidence interval for μ_1 is

$$43.5 - t_{.025,20}(5.64)\sqrt{\frac{1}{6}} \quad \text{to} \quad 43.5 + t_{.025,20}(5.64)\sqrt{\frac{1}{6}}$$

$$= 43.5 - (2.086)(5.64)(.408) \quad \text{to} \quad 43.5 + (2.086)(5.64)(.408)$$

$$= 43.5 - 4.80 \quad \text{to} \quad 43.5 + 4.80$$

$$= 38.7 \quad \text{to} \quad 48.3$$

As a result, we are 95% confident that the average lifetime of the brand 1 battery is between 38.7 and 48.3 hours.

The 95% confidence interval for $\mu_1 - \mu_3$ is

$$(\bar{X}_1 - \bar{X}_3) - t_{.025,20}s_p\sqrt{\frac{1}{n_1} + \frac{1}{n_3}} \quad \text{to} \quad (\bar{X}_1 - \bar{X}_3) + t_{.025,20}s_p\sqrt{\frac{1}{n_1} + \frac{1}{n_3}}$$

$$= (43.5 - 29.0) - (2.086)(5.64)\sqrt{\frac{1}{6} + \frac{1}{6}} \quad \text{to} \quad (43.5 - 29.0) + (2.086)(5.64)\sqrt{\frac{1}{6} + \frac{1}{6}}$$

$$= 14.5 - (2.086)(5.64)(.577) \quad \text{to} \quad 14.5 + (2.086)(5.64)(.577)$$

$$= 14.5 - 6.79 \quad \text{to} \quad 14.5 + 6.79$$

$$= 7.71 \quad \text{to} \quad 21.29$$

Based on this confidence interval, we are 95% confident that the average lifetime for brand 1 is between 7.71 and 21.29 hours *higher* than the average lifetime for brand 3. ∎

A Word of Warning

The procedure we used in Example 11.3 for determining confidence intervals is reliable, providing you decide which intervals you want computed *before* you observe your data. For example, constructing a confidence interval for the difference of two population means having the corresponding largest and smallest sample means is not an accurate procedure. If you do this, you let the data dictate which confidence interval you determine.

When using the procedure in Example 11.3 to construct confidence intervals for the difference of two population means, it is important to keep the number of such intervals as small as possible, because the probability of any one interval containing the true population difference is $1 - \alpha$, but the probability that *all* the intervals contain their respective population differences is not $1 - \alpha$. In other words, if $\alpha = .05$, the overall confidence level of this procedure is not 95%; it is something much less than 95%. *To compare all possible pairs of means effectively, you need to use a technique that will allow you to make all possible comparisons between population means while maintaining the Type I error rate at α. This is* called a **multiple comparison** procedure; one such procedure is discussed following Example 11.5.

EXAMPLE 11.4

Comptek, a computer software development firm, is interested in the effect of educational level on the job knowledge of the company's employees. They administer an exam to a randomly selected group of people having various educational backgrounds.

In the sample of 15 employees, 6 have only high school diplomas, 5 have only bachelor's degrees, and 4 have master's degrees. The exam scores are:

	HIGH SCHOOL DIPLOMA	BACHELOR'S DEGREE	MASTER'S DEGREE
	81	94	88
	84	83	89
	69	86	78
	85	81	85
	84	78	
	95		
Total (T)	498	422	340
Average ($\bar{X}$)	83.0	84.4	85.0

What would be your conclusion using a significance level of .10?

SOLUTION

Examining the sample means, you might be tempted to conclude that the higher a person's level of education, the higher their score on the exam. But is there a significant difference among these three means? An ANOVA analysis will clarify this.

The assumptions necessary here are:

1. The scores were obtained randomly and independently from each of the three populations.
2. The exam scores for each of the three populations follow a normal distribution, with means μ_1, μ_2, and μ_3. The scores in each of the samples are assumed to have the same amount of variation.

Because the sample sizes are not the same, the Hartley test for equal variances cannot be used here. As discussed earlier, we prefer not to use the same data for testing both the means and variances, and so a better procedure would be to obtain additional data (with equal sample sizes) for testing the equality of these three variances. We begin by calculating the necessary sum of squares:

$$\text{SS(factor)} = \left[\frac{T_1^2}{n_1} + \frac{T_2^2}{n_2} + \frac{T_3^2}{n_3} \right] - \frac{T^2}{n}$$

where

$$T = \Sigma x = T_1 + T_2 + T_3$$
$$= 498 + 422 + 340 = 1260$$
$$n = n_1 + n_2 + n_3$$
$$= 6 + 5 + 4 = 15$$

So,

$$\text{SS(factor)} = \frac{498^2}{6} + \frac{422^2}{5} + \frac{340^2}{4} - \frac{1260^2}{15}$$
$$= 105,850.8 - 105,840.0$$
$$= 10.8$$

$$SS(\text{total}) = \Sigma x^2 - \frac{T^2}{n}$$

$$= [81^2 + 84^2 + \cdots + 78^2 + 85^2] - \frac{1260^2}{15}$$

$$= 106{,}424 - 105{,}840$$

$$= 584$$

$$SS(\text{error}) = 584 - 10.8 = 573.2$$

Finally, because $k = 3$ and $n = 15$,

$$\text{df for factor} = k - 1 = 2$$
$$\text{df for error} = n - k = 12$$
$$\text{df for total} = n - 1 = 14$$

The resulting ANOVA table is

SOURCE	df	SS	MS	F
Factor	2	10.8	5.4	.11
Error	12	573.2	47.8	
Total	14	584		

The hypotheses are

$$H_0: \mu_1 = \mu_2 = \mu_3$$
$$H_a: \text{not all } \mu\text{'s are equal}$$

where each μ_i represents the average score of *all* employees having this particular educational level at Comptek.

We will reject H_0 if

$$F^* > F_{.10,2,12} = 2.81$$

Because $.11 < 2.81$, we fail to reject H_0.

We conclude that there is not sufficient evidence to indicate that the average performance on the exam is different among the three groups. As usual, we do not *accept H_0*; that is, we do *not* conclude that these three means *are* equal. There is simply not enough evidence to support the claim that employees with a higher educational level have greater job knowledge. ∎

The factor in Example 11.4 was the educational level of the employee; it had three levels. The results show that we are unable to demonstrate that this factor has a significant effect on exam performance.

EXAMPLE 11.5   In Example 11.4, before the exam was given, Comptek decided to construct a 95% confidence interval for the average exam score of all people holding a master's degree and the difference between the average exam scores for personnel with a master's degree and those with only a high school diploma. What are the confidence intervals?

SOLUTION   The point estimates are

$$\text{for } \mu_3: \bar{x}_3 = 85.0$$
$$\text{for } \mu_3 - \mu_1: \bar{x}_3 - \bar{x}_1 = 85.0 - 83.0 = 2.0$$

To construct the confidence intervals, you first need an estimate of the common variance of these three populations. Based on the results of Example 11.4, this is

$$s_p^2 = MS(\text{error}) = 47.8$$

so

$$s_p = \sqrt{47.8} = 6.91$$

Because $n = 15$, $k = 3$, and $n_3 = 4$, the 95% confidence interval for μ_3 is

$$\bar{X}_3 - t_{.025, 12} s_p \sqrt{\frac{1}{n_3}} \quad \text{to} \quad \bar{X}_3 + t_{.025, 12} s_p \sqrt{\frac{1}{n_3}}$$

$$= 85.0 - (2.179)(6.91)(.5) \quad \text{to} \quad 85.0 + (2.179)(6.91)(.5)$$

$$= 85.0 - 7.53 \quad \text{to} \quad 85.0 + 7.53$$

$$= 77.47 \quad \text{to} \quad 92.53$$

The 95% confidence interval for $\mu_3 - \mu_1$ is

$$(\bar{X}_3 - \bar{X}_1) - t_{.025, 12} s_p \sqrt{\frac{1}{n_3} + \frac{1}{n_1}} \quad \text{to} \quad (\bar{X}_3 - \bar{X}_1) + t_{.025, 12} s_p \sqrt{\frac{1}{n_3} + \frac{1}{n_1}}$$

$$= (85.0 - 83.0) - (2.179)(6.91)(.645) \quad \text{to} \quad (85.0 - 83.0) + (2.179)(6.91)(.645)$$

$$= 2.0 - 9.71 \quad \text{to} \quad 2.0 + 9.71$$

$$= -7.71 \quad \text{to} \quad 11.71$$

Consequently, we are 95% confident that the average exam score of all employees with master's degrees is between 7.71 *lower* to 11.71 *higher* than those with a high school diploma only. This implies that the data do *not* allow us to say that the employees with master's degrees performed better on the exam than those with high school degrees.

A MINITAB solution to this example is shown in Figure 11.5. The output contains summary information for each sample, the ANOVA table, and a graphical representation of the confidence interval for each population mean.   ∎

■ FIGURE 11.5
MINITAB solution for Examples 11.4 and 11.5.

```
MTB > READ INTO C1 C2
DATA> 81 1
DATA> 84 1
DATA> 69 1
DATA> 85 1
DATA> 84 1
DATA> 95 1
DATA> 94 2
DATA> 83 2
DATA> 86 2
DATA> 81 2
DATA> 78 2
DATA> 88 3
DATA> 89 3
DATA> 78 3
DATA> 85 3
DATA> END
      15 ROWS READ
MTB > NAME C1 = 'SCORE' C2 = 'EDLEVEL'
MTB > ONEWAY USING DATA IN C1, LEVELS IN C2

ANALYSIS OF VARIANCE ON SCORE
SOURCE      DF        SS        MS        F        p
EDLEVEL      2      10.8       5.4     0.11    0.894
ERROR       12     573.2      47.8
TOTAL       14     584.0
                                    INDIVIDUAL 95 PCT CI'S FOR MEAN
                                    BASED ON POOLED STDEV
    LEVEL    N      MEAN     STDEV   -------+---------+---------+---------
        1    6    83.000     8.367   (-----------*-----------)
        2    5    84.400     6.107    (-------------*-------------)
        3    4    85.000     4.967     (--------------*--------------)
                                    -------+---------+---------+---------
POOLED STDEV =     6.911               80.0      85.0      90.0
```

Multiple Comparisons: A Follow-up to the One-Factor ANOVA Procedure

If the one-factor ANOVA procedure leads to a rejection of H_0: all population means are equal, a logical question would be, which means do differ? In other words, rejecting the ANOVA null hypothesis informs us that the means are not all the same but provides no clue as to which of the population means are different. As we discussed prior to Example 11.4, performing a series of t tests to compare all possible pairs of means is not a good idea, since the chances of making at least one Type I error (concluding that a difference exists between two population means when in fact they are the same) using such a procedure is much larger than the predetermined α used for each of the t tests.

What is needed is a technique that compares all possible pairs of means in such a way that the probability of making **one or more** Type I errors is α. This is a *multiple comparisons procedure*. There are several methods available for making multiple comparisons; the one presented here is **Tukey's** test for multiple comparisons (Tukey is pronounced too'-key).

Tukey's procedure is based on a statistical test that uses the largest and smallest sample means. The form of this statistic is

$$Q = \frac{\text{maximum } (\bar{X}_i) - \text{minimum } (\bar{X}_i)}{\sqrt{\text{MS(error)}/n_r}} \qquad \textbf{(11.20)}$$

where

1. maximum $(\bar{X}_i)$ and minimum $(\bar{X}_i)$ are the largest and smallest sample means, respectively.
2. MS(error) is the sample variance.
3. n_r is the number of replicates in each sample.

Notice that Tukey's procedure assumes that each sample contains the same number (n_r) of replicates. Critical values of the Q statistic are contained in Table A.16. Define

$Q_{\alpha,k,v}$ = critical value of the Q statistic from Table A.16, using a significance level of α; k is the number of sample means (groups), and v is the df associated with MS(error)

Multiple Comparisons Procedure

1. Find $Q_{\alpha,k,v}$ using Table A.16.
2. Determine

$$D = Q_{\alpha,k,v} \cdot \sqrt{\frac{\text{MS(error)}}{n_r}}$$

where MS(error) is the sample variance and n_r is the number of replicates in each sample. For one-factor ANOVA, MS(error) is the same as s_p^2.
3. Place the sample means in order, from smallest to largest.
4. If two sample means differ by more than D, the conclusion is that the corresponding population means are unequal. In other words, if $|\bar{X}_i - \bar{X}_j| > D$, this implies that $\mu_i \neq \mu_j$.

To illustrate this procedure, reconsider Example 11.2. Here we concluded that the average lifetime on a test device was not the same for the four brands of batteries. The four sample means were

$$\text{Brand 1: } (\bar{x}_1 = 43.5)$$
$$\text{Brand 2: } (\bar{x}_2 = 42.0)$$
$$\text{Brand 3: } (\bar{x}_3 = 29.0)$$
$$\text{Brand 4: } (\bar{x}_4 = 30.5)$$

For this study, there were $n_r = 6$ replicates in each sample, with a resulting pooled variance of $s_p^2 = \text{MS(error)} = 31.85$. The study contained $k = 4$ groups and the df for the error sum of squares was $v = n - k = 24 - 4 = 20$. Using a significance level of .05, we begin by finding $Q_{.05,4,20}$ in Table A.16. This value is 3.96. Next we determine

$$D = Q_{.05,4,20} \cdot \sqrt{\frac{\text{MS(error)}}{n_r}}$$
$$= (3.96)\sqrt{31.85/6}$$
$$= 9.12$$

The sample means, in order, are

$$\overline{29.0, \ 30.5,} \quad \overline{42.0, 43.5}$$

Any two sample means are significantly different using the Tukey procedure if they differ by an amount greater than $D = 9.12$. Here there are four significant differences, namely,

$$\bar{x}_1 - \bar{x}_3 = 43.5 - 29.0 = 14.5 > 9.12$$
$$\bar{x}_2 - \bar{x}_3 = 42.0 - 29.0 = 13.0 > 9.12$$
$$\bar{x}_1 - \bar{x}_4 = 43.5 - 30.5 = 13.0 > 9.12$$
$$\bar{x}_2 - \bar{x}_4 = 42.0 - 30.5 = 11.5 > 9.12$$

The conclusion from the multiple comparisons analysis is that $\mu_1 \neq \mu_3$, $\mu_1 \neq \mu_4$, $\mu_2 \neq \mu_3$, and $\mu_2 \neq \mu_4$. There is no evidence of a difference between the brand 1 and the brand 2 populations or between the brand 3 and the brand 4 populations. This is indicated by the two overbars connecting these two pairs of sample means. In general, there is no evidence to indicate a difference in the population means for any group of sample means under such a bar.

One-Factor ANOVA Procedure

ASSUMPTIONS

The replicates are obtained *independently* and *randomly* from each of the populations. The value of one observation has no effect on any other replicates within the same sample or within the other samples.

The observations (replicates) from each population follow (approximately) a *normal* distribution.

The normal populations all have a *common variance*, σ^2. We expect the values in each sample to vary about the same amount. The ANOVA procedure will be much less sensitive to this assumption when we obtain samples of equal size from each population.

HYPOTHESES

$$H_0: \mu_1 = \mu_2 = \cdots = \mu_k$$

H_a: not all μ's are equal

Note that H_a is not the same as H_a': all μ's are unequal; H_a states that *at least two* of the μ's are different.

SUM OF SQUARES

$$SS(factor) = \left[\frac{T_1^2}{n_1} + \frac{T_2^2}{n_2} + \cdots + \frac{T_k^2}{n_k}\right] - \frac{T^2}{n}$$

where $n = n_1 + n_2 + \cdots + n_k$ and $T = \Sigma x = T_1 + T_2 + \cdots + T_k$.

$$SS(total) = \Sigma x^2 - \frac{T^2}{n}$$

$$SS(error) = SS(total) - SS(factor)$$

$$= \Sigma x^2 - \left[\frac{T_1^2}{n_1} + \frac{T_2^2}{n_2} + \cdots + \frac{T_k^2}{n_k}\right]$$

DEGREES OF FREEDOM

df for factor $= k - 1$

df for error $= n - k$

df for total $= n - 1$

Note that $(k - 1) + (n - k) = n - 1$.

ANOVA TABLE

SOURCE	df	SS	MS	F
Factor	$k - 1$	SS(factor)	$MS(factor) = \dfrac{SS(factor)}{k - 1}$	MS(factor)/MS(error)
Error	$n - k$	SS(error)	$MS(error) = \dfrac{SS(error)}{n - k}$	
Total	$n - 1$	SS(total)		

where MS = mean square = SS/df.

TESTING PROCEDURE

reject H_0 if $F^* > F_{\alpha,k-1,n-k}$

where $F_{\alpha,k-1,n-k}$ is obtained from Table A.7.

Exercises

11.1 A shoe manufacturer wanted to test whether there is a difference in the amount of wear on three different designs of rubber soles for a particular jogging shoe. Eighteen joggers were selected for the experiment. Each type of design was randomly assigned to six joggers. After running 200 miles, the joggers turned in their shoes. The manufacturer used an index to indicate the amount of rubber left on the sole. The measures obtained were:

Design 1	3.2	4.1	6.2	5.3	4.9	3.5
Design 2	4.7	6.3	4.0	5.4	7.1	4.5
Design 3	3.9	6.0	5.5	4.2	3.1	5.1

a. State the null and alternative hypotheses.

b. What assumptions are necessary to use the ANOVA procedure on these data?

c. What are the point estimates of the mean wear for each of the three designs?

d. What is the within-groups mean square?

e. Set up an ANOVA table and state the conclusion. Use a significance level of .05.

f. Using a significance level of .05, perform a multiple comparisons procedure, if appropriate.

11.2 In Exercise 11.1, subtract 3.0 from each of the observations in the table. Perform the ANOVA procedure. Are the sum of squares the same for the coded data as for the original data? Why or why not? What happens when any set of data is coded by adding or subtracting the same number to each observation value in terms of the sum of squares?

11.3 A manufacturer introduces a new car that gets 40 mpg with a standard deviation of 3. The manufacturer's competitor introduces a similar economy car and claims it also gets the same mpg with the same standard deviation. A random sample of 15 observations of mpg is taken for each manufacturer's car. Is there any difference in the mean mpg for these two cars? Use the ANOVA procedure and a significance level of .05.

Manufacturer	41	40	40	39	36	41	40	42	42	39	40	41	39	38	41
Competitor	38	39	37	40	42	43	41	39	38	37	37	38	39	39	38

11.4 Use the two-sample t test for the data in Exercise 11.3 to test for any difference in the mean mpg for these two cars. What is the relationship between the t test of this exercise and the F test of Exercise 11.3?

11.5 The science of ergonomics studies the influence of "human factors" in technology, i.e., how human beings relate to and work with machines. With the widespread use of computers for data processing, computer scientists and psychologists are getting together to study human factors. One typical study investigated the productivity of secretaries with different word processing programs. An identical task was given to 18 secretaries, randomly allocated to three groups. Group 1 used a primarily menu-driven program, Group 2 used a command-driven program, and Group 3 was a mixture of both approaches. The secretaries all had about the same level of experience, typing speed, and computer skills. The time (in minutes) taken to complete the task was observed. The results were as follows:

GROUP 1 (MENU-DRIVEN)	GROUP 2 (COMMAND-DRIVEN)	GROUP 3 (MIXED)
12	14	10
15	11	8
11	13	9
12	12	10
10	11	7
13	14	8

a. Do the necessary calculations to construct an ANOVA table, and test the hypothesis that there is no difference between the three types of word processing programs (i.e., on the average, the time taken to complete the task is about the same). Use $\alpha = .05$.

b. State the p-value for the test.

c. Does the type of word processing software used affect the performance of the secretaries?

d. If the secretaries had different levels of experience, typing speed, and computer skills, how would it affect the data? (Would it be an extraneous source of variation, or "noise"? Would it tend to increase the "within-sample" variation, the "between-sample" variation, both, or neither?)

e. Using a significance level of .05, perform a multiple comparisons procedure (if appropriate).

11.6 A small engine-repair shop can special-order parts from any one of three different warehouses and receive a substantial discount on the price. The manager of the shop is concerned with the length of time that it takes to special-order a part from one of the warehouses. The number of days it takes to special-order a part is recorded for 15 randomly selected orders from each of the three warehouses, as shown in the following table. Do the data indicate that there is a difference in the mean times that it takes to special-order a part from a warehouse? Use a .05 significance level. State the p-value.

WAREHOUSE

A	13	17	14	10	9	15	18	11	13	18	16	13	15	12	16
B	7	12	8	15	6	10	12	10	8	14	10	6	9	13	11
C	10	12	18	19	9	15	20	11	15	13	17	13	10	14	16

11.7 A sales manager wanted to know whether there was a significant difference in the monthly sales of three sales representatives. John is strictly on commission. Randy is on commission and a small salary, and Ted is on a small commission and a salary. Eight months were chosen at random. The data represent monthly sales.

John	969	905	801	850	910	1030	780	810
Randy	738	773	738	805	850	800	690	720
Ted	751	764	701	810	840	790	720	735

a. Using a significance level of .05, test the hypothesis that there is no difference in the mean monthly sales. (Coding the data may make the computations easier.)

b. What is the p-value?

c. Using a significance level of .05, perform a multiple comparisons procedure, if appropriate.

11.8 An instructor wanted to test whether there was a difference in the effectiveness of four different teaching techniques. Four groups of students were taught using one of the four teaching techniques. If the instructor examined the groups for mean differences one pair at a time, how many t tests would have to be performed? What is the advantage of using an ANOVA procedure instead?

11.9 Astral Airlines recently introduced a nonstop flight between Houston and Chicago. The vice president of marketing for Astral decided to run a test to see whether Astral's passenger load was similar to that of its two major competitors. Ten daytime flights were picked at random from each of the three airlines and the percent of unfilled seats on each flight was as follows:

Astral	10	14	12	10	8	13	11	8	12	9
Competitor 1	12	9	8	9	9	10	12	7	11	10
Competitor 2	15	10	15	8	14	9	8	11	10	12

Use a significance level of .05 and perform an ANOVA procedure. Find the p-value.

11.10 What assumptions do the data need to satisfy in Exercise 11.9 to ensure that the ANOVA procedure is valid?

11.11 After performing an ANOVA procedure on three groups, an analyst found that a significant difference existed at the .01 significance level. From the following statistics, perform a multiple comparisons procedure, using a significance level of .01:

$$\bar{x}_1 = 18.21, \ \bar{x}_2 = 19.14, \ \bar{x}_3 = 14.97, \ n_1 = 7, \ n_2 = 7, \ n_3 = 7, \ MSE = 2.8$$

11.12 The workers at a calculator assembly plant wish to bargain for more breaks during the workday. The manager believes that increasing the number of 15-minute breaks will affect productivity. The workers currently receive three breaks during the 8-hour work day. The manager decides to run a test by choosing four groups of five workers each and giving one group three breaks, the next group four breaks, and so on. The number of calculators assembled per day is recorded for five days. Test the manager's claim using an ANOVA procedure with a .10 significance level. Find the p-value. Using a significance level of .05, perform a multiple comparisons procedure, if appropriate.

3 breaks	200	205	197	210	205
4 breaks	210	203	201	197	199
5 breaks	198	190	185	188	180
6 breaks	197	180	190	192	175

11.13 Computer Aided Software Engineering (CASE) is a broad group of software technologies that support the automation of computer systems development. An experiment was conducted to evaluate first-time users' ease of use of CASE tools for drawing Data Flow Diagrams (DFDs). The null hypothesis (no difference between DFDs prepared manually and those prepared with CASE software by first-time users) was tested using two treatment groups. Twenty-six subjects were randomly assigned to each of two treatment groups: the manual treatment group and the treatment group with CASE software. Results from the subjects were scored on completeness. The two-group t test was calculated to be 3.58. What would the F statistic be if an ANOVA procedure was used? Find the p-value based on the F statistic.

(*Source:* G. Baram, G. Steinberg, and John Nosek, "Evaluation Of Ease of Use of CASE Tools By First-Time Users," *Proceedings of the 1990 Annual Meeting of the Decision Sciences Institute:* 934–936.)

11.14   Independent samples of size 16 are drawn from each of four normally distributed populations. The resulting sample standard deviations are: $s_1 = 2.0$, $s_2 = 2.5$, $s_3 = 2.2$, $s_4 = 2.5$. Do the data provide sufficient evidence that a significant difference exists in the population standard deviations of the four populations? Use a .05 significance level.

11.15   Three machines package 50-pound sacks of pinto beans. A preliminary test is performed using data from a pilot study to determine whether a significant difference exists among the variances of the amount of beans packaged for each machine. Use a 5% significance level in conducting a test for equality of variances for each of the machines from the following sample data (in pounds):

MACHINE 1	MACHINE 2	MACHINE 3	MACHINE 1	MACHINE 2	MACHINE 3
52	50	48			
51	49	46	49	50	52
48	51	51	56	49	50
50	50	50	51	51	51
46	52	52	56	50	49
55	53	50	45	49	50
53	55	51	50	48	50

11.16   A sales manager would like to determine whether there is a significant difference in the variance of the sales of three salespersons. Three independent samples of daily sales (in hundreds of dollars) are collected. Using a 5% significance level, determine whether the data indicate a difference in the variance of the sales of the three salespersons.

SALESPERSON 1	SALESPERSON 2	SALESPERSON 3	SALESPERSON 1	SALESPERSON 2	SALESPERSON 3
1.2	2.5	1.4	1.0	2.2	1.7
1.1	2.1	1.8	1.3	2.3	1.3
1.4	2.3	1.5	1.8	2.4	1.2
1.6	2.0	1.6	1.4	2.3	1.5
1.4	2.3	1.4	1.5	2.5	1.6

11.17   A manager wishes to know if there is a significant difference in the quality of a certain grade of plastic that is produced on each of three work shifts. Defects are recorded as the number of defects per 100 square feet. Ten inspection lots are sampled from each of the three shifts, and the data are presented below. Do the data provide sufficient evidence to indicate that a difference exists in the quality of the plastic rolls produced by the three work shifts? Use a .05 significance level.

SHIFT 1	SHIFT 2	SHIFT 3	SHIFT 1	SHIFT 2	SHIFT 3
8	7	13	9	16	16
12	6	11	20	8	23
15	13	8	14	19	10
10	15	17	13	10	13
13	5	21	18	11	17

11.18   In a small company, upper management wants to know if there is a difference in the three types of methods used to train its machine operators. One method uses a hands-on approach but is very expensive. A second method uses a combination of classroom instruction and some on-the-job training. The third method is the least expensive and is confined completely to the classroom. Eight trainees are assigned to each training technique. The following table gives the results of a test administered after completion of the training. Do the data provide sufficient evidence to indicate a difference in the methods of training at the .01 level of significance?

METHOD 1	METHOD 2	METHOD 3	METHOD 1	METHOD 2	METHOD 3
95	85	88	81	93	81
100	90	94	85	86	84
90	95	90	96	94	90
91	88	80	95	95	87

◢ 11.3
Designing an Experiment

The previous section introduced you to one-factor (or one-way) ANOVA. In this type of analysis, you randomly obtain samples from each of the k populations (levels) describing a single factor. The variable that is being measured (such as battery lifetime) is referred to as the **dependent** variable. Since replicates (repeat observations) are obtained in a completely random manner from each population, this type of sampling plan is called a **completely randomized design.** This section discusses other experimental designs, including the randomized block design and the two-way factorial design.

Suppose that the personnel director at Blackburn Industries is interested in examining the cost of dental claims filed by Blackburn employees. Let us consider using the amounts of these claims to examine various group differences; we will need to determine what type of design would be appropriate for each situation.

Situation 1: The Completely Randomized Design

One question of interest to the personnel director is whether the average annual amount claimed on the dental insurance plan differs among the four employee classifications. These classifications range from category 1 (consisting of production-line workers) to category 4 (consisting of upper-level management). Replicates are obtained randomly within each population (category), and the four samples are not related in any way. This illustration consists of one factor (employee classification) consisting of four levels. The question of interest here is, is there a difference in the average annual dental claims among the four types of employees? The corresponding null hypothesis is

$$H_0: \mu_1 = \mu_2 = \mu_3 = \mu_4$$

Essentially, this type of analysis (called one-way ANOVA) will fail to reject H_0 if the sample means are "close together" and reject this hypothesis otherwise.

Situation 2: The Randomized Block Design

Suppose instead that the personnel director at Blackburn Industries wished to investigate family dental claims. In particular, she wished to know if there was a difference in the amounts claimed (1) by the husband, (2) by the wife, and (3) per child in the family. For the study she randomly selected 15 (or however many) families having at least one child and recorded these three amounts for each family. This is an example of a **randomized block design.** The configuration of the sample results would resemble the following scheme, where each x represents a dollar amount.

FAMILY	HUSBAND	WIFE	PER CHILD	
1	x	x	x	(1st block)
2	x	x	x	(2nd block)
⋮				
15	x	x	x	(15th block)

This design consists of one *factor* with three levels (husband/wife/per child). Unlike the completely randomized design, the three samples *are not independent*, since the data are grouped (blocked) by family. For example, the first husband value is not independent of the first wife value, since they both belong to the same family. We encountered this very same design in Chapter 9, where we compared two population means using paired (that is, blocked) samples. When using the randomized block design, you can compare the means of more than two populations using a blocking strategy to gather your data.

The question of interest here is, is there a difference in the husband, wife, and per-child claims? In a situation similar to that of the completely randomized design, the question of interest is whether the factor of interest (family member type, here) has a significant effect on the value of the dependent variable (amount of the annual claim, here). *The difference between the randomized block design and the completely randomized situation is that here we use a blocking strategy rather than independent samples to obtain a more precise test for examining differences in the factor level means.* The null hypothesis for this illustration is that the group (factor level) means are identical, that is,

$$H_0: \mu_H = \mu_W = \mu_C$$

where μ_H is the average annual amount claimed by the husband, μ_W is the average amount for the wife, and μ_C is the average amount claimed per child.

The analysis for the randomized block design is discussed in the next section, but essentially this procedure removes the effects of the blocks (families, here) before testing for a difference between the factor level means. Consequently, this design removes the block effect from the error sum of squares in the completely randomized design. Several examples in the next section illustrate this technique.

Situation 3: The Two-Way Factorial Design

The **two-way factorial design** is very similar to situation 1, the one-way analysis of variance, except now *two* factors are of interest to the individual conducting the study. Suppose that the personnel director at Blackburn Industries decides to examine the dental claims for all the unmarried employees. She wants to investigate the effect of sex (factor A) and employee classification (factor B) on the amount of the annual claims. As before, employee classification ranges from category 1 (production-line workers) to category 4 (upper-level management). The previous one-way ANOVA illustration examined only the effect of employee classification on the amount of dental claims. The inclusion of the sex factor accomplishes two things: first, you can determine if the sex of the employee has an effect on the amount of the annual claim, and second, you can investigate whether the relationship between employee classification and the amount of the annual claim is different for male and female employees. In other words, whether factor B relates to the dependent variable (amount of annual dental claim) depends on the level of factor A. This type of effect is called **interaction** between factors A and B. *This differs from the randomized block design, where it is assumed that no interaction is present between the factor of interest and the blocks.*

Consequently, there are three sets of hypotheses that can be tested using the two-way factorial design. The corresponding null hypotheses are:

$H_{0,A}$: factor A (sex) is not significant

$H_{0,B}$: factor B (employee classification) is not significant

$H_{0,AB}$: there is no interaction between factor A and factor B

The first two hypotheses are similar to those tested in one-way ANOVA, but the third hypothesis is unique to the two-way (or higher) factorial design.

■ FIGURE 11.6
Illustration of a 2 × 4
factorial design using
three replicates.

		Factor B			
		Category 1	Category 2	Category 3	Category 4
Factor A	Male	x, x, x	x, x, x	x, x, x	x, x, x
	Female	x, x, x	x, x, x	x, x, x	x, x, x

To collect data for this design, the personnel director would obtain an amount for *every* combination of a factor A level and a factor B level. This particular illustration consists of two levels for factor A (male/female) and four levels for factor B (category 1/. . ./category 4) and is called a **2 × 4 factorial design.** Consequently, data are collected for eight possible factor level combinations, referred to as **treatments.** Data values within the same treatment are termed **replicates.** Furthermore, it is necessary when using this type of design *to obtain more than one replicate for each treatment.* An illustration using three replicates (two would be sufficient) is shown in Figure 11.6. Each *x* represents the amount of the annual dental claim. The actual two-way analysis of variance for this illustration is demonstrated in Section 11.5.

Exercises

11.19   **a.**   Name the three types of experimental design discussed in the preceding section.

b.   Which design or designs do not involve the necessity of having replicates?

c.   What is "interaction" between factors? Which design permits testing for interaction?

d.   In each of the three designs, how many dependent variables are there?

e.   If a one-way (one-factor) ANOVA design has four levels of the factor and six replicates at each level, how many treatments are being considered and how many total observations are made?

f.   If a 4 × 6 two-way factorial design is chosen for an ANOVA-based experiment, what are the number of treatments being considered and the *minimum* number of total observations necessary?

g.   What is the advantage of blocking and when is it necessary?

h.   What are some potential problems that can arise with the randomized block design?

11.20   An appraisal firm is interested in the amount of time it takes appraisers to appraise fairly new homes with market values between $100,000 and $150,000. An experiment is set up in which the factor is experience of the appraiser. Three levels are used: less than 1 year of experience, 1 year to 3 years of experience, and over 3 years of experience.

a.   Which design is appropriate to test whether there is a significant difference in the levels of experience with respect to the amount of time that it takes an appraiser to appraise a house? Construct a diagram to show how the data would look.

b.   Suppose that a second factor is added. Assume that the firm is interested in whether the price of the home affects the appraisal time. Two levels are considered: houses with market values between $50,000 and $100,000 and houses with market values between $100,000 and $150,000. Construct a diagram to show the setup of the data. What is the appropriate experimental design if interaction exists between the factors?

11.21   Suppose a supervisor is interested in the productivity of three different machines used to assemble electronics components.

a.   What design would be appropriate to test whether there is a significant difference in the productivity of the machines?

b.   Suppose that 20 operators use the machines over three shifts to assemble the components. How can a randomized block design be used?

c.   How can one be assured that observations within each block are randomized?

11.22   Explain what observations are independent in a completely randomized design. Are all observations independent in a randomized block design?

◢ *11.4*
Randomized Block Design

The previous section described the difference between the randomized block and completely randomized designs. Rather than obtaining independent samples from the k populations, the data for a randomized block design are organized into homogeneous units referred to as **blocks.** *Within* each block, any predictable difference in the observations is due to the effect of the factor of interest, such as sex or employee classification.

Consider the situation discussed in Section 9.5. Metalloy manufactures metal hinges, and as part of the quality inspection, these hinges are subjected to a hardness test in which a rod with a pointed tip is pressed into a hinge and the depth of the depression caused by the tip is measured. Two tips are available for the hardness tester. Twelve hinges were randomly selected, and both tips were used to test the hardness of each metal hinge. The coded data from this hardness test are as shown below:

HINGE	TIP #1	TIP #2	HINGE	TIP #1	TIP #2
1	39	35	7	45	41
2	32	34	8	48	47
3	42	38	9	38	35
4	49	48	10	48	46
5	45	47	11	41	37
6	47	43	12	47	44

Once again, there is a single factor of interest, namely, the tip used for the test. But are the 24 sample observations in fact replicates? Replicates (by definition) are obtained under (nearly) identical circumstances, so that any variation in the values within any one sample is due strictly to chance. In this situation, the hinge used in the test heavily influences each sample value. Also, each pair of values is obtained from the same hinge, and so these samples are *not independent,* violating a key assumption of the completely randomized design.

This situation fits the *paired sample* design discussed in Section 9.5, provided we assume that the population of differences (tip #1 minus tip #2) is normally distributed. We will now extend our discussion of Section 9.5 to consider more than two populations.

To determine whether there is a factor (tip) effect in these 24 observations, we must first account for the block (hinge) effect. If we ignore this effect, we could easily come to an incorrect conclusion regarding a difference in the effect of the two tips. This same point was made in Section 9.5, where a crucial question was whether to pair (block) the data. Figures 9.13 and 9.14 illustrated how one can arrive at an incorrect conclusion by failing to block the data.

Metalloy also makes a larger, softer hinge that requires a hardness test as part of the quality inspection. For this test, there are three tips available for inserting into the rod that is then pressed into the hinge with a specified force. The question of interest, as before, is whether there is a difference in the average reading for the three tips. Using ten hinges, an experiment was conducted in which all three tips were applied to each of the metal hinges. The results were:

HINGE	TIP #1	TIP #2	TIP #3	HINGE	TIP #1	TIP #2	TIP #3
1	68	72	65	6	80	91	86
2	40	43	42	7	47	58	50
3	82	89	84	8	55	68	52
4	56	60	50	9	78	77	75
5	70	75	68	10	53	65	60

The data from this situation constitute a randomized block design with a single factor (tip used in the rod) containing three levels (tip #1, tip #2, tip #3) as well

Block	FACTOR LEVEL (POPULATION)					Total
	1	2	3	$\cdots$	k	
1	x	x	x	$\cdots$	x	S_1
2	x	x	x	$\cdots$	x	S_2
3	x	x	x	$\cdots$	x	S_3
$\vdots$						$\vdots$
b	x	x	x	$\cdots$	x	S_b
Total	T_1	T_2	T_3	$\cdots$	T_k	T
Sample Mean	$\bar{X}_1$	$\bar{X}_2$	$\bar{X}_3$	$\cdots$	$\bar{X}_k$	

as 10 blocks (hinge 1, $\cdots$, hinge 10). The general appearance of such a design is shown in Table 11.1.

When using the randomized block design, the various levels should be applied in a *random* manner within each block. In the hardness testing, the test should *not* always use tip #1 first, tip #2 second, and tip #3 last. Instead, the three tips should be applied in a randomized order for each hinge—hence the name *randomized block design*.

The assumptions for the randomized block design are:

1. The observations within each factor level/block combination are obtained from a normal population.
2. These normal populations have a common variance, σ^2.

Furthermore, we assume that the factor effects are the same within each block; that is, there is no interaction effect between the factor and the blocks.

The analysis using the randomized block design is similar to that for the one-factor ANOVA, except that the total sum of squares (SS(total)) has an additional component. Now,

$$SS(\text{total}) = SS(\text{factor}) + SS(\text{blocks}) + SS(\text{error})$$

where SS(blocks) measures the variation due to the blocks. Consequently, this design extracts the block effect, as measured by SS(blocks), from the error of sum of squares in the completely randomized design.

If you use the randomized block design when blocking is not necessary, SS(blocks) will be very small in comparison to the other sums of squares. Referring to Table 11.1, this will occur when $S_1, S_2, \ldots, S_b$ are nearly the same. If all the S_i's are equal, then SS(blocks) = 0. The effect of the blocks will be significant whenever you observe a lot of variation in these block totals.

The sum of squares for the randomized block design is thus

$$SS(\text{factor}) = \frac{1}{b}\left[T_1^2 + T_2^2 + \cdots + T_k^2 \right] - \frac{T^2}{bk} \qquad (11.21)$$

where

$n =$ number of observations $= bk$

$T_1, T_2, \ldots, T_k$ represent the totals for the k factor levels

$S_1, S_2, \ldots, S_b$ are the totals for the b blocks

$T = T_1 + T_2 + \cdots + T_k$

$\quad = S_1 + S_2 + \cdots + S_b =$ total of all observations.

$$SS(blocks) = \frac{1}{k}\left[S_1^2 + S_2^2 + \cdots + S_b^2\right] - \frac{T^2}{bk} \qquad (11.22)$$

$$SS(total) = \Sigma x^2 - \frac{T^2}{bk} \qquad (11.23)$$

where $\Sigma x^2 =$ sum of the squares for each of the $n\,(=bk)$ observations.

$$SS(error) = SS(total) - SS(factor) - SS(blocks) \qquad (11.24)$$

The degrees of freedom are

$$df \text{ for factor} = k - 1$$
$$df \text{ for blocks} = b - 1$$
$$df \text{ for error} = (k - 1)(b - 1)$$
$$df \text{ for total} = bk - 1$$

The ANOVA table for a blocked design is very similar to the one-factor ANOVA table. There is one additional row because you now include the effect of the various blocks in your design:

SOURCE	df	SS	MS	F
Factor	$k - 1$	SS(factor)	$MS(factor) = \dfrac{SS(factor)}{k - 1}$	$F_1 = MS(factor)/MS(error)$
Blocks	$b - 1$	SS(blocks)	$MS(blocks) = \dfrac{SS(blocks)}{b - 1}$	$F_2 = MS(blocks)/MS(error)$
Error	$(k - 1)(b - 1)$	SS(error)	$MS(error) = \dfrac{SS(error)}{(k - 1)(b - 1)}$	
Total	$bk - 1$	SS(total)		

where

$$MS(factor) = \frac{SS(factor)}{(k - 1)}$$

$$MS(blocks) = \frac{SS(blocks)}{(b - 1)}$$

$$MS(error) = \frac{SS(error)}{(k - 1)(b - 1)}$$

Hypothesis Testing

Is there a difference in the average reading for the three tips in the previous illustration? In other words, does the tip used have a significant effect on the hardness reading? The hypotheses for this situation are $H_0: \mu_1 = \mu_2 = \mu_3$ and H_a: not all the means are equal. We determine the test statistic exactly as we did for the one-factor ANOVA:

$$F_1 = \frac{MS(factor)}{MS(error)} \qquad (11.25)$$

where the mean square values are obtained from the ANOVA table. Notice that the MS(factor) value is the *same* regardless of whether you block. However, when you use the block effect in the design, the SS(error) is smaller than the SS(error) value obtained using the completely randomized design. The df in the error term are different for the two designs, so it does not necessarily follow that the MS(error) is smaller in the randomized block design. If there is considerable variation among the block totals, then quite likely MS(error) *will* be smaller in the randomized block design. You are thus more likely to detect a difference in these k means when a difference does exist. Had you not included the block effect, the block variation would have been included in SS(error), resulting in a smaller F-value. This value often becomes small enough not to fall in the rejection region, leading you to conclude that no difference exists. *But perhaps there is a difference among these means (the factor does have a significant effect) that will go undetected if an incorrect experimental design is used.*

For the randomized block design, the test will be to

$$\text{reject } H_0 \text{ if } F_1 > F_{\alpha, v_1, v_2}$$

where $v_1 = k - 1$ and $v_2 = (k - 1)(b - 1)$. So, once again, we reject H_0 if the F-statistic falls in the right-tail rejection region, this time using Table A.7 with $v_1 = k - 1$, the df for factor, and $v_2 = (k - 1)(b - 1)$, the df for error.

Now suppose we wish to determine whether the effect of the hinge used for the hardness test is significant. We are attempting to determine whether there is a block effect, so the hypotheses are

H_0': there is no effect due to the hinges (blocks) (the block means are equal)

H_a': there is an effect due to the hinges (the block means are not all equal)

The corresponding test uses the "other" F statistic from the randomized block ANOVA table, namely,

$$F_2 = \frac{\text{MS(blocks)}}{\text{MS(error)}} \qquad (11.26)$$

and the test procedure is to

$$\text{reject } H_0' \text{ if } F_2 > F_{\alpha, v_1', v_2'}$$

where $v_1' = b - 1$ and $v_2' = (k - 1)(b - 1)$.

Let us reexamine our data. We will use $\alpha = .05$. Here, $k = 3$ levels (tips), $b = 10$ blocks (hinges), and $n = bk = 30$ observations.

HINGE	TIP #1	TIP #2	TIP #3	TOTALS
1	68	72	65	205
2	40	43	42	125
3	82	89	84	255
4	56	60	50	166
5	70	75	68	213
6	80	91	86	257
7	47	58	50	155
8	55	68	52	175
9	78	77	75	230
10	53	65	60	178
Total	629	698	632	1959
$\overline{X}$	62.9	69.8	63.2	

$$SS(\text{factor}) = \frac{1}{10}\left[629^2 + 698^2 + 632^2\right] - \frac{1959^2}{30}$$

$$= 128,226.9 - 127,922.7$$

$$= 304.2$$

$$SS(\text{blocks}) = \frac{1}{3}\left[205^2 + 125^2 + \cdots + 178^2\right] - \frac{1959^2}{30}$$

$$= 133,627.7 - 127,922.7$$

$$= 5705.0$$

$$SS(\text{total}) = \left[68^2 + 40^2 + \cdots + 75^2 + 60^2\right] - \frac{1959^2}{30}$$

$$= 134,107 - 127,922.7$$

$$= 6184.3$$

$$SS(\text{error}) = SS(\text{total}) - SS(\text{factor}) - SS(\text{blocks})$$

$$= 6184.3 - 304.2 - 5705.0$$

$$= 175.1$$

So

$$MS(\text{factor}) = \frac{SS(\text{factor})}{(k-1)}$$

$$= \frac{304.2}{2} = 152.1$$

$$MS(\text{blocks}) = \frac{SS(\text{blocks})}{(b-1)}$$

$$= \frac{5705.0}{9} = 633.9$$

$$MS(\text{error}) = \frac{SS(\text{error})}{(k-1)(b-1)}$$

$$= \frac{175.1}{18} = 9.73$$

The resulting ANOVA table is

SOURCE	df	SS	MS	F
Factor	2	304.2	152.1	$152.1/9.73 = 15.63\ (F_1)$
Blocks	9	5705.0	633.9	$633.9/9.73 = 65.15\ (F_2)$
Error	18	175.1	9.73	
Total	29	6184.3		

We first consider the hypotheses

$$H_0: \mu_1 = \mu_2 = \mu_3$$

$$H_a: \text{not all } \mu\text{'s are equal}$$

where

$$\mu_1 = \text{average reading using tip \#1 (estimate is } \bar{x}_1 = 62.9)$$

$$\mu_2 = \text{average reading using tip \#2 (estimate is } \bar{x}_2 = 69.8)$$

$$\mu_3 = \text{average reading using tip \#3 (estimate is } \bar{x}_3 = 63.2)$$

■ **FIGURE 11.7**
MINITAB solution for
hardness testing data.

```
MTB > READ INTO C1-C3
DATA> 68 1 1
DATA> 40 1 2
DATA> 82 1 3
DATA> 56 1 4
DATA> 70 1 5
DATA> 80 1 6
DATA> 47 1 7
DATA> 55 1 8
DATA> 78 1 9
DATA> 53 1 10
DATA> 72 2 1
DATA> 43 2 2
DATA> 89 2 3
DATA> 60 2 4
DATA> 75 2 5
DATA> 91 2 6
DATA> 58 2 7
DATA> 68 2 8
DATA> 77 2 9
DATA> 65 2 10
DATA> 65 3 1
DATA> 42 3 2
DATA> 84 3 3
DATA> 50 3 4
DATA> 68 3 5
DATA> 86 3 6
DATA> 50 3 7
DATA> 52 3 8
DATA> 75 3 9
DATA> 30 3 10
DATA> END
      30 ROWS READ

MTB > NAME C1 = 'HARDNESS' C2 = 'TIP' C3 = 'HINGE'
MTB > TWOWAY USING DATA IN C1, LEVELS IN C2, BLOCKS IN C3

ANALYSIS OF VARIANCE  HARDNESS

SOURCE      DF       SS        MS
TIP          2    304.20    152.10
HINGE        9   5704.97    633.89
ERROR       18    175.13      9.73
TOTAL       29   6184.30
```

C_2 contains factor (tip) levels

C_3 contains block values: 1 = hinge 1
                             2 = hinge 2

Because $F_1 = 15.63 > F_{.05,2,18} = 3.55$, we reject H_0 and conclude that there *is* a difference in the hardness readings for the three tips. This is not a surprising result, because it appears that the tip #2 readings were much higher than for tips #1 and #3. This means that the factor (tip used) *does* have a significant effect on the hardness reading.

We also wish to test

$$H_0': \text{there is no block effect}$$

$$H_a': \text{there is a block effect}$$

$S_1, S_2, \ldots, S_{10}$ appear to contain considerable variation, so our initial guess is that there is a block effect. Carrying out the statistical test, we see that

$$F_2 = 65.15 > F_{.05,9,18} = 2.46$$

Consequently, we strongly reject H_0' in favor of H_a'. The effect of the hinge used for the hardness test *is* significant.

A MINITAB solution for this problem is shown in Figure 11.7.

COMMENTS   Notice that the block (hinge) effect is highly significant here, indicating much variation in the totals for each row (hinge). This indicates that there is very little uniformity in the hardness of the hinges being produced, regardless of what tip is used for measurement. Consequently, Metalloy has a serious problem in the quality of the hinges being produced. As Chapter 12 ("Quality Control") will point out, one indicator of poor quality is too much variation in the quality characteristic being measured (hardness, here).

What would the result have been had we treated these 30 observations as replicates, ten from each of the three tips? In other words, what would happen if we failed to recognize that blocking was necessary and we incorrectly used the one-factor ANOVA? Because both SS(factor) and SS(total) do not change, the only difference is a new SS(error). Therefore,

$$SS(error) = SS(total) - SS(factor)$$
$$= 6184.3 - 304.2$$
$$= 5880.1$$

Also, in the one-factor ANOVA design,

$$df \text{ for total} = (df \text{ for factor}) + (df \text{ for error})$$

So,

$$df \text{ for error} = (df \text{ for total}) - (df \text{ for factor})$$
$$= 29 - 2 = 27$$

The resulting F-value will be

$$F = \frac{MS(factor)}{MS(error)} = \frac{304.2/2}{5880.1/27} = .70$$

Because $F^* = .70$ is *much less* than $F_{.05,2,27} = 3.35$, *we fail to detect a difference in the three means,* μ_1, μ_2, and μ_3. This is the effect of assuming independence among the samples when it does not exist. This example emphasizes that failing to recognize the need for a randomized block design can have serious consequences!

EXAMPLE 11.6 The personnel director at Blackburn Industries is investigating dental claims submitted by married employees having at least one child. Of interest is whether the average annual dollar amounts of dental work claimed by the husband, by the wife, and per child are the same. Data were collected by randomly selecting 15 families and recording these three dollar amounts (total claims for the year by the husband, by the wife, and per child). The results of the sample are shown below. Can the personnel director conclude that there is a difference in the three population means using a significance level of .05?

FAMILY	HUSBAND	WIFE	PER CHILD	TOTAL
1	78	84	112	274
2	105	80	274	459
3	95	184	305	584
4	85	158	280	523
5	148	180	263	591
6	284	208	145	637
7	124	145	340	609
8	118	75	130	323
9	153	112	239	504
10	143	204	262	609
11	84	110	182	376
12	106	172	248	526
13	218	185	320	723
14	145	90	226	461
15	175	304	152	631
Total	2061	2291	3478	7830
$\bar{X}$	$ 137.40	$ 152.73	$ 231.87	

SOLUTION   The hypotheses for this situation are

$$H_0: \mu_H = \mu_W = \mu_C$$

H_a: not all three means are equal

where μ_H is the average annual amount claimed by the husband, μ_W is the average for the wife, and μ_C is the average amount per child.

The various block and factor totals shown are obtained by summing across and down the array of data.

$$SS(\text{factor}) = \frac{1}{15}[(2061)^2 + (2291)^2 + (3478)^2] - \frac{7830^2}{45}$$

$$= 1{,}439{,}525.73 - 1{,}362{,}420 = 77{,}105.73$$

$$SS(\text{blocks}) = \frac{1}{3}[(274)^2 + (459)^2 + \cdots + (631)^2] - \frac{7830^2}{45}$$

$$= 1{,}435{,}654 - 1{,}362{,}420 = 73{,}234$$

$$SS(\text{total}) = [(78)^2 + (105)^2 + \cdots + (226)^2 + (152)^2] - \frac{7830^2}{45}$$

$$= 1{,}610{,}430 - 1{,}362{,}420 = 248{,}010$$

$$SS(\text{error}) = SS(\text{total}) - SS(\text{factor}) - SS(\text{blocks})$$

$$= 97{,}670.27$$

The corresponding ANOVA table follows. Note that the df for the factor are $3 - 1 = 2$, for blocks are $15 - 1 = 14$, and for total are $45 - 1 = 44$, leaving $44 - 2 - 14 = 28$ df for error.

SOURCE	df	SS	MS	F
Factor	2	77,105.73	38,552.87	$F_1 = 38552.87/3488.22 = 11.05$
Blocks	14	73,234	5,231	$F_2 = 5231/3488.22 = 1.50$
Error	28	97,670.27	3,488.22	
Total	44	248,010		

To test $H_0: \mu_H = \mu_W = \mu_C$, we use $F_1 = 11.05$. Since $11.05 > F_{.05,2,28} = 3.34$, we reject H_0 and conclude that the three average claim amounts are not equal. By observing the sample means, we notice that the claims per child are considerably higher than those for the husband and wife.

As a final note, the block (family) effect is *not* significant here, since $F_2 = 1.50 < F_{.05,14,28} \cong 2.04$ (using 15 and 28 df). This does not mean that including the block effect in the analysis was a mistake, since the samples were clearly *not* obtained independently. Furthermore, there is no guarantee that the block effect will once again turn out to be insignificant for the next set of data in this situation.   ∎

Constructing a Confidence Interval for the Difference Between Two Population Means

We can construct a confidence interval for the difference between any pair of means, $\mu_i - \mu_j$. Remember, however, that we must determine which confidence intervals we will construct *before* observing the data. Do not fall into the trap of letting the data dictate which confidence intervals you construct.

When using the randomized block design, our estimate of the common variance, σ^2, is now

$$s^2 = \text{estimate of } \sigma^2$$

$$= \text{MS(error)} = \frac{\text{SS(error)}}{(k-1)(b-1)} \qquad \textbf{(11.27)}$$

Thus, a $(1 - \alpha) \cdot 100\%$ confidence interval for $\mu_i - \mu_j$ is

$$(\bar{X}_i - \bar{X}_j) - t_{\alpha/2,\text{df}} \cdot s \cdot \sqrt{\frac{1}{b} + \frac{1}{b}} \quad \text{to}$$

$$(\bar{X}_i - \bar{X}_j) + t_{\alpha/2,\text{df}} \cdot s \cdot \sqrt{\frac{1}{b} + \frac{1}{b}} \qquad \textbf{(11.28)}$$

where df = degrees of freedom for the t-statistic (Table A.5) = $(k-1)(b-1)$; b = number of blocks; k = number of factor levels; and s is determined from equation 11.27.

EXAMPLE 11.7

Assume you have not yet observed the dental claim data in Example 11.6, and you decided to construct a 95% confidence interval for the difference between the average annual claim for the wife and the average annual claim per child. What does this confidence interval tell you?

SOLUTION

Using the ANOVA table for these data,

$$s^2 = \text{MS(error)} = 3488.22$$

and so $s = 59.06$. Also, $t_{.025,28} = 2.048$ using Table A.5. The resulting 95% confidence interval for $\mu_C - \mu_W$ is

$$(\bar{X}_C - \bar{X}_W) - t_{.025,28}\, s \sqrt{\frac{1}{15} + \frac{1}{15}} \quad \text{to} \quad (\bar{X}_C - \bar{X}_W) + t_{.025,28}\, s \sqrt{\frac{1}{15} + \frac{1}{15}}$$

$$= (231.87 - 152.73) - (2.048)(59.06)(.365) \quad \text{to}$$

$$(231.87 - 152.73) + (2.048)(59.06)(.365)$$

$$= 79.14 - 44.15 \quad \text{to} \quad 79.14 + 44.15$$

$$= 34.99 \quad \text{to} \quad 123.29$$

We are thus 95% confident that the average annual claim per child is between $34.99 and $123.29 *higher* than the average annual claim for the wife. ∎

Exercises

11.23 A real estate firm used two independent property appraisers. The firm wanted to know whether the two appraisers were consistent in determining the market value of local buildings. The appraisers each appraised 11 buildings and the following data were collected (values in dollars):

BUILDING	APPRAISER 1	APPRAISER 2	BUILDING	APPRAISER 1	APPRAISER 2
1	25,000	25,500	7	29,950	30,590
2	28,000	30,000	8	38,100	39,500
3	41,200	41,100	9	31,350	32,750
4	48,300	47,600	10	25,890	24,900
5	51,350	50,100	11	48,500	47,300
6	32,450	34,125			

a. Use a paired t test to test for differences due to the appraisers. Use a .05 significance level.

b. Use the F test in the randomized block design to test for no differences in appraisers. Use a .05 significance level.

c. What is the relationship between the t test in part a and the F test in part b?

11.24 A study compared the price of regular gas at Exgas stations and at Argas stations. Ten locations were randomly chosen in which both Exgas and Argas service stations were located. The price per gallon (in dollars) was:

LOCATION	1	2	3	4	5	6	7	8	9	10
Exgas	1.08	1.05	1.09	1.04	1.10	1.09	1.05	1.06	1.09	1.10
Argas	1.07	1.04	1.04	1.07	1.08	1.10	1.05	1.03	1.06	1.07

a. Use a paired t test to test for differences in the mean price of regular gas of the two companies. Use a .01 significance level.

b. Use the F test in the randomized block design to test for the difference in the mean price of regular gas at the Exgas and Argas service stations. Use a .01 significance level. Is the F-value equal to the square of the t-value in part a?

11.25 A particular application contains four blocks and four levels for the factor of interest. The totals for each of the four blocks are given as: $S_1 = 170$, $S_2 = 184$, $S_3 = 182$, $S_4 = 240$. The totals for each of the four levels of the factor are given as: $T_1 = 120$, $T_2 = 240$, $T_3 = 210$, $T_4 = 206$. Construct the ANOVA table for the randomized block design and assume that the total sum of squares is 2836.

11.26 Complete the following ANOVA table for a randomized block design. Find the p-value.

SOURCE	df	SS	MS	F
Factor	7			
Blocks	3	105.6		
Error		90.8		
Total		336.5		

11.27 The study in Exercise 11.5 was modified such that only six secretaries were used. Each secretary had a different typing speed. Each secretary tested all three word processing software packages. The same task could not be used for testing all three packages, since the "learning effect" would come into play, so each secretary performed three separate tasks. However, the tasks were of essentially the same length and difficulty level. Furthermore, which task was assigned to which word processor was randomly determined, and the order in which the three word processors were tested was also randomly decided. The secretaries relaxed between tasks to avoid "fatigue effects." Thus, a randomized block design was achieved, with secretaries constituting blocks and the three observations (levels of the factor) within each block being randomized. The following data were obtained (the secretary's typing speed in words per minute is given in parentheses for reference purposes, and the body of the table contains the time taken to complete the tasks):

SECRETARY	GROUP 1 (MENU)	GROUP 2 (COMMAND)	GROUP 3 (MIXED)
1 (75 wpm)	9	10	7
2 (65 wpm)	12	11	9
3 (55 wpm)	12	14	11
4 (50 wpm)	13	13	11
5 (45 wpm)	16	15	13
6 (30 wpm)	18	16	15

a. Compute the ANOVA table for the preceding data.

b. Conduct a hypothesis test to address the question, is there a significant difference between the three word processors (as measured by the performance of the secretaries)? Use $\alpha = .10$.

c. Determine the *p*-value. Does the conclusion change at $\alpha = .05$ and at $\alpha = .01$?

d. Is the block (secretary's) effect significant at $\alpha = .01$?

11.28 An analyst with a marketing firm wished to know if there was a difference in the number of responses from advertising a certain product at three different times on television. Ten days were randomly selected to run a commercial with a call-in phone number at each of the three times. The number of responses was recorded for 16 products:

PRODUCT	NOON	5:00 P.M.	10:30 P.M.	PRODUCT	NOON	5:00 P.M.	10:30 P.M.
1	12	18	14	9	7	5	8
2	12	30	22	10	35	39	37
3	5	4	3	11	17	15	16
4	21	20	24	12	31	45	40
5	13	19	14	13	15	25	21
6	17	16	15	14	18	29	18
7	35	37	33	15	7	10	6
8	20	29	20	16	20	17	18

a. At the .05 significance level, is there sufficient evidence to conclude that the mean number of responses at noon, 5:00 P.M., and 10:30 P.M. are different?

b. Find a 90% confidence interval for the difference in the mean number of responses for noon and 5:00 P.M.

11.29 In Exercise 11.28, subtract 3.0 from each of the observations in the table. Perform the ANOVA procedure at the .05 significance level and test the hypothesis that the three different times of day have no effect. Is the sum of squares the same for the coded data as for the original data? If any set of data is coded by adding (or subtracting) the same number to (or from) the value of each observation, how will the sum of squares be affected?

11.30 A quality-control engineer wishes to investigate the spray pattern delivered by three different windshield washer spray nozzles. The engineer uses eight different windshield designs with the three different spray nozzles. A score is given to each spray pattern to indicate the effectiveness of the spray pattern in cleaning the windshield. The experimental data are given below. Do the data provide sufficient evidence to indicate that there is a difference in the effectiveness of the three different windshield washer spray nozzles at the .05 significance level? Is there a significant difference due to blocks?

WINDSHIELD	SPRAY NOZZLES 1	2	3
1	67.2	75.3	70.1
2	63.5	72.1	68.7
3	50.8	65.1	62.5
4	71.3	78.8	71.8
5	78.1	79.9	79.0
6	69.5	74.8	64.3
7	74.6	79.6	70.8
8	70.1	76.8	69.8

11.31 In a randomized block design, if the df for the error sum of squares is given as 12 and the df for the factor sum of squares is 3, can you find the number of blocks used in the experiment? If yes, how many were used? If the total sum of squares is given as 520, the error sum of squares as 110, and the sum of squares due to blocks as 280, can you find the *F* test for this experiment? If yes, what is it?

11.32 A machine-shop supervisor is interested in knowing whether there is a significant difference among the production times of three machines running six different jobs.

	1	2	3	4	5	6
Machine 1	4.2	2.1	1.3	7.1	6.0	3.4
Machine 2	6.1	2.9	2.0	7.8	6.8	4.3
Machine 3	5.3	2.1	1.4	7.3	5.8	3.1

a. At the .01 level, test the hypothesis that there is no mean difference in the production time for each of the three machines.

b. Is the test for blocks at the .05 level significant?

c. Find a 90% confidence interval for the difference in mean production time for machine 1 and machine 2.

11.33 The Green Thumb lawn-care company is testing three different formulas for a fertilizer especially designed for lawns in Denton County. To adjust for variation in the soil, the formulas are tested in 13 locations. The growth rate of the grass is recorded. Do the coded results indicate a difference in the lawn growth due to the formulas at the .05 level of significance?

LOCATION	FORMULA 1	FORMULA 2	FORMULA 3	LOCATION	FORMULA 1	FORMULA 2	FORMULA 3
1	3.1	3.0	2.7	8	3.4	2.9	3.1
2	2.6	2.4	2.5	9	3.1	3.2	3.2
3	2.9	2.1	2.3	10	3.3	3.0	2.9
4	3.5	3.4	3.1	11	2.7	2.4	2.2
5	3.8	3.7	3.2	12	2.8	2.3	2.1
6	2.9	2.5	2.6	13	3.4	3.0	3.1
7	3.1	3.3	3.2				

11.34 Linoleum Unlimited is experimenting with three types of adhesives for laying linoleum. Each glue is tested on five different surfaces. The adhesiveness of the glue is measured, and the coded results are as follows. Construct the ANOVA table and test the hypothesis that there is no difference in the three types of adhesives at the .10 level of significance.

SURFACE	ADHESIVE 1	ADHESIVE 2	ADHESIVE 3
1	1.5	2.1	2.4
2	1.6	1.8	1.9
3	2.4	2.5	2.4
4	3.1	3.4	3.1
5	4.5	4.2	4.0

11.35 Suppose the following results were given on a computer printout: SS(factor) = 293.1, SS(blocks) = 5160.2, and SS(error) = 170.2. Assume four factor levels and ten blocks were used. Develop the ANOVA table and test for the effect of the factor and also for the effect of blocks. Use a significance level of .05.

11.36 What assumptions need to be made about the distribution of the population from which data are obtained in a randomized block design?

11.37 A consumer wished to know whether three leading brands of bread were priced approximately the same. Twelve supermarkets were selected as blocks. The following data were collected:

SUPERMARKET	BRAND 1	BRAND 2	BRAND 3	SUPERMARKET	BRAND 1	BRAND 2	BRAND 3
1	45	40	38	7	46	42	38
2	40	39	38	8	43	40	39
3	41	40	37	9	42	40	40
4	42	41	37	10	43	41	36
5	40	40	40	11	40	38	37
6	39	40	40	12	39	38	37

Test the hypothesis $H_0: \mu_1 = \mu_2 = \mu_3$ at the .01 significance level. Is the test for blocks significant at the .01 level?

◢ 11.5
The Two-Way Factorial Design

The two-way factorial design was introduced in Section 11.3. For this type of experiment, the researcher is considering two factors of interest, say, factor A and factor B. Of concern will be whether the individual factors have a significant effect on the observed variable (called the dependent variable) as well as the combined effect of the two factors.

Consider a simple example in which the dependent variable is the score on a test designed to measure assertiveness and managerial potential. The factors are sex

■ FIGURE 11.8
Scores on assertiveness/
managerial potential
exam.

■ FIGURE 11.8
Scores on assertiveness/
managerial potential
exam.

and marital status (single or married). Each of these two factors consists of two levels. Suppose we observed a significant difference between the male and female scores. Thus we would conclude that factor A (sex) is significant. The analysis procedure to investigate this hypothesis is described in this section. If a significant difference between the scores of the single and married subjects is observed, then we would conclude that factor B (marital status) is also significant.

Suppose that a closer look at the scores revealed that the married males and single females scored high, but the single males and married females scored low on the test, as illustrated in Figure 11.8.

Consequently, the relationship between sex and the dependent variable (exam score) *depends upon the marital status,* since this relationship is different for the single and married groups. Similarly, the relationship between marital status and the dependent variable depends upon the particular level of factor A (sex). This example illustrates **interaction** between factors A and B. A method of detecting interaction using a simple graph, along with a statistical test of hypothesis, will be explained in this section.

Degrees of Freedom

In a two-way factorial design, each level of factor A is combined with each level of factor B when obtaining the sample data. Suppose that factor A has a levels and factor B has b levels, as shown in Figure 11.9. Each x represents a test score.

If we record one observation for each factor A and factor B combination (referred to as a **treatment**), then we have $n = ab$ total observations. The df for each factor is one less than the number of levels and the df for the interaction term is the product of the factor A df and the factor B df. Consequently,

$$\text{df for factor A} = a - 1$$
$$\text{df for factor B} = b - 1$$
$$\text{df for interaction} = (a - 1)(b - 1)$$
$$\text{df for total} = n - 1 = ab - 1$$

We have a bit of a problem here. This design, like all experimental designs, must contain a source of variation due to error, that is, the unexplained variation. Suppose $a = 4$ and $b = 3$. Then the remaining df for error is (df for total) − (df for factor A) − (df for factor B) − (df for interaction), which in this case is $11 - 3 - 2 - 6 = 0$. It can be shown that the error df is zero *regardless* of the values of a and b. Since it will be necessary to measure this unexplained variation, this design requires that you obtain repeat observations (**replicates**) for each treatment. An illustration (using two replicates) including the various totals needed to

■ FIGURE 11.9
Layout for two-way
factorial design.

■ FIGURE 11.10
Illustration of two
replicates in a two-way
factorial design
($r = 2$).

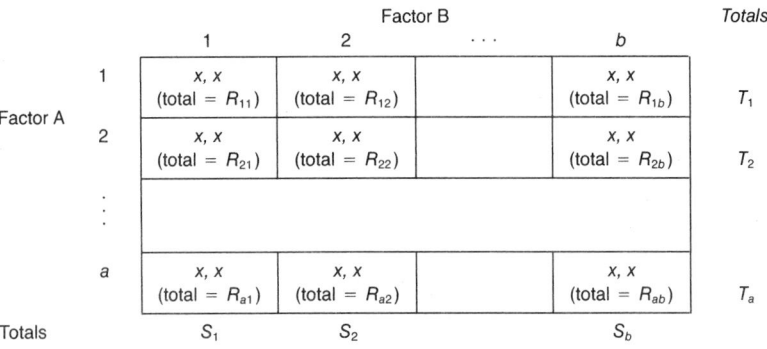

carry out the analysis is shown in Figure 11.10. *In general, you will need two or more replicates at each treatment.* The number of replicates at each treatment need not be the same, but we consider here only the case where there are r replicates at each treatment.

In the replicated design, the degrees of freedom are

$$\text{df for factor A} = a - 1$$
$$\text{df for factor B} = b - 1$$
$$\text{df for interaction} = (a - 1)(b - 1)$$
$$\text{df for total} = (\text{number of observations} - 1)$$
$$= abr - 1$$
$$\text{df for error} = (abr - 1) - (a - 1) - (b - 1) - (a - 1)(b - 1)$$
$$= ab(r - 1)$$

Sum of Squares and Mean Squares

The necessary sums of squares can be computed in a manner similar to that used in the previous designs. Using Figure 11.10, the following expressions give the corresponding sums of squares.

$$\text{factor A: SSA} = \frac{1}{br} [T_1^2 + T_2^2 + \cdots + T_a^2] - \frac{T^2}{abr} \quad \textbf{(11.29)}$$

where T = total of all $n = abr$ observations (that is, $T = T_1 + T_2 + \cdots + T_a$).

$$\text{factor B: SSB} = \frac{1}{ar} [S_1^2 + S_2^2 + \cdots + S_b^2] - \frac{T^2}{abr} \quad \textbf{(11.30)}$$

$$\text{interaction: SSAB} = \frac{1}{r} [\Sigma R^2] - \text{SSA} - \text{SSB} - \frac{T^2}{abr} \quad \textbf{(11.31)}$$

where the sum in the brackets is the sum of all the squares of the replicate totals, illustrated in Figure 11.10.

$$\text{total: SS(total)} = \Sigma x^2 - \frac{T^2}{abr} \quad \textbf{(11.32)}$$

where Σx^2 is the sum of the squares for each of the $n = abr$ observations. By subtraction,

$$SS(error) = SS(total) - SSA - SSB - SSAB \qquad \textbf{(11.33)}$$

The corresponding mean squares can be obtained by dividing each sum of squares by the corresponding degrees of freedom. Thus we have

$$MSA = \frac{SSA}{(a-1)} \qquad \textbf{(11.34)}$$

$$MSB = \frac{SSB}{(b-1)} \qquad \textbf{(11.35)}$$

$$MSAB = \frac{SSAB}{(a-1)(b-1)} \qquad \textbf{(11.36)}$$

$$MS(error) = \frac{SS(error)}{ab(r-1)} \qquad \textbf{(11.37)}$$

This analysis can be summarized in the following ANOVA table:

SOURCE	df	SS	MS	F
Factor A	$a-1$	SSA	$MSA = \dfrac{SSA}{a-1}$	$F_1 = MSA/MS(error)$
Factor B	$b-1$	SSB	$MSB = \dfrac{SSB}{b-1}$	$F_2 = MSB/MS(error)$
Interaction	$(a-1)(b-1)$	SSAB	$MSAB = \dfrac{SSAB}{(a-1)(b-1)}$	$F_3 = MSAB/MS(error)$
Error	$ab(r-1)$	SS(error)	$MS(error) = \dfrac{SS(error)}{ab(r-1)}$	
Total	$abr-1$	SS(total)		

The personnel director at Blackburn Industries is interested in examining the effect of sex and employee classification on the annual amount of dental claims for unmarried employees at Blackburn. Employee classifications range from category 1 (production-line workers) to category 4 (upper-level management). By utilizing a two-way factorial design, she can study the effect of sex (factor A) and employee classification (factor B), as well as the interaction effect between sex and employee classification, on the amount of the annual dental claims. These factors result in a 2×4 factorial design, since factor A consists of two levels and factor B has four levels. She decided to use three replicates for each of the eight treatment combinations, requiring annual claims from 24 different employees. The sample results are as follows, where the values in parentheses are the replicate totals for each of the treatments.

EMPLOYEE CLASSIFICATION (FACTOR B)

		CATEGORY 1	CATEGORY 2	CATEGORY 3	CATEGORY 4	TOTAL	AVERAGE
	Male	190, 225, 200 (615)	135, 180, 100 (415)	260, 330, 350 (940)	305, 275, 240 (820)	2790	232.50
SEX (FACTOR A)	Female	235, 190, 270 (695)	275, 305, 285 (865)	160, 205, 140 (505)	155, 110, 75 (340)	2405	200.42
	Total	1310	1280	1445	1160	5195	
	Average	218.33	213.33	240.83	193.33		

Using the previous discussion, the necessary sums of squares can be derived.

$$SSA = \frac{1}{(4)(3)} (2790^2 + 2405^2) - \frac{5195^2}{24} = 6176.04$$

$$SSB = \frac{1}{(2)(3)} (1310^2 + 1280^2 + 1445^2 + 1160^2) - \frac{5195^2}{24} = 6853.12$$

$$SSAB = \frac{1}{3} (615^2 + 415^2 + 940^2 + 820^2 + 695^2 + 865^2 + 505^2 + 340^2)$$

$$- 6176.04 - 6853.12 - \frac{5195^2}{24} = 98578.13$$

$$SS(total) = (190^2 + 225^2 + 200^2 + \cdots + 155^2 + 110^2 + 75^2) - \frac{5195^2}{24}$$

$$= 131173.96$$

Consequently,

$$SS(error) = 131173.96 - 6176.04 - 6853.12 - 98578.13 = 19566.67$$

The degrees of freedom here will be:

Sex factor: $a - 1 = 2 - 1 = 1$

Employee classification factor: $b - 1 = 4 - 1 = 3$

Interaction: $(a - 1)(b - 1) = (1)(3) = 3$

Error: $ab(r - 1) = (2)(4)(3 - 1) = 16$

Total: $abr - 1 = (2)(4)(3) - 1 = 24 - 1 = 23$

These calculations and the resulting mean squares can be summarized in the following ANOVA table.

SOURCE	df	SS	MS	F
Sex	1	6176.04	6176.04	$F_1 = 6176.04/1222.92 = 5.05$
Employee classification	3	6853.12	2284.37	$F_2 = 2284.37/1222.92 = 1.87$
Interaction	3	98578.13	32859.38	$F_3 = 32859.38/1222.92 = 26.87$
Error	16	19566.67	1222.92	
Total	23	131173.96		

Hypothesis Testing   When using a two-way factorial design, you can test for the significance of factor A, factor B, and the interaction of the two factors. For factor A, the null hypothesis is that the means are equal across the factor A levels. Written another way, we can define the following hypotheses:

$$H_{0,A}: \text{factor A is not significant } (\mu_M = \mu_F)$$

$$H_{a,A}: \text{factor A is significant } (\mu_m \neq \mu_F)$$

The corresponding test statistic is

$$F_1 = \frac{MSA}{MS(error)} \qquad (11.38)$$

and the testing procedure is to reject $H_{0,A}$ if

$$F_1 > F_{\alpha,v_1,v_2}$$

where F_{α,v_1,v_2} is from Table A.7, $v_1 = $ df for factor A $= a - 1$, and $v_2 = $ df for error $= ab(r - 1)$.

Similarly, to test for equal means of the factor B levels, we can define the hypotheses:

$$H_{0,B}: \text{factor B is not significant } (\mu_1 = \mu_2 = \mu_3 = \mu_4)$$

$$H_{a,B}: \text{factor B is significant (not all } \mu_i\text{'s are equal)}$$

The test statistic for determining the factor B effect is

$$F_2 = \frac{\text{MSB}}{\text{MS(error)}} \qquad (11.39)$$

and factor B is significant ($H_{0,B}$ is rejected) if

$$F_2 > F_{\alpha,v_1,v_2}$$

where v_1 = df for factor B = $b - 1$, and v_2 = df for error = $ab(r - 1)$.

The final set of hypotheses is concerned with the interaction effect between the two factors. The hypotheses for this procedure can be stated

$$H_{0,AB}: \text{there is no significant interaction between factor A and factor B}$$

$$H_{a,AB}: \text{there is significant interaction between factor A and factor B}$$

The test statistic is the remaining F statistic in the ANOVA table, namely,

$$F_3 = \frac{\text{MSAB}}{\text{MS(error)}} \qquad (11.40)$$

and the test procedure is to reject $H_{0,AB}$ if

$$F_3 > F_{\alpha,v_1,v_2}$$

where v_1 = df for interaction = $(a - 1)(b - 1)$ and v_2 = df for error = $ab(r - 1)$.

Multiple Comparisons

The method of multiple comparisons discussed in Section 11.2 for the one-factor ANOVA procedure (Tukey's method) can be used to examine pairwise differences between the various treatment means in two-way factorial designs. Since factor A has a levels and factor B has b levels, there are ab such means that can be compared, one pair at a time.

For the two-way factorial design, we use Table A.16 to find $Q_{\alpha,k,v}$, where α is the desired experimentwise significance level, $k = ab$ is the number of treatment means, and v is the degrees of freedom associated with MS(error). Any two (sample) treatment means differing by more than

$$D = Q_{\alpha,k,v} \cdot \sqrt{\frac{\text{MS(error)}}{r}}$$

will imply that the corresponding population means are unequal. This procedure is illustrated in the next example.

EXAMPLE 11.8   Using the previous ANOVA table constructed using the dental claims and the two factors, sex (factor A) and employee classification (factor B), determine whether (1) factor A is significant, (2) factor B is significant, (3) there is significant interaction between sex and employee classification, and (4) which pairs of the eight (population) treatment means are unequal. Use a significance level of .05.

SOLUTION 1   The df for the F statistic are $v_1 = 1$ and $v_2 = 16$. Using Table A.7, $F_{.05,1,16} = 4.49$, and the test is to reject $H_{0,A}$ if $F_1 > 4.49$. Since $F_1 = 5.05 > 4.49$, we conclude that the sex factor *is* significant. Examining the raw data, we observe that the sample mean for the males is $2790/12 = 232.50$, and the female average is $2405/12 = 200.42$. Thus we conclude that the difference between these sample means *is* significant, with higher dental claims occurring in the male population.

SOLUTION 2   For the employee classification factor, the df for the F statistic are $v_1 = 3$ and $v_2 = 16$, with a corresponding table value of $F_{.05,3,16} = 3.24$. Since $F_2 = 1.87 < 3.24$, the employee classification factor is *not* significant. Taking a closer look, we observe that the high (low) female values were balanced by the low (high) male values within each employee classification category. Consequently, there is insignificant variation in the means for the four employee classification groups, leading to the "fail to reject $H_{0,B}$" conclusion.

SOLUTION 3   The discussion in the solution to part 2 indicates the presence of interaction between the two factors. The four male means are $615/3 = 205$ (category 1), $415/3 = 138.33$ (category 2), $940/3 = 313.33$ (category 3), and $820/3 = 273.33$ (category 4). The corresponding means for the female sample are 231.67, 288.33, 168.33, and 113.33. These means are shown in Figure 11.11a, where interaction effect is

■ **FIGURE 11.11**   Illustration of interaction effect. (a) Interaction effect in Example 11.8. (b) Hypothetical situation containing no significant interaction between sex and employee classification.

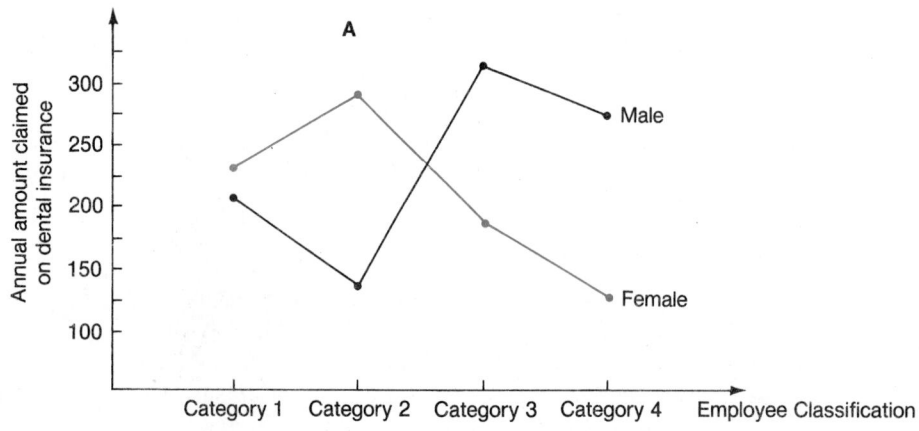

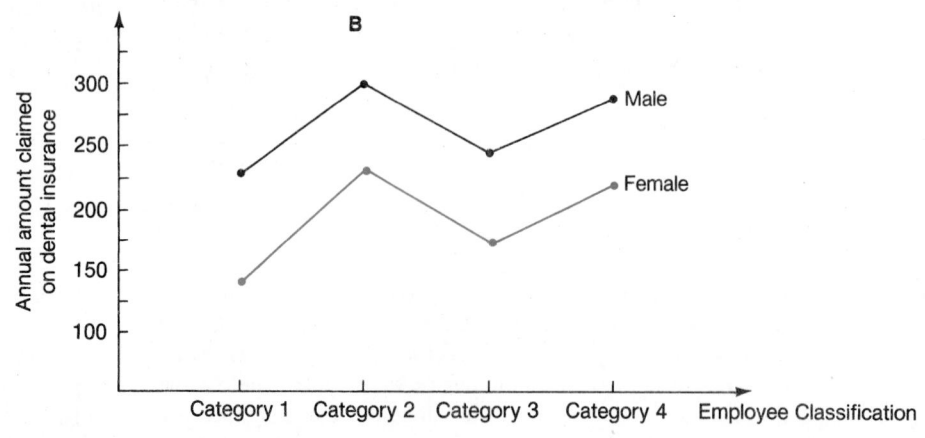

very apparent, since the male and female lines **are not parallel.** When no interaction exists between the two factors, such a graph should contain lines that are **nearly parallel,** as illustrated in Figure 11.11b.

The statistical test here supports this conclusion, since there is significant interaction provided F_3 is larger than $F_{.05,3,16} = 3.24$. Here, $F_3 = 26.87$, and so we once again conclude that there is significant interaction between sex and employee classification for this population.

Since the two lines in Figure 11.11a are not parallel, the relationship between employee classification and the amount of dental claims is not the same for males and females. In particular, the amount of dental claims is low for category 4 females and high for category 4 males. Also, the amount is high for category 2 females and low for category 2 males. *This type of discussion should always be included in the analysis whenever the interaction effect is significant.*

The MINITAB solution for this example is shown in Figure 11.12. Notice that the same TWOWAY command used for the randomized block design is used for the two-way factorial design. Due to the presence of replications, MINITAB assumes a possible interaction effect and includes this line in the ANOVA table.

SOLUTION 4   A multiple comparisons analysis will determine whether the average annual amount for category 1 males is the same as for category 4 males, the average for category 2 males is the same for category 3 females, and so forth. There are $2 \cdot 4 = 8$ means here, providing $_8C_2 = 28$ possible pairwise comparisons. In general, for a two-way factorial design there are $_{ab}C_2$ possible pairs of means that can be compared using the multiple comparisons procedure.

The critical value corresponding to $\alpha = .05$, $k = 8$, and $v = 16$ (the df associated with the error sum of squares) from Table A.16 is $Q_{.05,8,16} = 4.90$. Since

■ FIGURE 11.12
MINITAB solution for
two-way factorial design
in Example 11.8.

```
MTB > READ INTO C1-C3
DATA> 190 1 1
DATA> 225 1 1
DATA> 200 1 1
DATA> 135 1 2
DATA> 180 1 2
DATA> 100 1 2
DATA> 260 1 3
DATA> 330 1 3
DATA> 350 1 3
DATA> 305 1 4
DATA> 275 1 4
DATA> 240 1 4
DATA> 235 2 1
DATA> 190 2 1
DATA> 270 2 1
DATA> 275 2 2
DATA> 305 2 2
DATA> 285 2 2
DATA> 160 2 3
DATA> 205 2 3
DATA> 140 2 3
DATA> 155 2 4
DATA> 110 2 4
DATA> 75 2 4
DATA> END
      24 ROWS READ
MTB > NAME C1 = 'CLAIMS' C2 = 'GENDER' C3 = 'CLASSIF'
MTB > TWOWAY USING DATA IN C1, A LEVELS IN C2, B LEVELS IN C3

ANALYSIS OF VARIANCE   CLAIMS

SOURCE        DF       SS        MS
GENDER         1     6176      6176
CLASSIF        3     6853      2284
INTERACTION    3    98578     32859
ERROR         16    19567      1223
TOTAL         23   131174
```

MS(error) $= 1222.92$ and there are $r = 3$ replicates at each treatment level, we next determine

$$D = Q_{\alpha,k,v} \cdot \sqrt{\frac{MS(error)}{r}}$$

$$= (4.90) \sqrt{\frac{1222.92}{3}}$$

$$= 98.93$$

Consequently, any pair of sample treatment means differing by more than 98.93 will imply that the corresponding population means are unequal.

The eight sample means are obtained by dividing the corresponding replicate totals (R) by $r = 3$. Placing them in order, we obtain

113.33	138.33	168.33	205.00	231.67	273.33	288.33	313.33
(F,4)	(M, 2)	(F, 3)	(M, 1)	(F, 1)	(M, 4)	(F, 2)	(M, 3)

Here, M and F represent the sex (factor A) and 1, 2, 3, and 4 represent the employee classification (factor B).

Since $205.00 - 138.33 = 66.67 < 98.93$, we *cannot* conclude that $\mu_{M,1} \neq \mu_{M,2}$. This is represented by the overbar connecting these two sample means. Consider category 4 males and category 4 females. Here $273.33 - 113.33 = 160 > 98.93$, and so we conclude that there *is* a difference in the average amounts for these two groups; that is, $\mu_{M,4} \neq \mu_{F,4}$. This can be observed in the preceding sample means, since there is no overbar connecting these two means. Continuing this procedure, we arrive at the following summary for the multiple comparisons analysis:

males only: $\mu_{M,2} \neq \mu_{M,4}$, $\mu_{M,2} \neq \mu_{M,3}$, $\mu_{M,1} \neq \mu_{M,3}$

females only: $\mu_{F,4} \neq \mu_{F,1}$, $\mu_{F,4} \neq \mu_{F,2}$, $\mu_{F,3} \neq \mu_{F,2}$

males and females: $\mu_{F,4} \neq \mu_{M,4}$, $\mu_{F,4} \neq \mu_{M,3}$, $\mu_{F,2} \neq \mu_{M,2}$, $\mu_{F,3} \neq \mu_{M,4}$, $\mu_{F,3} \neq \mu_{M,3}$

Consequently, we observe three significant differences in each of the male and female populations and five significant differences in the amount of annual dental claims when comparing employee classifications across both sexes. ∎

Exercises

11.38 A two-way factorial experiment has five observations for each treatment. Three levels of factor A and two levels of factor B are used. The sample means for each treatment and the sum of squares are as follows.

		LEVELS (FACTOR A)		
		1	2	3
LEVELS	1	21.6	22.2	20.6
(FACTOR B)	2	23.0	23.6	22.8

$SSA = 7.20$, $SSB = 20.83$, $SSAB = 1.07$, $SS(total) = 236.30$

a. From viewing the sample means of each treatment, would you consider any of the factors or interaction to be significant?

b. Complete an ANOVA table. Find the corresponding p-values. Interpret the results.

c. Which pairs of the six treatment means are significantly different at the .05 significance level?

11.39 The comparative study of word processing software in Exercise 11.27 was modified to take into account different types of keyboards: enhanced keyboard, modified keyboard, and standard keyboard. Keyboard layout and type of software could not be assumed to be independent, because it was possible that a certain type of software might actually be enhanced by a certain type of keyboard (e.g., one with special function keys). In other words, interaction between factors was possible. Therefore, a 3×3 factorial design was implemented. The following table gives the completion time in minutes, with three observations for each treatment "cell."

	SOFTWARE TYPE		
KEYBOARD TYPE	GROUP 1 (MENU)	GROUP 2 (COMMAND)	GROUP 3 (MIXED)
Enhanced	9, 8, 10	8, 7, 7	8, 10, 10
Modified	14, 14, 13	10, 14, 12	12, 10, 14
Standard	15, 18, 17	18, 16, 15	15, 15, 14

a. Calculate the ANOVA table for the preceding experiment.

b. Assume that the assumptions of an ANOVA have been satisfied. Is there a significant difference in the three word processors, as measured by the productivity of the secretaries? Use $\alpha = .05$.

c. Do the different keyboards seem to affect productivity, as measured by completion times? Use $\alpha = .05$.

d. Is there a significant interaction between software type and keyboard type, at $\alpha = .05$?

e. For parts b, c, and d, find the corresponding p-value for each test.

11.40 A manufacturer is interested in reducing the number of defective components produced by its employees. A consultant recommends that each employee follow one of three proposed systematic procedures. To determine whether there was a difference in the three procedures, an experiment was conducted with one factor having three levels: less than a year of experience, 1 to 4 years experience, and over 4 years experience. The second factor was the systematic procedure, one level for each of the three proposed procedures. Data from the experiment, with three replications per cell, are given below and show the average number of nondefective components produced per day over a one-week period for each employee:

	PROCEDURE		
EXPERIENCE	1	2	3
Less than a year	12.6, 15.7, 10.5	8.6, 9.89, 11.2	13.6, 12.8, 10.2
Between 1 and 4 years	13.7, 14.2, 15.8	9.2, 12.6, 13.1	12.1, 11.8, 14.1
Over 4 years	17.5, 19.8, 20.4	16.4, 17.1, 14.2	16.5, 18.7, 17.0

At the 10% significance level, is there a difference in the results of the three systematic procedures?

11.41 The term *power base* can be used to refer to a type of relation in which party A may utilize some "force" to overcome the resistance of party B and alter B's behavior so that A achieves some desired objective. A 2×2 factorial design was used to analyze the effects of various levels of the type of power base of a principal and the effects of dependence of a food broker on the principal. The dependent variable was a measure of a food broker's readiness to respond to a request. Five levels of the "power base" factor (such as reward and coercion were used and two levels (high and low) were used with the second factor of dependence of the broker on the principal. The results of the experimental study are given below.

SOURCE	F RATIO
Power base	.76
Dependency	22.18
Interaction	1.30

(*Source:* J. Keith, D. Jackson, and L. Crosby, "Effects of Alternative Types of Influence Strategies Under Different Channel Dependence Structures," *Journal of Marketing* 54(1990): 30.)

Test for significance of each F ratio, assuming that the second degrees of freedom can be taken to be infinity. Use a .05 significance level. What are your conclusions?

11.42 Advertisers have often used celebrities to endorse a certain product or brand. An experiment was conducted in which a physically attractive celebrity (Tom Selleck) and a physically unattractive celebrity (Telly Savalas) were used as endorsers of a certain product. Graduate students were randomly assigned to one of four treatment groups. This 2×2 factorial experiment used physical attractiveness of celebrity (2 levels) as one factor and attractiveness of product as the second factor. The second factor had two levels—high attractiveness (luxury car) versus low attractiveness (home computer). The spokesman's believability was rated on a five-point scale. From the following F tests, what conclusions can be drawn? Use a .05 significance level and assume that the error degrees of freedom are 83.

SOURCE	F RATIO
Celebrity's physical attractiveness	10.535
Product's attractiveness	9.766
Interaction	1.555

(*Source:* Adapted from M. Kamis, "An Investigation Into the 'Match-Up' Hypothesis in Celebrity Advertising: When Beauty May Be Only Skin Deep," *Journal of Advertising* 19, no. 1 (1990): 4.)

11.43 A program to train middle-level managers is being experimented with to evaluate the overall training. One factor that is considered has two levels: computer-assisted training and no computer-assisted training. A second factor also has two levels: group instruction or self-paced program. A 2×2 factorial design is implemented. A training score on a 100-point scale is computed for each manager at the end of the training session.

	GROUP INSTRUCTION	SELF-PACED PROGRAM
Computer-assisted training	90, 93, 84, 94, 87	96, 92, 97, 99, 96
No computer used	87, 89, 98, 93, 94	97, 96, 91, 96, 95

a. How many replicates are there for each treatment combination?

b. Compute the ANOVA table for the preceding design.

c. Interpret the results of the ANOVA table. Are the factors or interaction significant at the .05 significance level?

d. Which pairs of the four treatment means are significantly different at the .05 level?

e. What assumptions are necessary to ensure that the ANOVA procedure is valid?

Summary

The **analysis of variance (ANOVA)** procedure is a method of detecting differences between the means of two or more normal populations. The various populations represent the *levels* of a *factor* under observation. The factor might consist of, for example, different locations (does the crime rate differ among five cities?), brand (does one brand outsell the others?), or time periods (is average attendance the same each day of the week?).

Samples for this analysis must be obtained independently of each other. The ANOVA technique measures sources of variation among the sample data by computing various **sums of squares.** The variation from one level (population) to the next is measured by the factor sum of squares (SS(factor)), which is large when there is great variation among the sample means. The variation *within* the samples is measured by the error sum of squares (SS(error)). Each of these SS has a corresponding df, which is divided into the SS to produce a **mean square,** MS.

The ratio of MS(factor) to MS(error) produces an F statistic that is used to test for equal means within the various populations. If the F-value is large (significant), we conclude that the means are not all the same, which implies that the factor of interest *does* have a significant effect on the variable under observation, called the **dependent** variable.

When we analyze the effect of a single factor, we perform a one-factor ANOVA and use a **completely randomized design.** The results of this analysis, including the various sums of squares, mean squares, and df, are summarized in an **ANOVA table.** If the ANOVA procedure concludes that the population means are not the same, a follow-up analysis can be conducted to determine which of the population means are unequal. This analysis is a **multiple comparisons** procedure and should be performed only in the event that the ANOVA null hypothesis of equal means is rejected.

When samples are not obtained in an independent manner, a **randomized block design** often can be used to test for differences in the population means. Again, there is a single factor of interest, but to determine the effect of this factor, the sample data are organized into *blocks.* For this situation, the samples are not independently obtained, but data within the same block may be gathered from the same city or person or at the same point in time. By including a block effect in the ANOVA procedure, we can analyze the factor of interest (the population means) using an F test. In addition, another F statistic can be used for determining whether there is a significant block effect within the sample data.

The other experimental design that was discussed was the **two-way factorial** design, where the effect of two factors can be investigated. Observations are obtained for each combination of factor levels, called **treatments.** For such a design, it is necessary to obtain two or more independent replicates for each treatment. The two-way factorial design allows the researcher to investigate the effect of each factor individually, as well as the combined effect of the two factors, which is referred to as the **interaction** effect.

Summary of Formulas

1. Completely randomized design (one-factor ANOVA):

$$SS(factor) = \left[\frac{T_1^2}{n_1} + \frac{T_2^2}{n_2} + \cdots + \frac{T_k^2}{n_k}\right] - \frac{T^2}{n}$$

$$SS(total) = \Sigma x^2 - \frac{T^2}{n}$$

$$SS(error) = SS(total) - SS(factor)$$

$$MS(factor) = \frac{SS(factor)}{k-1}$$

$$MS(error) = \frac{SS(error)}{n-k}$$

where

n = total number of observations

k = number of groups (populations) to be compared

T_i = total of sample values for the ith group

T = grand total = $T_1 + T_2 + \cdots + T_k$

test statistic:

$$F = \frac{MS(factor)}{MS(error)}$$

confidence interval for μ_i:

$$\bar{X}_i \pm t_{\alpha/2,n-k}s_p\sqrt{\frac{1}{n_i}}$$

confidence interval for $\mu_i - \mu_j$:

$$(\bar{X}_i - \bar{X}_j) \pm t_{\alpha/2,n-k}s_p\sqrt{\frac{1}{n_i} + \frac{1}{n_j}}$$

multiple comparisons:

$$\mu_i \neq \mu_j \text{ provided } |\bar{X}_i - \bar{X}_j| > D$$

where

$$D = Q_{\alpha,k,v}\sqrt{\frac{MS(error)}{n_r}}$$

v = df for error

n_r = number of replicates in each sample.

2. Randomized block design

$$SS(\text{factor}) = \frac{1}{b}\left[T_1^2 + T_2^2 + \cdots + T_k^2\right] - \frac{T^2}{bk}$$

$$SS(\text{blocks}) = \frac{1}{k}\left[S_1^2 + S_2^2 + \cdots + S_b^2\right] - \frac{T^2}{bk}$$

$$SS(\text{total}) = \Sigma x^2 - \frac{T^2}{bk}$$

$$SS(\text{error}) = SS(\text{total}) - SS(\text{factor}) - SS(\text{blocks})$$

$$MS(\text{factor}) = \frac{SS(\text{factor})}{(k-1)}$$

$$MS(\text{blocks}) = \frac{SS(\text{blocks})}{(b-1)}$$

$$MS(\text{error}) = \frac{SS(\text{error})}{(k-1)(b-1)}$$

where

b = number of blocks
k = number of factor levels
T = total of all bk observations

test statistic for factor effect:

$$F_1 = \frac{MS(\text{factor})}{MS(\text{error})}$$

test statistic for block effect:

$$F_2 = \frac{MS(\text{blocks})}{MS(\text{error})}$$

confidence interval for $\mu_i - \mu_j$:

$$(\bar{X}_i - \bar{X}_j) \pm t_{\alpha/2,df} \cdot s \cdot \sqrt{\frac{1}{b} + \frac{1}{b}}$$

where

$$df = (k-1)(b-1)$$
$$s = \sqrt{MS(\text{error})}$$

3. Two-way factorial design

$$SSA = \frac{1}{br}[T_1^2 + T_2^2 + \cdots + T_a^2] - \frac{T^2}{abr}$$

$$SSB = \frac{1}{ar}[S_1^2 + S_2^2 + \cdots + S_b^2] - \frac{T^2}{abr}$$

$$SSAB = \frac{1}{r}[\Sigma R^2] - SSA - SSB - \frac{T^2}{abr}$$

$$SS(\text{total}) = \Sigma x^2 - \frac{T^2}{abr}$$

$$SS(\text{error}) = SS(\text{total}) - SSA - SSB - SSAB$$

$$MSA = \frac{SSA}{(a-1)}$$

$$MSB = \frac{SSB}{(b-1)}$$

$$MSAB = \frac{SSAB}{(a-1)(b-1)}$$

$$MS(\text{error}) = \frac{SS(\text{error})}{ab(r-1)}$$

where

a = number of levels for factor A
b = number of levels for factor B
r = number of replicates for each factor A, factor B combination (treatment)
R = sum of replicates for each treatment
T = grand total of all observations

test statistic for factor A effect:

$$F_1 = \frac{MSA}{MS(\text{error})}$$

test statistic for factor B effect:

$$F_2 = \frac{MSB}{MS(\text{error})}$$

test statistic for interaction effect:

$$F_3 = \frac{MSAB}{MS(\text{error})}$$

Review Exercises

11.44 Research conducted by social scientists in the Northeast suggests that although older workers are often more productive than their younger counterparts, supervisors tend to rate the older workers lower. Consider the following experimental setup, where the rating is shown on a scale of 1 to 10. The observations are random and independent.

AGE GROUP	RATINGS BY SUPERVISORS						
< 30 years	7.2	5.5	8.0	7.5	6.3	9.0	6.6	7.1
31–45 years	6.1	7.9	5.8	8.0	6.8	7.3	8.2	7.7
>45 years	5.6	6.0	4.9	6.8	5.3	7.0	5.9	5.8

a.   State the number of factor levels, and the number of replicates at each level.

b.   If the assumptions of an ANOVA design are satisfied, what is the 95% confidence interval for:
(i)   The average rating for those less than 30 years of age?
(ii) The average difference between ratings for the "less than 30 years" and the "more than 45 years" groups?

c.   Construct an ANOVA table and test the hypothesis that there is no significant difference among the three age groups. At a 10% significance level, is there sufficient evidence to say that age level seems to affect the worker's rating?

d.   State the p-value for the test.

e.   Using a significance level of .05, perform a multiple comparisons procedure (if appropriate).

11.45   To compare the effectiveness of three motivational lectures, 21 employees hired in the past seven months were randomly divided into three groups. Each group heard one lecture. The increase in productivity of the 21 employees was measured over the two weeks following the lectures. The coded values for the increase in productivity were:

Lecture 1	5	6	7	4	6	5	6
Lecture 2	3	4	8	3	5	4	4
Lecture 3	1	2	6	7	2	4	3

a.   State the null and alternative hypotheses for this experiment.

b.   What is the between-sample variation?

c.   What is the within-sample variation?

d.   Test the null hypothesis at the .05 significance level and find the p-value.

11.46   The results of a one-factor ANOVA for four groups with six replicates per group yielded the following statistics:

$$\bar{x}_1 = 85, \ \bar{x}_2 = 90, \ \bar{x}_3 = 58, \ \bar{x}_4 = 53, \ MSE = 198.73$$

Perform a multiple comparisons procedure at the .05 significance level.

11.47   The following data give scores on an index that measures leadership ability, for respondents at three levels of management. It is assumed that the populations are normal and independent and the observations are randomly obtained. High scores represent greater leadership ability.

SUPERVISOR	MIDDLE-LEVEL MANAGER	UPPER-LEVEL MANAGER
18	36	55
21	60	42
16	21	68
45	31	33
20	40	48

a.   Compute the ANOVA table.

b.   Is there significant difference, on the average, between the leadership scores of the three groups? Use $\alpha = .05$. Find the p-value.

c.   What is the 95% confidence interval for the mean leadership score of middle-level managers?

d.   What is the 99% confidence interval for the difference between mean scores for the upper-level managers and the middle-level managers?

e.   Using a significance level of .05, perform a multiple comparisons procedure, if appropriate.

11.48   The study in Exercise 11.47 has been modified to cover the leadership scores of individual managers who advanced from the post of supervisor to upper-level manager. Fifteen persons were initially studied, but only ten actually went all the way to upper-level managerial positions. Since the same managers were used, a "randomized block" design was obtained, with three scores for each manager. Assume that the populations are normally distributed. The following table lists the leadership scores for the ten managers who completed the study.

RESPONDENT	SUPERVISOR	MIDDLE-LEVEL MANAGER	UPPER-LEVEL MANAGER
1	25	30	30
2	16	35	48
3	17	18	20
4	30	25	20
5	35	30	32
6	28	29	28
7	29	30	35
8	30	40	48
9	27	29	35
10	40	30	32

a. Compute the ANOVA table.

b. Test the hypothesis that the means for the three classes are equal, at $\alpha = .05$. Can you conclude that, on the average, leadership scores remain stable, or do they tend to change as the persons move up the managerial scale?

c. Is there a significant difference among the managers' mean leadership scores? Use $\alpha = .05$.

11.49 Exercise 11.47 was further modified to take into account the influence of sex. Thus, a 3×2 two-way factorial design was implemented. It was decided to have three replicates for each cell. The leadership ability scores are given in the following table.

SEX	SUPERVISOR	MIDDLE-LEVEL MANAGER	UPPER-LEVEL MANAGER
Male	16, 17, 25	18, 25, 30	20, 30, 42
Female	18, 20, 28	20, 28, 30	30, 41, 55

a. Compute the ANOVA table.

b. At $\alpha = .10$, test for a significant difference in leadership scores among the three groups of managers.

c. At $\alpha = .10$, is there a difference between the leadership scores of male and females?

d. Test for interaction between managerial level and sex at $\alpha = .10$.

e. Find the p-value for the three preceding hypothesis tests. Do the conclusions change at $\alpha = .05$ or $\alpha = .01$?

11.50 A factorial experiment with two levels for factor A and 4 levels for factor B is conducted. Five observations per treatment are used. The sample mean for each treatment and the sums of squares are given below:

		LEVELS (FACTOR B)			
		1	2	3	4
LEVELS	1	34.8	35.8	31.0	32.0
(FACTOR A)	2	34.8	32.0	31.2	31.0

SSA = 13.2, SSB = 97.9, SS(error) = 453.2, SS(total) = 589.80

a. Construct the ANOVA table with interaction.

b. What conclusions can be drawn from the ANOVA using a .05 significance level?

c. Which pairs of the eight treatment means are significantly different at the .05 significance level?

11.51 Assume four samples of size ten are taken from four normally distributed populations with a common variance. The total sum of squares is 221.6, and the error sum of squares is 3.7. Test the null hypothesis that there is no difference in the means of the populations at the .05 level. Find the p-value.

11.52 An engineer wishes to design the flashlight batteries that provide the brightest light but do not burn out the light bulb prematurely. An experiment is conducted to determine the life of flashlight light bulbs at voltage levels of 3.5, 4.0, and 4.5. Ten light bulbs were used at each of these levels, and the data collected on the life of the light bulbs are presented below in coded form.

VOLTAGE LEVEL 3.5	VOLTAGE LEVEL 4.0	VOLTAGE LEVEL 4.5
19	12	15
15	25	17
29	18	13
21	13	28
18	21	10
36	19	16
14	26	19
31	15	13
23	20	10
19	17	9

a. Compute the ANOVA table.

b. Is there a significant difference among the lifetimes of the light bulbs for the three groups? Use a .05 significance level.

c. What is a 95% confidence interval for the mean lifetime of the light bulbs in the group with the voltage level of 3.5?

d. What is the 99% confidence interval for the difference in mean lifetimes between the group with a voltage level of 3.5 and the group with a voltage level of 4.0?

e. Using a significance level of .05, perform a multiple comparisons procedure, if appropriate.

11.53 A quality engineer wishes to investigate the severity of cracks that develop on a casting design that is manufactured by three different processes. The engineer conducts an experiment in which three castings, one from each process, are subjected to extreme heat. A score is given to the severity of the crack. Various temperature levels are used, and the data are given below. Using a .05 significance level, what conclusions can be made from this experiment?

TEMPERATURE LEVEL	PROCESS 1	PROCESS 2	PROCESS 3
1	7.6	8.0	7.9
2	6.0	7.3	7.4
3	9.9	9.9	10.0
4	7.8	8.0	8.6
5	6.8	7.3	7.7
6	9.6	9.7	9.9
7	7.1	7.8	7.9
8	7.4	8.0	8.2

11.54 Suppose it is known in a randomized block design that the mean square for blocks is 75, the error mean square is 291, the total sum of squares is 5083, and five blocks and four factor levels are used in the experiment. What would be the value of the F-test for testing the hypothesis that there is no difference in the mean levels of the factor?

11.55 Complete the following ANOVA table for a two-way factorial design.

SOURCE	df	SS	MS	F
Factor A	1		3088	
Factor B		3400		
Interaction	3	49000		
Error				
Total	23	63000		

11.56 Fifteen university campuses of similar size were selected to determine which of three methods of advertising a blood-donation drive was most effective. Five randomly selected campuses advertised in the university newspaper (method 1). Another five advertised only by posters and signs around campus (method 2). The remaining five had each professor credit five points to the student's last test if the student contributed (method 3). The following table gives the percentage of the student body that contributed:

Method 1	.10	.15	.19	.21	.25
Method 2	.20	.18	.20	.23	.19
Method 3	.29	.20	.25	.30	.25

Do these data provide sufficient evidence at the .01 level of significance to reject the null hypothesis that there is no difference in the effectiveness of the three methods of advertising?

11.57 Sample data from three normally distributed populations were generated by a computer program for a simulation study. Do the data provide evidence to indicate that at least two of the population variances are not equal? Use a .05 significance level.

SAMPLE 1	SAMPLE 2	SAMPLE 3	SAMPLE 1	SAMPLE 2	SAMPLE 3
38	45	28	33	47	30
37	47	27	32	45	31
35	43	31	39	43	32
40	44	30	37	42	31
39	42	31	36	44	29
35	44	32	35	45	30
34	45	29			

11.58 Three different investment advisers were asked to give a performance rating from 0 to 100 on the risk-adjusted performances of 12 randomly selected aggressive growth mutual funds. From the following data, is there sufficient evidence to conclude that the three investment advisors differ significantly in their mean performance ratings? Use a .05 significance level.

MUTUAL FUNDS	INVESTMENT ADVISORS		
	A	B	C
1	81	76	70
2	83	72	75
3	51	51	43
4	96	92	90
5	67	70	68
6	71	64	71
7	88	75	83
8	51	55	53
9	41	37	45
10	88	90	87
11	59	61	60
12	78	73	74

11.59 In a 3 × 3 factorial experiment with two replicates per treatment, how many error df are there? If a one-way ANOVA was used with three levels and six observations per level, how many error df are there? If the error df are very small for a two-way factorial experiment, how can one increase the error df?

11.60 Many organizations believe that information systems (IS) planning can help them adopt new strategic postures by exploiting information-based competitive advantages. An experimental study used a 2 × 2 factorial design to investigate the significance of two factors, each at two levels (high and low), on resources provided to IS planning. One factor was the strategic impact of existing operating systems (current). The other factor was the strategic impact applications development portfolio (future). Questionnaires were mailed to organizations that could be classified into each of these treatment combinations. Based on the *p*-value given in the following ANOVA table, interpret the results of this experiment. Can an *F* ratio of less than 1 be considered nonsignificant at the .05 significance level regardless of the degrees of freedom?

SOURCE OF VARIATION	F RATIO	p-VALUE
Future portfolio	3.570	.036
Current portfolio	1.315	.253
Interaction	.611	.435

(*Source:* B. Raghunathan and T. Raghunathan, "Planning Implications of the Information Systems Strategic Grid: An Empirical Investigation," *Decision Sciences* (1990): 287.)

11.61 Suppose that an investment club is interested in the consistency of performance of growth and income mutual funds. The data below display the 1-year, 5-year, and 10-year annualized total return for ten growth and income mutual funds as of January 1, 1990. These ten growth and income mutual funds were randomly selected from mutual funds that have been in existence for at least ten years. Do these data provide sufficient evidence at the .01 level of significance to conclude that the mean annualized returns for the 1-year, 5-year, and 10-year periods differ for growth and income mutual funds?

MUTUAL FUND	1 YEAR	5 YEAR	10 YEAR
ABT Growth and Income	15.4	17.5	14.5
Colonial	20.0	17.4	15.8
Dodge and Cox Stock	26.9	21.5	18.8
Fidelity Fund	28.7	18.3	17.4
Fundamental Investors	28.6	19.7	18.1
New England Retirement Equity	22.6	17.7	16.4
Pioneer	23.4	16.5	14.3
Safeco Equity	35.8	19.5	15.1
Smith Barney Income & Growth	25.0	17.1	18.3
Windsor	15.0	18.2	20.1

(*Source:* "Mutual Fund Scoreboard," *Business Week,* (February 19, 1990): 78–105.)

11.62 A corporate educator was interested in whether there was a significant difference in the time that it takes an individual to complete each of three computer-aided instruction courses. Ten assistant managers were randomly selected to take each of the three courses. The time for completion of each course was recorded in hours.

PERSON	COURSE 1	COURSE 2	COURSE 3	PERSON	COURSE 1	COURSE 2	COURSE 3
1	2.5	2.8	2.0	6	2.5	2.4	2.1
2	2.9	2.7	2.6	7	2.9	2.8	2.5
3	3.5	3.0	2.9	8	3.8	4.0	3.5
4	2.4	2.8	2.4	9	2.6	2.2	2.3
5	3.8	3.5	3.1	10	2.7	2.4	2.2

Is there sufficient evidence that the mean times for completing each computer-aided instruction course differ? Use a .05 significance level.

11.63 The supervisor at an assembly plant can set a machine at three different speeds—fast, normal, and slow. The supervisor is interested in whether the mean number of times that the machine goes out of control differs at each of the speeds. The machine is run on eight randomly selected days at each of the three speeds. The MINITAB printout shows the data on the number of times the machine went out of control, with level 1 representing the fast speed, level 2 representing the normal speed, and level 3 representing the slow speed. From the printout, what statistical conclusions can the supervisor make, assuming a .10 significance level?

```
MTB > READ INTO C1 C2
DATA> 12 1
DATA> 10 1
DATA> 18 1
DATA>  7 1
DATA> 20 1
DATA> 14 1
DATA> 20 1
DATA> 15 1
DATA> 14 2
DATA>  7 2
DATA> 11 2
DATA> 15 2
DATA> 16 2
DATA> 18 2
DATA> 15 2
DATA> 13 2
DATA> 10 3
DATA> 11 3
DATA> 12 3
DATA>  6 3
DATA>  5 3
DATA> 13 3
DATA> 15 3
DATA> 14 3
DATA> END
      24 ROWS READ
MTB > ONEWAY ANOVA, DATA IN C1, LEVELS IN C2

ANALYSIS OF VARIANCE ON C1
SOURCE      DF       SS        MS         F
C2           2      61.6      30.8       1.98
ERROR       21     327.4      15.6
TOTAL       23     389.0
                                    INDIVIDUAL 95 PCT CI'S FOR MEAN
                                    BASED ON POOLED STDEV
LEVEL        N      MEAN      STDEV   ----+---------+---------+---------+--
    1        8    14.500     4.721                 (--------*---------)
    2        8    13.625     3.378              (--------*---------)
    3        8    10.750     3.615   (---------*---------)
                                    ----+---------+---------+---------+--
POOLED STDEV =     3.948              9.0      12.0      15.0      18.0
```

11.64   Workers at a production plant are paid three separate ways: (1) salary, (2) by the hour at $11 per hour, and (3) by the hour at $9 per hour but with a bonus at the end of each month for acceptable productivity levels (let A, B, and C represent these pay plans, respectively). Each of eight managers randomly selected three groups of ten workers for each pay plan. Each manager also recorded the total percentage increase or decrease in productivity for a six-month period for each group.

MANAGER	PLAN A	PLAN B	PLAN C
1	2.1	1.1	7.3
2	5.3	2.1	5.1
3	3.6	−1.6	4.3
4	−2.1	−3.5	1.1
5	−3.5	−4.1	.9
6	4.1	3.0	5.1
7	2.8	2.1	3.1
8	3.1	1.1	4.7

a.   From the MINITAB computer printout below, what conclusions can be drawn? Use a .01 significance level.

b.   Is the block effect, that is, the effect of different managers, significant at the .05 level?

c.   Construct a 95% confidence interval for the difference in the mean change in productivity between plan A and plan B.

```
MTB > name c2 'manager'
MTB > name c3 'pay-plan'
MTB > print c1 c2 c3
 ROW     C1   manager   pay-plan

   1    2.1      1          1
   2    1.1      1          2
   3    7.3      1          3
   4    5.3      2          1
   5    2.1      2          2
   6    5.1      2          3
   7    3.6      3          1
   8   -1.6      3          2
   9    4.3      3          3
  10   -2.1      4          1
  11   -3.5      4          2
  12    1.1      4          3
  13   -3.5      5          1
  14   -4.1      5          2
  15    0.9      5          3
  16    4.1      6          1
  17    3.0      6          2
  18    5.1      6          3
  19    2.8      7          1
  20    2.1      7          2
  21    3.1      7          3
  22    3.1      8          1
  23    1.1      8          2
  24    4.7      8          3

MTB > twoway anova,data in c1,pay-plan in c3,manager in c2

ANALYSIS OF VARIANCE   C1

SOURCE       DF       SS         MS
pay-plan      2     61.64      30.82
manager       7    128.30      18.33
ERROR        14     22.99       1.64
TOTAL        23    212.93
```

11.65 An experiment is set up to determine if there is a significant difference between three methods of debugging various programs. The first two methods are specific procedures for a programmer to follow, whereas the third method gives the programmer no instructions on how to debug the programs. Three types of difficulty levels are used for the programs needing debugging. Thus one factor is labeled "method," and the other factor is labeled "program." Two inexperienced programmers are randomly assigned to each combination of debugging method and program difficulty level. The time (in minutes) it takes to debug each program is recorded. From the MINITAB computer printout, what conclusions can be drawn? Use a .10 significance level.

```
MTB > name c1 'time'
MTB > name c2 'program'
MTB > name c3 'method'
MTB > print c1-c3
 ROW    time   program   method

   1     21       1         1
   2     19       1         1
   3     21       1         2
   4     17       1         2
   5     18       1         3
   6     20       1         3
   7     21       2         1
   8     17       2         1
   9     19       2         2
  10     20       2         2
  11     19       2         3
  12     19       2         3
  13     20       3         1
  14     18       3         1
  15     19       3         2
  16     20       3         2
  17     24       3         3
  18     26       3         3
```

```
MTB > twoway anova data in c1, levels in c2, c3

ANALYSIS OF VARIANCE  time

SOURCE        DF       SS       MS
program        2    14.78     7.39
method         2    11.11     5.56
INTERACTION    4    34.89     8.72
ERROR          9    25.00     2.78
TOTAL         17    85.78
```

Computer Exercises Using the Database

Exercise 1 -- Appendix I From the database, randomly select ten observations each from the NE sector, the NW sector, and SE sector (variable LOCATION). Using a .05 significance level, is there sufficient evidence to conclude that the mean house payment or apartment/house rent (variable HPAYRENT) is significantly different for the three locations? Include a multiple comparisons procedure, if appropriate.

Exercise 2 -- Appendix I Select at random 12 observations for each level of two factors from the database. Factor A has two levels: a non-zero income from the secondary wage earner, or no secondary income (variable INCOME2). Factor B has two levels: own or rent one's residence (variable OWNORENT). Determine the effect of these two factors on house payment or house/apartment rent (variable HPAYRENT). Set up an ANOVA table that includes interaction. What conclusions can be drawn at the .05 significance level?

Exercise 3 -- Appendix J From the database, select at random six observations from each bond rating and region combination. (Refer to variables BONDRATE and REGION.) Determine the effect of bond rating and region on the assets (variable ASSETS). Use a .05 significance level. Discuss the effect of interaction.

Exercise 4 -- Appendix J Repeat Exercise 3, but determine the effect of bond rating (BONDRATE) and region (REGION) on the net income (NETINC) instead of assets.

Case Study

The Packing of Ground Meat, or Variations on the Lean

At a ground-meat packing plant, rotary filling machines are used for packing ground meat. Batches of ground meat are directed consecutively to rotary filling machines for packing into cans. Each of the machines has six filling cylinders, with pistons pushing the meat through the cylinders. The quality assurance personnel are on the lookout for significant differences in the packed weight of the meat from batch to batch. They are also monitoring cylinder-to-cylinder differences and interaction effects between batch and cylinder.

To monitor these variables, a 6 × 5 factorial experiment was designed. Three filled cans were taken from each cylinder at random while each batch was being run. The weights of cans, coded by subtracting 12 oz., are shown in Table 11.2. The data are to be examined for significant batch-to-batch differences, cylinder-to-cylinder differences, or batch-cylinder interaction.

Since the filling machines work essentially on volume rather than weight, differences in batches might occur from changes in fat-to-lean meat ratio, air content, moisture content, and so forth, indicating the need for closer control of meat density. Differences between cylinders are likely to be mechanical in nature, mainly due to lack of standardization in piston stroke, showing the need for mechanical adjustment of the pistons.

An interaction would most likely indicate a change with time of the relative adjustment of the pistons or possibly a difference in their reaction to batches of different consistency.

Case Study Questions

1. Identify the dependent variable.
2. In the above example, a total of 90 observations was obtained. If we wished to maintain the 6 × 5 factorial design, what is the *minimum* number of observations required to conduct this type of experiment? Explain what problems occur if less than the minimum mumber of observations are obtained.
3. If we could assume that there was no batch-cylinder interaction and obtained one observation

■ TABLE 11.2
Coded Weights of
Filled Cans.

Cylinder	Batch 1	2	3	4	5	Row
1	1	4	6	3	1	
	1	3	3	1	3	46
	2	5	7	3	3	
2	−1	−2	3	2	1	
	3	1	1	0	0	14
	−1	0	5	1	1	
3	1	2	2	1	3	
	1	0	4	3	3	31
	1	1	3	3	3	
4	−2	−2	3	0	0	
	3	0	3	0	1	14
	0	1	4	2	1	
5	1	2	0	1	−2	
	1	1	1	0	3	14
	−1	5	2	−1	1	
6	0	0	3	3	3	
	1	0	3	0	1	26
	1	3	4	2	2	
Column	12	24	57	24	28	145

(*Source:* B.A. Griffith, A.E.R. Westman, and B.H. Lloyd, "Variance, the *F*-Test, and the Analysis of Variance Table," *Quality Engineering 2 (no. 2)* (1989–90): 195. Reprinted by permission, American Society for Quality Control.)

per treatment cell, what type of ANOVA design would we have?

4. Conduct hypothesis tests at 5% significance for:
a. A significant difference between the batches.
b. A significant difference between the cylinders.
c. Significant cylinder-batch interaction effects.
Report the *p*-value in each case.

5. Based on the results of these tests, which of the following actions would you recommend?
a. Adjust the density of the meat.
b. Adjust the cylinder piston strokes.
c. Do nothing.
d. Lay off the workers and close the plant.

6. Indicate whether each of the following would be reflected in the residual (error term):
a. Weighing errors.
b. Variation in empty can weights.
c. Within-batch variation in the meat.
d. Between-batch variation in the meat.
e. "Play" in the machine.

SPSS

☑ Solution
Example 11.4

Example 11.4 was concerned with using one-factor ANOVA procedures to test for a difference in two or more population means. The purpose was to test for a difference in job knowledge among three groups: (1) employees with a high school diploma only, (2) employees with a bachelor's degree only, and (3) employees with a master's degree. The SPSS program listing in Figure 11.13 was used to request the ANOVA table and in particular the *F*-value and the *p*-value for *F*. In this problem the SPSS commands are the same for both the mainframe and PC versions (remember to end each command line with a period when using the PC version).

The TITLE command names the SPSS run.

The DATA LIST command gives each variable a name and describes the data as being in free form.

The BEGIN DATA command indicates to SPSS that the input data immediately follow.

The next 15 lines contain the data values; each represents one employee's level of education and grade on the test. The first line indicates that the individual had a level 1 (high school) education and scored 81 on the exam.

The END DATA statement indicates the end of the data.

The ONEWAY command specifies a one-way analysis of variance model. GRADE is the dependent variable and EDUC is the independent variable. EDUC has minimum and maximum values of 1 and 3, respectively.

The RANGES = TUKEY command performs the Tukey multiple comparisons procedure.

Figure 11.14 shows the SPSS output obtained by executing the listing in Figure 11.13.

■ FIGURE 11.13
Input for SPSS
or SPSS/PC.
Remove the
periods for
SPSS input.

```
TITLE    JOB KNOWLEDGE.
DATA LIST FREE/EDUC GRADE.
BEGIN DATA.
1 81
1 84
1 69
1 85
1 84
1 95
2 94
2 83
2 86
2 81
2 78
3 88
3 89
3 78
3 85
END DATA.
ONEWAY GRADE BY EDUC(1,3)
/ RANGES = TUKEY.
```

■ FIGURE 11.14   SPSS output.

```
- - - - - - - - - - O N E W A Y - - - - - - - - - -

        Variable  GRADE
     By Variable  EDUC

                        Analysis of Variance

                        Sum of      Mean         F      F
        Source     D.F.  Squares    Squares     Ratio  Prob.

Between Groups      2    10.8000     5.4000     .1130  .8940

Within Groups      12   573.2000    47.7667

Total              14   584.0000                 F*    p-value

- - - - - - - - - - O N E W A Y - - - - - - - - - -

        Variable  GRADE
     By Variable  EDUC


Multiple Range Test


Tukey-HSD Procedure
Ranges for the  .050 level -

        3.77    3.77

The ranges above are table ranges.
The value actually compared with Mean(J)-Mean(I) is..
      4.8871 * Range * Sqrt(1/N(I) + 1/N(J))


No two groups are significantly different at the  .050 level
```

SPSS

Solution
Section 11.4

The hardness-testing example was based on a randomized block design. For this test, there were three tips available for inserting into the rod that was then pressed into the hinge with a specified force. The question of interest is whether there is a difference in the average reading for the three tips. Using ten hinges, an experiment was conducted in which all three tips were applied to each of the metal hinges. The three tips represented three factor levels; the ten hinges represented ten blocks.

The SPSS program listing in Figure 11.15 was used to request the ANOVA table and in particular the sum of squares for error, block, factor, and total. We were also interested in the *F*-value for the tips and the hinges and their respective *p*-values. In this problem the SPSS commands are the same for both the mainframe and PC versions (remember to end each command line with a period when using the PC version).

The TITLE command names the SPSS run.

The DATA LIST command gives each variable a name and describes the data as being in free form.

The BEGIN DATA command indicates to SPSS that the input data immediately follow.

The next 30 lines contain the data values. For example, in line 1, the first 1 represents the first hinge, the second 1 represents the first tip, and the 68 represents the hardness reading.

The END DATA statement indicates the end of the data.

FIGURE 11.15
Input for SPSS or SPSS/PC. Remove the periods for SPSS input.

```
TITLE   WARRANTY ASSESSMENTS.
DATA LIST FREE/HINGE TIP HARDNESS.
BEGIN DATA.
1 1 68
1 2 72
1 3 65
2 1 40
2 2 43
2 3 42
3 1 82
3 2 89
3 3 84
4 1 56
4 2 60
4 3 50
5 1 70
5 2 75
5 3 68
6 1 80
6 2 91
6 3 86
7 1 47
7 2 58
7 3 50
8 1 55
8 2 68
8 3 52
9 1 78
9 2 77
9 3 75
10 1 53
10 2 65
10 3 60
END DATA.
ANOVA HARDNESS BY TIP(1,3) HINGE(1,10)
/OPTIONS 3.
```

The ANOVA command specifies the ANOVA model. HARDNESS is the dependent variable, TIP is the factor of interest (ranging from 1 to 3), and HINGE represents the blocking variable (ranging from 1 to 10).

The /OPTIONS 3 statement specifies that there is no interaction between the factor and blocking variables (use OPTIONS 3 with the mainframe version).

Figure 11.16 shows the SPSS output obtained by executing the listing in Figure 11.15.

■ FIGURE 11.16   SPSS output.

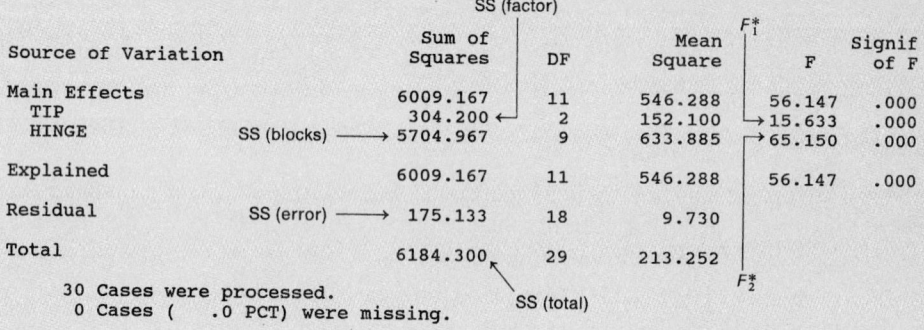

```
          WARRANTY ASSESSMENTS

          * * *  A N A L Y S I S   O F   V A R I A N C E  * * *

                   HARDNESS
             BY    TIP
                   HINGE

                                    SS (factor)
                              Sum of                    Mean              Signif
      Source of Variation     Squares     DF           Square      F      of F

      Main Effects           6009.167     11          546.288    56.147   .000
        TIP                   304.200 ←    2           152.100    15.633   .000
        HINGE    SS (blocks) → 5704.967    9           633.885    65.150   .000

      Explained              6009.167     11          546.288    56.147   .000

      Residual   SS (error) →  175.133    18            9.730

      Total                  6184.300     29          213.252

         30 Cases were processed.              SS (total)
          0 Cases (   .0 PCT) were missing.
```

F_1^*

F_2^*

SPSS

**Solution
Example 11.8**

The dental claims example was based on a two-way factorial design. The purpose was to determine the effects of sex and employee classification on level of dental claims by an employee. We wanted to determine the effects of these two factors as well as test for a possible interaction effect between these two variables. The SPSS program listing in Figure 11.17 was used to request the ANOVA table, *F*-values, and their respective *p*-values. In this problem the SPSS commands are the same for both the mainframe and PC versions (remember to end each command line with a period when using the PC version).

The TITLE command names the SPSS run.

The DATA LIST command gives each variable a name and describes the data as being in free form.

The BEGIN DATA command indicates to SPSS that the input data immediately follow.

The next 24 lines contain the data values. For example the first line of data represents one employee's sex (1 or 2), job classification (1, 2, 3, or 4), and level of dental claims (58).

The END DATA statement indicates the end of the data.

The ANOVA command specifies the ANOVA model. CLAIMS is the dependent variable, SEX (ranging from 1 to 2) and EMPCLASS (ranging from 1 to 4) are the two factors of interest.

Figure 11.18 shows the SPSS output obtained by executing the listing in Figure 11.17.

■ **FIGURE 11.17**
Input for SPSS
or SPSS/PC.
Remove the
periods for
SPSS input.

```
TITLE   DENTAL CLAIMS ANALYSIS.
DATA LIST FREE/SEX EMPCLASS CLAIMS.
BEGIN DATA.
1 1 190
1 1 225
1 1 200
1 2 135
1 2 180
1 2 100
1 3 260
1 3 330
1 3 350
1 4 305
1 4 275
1 4 240
2 1 235
2 1 190
2 1 270
2 2 275
2 2 305
2 2 285
2 3 160
2 3 205
2 3 140
2 4 155
2 4 110
2 4 075
END DATA.
ANOVA CLAIMS BY SEX(1,2) EMPCLASS(1,4).
```

■ **FIGURE 11.18** SPSS output.

DENTAL CLAIMS ANALYSIS

* * * A N A L Y S I S   O F   V A R I A N C E * * *

CLAIMS
BY   SEX
EMPCLASS

Source of Variation		Sum of Squares	DF	Mean Square	F	Signif of F
Main Effects		13029.167	4	3257.292	2.664	.071
SEX		6176.042	1	6176.042	5.050	.039
EMPCLASS		6853.125	3	2284.375	1.868	.176
2-way Interactions		98578.125	3	32859.375	26.870	.000
SEX EMPCLASS		98578.125	3	32859.375	26.870	.000
Explained		111607.292	7	15943.899	13.038	.000
Residual	SS (error) ⟶	19566.667	16	1222.917		
Total	SS (total) ⟶	131173.958	23	5703.216		

SSA — F_1^*
SSB — F_3^* F_2^*
SSAB

24 Cases were processed.
0 Cases (.0 PCT) were missing.

SAS

✔ Solution
Example 11.4

Example 11.4 was concerned with using one-factor ANOVA procedures to test for a difference in two or more population means. The purpose was to test for a difference in job knowledge among three groups: (1) employees with a high school diploma only, (2) employees with a bachelor's degree only, and (3) employees with a master's degree. The SAS program listing in Figure 11.19 was used to request the ANOVA table and in particular the F-value and the p-value for F. In this problem the SAS commands are the same for both the mainframe and PC versions.

■ **FIGURE 11.19**
   Input for SAS
   (mainframe or
   micro version).

```
TITLE    'JOB KNOWLEDGE';
DATA EDUCEXP;
INPUT EDUC GRADE;
CARDS;
1 81
1 84
1 69
1 85
1 84
1 95
2 94
2 83
2 86
2 81
2 78
3 88
3 89
3 78
3 85
PROC ANOVA;
  CLASS EDUC;
  MODEL GRADE=EDUC;
  MEANS EDUC / TUKEY;
```

■ **FIGURE 11.20**   SAS output.

JOB KNOWLEDGE

Analysis of Variance Procedure
Class Level Information

Class	Levels	Values
EDUC	3	1 2 3

Number of observations in data set = 15

Analysis of Variance Procedure

Dependent Variable: GRADE

Source	DF	Sum of Squares	Mean Square	F Value	Pr > F
Model	2	10.80000000	5.40000000	0.11	0.8940 ←
Error	12	573.20000000	47.76666667		
Corrected Total	14	584.00000000			

F^* p-value

R-Square	C.V.	Root MSE	GRADE Mean
0.018493	8.227790	6.911343	84.0000000

Source	DF	Anova SS	Mean Square	F Value	Pr > F
EDUC	2	10.80000000	5.40000000	0.11	0.8940

Tukey's Studentized Range (HSD) Test for variable: GRADE

NOTE: This test controls the type I experimentwise error rate.

Alpha= 0.05 Confidence= 0.95 df= 12 MSE= 47.76667
Critical Value of Studentized Range= 3.773

Comparisons significant at the 0.05 level are indicated by '***'.

EDUC Comparison	Simultaneous Lower Confidence Limit	Difference Between Means	Simultaneous Upper Confidence Limit
3 - 2	-11.769	0.600	12.969
3 - 1	-9.902	2.000	13.902
2 - 3	-12.969	-0.600	11.769
2 - 1	-9.765	1.400	12.565
1 - 3	-13.902	-2.000	9.902
1 - 2	-12.565	-1.400	9.765

The TITLE command names the SAS run.

The DATA command gives the data a name.

The INPUT command names and gives the correct order for the different fields on the data lines.

The CARDS command indicates to SAS that the input data immediately follow.

The next 15 lines contain the data values. Each line represents one employee's amount of education and grade on the test. The first line indicates that the individual had a level 1 (high school) education and scored 81 on the exam.

The PROC ANOVA command requests the ANOVA analysis procedure. The CLASS subcommand identifies EDUC as the factor of interest. The MODEL subcommand indicates that the exam grade is the dependent variable and the education level is the independent variable. The MEANS subcommand specifies that the TUKEY's test for multiple comparisons is to be performed.

Figure 11.20 shows the SAS output obtained by executing the listing in Figure 11.19.

SAS

Solution
Section 11.4

The hardness-testing example was based on a randomized block design. For this test, there were three tips available for inserting into the rod that was then pressed into the hinge with a specified force. The question of interest is whether there is a difference in the average reading for the three tips. Using ten hinges, an experiment was conducted in which all three tips were applied to each of the metal hinges. The three tips represented three factor levels; the ten hinges represented ten blocks.

The SAS program listing in Figure 11.21 was used to request the ANOVA table and in particular the sum of squares for error, block, factor, and total. We were also interested in the F-values for the tips and the hinges and their respective p-values. In this problem the SAS commands are the same for both the mainframe and PC versions.

The TITLE command names the SAS run.

The DATA command gives the data a name.

The INPUT command names the variables and specifies the correct order during input.

The CARDS command indicates to SAS that the input date immediately follow.

The next 30 lines contain the data values. For example, in line 1, the first 1 represents the first hinge, the second 1 represents the first tip, and the 68 represents the hardness reading.

The PROC ANOVA command requests the ANOVA analysis procedure. The CLASS subcommand identifies TIP and HINGE as variables to be classified in our study. The MODEL subcommand indicates that HARDNESS is the dependent variable while HINGE and TIP are the independent variables.

Figure 11.22 shows the SAS output obtained by executing the listing in Figure 11.21.

■ **FIGURE 11.21**
Input for SAS
(mainframe or
micro version).

```
                                 TITLE  'WARRANTY ASSESSMENTS';
DATA WARRANTY;
INPUT HINGE TIP HARDNESS;
CARDS;
1  1  68
1  2  72
1  3  65
2  1  40
2  2  43
2  3  42
3  1  82
3  2  89
3  3  84
4  1  56
4  2  60
4  3  50
5  1  70
5  2  75
5  3  68
6  1  80
6  2  91
6  3  86
7  1  47
7  2  58
7  3  50
8  1  55
8  2  68
8  3  52
9  1  78
9  2  77
9  3  75
10 1  53
10 2  65
10 3  60
PROC ANOVA;
  CLASS TIP HINGE;
  MODEL HARDNESS = TIP HINGE;
```

■ **FIGURE 11.22** SAS output.

WARRANTY ASSESSMENTS

Analysis of Variance Procedure
Class Level Information

Class	Levels	Values
TIP	3	1 2 3
HINGE	10	1 2 3 4 5 6 7 8 9 10

Number of observations in data set = 30

WARRANTY ASSESSMENTS

Analysis of Variance Procedure

Dependent Variable: HARDNESS

Source	DF	Sum of Squares	Mean Square	F Value	Pr > F
Model	11	6009.166667	546.287879	56.15	0.0001
Error	18	175.133333 ← SS (error)	9.729630		
Corrected Total	29	6184.300000 ← SS (total)			

R-Square	C.V.	Root MSE	HARDNESS Mean
0.971681	4.776777	3.119235	65.3000000

Source	DF	Anova SS	Mean Square	F Value	Pr > F
TIP	2	304.200000 ←	152.100000	15.63 ←	0.0001 ⌐
HINGE	9	5704.966667	633.885185	65.15	0.0001 ⌐

SS (blocks) SS (factor) F_2^* F_1^* p-values ⌐

SAS

Solution
Example 11.8

The dental claims example was based on a two-way factorial design. The purpose was to determine the effects of sex and employee classification on level of dental claims by an employee. We wanted to determine the effects of these two factors as well as test for a possible interaction effect between these two variables. The SAS program listing in Figure 11.23 was used to request the ANOVA table, F-values, and their respective p-values. In this problem the SAS commands are the same for both the mainframe and PC versions.

The TITLE command names the SAS run.

The DATA command gives the data a name.

The INPUT command names the variables and specifies the correct order during input.

The CARDS command indicates to SAS that the input data immediately follow.

The next 24 lines contain the data values. For example the first line of data represents one employee's sex (1 or 2), job classification (1, 2, 3, or 4), and level of dental claims (58).

The PROC ANOVA command requests the ANOVA analysis procedure. The CLASS subcommand identifies SEX and EMPCLASS as variables to be classified in our study. The MODEL subcommand indicates that CLAIMS is the dependent variable and that the analysis will consider the effects of SEX, EMPCLASS, and the interaction between them (SEX*EMPCLASS). The MEANS subcommand specifies that the TUKEY's test for multiple comparisons is to be performed.

Figure 11.24 shows the SAS output obtained by executing the listing in Figure 11.23.

FIGURE 11.23
Input for SAS
(mainframe or
micro version).

```
TITLE  'DENTAL CLAIMS ANALYSIS';
DATA DENTAL;
INPUT SEX EMPCLASS CLAIMS;
CARDS;
1 1 190
1 1 225
1 1 200
1 2 135
1 2 180
1 2 100
1 3 260
1 3 330
1 3 350
1 4 305
1 4 275
1 4 240
2 1 235
2 1 190
2 1 270
2 2 275
2 2 305
2 2 285
2 3 160
2 3 205
2 3 140
2 4 155
2 4 110
2 4 075
PROC ANOVA;
  CLASS SEX EMPCLASS;
  MODEL CLAIMS = SEX EMPCLASS SEX*EMPCLASS;
  MEANS SEX EMPCLASS SEX*EMPCLASS / TUKEY;
```

■ **FIGURE 11.24** SAS output.

DENTAL CLAIMS ANALYSIS

Analysis of Variance Procedure

Dependent Variable: CLAIMS

Source	DF	Sum of Squares	Mean Square	F Value	Pr > F
Model	7	111607.2917	15943.8988	13.04	0.0001
Error	16	19566.6667	1222.9167		
			SS (error)		
Corrected Total	23	131173.9583	SS (total)		

	R-Square	C.V.	Root MSE	CLAIMS Mean
	0.850834	16.15564	34.97023	216.458333

Source	DF	Anova SS	Mean Square	F Value	Pr > F
SEX	1	6176.04167	6176.04167	5.05	0.0391
EMPCLASS	3	6853.12500	2284.37500	1.87	0.1757
SEX*EMPCLASS	3	98578.12500	32859.37500	26.87	0.0001

SSAB   SSB   SSA

p-value

CHAPTER

Quality Control

A LOOK BACK/INTRODUCTION

Previous chapters have introduced you to various statistical measures, such as the sample mean (a measure of *location*) and the sample range (a measure of *dispersion* or *variation*). When examining a population, knowledge of these measures can tell you a great deal about the population. For example, if your company advertises 10-pound bags of dog food but a large random sample of bags produces an average ($\bar{x}$) of 8.7 pounds, the company will undoubtedly encounter many angry consumers and may well be out of the dog food business if the sample is representative of all bags being produced.

Similarly, too much variation in the bag weights would indicate that the *production process* is too erratic. The company would then lose money due to dissatisfied customers (whose bags weigh under 10 pounds) or excess product being packaged (in bags that weigh over 10 pounds). Ideally, this production process could be fine-tuned to always produce 10-pound bags with no variation, but realistically, such a goal is nearly impossible. Our example, however, points out the need to be *on target* and *consistent.*

In part, **quality control is the study of variability.** Recently, it has received a great deal of national attention, as more and more companies are facing competitors who focus on quality and offer less expensive, more reliable products. This chapter will examine many of the popular tools for implementing quality standards in manufacturing and service industries. Many of the statistical measures and procedures from the earlier chapters will be used to develop simple, yet powerful, methods of monitoring and controlling process location and variation.

12.1 Quality Control: What Is It?

Rarely a day goes by in which we're not bombarded by such statements as, "Quality is our most important job at Company A," or "Company B is a leader in quality." It is safe to say that a focus on quality is taking over American business; however, using the word "quality" in one's advertising is not what this movement is all about. Rather, the current focus on quality represents a change in the entire corporate culture from the top on down, ranging from a dramatic change in how we manage and treat company employees all the way to methods of inspecting products to make sure they conform to quality standards.

A **quality** product or service is one that meets or surpasses the needs and expectations of the customer; that is, quality means general excellence. The quality characteristics for a product might include strength, durability, appearance, and lifetime. For a service organization, quality characteristics might include promptness, attitude, and cleanliness. In this chapter, quality is viewed as some attribute or measurable characteristic of a process or product for which a standard has been established. Consequently, a quality product or process is one that *conforms* to the standards established for it; that is, is fit for its intended use.

This chapter will focus on the inspection aspect of the relatively new world of applying quality-control procedures in the American workplace. Such methods have achieved dramatic results outside the United States, most notably in Japan. An interesting point about the Japanese quality revolution is that it was led by an American statistician, W. Edwards Deming, whose message was largely ignored in the United States until recently. Now his message and the advice of other quality experts are in great demand in the United States and Canada, since businesses in these countries have come to realize that they *must* focus on quality if they hope to survive and thrive in an increasingly competitive and quality-conscious marketplace.

What Is Variation, and Is It So Bad?

Variation exists in every process and in almost every sample of measurements. No two units of product are identical. Consequently, *the measured quality of a manu-factured product is always subject to a certain amount of variation as a result of chance.* We refer to this situation as a *stable* system of *chance causes* of variation. Variation within this stable pattern is inevitable. Variations *outside* of this stable pattern is another matter. Once discovered, causes for such variation (called *assignable causes*) should be searched out and corrected. Detection of such variation is a key aspect of quality control and will be discussed in the next section, which deals with *control charts*.

By keeping a close eye on process variation and not overreacting to chance variation, a company can bring about dramatic improvements in product quality and a reduction of spoilage and rework. A major portion of the quality-control process is an effort to constantly improve the product (process) quality. A goal of such efforts is to reduce the variability in the process with an eye toward eliminating this process variation. Such efforts result in reduced costs and improved quality.

It is important to note that the examination of variation and the improvement of quality is an important area of concern to service industries or departments. For example, the accounting department of your firm might want to monitor the time required to collect on their accounts receivable and take appropriate action whenever the process appears to be "out of control."

Each year the United States Department of Commerce awards one or more **Malcolm Baldrige National Quality Awards** to companies that have demonstrated and documented significant strides in the area of quality improvement. Past winners include Motorola (1988), Xerox (1989), Cadillac Motor Car Division (1990), and Federal Express Corporation (1990). Note that Federal Express is not a manufacturing company: we emphasize again that the concepts introduced in this chapter apply to *any* organization that provides a product or service.

The remainder of this chapter will focus on many of the *tools* for measuring, monitoring, and improving process quality. The following key terms will be used:

1. **Process** Any combination of people, machinery, material, and methods that is intended to produce a product or service.
2. **Quality characteristics** Features of a product that describe its fitness for use, such as length, weight, taste, appearance, reliability.
3. **Statistical process control (SPC)** The application of statistical quality control to measuring and analyzing the variation found in processes.
4. **Acceptance sampling** The procedure of accepting or rejecting a large lot (batch) based on the results of a sample from this lot (see Figure 12.1).
5. **Control chart** A statistical chart used to monitor various aspects of a process (such as the process average) and to determine if the process is in control (stable) or out of control (unstable).

■ **FIGURE 12.1**
Acceptance sampling.

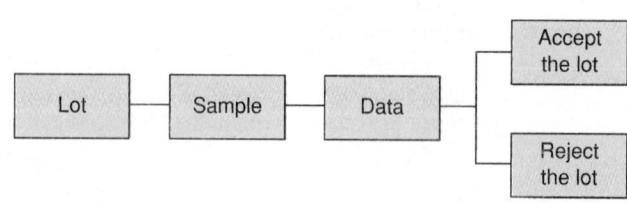

FIGURE 12.2
The application of
control charts.

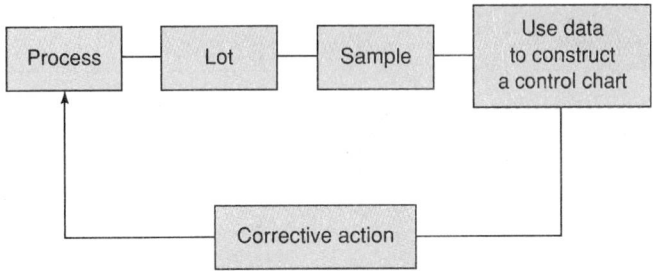

The process of using a control chart to monitor and improve process quality is shown in Figure 12.2.

12.2
Control Charts

In a sense, control charts form the foundation of the inspection side of statistical quality control. These charts were first introduced by W. A. Shewhart at Bell Telephone Laboratories in 1924. Essentially, control charts allow you to monitor a process (usually, but not necessarily, a manufacturing process) to determine if the process is "in control" or "out of control."

Definition

A process is **in control** if the observed variation is due to inherent or natural variability. This variability is the cumulative effect of many small, essentially uncontrollable, causes. It is important to note that a process may be in control yet entirely unsatisfactory. For example, your process of producing 10-pound bags of dog food may be perfectly in control with little variation, yet a closer look may reveal that the process appears to be centered at 8.7 pounds, which would certainly be unacceptable to dog owners who check the weight of the bags.

Definition

A process is **out of control** if a relatively large variation is introduced that can be traced to an *assignable cause*. Such causes are generally the result of an improperly adjusted machine, an operator error, or defective raw material.

During the in-control state, we say that the variation is due to **chance cause.** Such variation can be reduced by careful analysis of the process, but it can never be completely eliminated. On the other hand, variation during an out-of-control state is due to one or more assignable causes. Such causes *are* avoidable and cannot be overlooked.

The purpose of the control chart is to detect the presence of an out-of-control state, during which a large portion of the process output will not be conforming to requirements. If an assignable cause can be determined, corrective action is taken and an attempt is made to return the process to an in-control state. An important point to be made here is that while the process is believed to be in control, it should not be adjusted or tampered with, since such well-meaning adjustments generally result in an *increase* in process variation.

The general form of a control chart is shown in Figure 12.3. The chart contains a **center line (CL),** which represents the average value of the quality characteristic corresponding to the in-control state. The other two horizontal lines are the **control limits;** the *upper control limit* (UCL) and the *lower control limit* (LCL) are the

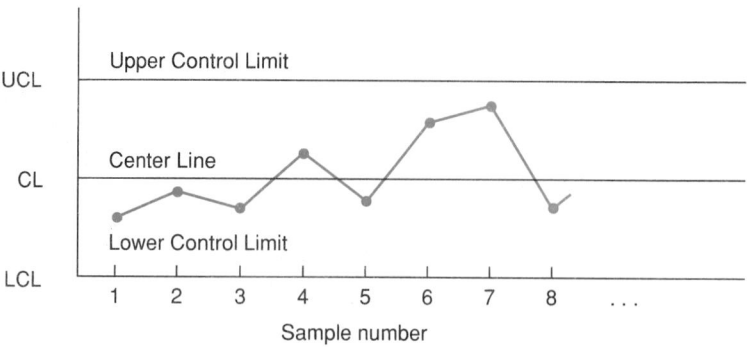

same distance from the center line. *These limits are chosen so that if the process is in control, nearly all the plotted points will be between the two control limits.* For example, a plotted point might represent the average weight of a sample of five filled coffee cans. As long as the plotted averages fall within the control limits, the process is determined to be in control, with only chance variation present.

When measuring a quality characteristic such as the weight of a filled coffee can, the resulting data are referred to as **variables data.** Conversely, when counting the number of solder defects in the coffee can, such data are referred to as **attribute data.** Chapter 1 referred to variables data as *continuous* data and to attribute data as *discrete* data. The type of control chart that is needed for a particular situation depends on what quality characteristic is being measured and what type of data is being collected. For the filled weight of the coffee cans, we would undoubtedly be interested in the average weight of each sample of five cans and would construct a control chart referred to as an $\bar{X}$ chart ($\bar{X}$ is read as "X bar"). We would also want to maintain a control chart for the *variation* within each sample. The control chart to use for this situation would be an R chart, where R represents the sample range.

Since control charts are typically used to monitor *small* samples, the sample range is generally used to measure sample variation. The range is easier than the sample standard deviation (s) to calculate and provides nearly as much information about the sample variation for sample sizes under 10. For sample sizes of 10 or more, the sample standard deviation should be used to measure sample variation, producing $\bar{X}$ and s charts. For a more thorough discussion of these charts, refer to the book by D. C. Montgomery in the list of extra readings at the end of the chapter.

Control Charts for Variables Data: The $\bar{X}$ and R Charts

Variables data are obtained by observing a continuous variable, which has *no gaps* in the values it can have, over the range of possible values. A corresponding control chart could be constructed for controlling and analyzing a process whose quality characteristic is a continuous variable, such as weight, length, or concentration.

The first step is to obtain data for construction of the control chart. A number of preliminary samples (say, m of these) is obtained, each of size n, when the process is thought to be in control. Typically, m ranges from 20 to 25 and the sample size, n, is 4, 5, or 6. Define $\bar{X}_1, \bar{X}_2, \ldots, \bar{X}_m$ to be the m sample averages. Also define $R_1, R_2, \ldots, R_m$ to be the m sample ranges.

For example, suppose the quality characteristic is the weight of a filled coffee can and the first three samples are (data in ounces):

SAMPLE	DATA	$\bar{X}$	R
1	19.8, 20.1, 20.2, 19.9, 20.0	20.00	20.2 − 19.8 = .4
2	20.3, 19.9, 19.8, 19.8, 20.1	19.98	20.3 − 19.8 = .5
3	20.0, 19.7, 20.2, 19.8, 19.7	19.88	20.2 − 19.7 = .5

■ TABLE 12.1	SAMPLE	1	2	3	4	5	6	7	8	9	10
Preliminary sample results.	$\bar{X}$	20.00	19.98	19.88	19.94	20.04	20.06	20.02	19.82	20.02	20.06
	R	.4	.5	.5	.4	.6	.3	.4	.4	.5	.7

	SAMPLE	11	12	13	14	15	16	17	18	19	20
	$\bar{X}$	19.94	19.86	19.90	20.12	19.92	20.04	20.06	19.98	19.88	20.08
	R	.4	.3	.2	.5	.5	.4	.3	.5	.6	.4

For these samples, $\bar{X}_1 = 20.00$, $\bar{X}_2 = 19.98$, $\bar{X}_3 = 19.88$, $R_1 = .4$, $R_2 = .5$, and $R_3 = .5$. These results and those for the next 17 samples are shown in Table 12.1.

In the quality-control context, the population consists of the *process* values. Thus we speak of the process mean, the process standard deviation, and so on. The best estimator of μ, the process mean, is

$$\bar{\bar{X}} = \frac{\bar{X}_1 + \bar{X}_2 + \cdots + \bar{X}_m}{m} \tag{12.1}$$

For our coffee-can filling example, the process consists of taking empty coffee cans and filling them with a prescribed amount of coffee. The quality characteristic is the filled weight of the can. The estimate of the process average is

$$\bar{\bar{X}} = \frac{20.00 + 19.98 + \cdots + 20.08}{20}$$

$$= \frac{399.60}{20}$$

$$= 19.98 \; ounces$$

We also need an estimate of the process standard deviation (σ). As mentioned earlier, there is more than one way to estimate this parameter, but since the sample sizes are small (under 10), the best way to proceed is to use the sample ranges (R) and carry out the following steps:

Determine the average of the m values of R. Call this $\bar{R}$.
Select the value of d_2 from Table 12.2 using the corresponding sample size, n. Estimate σ using

$$\hat{\sigma} = \frac{\bar{R}}{d_2} \tag{12.2}$$

■ TABLE 12.2	n	d_2	d_3	D_3	D_4
Factors for constructing an R chart.	2	1.128	.853	0	3.267
	3	1.693	.888	0	2.574
	4	2.059	.880	0	2.282
	5	2.326	.864	0	2.114
	6	2.534	.848	0	2.004
	7	2.704	.833	.076	1.924
	8	2.847	.820	.136	1.864
	9	2.970	.808	.184	1.816
	10	3.078	.797	.223	1.777

For the coffee-can illustration,

$$\bar{R} = \frac{.4 + .5 + \cdots + .4}{20}$$

$$= \frac{8.8}{20}$$

$$= .44$$

Using Table 12.2, the estimate of the process standard deviation is

$$\hat{\sigma} = \frac{\bar{R}}{d_2} = \frac{.44}{2.326} = .189 \; ounce$$

In Chapter 7, a procedure was outlined for constructing a confidence interval for the population mean using the results of a *single* sample. When constructing a control chart for the process average, a similar procedure is followed using the results of *multiple* samples (*m* of them). The center line and control limits are defined below:

$$UCL = \bar{\bar{X}} + 3\,\frac{\hat{\sigma}}{\sqrt{n}} = \bar{\bar{X}} + 3\,\frac{(\bar{R}/d_2)}{\sqrt{n}}$$

$$Center\ Line = \bar{\bar{X}} \tag{12.3}$$

$$LCL = \bar{\bar{X}} - 3\,\frac{\hat{\sigma}}{\sqrt{n}} = \bar{\bar{X}} - 3\,\frac{(\bar{R}/d_2)}{\sqrt{n}}$$

Notice that the control limits take on the appearance of a confidence interval where the value 3 replaces the value previously obtained from the standard normal (*Z*) table or the *t* table. The resulting control limits are referred to as the **3-sigma control limits.** It should be mentioned that the value 3 can be changed to fit the quality requirements of the process. Essentially, using this value produces control limits that will be exceeded approximately 27 times in ten thousand (.0027) if the process average and variation remain stable. Assuming a normal (or nearly normal) process, the value .0027 is obtained by finding the combined tail area under a standard normal curve outside ± 3 (see Figure 12.4).

For the coffee can illustration, $\bar{\bar{X}} = 19.98$ and $\hat{\sigma} = .189$, so the control limits would be

$$UCL = 19.98 + 3\,\frac{.189}{\sqrt{5}} = 19.98 + .25 = 20.23$$

$$Center\ Line = 19.98$$

$$LCL = 19.98 - 3\,\frac{.189}{\sqrt{5}} = 19.98 - .25 = 19.73$$

■ FIGURE 12.4
Tail area outside ± 3 for the standard normal curve.

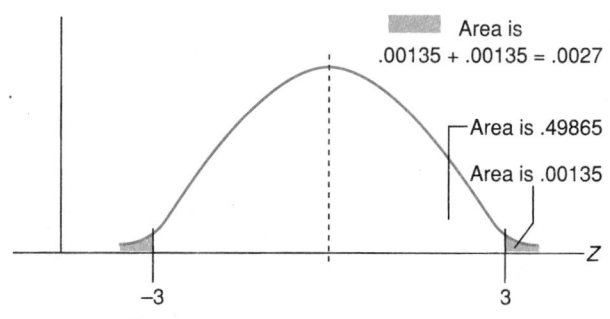

Area is
.00135 + .00135 = .0027

Area is .49865

Area is .00135

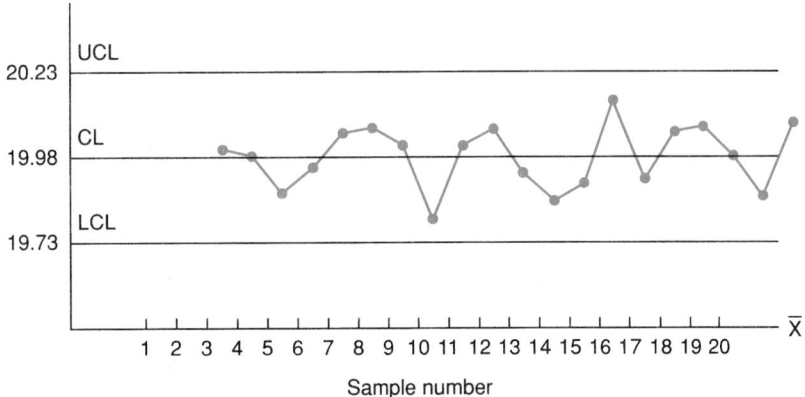

$\bar{X}$ chart for coffee-can example.

The control chart is shown in Figure 12.5. Notice that all 20 of the sample means are within the control limits, indicating that the control chart is ready for use. If one or more of the sample means falls outside the control limits, a search should be made to determine whether there is an assignable cause behind these extreme sample values. If such a cause is found, this sample should be removed and the control limits (including the center line) should be rederived.

There are many excellent computer packages available for constructing quality-control charts. MINITAB offers a variety of control chart commands for simple chart construction. The data from Table 12.1 and the corresponding MINITAB control chart are shown in Figure 12.6. Notice that the previously determined estimate of the standard deviation ($\hat{\sigma} = .189$) was specified using a MINITAB subcommand. The default MINITAB procedure estimates this standard deviation differently, but this estimated standard deviation is provided so that the results will agree with those in the previous discussion.

■ FIGURE 12.6
MINITAB $\bar{X}$ chart for the coffee-can example, using a specified standard deviation.

```
MTB > SET INTO C1
DATA> 19.8    20.1    20.2    19.9    20.0    20.3    19.9    19.8    19.8    20.1
DATA> 20.0    19.7    20.2    19.8    19.7    20.2    19.8    19.8    20.0    19.9
DATA> 20.3    19.9    20.1    20.2    19.7    20.2    20.1    20.0    19.9    20.1
DATA> 20.0    20.2    19.8    20.2    19.9    19.9    19.6    20.0    19.8    19.8
DATA> 20.0    20.3    19.8    20.1    19.9    20.4    20.1    20.0    19.7    20.1
DATA> 20.0    19.8    19.9    20.2    19.8    19.8    19.9    19.7    19.9    20.0
DATA> 19.8    20.0    19.8    19.9    20.0    19.9    19.9    20.2    20.2    20.4
DATA> 20.0    19.9    19.8    19.7    20.2    20.3    19.9    20.0    20.1    19.9
DATA> 20.1    20.1    20.2    19.9    20.0    19.8    20.2    20.2    19.7    20.0
DATA> 20.2    19.6    19.9    20.0    19.7    20.3    19.9    20.1    20.2    19.9

MTB > XBARCHART USING DATA IN C1, SAMPLE SIZE IS 5
SUBC> SIGMA = .189.

                X-bar Chart for C1

         -
         -----------------------------------------------UCL=20.23
         -
   S     -
   a  20.125+                          +
   m     -
   p     -        + +         +              + +
   l  •  -              +   +                         =
   e    --+-+-------------------------------------+----X=19.98
      19.950+      +                 +
   M     -
   e     -      +                  + +
   a     -                     +              +
   n     -
      19.775+             +
         -----------------------------------------------LCL=19.73
         -
         -
         +---------+---------+---------+---------+
         0         5        10        15        20
                  Sample Number
```

The R Chart   To monitor the process variability, we use an R chart to plot values of the sample range, R. As we do with the $\bar{X}$ chart, we conclude that the sample *variability* is out of control when a sample range falls outside the control limits of the R chart. If the sample range falls within the control limits, the process variation is in control, that is, stable. Note that it is possible for a sample range to be out of control while the corresponding sample mean ($\bar{X}$) is well in control; that is, a sample can contain extreme variation but be centered properly.

The center line for the R chart is the average ($\bar{R}$) of the m ranges. For the summary data in Table 12.1, we have already determined that $\bar{R} = .44$. The 3-sigma control limits are derived by again adding and subtracting three times the estimated standard deviation of $\bar{R}$, say $s_{\bar{R}}$. The value of $s_{\bar{R}}$ can be derived using

$$s_{\bar{R}} = \bar{R}(d_3/d_2) \tag{12.4}$$

where the values of d_2 and d_3 are provided in Table 12.2.

The control limits for the R chart are

$$\text{UCL} = \bar{R} + 3s_{\bar{R}} = \bar{R} + 3\bar{R}(d_3/d_2) = \left(1 + 3\frac{d_3}{d_2}\right)\bar{R}$$

$$\text{LCL} = \bar{R} - 3s_{\bar{R}} = \bar{R} - 3\bar{R}(d_3/d_2) = \left(1 - 3\frac{d_3}{d_2}\right)\bar{R}$$

By defining

$$D_3 = 1 - 3\frac{d_3}{d_2}$$

$$\text{and } D_4 = 1 + 3\frac{d_3}{d_2} \tag{12.5}$$

the R chart can be defined using

$$\text{UCL} = D_4\bar{R}$$
$$\text{Center Line} = \bar{R} \tag{12.6}$$
$$\text{LCL} = D_3\bar{R}$$

COMMENTS

1. Values of D_3 and D_4 are provided in Table 12.2.
2. Since a sample range is never negative, D_3 is defined to be zero whenever the expression in (12.5) is negative; that is, for $n = 2, 3, 4, 5$, and 6.

For our previous example, in which the quality characteristic is the filled weight of a coffee can, the limits for the R chart are easily found:

$$\text{UCL} = (2.114)(.44) = .93$$
$$\text{Center Line} = .44$$
$$\text{LCL} = 0$$

This R chart is shown in Figure 12.7. Note that all the sample ranges appear to be in control, so the R chart is ready for use. If future range values fall within these

■ **FIGURE 12.7**
MINITAB R chart for
the coffee-can example,
using a specified
standard deviation.

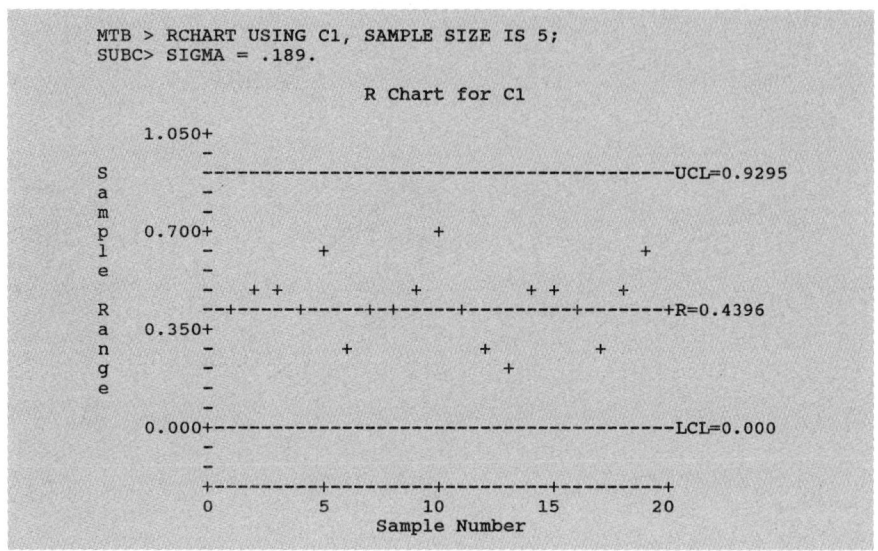

```
MTB > RCHART USING C1, SAMPLE SIZE IS 5;
SUBC> SIGMA = .189.

                         R Chart for C1

      1.050+
    S     -
    a     -----------------------------------------------UCL=0.9295
    m     -
    p 0.700+              +
    l     -        +                                    +
    e     -
      -      + +            +              + +       +
    R   --+-----+-----+-+-----+---------+-------+R=0.4396
    a 0.350+
    n     -        +          +          +
    g     -                        +
    e     -
      -
      0.000+-------------------------------------------LCL=0.000
          -
          -
      +---------+---------+---------+---------+
          0         5        10        15        20
                      Sample Number
```

limits, we conclude that the process variation is in control. Our procedure is the
same as for the $\bar{X}$ chart: if any of the values in Figure 12.7 had fallen outside the
control limits, a search would have been made for an assignable cause (or causes)
for these sample points. Samples for which such a cause is found should be removed
and the R chart derived again.

EXAMPLE 12.1

Following the construction of the control charts using the data in Table 12.1 (shown
in Figures 12.5 and 12.7), samples of five filled coffee cans were obtained every
half hour over a three-hour period. The data for these samples are shown below (in
ounces).

SAMPLE	DATA
1	19.9, 19.7, 19.9, 20.2, 20.3
2	20.1, 20.3, 19.6, 19.8, 19.5
3	19.9, 20.1, 20.3, 19.9, 19.9
4	20.1, 19.9, 20.0, 20.1, 20.3
5	20.0, 19.5, 19.5, 20.1, 20.2
6	19.7, 19.8, 20.3, 19.7, 20.1

Using the proper control charts, determine whether the process location and
variability are in control during this three-hour period.

SOLUTION

The first step is to find the sample averages and ranges.

SAMPLE	1	2	3	4	5	6
$\bar{X}$	20.00	19.86	20.02	20.08	19.86	19.92
R	.6	.8	.4	.4	.7	.6

Each sample mean is plotted in the $\bar{X}$ chart and each sample range in the R chart.
Both charts are shown in Figure 12.8. Since all points in the two charts are within
the control limits, we conclude that the process is in control (stable) and that no
adjustments to the process are necessary.

Steps for Making $\bar{X}$ and R charts

1. Collect m samples of data, each of size n (In our first example, $m = 20$ samples
of $n = 5$ observations each).
2. Compute the average of each subgroup ($\bar{X}_1, \bar{X}_2, \bar{X}_3, \ldots, \bar{X}_m$).
3. Compute the range for each subgroup ($R_1, R_2, R_3, \ldots, R_m$).

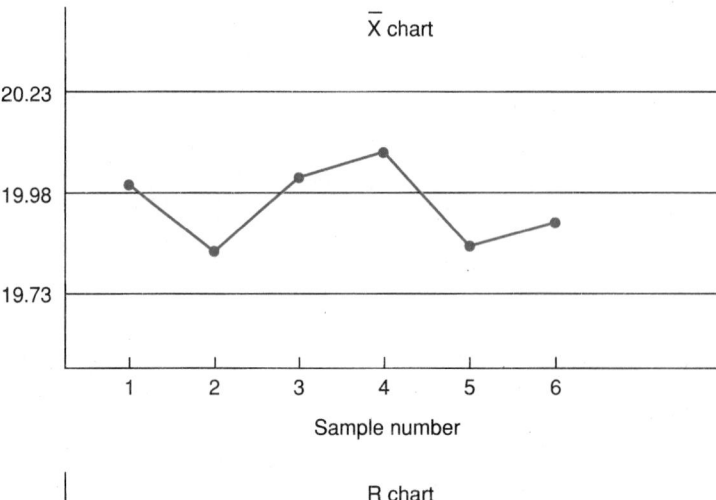

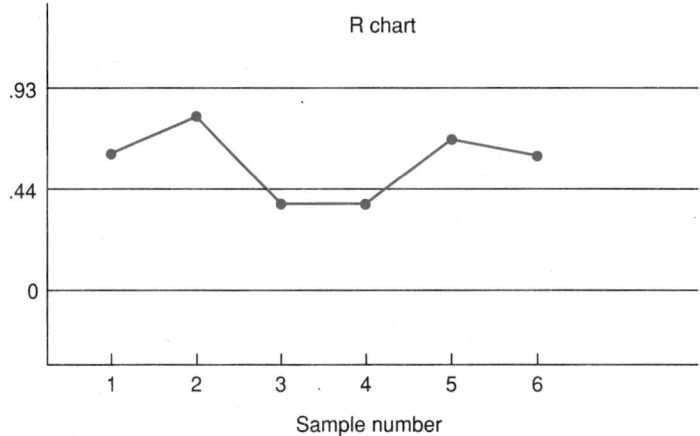

4. Find the overall mean, $\bar{\bar{X}}$, where $\bar{\bar{X}}$ is the average of the m values of $\bar{X}$.

5. Find the average range, $\bar{R}$, where $\bar{R}$ is the average of the m values of R.

6. To estimate σ (say, $\hat{\sigma}$), compute $\bar{R}/d_2$, where d_2 can be found using Table 12.2.

7. Compute the 3-sigma control limits for the $\bar{X}$ control chart:

$$\text{UCL} = \bar{\bar{X}} + 3\,\frac{\hat{\sigma}}{\sqrt{n}}$$

$$CL = \bar{\bar{X}}$$

$$\text{LCL} = \bar{\bar{X}} - 3\,\frac{\hat{\sigma}}{\sqrt{n}}$$

8. Compute the 3-sigma control limits for the R control chart.

$$\text{UCL} = D_4\bar{R} \qquad \text{(using Table 12.2)}$$
$$CL = \bar{R}$$
$$\text{LCL} = D_3\bar{R} \qquad \text{(using Table 12.2)}$$

9. Construct the control charts by plotting the $\bar{X}$ and R points for each subgroup on the same vertical line.

Pattern Analysis for $\bar{X}$ Charts   So far, our discussion of control charts has focused on determining out-of-control conditions by identifying a point beyond the 3-sigma control limits on the $\bar{X}$ or R charts. For $\bar{X}$ charts, a closer look at the pattern of the control chart points (all of which may be within the control limits) may also reveal a process that is out of control and requiring attention. For example, six points in a row, all increasing or decreasing, indicates an out-of-control process, possibly due to a gradual wearing down of a machine part or due to operator fatigue. While the process may be not outside the control limits, it can be improved by identifying and eliminating this source of variation (an assignable cause).

Pattern analysis is concerned with recognizing systematic or nonrandom patterns in a $\bar{X}$ control chart and identifying the source of such process variation. To help us detect nonrandom patterns in $\bar{X}$ charts, we divide each chart into **zones:**

Zone A contains the area between the 2- and the 3-sigma limits, both above and below the center line.

Zone B contains the area between the 1- and the 2-sigma limits, both above and below the center line.

Zone C contains the area between the center line and the 1-sigma limit, both above and below the center line.

Specific patterns indicating nonrandom variation can be summarized as follows:*

PATTERN	DESCRIPTION
1	One point beyond zone A
2	Nine points in a row in zone C or beyond, all on one side of center line
3	Six points in a row, all increasing or all decreasing
4	Fourteen points in a row, alternating up and down
5	Two out of three points in a row in zone A or beyond
6	Four out of five points in a row in zone B or beyond (on one side of center line)
7	Fifteen points in a row in zones C (above or below center line)
8	Eight points in a row beyond zones C (above or below center line)

These patterns are illustrated in Figure 12.9, where A, B, and C refer to zones A, B, and C. Note that pattern 1 illustrates what we have, up to this point, identified as an out-of-control state.

We will not attempt in this discussion to interpret the causes behind these nonrandom patterns; rather, these patterns should point out to you that *there is more to control chart inspection than looking for points outside the control limits.* The ability to interpret a particular pattern in terms of assignable causes requires a great deal of experience and knowledge of the process. For more information on pattern interpretation, see the article by Nelson and the textbook by Montgomery listed in the extra readings section.

Using the appropriate subcommands, the MINITAB $\bar{X}$ chart in Figure 12.6 is reproduced in Figure 12.10, where zones A, B, and C are included (using subcommand SLIMITS 1 2 3) and MINITAB is instructed to search for patterns 1 through 8 (using subcommand TEST 1:8). No patterns were identified. If MINITAB does detect a pattern, the sample at which the pattern is completed is flagged by placing a value (1 for pattern 1, 2 for pattern 2, and so forth) above or below the plotted point.

 *Exercises*

12.1 What is the difference between an $\bar{X}$ and an R chart (i.e., what is each useful for controlling)? What type of data is required for their use?

12.2 A marketing manager oversees a district with 15 sales representatives, each of whom travels a territory of roughly the same geographic size. The manager randomly selects five

---

*There is no general agreement about the set of nonrandom patterns. This set, which is used by MINITAB, is taken from the article by L. Nelson cited in the extra readings section at the end of this chapter.

■ FIGURE 12.9
Eight patterns requiring
a search for assignable
causes due to
nonrandom variation.
The dotted lines indicate
the nonrandom pattern.

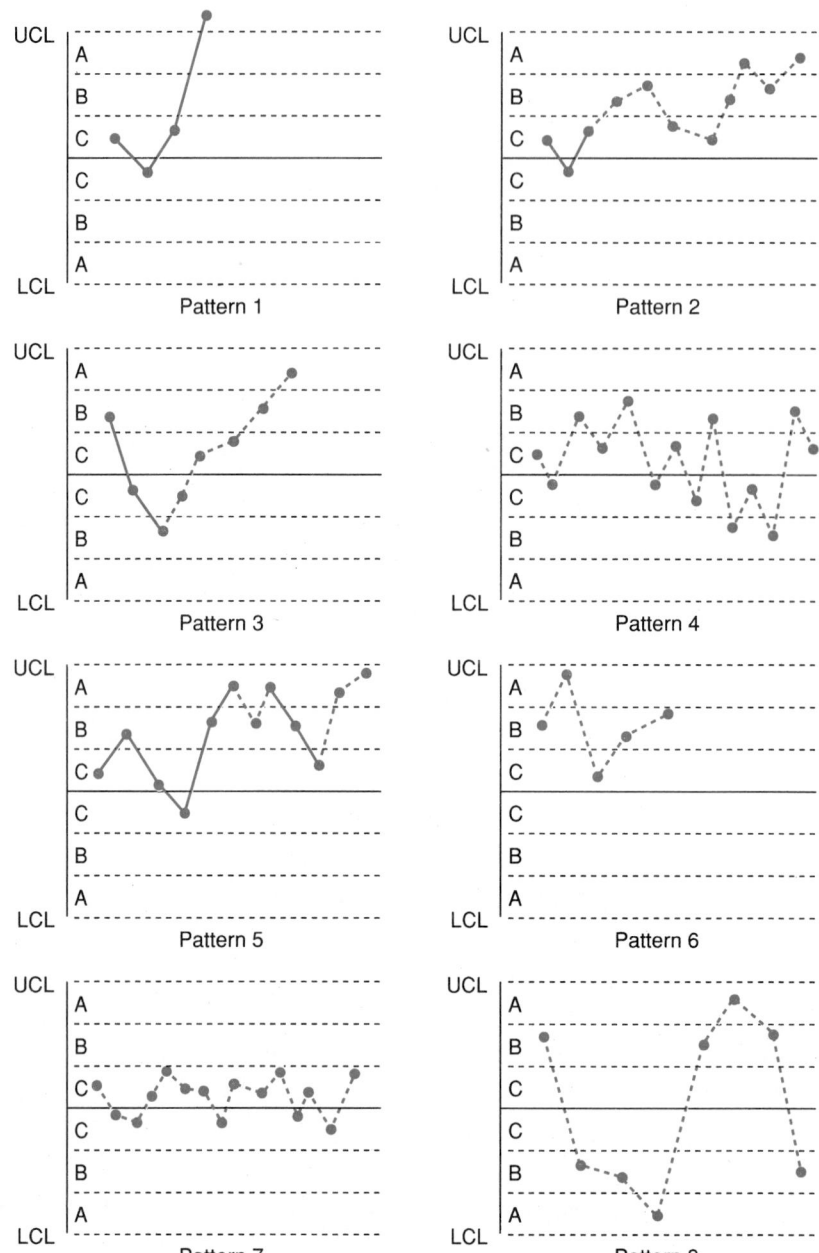

travel expense reports from the district for each of the past 20 weeks. The following are the
expense data in dollars:

WEEK	EXPENSES	WEEK	EXPENSES
1	350, 278, 319, 336, 299	11	298, 365, 312, 308, 334
2	366, 323, 297, 312, 279	12	318, 287, 312, 345, 298
3	346, 335, 288, 314, 351	13	299, 289, 356, 357, 333
4	321, 334, 357, 288, 299	14	279, 312, 322, 345, 329
5	354, 348, 311, 332, 280	15	279, 346, 289, 312, 344
6	277, 335, 328, 340, 301	16	288, 312, 338, 347, 311
7	316, 305, 355, 327, 296	17	344, 284, 295, 314, 356
8	276, 322, 345, 291, 361	18	345, 295, 312, 344, 303
9	288, 312, 335, 322, 299	19	312, 322, 346, 287, 301
10	335, 306, 317, 281, 319	20	269, 355, 302, 298, 331

a.   Construct an $\bar{X}$ chart that the manager can use to control travel expenses in the district.

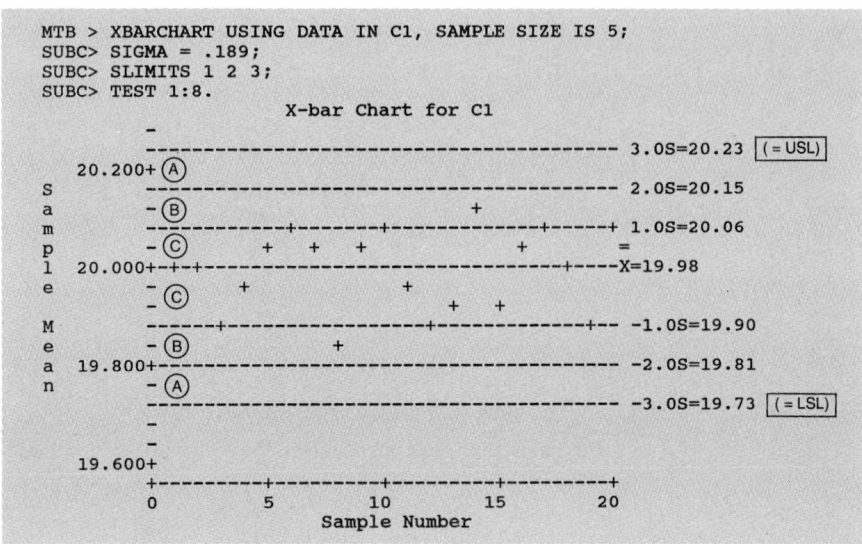

■ **FIGURE 12.10**
MINITAB $\bar{X}$ chart for the coffee-can example with specified zones showing 1-, 2-, and 3-sigma limits. MINITAB also searches for eight nonrandom patterns.

b. Perform a pattern analysis of the $\bar{X}$ chart in part a by checking for the eight nonrandom patterns discussed in this section.

c. Construct an R chart that the manager can use to control travel expenses in the district.

12.3 Assume that the Office of Thrift Supervision (OTS) has concluded that both hasty and prolonged examinations produce an inferior regulatory product. The supervisory agent for the IXth District instructs you to formulate a district standard for the timely and thorough completion of examination reports. You audit and select four accurate and thorough examination reports for each of the past twenty months and find the following completion times:

MONTH	TIME REQUIRED (DAYS)	MONTH	TIME REQUIRED (DAYS)
1	23, 21, 15, 19	11	15, 25, 13, 29
2	9, 19, 18, 25	12	13, 30, 23, 14
3	31, 19, 15, 22	13	26, 25, 12, 14
4	16, 22, 21, 13	14	8, 6, 13, 30
5	23, 13, 22, 9	15	15, 12, 29, 7
6	7, 18, 26, 25	16	23, 11, 29, 28
7	17, 11, 22, 12	17	28, 27, 23, 19
8	18, 26, 20, 23	18	7, 12, 5, 18
9	28, 26, 25, 20	19	16, 11, 12, 19
10	7, 5, 22, 11	20	16, 30, 29, 23

a. Construct an $\bar{X}$ chart that can be used to control examination completion times.

b. Perform a pattern analysis of the $\bar{X}$ chart in part a by checking for nonrandom patterns.

c. Construct an R chart that can be used to control examination completion times.

12.4 Samples of seven 4-inch pieces of strapping tape are taken at 30-minute intervals during an 8-hour production run. The tape is stretched to the breaking point, and the mean strength (in ft-lbs) is recorded for each sample. The mean range for the sixteen samples is 19.67 ($\bar{R} = 19.67$).

SAMPLE	MEAN TENSILE STRENGTH	SAMPLE	MEAN TENSILE STRENGTH
1	106.26	9	99.62
2	105.83	10	101.23
3	103.69	11	96.89
4	101.20	12	95.72
5	99.80	13	96.34
6	114.19	14	93.48
7	102.56	15	86.09
8	97.88	16	87.50

a. Construct an $\bar{X}$ chart using the above data.

b. Do you have any reservations about this $\bar{X}$ chart? What course of action do you recommend?

12.5   Samples of five light bulbs are taken from a production run at 15-minute intervals, and the intensity of each bulb is measured. The sample means for the first ten samples are reported below. The mean range for the ten samples is 97.3 ($\bar{R}$ = 97.3).

SAMPLE	SAMPLE MEAN INTENSITY (LUMENS)	SAMPLE	SAMPLE MEAN INTENSITY (LUMENS)
1	1545.6	6	1558.3
2	1489.4	7	1492.3
3	1495.2	8	1565.3
4	1568.4	9	1486.5
5	1571.9	10	1524.6

a. Construct an $\bar{X}$ chart using the above data.

The production run continues and the following are the mean intensities for the next ten samples:

SAMPLE	SAMPLE MEAN INTENSITY (LUMENS)	SAMPLE	SAMPLE MEAN INTENSITY (LUMENS)
1	1545.6	6	1479.00
2	1514.9	7	1542.10
3	1499.2	8	1482.70
4	1489.6	9	1495.80
5	1569.8	10	1522.40

b. Is the process in or out of control at this point? (Use the control chart from part a.)

12.6   A drill press operator selects samples of eight metal bushings and measures and records the internal diameter of the bushings at 15-minute intervals for the first two hours of his shift. The following are the sample ranges of the internal diameter measurements:

SAMPLE	RANGE (mm)	SAMPLE	RANGE (mm)
1	1.26	5	1.42
2	1.20	6	1.44
3	1.39	7	0.99
4	1.62	8	1.21

a. Construct an R chart using the above data.

b. What are your observations concerning the pattern of these data?

12.7   The drill press operator in Exercise 12.6 continues to sample for the next two hours of his shift and gathers the following internal diameter sample ranges:

SAMPLE	RANGE (mm)	SAMPLE	RANGE (mm)
1	1.26	5	1.90
2	1.49	6	2.06
3	1.54	7	2.19
4	1.75	8	2.34

a. Plot the above data using the R chart developed in Exercise 12.6.

b. What are your observations concerning the pattern of the data now?

c. What are the likely consequences if the process proceeds undisturbed?

d. What are some likely causes for the pattern of variation that you have observed?

12.8   Samples of eight boxes of cereal taken periodically from a production run yield the following mean weight measurements and sample ranges (in ounces):

SAMPLE	MEAN WEIGHT	RANGE	SAMPLE	MEAN WEIGHT	RANGE
1	18.9	3.1	6	17.2	4.1
2	19.3	2.7	7	19.3	3.3
3	17.5	3.6	8	17.0	3.8
4	18.4	3.8	9	18.1	3.4
5	19.0	2.9	10	18.2	4.3

a. Calculate the control limits for an $\bar{X}$ chart using the above data.

b. Calculate the control limits for an R chart using the above data.

12.9 In a process evaluation study, the lower glue joint gaps of corrugated boxes are measured in units of 1/128 inch. The standard is 48 units. In a first attempt at process evaluation, initial measurements on 80 boxes in groups of five produced the following sample means and ranges:

SAMPLE	$\bar{X}$	R	SAMPLE	$\bar{X}$	R
1	19	49	9	25	40
2	16	30	10	32	17
3	21	34	11	30	12
4	51	37	12	31	25
5	46	34	13	28	28
6	42	32	14	27	20
7	24	24	15	28	18
8	14	25	16	21	15

(*Source:* Boris Iglewicz and David Hoaglin, "Use of Box-plots for Process Evaluation," *Journal of Quality Technology* 19, no. 4 (October, 1987): 180.)

a. Construct an $\bar{X}$ chart using the above sample data.

b. Construct an R chart using the above sample data.

c. Is the process in control for these two charts?

12.10 The process in Exercise 12.9 was modified to improve the product. The following sample data were recorded after the modification.

SAMPLE	$\bar{X}$	R	SAMPLE	$\bar{X}$	R
17	53	40	25	44	20
18	48	13	26	48	17
19	45	16	27	53	7
20	56	22	28	48	15
21	46	19	29	45	19
22	53	22	30	57	49
23	53	23	31	46	26

a. Construct an $\bar{X}$ chart for the process after the modification.

b. Construct an R chart for the process after the modification.

c. Discuss the precision of the process after the modification.

12.11 a. Using MINITAB or another software package, construct $\bar{X}$ and R charts for the following data:

SAMPLE	OBSERVATIONS	SAMPLE	OBSERVATIONS
1	56, 78, 66, 65, 61, 59	21	75, 56, 66, 55, 49, 86
2	83, 56, 78, 65, 77, 55	22	74, 74, 56, 58, 89, 54
3	66, 78, 72, 55, 58, 61	23	78, 77, 62, 83, 74, 59
4	59, 77, 56, 82, 74, 66	24	77, 64, 62, 78, 55, 68
5	74, 56, 58, 82, 78, 61	25	67, 61, 81, 56, 74, 83
6	52, 81, 68, 88, 56, 79	26	69, 56, 82, 69, 61, 73
7	69, 52, 75, 66, 79, 89	27	71, 53, 47, 79, 51, 58
8	78, 76, 85, 78, 56, 63	28	65, 71, 85, 72, 65, 54
9	79, 69, 56, 68, 84, 51	29	81, 55, 80, 50, 69, 74
10	59, 68, 59, 86, 78, 62	30	59, 60, 89, 53, 64, 85
11	56, 81, 82, 59, 78, 74	31	69, 62, 55, 87, 84, 53
12	81, 72, 69, 66, 56, 58	32	78, 71, 54, 66, 52, 74
13	80, 64, 78, 73, 74, 60	33	73, 54, 76, 62, 61, 80
14	78, 59, 82, 56, 49, 53	34	76, 59, 76, 78, 83, 54
15	56, 66, 66, 52, 87, 73	35	72, 69, 47, 80, 54, 56
16	65, 58, 88, 75, 83, 49	36	69, 67, 88, 80, 74, 60
17	58, 69, 82, 75, 64, 54	37	85, 71, 60, 81, 66, 46
18	64, 78, 82, 75, 56, 49	38	62, 74, 77, 84, 52, 63
19	81, 73, 72, 61, 83, 48	39	80, 64, 85, 72, 46, 58
20	62, 75, 59, 56, 73, 81	40	67, 68, 77, 76, 85, 59

b. Perform a pattern analysis of the $\bar{X}$ chart in part a by checking for nonrandom patterns.

Control Charts for Attribute Data: The p and c Chart

Many quality characteristics cannot be measured. In these situations, an item is typically classified as either *conforming* or *nonconforming,* where the word *conforming* implies that the item conforms to specifications imposed upon the process (such as no surface scratches). Such quality characteristics are called **attributes.** The data gathered on these characteristics consist of counts or values based on counts, such as proportions. Examples of such data would be the number of blemishes in a square yard of sheet metal (monitored using a c chart) or the proportion of nonconforming computer chips in a sample of size 200 (monitored using a p chart).

Control Chart for the Proportion Nonconforming: The p Chart   The proportion nonconforming is defined as the number of nonconforming items in a population divided by the population size. It is denoted by p, and it corresponds to the binomial parameter p discussed in Chapters 5 and 10, where (as in Chapter 10) the value of p is unknown. Consequently, the process involved must satisfy the assumptions behind a binomial situation, described in Chapter 5, section 4 (page 152). When using a p chart, we concentrate on the parameter p and observe when this proportion appears to be out of control. If an item is judged on more than one quality characteristic, it is said to be nonconforming if the item does not conform to standard on one or more of these characteristics.

The reasons for using a p chart include the following:

1. Quality measurements are not possible.
2. Quality measurements are possible, but not practical (such as determining the atmospheric pressure at which an electrical component is destroyed).
3. Many characteristics on each part are being judged during inspection.
4. The main question of interest is, "will the process be able to produce conforming products over time?"

The construction of a p chart is done in basically the same way as for an $\bar{X}$ or R chart. A collection of samples (usually 20 to 25) is obtained while the process is believed to be in control. Let T_i be the number of nonconforming items in the i-th sample, and let n be the sample size. The resulting proportion nonconforming is $p_i = T_i/n$. For example, if sample 4 contains 150 items, 3 of which are nonconforming, then $n = 150$, $T_4 = 3$ and $p_4 = 3/150 = .02$.

The five-step procedure for constructing a p chart is given below and illustrated in the next example.

Steps for Making p Charts (Constant Sample Size)
1. Collect m samples of data (typically, 20 to 25), each of size, n.
2. Determine the proportion nonconforming for each sample. Call this value p_i.

$$p_i = T_i/n$$

where T_i is the number of nonconforming items in sample i and n is the sample size.
3. Find $\bar{p}$, the overall proportion nonconforming.
   $\bar{p} = $ (total number of nonconforming units)/(total sample size),
   that is, $\bar{p} = (\Sigma T_i)/(mn)$
   Note: $\bar{p}$ is merely the average of the m values of p_i.
4. Compute the 3-sigma control limits

$$\text{UCL} = \bar{p} + 3\sqrt{\frac{\bar{p}(1 - \bar{p})}{n}}$$

$$\text{CL} = \bar{p}$$ (12.7)

$$\text{LCL} = \bar{p} - 3\sqrt{\frac{\bar{p}(1 - \bar{p})}{n}}$$

5. Draw in the control lines and plot the values of p_i.

EXAMPLE 12.2 Repeated samples of 150 coffee cans are inspected to determine whether a can is out of round (the cylindrical shape of the can has been distorted) or whether it contains leaks due to improper construction. Such a can is said to be nonconforming, and p represents the proportion of nonconforming cans in the population. Twenty preliminary samples (150 cans each) are obtained.

SAMPLE NUMBER	NUMBER OF NONCONFORMING CANS	p_i
1	7	.047
2	4	.027
3	1	.007
4	3	.020
5	4	.027
6	8	.053
7	10	.067
8	5	.033
9	2	.013
10	7	.047
11	6	.040
12	8	.053
13	0	.000
14	9	.060
15	3	.020
16	1	.007
17	4	.027
18	5	.033
19	7	.047
20	2	.013
	96	

Construct the p chart for these data.

SOLUTION Steps 1 and 2 have been completed. The next step is to determine the overall proportion nonconforming, $\bar{p}$, which is found by dividing the total number of nonconforming items (96) by the total sample size (20 · 150 = 3000). Consequently,

$$\bar{p} = \frac{96}{3000} = .032$$

The control limits are then easily derived:

$$\text{UCL} = .032 + 3\sqrt{\frac{(.032)(.968)}{150}} = .075$$

$$\text{Center Line} = .032$$

$$\text{LCL} = .032 - 3\sqrt{\frac{(.032)(.968)}{150}} = -.011 \text{ (set LCL} = 0)$$

■ FIGURE 12.11
p chart for
Example 12.2.

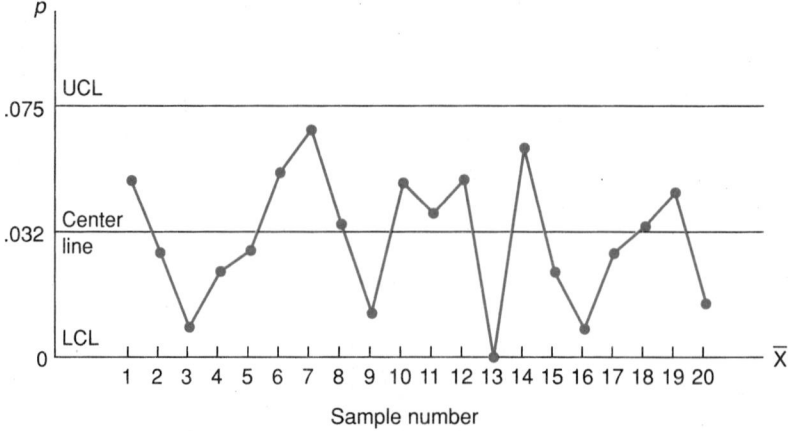

■ FIGURE 12.12
MINITAB *p* chart
for Example 12.2.

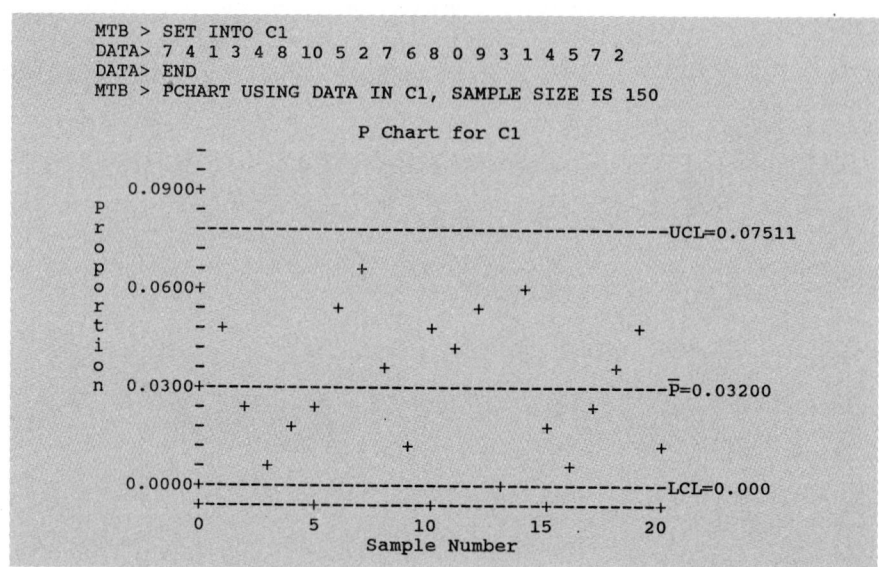

Since a negative proportion is impossible, the LCL is set equal to zero and is inactive. The resulting *p* chart is shown in Figure 12.11. Since none of the 20 sample proportions are outside the control limits, the *p* chart has been established and is ready for use.   ■

MINITAB provides a very simple method of constructing a *p* chart. The input and output for the previous example are shown in Figure 12.12 and agree with the results obtained in Example 12.2.

EXAMPLE 12.3   Continuing with Example 12.2, the next five coffee can samples (each of size 150) produced 5, 2, 11, 6, and 8 nonconforming cans, respectively. Do the sample proportions indicate that the process is in control for each of these samples?

SOLUTION   The five sample proportions are

$$5/150 = .033$$
$$2/150 = .013$$
$$11/150 = .073$$
$$6/150 = .040$$
$$8/150 = .053$$

Each of these proportions fall within the control limits in Figure 12.11, so the conclusion would be that the process is in control during this period. ∎

Notice that the third sample came close to exceeding the upper control limit of .075, and it may be tempting to react and start tampering with the process to "fix it." But, this action would defeat the purpose of a control chart, since we conclude that the occurrence of 11 nonconforming cans is simply due to chance variation and no process adjustment is called for. As was pointed out earlier, such well-meaning overadjustments result in *increased* process variation rather than improved product quality.

Control Chart for Number of Nonconformities: The c Chart   The p chart dealt with monitoring the proportion (or in a sense, the number) of *nonconforming units* in a sample of n units. Since it is quite possible for a unit to have more than one nonconformity, we must also consider the *number of nonconformities per unit*. The c chart can be used for controlling a single type of nonconformity or for controlling all types of nonconformities without distinguishing between types. Situations in which a c chart could be used would include monitoring the number of scratches on a CRT casing, the number of minor blemishes on a rubber tire, the number of loose bolts on a manufactured assembly, and so forth.

The assumption behind the process is that the number of nonconformities occurring in a unit satisfies the assumptions behind the Poisson process, described in Chapter 5, Section 6, page 162. An important characteristic of the Poisson random variable is that mean and variance are identical, and so the standard deviation is the square root of the mean. Therefore, the control chart for the number of nonconformities per unit (say, c) is very easy to construct. The method for constructing a c chart is described in the following five-step procedure.

Steps for Making a c Chart
1. Collect m samples of data (typically, 20 to 25 units), where each sample is obtained by observing a single unit.
2. Determine the number of nonconformities for the i-th unit. Call this value c_i.
3. Find the *average* number of nonconformities per unit, $\bar{c}$, where

$$\bar{c} = \Sigma c_i / m$$

4. Compute the 3-sigma control limits

$$\text{UCL} = \bar{c} + 3\sqrt{\bar{c}}$$
$$\text{Center Line} = \bar{c} \qquad (12.8)$$
$$\text{LCL} = \bar{c} - 3\sqrt{\bar{c}}$$

NOTE: If the mean of the process is known, its value may be substituted for $\bar{c}$ in the above control limits and center line.

5. Construct the control chart by drawing in the control lines and plotting the values of c_i.

EXAMPLE 12.4   An automobile assembly worker is interested in monitoring and controlling the number of minor paint blemishes appearing on the outside door panel on the driver's side of a certain make of automobile. The following data were obtained, using a sample of 25 door panels.

Panel	1	2	3	4	5	6	7	8	9	10	11	12	13	14
Number of paint blemishes	1	0	3	3	1	2	5	0	2	1	2	0	8	0

Panel	15	16	17	18	19	20	21	22	23	24	25
Number of paint blemishes	2	1	4	0	2	4	1	1	0	2	3

Construct the control chart for this situation and determine whether all the plotted points are in control.

SOLUTION   The average number of nonconformities (minor paint blemishes) for the sample of 25 door panels is $\bar{c}$, where

$$\bar{c} = \frac{1 + 0 + 3 + \cdots + 2 + 3}{25}$$

$$= \frac{48}{25}$$

$$= 1.92$$

The limits and center line for the resulting c chart are

$$\text{UCL} = 1.92 + 3\sqrt{1.92} = 6.08$$

$$\text{Center Line} = 1.92$$

$$\text{LCL} = 1.92 - 3\sqrt{1.92} = -2.24 \text{ (set LCL} = 0)$$

Since the Poisson variable is never negative, the LCL is set equal to zero here and is inactive.

The c chart constructed using MINITAB is shown in Figure 12.13. Notice that the 13th sample contains an out-of-control observation. If control limits for current or future production are to be meaningful, they must be based on data from a process that is *in control*.

The next step here would be to examine the 13th observation to determine if an assignable cause can be located. If it can, the observation should be removed, the

■ **FIGURE 12.13**
MINITAB c chart
for Example 12.4.

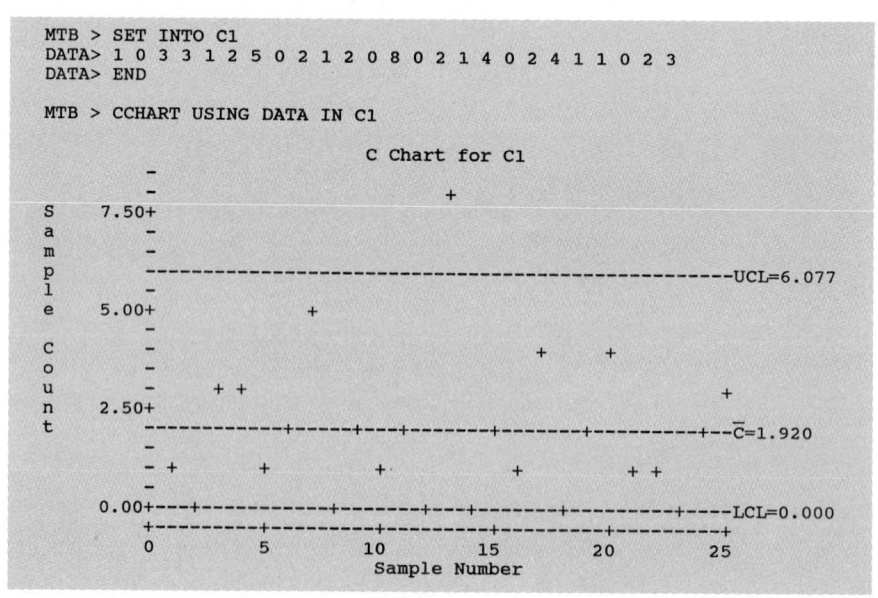

control limits recomputed (using 24 samples), and the procedure continued until all observations are in control. If no assignable cause is found, two choices exist: (1) retain this observation and keep the present control limits or (2) drop this observation assuming that an assignable cause exists, but was not identified. ∎

EXAMPLE 12.5 It was discovered that defective paint was used during sample number 13 in Example 12.4. Since this is a legitimate assignable cause, the observation was removed from the sample. Construct the modified control limits for the c chart and determine if the control chart is ready for use.

SOLUTION Removing the 13th observation produces a new total of $48 - 8 = 40$ and an average of

$$\bar{c} = \frac{40}{24} = 1.67$$

The resulting control limits are

$$\text{UCL} = 1.67 + 3\sqrt{1.67} = 5.55$$

$$\text{Center Line} = 1.67$$

$$\text{LCL} = 1.67 - 3\sqrt{1.67} = -2.21 \text{ (set LCL} = 0)$$

Reviewing the 24 samples, all of the number of nonconformities fall within the revised control limits, so the c chart has been established. ∎

COMMENTS Any control chart should be modified when it appears that a substantial decrease in variability has produced a process that is more consistent. As product quality improves, the definition of what is "in control" should also change to reflect the new standards resulting from the improved process. One always hopes that *this year's control limits will not be acceptable on next year's control chart.*

 *Exercises*

12.12 What is the difference between a p and a c chart (i.e., what does each control)? What type of data is required for their use?

12.13 Indicate whether a p or a c chart would be appropriate for controlling:

a. The number of defective RAM chips in a sample of 50.

b. The number of knots in a linear foot of a $2'' \times 4''$ stud.

c. The number of hairline cracks in an engine block.

d. The number of cartons containing broken eggs in a crate of 20 cartons.

e. The number of flaws in a cashmere sweater.

12.14 Samples of 200 fuses are subjected to a pulse of current for which they are rated. The following are the number of fuses that failed in each of twenty samples. Construct a p chart to control the proportion of fuses that fail.

SAMPLE	FAILED FUSES	SAMPLE	FAILED FUSES
1	13	11	7
2	5	12	6
3	8	13	12
4	7	14	10
5	6	15	5
6	9	16	14
7	10	17	8
8	13	18	6
9	6	19	9
10	9	20	9

12.15 A firm that produces desks inspects them for finish blemishes. The following data represent the number of blemishes found on each of 20 desks inspected. Construct a *c* chart for controlling the number of finish blemishes.

TABLE	BLEMISHES	TABLE	BLEMISHES
1	13	11	8
2	8	12	16
3	11	13	11
4	10	14	10
5	9	15	9
6	5	16	18
7	15	17	9
8	12	18	8
9	9	19	8
10	10	20	14

12.16 Assume that the historical average number of nonconformities per unit for a proven and stable process is 1.8 ($\bar{c} = 1.8$). What are the control limits for the *c* chart?

12.17 Assume that for a stable process producing integrated circuits, the historical population proportion nonconforming (p) is known to be .03.

a.   What are the control limits if samples of 500 integrated circuits are taken?

b.   What are the control limits if samples of 1,500 integrated circuits are taken?

c.   What effect does sample size have on the *p* chart control limits?

12.18 Benworth Motors builds heavy trucks. Dashboard assemblies are subjected to inspection and the number of total assembly defects are recorded. The following are the results for the inspection of sixteen dashboards.

DASHBOARD	DEFECTS	DASHBOARD	DEFECTS
1	12	9	11
2	9	10	9
3	12	11	16
4	10	12	10
5	13	13	25
6	15	14	10
7	12	15	13
8	10	16	17

a.   What type of chart is appropriate for controlling the number of dashboard defects?

b.   Construct the appropriate control chart.

c.   Do you have any recommendations concerning this chart?

12.19 Samples of 200 floppy diskettes are taken from each production run, and the diskettes are checked for bad sectors. The following are the number of diskettes with bad sectors for the past eight production runs:

RUN	BAD DISKETTES	RUN	BAD DISKETTES
1	13	5	23
2	15	6	26
3	14	7	29
4	19	8	37

a.   What type of chart is appropriate to use in this instance?

b.   Construct the appropriate control chart.

c.   Is this process in or out of control?

12.20 Samples of 150 carbon-fiber mounting struts are x-rayed for structural defects. The number of defective struts found in samples taken from the past forty production batches are shown below. Using MINITAB or another software package, construct the appropriate control chart for this data.

BATCH	DEFECTIVE STRUTS	BATCH	DEFECTIVE STRUTS
1	8	21	10
2	5	22	6
3	1	23	5
4	0	24	7
5	7	25	4
6	11	26	2
7	6	27	8
8	5	28	1
9	7	29	9
10	3	30	2
11	2	31	5
12	0	32	8
13	9	33	3
14	12	34	4
15	11	35	1
16	0	36	1
17	4	37	10
18	5	38	12
19	8	39	3
20	9	40	5

12.21 In some practical situations where inspection might destroy or deteriorate a product, acceptance sampling is still desirable. The main question is how to estimate the percent nonconforming in an accepted product after using a defect sampling plan. The data summarized in the following table pertains to a manufacturing facility that produces lots of approximately 5000 units. A random sample of 125 units is selected from each lot. In a recent month, 305 lots were submitted and the number nonconforming in a sample of 125 units from each lot was recorded.

Number nonconforming in sample of 125	0	1	2	3	4	5	8	10	11	14	15	34	106
Number of lots	227	48	10	4	5	1	3	1	1	2	1	1	1

(*Source:* Gerald J. Hahn, "Estimating the Percent Nonconforming in the Accepted Product After Zero Defect Sampling," *Journal of Quality Technology* 18, no. 3 (July 1986): 182.)

a.   Find the overall proportion nonconforming.

b.   Compute the control limits.

c.   Is the process in control during this sampling period?

Process Capability

The previous section dealt with control charts that monitor a process to determine whether it is operating in control. But the term "in control" merely indicates that the process is performing within natural variation *as measured by past performance*. It is entirely possible that the process might be in control but not be *capable* of meeting the process requirements, referred to as **specification limits.**

For example, suppose the process that produces piston rings is operating in control, with the inside diameter of the rings centered at 12.1 cm with a standard deviation of .03 cm. However, the product specifications state that in order to be of acceptable quality, the piston rings must have an inside diameter between 11.95 and 12.05 cm. The value 11.95 is called the **lower spec limit** (LSL) and 12.05 is the **upper spec limit** (USL). From this information, we would have to conclude that the piston rings process is incapable of meeting the required specifications (specs) (see Figure 12.14).

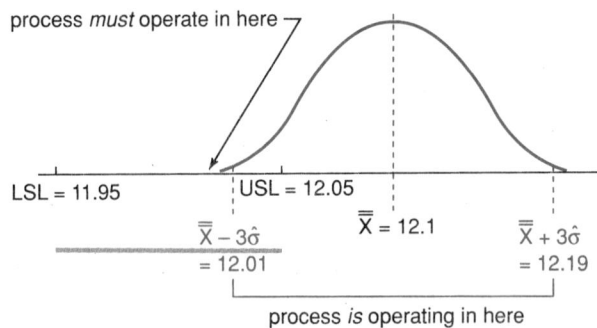

In Figure 12.14, the difference between 12.19 and 12.01 is six standard deviations (estimated) and is referred to as the **process spread:**

$$\text{process spread} = 6\hat{\sigma} \qquad\qquad (12.9)$$

For this illustration, $6\hat{\sigma} = 12.19 - 12.01 = .18$ cm.

A visual method of checking process capability is to obtain a fairly large sample and plot a histogram against the required spec limits. In the preceding example, the quality characteristic of interest is the inside diameter of a piston ring; suppose that the histogram of inside diameters looks like the one in Figure 12.15. Visually, it is clear that the process is centered at a value much too large and that the process is not capable of meeting the required specifications.

This section will introduce two descriptors of process capability that *measure* how well the process is conforming to the required specifications. We will move away from a subjective visual assessment of process capability toward an objective measure that is based on facts (sample information) rather than opinion.

Process Capability Ratios, C_p and C_{pk}

When computing the first measure (C_p), the assumptions are:

1. The process output is centered within specification.
2. The process is normally distributed.
3. The process is stable (in control).

The measure C_p is simply a comparison of the process capability with the specifications, and it is a valid indicator *only* if the above assumptions are true. It is clear from Figure 12.15 that assumption 1 is violated for the piston ring illustration, since the process is centered at a value much larger than 12 cm (the center of the spec limits).

Determining the ratio C_p is similar to comparing the width of a car driving down the center of a road to the width of the road, where the width of the car is

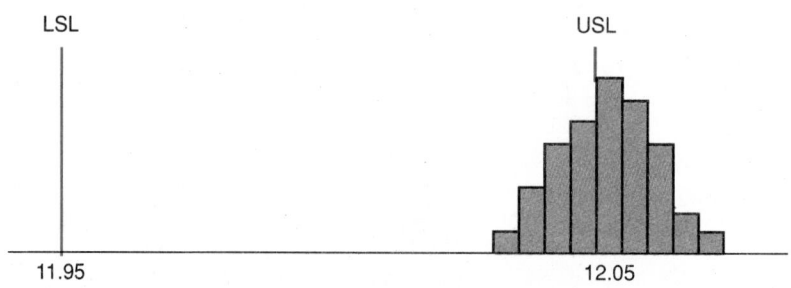

the process spread, $6\hat{\sigma}$, and the width of the road is the width of the process specs (USL − LSL). The process capability ratio C_p for a two-sided spec limit is found by dividing these two "widths":

$$C_p = \frac{USL - LSL}{6\hat{\sigma}} \qquad (12.10)$$

Very often process specifications are one-sided, specifying only a lower spec limit (for example, the bursting strength of a glass bottle) or an upper spec limit (for example, the number of missing rivets in an aircraft assembly). For these situations, we compare the distance between the sample mean and the spec limit with three standard deviations:

$$C_p = \frac{USL - \bar{X}}{3\hat{\sigma}} \qquad \text{(upper spec limit only)} \qquad (12.11)$$

$$C_p = \frac{\bar{X} - LSL}{3\hat{\sigma}} \qquad \text{(lower spec limit only)} \qquad (12.12)$$

EXAMPLE 12.6   The specification limits for the filled weight of a coffee can are 20 ± 1 (ounces), that is, the USL is 21 oz. and the LSL is 19 oz. A sample of 100 cans provides a mean of 19.98 oz. and an estimated standard deviation of .189 oz. Determine the capability ratio, C_p.

SOLUTION   The process spread is $(6)(.189) = 1.134$ oz. The value of C_p can be found from equation 12.10:

$$C_p = \frac{21 - 19}{1.134} = 1.76 \qquad \blacksquare$$

EXAMPLE 12.7   The bursting strength of a particular soft drink container has a lower spec limit of 200 psi (pounds per square inch). A sample of 100 containers produced a sample mean of 226 psi and a standard deviation of 11.3 psi. What is the capability ratio, C_p?

SOLUTION   Using equation 12.12,

$$C_p = \frac{226 - 200}{3(11.3)} = \frac{26}{33.9} = .77 \qquad \blacksquare$$

Interpreting C_p

A general rule for interpreting C_p is as follows:*

$$C_p \geq 1.33 \qquad \text{good}$$
$$1 \leq C_p < 1.33 \qquad \text{adequate}$$
$$C_p < 1 \qquad \text{inadequate}$$

---

*A more precise interpretation of C_p should consider whether the process is new or existing and whether the quality characteristic is of critical importance (such as one related to consumer safety). See the reference by Montgomery (page 373) in the extra readings section for more detail.

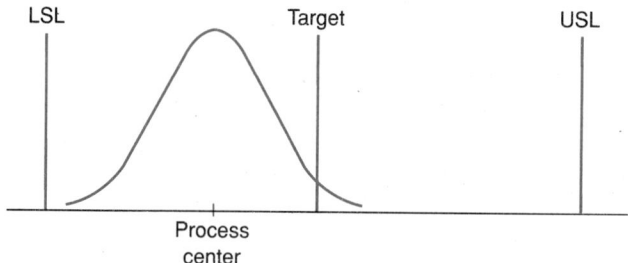

For Example 12.6, the value of $C_p = 1.76$ would be in the "good" category, and we would conclude that the *potential* capability of the coffee-can process to meet the product specs is very good, with little chance of producing a nonconforming (out-of-spec) product. The term *potential* is needed here, since *no attention has been paid to where the process is centered when determining this capability ratio.* The C_{pk} capability ratio (discussed next) considers this very important aspect of the process performance.

A value of $C_p = .77$ was determined for Example 12.7, indicating that this process has a tendency to operate dangerously close to the lower spec limit. We can expect that an unsatisfactory number of containers will have a bursting strength below the lower spec limit of 200 psi.

Consideration of Process Location: Use of C_{pk}

When determining the C_p ratio, we used the analogy of comparing the width of a car (the process spread) to the width of the road (the difference between the spec limits), while *assuming* that the car is traveling down the center of the road. The C_p ratio is a measure of potential capability.

The C_{pk} process capability ratio not only compares the width of the car to the width of the road but also questions whether the car is on or off the center stripe. This ratio examines the distance from the process center to the *nearest* spec limit (assuming both an upper and a lower spec limit), as illustrated in Figure 12.16.

The assumptions behind the use of the C_{pk} ratio are:

1. The process may or may not be centered in spec.
2. The process is normally distributed.
3. The process is stable.
4. Control charts will be used to monitor the process over time.

Procedure for Finding C_{pk}

1. Determine $R_L = \dfrac{\bar{X} - \text{LSL}}{3\hat{\sigma}}$

2. Determine $R_U = \dfrac{\text{USL} - \bar{X}}{3\hat{\sigma}}$

3. C_{pk} = Minimum of R_L and R_U

Referring to Example 12.6, we have $\bar{x} = 19.98$ oz., $\hat{\sigma} = .189$ oz., LSL $= 19$ oz., and USL $= 21$ oz. Consequently,

$$R_L = \frac{19.98 - 19}{3(.189)} = 1.73$$

$$R_U = \frac{21 - 19.98}{3(.189)} = 1.80$$

and so C_{pk} is the minimum of 1.73 and 1.80, that is, $C_{pk} = 1.73$.

The value of C_{pk} will always be less than or equal to the corresponding value of C_p. **A generally acceptable value of C_{pk} is 1.** Consequently, the coffee-can process in Example 12.6 is well centered and operating well within the upper and lower spec limits.

Taking C_{pk} One Step Further   If the process is capable ($C_{pk} > 1$)

- Monitor the process.
- Pursue continuous improvement.

If the process is not capable ($C_{pk} \leq 1$)

- Monitor the process.
- Pursue continuous improvement.
- Invest time, money, and resources to reduce process variation.
- Consider removing this product from production.

Although the fourth statement under $C_{pk} \leq 1$ may appear to be a bit drastic, it is worthy of serious consideration, given the present world of increasing product quality requirements and consumer quality demands.

Determining the Percent Nonconforming

Another method of measuring process capability is to estimate the number of nonconforming units, that is, those outside the spec limits. (The basic assumption behind this procedure is that we have reason to believe that the process is *normally distributed*. If there is reason to doubt this assumption, then the results of this section are unreliable.) There is nothing new about the procedure we will use here, since we learned in Chapter 6 that the percent nonconforming can be determined by finding the area under a normal curve. The mean and standard deviation are estimated using the sample statistics, again assuming that the process is in control during the collection of this sample. This procedure is illustrated in the following example.

EXAMPLE 12.8   A machine is used to fill plastic containers of motor oil, each of which is supposed to contain 32 fluid ounces. The process specs are $32 \pm .5$ fluid ounces. A sample of 75 containers produces a mean of $\bar{x} = 31.92$ fluid ounces and a standard deviation of $s = .16$. Describe the process capability.

SOLUTION   The sample standard deviation is the estimated process standard deviation, that is, $\hat{\sigma}$. First we determine the process capability ratio C_{pk}.

$$R_L = \frac{31.92 - 31.5}{3(.16)} = \frac{.42}{.48} = .88$$

$$R_U = \frac{32.5 - 31.92}{3(.16)} = \frac{.58}{.48} = 1.21$$

Consequently,

$$C_{pk} = \text{minimum of } .88 \text{ and } 1.21$$
$$= .88,$$

an unacceptable value.

To estimate the percent nonconforming, we examine the tails of a normal curve outside the lower and upper spec limits of 31.5 and 32.5. These tail areas are shaded

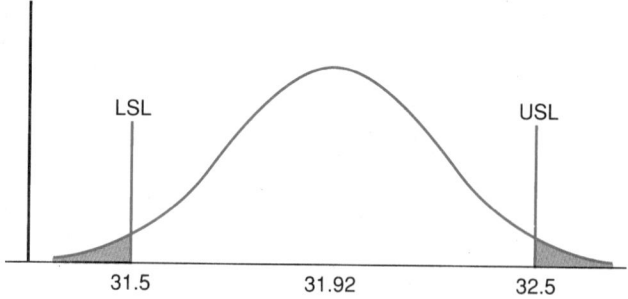

in Figure 12.17, where the curve is centered at $\bar{x} = 31.92$ with a standard deviation of $\hat{\sigma} = .16$.

The standardized values for the spec limits are

$$\text{standardized LSL} = \frac{31.5 - 31.92}{.16} = -2.62$$

$$\text{standardized USL} = \frac{32.5 - 31.92}{.16} = 3.62$$

The proportion nonconforming will be

(the proportion to the left of -2.62 under a Z curve)

+ (the proportion to the right of 3.62 under a Z curve)

= .0044 + .0002 (approximately)

= .0046

So, an estimated .46% of the oil containers will be nonconforming. At first glance, this appears to be a reasonably "small" number. However, when discussing the proportion of units nonconforming, it is common to talk in terms of the number of units nonconforming *per million* units produced. For this example, the estimated number of nonconforming oil containers per million produced is (.0046)(1,000,000) or 4,600, a number that should be large enough to get management's attention.
■

The expected number of nonconforming units per million produced *provided the process is centered in the specs* can be roughly estimated from the value of C_{pk} using Table 12.3. This table is constructed using the procedure discussed in the previous example.

Since Table 12.3 assumes the process is exactly centered on target (midway between the spec limits), it provides a measure of the *potential* number of nonconforming units per million produced, given the variation of the process. Since the process in the previous example was very nearly centered on target (32 fluid ounces) with a C_{pk} of .88, we would expect Table 12.3 to provide a crude estimate of the

■ **TABLE 12.3**
Determining the number
of nonconforming units
per million produced
using C_{pk}.

C_{pk}	NUMBER OF NONCONFORMING UNITS PER MILLION PRODUCED
.5	133,614
.75	24,448
1.00	2,700
1.30	96
1.50	6.8
2.00	.002

number nonconforming, namely, between 2,700 and 24,448. The actual number of nonconforming units per million produced was estimated to be 4,600.

Exercises

12.22 What phenomena does process capability measure? Explain the primary difference between the measures of process capability C_p and C_{pk}.

12.23 For each of the following processes (i.e., A–D), a sample of size 400 is collected in order to estimate the process mean. Indicate which measure of process capability appears the most appropriate to use for each process.

PROCESS	SPECIFICATIONS	SAMPLE MEAN
A	124.0 ± 8.5	117.6
B	0.66 ± 0.08	0.67
C	89.2 ± 2.0	90.9
D	0.13 ± 0.02	0.132

12.24 Use the data from Exercise 12.8 and assume that the process specification limits are 17.5 ± 2.5 (ounces).

a. Calculate C_p.

b. What is your assessment of the capability of this process?

c. Would the measure of process capability C_{pk} be more appropriate to use? Why?

12.25 Replicate Exercise 12.24, but calculate C_{pk} instead. Is this process centered nearer the upper or lower specification limit?

12.26 Refer to Exercise 12.9. The specification limits for the glue joints are 48 ± 16 units.

a. Calculate C_p.

b. What is your assessment of the capability of this process?

c. Calculate C_{pk}. Would this index be more appropriate to use? Why?

12.27 For each of the following centered processes (i.e., A–D), a sample of size 325 is collected and the standard deviation of the process is estimated. Calculate C_p and indicate whether you believe the process capability is "good," "adequate," or "inadequate."

PROCESS	SPECIFICATIONS	SAMPLE STANDARD DEVIATION
A	0.65 ± 0.07	0.03
B	29.3 ± 2.5	0.74
C	99.5 ± 7.2	1.11
D	0.25 ± 0.02	0.006

12.28 The specification limits for the crossover frequency of a given capacitor are 2500 ± 200 hertz (hz). A normally distributed sample of 250 of these capacitors yields a mean of 2,409 hz and a standard deviation of 105 hz.

a. Calculate C_{pk} for this process.

b. What proportion of capacitor crossover frequencies fall below the LSL?

c. What proportion of capacitor crossover frequencies fall above the USL?

12.29 The specification limits for the length of a specific piston rod are 171.0 ± 0.9 mm. A normally distributed sample of 180 piston rods yields a mean length of 171.2 mm and a standard deviation of 0.47 mm.

a. Calculate C_{pk} for this process.

b. Estimate the number of nonconforming pistons per million using Table 12.3.

c. Calculate the actual number nonconforming per million.

12.30 A decision must be made between two experimental processes for producing silicon chips. The specification limits on the thickness of the chips are 0.125 ± 0.003 mm. Samples of 500 chips are drawn from experimental runs for each process and means and standard deviations are recorded below.

PROCESS	SAMPLE MEAN	SAMPLE STANDARD DEVIATION
A	0.1254	0.0008
B	0.1231	0.0003

a. Which of the two processes is better centered?

b. Calculate C_{pk} for both processes.

c. Which process do you recommend, and why?

12.31 Three processes in your manufacturing facility have been producing unacceptably high numbers of nonconforming parts. You have drawn normally distributed samples of 375 from each process. The sample means and standard deviations are shown below:

PROCESS	SPECIFICATIONS	SAMPLE MEAN	SAMPLE STANDARD DEVIATION
A	51.0 ± 1.2	51.2	1.1
B	102.5 ± 8.9	108.1	2.3
C	3.4 ± 0.5	3.7	0.4

a. Calculate the Z scores for the specification limits and the spread between specification limits (in standard deviations) for each process.

b. The plant manager wants to know your thoughts regarding this predicament. Considering the Z scores and spreads that you have calculated, do you believe that the problems are attributable to improper centering of the processes, excessive process variability, or both? Formulate your answer for each process individually.

12.4 Acceptance Sampling

Acceptance sampling was briefly introduced in Chapter 5 (Example 5.11). In acceptance sampling, we decide whether to accept a lot (batch) of say, electrical components or machine parts, based on the results of a sample from this lot. Such a sampling strategy might resemble the following:

The lot of 8,000 parts will be accepted if a sample of 100 parts contains 2 or fewer parts that are defective.

Alternatives to acceptance sampling include: (1) accept the lot with no sampling and (2) 100% inspection. It is important to note that even with 100% inspection, there is no guarantee that all defective (or nonconforming) units will be detected, since inspector fatigue and normal human error will result in defective units passing through inspection. Such a sampling plan is useful, however, when the component is extremely critical or when the vendor (supplier of the part) has a process capability that is unable to adequately meet the product specifications.

One aspect of acceptance sampling that has come under fire is that such a plan will not *improve* product quality, since it is only a mechanism for separating the conforming units from the nonconforming. Nevertheless, when the sampling plan is applied to a stream of lots from a vendor, it offers protection for both the producer of the lot and the consumer. It generally has the effect of pressuring the vendor to improve the production process, since, in these days of higher quality awareness, vendors realize that better quality is required to stay in business.

The type of sampling plan discussed in this section is referred to as a **single-sampling plan for attributes.** For such a plan, we need to specify

N = the lot size

n = the sample size

c = the cutoff point, above which the lot is rejected

■ FIGURE 12.18
Single-sampling plan.

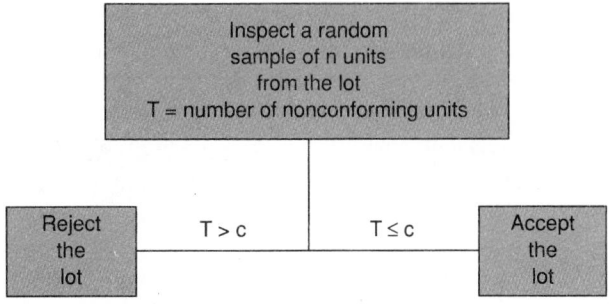

For the illustration at the start of this section, $N = 8000$, $n = 100$, and $c = 2$. This sampling plan is illustrated in Figure 12.18.

EXAMPLE 12.9

A lot contains 8000 machine parts that must have an outside diameter between 2.65 centimeters and 2.75 centimeters. A part will be nonconforming if its diameter is less than 2.65 centimeters or greater than 2.75 centimeters. A random sample of 100 parts is selected and measured. The lot will be accepted if 2 or fewer of the parts in the sample are nonconforming. If 3% of the parts in the lot are, in fact, nonconforming, determine the probability that the lot will be accepted.

SOLUTION

This is a sampling plan with $N = 8000$, $n = 100$, and $c = 2$. Such a plan fits the *hypergeometric* situation defined in Chapter 5, where a "success" is defined to be that a machine part is nonconforming. As described in Chapter 5, hypergeometric probabilities can be approximated by the *binomial* random variable, provided n/N is less than .05. Here, $n/N = 100/8000 = .0125$, so the binomial provides an excellent hypergeometric approximation.

The binomial parameter, p, was specified as $p = .03$. In fact, the value of p is generally unknown, so the probability of accepting the lot should be determined for a variety of values of p. Such an analysis leads to a set of corresponding probabilities, one for each value of p, that can be connected to form a curve. This curve is called an **operating characteristic (OC)** curve and will be described in the next section.

The probability of accepting the lot using the binomial probabilities is

$$P(\text{accept lot}) = {}_{100}C_0(.03)^0(.97)^{100} + {}_{100}C_1(.03)^1(.97)^{99} + {}_{100}C_2(.03)^2(.97)^{98}$$
$$= (.97)^{100} + (100)(.03)(.97)^{99} + (4950)(.0009)(.97)^{98}$$
$$= .0476 + .1471 + .2252$$
$$= .420$$

Consequently, there is a 42% chance of accepting the lot of machine parts using this sampling plan if 3% of the lot is nonconforming. ■

This binomial probability could also have been approximated using the Poisson distribution. By using a Poisson variable with mean $= \mu = np = (100)(.03) = 3$, the probability of two or fewer nonconformities can be obtained from Table A.2 with $\mu = 3$; namely

$$P(\text{accept lot}) = P(2 \text{ or fewer})$$
$$= .0498 + .1494 + .2240$$
$$= .4232$$

Other Acceptance Sampling Plans

Double-Sampling Plans   A double-sampling plan provides more flexibility than the single-stage plan. With this type of plan, both very good and very poor lots can be detected with smaller sample sizes. To completely specify a double-sampling plan, values of n_1, n_2, c_1, and c_2 must be determined, where the sampling plan is:

1. Accept the lot if the first sample of size n_1 contains c_1 or fewer nonconforming units.
2. Reject the lot if the first sample contains more than c_2 nonconforming units.
3. If the number of nonconforming units in the first sample is greater than c_1 and less than or equal to c_2, obtain a second sample of size n_2.
4. Accept the lot if the *combined* sample of size $n_1 + n_2$ contains c_2 or fewer nonconforming units.
5. Reject the lot if the *combined* sample contains more than c_2 nonconforming units.

The double-sampling plan is shown in Figure 12.19.

   A double-sampling plan may reduce the total amount of required inspection since the value of n_1 is typically less than the sample size for a comparable single-stage plan. If the lot is extremely good, then there should be c_1 or fewer nonconformities, thereby saving inspection effort. Conversely, if the lot is of very poor quality, then the number of nonconformities will be greater than c_2 and will be detected with the small sample of size n_1. On the other hand, a double-sampling plan offers a second chance to a lot of questionable quality and provides the vendor two opportunities to have the lot accepted.

   Tables exist that outline a variety of single-sampling and double-sampling plans. The most commonly used set of tables was developed during World War II and is referred to as MIL-STD-105D (MIL-STD stands for "military standard"). To use these tables, the consumer must specify the **acceptable quality level (AQL)** where

   AQL is the poorest level of quality for the vendor's process that the consumer would consider to be acceptable as a process average.

Based upon this specified value, a single-, double- (or multiple-) stage sampling plan can be found by merely consulting the corresponding MIL-STD-105D table.

■ **FIGURE 12.19**
Double-sampling plan.

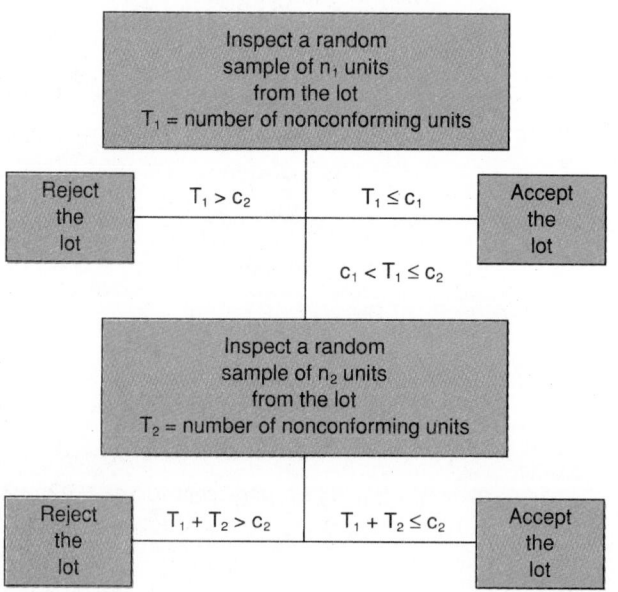

Lot Acceptance Plan for Variables Another set of military standard tables exists for constructing sampling plans for use with variables sampling. For this situation, the quality characteristic is a *measurable* variable, such as length or weight. The sampling plans are based (as in MIL-STD-105D) on a specified value for the AQL and are summarized in a set of tables called the MIL-STD-414 tables.

The MIL-STD-414 sampling procedures operate in much the same way that the process capability ratio C_{pk} was calculated. The value of C_{pk} is unacceptable (generally, if it is less than 1) if the process mean is "pushing" either the lower spec limit (LSL) or the upper spec limit (USL). By the term *pushing* (when using the MIL-STD-414 tables), we mean that $(\bar{X} - LSL)/s$ or $(USL - \bar{X})/s$ is less than a specified cutoff value provided by the MIL-STD-414 tables. The lot is accepted provided the sample mean is at an acceptable distance from the LSL *and* the USL.

Operating Characteristic (OC) Curves

When using acceptance sampling plans, the user of the sampling plan is really conducting a statistical test, with hypotheses

H_0: the lot is of acceptable quality

H_a: the lot is not of acceptable quality

So in a sense, rejecting the lot is the same as rejecting the null hypotheses (H_0). In Chapter 8 (section 8.1) we discussed the **power** of a statistical test, where

$$\text{power} = 1 - \beta = P(\text{rejecting } H_0)$$

and β = probability of a Type II error.

In this section, we focus on β, the chance of *accepting* the lot given the nonconforming proportion (p) of the lot. In Example 12.9, for a single-stage sampling plan with $N = 8000$, $n = 100$, and $c = 2$, there was a 42% chance of accepting the lot given that the proportion of nonconforming units in the lot was $p = .03$. For this situation, $\beta = .42$.

EXAMPLE 12.10 For a single-stage sampling plan with $N = 8000$, $n = 100$, and $c = 2$, determine the probability of accepting the lot if (1) $p = .05$ and (2) $p = .10$.

SOLUTION 1 Using the binomial approximation,

$$\begin{aligned}
\beta = P(\text{accept lot}) &= {}_{100}C_0(.05)^0(.95)^{100} + {}_{100}C_1(.05)^1(.95)^{99} \\
&\quad + {}_{100}C_2(.05)^2(.95)^{98} \\
&= (.95)^{100} + (100)(.05)(.95)^{99} \\
&\quad + (4950)(.0025)(.95)^{98} \\
&= .0059 + .0312 + .0812 \\
&= .1183
\end{aligned}$$

SOLUTION 2
$$\begin{aligned}
\beta = P(\text{accept lot}) &= (.9)^{100} + (100)(.1)(.9)^{99} + (4950)(.01)(.9)^{98} \\
&= .0000 + .0003 + .0016 \\
&= .0019
\end{aligned}$$

If we plot β against the proportion nonconforming, p, the resulting curve is called an **operating characteristic (OC)** curve. An example is shown in Figure 12.20, where the three points determined in Examples 12.9 and 12.10 are labeled 1 ($p = .03$), 2 ($p = .05$), and 3 ($p = .10$). Once the curve is constructed, you

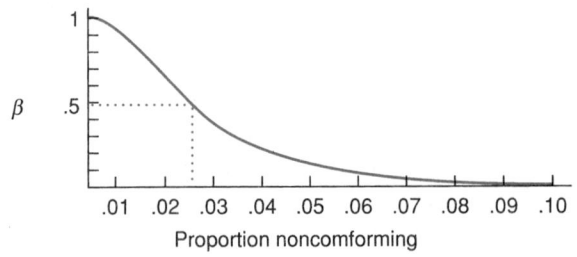

can easily interpolate for other values of p. For example, when the proportion of nonconforming units is approximately .025, there is a 50-50 chance of accepting the lot using this particular sampling plan. *If the consumer is dissatisfied with a particular OC curve, then one option would be to change the sampling plan to one that would offer better protection.*

Protection   A key aspect of an acceptance sampling plan is to offer protection, both to the consumer (who doesn't want to accept a bad lot) and to the producer/vendor (who doesn't want a good lot to be rejected). From the viewpoint of the hypotheses introduced at the start of this section, we can define

α = **producer's risk** = probability of rejecting
product of acceptable quality

β = **consumer's risk** = probability of accepting
product of unacceptable quality

For a given sampling plan, the corresponding OC curve provides a value for the *consumer's risk* for each specified value of product quality (the proportion nonconforming). In Example 12.10, the consumer will have an 11.8% risk of accepting a lot containing 5% nonconforming parts using that particular sampling plan. One option for the producer is to negotiate with the consumer for an *acceptable quality level* (AQL) that allows for a high chance of accepting a lot containing a low percentage of nonconforming parts.

It becomes clear that there is no one sampling plan that best fits all situations. Both the producer and the consumer want a plan that assures good quality yet preserves their individual interests. It is therefore appropriate to give careful consideration to the family of OC curves in the selection of a sampling plan.

There is no shortage of OC curves available in the literature, both for attribute sampling (single-stage and double) and for variable sampling. The complete MIL-STD-105D and MIL-STD-414 tables contain a very comprehensive set of OC curves for the sampling plans discussed in those documents. Note that double-sampling plans do *not* offer additional consumer protection, since single-stage and double-sampling plans can be chosen so that they have identical OC curves. Both plans would then offer the same protection against accepting or rejecting lots of specified quality.

◢ *Exercises*

12.32   What is the difference between acceptance sampling and process control (what are the objectives of each)? What is meant by the statement, "You can't inspect quality into a product"?

12.33   What is the difference between MIL-STD-105D and MIL-STD-414? When is each appropriate to use (what type of data is required)?

12.34   A single-stage sampling plan uses a sample size of 30 and a cutoff point of 3 (i.e., $n = 30$, $c = 3$). Using the Poisson approximation, what is the probability of accepting a lot of size 700 ($N = 700$) if:

a.   the proportion of the lot nonconforming is known to be 5% ($p = 0.05$)?

b. $p = 0.10$?

c. $p = 0.15$?

12.35 A loudspeaker manufacturer buys polypropylene woofers from a British vendor. Samples of 20 woofers are selected from incoming lots of 1,000 woofers, and the cutoff point is one nonconforming woofer ($c = 1$). What is the probability of accepting a lot if the vendor's process historically produces 5% of product nonconforming ($p = 0.05$)?

a.   Calculate this probability using the binomial probability (Table A.1 in appendix A).

b.   Calculate this probability using the Poisson probability (Table A.3 in appendix A).

12.36 A firm that provides on-site inventory services claims a historical accuracy rate of 98% (i.e., $p = 0.02$). A store containing 10,000 items is inventoried. Following the inventory, a sample of 100 items is taken, and the corresponding inventory information is audited. If the cutoff point is two items nonconforming ($c = 2$), what is the probability of accepting this job?

a.   Calculate this probability using the binomial probability mass function.

b.   Calculate this probability using the Poisson probability (Table A.3 in appendix A).

12.37 Construct a flowchart depicting the double-sampling plan process.

12.38 Using a double-sampling plan with $n_1 = 15$, $n_2 = 45$, $c_1 = 1$, and $c_2 = 4$, determine the appropriate course of action if a sample of size n_1 is taken and results in:

a.   0 units nonconforming

b.   1 unit nonconforming

c.   2 units nonconforming

d.   3 units nonconforming

e.   4 units nonconforming

f.   5 units nonconforming

12.39 Using a double-sampling plan with $n_1 = 5$, $n_2 = 20$, $c_1 = 0$, and $c_2 = 3$, a sample of size n_1 was taken and resulted in one nonconforming unit. Subsequently, a second sample of size n_2 was taken. What is the appropriate course of action if the second sample contained:

a.   0 units nonconforming

b.   1 unit nonconforming

c.   2 units nonconforming

d.   3 units nonconforming

e.   4 units nonconforming

12.40 What is the difference between producer's risk and consumer's risk? What type of error does each represent (Type I or II)?

Use the Poisson approximation for Exercises 12.41–12.47.

12.41 For a single-stage sampling plan, for samples drawn from lots of 2,000, with $n = 80$, and $c = 3$, calculate β when the proportion of nonconforming units in the lot is known to be:

a. $p = 0.005$

b. $p = 0.03$

c. $p = 0.05$

d. $p = 0.07$

e. $p = 0.10$

12.42 Formulate the operating characteristic curve for the sampling plan of Exercise 12.41.

12.43 Formulate the OC curve for a single-stage sampling plan with $N = 1,500$, $n = 60$, and $c = 2$. Calculate β using at least five different values of p, of your choosing, in order to plot the curve.

12.44 Replicate Exercise 12.43 for a single-stage sampling plan with $N = 1,500$, $n = 60$, and $c = 4$.

a. What is the value of β when $p = 0.04$? What is it using the OC curve of Exercise 12.43?

b. What effect does increasing the cutoff point have on consumer's risk (β), if all other conditions are held constant?

c. What effect do you think reducing c will have on β, if all other conditions are held constant?

12.45 Formulate the OC curve for a single-stage sampling plan with $N = 2,500$, $n = 100$, and $c = 2$. Calculate β using at least five different values of p, of your choosing, in order to plot the curve.

12.46 Replicate Exercise 12.45 for a single-stage sampling plan with $N = 2,500$, $n = 50$, and $c = 2$.

a. What is the value of β when $p = 0.05$? What is it using the OC curve of Exercise 12.45?

b. What effect does decreasing sample size have on consumer's risk (β), if all other conditions are held constant?

c. What effect do you think increasing n will have on β, if all other conditions are held constant?

12.47 You have negotiated two possible single-stage sampling plans by which to inspect lots of 2,500 resistors, both of which are acceptable to the vendor (producer). The two plans have the following parameters:

PLAN	n	c
A	75	2
B	125	3

Assuming that $p = 0.04$, which of the two plans do you, the consumer, desire? (*Hint:* Calculate β for both plans.)

Summary American manufacturing and service industries are very definitely undergoing an evolution that emphasizes quality. The need for tools to monitor quality has sparked a renewed interest in the everyday application of statistical thinking. Decisions are routinely based on *facts* gathered from sample data.

This chapter has focused primarily on the **inspection** side of statistical quality control. It is important to keep in mind that the world of quality emphasis extends into changing the management philosophy (see the book by Mary Walton in the extra readings section) as well as concentrating efforts into building quality into a product or process. Quality does not simply mean inspection. Nevertheless, the statistical quality tools introduced in this chapter are a key part of the quality improvement process.

A **process** can be any combination of resources, such as people, machines, and/or material that lead to a product or service. **Statistical process control** (SPC) is the application of statistical procedures intended to measure, analyze, and control this process.

To monitor a process, you must first define what it is you wish to measure or count. For measurement data, what you are measuring is the **quality characteristic** and such measurements result in **variables data.** Examples of variables data include the weight or length of a manufactured unit. When counting, you obtain **attribute data,** such as the number of nonconformities per unit.

For each type of data, **control charts** can be set up to monitor a process to determine whether it is in control (stable) or out of control (unstable). Provided the plotted points for a control chart stay within the **upper control limit** (UCL) and the

lower control limit (LCL), the process is exhibiting natural variation and is said to be in control. If a plotted point exceeds either of the limits, a search is made for an **assignable cause,** that is, an explanation (such as defective raw material) for this extreme value in the control chart. Control charts for variables data consist of the $\bar{X}$ **chart** for monitoring the process center and the R **chart** for monitoring the process variation. For attribute data, the most commonly used control charts are the p **chart** for observing and controlling the proportion of nonconforming units and the $\dot{c}$ **chart** for the number of nonconformities per unit.

For most production situations, a process must conform to certain requirements referred to as **specification (spec) limits.** These may be imposed by an outside purchaser or internally by the company's engineering staff. The ability of the process to conform to these specifications is the **process capability.** If a "large" percentage of the units produced can be expected to lie outside the spec limits, we would conclude that the process is not capable of performing to specification. Measures of process capability consist of the C_p ratio, which does not consider where the process is centered, and the C_{pk} ratio, which does. The C_p ratio measures how well the process *should* perform (the difference of the spec limits) compared with how well the process *is* performing (the **process spread,** defined as six standard deviations).

The procedure of inspecting large lots (batches) and accepting or rejecting them based on sample results is **acceptance sampling.** Sampling plans must take into consideration the chances of accepting a "bad" lot (the **consumer's risk**) and rejecting a "good" lot (the **producer's risk**). One option is for both parties to agree on the poorest level of quality that would be considered to be acceptable as a process average and choose the sampling plan accordingly. This level is the **acceptable quality level (AQL)** and is a value that must be specified in order to use the military standard tables such as MIL-STD-105D (used for attribute sampling) and MIL-STD-414 (used for variables sampling).

If the quality of the lot can be measured by the proportion (p) of nonconforming units, a graph called the **operating characteristic (OC)** curve can be constructed for any sampling plan; it plots the probability of accepting the lot against the proportion, p. For example, with a particular sampling plan, one of the points on the corresponding OC curve might indicate that there is a 62% chance of accepting the lot if 3% of the units are nonconforming. OC curves can be constructed for *any* sampling plan, using either variables or attribute data.

Summary of Formulas

Control Charts

1. $\bar{X}$ Chart:

$$UCL = \bar{\bar{X}} + 3\frac{\hat{\sigma}}{\sqrt{n}}$$

$$CL = \bar{\bar{X}}$$

$$LCL = \bar{\bar{X}} - 3\frac{\hat{\sigma}}{\sqrt{n}}$$

where $\bar{\bar{X}} = \Sigma \bar{X}_i/m$ and $\hat{\sigma} = \bar{R}/d_2$
(values of d_2 provided in Table 12.2)

2. R chart:

$$UCL = D_4\bar{R}$$

$$CL = \bar{R}$$

$$LCL = D_3\bar{R}$$

(values of D_3 and D_4 provided in Table 12.2)

3. p chart:

$$UCL = \bar{p} + 3\sqrt{\frac{\bar{p}(1 - \bar{p})}{n}}$$

$$CL = \bar{p}$$

$$LCL = \bar{p} - 3\sqrt{\frac{\bar{p}(1 - \bar{p})}{n}}$$

4. *c* chart:

$$UCL = \bar{c} + 3\sqrt{\bar{c}}$$

$$CL = \bar{c}$$

$$LCL = \bar{c} - 3\sqrt{\bar{c}}$$

where $\bar{c}$ = average number of nonconformities per unit

Process Capability

5. Process spread = $6\hat{\sigma}$

6.

$$C_p = \frac{USL - LSL}{6\hat{\sigma}}$$ (two-sided spec limit)

$$C_p = \frac{USL - \bar{X}}{3\hat{\sigma}}$$ (upper spec limit only)

$$C_p = \frac{\bar{X} - LSL}{3\hat{\sigma}}$$ (lower spec limit only)

7. C_{pk} = minimum of R_L and R_U

where

$$R_L = \frac{\bar{X} - LSL}{3\hat{\sigma}}$$

$$R_U = \frac{USL - \bar{X}}{3\hat{\sigma}}$$

Acceptance Sampling (attribute)
Probability of accepting the lot:

$$_nC_0 p^0 (1 - p)^n + {_nC_1} p^1 (1 - p)^{n-1}$$
$$+ \cdots + {_nC_c} p^c (1 - p)^{n-c}$$

where

n = sample size
p = proportion of the lot that is nonconforming
c = cutoff, above which the lot is rejected

Further Reading

1. Duncan, A. J. *Industrial Quality Control.* 5th ed. Homewood, Ill.: Irwin, 1986.
2. Duncan, A. J. *Quality Control and Industrial Statistics.* 5th ed. Homewood, Ill.: Irwin, 1986.
3. Grant, E. L., and R. S. Leavenworth. *Statistical Quality Control.* 6th ed. New York: McGraw-Hill, 1988.
4. Juran, J. M. *Quality Control Handbook.* 4th ed. New York: McGraw-Hill, 1988.
5. Kane, V. E. "Process Capability Indices." *Journal of Quality Technology* 18, no. 1 (1986): 41.
6. Montgomery, D. C. *Introduction to Statistical Quality Control.* 2nd ed. New York: Wiley, 1991.
7. Nelson, L. "The Shewhart Control Chart—Tests for Special Causes." *Journal of Quality Technology* 16 (1984): 237.
8. Shewhart, W. A. "Quality Control Charts." *Bell System Technical Journal* (1926).
9. United States Department of Defense. *Sampling Procedures and Tables for Inspection by Attributes* MIL STD 105D. Washington, D.C.: Government Printing Office, 1963.
10. United States Department of Defense. *Sampling Procedures and Tables for Inspection by Variables for Percent Defective* MIL STD 414. Washington, D.C.: Government Printing Office, 1957.
11. Walton, M. *The Deming Management Method.* New York: Putnam Publishing Group, 1986.

Review Exercises

12.48 What are the different types of variation? Are there differences in our ability to diagnose and control the different types of variation? Explain your answer.

12.49 Explain what is meant when a process is said to be "in statistical control."

12.50 Differentiate between specification limits and control limits. Is there a measure that conveys the relationship between the two? Explain.

12.51 Classify the following as attribute or variables data:

a. the refractive index of a lens

b. the number of typographical errors in a legal document

c. the number of defective tires in a production batch

d. the voltage output of a transformer

e. the torque of an electric motor

12.52 Motorola, a winner of the Malcolm Baldrige National Quality Award in 1988, is planning to achieve six sigma quality (99.9997% defect-free products) in its product line by 1992.

(*Source:* Brian M. Cook, "In Search of Six Sigma: 99.9997% Defect-Free," *Industry Week,* October 1, 1990.)

a. Comment on the term *six sigma*. What does it mean?

b. If six sigma is to be achieved by Motorola, about how many defects per million are allowed in the production process?

c. Suppose that in a sample of 20,000 memory chips taken each day from the process line for a period of one month, the overall defective rate is .00003. Construct a p chart for this process.

12.53 Samples of 100 hard-drive controllers are tested following a soldering process. The following are the number of nonconforming controllers out of each of the last fourteen samples:

SAMPLE	NUMBER NONCONFORMING	SAMPLE	NUMBER NONCONFORMING
1	7	8	8
2	10	9	7
3	9	10	10
4	8	11	8
5	7	12	7
6	9	13	7
7	10	14	9

a. What type of data is this?

b. What type of chart is appropriate for controlling the number of nonconforming hard-drive controllers?

c. Construct the appropriate control chart.

12.54 A single-stage sampling plan is used in the inspection of lots of 5,000 LEDs. The sample size used is 90 ($n = 90$), and the cutoff point is four ($c = 4$). Using the Poisson approximation, what is the probability of accepting a lot when:

a. $p = .01$

b. $p = .05$

c. $p = .08$

d. $p = .11$

12.55 Samples of nine bottles are taken from a capping process at 15-minute intervals and the seal strength tested. The following are the sample mean seal strengths and ranges (in ft-lbs) for the last 12 samples.

SAMPLE	MEAN	RANGE	SAMPLE	MEAN	RANGE
1	82.8	6.9	7	83.9	8.2
2	83.9	8.9	8	86.9	7.9
3	84.7	9.6	9	82.7	7.3
4	86.9	7.1	10	81.9	8.5
5	83.6	7.5	11	90.2	17.5
6	86.3	8.5	12	80.2	9.0

a. Construct an $\bar{X}$ chart using the above data.

b. Construct an R chart using the above data.

c. Perform a pattern analysis of the $\bar{X}$ chart in part a by checking for the eight nonrandom patterns discussed in Section 12.2.

d. Do you have any recommendations concerning these charts?

12.56 The specification limits for the tear strength of a cardboard container are 150 ± 2 ft-lbs. A normally distributed sample of 100 containers yields a mean tear strength of 150.3 ft-lbs with a standard deviation of .7 ft-lbs.

a. Calculate C_{pk}.

b. Estimate the number of containers nonconforming per million by using Table 12.3.

c. Calculate the number of containers nonconforming per million.

12.57 An unsophisticated vendor is agreeable to any of three sampling plans. The specifics of these sampling plans are as follows:

PLAN	n	c
A	50	0
B	70	2
C	40	1

Assume that the historical proportion nonconforming for the vendor's product is 6% ($p =$ .06). Which of the plans appears the most attractive to you, the consumer?

12.58 Samples of five 100-pound bags of pinto beans are taken hourly and weighed. The mean weights for the last sixteen samples are shown below. The mean range for the last sixteen samples is 12.7 pounds. ($\bar{R} = 12.7$).

SAMPLE	WEIGHT (LB)	SAMPLE	WEIGHT (LB)
1	102.3	9	100.8
2	99.8	10	100.2
3	101.8	11	103.2
4	99.2	12	100.5
5	103.1	13	98.2
6	102.5	14	96.4
7	98.7	15	92.3
8	103.1	16	89.9

a. Construct an $\bar{X}$ chart using the above data.

b. Perform a pattern analysis of the $\bar{X}$ chart in part a by checking for the eight nonrandom patterns discussed in section 12.2.

c. Do you have any recommendations concerning this control chart?

12.59 Indicate whether an $\bar{X}$, R, p, or c chart would be the most appropriate for controlling:

a. the number of defective resistors in a sample of 500 resistors

b. the mean cranking amperage of a sample of car batteries

c. the discrepancy in the time it takes to perform identical oil changes

d. the number of scratches and other imperfections in an $8' \times 4'$ sheet of paneling

e. the mean length of a sample of bolts

f. the number of nonconforming sprocket sets in a sample of 250

12.60 Twenty automobiles are sampled and the number of paint imperfections are observed. Construct a c chart for the number of paint imperfections using the following data:

AUTOMOBILE	IMPERFECTIONS	AUTOMOBILE	IMPERFECTIONS
1	15	11	19
2	17	12	14
3	11	13	23
4	18	14	13
5	15	15	15
6	19	16	18
7	9	17	16
8	22	18	10
9	16	19	13
10	10	20	21

12.61 A double-sampling plan is used with $n_1 = 7$, $n_2 = 35$, $c_1 = 1$, and $c_2 = 4$. A sample of size n_1 is taken and yields two units nonconforming. Should you take a second sample? If so, how many units nonconforming will have to be found to:

a. accept the lot?

b. reject the lot?

12.62 Formulate the OC curve for a single-stage sampling plan with $N = 2,000$, $n = 90$, and $c = 5$.

a. Calculate β using $p = .02$, $p = .04$, $p = .06$, $p = .08$, and $p = .11$; use these data to plot the curve.

b. Using the curve you have drawn, what is β when $p = .09$?

12.63 MINITAB was used to construct the following $\bar{X}$ chart to monitor a production process:

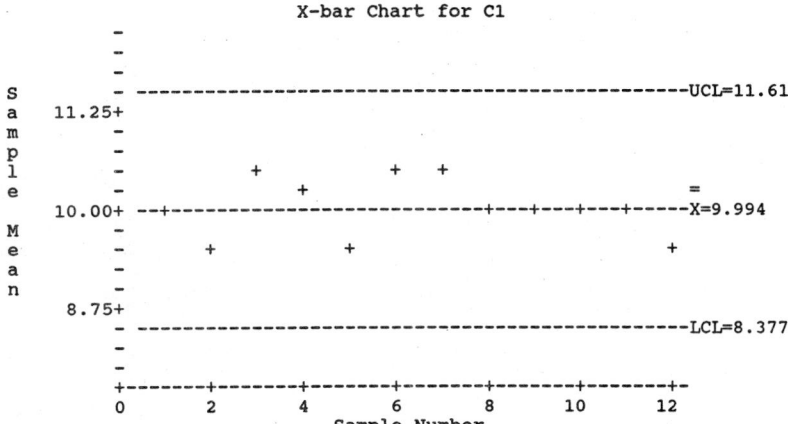

```
MTB > SET INTO C1
DATA> 9.5 9.1 9.6 11.5 9.7 11.2 8.9 7.9 12.7 10.8 9.9
DATA> 8.8 9.8 9.9 10.5 10.7 9.9 9.8 10.5 7.4 11.2 9.8
DATA> 11.4 9.5 10.8 9.9 11.4 9.5 10.6 9.7 11.5 8.5 9.6
DATA> 10.3 10.3 10.1 9.9 9.5 10.4 9.8 11.2 8.7 10.5 9.8
DATA> 10.4 9.5 7.9 9.9
DATA> END
MTB > XBARCHART USING C1, SAMPLE SIZE IS 4
```

a. Is the process in control?

b. Perform a pattern analysis of the chart by checking for the eight nonrandom patterns discussed in section 12.2.

12.64 Using MINITAB and the data in Exercise 12.63, the following R chart has been constructed to monitor a production process.

```
MTB > RCHART USING C1, SAMPLE SIZE IS 4
```

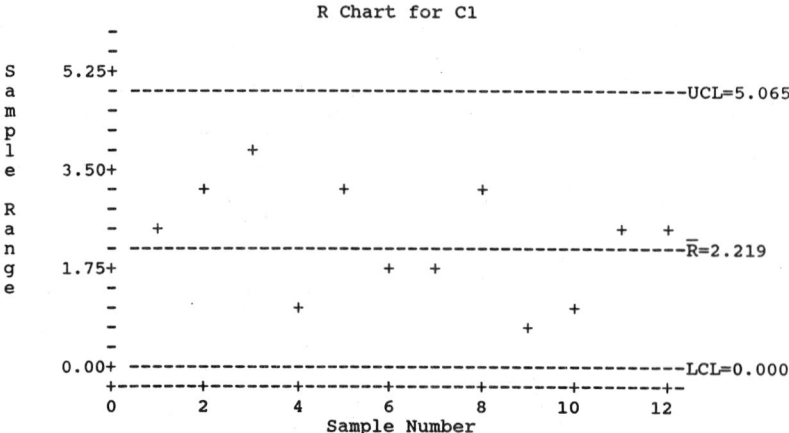

a. Is the process in control?

b. Explain how the R chart will be used to determine whether the process is in control?

12.65 To examine the fraction of nonconforming memory chips, an electronics company randomly selected 200 chips each day for a period of 15 days and observed the number of nonconforming chips in each sample. The following MINITAB output is a p chart for this situation:

```
MTB > SET INTO C1
DATA> 12 14 23 21 15 22 34 31 12 21 21 20 17 18 14
DATA> END
MTB > PCHART USING C1, SAMPLE SIZE IS 200

                      P Chart for C1

            -
            -
            -
            -                          +
    P       - ----------------------+-------------------UCL=0.1615
    r   0.150+
    o       -
    p       -
    o       -          +
    r       -                    +
    t   0.100+ ----------+-------------------+--+--+---------P=0.09833
    i       -                                   +   +
    o       -                    +
    n       -          +                                   +
            -   +                          +
       0.050+
            - ------------------------------------------LCL=0.03517
            -
            -
            +-----+-----+-----+-----+-----+-----+-----+-----+-
            0     2     4     6     8    10    12    14    16
                           Sample Number
```

a. Is the process in control?

b. If the process is not in control, what would be your next step?

Case Study

Don't Forget the Target—An Alternative Measure of Process Capability

This chapter has examined various measures of process capability. As we have discussed, the topic of quality control is getting increased attention due to increasing consumer quality awareness and the in-flux of better and less expensive products from foreign markets. Slogans about doing things right the first time and zero defects are noble ideas, but if a process is not capable of meeting requirements, resources are wasted and frustration is sure to follow.

The process spread, defined in this chapter as six (estimated) standard deviations, is a simple measure of the process capability. However, this measure depends on the units of measurement (inches, pounds, and so forth) and does not permit comparison between two processes with different units of measurement. Two measures of process capability that relate the actual process spread to the required specification (spec) limits are the C_p and the C_{pk} ratios. The C_p index divides the allowable process spread (the difference between the upper and lower spec limits) by the actual process spread. The C_{pk} index determines the distance from the process *center* (not necessarily the process target) to the nearest spec limit divided by half the actual process spread. The advantage of C_{pk} over C_p is that C_{pk} considers the process center, whereas C_p does not.

Here we suggest an alternative to these two measures that not only considers the process center but also the process *target*. The estimated process standard deviation (s) is a measure of the variation about the process mean (center). We replace this measure with a measure of the variation about the process target, T, assumed here to be midway between the upper spec limit (USL) and the lower spec limit (LSL). This measure is written C_{pm} and is defined as

$$C_{pm} = \frac{USL - LSL}{6s'}$$

where s' measures the variation about the process target and is defined as

$$s' = \sqrt{\frac{\Sigma (x - T)^2}{n - 1}}$$

where the summation is over the n sample values. Note that replacing T in this expression with $\bar{x}$ produces the sample standard deviation, s. This is a small, yet powerful, change to the usual C_p index.

The C_{pm} index will decrease as s' increases due to a shift from the process target. When the process variance changes and the process mean drifts from T concurrently, the C_{pm} index has the ability to reflect these changes. To illustrate the difference among these three different measures of process capability, case study question 2 will ask you to determine these measures for the five processes illustrated in Figure 12.21.

In another application concerning a machine qualification run, the measured quality characteristic was the radial length of machined holes. Radial length is defined as the distance between the hole center and the centerline of the part. The upper and lower spec limits are 20 and -20 units, respectively, with a target of $T = 0$. Table 12.4 shows the results from three stages of development. In the first stage, a machine compensator was used, resulting in poor process capability. In the second stage, the compensator was disconnected and the machine cutting tool feed rate was increased, resulting in even less process capability. In the third stage, the compensator remained disconnected but the feed rate was reduced.

Case Study Questions

1. The summation $\Sigma (x - T)^2$ can be written as $\Sigma (x - \bar{x})^2 + n(\bar{x} - T)^2$. Why is this true? As a result,

$$s' = \sqrt{\Sigma \frac{(x - \bar{x})^2}{n - 1} + \frac{n(\bar{x} - T)^2}{n - 1}}$$

$$= \sqrt{s^2 + \frac{n(\bar{x} - T)^2}{n - 1}}$$

Use this information when answering the following questions.

2. Five different processes are illustrated in Figure 12.21, where the process target is $T = 14$, spec limits are 10 and 18, sample size is $n = 100$, and the process standard deviation is $s = 2/3$. For each of the five processes, determine C_p, C_{pk}, and C_{pm}. Discuss these three values.

The following four questions refer to the radial measurements application.

3. For each stage, determine C_p. Interpret these values.

4. For each stage, determine C_{pk}. Interpret these values.

5. For each stage, determine C_{pm}. Interpret these values.

6. What could be done to further improve the value of C_{pm} in stage 3?

7. Which of the three measures do you think best measures the capability of a process that has a high likelihood of being off-target? Why?

■ TABLE 12.4 Summary of radial length data.	STAGE	n	$\bar{x}$	s
	1	201	4.7	8.7
	2	96	10.4	21.1
	3	316	5.0	5.4

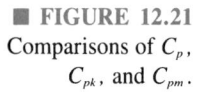
■ **FIGURE 12.21**
Comparisons of C_p,
C_{pk}, and C_{pm}.

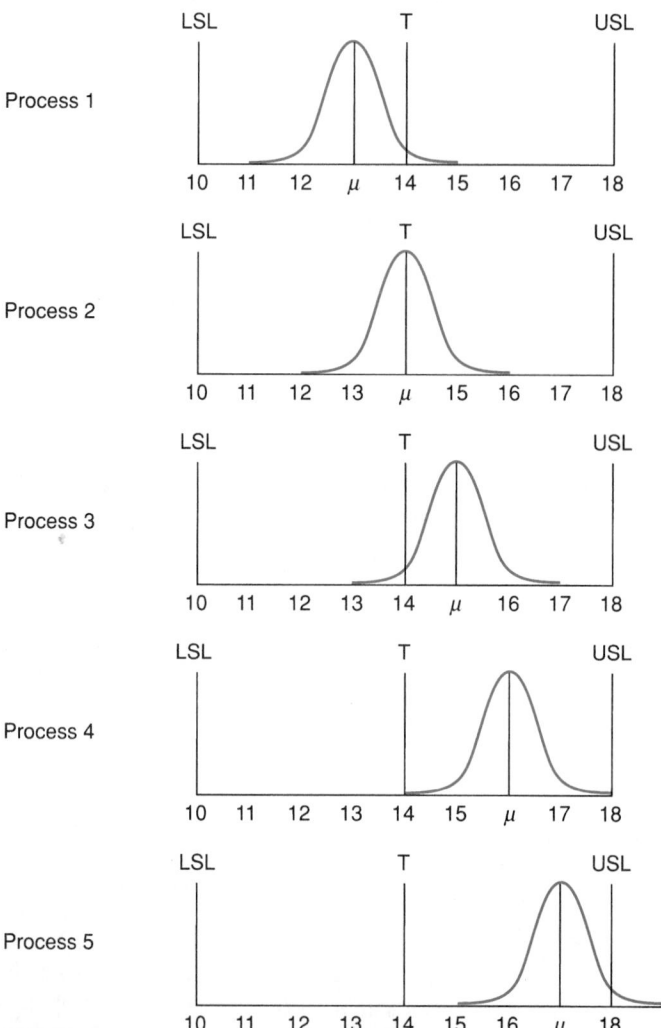

(*Source:* Fred A. Spiring, "The C_{pm} Index: An Alternate Measure of Process Capability,"
Quality Progress (February 1991): 57. © 1991 American Society for Quality Control. Reprinted by permission.)

SAS

⧄ **$\bar{X}$ Chart**   SAS offers the ability to construct control charts. To use this feature, you need
Figure 12.6   SAS/QC software. The SAS commands to construct an $\bar{X}$ control chart using the
raw data from the coffee-can illustration are shown below. The first set of commands enters the data in a SAS data file, and the second set of commands constructs the control chart.

The TITLE command names the SAS run.

The DATA command gives the data a name.

The INPUT command names and gives the correct order for the different
fields on the data lines.

The LABEL command provides a descriptor for the data fields.

The CARDS command indicates to SAS that the input data immediately follow.

The next 100 lines contain the subgroup samples for the coffee-can data shown in Figure 12.6. There are 20 subgroups, each of sample size 5.

The RUN command indicates the completion of the data commands and signals the end of this SAS step.

The TITLE1 and TITLE2 commands provide a title for the output.

The PROC SHEWHART command indicates the broad classification of Shewhart charts.

GRAPHICS creates charts for a graphics device. If you do not use this option, the procedure provides line printer charts. To use GRAPHICS you need SAS/GRAPH software in addition to SAS/QC software.

DATA = WEIGHTS indicates that the data set WEIGHTS will be used.

XCHART indicates an $\bar{X}$ chart for subgroup averages.

WEIGHT * CAN = '*' indicates that WEIGHT and CAN data will be used to construct the chart. The symbol * will be used to show the data points on the chart.

Figure 12.23 shows the SAS output obtained by executing the listing in Figure 12.22.

■ FIGURE 12.22
Input for SAS
(mainframe or
micro).

```
/********************************/
/*                              */
/*         X BAR CHART          */
/*                              */
/********************************/

DATA WEIGHTS;
  INPUT CAN WEIGHT;
  LABEL CAN = 'SUBGROUP CAN SAMPLE'
        WEIGHT = 'WEIGHT OF CANS IN SAMPLE';
CARDS;
1    19.8
1    20.1
1    20.2
1    19.9
1    20.0
2    20.3
2    19.9
2    19.8
2    19.8
2    20.1
       .
       .
       .
19   20.2
19   19.6
19   19.9
19   20.0
19   19.7
20   20.3
20   19.9
20   20.1
20   20.2
20   19.9
;
RUN;
TITLE1 '                              ';
TITLE2 'X BAR CHART FOR CAN WEIGHTS';
PROC SHEWHART GRAPHICS DATA=WEIGHTS;
     XCHART WEIGHT*CAN='*';
RUN;
```

■ FIGURE 12.23   SAS output.

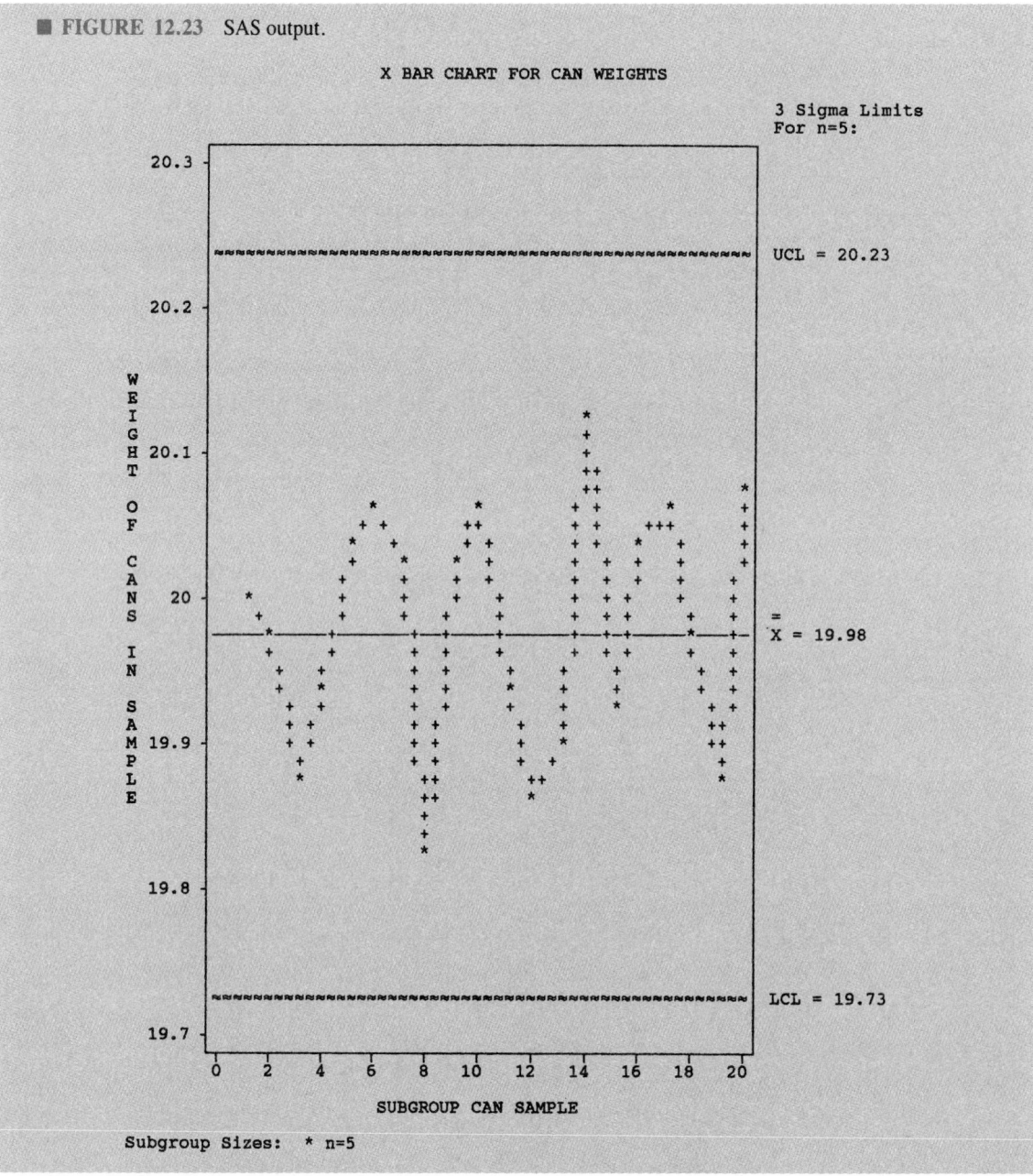

SAS

◢ **R Chart**
Figure 12.7

You can construct control charts using SAS if you have SAS/QC software. The SAS commands to construct an *R* control chart using the raw data from the coffee-can illustration are shown below. The first set of commands enters the data in a SAS data file, and the second set of commands constructs the control chart.

The TITLE command names the SAS run.

The DATA command gives the data a name.

The INPUT command names and gives the correct order for the different fields on the data lines.

The LABEL command provides a descriptor for the data fields.

The CARDS command indicates to SAS that the input data immediately follow.

The next 100 lines contain the subgroup samples for the coffee-can data shown in Figure 12.6. There are 20 subgroups, each of sample size 5.

The RUN command indicates the completion of the data commands and signals the end of this SAS step.

The TITLE1 and TITLE2 commands provide a title for the output.

The PROC SHEWHART command indicates the broad classification of Shewhart charts.

GRAPHICS creates charts for a graphics device. If you do not use this option, the procedure provides line printer charts. To use GRAPHICS, you need SAS/GRAPH software in addition to SAS/QC software.

DATA = WEIGHTS indicates that the data set WEIGHTS will be used.

RCHART indicates an R chart for subgroup ranges.

WEIGHT * CAN = '*' indicates that WEIGHT and CAN data will be used to construct the chart. The symbol * will be used to show the data points on the chart.

Figure 12.25 shows the SAS output obtained by executing the listing in Figure 12.24.

■ FIGURE 12.24
Input for SAS
(mainframe or
micro).

```
/*******************************/
/*                             */
/*          R  CHART           */
/*                             */
/*******************************/

DATA WEIGHTS;
  INPUT CAN WEIGHT;
  LABEL CAN = 'SUBGROUP CAN SAMPLE'
        WEIGHT = 'WEIGHT OF CANS IN SAMPLE';
CARDS;
1    19.8
1    20.1
1    20.2
1    19.9
1    20.0
2    20.3
2    19.9
2    19.8
2    19.8
2    20.1
        .
        .
        .
19   20.2
19   19.6
19   19.9
19   20.0
19   19.7
20   20.3
20   19.9
20   20.1
20   20.2
20   19.9
;
RUN;
TITLE1 '                        ';
TITLE2 'R CHART FOR CAN WEIGHTS';
PROC SHEWHART GRAPHICS DATA=WEIGHTS;
    RCHART WEIGHT*CAN='*';
RUN;
```

■ FIGURE 12.25   SAS output.

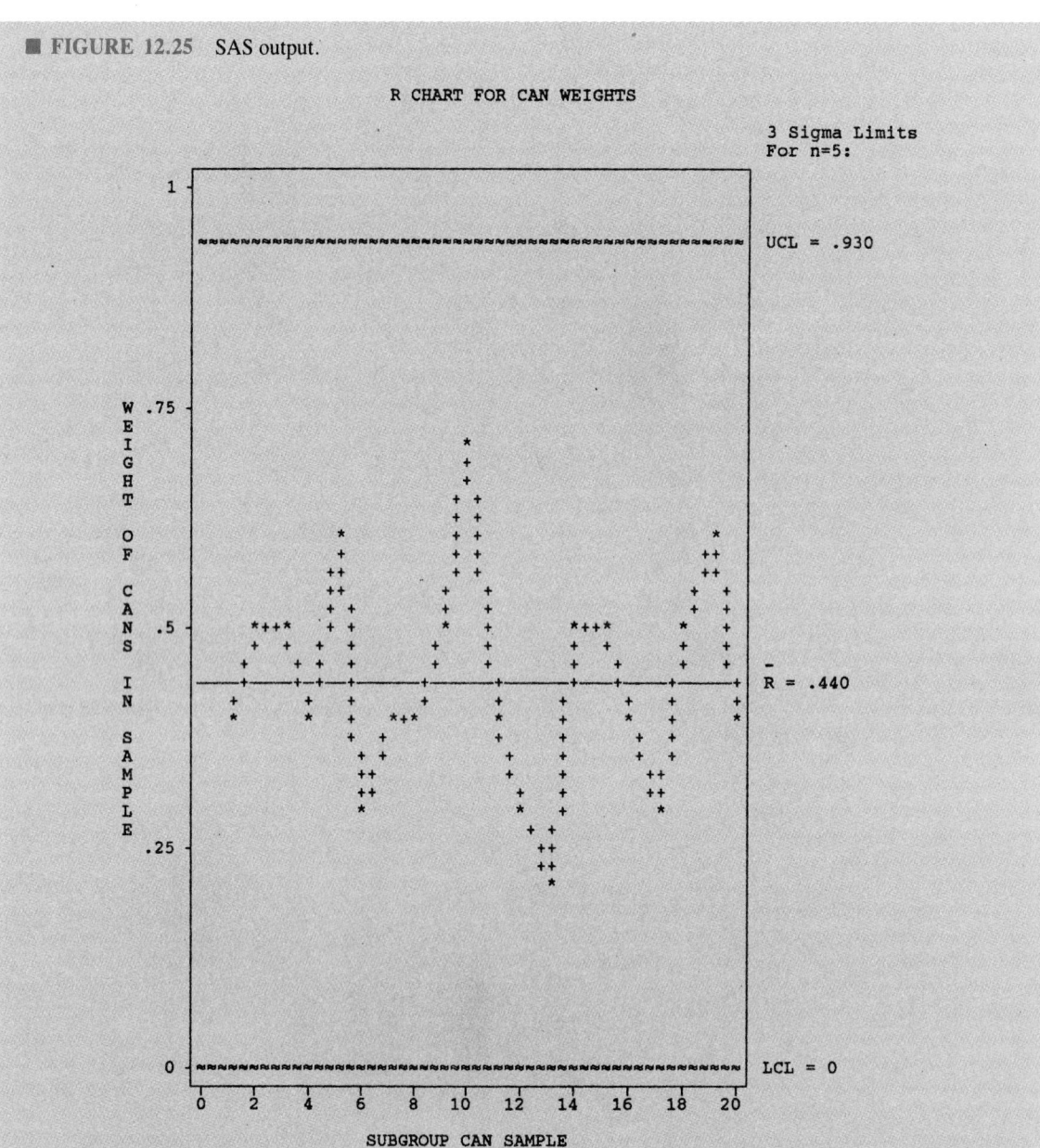

SAS

◢ *p* **Chart**
Figure 12.11,
Example 12.2

Example 12.2 was concerned with constructing a *p* chart. SAS offers the ability to construct *p* charts. To use this feature, you need SAS/QC software. The SAS commands to construct a *p* control chart using the coffee-can data are shown below. The first set of commands enters the data in a SAS data file, and the second set of commands constructs the control chart.

The TITLE command names the SAS run.

The DATA command gives the data a name.

The INPUT command names and gives the correct order for the different fields on the data lines.

The LABEL command provides a descriptor for the data fields.

The CARDS command indicates to SAS that the input data immediately follow.

The next 20 lines contain the subgroup samples for the coffee-can data. There are 20 subgroups, each of sample size 150. Each line contains the sample number and the number nonconforming.

The RUN command indicates the completion of the data commands and signals the end of this SAS step.

The TITLE1 and TITLE2 commands provide a title for the output.

The PROC SHEWHART command indicates the broad classification of Shewhart charts.

GRAPHICS creates charts for a graphics device. If you do not use this option, the procedure provides line printer charts. To use GRAPHICS, you need SAS/GRAPH software in addition to SAS/QC software.

DATA = CANS indicates that the data set CANS will be used.

PCHART indicates a p control chart.

NUMBER * SAMPLE = '*' indicates that NUMBER and SAMPLE data will be used to construct the chart. The symbol * will be used to show the data points on the chart.

/SUBGROUPN = 150 indicates the subgroup sample size.

Figure 12.27 shows the SAS output obtained by executing the listing in Figure 12.26.

■ FIGURE 12.26
Input for SAS
(mainframe or micro).

```
/****************************************************************/
/*                                                              */
/*    P CHART                                                   */
/*                                                              */
/****************************************************************/
DATA CANS;
   INPUT SAMPLE NUMBER;
   LABEL SAMPLE = 'SUBGROUP SAMPLE'
         NUMBER = 'NUMBER OF NONCONFORMITIES';
   CARDS;
1    7
2    4
3    1
4    3
5    4
6    8
7    10
8    5
9    2
10   7
11   6
12   8
13   0
14   9
15   3
16   1
17   4
18   5
19   7
20   2
;
RUN;

TITLE1 '                              ';
TITLE2 'P CHART FOR COFFEE CANS';
PROC SHEWHART GRAPHICS DATA=CANS;
     PCHART NUMBER*SAMPLE='*'/SUBGROUPN=150;
RUN;
```

 FIGURE 12.27   SAS output.

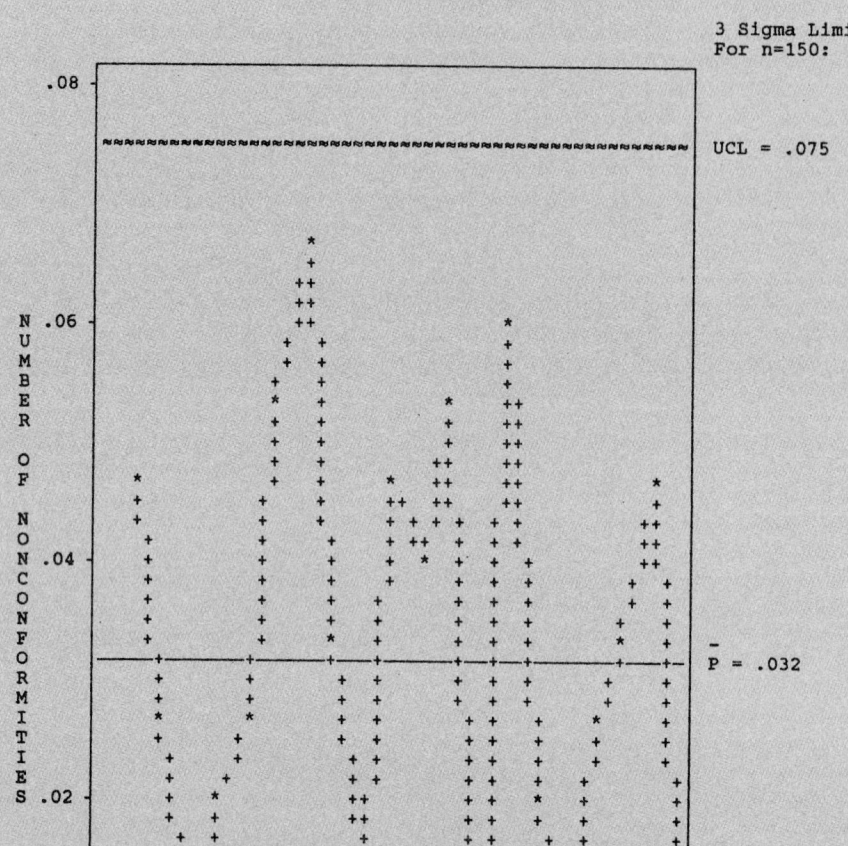

Subgroup Sizes: * n=150

SAS

☑ c Chart
Figure 12.13,
Example 12.4

Example 12.4 was concerned with constructing a *c* chart. You can construct control charts using SAS if you have SAS/QC software. The SAS commands to construct a *c* control chart using the door panel data are shown below. The first set of commands enters the data in a SAS data file, and the second set of commands constructs the control chart.

The TITLE command names the SAS run.

The DATA command gives the data a name.

The INPUT command names and gives the correct order for the different fields on the data lines.

The LABEL command provides a descriptor for the data fields.

The CARDS command indicates to SAS that the input data immediately follow.

The next 25 lines contain the door panel data. Each line contains the sample number and the number of nonconformities.

The RUN command indicates the completion of the data commands and signals the end of this SAS step.

The TITLE1 and TITLE2 commands provide a title for the output.

The PROC SHEWHART command indicates the broad classification of Shewhart charts.

GRAPHICS creates charts for a graphics device. If you do not use this option, the procedure provides line printer charts. To use GRAPHICS, you need SAS/GRAPH software in addition to SAS/QC software.

DATA = DOORS indicates that the data set DOORS will be used.

CCHART indicates a c control chart.

NUMBER * PANEL = '*' indicates that NUMBER and PANEL data will be used to construct the chart. The symbol * will be used to show the data points on the chart.

Figure 12.29 shows the SAS output obtained by executing the listing in Figure 12.28.

■ FIGURE 12.28
Input for SAS
(mainframe or micro).

```
/******************************************************************/
/*                                                                */
/*   C CHART                                                      */
/*                                                                */
/******************************************************************/
DATA DOORS;
   INPUT PANEL NUMBER;
   LABEL PANEL = 'DOOR PANEL'
         NUMBER = 'NUMBER OF NONCONFORMITIES';
   CARDS;
1    1
2    0
3    3
4    3
5    1
6    2
7    5
8    0
9    2
10   1
11   2
12   0
13   8
14   0
15   2
16   1
17   4
18   0
19   2
20   4
21   1
22   1
23   0
24   2
25   3
;
RUN;
```

■ **FIGURE 12.28**
(*continued*)

```
TITLE1 '                        ';
TITLE2 'C CHART FOR DOOR PANELS';
PROC SHEWHART GRAPHICS DATA=DOORS;
    CCHART NUMBER*PANEL='*';
RUN;
```

■ **FIGURE 12.29** SAS output.

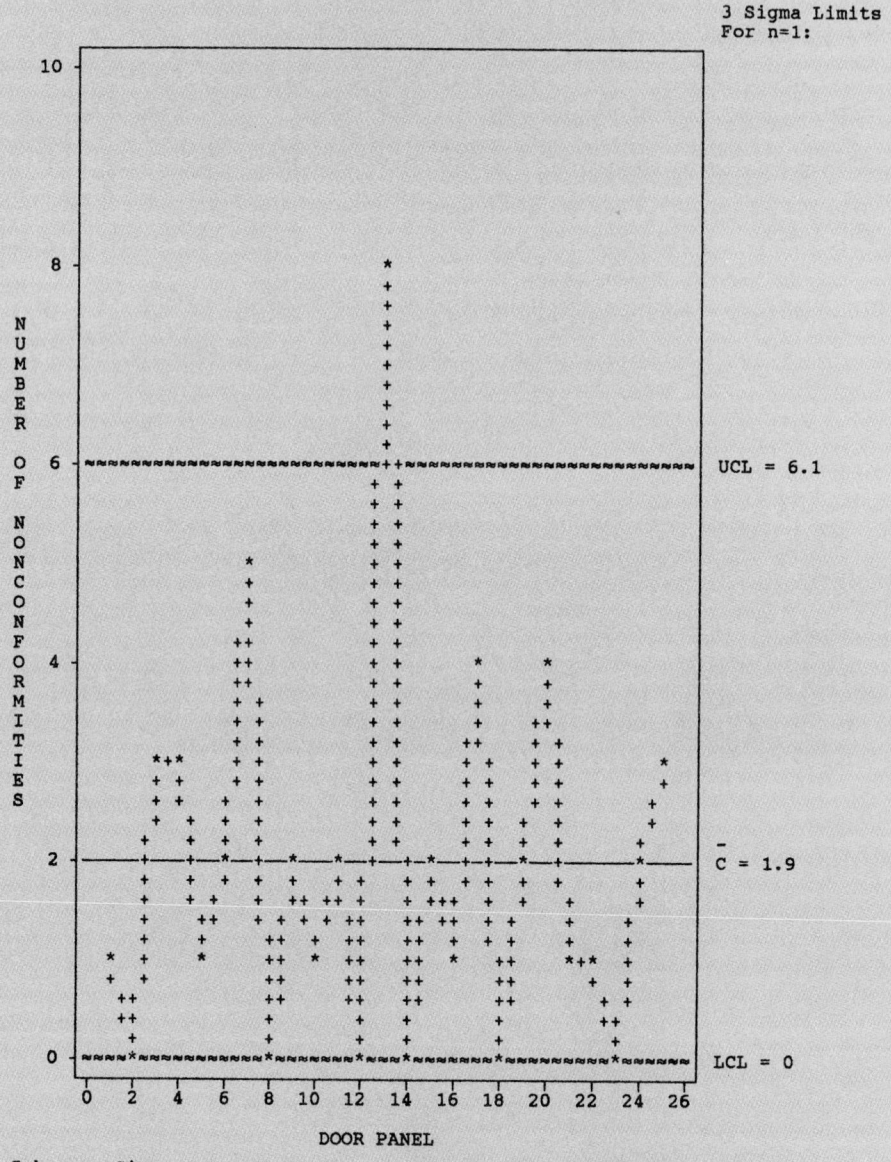

CHAPTER

13 Applications of the Chi-Square Statistic

A LOOK BACK/INTRODUCTION

We have now examined several topics in *descriptive* and *inferential* statistics. The descriptive area introduced you to both the numeric methods (for example, mean, median, and variance) and the graphic methods (for example, histogram and scatter diagram) of describing data. In inferential statistics, we discussed point estimation, confidence intervals, and tests of hypothesis. In the remaining chapters, we turn our attention to other applications of the material from these earlier chapters.

In Chapter 8, we introduced the chi-square (χ^2) distribution. We used this distribution to test that the variance of a normal distribution was equal to a specified value. The shape of the chi-square distribution is skewed right (with a right tail) and is nonnegative. The shape of a chi-square curve and areas (probabilities) under such a curve are contained in Table A.6. The test statistic in Chapter 8 had a chi-square distribution. This chapter introduces you to additional applications of statistics by using the chi-square distribution to answer such questions as:

Do reported percentages of market share accurately describe the product mix for the new cars sold this past year in Minneapolis, Minnesota?

Does a person's age have an influence on buying behavior?

13.1 Chi-Square Goodness-of-Fit Tests

The Binomial Situation

In the binomial situation (introduced in Chapter 5) the following four conditions must be satisfied:

1. The experiment consists of n repetitions, called *trials*.
2. The trials are *independent*.
3. Each trial has two (and only two) possible outcomes, referred to as *success* and *failure*.
4. The probability of a success for each trial is p, where p remains the same for each trial. For a large finite population, p is the *proportion* of successes in this population.

Consequently, the binomial distribution applies to applications where there are only two possible outcomes, such as:

The person selected is a male or a female.

The product tested is either defective or not defective.

A new-car buyer buys either an American-made car or a foreign-made car.

Inferences for the Binomial Situation

Estimating the binomial parameter, p, was covered in Chapter 10. We obtained a random sample of size n and observed the number of successes, x. The estimator of p, the proportion of successes in the *population*, was $\hat{p} = x/n =$ the proportion of successes in the *sample*. We also discussed hypothesis testing for p. For example, we discussed a binomial situation in which a calculator was either defective (with probability p) or not defective. The hypothetical value of p was .04, and we

determined whether the results of the sample (13 defectives out of 150) indicated a departure from this percentage. Here $\hat{p} = 13/150 = .0867$. So, 8.67% of the sampled calculators were defective. Is this percentage large enough for us to conclude that p is different from .04, or is this large value of $\hat{p}$ just due to the fact that we tested a sample and not the entire population—that is, is this sampling error?

The resulting value of the test statistic was

$$Z^* = \frac{\hat{p} - .04}{\sqrt{\dfrac{(.04)(.96)}{150}}} = \frac{.0867 - .04}{.016} = 2.92$$

By comparing $Z^* = 2.92$ with the value 1.96 in Table A.4, we rejected H_0 using $\alpha = .05$; that is, the proportion of defective calculators was not 4%. The corresponding p-value was .0036.

Another Test for H_0: $p = p_0$ versus H_a: $p \neq p_0$

There is another test for a *two-tailed* test on p. This new test extends easily to a situation in which there are *more than two possible outcomes* for each trial: the **multinomial situation.**

To demonstrate this new testing procedure, the **chi-square goodness-of-fit** test, let's look at the lot sampling example. Note that the population consists of two **categories**—defective (category 1) and nondefective (category 2). Let $p_1 = $ the proportion of defectives in the population and $p_2 = $ the proportion of nondefectives in the population. In the previous solution, $p_1 = p$ and $p_2 = 1 - p$.

We *observed* 13 sample values in category 1 (defective) and 137 in category 2. So define

$$O_1 = 13$$
$$O_2 = 137$$

How many units do we *expect* to see in each category if H_0 is *true?* The hypotheses here can be written

$$H_0: p_1 = .04, \ p_2 = .96$$
$$H_a: p_1 \neq .04, \ p_2 \neq .96$$

This means that if H_0 is true, then, on the average, 4% of the sample values should be defective (category 1) and 96% should be nondefective (category 2). Define

$E_1 = $ expected number of sample values in category 1 if H_0 is true

$\quad = (150)(.04) = 6.0$

$E_2 = $ expected number of sample values in category 2 if H_0 is true

$\quad = (150)(.96) = 144.0$

We next define a test statistic that has an approximate chi-square distribution:

$$\chi^2 = \sum \frac{(O - E)^2}{E} \qquad \qquad (13.1)$$

where the summation is over all categories (two here). In previous uses of this distribution, its shape depended on the sample size, specified by the degrees of freedom (df). Now the shape depends on the number of categories, and

$$df = \text{number of categories} - 1$$

For the binomial situation,

$$df = 2 - 1 = 1$$

Therefore, for any *binomial* application, the test statistic in equation 13.1 has a *chi-square distribution* with 1 *df*.

EXAMPLE 13.1   Analyze the lot sampling data using the chi-square test statistic and a significance level of $\alpha = .05$.

SOLUTION   **Step 1.** The hypotheses are

$$H_0: p_1 = .04, \, p_2 = .96$$
$$H_a: p_1 \neq .04, \, p_2 \neq .96$$

Step 2. The test statistic is

$$\chi^2 = \sum \frac{(O - E)^2}{E}$$
$$= \frac{(O_1 - E_1)^2}{E_1} + \frac{(O_2 - E_2)^2}{E_2}$$

Step 3. If H_0 is not true (H_a is true), we expect the observed values to be different from the expected values, resulting in a *large* value for χ^2. So the procedure is to reject H_0 if the chi-square test statistic lies in the *right* tail. Consequently, we

$$\text{reject } H_0 \text{ if } \chi^2 > \chi^2_{.05,1}$$

where $\chi^2_{.05,1}$ is the χ^2 value having a right-tail area of .05 with 1 df. Using Table A.6, this value is 3.84. Therefore, we

$$\text{reject } H_0 \text{ if } \chi^2 > 3.84$$

Step 4. We have

$$O_1 = 13 \qquad E_1 = 6$$
$$O_2 = 137 \qquad E_2 = 144$$

(Note that $O_1 + O_2 = E_1 + E_2 = n = 150$.) The calculated value of the test statistic is

$$\chi^{2*} = \frac{(13 - 6)^2}{6} + \frac{(137 - 144)^2}{144}$$
$$= 8.17 + .34$$
$$= 8.51$$

This value is larger than 3.84, so we reject H_0.

Step 5. We conclude, as before, that the proportion of defectives (p_1) is not .04.

The *p*-value for Example 13.1 using the chi-square analysis is shown in Figure 13.1; it is the shaded area to the right of 8.51. Using Table A.6, all we can say is that this value is less than .005. The actual value is .0036 (calculated using a statistical software package). This is the *same p*-value we obtained when Z^* was used to perform this test of hypothesis.

In the lot sampling examples, we observe some quite fascinating (would you believe mildly interesting?) parallels with Chapter 11. In Chapter 11, we noted that when using the F test from the ANOVA procedure to test $H_0: \mu_1 = \mu_2$, we obtained

■ **FIGURE 13.1**
p-value for Example
13.1 using the chi-
square analysis.

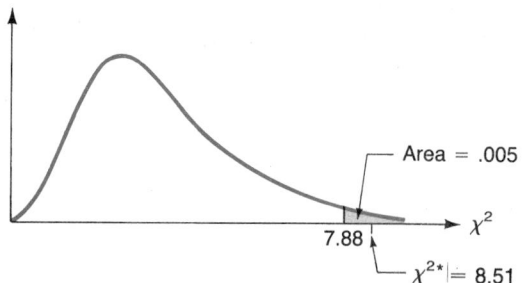

an *F*-value that was the *square* of the *t* value obtained using the corresponding *t* test. Also, the value from the *F* table used to define the rejection region was the square of the corresponding *t* value. This relationship held only when testing the equality of two means. Finally, the *p*-values from the two tests were identical; it made no difference which test we used for a two-tailed test on two means because the results were the same for both procedures. However, the ANOVA technique also could be used for comparing the means of more than two populations.

Using the results from the lot-sampling example in this chapter, we again find that:

1. $\chi^{2*} = 8.51 = (2.92)^2 = (Z^*)^2$.
2. The table values for the rejection region are 1.96 for the *Z* test and 3.84 = $(1.96)^2$ for the χ^2 test.
3. The *p*-value for each test was the same.

So again we have two testing procedures that produce identical conclusions. *The chi-square test, however, extends easily to the multinomial situation.* The chi-square goodness-of-fit test is an *extension* of the *Z* test used to test a binomial parameter. Furthermore, there is a definite relationship between the standard normal distribution (*Z*) and the chi-square distribution: the *square* of *Z* always is a chi-square random variable with 1 df.

Testing $H_0: p = p_0$ versus $H_a: p \neq p_0$

USING Z TEST

Test statistic:

$$Z = \frac{\hat{p} - p_0}{\sqrt{\dfrac{p_0(1 - p_0)}{n}}}$$

Rejection region:

reject H_0 if $|Z| > Z_{\alpha/2}$

(Use Table A.4.)

USING χ^2 TEST

Test statistic:

$$\chi^2 = \frac{(O_1 - E_1)^2}{E_1} + \frac{(O_2 - E_2)^2}{E_2}$$

$$= \sum \frac{(O - E)^2}{E}$$

Rejection region:

reject H_0 if $\chi^2 > \chi^2_{\alpha,1}$

(Use Table A.6.)

The Multinomial Situation

The multinomial situation is identical to the binomial situation, except that there are *k* possible outcomes on each trial rather than two. Here *k* is any integer that is at least 2.

Suppose that a recent survey indicated the following percent of market share for U.S. auto manufacturers:

GA	62.5
K	22.6
L	11.5
M	1.4
Other	2.0
	100

An executive at GA questions whether these percentages apply to the new cars sold during the past year in Minneapolis, Minnesota. Obtaining a random sample of 500 new Minneapolis car registrations from that year, she observes the following frequencies of cars sold:

GA	290
K	125
L	65
M	10
Other	10
	500

The assumptions necessary for a multinomial experiment are:

1. The experiment consists of n independent repetitions (trials).
2. Each trial outcome falls in exactly one of k categories.
3. The probabilities of the k outcomes are denoted by $p_1, p_2, \ldots, p_k$, where these probabilities (proportions) remain the same on each trial. Also, $p_1 + p_2 + \cdots + p_k = 1$.

For this situation, we can define k random variables as the k observed values, where

$$O_1 = \text{the observed number of sample values in category 1}$$

$$O_2 = \text{the observed number of sample values in category 2}$$

$$\vdots$$

$$O_k = \text{the observed number of sample values in category } k$$

For our example, $n = 500$ trials, where each trial consists of obtaining a new car registration and observing in which of the $k = 5$ categories this new car lies. Assuming these registrations are obtained in a *random* manner (not the first 500 cars in May, for example), then these trials are independent. Also, let

$p_1 = $ the proportion of cars sold in Minneapolis that were GA cars for that year

$p_2 = $ the proportion of cars sold in Minneapolis that were K cars for that year

$$\vdots$$

The five random variables in our example are

$$O_1 = \text{the number of GA cars in the sample}$$

$$O_2 = \text{the number of K cars in the sample}$$

$$\vdots$$

Thus, this example fits the assumptions for the multinomial situation.

Hypothesis Testing for a Multinomial Situation

The hypotheses for the Minneapolis market share example are

$$H_0: p_1 = .625, \; p_2 = .226, \; p_3 = .115, \; p_4 = .014, \; p_5 = .02$$

H_a: at least one of the p_i's is incorrect

Notice that H_a is *not* $p_1 \neq .625, \; p_2 \neq .226, \ldots, \; p_5 \neq .02$. This hypothesis is too strong and is not the opposite of H_0.

Let $p_{1,0}$ be any specified value of p_1, $p_{2,0}$ any specified value of p_2, and so on. The hypotheses to test the multinomial parameters are

$$H_0: p_1 = p_{1,0}, \; p_2 = p_{2,0}, \ldots, \; p_k = p_{k,0}$$

H_a: at least one of the p_i's is incorrect

Using the observed values $(O_1, O_2, \ldots)$, the point estimates here are

$$\hat{p}_1 = \text{estimate of } p_1 = O_1/n$$
$$\hat{p}_2 = \text{estimate of } p_2 = O_2/n$$
$$\vdots$$

To test H_0 versus H_a, we use the previously stated chi-square statistic. To define the rejection region, notice that when H_a is true, we would expect the O's and E's to be "far apart," because the E's are determined by assuming that H_0 is true. In other words, if H_a is true, the chi-square test statistic should be large. Consequently, we always reject H_0 when χ^{2*} lies in the *right tail* when using this particular statistic.

To test H_0 versus H_a, compute

$$\chi^2 = \sum \frac{(O - E)^2}{E} \tag{13.2}$$

where

1. The summation is across all categories (outcomes).
2. The O's are the *observed* frequencies in each category using the sample.
3. The E's are the *expected* frequencies in each category if H_0 is true, so

$$E_1 = np_{1,0}$$
$$E_2 = np_{2,0}$$
$$E_3 = np_{3,0}$$
$$\vdots$$

4. The df for the chi-square statistic are $k - 1$, where k is the number of categories.

To carry out the test,

$$\text{reject } H_0 \text{ if } \chi^2 > \chi^2_{\alpha, \text{df}}$$

Notice that the hypothetical proportions (probabilities) for each of the categories are specified in H_0. Consequently, we will complete the analysis by concluding that at least one of the proportions is incorrect (we reject H_0) or that there is not enough evidence to conclude that these proportions are incorrect (we fail to reject H_0). We do not *accept* H_0; we never conclude that these specified proportions *are* correct. We act like the juror who acquits a defendant not because he or she is convinced that this person is innocent but rather because there was not sufficient evidence for conviction.

When we introduced the ANOVA technique, we mentioned that this procedure allowed us to determine whether many population means were equal using a *single* test. This technique was preferable to using many *t* tests to test the equality of two means, one pair at a time, because these tests would not be independent, and the overall significance level would be difficult to determine. We encounter the same situation here. *It is much better to use a chi-square goodness-to-fit test to test* all *of the proportions at once rather than using many Z tests to test the individual proportions.*

EXAMPLE 13.2

What do the observed number of cars sold in our market share example tell us about the mix of new car sales in Minneapolis for that year? Do they conform to the percentages for all U.S. auto sales? Use a significance level of $\alpha = .05$.

SOLUTION

Step 1. Let p_1 = proportion of all Minneapolis new car sales that are GA, p_2 are K, p_3 are L, p_4 are M, and p_5 are all other new (U.S. made) cars. The hypotheses under investigation are

$$H_0: p_1 = .625, p_2 = .226, p_3 = .115, p_4 = .014, p_5 = .02$$

$$H_a: \text{at least one of these } p_i\text{'s is incorrect}$$

Step 2. The test statistic is

$$\chi^2 = \sum \frac{(O - E)^2}{E}$$

where the summation is over the five categories.

Step 3. Your test procedure here is to

$$\text{reject } H_0 \text{ if } \chi^2 > \chi^2_{\alpha, df}$$

The df is (number of categories) $- 1$, so df $= 5 - 1 = 4$. The chi-square value from Table A.6 is $\chi^2_{.05, 4} = 9.49$, and we

$$\text{reject } H_0 \text{ if } \chi^2 > 9.49$$

Step 4. The observed values are

$$O_1 = 290, O_2 = 125, O_3 = 65, O_4 = 10, O_5 = 10$$

The expected values when H_0 is true are obtained by multiplying $n = 500$ by each of the proportions in H_0. So,

$$E_1 = (500)(.625) = 312.5$$
$$E_2 = (500)(.226) = 113$$
$$E_3 = (500)(.115) = 57.5$$
$$E_4 = (500)(.014) = 7$$
$$E_5 = (500)(.02) = \underline{10}$$
$$500$$

Note that we do not round the expected values because they are *averages*.

The computed value of the chi-square test statistic is

$$\chi^{2*} = \frac{(290 - 312.5)^2}{312.5} + \frac{(125 - 113)^2}{113} + \frac{(65 - 57.5)^2}{57.5}$$
$$+ \frac{(10 - 7)^2}{7} + \frac{(10 - 10)^2}{10}$$
$$= 5.16$$

Because 5.16 does not exceed 9.49, we fail to reject H_0.

Step 5. There is insufficient evidence to suggest that the Minneapolis car sales differ from the U.S. mixture. In other words, the observed values were "close enough" to the expected values under H_0 to let this hypothesis stand.   ∎

In Example 13.2, the proportions under investigation were directly specified. We can also use the chi-square statistic when the proportions are implied.

EXAMPLE 13.3

Allied Health Corporation owns and operates hospitals in the southeast. Three of their hospitals were recently audited by federal auditors to determine if the three hospitals were in compliance with Medicare billing regulations. According to the auditors, a billing selected for audit had an equal probability of being from each of the three hospitals. A random sample of the audited billings revealed the following number of billings selected from each hospital: hospital 1: 485, hospital 2: 405, hospital 3: 310; total = 1200. Using a significance level of .01, what can you conclude about the auditors' selection of billings?

SOLUTION

Step 1. Let p_1 = proportion of audited billings from hospital 1, p_2 = proportion from hospital 2, and p_3 = proportion from hospital 3. If an audited billing has an equal probability of belonging to each hospital, then each of these proportions will be 1/3, providing the values for H_0:

$$H_0: p_1 = 1/3, \ p_2 = 1/3, \ p_3 = 1/3$$

$$H_a: \text{at least one of these proportions is incorrect}$$

Steps 2, 3. The test procedure will be to

$$\text{reject } H_0 \text{ if } \chi^2 > \chi^2_{.01,2} = 9.21$$

because df = $k - 1 = 3 - 1 = 2$. The value of χ^2 is determined from equation 13.2.

Step 4. The observed and expected values are

O	E (if H_0 is true)	$\hat{p}$
$O_1 = 485$	$E_1 = (1200)(1/3) = 400$	$485/1200 = .404$
$O_2 = 405$	$E_2 = (1200)(1/3) = 400$	$405/1200 = .338$
$O_3 = 310$	$E_3 = (1200)(1/3) = 400$	$310/1200 = .258$
1200	1200	1.0

$$\chi^{2*} = \frac{(485 - 400)^2}{400} + \frac{(405 - 400)^2}{400} + \frac{(310 - 400)^2}{400}$$
$$= 18.06 + .06 + 20.25$$
$$= 38.4$$

So we reject H_0 because 38.4 > 9.21.

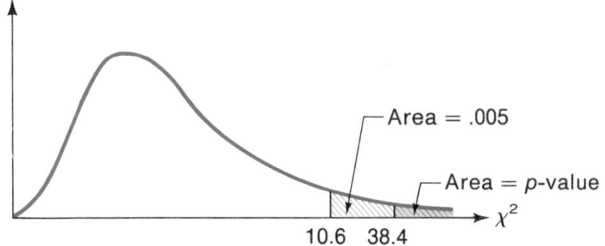

■ FIGURE 13.2
Shaded area is *p*-value
for Example 13.3.

Area = .005

Area = *p*-value

χ^2

10.6 38.4

Step 5. There *is* an unequal distribution of hospital selection here. Your next step should be to examine the three values making up this large χ^2 value. This large value is due to the 18.06 (the number of billings from hospital 1 was larger than expected under H_0) and the 20.25 (the number of billings from hospital 3 was smaller than expected under H_0). This discrepancy can also be seen in the values of $\hat{p}$ from step 4. ■

The *p*-value for the results in Example 13.3 is shown in Figure 13.2. Using Table A.6 and 2 df, the largest value here is 10.6, with a corresponding right-tail area of .005. All you can say using this table is that the *p*-value is less than .005. At any rate, it is small and would lead you to reject H_0 for the most common values of α.

Pooling Categories

When using the chi-square procedure of comparing observed and expected values, we determine the difference between these two values for each category, square it, and *divide by the expected value, E.* If one value of *E* is very small (say, less than 5), then this computation produces an extremely *large* contribution to the final χ^2 value from this category. In other words, this small expected value produces an inflated chi-square value, with the result that we reject H_0 when perhaps we should not have. To prevent this from occurring, we use the following rule: When using equation 13.2, each expected value, *E*, should be at least 5.*

If you encounter an application where one or more of the expected values is less than 5, you can handle this situation by *pooling* your categories such that each of the new categories has an expected value that is at least 5.

EXAMPLE 13.4   The analysis in Example 13.2 was repeated for a much smaller community using 200 observations rather than $n = 500$. The following mixture was observed (number of cars sold):

GA	95
K	50
L	41
M	5
Other	9
	200

Do the data from this community appear to fit the U.S. proportions of 62.5% for GA, 22.6% for K, and so on? Use $\alpha = .05$.

---

*This rule is somewhat arbitrary but commonly used. Another procedure used for pooling requires that all the expected values be 3 or more, while yet another procedure requires that no more than 20% of all the expected values be less than 5 with none less than 1.

SOLUTION

CATEGORY	OBSERVED (O)	EXPECTED (E), IF H_0 IS TRUE
GA	$O_1 = 95$	$E_1 = (200)(.625) = 125$
K	$O_2 = 50$	$E_2 = (200)(.226) = 45.2$
L	$O_3 = 41$	$E_3 = (200)(.115) = 23$
M	$O_4 = 5$	$E_4 = (200)(.014) = 2.8$
Other	$O_5 = 9$	$E_5 = (200)(.02) = 4$
	200	200

Notice that the last two expected values are less than 5. These two categories need to be pooled (combined) into a new category, which we will label category 4: other. The new summary is

SOLUTION

CATEGORY	OBSERVED (O)	EXPECTED (E), IF H_0 IS TRUE
GA	95	125
K	50	45.2
L	41	23
Other	14 $(= 5 + 9)$	6.8 $(= 2.8 + 4)$

Now each of the expected values is at least 5, and we can continue the analysis. The hypotheses using the four categories are

$$H_0: p_1 = .625, \ p_2 = .226, \ p_3 = .115, \ p_4 = .034$$

$$H_a: \text{at least one of the proportions is incorrect}$$

The value of p_4 represents the proportion of all other cars (including Ms) sold in this community. The hypothetical value of p_4 becomes $.014 + .020 = .034$.

The computed chi-square value is now

$$\chi^{2*} = \frac{(95 - 125)^2}{125} + \frac{(50 - 45.2)^2}{45.2} + \frac{(41 - 23)^2}{23} + \frac{(14 - 6.8)^2}{6.8}$$

$$= 7.2 + .51 + 14.09 + 7.62 = 29.4$$

This value exceeds the Table A.6 value of $\chi^2_{.05,3} = 7.81$, so we reject H_0 and conclude that we do have a significant departure from the U.S. percentages in this community. The largest contributor to this chi-square value is from the L category, which exceeds expectations by 18 cars (78% more than expected if H_0 is true). ∎

Testing a Hypothesis about a Distributional Form

In this discussion of the goodness-of-fit test, we examine such questions as:

Is it true that these data came from a binomial distribution with probability of success, p, equal to .2?

Does this particular set of data violate the assumption that the number of defects in this product follow a Poisson distribution?

Is there any reason to doubt the assumption that the weights of all Rice Krinkle cereal boxes follow a normal distribution using a recently obtained sample of boxes?

The first two questions concern *discrete* distributions (binomial and Poisson). The final question is concerned with whether the data came from a particular *continuous* (in this case, normal) distribution. We illustrate the chi-square technique using a goodness-of-fit test for a discrete situation. (Goodness-of-fit tests for the normal distribution are illustrated in Exercises 13.13, 13.14, and 13.15.)

Suppose that Blitz laundry detergent is well known for its obnoxious commercials, which advertise that 20% of all Blitz boxes contain a valuable discount coupon. The commercials also claim that the boxes containing coupons are randomly distributed across all stores carrying the product. A recent study obtained a random sample of ten Blitz boxes from each of 100 different stores. The results were

OF 10 BOXES, NUMBER CONTAINING COUPONS	NUMBER OF STORES
0	9
1	31
2	29
3	18
>3	13
	100

We wish to know whether these data appear to come from a binomial distribution with $p = .2$, using $\alpha = .05$.

Your immediate reaction may well be that this problem does not fit a multinomial situation. However, there are 100 independent trials, each trial consisting of randomly selecting ten boxes of Blitz detergent. Also, we can set up five categories here, namely:

Category 1: Observe 0 coupons in the ten boxes (probability p_1)

Category 2: Observe 1 coupon in the ten boxes (probability p_2)

Category 3: Observe 2 coupons in the ten boxes (probability p_3)

Category 4: Observe 3 coupons in the ten boxes (probability p_4)

Category 5: Observe > 3 coupons in the ten boxes (probability p_5)

So we *do* have a multinomial situation here. The hypotheses can be stated as

H_0: the data follow a binomial distribution with $p = .2$

H_a: the data do not follow a binomial distribution with $p = .2$

If H_0 is true, how often should we observe zero coupons in ten boxes? Each multinomial trial fits a binomial situation, where a success consists of a box containing a coupon. We repeat the trial ten times and count the number of successes (coupons). According to Table A.1, with $n = 10$ and probability of success $p = .2$, you should observe zero coupons out of ten boxes 10.7% of the time. So, if H_0 is true, $p_1 = .107$. Similarly, if H_0 is true, we should see one success out of ten trials 26.8% of the time. In other words, $p_2 = .268$. Therefore, another way to state your hypotheses is

H_0: $p_1 = .107$, $p_2 = .268$, $p_3 = .302$, $p_4 = .201$, $p_5 = .122$

H_a: at least one of these p_i's is incorrect

We obtain p_5 by finding the probability of more than three successes in ten trials:

$$1 - (\text{probability of 3 or less}) = 1 - (p_1 + p_2 + p_3 + p_4)$$
$$= 1 - .878 = .122$$

So this is a multinomial test of hypothesis in disguise. Next, we compute the expected values, E.

CATEGORY	OBSERVED (O)	EXPECTED (E), IF H_0 IS TRUE
0 boxes	9	$(100)(.107) = 10.7$
1 box	31	$(100)(.268) = 26.8$
2 boxes	29	$(100)(.302) = 30.2$
3 boxes	18	$(100)(.201) = 20.1$
>3 boxes	13	$(100)(.122) = 12.2$
	100	100

Make sure that all the E's (do not worry about the O's) are at least 5. In this case, all E's are greater than 5, so no pooling of categories is necessary.

To define the rejection region, we notice that there are $k =$ five categories. So the df in the chi-square statistic is df $= k - 1 = 4$. Also, $\chi^2_{.05,4} = 9.49$, and so the test procedure will be to

$$\text{reject } H_0 \text{ if } \chi^2 > 9.49$$

The value of our test statistic is

$$\chi^{2*} = \frac{(9 - 10.7)^2}{10.7} + \frac{(31 - 26.8)^2}{26.8} + \cdots + \frac{(13 - 12.2)^2}{12.2}$$
$$= 1.25$$

Because 1.25 is less than 9.49, we fail to reject H_0 and conclude that there is no evidence to suggest that these data have violated the binomial assumption. These 100 observations suggest that we have no reason to accuse Blitz of false advertising for their claim that 20% of their boxes contain coupons.

In summary, you can refer to the following guidelines if you are trying to perform a test of hypothesis on a binomial parameter, p (for example, $H_0: p = .2$).

1. If you have a *single* sample of size n, then the results of Chapter 10 apply; you have a binomial experiment.
2. If you have *many* samples of size n, then the results of this section apply; this problem can be expressed as a multinomial experiment. The chi-square goodness-of-fit procedure allows you to perform a test on the population proportion, p, as well as determine if the population follows a binomial distribution. The previous example illustrates this type of situation.

Distributional Form with Unknown Parameters

In the Blitz cereal example, H_0 not only stated that the data followed a binomial distribution, it also specified a value of the binomial parameter p (namely, $p = .2$). Often your only concern is whether the data follow a particular distribution (such as binomial, Poisson, or normal), and the values of the corresponding parameters are not preset.

For example, suppose that the manager of Case Electronics has always assumed that the weekly sales of his top-of-the-line telephone answering machine followed a Poisson distribution. Data from a 50-week period were gathered, with the following results:

UNITS SOLD	NUMBER OF WEEKS	UNITS SOLD	NUMBER OF WEEKS
0	1	5	7
1	3	6	5
2	6	7	3
3	11	8	4
4	10	>8	0
			‾‾
			50

How can we test the hypothesis that the number of units sold follows a Poisson distribution, using $\alpha = .1$?

The correct hypotheses here are

$$H_0: \text{weekly sales follow a Poisson distribution}$$

$$H_a: \text{weekly sales do not follow a Poisson distribution}$$

The probability function for the Poisson distribution has one parameter, μ, because this function (from equation 5.12) is given by

$$P(X = x) = \frac{\mu^x e^{-\mu}}{x!}$$

for $x = 0, 1, 2, \ldots$, where $x =$ the number of units sold during a particular week.

However, the value of μ was not specified in H_0. In this case, we estimate any unknown parameter (μ, here) from the sample information and replace each parameter by its estimate in the probability function. In this way, we can estimate all the expected frequencies ($E_1, E_2, E_3, \ldots$).

Whenever you estimate unknown parameters for use with the chi-square test, you need to *adjust the corresponding degrees of freedom*, df. In general, the df for the chi-square goodness-of-fit statistic is given by

$$df = (\text{number of classes}) - 1 - (\text{number of estimated parameters})$$

For the Poisson situation, you are estimating only one parameter, μ, and so the df = (number of classes) $- 1 - 1 =$ (number of classes) $- 2$. The same argument holds true for a test of hypothesis on a binomial distribution where the single parameter, p, is unspecified in H_0 and is instead estimated from the sample information.

Estimating μ

Because μ is the mean of the telephone answering machine sales population, we estimate it using the average (mean) of the sample. In the sample, we observe 1 value of zero, 3 values of one, 6 values of two, and so on for all 50 values. The sample average, our estimate of μ, will be

$$\hat{\mu} = \frac{(0)(1) + (1)(3) + (2)(6) + \cdots + (8)(4)}{50}$$

$$= \frac{206}{50} = 4.12$$

Rounding this to $\hat{\mu} = 4.1$, the estimated probability function is

$$P(X = x) = \frac{(4.1)^x e^{-4.1}}{x!}$$

for $x = 0, 1, 2, \ldots$.

We can now use Table A.3 (the Poisson table) to estimate the expected number of weeks with zero sales, with one sale, and so on. We are *estimating* each expected value, so we denote each of them as $\hat{E}$.

X	$P(X = x)$	$\hat{E}$	O
0	.0166	$(.0166)(50) = $ .83	1
1	.0679	$(.0679)(50) = $ 3.39	3
2	.1393	$(.1393)(50) = $ 6.97	6
3	.1904	$(.1904)(50) = $ 9.52	11
4	.1951	$(.1951)(50) = $ 9.76	10
5	.1600	$(.1600)(50) = $ 8.00	7
6	.1093	$(.1093)(50) = $ 5.46	5
7	.0640	$(.0640)(50) = $ 3.20	3
8	.0328	$(.0328)(50) = $ 1.64	4
>8	.0246	$(.0246)(50) = $ 1.23	0
	1	50	50

Notice that, for the category $X > 8$, the corresponding probability is $1 - (.0166 + .0679 + \cdots + .0640 + .0328) = .0246$.

The next step always is to check your expected frequencies $(\hat{E})$ to see if pooling is necessary. Each $\hat{E}$ value must be at least 5, so it is necessary to pool the first three classes $(.83 + 3.39 + 6.97 = 11.19)$ and the last three classes $(3.20 + 1.64 + 1.23 = 6.07)$. Now you can evaluate the chi-square statistic.

X	$\hat{E}$	O	$(O - \hat{E})$	$(O - \hat{E})^2/\hat{E}$
≤ 2	11.19	10	-1.19	.127
3	9.52	11	1.48	.230
4	9.76	10	.24	.006
5	8.00	7	-1.00	.125
6	5.46	5	$-.46$	.039
≥ 7	6.07	7	.93	.142
	50	50	0	.669

(check)

$$\chi^2 = \sum \frac{(O - \hat{E})^2}{\hat{E}} = .669$$

The degrees of freedom for the corresponding test are

$$\text{df} = (\text{number of classes}) - 1 - (\text{number of estimated parameters})$$
$$= 6 - 1 - 1 = 4$$

The resulting test, using Table A.6 and $\alpha = .1$, is

$$\text{reject } H_0 \text{ if } \chi^2 > 7.779$$

Because $.669 < 7.779$, we fail to reject H_0 and conclude that there is not enough evidence to indicate that the Poisson distribution assumption is incorrect.

Exercises

13.1   A researcher in the marketing department of an investment firm believes that the proportion of full-time workers who are over 40 years of age in a certain locality is equal to 35%. A random sample of 100 observations was taken, and the estimate of the proportion of full-time workers who are over 40 years of age was .31.

a.   Is there sufficient evidence that the proportion of full-time workers who are over 40 years of age is not equal to 35%? Use the Z-test from Section 13.1. Let the significance level be .05.

b.   Use the chi-square goodness-of-fit test to test the hypothesis in part a. Let the significance level be .05.

c.   What is the relationship between the test statistics in parts a and b?

13.2   The owner of a car-insurance company believes that 20% of drivers under 25 years of age have been in exactly one automobile accident in the past two years. She also believes that 15% of the drivers under 25 have been in exactly two automobile accidents in the past two years. Finally, she believes that 10% of the drivers under 25 have been in more than two automobile accidents in the past two years. A survey of 300 randomly selected drivers under 25 years of age was taken. Test the beliefs using the following data and letting the significance level be .10:

NO ACCIDENT	1 ACCIDENT	2 ACCIDENTS	>2 ACCIDENTS
153	68	51	28

13.3   In Exercise 13.2, the equality of proportions for any two categories can be tested using the Z test in Section 13.1. Is there any difference in testing the equality of two proportions one pair at a time and testing all the proportions at once using the chi-square goodness-of-fit test? Is the overall significance level the same in both cases?

13.4 Barton's Food Store carries three brands of milk. Recently, Barton has been getting numerous complaints from customers about the milk being spoiled. Barton decided to categorize 34 randomly selected complaints to see whether they were equally divided among the three brands of milk. Using the accompanying data, test that the complaints are equally divided among the three brands. Let the significance level be .05.

BRAND A	BRAND B	BRAND C
7	13	14

13.5 A stockbroker believes that when too many of the stockmarket newsletters are bullish on the market (that is, they predict that stock prices will go higher), the stockmarket will most likely fall. Thirty-two randomly selected stockmarket newsletters were each placed in one of three categories:

BEARISH ON STOCKMARKET	NEUTRAL ON STOCKMARKET	BULLISH ON STOCKMARKET
9	10	13

Test the null hypothesis that the newsletters are equally divided among the three categories. Use a .05 significance level.

13.6 In order to understand what motivates students to enter a sales career, a questionnaire was mailed to college students who indicated interest in going into sales. The data below show the responses of 965 students. Each student listed the most important factor that motivated him or her to consider a sales career.

FACTOR	NUMBER SAYING THIS IS THE MOST IMPORTANT FACTOR
Job itself	395
Pay	231
Advancement & sense of achievement	231
Other reasons	108

(*Source:* Adapted from S. Castleberry, "The Importance of Various Motivational Factors to College Students Interested in Sales Positions," *Journal of Personal Selling and Sales Management* 10, no. 2, (1990): 65.)

A marketing sales manager believes that there is no difference among the respondents' selection of the first three factors and that the last factor (other reasons) accounts for 10% of the students' selection. Do the data support this belief? Use a .10 significance level.

13.7 A survey of 158 small firms was conducted to determine perceived barriers to small business participation in the Federal marketplace. The results of the survey given below show these firms reported the following types of major barriers to obtaining Federal contracts. Test the null hypothesis that the major barriers are equally divided among the categories listed below. Use a .10 significance level.

TYPE OF MAJOR BARRIER	NUMBER OF SMALL FIRMS REPORTING THIS CATEGORY AS THEIR MAJOR BARRIER
Knowing about solicitation and/or lacking sales and marketing resources	41
Bid and proposal costs and/or efforts to be certified as a legitimate offeror	30
No barriers	51
Other reasons	36

(*Source:* Adapted from J. Cantor, "The Small Business And The Federal Marketplace: A Study Of Barriers To Successful Access," *Business Journal* (Fall 1989): 42.)

13.8 Electrical fuses are packaged in lots of 20. The quality-control department claims that on the average only about 10% of the fuses in each package of 20 fuses are defective. A random sample of 40 packages was selected and the results were:

DEFECTIVE FUSES	PACKAGES
0	7
1	12
2	10
3	7
4	1
5	1
>5	2

Do these data appear to have come from a binomial distribution with $p = .10$? Use a significance level of .05.

13.9 A quality-control engineer samples 100 crates, each of which contains 20 strips of low-tension electrical insulators. The engineer counts the number of strips that are nonconforming (defective). The engineer believes that the number of nonconforming strips per crate follows a binomial distribution with $p = .05$. Seven classes are used to summarize the frequency of different numbers of defective items per crate. Test whether the data below follow a binomial distribution with $p = .05$ at a .10 significance level.

CLASS	NUMBER OF NONCONFORMING STRIPS PER CRATE	OBSERVED FREQUENCIES
1	0	30
2	1	41
3	2	16
4	3	10
5	4	2
6	5	1
7	6 or more	0

13.10 An auditor believes that the number of errors per 25 invoices in the records of a discount furniture store chain follow a Poisson distribution with a mean of 2.2. To test the auditor's belief, 25 stores were randomly selected and the number of errors were tabulated.

ERRORS PER 25 INVOICES	STORES	ERRORS PER 25 INVOICES	STORES
0	3	4	2
1	5	5	2
2	7	6	1
3	4	>6	1

Do these data appear to have come from a Poisson distribution with a mean of 2.2? Use a .10 significance level.

13.11 On each flight of Astral Airways, 12 randomly selected passengers are asked if they would be willing to pay a 5% airfare increase to fly on an airline that had an open bar in the airplane. Results of the survey from 50 different flights were as follows:

YES ANSWERS	FLIGHTS	YES ANSWERS	FLIGHTS
0	3	4	7
1	10	5	3
2	13	6	2
3	12	>6	0

Use a chi-square goodness-of-fit test to determine whether these data came from a binomial distribution. Let the significance level be .05. [*Hint: p* must be estimated using (total number of people who said yes)/(total number asked).]

13.12 An assembly-line operation coats moving sheets of steel with a plastic film. If the thickness of the plastic coating at any point is less than 5mm, then a nonconformity is present. A quality inspector samples 150 lots of 1 meter by 10 meter sheets that have been coated. The number of nonconformities is believed to follow a Poisson distribution. Eight classes are used to classify the frequencies of each number of nonconformities per lot. Test the hypothesis that the data follow a Poisson distribution. Use a .05 significance level.

CLASS	NUMBER OF NONCONFORMITIES PER LOT	OBSERVED FREQUENCY
1	0	21
2	1	43
3	2	37
4	3	25
5	4	12
6	5	7
7	6	5
8	7 or more	0

13.13   To perform certain statistical tests on a set of data, the assumption of normality is required. It is thought that the percentage gain over the past three years in mutual funds that have balanced portfolios of both long-term growth stocks and income-oriented stocks is normally distributed, with a mean of 35% and a standard deviation of 10%. A sample of 75 mutual funds of this type is selected to test this assumption of normality using a chi-square goodness-of-fit test. For the intervals listed, probabilities can be found from the normal table (Table A.4). To find the expected frequencies in the third column, the sample size is multiplied by each probability. If the differences between the observed and expected frequencies are large, then the chi-square statistic based on the observed and expected frequencies would be large and would cause the null hypothesis (that the data was sampled from a normally distributed population with mean = 35 and standard deviation = 10) to be rejected.

PERCENTAGE GAIN INTERVAL	PROBABILITY	EXPECTED FREQUENCY	OBSERVED FREQUENCY
less than 20	.0668	5.01	7
20 and less than 30	.2417	18.1275	15
30 and less than 40	.3830	28.725	26
40 and less than 50	.2417	18.1275	21
50 or more	.0668	5.01	6

At the 5% significance level, complete the chi-square goodness-of-fit test by calculating the chi-square statistic presented in this chapter and by using a tabulated chi-square value for the critical value of the rejection region. The degrees of freedom is taken to be equal to the number of intervals minus one.

13.14   The monthly maintenance time on a particular machine at a manufacturing plant is believed to be normally distributed with a mean of 6 hours and a standard deviation of 1.5 hours. A sample of 45 months is selected and the maintenance time is recorded (in hours). From the following data, use the chi-square goodness-of-fit procedure in Exercise 13.13 to test whether there is enough evidence to support the hypothesis that the maintenance time is not normally distributed with a mean of 6 and a standard deviation of 1.5. Use a 10% significance level.

MAINTENANCE TIME INTERVAL	OBSERVED FREQUENCY
less than 5	10
5 hours but less than 6	6
6 hours but less than 6.5	5
6.5 hours but less than 7	7
7 hours but less than 8	6
8 hours or more	11

13.15   The weekly traffic flow between 8 A.M. and 6 P.M. at a certain intersection in Oklahoma City is believed to be normally distributed, with a mean of 65,000 cars and a standard deviation of 12,000. Fifty weeks are randomly selected over the past three years. Using the following data (in thousands of cars) and the chi-square goodness-of-fit test procedure in Exercise 13.13, is there sufficient evidence to conclude that the traffic flow is not normally distributed with a mean of 65 (thousand) and a standard deviation of 12 (thousand)? Use a .10 significance level.

TRAFFIC FLOW	OBSERVED FREQUENCY
less than 50	9
50 and less than 60	10
60 and less than 70	11
70 and less than 80	10
80 or more	10

13.2
Chi-Square Tests of Independence

In the previous section, we classified each member of a population into one of many categories. This classification was one-dimensional, because each member was classified using only *one* criterion (brand, color, and so on). In this section, we extend classification to a two-dimensional situation, in which each element in the population is classified according to two criteria, such as sex and income level (high, medium, or low). The question of interest is, are these two variables (classifications) *independent?* For example, if sex and income level are not independent, perhaps sex discrimination is present in the salary structure of a company. If a peron's salary is not related to sex, these two classifications *would* be independent.

In Chapter 4, we examined a survey concerned with the age and sex of the purchasers of a recently released microcomputer. The results were summarized in a *contingency* (or *cross-tab*) table. This table consisted of **cells,** where each cell contains the **frequency** of people in the sample who satisfy each of the various cross-classifications:

SEX	AGE <30	AGE 30–45	AGE >45	TOTAL
Male	60	20	40	120
Female	40	30	10	80
Total	100	50	50	200

This 2 × 3 contingency table shows that there were 60 people who were both male *and* under 30. In Chapter 4, we determined various probabilities for a person selected at random from this group of 200, such as the probability that this person is both a male and over 45 years. Now we will view these data as the results of a particular experiment (survey) and attempt to determine whether the variables—age and sex—are independent for this application. Put another way, is the age structure of the male buyers the same as that of the female purchasers? The hypotheses are

H_0: the classifications (age and sex) are independent

H_a: the classifications are dependent

This problem can also be viewed as a multinomial experiment containing 200 trials and (2)(3) = 6 possible categories for each trial outcome.

Deriving a Test of Hypothesis for Independent Classifications

Calculating the Expected Values   We want to decide whether the data about the purchasers exhibit random variation or a pattern of some type due to a dependency between age and sex. If these classifications *are* independent (H_0 is true), how many people would you expect in each cell? Consider the upper right cell, which shows males over 45 years. The expected number of sample observations in this cell is $200 \cdot P$(sampled purchaser is a male and over 45). Assuming independence, this is $200 \cdot P$(sampled purchaser is a male) $\cdot$ P(sampled purchaser is over 45), using the multiplicative rule for independent events discussed in Chapter 4.

What is P(sampled purchaser is a male)? We do not know, because we do not have enough information to determine what percentage of *all* purchasers are male. However, we can *estimate* this probability using the percentage of males in the sample: $120/200 = .6$.

Similarly, P(sampled purchaser is over 45) can be estimated by the fraction of people over 45 in the sample—namely, $50/200$. So, our estimate of the expected number of observations for this cell is

$$\hat{E} = 200 \cdot \frac{120}{200} \cdot \frac{50}{200} = \frac{(120)(50)}{200} = 30$$

So, for this cell, the observed frequency is $O = 40$, and our estimate of the expected frequency (if H_0 is true) is $\hat{E} = 30$. In general,

$$\hat{E} = \frac{\text{(row total for this cell)} \cdot \text{(column total for this cell)}}{n}$$

where n = total sample size. A summary of the calculations can be tabulated as follows.

SEX	AGE	OBSERVED (O)	EXPECTED ($\hat{E}$), IF H_0 IS TRUE
Male	< 30	60	$(120)(100)/200 = 60$
	30–45	20	$(120)(50)/200 = 30$
	>45	40	$(120)(50)/200 = 30$
Female	<30	40	$(80)(100)/200 = 40$
	30–45	30	$(80)(50)/200 = 20$
	>45	10	$(80)(50)/200 = 20$
		200	200

The easiest way to represent these 12 values is to place the expected value in parentheses alongside the observed value in each cell:

SEX	AGE <30	AGE 30–45	AGE >45	TOTAL
Male	60 (60)	20 (30)	40 (30)	120
Female	40 (40)	30 (20)	10 (20)	80
Total	100	50	50	200

Pooling

At this point, you need to check your expected values. If any one of them is less than 5, you need to combine the column (or row) in which this small value occurs with another column (or row) to comply with our earlier requirement that all expected values in the chi-square statistic are at least 5. The observed and expected values for this new column (row) are obtained by summing the values for the two columns (rows).

The Test Statistic

The test statistic for testing H_0: the classifications are independent versus H_a: the classifications are dependent is the usual chi-square statistic, which in this case compares each *observed* frequency with the corresponding *expected* frequency estimate.

$$\chi^2 = \sum \frac{(O - \hat{E})^2}{\hat{E}} \qquad (13.3)$$

where the summation is over all cells of the contingency table.

Degrees of Freedom

In the multinomial situation, the degrees of freedom for the chi-square statistic were $k - 1$, where $k =$ the number of categories (outcomes). In this situation, there were k values of $(O - \hat{E})$. However, because the sum of the observed frequencies is the same as the sum of the expected frequencies, the sum of the k values of $(O - \hat{E})$ is *always zero*. This means that, of these k values, only $k - 1$ are free to vary, resulting in $k - 1$ df for the chi-square statistic.

Take a close look at the observed and expected frequencies in the contingency table for age and sex of purchasers. Notice that (1) for each row, sum of O's = sum of $\hat{E}$'s and (2) for each column, sum of O's = sum of $\hat{E}$'s. In general, if classification 1 has c categories and classification 2 has r categories, you construct an **$r \times c$ contingency table** (Figure 13.3). Of the c values of $(O - \hat{E})$ in each row, only $c - 1$ are free to vary. Similarly, only $r - 1$ of the values in each column are free to assume any value. So, for this contingency table, only $(r - 1)(c - 1)$ values are free to vary. Therefore, for the chi-square test of independence,

$$df = (r - 1)(c - 1) \qquad (13.4)$$

Testing Procedure

When H_0 is not true, the expected frequencies and observed frequencies will be very different, producing a large χ^2 value. We again reject H_0 if the value of the test statistic falls in the *right-tail* rejection region, so we

$$\text{reject } H_0 \text{ if } \chi^2 > \chi^2_{\alpha,df}$$

where df $= (r - 1)(c - 1)$.

■ **FIGURE 13.3**
Expected value estimates for an $r \times c$ contingency table.

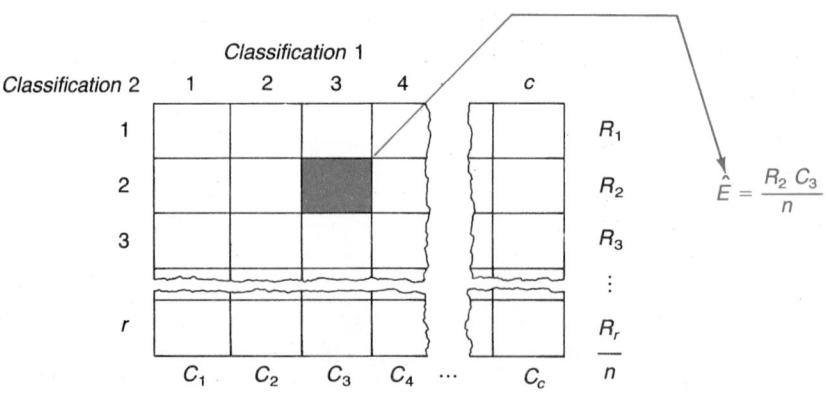

In summary, the chi-square test for independence hypotheses is

H_0: the row and column classifications are independent (not related)

H_a: the classifications are dependent (related or associated in some way)

The test statistic is

$$\chi^2 = \sum \frac{(O - \hat{E})^2}{\hat{E}}$$

where

1. The summation is over all cells of the contingency table consisting of r rows and c columns.
2. O is the observed frequency in this cell.
3. $\hat{E}$ is the estimated expected frequency for this cell.

$$\hat{E} = \frac{\left(\begin{array}{c}\text{total of row in} \\ \text{which the cell lies}\end{array}\right) \cdot \left(\begin{array}{c}\text{total of column in} \\ \text{which the cell lies}\end{array}\right)}{(\text{total of all cells})}$$

4. The degrees of freedom for the chi-square statistic are df $= (r - 1)(c - 1)$.

The test procedure is (using Table A.6):

$$\text{reject } H_0 \text{ if } \chi^2 > \chi^2_{\alpha,\text{df}}$$

We can now return to our question of whether the age and sex of microcomputer purchasers are independent. Step 1 (statement of hypotheses) and step 2 (definition of test statistic) of our five-step procedure have been discussed already. Assume that a significance level of $\alpha = .1$ was specified. For step 3, the df are $(2 - 1)(3 - 1) = 2$. Using Table A.6, $\chi^2_{.1,2} = 4.61$. So we will reject H_0 if $\chi^2 > 4.61$. For step 4, referring to the contingency table,

$$\chi^{2*} = \frac{(60 - 60)^2}{60} + \frac{(20 - 30)^2}{30} + \frac{(40 - 30)^2}{30} + \frac{(40 - 40)^2}{40}$$

$$+ \frac{(30 - 20)^2}{20} + \frac{(10 - 20)^2}{20}$$

$$= 0 + 3.33 + 3.33 + 0 + 5 + 5$$

$$= 16.66$$

This exceeds the table value of 4.61, so we reject H_0. We thus conclude that the age and sex classifications are *not* independent (step 5).

If the results of the chi-square test lead to a conclusion that the classifications are not independent, a closer look at the individual terms in the chi-square statistic can often reveal what the relationship is between these two variables. Examining the six terms, we observe four large values, namely, 3.33 (male/age 30–45), 3.33 (male/age over 45), 5 (female/age 30–45), and 5 (female/age over 45). We obtained more men (and fewer women) over 45 years than we would expect if there was no dependency. Similarly, there were fewer men (and more women) between 30 and 45 years.

We can find the *p*-value for this situation also, given $\chi^{2*} = 16.66$. Using a χ^2 curve with 2 df, the area to the right of 16.66, using Table A.6, is $<.005$. The *p*-value indicates the **strength** of the dependency between two classifications. *The smaller the p-value is, the more you tend to support the alternative hypothesis, which indicates a stronger dependency between the two variables.* For the age and sex illustration, $p < .005$, wo we conclude that the age and sex of these purchasers are strongly related.

It is worth mentioning at this point that it is possible that examining one category (such as sex) can fail to show any differences among subcategories (male versus female), but when the category is examined along with another category (such as age classification), patterns can emerge. Such a technique is often useful in detecting job discrimination within companies. For example, no sex discrimination may be evident in a sample, but when it is examined along with race or age categories, certain discriminatory practices can be identified.

EXAMPLE 13.5

In Example 11.4, a personnel director attempted to determine whether an employee's educational level had an effect on his or her job performance. An exam was given to a sample of the employees, and we used the ANOVA procedure to test for a difference among the three groups: (1) those with a high school diploma only, (2) those with a bachelor's degree only, and (3) those with a master's degree.

The director decided to expand this procedure by testing 120 employees; rather than recording the exam scores, she rated each person's exam performance as high, average, or low. The results of this study are:

	HIGH	AVERAGE	LOW	TOTAL
Master's degree	4	20	11	35
Bachelor's degree	12	18	15	45
High school diploma	9	22	9	40
Total	25	60	35	120

Does job performance as measured by the exam appear to be related to the level of an employee's education, at this particular firm? Use $\alpha = .05$.

SOLUTION

Step 1. This problem calls for a chi-square test of independence, with hypotheses

H_0: exam performance is independent of educational level

H_a: these classifications are dependent

Steps 2, 3. Your test statistic is the chi-square statistic in equation 13.3. The table of frequencies here is a 3×3 contingency table, which means that the degrees of freedom are df $= (3 - 1)(3 - 1) = 4$. From Table A.6, we determine that $\chi^2_{.05,4} = 9.49$, so the testing procedure is to

reject H_0 if $\chi^2 > 9.49$

Step 4. Computing the expected frequency estimates in the usual way, we arrive at the following table:

	HIGH	AVERAGE	LOW	TOTAL
Master's degree	4 (7.29)	20 (17.5)	11 (10.21)	35
Bachelor's degree	12 (9.38)	18 (22.5)	15 (13.12)	45
High school diploma	9 (8.33)	22 (20.0)	9 (11.67)	40
Total	25	60	35	120

To illustrate the calculations, the 11.67 in the lower right cell is $(40 \cdot 35)/120$. The computed chi-square value is

$$\chi^{2*} = \frac{(4 - 7.29)^2}{7.29} + \frac{(20 - 17.5)^2}{17.5} + \cdots + \frac{(9 - 11.67)^2}{11.67} = 4.67$$

This value is <9.49, and so we fail to reject H_0.

```
MTB > READ INTO C1-C3
DATA> 4 20 11
DATA> 12 18 15
DATA> 9 22 9
DATA> END
      3 ROWS READ
MTB > NAME C1 = 'HIGH' C2 = 'AVERAGE' C3 = 'LOW'
MTB > CHISQUARE USING C1-C3

Expected counts are printed below observed counts

            HIGH  AVERAGE     LOW   Total
     1         4       20      11       35
            7.29    17.50   10.21              Observed (O)
                                               Expected (E)
     2        12       18      15       45
            9.37    22.50   13.12

     3         9       22       9       40
            8.33    20.00   11.67

  Total       25       60      35      120

  ChiSq =  1.486 +  0.357 +  0.061 +
           0.735 +  0.900 +  0.268 +
           0.053 +  0.200 +  0.610 = 4.670
  df = 4
```

Step 5. We see no evidence of a relationship between job performance as measured by the exam and level of education.

We do not conclude that these data demonstrate that the two classifications are clearly *independent*, because this would amount to accepting H_0. We are simply unable to demonstrate that a relationship exists. ■

A MINITAB solution to Example 13.5 is contained in Figure 13.4. Notice that the format of this table is similar to that of the one we constructed, with the expected value (assuming H_0) and the observed value shown for each cell.

COMMENTS In Example 13.5, the personnel director recorded the exam performance as high, average, or low rather than listing the actual exam score. Why would anyone take *interval/ratio* data (the exam scores) and convert them to seemingly weaker *ordinal* data (the exam performance classifications)? Do you lose useful information by doing this? When using the ANOVA procedure, we were forced to assume that these data came from *normal* populations with equal variances. In this chapter, *no* assumptions regarding the populations (aside from the randomness of the sample) were necessary. So, by converting the exam scores to a form suitable for a contingency table and using the chi-square test of independence, we can avoid the assumptions of normality and equal variances.

This question introduces **nonparametric statistics,** often called *distribution-free* statistics. The beauty of these procedures is that they require only very weak assumptions regarding the populations. However, if the data *do* satisfy the requirements of the ANOVA procedure (or nearly so), the nonparametric test is less sensitive to differences among the populations (such as educational level) and so is less *powerful* than the ANOVA F test. Additional nonparametric tests of hypothesis are discussed in Chapter 19.

Test of Independence with Fixed Marginal Totals (Test of Homogeneity)

We use a slightly different interpretation of the previous chi-square procedure when we determine *in advance* the number of observations to be sampled within each column (or row). In the previous discussion, the row and column totals were random variables because we had no way of knowing what they would be before the sample was obtained. In this discussion, the contingency table is the same, except that the column (or row) totals are predetermined.

Assume Lextron International, a manufacturer of electronic components, has facilities located in Dallas, Boston, Seattle, and Denver. Over the years, Lextron has gone to great lengths to discourage the formation of labor unions at these plants, including constructing employee recreational centers and offering better-than-average employee benefits. Management suspects, however, that there is growing interest among the employees in forming a union. Of particular interest is whether employee interest in a union differs among the four plants.

The Dallas and Denver plants are considerably larger than the other two, so Lextron obtains a random sample of 200 employees from each of these two plants and 100 from each of the two smaller facilities. The results of the survey are:

	DALLAS	BOSTON	SEATTLE	DENVER	TOTAL
Interested	120	41	45	112	318
Not interested	35	38	40	36	149
Indifferent	45	21	15	52	133
Total	200	100	100	200	600

In the previous tests of independence, we had a *single* population, where each member was classified according to two criteria, such as age and sex. Now we have four distinct populations, namely, the Lextron employees in each of the four cities. Consequently, we obtained a random sample from each one. The column totals (sample sizes) were determined in advance. This example differs from our previous ones, where we had no idea what the row or column totals would be before the sample was obtained.

The question of interest here becomes, is interest in a labor union the same in each of the four cities? In other words, we are trying to determine whether these four populations can be viewed as belonging to the *same* population (in terms of this criterion). Identical populations are said to be **homogeneous.** Consequently, the test of hypothesis here is a **test of homogeneity** as well as a test for independence. The null hypothesis can be written as

H_0: the four populations are homogeneous in their interest in a union

or as

H_0: plant location and employee interest are independent classifications

The procedure for analyzing a contingency table is the *same* whether or not the column (or row) totals are fixed in advance.

A MINITAB solution using $\alpha = .05$ is provided in Figure 13.5. The expected cell frequencies are computed by finding

$$\hat{E} = \frac{(\text{row total})(\text{column total})}{600}$$

The computed chi-square value is

$$\chi^{2*} = \frac{(120 - 106.0)^2}{106} + \frac{(41 - 53.0)^2}{53} + \cdots + \frac{(52 - 44.3)^2}{44.3} = 34.16$$

The degrees of freedom here are $(3 - 1)(4 - 1) = 6$, so we reject H_0 if $\chi^2 > 12.59$, where 12.59 is $\chi^2_{.05,6}$. The computed value (34.16) exceeds the tabled value, so we reject H_0. We conclude that these four populations are *not* homogeneous. The employee interest in a labor union is not identical at each of the four locations. We can also say that the location and union interest classifications are not independent.

Examining the individual terms of the chi-square value in Figure 13.5, we note that the larger plants (Dallas and Denver) had a higher proportion of employees interested in forming a union. In Dallas, for example, if these classifications were independent, we would expect 106 employees to be interested; instead, we observed 120. The same argument applies to the Denver plant.

■ FIGURE 13.5
MINITAB solution to
test H_0: Plant location
and employee interest
are independent
classifications (test
of homogeneity).

```
MTB > READ INTO C1-C4
DATA> 120 41 45 112
DATA> 35 38 40 36
DATA> 45 21 15 52
DATA> END
     3 ROWS READ
MTB > NAME C1 = 'DALLAS' C2 = 'BOSTON' C3 = 'SEATTLE' C4 = 'DENVER'
MTB > CHISQUARE USING C1-C4

Expected counts are printed below observed counts

           DALLAS    BOSTON   SEATTLE    DENVER     Total
      1       120        41        45       112       318
           106.00     53.00     53.00    106.00

      2        35        38        40        36       149
            49.67     24.83     24.83     49.67

      3        45        21        15        52       133
            44.33     22.17     22.17     44.33

Total        200       100       100       200       600

ChiSq =    1.849 +   2.717 +   1.208 +   0.340 +
           4.331 +   6.981 +   9.263 +   3.761 +
           0.010 +   0.061 +   2.317 +   1.326 = 34.163
      df = 6
```

Using Table A.6, we find that the p-value is less than .005. Because of this extremely small value, we can conclude that employees at these four plants have considerably different views in regard to the formation of a labor union. The small p-value also implies that there is an extremely strong dependence between the two classifications.

EXAMPLE 13.6

Masuturi, a manufacturer of printed circuit boards, has determined that boards classified as nonconforming nearly always have one of three defects: a component on the board is either missing, damaged, or raised (installed improperly). The boards are produced on three machines (A, B, and C). Machine A is 90% computer-controlled and produces twice as many boards per day as machines B and C, which are 50% computer-controlled. To determine whether there is a relationship between the type of nonconformity and the machine, a sample of 500 nonconforming boards was obtained, 250 from machine A and 125 each from machines B and C, since machine A produces twice as many boards as machines B and C. The following data were obtained:

MACHINE	TYPE OF NONCONFORMITY			TOTAL
	MISSING	DAMAGED	RAISED	
A	50	80	120	250
B	60	55	10	125
C	65	45	15	125
Total	175	180	145	500

Is the type of nonconformity related to the machine used for production? Use a significance level of .10.

SOLUTION **Step 1.** The hypotheses for this test of homogeneity are

H_0: the three machine populations are homogeneous in the types of nonconformity produced

H_a: the three machine populations are not homogeneous in the types of nonconformity produced

Steps 2, 3. Your test statistic is the chi-square statistic in equation 13.3 with degrees of freedom $(3 - 1)(3 - 1) = 4$. Using Table A.6, the testing procedure is to

$$\text{reject } H_0 \text{ if } \chi^2 > 7.78$$

Step 4. Computing the expected frequencies in the usual way produces the following table:

MACHINE	TYPE OF NONCONFORMITY			TOTAL
	MISSING	DAMAGED	RAISED	
A	50 (87.5)	80 (90.0)	120 (72.5)	250
B	60 (43.75)	55 (45.0)	10 (36.25)	125
C	65 (43.75)	45 (45.0)	15 (36.25)	125
Total	175	180	145	500

The computed chi-square value is

$$\chi^{2*} = \frac{(50 - 87.5)^2}{87.5} + \frac{(80 - 90)^2}{90} + \cdots + \frac{(15 - 36.25)^2}{36.25}$$

$$= 16.071 + 1.111 + 31.121$$
$$+ 6.036 + 2.222 + 19.009$$
$$+ 10.321 + 0.000 + 12.457 = 98.35$$

This value is >7.78, so we reject H_0.

Step 5. There is evidence to indicate that the three machine populations are not homogeneous; that is, the type of nonconformity *is* related to the machine used for production of the circuit boards. Comparing the observed and expected values, we observed more raised component nonconformities than expected on machine A and fewer than expected on machines B and C. Also, we observed fewer missing-component nonconformities than expected on machine A and more than expected on machines B and C.

 *Exercises*

13.16 Suppose you are interested in determining whether there is a relationship between one's educational preference (major) and one's sex. A random sample of 172 students at Hamilton College yields the following data:

MAJOR	FEMALE	MALE	TOTAL
Liberal arts	35	25	60
Home economics	6	9	15
Physics	18	21	39
Business	26	32	58
Total	85	87	172

a. Formulate the necessary hypotheses.

b. Using the chi-square test, test the hypotheses to determine whether educational preference is independent of sex, using a significance level of .10.

c. Find and interpret the p-value for the chi-square test.

13.17 A lawn-equipment shop is considering adding a brand of lawnmowers to its merchandise. The manager of the shop believes that the highest-quality lawnmowers are Trooper, Lawneater, and Nipper, and he needs to decide whether it makes a difference which of these three the shop adds to its existing merchandise. Twenty owners of each of these three types of lawnmowers are randomly sampled and asked how satisfied they are with their lawnmowers.

LAWNMOWER	VERY SATISFIED	SATISFIED	NOT SATISFIED	TOTAL
Trooper	11	6	3	20
Lawneater	13	4	3	20
Nipper	13	6	1	20

Are the owners of the lawnmowers homogeneous in their response to the survey? Use a 5% significance level.

13.18 A real-estate firm wanted to know whether the type of house purchased is associated with the amount of education of the head of the household. Fox and Jones Construction builds four styles of homes. A random sample of 175 homeowners who own a Fox and Jones house was taken, and the education level of the household head was noted.

TYPE OF HOUSE	NO COLLEGE DEGREE	BACHELOR'S DEGREE	MASTER'S DEGREE	DOCTORAL DEGREE	TOTAL
1	12	5	1	0	18
2	13	10	8	2	33
3	10	20	25	10	65
4	2	18	30	9	59

Do the data provide sufficient evidence to indicate that the type of house owned is related to the educational level of the head of the household? Use a .05 level of significance.

13.19 An experimental project was undertaken to verify the belief that novice computer users prefer a menu-oriented interface and that experienced users prefer a command-oriented interface. Lotus 1-2-3, a software product that can be used with both interfaces, was selected for this experiment. Novice, trained, and experienced computer users were given various tasks to perform, and a tally was kept on whether the computer user selected a menu-oriented or a command-oriented interface. Do the data given below provide sufficient evidence to indicate a relationship between experience level and interface preference? Use a .05 significance level.

EXPERIENCE LEVEL	PREFERENCE	
	COMMANDS	MENU
Novice	8	24
Trained	8	6
Experienced	7	4

(*Source:* William Remington, "Interface Preference Versus User Experience Level: An Empirical Investigation," *Proceedings of the 1990 Annual Meeting of the Decision Sciences Institute:* 1253.)

13.20 An insurance company claims that full-size cars are more prone to automobile accidents and hence should be subject to a higher insurance premium. To test the validity of the claim, an auto firm gathered a random sample:

CAR SIZE	AT LEAST ONE ACCIDENT	NO ACCIDENTS
Full size	24	13
Compact	36	117
Small	108	214

Formulate the necessary hypotheses and test at the 10% significance level. Also calculate the *p*-value. Based on this value, would you reject the null hypothesis that car size and occurrence of accidents are independent at the 1% level?

13.21 The Meyers-Briggs Type Indicator (MBTI) is a personality scale that can be used to classify qualities like Extrovert (E), Introvert (I), Intuitive (N), Feeling (F), Sensing (S), Thinking (T), Judging (J), Perceptive (P). Thus, EN means extrovert-intuitive, SP means sensing-perceptive. Consider the following hypothetical frequencies for a cross-tabulation of four types of personality using the MBTI against profession.

PROFESSION	PERSONALITY (MBTI)				
	EN	IF	SP	JT	TOTAL
Computer programmer	4	6	5	6	21
Accountant	3	7	5	5	20
Marketer	9	3	7	4	23
Educator	5	5	5	5	20
Total	21	21	22	20	84

a. From the above table, can you conclude at a 1% significance level, whether personality type and profession are related?

b. State the *p*-value for your test.

13.22 A study was conducted to determine whether there is a relationship between firm size and the use of financial concepts. In a survey, large and small firms were asked the question, "Do you use financial ratio analysis to assist you in making decisions?" The data below summarize the results of the survey. Is there sufficient evidence that the response to the question depends on whether the firm is small or large? Use a 10% significance level.

RESPONSE	SMALL	LARGE
Yes	87	126
No	15	5

(*Source:* Adapted from M. Lamberson, "Financial Analysis and Working Capital Management Techniques Used By Manufacturers: A Comparison of Small And Large Firms," *Journal of Business And Economic Perspectives* 16, no. 1, (1990): 1.)

13.23 A research team conducts a study to see whether the voting for candidates in the recent local election is homogeneous across age groups (given in years). One hundred voters were randomly selected from each of five age classifications. Do the data indicate that voting is homogeneous with respect to age group? Use a 5% significance level.

AGE	CANDIDATE A	CANDIDATE B	CANDIDATE C	TOTAL
Less than 25	48	22	30	100
25 and less than 35	55	20	25	100
35 and less than 45	50	28	22	100
45 and less than 55	45	21	34	100
Over 55	49	21	30	100

 *Summary*

When performing a two-tailed test of hypothesis on a binomial parameter (for example, $p = .75$) we can use a chi-square test statistic. The advantage of this approach is that it extends easily to the **multinomial situation,** where each trial can result in any specified number of outcomes. For example, the roll of a single die has six possible outcomes on each roll.

In the multinomial situation, the probability of observing each possible outcome may be specified (such as 1/6 for each outcome in the single die illustration). To test the hypothesis, a random sample of observations is obtained, and a chi-square test statistic is evaluated either to reject or to fail to reject this set of probabilities (percentages). Such a test is referred to as a **chi-square goodness-of-fit test.** The form of this chi-square test statistic is

$$\chi^2 = \sum \frac{(O - E)^2}{E}$$

where

1. *O* represents the *observed* frequency of observations in a particular category (such as the observed number of 3s in 60 rolls of a single die).

2. E is the *expected* frequency for this category. For example, we would expect to see $60 \cdot \frac{1}{6} = 10$ values of 3 in the die illustration.
3. The chi-square value is obtained by summing over all categories of the multinomial random variable.
4. Categories must be combined (pooled) together whenever an expected value (E) for a particular category is less than 5.

The chi-square goodness-of-fit procedure can be used to determine whether a certain set of sample data came from a specified probability distribution. For example, you might attempt to determine whether the number of nonconformities in a particular product follows a Poisson distribution. By collecting a random sample and counting the number of nonconformities in each product, you can compare the observed values (how many 0s, how many 1s, and so on) with what you would expect if the null hypothesis (H_0: the data are from a Poisson distribution) is true. If the calculated chi-square value is significantly large (in the right tail), this hypothesis will be rejected. Whenever any of the parameters for this distribution are unknown (such as μ for the Poisson illustration), they can be estimated using the sample data. The degrees of freedom of the chi-square test statistic are reduced by one for *each* estimated parameter.

Finally, this chi-square statistic can be used to test whether two classifications (such as age and performance) used to define a contingency table are independent. This is the **chi-square test of independence.** The expected value within each cell of the contingency table is determined under the assumption that H_0 is true, for H_0: the row and column classifications are independent. This also leads to a right-tailed rejection region using the chi-square statistic.

This procedure can be used as a **test for homogeneity** when fixed sample sizes are used for each row or column of the table. If the column totals are fixed in advance, this test will determine whether the populations defined by the column categories are homogeneous (identical) with respect to the variable defining the rows. A similar argument applies when the row totals are predetermined. The test statistic used for a test of homogeneity is the same chi-square statistic used in the test of independence.

Summary of Formulas

Chi-square test statistic:

$$\chi^2 = \sum \frac{(O - E)^2}{E}$$

summed over all categories for a goodness-of-fit test and over all cells for a test of independence or homogeneity. In the latter case, each expected value (E) is replaced by its estimate, $\hat{E}$.

Estimated expected value for a test of independence/homogeneity:

$$\hat{E} = \frac{(\text{row total for cell})(\text{column total for cell})}{n}$$

where n is the total of all cells in the contingency table.

Review Exercises

13.24 The manager of the Grandiose Hotel guarantees that a customer's room will be ready at 6:00 P.M. if a reservation is made. Otherwise, the customer stays at the hotel for free. The manager believes that this policy should be continued; he believes that a customer's room is not available on time only 5% of the time. A random sample of 200 past reservations was selected and yielded an estimate of .065 for the proportion of customers whose room was not available on time.

a. Letting the significance level be .05, test the belief that the proportion of customers whose room is not available on time is .05. Use the Z-test from Section 10.2.

b. Use the chi-square goodness-of-fit test instead of the Z-test in part a.

c. What is the relationship between the test statistics in parts a and b?

13.25 A quality-control inspector believes that the number of nonconformities related to the manufacturing of certain containers is equally divided among the four categories given below. A sample of 50 nonconforming items yielded the following results:

CATEGORY OF NONCONFORMITY	NUMBER OF TIMES THIS NONCONFORMITY OCCURRED
Sides of the container not sealed properly	11
Dent in the side of the container	14
Improper label	15
Smeared print	10

Test the belief that all nonconformities occur equally often. Use a significance level of 10 percent.

13.26 A manufacturer of clothes dryers believes that historically, 40% of its sales are for the basic 18-pound-capacity clothes dryer, 35% are for the 20-pound-capacity dryer, and 25% are for the 22-pound-capacity dryer. A random sample of 200 clothes dryers sold during the past 6 months was obtained, with the following results (capacity in pounds):

CAPACITY	NUMBER SOLD
18	72
20	68
22	60

Using a significance level of .10, is there sufficient evidence to indicate that the manufacturer was wrong in its statement of these percentages?

13.27 A construction company sells and installs solid vinyl siding. Siding comes in five basic colors: white, brown, avocado, reddish-tan, and yellow. Historically, the percentages of sales for each color are 50%, 27%, 12%, 8% and 3%, respectively. A random sample of 100 recent sales gives the following data:

COLOR	NUMBER SOLD
White	43
Brown	29
Avocado	14
Reddish-tan	13
Yellow	1

Does the sales distribution of these colors appear to be the same as the historical distribution? Test at the 1% significance level.

13.28 A large department store in New York City has five entrances and exits. It is believed that the proportion of shoppers entering or leaving the store is approximately the same for each of the five doorways on any single day. The number of customers entering or leaving the store is tallied at each doorway for three randomly selected days:

DOORWAYS	CUSTOMERS
1	150
2	123
3	126
4	163
5	152

Do the data justify the statement that all five entrances and exits are used equally often? Use a 5% significance level.

13.29 The manager of a news and magazine store in a metropolitan area with four newspapers believes that the proportions of customers preferring newspapers A, B, C, and D are

30%, 40%, 15%, and 15%. One hundred randomly selected customers were chosen, and the newspaper they bought was recorded as shown below:

NEWSPAPER	NUMBER OF CUSTOMERS BUYING THE NEWSPAPER
A	26
B	44
C	12
D	18

Do the data justify the percentages used by the manager of the news and magazine store? Use a 1% significance level.

13.30 A car-rental company has 15 cars to rent. The owner believes that the number of cars rented daily is binomially distributed. He also believes that each car has a 30% chance of being rented each day. Forty-five randomly selected days are chosen and the number of cars rented are recorded. From the data, test the hypothesis that the daily rental of cars is binomially distributed with $p = .30$. Use a 5% significance level.

CARS RENTED	DAYS OCCURRED	CARS RENTED	DAYS OCCURRED
0	0	5	12
1	3	6	6
2	3	7	3
3	6	≥ 8	3
4	9		

13.31 An advertising firm believes that the number of daily responses to an advertisement in the *Wall Street Journal* follows a Poisson distribution. Forty days were randomly selected and the following data were collected:

RESPONSES	DAYS	RESPONSES	DAYS
0	0	4	6
1	8	5	6
2	8	6	2
3	10		

Can you conclude that the data did come from a Poisson distribution? Use a 1% level of significance.

13.32 The assistant dean of the College of Business at Oceanside University believes that the number of students dropping a class is Poisson distributed. Fifty classes, all containing the same number of students, were randomly selected. The number of withdrawals from the classes was recorded. Based on the following data, what conclusion can be drawn about whether these data come from a Poisson distribution? Use a significance level of .05 to justify your conclusion.

DROPS	CLASSES	DROPS	CLASSES
0	0	5	4
1	2	6	6
2	6	7	4
3	10	>7	0
4	18		

13.33 Favin Copiers, Inc. has repairpersons who are used to traveling to sites where a Favin copier needs repair. For a certain metropolitan area, the company has ten repairpersons who stand by for calls to repair a copier. A manager believes that the number of repairpersons used each day is binomially distributed with $p = .3$, where p is the probability that a repairman is sent out on any given day. One hundred days of operation yielded the following results:

NUMBER OF REPAIRPERSONS SENT OUT	DAYS WITH THIS NUMBER OF REPAIRPERSONS SENT OUT
0	4
1	10
2	21
3	27
4	18
5	12
6	5
7	3
8 or more	0

Test the goodness-of-fit of the data to a binomial distribution with $p = .3$. Use a significance level of .05.

13.34   A computer generates 100 observations from a normally distributed population with mean 35 and standard deviation 2. The results of the 100 observations generated are:

INTERVAL	OBSERVED FREQUENCY	INTERVAL	OBSERVED FREQUENCY
less than 32	6	35 but less than 36	19
32 but less than 33	9	36 but less than 37	15
33 but less than 34	12	37 but less than 38	11
34 but less than 35	23	38 or more	5

Use the chi-square goodness-of-fit procedure in Exercise 13.13 to test whether there is enough evidence to support the conclusion that the generated numbers did not come from a normally distributed population with mean = 35 and standard deviation = 2. Use a 1% significance level.

13.35   The vice president of a national firm wants to know the response of workers at a certain plant to a proposal to relocate the plant. Forty workers were randomly selected from each of the five divisions at the plant and asked if they favored a relocation of the plant.

DIVISION	FAVORED	DO NOT FAVOR	TOTAL
A	15	25	40
B	18	22	40
C	24	16	40
D	17	23	40
E	20	20	40

Do the data indicate that the divisions are not homogeneous with respect to the proportion of workers who favor a relocation of the plant? Use a .05 significance level.

13.36   A manufacturer of bleach sold the same bleach to three different companies, which placed their own brand names on the bottle. A marketing-research firm wanted to know whether the same amount of each brand of bleach would be sold, provided the price was the same for each brand. Supplies of the three brands of bleach were placed on a supermarket shelf. By the end of the month, 335 bottles of bleach had been sold, as follows:

BLEACH X	BLEACH Y	BLEACH Z
105	133	97

Test the null hypothesis that each brand sells equally well. Use a significance level of .10.

13.37   An immigration attorney was investigating which industries to target for obtaining new clients who might have problems with changes in the immigration laws. Five industries were selected. Twenty workers were chosen in each industry, and their visa status was verified. The data are summarized as follows:

	INDUSTRY					
VISA STATUS	**A**	**B**	**C**	**D**	**E**	**TOTAL**
Illegal alien	8	10	5	10	1	34
Legal resident	4	2	6	4	9	25
U.S. citizen	8	8	9	6	10	41
Total	20	20	20	20	20	100

a. Are the five industries homogeneous with respect to the visa status of their workers? Use $\alpha = .05$.

b. State the p-value.

13.38 An analyst for a marketing firm believes that the number of shopping mall customers who have come a distance of over 10 miles is equal to 25%. A random sample of 200 shoppers was selected, and the estimate of the proportion of shoppers who traveled over 10 miles was .22.

a. Is there sufficient evidence to conclude that the percentage of shoppers who come a distance over 10 miles is not equal to 25%? Use a significance level of .05 and the Z test from Section 13.1.

b. Use the chi-square goodness-of-fit test instead of the Z test in part a. Let the significance level be .05.

c. What is the relationship between the test statistic in parts a and b?

13.39 Microtron, a maker of semiconductors, has plants in four states, Texas, Georgia, California, and New York. Management would like to evaluate the effect of a mandatory retirement age of 60. The largest plants are in California and New York, so they sample 300 employees from each of these states and 200 employees each in Texas and Georgia.

RESPONSE	**TEXAS**	**GEORGIA**	**CALIFORNIA**	**NEW YORK**
Favor	81	85	122	128
Against	119	115	178	172

Do the data support the hypothesis that employees in different states have different opinions about the mandatory retirement age? Use a 5% significance level.

13.40 A marketing survey was taken of 277 frequent flyers by recording their socioeconomic class and airline preference. Do the sample data provide sufficient evidence to reject the null hypothesis of independence of airline preference and social class at the .05 significance level?

SOCIOECONOMIC CLASS	**DELTA**	**SOUTHWEST**	**AMERICAN**	**OTHER**	**TOTAL**
Low	20	45	23	4	92
Middle	25	40	20	20	105
Upper	18	15	30	17	80

13.41 A record company wanted to survey its customers regarding music preferences. A random sample of 258 frequent customers of the record company was selected, and information was gathered on their music preference and job classification. From the following data, can the null hypothesis of independence between type of music preferred and working status be rejected at the 10% significance level?

JOB CLASSIFICATION	**COUNTRY AND WESTERN**	**ROCK**	**CLASSICAL**	**JAZZ**	**TOTAL**
Clerical	25	40	17	5	87
Managerial	21	25	29	15	90
Blue collar	27	33	14	7	81
Total	73	98	60	27	258

13.42 The personnel department of a particular firm wants to know if an employee's age is associated with productivity (given in items per hour). The manager of the personnel department draws a random sample of 60 employees from each of the age classifications listed. Do the data support the hypothesis that the five age categories are not homogeneous with respect to productivity? Use a 10% significance level.

AGE	4–5 ITEMS	6–7 ITEMS	≥ 8 ITEMS	TOTAL
20 and under 30	15	25	20	60
30 and under 40	13	29	18	60
40 and under 50	16	26	18	60
50 and under 60	19	26	15	60
60 and under 70	22	24	14	60

13.43 An aspiring politician decided to sample 300 citizens from each of two major cities to find out whether the two populations were homogeneous with regard to their opinion on gun control. Do the following data indicate a lack of homogeneity? Use a 10% significance level.

CITY	FAVOR GUN CONTROL	AGAINST GUN CONTROL
A	126	174
B	148	152

13.44 Axiom Market Research published the following data concerning education level and attendance at "regular" theater performances. A sample size of 950 was selected. Do the data indicate, at the .05 significance level, a relationship between level of education and regular theater attendance?

EDUCATION	ATTEND MORE THAN ONCE PER YEAR	ATTEND NO MORE THAN ONCE PER YEAR	TOTAL
College graduate	82	120	202
Some college	75	131	206
High school graduate	106	215	321
Not a high school graduate	51	170	221

13.45 A sample of households classified as having incomes below the poverty level revealed the following distribution of persons by age:

AGE	FREQUENCY
0 to <5	27
5 to <18	53
18 to <22	16
22 to <45	60
45 to <65	26
65 to <72	18
Total	200

a. Compute the mean and standard deviation for the above distribution.

b. Using a 10% significance level, determine with a chi-square test whether the data fit a normal population.

c. Find the p-value for the test.

d. Does your conclusion change if $\alpha = .05$ or $\alpha = .01$?

13.46 In an effort to monitor the service of its employees, a parcel-delivery firm keeps a tally of the number of packages misrouted each week at each of its 25 distribution centers, with the following results:

MISROUTED PACKAGES	DISTRIBUTION CENTERS
0	5
1	6
2	8
3	6
>3	0

Do these data appear to come from a Poisson distribution? Use a .05 significance level.

13.47 In the field of organizational psychology, extensive study has been made of different leadership styles. One researcher refers to two extremes as authoritarian versus democratic; another refers to task-oriented versus people-oriented; yet others have their own labels for these qualities. Whatever the label, do these different styles affect the morale of the subordinates? To address this issue, a researcher established a ranking scale for worker morale, based on interviews, and grouped the workers into low, acceptable, and high morale categories. These were cross-classified against the leadership style of the supervisor. The following contingency table summarizes the results.

	LEADERSHIP STYLE		
WORKER MORALE	AUTHORITARIAN	DEMOCRATIC	TOTAL
Low	10	5	15
Acceptable	8	12	20
High	6	9	15
Total	24	26	50

a. Apply the chi-square test of independence to these data, at a 5% significance level.

b. State the p-value for your test.

c. Is worker morale related to the supervisor's leadership style, or are these qualities independent?

13.48 Kingston Pencils is considering a new bonus plan. Under the current bonus plan, the amount of bonus is not linked to production but only linked to profits. According to the proposed bonus plan, the amount of bonus will be linked to the quantity produced but will be subject to the amount of profits. The controller of Kingston is interested in examining whether employee opinion of the bonus plan is independent of job classification.

EMPLOYEE	FAVORABLE	UNFAVORABLE
White collar	67	28
Blue collar	43	19

Calculate the p-value and interpret it.

13.49 Nonresponse bias is a problem for many sample surveys. One way to minimize the "refusal to respond" is to include an incentive as a reward for participating in the survey. An experiment was carried out in which a survey instrument was sent to a control group, which did not receive any reward for participating in the survey, and also to a group in which each participating member received a digital clock pen. The results of the experiment are given below. Do the data provide sufficient evidence to indicate that the incentive and failure to respond are related? Base your conclusion on the p-value.

	CONTROL	PEN
Completed questionnaire	213	227
Refusal to complete questionnaire	66	77

(*Source:* Adapted from S. Pharr, R. Stuefen, and M. Wilber, "The Effects of Nonmonetary Incentives Upon Survey Refusal Tendencies of the Affluent Consumer Population," *The Journal of Applied Business Research* 6, no. 3 (1990): 88.)

13.50 Marketing research firms have been using computer administered questionnaires (CAQ) to collect data from certain populations. CAQs eliminate human intervention and usually include a short pictorial-type tutorial to facilitate ease of use. An experiment was conducted to determine whether computer experience was related to a person's willingness to participate in such surveys. In the experiment, a traditional sample was taken to estimate characteristics about the population of people visiting a local attraction. An unattended kiosk was also present with a large "Visitor Survey" sign. A CAQ was administered to people at the kiosk. The results of the experiment are summarized below. What conclusion can be drawn from the data below with respect to a biased sample from the CAQ because of computer experience? Use a .01 significance level.

COMPUTER EXPERIENCE	TRADITIONAL SAMPLE SURVEY	CAQ (KIOSK)
None	177	75
Once/Twice	7	71
Occasional	114	98
Regular	85	104
Extensive	106	141

(*Source:* Adapted from G. Bratton and P. Newsted, "An Investigation of Sampling Bias In Unattended Computer Administered Questionnaires," *Proceedings of the 1990 Annual Meeting of Decision Sciences Institute:* 1212.)

13.51 A bank offers three types of money market accounts. The vice president is interested in whether there is a relationship between the account balances and the type of money market account. A random sample of 2500 accounts are selected, with the results given in the following table. What conclusions can be drawn from the MINITAB computer printout? Base your decision on the *p*-value.

```
MTB > CHISQUARE USING C1-C3

Expected counts are printed below observed counts

            ACCT I  ACCT II ACCT III    Total
       1       194      190      215      599
             199.1    192.2    207.7

       2       220      225      231      676
             224.7    216.9    234.4

       3       239      215      250      704
             234.0    225.8    244.1

       4       178      172      171      521
             173.2    167.1    180.7

  Total        831      802      867     2500

ChiSq =     0.13 +   0.02 +   0.25 +
            0.10 +   0.31 +   0.05 +
            0.11 +   0.52 +   0.14 +
            0.13 +   0.14 +   0.52 = 2.43
    df = 6
```

ACCOUNT BALANCE	MONEY MARKET ACCOUNT I	MONEY MARKET ACCOUNT II	MONEY MARKET ACCOUNT III
<500	194	190	215
500 and <1000	220	225	231
1000 and <5000	239	215	250
≥ 5000	178	172	171

13.52 In a survey, consumers are questioned about how often they purchase each of three products. One hundred consumers are randomly selected to participate in the survey. From the MINITAB computer printout, is there sufficient evidence to indicate that there is a relationship between the products purchased and how often the products are purchased at a significance level of .05? At a significance level of .01? Each row represents how often a product is purchased (frequently, occasionally, or never) and each column represents one of the three products.

```
MTB > NAME C1 'ITEM A'
MTB > NAME C2 'ITEM B'
MTB > NAME C3 'ITEM C'
MTB > PRINT C1-C3
 ROW   ITEM A   ITEM B   ITEM C

  1       30       10        8
  2        8        6       11
  3        9        7       11
```

```
MTB > CHISQUARE USING C1-C3

Expected counts are printed below observed counts

                ITEM A    ITEM B    ITEM C    Total
        1         30        10         8        48
                 22.6      11.0      14.4

        2          8         6        11        25
                 11.8,      5.7       7.5

        3          9         7        11        27
                 12.7       6.2       8.1

     Total        47        23        30       100

   ChiSq =      2.45 +     0.10 +    2.84 +
                1.20 +     0.01 +    1.63 +
                1.07 +     0.10 +    1.04 = 10.45
   df = 4
```

Computer Exercises
Using the Database

Exercise 1 -- Appendix I Randomly select 100 observations from the database. Determine whether the variable family size (FAMLSIZE) has a distribution that is significantly different from the binomial distribution. Use a .05 significance level.

Exercise 2 -- Appendix I Randomly select 50 observations from the database. Are the categories own or rent one's residence (variable OWNORENT) and family size (variable FAMLSIZE) independent? Use a .05 significance level.

Exercise 3 -- Appendix J Randomly select 100 observations from the database. Determine whether the total asset value (variable TOTAL) has a distribution that is significantly different from the normal distribution. Use a .05 significance level. (*Hint:* See Exercise 13.13.)

Exercise 4 -- Appendix J Randomly select 100 observations from the database. Are the categories bond rating (BONDRATE) and positive or negative net income (NETINC) independent? Use a .05 significance level.

 Case Study

Are Small Businesses Headed by Women Less Likely to Survive and Succeed?

In recent years, more and more women have been heading their own businesses. The rate of growth in self-employment is greater among women than among men; in recent years, the rate of increase has been 35% for women compared to 12% for men. Overall, however, more businesses are still headed by men. According to the U.S. Small Business Administration, recent statistics indicated that the average male-owned business had seven times the revenues of the average female-owned business. It is not surprising, then, that a widespread perception exists among scholars and laypersons alike that businesses owned by men are more successful than those owned by women.

Kalleberg and Leicht (1991) point out that there is very little empirical evidence on how differences in sex affect organizational survival and performance. Their study, conducted under the auspices of the University of North Carolina's Institute for Private Enterprise and published in the *Academy of Management Journal,* seeks to address this issue (among others). Their analysis did extend into other areas, and they used techniques such as regression (to be covered in a later chapter). We shall limit our attention to those areas of the study where the chi-square test could be applied.

The data in Table 13.1 is based on a group of small businesses in three industries—food and drink establishments, computer sales and software companies, and health-related businesses—in south central Indiana. The data show the survival pattern for the businesses, which are classified by gender of owner-operator and by industry.

The authors of the study theorized that organizational performance could be analyzed at two levels: At the *macro* level of analysis, differences in survival and success of organizations can be explained by institutional and environmental characteristics (such as market structures, intensity of competition, industry size, and growth patterns). At the *micro* level of analysis, personal characteristics of the entrepreneur (such as gender and experience),

■ TABLE 13.1
Summary of Businesses
Studied.

INDUSTRY	BUSINESSES HEADED BY MEN			BUSINESSES HEADED BY WOMEN		
	N	Number of Failures	% Failures	N	Number of Failures	% Failures
Food and drink	106	15	14	42	7	17
Computers	122	23	19	14	1	7
Health	84	12	14	43	7	16
Totals	312	50	16	99	15	15

(*Source:* Arne L. Kalleberg and Kevin T. Leicht, "Gender and Organizational Performance: Determinants of Small Business Survival and Success," *Academy of Management Journal* 34, no. 1 (1991): 136.)

or individual firm characteristics (such as the age of the firm), could be related to success or failure.

Simplifying the analysis a little, we shall pick three hypotheses considered by the authors. From a micro viewpoint (focusing on gender of the owner-operator), we may formulate the following claims:

Claim 1: For the food and drink industry, small business success is related to the gender of the owner-operator.

Claim 2: For the computer industry, small business success is related to the gender of the owner-operator.

Claim 3: For the health industry, small business success is related to the gender of the owner-operator.

Case Study Questions

1. a. Do the data in Table 13.1 support claim 1 using a significance level of 10 percent?
b. What is the smallest significance level at which the null hypothesis could be rejected for testing claim 1?
2. a. Do the data in Table 13.1 support claim 2 using a significance level of 10 percent?
b. What is the smallest significance level at which the null hypothesis could be rejected for testing claim 2?
3. a. Do the data in Table 13.1 support claim 3 using a significance level of 10 percent?
b. What is the smallest significance level at which the null hypothesis could be rejected for testing claim 3?

SPSS

Solution
Example 13.4

Example 13.4 was concerned with a multinomial goodness-of-fit test. One of the expected values was less than 5, so it was necessary to pool the categories and create a new category with an expected value of at least 5. The problem was to determine whether the percentage of sales in the sample community fit the U.S. proportions. The percentages in the null hypothesis were .625, .226, .115, and .034. The SPSS program listing in Figure 13.6 was used to request a computed chi-square value when the categories were pooled. In this problem the SPSS commands are the same for both the mainframe and the PC versions (remember to end each command line with a period when using the PC version).

The TITLE command names the SPSS run.

The DATA LIST command gives each variable a name and describes the data as being in free form.

The BEGIN DATA command indicates to SPSS that the input data immediately follow.

The next four lines contain the data values, with each line representing a type (1 = GA, etc.) and the observed number of cars sold.

The END DATA statement indicates the end of the data.

The VALUE LABELS statement assigns the labels GA to type 1 records, K to type 2 records, L to type 3 records, and OTHER to type 4 records.

The WEIGHT command is used to weight our cases by the number of observed cars sold.

The NPAR TESTS CHISQUARE= statement requests a chi-square test between the observed and expected sales. For instance, 0.625 is the expected market share for the variable labeled GA.

Figure 13.7 shows the SPSS output obtained by executing the listing in Figure 13.6.

■ **FIGURE 13.6**
Input for SPSS
or SPSS/PC.
Remove the periods
for SPSS input.

```
TITLE    NEW CAR SALES MIXTURE.
DATA LIST FREE/TYPE CARSSOLD.
BEGIN DATA.
1 95
2 50
3 41
4 14
END DATA.
VALUE LABELS TYPE 1 'GA' 2 'K' 3 'L' 4 'OTHER'.
WEIGHT BY CARSSOLD.
NPAR TESTS CHISQUARE=TYPE/EXPECTED=0.625,0.226,0.115,0.034.
```

■ **FIGURE 13.7**
SPSS output.

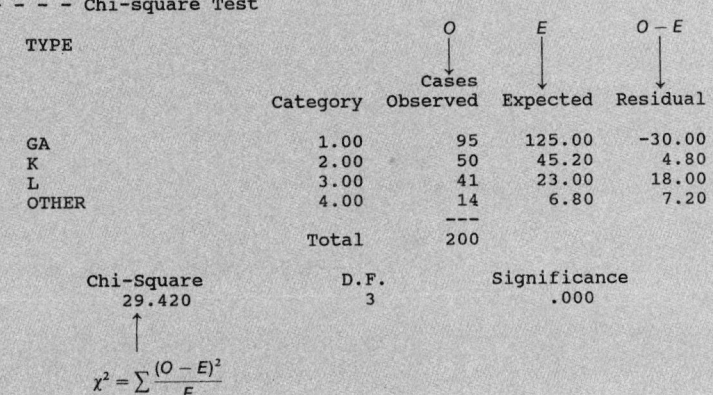

```
                    NEW CAR SALES MIXTURE

- - - - - Chi-square Test

                                      O         E        O - E

     TYPE
                                    Cases
                       Category   Observed  Expected  Residual

     GA                 1.00         95      125.00    -30.00
     K                  2.00         50       45.20      4.80
     L                  3.00         41       23.00     18.00
     OTHER              4.00         14        6.80      7.20
                                    ---
                       Total        200


          Chi-Square             D.F.          Significance
           29.420                  3               .000
```

$$\chi^2 = \sum \frac{(O - E)^2}{E}$$

SPSS

Solution
Example 13.5

Example 13.5 was concerned with a chi-square test of independence. The purpose was to determine whether an employee's educational level had an effect on job performance as measured by an exam. The null hypothesis was that there was no relationship between educational level and job performance. The SPSS program listing in Figure 13.8 requests the chi-square and p-value statistics from the data obtained by testing 120 employees. In this problem the SPSS commands are the same for both the mainframe and PC versions (remember to end each command line with a period when using the PC version).

The TITLE command names the SPSS run.

The DATA LIST command gives each variable a name and describes the data as being in free form.

The BEGIN DATA command indicates to SPSS that the input data immediately follow.

The next nine lines contain the data values. Each line contains an educational level code, a performance level code, and the number of observations that comprise the two adjacent categories.

The END DATA statement indicates the end of the data.

The VALUE LABELS statement assigns codes to different categories (LEVEL and PERFORM) of data. These are positional categories. The first position of the input data stream is the LEVEL category, while the second position is the PERFORM category. A 1 in the first position of the data stream indicates that the employee has a master's degree, a 2 indicates a bachelor's degree, and a 3 indicates a high school degree. For the second position of the data stream, a 1 indicates that the employee scored HIGH on the test for job knowledge, 2 indicates AVERAGE, and 3 indicates LOW.

The WEIGHT command requests that the data be weighted by the variable COUNT. This variable indicates the total number of employees who satisfy the criteria in the adjacent LEVEL and PERFORM categories. Looking at the first row of data, four employees had master's degrees and scored HIGH on the exam.

The CROSSTABS command produces a cross-tabulation of the variables LEVEL and PERFORM.

The STATISTICS 1 command requests the chi-square test (use /STATISTICS 1 for the PC version).

The OPTIONS 14 command requests the expected frequencies to be printed (use /OPTIONS 14 for the PC version).

Figure 13.9 shows the SPSS output obtained by executing the listing in Figure 13.8.

■ FIGURE 13.8
Input for SPSS
or SPSS/PC.
Remove the periods
for SPSS input.

```
TITLE    EDUCATIONAL LEVEL VERSUS JOB KNOWLEDGE.
DATA LIST FREE/LEVEL PERFORM COUNT.
BEGIN DATA.
1  1  4
1  2  20
1  3  11
2  1  12
2  2  18
2  3  15
3  1  9
3  2  22
3  3  9
END DATA.
VALUE LABELS LEVEL 1 'MASTERS' 2 'BACHELOR' 3 'HIGH SCHOOL'/
             PERFORM 1 'HIGH' 2 'AVERAGE' 3 'LOW'.
WEIGHT BY COUNT.
CROSSTABS TABLES=LEVEL BY PERFORM
/STATISTICS 1
/OPTIONS 14.
```

■ FIGURE 13.9 SPSS output.

EDUCATIONAL LEVEL VERSUS JOB KNOWLEDGE

Crosstabulation: LEVEL
                   By PERFORM

PERFORM->	Count Exp Val	HIGH 1.00	AVERAGE 2.00	LOW 3.00	Row Total
LEVEL					
MASTERS	1.00	4 7.3	20 17.5	11 10.2	35 29.2%
BACHELOR	2.00	12 9.4	18 22.5	15 13.1	45 37.5%
HIGH SCHOOL	3.00	9 8.3	22 20.0	9 11.7	40 33.3%
	Column Total	25 20.8%	60 50.0%	35 29.2%	120 100.0%

■ FIGURE 13.9   (continued))

EDUCATIONAL LEVEL VERSUS JOB KNOWLEDGE

Chi-Square	D.F.	Significance	Min E.F.	Cells with E.F.< 5
4.67020	4	.3228 ←	7.292	None

Number of Missing Observations =   0

p-value

$$\chi^2 = \sum \frac{(O-E)^2}{E}$$

SAS

**Solution
Example 13.5**

Example 13.5 was concerned with a chi-square test of independence. The purpose was to determine whether an employee's educational level had an effect on job performance as measured by an exam. The null hypothesis was that there is no relationship between educational level and job performance. The SAS program listing in Figure 13.10 requests the chi-square and *p*-value statistics from the data obtained by testing 120 employees.

In this problem the SAS commands are the same for both the mainframe and PC versions.

The TITLE command names the SAS run.

The DATA command gives the data a name.

The INPUT command names and gives the correct order for the different fields on the data lines. The $ implies that both LEVEL and PERFORM are character data. The @@ signs indicate that each data line contains two additional sets of data.

The CARDS command indicates to SAS that the input data immediately follow.

The next three lines are data values. The first line, for example, indicates that 4 employees have a master's degree and scored HIGH on the exam, 20 with master's degrees scored AVERAGE, and 11 with master's degrees scored LOW.

The PROC FREQ command and WEIGHT COUNT subcommand specify that the values of the variable COUNT are relative weights for the observations.

The TABLES subcommand produces a cross-tabulation of the variables LEVEL and PERFORM.

The CHISQ command generates chi-square statistics.

■ FIGURE 13.10   Input for SAS (mainframe or micro version).

```
TITLE    'EDUCATIONAL LEVEL VERSUS JOB KNOWLEDGE
DATA EXAM PERFORM;
 INPUT LEVEL $ PERFORM $ COUNT@@;
CARDS;
MASTERS HIGH 4 MASTERS AVERAGE 20 MASTERS LOW 11
BACHELORS HIGH 12 BACHELORS AVERAGE 18 BACHELORS LOW 15
HIGHSCHOOL HIGH 9 HIGHSCHOOL AVERAGE 22 HIGHSCHOOL LOW 9
PROC FREQ;
 WEIGHT COUNT;
 TABLES LEVEL*PERFORM/CHISQ;
```

Figure 13.11 shows the SAS output obtained by executing the listing in Figure 13.10.

■ FIGURE 13.11
SAS output.

EDUCATIONAL LEVEL VERSUS JOB KNOWLEDGE

TABLE OF LEVEL BY PERFORM

LEVEL   PERFORM

Frequency Percent Row Pct Col Pct	AVERAGE	HIGH	LOW	Total	observed frequency (O)
BACHELOR	18 15.00 40.00 30.00	12 10.00 26.67 48.00	15 12.50 33.33 42.86	45 37.50	
HIGHSCHO	22 18.33 55.00 36.67	9 7.50 22.50 36.00	9 7.50 22.50 25.71	40 33.33	
MASTERS	20 16.67 57.14 33.33	4 3.33 11.43 16.00	11 9.17 31.43 31.43	35 29.17	
Total	60 50.00	25 20.83	35 29.17	120 100.00	

STATISTICS FOR TABLE OF LEVEL BY PERFORM

Statistic	DF	Value	Prob	
Chi-Square	4	4.670	0.323	←── p-value
Likelihood Ratio Chi-Square	4	4.986	0.289	
Mantel-Haenszel Chi-Square	1	1.094	0.296	
Phi Coefficient		0.197		
Contingency Coefficient		0.194		
Cramer's V		0.139		

Sample Size = 120   computed value of
                            chi-square statistic

CHAPTER

14 Correlation and Simple Linear Regression

A LOOK BACK/INTRODUCTION

The early chapters discussed methods of reducing a set of values for one variable to a graph (such as a histogram) or a numerical measure (such as a mean). A **variable** here is a characteristic of the population being measured or observed. For example, the variable of interest might be an individual's height or income. The sample then consists of random observations of the variable describing a given population.

In this chapter we discuss the situation in which the population and sample consist of measurements not of *one* variable but of *two.* As a result, we not only can describe each variable individually—we can also describe how the two variables are related. The relationship between the two variables can be described using a simple graph or a numerical measure (statistic). We can then use the sample results to form a conclusion about the population from which the sample was obtained. If we believe that a linear relationship between the two variables exists, the next step is to construct the "best-fitting" line through the points defined by the sample bivariate data.

Finally, we turn our attention to the question of what we are estimating when sampling from a population of bivariate data. How can we determine whether a significant linear relationship exists? To answer this question, we introduce the concept of a **statistical model** and the assumptions behind it. Various tests of hypothesis examine the adequacy of this model (is it a good one?), and an assortment of confidence intervals measure the reliability of the corresponding estimates using this model.

14.1
Bivariate Data and Correlation

With bivariate data, each observation consists of data on two variables. For example, you obtain a sample of people and record their ages (X) and liquid assets (Y). Or, for each month, you record the average interest rate (X) and the number of new housing starts (Y). These data are *paired.*

Suppose that a real-estate developer is interested in determining the relationship between family income (X, in thousands of dollars) of the local residents and the square footage of their homes (Y, in hundreds of square feet). A random sample of ten families is obtained with the following results:

Income (X)	22	26	45	37	28	50	56	34	60	40
Square footage (Y)	16	17	26	24	22	21	32	18	30	20

Bivariate data can be represented graphically using a **scatter diagram.** In this graph, each observation is represented by a point, where the X axis is always horizontal and the Y axis is vertical. A scatter diagram of the real-estate data is shown in Figure 14.1a. The underlying pattern here appears to be that larger incomes (X) are associated with larger home sizes (Y). In this example X and Y have a **positive (direct) relationship.** A **negative (inverse) relationship** occurs when Y decreases as X increases—for example, when Y is the demand for a particular consumer product and X is the selling price.

We next try to determine whether we can estimate this relationship by means of a straight line. One possible line is sketched in Figure 14.1b; it passes among these points and has a positive slope. To measure the strength of the linear relationship between these two variables, we determine the coefficient of correlation.

■ **FIGURE 14.1**   Scatter diagram of real-estate data. (a) Scatter diagram of sample data. (b) Line through sample data.

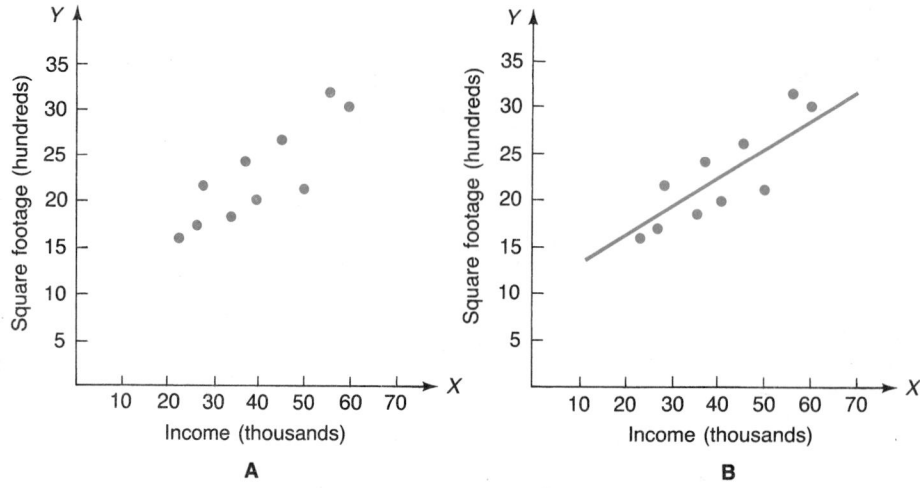

Coefficient of Correlation

It is often difficult to determine whether a *significant* linear relationship exists between X and Y by inspecting a scatter diagram of the data. A second procedure is to include a *measure* of this linearity—the sample coefficient of correlation. It is computed from the sample data by combining these pairs of values into a single number, written as r. The sample **coefficient of correlation,** r, measures the strength of the linear relationship that exists within a sample of n bivariate data. Its value is given by

$$r = \frac{\Sigma(x - \bar{x})(y - \bar{y})}{\sqrt{\Sigma(x - \bar{x})^2}\sqrt{\Sigma(y - \bar{y})^2}} \qquad (14.1)$$

$$= \frac{\Sigma xy - (\Sigma x)(\Sigma y)/n}{\sqrt{\Sigma x^2 - (\Sigma x)^2/n}\sqrt{\Sigma y^2 - (\Sigma y)^2/n}} \qquad (14.2)$$

where Σx = sum of X values, Σx^2 = sum of X^2 values, Σy = sum of Y values, Σy^2 = sum of Y^2 values, Σxy = sum of XY values, $\bar{x} = \Sigma x/n$, and $\bar{y} = \Sigma y/n$. When using a calculator to determine a coefficient of correlation, equation 14.2 provides a computationally easier procedure. Notice that the summations in the denominator of equation 14.1 are the numerators for the sample variances of X and Y.

Sum of Squares

We will introduce a shorthand notation at this point, related to the notation in Chapter 11 for ANOVA. Let

$$\begin{aligned} SS_X &= \text{sum of squares for } X \\ &= \Sigma(x - \bar{x})^2 \\ &= \Sigma x^2 - (\Sigma x)^2/n \qquad (14.3) \\ SS_Y &= \text{sum of squares for } Y \\ &= \Sigma(y - \bar{y})^2 \\ &= \Sigma y^2 - (\Sigma y)^2/n \qquad (14.4) \end{aligned}$$

$$SCP_{XY} = \text{sum of cross products for } XY$$
$$= \Sigma(x - \bar{x})(y - \bar{y})$$
$$= \Sigma xy - (\Sigma x)(\Sigma y)/n \qquad \textbf{(14.5)}$$

Using this notation, we can write r as

$$r = \frac{SCP_{XY}}{\sqrt{SS_X}\ \sqrt{SS_Y}} \qquad \textbf{(14.6)}$$

The following are some important properties of the sample correlation coefficient, r.

1. r ranges from -1.0 to 1.0.
2. The larger $|r|$ (absolute value of r) is, the stronger is the linear relationship.
3. r near zero indicates that there is no linear relationship between X and Y, and the scatter diagram *typically* (although not necessarily) appears to have a shotgun effect (Figure 14.2a). Here, X and Y are uncorrelated.
4. $r = 1$ or $r = -1$ implies that a perfect linear pattern exists between the two variables in the sample, that is, a single line will go *through* each point. Here we say that X and Y are **perfectly correlated** (Figure 14.2b and c).

■ **FIGURE 14.2**
Scatter diagrams for various values of the sample correlation coefficient.

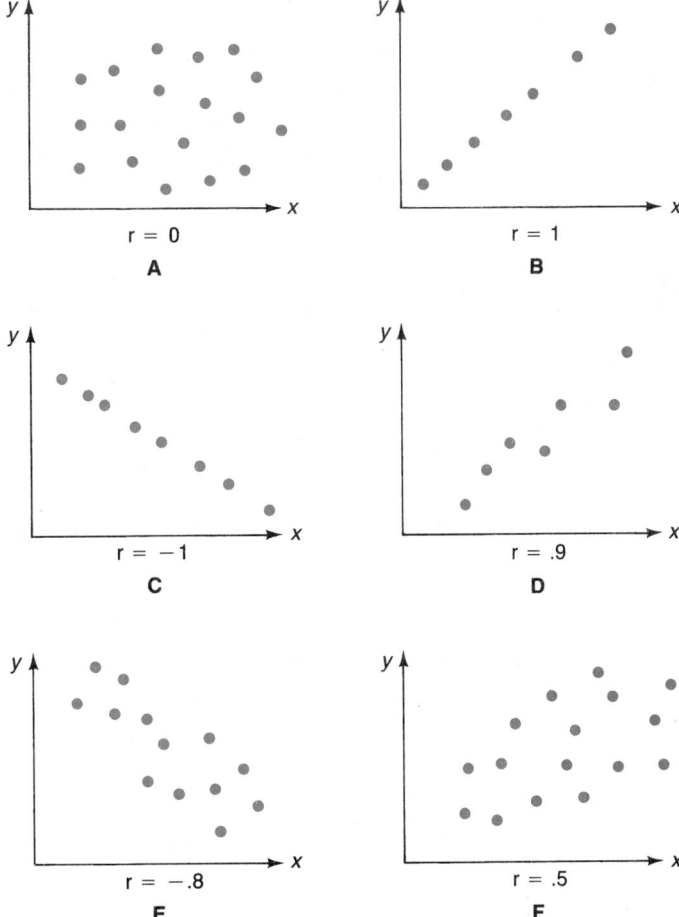

■ **FIGURE 14.3**
Although (a) has a large
slope and (b) has a
small slope, both are
scatter diagrams for
$r = .9$.

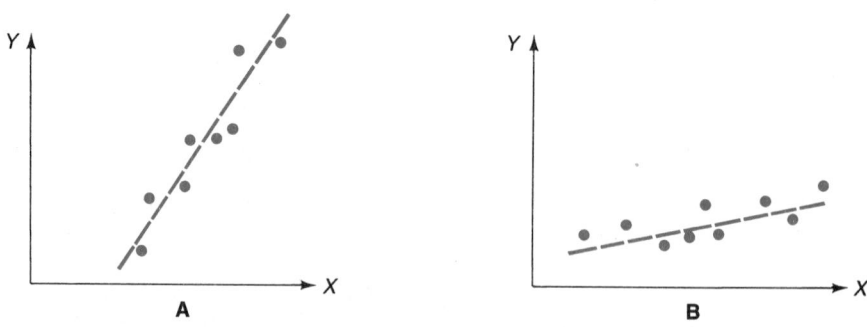

5. Values of $r = 0$, 1, or -1 are rare in practice. Several other values of the correlation coefficient are illustrated in Figure 14.2d, e, and f.
6. The sign of r tells you whether the relationship between X and Y is a positive (direct) or a negative (inverse) one.
7. The value of r tells you nothing about the slope of the line through these points (except for the sign of r). If r is positive, the line through these points has positive slope, and similarly, this line will have negative slope if r is negative. However, a set of data with $r = .9$ will not necessarily have a steeper line passing through it than will a set of data with $r = .4$. All you will observe in the first data set is a set of points that is very close to some straight line with positive slope, but you know nothing (except for the sign) about the slope of this line. See Figure 14.3, where both sets of data have an r value of .9.

EXAMPLE 14.1   Determine the sample correlation coefficient for the real-estate data in Figure 14.1.

SOLUTION   Your calculations can be organized as follows:

FAMILY	X (INCOME)	Y (SQUARE FOOTAGE)	XY	X²	Y²
1	22	16	352	484	256
2	26	17	442	676	289
3	45	26	1,170	2,025	676
4	37	24	888	1,369	576
5	28	22	616	784	484
6	50	21	1,050	2,500	441
7	56	32	1,792	3,136	1,024
8	34	18	612	1,156	324
9	60	30	1,800	3,600	900
10	40	20	800	1,600	400
	398	226	9,522	17,330	5,370

Using the totals from this table,

$$SS_X = 17{,}330 - (398)^2/10 = 1489.6$$

$$SS_Y = 5370 - (226)^2/10 = 262.4$$

$$SCP_{XY} = 9522 - (398)(226)/10 = 527.2$$

This value of the sample correlation coefficient is

$$r = \frac{SCP_{XY}}{\sqrt{SS_X}\,\sqrt{SS_Y}}$$

$$= \frac{527.2}{\sqrt{1489.6}\,\sqrt{262.4}} = \frac{527.2}{625.2}$$

$$= .843$$

■

■ **FIGURE 14.4**   Computer-generated correlation coefficient and scatter diagram for real-estate data using MINITAB.

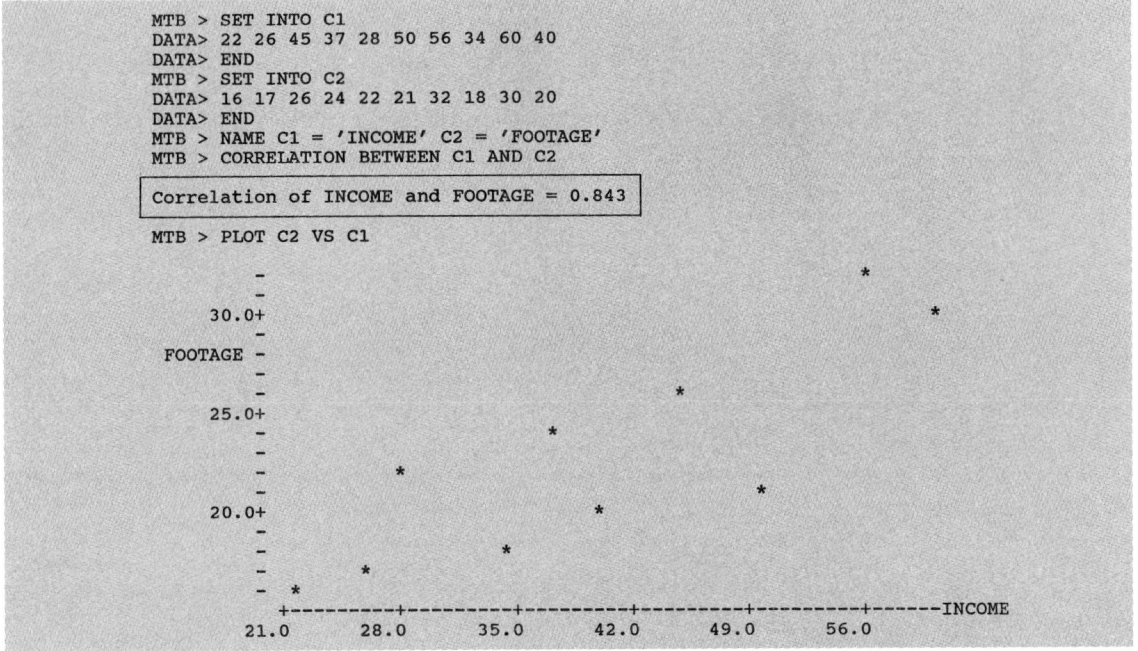

Using the Computer   For large data sets, the only reasonable way to obtain a scatter diagram and calculate r is to use a computer. At the end of the chapter, we will show you how to do this using SAS and SPSS. A computer-generated scatter diagram of the real-estate data using MINITAB is contained in Figure 14.4.

Covariance

Another commonly used measure of the association between two variables, X and Y, is the sample covariance, written cov(X, Y). It is similar to the sample correlation between these two variables. For one thing, the covariance and correlation always have the *same sign*. Consequently, if large values of X are associated with large values of Y, then both the covariance and correlation are positive. Similarly, both values are negative whenever large values of X are associated with small values of Y. For any two variables, X and Y, the sample **covariance** between these variables is defined in the following box.

Definition
The sample covariance between two variables, cov(X, Y), is a measure of the joint variation of the two variables, X and Y, and is defined to be

$$\text{cov}(X, Y) = \frac{1}{n-1} \Sigma(x - \bar{x})(y - \bar{y}) \tag{14.7}$$

$$= \frac{1}{n-1} \text{SCP}_{XY} \tag{14.8}$$

In Example 14.1, the sample covariance between income (X) and home size (Y) is

$$\text{cov}(X, Y) = \frac{1}{n - 1} \text{SCP}_{XY} = \frac{1}{9} (527.2) = 58.58$$

To see how the sample covariance and sample correlation (r) are related, let

$$s_X = \text{standard deviation of the } X \text{ values} = \sqrt{\frac{\text{SS}_X}{n - 1}}$$

and

$$s_Y = \text{standard deviation of the } Y \text{ values} = \sqrt{\frac{\text{SS}_Y}{n - 1}}$$

Then

$$
\begin{aligned}
r &= \text{sample correlation between } X \text{ and } Y \\
&= \frac{\text{cov}(X, Y)}{s_X s_Y}
\end{aligned}
\tag{14.9}
$$

In Example 14.1,

$$s_X = \sqrt{\frac{1489.6}{9}} = 12.865$$

$$s_Y = \sqrt{\frac{262.4}{9}} = 5.400$$

and so

$$r = \frac{58.58}{(12.865)(5.400)} = .843 \qquad \text{(as before)}$$

The correlation between two variables is used more often than the covariance because r always ranges from -1 to 1. The covariance, on the other hand, has no limits and can assume any value. Furthermore, the units of measurement for a covariance are difficult to interpret. For example, the previously calculated covariance is 58.58 (thousands of dollars) · (hundreds of square feet)—a somewhat meaningless unit of measurement. So, in a sense, the correlation is a scaled version of the covariance and has no units of measurement (a nice feature). To illustrate, the sample correlation between body weight and height will be the same whether you use the metric or the English systems to obtain the sample data. The covariance, however, will *not* be the same for these two situations. The covariance does have its applications, however, particularly in financial analyses, such as determining the risk associated with a number of interrelated investment opportunities.

As a final look at these two measures, you can consider the correlation between two variables to be the covariance between the **standardized** variables. By defining

$$X' = \frac{X - \bar{X}}{s_X}$$

■ **FIGURE 14.5**
Vertical distances from
line L to real-estate data
(Example 14.1),
represented by
$d_1, d_2, \ldots, d_{10}$.

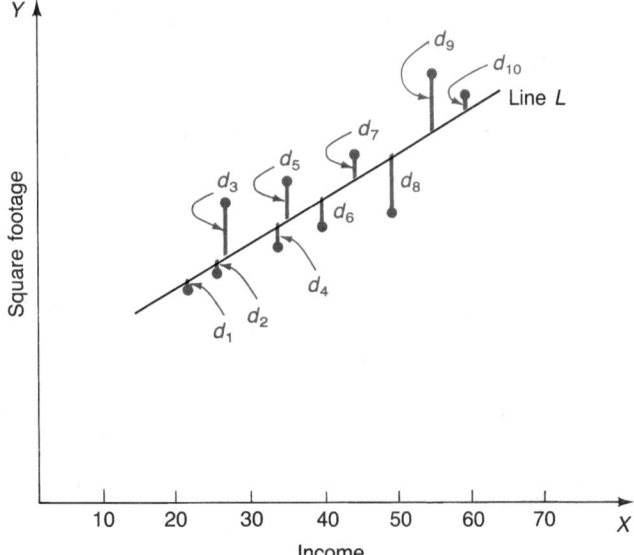

and

$$Y' = \frac{Y - \bar{Y}}{s_Y}$$

then

$$\mathrm{cov}(X', Y') = \text{correlation between } X \text{ and } Y = r$$

Least Squares Line

If we believe that two variables do exhibit an underlying linear pattern, how can we determine a straight line that best passes through these points? So far, we have demonstrated only the calculations necessary to compute a correlation coefficient. We next illustrate how to construct a line through a set of points exhibiting a linear pattern; we look at the assumptions behind this procedure in the next section.

Look at the scatter diagram in Figure 14.1b, which shows one possible line through these points. The scatter diagram and line are repeated in Figure 14.5, which also shows the vertical distances from each point to the line ($d_1, d_2, \ldots$).

Is line L the best line through these points? Because we would like the distances $d_1, d_2, \ldots, d_{10}$ to be *small*, we define the best line to be the one that minimizes

$$\Sigma d^2 = d_1^2 + d_2^2 + d_3^2 + \cdots + d_{10}^2 \qquad \textbf{(14.10)}$$

We square each distance because some of these distances are positive (the point lies *above* line L) and some are negative (the point lies *below* line L). If we did not square each distance, d, the positive d's might cancel out the negative ones. This means that using $(d_1 + d_2 + \cdots + d_{10})$ as a *measure of fit* is *not* a good idea. A better method is to determine which line makes equation 14.10 as small as possible;

■ **FIGURE 14.6**
The least squares line for Example 14.1. Each $d = Y - \hat{Y}$, the error encountered by using the straight line to estimate the value of Y at the corresponding point.

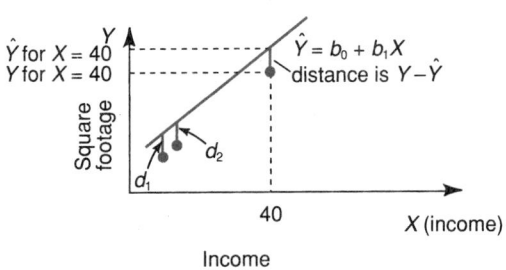

this line is called the **least squares line.** Deriving this line in general requires the use of calculus (derivatives, in particular).*

Because we intend to use this line to predict Y for a particular value of X, we use the notation $\hat{Y}$ (Y-hat) to describe the equation of the line. We can now define, for the least squares line, the b_0 and b_1 that minimize $(d_1^2 + d_2^2 + \cdots + d_n^2)$, given by

$$b_1 = \frac{SCP_{XY}}{SS_X} \tag{14.11}$$

$$b_0 = \bar{y} - b_1\bar{x} \tag{14.12}$$

where SS_X and SCP_{XY} are as defined in equations 14.3 and 14.5. Also, $\bar{x} = \Sigma x/n$ and $\bar{y} = \Sigma y/n$. The resulting least squares line is

$$\hat{Y} = b_0 + b_1 X$$

In Figure 14.6, notice that each distance, d, is actually $Y - \hat{Y}$ and consists of the **residual,** encountered by using the straight line to estimate the value of Y at this point. So

$$\Sigma d^2 = \Sigma(y - \hat{y})^2$$

This term is the **sum of squares of error** (or *residual sum of squares*) and is written **SSE.** Consequently, the least squares line is the one that makes SSE as small as possible.

$$SSE = \Sigma d^2 = \Sigma(y - \hat{y})^2 \tag{14.13}$$

---

*For the mathematically curious, we provide a condensed derivation of these coefficients. To minimize Σd^2, first write this expression as

$$f(b_0, b_1) = \Sigma d^2 = \Sigma(y - \hat{y})^2$$
$$= \Sigma(y - b_0 - b_1 x)^2$$

because $\hat{y} = b_0 + b_1 x$.

To minimize this function, determine the partial derivatives with respect to b_0 (written as f_{b_0} and with respect to b_1 (written as f_{b_1}). These are

$$f_{b_0} = 2\Sigma(y - b_0 - b_1 x)(-1) = -2[\Sigma y - nb_0 - b_1\Sigma x]$$
$$f_{b_1} = 2\Sigma(y - b_0 - b_1 x)(-x) = -2[\Sigma xy - b_0\Sigma x - b_1\Sigma x^2]$$

Setting $f_{b_0} = f_{b_1} = 0$ and solving for b_0 and b_1 results in equations 14.11 and 14.12.

There is another method of determining SSE when using the least squares line, which avoids having to determine the value of $\hat{Y}$ at each point:

$$SSE = SS_Y - \frac{(SCP_{XY})^2}{SS_X} \qquad (14.14)$$

EXAMPLE 14.2   Determine the least squares line for the real-estate data we used in Example 14.1. What is the SSE?

SOLUTION   Using the calculations from Example 14.1, $SCP_{XY} = 527.2$, $SS_X = 1489.6$, and $SS_Y = 262.4$, leading to

$$b_1 = \frac{SCP_{XY}}{SS_X}$$

$$= \frac{527.2}{1489.6} = .354$$

and

$$b_0 = \bar{y} - b_1\bar{x}$$

$$= 22.6 - (.354)(39.8) = 8.51$$

because

$$\bar{y} = \Sigma y/n$$

$$= 226/10 = 22.6$$

and

$$\bar{x} = \Sigma x/n$$

$$= 398/10 = 39.8$$

So the equation of the best (least squares) line through these points is

$$\hat{Y} = 8.51 + .354X$$

This equation tells us that in the sample data an increase of $1000 in income ($X$ increases by 1) is accompanied by an increase of 35.4 square feet in home size (Y increases by .354), on the average. For this illustration (and many others in practice), the *intercept*, b_0, has no real meaning because it corresponds to an income of zero dollars. Furthermore, an income of zero is considerably outside the range of the incomes in the sample. It is unsafe to assume that the linear relationship between X and Y present over the range of sample incomes ($22,000 to $60,000) exists outside this range—in particular, all the way to an income of zero. The *slope, b_1*, generally is the more informative value.

In Figure 14.7, the actual value of Y (in the sample data) for $X = 40$ is $Y = 20$ (the last pair of X, Y values). The predicted value of Y using the least squares line is

$$\hat{Y} = 8.51 + .354(40) = 22.67$$

The residual at this point is

$$\text{residual} = Y - \hat{Y} = 20 - 22.67 = -2.67$$

■ FIGURE 14.7
Least squares line for
real-estate data
(Example 14.2).

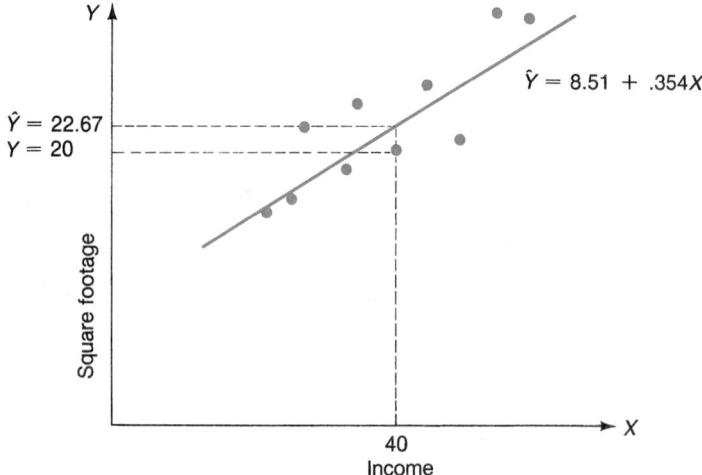

Repeating this for the other nine points leads to the following results. Notice that the sum of the residuals when using the least squares line is zero. This is always true.

X	Y	$\hat{Y}$	$Y - \hat{Y}$	$(Y - \hat{Y})^2$
22	16	16.30	−.30	.090
26	17	17.71	−.71	.504
45	26	24.44	1.56	2.434
37	24	21.61	2.39	5.712
28	22	18.42	3.58	12.816
50	21	26.21	−5.21	27.144
56	32	28.33	3.67	13.469
34	18	20.54	−2.54	6.452
60	30	29.75	.25	.063
40	20	22.67	−2.67	7.129
			0	75.81

As you can see, calculating the SSE (=75.81) using the table and equation 14.13 is tedious. Using equation 14.14 instead leads to

$$SSE = 262.4 - \frac{(527.2)^2}{1489.6}$$

$$= 262.4 - 186.59$$

$$= 75.81 \quad \text{(the same as before)}$$

Remember, however, that equation 14.13 applies to *any* line that you choose to construct through these points, whereas equation 14.14 applies only to the SSE for the least squares line.   ■

In Example 14.2, we attempted to predict the size of a home (Y) using the corresponding income (X). The variable Y is the **dependent variable,** and X is the **independent variable.** By passing a straight line through the sample points with Y as the dependent variable, we are **regressing** Y on X. *In linear regression, you regress the dependent variable, Y, which you are trying to predict, on the indepen-dent (or predictor or explanatory) variable, X.*

Exercises   **14.1**   The tensile strength of a paper product is related to the amount of hardwood in the pulp. Ten samples are produced in a pilot plant and the following data are obtained:

SAMPLE	STRENGTH (Y)	PERCENT HARDWOOD (X)
1	160	14
2	171	15
3	175	20
4	182	16
5	184	20
6	181	12
7	188	22
8	193	25
9	195	24
10	200	35

a.   Draw a scatter diagram of the X and Y values. What would you estimate the coefficient of correlation to be (without calculating it)?

b.   Calculate the coefficient of correlation.

14.2   Mr. Smart Fellow, the president of Well-Run Car, is examining the nature of the relationship between new car sales and annual advertising expenditure. His administrative assistant gathered the following information from the company's records.

YEAR	NEW CAR SALES (Y)	ADVERTISING DOLLARS (X)	YEAR	NEW CAR SALES (Y)	ADVERTISING DOLLARS (X)
1979	4,000	120,000	1985	6,000	144,000
1980	4,500	127,000	1986	6,100	147,000
1981	4,200	131,000	1987	6,800	152,000
1982	4,800	134,000	1988	7,200	160,000
1983	5,400	140,000	1989	7,800	165,000
1984	5,750	139,000	1990	9,100	170,000

a.   Calculate the coefficient of correlation.

b.   Find the least squares line.

c.   Graph the data and the least squares line, as a check on your calculations.

d.   Verify that the sum of the deviations from the least squares line is zero.

e.   Interpret the coefficients of the least squares line.

14.3   Tony's used-car lot has been paying car salespeople the highest commission in town. Tony decides to compile data to substantiate his belief that monthly net earnings increase when the car salespeople are highly paid. Fifteen months are chosen:

NET EARNINGS (Y)	TOTAL COMMISSIONS PAID (X)	NET EARNINGS (Y)	TOTAL COMMISSIONS PAID (X)
10,780	3,680	11,915	3,161
15,120	5,160	25,160	7,540
18,195	5,180	26,151	8,216
21,690	7,150	18,630	6,051
14,691	5,030	15,551	4,980
16,151	5,210	16,980	5,801
11,015	2,991	24,130	7,160
10,151	3,151		

a.   Graph the data and draw a line through them, using the "eyeball" method.

b.   Calculate the least squares line. How does it compare to the line in part a?

14.4   The supervisor of a group of assembly-line workers wanted to compare last year's productivity (X) to this year's productivity (Y) for each of the 20 employees that she supervises. In the past, an approximate linear relationship has existed between these two variables. Last year the average productivity per worker was 9.5 items per hour. This year, the average productivity per worker is 12.1 items per hour. The supervisor found the following sums for her 20 employees:

$$SCP_{XY} = 0.4$$
$$SS_X = 0.3$$
$$SS_Y = 0.8$$

a.   Calculate the correlation coefficient.

b.   Calculate the least squares line.

c.   Calculate the sum of squares for error.

14.5   Because $b_0 = \bar{y} - b_1\bar{x}$, we can replace b_0 in $\hat{Y} = b_0 + b_1X$ by $\bar{y} - b_1\bar{x}$. Hence, we have $\hat{Y} = \bar{y} + b_1(X - \bar{x})$. From this equation, show that the point $(\bar{x}, \bar{y})$ always falls on the least squares line.

14.6   Compare the formulas for the sample correlation, r, and the slope of the least squares line, b_1, and verify that $b_1 = r\sqrt{SS_Y/SS_X}$. What can we say about the sign of r and b_1?

14.7   A recent study identified factors that influence overall user information satisfaction (UIS) with management information systems (MIS). The following factors were identified: equity in the allocation of MIS (EAMIS), user's knowledge and involvement level (KIL), and quality of information products (QIP). A survey of 226 users yielded the following correlations:

VARIABLE	CORRELATION WITH UIS
EAMIS	.6653
KIL	.5158
QIP	.5301

(*Source:* Adapted from K. Joshi, "An Investigation of Equity as a Determinant of User Information Satisfaction," *Decision Sciences* 21, no. 4 (1990): 786.)

In your opinion, which variable is the most influential with UIS? How good do you think this variable will be in predicting UIS?

14.8   The owner of Grandmother's Cake Shop would like to predict the quantity of cakes sold when they are marked at low prices. There are no restrictions on the quantity, because the shop can easily bake several cakes in an hour if the demand is stronger than predicted. Past data show the following results.

NUMBER OF CAKES SOLD (Y)	PRICE OF CAKE (X)	NUMBER OF CAKES SOLD (Y)	PRICE OF CAKE (X)
14	2.30	16	1.99
16	2.10	17	1.90
17	1.80	15	2.25
17	1.89	14	2.39
13	2.50	13	2.70
12	2.80		

a.   Find the least squares line for X and Y.

b.   Graph the data and the least squares line.

c.   Suppose that the manager believes that there is a strong linear relationship between Y and X^2. Find the prediction equation for Y using X^2 only.

d. Compare the SSE for the least squares line found in question a with the least squares line found in part c.

14.9 The owner of an ice-cream stand believes that there is a linear relationship between the noon temperature (X) and the number of ice creams sold (Y). Data were collected during the noon hour every day for 20 days. The average number of ice creams sold during this hour is 35.6, and the average noon temperature over the 20 days is 87.4. The following sample statistics were calculated:

$$SCP_{XY} = 8.4$$
$$SS_X = 28.1$$
$$SS_Y = 3.9$$

a. Calculate the correlation coefficient.

b. Calculate the least squares line.

c. Calculate the error sum of squares.

14.10 The manager of a city zoo would like to use his staff more efficiently to accommodate large crowds. He randomly selected fifteen days on which attendance and high temperature for the day were recorded. Do the data indicate a significant correlation between attendance and daily high temperature?

ATTENDANCE, 1000s (Y)	HIGH TEMPERATURE, °F (X)	ATTENDANCE, 1000s (Y)	HIGH TEMPERATURE, °F (X)
1.9	82	2.6	76
0.8	104	0.7	105
1.2	90	1.3	90
1.4	92	1.6	85
2.4	75	2.1	78
2.8	70	1.8	83
1.5	86	2.3	77
1.4	87		

a. Calculate the least squares line. Interpret the coefficients of the least squares line in the context of the problem.

b. Calculate the error sum of squares.

c. Graph the data and draw the least squares line through them.

14.11 It is well known that the federal funds rate influences the yield on 13-week treasury bills. (The federal funds rate is the rate at which reserves are traded among commercial banks for overnight use. Treasury bills are short-term government bills sold at an auction at a discount from the face value.) The following data were collected:

TREASURY BILL YIELD (Y)	FEDERAL FUNDS RATE (X)	TREASURY BILL YIELD (Y)	FEDERAL FUNDS RATE (X)
12.89	14.23	8.79	9.43
12.36	14.51	9.39	9.56
9.71	11.01	9.05	9.45
7.93	9.29	8.71	9.48
8.08	8.65	8.71	9.34
8.42	8.80	8.96	9.47
9.19	9.46		

Find the least squares line and predict what the treasury bill rate would be if the federal funds rate was 9.67. Interpret the coefficient of the X variable in the least squares line.

14.12 The average weekly earnings of production workers in the construction and manufacturing industries in the United States are given below for 1980–1989:

	AVERAGE WEEKLY EARNINGS	
YEAR	CONSTRUCTION	MANUFACTURING
1980	$386.78	$288.62
1981	399.26	318.00
1982	426.82	330.26
1983	442.97	354.08
1984	458.51	374.03
1985	464.46	386.37
1986	466.75	396.01
1987	480.44	406.31
1988	495.73	418.81
1989	512.41	430.09

(*Source:* Adapted from *The World Almanac and Book of Facts* (1991): 142.)

a. Find the correlation coefficient between the average weekly earnings of the construction and manufacturing industries.

b. Calculate the least squares line that regresses the average weekly earnings of the construction industry on the average weekly earnings of the manufacturing industry.

c. Calculate the error sum of squares for the least squares line in part b.

▰ 14.2
The Simple Linear Regression Model

When we construct a straight line through a set of data points, we are attempting to predict the behavior of a dependent variable, Y, using a straight line equation with one predictor (independent) variable, X. Examples 14.1 and 14.2 examined the relationship in a particular community between the square footage (Y) of a particular home and the income of the owner (X).

Another application is attempting to predict the sales (Y) of a certain brand of shampoo using the amount of advertising expenditure (X) as the independent variable. We expect that as more advertising dollars are spent, the sales will increase. In other words, we expect a *positive* relationship for this situation.

Regression analysis is a method of studying the relationship between two (or more) variables, one purpose being to arrive at a method for predicting a value of the dependent variable. In **simple linear regression,** we use only *one* predictor variable, X, to describe the behavior of the dependent variable, Y. Also, the relationship between X and Y is assumed to be basically linear.

We have learned the mechanics of constructing a line through a set of bivariate sample values. We are now ready to introduce the concept of a statistical model.

Defining the Model

Return to Example 14.2 and Figure 14.4. This set of sample data contained a value of $X = 40$ and $Y = 20$. Consider the population of *all* houses in this community where the owner's income is 40 (that is, $40,000). Will they all have the same square footage? Unless this is a very boring-looking neighborhood, certainly not. Does this mean that the straight line predictor is of no use? The answer again is no; we do not expect things in this world to be that perfectly predictable. When you use a straight line to predict the square footage, you should be aware that there will be a certain amount of *error* present in this estimate. This is similar to the situation in which we estimate the mean, μ, of a population and the sample mean, $\bar{X}$, always estimates this parameter with a certain amount of inherent error.

When we elect to use a straight-line predictor, we employ a **statistical model** of the form

$$Y = \beta_0 + \beta_1 X + e \qquad (14.15)$$

where (1) $\beta_0 + \beta_1 X$ is the *assumed* line about which *all* values of X and Y will fall, called the **deterministic** portion of the model, and (2) e is the error component, referred to as the **random** part of the model.

In other words, there exists some (unknown) line about which all X, Y values can be expected to fall. Notice that we said "about which," not "on which"— hence the necessity of the error term, e, which is the unexplained error that is part of the simple linear model. Because this model considers only one independent variable, the effect of other predictor variables (perhaps unknown to the analyst) is contained in this error term.

We emphasize that the deterministic portion, $\beta_0 + \beta_1 X$, refers to the straight line for the *population* and will remain unknown. However, by obtaining a random sample of bivariate data from this population, we are able to estimate the unknown parameters, β_0 and β_1. Thus b_0 is the **intercept** of the sample regression line and is the estimate of the population intercept, β_0. The value of b_0 can be calculated using equation 14.12. Similarly, b_1 is the **slope** of the sample regression line and is the estimate of the population slope, β_1. The value of b_1 can be calculated using equation 14.11.

Assumptions for the Simple Linear Regression Model

We can construct a least squares line through *any* set of sample points, whether or not the pattern is linear. We could construct a least squares line through a set of sample data exhibiting no linear pattern at all. However, to have an effective predictor and a model that will enable us to make statistical decisions, certain assumptions are necessary.

We treat the values of X as fixed (nonrandom) quantities when using the simple linear regression model. For any given value of X, the only source of variation comes from the error component, e, which is a random variable. In fact, there are many random variables here, one for each possible value of X. The assumptions used with this model are concerned with the nature of these random variables.

The first three assumptions are concerned with the behavior of the error component for a fixed value of X. The fourth assumption deals with the manner in which the error components (random variables) affect each other.

Assumption 1   *The mean of each error component is zero.* This is the key assumption behind simple linear regression. Look at Figure 14.8, where we once

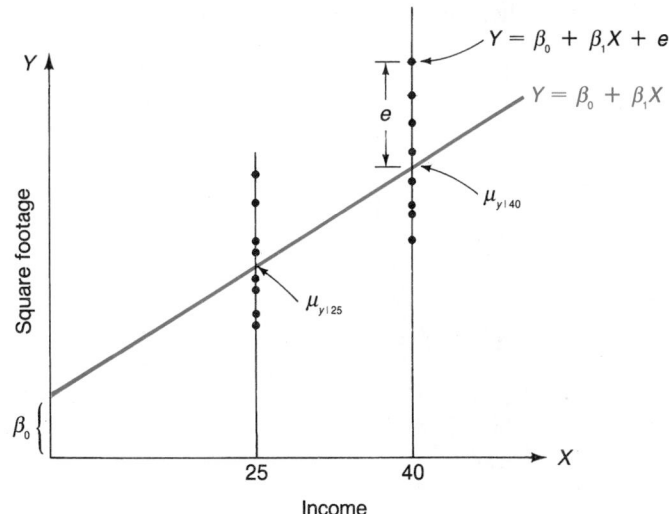

■ **FIGURE 14.8**
Illustration of assumption 1; see text.

again examine a value of $X = 40$. Considering all homes (in this community) whose owners have an income of \$40,000 ($X = 40$), we have already decided that these homes do not all have the same square footage, Y. In fact, the square-footage values will be scattered about the (unknown) line $Y = \beta_0 + \beta_1 X$, with some values lying above the line (e is positive) and some falling below it (e is negative). Consider the average of *all* Y values with $X = 40$. This is written as

$$\mu_{Y|40}$$

which is the mean of Y *given* $X = 40$. Our assumption here is that $\mu_{Y|40}$ *lies on this line;* that is, for *any* value of X, $\mu_{Y|X}$ *lies on the line* $Y = \beta_0 + \beta_1 X$ (such as $\mu_{Y|25}$ in Figure 14.8). Put another way, the error is zero, *on the average.*

Assumption 2 *Each error component (random variable) follows an approximate normal distribution.* In our sample of ten homes and incomes, we had one family with $X = 40$ and $Y = 20$. Figure 14.8 illustrates what we might expect if we were to examine other homes whose owners had an income of \$40,000. We assume here that if we were to obtain 100 homes, for example, whose owners had this income, a histogram of the resulting errors (e) would be bell-shaped in appearance. So we would expect a concentration of errors near zero (from assumption 1), with half of them positive and half of them negative.

Assumption 3 *The variance of the error component, σ_e^2, is the same for each value of X.* For each value of X, the errors illustrated in Figure 14.8 have so far been assumed to follow a normal distribution with a mean of zero. So each error, e, is from such a normal population. The variance of this population is σ_e^2. The assumption here is that σ_e^2 *does not change* as the value of X changes. This is the assumption of **homoscedasticity.** A situation where this assumption is violated is illustrated in Figure 14.9, where we once again consider what might occur if we *were* to obtain (we will not, actually) many values of Y for $X = 25$ and also for $X = 50$. If Figure 14.9 were the result, assumption 3 would be violated because the errors would be much larger (in absolute value) for the \$50,000-income homes than they would for the \$25,000-income homes. Figure 14.9 illustrates **heteroscedasticity,** which does pose a problem when we try to infer results from a linear regression equation.

You might argue that, proportionally, the errors for $X = 50$ seem about the same as those for $X = 25$, which means that you would expect larger errors for larger values of X here. If this is the case, the confidence intervals and tests of hypothesis that we are about to develop for the simple linear regression model are

■ FIGURE 14.9
A violation of
assumption 3; see text.

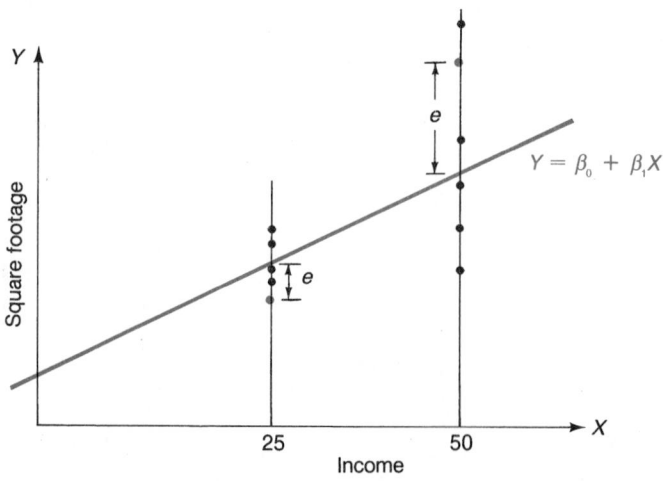

■ **FIGURE 14.10**
Illustration of
assumptions 1, 2, 3;
see text.

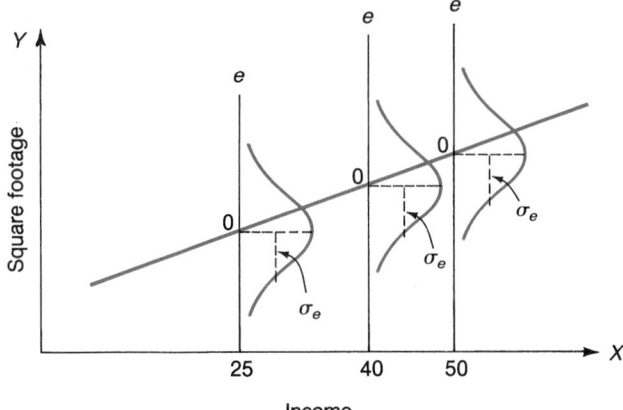

not appropriate. There are methods of "repairing" this situation, by applying a *transformation* to the dependent variable, Y, such as $\sqrt{Y}$ or $\log(Y)$. By using this "new" dependent variable rather than the original Y, the resulting errors often will exhibit a nearly constant variance. Such transformations, however, are beyond the scope of this text.

A summary of the first three assumptions is shown in Figure 14.10. Note that the distribution of errors is *identical* for each illustrated value of X; namely, it is a normal distribution with mean $= 0$ and variance σ_e^2.

Assumption 4 *The errors are independent of each other.* This implies that the error encountered for one value of Y is unaffected by the error for any other value of Y. To illustrate, consider the real-estate data and suppose that the sample is *not* random but that instead the sampled houses are all located on a certain street. The first house has a positive error when predicting the square footage. If the probability is greater than .5 that the next house in the sample also has a positive error (that is, if its location makes it probable that it will be a certain size), then the assumption of independence is violated. In other words, the sample was poorly chosen because the houses on one street are likely to be more or less the same size and their owners are likely to have similar incomes. The nonrandom sample led to a violation of assumption 4.

We can draw two conclusions from these assumptions. First, each value of the dependent variable, Y, is a normal random variable with mean $= \beta_0 + \beta_1 X$ and variance σ_e^2. Second, the error components come from the same normal population, *regardless of the value* of X. In other words, it makes sense to examine the residuals resulting from each value of X in the sample, to construct a histogram of these residuals, and to determine whether its appearance is bell-shaped (normal), centered at zero. A key assumption when using simple linear regression is that the errors follow a normal distribution with a mean of zero. Constructing a histogram of the sample residuals provides a convenient method of determining whether this assumption is reasonable for a particular application.

We further discuss methods of analyzing the validity of each of these assumptions in Chapter 15, where we learn how to use more than one independent variable in a linear regression equation.

Estimating the Error Variance, σ_e^2

The variance of the error components, σ_e^2, measures the variation of the error terms resulting from the simple linear regression model. The value of σ_e^2 severely affects our ability to use this model as an effective predictor for a given situation. Suppose, for example, that σ_e^2 is very large in Figure 14.10. This means that if we were to obtain many observations (square footage values, Y) for a *fixed* value of X (say, income = \$40,000), these Y values would vary a great deal, decreasing the accuracy of our model; we would prefer that these values were grouped closely about the mean, $\mu_{Y|40}$.

In practice, σ_e^2 typically is unknown and must be estimated from the sample. To estimate this variance, we first determine the sum of squares of error, SSE, using SSE = $\Sigma(y - \hat{y})^2$ or equation 14.14. Estimating β_0 and β_1 for the simple regression model results in a loss of 2 df, leaving $n - 2$ df for estimating the error variance. Consequently,

$$s^2 = \hat{\sigma}_e^2 = \text{estimate of } \sigma_e^2 = \frac{\text{SSE}}{n - 2} \qquad (14.16)$$

where

$$\text{SSE} = \Sigma(y - \hat{y})^2 = \text{SS}_Y - \frac{(\text{SCP}_{XY})^2}{\text{SS}_X}$$

Note that SSE is the expression that was minimized in Figure 14.5 when determining the "best" line through the sample points. In Figure 14.5, each distance d corresponds to $y - \hat{y}$.

We can determine the estimate of σ_e^2 and σ_e for the real-estate data in Example 14.2, where we calculated the value of SSE to be 75.81. Our estimate of σ_e^2 is

$$s^2 = \frac{\text{SSE}}{n - 2} = \frac{75.81}{8} = 9.476$$

and so $s = \sqrt{9.476} = 3.078$ provides an estimate of σ_e. The values of s^2 and s are a measure of the variation of the Y values about the least squares line.

COMMENTS   We know from the empirical rule that approximately 95% of the data from a normal population should lie within two standard deviations of the mean. For this example, this rule implies that approximately 95% of the residuals should lie within 2(3.078) = 6.16 of the mean. In the table in Example 14.2, the sample residuals are in the fourth column. Their sum is *always* zero, when using the least squares line; therefore, their mean is *zero*. So, approximately 95% of the residuals should be no larger (in absolute value) than 6.16. In fact, all of them are less than 6.16—not a surprising result, given that we had only ten values in the sample.

Exercises   **14.13**   A stock broker collected data on company XYZ's quarterly earnings (X) and also on the company's closing price (Y) on the day that the quarterly earnings were reported.

CLOSING PRICE (Y)	QUARTERLY EARNINGS (X)	CLOSING PRICE (Y)	QUARTERLY EARNINGS (X)
10.125	1.09	14.0	2.0
10.0	1.10	14.25	2.10
10.25	1.12	14.37	2.50
10.75	1.80	15.0	2.85
10.5	1.95	14.55	2.65

a.   Find the least squares line. Then graph the data and the least squares line.

b.   Find the residual ($Y - \hat{Y}$) for each value of Y.

c.   Is there any indication that the error terms may be correlated?

14.14   The following data were collected for labor hours (X, in hundreds) spent on maintenance and total cost (Y in thousands of dollars) of maintenance.

LABOR HOURS (X)	TOTAL COST (Y)	LABOR HOURS (X)	TOTAL COST (Y)
2.1	5.5	4.1	9.4
2.9	6.4	2.3	4.7
4.9	11.2	6.7	14.9
3.8	7.9	7.2	13.3
2.8	6.3	4.8	13.0
1.4	6.2	5.3	12.9
6.1	12.9	5.2	12.2
5.0	13.5	1.2	2.5
6.2	12.8	4.5	8.6
4.3	10.7	3.8	8.4

a.   Calculate the least squares line. Interpret the coefficients of the least squares line in the context of the problem.

b.   Find the residuals ($Y - \hat{Y}$) for each value of Y.

c.   Construct a histogram for the residuals. Do they appear to follow a normal distribution?

14.15   The following data show the number of total annual bankruptcy petitions filed in the northern district of a southern state (in thousands) and the size of the permanent staff at the U.S. Bankruptcy Court for that district.

BANKRUPTCIES, IN THOUSANDS (X)	PERMANENT STAFF AT U.S. BANKRUPTCY COURT (Y)
2.1	15
3.8	18
4.1	18
10.0	59
3.2	14
3.9	18
6.1	24

a.   Compute the least squares line.

b.   Identify the values of the slope and the intercept for the simple linear regression model.

c.   Estimate the variance of the error for the model.

d.   Find the residuals ($Y - \hat{Y}$) for all the Y values.

14.16   What assumptions need to be made about the error component of a linear model in order that statistical inference can be used?

14.17   The following is a list of sample errors ($Y - \hat{Y}$) from a linear regression application:

$$2.1, \ -.3, \ 1.4, \ -2.8, \ -3.9, \ 4.2, \ 3.6, \ 4.3, \ 1.8, \ -2.7, \ -.8, \ 1.2, \ .9,$$
$$-1.1, \ -4.5, \ -5.2, \ -1.3, \ .5, \ .9, \ -.6, \ 1.5, \ 2.1, \ -2.2, \ .9$$

Do the data appear to conform to the empirical rule that approximately 95% of the errors should lie within two standard deviations of the mean? Construct a histogram for the residuals.

14.18   Let X be the distance an employee lives from his or her job. Let Y be the average time that it takes the employee to drive to work. Data from 30 employees gave the following sample statistics.

$$SCP_{XY} = 8.4 \quad SS_X = 9.4 \quad SS_Y = 12.2$$

a.   Find the estimate of the error variance.

b.   Find the interval in which approximately 68% of the error values should fall.

14.19   The following are residuals resulting from a regression analysis:

$Y - \hat{Y}$	X	$Y - \hat{Y}$	X
.2	1	1.0	4.00
−.2	1.5	−1.5	5.00
−.5	1.75	−1.7	6.00
.6	2.00	2.1	7.00
−.5	2.50	−2.5	8.00
−.8	3.00	3.8	9.00

From these data, where Y is the dependent variable and X is the independent variable, does it appear that any of the standard assumptions of regression analysis are violated?

14.20   Why is $\Sigma(y - \hat{y})^2$ used in estimating the variance of the error term instead of $\Sigma(y - \hat{y})$?

14.21   In the statistical model $Y = \beta_0 + \beta_1 X + e$, is X a random variable? Comment on your answer.

14.22   Let X be a person's income. Let Y be the amount of life insurance (in thousands of dollars) that this person has. Data from 20 people were collected.

INCOME (X)	LIFE INSURANCE (Y)	INCOME (X)	LIFE INSURANCE (Y)
15.4	33.2	28.6	53.7
19.8	39.5	38.7	67.6
20.6	42.2	41.5	75.4
29.4	52.5	40.1	68.3
22.3	44.3	36.5	65.2
19.5	42.3	27.4	51.2
30.8	57.6	28.6	54.9
25.5	49.2	21.4	41.0
20.4	41.6	19.8	40.9
18.4	36.7	20.1	40.2

a.   Calculate the least squares line.

b.   Find the residual $Y - \hat{Y}$ for each value of Y.

c.   Construct a histogram of the residuals.

d.   Do the residuals appear to follow a normal distribution?

◢ 14.3
Inference on the Slope, β_1

Performing a Test of Hypothesis on the Slope of the Regression Line

Under the assumptions of the simple linear regression model outlined in the previous section, we are now in a position to determine whether a linear relationship exists between the variables X and Y. Examining the estimate of the slope, b_1, will provide information as to the nature of this relationship.

Consider the *population* slope, β_1. Three possible situations are demonstrated in Figure 14.11. What can you say about using X as a predictor of Y in Figure 14.11a? When $\beta_1 = 0$, the population line is perfectly horizontal. As a result, the value of Y is the *same* for each value of X, and so X is not a good predictor of Y; the value of X provides no information regarding the value of Y. In the event $\beta_1 = 0$, the best predictor of Y is given by $\hat{Y} = \bar{y}$, and so $\beta_1 \neq 0$ is equivalent to saying that $\hat{Y}$ (using X as a predictor) is superior to using the sample mean ($\hat{Y} = \bar{y}$) as a predictor.

To determine whether X provides information in predicting Y, the hypotheses are

$$H_0: \beta_1 = 0 \qquad (X \text{ provides no information})$$
$$H_a: \beta_1 \neq 0 \qquad (X \text{ does provide information})$$

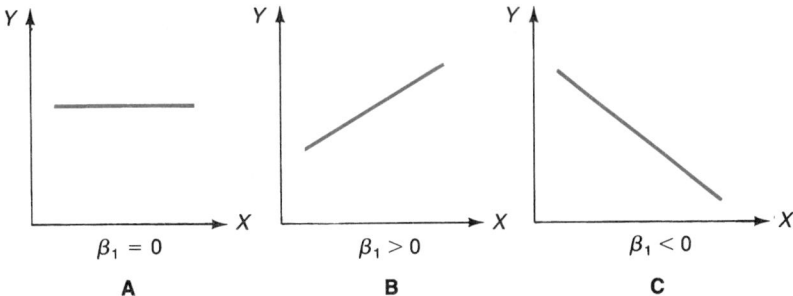

A $\beta_1 = 0$ **B** $\beta_1 > 0$ **C** $\beta_1 < 0$

Other Alternative Hypotheses If we are attempting to demonstrate that a significant *positive* linear relationship exists between X and Y, the appropriate alternative hypothesis would be $H_a: \beta_1 > 0$. For example, do the data in Example 14.1 support the hypothesis that owners with large incomes have larger homes?

When the purpose of the analysis is to determine whether a *negative* linear relationship exists between X and Y, the alternative hypothesis should be $H_a: \beta_1 < 0$. For example, you would expect such a relationship between the number of new housing starts (Y) and the interest rate (X) (as the interest rate increases, you would expect the number of new houses under construction to decrease).

The Test Statistic We use the point estimate of β_1 (that is, b_1) in the test statistic to determine the nature of β_1. What is b_1? A constant? A variable? Suppose that we obtained a different set of data and recalculated b_1. The new value would not be exactly the same as the previous value, implying that b_1 is actually a variable. To be more precise, under the assumptions of the previous section, b_1 is a *normal* random variable with mean = β_1 and variance = $\sigma_{b_1}^2 = (\sigma_e^2)/(SS_X)$. Notice that b_1 is, on the average, equal to β_1; that is, b_1 is an *unbiased* estimator of β_1. The variance $\sigma_{b_1}^2$ is a parameter describing the variation in the b_1 values if we were to obtain random samples of n observations indefinitely.

If we replace the unknown σ_e^2 by its estimate, s^2, then the *estimated* variance of b_1 is $s_{b_1}^2 = s^2/SS_X$. As a result

$$t = \frac{b_1 - \beta_1}{s/\sqrt{SS_X}} = \frac{b_1 - \beta_1}{s_{b_1}} \tag{14.17}$$

has a t distribution with $n - 2$ df. If the null hypothesis is $H_0: \beta_1 = 0$, the test statistic becomes

$$t = \frac{b_1}{s/\sqrt{SS_X}} \tag{14.18}$$

A summary of the testing procedure is shown in the accompanying box.

Test of Hypothesis on the Slope of the Regression Line

TWO-TAILED TEST
$H_0: \beta_1 = 0$

$H_a: \beta_1 \neq 0$

Test statistic:

$$t = \frac{b_1}{s_{b_1}}$$

where $s_{b_1} = s/\sqrt{SS_x}$ and df $= n - 2$.

Test:

reject H_0 if $|t| > t_{\alpha/2, n-2}$

ONE-TAILED TEST

$H_0: \beta_1 \leq 0$ $H_0: \beta_1 \geq 0$
$H_a: \beta_1 > 0$ $H_a: \beta_1 < 0$

Test statistic: Test statistic:

$$t = \frac{b_1}{s_{b_1}}$$ $$t = \frac{b_1}{s_{b_1}}$$

where $s_{b_1} = s/\sqrt{SS_x}$ and   where $s_{b_1} = s/\sqrt{SS_x}$ and
df $= n - 2$.   df $= n - 2$.

Test: Test:

reject H_0 if $t > t_{\alpha, n-2}$ reject H_0 if $t < -t_{\alpha, n-2}$

EXAMPLE 14.3

Is there sufficient evidence, using the real-estate data in Example 14.1, to conclude that a significant positive relationship exists between income (X) and home size (Y)? Use $\alpha = .05$.

SOLUTION

Step 1. The hypotheses indicated here are

$$H_0: \beta_1 \leq 0$$
$$H_a: \beta_1 > 0$$

Step 2. The test statistic is

$$t = \frac{b_1}{s_{b_1}}$$

which has a t distribution with $n - 2 = 8$ df.

Step 3. The testing procedure is to

reject H_0 if $t > t_{.05,8} \doteq 1.860$

The t curve is shown in Figure 14.12.

■ **FIGURE 14.12**
t curve with 8 df
showing rejection region
(shaded) for
Example 14.3.

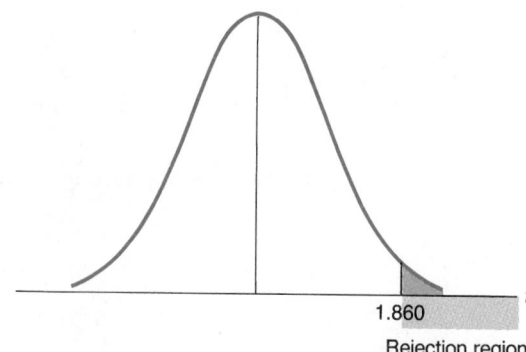

1.860

Rejection region

■ **FIGURE 14.13** MINITAB solution to Example 14.3.

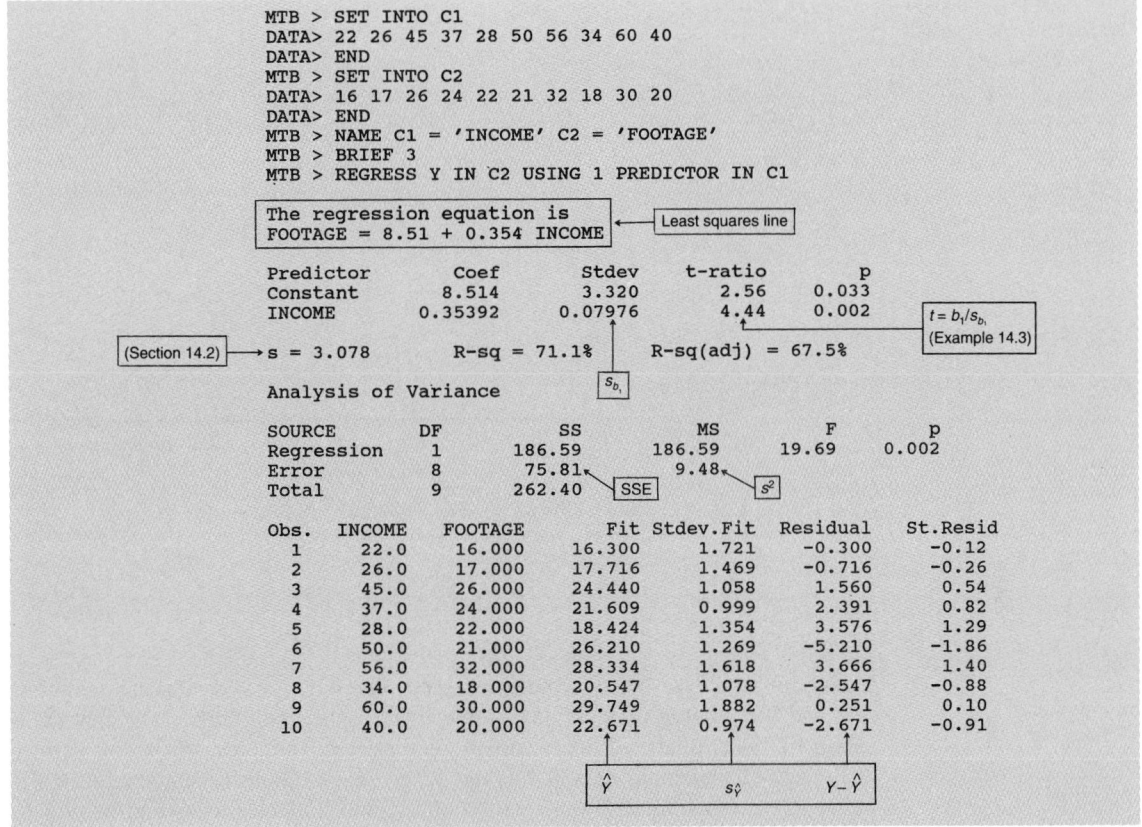

Step 4. We previously determined that $SS_x = 1489.6$, $b_1 = .354$, and $s = 3.078$. The calculated test statistic is then

$$t^* = \frac{.354}{3.078/\sqrt{1489.6}} = \frac{.354}{.0797} = 4.44$$

where $s_{b_1} = .0797$. Because $4.44 > 1.86$, we reject H_0.

Step 5. Based on these ten observations, we conclude that a positive linear relationship does exist between income and home size. ■

A MINITAB solution using the real-estate data is shown in Figure 14.13. This output contains nearly all the calculations performed so far. In particular, note that:

1. The least squares equation is $\hat{Y} = 8.51 + .354X$.
2. The standard deviation of b_1 is $s_{b_1} = .07976$.
3. The value of the test statistic is $t^* = b_1/s_{b_1} = 4.44$.
4. The estimated standard deviation of the error components is $s = 3.078$.
5. The value of SSE is 75.81, contained in the ANOVA table (construction of this table is discussed in Chapter 15).
6. The column of estimated Y's ($\hat{Y}$'s) and the corresponding residuals are in the column labeled $Y - \hat{Y}$.

EXAMPLE 14.4 The firm of Smithson Financial Consultants has been hired by Blackburn Industries to determine whether a relationship exists between the age of unmarried male

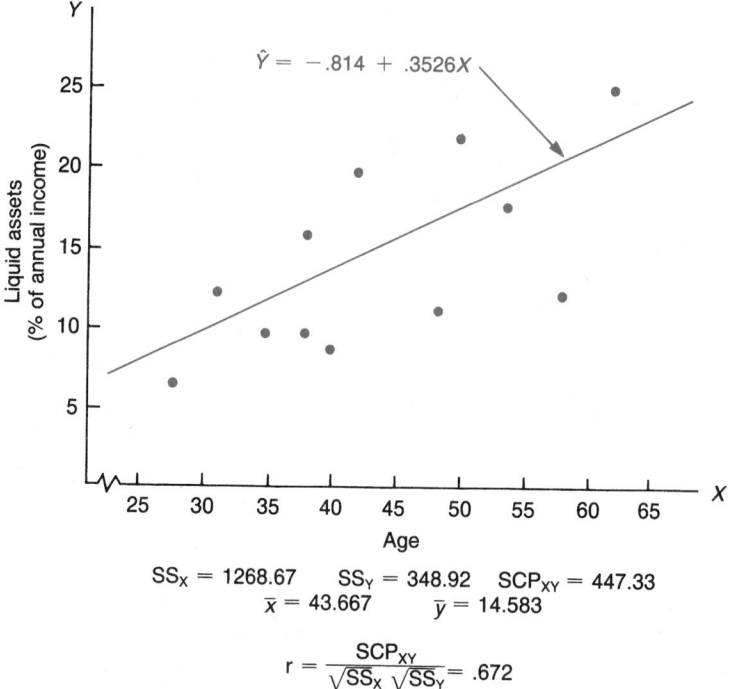

$$SS_X = 1268.67 \qquad SS_Y = 348.92 \qquad SCP_{XY} = 447.33$$
$$\bar{x} = 43.667 \qquad \bar{y} = 14.583$$

$$r = \frac{SCP_{XY}}{\sqrt{SS_X}\ \sqrt{SS_Y}} = .672$$

Blackburn employees (that is, never married, divorced, or widowed male employees) and the amount of individual liquid assets. The main question of interest is whether a linear relationship exists between these two variables, where X is defined as the age of the employee and Y is the *percentage* of annual income allocated to liquid assets (such as cash, savings accounts, and tradable stocks and bonds). A random sample of 12 unmarried male employees is selected, and the following data are obtained:

AGE (X)	LIQUID ASSETS (Y, PERCENTAGE OF ANNUAL INCOME)	AGE (X)	LIQUID ASSETS (Y, PERCENTAGE OF ANNUAL INCOME)
38	16	58	13
48	12	31	13
38	10	42	20
28	7	35	10
40	9	54	18
50	22	62	25

A scatter diagram of these 12 observations is provided in Figure 14.14, with a summary of the calculations. Using $\alpha = .10$, do you think that an employee's age provides useful information for predicting the percentage of total income allocated to liquid assets?

SOLUTION   To derive the least squares regression line, we determine

$$b_1 = \frac{SCP_{XY}}{SS_X} = \frac{447.33}{1268.67} = .3526$$

and

$$b_0 = \bar{y} - b_1\bar{x}$$
$$= 14.583 - (.3526)(43.667) = -.814$$

Consequently, the least squares line is

$$\hat{Y} = -.814 + .3526X$$

Notice that the slope of this line is positive. As the following test of hypothesis will conclude, this slope is significant. Consequently, a higher percentage invested in liquid assets is associated with the *older* employees. According to these data, each additional year of age is accompanied by an increase of .35 percent of income allocated to liquid assets, on the average, for the unmarried male population at Blackburn.

To carry out a test of hypothesis, we follow the usual five-step procedure.

Step 1. Because the suspected direction of the relationship between these two variables (positive or negative) is unknown before the data are obtained, a two-tailed test is appropriate. The hypotheses are

$$H_0: \beta_1 = 0$$

$$H_a: \beta_1 \neq 0$$

Step 2. The test statistic is $t = b_1/s_{b_1}$, which has $n - 2 = 10$ df.

Step 3. The test procedure is to

$$\text{reject } H_0 \text{ if } |t| > t_{.10/2,10} = t_{.05,10} = 1.812$$

Step 4. Based on the data summary in Figure 14.14 and using equation 14.14,

$$\text{SSE} = \text{SS}_Y - \frac{(\text{SCP}_{XY})^2}{\text{SS}_X}$$

$$= 348.92 - \frac{(447.33)^2}{1268.67}$$

$$= 348.92 - 157.73$$

$$= 191.19$$

Consequently,

$$s^2 = \frac{\text{SSE}}{n - 2} = \frac{191.19}{10} = 19.12$$

and so

$$s_{b_1} = \frac{s}{\sqrt{\text{SS}_X}} = \frac{\sqrt{19.12}}{\sqrt{1268.67}} = .1228$$

The computed value of the test statistic is therefore:

$$t^* = \frac{b_1}{s_{b_1}}$$

$$= \frac{.3526}{.1228} = 2.87$$

Because $t^* = 2.87$ exceeds the table value of 1.812, we reject H_0 in support of H_a.

Step 5. Our conclusion is that age is a useful (although imperfect) predictor of percentage of income invested in liquid assets for this particular population.

One thing to keep in mind is that *statistical* significance does not always imply *practical* significance. In other words, rejection of $H_0: \beta_1 = 0$ (statistical significance) does not mean that precise prediction (practical significance) follows. It *does* demonstrate to the researcher that, within the sample data at least, this particular independent variable has an association with the dependent variable.   ■

Confidence Interval for β_1

Following our usual procedure of providing a confidence interval with a point estimate, we use the t distribution of the previous test statistic and equation 14.17 to define a confidence interval for β_1. The narrower this confidence interval is, the more faith we have in our estimate of β_1 and in our model as an accurate, reliable predictor of the dependent variable. A $(1 - \alpha) \cdot 100\%$ confidence interval for β_1 is

$$b_1 - t_{\alpha/2, n-2} s_{b_1} \quad \text{to} \quad b_1 + t_{\alpha/2, n-2} s_{b_1}$$

EXAMPLE 14.5   Construct a 90% confidence interval for the population slope, β_1, using the real-estate data in Example 14.1 and Example 14.3.

SOLUTION   All the necessary calculations have been completed; $b_1 = .354$ and $s_{b_1} = .0797$ (from Example 14.3). Using $t_{.05,8} = 1.860$, the resulting confidence interval is

$$.354 - (1.860)(.0797) \quad \text{to} \quad .354 + (1.860)(.0797)$$
$$= .354 - .148 \quad \text{to} \quad .354 + .148$$
$$= .206 \quad \text{to} \quad .502$$

So we are 90% confident that the value of the estimated slope ($b_1 = .354$) is within .148 of the actual slope, β_1. The large width of this interval is due in part to the lack of information (small sample size) used to derive the estimates; a larger sample would decrease the width of this confidence interval.   ■

COMMENTS   A failure to reject H_0 when performing a hypothesis test on β_1 does not always indicate that no relationship exists between the two variables. Some form of nonlinear relationship may exist between these variables. For example, in Figure 14.15, there is clearly a strong curved (**curvilinear**) relationship between X and Y. However, the least squares line through these points is horizontal, leading to a t value equal to zero and a failure to reject H_0. Furthermore, the sample correlation coefficient, r, for these data is zero.

Of course, you may fail to reject H_0 as the result of a type II error. In other words, you failed to reject H_0 when in fact a significant linear relationship does exist. This situation is more apt to occur when using a small sample to test the null hypothesis.

More often, a failure to reject H_0 occurs when there is no visible relationship between the two variables within the sample data. To determine whether there is no relationship or that there is a nonlinear one, you should inspect either a scatter diagram of the data, a scatter diagram of the residuals, or, better yet, both. The latter diagram is a picture of the residuals $(Y - \hat{Y})$ plotted against the independent variable, X. Residual plots are discussed further in Chapter 15.

In many situations, a business analyst has the opportunity to select the values of the independent variable, X, *before* the sample is obtained. At first glance, it might appear that the precision of b_1 (as an estimator of β_1) is unaffected by the X values. This is partially but not completely true. Because a narrow confidence interval for β_1 lends credibility to our model, we may choose to decrease the width of this confidence interval by decreasing s_{b_1}. Now $s_{b_1} = s/\sqrt{SS_X}$, so if we make SS_X large, the resulting s_{b_1} will be small. Therefore, given the opportunity, select a set of X values having a *large variance*. You can accomplish this by choosing a great many X values on the lower end of your range of interest, a large number of values at the upper end, and some values in between to detect any curvature that exists (as in Figure 14.15).

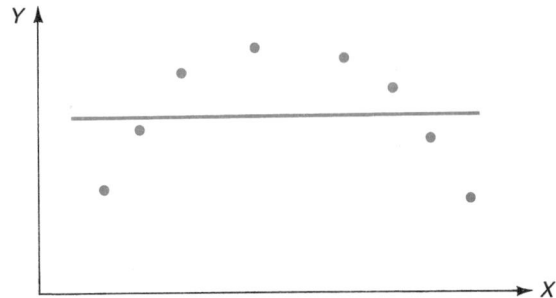

■ **FIGURE 14.15**
Curvilinear relationship.
The horizontal line is
the least squares line.

■ *Exercises*   **14.23**   A banker is interested in the relationship between a person's income and the amount of money the person has in tax-free investment instruments (such as municipal bonds or IRAs). Data on 20 working individuals were collected. The results are as follows (in thousands of dollars):

INCOME (X)	MONEY IN TAX-FREE INVESTMENTS (Y)	INCOME (X)	MONEY IN TAX-FREE INVESTMENTS (Y)
20.2	5.1	24.1	5.2
33.2	7.5	25.1	4.4
35.1	8.1	34.2	7.1
29.4	6.7	33.0	5.1
33.0	7.4	45.1	12.4
40.1	9.4	41.0	9.8
41.0	9.7	45.1	8.9
45.1	10.1	40.1	8.8
42.3	9.1	31.2	6.9
45.3	11.4	24.0	4.2

a.   Is there sufficient evidence using the observed data to conclude that a positive relationship exists between X and Y? Use a 5% significance level.

b.   Find the p-value for the test statistic in question a. What is your conclusion based on this value?

14.24   The regression equation $\hat{Y} = 2.3 + 1.5X$ was arrived at by fitting a least squares line to 25 data points. The standard deviation (error) of the estimate of the slope was found to be 0.812. Test the null hypothesis that the slope of the line is equal to zero at the .01 level of significance.

14.25   It is believed that the size of the U.S. population (X) is a variable that influences personal consumption expenditure for housing (Y). However, the relationship historically does not appear to be linear. Therefore, a log transformation of housing expenditure is used. Fifteen observations are taken over previous years. The units of Y are millions and the units of X are billions.

X	LOG Y	X	LOG Y	X	LOG Y
183.69	3.935	196.56	4.241	207.66	4.631
186.54	4.001	198.71	4.305	209.90	4.722
189.24	4.060	200.71	4.379	211.91	4.818
191.89	4.117	202.68	4.465	213.85	4.923
194.30	4.182	205.05	4.542	215.97	5.009

From the data, does there appear to be a significant positive relationship between X and log Y? Use a significance level of .05.

14.26   The life of a lawn-mower engine can be extended by frequent oil changes. An experiment was conducted in which 20 lawn mowers were used over many years with different time intervals between oil changes. Let X be the number of hours of operation between oil changes. Let Y be the number of years that the engine was able to perform adequately.

X	Y	X	Y	X	Y
11.25	12.1	22.0	9.5	25.5	6.1
15.5	11.8	22.5	9.2	26.0	5.4
17.5	11.5	23.0	8.4	26.5	4.8
20.5	10.1	23.5	8.8	27.0	4.6
19.5	9.9	24.0	7.1	28.0	4.8
18.5	9.7	24.5	7.2	30.0	4.1
21.5	10.1	25.0	5.8		

a.   Graph the data and the least squares line.

b.   Is there sufficient evidence to conclude, at the 10% significance level, that a negative relationship exists between Y and X? What is the critical region?

14.27   The Y fares, the base coach fares from which discounts are subtracted by the airline industry, have been rising at an alarming rate. Nine top U.S. markets, in terms of passengers flown, are listed below with the average Y fares on November 1, 1978, and on March 28, 1988. From the data, does there appear to be a significant positive relationship between the X and Y variables given below? Use a significance level of .05.

ROUTE	NOVEMBER 1, 1978 AVERAGE "Y" FARE X	MARCH 28, 1988 AVERAGE "Y" FARE Y
Boston to New York	76	320.75
Newark to D.C.	82	316.80
Chicago to New York	176	614.60
Los Angeles to New York	440	1,068.73
Miami to New York	232	702.91
Dallas/Fort Worth to Houston	84	420.00
Ft. Lauderdale to New York	232	694.60
Newark to San Francisco	458	1,163.33
Atlanta to New York	180	649.56

(*Source:* Adapted from R. Fox and F. Stephenson, "How to Control Corporate Air Travel Costs," *Business* (July–September 1990): 3–9.)

14.28   A medical researcher was interested in the amount of weight loss caused by a particular diuretic. In a controlled experiment with 18 rats, the amount of weight loss was recorded after 1 month of daily dose of the diuretic. Let X be the amount, in milligrams, of diuretic given. Let Y be the weight loss in pounds.

X	Y	X	Y	X	Y
.10	.05	.25	.35	.40	.44
.10	.08	.25	.31	.40	.47
.15	.11	.30	.41	.45	.51
.15	.13	.30	.42	.45	.52
.20	.19	.35	.43	.50	.54
.20	.21	.35	.42	.50	.53

Is there sufficient evidence to conclude that a significant positive relationship exists between the amount of diuretic given and the amount of weight loss? Use a significance level of 10%. Find a 90% confidence interval for the slope of the regression equation used to predict Y.

14.29   Using the data in Exercise 14.4, find a 95% confidence interval for the slope of the regression equation used to predict the current year's productivity from the previous year's productivity for each employee.

14.30   An investment counselor wanted to know the relationship between the price/earnings ratio (Y) and the yield (X) for high-yield stocks. If a stock yielded over 5.5%, it was considered to be a high-yield stock. Twenty-five high-yield stocks were randomly selected. The following sample statistics were found:

$$SCP_{XY} = -10.4$$

$$SS_X = 11.4$$

$$SS_Y = 21.4$$

a. Test that the slope of the regression equation used to predict the price/earnings ratio from the yield of a stock is negative. Use a 10% significance level.

b. Find a 95% confidence interval for the slope in question a.

14.31 In Exercise 14.9, find the 90% confidence interval for the slope of the regression equation used to predict the number of ice creams sold by using the independent variable, temperature.

14.32 A survey of the students of Highpoint College gathered the following information with regard to their study time (hours per week) and grade point averages.

Y (STUDY TIME)	X (GRADE-POINT AVERAGE)	Y (STUDY TIME)	X (GRADE-POINT AVERAGE)
16	4.0	8	2.2
15	3.8	6	1.5
14	3.5	4	1.0
12	3.0	2	0.5
10	2.8	0	0.2

Find the 95% confidence interval for the slope of the regression equation used to predict study time.

 **14.4**

Measuring the Strength of the Model

We have already used the sample coefficient of correlation, r, as a measure of the amount of linear association within a sample of bivariate data. The value of r is given by

$$r = \frac{SCP_{XY}}{\sqrt{SS_X}\,\sqrt{SS_Y}} \qquad \textbf{(14.19)}$$

The possible range for r is -1 to 1.

Comparing the equations for r and b_1, we see that

$$r = b_1 \sqrt{\frac{SS_X}{SS_Y}}$$

Because SS_X and SS_Y are *always greater than zero*, r and b_1 have the same sign. Thus, if a positive relationship exists between X and Y, then both r and b_1 will be greater than zero. Similarly, they are both less than zero if the relationship is negative.

When you determine r, you use a sample of observations; r is a *statistic*. What does r estimate? It is actually an estimate of ρ (rho, pronounced "roe"), the **population correlation coefficient.** To grasp what ρ is, imagine obtaining *all* possible X, Y values and using equation 14.19 to determine a correlation. The resulting value is ρ.

The population slope, β_1, and ρ are closely related. In particular, $\beta_1 = 0$ if and only if $\rho = 0$. This leads to another method of determining whether the simple linear regression model (using X to predict Y) is satisfactory. The hypotheses are

$H_0: \rho = 0$ (no linear relationship exists between X and Y)

$H_a: \rho \neq 0$ (linear relationship does exist)

In a similar manner, alternative hypotheses can be set up to demonstrate a positive relationship ($H_a: \rho > 0$) or a negative relationship ($H_a: \rho < 0$). The test statistic uses the point estimate of ρ (that is, r) and is defined by

$$t = \frac{r}{\sqrt{\dfrac{1 - r^2}{n - 2}}} \qquad \textbf{(14.20)}$$

where n = the number of observations in the sample. It is also a t statistic with $n - 2$ df. Although equations 14.18 and 14.20 appear to be unrelated, the two are algebraically equivalent and *their values for t are always the same.*

Thus, the t test for H_0: $\beta_1 = 0$ and H_0: $\rho = 0$ produce identical results, provided both tests use the same level of significance. These tests are therefore redundant; they both produce the same conclusion. Remember, if you have already computed the sample correlation coefficient, r, equation 14.20 offers a much easier method of determining whether the simple linear model is statistically significant. Notice also in equation 14.20 that the significance of the t value depends on the sample size, n. *As a result, if the sample size is large enough, then virtually any value of r can produce a "significantly large" value of t.*

EXAMPLE 14.6

Use equation 14.20 to determine whether a positive linear relationship exists between X = income and Y = home square footage, based on the real-estate data from Example 14.1. Use $\alpha = .05$.

SOLUTION

The hypotheses to be used here are H_0: $\rho \le 0$ versus H_a: $\rho > 0$. In Example 14.1, we found that $r = .843$. This leads to a computed test statistic value of

$$t^* = \frac{r}{\sqrt{\dfrac{1 - r^2}{n - 2}}} = \frac{.843}{\sqrt{\dfrac{1 - (.843)^2}{8}}}$$

$$= \frac{.843}{.190} = 4.44$$

Because this value is the same as the one obtained in Example 14.3 (testing H_0: $\beta_1 \le 0$ versus H_a: $\beta_1 > 0$), we draw the same conclusion. A positive linear relationship *does* exist between these two variables. In other words, r is large enough to justify this conclusion. ∎

Remember, there is no harm in using equation 14.20 as a substitute for equation 14.18 with H_0: $\beta_1 = 0$ (or ≤ 0, or ≥ 0), particularly if you have already determined the value of r.

Danger of Assuming Causality

A word of warning is in order here—namely, that high statistical correlation does not imply *causality*. Even if the correlation between X and Y is extremely high (say, $r = .95$), a unit increase in X does not necessarily *cause* an increase in Y. All we know is that in the sample data, as X increased, so did Y. As a simple example, consider X = percentage of gray hairs and Y = blood pressure. One might expect to observe a high correlation between these two variables, but it is probably absurd to say that an additional gray hair will *cause* a person's blood pressure to increase. What is actually happening is that there is another variable, in this case age, that is causing both percentage of gray hair and blood pressure to increase.

In many business and economics applications, we observe highly correlated variables when each pair of observations corresponds to a particular time period. For example, we would expect a high correlation between average annual wages (X) and the U.S. gross national product (GNP; Y) when measured over time. Even though wages may be a good predictor of GNP, this correlation does not imply that an increase in wages *causes* an increase in GNP. It is much more likely that a third factor—inflation—caused both wages and GNP to increase.

Coefficient of Determination

In our earlier discussion of ANOVA techniques, we used the expression SS(total) $= \Sigma(y - \bar{y})^2$ to measure the tendency of a set of observations to group about the mean. If this value was large, then the observations (data) contained much variation and were *not* all clustered about the mean, $\bar{y}$.

In the simple linear regression model, $SS_y = \Sigma(y - \bar{y})^2$ is computed in the same way and (as before) measures the total variation in the values of the dependent variable.

$$SS_Y = \text{total variation of the dependent variable observations}$$

When comparing the sum of squares of error, SSE, to the total variation, SS_Y, we use the ratio SSE/SS_Y. If all $\hat{Y}$ values are equal to their respective Y values, there is a perfect fit, with SSE = 0 and $r = 1$ or -1. Our model explains 100% of this total variation, and the unexplained variation is zero.

In general, SSE/SS_Y (expressed as a percentage) is the **percentage of unexplained variation.** Recall from equations 14.14 and 14.19 that

$$SSE = SS_Y - \frac{(SCP_{XY})^2}{SS_X} \quad \text{and} \quad r^2 = \frac{(SCP_{XY})^2}{SS_X SS_Y}$$

Thus

$$r^2 = 1 - \frac{SSE}{SS_Y}$$

As a result, r^2 may be interpreted as a measure of the *explained variation* in the dependent variable using the simple linear model; r^2 is the **coefficient of determination.**

r^2 = coefficient of determination

$$= 1 - \frac{SSE}{SSY} \tag{14.21}$$

= the percentage of explained variation in the dependent variable using the simple linear regression model

For this model, we can determine r^2 simply by squaring the coefficient of correlation. In Chapter 15, we will predict the dependent variable, Y, using *more than one* predictor (independent) variable. To derive the coefficient of determination in this case, we must first calculate SSE and then use equation 14.21. So, although this definition may appear to be unnecessary, it will enable us to compute this value when we use a multiple linear regression model.

EXAMPLE 14.7

What percentage of the total variation of the home sizes is explained by means of the single predictor, income, using the real-estate data from Example 14.1?

SOLUTION We previously calculated r to be .843, so the coefficient of determination is

$$r^2 = (.843)^2 = .71$$

Therefore, we have accounted for 71% of the total variation in the home sizes by using income as a predictor of home size.

■ **FIGURE 14.16**
Splitting $(y - \bar{y})$ into
two deviations, $(\hat{y} - \bar{y})$
$+ (y - \hat{y})$.

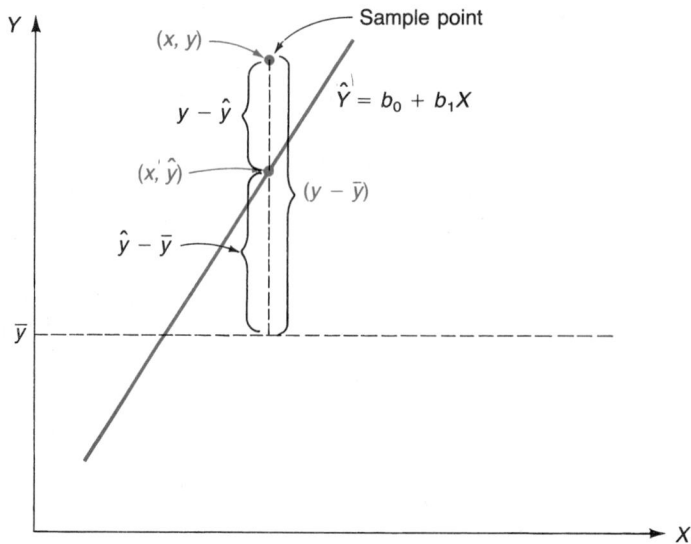

Notice that we could have determined this value by using the calculations from Examples 14.1 and 14.2, where

$$r^2 = 1 - \frac{\text{SSE}}{\text{SS}_Y}$$

$$= 1 - \frac{75.81}{262.4} = .71$$

■

Total Variation, SS_Y

In Chapter 11, when discussing the ANOVA procedure, we partitioned the total variation in the observations, measured by SS(total), into two sums of squares, namely, SS(factor) and SS(error). The resulting equation was

$$\text{SS(total)} = \text{SS(factor)} + \text{SS(error)}$$

In a similar fashion, we can partition the total variation of the Y values in linear regression, measured by SS_Y, into two other sums of squares. In Figure 14.16, notice that the value of $y - \bar{y}$ can be written as the sum of two deviations, namely,

$$y - \bar{y} = (\hat{y} - \bar{y}) + (y - \hat{y})$$

By squaring and summing over *all* the data points in the sample, we can show that*

$$\Sigma(y - \bar{y})^2 = \Sigma(\hat{y} - \bar{y})^2 + \Sigma(y - \hat{y})^2$$

The summation on the left of the equal sign is SS_Y. The second summation on the right is the sum of squares of error, SSE. The first summation on the right is defined to be the **sum of squares of regression, SSR.**

$$\Sigma(\hat{y} - \bar{y})^2 = \text{SSR}$$

---

*This result follows since it can be shown that $\Sigma(\hat{y} - \bar{y})(y - \hat{y}) = 0$ when using the least squares line.

As a result, we have

$$SS_Y = SSR + SSE \qquad (14.22)$$

By comparing equations 14.14 and 14.22, we see that a simple way to calculate the sum of squares of regression is

$$SSR = \frac{(SCP_{XY})^2}{SS_X} \qquad (14.23)$$

The regression sum of squares, SSR, measures the variation in the Y values that would exist if differences in X were the *only* cause of differences among the Y's. If this were the case, then all the (X, Y) points would lie exactly on the regression line. In practice, this does not happen when using a simple linear regression model. Otherwise, we would have a deterministic phenomenon, not an object of statistical investigation. Consequently, the sample points can be assumed to lie about the regression line rather than on this line. This variation *about* the regression line is measured by the error sum of squares, SSE.

Exercises

14.33 The sales manager of a real-estate firm believes that experience is the best predictor for determining the yearly sales of the various salespeople in the real-estate industry. Data were collected from 15 salespeople. Let X be the number of years of prior experience. Let Y be the annual sales (in thousands).

SALES (Y)	EXPERIENCE (X)	SALES (Y)	EXPERIENCE (X)
50	1.3	78	2.2
161	5.1	124	3.4
195	6.2	131	7.1
172	5.4	64	2.1
132	3.9	80	4.5
133	4.1	110	3.8
181	6.1	127	4.4
69	1.9		

Using a 10% significance level, test whether the population correlation coefficient between the variables X and Y is zero.

14.34 The manager of a company that relies on traveling salespersons to sell the company's products wants to examine the relationship between sales and the amount of time a salesperson spends with each established customer who regularly orders the company's products. The manager collects data on 12 salespersons. Let Y represent sales per month and X represent hours spent with customers per month.

X	Y	X	Y	X	Y
3.2	412	6.1	715	5.1	570
4.6	500	4.2	500	7.1	800
3.9	450	5.6	610	6.5	725
5.3	610	5.3	600	7.8	850

Can one conclude that the population correlation coefficient between X and Y is positive? Use a 10% significance level. Can one conclude that spending more time with customers increases sales?

14.35 Using the data in Exercise 14.3, test that there is no linear relationship between the total commissions paid and the net earnings of Tony's used-car lot. Use a 5% significance level.

14.36   Refer to Exercise 14.23. Use equation 14.20 to test whether there is a positive relationship between a person's investment in tax-free investments and a person's income. Use a significance level of .05. Is the conclusion the same as that in Exercise 14.23?

14.37   Ten cards numbered 1 through 10 are shuffled and a person is asked to pick one card. The card is replaced and the deck is reshuffled. Then the person is asked to draw a second card. If the second card is higher than the first, the dealer gives $.85 to the player. If the second card is not higher than the first, the player pays $1.15 to the dealer. A sample of 15 pairs of draws is taken to see whether there is any correlation between the first and the second cards.

a.   Would you expect to observe significant correlation here? Why or why not?

b.   Find the coefficient of determination for the following data and test using a 5% significance level that there is no correlation between the first and second cards. Interpret the value of the coefficient of determination.

FIRST CARD (X)	SECOND CARD (Y)	FIRST CARD (X)	SECOND CARD (Y)
7	3	10	5
3	10	3	6
8	2	4	3
5	8	6	1
2	7	7	8
7	9	8	4
9	4	2	6
1	1		

14.38   Refer to Exercise 14.25. Use equation 14.20 to test that there is no linear relationship between the size of the U.S. population and the logarithm of personal consumption expenditure on housing. Use a significance level of 5%.

14.39   For the data in Exercise 14.26, test that there is no linear relationship between the number of hours of operation between oil changes and the number of years that the engine was able to perform adequately. Use equation 14.20 and test with a 10% significance level. Is the result the same as in Exercise 14.26?

14.40   A sample of 35 pairs of observations is taken and a sample correlation coefficient is computed to be $r = .48$. Do the data provide sufficient evidence to reject the null hypothesis of no correlation? Use a 1% significance level.

14.41   Fifty people were asked to record their yearly income and their expenditures on vacations during the year. A correlation value of .39 was found. Do the data provide sufficient evidence to reject the null hypothesis of no correlation between the two variables? Use a 5% significance level.

14.42   The following pairs of observations represent the scores of a test given by a psychologist to a group before an experiment (X) and then to the same group after the experiment (Y).

X	Y	X	Y
2.1	9.4	2.4	12.9
3.4	35.6	1.9	5.2
1.6	3.5	1.3	3.4
2.7	15.4	0.2	0.1
3.2	30.1	1.5	4.6
4.5	52.7	2.1	9.3
1.8	17.4		

a.   Calculate the correlation coefficient, r, between X and Y.

b.   Take the log to the base 10 of Y. Calculate the correlation coefficient between X and log Y.

c.   In parts a and b, is there sufficient evidence to conclude that nonzero correlation exists? Use a 5% significance level.

14.5
Estimation and Prediction Using the Simple Linear Model

We have concentrated on predicting a value of the dependent variable (Y) for a given value of X. In the previous examples, we used a person's income, X, to predict the size of that person's home (Y). Notice in Figure 14.8 that we can also use the least squares line to estimate the *average (mean)* value of Y for a specified value of X. So we can use this line in two different situations.

Situation 1   The regression equation $\hat{Y} = b_0 + b_1 X$ estimates the **mean** value of Y for a specified value of the independent variable, X. For $X = x_0$, this value would be written $\mu_{Y|x_0}$ (the mean of Y given $X = x_0$).

For example, the least squares line passing through the real-estate data in Example 14.1 is $\hat{Y} = 8.51 + .354X$. The average square footage for *all* homes in the population with an income of \$40,000 ($X = 40$) is $\mu_{Y|40}$. Its estimate is provided by the corresponding value on the least squares line, namely,

$$\hat{Y} = 8.51 + .354(40) = 22.67$$

So the estimate of the average square footage of all such homes is 2267 square feet (Figure 14.7).

Situation 2   An **individual** predicted value of Y also uses the regression equation $\hat{Y} = b_0 + b_1 X$ for a specified value of X. This value of Y is denoted by Y_{x_0} for $X = x_0$. This application is the more common one in business, because a regression equation is generally used for individual forecasts.

For example, assume the Jenkins family resides in our sample community and has an income of \$40,000. A prediction of their home size (Y_{40}) is also

$$\hat{Y} = 8.51 + .354(40) = 22.67$$

which is 2267 square feet (Figure 14.7).

We see that the least squares line can be used to estimate average values (situation 1) or predict individual values (situation 2). Since $\mu_{Y|40}$ is a parameter, we use $\hat{Y}$ to estimate this value. On the other hand, Y_{40} represents a particular value of a dependent (random) variable, and so $\hat{Y}$ is used to predict this value. In the first situation, we can determine a confidence interval for $\mu_{Y|40}$; in the second situation, we determine a prediction interval for Y_{40}.

Confidence Interval for $\mu_{Y|x_0}$ (Situation 1)

We have already established that the point estimate of $\mu_{Y|x_0}$ is the corresponding value of $\hat{Y}$. The reliability of this estimate depends on (1) the number of observations in the sample, (2) the amount of variation in the sample, and (3) the value of $X = x_0$. A confidence interval for $\mu_{Y|x_0}$ takes all three factors into consideration.

A $(1 - \alpha) \cdot 100\%$ confidence interval for $\mu_{Y|x_0}$ is

$$\hat{Y} - t_{\alpha/2, n-2} s \sqrt{\frac{1}{n} + \frac{(x_0 - \bar{x})^2}{SS_X}} \quad \text{to}$$

$$\hat{Y} + t_{\alpha/2, n-2} s \sqrt{\frac{1}{n} + \frac{(x_0 - \bar{x})^2}{SS_X}} \quad \textbf{(14.24)}$$

EXAMPLE 14.8   Determine a 95% confidence interval for the average home size of families with an income of \$35,000, using the real-estate data from Example 14.1.

SOLUTION   We previously determined that $n = 10$, $\bar{x} = 39.8$, $SS_x = 1489.6$, and $s = 3.078$. The point estimate for the average square footage, $\mu_{Y|35}$, is

$$\hat{Y} = 8.51 + .354(35)$$
$$= 20.90(2090 \text{ square feet})$$

Obtaining $t_{.025,8} = 2.306$ from Table A.5, the 95% confidence interval for $\mu_{Y|35}$ is

$$20.90 - (2.306)(3.078)\sqrt{\frac{1}{10} + \frac{(35 - 39.8)^2}{1489.6}} \quad \text{to} \quad 20.90 + (2.306)(3.078)\sqrt{\frac{1}{10} + \frac{(35 - 39.8)^2}{1489.6}}$$

$$= 20.90 - (2.306)(3.078)(.340) \quad \text{to} \quad 20.90 + (2.306)(3.078)(.340)$$
$$= 20.90 - 2.41 \quad \text{to} \quad 20.90 + 2.41$$
$$= 18.49 \quad \text{to} \quad 23.31$$

We are thus 95% confident that the average home size for families earning $35,000 is between 1849 and 2331 square feet.   ∎

Using MINITAB to Construct Confidence Intervals   The MINITAB solution for the real-estate problem is contained in Figure 14.13. To construct confidence intervals, the column of interest is labeled as Stdev. Fit, which, when translated, means the standard deviation of the predicted Y. Writing this as $s_{\hat{Y}}$,

$$s_{\hat{Y}} = s\sqrt{\frac{1}{n} + \frac{(x_0 - \bar{x})^2}{SS_X}} \qquad (14.25)$$

For each value of X in the sample (say, x_0), the corresponding confidence interval for $\mu_{Y|x_0}$ is

$$\hat{Y} - t \cdot s_{\hat{Y}} \quad \text{to} \quad \hat{Y} + t \cdot s_{\hat{Y}}$$

where $t = t_{\alpha/2, n-2}$, as before, and $\hat{Y}$ is contained in the column to the left of the standard deviations.

Using the MINITAB output in Figure 14.13, we can find the confidence intervals corresponding to X values of 22, 40, and 60. The remaining seven confidence intervals are constructed in a similar manner.

For $X = 22$, the confidence interval is

$$16.300 - (2.306)(1.721) \quad \text{to} \quad 16.300 + (2.306)(1.721) = 12.33 \quad \text{to} \quad 20.27$$

For $X = 40$, the confidence interval is

$$22.671 - (2.306)(.974) \quad \text{to} \quad 22.671 + (2.306)(.974) = 20.42 \quad \text{to} \quad 24.92$$

For $X = 60$, the confidence interval is

$$29.749 - (2.306)(1.882) \quad \text{to} \quad 29.749 + (2.306)(1.882) = 25.41 \quad \text{to} \quad 34.09$$

Notice that the confidence intervals are much wider for $X = 22$ and $X = 60$ than for $X = 40$.

An easier method of determining confidence intervals using MINITAB is to use the PREDICT subcommand, as illustrated in Figure 14.17. Notice that the confidence intervals (labeled **95% C.I.**) agree with the previous results for $X = 22, 40$, and 60. The word PREDICT can be followed by a single number or a column (as in Figure 14.17). For values of X not in the sample, use the PREDICT subcommand, followed by this value (such as PREDICT 38.). For multiple values of X not in the sample, we suggest you store these values in a new column (say, C3) and

■ **FIGURE 14.17**
95% confidence
and prediction intervals
for data entered in
Figure 14.13.

```
NAME C1 = 'INCOME' C2 = 'FOOTAGE'
MTB > BRIEF 1
MTB > REGRESS Y IN C2 USING 1 PREDICTOR IN C1;
SUBC> PREDICT C1.

The regression equation is
FOOTAGE = 8.51 + 0.354 INCOME

Predictor        Coef        Stdev      t-ratio          p
Constant        8.514        3.320         2.56      0.033
INCOME        0.35392      0.07976         4.44      0.002

s = 3.078        R-sq = 71.1%      R-sq(adj) = 67.5%

Analysis of Variance

SOURCE        DF          SS          MS          F          p
Regression     1       186.59      186.59      19.69      0.002
Error          8        75.81        9.48
Total          9       262.40


    Fit   Stdev.Fit         95% C.I.             95% P.I.
 16.300      1.721   ( 12.329, 20.271)   (  8.165, 24.436)  ←⎡X = 22⎤
 17.716      1.469   ( 14.326, 21.105)   (  9.848, 25.584)
 24.440      1.058   ( 22.000, 26.881)   ( 16.932, 31.949)
 21.609      0.999   ( 19.305, 23.913)   ( 14.144, 29.074)
 18.424      1.354   ( 15.300, 21.547)   ( 10.666, 26.181)
 26.210      1.269   ( 23.284, 29.136)   ( 18.530, 33.890)
 28.334      1.618   ( 24.602, 32.065)   ( 20.312, 36.355)
 20.547      1.078   ( 18.061, 23.033)   ( 13.024, 28.071)
 29.749      1.882   ( 25.407, 34.091)   ( 21.426, 38.072)  ←⎡X = 60⎤
 22.671      0.974   ( 20.425, 24.917)   ( 15.223, 30.118)  ←⎡X = 40⎤
```

use the PREDICT subcommand, followed by this new column (as is done in Figure 14.17).

By connecting the upper end of the confidence intervals for all ten data points and connecting the lower limits, we obtain Figure 14.18. Equation 14.23 indicates that the confidence interval is narrowest when $(x_0 - \bar{x})^2 = 0$, that is, at $X = x_0 = \bar{x}$. For values of X to the left or right of $\bar{x}$, the confidence interval is wider. *In other words, the farther x_0 is from $\bar{x}$, the less reliable is the estimate.*

The Danger of Extrapolation   Extrapolation is the process of calculating an estimate corresponding to a value of X outside the range of the data used to derive the prediction equation (the least squares line). For example, in Figure 14.18 the least

■ **FIGURE 14.18**
95% confidence
intervals for the real-
estate data derived from
MINITAB output shown
in Figure 14.13.

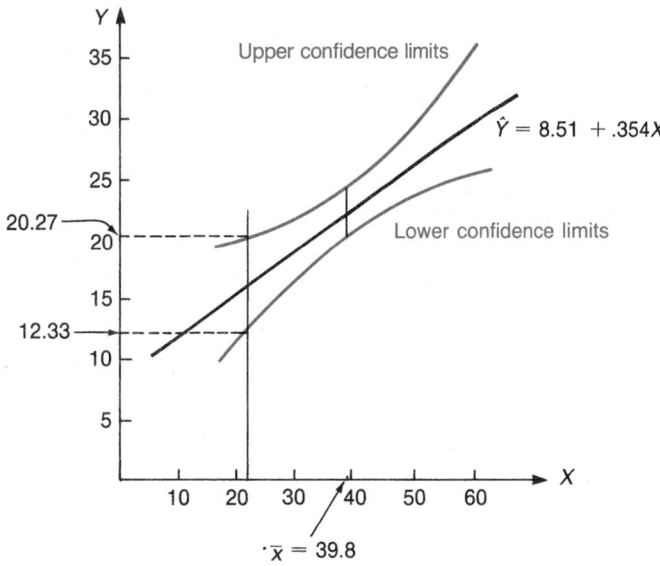

squares line could be used to estimate the average home size for families with an income of \$100,000. Although we *can* estimate $\mu_{Y|100}$, the corresponding confidence interval for this parameter will be extremely wide, so the point estimate, $\hat{Y}$, will have little practical value.

To use the simple regression model effectively for estimation, you need to stay within the range of the sampled values for the independent variable, X. This process is called **interpolation.** If you use values outside this range, you need to be aware that given *another* set of data, you would quite likely obtain a considerably different estimate. Furthermore, you have no assurance that the linear relationship still holds outside the range of your sample data.

Prediction Interval for Y_{x_0} (Situation 2)

The procedure of predicting individual values is used more often in business applications. The regression equation is generally used to **forecast** (predict) a future value of the dependent variable for a particular value of the independent variable. When attempting to predict a single value of the dependent variable, Y, using the simple linear regression model, we begin, as before, with $\hat{Y}$. Substituting $X = x_0$ into the regression equation provides the best prediction of Y_{x_0}. For example, if the Johnson family has an income of \$35,000, our best guess as to their home size (using this particular model) is $\hat{Y}$ for $X = 35$. From the results of Example 14.8, this is 20.90, or 2090 square feet.

We do not use the term *confidence interval* for this procedure because what we are estimating (Y_{x_0}) is not a parameter. It is a value of a random variable, so we use the term **prediction interval.**

The variability of the error in predicting a single value of Y is more than that for estimating the average value of Y (situation 1). It can be shown that an estimate of the variance of the error ($Y - \hat{Y}$), when using $\hat{Y}$ to predict an individual Y for $X = x_0$, is*

$$s_{\hat{Y}}^2 = s^2 \left(1 + \frac{1}{n} + \frac{(x_0 - \bar{x})^2}{SS_X} \right) \qquad \textbf{(14.26)}$$

This result can be used to construct a $(1 - \alpha) \cdot 100\%$ prediction interval for Y_{x_0}, as follows:

$$\hat{Y} - t_{\alpha/2, n-2} s \sqrt{1 + \frac{1}{n} + \frac{(x_0 - \bar{x})^2}{SS_X}} \quad \text{to}$$

$$\hat{Y} + t_{\alpha/2, n-2} s \sqrt{1 + \frac{1}{n} + \frac{(x_0 - \bar{x})^2}{SS_X}} \qquad \textbf{(14.27)}$$

Notice that the only difference between this prediction interval and the confidence interval in equation 14.24 is the inclusion of "$1 +$" under the square root sign. The other two terms under the square root are usually quite small, so this "$1 +$" has a large effect on the width of the resulting interval. Be aware that our warning about extrapolating outside the range of the data applies here as well. In equations 14.26 and 14.27, the distance from the mean ($x_0 - \bar{x}$) is squared, which increases the risk of predicting beyond the range of the sampled data.

---

*This follows since $Y = \hat{Y} + e$ and, as a result, $s_{\hat{Y}}^2 = s_{\hat{y}}^2 + s_e^2$. Substituting equation 14.25 for $s_{\hat{y}}^2$ and s^2 for s_e^2 produces the desired result.

EXAMPLE 14.9

We previously determined that if the Johnson family has an income of $35,000, the best prediction of their home size is $\hat{Y} = 20.90$. Determine a 95% prediction interval for this situation.

SOLUTION

We can use the calculations from Example 14.8 to derive the prediction interval for Y_{35}. The result is

$$20.90 - (2.306)(3.078)\sqrt{1 + \frac{1}{10} + \frac{(35 - 39.8)^2}{1489.6}} \quad \text{to}$$

$$20.90 + (2.306)(3.078)\sqrt{1 + \frac{1}{10} + \frac{(35 - 39.8)^2}{1489.6}}$$

$$= 20.90 - (2.306)(3.078)(1.056) \quad \text{to} \quad 20.90 + (2.306)(3.078)(1.056)$$

$$= 20.90 - 7.49 \quad \text{to} \quad 20.90 + 7.49$$

$$= 13.41 \quad \text{to} \quad 28.39$$

Comparing this interval to the confidence interval for $\mu_{Y|35}$ in Example 14.8, we see that individual predictions are considerably less accurate than estimations for the mean home size. Of course, we could reduce the width of this interval by obtaining additional data. Expecting accurate results from a sample of ten observations is being a bit optimistic. ∎

Using MINITAB for Constructing Prediction Intervals Prediction intervals can be determined without any calculations by using the PREDICT subcommand, as illustrated in Figure 14.17. From Figure 14.17, we observe the following prediction intervals:

$X = 22$: prediction interval for Y_{22} is from 8.165 to 24.436

$X = 40$: prediction interval for Y_{40} is from 15.223 to 30.118

$X = 60$: prediction interval for Y_{60} is from 21.426 to 38.072

Notice that these intervals are considerably wider than the corresponding confidence intervals.

Figure 14.19 shows the prediction intervals for all ten data points; the upper and lower limits have been connected. The increased width of a prediction interval

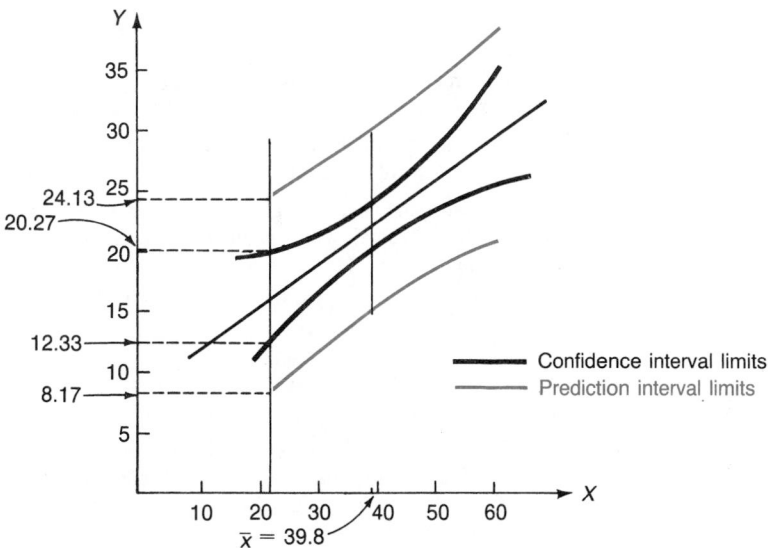

■ FIGURE 14.19

The 95% prediction and confidence intervals for the real-estate data.

versus a confidence interval is quite apparent from this graph. Also, like the width of a confidence interval, the width of a prediction interval increases as the value of X strays from $\bar{x}$.

Exercises

14.43 The marketing division of Astral Airlines wants to determine the relationship between the amount of money a person spends yearly on air transportation and the yearly income of the person. They randomly selected 20 airline passengers. The data (in thousands of dollars) are:

YEARLY EXPENDITURE (Y)	YEARLY INCOME (X)	YEARLY EXPENDITURE (Y)	YEARLY INCOME (X)
423	22.5	640	29.8
396	31.4	675	33.4
120	18.1	745	38.4
140	19.1	425	21.8
550	26.4	380	22.5
690	44.5	725	30.4
740	37.1	950	38.7
320	16.8	925	40.6
1200	50.5	210	18.8
470	21.8	425	21.7

a. Determine a 95% confidence interval for the average yearly airline expenditure for people with an annual income of $30,000.

b. Find a 95% prediction interval for the yearly airline expenditure of a person with an annual income of $30,000.

c. Interpret and compare the intervals in parts a and b.

14.44 A statistics instructor believes that there is a strong correlation between a student's grade in college algebra and a student's grade in introductory statistics. The following statistics were collected from 20 randomly selected students. Let X be the student's grade in college algebra. Let Y be the student's grade in introductory statistics (A = 4.0, B = 3.0, and so on).

$$SCP_{XY} = 31.8$$
$$SS_X = 32.3$$
$$SS_Y = 45.4$$
$$\bar{x} = 2.3$$
$$\bar{y} = 2.5$$
$$n = 20$$

a. Determine a 95% confidence interval for the average grade in introductory statistics for students who get a B (equal to 3.0) in college algebra.

b. Find the 95% prediction interval for the grade in introductory statistics for a student who obtained a B in college algebra.

14.45 Using the data in Exercise 14.23, find a 90% prediction interval for the amount of money an individual with an annual income of $40,000 places in tax-free investment instruments.

14.46 For the data in Exercise 14.26, find the 99% confidence interval for the average number of years that a lawn mower will be able to function properly if the number of hours of operation between oil changes is 23 hours. What is the standard deviation for the predicted value of the number of years that a single lawn mower will be able to perform adequately if the number of hours of operation between oil changes is 23?

14.47 For a fixed value of X, which interval is larger, the confidence interval for the mean value of Y at X (equation 14.24) or the prediction interval for a predicted value of Y at X (equation 14.27)? What value can you assign to X to achieve the smallest confidence interval for the mean value of Y at X or for the predicted value of Y at X?

14.48 A sample of 200 executives who work in Chicago was taken to find out how much of their own money the executives invest each year in stock of the company that they work for. The following regression equation was developed, where X is the income (in thousands) of an executive and Y is the amount of money (in thousands) he or she invests each year in the company. The prediction equation is

$$\hat{Y} = 9.5 + 0.05X$$

Based on the regression equation, can the following statement be made? A Chicago-area executive who earns $15,000 a year would invest about $10,250 in company stock. Comment.

14.49 The manager of an engineering firm believes that an employee's overall performance is related to the employee's score on Randall and Cantrell's job-aptitude test. Fourteen employees were selected randomly and the following data were collected. Job performance was rated by the manager on a one to ten scale, with ten being the score of a perfect employee.

EVALUATION OF JOB PERFORMANCE (Y)	SCORE ON APTITUDE TEST (X)	EVALUATION OF JOB PERFORMANCE (Y)	SCORE ON APTITUDE TEST (X)
9.5	90	4.5	59
6.0	71	5.5	63
2.0	33	7.0	84
6.5	49	9.0	94
7.5	82	8.5	83
5.0	61	6.0	74
7.5	73	7.5	51

a. Determine a 95% confidence interval for the average job performance evaluation of employees with a score of 75.

b. Find a 95% prediction interval for the job performance of an employee with a score of 75.

c. What score would give the smallest 95% prediction interval for an employee?

14.50 Using the following statistics, find a 95% confidence interval for the mean value of Y at $X = 6.0$.

$$\Sigma x = 137 \qquad \Sigma y = 253 \qquad n = 25$$
$$\Sigma xy = 1,609 \qquad \Sigma x^2 = 895 \qquad \Sigma y^2 = 2,943$$

14.51 For the data in Exercise 14.28, find a 99% prediction interval for the monthly weight loss of an individual rat that receives a daily dose of .30 milligrams of the diuretic.

14.52 The owner of a used-car lot would like to explain to potential car buyers the relationship between the horsepower rating of a car and the gasoline mileage. The owner collects the following data on twenty 1985 automobiles, with the objective of showing that horsepower rating gives a good indication of gasoline mileage.

GASOLINE MILEAGE (Y)	HORSEPOWER RATING (X)	GASOLINE MILEAGE (Y)	HORSEPOWER RATING (X)
16.0	180	16.8	195
14.9	195	20.0	290
14.1	160	14.8	150
18.5	235	13.0	100
15.3	175	20.3	290
21.9	285	18.2	255
15.2	210	17.1	220
18.2	230	19.8	235
17.0	235	17.5	260
16.9	200	21.4	275

a. Construct a scatter diagram of the data.

b.   Find the least squares line for the data.

c.   Test the hypothesis that there is no linear relationship between the variables Y and X at the .10 significance level.

 *Summary*

When dealing with a pair of variables (say, X and Y), we generally are interested in determining whether the variables are related in some manner. If a relationship does exist, perhaps the **independent** variable (X) can be used to predict values of the **dependent** variable (Y). If a significant linear relationship exists within the sample data, both the direction (positive or negative) and the strength of this linear relationship can be measured using the sample **coefficient of correlation,** r. The sample correlation coefficient is an estimate of the population coefficient of correlation, ρ. Another commonly used measure of association between two variables is the sample **covariance.**

Whenever a sample of bivariate data contains a significant linear pattern, we determine the **least squares line** through the data points and generate an equation that can be used to predict values of the dependent variable. To describe accurately the assumptions behind this procedure, we introduced the concept of a **statistical model** consisting of a **deterministic** portion (the straight line) and a **random error** component. This model can be written as $Y = \beta_0 + \beta_1 X + e$, where $\beta_0 + \beta_1 X$ is the deterministic component and e represents the error component. When we perform any test of hypothesis regarding the underlying bivariate population, we must be careful to satisfy the necessary assumptions behind this procedure. These assumptions will be examined more closely in Chapter 15.

By regressing Y on X we are able to determine the least squares line, $\hat{Y} = b_0 + b_1 X$. The value of b_0 is the **intercept** of the least squares line and estimates the population intercept β_0. The **slope** of the least squares line, b_1, estimates the population slope β_1. One question of interest is, if we regress X on Y (that is, switch the independent and dependent variables), can we rearrange the previous equation and say that $\hat{X} = (-b_0/b_1) + (1/b_1)Y$? The answer is no, although the coefficient of correlation, r, is the same in either case. Consequently, constructing a least squares line is not a good idea when it is not obvious which variable is the dependent variable.

Various methods for determining the utility of the model as a predictor of the dependent variable include: (1) a t test for detecting a significant slope, b_1 (a value of $\beta_1 = 0$ indicates that X has no predictive ability); (2) a t test for determining whether the sample correlation, r, is significantly large (a value of $\rho = 0$ indicates that there is no linear relationship between the two variables); and (3) a confidence interval for the slope, β_1. The two t tests appear to be quite different, but their computed values (and df) are *identical;* there is no point in performing both tests.

Another measure of how well the model provides estimates that fit the sample data is given by the **coefficient of determination,** r^2. For simple linear regression (one independent variable), this is the square of the correlation coefficient. Another definition of the coefficient of determination can also be used to examine more than one independent variable (called multiple linear regression), namely, $r^2 = 1 - (\text{SSE}/\text{SS}_Y)$. Here, SSE is the sum of squared errors and SS_Y represents the total variation in the sample Y values. For example, if $r^2 = .85$, then 85% of the variation in the sample Y values has been explained using this model.

The value of $\hat{Y}$ from the least squares regression line at a specific value of X (say, $X = x_0$), can be used to estimate an *average* value of Y, given this value of X (written $\mu_{Y|x_0}$). The value of $\hat{Y}$ centers a *confidence interval* for $\mu_{Y|x_0}$. Similarly, we can use the $\hat{Y}$ value to center a *prediction interval* for an individual value of the dependent variable, given this specific value of X (written Y_{x_0}). The value of $\hat{Y}$ can be used to *estimate* the value of $\mu_{Y|x_0}$ or *predict* the value of Y_{x_0}.

Summary of Formulas

Correlation between two variables

$$r = \frac{\text{SCP}_{XY}}{\sqrt{\text{SS}_x}\sqrt{\text{SS}_Y}}$$

where

$$\text{SCP}_{XY} = \Sigma xy - (\Sigma x)(\Sigma y)/n$$
$$\text{SS}_x = \Sigma x^2 - (\Sigma x)^2/n$$
$$\text{SS}_Y = \Sigma y^2 - (\Sigma y)^2/n$$

Least squares line

$$\hat{Y} = b_0 + b_1 X$$

where

$$b_1 = \text{SCP}_{XY}/\text{SS}_x$$

and

$$b_0 = \bar{y} - b_1 \bar{x}$$

Estimate of the residual variance

$$\hat{\sigma}_e^2 = s^2 = \frac{\text{SSE}}{n-2}$$

where

$$\text{SSE} = \Sigma(y - \hat{y})^2$$
$$= \text{SS}_Y - \frac{(\text{SCP}_{XY})^2}{\text{SS}_x}$$

t statistic for detecting a significant slope

$$t = \frac{b_1}{s_{b_1}}$$

(df $= n - 2$) where

$$s_{b_1} = s/\sqrt{\text{SS}_x}$$

Confidence interval for the slope, β_1

$$b_1 - t_{\alpha/2,n-2}s_{b_1} \quad \text{to} \quad b_1 + t_{\alpha/2,n-2}s_{b_1}$$

t statistic for detecting a significant correlation

$$t = \frac{r}{\sqrt{\dfrac{1 - r^2}{n - 2}}}$$

(df $= n - 2$)

Coefficient of determination

$$r^2 = \text{square of correlation coefficient}$$
$$= 1 - \frac{\text{SSE}}{\text{SS}_Y}$$

Confidence interval for the average value of Y at a specific value of X (say, x_0)

$$\hat{Y} \pm t_{\alpha/2,n-2}s\sqrt{\frac{1}{n} + \frac{(x_0 - \bar{x})^2}{\text{SS}_x}}$$

Prediction interval for a particular value of Y at a specific value of X (say, x_0)

$$\hat{Y} \pm t_{\alpha/2,n-2}s\sqrt{1 + \frac{1}{n} + \frac{(x_0 - \bar{x})^2}{\text{SS}_x}}$$

Review Exercises

14.53 Let X represent the sale price of a certain tool sold at a hardware store. Let Y represent the number of tools sold over a one-month period of time. Eight months were randomly selected over a two-year time frame to advertise the tool at various sale prices. The following statistics were compiled.

$$\Sigma x = 1022 \qquad \Sigma y = 157 \qquad \Sigma xy = 19,064$$
$$\Sigma x^2 = 136,938 \qquad \Sigma y^2 = 3,259 \qquad n = 8$$
$$\bar{x} = 127.75 \qquad y = 19.625$$

a. Obtain the least squares line for the data by regressing the number of tools sold on the sale price.

b. Find the coefficient of determination.

c. Find a 90% confidence interval for the slope of the regression line.

d. Is there sufficient evidence to conclude, at the .10 significance level, that there is a linear relationship between X and Y?

e. Find a 90% confidence interval for the mean of Y if the sale price is $120.

f. Find a 90% prediction interval for the number of tools sold if the sale price is $120.

g. What assumptions are necessary for the results of parts d, e, and f to be valid?

14.54 A manager at an electronics manufacturing plant wished to know if a salary pay structure would increase the productivity of the workers. The workers had been on an hourly pay scale. Eighteen workers were randomly selected and were placed on a salary pay structure. Times were recorded for these eighteen workers for an assembly-line task while the workers were under each pay structure.

WORKERS	TIME (IN HOURS) UNDER HOURLY PAY STRUCTURE (X)	TIME (IN HOURS) UNDER SALARY PAY STRUCTURE (Y)
1	3.3	2.8
2	2.7	2.3
3	3.8	3.4
4	3.5	3.4
5	3.9	3.4
6	4.1	3.6
7	3.2	2.9
8	3.5	3.2
9	4.0	3.6
10	3.2	3.0
11	4.3	4.3
12	3.8	3.3
13	3.4	3.0
14	3.1	2.9
15	3.0	3.0
16	4.3	3.7
17	4.5	4.1
18	3.2	3.2

a.   Find the least squares line for the data, with X as the independent variable and Y as the dependent variable.

b.   Find the coefficient of determination.

c.   Find the 90% confidence interval for the slope of the regression line.

d.   Find the 90% confidence interval for the mean value of Y if X is equal to 3.5.

e.   Find the 90% prediction interval for Y if X is equal to 3.5.

14.55 Dolls-R-Us believes that television advertising is the most effective way to market their new line of dolls. The sales manager recorded the amount of money spent on advertising and the amount of sales for 20 randomly selected months. The average cost for television advertising for the 20 months was $110,000. The average sales volume for the 20 months was $675,000. The following sample statistics were found from the data for the 20 months.

$$SCP_{XY} = 198.4$$

$$SS_X = 205.3$$

$$SS_Y = 341.6$$

where Y represents the sales volume (in thousands) and X represents the television advertising costs (in thousands of dollars).

a.   Calculate the least squares line.

b.   Calculate the coefficient of determination.

c.   Calculate the sum of squares of error.

d.   What is the estimate of the variance of the error component for the model?

e.   Is there sufficient evidence from the data to conclude at the .01 significance level that a positive relationship exists between X and Y?

f.   Find a 95% prediction interval for the monthly sales volume if the television advertising expenditure during one particular month is $120,000.

14.56 A car rental agency has a fleet of 200 cars available for rent at Kennedy airport in New York City. The owner of the agency uses a regression equation for estimating the company's daily revenue based on the number of incoming flights that day. The regression equation is $\hat{Y} = 2500 + 21.4X$, where X is the number of daily incoming flights and Y is the daily revenue in dollars. The data used to find the least squares line are based on a sample of 100 randomly selected days in 1984. Can the following statement be made based on regression analysis? If Kennedy airport increases its daily incoming flights by 50 flights next year, then the car agency can expect to make an additional daily revenue of $1,070. Comment.

14.57 Each week, a realtor advertises the houses she manages that are available for rent. The number of telephone calls from people inquiring about the advertisement were recorded for several weeks, during which various sizes of the advertisement were used. Is there sufficient evidence from the data below to conclude, at the .10 significance level, that a nonzero correlation exists?

X HEIGHT OF AD, INCHES)	Y (NUMBER OF INQUIRIES)	X (HEIGHT OF AD, INCHES)	Y (NUMBER OF INQUIRIES)
0.5	3	2.5	10
1.0	4	3.0	14
1.5	6	3.5	12
2.0	5	4.0	18

14.58 The manager of a firm that specializes in assisting individuals in filling out federal income tax forms obtained data from the Internal Revenue Service pertaining to deductions for charitable contributions. The following table provides a distribution of charitable contributions for eight groups with different adjusted gross incomes.

MEDIAN ADJUSTED GROSS INCOME (IN THOUSANDS OF DOLLARS) (X)	PERCENTAGE IN GROUP MAKING CHARITABLE CONTRIBUTIONS (CLAIMING ITEMIZED DEDUCTIONS) (Y)
5.0	17.0
7.5	36.0
12.5	40.5
17.5	38.5
25.0	29.2
40.0	14.0
75.0	4.2
100.0	1.5

a. Obtain the least squares line for these data.

b. Identify the values of the intercept, the slope, and the variance of the error for the simple linear regression model.

c. Find the residuals for all the Y values.

d. If the correlation between X and Y above was very strong, would it then be correct to conclude that an increase in income causes people to become less charitable?

14.59 The following data were collected for a certain regression analysis.

X	Y
.2	1.03
1.1	2.21
.5	1.20
.3	1.15
−3.5	13.1
2.4	7.5
.9	1.90
5.0	27.0
−4.2	16.5
8.0	64.1
6.1	38.1
−2.0	5.1
7.4	53.5
6.5	44.3
9.5	91.0

a. Obtain the least squares line by regressing Y on X.

b. Obtain the least squares line of Y regressed on X^2.

c. Find the coefficient of determination for the regression equations in parts a and b.

d. Plot the regression lines from parts a and b with the data. Comment on the adequacy of the two regression lines.

14.60 One management policy is based on the hypothesis that the more productive a worker is, the more satisfied the worker will be. A scale from one to ten is used to measure productivity, with ten being assigned to an extremely productive worker. A second scale from one to ten is used to measure satisfaction. The worker assigns him- or herself a ten if he or she is satisfied in every aspect of the job. Twenty employees were selected randomly from the production-and-research department of Tellon Oil. The results of the data collection are as follows:

SATISFACTION (Y)	PRODUCTIVITY (X)	SATISFACTION (Y)	PRODUCTIVITY (X)
5	4	9	7
2	3	7	5
9	8	4	4
9	9	8	7
5	6	9	8
3	5	10	9
5	4	5	6
7	7	1	2
9	8	7	8
2	3	9	9

a. Draw a scatter diagram of the data.

b. Test the hypothesis, at a 5% significance level, that productivity does not positively influence a worker's satisfaction.

c. Find a 99% confidence interval for the slope of the regression equation.

d. Calculate the coefficient of determination.

e. Find a 99% prediction interval for the satisfaction of a particular worker whose measure of productivity is 7.

14.61 A certain risk-averse investor calculates the beta for a stock before investing in the stock. By regressing the weekly percent return of, say, stock XYZ on the weekly percent return of the Standard and Poor's 500 Index (S&P 500), the investor can determine the stock's beta, which is equal to the slope of the regression line. The following data represent the weekly return over 20 weeks for both the S&P 500 and stock XYZ.

WEEK	S&P 500 (IN PERCENT) (X)	STOCK XYZ (IN PERCENT) (Y)	WEEK	S&P 500 (IN PERCENT) (X)	STOCK XYZ (IN PERCENT) (Y)
1	.51	.95	11	−1.12	−1.13
2	.22	.66	12	−.80	−.74
3	−.43	−.21	13	1.55	2.32
4	−2.51	−3.00	14	2.34	3.34
5	3.05	4.11	15	−.50	−.40
6	.40	.75	16	2.81	4.00
7	−.21	.01	17	3.33	4.63
8	1.80	2.64	18	−1.64	−1.83
9	2.55	4.51	19	1.75	2.58
10	3.80	6.12	20	2.20	3.11

a. Calculate the least squares line.

b. A slope greater than one indicates that the stock is more volatile than the S&P market index. Interpret the coefficients of the regression equation.

c. Graph the data and the least squares line. Comment on the fit.

14.62 Diane's Beauty Salon is currently hiring beauticians at its new location in a popular mall. Diane wants to know what percentage of commission to pay the beauticians based on experience. A survey of 12 licensed beauticians was taken with the following results.

PERCENTAGE OF COMMISSION (Y)	YEARS OF EXPERIENCE (X)	PERCENTAGE OF COMMISSION (Y)	YEARS OF EXPERIENCE (X)
24	2	25	4
18	1	44	12
30	5	33	8
41	10	24	3
35	8	20	1
35	7	40	10

a. Find the least squares line.

b. Calculate the sum of squares due to error.

c. Test the null hypothesis that there is no linear relationship between years of experience and percentage of commissions paid. Use a significance level of .05.

d. Find a 90% confidence interval for the slope of the least squares line.

14.63 A regression line is fitted to a set of data and the values of $Y - \hat{Y}$ are calculated. Does the following set of sample errors appear to conform to the empirical rule that 95% of the data should lie within two standard deviations of the mean? Construct a histogram for the residuals. Comment on the slope of the histogram.

1.1, -0.8, 2.6, 1.5, 0.2, -0.4, 0.8, -1.8, -2.3, 0.9, -2.7, 3.1, -1.0, 0.9, 4.5, -3.4, -0.1, -0.2, 2.4, -1.7, -0.7

14.64 The average list price of starter homes and the average list price of executive homes are given below for several cities across the United States.

CITY	STARTER HOME LIST PRICE (Y)	EXECUTIVE HOME LIST PRICE (X)
Atlanta	$ 74,000	$190,000
Boston	125,000	300,000
Chicago	79,200	421,000
Concord, New Hampshire	101,740	177,240
Dallas	80,000	180,000
Denver	84,000	155,000
Indianapolis	49,800	135,800
Los Angeles	223,125	498,750
New York	220,000	600,000
Orlando	72,000	155,000
Philadelphia	85,000	200,000
St. Louis	58,900	159,000
San Diego	180,000	291,800
San Francisco	250,000	430,000
Seattle	120,000	225,000

(*Source:* Adapted from "Your Home," *Changing Times* (February 1991): 22–24.)

a. Plot the values of X and Y. Does a least squares line appear to be a good fit, from eyeballing the scatter plot?

b. Calculate the correlation coefficient.

c. Find the least squares line.

14.65 Total personal taxes (in billions) and total personal income (in billions) are listed below for the years 1975 to 1989 for the United States.

YEAR	PERSONAL INCOME (X)	PERSONAL TAXES (Y)
1975	1,265.0	168.9
1976	1,391.2	196.8
1977	1,540.4	226.4
1978	1,732.7	258.7
1979	1,951.2	301.0
1980	2,165.3	336.5
1981	2,429.5	387.7
1982	2,584.6	404.1
1983	2,838.6	410.5
1984	3,108.7	440.2
1985	3,325.3	486.6
1986	3,526.2	512.9
1987	3,776.6	571.7
1988	4,070.8	591.6
1989	4,384.3	658.8

(Source: The World Almanac and Book of Facts (1991): 111.)

a. Find the least squares equation for predicting personal taxes from personal income.

b. Is there sufficient evidence to conclude that personal taxes are positively correlated with personal income? Use a .10 significance level.

c. Find a 90% prediction interval for personal taxes when personal income is equal to 3,000.

d. Find a 90% confidence interval for the mean amount of tax when personal income is 3,000.

14.66 Twenty employees were asked to rate on a continuous scale from 1 to 7 (1 = strongly disagree to 7 = strongly agree) their feelings about several statements that measured job satisfaction, variety (the number of different and new tasks required of an employee), autonomy (freedom to make decisions), and task significance (the meaningfulness and importance of the tasks required of an employee). The results are shown in the MINITAB output below.

EMPLOYEE	JOB SATISFACTION	VARIETY	AUTONOMY	TASK SIGNIFICANCE
1	4.6	3.5	4.7	5.2
2	3.3	4.8	3.1	4.7
3	6.8	5.0	7.0	7.0
4	6.2	4.5	6.7	7.0
5	5.5	6.5	6.0	6.4
6	5.0	4.5	5.4	7.0
7	3.4	2.5	3.2	3.4
8	5.7	6.8	6.5	6.4
9	5.5	4.0	6.0	6.3
10	4.0	5.5	4.0	4.3
11	5.0	4.0	5.5	5.3
12	3.5	4.8	3.5	5.0
13	6.0	5.0	6.5	7.0
14	4.0	5.6	3.9	4.2
15	5.0	3.0	5.4	5.5
16	4.5	5.5	4.8	5.0
17	5.5	4.0	5.8	6.4
18	5.5	6.5	6.5	6.0
19	5.7	4.0	6.0	6.7
20	4.6	4.5	5.0	6.2

```
MTB > name c1 'person'
MTB > name c2 'satisfac'
MTB > name c3 'variety'
MTB > name c4 'autonomy'
MTB > name c5 'task sig'
```

```
MTB > correlations for c2-c5

             satisfac  variety autonomy
     variety   0.208
     autonomy  0.983     0.276
     task sig  0.873     0.211     0.874
```

a. Interpret the values of the correlation coefficients shown in the above printout.

b. Which single variable would be the best predictor of job satisfaction?

c. Which correlations are significantly different from zero? Use a .05 significance level.

14.67 A realtor is interested in the relationship between the initial listing price of a single-family house in a large residential subdivision and the final selling price. A random selection of 16 houses is chosen from a list of homes that have been sold in the past six months. The units are in thousands of dollars. C1 contains the final selling price, and C2 contains the initial listing price in the following MINITAB printout.

```
MTB > name c1 'final'
MTB > name c2 'initial'
MTB > print c1 c2
 ROW   final  initial

   1   125.5   130.1
   2   115.8   121.5
   3   105.0   110.3
   4   122.9   127.5
   5    92.0    95.0
   6   101.5   105.9
   7    88.1    99.5
   8    92.1    93.0
   9   117.0   121.3
  10   112.5   118.0
  11   101.0   105.0
  12   107.5   110.5
  13   103.0   106.9
  14   113.4   117.0
  15   121.0   125.5
  16   115.0   119.0

MTB > regress y in c1 using 1 predictor in c2

The regression equation is
final = - 3.04 + 0.987 initial

Predictor      Coef      Stdev     t-ratio
Constant     -3.038      5.725      -0.53
initial      0.98666    0.05048     19.54

s = 2.228      R-sq = 96.5%      R-sq(adj) = 96.2%

Analysis of Variance

SOURCE       DF         SS          MS
Regression    1      1896.6      1896.6
Error        14        69.5         5.0
Total        15      1966.1

Unusual Observations
Obs. initial    final     Fit Stdev.Fit  Residual  St.Resid
  7      100    88.100  95.135     0.875    -7.035    -3.43R

R denotes an obs. with a large st. resid.
```

a. What is the predicted selling price of a house that is listed initially for $115 thousand?

b. Does the initial listing price contribute to the prediction of the final selling price? Use a .05 significance level.

c. Find a 95% confidence interval for the slope of the regression equation.

d. If observation 7 were removed from the data, do you think the coefficient of determination would improve? Try it.

14.68 An economist was analyzing the relationship between the price (Y) and the supply (X) of a certain product. As the supply diminished, the price increased rapidly. Twelve data points were collected from historical data. A plot of Y and X is given, followed by a plot of Y and $1/X$. The variable Y is in dollars and the variable X has been coded. Comment on the adequacy of the relationship between $1/X$ and Y. Interpret the 95% prediction interval for the price at $X = 3.0$ and the 95% confidence interval for the mean price at $X = 3.0$.

```
MTB > name c1 'supply'
MTB > name c2 'price'
MTB > plot c2  c1

price   -     *
        -
        -
   3.00+     *
        -       *
        -          *
        -
        -
   2.40+
        -
        -
        -             2
        -               *
   1.80+                  *    *
        -
        -                           *
        -                              *
        -                                   *
   1.20+
        ----+---------+---------+---------+---------+---------+---supply
          1.5       3.0       4.5       6.0       7.5       9.0
```

```
MTB > let c3 = 1/c1
MTB > name c3 '1/supply'
MTB > plot c2. c3

price   -                                              *
        -
        -
   3.00+                                           *
        -                                    *
        -                                 *
        -
        -
   2.40+
        -
        -
        -                          **
        -                       *
   1.80+                  *    *
        -
        -              *
        -           *
        -         *
   1.20+
        +---------+---------+---------+---------+---------+------1/supply
      -0.00      0.15      0.30      0.45      0.60      0.75
```

```
MTB > regress c2 on 1 predictor in c3;
SUBC> predict y value for reciprocal of x equal to .333.

The regression equation is
price = 1.05 + 3.12 1/supply

Predictor       Coef        Stdev       t-ratio
Constant      1.0494       0.1052          9.98
1/supply      3.1207       0.2585         12.07

s = 0.1808      R-sq = 93.6%      R-sq(adj) = 92.9%

Analysis of Variance

SOURCE         DF          SS           MS
Regression      1       4.7645       4.7645
Error          10       0.3269       0.0327
Total          11       5.0914
```

```
Obs. 1/supply      price    Fit Stdev.Fit  Residual   St.Resid
  1    0.769      3.3000   3.4499   0.1195    -0.1499     -1.11
  2    0.667      3.0000   3.1299   0.0964    -0.1299     -0.85
  3    0.526      2.9000   2.6919   0.0687     0.2081      1.24
  4    0.417      2.8000   2.3497   0.0547     0.4503      2.61R
  5    0.357      2.0700   2.1639   0.0522    -0.0939     -0.54
  6    0.345      2.1000   2.1255   0.0522    -0.0255     -0.15
  7    0.303      1.9000   1.9951   0.0538    -0.0951     -0.55
  8    0.263      1.8000   1.8706   0.0571    -0.0706     -0.41
  9    0.200      1.7500   1.6735   0.0655     0.0765      0.45
 10    0.154      1.5000   1.5295   0.0733    -0.0295     -0.18
 11    0.127      1.4000   1.4444   0.0785    -0.0444     -0.27
 12    0.111      1.3000   1.3961   0.0815    -0.0961     -0.60

R denotes an obs. with a large st. resid.

    Fit   Stdev.Fit        95% C.I.            95% P.I.
  2.0886     0.0525    ( 1.9717, 2.2055)   ( 1.6690, 2.5081)
```

Computer Exercises Using the Database

Exercise 1 -- Appendix I From the database, select 50 random observations. Compute the sample correlation between the variable HPAYRENT (house payment or rent) and the variable UTILITY (monthly utility expenditure). Is there sufficient evidence to conclude that a positive correlation exists? Use a .05 significance level.

Exercise 2 -- Appendix I Select 50 random observations from the database. Plot the values of total indebtedness (TOTLDEBT) and total income (INCOME1 + INCOME2). Also plot the value of total indebtedness with house payment or apartment/house rent (HPAYRENT). Choose the graph that appears to have a more linear relationship between the variables graphed. Fit a least squares line to the data. Test if the predictor variable significantly contributes to the prediction of total indebtedness. Use a .10 significance level.

Exercise 3 -- Appendix J Select 50 random observations from the database. Compute the sample correlation between the variable NETINC (net income) and each of the variables SALES (gross sales) and COSTSALE (cost of sales). Select the variable from these two that has the higher correlation with net income. Regress net income on this variable and test if this variable significantly contributes to the prediction of net income. Use a .05 significance level.

Exercise 4 -- Appendix J Select 50 random observations from the database. Compute the regression line for predicting total assets (TOTAL) from current assets (ASSETS). Also compute the regression line for predicting total assets from current liabilities (LIABIL). Test for the adequacy of the fit of these two regression lines to the data. Use a .10 significance level. Which of these two regression lines has a higher value for the coefficient of determination, and what does the higher value indicate?

Case Study

The 1991 EPA Mileage Estimates

The United States Environmental Protection Agency (EPA) releases an annual ranking of fuel efficiency in automobiles. The EPA estimates these fuel economy ratings by testing the exhaust emissions of new cars with computerized equipment. The EPA is charged with this responsibility primarily to enable the federal government to monitor auto manufacturers' compliance with legislation mandating a minimum average fuel economy that a manufacturer's fleet of cars must reach. These fuel economies are known as the CAFE (Corporate Average Fuel Economy) standards. For the year 1991, the CAFE standards required each automaker to meet an aver-age fuel-efficiency rating of 27.5 miles per gallon (mpg).

The following table was adapted from EPA data published in *Consumer's Research* magazine. It shows some randomly selected models, along with their annual fuel cost in dollars based on 15,000 miles of driving, an estimated MPG rating based on highway driving, and engine capacity in liters. All figures apply to models with manual transmission.

Case Study Questions

1. Obtain a correlation matrix for annual fuel cost, estimated MPG, and engine capacity. Which two variables have the strongest relationship? Is this surprising?

2. a. Find the least squares regression equation to

CAR MODEL	ANNUAL FUEL COST	EST. MPG (HWY)	ENGINE CAPACITY
Ferrari Testarossa	1704	15	4.9
Ford Escort	669	31	1.8
Mazda 626	694	31	2.2
Nissan 300ZX	1088	24	3.0
Pontiac Grand Am	722	33	2.5
Geo Metro LSI	399	50	1.0
Toyota Camry	647	34	2.0
Porsche 944 S2	1088	26	3.0
BMW 850i	1339	19	5.0
Audi 80 Quatro	938	24	2.3
Chevrolet Beretta	805	33	2.3
Saturn SL	606	37	1.9
Eagle Summit	606	35	1.5
Chrysler LeBaron	816	29	3.0
Ford Taurus	1035	26	3.0
Honda Accord	722	29	2.2
Acura Integra	722	28	1.8
Hyundai Sonata	782	28	2.4
Lexus ES250	892	25	2.5
Toyota Tercel	606	35	1.5
Rover Sterling 827	938	23	2.7
Peugeot 405 Sedan	816	28	1.9
Oldsmobile Cutlass Calais	722	34	2.5
Mercedes-Benz 190E	990	27	2.6
Suzuki Swift	469	43	1.3
Lamborghini DB132/Diablo	1977	14	5.7
Cadillac Seville	1088	26	4.9

(*Source:* "Fuel Economy Comparisons: The 1991 EPA Mileage Estimates," *Consumer's Research* (November 1990): 15–22. Reprinted, with permission, from *Consumers' Research* magazine, Washington, D.C.)

predict average annual fuel cost (Y) using estimated MPG as the independent variable, X. Plot the two variables.

b. Interpret what r^2 tells us.

c. Determine whether this regression model represents a significant linear relationship between Y and X at a 5% significance level.

d. State the p-value for the test, and interpret what it means.

e. What is the variance of the error in the model?

3. Repeat all five parts of question 2 using engine capacity as the predictor of average annual fuel cost.

4. a. Which of the two models do you think is better at predicting annual fuel cost?

b. Would any of the following criteria indicate one model is better than another:

i. lower vs higher MSE?

ii. lower vs. higher MSR?

iii. lower vs. higher correlation?

iv. lower vs. higher coefficient of determination?

v. lower vs. higher regression coefficient (b_1)?

5. Suppose you had an automobile with an engine capacity of 1.6 liters and an EPA rating of 32 MPG. Using the better of the two models from the above analysis, find:

a. The point estimate for the predicted annual fuel cost for this car.

b. The 95% interval estimate of annual fuel cost for any individual car of this type.

c. The 95% interval estimate of mean annual fuel cost for all cars of this type.

SPSS

☐ **Solution**
Example 14.2

Example 14.2 was concerned with computing the regression equation $\hat{Y} = b_0 + b_1X$. The purpose of this problem was to determine the relationship between family incomes and the square footage of their homes. The SPSS program listing in Figure 14.20 was used to request the SSE, b_0 and b_1 values, as well as other statistics. The regression analysis should always include a scatter diagram of the sample data. In this problem the SPSS commands are the same for both the mainframe and PC versions (remember to end each command line with a period when using the PC version).

The TITLE command names the SPSS run.

The DATA LIST command gives each variable a name and describes the data as being in free form.

The BEGIN DATA command indicates to SPSS that the input data immediately follow.

The next 10 lines contain the data values, which represent square footage in hundreds of square feet and income in thousands of dollars. For example, the first card image (16 22) represents a family living in a 1,600 square foot home with an income of $22,000 per year.

The END DATA statement indicates the end of the data.

The REGRESSION statement defines the variables, SQFOOT and INCOME, as the regression variables, and specifies that SQFOOT is to be the dependent variable. The ENTER subcommand means that all independent variables are to be entered into the regression analysis. In this example, however, there is only one independent variable, INCOME.

The SCATTERPLOT subcommand requests a scatter diagram of the sample data with SQFOOT on the vertical axis and INCOME on the horizontal axis. Each variable in this plot is standardized; that is, the mean is subtracted from each value and the result is divided by the standard deviation. Standardized values (referred to as Z scores in Chapter 3) generally range from -3 to 3.

Figure 14.21 shows the SPSS output obtained by executing the listing in Figure 14.20.

■ **FIGURE 14.20**
Input for SPSS
or SPSS/PC.
Remove the periods
for SPSS input.

```
TITLE    REAL ESTATE EXAMPLE USING ONE PREDICTOR.
DATA LIST FREE/SQFOOT INCOME.
BEGIN DATA.
16 22
17 26
26 45
24 37
22 28
21 50
32 56
18 34
30 60
20 40
END DATA.
REGRESSION VARIABLES=SQFOOT,INCOME/DEPENDENT=SQFOOT/ENTER/
   SCATTERPLOT (SQFOOT,INCOME).
```

■ **FIGURE 14.21**
SPSS output.

```
         REAL ESTATE EXAMPLE USING ONE PREDICTOR

   * * * *   M U L T I P L E   R E G R E S S I O N   * * * *


Listwise Deletion of Missing Data

Equation Number 1    Dependent Variable..    SQFOOT

Beginning Block Number 1.  Method:  Enter

Variable(s) Entered on Step Number
   1..    INCOME
```

■ **FIGURE 14.21**
(continued)

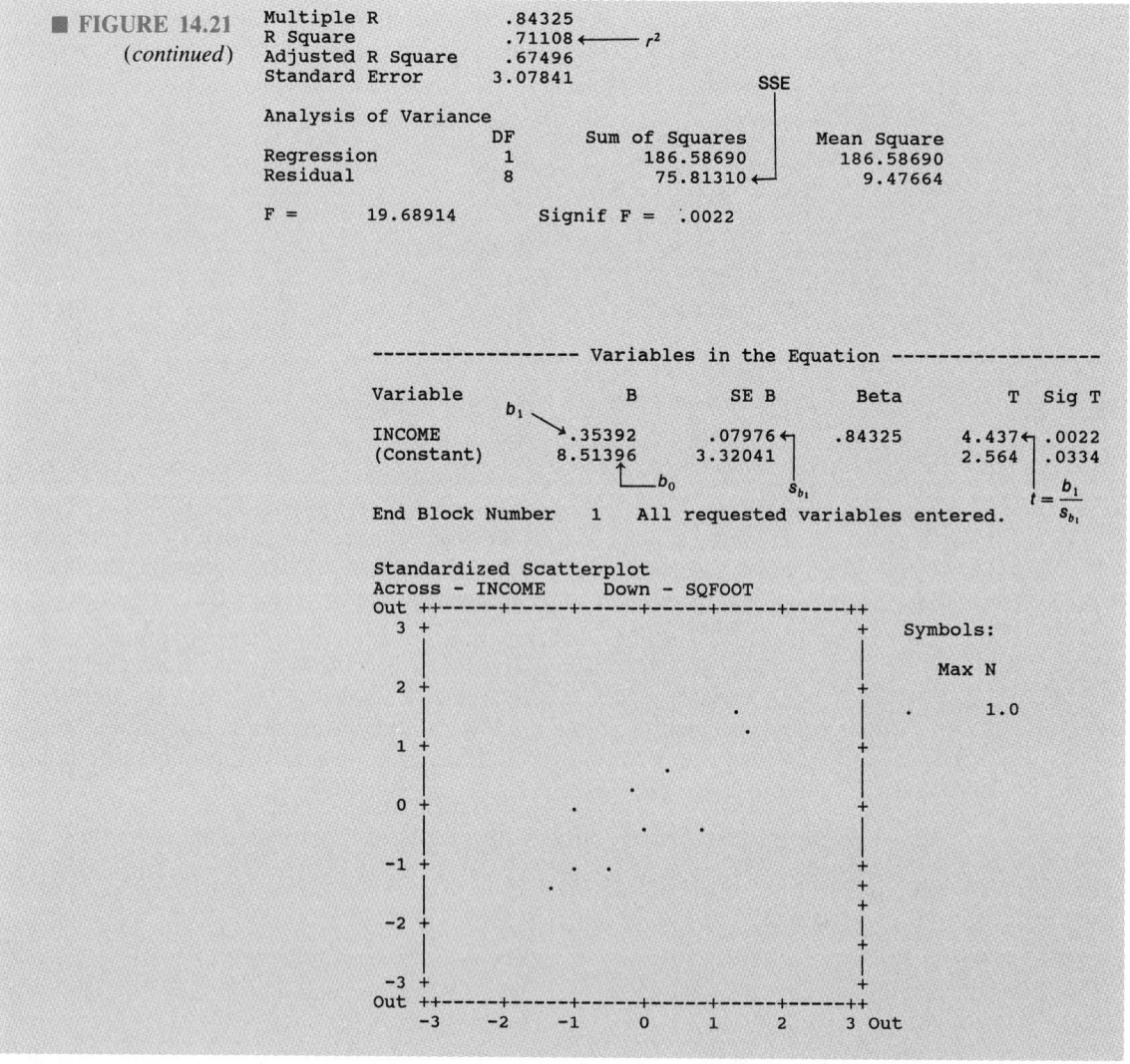

SAS

🖊 **Solution**
Example 14.2

Example 14.2 was concerned with computing the regression equation $\hat{Y} = b_0 + b_1 X$. The purpose of this problem was to determine the relationship between families' incomes and the square footage of their homes. The SAS program listing in Figure 14.22 was used to request the SSE, b_0 and b_1 values, as well as other statistics. The regression analysis should always include a scatter diagram of the sample data. In this problem the SAS commands are the same for both the mainframe and PC versions.

The TITLE command names the SAS run.

The DATA LIST command gives the data a name.

The INPUT command names and gives the correct order for the different fields on the data lines.

The CARDS command indicates to SAS that the input data immediately follow.

The next 10 lines contain the data values, which represent square footage in hundreds of square feet and income in thousands of dollars. For example, the first card image (16 22) represents a family living in a 1,600 square foot home with an income of $22,000 per year.

The PROC PLOT command requests a scatter diagram of the sample data, named in the previous DATA statement. The PLOT SQFOOT*INCOME statement will plot the SQFOOT values on the vertical axis and the IN-COME values on the horizontal axis.

The PROC REG command and MODEL subcommand indicate that SQFOOT and INCOME are the regression variables, with SQFOOT being the dependent variable and INCOME the independent variable.

Figure 14.23 shows the SAS output obtained by executing the listing in Figure 14.22.

FIGURE 14.22
Input for SAS
(mainframe or
micro version).

```
TITLE    'REAL ESTATE EXAMPLE USING ONE PREDICTOR';
DATA REALEST;
 INPUT SQFOOT INCOME;
CARDS;
16 22
17 26
26 45
24 37
22 28
21 50
32 56
18 34
30 60
20 40
PROC PLOT DATA = REALEST;
 PLOT SQFOOT*INCOME;
PROC REG;
 MODEL SQFOOT=INCOME;
```

FIGURE 14.23 SAS output.

REAL ESTATE EXAMPLE USING ONE PREDICTOR

Model: MODEL1
Dependent Variable: SQFOOT

Analysis of Variance

Source	DF	Sum of Squares	Mean Square	F Value	Prob>F
Model	1	186.58690	186.58690	19.689	0.0022
Error	8	75.81310	9.47664		
C Total	9	262.40000			

SSE $\rightarrow$ (pointing to Sum of Squares column for Error row 75.81310)

Root MSE	3.07841	R-square	0.7111 $\leftarrow r^2$
Dep Mean	22.60000	Adj R-sq	0.6750
C.V.	13.62130		

Parameter Estimates

Variable	DF	Parameter Estimate	Standard Error	T for H0: Parameter=0	Prob > |T|
INTERCEP	1	8.513963	3.32040910	2.564	0.0334
INCOME	1	0.353921	0.07976132	4.437	0.0022

b_0 b_1 s_{b_1} $t = \dfrac{b_1}{s_{b_1}}$ p-value (two-tailed)

■ FIGURE 14.23   (*continued*)

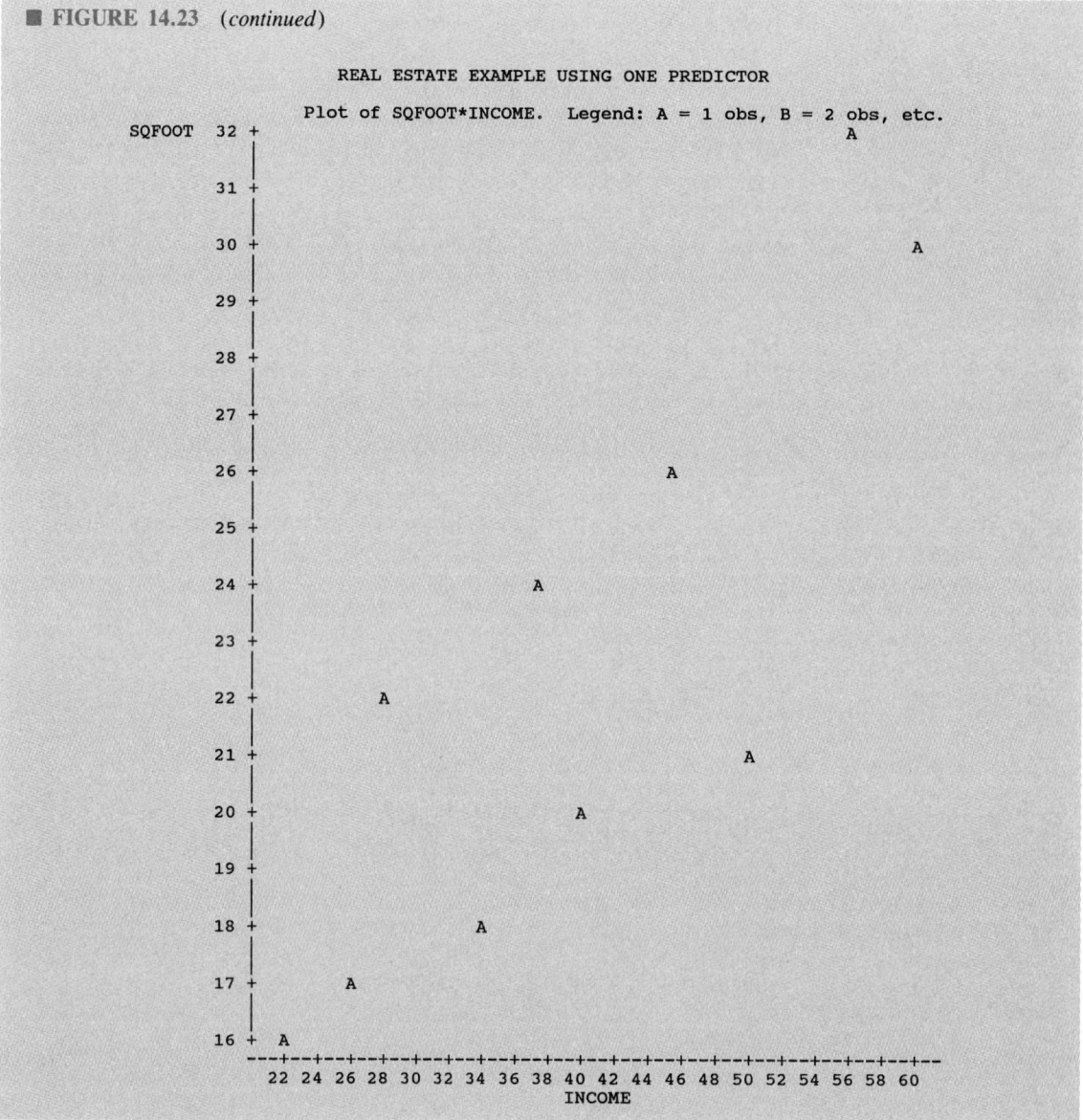

Multiple Linear Regression

We used the technique of simple linear regression in Chapter 14 to explain the behavior of a dependent variable using a single predictor (independent) variable. For example, we can attempt to explain the amount of new housing construction using the interest rate as a predictor variable.

To define this procedure in statistical terms, we introduced the concept of a statistical model. This model consists of two parts. The first part, the deterministic component, is assumed to be $Y = \beta_0 + \beta_1 X$ (a straight line), implying that the underlying pattern for the X and Y variables is linear. If a simple linear regression model is appropriate for the construction illustration, a scatter diagram of the new housing starts (Y) and the corresponding interest rates (X) should reveal a basic linear pattern. We never expect all the sample data to lie *exactly* on a straight line; we realize that with any statistical model there is error involved. This error makes up the random component. The actual model used for simple linear regression is $Y = \beta_0 + \beta_1 X + e$, where e represents the distance from the actual Y value to the line passing

through all X, Y values. The value of e is the error and represents the error component of the model. The assumptions behind the use of this model deal with the behavior of these error terms—are they normally distributed, centered at zero, with the same variance? Are they independent?

In the construction example, it seems reasonable to assume that the volume of housing construction is affected not only by the interest rate but also by many other factors (variables) as well, including cost of materials, geographic location, and unemployment rate in the area. We next look at statistical models that predict the dependent variable (such as Y = the number of new housing starts) as a function of *more than one* independent variable. The concept and assumptions are the same as before—now we are merely concerned with more than one predictor variable. When we include these additional variables, the predictive ability of the model should be significantly improved. This procedure is called multiple linear regression, and it is a very useful statistical technique.

▲ 15.1
The Multiple Linear Regression Model

Prediction Using More Than One Variable

To explain or predict the behavior of a certain dependent variable using more than one predictor variable, we use a **multiple linear regression** model. The form of this model is

$$Y = \beta_0 + \beta_1 X_1 + \beta_2 X_2 + \cdots + \beta_k X_k + e \qquad (15.1)$$

where $X_1, X_2, \ldots, X_k$ are the k independent (predictor) variables and e is the error associated with this model.

Notice that equation 15.1 is similar to the equation for the simple linear regression model, except that the *deterministic component* is now

$$\beta_0 + \beta_1 X_1 + \cdots + \beta_k X_k \qquad (15.2)$$

■ **FIGURE 15.1**
The multiple linear
regression model (two
independent variables).

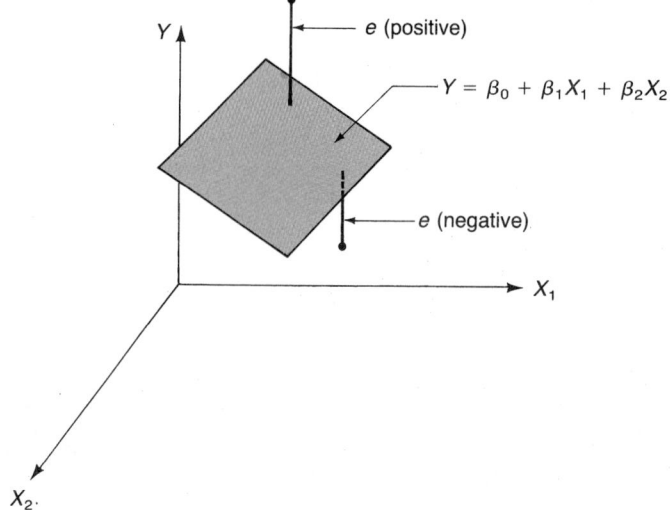

rather than $\beta_0 + \beta_1 X$. Once again the error term, e, is included to provide for deviations about the deterministic component.

What is the appearance of the deterministic portion in equation 15.2? In Chapter 14, where we discussed simple linear regression, the deterministic portion was a straight line. In the case of multiple linear regression, the deterministic portion is more difficult (usually impossible) to represent graphically. If your model contains two predictor variables, X_1 and X_2, the deterministic component becomes a plane, as shown in Figure 15.1. Consequently, the key assumption behind the use of this particular model is that the Y values will lie in this plane, *on the average*, for any particular values of X_1 and X_2.

In Chapter 14, we examined the relationship between the square footage of a home (Y) and the corresponding household income (X). The results were:

least squares line: $\hat{Y} = 8.51 + .354X$

correlation between X and Y: $r = .843$

coefficient of determination: $r^2 = .711$

significant linear relationship exists

We now want to include two additional variables in this model. The real-estate developer performing the study suspects that (1) larger families have larger homes and (2) the size of the home is affected by the amount of formal education (years of college) of the wage earner(s) in the home. We now have three independent variables:

$X_1 =$ annual income (thousands of dollars)

$X_2 =$ family size

$X_3 =$ combined years of formal education (beyond high school) for all household wage earners

The same ten families were used in the study, but data were collected on the two additional variables, X_2 and X_3.*

---

*A sample of size ten is unrealistically small in practice.

FAMILY	Y (HOME SQUARE FOOTAGE)	X_1 (INCOME)	X_2 (FAMILY SIZE)	X_3 (YEARS OF FORMAL EDUCATION)
1	16	22	2	4
2	17	26	2	8
3	26	45	3	7
4	24	37	4	0
5	22	28	4	2
6	21	50	3	10
7	32	56	6	8
8	18	34	3	8
9	30	60	5	2
10	20	40	3	6

The data configuration now has four columns (including Y) and ten rows (called *observations*). Our task is to use the data on all *three* variables (X_1, X_2, and X_3) to provide a better estimate of home size (Y).

The Least Squares Estimate   Using Figure 15.1, we proceed as we did for simple regression and determine an estimate of the β's that will make the sum of squares of the residuals as small as possible. A **residual** is defined as the difference between the actual Y value and its estimate; that is, $Y - \hat{Y}$. In other words, we attempt to find the b_0, b_1, . . . , b_k that minimize the sum of squares of error,

$$SSE = \Sigma(Y - \hat{Y})^2 \qquad \textbf{(15.3)}$$

where now $\hat{Y} = b_0 + b_1X_1 + b_2X_2 + \cdots + b_kX_k$ and b_0, b_1, . . . , b_k are called the **least squares estimates** of β_0, β_1, . . . , β_k.

By determining the estimated *regression coefficients* (b_0, b_1, . . . , b_k) that minimize SSE rather than $\Sigma(Y - \hat{Y})$, we once again avoid the problem of positive errors canceling out negative ones. Another advantage of this procedure is that, by means of a little calculus, we can show that a fairly simple expression exists for these sample regression coefficients. Because this expression involves the use of *matrix notation,* we omit this result.*

There is only one way to solve a multiple regression problem in practice, and that is with the help of a computer. All computer packages determine the values of b_0, b_1, . . . , b_k in the same way—namely, by minimizing SSE. As a result, these values will be identical (except for numerical rounding errors), regardless of which computer program you use.

In the example where we attempt to predict home size using the three predictor variables, the prediction equation is

$$\hat{Y} = b_0 + b_1X_1 + b_2X_2 + b_3X_3$$

where

$$\hat{Y} = \text{predicted home size}$$

$$X_1 = \text{income}$$

$$X_2 = \text{family size}$$

$$X_3 = \text{years of education}$$

and b_0, b_1, b_2 and b_3 are the least squares estimates of β_0, β_1, β_2, and β_3.

---

*Information on this expression is presented in W. Mendenhall and T. Sincich, *A Second Course in Business Statistics: Regression Analysis* 3d ed. (San Francisco: Dellen, 1989); J. Neter, W. Wasserman, and M. Kutner, *Applied Linear Regression Models* 2d ed. (Homewood, Ill.: Richard D. Irwin, 1989).

■ **FIGURE 15.2**
MINITAB multiple
regression solution to
house size using three
predictor variables.
See text.

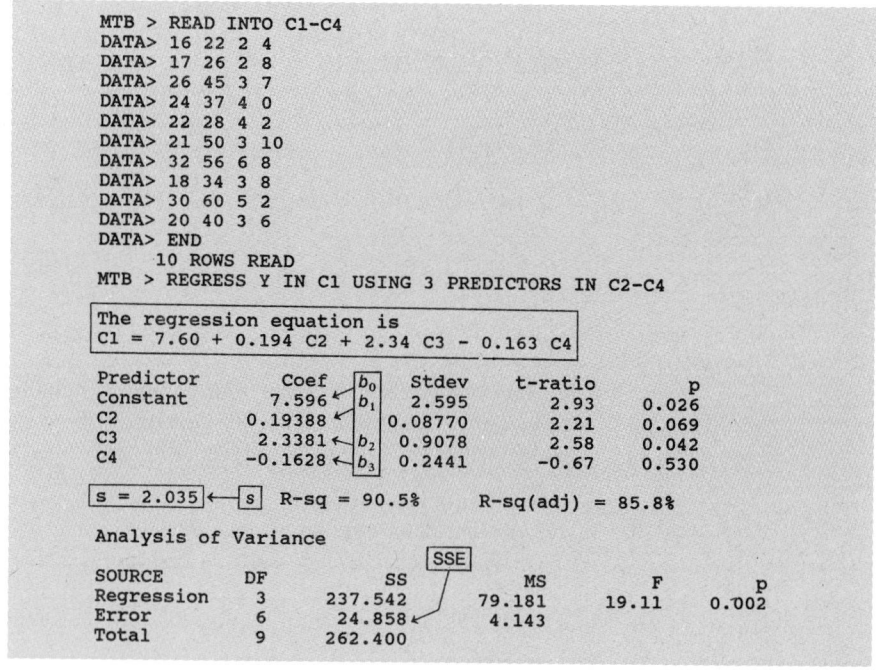

```
MTB > READ INTO C1-C4
DATA> 16 22 2 4
DATA> 17 26 2 8
DATA> 26 45 3 7
DATA> 24 37 4 0
DATA> 22 28 4 2
DATA> 21 50 3 10
DATA> 32 56 6 8
DATA> 18 34 3 8
DATA> 30 60 5 2
DATA> 20 40 3 6
DATA> END
      10 ROWS READ
MTB > REGRESS Y IN C1 USING 3 PREDICTORS IN C2-C4

The regression equation is
C1 = 7.60 + 0.194 C2 + 2.34 C3 - 0.163 C4

Predictor      Coef   b0    Stdev    t-ratio        p
Constant      7.596   b1    2.595       2.93    0.026
C2           0.19388        0.08770     2.21    0.069
C3            2.3381  b2    0.9078      2.58    0.042
C4           -0.1628  b3    0.2441     -0.67    0.530

s = 2.035   s   R-sq = 90.5%     R-sq(adj) = 85.8%

Analysis of Variance
                                SSE
SOURCE         DF       SS          MS        F        p
Regression      3   237.542     79.181    19.11    0.002
Error           6    24.858      4.143
Total           9   262.400
```

Figure 15.2 contains the MINITAB solution using the data we presented. According to this output, the best prediction equation (in the least squares sense) for home size is

$$\hat{Y} = 7.60 + .194X_1 + 2.34X_2 - .163X_3$$

So this solution minimizes SSE. But what is the SSE here? We need to determine how well this equation "fits" the ten observations in the data set. Consider the first family, where $X_1 = 22$ (income = \$22,000), $X_2 = 2$ (family size = 2, such as an adult couple with no children), and $X_3 = 4$ (combined years of college = 4). The predicted home size here is

$$\hat{Y} = 7.60 + .194(22) + 2.34(2) - .163(4) = 15.89$$

Consequently, the predicted home size is 1589 square feet. The actual square footage for this observation is 1600 ($Y = 16$), so the sample residual here is $Y - \hat{Y} = 16 - 15.89 = .11$.

Using this procedure on the remaining nine observations, we get the following results:

Y	$\hat{Y}$	$Y - \hat{Y}$	$(Y - \hat{Y})^2$
16	15.89	0.11	.0121
17	16.01	0.99	.9801
26	22.19	3.81	14.5161
24	24.12	−0.12	.0144
22	22.05	−0.05	.0025
21	22.68	−1.68	2.8224
32	31.18	0.82	.6724
18	19.90	−1.90	3.6100
30	30.59	−0.59	.3481
20	21.39	−1.39	1.9321
		0	24.91 ≈ SSE

The computed value for the error sum of squares is SSE = 24.91. This value also is contained in the MINITAB output in Figure 15.2. The MINITAB value for SSE is 24.86, which differs slightly from the result in the table because the computer uses much more accurate calculations than those in the table. Since SSE = 24.86 is more accurate, we will use this value in the remaining discussion.

This result implies that for *any* other values of b_0, b_1, b_2, and b_3, if we were to find the corresponding $\hat{Y}$'s and the resulting SSE $= \Sigma(Y - \hat{Y})^2$ using these values, this new SSE would be *larger* than 24.86. Thus, $b_0 = 7.60$, $b_1 = .194$, $b_2 = 2.34$, and $b_3 = -.163$ minimize the error sum of squares, SSE. Put still another way, these values of b_0, b_1, b_2, and b_3 provide the **best fit** to our data.

Using only income (X_1) as a predictor in Chapter 14, we found the SSE to be 75.81 in our table. By including the additional two variables, the SSE has been reduced from 75.81 to 24.86 (a 67% reduction). It appears that either family size (X_2), years of education (X_3), or both contribute, perhaps significantly, to the prediction of Y.

Interpreting the Regression Coefficients When using a multiple linear regression equation, such as $Y = \beta_0 + \beta_1 X_1 + \beta_2 X_2 + \beta_3 X_3 + e$, what does β_2 represent? Very simply, it reflects the change in Y that can be expected to accompany a change of one unit in X_2, *provided all other variables* (namely, X_1 and X_3) *are held constant*.

In the previous example, the sample estimate of β_2 was $b_2 = 2.34$. Can we expect an increase of 2.34 on the average as X_2 (the family size) increases by one if X_1 and X_3 are held constant? This type of argument is filled with problems, as we demonstrate later. The primary problem is that a change in one of the predictor variables (such as X_2) always (or almost always) is accompanied by a change in one of the other predictors (say, X_1) in the sample observations. Consequently, variables X_1 and X_2 are related in some manner, such as $X_1 \cong 1 + 5X_2$. In other words, a situation in which X_2, for instance, changed and the others remained constant would not be observed in the sample data.

In the case (typically not observed in business applications) where the predictor variables *are* totally unrelated, a unit change in X_2, for example, can be expected to be accompanied by a change of β_2 in the dependent variable.

In general, it is not safe to assume that the predictor variables are unrelated. As a result, the b's usually do not reflect the true "partial effects" of the predictor variables, and you should avoid such conclusions. Section 15.4 discusses methods of dealing with this type of situation.

The Assumptions Behind the Multiple Linear Regression Model

The form of the multiple linear regression model is given by equation 15.1, which contains a linear combination of the k predictor (independent) variables as well as the error component, e. The assumptions for the case of $k > 1$ predictors are exactly the same as for $k = 1$ independent variable (simple linear regression). These assumptions, discussed in Chapter 14, are:

1. The errors follow a normal distribution, centered at zero, with common variance, σ_e^2.
2. The errors are (statistically) independent.

The case for $k = 2$ predictor variables can be represented graphically, as shown in Figures 15.1 and 15.3. Using Figure 15.3, consider the situation in which $X_1 = 20$ and $X_2 = 15$. If you *were* to obtain repeated values of Y for these values for X_1 and X_2, you would obtain some Y's above the plane and some below. The assumptions are that the *average* value of Y with $X_1 = 20$ and $X_2 = 15$ lies *on* the plane and that, moreover, these errors are normally distributed.

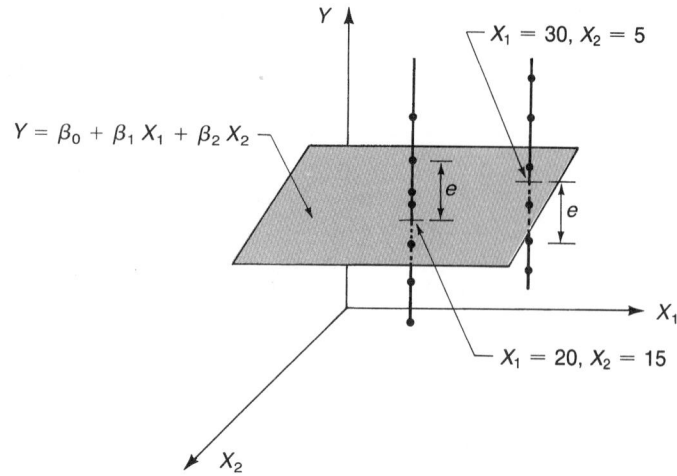

The final part of assumption 1 is that the variation about this plane does not depend on the values of X_1 and X_2. You should see roughly the same amount of variation if you obtain repeated values of Y corresponding to $X_1 = 30$ and $X_2 = 5$ as you observed for $X_1 = 20$ and $X_2 = 15$. The variance of these errors, if you could observe Y indefinitely, is σ_e^2.

Finally, assumption 2 means that the error encountered at $X_1 = 30$ and $X_2 = 5$, for instance, is not affected by a known error at any other point, such as $X_1 = 20$ and $X_2 = 15$. The error associated with one pair of X_1, X_2 values has no effect on any other error.

An Estimate of σ_e^2   When using a straight line to model a relationship between Y and a single predictor, the estimate of σ_e^2 was given by equation 14.16, where

$$s^2 = \hat{\sigma}_e^2 = \frac{\text{SSE}}{n - 2}$$

In general, for k predictors and n observations, the estimate of this variance is

$$s^2 = \hat{\sigma}_e^2 = \frac{\text{SSE}}{n - (k + 1)} = \frac{\text{SSE}}{n - k - 1} \tag{15.4}$$

The value of s^2 is critical in determining the reliability and usefulness of the model as a predictor. If $s^2 = 0$, then SSE $= 0$, implying that $Y = \hat{Y}$ for each of the observations in the sample data. This rarely happens in practice, but it does point out that a small s^2 is desirable. As s^2 increases, you can expect more error when predicting a value of Y for specified values of $X_1, X_2, \ldots, X_k$. In the next section, we use s^2 as a key to determining whether the model is satisfactory and which of the independent variables are useful in the prediction of the dependent variable.

The square root of this estimated variance is the **residual standard deviation.**

$$s = \sqrt{\frac{\text{SSE}}{n - k - 1}} \tag{15.5}$$

In the MINITAB solution in Figure 15.2, the value of s is shown in the box containing $s = 2.035$.

EXAMPLE 15.1 Determine the estimate of σ_e^2 and the residual standard deviation for the real-estate data on page 607.

SOLUTION This example contained $n = 10$ observations and $k = 3$ predictor variables. The resulting error sum of squares was SSE $= 24.86$ (from Figure 15.2).

Therefore,

$$s^2 = \hat{\sigma}_e^2$$

$$= \frac{24.86}{10 - 3 - 1} = \frac{24.86}{6} = 4.14$$

and

$$s = \sqrt{4.14} = 2.035$$

That is, the residual standard deviation is 203.5 square feet. ■

If a particular regression model meets all the required assumptions, then the next question of interest is whether this set of independent variables provides an accurate method of predicting the dependent variable, Y. The next section shows how to calculate the predictive ability of your model and determine which variables contribute significantly to an accurate prediction of Y.

Exercises

15.1 A management-consulting firm uses a regression model where variable X_1 stands for previous experience, variable X_2 for number of years at current job, and variable X_3 for score on a job-aptitude test. These variables are used in a regression model to predict job satisfaction. Job satisfaction ranges from 1 to 20, with 20 indicating that an employee is satisfied with every aspect of his or her job. The prediction equation is:

$$\hat{Y} = 1.7 - 0.15X_1 + 0.25X_2 + 0.14X_3$$

a. What would the consulting firm predict for the job satisfaction of an employee who has 15 years of prior experience, 10 years of employment at the present job, and an aptitude test score of 85?

b. If an employee's score on the aptitude test is increased by 20, with the years of prior experience and years of employment at the current job remaining constant, what would be the net change in the employee's predicted satisfaction score?

15.2 The marketing department at Computeron would like to predict the sales volume of its software for its personal computers. They believe that the sales volume of software (given in thousands) increases when the number of units of personal computers increases and when advertising expenditure (given in thousands of dollars) increases. The following data were collected for 14 months:

SALES OF SOFTWARE (Y)	UNITS OF PERSONAL COMPUTERS SOLD (X_1)	ADVERTISING EXPENDITURE (X_2)	SALES OF SOFTWARE (Y)	UNITS OF PERSONAL COMPUTERS SOLD (X_1)	ADVERTISING EXPENDITURES (X_2)
7.2	12	4.2	3.4	10	2.1
5.4	11	3.1	7.0	12	3.8
7.7	14	5.1	12.1	22	5.8
5.6	11	3.5	8.4	14	4.9
9.1	17	5.4	9.7	19	5.5
8.8	17	4.4	8.3	15	4.6
6.2	11	3.5	7.1	13	4.0

a. Using the least squares line $\hat{Y} = -1.21768 + 0.3141X_1 + 1.016X_2$, find the estimate of the variance of the error component.

b. Interpret the coefficients of the regression equation in the context of the problem. Would it be reasonable to expect sales of software to increase by 1.016 thousand if X_2 is increased by one?

15.3 An oil-service company decided to fit a least squares equation to a set of data to predict the total cost of building a well. The independent variables are X_1 = drilling days, X_2 = total depth, and X_3 = intermediate casing depth. After calculating the least squares equation, the residuals were calculated to find out whether the assumptions of regression analysis are satisfied. The following are the residuals from 20 observations:

$$-0.8,\ 1.5,\ -3.7,\ 4.1,\ -3.1,\ -5.2,\ 4.3,\ -2.1,\ -1.6,\ 4.1,\ 0.9,\ -0.3,$$
$$4.5,\ -4.2,\ 3.2,\ -2.7,\ 1.7,\ -2.2,\ 3.4,\ -1.8$$

Do the residuals $Y - \hat{Y}$ appear to conform to the empirical rule that approximately 95% of the data should lie within two standard deviations of the mean?

15.4 What assumptions need to be made about the error component of a multiple linear regression model in order that the results of statistical inference can be used?

15.5 Tony owns a used-car lot. He would like to predict monthly sales volume. Tony believes that sales volume (given in thousands) is directly related to the number of sales-people employed and the number of cars on the lot for sale. The following data were collected over a period of ten months:

MONTHLY SALES VOLUME (Y)	SALESPEOPLE (X_1)	CARS (X_2)	MONTHLY SALES VOLUME (Y)	SALESPEOPLE (X_1)	CARS (X_2)
5.8	4	20	8.1	4	25
7.5	5	15	13.3	8	30
11.4	7	25	15.0	9	35
7.0	3	17	8.3	5	20
5.1	2	18	6.8	4	23

a. Using a computerized statistical package, determine the least squares prediction equation for these data.

b. Find the value of SSE.

15.6 Using the multiple regression model $\hat{Y} = b_0 + b_1X_1 + b_2X_2 + b_3X_3$, where do you expect the average value of Y to fall for X_1 = 3, X_2 = 4.1, and X_3 = 5.6?

15.7 A regression analysis was performed for data with three independent variables. The following residuals were found for the 20 observations of the dependent variable:

$$5.4,\ 8.1,\ -7.4,\ 2.5,\ -3.5,\ -4.1,\ -8.1,\ 6.5,\ 4.3,\ -7.8,\ 2.8,\ 7.1,\ -6.2,$$
$$5.6,\ -5.1,\ 2.9,\ 2.8,\ -7.2,\ 8.3,\ -6.9$$

Find the residual standard deviation.

15.8 A real-estate broker uses four independent variables to predict the appraised value of homes in a certain subdivision. The variables are X_1 = lot size in square feet, X_2 = house size in square feet, X_3 = age of the house in years, and $X_4 = X_2X_3$. Twenty-seven observations were used to find the least squares prediction equation. The real-estate broker found the value of the residual standard deviation to be 650. What is the value of SSE?

▌ *15.2*
Hypothesis Testing and Confidence Intervals for the β Parameters

Multiple linear regression is a popular tool in the application of statistical techniques to business decisions. However, this modeling procedure does not always result in an accurate and reliable predictor. When the independent variables that you have selected account for very little of the variation in the values of the dependent variable, the model (as is) serves no useful purpose.

The first thing we demonstrate is how to determine whether your overall model is satisfactory. We begin by summarizing a regression analysis in an ANOVA table, much as we did in Chapter 11.

The ANOVA Table

The summary ANOVA table contains the usual headings.

SOURCE	df	SS	MS	F
Regression	k	SSR	MSR	MSR/MSE
Residual	$n - k - 1$	SSE	MSE	
Total	$n - 1$	SST		

where n = number of observations and k = number of independent variables.

$$\text{SST} = \text{total sum of squares}$$
$$= \text{SS}_Y$$
$$= \Sigma(Y - \bar{Y})^2 = \Sigma Y^2 - (\Sigma Y)^2/n \qquad (15.6)$$
$$\text{SSE} = \text{sum of squares for error}$$
$$= \Sigma(Y - \hat{Y})^2 \qquad (15.7)$$
$$\text{SSR} = \text{sum of squares for regression}$$
$$= \Sigma(\hat{Y} - \bar{Y})^2$$
$$= \text{SST} - \text{SSE} \qquad (15.8)$$
$$\text{MSR} = \text{mean square for regression}$$
$$= \text{SSR}/k \qquad (15.9)$$
$$\text{MSE} = \text{mean square for error}$$
$$= \text{SSE}/(n - k - 1) \qquad (15.10)$$

Practically all computer packages provide you with this ANOVA summary as part of the standard output. The ANOVA section of the MINITAB solution for the real-estate model is highlighted in Figure 15.4.

Notice that

$$\text{SST} = \text{SS}_Y$$
$$= (16^2 + 17^2 + \cdots + 20^2) - (16 + 17 + \cdots + 20)^2/10$$
$$= 262.4$$

This is the same value of SS_Y we obtained for the same example in Chapter 14, when we used only income (X_1) as the predictor variable. This is hardly surprising because *this value is strictly a function of the Y values* and is unaffected by the model that we are using to predict Y. The total sum of squares (SST) measures the total variation in the values of the dependent variable. Its value is the same, regardless of which predictor variables are included in the model.

The df for the regression source of variation is k = the number of predictor variables in the analysis. The df for the error sum of squares is $n - k - 1$, where n = the number of observations in the sample data.

As in the case of simple linear regression, the sum of squares of regression (SSR) measures the variation *explained* by the model—the variation in the Y values that would exist if differences in the values of the predictor variables were the only cause of differences among the Y's. On the other hand, the sum of squares of error (SSE) represents the variation *unexplained* by the model. The easiest way to determine the sum of squares of regression is to subtract:

$$\text{SSR} = \text{SST} - \text{SSE}$$

■ **FIGURE 15.4**
MINITAB output
(see Figure 15.2).
(a) Prediction equation
and ANOVA table using
X_1, X_2, and X_3.
(b) Prediction equation
and ANOVA table using
X_1 and X_2.

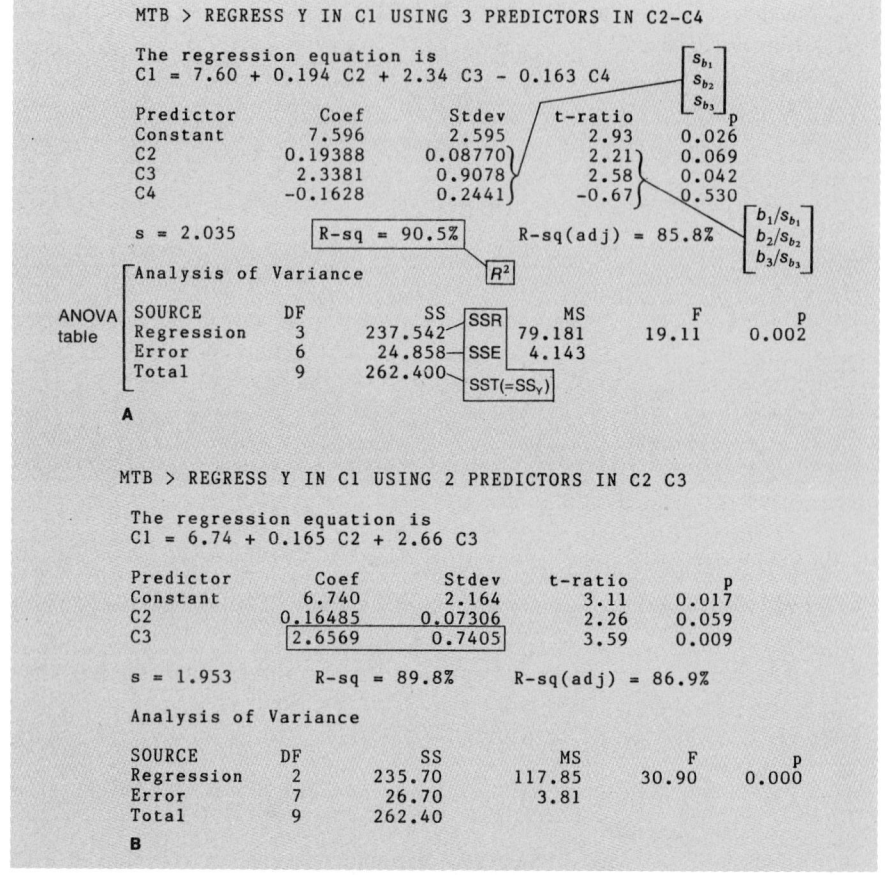

The error mean square is MSE = SSE/$(n - k - 1)$ = 24.858/(10 –
= 4.14. This is the same as the *estimate* of σ_e^2 determined in Example 15.

$$s^2 = \hat{\sigma}_e^2 = \text{MSE}$$

A Test for H_0: All β's = 0

We have yet to make use of the F-value calculated in the ANOVA table, where

$$F = \frac{\text{MSR}}{\text{MSE}} \qquad\qquad \textbf{(15.11)}$$

When using the simple regression model, we previously argued that one way
to determine whether X is a significant predictor of Y is to test H_0: $\beta_1 = 0$, where
β_1 is the coefficient of X in the model $Y = \beta_0 + \beta_1 X + e$. If you reject H_0, the
conclusion is that the independent variable X *does* contribute significantly to the
prediction of Y. For example, in Example 14.3, by rejecting H_0: $\beta_1 = 0$, we concluded that income (X_1) was a useful predictor of home size (Y) using the simple
linear model.

We use a similar test as the first step in the multiple regression analysis, where
we examine the hypotheses

$$H_0: \beta_1 = \beta_2 = \cdots = \beta_k = 0$$

$$H_a: \text{at least one of the } \beta\text{'s} \neq 0$$

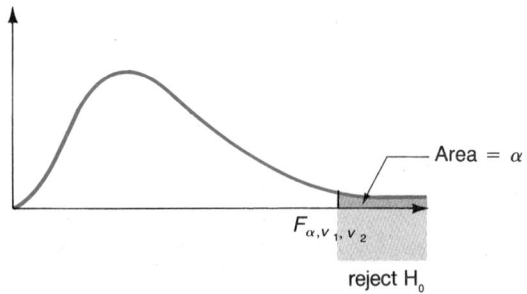

■ **FIGURE 15.5**
F curve with k and $n - k - 1$ df. The lightly shaded area is the rejection region.

Area = α

F_{α, v_1, v_2}

reject H_0

If we *reject* H_0, we can conclude that at least one (but maybe not all) of the independent variables contributes significantly to the prediction of Y. If we *fail to reject* H_0, we are unable to demonstrate that any of the independent variables (or combination of them) helps explain the behavior of the dependent variable, Y. For example, in our housing example, if we were to fail to reject H_0, this would imply that we are unable to demonstrate that the variation in the home sizes (Y) can be explained by the effect of income, family size, and years of education.

Test Statistic for H_0 Versus H_a The test statistic used to determine whether our multiple regression model contains at least one explanatory variable is the F statistic from the preceding ANOVA table.

When testing H_0: all β's $= 0$ (this set of predictor variables is no good at all) versus H_a: at least one $\beta \neq 0$ (at least one of these variables is a good predictor), the test statistic is

$$F = \frac{\text{MSR}}{\text{MSE}}$$

which has an F distribution with k and $n - k - 1$ df. The expression $n - k - 1$ can be written as $n - (k + 1)$, where $k + 1$ is the number of coefficients (β's) estimated including the constant term.

Notice that the df for the F statistic comes directly from the ANOVA table. The testing procedure is to

$$\text{reject } H_0 \text{ if } F > F_{\alpha, v_1, v_2}$$

where (1) $v_1 = k$, $v_2 = n - k - 1$ and (2) F_{α, v_1, v_2} is the corresponding F-value in Table A.7, having a *right-tail area* $= \alpha$ (Figure 15.5).

EXAMPLE 15.2 Using the real-estate data and the model we developed, what can you say about the predictive ability of the independent variables, income (X_1), family size (X_2), and years of education (X_3)? Use $\alpha = .10$.

SOLUTION **Step 1.** The hypotheses are

$$H_0: \beta_1 = \beta_2 = \beta_3 = 0$$

$$H_a: \text{at least one } \beta \neq 0$$

Remember that our hope here is to reject H_0. If you are unable to demonstrate that any of your independent variables have any predictive ability, then you will fail to reject H_0.

Step 2. The test statistic is

$$F = \frac{\text{MSR}}{\text{MSE}}$$

The mean squares are obtained from the ANOVA summary of the regression analysis (see Figure 15.4).

Step 3. The df for the F statistic are $k = 3$ and $n - k - 1 = 10 - 3 - 1 = 6$. So we will

$$\text{reject } H_0 \text{ if } F > F_{.10,3,6} = 3.29$$

Step 4. Using the results in Figure 15.4, the computed F-value is

$$F^* = \frac{79.18}{4.14} = 19.1$$

Because $F^* > 3.29$, we reject H_0.

Step 5. The three independent variables *as a group* constitute a good predictor of home size. This does *not* imply that all three variables have significant predictive ability; however, at least one of them does. The next section shows how you can tell *which* of these predictor variables significantly contributes to the prediction of home size.   ∎

A Test for H_0: $\beta_i = 0$

Assuming that you rejected the null hypothesis that all of the β's are zero, the next logical question would be, which of the independent variables contributes to the prediction of Y?

In Example 15.2, we rejected the null hypothesis, so at least one of these three independent variables affects the variation of the ten home sizes in the sample. To determine the contribution of each variable, we perform three separate t tests:

$$H_0: \beta_1 = 0 \ (X_1 \text{ does not contribute})$$

$$H_a: \beta_1 \neq 0 \ (X_1 \text{ does contribute})$$

$$H_0: \beta_2 = 0 \ (X_2 \text{ does not contribute})$$

$$H_a: \beta_2 \neq 0 \ (X_2 \text{ does contribute})$$

$$H_0: \beta_3 = 0 \ (X_3 \text{ does not contribute})$$

$$H_a: \beta_3 \neq 0 \ (X_3 \text{ does contribute})$$

One-tailed tests also can be used here, but we will demonstrate this procedure using two-tailed tests. This means that we are testing to see whether this particular X contributes to the prediction of Y, but we are not concerned about the direction (positive or negative) of this relationship.

When income (X_1) was the only predictor of home size (Y), we used a t test to determine whether the simple linear regression model was adequate. In Example 14.3, the value of the test statistic was derived, where

$$t = \frac{b_1}{s_{b_1}} \tag{15.12}$$

Also, b_1 is the estimate of β_1 in the simple regression model, and s_{b_1} is the (estimated) standard deviation of b_1.

All computer packages provide both the estimated coefficient (b_1) and its standard deviation (s_{b_1}). In Example 14.3, the computed value of this t statistic was $t^* = 4.44$. This result led us to conclude that income was a good predictor of home size because a significant positive relationship existed between these two variables.

We use the same t statistic procedure to test the effect of the individual variables in a multiple regression model. When examining the effect of an individual independent variable, X_i, on the prediction of a dependent variable, the hypotheses are

$$H_0: \beta_i = 0$$
$$H_a: \beta_i \neq 0$$

The test statistic is

$$t = \frac{b_i}{s_{b_i}}$$

where (1) b_i is the estimate of β_i, (2) s_{b_i} is the (estimated) standard deviation of b_i, and (3) the df for the t statistic is $n - k - 1$.

The test of H_0 versus H_a is to

$$\text{reject } H_0 \text{ if } |t| > t_{\alpha/2, n-k-1}$$

where $t_{\alpha/2, n-k-1}$ is obtained from Table A.5.

We can now reexamine the real-estate data in Example 15.2.

X_1 = Income   Consider the hypotheses

$$H_0: \beta_1 = 0$$
$$H_a: \beta_1 \neq 0$$

As in Example 15.2, we use $\alpha = .10$.

According to Figure 15.4, $b_1 = .194$ and $s_{b_1} = .0877$. Also contained in the output is the computed value of

$$t^* = \frac{b_1}{s_{b_1}} = \frac{.194}{.0877} = 2.21$$

Why is this value of t^* *not* the same as the value of t calculated for this variable in Chapter 14, when income was the only predictor of Y? When there are three predictors in the model, t^* for income is 2.21. When income is the only predictor in the model, $t^* = 4.44$. The difference in the two values is simply that $t^* = 2.21$ provides a measure of the contribution of $X_1 = $ income, *given that X_2 and X_3 already have been included in the model.* A large value of t^* indicates that X_1 contributes significantly to the prediction of Y, even if X_2 and X_3 have been included previously as predictors.

The hypotheses can better be stated as

H_0: income *does not* contribute to the prediction of home size, *given* that family size and years of education already have been included in the model

H_a: income *does* contribute to this prediction, given that family size and years of education already have been included in the model

or as

$$H_0: \beta_1 = 0 \quad \text{(if } X_2 \text{ and } X_3 \text{ are included)}$$
$$H_a: \beta_1 \neq 0$$

Because $t^* = 2.21$ exceeds the table value of $t_{\alpha/2, n-k-1} = t_{.05, 10-3-1} = t_{.05, 6} = 1.943$, we conclude that income contributes significantly to the prediction of home size and should be kept in the model.

X_2 = Family Size   Using a similar argument, the following test of hypothesis will determine the contribution of family size, X_2, as a predictor of the home square footage, given that X_1 and X_3 already have been included. The hypotheses here are

$$H_0: \beta_2 = 0 \quad \text{(if } X_1 \text{ and } X_3 \text{ are included)}$$

$$H_a: \beta_2 \neq 0$$

According to Figure 15.4, the computed t statistic here is

$$t^* = \frac{b_2}{s_{b_2}} = \frac{2.34}{.9078} = 2.58$$

This value also exceeds $t_{.05,6} = 1.943$, and so family size provides useful information in predicting the square footage of a home. We conclude that we should keep X_2 in the model.

X_3 = Years of Education   To test

$$H_0: \beta_3 = 0 \quad \text{(if } X_1 \text{ and } X_2 \text{ are included)}$$

$$H_a: \beta_3 \neq 0$$

we once again use the t statistic.

$$t = \frac{b_3}{s_{b_3}}$$

Using Figure 15.4, the computed value of this statistic is

$$t^* = \frac{-.163}{.2441} = -.67$$

Because $|t^*| = .67$, which does *not* exceed $t_{.05,6} = 1.943$, we fail to reject H_0. We conclude that, given the values of X_1 = income and X_2 = family size, the level of a family's education appears not to contribute to the prediction of the size of their home. This means that X_3 can be ignored in the final prediction equation, leaving only X_1 and X_2.

COMMENTS   As a word of warning, you should *not* simply remove this term from the equation containing all three variables. Since the predictor variables are typically related in some manner, the sample regression coefficients ($b_0, b_1, \ldots$) change as variables are added to or deleted from the model. Referring to Figure 15.4a, the final prediction equation is not $\hat{Y} = 7.60 + .194X_1 + 2.34X_2$. Instead, the coefficients of X_1 and X_2 should be derived by repeating the analysis using only these two variables. According to Figure 15.4b, this prediction equation is $\hat{Y} = 6.74 + .165X_1 + 2.66X_2$.

A Confidence Interval for β_i

Using what you believe to be the "best" model, you can easily construct a $(1 - \alpha) \cdot 100\%$ confidence interval for β_i based on the previous t statistic:

$$b_i - t_{\alpha/2, n-k-1} s_{b_i} \quad \text{to} \quad b_i + t_{\alpha/2, n-k-1} s_{b_i} \tag{15.13}$$

Once again, k represents the number of predictor variables used to estimate β_i.

EXAMPLE 15.3   Suppose you decide to retain only X_1 = income and X_2 = family size in the prediction equation. Referring to Figure 15.4b, construct a 90% confidence interval for β_2, the coefficient for X_2.

SOLUTION   Since this model contains $k = 2$ predictor variables, we first find $t_{\alpha/2, n-k-1} = t_{.05,7} = 1.895$. Based upon the MINITAB results in Figure 15.4b, the confidence interval for β_2 is

$$2.6569 \ - \ (1.895)(.7405) \quad \text{to} \quad 2.6569 \ + \ (1.895)(.7405)$$
$$= \ 2.6569 \ - \ 1.4032 \quad \text{to} \quad 2.6569 \ + \ 1.4032$$
$$= \ 1.25 \quad \text{to} \quad 4.06$$

Therefore, we are 90% confident that the estimate of β_2 (that is, $b_2 = 2.6569$) is within 1.4032 of the actual value of β_2. Notice that this is an extremely wide confidence interval. As usual, increasing the sample size would help to reduce the width of this confidence interval.   ∎

EXAMPLE 15.4   The management of BB Investments decided to develop a model to predict the amount of money invested by various clients in their portfolio of high-risk securities. It was generally agreed that the income of the investor should be a major factor in predicting his or her annual investment and would explain a major portion of the variability in the amount invested. In addition, the investor's willingness to assume risk also was influenced by the investor's view of present and future economic conditions. On the assumption that the investors would use economic forecasts and economists' indices of future expectations, the financial group at BB Investments constructed an economic index that ranged from 0 to 100. When applied to any particular point in time, this index was tied to the expected increase in interest rates and borrowing levels, the expected increase in manufacturing costs because of the rate of inflation, and the expected level of price inflation at the retail level. This meant that the *lower* the index, the *better* the future economic conditions were expected to be.

Data were obtained by randomly selecting 50 high-risk portfolio customers and recording their incomes and the amounts of their investments. The income figures represent annual incomes and the economic index values are the index values at the time the investment was made.

INVESTOR	INVESTMENT (Y)	ECONOMIC INDEX (X_1)	INCOME (X_2)	INVESTOR	INVESTMENT (Y)	ECONOMIC INDEX (X_1)	INCOME (X_2)
1	2,500	86	55,800	26	3,600	40	61,600
2	3,700	54	60,400	27	2,800	81	60,000
3	3,900	21	72,700	28	2,200	44	50,600
4	1,700	91	41,700	29	3,800	36	66,300
5	1,000	72	35,200	30	4,300	50	70,900
6	1,700	16	41,800	31	3,300	95	66,600
7	2,500	81	43,700	32	3,300	47	64,400
8	3,400	32	67,900	33	2,100	3	52,900
9	2,500	37	53,700	34	3,800	55	68,500
10	2,900	89	57,400	35	1,700	65	36,400
11	2,100	48	47,100	36	3,800	28	69,100
12	2,600	61	55,300	37	2,000	47	44,400
13	1,700	33	40,000	38	1,900	50	47,900
14	2,100	82	40,200	39	2,400	15	57,100
15	1,500	95	36,900	40	3,800	84	67,500
16	1,700	73	40,700	41	3,700	44	61,900
17	1,400	9	35,100	42	1,500	28	36,100
18	2,400	42	50,900	43	3,000	36	58,900
19	1,000	74	36,300	44	2,200	69	50,700
20	3,200	31	63,700	45	2,400	69	49,800
21	2,500	12	46,800	46	3,200	3	62,500
22	4,500	25	75,200	47	2,600	70	51,100
23	2,400	24	42,400	48	3,000	17	52,600
24	2,000	88	42,000	49	1,900	15	43,700
25	2,900	53	54,600	50	1,700	63	40,100

Determine the predicted investment for an investor with an income of \$48,500 at a time when the economic index has a value of 72.

SOLUTION   From Figure 15.6, the least squares equation is

$$\hat{Y} = -1183.3 - .127X_1 + .072X_2$$

The predicted investment is

$$\hat{Y} = -1183.3 - .127(72) + .072(48,500)$$
$$= -1183.3 - 9.1 + 3492.0$$
$$= 2299.6$$

that is, approximately \$2300. Note that, following the same argument used in Chapter 14, \$2300 also serves as an estimate of the average investment whenever $X_1 = 72$ and $X_2 = 48,500$. This topic is explored further in Section 15.5, where we discuss the construction of a confidence interval for an *average* investment or a prediction interval for an *individual* investment.

The first test of hypothesis determines whether these two variables *as a group* provide a useful model for predicting the amount of an investment:

$$H_0: \beta_1 = \beta_2 = 0$$

$$H_a: \beta_1 \neq 0, \beta_2 \neq 0, \text{ or both} \neq 0$$

Using the ANOVA table in Figure 15.6, the value of the F statistic is

$$F^* = \frac{\text{MSR}}{\text{MSE}} = \frac{16,647,068}{81,108} = 207.8$$

The df here is $v_1 = k = 2$ and $v_2 = n - k - 1 = 50 - 2 - 1 = 47$. Because $F_{.10,2,47}$ is not in Table A.7, we use the nearest value, $F_{.10,2,40} = 2.44$. The computed F^* exceeds this value, so we reject H_0 and conclude that at least one of these two independent variables is a significant predictor of investment amounts.   ∎

The t Tests   Because we rejected H_0 in Example 15.4, the next step is to examine the t tests to determine which of the two independent variables are useful predictors. The t value from Table A.5 is $t_{\alpha/2, n-k-1} = t_{.05,47} \cong 1.684$. The computed t values

■ FIGURE 15.6
MINITAB output for
Example 15.4.

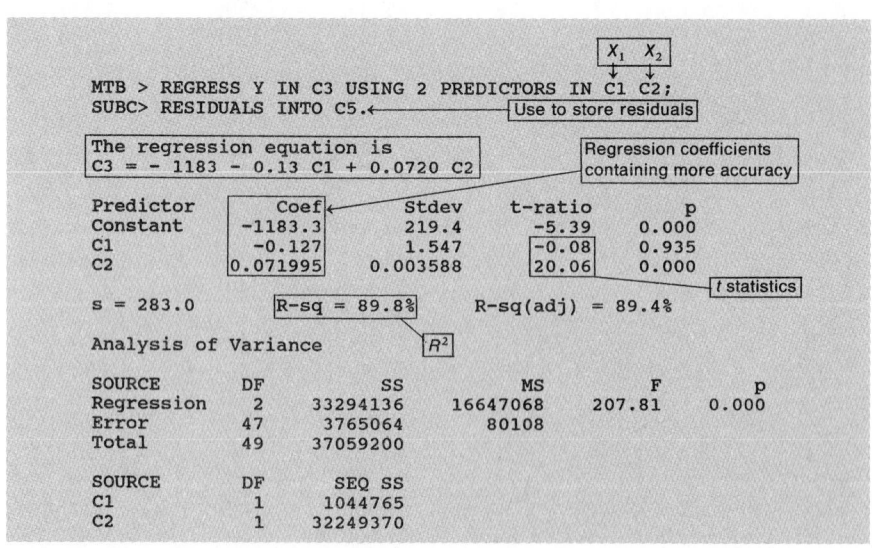

in Figure 15.6 lead to the following conclusions: First, the t value for $X_1 =$ economic index is $t^* = -.08$. The absolute value of t^* is *less than* 1.684, which means that, given the presence of X_2 in the model, X_1 does not contribute useful information to the prediction of the amount of an investment. It can be removed from the model without seriously affecting the accuracy of the resulting prediction. Second, for $X_2 =$ income, the computed t value is $t^* = 20.06$. Since this value exceeds 1.684, the investor's income *is* an excellent predictor of the amount of an investment. It was the contribution of this variable and not of X_1 that produced the extremely large F-value we obtained.

As we have seen, a quick glance at the computer output allows you to determine whether your model is useful as a whole and, furthermore, which variables are useful predictors. But beware—the analysis is not over! *Before you form your conclusions from this analysis and make critical decisions based on several tests of hypotheses, you need be sure that none of the assumptions of the multiple linear regression model (discussed earlier) have been violated.* We will discuss this problem in the final section of this chapter, where we conclude the analysis by examining the sample *residuals, $Y - \hat{Y}$.*

The use of t tests allows you to determine the predictive contribution of each independent variable, provided you want to examine the contribution of *one* such variable while assuming that the remaining variables are included in the equation. The next section shows you how to extend this procedure to a situation in which you wish to determine the contribution of any *set* of predictor variables by using a single test.

Exercises

15.9 The following information is selected from a computer printout of a multiple regression analysis:

PREDICTOR	COEFFICIENT	S.D.	t RATIO
Constant	$-.50$	.198	-2.525
X_1	2.40	.256	9.375
X_2	2.95	.210	14.048

R Square $= 0.9381$

ANALYSIS OF VARIANCE

SOURCE	df	SS	MS	F
Model	2	295.30	147.65	128.39
Residual	17	19.50	1.15	
Total	19	314.80		

Answer the following questions:

a. Write the multiple regression equation.

b. What percentage of variation in Y is explained by the model?

c. What is the sample size n used for the above regression analysis?

d. Find the value of:

    i. The total sum of squares
    ii. The error sum of squares
    iii. The regression sum of squares
    iv. The F statistic to test $H_0: \beta_1 = \beta_2 = 0$
    v. The rejection region for the test in part iv at the .05 significance level
    vi. The t statistic to test $H_0: \beta_2 = 0$
    vii. The rejection region for the test in part vi at the .05 significance level
    viii. The estimated variance of the error in the model

15.10 Complete the following ANOVA table to test the usefulness of a model with five independent variables that attempts to explain the variation in the dependent variable:

SOURCE	df	SS	MS	F
Regression				
Error		180		
Total	50	215		

15.11 Many chief executives (CEOs) have been under serious criticism from organized labor for the fat paychecks CEOs take home. Many of the CEOs have advocated sacrifice and leaner paychecks for large groups of employees of their companies. An experiment was set up in which 15 observations were taken on the variables:

$$Y = \text{CEOs pay (in thousands of dollars)}$$

$$X_1 = \text{company's net profit (in millions of dollars)}$$

$$X_2 = \text{number of employees (in thousands)}$$

A computer package gave the following sample statistics:

$$b_1 = .1336 \qquad b_2 = -.86$$

$$s_{b_1} = .0424 \qquad s_{b_2} = .39$$

a. Given that X_2 is in the model, does X_1 contribute to predicting the dependent variable at the .05 significance level?

b. Given that X_1 is in the model, does X_2 contribute to predicting the dependent variable at the .05 significance level?

15.12 Brown and Gilbert's law firm would like to predict the salary for a legal secretary based on years of college education (X_1), typing speed in words per minute (X_2), and years of experience (X_3). The following data were collected:

Y	X_1	X_2	X_3	Y	X_1	X_2	X_3
15,120	2	65	2	12,500	0	45	.5
12,500	1	45	2	15,800	2.5	60	2
26,000	3.5	85	9	19,600	1	70	3
19,000	0	55	11	21,800	3	75	6
16,000	4	85	1	12,400	0	60	.5
15,000	0	65	1	22,500	2	75	7

a. Using a computerized statistical package, determine the least squares prediction equation.

b. What is the value of the residual standard deviation?

c. Do the variables X_1, X_2, and X_3 contribute to predicting salaries at the .10 significance level?

d. Find a 90% confidence interval for β_1.

e. Test the null hypothesis that $\beta_1 = 0$ at the 10% significance level.

f. Interpret the results of the hypothesis test in question e.

15.13 The following sample statistics were computed for a regression analysis:

$$b_0 = 10.2 \quad b_1 = 5.6 \quad b_2 = 100.4 \quad s_{b_1} = 1.04 \quad s_{b_2} = 17.95$$

Assume that 20 observations were taken.

a. Test that X_2 significantly contributes to the prediction of Y given that X_1 is in the model. Use a 5% significance level.

b. Find a 95% confidence interval for β_1.

15.14 The model $\hat{Y} = 3.2 + 6.1X_1 + 5.2X_2$ was calculated to fit 20 data points pertaining to the growth rate of a hog. The variable X_1 represents the daily food consumption of the hog and X_2 represents the age of the hog. If the standard deviation of the estimate of β_1 is 2.5, what is a 95% confidence interval for the parameter β_1?

15.15 The tensile strength (Y) of a paper product is related both to the amount of hardwood in the pulp (X_1) and to the amount of time the paper spent soaking in a preparatory solution prior to cutting (X_2). A quality engineer collected ten samples of the variables Y, X_1, and X_2. Complete the following ANOVA table to test the null hypothesis that the independent variables X_1 and X_2 are not useful predictors of the tensile strength of the paper product.

SOURCE	df	SS	MS	F
Regression			582.83	
Error				
Total		1300.90		

15.16 Datamatics Equipment, a Seattle-based electronics firm, is interested in identifying variables in the manufacturing environment that have a linear relationship with the number of line shortages on the manufacturing floor. The sample data used in a regression analysis are as follows:

WEEK	Y	X_1	X_2	X_3	WEEK	Y	X_1	X_2	X_3
1	293	205	5.936	343	9	420	365	4.780	453
2	348	215	5.815	259	10	407	329	4.905	460
3	416	227	4.983	250	11	397	345	5.009	426
4	445	301	4.841	236	12	430	249	4.869	408
5	453	362	4.755	243	13	497	356	4.791	324
6	392	358	4.775	303	14	534	424	4.754	330
7	382	302	4.813	411	15	547	430	4.598	283
8	365	246	4.909	420					

where

Y = number of line shortages with back-order status for a given week

X_1 = number of delinquent purchase orders for a given week

X_2 = inventory level (in millions of dollars) for prior weeks

X_3 = number of purchased items for prior weeks

The least squares regression equation was found to be:

$$\hat{Y} = 710.9 + 0.4767X_1 - 70.90X_2 - 0.2525X_3$$

a. Does the complete model significantly contribute to predicting the dependent variable? Use a 10% significance level.

b. If s_{b_2} is 36.886, find a 95% confidence interval for β_2.

c. Interpret the results of the hypothesis test in question a, and interpret the confidence interval in part b.

15.17 The least squares equation of $\hat{Y} = 3.4 + 1.2X_1 + 4.3X_2$ was obtained. The sample residuals of the 20 observations used in fitting the regression equation are:

4.1, -3.2, 1.5, 6.7, 6.4, 3.8, -4.2, -2.4, 1.6, -8.7, -3.1, 1.2, -5.1, 2.1, 0.6, 5.4, 3.4, -7.1, -6.2, 3.2

Given that the value of SST is 510, test the null hypothesis that the variables X_1 and X_2 do not contribute to predicting the variation in the dependent variable. Use a 5% significance level.

15.18 Do the number of units of personal computers sold and the advertising expenditures contribute significantly to predicting the variation in the sales of software in Exercise 15.2? Use a 5% level of significance.

15.19 The job placement center at Ozark Technological University would like to predict the starting salaries (given in thousands of dollars) for the college graduates in the engineering department. Two variables are used. The variable X_1 represents the student's overall grade point average (GPA). The variable X_2 represents the number of years of prior job-related experience. Data for fifteen randomly selected graduating students are:

STARTING SALARY (Y)	OVERALL GPA (X_1)	YEARS OF JOB-RELATED EXPERIENCE (X_2)	STARTING SALARY (Y)	OVERALL GPA (X_1)	YEARS OF JOB-RELATED EXPERIENCE (X_2)
27.1	3.7	0	32.3	3.8	2.5
23.3	2.9	1.1	18.1	2.1	1.4
21.4	2.4	1.5	22.5	3.0	0.3
24.2	3.2	0.5	23.8	3.4	0.5
26.1	3.6	0.8	20.9	2.8	0
19.8	2.7	0	20.0	2.5	1.0
22.8	3.1	0	27.8	3.3	2.1
20.5	2.2	2.1			

The least squares equation for the data is:

$$\hat{Y} = 2.189 + 6.5144X_1 + 1.9259X_2$$

a. Find the F-value using an ANOVA table to test the hypothesis that $\beta_1 = 0$ and $\beta_2 = 0$ at the 5% level of significance.

b. Interpret the coefficients in the context of the problem.

15.20   If the residual standard deviation for a set of data is 3.82 and the total sum of squares is 269, what is the F test for testing that a model with five independent variables does not contribute to predicting the variation in the dependent variable? Assume that 15 observations of the dependent variable were taken.

▌ 15.3
Determining the Predictive Ability of Certain Independent Variables

We can extend the procedure we used to examine the contribution of each independent variable, one at a time, using a t-test.

Assume that the personnel director of an accounting firm has developed a regression model to predict an individual's performance on the CPA exam. The multiple linear regression model contains eight independent variables, three of which (say, X_6, X_7, X_8) describe the physical attributes of each individual (say, height, weight, and age). Can all three of these variables be removed from the analysis without seriously affecting the predictive ability of the model?

To answer this question, we return to a statistic described in Chapter 14 that measures how well the model captures the variation in the values of the dependent variable.

Coefficient of Determination

The total variation of the sampled dependent variable is determined by

$$SST = \text{total sum of squares}$$
$$= SS_Y$$
$$= \Sigma(Y - \bar{Y})^2$$
$$= \Sigma Y^2 - (\Sigma Y)^2/n$$

where n = number of observations. To determine what percentage of this variation has been explained by the predictor variables in the regression equation, we determine the **coefficient of determination, R^2.**

$$R^2 = 1 - \frac{SSE}{SST} \qquad \textbf{(15.14)}$$

The range for R^2 is 0 to 1. If $R^2 = 1$, then 100% of the total variation has been explained, because in this case SSE $= \Sigma(Y - \hat{Y})^2 = 0$, and so $Y = \hat{Y}$ for each observation in the sample; that is, the model provides a *perfect predictor*. This does not occur in practice, but the main point is that a large value of R^2 is generally desirable for a regression application. It should be mentioned that $R^2 = 1$ whenever the number of observations (n) is equal to the number of estimated coefficients ($k + 1$). This does not mean that you have a "wonderful" model; rather, you have inadequate data. As a result, you need to guard against using too small a sample in your regression analysis. *A general rule of thumb is to use a sample containing at least three times as many (unique) observations as the number of predictor variables (k) in the model.*

H_0: All β's $= 0$ A test statistic for testing H_0: all β's $= 0$ was introduced in equation 15.11, which used the ratio of two mean squares from the ANOVA table. Another way to calculate this F-value is to use

$$F = \frac{R^2/k}{(1 - R^2)/(n - k - 1)} \qquad (15.15)$$

This version of the F statistic is used to answer the question, is the value of R^2 significantly large? If H_0 is rejected, then the answer is yes, and so this group of predictor variables has at least some predictive ability for predicting Y.

The F-value computed in this way will be exactly the *same* as the one computed using $F = $ MSR/MSE, except for possible rounding error (see Example 15.5).

Once again, remember that *statistical* significance does not always imply *practical* significance. A large value of R^2 (rejecting H_0) does not imply that precise prediction (practical significance) will follow. However, it does inform the researcher that these predictor variables, as a group, are associated with the dependent variable.

EXAMPLE 15.5   In Chapter 14, we determined that $X = $ income explained 71% of the total variation of the home sizes (Y) in the sample, since the computed value of r^2 was .711. What percentage is explained using all three predictors (income, family size, and years of education)?

SOLUTION   The coefficient of determination using X_1 only is .711. Using the MINITAB solution in Figure 15.4a, the coefficient of determination using X_1, X_2, and X_3 is

$$R^2 = 1 - \frac{SSE}{SST}$$

$$= 1 - \frac{24.858}{262.4} = .905$$

Consequently, 90.5% of this variation has been explained using the three independent variables.

The F-value determined in Example 15.2 for testing H_0: $\beta_1 = \beta_2 = \beta_3 = 0$ can be duplicated using equation 15.15 because

$$F = \frac{.905/3}{(1 - .905)/(10 - 3 - 1)} = \frac{.905/3}{.095/6}$$

$$= 19.1 \text{ (as before)} \qquad \blacksquare$$

COMMENTS

1. In Example 15.5, notice that the value of R^2 *increased* when we went from using one independent variable to using three. As you add variables to your regression model, R^2 *never decreases*. However, the increase

may not be a significant one. If adding ten more predictor variables to your model causes R^2 to increase from .91 to .92, this is not a *significant* increase. Therefore, do not include these ten variables; they clutter up your model and are likely to add spurious predictive ability to it.

2. Nearly every computer package (including MINITAB, SPSS, and SAS) will provide a value in the output, referred to as the **adjusted R^2**. This particular statistic does *not* necessarily increase as additional predictor variables are added to the model, and many researchers use this value to determine the predictive contribution of a variable added to the model. The adjusted R^2 is found by dividing SSE and SST by their respective degrees of freedom.

$$R^2(\text{adj}) = 1 - \frac{\text{SSE}/(n - k - 1)}{\text{SST}/(n - 1)} \qquad \textbf{(15.16)}$$

Referring to Example 15.5 and Figure 15.4a, the adjusted R^2 value is

$$R^2(\text{adj}) = 1 - \frac{24.858/6}{262.4/9} = 1 - \frac{4.143}{29.156} = .858$$

which is also given in the MINITAB output in Figure 15.4a.

How can we tell if adding (or removing) a certain set of X variables causes a *significant* increase (or decrease) in R^2?

The Partial F Test

Consider the situation in which the personnel director is trying to determine whether to retain three variables (X_6 = height, X_7 = weight, X_8 = age) as predictors of a person's performance on a CPA exam. We know one thing—R^2 *will* be higher with these three variables included in the model. If we do not observe a *significant* increase, however, our advice would be to remove these variables from the analysis. To determine the extent of this increase, we use another F test.

We define two models—one contains X_6, X_7, and X_8, and one does not:

complete model: uses all predictor variables, including X_6, X_7, and X_8

reduced model: uses the same predictor variables as the complete model except X_6, X_7, and X_8

Also, let

$$R_c^2 = \text{the value of } R^2 \text{ for the complete model}$$

$$R_r^2 = \text{the value of } R^2 \text{ for the reduced model}$$

Do X_6, X_7, and X_8 contribute to the prediction of Y? We will test

$$H_0: \beta_6 = \beta_7 = \beta_8 = 0 \text{ (they do not contribute)}$$

$$H_a: \text{at least one of the } \beta\text{'s} \neq 0 \text{ (at least one of them does contribute)}$$

The test statistic here is

$$F = \frac{(R_c^2 - R_r^2)/v_1}{(1 - R_c^2)/v_2} \qquad \textbf{(15.17)}$$

where v_1 = number of β's in H_0 and $v_2 = n - 1 - $ (number of X's in the complete model).

For this illustration, $v_1 = 3$ because there are three β's in H_0. Assuming that there are eight predictor variables in the complete model, then $v_2 = n - 1 - 8 = n - 9$. Here, n is the total number of observations (rows) in the data. This F statistic measures the *partial* effect of these three variables; it is a **partial F statistic.**

Equation 15.17 resembles the F statistic given in equation 15.15, which we used to test H_0: all β's $= 0$. If all the β's are zero, then the reduced model consists of only a constant term, and the resulting R^2 will be zero; that is, $R_r^2 = 0$. Setting $R_r^2 = 0$ in equation 15.17, produces equation 15.15, where $v_1 = k$ and $v_2 = n - k - 1$.

These variables (as a group) contribute significantly if the computed partial F-value in equation 15.17 exceeds F_{α,v_1,v_2} from Table A.7.

EXAMPLE 15.6

The personnel director gathered data from 30 individuals using all eight independent variables. These data were entered into a computer, and a multiple linear regression analysis was performed. The resulting R^2 was .857.

Next, variables X_6, X_7, and X_8 were omitted, and a second regression analysis was performed. The resulting R^2 was .824. Do the variables X_6, X_7, and X_8 (height, weight, and age) appear to have any predictive ability? Use $\alpha = .10$.

SOLUTION

Here, $n = 30$ and

$$R_c^2 = .857 \text{ (complete model)}$$

$$R_r^2 = .824 \text{ (reduced model)}$$

Based on the previous discussion, the value of the partial F statistic is

$$F^* = \frac{(.857 - .824)/3}{(1 - .857)/(30 - 1 - 8)} = \frac{.033/3}{.143/21} = 1.61$$

The procedure is to reject H_0: $\beta_6 = \beta_7 = \beta_8 = 0$ if $F^* > F_{.10,3,21} = 2.36$. The computed F-value does not exceed the table value, so we fail to reject H_0. We conclude that these variables should be removed from the analysis because including them in the model fails to produce a significantly larger R^2. ∎

The partial F test also can be used to determine the effect of adding a *single* variable to the model.

EXAMPLE 15.7

Using the real-estate data analyzed in Example 15.2, determine whether $X_2 =$ family size contributes to the prediction of home size, given that $X_1 =$ income and $X_3 =$ years of education are included in the model. Use a significance level of $\alpha = .10$.

SOLUTION

We test the hypotheses

$$H_0: \beta_2 = 0 \quad \text{(if } X_1 \text{ and } X_3 \text{ are included)}$$

$$H_a: \beta_2 \neq 0.$$

The complete model uses X_1, X_2, and X_3. Using Example 15.5,

$$R_c^2 = .905$$

The reduced model uses only X_1 and X_3. Figure 15.7 shows the MINITAB output for this, and

$$R_r^2 = .801$$

■ **FIGURE 15.7**
MINITAB output using
X_1 = income and
X_3 = years of education
as predictors.

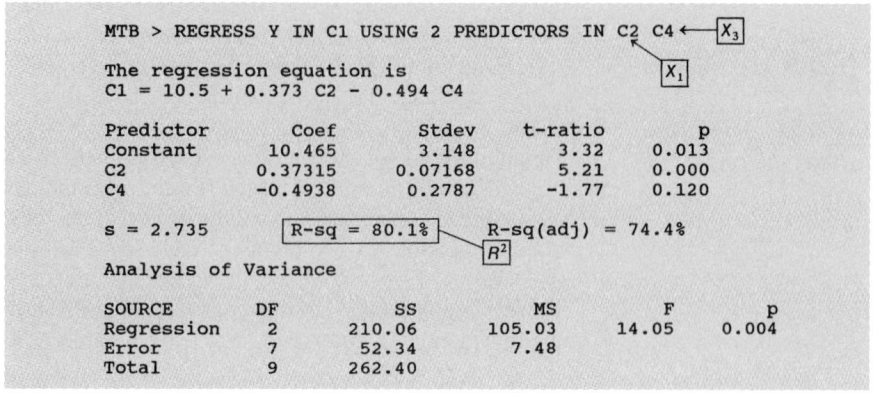

```
MTB > REGRESS Y IN C1 USING 2 PREDICTORS IN C2 C4 ←— X₃

The regression equation is                          X₁
C1 = 10.5 + 0.373 C2 - 0.494 C4

Predictor      Coef       Stdev     t-ratio         p
Constant     10.465       3.148        3.32     0.013
C2          0.37315     0.07168        5.21     0.000
C4          -0.4938       0.2787       -1.77     0.120

s = 2.735        R-sq = 80.1%     R-sq(adj) = 74.4%
                                 R²
Analysis of Variance

SOURCE         DF          SS          MS         F        p
Regression      2      210.06      105.03     14.05    0.004
Error           7       52.34        7.48
Total           9      262.40
```

The value of the partial F statistic is

$$F^* = \frac{(.905 - .801)/1}{(1 - .905)/(10 - 1 - 3)} = \frac{.104/1}{.095/6} = 6.6$$

The 1 in the numerator indicates that there is one β in H_0; subtracting the 3 in the denominator indicates that there are three X's in the complete model.

This value does exceed $F_{.10,1,6} = 3.78$, and so X_2 = family size does (as suspected from the earlier t test) significantly improve the model's predictive ability when included with X_1 and X_3. In other words, there is a significant increase in R^2 (from .801 to .905) when X_2 is added to the model, and as a result, our conclusion is to retain this variable in the model. ■

COMMENTS   Both Example 15.7 and the t test for X_2 discussed on page 616 dealt with testing $H_0: \beta_2 = 0$ versus $H_a: \beta_2 \neq 0$. Both tests attempted to determine whether X_2 should be included as a predictor given that X_1 and X_3 were already included as predictor variables. Note: (1) the partial F-value = 6.6 = $(2.578)^2$ = (t value)², and (2) the p-value using the t test (.042) = the p-value using the F test (not shown).

We can see that these tests are *identical:* they result in exactly the same p-value and the same conclusion. This result demonstrates that to determine the predictive ability of an individual independent variable, we can compute the partial F statistic or the somewhat simpler t statistic. Some computer packages use the F statistics to summarize the individual predictors, whereas others (such as MINITAB) use the t values to measure the influence of each predictor. You should use whatever is provided (the F statistic or t statistic) to measure the partial effect of each variable; both sets of statistics accomplish the same thing.

Using Curvilinear Models: Polynomial Regression

Motormax produces electric motors for use in home furnaces. The company has formed a team of employees to examine the relationship between the dollars spent per week in inspecting finished products (X) and the number of motors produced during that week that were returned to the factory by the customer (Y). Motormax suspects that the number of returned motors will decrease as the amount spent on inspection increases but that after a certain point the decrease will slow down, that is, the number of returned motors will continue to decrease but at a slower rate. In other words, after spending a certain amount on inspecting finished product, they will reach a point where there will be little decrease in the number of returned motors, even though they spend a much larger amount on inspection.

The following data were gathered from company records covering 15 (nonconsecutive) weeks (the inspection expenditure (X) is in thousands of dollars):

■ FIGURE 15.8

MINITAB scatter
diagram of data for
inspection expense
example.

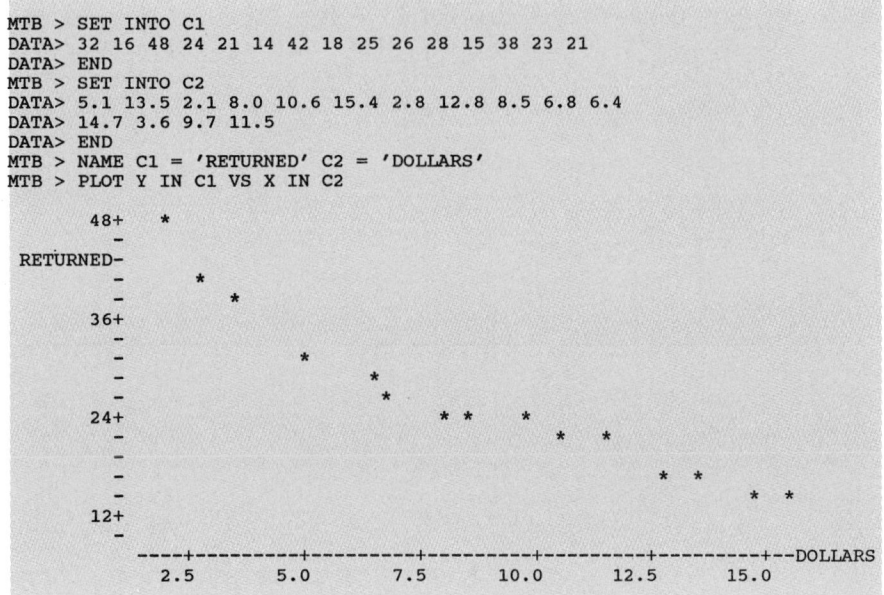

```
MTB > SET INTO C1
DATA> 32 16 48 24 21 14 42 18 25 26 28 15 38 23 21
DATA> END
MTB > SET INTO C2
DATA> 5.1 13.5 2.1 8.0 10.6 15.4 2.8 12.8 8.5 6.8 6.4
DATA> 14.7 3.6 9.7 11.5
DATA> END
MTB > NAME C1 = 'RETURNED' C2 = 'DOLLARS'
MTB > PLOT Y IN C1 VS X IN C2
```

WEEK	UNITS RETURNED (Y)	INSPECTION EXPENDITURES (X)
1	32	5.1
2	16	13.5
3	48	2.1
4	24	8.0
5	21	10.6
6	14	15.4
7	42	2.8
8	18	12.8
9	25	8.5
10	26	6.8
11	28	6.4
12	15	14.7
13	38	3.6
14	23	9.7
15	21	11.5

The scatter diagram is shown in Figure 15.8. Motormax seems to have a point—the number of returned motors does appear to level off after a certain amount of inspection expense.

Does the simple linear model $Y = \beta_0 + \beta_1 X + e$ capture the relationship between inspection expenditure (X) and number of units returned (Y)? Although Y does decrease as X increases here, the linear model does not capture the "slowing down" of Y for larger values of X. The least squares line (sketched in Figure 15.8) overpredicts Y for the middle range of X but underpredicts Y for small or large values of X.

Figure 15.9 shows **quadratic curves** rather than straight lines. If we include X^2 in the model, we can describe the curved relationship that seems to exist between the number of returned motors and inspection expense. More specifically, the left half of Figure 15.9a resembles the shape of the scatter diagram in Figure 15.8.

■ **FIGURE 15.9**   Quadratic curves. (a) Graph of $Y = 34 - 12X + 2X^2$. In general, this is the shape of $Y = \beta_0 + \beta_1 X + \beta_2 X^2$, where $\beta_2 > 0$. (b) Graph of $Y = 6 + 12X - 2X^2$. In general, this is the shape of $Y = \beta_0 + \beta_1 X + \beta_2 X^2$, where $\beta_2 < 0$.

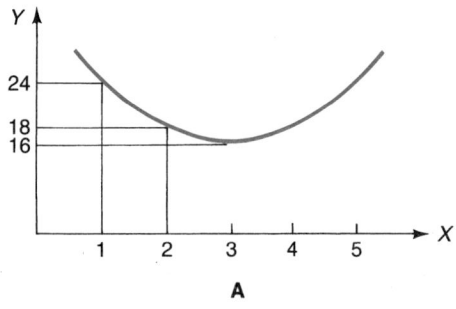

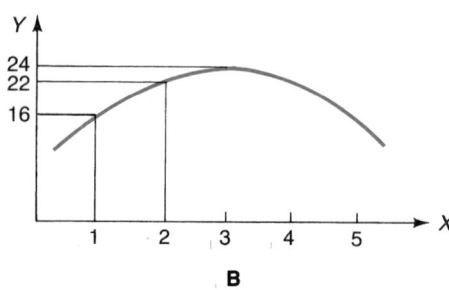

Consider the model

$$Y = \beta_0 + \beta_1 X + \beta_2 X^2 + e \qquad (15.18)$$

Is this a linear regression model? At first glance, it would appear not to be. However, by the word *linear* we really mean that the model is **linear in the unknown** β**'s**, not in X. In equation 15.18, there are no terms such as β_1^2, $\beta_1 \beta_2$, $\sqrt{\beta_0}$, and so on. So the model is linear in the β's, and this is a (multiple) linear regression application.

The model in equation 15.18 is a **curvilinear model** and is an example of **polynomial regression.** Such models are very useful when a particular independent variable and dependent variable exhibit a definite increasing and/or decreasing relationship that is nonlinear.

Solving for β_0, β_1, **and** β_2 Equation 15.18 represents a multiple regression model containing two predictors, namely, $X_1 = X$ and $X_2 = X^2$. The data for the model then are

Y	X_1	X_2
32	5.1	26.01 $(= 5.1^2)$
16	13.5	182.25 $(= 13.5^2)$
48	2.1	4.41
⋮	⋮	⋮
23	9.7	94.09
21	11.5	132.25

These data for Y, X_1, and X_2 are your input to the multiple linear regression computer program. You can simplify this task by letting the computer build the $X_2 = X^2$ column of data by squaring the entries in the $X_1 = X$ column.

EXAMPLE 15.8   Look at the MINITAB solution using the model $Y = \beta_0 + \beta_1 X + \beta_2 X^2 + e$ for the Motormax data shown in Figure 15.10.

1. Predict the number of returned motors for a week in which Motormax spends $20,000 on inspecting final product.
2. What do the F and t tests tell you about this model? Use $\alpha = .10$.

■ **FIGURE 15.10**
MINITAB solution
using $Y = \beta_0 + \beta_1 X + \beta_2 X^2$.

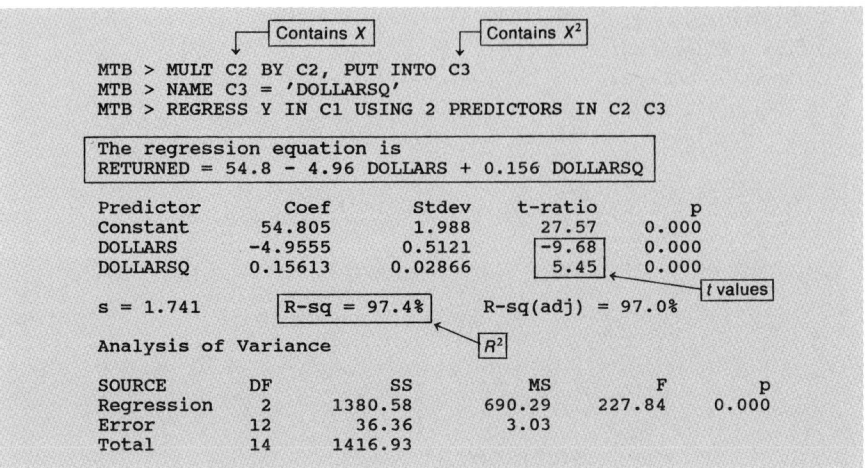

SOLUTION 1 The predicted number returned for $X_1 = 20$ (thousand) is

$$\hat{Y} = 54.8 - 4.96(20) + .156(20)^2 = 18.0$$

That is, 18 units.

SOLUTION 2 We first examine the F test. Our first test of hypothesis determines whether the overall model has predictive ability.

$$H_0: \beta_1 = \beta_2 = 0$$

$$H_a: \text{at least one of the } \beta\text{'s} \neq 0$$

Using the R^2 value from Figure 15.10 and equation 15.15,

$$F^* = \frac{.974/2}{.026/(15 - 2 - 1)} = \frac{.974/2}{.026/12} = 224.8*$$

As we might have suspected, this model does have significant predictive ability; $F^* = 224.8$ exceeds $F_{\alpha,k,n-k-1} = F_{.10,2,12} = 2.81$ from Table A.7.

Now we want to look at the t tests (same as partial F tests). Here, we examine each variable in the model, namely, X and X^2. The t value from Table A.5 is $t_{.10,12} = 1.356$ for a one-tailed test. We want to determine first whether $X_1 =$ inspection expenditure should be included in the model. Increased expenditure should be associated with decreased returns, so β_1 should be less than zero. We will therefore use a one-tailed procedure to test $H_0: \beta_1 \geq 0$ versus $H_a: \beta_1 < 0$.

According to Figure 15.10, the computed t statistic is $t^* = b_1/(\text{standard deviation of } b_1) = -9.68$. Now, $t^* = -9.68 < -1.356$, which means that the expenditure variable should be retained as a predictor of returns.

Next, we want to determine whether $X_2 = (\text{inspection expenditures})^2$ contributes significantly to the prediction of number returned. We are asking whether including the *quadratic term* was necessary. If this model is the correct one, then according to Figure 15.9a, β_2 should not only be unequal to zero but also, more specifically, should be greater than zero. This follows since if the number of returned motors does, in fact, level off after a certain amount of inspection expenditures, the curve should resemble the left half of the quadratic curve in Figure 15.9a.

---

*This differs from the MINITAB F value of 227.84 due to rounding.

■ **FIGURE 15.11**
Error resulting from
extrapolation. See text
and Figure 15.9b.

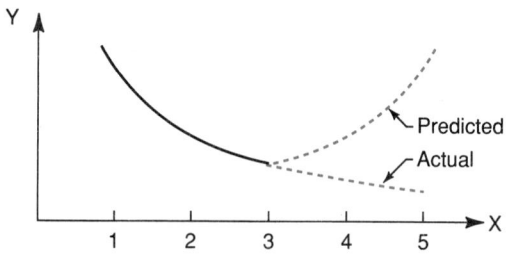

The appropriate hypotheses are

$$H_0: \beta_2 \leq 0$$
$$H_a: \beta_2 > 0$$

We reject H_0 if $t > t_{.10, 12} = 1.356$.

From Figure 15.10, we see that $t^* = b_2 /(\text{standard deviation of } b_2) = 5.45$. This value lies in the rejection region, so we conclude that $\beta_2 > 0$, which means that the quadratic term, X^2, contributes significantly and in the correct direction.

■

COMMENTS   There are three points you should note about the curvilinear model.

1. Curvilinear models often are used for situations in which the rate of increase or decrease in the dependent variable is not constant when plotted against a particular independent variable. The use of X^2 (and in some cases, X^3) in your model allows you to capture this nonlinear relationship between your variables.
2. There are other methods available for modeling a nonlinear relationship, including

$$Y = \beta_0 + \beta_1(1/X) + \text{error}$$

and

$$Y = \beta_0 + \beta_1 e^{-x} + \text{error}$$

These models also are (simple, here) linear regressions; they are linear in the unknown parameters. Unlike the quadratic model discussed previously, these models involve a **transformation** of the independent variable, X. When replacing X by the transformed X (such as $1/X$ or e^{-x}) in the model, one obtains many other curvilinear models that may better fit a set of sample data displaying a nonlinear pattern.
3. Avoid using the model $Y = \beta_0 + \beta_1 X + \beta_2 X^2 + e$ for values of X outside the range of data used in the analysis. Extrapolation is extremely dangerous when using this modeling technique. Consider Figure 15.9a, and suppose that values of X between 1 and 3 were used to derive the estimate of β_0, β_1, and β_2. Figure 15.11 shows the results. For values of X larger than 3, the predicting equation will turn up, whereas the actual relationship will probably continue to level off. So this model works for interpolation (for values of X between 1 and 3, here) but is extremely unreliable for extrapolation.

Exercises

15.21   If all the values of the dependent variable Y fell on the plane $\hat{Y} = 2 + 5.2X_1 + 10X_2$ and the variables X_1 and X_2 were used in a least squares fit to the data, what would be the value of the coefficient of determination?

15.22   If the residual deviation is 10.0 for a regression model with two independent variables and ten observations, what is the value of the adjusted R^2? What would it mean if the adjusted R^2 value dropped when another independent variable was added to this model? Assume that SST $= 5000$.

15.23   The manager of the personnel department of a computer firm is interested in knowing the relationship between the pay raise (Y) given to an employee of the firm and the following variables: yearly performance evaluation (X_1), years with the company (X_2), and number of credit hours of computer courses that the employee has taken in college (X_3). After observing

50 employees under different values of X_1, X_2, and X_3, the manager wishes to test that X_2 and X_3 contribute to predicting the variation in pay raises. The coefficient of determination for the model involving just Y and X_1 is .71. The coefficient of determination for the model with X_1, X_2, and X_3 is .82. Do the additional independent variables contribute significantly to the model? Use a 5% significance level.

15.24 A recent study found several good predictors of computer proficiency for female undergraduate students. The independent variables were GPA (grade point average), MEIND (major life event indicator), IMPORT (perceived importance of course), EFFORT (self-evaluated effort in class), and CCHS (computer courses in high school). The dependent variable was the score on a testing instrument that measures computer proficiency. Data collected from 83 females yielded an R^2 value of .64. At what significance level would the overall F test indicate that the model contributes to the prediction of computer proficiency for female undergraduate students? What is the value of the adjusted R^2?

(*Source:* Adapted from M. Simkin and B. Bowman, "Predictors of Computer Proficiency: Some Additional Evidence On Process Variables," *Proceedings of the 1990 Annual Meeting of the Decision Sciences Institute* (1990): 685.)

15.25 The dean of the college of business at Fargo University would like to see whether several variables affect a student's grade point average. Thirty first-year students were randomly selected and data were collected on the following variables:

Y = grade point average for the first year

X_1 = average time spent per month at fraternity or sorority functions

X_2 = average time spent per month working part time

X_3 = total number of hours of coursework attempted

The SSE for the least squares line involving only Y and X_1 was found to be 5.21. The SSE for the complete model was found to be 4.31. The SST is 24.1. At the 5% significance level, test the null hypothesis that the independent variables X_2 and X_3 do not contribute to predicting the variation in Y given that X_1 is already in the model.

15.26 A regression model was used to predict the total world production of crude oil (in millions of barrels per day). Two independent variables were used. The variable X_1 represented the crude oil production from OPEC and the variable X_2 represented the crude oil production from the USSR. Annual data were used for the years 1975–1989. The regression equation $\hat{Y} = 1.9 + .661X_1 + 3.52X_2$ had an R^2 of 80.0%. The regression equation $\hat{Y} = 47.9 + .3761X_1$ had an R^2 value of 45.7%. At the .01 significance level, does the variable X_2 contribute to the prediction of total world production of crude oil given that X_1 is in the model? What is the value of the adjusted R^2?

(*Source:* Adapted from *The World Almanac and Book of Facts* (1991): 181.)

15.27 In an effort to control a manufacturing plant's operating costs, a supervisor wishes to know the relationship between the time that it takes an employee to perform a task and the employee's mechanical aptitude and years of experience. The variables used in fitting a regression line were as follows:

Y = time it takes to perform the task

X_1 = mechanical aptitude level

X_2 = number of years of experience

$X_3 = X_1 X_2$

$X_4 = X_1^2$

$X_5 = X_2^2$

Data were collected on 15 randomly selected employees. The coefficient of determination for the model involving only the independent variables X_1 and X_2 is .53. The coefficient of determination for the model involving all five independent variables is .75. Do the variables X_3, X_4, and X_5 contribute to predicting the variation in the time that it takes to perform the task?

15.28   An economist would like to examine the relationship between personal savings and the following independent variables:

$$X_1 = \text{total personal income}$$

$$X_2 = \text{yield on U.S. Government securities}$$

$$X_3 = \text{consumer price index}$$

The following data were collected for 14 randomly selected months:

Y	X_1	X_2	X_3	Y	X_1	X_2	X_3
80.2	2077.2	12.036	233.2	107.4	2179.4	9.259	249.4
91.6	2086.4	12.814	236.4	116.8	2205.7	10.321	252.7
87.4	2101.0	15.526	239.8	102.1	2234.3	11.580	253.9
104.9	2102.1	14.003	242.5	97.9	2257.6	13.888	256.2
116.2	2114.1	9.150	244.9	93.3	2276.6	15.661	258.4
109.1	2127.1	6.995	247.6	83.6	2300.7	14.724	260.5
110.1	2161.2	8.126	247.8	91.0	2318.2	14.905	263.2

a.   Using a computerized statistical package, determine the least squares equation for these data.

b.   Use only the variables X_1 and X_2. What is the new prediction equation?

c.   Does the variable X_3 contribute to predicting the variation in personal savings, given that X_1 and X_2 are in the model? Use a 10% significance level.

15.29   The yields on tax-exempt bond funds are influenced by the average maturity of the bonds in the portfolio and by the percentage of cash that the portfolio manager is holding. Fifteen tax-exempt bond funds were selected, and the following data were recorded:

FUND NAME	YIELD (Y)	AVERAGE MATURITY (IN YEARS) (X_1)	PERCENTAGE OF ASSETS HELD IN CASH (X_2)
Cigna Municipal Bond	7.0	26.1	11.1
Composite Tax-Exempt Bond	6.6	8.4	4.9
Dreyfus Short-Intermediate Tax-Exempt	6.2	2.6	11.7
Eaton Vance Muni Bond	7.2	23.9	.4
Fidelity Insured Tax-Free	6.4	21.1	13.5
Fidelity Municipal Bond	6.6	19.6	21.1
Franklin Federal Tax-Free Income	7.6	25.0	3.0
IDS Tax-Exempt Bond	6.5	24.2	12.0
Keystone Tax-Exempt T Bond	6.6	22.6	7.0
Merrill Lynch Muni-Limited Maturity	6.2	1.2	12.0
Pru-Bache Muni-Insured Series B	6.0	24.7	10.0
Smith Barney Muni-National Portfolio	7.4	26.8	3.0
SteinRoe Intermediate Municipals	5.9	7.3	16.8
Tax-Exempt Bond of America	6.6	20.9	14.2
Vanguard Muni-Long-Term Portfolio	6.9	20.8	2.0

Complete the following ANOVA table and determine the significance of the contribution of the regression model with X_1 and X_2 in predicting the yields on tax-exempt bond funds. Use the p-value to determine your conclusions. Calculate the value of R^2 and the value of the adjusted R^2.

SOURCE	df	SS	MS	F
Regression		2.0834		
Error				
Total		3.4373		

(*Source:* Adapted from "The MONEY Rankings: Tax-Exempt Bond Funds," *Money* (February 1991): 162–168.)

15.30   The amount of money that a family spends monthly on food is believed to be related to the number of family members (X_1), the joint income of the husband and wife (X_2), and the age of the oldest child (X_3). A regression procedure was run with 50 observations on the model with X_1, X_2, and X_3. The SSE was 121,580 and the SST was 486,321. Another computer run also included the variables $\sqrt{X_1}$ and $\sqrt{X_3}$. The SSE for this complete model was 77,811. Do the variables $\sqrt{X_1}$ and $\sqrt{X_3}$ contribute to predicting the amount that a family spends on food? Use a 5% level of significance.

◢ 15.4

The Problem of Multicollinearity

Another possible title for this section is: What do the individual b_i's tell you? We discuss one of the common problems in the use (or misuse) of multiple linear regression—namely, trying to extract more information from the results than they actually contain.

We examine the validity of such statements as: Because $b_1 = 10$, increasing X_1 by 1 while *holding X_2 constant* will result in an increase of 10 in Y.

Assume that a sample of ten employees at Bellaire Industries was examined in an effort to determine the ability of age (X_1) and years of experience (X_2) to predict an employee's salary (Y). The following data were obtained:

EMPLOYEE	Y (SALARY)	X_1 (AGE)	X_2 (YEARS OF EXPERIENCE)	EMPLOYEE	Y (SALARY)	X_1 (AGE)	X_2 (YEARS OF EXPERIENCE)
1	52	52	33	6	60	55	30
2	35	47	21	7	31	36	8
3	45	38	14	8	38	40	15
4	28	25	3	9	33	32	7
5	42	44	18	10	48	50	27

First, we can ask how well X_1 (age) predicts Y (salary).

A MINITAB solution using the model $Y = \beta_0 + \beta_1 \cdot \text{(age)} + e$ is shown in Figure 15.12. Notice the computed t value. Now, $k = 1$ because this model consid-

■ **FIGURE 15.12**
MINITAB solution
to $Y(\text{salary}) = b_0 + b_1 \cdot \text{(age)}$.

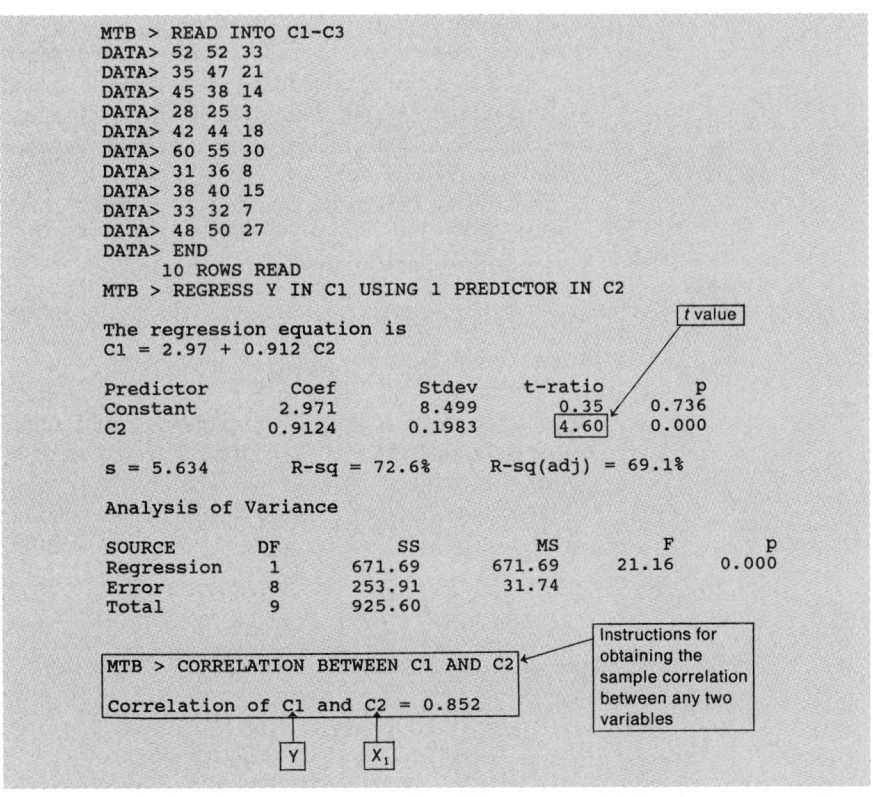

```
MTB > READ INTO C1-C3
DATA> 52 52 33
DATA> 35 47 21
DATA> 45 38 14
DATA> 28 25 3
DATA> 42 44 18
DATA> 60 55 30
DATA> 31 36 8
DATA> 38 40 15
DATA> 33 32 7
DATA> 48 50 27
DATA> END
     10 ROWS READ
MTB > REGRESS Y IN C1 USING 1 PREDICTOR IN C2

The regression equation is
C1 = 2.97 + 0.912 C2

Predictor      Coef      Stdev    t-ratio        p
Constant      2.971      8.499       0.35    0.736
C2           0.9124     0.1983       4.60    0.000

s = 5.634      R-sq = 72.6%     R-sq(adj) = 69.1%

Analysis of Variance

SOURCE        DF          SS         MS         F        p
Regression     1      671.69     671.69     21.16    0.000
Error          8      253.91      31.74
Total          9      925.60

MTB > CORRELATION BETWEEN C1 AND C2

Correlation of C1 and C2 = 0.852
```

t value

Instructions for obtaining the sample correlation between any two variables

Y X_1

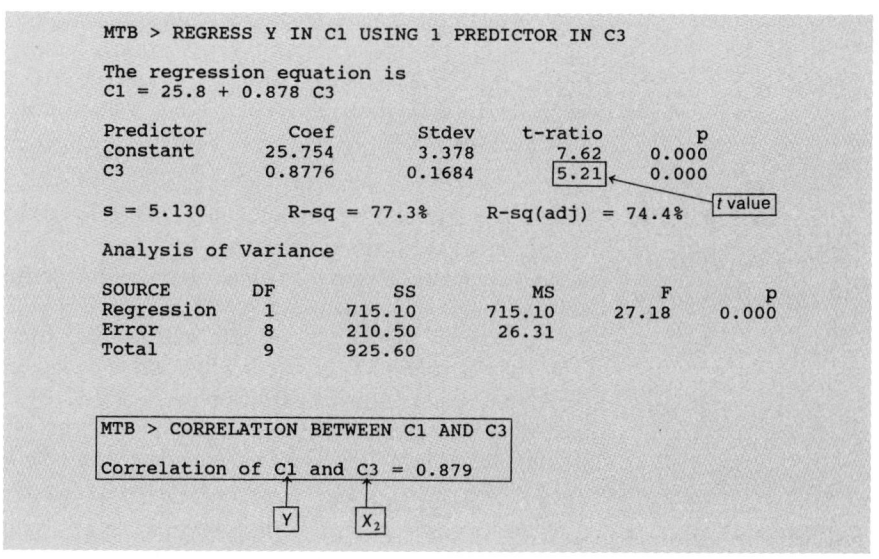

ers only one independent variable, so the tabulated value for comparison (using $\alpha = .10$) is $t_{\alpha/2, n-k-1} = t_{.05, 10-1-1} = t_{.05, 8} = 1.860$. The value of $t^* = 4.60$ is considerably larger than 1.86, so X_1 (age) is an excellent predictor of Y (salary).

What is the *correlation* between X_1 and Y? It seems reasonable that it would be quite large, because age has been shown to be a good predictor. In fact, according to Figure 15.12, this value is .852. So there is a *positive* relationship between age and salary, as one would expect.

Next, we determine how well X_2 (years of experience) predicts Y (salary). The solution using $Y = \beta_0 + \beta_1 \cdot$ (years of experience) $+ e$ is shown in Figure 15.13. Once again, the computed t value $= t^* = 5.21$ is much larger than $t_{.05, 8} = 1.860$, and the correlation between these two variables is .879. This result is not surprising; we might expect people with more years of experience to have higher salaries. Consequently, a significant positive relationship appears to exist between these two variables.

Finally, we turn to the question, how well do both X_1 (age) and X_2 (years of experience) predict salary? The model here is $Y = \beta_0 + \beta_1 X_1 + \beta_2 X_2 + e$. The least squares solution is shown in Figure 15.14.

$$\hat{Y} = 26.1 - .014 X_1 + .890 X_2$$

A few seemingly bizarre things show up here.

The coefficient of X_1 is $b_1 = -.014$. This result would appear to indicate that larger values of X_1 (older people) produce smaller salaries. But we know from our first analysis that the *opposite* is true. We would have expected a *positive* value of b_1 here, and so the coefficient of X_1 appears to have the wrong sign.

The small t values also are puzzling. The value of the F statistic (using Figure 15.14) is

$$F^* = \frac{R^2/2}{(1 - R^2)/(10 - 1 - 2)} = \frac{.773/2}{.227/7} = 11.9$$

As before, you can compute this value using the ANOVA table, where

$$F^* = \frac{\text{MSR}}{\text{MSE}} = \frac{357.55}{30.07} = 11.9$$

Using $\alpha = .10$, this value is much larger than $F_{.10,2,7} = 3.26$, and so the model does provide a very good predictor of Y. The coefficient of determination is $R^2 = .77$; these two predictor variables explain 77% of the total variation in the ten salary values.

The t values are very small; both are smaller in absolute value than $t_{\alpha/2, n-k-1} = t_{.05,10-2-1} = t_{.05,7} = 1.895$. Does this imply that both predictors are weak and should be removed from the model? Certainly not, as our previous analyses made clear.

This example demonstrates the problem of **multicollinearity.** In multiple regression models, it is desirable for each independent variable, X, to be highly correlated with Y, but it is *not* desirable for the X's to be highly correlated *with each other.* In business applications of multiple linear regression, the independent variables typically have a certain amount of pairwise correlation (usually positive). Extremely high correlation between any pair of variables can cause a variety of problems, as we will show.

The (sample) correlation between X_1 and X_2 is

$$r = \frac{\Sigma X_1 X_2 - (\Sigma X_1)(\Sigma X_2)/n}{\sqrt{\Sigma X_1^2 - (\Sigma X_1)^2/n}\ \sqrt{\Sigma X_2^2 - (\Sigma X_2)^2/n}}$$

This value, using Figure 15.14, is $r = .970$. Notice in the data set that nearly every time X_1 increases, so does X_2; X_1 and X_2 are highly correlated. As a result, these data contain a great deal of multicollinearity.

Implications

First of all, the correlation of X_1 and X_2 explains the small t values. Remember that each t value describes the contribution of that particular independent variable *after* all other independent variables have been included in the model. X_1 is very nearly a linear function of X_2 (as evidenced by $r = .970$), so it contributes very little to the prediction of Y, given that X_2 is in the model. The same argument applies to X_2. This means that neither X_1 nor X_2 is a strong predictor given that the other variable is included—not that each one is a weak predictor by itself.

The second implication of the multicollinearity is that the situation in which X_1

■ **FIGURE 15.14**
MINITAB solution to
Y (salary) $= b_0 + b_1 \cdot$ (age) $+ b_2 \cdot$ (years
of experience).

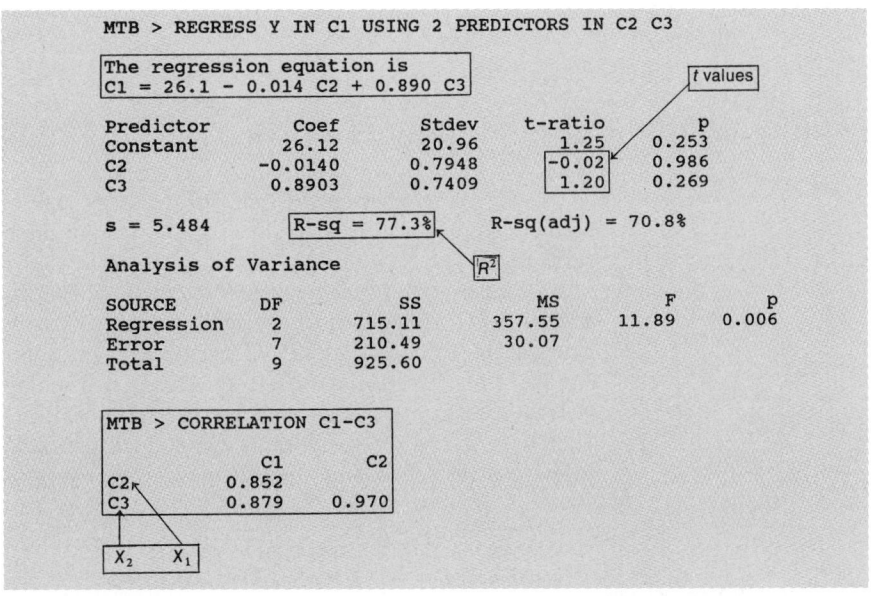

increases by 1 while X_2 remains constant never occurred in the sample data—as X_1 increased by 1, X_2 always changed also, because X_1 and X_2 are so highly correlated.

Finally, the sample coefficients (b_1 and b_2) of our independent variables have very large variances. If we took another sample from this population, the values of b_1 and b_2 probably would change dramatically—this is not a good situation. In fact, as this example has demonstrated, these coefficients can even have the "wrong" sign, a sign different from that obtained when regressing X_1 or X_2 alone on Y.

Eliminating the Effects of Multicollinearity

One way out of this dilemma is to remove some of the correlated predictors from the model. For this illustration, we should remove either X_1 or X_2 (but not both). Our best bet would be to retain X_2 = years of experience because it has the highest correlation with Y.

One method of eliminating correlated predictor variables is to use a **stepwise** selection procedure. This technique of selecting the variables to be used in a multiple linear regression equation is discussed in the next section. Essentially, it selects variables one at a time and generally (although not always) does not insert into the regression equation a variable that is highly correlated with a variable already in the equation. In the previous example, a stepwise procedure would have selected variable X_2 (the single best predictor of Y) and then informed the user that X_1 did not significantly improve the prediction of Y, given that X_2 was already included in the prediction equation.

Whenever you perform a multiple regression analysis, it is always a good idea to examine the pairwise correlations between all your variables, including the dependent variable. In this way, you often can easily detect the two independent variables that are contributing to the multicollinearity problem. These correlations can be obtained using a single command with most computer packages. The MINITAB command to generate a table (often called a **correlation matrix**) of pairwise correlations is shown in the bottom box in Figure 15.14. This output indicates that the correlation between Y and X_1 is .852, between Y and X_2 is .879, and between X_1 and X_2 is .970. Since the correlation of any variable with itself is 1, this particular correlation matrix is generally written as

$$
\begin{array}{c c}
 & \begin{array}{ccc} Y & X_1 & X_2 \end{array} \\
\begin{array}{c} Y \\ X_1 \\ X_2 \end{array} &
\left[\begin{array}{ccc}
1.0 & .852 & .879 \\
.852 & 1.0 & .970 \\
.879 & .970 & 1.0
\end{array} \right]
\end{array}
$$

Other, more advanced methods of detecting and treating the multicollinearity problem are beyond the scope of this text. One of the more popular procedures is *ridge regression*.*

We have seen that the problem of multicollinearity enters into our regression analysis when an independent variable is highly correlated with one or more other independent variables. Multicollinearity produces inflated regression coefficients that can even have the wrong sign. Also, the resulting t statistics can be small, making it difficult to determine the predictive ability of an individual variable. Therefore, b_1, b_2, . . . tell us nothing about the partial effect of each variable, unless we can demonstrate that there is no correlation among our predictor vari-

---

*For an excellent discussion of this topic, see J. Neter, W. Wasserman, and M. Kutner, *Applied Linear Regression Models* 2d ed. (Homewood, Ill.: Richard D. Irwin, 1989).

ables. In business applications, correlation (in particular, *positive* correlation) among the independent variables is far from unusual.

As a final note care should be taken in the selection of a model not to include variables that will be likely to produce multicollinearity. The detection and correction of the multicollinearity problem is often difficult to accomplish, and the methods discussed in this section are open to debate. The treatments of multicollinearity discussed so far are highly data dependent; that is, a new set of data could very well produce different results. Also, examining pairwise correlations may well miss the presence of multicollinearity, since multicollinearity will exist when one predictor variable is nearly a linear combination of two or more predictor variables and is not highly correlated with either of them. *In short, there is no easy way out of the multicollinearity problem.*

Exercises

15.31 What might cause the following situation to occur for a regression model with two independent variables? The t values for both β_1 and β_2 are nonsignificant. However, the F test for both $\beta_1 = \beta_2 = 0$ is highly significant.

15.32 If it is known that multicollinearity exists between three independent variables, how would you choose the independent variables that should remain in the model?

15.33 Refer to Exercise 15.2. Find the correlation between the number of units of personal computers sold and the advertising expenditure. Is multicollinearity a concern?

15.34 A least squares equation was fit to a set of data for an experiment and was found to be $\hat{Y} = 30 - 501X_1 + 300X_2$. The experiment was repeated and a new set of data from the same population was fit with the least squares line $\hat{Y} = -20 + 309X_1 - 151X_2$. Is there any explanation for these two different prediction equations?

15.35 The following set of data was collected:

Y	X_1	X_2	Y	X_1	X_2
2.02	1.01	.97	4.20	1.61	2.62
7.95	2.34	5.50	2.62	1.19	1.42
2.61	1.21	1.49	.07	.07	.01
.31	.23	.05	1.53	.80	.67
1.63	.85	.72	6.19	2.03	4.17

a. Construct the correlation matrix for the variables. Does multicollinearity appear to be a problem?

b. Find the coefficient of determination for the model using only X_1. Then find it using only X_2.

c. The coefficient of determination for the complete model is .9996. Does it appear that both variables, X_1 and X_2, should stay in the model?

15.36 Consider the following set of data for 12 emerging growth-oriented companies. Y represents the growth rate of a company for the current year, X_1 represents the growth rate of the company for the previous year, and X_2 represents the percent of the market that does not use the company's product or a similar product. All values are percentages.

Y	X_1	X_2	Y	X_1	X_2
20	10	30	30	15	60
24	12	35	36	42	38
18	15	25	47	45	40
33	30	40	35	32	32
27	19	32	28	24	31
20	24	20	32	⁻20	50

a. Construct the correlation matrix for the variables. Does multicollinearity appear to be a problem?

b. Find the coefficient of determination for the model with only X_1 included in the model.

c. Find the coefficient of determination for the model with only X_2 in the model.

d. The coefficient of determination for the complete model is .896. Does it appear from observing the values of the coefficient of determination in questions a and b that both variables X_1 and X_2 should stay in the model?

15.37 The marketing department of a local industry used a regression equation to predict monthly sales based on total advertising expenditure and television advertising expenditure (both in thousands of dollars). The least squares equation used to predict monthly sales is $\hat{Y} = 103.2 - .20X_1 + 3.4X_2$, where X_1 = total advertising expenditure and X_2 = television advertising expenditure. Can you assume that if television advertising expenditure stays constant and total advertising increases, monthly sales will decrease?

15.38 Looking at the regression equation,

$$\hat{Y} = 25 + 3.5X_1 - 1.8X_2$$

a business student interpreted the values as follows: $b_1 = 3.5$ means that Y increases by 3.5 for each unit increase in X_1 while X_2 is held constant. Similarly, $b_2 = -1.8$ means that Y decreases by 1.8 for each unit increase in X_2 while X_1 is held constant. Although this sounds very good in theory, what is the practical problem in trying to interpret the regression coefficients in this fashion?

15.5   The Use of Dummy Variables

The use of **dummy,** or **indicator,** variables in regression analysis allows you to include *qualitative* variables in the model. For example, if you wanted to include an employee's sex as a predictor variable in a regression model, define

$$X_1 = \begin{cases} 1 & \text{if female} \\ 0 & \text{if male} \end{cases}$$

Note that the choice of which sex is assigned the value of 1, male or female, is arbitrary. The estimated value of Y will be the same, regardless of which coding procedure is used.

Returning to the data we used in Example 15.2, the real-estate developer noticed that all the houses in the population were from three neighborhoods, A, B, and C. Taking note of which neighborhood each of the sampled houses was from led to the following data (in the discussion following Example 15.2, X_3 = years of education was shown to be a weak predictor and so is removed from the model here):

FAMILY	HOME SQUARE FOOTAGE (Y)	INCOME (X₁)	FAMILY SIZE (X₂)	NEIGHBORHOOD
1	16	22	2	B
2	17	26	2	C
3	26	45	3	A
4	24	37	4	C
5	22	28	4	B
6	21	50	3	C
7	32	56	6	B
8	18	34	3	B
9	30	60	5	A
10	20	40	3	A

Using these data, we can construct the necessary dummy variables and determine whether they contribute significantly to the prediction of home size (Y).

One way to code neighborhoods would be to define

$$X_3 = \begin{cases} 0 & \text{if neighborhood A} \\ 1 & \text{if neighborhood B} \\ 2 & \text{if neighborhood C} \end{cases}$$

However, this type of coding has many problems. First, because $0 < 1 < 2$, the codes imply that neighborhood A is smaller than neighborhood B, which is smaller than neighborhood C. Furthermore, any difference between neighborhoods A and C receives twice the weight (because $2 - 0 = 2$) of any difference between neighborhoods A and B or B and C. So this coding transforms data that are actually *nominal* to data that are *interval*, a much stronger type.

A better procedure is to use the necessary number of dummy variables (coded 0 or 1) to represent the neighborhoods. We needed one dummy variable with two categories (male and female) to specify a person's sex. To represent the three neighborhoods, we use two dummy variables by letting

$$X_3 = \begin{cases} 1 & \text{if house is in A} \\ 0 & \text{otherwise} \end{cases}$$

and

$$X_4 = \begin{cases} 1 & \text{if house is in B} \\ 0 & \text{otherwise} \end{cases}$$

Note that as for the male/female dummy variable, this coding is arbitrary as far as the prediction, $\hat{Y}$, is concerned. We could have assigned $X_3 = 0$ and $X_4 = 0$ to neighborhood A, with $X_3 = 1$ for B and $X_4 = 1$ for C.

What happened to neighborhood C? It is not necessary to develop a third dummy variable here because we have the following scheme:

HOUSE IS IN NEIGHBORHOOD	X_3	X_4
A	1	0
B	0	1
C	0	0

In fact, it can be shown that a third dummy variable is not only unnecessary, it is very important that you not include it. If you attempted to use three such dummy variables in your model, you would receive a message in your computer output informing you that "no solution exists" for this model. Suppose we had introduced a third dummy variable (say, X_5) that was equal to 1 if the house was in neighborhood C. For each observation in the sample, we would have

$$X_5 = 1 - X_3 - X_4$$

Whenever any one predictor variable is a linear function (including a constant term) of one or more other predictors, then mathematically *no solution exists* for the least squares coefficients, since you have multicollinearity at its worst. To arrive at a usable equation, any such predictor variable must not be included.

The resulting model here is*

$$Y = \beta_0 + \beta_1 X_1 + \beta_2 X_2 + \beta_3 X_3 + \beta_4 X_4 + e$$

---

*Models that include dummy variables typically contain terms that reflect any interaction between the dummy variables and the other quantitative variables. For this model, this would amount to adding four additional terms to the model, namely, $X_1 X_3$, $X_1 X_4$, $X_2 X_3$, and $X_2 X_4$. Such a model would require a larger sample size (n) than that used in this illustration, since the model would then contain $k = 8$ predictor variables.

The final array of data (ready for input into a computer program) is

ROW	Y	X_1	X_2	X_3	X_4	ROW	Y	X_1	X_2	X_3	X_4
1	16	22	2	0	1	6	21	50	3	0	0
2	17	26	2	0	0	7	32	56	6	0	1
3	26	45	3	1	0	8	18	34	3	0	1
4	24	37	4	0	0	9	30	60	5	1	0
5	22	28	4	0	1	10	20	40	3	1	0

where Y = square footage of home, X_1 = income, X_2 = family size, X_3 = 1 if neighborhood A, and X_4 = 1 if neighborhood B.

A MINITAB solution is shown in Figure 15.15. To determine whether the particular neighborhood has any effect on the prediction of home size, we test

$$H_0: \beta_3 = \beta_4 = 0 \quad \text{(if } X_1 \text{ and } X_2 \text{ are included)}$$

$$H_a: \beta_3 \text{ or } \beta_4 \text{ (or both)} \neq 0$$

In the complete model, the variables are X_1, X_2, X_3, and X_4, and from Figure 15.15(a),

$$R_c^2 = .921$$

■ FIGURE 15.15
MINITAB solution
to real-estate dummy
variable problem. (a)
Solution using variables
X_1, X_2, X_3, and X_4.
(b) Solution using
variables X_1, X_2.

```
MTB > READ INTO C1-C5
DATA> 16 22 2 0 1
DATA> 17 26 2 0 0
DATA> 26 45 3 1 0
DATA> 24 37 4 0 0
DATA> 22 28 4 0 1
DATA> 21 50 3 0 0
DATA> 32 56 6 0 1
DATA> 18 34 3 0 1
DATA> 30 60 5 1 0
DATA> 20 40 3 1 0
DATA> END
        10 ROWS READ
MTB > REGRESS Y IN C1 USING 4 PREDICTORS IN C2-C5

The regression equation is
C1 = 7.77 + 0.082 C2 + 3.27 C3 + 1.61 C4 - 0.90 C5

Predictor      Coef       Stdev     t-ratio       p
Constant       7.772      2.557       3.04      0.029
C2             0.0819     0.1059      0.77      0.474
C3             3.2696     0.9870      3.31      0.021
C4             1.613      1.801       0.90      0.411
C5            -0.900      1.841      -0.49      0.646

s = 2.036      R-sq = 92.1%    R-sq(adj) = 85.8%

A                   ↑
                  $R_c^2$

MTB > REGRESS Y IN C1 USING 2 PREDICTORS IN C2 C3

The regression equation is
C1 = 6.74 + 0.165 C2 + 2.66 C3

Predictor      Coef       Stdev     t-ratio       p
Constant       6.740      2.164       3.11      0.017
C2             0.16485    0.07306     2.26      0.059
C3             2.6569     0.7405      3.59      0.009

s = 1.953      R-sq = 89.8%    R-sq(adj) = 86.9%

B                   ↑
                  $R_r^2$
```

In the reduced model, the variables are X_1 and X_2 only, and from Figure 15.15(b),

$$R_r^2 = .898$$

At first glance, it does not appear that X_3 and X_4 produced a significant increase in R^2. The partial F test will determine whether this is true.

$$F = \frac{(R_c^2 - R_r^2)/(\text{number of } \beta\text{'s in } H_0)}{(1 - R_c^2)/[n - 1 - (\text{number of } X\text{'s in the complete model})]}$$

$$= \frac{(.921 - .898)/2}{(1 - .921)/(10 - 1 - 4)} = \frac{.023/2}{.079/5} = .73$$

Using $\alpha = .10$, this result is considerably less than $F_{.10,2,5} = 3.78$, so there is no evidence that the neighborhood dummy variables significantly improve the prediction of home size.

In this example, the dummy variables were not significant predictors in the model. However, do not let this mislead you. In many business applications, dummy variables representing location, weather conditions, yes/no situations, time, and many other variables can have a tremendous effect on improving the results of a multiple regression model.

Stepwise Procedures

Assume you wish to predict annual divisional profits for a large corporation using, among other techniques, a multiple linear regression model. Your strategy is to consider any variable that you think *could* have an effect on these profits. You have identified twelve such variables.

One possibility is to include all these variables in your model and to use the *t* tests to decide which variables are significant predictors. However, this procedure invites multicollinearity, because your model is more apt to include correlated predictors, severely hindering the interpretation of your model. In particular, two independent variables that are very highly correlated may both have small *t* values (as we saw in the employee example), causing you possibly to discard both of them from the model—this is *not* the right thing to do because you possibly should have retained one of them.

A better way to proceed here is to use one of the several stepwise selection procedures. These techniques either choose or eliminate variables, one at a time, in an effort not to include those variables that either have no predictive ability or are highly correlated with other predictor variables. A word of caution—these procedures do not provide a guarantee against multicollinearity; however, they greatly reduce the chances of including a large set of correlated independent variables.

These procedures consist of three different selection techniques: (1) forward regression, (2) backward regression, and (3) stepwise regression.

Forward Regression The forward regression method of model selection puts variables into the equation, one at a time, beginning with that variable having the highest correlation (or R^2) with Y. For sake of argument, call this variable X_1.

Next, it examines the remaining variables for the variable that, when included with X_1, has the highest R^2. That predictor (with X_1) is inserted into the model. This procedure continues until adding the "best" remaining variable at that stage results in an insignificant increase in R^2 according to the partial F test.

Backward Regression Backward regression is the opposite of forward regression: it begins with *all* variables in the model and, one by one, removes them. It begins by finding the "worst" variable—the one that causes the smallest decrease in R^2 when removed from the complete model. If the decrease is insignificant, this variable is removed, and the process continues.

The variable among those remaining in the model that causes the smallest decrease in the new R^2 is considered next. You continue this procedure of removing variables until a significant drop in R^2 is obtained, at which point you replace this significant predictor and terminate the selection.

Will the model resulting from a backward regression be the same as that obtained using forward regression? Not necessarily; usually, however, the resulting models are very similar. Of course, if two variables are highly correlated, the forward procedure could choose one of the correlated predictors, whereas the backward procedure could choose the other.

Stepwise Regression   Stepwise regression is a modification of forward regression. *It is the most popular and flexible of the three selection techniques.* It proceeds exactly as does forward regression, except that at each stage it can *remove* any variable whose partial F-value indicates that this variable does not contribute, given the present set of independent variables in the model. Like forward regression, it stops when the "best" variable among those remaining produces an insignificant increase in R^2.

Figure 15.16 illustrates this procedure for the example on predicting divisional profits (the data are not shown). Data from all 12 independent variables, as well as from Y, are used as input to a stepwise regression program. One possible outcome from this analysis is shown in Figure 15.16.

The stepwise solution for the data we used to predict home size is contained in the end-of-chapter MINITAB appendix. As we previously determined, X_3 = educational level does not contribute significantly, and so the resulting prediction equation includes only X_1 = income and X_2 = family size. This equation is

$$\hat{Y} = 6.74 + .165X_1 + 2.66X_2$$

■ **FIGURE 15.16**
Possible solution using stepwise regression on divisional profits data.

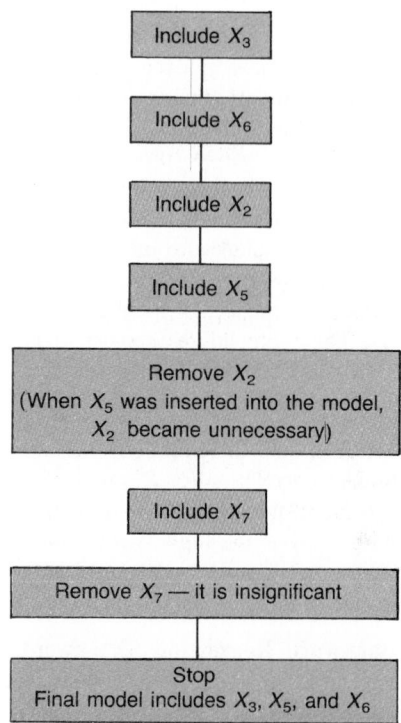

Using Dummy Variables in Forward or Stepwise Regression We emphasized that $C - 1$ dummy variables should be used to represent C categories if *all* the dummy variables were to be inserted into the regression equation. When using a forward or stepwise regression procedure, this may not be the best way to proceed, as the following illustration shows.

Suppose you are using nine dummy (indicator) variables to represent ten cities. The dependent variable is monthly sales, and the purpose is to determine which city (or cities) exhibits very large or very small sales. If a forward or stepwise selection procedure is used, then including one of these dummy variables indicates that specifying this particular city significantly improves the prediction of sales. In other words, it indicates that sales for this city are not just average but are much higher (its coefficient will be positive) or lower (its coefficient will be negative) than average.

When you use the forward or stepwise techniques, you probably will not include all nine dummy variables in the model. Your ability to predict sales (Y) is unaffected by not defining a tenth dummy variable and, in fact, as pointed out earlier, the regression analysis will not accept all ten dummy variables.

For this situation, however, there is the danger of not detecting extremely high or low sales in the tenth city, which did not receive a dummy variable. When including these variables one at a time in the regression equation using a forward or stepwise procedure, we can allow the regression model to examine the effect of all ten cities. We do this by defining ten such dummy variables, one for each city.*

Because a forward regression procedure generally will not attempt to include all ten dummy variables, you are able to investigate the existence of high or low sales in each of the ten cities. When using dummy variables in a forward or stepwise regression procedure, it is perfectly acceptable to use C such variables to represent C categories.

Checking the Assumptions: Examination of the Residuals

When you use a multiple linear regression model, you should keep two points in mind. First, no assumptions are necessary to derive the least squares estimates of β_1, β_2, β_3, ... The regression coefficients b_1, b_2, b_3, . . . determined by a computer solution are the "best" estimates, in the least squares sense.

Second, several key assumptions *are* required to construct confidence intervals and perform any test of hypothesis. If these assumptions are violated, you may still have an accurate prediction, $\hat{Y}$, but the validity of these inference procedures will be very questionable.

Your final step in any regression analysis should be to verify your assumptions.

Assumption 1 The errors are normally distributed, with a mean of zero.

An easy method to determine whether the errors follow a normal distribution, centered at zero, is to let the computer construct a histogram of the sample residuals ($Y - \hat{Y}$). Since the residuals *always* sum to zero, the residual histogram is typically centered at zero. This plot should reveal whether the distribution of residuals is severely skewed.

Consider the 50 residuals resulting from the analysis in Example 15.4. The computer solution for this problem is shown in Figure 15.6. Notice that the RESIDUALS subcommand following the REGRESS command can be used to store the residuals in any column (such as column C5 in Figure 15.6).

---

*This problem is discussed in D. Dorsett and J. T. Webster, "Guidelines for Variable Selection Problems When Dummy Variables Are Used," *The American Statistician* 37, no. 4 (1983): 337.

```
MTB > HISTOGRAM OF C5, FIRST MIDPOINT AT -500, CLASS WIDTH OF 125

Histogram of C5    N = 50

Midpoint     Count
    -500         2    **
    -375         5    *****
    -250         6    ******
    -125         8    ********
       0        12    ************
     125         5    *****
     250         3    ***
     375         6    ******
     500         3    ***
```

A MINITAB histogram of these values is shown in Figure 15.17. The distribution of these residuals appears to be centered at zero and, except for a slight left skew, is bell-shaped (normally distributed) in appearance. Remember that an exact normal distribution is not necessary here; problems arise only when the distribution is severely skewed and does not resemble a normal distribution.

More sophisticated methods of checking this assumption involve the use of a *probability plot*, or a *chi-square goodness-of-fit test*. We do not discuss the probability plot technique here, except to say that you plot the residuals in a specialized type of graph. If the resulting graph is basically linear in appearance, the normality assumption has been verified. The goodness-of-fit test was discussed in Chapter 13, where we used a chi-square statistic to test the hypothesis that a particular set of data (in this case, the regression residuals) came from a specific distribution. The end-of-chapter exercises in Chapter 13 discuss how to use the chi-square test for a suspected *normal* population.

If you have reason to believe that this assumption of your model has been violated, then you need to search for another model. This new model may include additional predictor variables that have been overlooked. Another possibility is to transform the dependent variable (for example, use $\sqrt{Y}$ rather than Y) or transform one or more of the predictor variables. As your model tends to "improve," you should observe the residuals tending toward a normal distribution.

Assumption 2   The variance of the errors remains constant. For example, you should not observe larger errors associated with larger values of $\hat{Y}$.

When the residuals $(Y - \hat{Y})$ are plotted against the predicted values $(\hat{Y})$, we hope to observe *no pattern* (a "shotgun blast" appearance) in this graph, as in Figure 15.18a. Remember—the assumptions are essentially that the errors consist of what engineers call *noise*, with no observable pattern.

A common violation of the assumption of equal variances occurs when the value of the residual increases as $\hat{Y}$ or an individual predictor increases, as is illustrated in Figure 15.18b. In this figure, the variance of the residual increases with $\hat{Y}$. This has a serious effect on the validity of the hypothesis tests developed in this

■ FIGURE 15.18
Examination of the residuals. (a) The shotgun effect (no violation of assumptions 1 and 2). (b) A violation of the equal variance assumption (assumption 2).

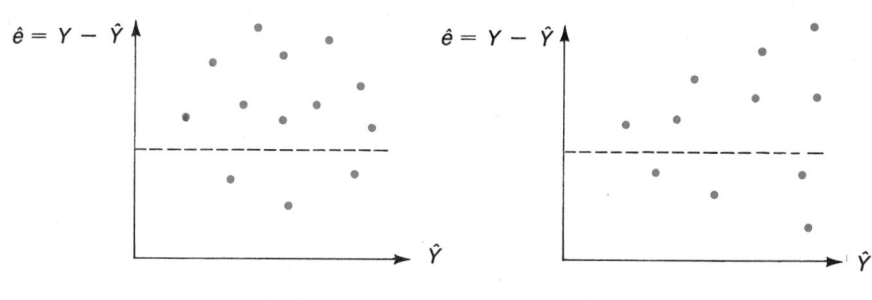

A   B

■ **FIGURE 15.19**
Autocorrelated errors.

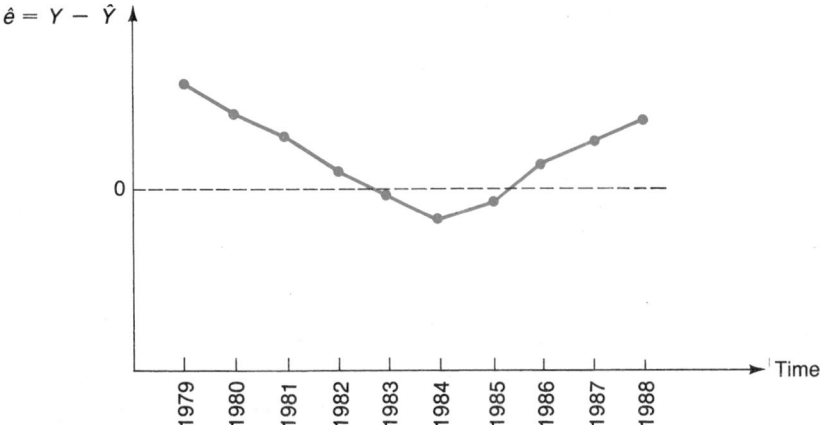

chapter, which determine the strength of the regression model and the individual predictors.

When you encounter a violation of this type, you need to resort to more advanced modeling techniques, such as *weighted least squares* or *transformations* of your dependent variable.*

Assumption 3 The errors are independent.

To examine this assumption after the regression equation has been determined involves using the residual from each of the sample observations. For given values of $X_1, X_2, \ldots, X_k$, the actual error is

$$e = Y - (\beta_0 + \beta_1 X_1 + \beta_2 X_2 + \cdots + \beta_k X_k)$$

The β's are unknown, so we estimate the error by using the residual for this particular observation,

$$Y - \hat{Y} = Y - (b_0 + b_1 X_1 + b_2 X_2 + \cdots + b_k X_k)$$

The residuals for the real-estate data are shown in the third column, labeled $Y - \hat{Y}$, of the table on page 608. In general, a close examination of these values will reveal any departures from the regression assumptions.

When your regression data consist of *time series* data, your errors often are not independent. This type of data has the appearance

TIME	Y	X_1	X_2	$\cdots$	X_k
1979	x	x	x		x
1980	x	x	x		x
1981	x	x	x		x
⋮	⋮	⋮	⋮		⋮

Also remember that the error component *includes the effect of missing variables* in your model. In many business applications, there is a positive relationship between time-related predictor variables, such as prices and wages, because they both increase over time. This relationship can produce a set of residuals in your regression analysis that are not independent of one another but instead display a pattern similar to the one in Figure 15.19. This plot contains the sample residuals on the vertical axis and time on the horizontal axis. If this assumption were *not*

*See J. Neter, W. Wasserman, and M. Kutner, *Applied Linear Regression Models* 2d ed. (Homewood, Ill.: Richard D. Irwin, 1989).

violated here, we would observe the shotgun appearance. Instead we notice that adjacent residuals have roughly the same value and so are correlated with each other. This is **autocorrelation.** To be more specific, the pattern in Figure 15.19 is one of *positive* autocorrelation. Negative autocorrelation exists when most of the neighboring residuals are very unequal in size.

The amount of autocorrelation that exists in residuals is measured by the **Durbin-Watson statistic.** This statistic ranges from 0 to 4, with a value near 0 indicating strong *positive* autocorrelation and a value close to 4 meaning that there is a significant *negative* autocorrelation. A value near 2 indicates that there is no (or very little) autocorrelation—the ideal situation. Chapter 17 discusses the calculation of this statistic and its use in detecting autocorrelated residuals. Also, Chapter 19 (in particular, Example 19.3) discusses a nonparametric procedure, the runs test, that can be used to determine whether the residuals are significantly autocorrelated. The runs test does not assume (unlike the Durbin-Watson test) that the residuals are from a normal population.

The problem of autocorrelated errors is the most difficult of the three assumptions to correct. The error term is not noise, as we originally assumed, but instead has a definite pattern (as in Figure 15.19). Several ways of treating this problem are discussed in Chapter 17.

Prediction Using Multiple Regression

Once a regression equation has been derived, its primary application generally is to derive predicted values of the dependent variable. Computer packages provide an easy method of deriving such an estimate. To illustrate, consider the regression equation we developed for the real-estate data. For this illustration we include X_3 = years of formal education, although, as we demonstrated in Section 15.2, this variable could be dropped without any significant loss in the prediction of home size. The resulting prediction equation was

$$\hat{Y} = 7.60 + .194X_1 + 2.34X_2 - .163X_3$$

Consider a situation in which

$$X_1 = \text{income} = 36 \text{ (thousands of dollars)}$$

$$X_2 = \text{family size} = 4$$

$$X_3 = \text{years of formal education} = 8 \text{ (years)}$$

The predicted home size (Y) here is

$$\hat{Y} = 7.60 + .194(36) + 2.34(4) - .163(8)$$
$$= 22.64 \quad (2264 \text{ square feet})$$

We can derive this predicted value using MINITAB, SAS, or SPSS. When using the latter two packages, we can derive a predicted value by adding one additional row to our input data containing these specific values of the predictor variables and a *missing value* for the dependent variable value. The computer routine ignores this row when deriving the regression equation (and all subsequent tests of hypotheses) but attaches this row when listing the predicted values. The procedure when using MINITAB is slightly different, as described next.

Using MINITAB   The MINITAB solution for the preceding illustration is shown in Figure 15.20. By using the subcommand PREDICT following the REGRESS command, you can easily derive a predicted value for the input values of the independent variables. The output will contain the predicted (fitted) Y value, the standard deviation of the predicted Y, a 95% confidence interval, and a 95% prediction

■ **FIGURE 15.20**
Prediction for new
data using MINITAB.
For the input data,
See Figure 15.2.

```
MTB > BRIEF 3
MTB > REGRESS Y IN C1 USING 3 PREDICTORS IN C2-C4;
SUBC> PREDICT FOR 36 4 8.

The regression equation is
C1 = 7.60 + 0.194 C2 + 2.34 C3 - 0.163 C4

Predictor       Coef      Stdev    t-ratio         p
Constant       7.596      2.595       2.93     0.026
C2           0.19388    0.08770       2.21     0.069
C3            2.3381     0.9078       2.58     0.042
C4           -0.1628     0.2441      -0.67     0.530

s = 2.035        R-sq = 90.5%      R-sq(adj) = 85.8%

Analysis of Variance

SOURCE        DF         SS         MS        F        p
Regression     3    237.542     79.181    19.11    0.002
Error          6     24.858      4.143
Total          9    262.400


Obs.      C2         C1     Fit Stdev.Fit  Residual   St.Resid
  1     22.0     16.000  15.886    1.193     0.114       0.07
  2     26.0     17.000  16.010    1.161     0.990       0.59
  3     45.0     26.000  22.195    0.975     3.805       2.13R
  4     37.0     24.000  24.121    1.300    -0.121      -0.08
  5     28.0     22.000  22.051    1.371    -0.051      -0.03
  6     50.0     21.000  22.676    1.331    -1.676      -1.09
  7     56.0     32.000  31.179    1.854     0.821       0.98
  8     34.0     18.000  19.900    0.949    -1.900      -1.05
  9     60.0     30.000  30.593    1.608    -0.593      -0.48
 10     40.0     20.000  21.388    0.767    -1.388      -0.74

R denotes an obs. with a large st. resid.

   Fit  Stdev.Fit        95% C.I.             95% P.I.
22.625      1.355   ( 19.308, 25.943)   ( 16.640, 28.611)
```
$\hat{Y}$ using $X_1 = 36$, $X_2 = 4$, $X_3 = 8$

interval. The resulting predicted value for $X_1 = 36$, $X_2 = 4$, and $X_3 = 8$ is $\hat{Y} = 22.625$, which is more accurate than the previously derived value of 22.64.

Using SPSS   A solution for this illustration using SPSS is presented at the end of the chapter. A numeric value (such as -9) is used for the missing Y value, and then SPSS is informed that such a value represents a missing value using the MISSING VALUES command. This command instructs SPSS that a value of -9 appearing as a value for SQFOOT (the Y variable) should be interpreted as a missing value. The resulting predicted value of $\hat{Y} = 22.625$ appears at the end of the list of predicted values.

Using SAS   Much like SPSS, SAS predicts values by including an additional row containing a missing value for the dependent variable. Any single *letter* (A to Z) can be used to represent the missing value; SAS is informed that such a character represents a missing value by using the MISSING statement after the DATA statement. This row is then automatically ignored during subsequent calculations, but, once again, the predicted value of $\hat{Y} = 22.625$ is generated for this set of X values at the end of the list of predicted values.

Confidence and Prediction Intervals   In the preceding illustration, what does $\hat{Y} = 22.625$ estimate? For ease of notation, let X_0 represent the set of X values used for this estimate; that is, $X_0 = (36, 4, 8)$, where $X_1 = 36$, $X_2 = 4$, and $X_3 = 8$. This value of $\hat{Y}$ estimates (1) the *average* home size of all families with this specific set of X values, written $\mu_{Y|X_0}$ and (2) the home size for an *individual* family having this specific set of X values, written Y_{X_0}.

Using the notation from Chapter 14, let

$$s_{\hat{Y}} = \text{the standard deviation of the predicted } Y \text{ mean}$$

These values can be computed and included in the output by each of the computer packages. To determine the reliability of this particular point estimate, $\hat{Y}$, you can (1) derive a *confidence interval* for $\mu_{Y|X_0}$ if your intent is to estimate the *average* value of Y given X_0 (not the usual situation) or (2) derive a *prediction interval* for Y_{X_0} if the purpose is to forecast an *individual* value of Y given this specific set of values for the predictor variables. In business applications, deriving a specific forecast is, by far, the more popular use of linear regression.

These intervals are summarized as follows. A $(1 - \alpha) \cdot 100\%$ confidence interval for $\mu_{Y|X_0}$ is

$$\hat{Y} - t_{\alpha/2, n-k-1} s_{\hat{Y}} \quad \text{to} \quad \hat{Y} + t_{\alpha/2, n-k-1} s_{\hat{Y}} \qquad \textbf{(15.19)}$$

A $(1 - \alpha) \cdot 100\%$ prediction interval for Y_{X_0} is

$$\hat{Y} - t_{\alpha/2, n-k-1} \sqrt{s^2 + s_{\hat{Y}}^2} \quad \text{to} \quad \hat{Y} + t_{\alpha/2, n-k-1} \sqrt{s^2 + s_{\hat{Y}}^2} \qquad \textbf{(15.20)}$$

Using MINITAB $[X_0 = (36, 4, 8)]$ Prediction and confidence intervals are easy to derive using MINITAB by using the PREDICT subcommand, as illustrated in Figure 15.20. The resulting confidence interval for the average home size, given X_0, is (19.308, 25.943). As usual, the 95% prediction interval for an individual family with this set of X values, namely, (16.640, 28.611), is wider than the corresponding confidence interval.

Using SPSS $[X_0 = (36, 4, 8)]$ These intervals are not directly available on this package but can be obtained easily using equations 15.19 and 15.20 and the regression output. SPSS provides the estimated value, $\hat{Y}$, and the standard deviation of the predicted value, $s_{\hat{Y}}$. Referring to the discussion at the end of the chapter, these values are $\hat{Y} = 22.625$ and $s_{\hat{Y}} = 1.355$.

Using equation 15.19, a 95% confidence interval for $\mu_{Y|X_0}$ is derived by first using Table A.5 to obtain $t_{\alpha/2, n-k-1} = t_{.025, 10-3-1} = t_{.025, 6} = 2.447$. The resulting confidence interval is

$$22.625 - (2.447)(1.355) \quad \text{to} \quad 22.625 + (2.447)(1.355)$$
$$= 22.625 - 3.316 \quad \text{to} \quad 22.625 + 3.316$$
$$= 19.309 \quad \text{to} \quad 25.941$$

Consequently, we have estimated the average home size for families with $X_1 = 36$, $X_2 = 4$, $X_3 = 8$ to within 331.6 square feet of the actual mean with 95% confidence. This confidence interval is slightly different from that obtained with MINITAB due to rounding error.

The prediction interval from equation 15.20 is derived by using MSE = 4.143 from the computer solution to obtain

$$22.625 - 2.447 \sqrt{4.143 + (1.355)^2} \quad \text{to}$$
$$22.625 + 2.447 \sqrt{4.143 + (1.355)^2}$$
$$= 22.625 - 5.983 \quad \text{to} \quad 22.625 + 5.983$$
$$= 16.642 \quad \text{to} \quad 28.608$$

This means that we have predicted the home size of an individual family with $X_1 = 36$, $X_2 = 4$, and $X_3 = 8$ to within 598.3 square feet of the actual value with 95% confidence. Once again, this result differs slightly from that obtained with MINITAB due to rounding error.

Using SAS [$X_0 = (36, 4, 8)$] SAS provides an easy method of determining these intervals by including CLI (for an individual Y_{X_0}), CLM (for a mean $\mu_{Y|X_0}$), or both in the final SAS statement. According to the output at the end of the chapter, the 95% confidence interval for the average home size, given X_0, is (19.309, 25.942), with a corresponding prediction interval of (16.641, 28.609).

Exercises

15.39 A real-estate agency was interested in the amount of rent paid (monthly) for commercial buildings near downtown Houston. A broker at the real-estate agency used the following independent variables:

$$X_1 = \text{age of building}$$

$$X_2 = \begin{cases} 1 & \text{if low-rise building} \\ 0 & \text{if not} \end{cases}$$

$$X_3 = \begin{cases} 1 & \text{if mid-rise building} \\ 0 & \text{if not} \end{cases}$$

$$X_4 = \text{square footage of building}$$

The data from 20 commercial buildings yielded the least squares equation $\hat{Y} = 130 - 50X_1 + 70X_2 - 20X_3 + 0.31X_4$.

a. What is the predicted rent for a five-year-old high-rise commercial building that has 25,000 square feet of space?

b. Test the hypothesis that the dummy variables significantly improve the prediction of monthly rent for commercial buildings. Assume that R^2 for the model including only X_1 and X_4 is .65 and that R^2 for the complete model including X_1, X_2, X_3, and X_4 is .85.

15.40 If an economist is interested in examining the relationship between household income and household recreational expenses over time, then the economist would use *time series* data. However, if an economist is interested in estimating household recreational expenses as a function of household income, then he or she would use *cross-sectional* data. A set of cross-sectional data were collected from a sample of 30 households in a large metropolitan area. The independent variables are yearly household income (in thousands), X_1, and house payment (either rent or mortgage), X_2. The dependent variable was annual household recreational expenses. The least squares line is $\hat{Y} = 51.3 + 12.3X_1 + .11X_2$. Given that the standard deviations of the estimates of the coefficients of X_1 and X_2 are 5.59 and .048, test the hypothesis that the variable X_1 contributes to the prediction of Y, given that X_2 is in the model. Also, test that X_2 contributes to the prediction of Y, given that X_1 is in the model. Use a .05 significance level.

15.41 Nebraska Associated Insurance handles workers' compensation insurance for three large manufacturing firms. The insurance company believes that the following independent variables are important in determining the total amount of compensation paid for each claim from the three manufacturers:

age

sex

marital status

length of employment

type of injury (to a limb, to the head, or to other parts of the body)

manufacturer employing the worker

Set up an appropriate regression model to predict total amount of compensation paid based on the independent variables. Define your variables.

15.42   Data are collected for the variables Y, X_1, and X_2. A computer printout of the correlation matrix is:

	Y	X_1	X_2
Y	1	.49	.30
X_1	.49	1	.12
X_2	.30	.12	1

a.   Which independent variable, X_1 or X_2, would be selected first in a forward regression procedure?

b.   Which independent variable, X_1 or X_2, would be a better predictor of Y? Why?

15.43   The following is a correlation matrix for three independent variables and one dependent variable:

	Y	X_1	X_2	X_3
Y	1	.25	.36	.59
X_1	.25	1	.54	.22
X_2	.36	.54	1	.31
X_3	.59	.22	.31	1

a.   Which independent variable would be chosen for the first stage of a forward regression procedure?

b.   Which independent variable would be chosen for the first step of a stepwise regression procedure?

15.44   The least squares regression equation

$$\hat{Y} = 1.5 + 3.5X_1 + 7.5X_2 - 150X_3$$

has the following t values for the independent variables:

NULL HYPOTHESIS	t STATISTIC
$\beta_1 = 0$	4.5
$\beta_2 = 0$	1.89
$\beta_3 = 0$	1.52

Twenty observations were used in calculating the least squares equation. In the first stage of a backward selection procedure, which independent variable would be eliminated first? Use a 5% level of significance.

15.45   Describe the main difference between the forward selection procedure and the stepwise selection procedure in regression analysis.

15.46   If a statistician would like to include dummy variables to indicate one of three cities and also one of four salespeople in a regression model, how many dummy variables would be needed in the model?

15.47   Which of the standard assumptions of regression appear to have been violated in the data from the following table, which lists the dependent variable and the residual values?

$\hat{Y}$	$Y - \hat{Y}$	$\hat{Y}$	$Y - \hat{Y}$
1.5	0.12	5.0	-1.45
2.1	$-.70$	5.5	1.61
3.5	$-.91$	6.0	1.79
4.0	1.02	7.0	-2.40
4.5	-1.18	7.5	2.10

15.48   How should a plot of the residuals $(Y - \hat{Y})$ plotted against the predicted values $\hat{Y}$ look if the standard assumptions of regression are satisfied?

15.49   A set of 20 observations is used to obtain the least squares line

$$\hat{Y} = 1.5 + 3.6X_1 + 4.9X_2$$

a.   Given that the estimated standard deviation of Y at $X_1 = 1.0$ and $X_2 = 2.0$ is 3.4, find a 90% confidence interval for the mean value of Y at $X_1 = 1.0$ and $X_2 = 2.0$.

b. Given that the MSE from this analysis is 21.5, then, using the information in part a, find a 90% prediction interval for an individual value of Y at $X_1 = 1.0$ and $X_2 = 2.0$.

15.50 Fifteen months are randomly selected to estimate the monthly sales, Y (in thousands of dollars), of a retail store based on monthly advertising expenditure, X (in thousands of dollars). The prediction equation is found to be

$$\hat{Y} = .2 + 1.5X + .4X^2$$

a. Find a 95% confidence interval for the mean monthly sales with a monthly advertising expenditure of $1.4 thousand if the estimated standard deviation of the monthly sales for $X = 1.4$ is .8.

b. Using the information in part a, find a 95% prediction interval for the monthly sales of a month that has an advertising expenditure of $1.4 thousand if the MSE from the analysis is 1.55.

15.51 Explain the difference between a confidence interval for the mean value of Y at particular values of the independent variables and a prediction interval for a future value of Y at particular values of the independent variables. Will the prediction interval for Y always be larger than the corresponding confidence interval for particular values of the independent variables?

15.52 A real-estate firm would like to determine the monthly income (Y) of homeowners in a certain section of town by using the monthly mortgage payment (X_1), the market value of the homeowner's car(s) (X_2), and the age of the homeowner (X_3). The following data are collected from 15 randomly selected households (Y, X_1, and X_2 are in dollars; X_3 in years):

Y	X₁	X₂	X₃	Y	X₁	X₂	X₃	Y	X₁	X₂	X₃
2963	820	7,800	32	2225	725	4,380	30	3180	635	9,450	36
2100	710	5,100	33	1630	538	3,760	27	3350	758	12,600	31
2820	520	10,500	26	3070	679	7,350	37	3267	810	10,630	29
3350	630	9,500	30	2950	975	6,580	34	2120	710	5,340	28
2640	925	6,260	35	3460	1120	7,900	33	2280	504	4,690	32

Use a computerized statistical package to answer the following questions:

a. What is the mean income for a homeowner with $X_1 = 800$, $X_2 = 7000$, and $X_3 = 30$?

b. Find a 95% confidence interval for the mean monthly income of a homeowner with $X_1 = 800$, $X_2 = 7000$, and $X_3 = 30$.

c. Find a 95% prediction interval for the income of a homeowner with $X_1 = 800$, $X_2 = 7000$, and $X_3 = 30$.

15.53 The operations manager in charge of a production process is interested in the amount of time in minutes, Y, that it takes an assembly-line worker to perform a certain task relative to his or her score, X, on an aptitude test. The proposed model is

$$Y = \beta_0 + \beta_1 X + \beta_2 X^2 + e$$

Twelve assembly line workers were randomly selected with the following results:

Y	X	Y	X
49	58	19	85
37	67	17	89
12	95	50	52
60	43	67	41
33	72	39	67
22	83	35	70

a. Using a computerized statistical package and the proposed model, construct a 99% confidence interval for the mean time that it takes assembly-line workers whose aptitude score is 80 to complete the task.

b. Using a computerized statistical package and the proposed model, construct a 99% prediction interval for the time it takes a worker to complete the task if the aptitude score of that worker is 80.

c. Compare the answers in parts a and b.

15.54   The vice president of a computer firm is interested in funding research proposals by graduate students who wish to perform experiments in the firm's advanced technology laboratory during the summer months. The vice president receives 18 proposals and sends these proposals to the director of the laboratory for evaluation. The director rates the proposals on two criteria and gives a score between zero and ten to each criterion, with 10 representing the best score possible. The variables X_1 and X_2 are used to represent these two scores. The dependent variable Y is the level of funding that the director of the laboratory would be willing to grant for the proposal. The variable Y is given in units of 1000's. The collected data are given below:

Y	X_1	X_2	Y	X_1	X_2	Y	X_1	X_2
9.5	8.7	9.2	7.0	7.9	7.9	8.5	8.6	8.8
7.3	8.1	8.0	7.4	8.2	8.0	7.2	8.3	7.8
6.5	7.4	7.7	8.3	8.5	8.4	5.8	6.7	7.0
8.4	8.4	8.6	8.2	8.6	7.9	6.3	7.3	7.5
8.0	8.3	8.0	5.3	6.6	6.9	9.0	8.6	9.0
6.1	7.0	7.3	6.7	7.8	7.5	6.4	7.7	7.5

a.   Find a 90% confidence interval for the mean value of Y at $X_1 = 8.0$ and $X_2 = 7.8$.

b.   Interpret the confidence interval in part a in the context of this exercise.

c.   Find a 90% prediction interval for the value of Y at $X_1 = 8.0$ and $X_2 = 7.8$.

d.   Interpret the prediction interval in part c in the context of this exercise.

Summary

Multiple linear regression offers a method of predicting (or modeling) the behavior of a particular **dependent** variable (Y) using two or more **independent (predictor)** variables. As in the case of simple linear regression, which uses one predictor variable, the regression coefficients are those that minimize

$$\text{SSE} = \text{sum of squares of error} = \Sigma(Y - \hat{Y})^2$$

To use this technique properly, you must pay special attention to the assumptions behind it: (1) the regression errors follow a normal distribution, centered at zero, with a common variance and (2) the errors are statistically independent. An estimate of this common variance is

$$\hat{\sigma}_e^2 = s^2 = \text{MSE} = \frac{\text{SSE}}{n - k - 1}$$

To determine the adequacy of the regression model, you can test the entire set of predictor variables using an F test with k and $n - k - 1$ degrees of freedom:

$$F = \frac{\text{MSR}}{\text{MSE}} = \frac{R^2/k}{(1 - R^2)/(n - k - 1)}$$

The contribution of an individual predictor variable (say, X_i) can be tested using a t statistic with $n - k - 1$ df:

$$t = \frac{b_i}{s_{b_i}}$$

where s_{b_i} represents the estimated standard deviation of b_i. Here b_i is the least squares estimate of the population parameter, β_i, and centers the confidence interval for this parameter.

The **coefficient of determination, R^2,** describes the percentage of the total variation in the sample Y values explained by this set of predictor variables. To determine the contribution of a particular subset of the predictor variables—such as X_2 and X_4—R^2 is computed with X_2 and X_4 included and then with X_2 and X_4 excluded from the regression equation. A **partial F test** is then used to determine whether the resulting decrease in R^2 is significant. The **adjusted value of R^2,** R^2(adj), is a statistic that, unlike R^2, does not necessarily increase as variables are added to the model. R^2(adj) helps distinguish variables that significantly improve the model from those that do not.

When a **curvilinear** pattern exists between two variables, X and Y, this non-linear relationship often can be modeled by including an X^2 term in the regression equation. The resulting equation is

$$\hat{Y} = b_0 + b_1X + b_2X^2$$

This type of model often works well in situations where Y (for example, sales) appears to increase more slowly as the independent variable, X (for example, the amount of shelf space devoted to this product) continues to increase.

The problem of **multicollinearity** arises in the application of multiple linear regression when two or more independent variables are highly correlated. The resulting regression equation contains coefficients that are highly inflated (have a large variance) with t statistics that are extremely small, despite the fact that one or more of these seemingly insignificant variables are very useful predictors.

Stepwise techniques allow you to insert variables one at a time into the equation (**forward regression**), remove them one at a time after initially including all variables in the equation (**backward regression**), or perform a combination of the two by inserting variables one at a time but removing a variable that has become redundant at any stage (**stepwise regression**). Once the variables for the model have been selected, **residual plots** should be obtained to examine the underlying assumptions that are necessary in a regression analysis.

Dummy variables can be used in a regression application to represent the categories of a qualitative variable (such as city). If all dummy variables are to be inserted into the equation, then $C - 1$ such variables should be defined to represent C categories. If a forward or stepwise selection procedure is used to define the final regression equation, then a better procedure is to define C dummy variables to represent this situation.

Use of a computer package is essential in the derivation of a multiple regression equation. In this chapter, we used MINITAB, SPSS, and SAS. They provide the sampling coefficients (b_0, b_1, b_2, . . .), the statistics necessary to perform any test of hypothesis, and those needed for the prediction and confidence intervals for any specific set of predictor variable values. A **confidence interval** is derived whenever the predicted Y value is used to estimate the average value of the dependent variable for a specific set of X values. When the predicted Y value is used to predict an individual value of Y for a specific set of X values, a **prediction interval** can be used to place bounds on the actual Y value.

Summary of Formulas

H_0: all β's $= 0$

H_a: at least one $\beta \neq 0$

$$F = \frac{MSR}{MSE} = \frac{R^2/k}{(1 - R^2)/(n - k - 1)}$$

(df $= k$ and $n - k - 1$)

H_0: $\beta_i = 0$

H_a: $\beta_i \neq 0$

(or H_a: $\beta_i > 0$)

(or H_a: $\beta_i < 0$)

$$t = \frac{b_i}{s_{b_i}}$$

(df $= n - k - 1$)

Confidence interval for β_i

$$b_i - t_{\alpha/2, n-k-1}s_{b_i} \quad \text{to} \quad b_i + t_{\alpha/2, n-k-1}s_{b_i}$$

Coefficient of determination

$$R^2 = 1 - (SSE/SST)$$

where

$$SST = \Sigma(Y - \bar{Y})^2 = \Sigma Y^2 - (\Sigma Y)^2/n$$

and

$$SSE = \Sigma(Y - \hat{Y})^2$$

Coefficient of determination (adjusted)	where (1) R_c^2 is the R^2 including the variables in H_0 (the complete model), (2) R_r^2 is the R^2 excluding the variables in H_0 (the reduced model), (3) $v_1 =$ the number of β's in H_0, (4) $v_2 = n - 1 -$ (the number of X's in the complete model), and (5) the degrees of freedom for the F statistic are v_1 and v_2

$$R^2(\text{adj}) = 1 - \frac{\text{SSE}/(n - k - 1)}{\text{SST}/(n - 1)}$$

H_0: X_i, X_{i+1}, $\cdots$, X_j do not contribute

H_a: at least one of them contributes

$$F = \frac{(R_c^2 - R_r^2)/v_1}{(1 - R_c^2)/v_2}$$

Review Exercises

15.55 Given below is selected information from a computer printout of a multiple regression analysis:

ANALYSIS OF VARIANCE

	df	SUM OF SQUARES	MEAN SQUARES	F RATIO
Regression	2	86.091	43.0455	8.649
Error	17	84.609	4.9770	

VARIABLE	COEFFICIENT	ST. DEV. COEFF.	t RATIO
Intercept	.10255		
X_1	.68808	.195	3.529
X_2	.39844	.295	1.351

Answer the following questions.

a. Write down the multiple regression equation.

b. What percentage of variation in Y is explained by the model?

c. What is the size of the sample (n) used in the above regression analysis?

d. Find the value of

    i. The total sum of squares
    ii. The error sum of squares
    iii. The estimated variance of the error in the model

e. Is the model as a whole significant? Use a .05 significance level.

f. Does X_1 contribute to this model given that X_2 is in the model? Use a .05 significance level.

g. Does X_2 contribute to this model given that X_1 is in the model? Use a .05 significance level.

15.56 A company has opened several outdoor ice-skating rinks and would like to know what factors affect the attendance at the rinks. The manager believes that the following variables affect attendance:

$$X_1 = \text{temperature (forecasted high)}$$
$$X_2 = \text{wind speed (forecasted high)}$$
$$X_3 = 1 \text{ if weekend and 0 otherwise}$$
$$X_4 = X_1 X_2$$

The following least squares model was found from 30 days of data:

$$\hat{Y} = 250 + 4.8X_1 - 30X_2 + 1.3X_3 + 35X_4$$

a. What is the predicted attendance on a weekend if the forecasted high temperature is 28°F and the forecasted high wind speed is 12 miles per hour?

b. If the coefficient of determination for the model is .67, test that the overall model contributes to predicting the attendance at the ice-skating rinks. Use a .05 significance level. What is the value of the adjusted R^2?

c. If the standard deviation of the estimate of the coefficient of X_2 is 2.01, does the variable wind speed contribute to predicting the variation in attendance, assuming that the variables X_1, X_3, and X_4 are in the model? Use a .05 significance level.

15.57 An automobile dealer decided to collect data to predict the demand for automobiles using regression analysis. Using historical data, the multiple regression method gave the least squares equation:

$$\hat{Y} = -307.2 + 1.994X_1 + 0.0207X_2 + 0.00876X_3 - 10.48X_4$$

where

Y = amount spent on new automobiles (in billions of dollars)

X_1 = U.S. population (in millions)

X_2 = disposable personal income (in billions of dollars)

X_3 = number of marriages (in thousands)

X_4 = financial interest rate (in percent) for automobile loans

a. Would you expect any multicollinearity to be present in these variables? Discuss.

b. If disposable personal income increased by \$200 billion and the value of the other independent variables remained constant, how much would you expect the demand for automobiles to increase? Is your conclusion valid if multicollinearity exists?

15.58 A real-estate agent wanted to explore the feasibility of using multiple regression analysis in appraising the value of single-family homes within a certain community. The following variables were used:

Y = selling price of a house (in dollars)

X_1 = total living area (in square feet)

$$X_2 = \begin{cases} 1 & \text{if in neighborhood 1} \\ 0 & \text{if not} \end{cases}$$

$$X_3 = \begin{cases} 1 & \text{if in neighborhood 2} \\ 0 & \text{if not} \end{cases}$$

$$X_4 = \begin{cases} 1 & \text{if lot size is larger than the typical house lot} \\ 0 & \text{if not} \end{cases}$$

The data are as follows:

Y	X_1	X_2	X_3	X_4	Y	X_1	X_2	X_3	X_4
63,000	2020	1	0	1	31,350	640	0	1	0
36,000	980	1	0	0	49,400	1910	0	0	1
44,000	1230	0	0	1	31,000	900	1	0	0
37,000	980	0	1	0	56,000	1890	1	0	0
28,000	640	0	1	0	63,500	1900	0	0	1
28,000	720	0	1	0	49,000	2080	1	0	1
56,000	2400	1	0	1	63,000	1900	0	0	1
28,600	670	0	1	0					

Using a computerized statistical package, find the following:

a. The least squares equation

b. The 95% confidence interval for the coefficient of total living area

c. The 95% prediction interval for selling price given that $X_1 = 1800$, $X_2 = 1$, $X_3 = 0$, and $X_4 = 0$

d. The overall F test for the model and the resulting conclusion using a 5% significance level

15.59 To predict the asking price of a used Chevrolet Camaro, the following data were collected on the car's age, condition, and mileage and on whether the seller is an individual or a dealer. The data are as follows:

ASKING PRICE (Y)	AGE (IN YEARS) (X_1)	MILEAGE (IN THOUSANDS) (X_2)	CONDITION (EXCELLENT, AVERAGE, POOR) (X_3)	(X_4)	DEALER OR INDIVIDUAL (X_5)
3,000	9	70	1	0	0
2,700	9	99	0	1	0
2,995	8	120	0	1	0
5,500	7	56	1	0	1
3,988	7	50	0	1	0
3,900	7	83	0	1	0
2,800	7	106	0	1	0
6,800	6	70	0	0	1
6,295	6	66	1	0	1
3,700	6	60	0	1	0
7,450	5	55	1	0	1
6,800	5	67	0	1	0
6,795	5	62	1	0	0
6,476	5	60	0	1	0
6,450	5	55	0	1	0
4,800	5	75	0	1	0
9,695	4	44	1	0	1
9,675	4	37	0	0	1
9,595	4	44	1	0	0
8,500	4	55	1	0	0
7,995	4	46	0	1	0
6,995	4	56	0	1	0
6,450	4	65	0	1	0
14,350	3	29	0	0	1
11,965	3	23	0	1	1
11,850	3	27	0	0	1
11,000	3	31	1	0	1
7,600	3	45	0	1	0
19,888	2	18	0	0	1
16,000	2	19	0	0	1
17,650	1	9	0	0	1

The dummy variable X_3 is equal to 1 if the car is in average condition, 0 if not. The variable X_4 is equal to 1 if the car is in poor condition, 0 if not. The dummy variable X_5 is equal to 1 if the seller is a dealer and is equal to zero if the seller is an individual. Use a computerized statistical package to answer the following questions.

a. Find the least squares equation.

b. Does the overall model contribute significantly to predicting the asking price of a used Chevrolet Camaro? Use a .01 significance level.

c. Find a 95% prediction interval for the asking price of a 5-year-old Camaro in average condition with 70,000 miles, sold by an individual.

d. Calculate the correlation matrix of all the variables. Would you suspect any multicollinearity problems by observing the correlations in this matrix?

e. Do a forward regression analysis using a significance level of .10.

15.60 The owner of a photographic laboratory would like to explore the relationship between her weekly profits (Y) and

X_1 = number of rolls of film sold

X_2 = number of enlargements given out free for advertising purposes

X_3 = number of prints

X_4 = number of reprints

Several weeks were selected randomly to collect the following data:

Y	X_1	X_2	X_3	X_4	Y	X_1	X_2	X_3	X_4
350	50	15	130	50	358	62	17	125	35
414	61	18	150	39	392	55	19	150	36
385	71	12	125	45	415	59	24	157	44
429	86	21	141	36	380	63	28	140	38
415	90	22	133	40					

Use a computerized statistical package.

a. Find the least squares prediction equation.

b. Test the null hypothesis that X_4 does not contribute to predicting the variation in Y given that X_1, X_2, and X_3 are already in the model. Use a .05 significance level.

c. Find the 90% confidence interval for the mean value of Y given $X_1 = 85$, $X_2 = 20$, $X_3 = 135$, and $X_4 = 37$.

d. Find the coefficient of determination for the complete model and interpret its value.

15.61 Use a computerized statistical package to analyze the following data:

Y	X_1	X_2	X_3	Y	X_1	X_2	X_3
154	30	1	1	220	34	5	25
223	41	3	9	210	38	4	16
201	33	5	25	230	44	3	9
177	31	4	16	265	51	2	4
143	25	3	9	306	55	5	25
155	29	2	4	170	31	4	16

a. Find the least squares prediction equation.

b. Find the coefficient of determination for the model.

c. Test at the .10 significance level that X_2 and X_3 contribute to the prediction of Y, given that X_1 is in the model.

d. Test at the .10 significance level that X_1 contributes to the prediction of Y, given that X_2 and X_3 are in the model.

e. Plot the residuals of the complete model versus the predicted values. Do the residuals appear to be random?

15.62 Complete the following ANOVA table for testing whether a model with five independent variables contributes significantly to the prediction of the dependent variable:

SOURCE	df	SS	MS	F
Regression		95.6		
Error	20	159.0		
Total				

15.63 The manager of Stay Trim Health Studios would like to determine the average number of times per month a member attends the health studio (Y). The following independent variables were used in the analysis:

$$X_1 = \text{weight at initial visit (in pounds)}$$

$$X_2 = X_1^2$$

$$X_3 = \text{age at initial visit}$$

$$X_4 = \text{length of membership (in years)}$$

$$X_5 = \begin{cases} 1 & \text{if employed} \\ 0 & \text{if not} \end{cases}$$

The manager collected the following data:

Y	X_1	X_3	X_4	X_5	Y	X_1	X_3	X_4	X_5
11	202	30	1	1	13	245	35	1	0
9	180	22	2	1	15	215	24	3	0
7	130	19	1	0	11	185	43	2	1
14	175	32	4	1	12	165	27	3	1
12	225	41	2	1	12	195	38	1	0
19	191	52	5	1	11	217	42	1	1
7	142	40	1	1	10	205	40	1	1
11	208	33	2	1					

The least squares equation was found to be:

$$\hat{Y} = -11.218 + 0.15178X_1 - 0.0003X_2 + 0.08286X_3 + 1.9138X_4 - 2.299X_5$$

a. Does the overall model contribute significantly to predicting the monthly attendance at Stay Trim Health Studios? Use a 10% significance level.

b. Does weight squared contribute significantly to predicting the monthly attendance, assuming that the variables X_1, X_3, X_4, and X_5 are in the model? Use a 10% significance level.

c. Find a 95% confidence interval for the coefficient of age.

d. Find a 95% confidence interval for the coefficient of length of membership.

e. Use the model to predict the monthly attendance of a 35-year-old member who weighs 200 pounds, has a 2-year membership, and is currently employed.

f. Construct a histogram for the residuals of the complete model.

15.64 By using many independent variables in a model, a stepwise procedure can sometimes produce a spurious "significant" model. Using MINITAB, type in the following two commands.

RAND 50 C1-C30

STEPWISE C1 C2-C30

What do the results of this procedure imply in the interpretation of the stepwise selection procedure? Would a second set of data from this population necessarily produce the same set of significant predictors? Try it.

15.65 The U.S. domestic coin production of cents and nickels is given below along with the total production of all coins for the years 1980 to 1989. Units are in millions of coins.

YEAR	CENTS (X_1)	NICKELS (X_2)	TOTAL (Y)
1980	12,555	1,095	16,426
1981	12,865	1,022	16,521
1982	16,726	666	19,459
1983	14,220	1,098	18,053
1984	13,720	1,264	17,822
1985	10,936	1,107	14,670
1986	8,934	899	12,073
1987	9,562	782	12,998
1988	11,347	1,435	15,959
1989	12,837	1,498	18,034

(*Source:* Adapted from *The World Almanac and Book of Facts* (1991): 123.)

a. Find the ANOVA table for testing the regression equation $\hat{Y} = b_0 + b_1X_1 + b_2X_2$. Use a computerized package. Test the model at the .05 significance level.

b. What is the value of R^2 and R^2 (adj)? Should these values be equal? Explain.

c. Is there a problem with multicollinearity?

d. Find a 95% prediction interval for the total coin production given that $X_1 = 13,000$ and $X_2 = 1,000$.

15.66 Forty of the largest companies traded on the U.S. stock exchange are listed with each company's earnings per share (EPS), sales, and profit for the 12-month period ending April 1, 1988. The sales and profits are given in millions of dollars. The data are as follows:

COMPANY	EPS (Y)	SALES (X_1)	PROFITS (X_2)
1. IBM	8.72	54,217	5,253
2. EXXON	3.43	84,116	4,841
3. GE	3.38	39,310	3,089
4. AT&T	1.88	33,598	2,044
5. FORD MOTOR	9.05	71,640	4,625
6. GM	10.06	101,782	3,551
7. AMOCO	5.77	22,390	1,360
8. MOBIL	3.08	56,446	1,264
9. DOW CHEMICAL	6.50	13,377	1,245
10. HEWLETT-PACKARD	2.76	8,540	644
11. WAL-MART	1.16	16,064	656
12. CHEVRON	2.27	29,100	773
13. SEARS	4.35	48,440	1,650
14. EASTMAN KODAK	3.52	13,305	1,178
15. COCA-COLA	2.43	7,660	916
16. JOHNSON & JOHNSON	4.83	8,010	833
17. PROCTER & GAMBLE	5.48	17,892	786
18. 3M	4.02	9,429	918
19. ARCO	6.68	16,829	1,224
20. NABISCO	4.19	15,766	1,081
21. BRISTOL-MYERS	2.47	5,400	710
22. GTE	3.29	15,421	1,119
23. SOUTHWESTERN BELL	3.48	8,003	1,047
24. AMERICAN EXPRESS	1.20	16,141	533
25. TEXACO	2.11	35,300	493
26. NISSAN	.07	33,461	55
27. ANHEUSER-BUSCH	2.04	8,258	615
28. PEPSI CO	2.30	11,485	605
29. McDONALD'S	2.89	4,894	549
30. WALT DISNEY	3.05	2,951	392
31. KRAFT	2.87	9,876	390
32. NEC	.42	17,338	103
33. PHILIP MORRIS	7.75	27,694	1,840
34. BELL SOUTH	3.46	12,269	1,665
35. BELL ATLANTIC	6.24	10,303	1,240
36. AMERITECH	8.47	9,536	1,188
37. FUJI PHOTO FILM	2.22	7,138	404
38. ABBOTT LABORATORIES	2.78	4,390	633
39. DUN & BRADSTREET	2.58	3,359	393
40. AMER. INTERNAT'L	5.37	10,679	657

(*Source: Financial World,* April 5, 1988.)

Using a computerized statistical package, find the following:

a. The least squares equation for predicting EPS.

b. The overall F-test for the model and the resulting conclusion using a 10% significance level.

c. The individual t tests for testing the significance of each of the two independent variables and the resulting conclusions using a 10% significance level.

d. The coefficient of determination and its interpretation.

15.67 Twelve employees each take two written tests that are used to indicate how well the employee will perform on the job. In the MINITAB printout, C2 and C3 are the scores from the two tests. In C1, an evaluation is given of the employee's performance on the job. Using a significance level of .01, what conclusion can be drawn from the ANOVA table with regard to the prediction of job performance by the two prediction variables? Is there a conflict between the significance of the F statistic and significance of the two t tests? How can this inconsistency be corrected?

```
MTB > NAME C1 'EVAL'
MTB > NAME C2 'SCORE1'
MTB > NAME C3 'SCORE2'
MTB > PRINT C1-C3
 ROW    EVAL   SCORE1   SCORE2

   1      99      99       66
   2      72      93       42
   3      90      75       29
   4      55      50        6
   5      95     100       54
   6      65      64       14
   7      75      80       35
   8      60      72       16
   9      83      88       36
  10      88      75       28
  11      59      73       18
  12     100     100       67

MTB > REGRESS Y IN C1 USING 2 PREDICTORS IN C2 , C3

The regression equation is
EVAL = 71.8 - 0.317 SCORE1 + 0.939 SCORE2

Predictor        Coef        Stdev      t-ratio
Constant        71.82        28.48        2.52
SCORE1         -0.3168       0.5052      -0.63
SCORE2          0.9394       0.3959       2.37

s = 8.737      R-sq = 76.4%      R-sq(adj) = 71.1%

Analysis of Variance

SOURCE        DF          SS          MS
Regression     2        2221.9      1111.0
Error          9         687.0        76.3
Total         11        2908.9

SOURCE        DF       SEQ SS
SCORE1         1       1792.1
SCORE2         1        429.8
```

15.68 An independent research firm is investigating sex discrimination in the salaries of managers of a small firm. Fourteen supervisors and midlevel managers were chosen at random and the salary, years of experience, and sex of each were recorded. The independent research firm believes that the salary compensation may increase with years of experience differently for males and females. Data are shown in the MINITAB printout. Test that sex significantly contributes to the prediction of salaries in the model

$$Y = \beta_0 + \beta_1 X_1 + \beta_2 X_2 + \beta_3 X_3 + e$$

where X_1 is years of experiences, X_2 is 1 for male and 0 for females, and $X_3 = X_1 X_2$. Use the MINITAB printout and let the significance level be .05. Note that the variable $X_3 = X_1 X_2$

```
MTB > LET C4 = C2*C3
MTB > NAME C1 'SALARY'
MTB > NAME C2 'EXPER'
MTB > NAME C3 'SEX'
MTB > NAME C4 'INTERACT'
MTB > PRINT C1-C4
 ROW   SALARY   EXPER   SEX   INTERACT

   1     17.1     2.3     0      0.0
   2     26.0     3.7     0      0.0
   3     30.8     4.1     1      4.1
   4     19.1     2.5     1      2.5
   5     71.2    10.3     1     10.3
   6     47.3     6.4     1      6.4
   7     25.5     3.4     1      3.4
   8     33.6     4.7     0      0.0
   9     60.3     8.3     0      0.0
  10     36.7     5.1     1      5.1
  11     26.5     3.7     0      0.0
  12     37.3     5.0     1      5.0
  13     27.7     3.8     1      3.8
  14     32.5     4.6     0      0.0
```

```
MTB > REGRESS Y IN C1 USING 3 PREDICTORS IN C2-C4

The regression equation is
SALARY = - 0.404 + 7.27 EXPER + 3.31 SEX - 0.559 INTERACT

Predictor        Coef        Stdev      t-ratio
Constant       -0.4040       0.8386      -0.48
EXPER           7.2683       0.1707      42.58
SEX             3.306        1.075        3.08
INTERACT       -0.5593       0.2092      -2.67

s = 0.7747      R-sq = 99.8%      R-sq(adj) = 99.7%

Analysis of Variance

SOURCE        DF          SS           MS
Regression     3       2999.68       999.89
Error         10          6.00         0.60
Total         13       3005.68

SOURCE        DF       SEQ SS
EXPER          1       2993.90
SEX            1          1.48
INTERACT       1          4.29

MTB > REGRESS Y IN C1 USING 1 PREDICTOR IN C2

The regression equation is
SALARY = 1.58 + 6.91 EXPER

Predictor        Coef        Stdev      t-ratio
Constant        1.5775       0.6624       2.38
EXPER           6.9148       0.1252      55.23

s = 0.9907      R-sq = 99.6%      R-sq(adj) = 99.6%

Analysis of Variance

SOURCE        DF          SS           MS
Regression     1       2993.9        2993.9
Error         12         11.8           1.0
Total         13       3005.7
```

(called the interaction of sex and experience) allows for the slope of the regression line for males to be different from the slope of the regression line for females. Salary is in C1 in units of thousands. Years of experience are in C2. A 1 in C3 represents a male and a 0 represents a female. (*Hint:* The reduced model contains only the predictor variable X_1.)

15.69   Nine similar machines are used in a manufacturing process at an assembly plant. The operations manager believes that the repair costs for the machines are influenced by both the age of the machines and the operator of the machine. The MINITAB printout shows the costs of repair of the machines over the past six months (C1), the age of each machine at the beginning of the six month period (C2), and which of three operators used the machines (C3) and (C4). Each machine was used by only one operator during this time. What conclusions can be drawn from analyzing the MINITAB printout? Use a .05 significance level. Comment on the residual plot. Also interpret the three prediction intervals for the repair cost if a machine is 6 years old and is operated by one of the three operators.

```
MTB > NAME C1 'COST'
MTB > NAME C2 'AGE'
MTB > NAME C3 'DUMMY1'
MTB > NAME C4 'DUMMY2'
MTB > PRINT C1-C4
ROW    COST    AGE   DUMMY1   DUMMY2

  1     310     4      1        0
  2     300     8      0        1
  3     175     3      0        0
  4     200     2      1        0
  5     620    15      0        1
  6     365     9      0        0
  7     370     6      1        0
  8     175     5      0        1
  9     365     8      0        0
```

```
MTB > BRIEF 3
MTB > REGRESS Y IN C1 USING 3 PREDICTORS IN C2-C4, RES IN C5, YHATS IN C6;
SUBC> PREDICT FOR 6 1 0;
SUBC> PREDICT FOR 6 0 1;
SUBC> PREDICT FOR 6 0 0.

The regression equation is
COST = 23.7 + 41.7 AGE + 103 DUMMY1 - 47.9 DUMMY2

Predictor        Coef         Stdev       t-ratio
Constant         23.66        22.39          1.06
AGE              41.701        2.646        15.76
DUMMY1          102.87        20.73          4.96
DUMMY2          -47.87        20.73         -2.31

s = 23.87        R-sq = 98.1%      R-sq(adj) = 97.0%

Analysis of Variance

SOURCE          DF           SS            MS
Regression       3        150652         50217
Error            5          2848           570
Total            8        153500

SOURCE          DF        SEQ SS
AGE              1        126784
DUMMY1           1         20829
DUMMY2           1          3039

Obs.      AGE       COST      Fit Stdev.Fit   Residual   St.Resid
  1       4.0     310.00    293.33    13.78      16.67       0.86
  2       8.0     300.00    309.40    14.22      -9.40      -0.49
  3       3.0     175.00    148.76    16.85      26.24       1.55
  4       2.0     200.00    209.93    14.76      -9.93      -0.53
  5      15.0     620.00    601.30    20.37      18.70       1.50
  6       9.0     365.00    398.97    15.10     -33.97      -1.84
  7       6.0     370.00    376.73    14.76      -6.73      -0.36
  8       5.0     175.00    184.30    17.93      -9.30      -0.59
  9       8.0     365.00    357.27    14.22       7.73       0.40


    Fit   Stdev.Fit        95% C.I.            95% P.I.
 376.73      14.76    ( 338.78, 414.69)    ( 304.58, 448.89)

 226.00      16.36    ( 183.93, 268.07)    ( 151.59, 300.40)

 273.87      13.89    ( 238.15, 309.59)    ( 202.86, 344.87)


MTB > NAME C5 'RESID'
MTB > NAME C6 'PRED'
MTB > PLOT C5 C6

RESID   -        *                                                 *
        -
        -
   1.0+ -
        -                       *
        -                          *
        -
  -0.0+ -
        -
        -              *       *
        -         *  *
  -1.0+ -
        -
        -
        -                        *
  -2.0+ -
        +---------+---------+---------+---------+---------+------PRED
       100       200       300       400       500       600
```

15.70 The MINITAB computer printout shows the execution of the stepwise procedure with FENTER = 0 and FREMOVE = 0. For this situation, all independent variables are put in the model, starting with the most significant independent variable. Using a .05 significance level, determine the model that the forward regression method would choose.

```
MTB > NAME C1 'Y'
MTB > PRINT 'Y' C2-C7
ROW        Y        C2        C3        C4       C5       C6        C7

  1     1778.47    114    468.000    113.3     57.6     26.5    15.7663
  2     1628.18    114    341.759     87.4     57.6     56.0    11.7664
  3     1465.00     26    266.034     72.6     57.6     79.6    10.4331
  4     1864.35     81    547.724    128.1     72.1     73.7     2.4333
  5      610.47    125     84.172     28.2    104.0     14.7     7.7665
  6     2031.82     81    575.827    124.4     51.8     56.0    10.4331
  7      863.82    114    123.966     43.0    101.1     14.7     9.0998
  8      447.29     26     65.655     17.1    104.0     14.7     2.4333
  9     1752.71     15    532.172    131.8    104.0     14.7     9.0998
 10     1387.71     59    243.759     68.9     69.2     67.8     6.4332
 11      245.47     15     56.931      9.7     83.7     38.3     3.7666
 12     1022.71     26    128.552     39.3     98.2     20.6    14.4330
 13     1237.41    125    262.827     83.7     92.4     32.4     2.4333
 14      777.94     37    110.862     39.3     80.8     26.5    11.7664
 15      876.71     15    202.000     83.7    101.1     20.6     2.4333
 16     1336.18     26    215.345     61.5     63.4     44.2    15.7663
 17     1726.94    125    483.655    120.7     80.8     56.0     3.7666
 18      404.35     26     58.207      9.7     98.2     14.7    10.4331

MTB > STEPWISE 'Y' C2-C7;
SUBC> FENTER = 0.0;
SUBC> FREMOVE = 0.0.

STEPWISE REGRESSION OF  Y ON 6 PREDICTORS, WITH N =   18
```

STEP	1	2	3	4	5	6
CONSTANT	437.28	255.65	129.29	-17.69	-59.78	-532.32
C3	2.85	2.84	2.54	0.51	0.48	0.63
T-RATIO	11.02	12.66	12.08	0.88	0.88	1.16
C7		22.2	24.4	28.4	28.9	35.8
T-RATIO		2.52	3.37	5.24	5.67	4.89
C6			5.0	5.5	5.6	8.0
T-RATIO			2.88	4.33	4.67	3.63
C4				9.0	8.8	8.3
T-RATIO				3.62	3.76	3.60
C2					0.90	1.04
T-RATIO					1.65	1.91
C5						3.8
T-RATIO						1.28
S	195	169	139	101	95.4	92.9
R-SQ	88.35	91.82	94.87	97.45	97.92	98.19

Computer Exercises Using the Database

Exercise 1 -- Appendix I From the database, randomly select 40 observations. Regress the variable HPAYRENT (house payment or apartment/house rent) on the prediction variables INCOME1 (primary income), INCOME2 (secondary income), and FAMLSIZE (size of family). Find the coefficient of determination for the complete model. Find a 90% confidence interval on the mean value of HPAYRENT for a family that has a principal income of $33,000, a secondary income of 18,000 and a family size equal to three.

Exercise 2 -- Appendix I Using the data from the previous problem along with dummy variables representing the LOCATION of the residences, do both a forward regression analysis and a backward regression analysis with a significance level of .10. Compare the two resulting models.

Exercise 3 -- Appendix J From the database, randomly select 50 observations. Consider a multiple regression model, where the dependent variable is SALES and predictor variables are COSTSALE (sales cost), EMPLOYEE (number of employees), NETINC (net income), ASSETS, and TOTAL. Using these predictor variables, what percentage of the variation in the SALES has been explained? Construct a histogram of the residuals. Do the regression assumptions appear to be satisfied?

Exercise 4 -- Appendix J Using the data from the previous problem, perform both a forward regression analysis and backward regression analysis with a significance level of .10. Compare the resulting models.

Case Study

Mega Bucks for the Top Executive Who Can't Pass the Buck

Compensation paid to chief executive officers (CEOs) of large corporations has always generated much interest. The salaries and bonuses of high profile CEOs like Lee Iacocca of Chrysler Corporation, Jim Manzi of Lotus Development, and Michael Eisner of Walt Disney run into several millions of dollars each year. CEO compensation has historically been high in comparison with other company salaries. It is therefore useful to investigate how CEO compensation is determined and also what factors might influence compensation.

Forbes magazine publishes an annual survey of the executive compensation of the 800 most highly paid CEOs of major American companies. Their May 28, 1990, issue reports that a total of a billion dollars in executive compensation was split among these 800 individuals in 1989. Of these 800 CEOs, 347 earned $1 million or more. While these figures seem astonishing, and not all CEOs necessarily deserved what they got, Forbes points out that these 800 companies had combined profits of $175 billion and revenues of $3.46 trillion. So the CEOs earned a penny on 2.5 dollars of profit, or a penny on every 49 cents of revenue.

The Forbes CEO profiles contained, among other things, the annual compensation, the actual size of the company, tenure (the number of years the CEO had spent with the company), and other variables. The table below represents 35 companies selected from the 800 reported by Forbes, with data on the following variables for each company:

	1	2	3	4	5	6
COMPANY	SALES	PROFIT	TENURE	AGE	SHARES OWNED	COMPEN-SATION
Aetna Life	19671	639.4	12	63	1.3	2123
IBM	62710	3758.0	30	55	4.3	1886
Amdahl	2101	153.0	13	54	1.1	1486
American Express	25047	1157.0	19	54	14.9	3537
AT&T	36112	2697.0	33	55	2.3	2151
AMR Corp	10480	454.8	17	55	8.2	2081
BankAmerica	11389	820.0	4	60	0.8	1292
K Mart	29792	322.7	26	48	1.0	928
Black & Decker	3615	−11.5	5	46	0.1	1877
H&R Block	975	107.2	35	67	117.4	983
Boeing	20276	675.0	32	58	0.8	1167
Lotus	556	67.7	7	38	34.8	16412
Maytag	3089	131.5	37	63	2.8	1239
CBS	2962	297.1	4	67	0.1	1360
Chase Manhattan	13904	−665.0	43	63	1.1	1467
Chevron	29443	251.0	30	53	3.9	1034
Chrysler	34922	359.0	12	65	3.0	3322
McDonald's	6066	726.7	24	45	7.1	1201
Citicorp	37970	498.0	25	51	11.8	1617
Coca-Cola	8966	1723.8	36	58	81.1	10814
Compaq Computer	2876	333.3	8	45	8.1	2115
Delta Airlines	8572	473.2	27	48	0.6	1414
Digital Equipment	12937	875.8	33	64	204.0	950
New York Times	1769	266.6	36	64	10.2	899
Walt Disney	4839	729.4	6	48	42.6	9595
Dow Chemical	17600	2487.0	31	54	4.8	2246
Du Pont	35099	2480.0	32	56	20.0	2569
Eastman Kodak	18398	529.0	40	65	1.1	1151
Exxon	86656	2975.0	38	62	6.5	1458
Federal Express	6769	113.6	19	45	235.4	880
NIKE	2040	212.2	26	52	943.6	274
Ford Motor Co	96146	3835.0	39	64	7.8	2199
General Motors	126932	4224.3	41	64	4.6	2109
Hilton Hotels	954	110.1	36	62	556.8	971
Zenith Electronics	1549	−68.4	19	51	0.5	463

(*Source:* "Corporate America's Most Powerful People: The Pay," *Forbes* (May 28, 1990): 266–316.) Reprinted by permission of *Forbes* magazine. © Forbes, Inc., 1990.

1. Sales—the annual sales (in millions of dollars) of the corporation.
2. Profit—the annual after-tax profit of the corporation (in millions of dollars).
3. Tenure—the number of years the CEO has worked for the corporation.
4. Age—the chronological age of the CEO in years.
5. Shares owned—the market value of shares of stock owned by the CEO in the corporation, in millions of dollars.
6. Compensation—the total compensation of the CEO, including salary, bonus, and stock options, in units of hundreds of thousands of dollars.

Case Study Questions

1. Perform a multiple regression analysis on the above data, using the five independent variables in columns 1 to 5 to predict compensation of CEO (column 6).
a. What percentage of the variation in CEO compensation is explained by these predictors?
b. Is the model as a whole significant at a .05 significance level?
c. Comment on how significantly each predictor variable contributes individually, given the other variables in the model.
2. Obtain a correlation matrix of all the variables. If you wished to use only *one* independent variable to predict CEO compensation, which variable would you choose? Conduct a simple regression procedure using just this one variable to predict CEO compensation.

3. Examine the correlation matrix to detect any potential for multicollinearity. What would you look for? If you detect multicollinearity, what action would you take?
4. Using the partial F test, determine whether the five-predictor model (question 1) is better at explaining CEO compensation than the simple regression model (question 2). Which is the "better" model of the two? State the p-value for the test.
5. Obtain a plot of the residuals $(Y - \hat{Y})$ against the predicted Y values $(\hat{Y})$. What do you look for? How would you detect heteroscedasticity (a violation of the assumption of constant variance)?
6. Print a histogram of the residuals. What shape would you like to see? If the histogram does not have this shape, what is the problem?
7. Suppose you have a large corporation with sales of $750 million, profits of $62 million, and a 54-year-old CEO with a tenure of 15 years who owns a half million dollars worth of shares in the company.
a. What is the predicted value (point estimate) of the CEO's compensation?
b. Find the 95% prediction interval and the 95% confidence interval of CEO compensation for this situation.
c. What is the difference between the prediction and confidence intervals, i.e., what different kinds of information do they provide?
8. Would the above regression model be applicable to all United States companies in general? Justify your answer. Also, comment on how the above model might be adversely influenced by those companies that suffered a loss.

SPSS

Solution
Example 15.2

At the beginning of the chapter, we computed the regression equation for predicting the estimate of home size, based on the three predictor variables: income, family size, and level of education. The SPSS program listing in Figure 15.21 requests a multiple regression solution. In this problem the SPSS commands are the same for both the mainframe and PC versions (remember to end each command line with a period when using the PC version).

The TITLE command names the SPSS run.

The DATA LIST command gives each variable a name and describes the data as being in free form.

The BEGIN DATA command indicates to SPSS that the input data immediately follow.

The next ten lines contain the data values representing the four variables to be considered in the regression analysis. The first card image represents a home with 1,600 square feet, an income of $22,000, a family of two people, and four years of educational experience at the college level.

The END DATA statement indicates the end of the data.

The CORRELATION statement requests a correlation matrix of the four variables SQFOOT, INCOME, SIZE, and EDUC.

The REGRESSION statement requests that the independent variables INCOME, SIZE, and EDUC be entered in the regression equation to predict the dependent variable SQFOOT.

Figure 15.22 shows the SPSS output obtained by executing the listing in Figure 15.21.

■ **FIGURE 15.21**
Input for SPSS or SPSS/PC. Remove the periods for SPSS input.

```
TITLE    REAL ESTATE EXAMPLE USING THREE PREDICTORS.
DATA LIST FREE/SQFOOT INCOME SIZE EDUC.
BEGIN DATA.
16 22  2  4
17 26  2  8
26 45  3  7
24 37  4  0
22 28  4  2
21 50  3  10
32 56  6  8
18 34  3  8
30 60  5  2
20 40  3  6
END DATA.
CORRELATION SQFOOT,INCOME,SIZE,EDUC.
REGRESSION VARIABLES=SQFOOT,INCOME,SIZE,EDUC/
         DEPENDENT=SQFOOT/ENTER.
```

■ **FIGURE 15.22** SPSS output.

```
                 REAL ESTATE EXAMPLE USING THREE PREDICTORS

    Correlations:  SQFOOT      INCOME      SIZE        EDUC

       SQFOOT      1.0000      .8433*      .9079**    -.1679
       INCOME       .8433*     1.0000      .7213*      .1514
       SIZE         .9079**     .7213*     1.0000     -.2514
       EDUC        -.1679       .1514     -.2514      1.0000

    N of cases:   10           1-tailed Signif:  * - .01  ** - .001

    " . " is printed if a coefficient cannot be computed

          * * * *   M U L T I P L E   R E G R E S S I O N   * * * *

    Listwise Deletion of Missing Data

    Equation Number 1    Dependent Variable..   SQFOOT

    Beginning Block Number  1.  Method:  Enter

    Variable(s) Entered on Step Number
        1..     EDUC
        2..     INCOME
        3..     SIZE

    Multiple R           .95145
    R Square             .90527  ◄——— R²
    Adjusted R Square    .85790                          SSE
    Standard Error      2.03545

    Analysis of Variance
                        DF      Sum of Squares      Mean Square
    Regression           3          237.54155         79.18052
    Residual             6           24.85845          4.14307  ◄---- MSE = s²

    F =      19.11154       Signif F =  .0018  ◄——— p value for F test
```

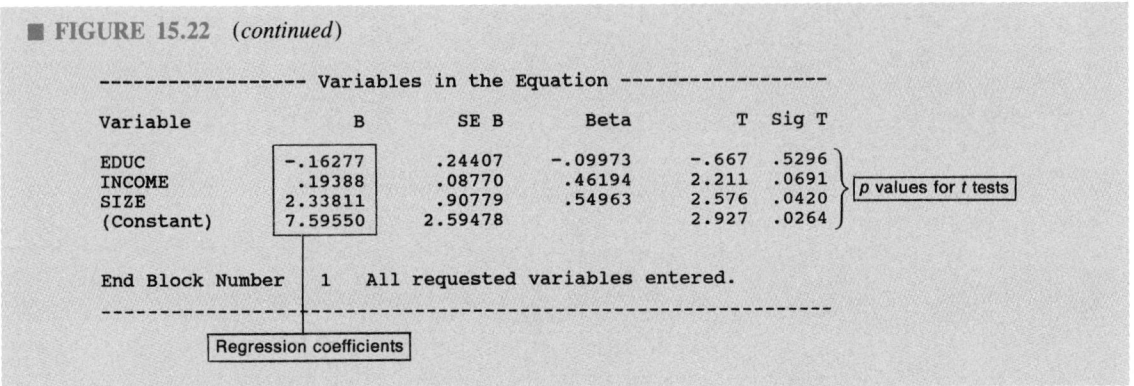

■ FIGURE 15.22 (*continued*)

```
------------------ Variables in the Equation ------------------

Variable              B        SE B      Beta       T   Sig T

EDUC              -.16277     .24407    -.09973    -.667  .5296
INCOME             .19388     .08770     .46194    2.211  .0691
SIZE              2.33811     .90779     .54963    2.576  .0420
(Constant)        7.59550    2.59478               2.927  .0264

End Block Number    1    All requested variables entered.
```

p values for *t* tests

Regression coefficients

SPSS

▟ Example 15.2
Stepwise Method

SPSS can be used for determining the stepwise solution for estimating home size based on the three predictor variables: income, family size, and level of education. The SPSS program listing in Figure 15.23 requests a stepwise regression solution. Notice that the word STEPWISE is substituted for ENTER and that a new line has been added. Instead of forcing all variables into the equation with an ENTER command, STEPWISE selects the variables that meet the entry criteria. The CRITERIA = PIN (0.1) statement specifies that each independent variable must produce a significant increase in R at a significance level of .1 before it is allowed to enter into the regression equation. The POUT (.15) statement dictates that any variable in the model not producing a significant decrease in R at a significance level of .15 when dropped from the model should be removed. The value contained in POUT must exceed the value contained in PIN. In this problem the SPSS commands are the same for both the mainframe and PC versions (remember to end each command line with a period when using the PC version).

The TITLE command names the SPSS run.

The DATA LIST command gives each variable a name and describes the data as being in free form.

The BEGIN DATA command indicates to SPSS that the input data immediately follow.

The next ten lines contain the data values representing the four variables to be considered in the regression analysis. The first card image represents a home with 1,600 square feet, an income of $22,000, a family of two people, and four years of educational experience at the college level.

The END DATA statement indicates the end of the data.

The REGRESSION statement requests that the independent variables INCOME, SIZE, and EDUC be entered in the regression equation to predict the dependent variable SQFOOT. The regression subcommands were discussed above.

Figure 15.24 shows the SPSS output obtained by executing the listing in Figure 15.23.

■ **FIGURE 15.23**
Input for SPSS or
SPSS/PC. Remove
the periods for
SPSS input.

```
TITLE    REAL ESTATE EXAMPLE USING STEPWISE.
DATA LIST FREE/SQFOOT INCOME SIZE EDUC.
BEGIN DATA.
16 22  2  4
17 26  2  8
26 45  3  7
24 37  4  0
22 28  4  2
21 50  3  10
32 56  6  8
18 34  3  8
30 60  5  2
20 40  3  6
END DATA.
REGRESSION VARIABLES=SQFOOT,INCOME,SIZE,EDUC/
           CRITERIA=PIN(0.1)/
           DEPENDENT=SQFOOT/STEPWISE.
```

■ **FIGURE 15.24** SPSS output.

```
            REAL ESTATE EXAMPLE USING STEPWISE

        * * * *   M U L T I P L E   R E G R E S S I O N   * * * *


Listwise Deletion of Missing Data

Equation Number 1    Dependent Variable..   SQFOOT

Beginning Block Number  1.  Method:  Stepwise

Variable(s) Entered on Step Number
    1..    SIZE ◄──────────────┤ 1st variable entered │

Multiple R          .90787
R Square            .82422
Adjusted R Square   .80225
Standard Error     2.40115

Analysis of Variance
                  DF     Sum of Squares      Mean Square
Regression         1        216.27586         216.27586
Residual           8         46.12414           5.76552

F =     37.51196      Signif F =  .0003


------------------ Variables in the Equation ------------------

Variable           B          SE B        Beta         T    Sig T

SIZE            3.86207      .63057      .90787      6.125   .0003
(Constant)      9.08276     2.33397                  3.892   .0046


------------- Variables not in the Equation -------------

Variable      Beta In  Partial  Min Toler       T   Sig T

INCOME         .39278   .64892    .47980     2.257 ►.0586
EDUC           .06434   .14853    .93681      .397   .7029

                                    │ Enter income into the equation because
                                    │ the value is < PIN, where PIN = .1


            REAL ESTATE EXAMPLE USING STEPWISE

        * * * *   M U L T I P L E   R E G R E S S I O N   * * * *

Equation Number 1    Dependent Variable..   SQFOOT


Variable(s) Entered on Step Number
    2..    INCOME ◄──────────────┤ 2nd variable entered │
```

■ **FIGURE 15.24** (*continued*)

```
Multiple R            .94776
R Square              .89824
Adjusted R Square     .86917
Standard Error       1.95306

Analysis of Variance
                    DF     Sum of Squares     Mean Square
Regression           2         235.69890       117.84945
Residual             7          26.70110         3.81444

F =      30.89559      Signif F =   .0003
```

```
                REAL ESTATE EXAMPLE USING STEPWISE

        * * * *   M U L T I P L E   R E G R E S S I O N   * * * *

Equation Number 1    Dependent Variable..   SQFOOT


------------------ Variables in the Equation ------------------

Variable              B         SE B        Beta        T   Sig T

SIZE            2.65694       .74046      .62457    3.588   .0089
INCOME           .16485       .07306      .39278    2.257   .0586
(Constant)      6.73958      2.16385                3.115   .0170


------------- Variables not in the Equation -------------

Variable      Beta In  Partial  Min Toler        T  Sig T

EDUC         -.09973  -.26270     .34672     -.667  .5296


End Block Number   1    PIN =      .100 Limits reached.
```

Final Equation: $\hat{Y} = 6.74 + .16(\text{income}) + 2.66(\text{size})$

Do not enter education because this value is > PIN, where PIN = .1

SPSS

▨ **Example** **Section 15.5** In the real-estate example in section 15.5, we wished to determine the predicted square footage ($\hat{Y}$) for values $X_1 = 36$, $X_2 = 4$, and $X_3 = 8$. Of course, one way to do this is to insert them manually into the regression equation resulting from the ten observations in the previous SPSS example. An easier way is to attach these values at the end of the input data along with a numeric value for Y (we used -9 here) that will be identified as a *missing value* using the MISSING VALUES SQFOOT(-9) statement. SPSS ignores this row of data and computes the regression equation but then includes this row when it summarizes the predicted values. The predicted values as well as their standard errors ($s_{\hat{Y}}$) are calculated and included in the output using the CASEWISE = ALL PRED SEPRED statement. This statement informs SPSS that you would like to see the predicted values (and their standard errors) for *all* of the cases in the input data, including the row(s) with the missing Y value. The listing in Figure 15.25 is used to obtain the predicted values and corresponding standard errors. In this problem the SPSS commands are the same for both the mainframe and PC versions (remember to end each command line with a period when using the PC version).

The TITLE command names the SPSS run.

The DATA LIST command gives each variable a name and describes the data as being in free form.

The BEGIN DATA command indicates to SPSS that the input data immediately follow.

The next ten lines contain the data values representing the four variables to be considered in the regression analysis. The first card image represents a home with 1,600 square feet, an income of $22,000, a family of two people, and four years of educational experience at the college level.

The END DATA statement indicates the end of the data.

The REGRESSION statement requests that the independent variables INCOME, SIZE, and EDUC be entered in the regression equation to predict the dependent variable SQFOOT.

Figure 15.26 shows the SPSS output obtained by executing the listing in Figure 15.25.

■ **FIGURE 15.25**
Input for SPSS or
SPSS/PC. Remove
the periods for
SPSS input.

```
TITLE    PREDICTION FOR REAL ESTATE EXAMPLE.
DATA LIST FREE/SQFOOT INCOME SIZE EDUC.
BEGIN DATA.
16 22 2 4
17 26 2 8
26 45 3 7
24 37 4 0
22 28 4 2
21 50 3 10
32 56 6 8
18 34 3 8
30 60 5 2
20 40 3 6
-9 36 4 8
END DATA.
MISSING VALUES SQFOOT(-9).
REGRESSION VARIABLES=SQFOOT,INCOME,SIZE,EDUC/
          DEPENDENT=SQFOOT/ENTER/
          CASEWISE=ALL PRED SEPRED.
```

■ **FIGURE 15.26**
SPSS output.

```
                PREDICTION FOR REAL ESTATE EXAMPLE

      * * * *   M U L T I P L E   R E G R E S S I O N   * * * *

Listwise Deletion of Missing Data

Equation Number 1    Dependent Variable..   SQFOOT

Beginning Block Number  1.  Method: Enter


Variable(s) Entered on Step Number
     1..    EDUC
     2..    INCOME
     3..    SIZE

Multiple R            .95145
R Square              .90527
Adjusted R Square     .85790
Standard Error       2.03545

Analysis of Variance
                    DF      Sum of Squares      Mean Square
Regression           3          237.54155         79.18052
Residual             6           24.85845          4.14307

F =      19.11154        Signif F =   .0018
```

FIGURE 15.26
(*continued*)

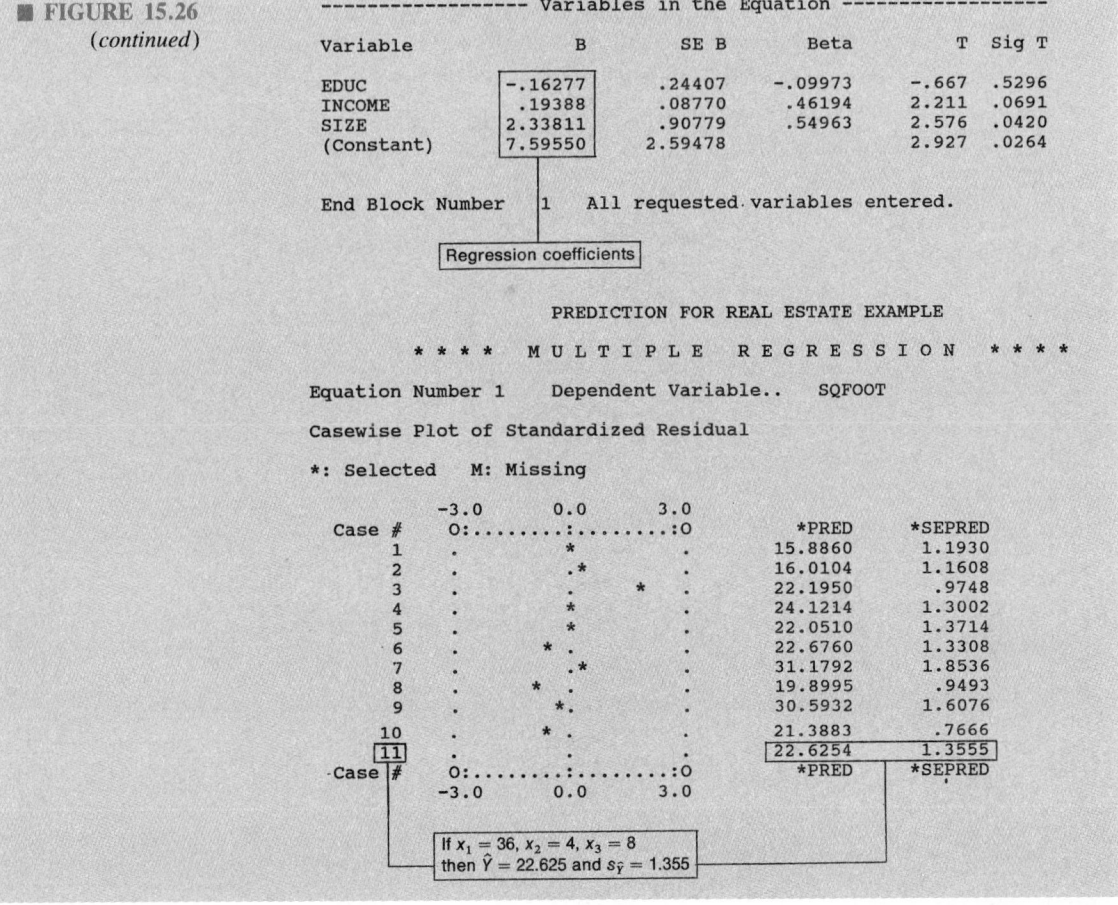

```
------------------ Variables in the Equation ------------------

Variable              B          SE B        Beta        T    Sig T

EDUC              -.16277       .24407     -.09973     -.667   .5296
INCOME             .19388       .08770      .46194     2.211   .0691
SIZE              2.33811       .90779      .54963     2.576   .0420
(Constant)        7.59550      2.59478                 2.927   .0264


End Block Number   1   All requested variables entered.
```

Regression coefficients

```
                  PREDICTION FOR REAL ESTATE EXAMPLE

            * * * *   M U L T I P L E   R E G R E S S I O N   * * * *

Equation Number 1    Dependent Variable..   SQFOOT

Casewise Plot of Standardized Residual

*: Selected   M: Missing

              -3.0      0.0       3.0
   Case  #    O:........:........:O        *PRED      *SEPRED
      1      .         *         .        15.8860     1.1930
      2      .        .*         .        16.0104     1.1608
      3      .         .    *    .        22.1950      .9748
      4      .         *         .        24.1214     1.3002
      5      .         *         .        22.0510     1.3714
      6      .      *  .         .        22.6760     1.3308
      7      .         .*        .        31.1792     1.8536
      8      .     *   .         .        19.8995      .9493
      9      .        *.         .        30.5932     1.6076
     10      .      *  .         .        21.3883      .7666
     11      .         .         .        22.6254     1.3555
   Case  #   O:........:........:O        *PRED      *SEPRED
              -3.0      0.0       3.0
```

If $x_1 = 36$, $x_2 = 4$, $x_3 = 8$
then $\hat{Y} = 22.625$ and $s_{\hat{Y}} = 1.355$

SAS

Solution
Example 15.2

At the beginning of the chapter, we computed the regression equation for the estimate of home size based on the three predictor variables, income, family size, and level of education. The SAS program listing in Figure 15.27 requests a multiple regression solution. In this problem the SAS commands are the same for both the mainframe and PC versions.

The TITLE command names the SAS run.

The DATA command gives the data a name.

The INPUT command names and gives the correct order for the different fields on the data lines.

The CARDS command indicates to SAS that the input data immediately follow.

The next ten lines contain the data values and represents the four variables to be considered in the regression analysis. The first card image represents a home with 1,600 square feet, an income of $22,000, a family of two people, and four years of educational experience at the college level.

The PROC CORR command and VAR subcommand request a correlation matrix of the variables SQFOOT, INCOME, SIZE, and EDUC.

The PROC REG command and MODEL subcommand indicate that the independent variables INCOME, SIZE, and EDUC are to be entered in the regression equation to predict the dependent variable SQFOOT.

Figure 15.28 shows the SAS output obtained by executing the listing in Figure 15.27.

■ FIGURE 15.27
Input for SAS (mainframe or micro version).

```
TITLE  'REAL ESTATE EXAMPLE USING THREE PREDICTORS';
DATA REAL ESTATE;
  INPUT SQFOOT INCOME SIZE EDUC;
CARDS;
16 22 2 4
17 26 2 8
26 45 3 7
24 37 4 0
22 28 4 2
21 50 3 10
32 56 6 8
18 34 3 8
30 60 5 2
20 40 3 6
PROC CORR;
  VAR SQFOOT INCOME SIZE EDUC;
PROC REG;
  MODEL SQFOOT=INCOME SIZE EDUC;
```

■ FIGURE 15.28   SAS output.

REAL ESTATE EXAMPLE USING THREE PREDICTORS

CORRELATION ANALYSIS

Pearson Correlation Coefficients / Prob > |R| under Ho: Rho=0 / N = 10

	SQFOOT	INCOME	SIZE	EDUC
SQFOOT	1.00000 0.0	0.84325 0.0022	0.90787 0.0003	-0.16794 0.6428
INCOME	0.84325 0.0022	1.00000 0.0	0.72125 0.0186	0.15142 0.6763
SIZE	0.90787 0.0003	0.72125 0.0186	1.00000 0.0	-0.25137 0.4836
EDUC	-0.16794 0.6428	0.15142 0.6763	-0.25137 0.4836	1.00000 0.0

Model: MODEL1
Dependent Variable: SQFOOT

Analysis of Variance

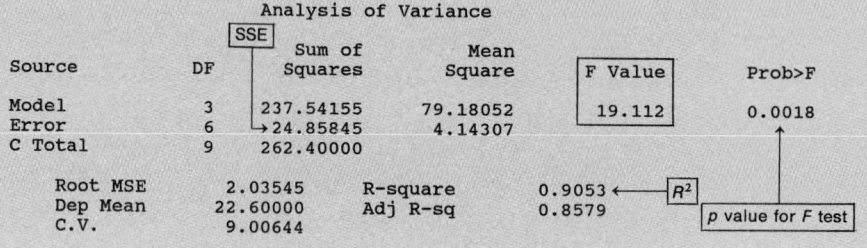

Source	DF	Sum of Squares	Mean Square	F Value	Prob>F
Model	3	237.54155	79.18052	19.112	0.0018
Error	6	24.85845	4.14307		
C Total	9	262.40000			

Root MSE	2.03545	R-square	0.9053	
Dep Mean	22.60000	Adj R-sq	0.8579	
C.V.	9.00644			

R^2

p value for F test

Parameter Estimates

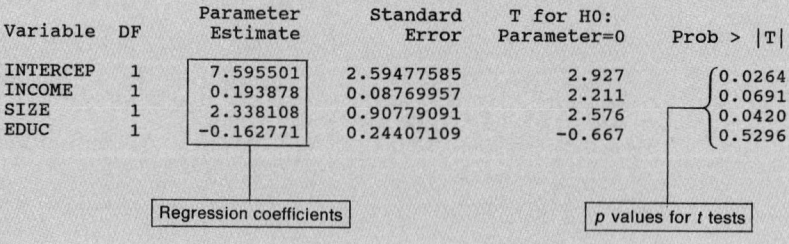

Variable	DF	Parameter Estimate	Standard Error	T for H0: Parameter=0	Prob > |T|
INTERCEP	1	7.595501	2.59477585	2.927	0.0264
INCOME	1	0.193878	0.08769957	2.211	0.0691
SIZE	1	2.338108	0.90779091	2.576	0.0420
EDUC	1	-0.162771	0.24407109	-0.667	0.5296

Regression coefficients

p values for t tests

SAS

Example 15.2
Stepwise Method

SAS can be used for determining the stepwise solution used in predicting home size based on the three predictor variables, income, family size, and level of education. Notice that the word STEPWISE is substituted for REG and that a new line has been added. Instead of forcing all variables into the equation with the REG command, STEPWISE selects the variables that meet the entry criteria. The SLENTRY = 0.1 statement specifies the significance level for entering a variable into the regression equation. The SAS program listing in Figure 15.29 requests a stepwise regression solution. In this problem the SAS commands are the same for both the mainframe and PC versions.

The TITLE command names the SAS run.

The DATA command gives the data a name.

The INPUT command names and gives the correct order for the different fields on the data lines.

The CARDS command indicates to SAS that the input data immediately follow.

The next ten lines contain the data values, which represent the four variables to be considered in the regression analysis. The first card image represents a home with 1,600 square feet, an income of $22,000, a family of two people, and four years of educational experience at the college level.

The PROC STEPWISE command and MODEL subcommand indicate that the independent variables INCOME, SIZE, and EDUC are to be entered in the regression equation to predict the dependent variable SQFOOT. The SLENTRY subcommand was discussed above.

Figure 15.30 shows the SAS output obtained by executing the listing in Figure 15.29.

■ **FIGURE 15.29**
Input for SAS
(mainframe or
micro version).

```
TITLE  'REAL ESTATE EXAMPLE USING STEPWISE';
DATA REAL ESTATE;
 INPUT SQFOOT INCOME SIZE EDUC;
CARDS;
16 22 2 4
17 26 2 8
26 45 3 7
24 37 4 0
22 28 4 2
21 50 3 10
32 56 6 8
18 34 3 8
30 60 5 2
20 40 3 6
PROC STEPWISE;
 MODEL SQFOOT=INCOME SIZE EDUC/
         SLENTRY=0.1;
```

■ **FIGURE 15.30** SAS output.

REAL ESTATE EXAMPLE USING STEPWISE

Stepwise Procedure for Dependent Variable SQFOOT

Step 1 | Variable SIZE Entered | | R-square = 0.82422204 | C(p) = 5.13282842

	DF	Sum of Squares	Mean Square	F	Prob>F
Regression	1	216.27586207	216.27586207	37.51	0.0003
Error	8	46.12413793	5.76551724		
Total	9	262.40000000			

■ **FIGURE 15.30** (*continued*)

```
                   Parameter        Standard        Type II
Variable           Estimate         Error       Sum of Squares        F      Prob>F

INTERCEP          9.08275862       2.33397081     87.31381827        15.14    0.0046
SIZE              3.86206897       0.63057266    216.27586207        37.51    0.0003

Bounds on condition number:              1,               1
--------------------------------------------------------------------------------

Step 2  | Variable INCOME Entered |    | R-square = 0.89824277 |  C(p) = 2.44475400

                   DF         Sum of Squares      Mean Square         F      Prob>F

Regression          2          235.69890381      117.84945191       30.90    0.0003
Error               7           26.70109619        3.81444231
Total               9          262.40000000

                   Parameter        Standard        Type II
Variable           Estimate         Error       Sum of Squares        F      Prob>F

INTERCEP         | 6.73957851 |     2.16385176     37.00339013        9.70    0.0170
INCOME           | 0.16485256 |—    0.07305544     19.42304174        5.09    0.0586
SIZE             | 2.65693994 |     0.74046309     49.11200800       12.88    0.0089

Bounds on condition number:      2.084221,       8.336884
--------------------------------------------------------------------------------
             |
             | Final equation: Ŷ = 6.74 + .16(income) + 2.66(size) |

All variables in the model are significant at the 0.1500 level.
No other variable met the 0.1000 significance level for entry into the model.

      Summary of Stepwise Procedure for Dependent Variable SQFOOT

        Variable          Number    Partial    Model
Step    Entered Removed      In      R**2       R**2       C(p)          F      Prob>F

  1     SIZE                  1      0.8242     0.8242     5.1328      37.5120    0.0003
  2     INCOME                2      0.0740     0.8982     2.4448       5.0920    0.0586
```

SAS

**✓ Example
Section 15.5**

In the real-estate example in section 15.5, we wished to determine the predicted square footage ($\hat{Y}$) for values $X_1 = 36$, $X_2 = 4$, and $X_3 = 8$. Of course, one way to do this is to insert them manually into the regression equation resulting from the ten observations in the previous SAS example. An easier way is to attach these values at the end of the input data along with any single character from A to Z for the value of the dependent variable to indicate to SAS that this value is missing. Which character you use is arbitrary, but it should be specified in the MISSING statement immediately following the DATA statement. SAS ignores this row of data when it computes the regression equation but includes this row when it summarizes the predicted values.

The predicted values as well as the corresponding confidence intervals and prediction intervals are calculated and included in the output by inserting /CLI and CLM; in the final MODEL statement. The CLI command generates the prediction intervals for an *individual* whereas CLM will produce confidence intervals for the *mean*. The row(s) containing the missing value(s) will be included in this summary.

The SAS program listing in Figure 15.31 is used to generate the predicted value, confidence intervals, and prediction intervals. In this problem the SAS commands are the same for both the mainframe and PC versions.

The TITLE command names the SAS run.

The DATA command gives the data a name.

The INPUT command names and gives the correct order for the different fields on the data lines.

The CARDS command indicates to SAS that the input data immediately follow.

The next ten lines contain the data values representing the four variables to be considered in the regression analysis. The first card image represents a home with 1,600 square feet, an income of $22,000, a family of two people, and four years of educational experience at the college level.

The PROC REG command and MODEL subcommand indicate that the independent variables INCOME, SIZE, and EDUC are to be entered in the regression equation to predict the dependent variable SQFOOT. The CLI and CLM subcommands were discussed above.

Figure 15.32 shows the SAS output obtained by executing the listing in Figure 15.31.

■ FIGURE 15.31
Input for SAS
(mainframe or
micro version).

```
TITLE  'PREDICTION FOR REAL ESTATE EXAMPLE';
DATA REAL ESTATE;
    MISSING A;
  INPUT SQFOOT INCOME SIZE EDUC;
CARDS;
16 22  2  4
17 26  2  8
26 45  3  7
24 37  4  0
22 28  4  2
21 50  3 10
32 56  6  8
18 34  3  8
30 60  5  2
20 40  3  6
A 36  4  8
PROC REG;
    MODEL SQFOOT=INCOME SIZE EDUC/ CLI CLM;
```

■ FIGURE 15.32   SAS output.

PREDICTION FOR REAL ESTATE EXAMPLE

Model: MODEL1
Dependent Variable: SQFOOT

Analysis of Variance

Source	DF	Sum of Squares	Mean Square	F Value	Prob>F
Model	3	237.54155	79.18052	19.112	0.0018
Error	6	24.85845	4.14307		
C Total	9	262.40000			

Root MSE	2.03545	R-square	0.9053
Dep Mean	22.60000	Adj R-sq	0.8579
C.V.	9.00644		

Parameter Estimates

Variable	DF	Parameter Estimate	Standard Error	T for H0: Parameter=0	Prob > |T|
INTERCEP	1	7.595501	2.59477585	2.927	0.0264
INCOME	1	0.193878	0.08769957	2.211	0.0691
SIZE	1	2.338108	0.90779091	2.576	0.0420
EDUC	1	-0.162771	0.24407109	-0.667	0.5296

■ FIGURE 15.32 (*continued*)

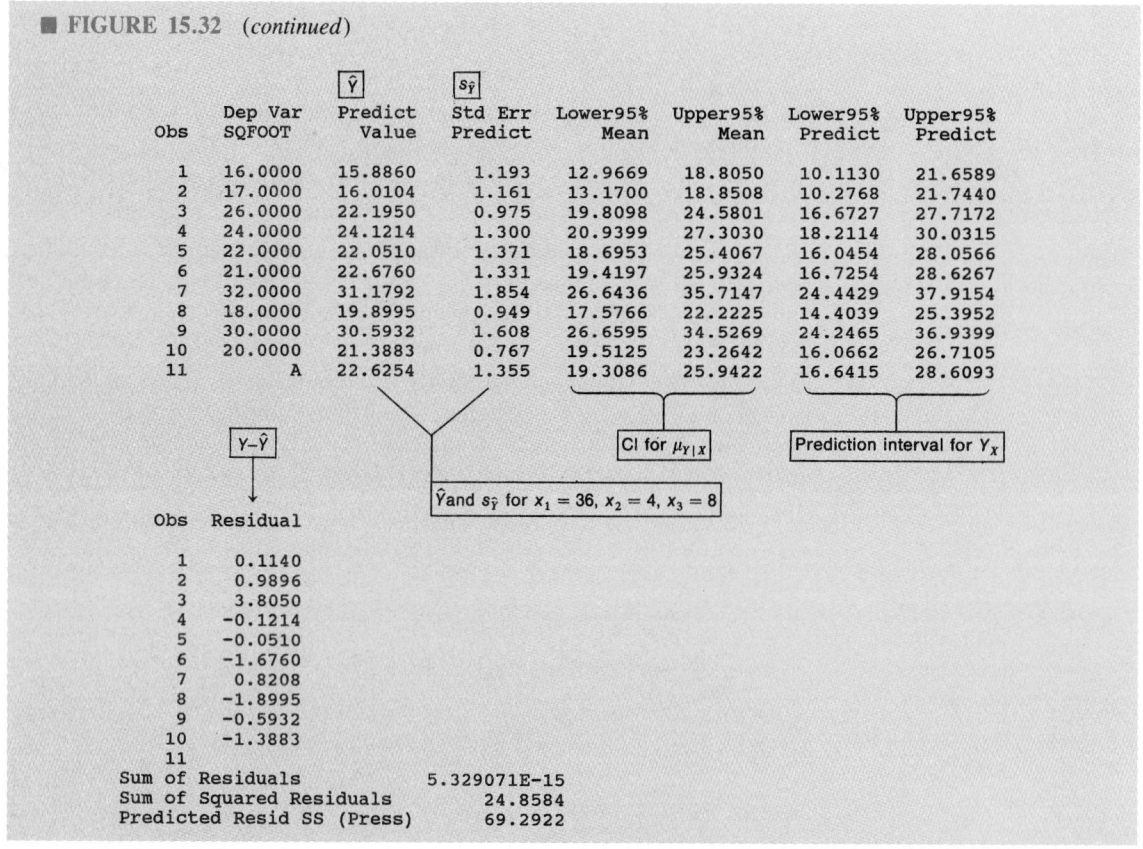

Obs	Dep Var SQFOOT	$\hat{Y}$ Predict Value	$s_{\hat{Y}}$ Std Err Predict	Lower95% Mean	Upper95% Mean	Lower95% Predict	Upper95% Predict
1	16.0000	15.8860	1.193	12.9669	18.8050	10.1130	21.6589
2	17.0000	16.0104	1.161	13.1700	18.8508	10.2768	21.7440
3	26.0000	22.1950	0.975	19.8098	24.5801	16.6727	27.7172
4	24.0000	24.1214	1.300	20.9399	27.3030	18.2114	30.0315
5	22.0000	22.0510	1.371	18.6953	25.4067	16.0454	28.0566
6	21.0000	22.6760	1.331	19.4197	25.9324	16.7254	28.6267
7	32.0000	31.1792	1.854	26.6436	35.7147	24.4429	37.9154
8	18.0000	19.8995	0.949	17.5766	22.2225	14.4039	25.3952
9	30.0000	30.5932	1.608	26.6595	34.5269	24.2465	36.9399
10	20.0000	21.3883	0.767	19.5125	23.2642	16.0662	26.7105
11	A	22.6254	1.355	19.3086	25.9422	16.6415	28.6093

$Y - \hat{Y}$

CI for $\mu_{Y|x}$ Prediction interval for Y_x

$\hat{Y}$ and $s_{\hat{Y}}$ for $x_1 = 36$, $x_2 = 4$, $x_3 = 8$

Obs	Residual
1	0.1140
2	0.9896
3	3.8050
4	-0.1214
5	-0.0510
6	-1.6760
7	0.8208
8	-1.8995
9	-0.5932
10	-1.3883
11	

Sum of Residuals   5.329071E-15
Sum of Squared Residuals   24.8584
Predicted Resid SS (Press)   69.2922

MINITAB

■ Instructions for Multiple and Stepwise Regression

The Regress Command

The MINITAB REGRESS command is illustrated in many of the examples contained in this chapter. By using various subcommands you can store the residuals and the values of $b_0, b_1 \ldots, b_k$.

For example, consider the sequence

REGRESS Y IN C1 USING 2 PREDICTORS IN C2 C3;

RESIDUALS IN C4;

COEF IN C5.

This sequence stores the residuals in column C4 and the estimated regression coefficients in column C5. Also, the abbreviated form of the REGRESS command here would be

REGRESS C1 2 C2 C3;

If you wish merely to examine the residuals but not to store them, type the command BRIEF 3 at some point before performing the regression analysis. This will provide additional output (including the residuals) when the REGRESS command is used, as illustrated in Figure 15.33).

Stepwise Regression

MINITAB also performs stepwise regression using the command

STEPWISE REGRESSION OF Y IN C1 USING PREDICTORS C2 C3 C4 . . .

```
MTB > BRIEF 3
MTB > REGRESS Y IN C1 USING 3 PREDICTORS IN C2-C4

The regression equation is
C1 = 7.60 + 0.194 C2 + 2.34 C3 - 0.163 C4

Predictor         Coef        Stdev       t-ratio          p
Constant         7.596        2.595          2.93      0.026
C2             0.19388      0.08770          2.21      0.069
C3              2.3381       0.9078          2.58      0.042
C4             -0.1628       0.2441         -0.67      0.530

s = 2.035      R-sq = 90.5%      R-sq(adj) = 85.8%

Analysis of Variance

SOURCE          DF           SS           MS          F          p
Regression       3      237.542       79.181      19.11      0.002
Error            6       24.858        4.143  ← MSE
Total            9      262.400

SOURCE          DF       SEQ SS
C2               1      186.587
C3               1       49.112
C4               1        1.843

Obs.        C2          C1        Fit  Stdev.Fit   Residual   St.Resid
  1       22.0      16.000     15.886      1.193      0.114       0.07
  2       26.0      17.000     16.010      1.161      0.990       0.59
  3       45.0      26.000     22.195      0.975      3.805       2.13R
  4       37.0      24.000     24.121      1.300     -0.121      -0.08
  5       28.0      22.000     22.051      1.371     -0.051      -0.03
  6       50.0      21.000     22.676      1.331     -1.676      -1.09
  7       56.0      32.000     31.179      1.854      0.821       0.98
  8       34.0      18.000     19.900      0.949     -1.900      -1.05
  9       60.0      30.000     30.593      1.608     -0.593      -0.48
 10       40.0      20.000     21.388      0.767     -1.388      -0.74

R denotes an obs. with a large st. resid.
```

The abbreviated statement is:

STEP C1 C2 C3 C4 . . .

where C2, C3, C4, . . . are the predictor variables.

This procedure enters a variable if its corresponding (partial) F-value exceeds FENTER $= 4$ and removes any variable whose (partial) F-value falls below FREMOVE $= 4$. An illustration of this procedure using the real-estate data is shown in Figure 15.34. Only two of the three variables being considered were selected (using the default values of FREMOVE $=$ FENTER $= 4$). The resulting equation is

$$\hat{Y} = 6.74 + .165X_1 + 2.66X_2$$

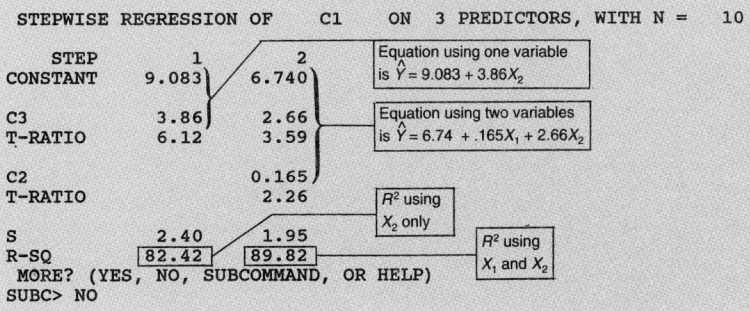

To change the values of FENTER and/or FREMOVE, use the following sequence of commands. The semicolon (;) at the end of the REGRESS command informs MINITAB that subcommands are needed. The period following the final subcommand indicates that there are no further subcommands.

MTB > STEPWISE REGRESSION OF Y IN C1 USING C2, C3, C4, C5, C6;

SUBC > FENTER = 3.5;

SUBC > FREMOVE = 3.5.

To perform a forward selection, you simply do not allow any variable to be removed once it is included in the model. Setting FREMOVE = 0 will accomplish this. The procedure ends when the (partial) F statistic for an entering variable is below FENTER.

Similarly, you can perform backwards regression by first using the subcommand ENTER:

SUBC > ENTER C2-C6;

where C2–C6 are all of your predictor variables. This subcommand enters all of your predictor variables into the model. Next, use

SUBC > FENTER = 10000.

or any large value. This procedure stops when no variable in the model has an F-value less than FREMOVE.

CHAPTER

16

Time Series Analysis and Index Numbers

The previous two chapters introduced you to a method of predicting the value of a dependent variable using the technique of linear regression. You determined a set of one or more predictor (independent) variables (X_1, X_2, . . .) that could be used to model the behavior of the dependent variable, Y.

When the dependent variable is measured over *time*, there is another method of describing the behavior of this variable—**time series analysis.** For example, consider the following data, where Y is the amount of electrical power consumed in Pine Bluff over a ten-year period:

YEAR	POWER CONSUMPTION (MILLION kwh)
1981	95
1982	145
1983	174
1984	200
1985	224
1986	245
1987	263
1988	275
1989	283
1990	288

This is an example of a (very short) time series. Typically, a time series covers many more periods, especially when measured for each month, week, or even day. To describe the behavior of the variable Y, we examine the past data and, rather than searching for a number of predictor variables, we try to capture the patterns that exist in the Y observations over a

period of time. In other words, we assume that *time-related patterns can serve as predictors.* In this illustration, one pattern is clear—the power consumption values increase from one year to the next.

The process of using the patterns contained in the past data to predict future values is referred to as **forecasting.** Forecasting using time series data has both advantages and disadvantages. The primary advantage of using time series analysis is that often you can describe your variable of interest, Y, by using only a sample of past observations. Inherent to this type of forecasting procedure is the assumption that past patterns will continue into the future. The disadvantage of time series forecasting is that the past observations often contain patterns that are difficult to extract and, as a result, the models can become very complex.

In this chapter, we will concentrate on methods of *describing* a time series by isolating its various *components* (for example, sales in December are always much higher than the yearly average). In the next chapter, methods of forecasting are discussed. We should note at this point that in general, there is no single best forecasting technique. Instead, the forecaster should attempt to match the forecasting technique to patterns observed in the time series data. Consequently, this chapter and the next chapter are highly intertwined, since by describing the nature of the time series (Chapter 16), you will have a better idea as to which forecasting technique to employ (Chapter 17).

16.1
Components of a Time Series

A **time series** represents a variable observed across time. The time increment can be years, quarters, months, or even days. The values of the time series can be presented in a table or illustrated using a scatter diagram. Usually, the points in the graph are connected by straight lines, making it easier to detect any existing patterns; such a graph is called a **line graph.**

The time series for the power-consumption data is shown in Figure 16.1. As we noted, the power-consumption values increase steadily from one year to the next. This long-term movement in the time series is called a *trend*. These values exhibit a definite increasing trend (or growth). Trend is only one of several com-

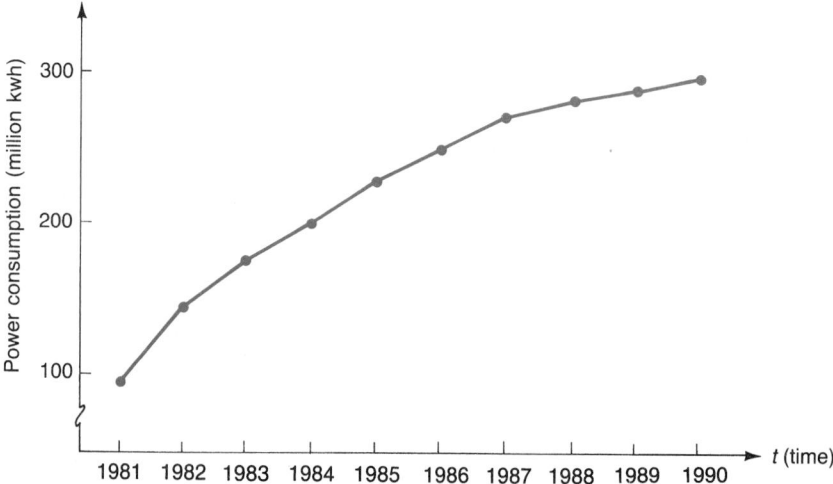

ponents that describe the behavior of any time series. The **components** of a time series are:

trend (TR)

seasonal variation (S)

cyclical variation (C)

irregular activity (I)

The purpose of time series analysis is to describe a particular data set by estimating the various components that make up this time series. We examine each of these components individually, although time series data usually contain a mixture of all four. This section will *not* attempt to measure these components, but rather will introduce you to the nature of each component. The remainder of this chapter demonstrates methods of capturing and measuring these individual components.

Trend (TR)

The **trend** is a steady increase or decrease in the time series. If a particular time series is neither increasing nor decreasing over its range of time, it contains *no trend*. The trend reflects any long-term growth or decline in the observations. For example, a trend may be due to inflation, increases in the population, increases in personal income, market growth or decline, or changes in technology. Each of these factors could have a long-term effect on the variable of interest and would be reflected in the trend in the corresponding time series.

This long-term growth or decay pattern can take a variety of shapes. If the rate of change in Y from one time period to the next is relatively constant, the trend is a **linear trend:**

$$TR = b_0 + b_1 t$$

(for some b_0 and b_1), where the predictor variable is time t.

When the time series appears to be slowing down or accelerating as time increases, then a nonlinear trend may be present. It may be a **quadratic trend**

$$TR = b_0 + b_1 t + b_2 t^2$$

or a **decaying trend**

$$TR = b_0 + b_1(1/t)$$

or

$$TR = b_0 + b_1 e^{-t}$$

These trend equations can be derived from the linear regression equations developed in Chapter 14 (for linear trend) and Chapter 15 (for quadratic trend). The linear trend equation is an application of *simple* linear regression, whereas the quadratic trend uses a *multiple* regression equation using two predictors, t and t^2. Simple linear regression techniques also can be used to derive b_0 and b_1 for the decaying trend equations, where values of t are replaced by the values of $1/t$ or e^{-t} in the data input.

The number of employees from 1983 to 1990 at Video-Comp, an expanding microcomputer-software firm, are recorded in the following table and illustrated in Figure 16.2.

YEAR	NUMBER OF EMPLOYEES (thousands)
1983	1.1
1984	2.4
1985	4.6
1986	5.4
1987	5.9
1988	8.0
1989	9.7
1990	11.2

The underlying long-term growth trend in this time series appears to be nearly *linear,* as represented by the dotted line in Figure 16.2. To determine the equation of this line, we use the technique of simple linear regression, where X = the predictor variable = time and Y = the number of employees. We can estimate the existing trend using

$$\hat{y}_t = b_0 + b_1 t$$

■ **FIGURE 16.2**
Number of employees at Video-Comp (an example of linear trend).

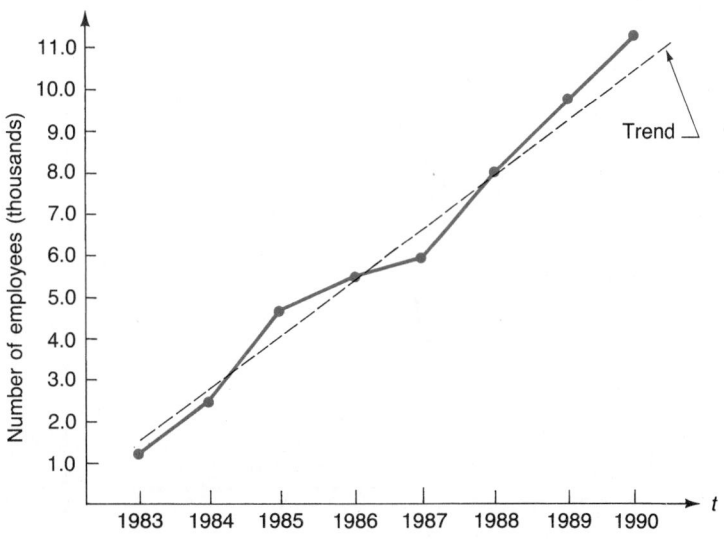

■ FIGURE 16.3
(a) Increasing linear trend: $TR = b_0 + b_1 t$ $(b_1 > 0)$. (b) Decreasing linear trend: $TR = b_0 + b_1 t (b_1 < 0)$.

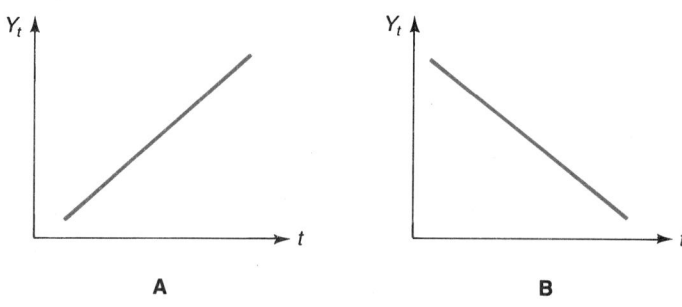

where t represents the time variable and y_t is the value of Y at time period t. Here b_0 and b_1 are the least squares regression coefficients for a straight line predictor. The procedure of deriving these least squares estimates is developed later in the chapter. Figure 16.3 shows an *increasing* linear trend (y_t increases over time) and a decreasing linear trend (y_t decreases over time).

EXAMPLE 16.1

What type of trend exists in the power-consumption data (Figure 16.1)?

SOLUTION   Although this time series increases steadily, *it increases at a decreasing rate:* it starts off with large increases from one time period to the next, but these increments gradually become smaller. When the growth is linear, the values increase at a nearly constant rate. Figure 16.1 is an illustration of **quadratic trend,** where the time series randomly fluctuates about a quadratic (or curvilinear) level over time. This trend is captured by the equation

$$\hat{y}_t = b_0 + b_1 t + b_2 t^2$$

To derive these estimates, we use the multiple linear regression approach discussed in Chapter 15 (curvilinear models). Section 16.2 demonstrates this technique.

The four types of quadratic trend are summarized in Figure 16.4.   ■

■ FIGURE 16.4
Quadratic trend. (a) Y increases at a decreasing rate. (b) Y decreases at an increasing rate. (c) Y decreases at a decreasing rate. (d) Y increases at an increasing rate.

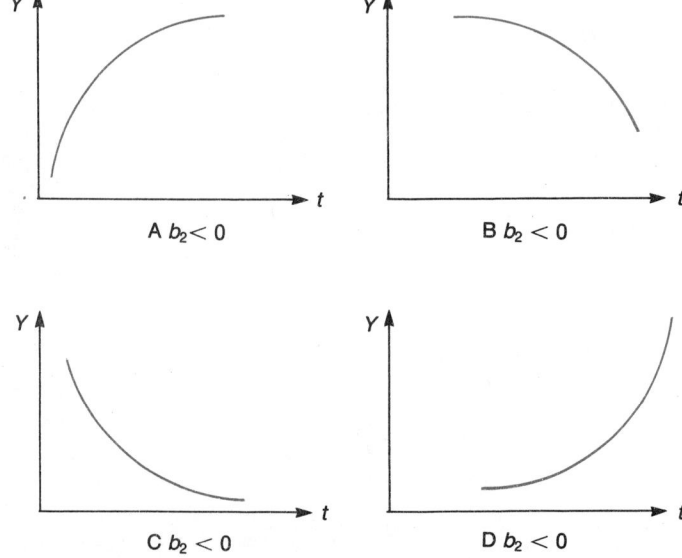

Seasonality (*S*)

Seasonal variation, or **seasonality,** refers to periodic increases or decreases that occur *within a calendar year* in a time series. They are very predictable because they occur every year. When a time series consists of annual data (as in Figure 16.1), you cannot see what is going on within each year. Data reported in annual increments therefore cannot be used to examine seasonality. Seasonality may or may not exist; annual increment data are not in a form that will show whether it does.

When time series data are quarterly or monthly, seasonal variation may be evident. For example, if the power-consumption data were available for each month over these 10 years, then the resulting time series would contain $12 \cdot 10 = 120$ observations. A plot of monthly data for the last 3 years (36 observations) is shown in Figure 16.5. The seasonal effects here consist of

Extremely high power consumption during the hot summer months (July and August)

Very high consumption during the coldest part of the winter (December and January)

Gradually declining consumption during the spring, reaching a low level in April and then increasing until July

Gradually declining power consumption during the fall, but beginning to increase in November

The key is that these movements in the time series follow the same pattern each year and so probably are due to seasonality. An analysis of seasonal variation is often a crucial step in planning sales and production. Just because your sales drop from one month to the next does not necessarily mean that it is time to panic. If a review of past observations indicates that sales *always* drop between these two months, then quite likely there is no cause for concern. On the production side, if sales always are extremely high in December, then you will need to increase pro-

■ FIGURE 16.5
Illustration of seasonal variation. These are monthly observations; compare with annual data in Figure 16.1.

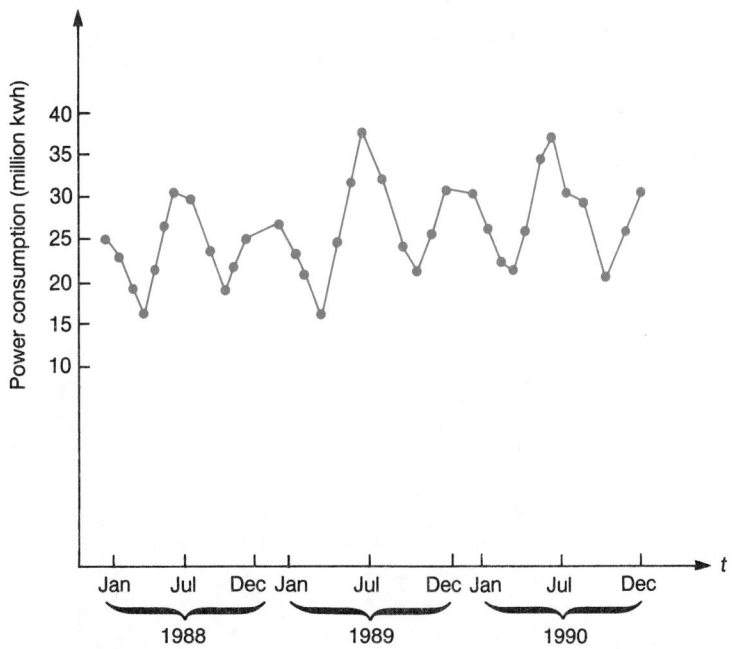

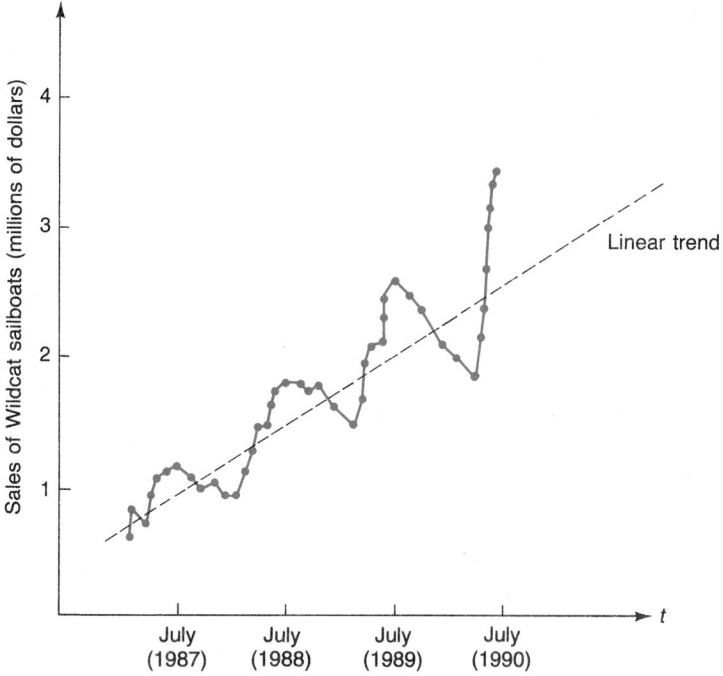

duction in the months prior to December so that you will have the necessary inventory level for this peak month. Measurement of this seasonal component is discussed later.

As mentioned earlier, a time series often contains the effect of trend and seasonality (as well as cyclical and irregular activity). The sales of Wildcat sailboats, illustrated in Figure 16.6, contain a strong linear trend as well as definite seasonal variation. In particular, the highest sales occur in the summer months of each year.

As manager of Wildcat Enterprises, would you be concerned that the sales of these boats in December 1990 were lower than those in July 1990? There may or may not be a problem; this seasonal pattern exists in Figure 16.6 despite an overall growth. More data would be required to determine whether the December sales were lower than expected for that month. What would you think if sales in July 1991 were lower than those in July 1990? This event should definitely concern you. This is a year-to-year comparison, and seasonal variation or not, we would expect the sales for July 1991 to be larger than for July 1990 if the long-term growth trend in Figure 16.6 is still present. Lower sales in July 1991 would indicate a possible leveling off or a drop in boat sales in 1991.

Cyclical Variation (*C*)

Cyclical variation describes a gradual cyclical movement about the trend; it is generally attributable to business and economic conditions. The length of a cycle is the **period** of that cycle. The period of a cycle can be measured from one **peak** to the next, one **trough** (valley) to the next, or from the time value at which the cycle crosses the horizontal line (where no cyclic activity exists) to the value where it completes the cycle and returns to this point. Figure 16.7 shows that the cycle length can be measured from P_1 to P_2, from V_1 to V_2, or from Z_1 to Z_2. In the illustrations to follow, we use the Z_1 to Z_2 approach.

In business applications, cycles typically are long-term movements, with periods ranging from two to ten years. The primary difference between the cyclical

■ **FIGURE 16.7**
The cycle can be
measured from P_1 to P_2,
from V_1 to V_2, or from
Z_1 to Z_2.

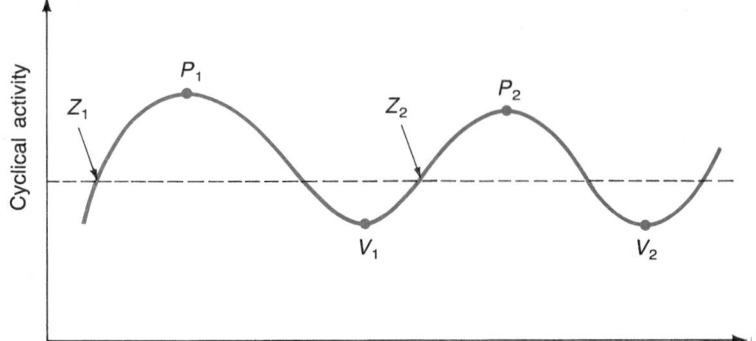

and seasonal factors is the period length. Seasonal effects take place *within* one year, whereas the period for cyclical activity is usually *more than* one year.

Cyclical activity need not follow a definite, recurrent pattern. The cycles generally represent conditions within the economy, where a peak occurs at the height of an expansion (prosperity) period and is generally followed by a period of contraction in economic activity. The low point (trough) of each cycle usually takes place at the low point of an economic recession or depression. This low point is then followed by a gradual increase during the recovery period.

EXAMPLE 16.2   The annual corporate taxes paid by Lindale (a clothing manufacturer) over a 25-year period are shown in Figure 16.8. How many cycles do you observe?

SOLUTION   The year 1967 began a cycle lasting approximately eight years. There are three cycles contained within the time series, which ends in the midst of an "up cycle." Notice that the cycle lengths are not the same.   ■

Irregular Activity (*I*)

Irregular activity consists of what is "left over" after accounting for the effect of any trend, seasonality, or cyclical activity. These values should consist of noise, much like the error term in the linear regression models discussed in the previous chapters. *The irregular activity should contain no observable or predictable pattern.* An extremely large irregular component can be caused by a measurement error in the variable. Such an outlier should always be checked to ensure its accuracy.

The irregular component (1) measures the random movement in your time se-

■ **FIGURE 16.8**
Annual taxes paid by
Lindale (illustration of
cyclical activity;
Example 16.2).

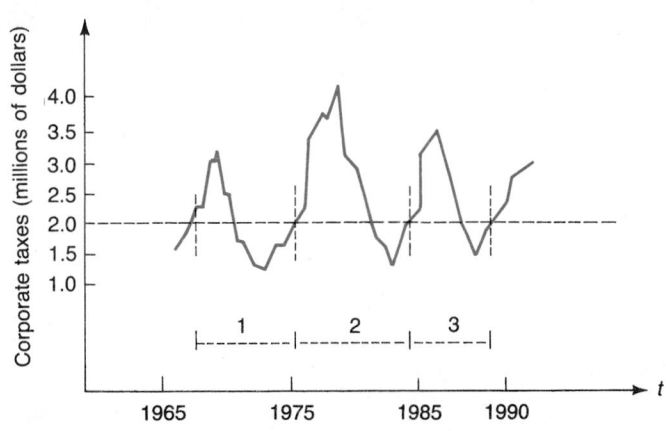

ries and (2) represents the effect introduced by unpredictable rare events, such as earthquakes, oil embargoes, or strikes.

If a noticeable jump in the resulting irregular components (when plotted across time) can be attributed to a particular rare event, you may wish to eliminate such data from the time series. You can then examine the remaining data to measure more accurately the other time series components.

Combining the Components

The time series components can be combined in various ways to describe the behavior of a particular time series. One method is to describe the time series variable, y_t, as a *sum* of these four components

$$y_t = TR_t + S_t + C_t + I_t$$

This is called the **additive structure.** The implication here is that any seasonal effects are additive from one year to the next. For example, if the seasonal effect of December for a time series representing sales is an increase of 250 units over the average yearly sales, then this same increase will occur each year regardless of the sales volume. Whether the average yearly sales are 1000 units or 10,000 units, December should show a sales volume of approximately 1250 (the first case) or 10,250 (the latter case).

Better success has been achieved by describing a time series using the **multiplicative structure,** where

$$y_t = TR_t \cdot S_t \cdot C_t \cdot I_t$$

Here, the seasonal effect increases or decreases according to the underlying trend and cyclical effect. Using the previous illustration, the difference between the December sales and the yearly average will be *higher* for the latter case (where the yearly average is 10,000 units). For example, for the first case, the December sales might be 1250 (a 25% increase over the yearly average) and, for the latter situation, it might be 12,500 (also a 25% increase). This result follows from the implication in the multiplicative structure that as the sales increase from one year to the next, the changes in volume due to seasonality also increase. For our illustration, this shift was 250 units for the first case and 2500 units for the second case.

▟ *Exercises*

16.1   The management of a pharmaceutical firm would like to predict the effects of a technological breakthrough in an antiulcer drug in order to plan for company growth and capital expenditure. Would a time series analysis be appropriate? Why or why not?

16.2   Describe in words the trends for the quarterly sales figures (in thousands of dollars) of the companies A, B, and C. Graph the data over time.

YEAR	QUARTER	COMPANY A	COMPANY B	COMPANY C
1989	1	13.1	8.3	5.1
	2	10.2	7.3	7.1
	3	11.1	8.0	6.4
	4	16.5	7.3	5.3
1990	1	14.1	7.1	5.4
	2	11.3	6.4	7.2
	3	12.5	7.0	6.3
	4	17.8	6.4	5.1
1991	1	15.1	6.2	5.0
	2	12.2	5.3	7.0
	3	13.4	6.0	6.1
	4	18.3	5.2	5.3

16.3   Construction in the housing industry usually appears to peak in the middle of the summer and to bottom out around January. If the number of new housing starts are the same for the month of March and the month of July in a particular year, of what concern would these figures be to housing construction companies? Would they be pleased, worried, or indifferent? Why?

16.4   The end-of-year inventory levels, in dollars, of West Coast Distributing are given in the following table. Estimate the period of the cyclical component by graphing the data.

YEAR	INVENTORY	YEAR	INVENTORY
1979	80	1986	80
1980	75	1987	83
1981	71	1988	80
1982	73	1989	77
1983	82	1990	79
1984	76	1991	84
1985	78		

16.5   To which of the four components of a time series would each of the following influences on housing starts contribute?

a.   Presidential election year

b.   Start of the school year in September

c.   Long-term growth of the housing industry

d.   Shortage of lumber because of a strike

16.6   Describe in words both the trend and the seasonal components for the following sales figures (in thousands of dollars). (*Hint:* draw a graph for each.)

MONTH	1990	1991	MONTH	1990	1991
Jan	1.2	2.2	July	2.9	3.8
Feb	1.4	2.4	Aug	3.2	4.0
Mar	1.3	2.3	Sept	2.5	3.5
Apr	1.5	2.4	Oct	2.4	3.0
May	1.5	2.5	Nov	2.3	2.8
June	2.3	3.5	Dec	2.1	2.5

▌ *16.2*
Measuring Trend: No Seasonality

Suppose that you have a time series containing trend and cyclical activity but no seasonality. For example, the employment data in Figure 16.2 are annual and so contain no seasonality. The same is true for the annual power-consumption data in Figure 16.1. When data are collected on a yearly basis, we are not concerned with any seasonality in the data; we need data from quarterly or shorter intervals to identify any seasonality. Yearly data may have trend (TR), cyclical activity (C), or irregular activity (I). If we observe a strong linear trend (as in Figure 16.2) or a quadratic trend (Figure 16.4), we can estimate it using the least squares technique developed in Chapters 14 and 15. We use simple linear regression for linear trends and multiple linear regression for quadratic trends.

Linear Trend

We begin by **coding** the time variable to make the calculations (or computer input) easier.

In Figure 16.2, we can find an equation for the trend line passing through the eight observations in the time series. The least squares trend line through these eight values is sketched in Figure 16.9. The equation of the trend line is

$$\hat{y}_t = TR_t = b_0 + b_1 t$$

where t represents the time variable. For this equation, TR_t represents the trend component of the sample observation at time period t and is simply a new name for the trend effect that this equation allows us to estimate.

We could use $t = 1983, 1984, \ldots$ to represent time, but a much simpler

■ FIGURE 16.9

Least squares trend line using coded time data (compare with Figure 16.2). $\hat{y}_t = b_0 + b_1 t$.

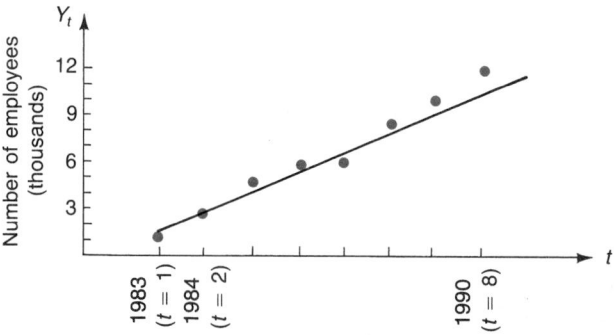

method is to *code* the data, as illustrated in Figure 16.9. By using $t = 1, 2, \ldots$, the estimate, $\hat{y}_t$, is not affected and the calculations are easier. You are able to code the predictor variable, t, because the sample values are equally spaced—they are all one year apart. As we saw in Chapter 14, this equal spacing does not occur in all simple regression applications. Continuing the scheme in Figure 16.9, $t = 9$ represents the year 1991 and the estimated number of employees for 1991 (using trend only) is

$$\hat{y}_9 = b_0 + b_1(9)$$

To derive the "best" line through the time series data, we use the least squares estimates discussed in Chapter 14; the independent variable here is time, t. The data are

t	y_t
1	1.1 $(= y_1)$
2	2.4 $(= y_2)$
3	4.6 :
4	5.4 :
5	5.9
6	8.0
7	9.7
8	11.2

The calculations are

$$\Sigma t = 1 + 2 + \cdots + 8 = 36$$
$$\Sigma t^2 = 1 + 4 + \cdots + 64 = 204$$
$$\Sigma y_t = 1.1 + 2.4 + \cdots + 11.2 = 48.3$$
$$\Sigma y_t^2 = (1.1)^2 + (2.4)^2 + \cdots + (11.2)^2 = 375.63$$
$$\Sigma t y_t = (1)(1.1) + (2)(2.4) + \cdots + (8)(11.2) = 276.3$$

In Chapter 14, when we regressed the variable Y on a single variable X, the estimate for the slope of the least squares line (from equation 14.11) was given by

$$b_1 = \frac{\text{SCP}_{XY}}{\text{SS}_X} = \frac{\Sigma xy - (\Sigma x)(\Sigma y)/n}{\Sigma x^2 - (\Sigma x)^2/n}$$

where n was the number of sample observations. To determine a linear trend line for a time series, this equation becomes

$$b_1 = \frac{\Sigma t y_t - (\Sigma t)(\Sigma y_t)/T}{\Sigma t^2 - (\Sigma t)^2/T} \qquad \textbf{(16.1)}$$

where T = the number of observations in the time series.

The sample estimate of the intercept is

$$b_0 = \bar{y} - b_1\bar{x} = \bar{y}_t - b_1\bar{t} \qquad \textbf{(16.2)}$$

where $\bar{y}_t = (y_1 + y_2 + \cdots + y_t)/T$.

Because the time variable, t, *always* is 1, 2, . . . , T, there is an easier way to calculate Σt, Σt^2, and $\bar{t}$.

$$\Sigma t = 1 + 2 + \cdots + T$$
$$= \frac{T(T + 1)}{2} \qquad \textbf{(16.3)}$$
$$\Sigma t^2 = 1 + 4 + \cdots + T^2$$
$$= \frac{T(T + 1)(2T + 1)}{6} \qquad \textbf{(16.4)}$$
$$\bar{t} = \frac{\Sigma t}{T} = \frac{T + 1}{2} \qquad \textbf{(16.5)}$$

We use these equations to derive the least squares line in Figure 16.9. Using equations 16.3 and 16.4,

$$\Sigma t = \frac{T(T + 1)}{2} = \frac{(8)(9)}{2} = 36$$

and

$$\Sigma t^2 = \frac{T(T + 1)(2T + 1)}{6} = \frac{(8)(9)(17)}{6} = 204$$

Also,

$$\bar{t} = \frac{\Sigma t}{T} = \frac{36}{8} = 4.5$$

This value can also be found using equation 16.5:

$$\bar{t} = \frac{T + 1}{2} = \frac{9}{2} = 4.5$$

So we can now calculate

$$b_1 = \frac{\Sigma ty_t - (\Sigma t)(\Sigma y_t)/T}{\Sigma t^2 - (\Sigma t)^2/T}$$
$$= \frac{276.3 - (36)(48.3)/8}{204 - (36)^2/8} = \frac{58.95}{42} = 1.4036$$

and

$$b_0 = \bar{y}_t - b_1\bar{t}$$
$$= \frac{48.3}{8} - (1.4036)(4.5) = 6.0375 - 6.3162 = -.279$$

The trend line for this time series is

$$\hat{y}_t = -.279 + 1.404t$$

■ **FIGURE 16.10**
MINITAB solution of
least squares trend line.

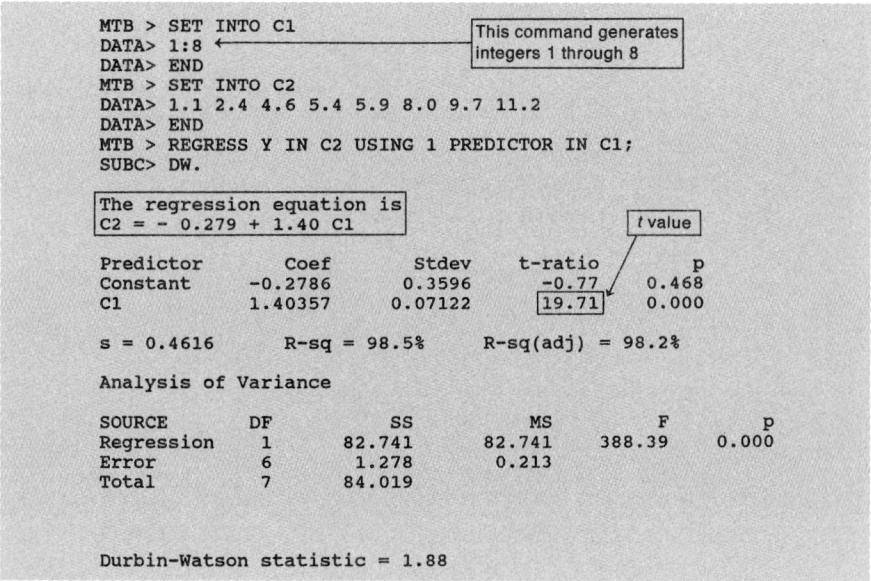

We conclude that the number of employees appears to increase at the rate of 1404 per year, on the average.

The trend line is derived using the same least squares procedure we discussed in Chapter 14—you can use the computer instructions contained in the simple linear regression illustrations. A computer solution (using MINITAB) is shown in Figure 16.10. It produces the same results we obtained.

Figure 16.10 contains the t statistic; you may be tempted to use it to determine whether time is a significant predictor of Y = number of employees. However, to use this statistic, you must assume that the errors about the trend line are completely *independent* of one another and contain *no observable pattern*. Do not forget that there may well be considerable cyclical activity about the trend line, and this cyclical activity will be contained in the residuals of the regression analysis. Thus there probably will be a cyclical pattern to these residuals, so the assumption of complete independence is not met. The errors are therefore *autocorrelated* and any test of hypothesis is invalid.

This situation poses no serious problems at this point, however, because *our intent is simply to describe the time series by measuring the various components, and not to perform a statistical test of hypothesis*. If, however, the residuals about the trend line appear to be extremely large, it suggests that a linear trend component is not appropriate.

Quadratic Trend

The nature of a quadratic trend is illustrated in Figure 16.4. This type of trend is common for a time series that increases or decreases rapidly and then gradually levels off over the observed values. We discussed a similar situation in Chapter 15, where a quadratic model of the form

$$\hat{Y} = b_0 + b_1 X + b_2 X^2$$

was used to capture a curvilinear relationship between two variables. We use exactly the same technique to describe a quadratic trend; now X is replaced by time, t.

The power-consumption time series in Figure 16.1 indicates that as time increases, the amount of power consumption (y_t) also increases, but at a decreasing rate. More specifically, the increase for 1986 to 1987 is 18; for 1987 to 1988 is 12 (12 < 18); for 1988 to 1989 is 8(8 < 12); and for 1989 to 1990 is 5 (5 < 8).

When you observe a series where the *changes* from one year to the next are not (approximately) constant but seem to be either increasing or decreasing with time, these changes indicate a quadratic trend.

The equation of this curvilinear (quadratic) trend is

$$\hat{y}_t = b_0 + b_1 t + b_2 t^2$$

To derive the least squares estimates b_0, b_1, and b_2, we use the multiple linear regression procedure of Chapter 15.

What would be the input to a computer program (such as MINITAB, SAS, or SPSS) for the power-consumption data? For the regression program, you have two predictor variables, $X_1 = t$ and $X_2 = t^2$. The resulting data configuration is

y_t	t	t^2	
95	1	1	(for 1981)
145	2	4	(for 1982)
174	3	9	(for 1983)
200	4	16	⋮
224	5	25	⋮
245	6	36	
263	7	49	
275	8	64	
283	9	81	(for 1989)
288	10	100	(for 1990)

[Note that here the time series data (y_t) are put in the first column of the input data. This placement is arbitrary.]

The solution for these data (shown in Figure 16.11) is

$$\hat{y}_t = 58.6 + 44.048t - 2.1212t^2$$

To illustrate the use of this equation, the actual value for the second time period is $y_2 = 145$ and the predicted value is

$$\hat{y}_2 = 58.6 + 44.048(2) - 2.1212(2)^2 = 138.21$$

■ **FIGURE 16.11**
MINITAB solution for quadratic trend (power-consumption data).

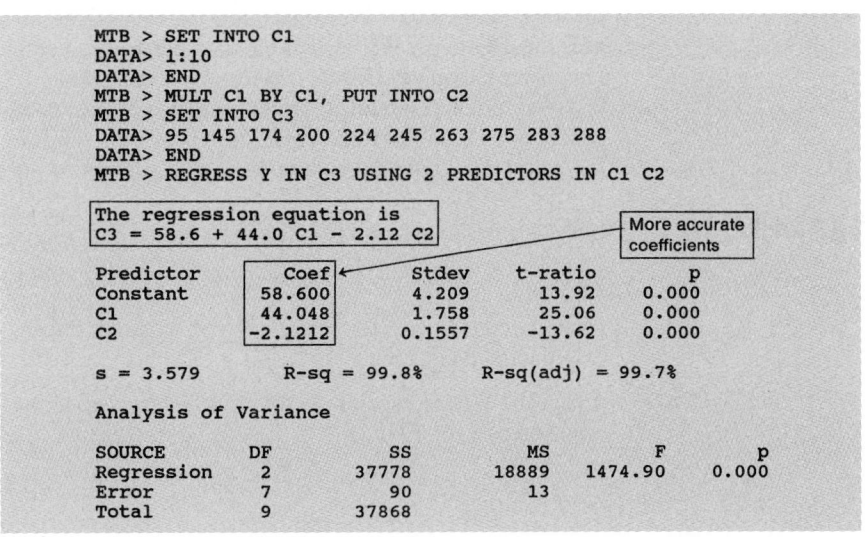

```
MTB > SET INTO C1
DATA> 1:10
DATA> END
MTB > MULT C1 BY C1, PUT INTO C2
MTB > SET INTO C3
DATA> 95 145 174 200 224 245 263 275 283 288
DATA> END
MTB > REGRESS Y IN C3 USING 2 PREDICTORS IN C1 C2

The regression equation is
C3 = 58.6 + 44.0 C1 - 2.12 C2                    More accurate
                                                coefficients

Predictor       Coef        Stdev      t-ratio         p
Constant       58.600       4.209       13.92      0.000
C1             44.048       1.758       25.06      0.000
C2            -2.1212       0.1557     -13.62      0.000

s = 3.579       R-sq = 99.8%      R-sq(adj) = 99.7%

Analysis of Variance

SOURCE          DF          SS          MS          F          p
Regression      2         37778       18889     1474.90     0.000
Error           7            90          13
Total           9         37868
```

A First Look at Forecasting: Extending the Trend

Whenever a time series contains very little seasonality (such as *annual* data, which have *no* seasonality) and a strong trend, an easy method of providing future forecasts is to project the observed growth pattern, as measured by the trend equation, into the future. For example, if a city's tax revenues have increased steadily by approximately $15,000 per year over the past 10 years, it seems reasonable to expect that this pattern will continue, at least for a short time. (Of course, assuming that such a growth will continue indefinitely is a hazardous gamble at best!)

The process of extending a trend equation is called **forecasting,** or **extrapolation.** The following examples illustrate that extending a straight-line trend equation can provide useful estimates of future values. A quadratic trend equation is, however, useful only *within* the range of the sample data, that is, for **interpolation.**

This method of forecasting is but one of many possible ways of predicting the future time series values by capturing patterns present in the past observations. Chapter 17 examines other methods of using the past observations to forecast future values.

EXAMPLE 16.3   Using the trend line from Figure 16.9, estimate the number of employees in 1991.

SOLUTION $t = 9$ corresponds to the year 1991, so the *forecast* for this year is

$$\hat{y}_9 = -.279 + 1.404(9) = 12.357$$

that is, 12,357 employees.   ∎

As mentioned earlier, the basic assumption when using the trend line to determine a forecast is that this same pattern *will continue* into the future. This may or may not be true. Very often a time series will increase at a more or less constant rate and then begin to level off. One example is the sales of an innovative product. Such a time series will grow from one year to the next as people think that they just have to have this product, but eventually a saturation point is reached and the sales grow at a much smaller rate. If the historical data used to determine the trend line are collected during the growth stage, then you will stop short of and miss the "slowing down" of the time series, severely overestimating the sales. This problem is not a flaw in the technique; any time series model makes predictions by capturing the pattern(s) in the past observations and extending this pattern beyond the last year of the data. It does, however, place a great deal of responsibility on the person who uses the data to predict beyond the data range. If you do not know what underlying factors are driving the trend, serious errors can result.

Very often, a nonlinear growth rate can be described accurately by including a quadratic term in the trend equation. However, using such an equation to forecast *future* values is not a reliable procedure, as the following section demonstrates.

EXAMPLE 16.4   Using the trend equation from Figure 16.11

$$\hat{y}_t = TR_t = 58.6 + 44.048t - 2.1212t^2$$

what is your forecast for the power consumption during 1991? During 1992? Use only the trend equation.

SOLUTION   The year 1991 corresponds to $t = 11$ (the last year of your data is $t = 10$ for 1990). Your forecast for 1991 is

$$\hat{y}_{11} = 58.6 + 44.048(11) - 2.1212(11)^2$$

$$= 286.46$$

that is, 2,864,600 kilowatt-hours. For 1992, your forecast is

$$\hat{y}_{12} = 58.6 + 44.048(12) - 2.1212(12)^2$$
$$= 281.72 \qquad \blacksquare$$

The sermon we delivered about projecting a trend line beyond the range of the data applies to a quadratic trend as well: by forecasting with such an equation, you assume that this quadratic (curved) pattern observed in the time series observations will continue.

In addition, there is another danger when forecasting with a quadratic trend equation. Every such equation looks like

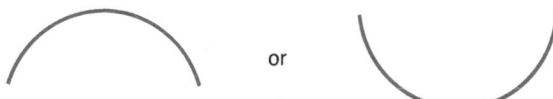

or

In other words, the curve reaches a peak (or trough) and then reverses. The problem is that you really do not know what your equation will do outside the range of your data.

The forecasts for power consumption (Example 16.4) for 1991 and 1992 provide a good illustration of this problem. Notice that the predicted value for 1991 is less than the actual value for 1990, despite a steadily increasing pattern in the time series data. Even worse, the 1992 estimate is less than that for 1991. These values imply that the trend equation is decreasing during the years after 1990. We see that the trend equation forecasts appear to be poor estimates—we have no reason to suspect a downturn in the amount of power consumption in these future years. The trend is appropriately described by the quadratic curve, but only within the range of the data. Beyond this range, the curve turns down. Because we have no reason to believe that the demand for electrical power will decrease, the quadratic equation is no longer appropriate.

To describe the trend *within* the years of your time series data, the quadratic trend equation may work well. However, remember that, *as a forecasting procedure, it is very dangerous; do not use it for this purpose.*

This section has demonstrated how you can derive linear or quadratic trend equations by using linear regression techniques. Extending such a trend equation into the future is a method of statistical forecasting, a subject that is discussed at length in Chapter 17.

Exercises

16.7 A company that supplies quotation machines to brokerage firms has had increased sales volume every year for the past ten years. Given in thousands of dollars, the sales figures are:

YEAR	SALES	YEAR	SALES
1982	60	1987	170
1983	80	1988	190
1984	110	1989	230
1985	130	1990	240
1986	160	1991	270

a. Using the simple regression formula, find the least squares line to describe sales from time, t, where t is equal to 1982, 1983, . . . , 1991.

b. Do the calculations in part a with $t = 1, 2, \ldots, 10$.

c. Compare the prediction equations given in parts a and b. Do these equations give the same predicted values?

16.8 The community of Farlington has seen its population grow dramatically over the past eight years. Data on the community's population is given in the following table (units are in thousands):

YEAR	POPULATION	YEAR	POPULATION
1984	3.1	1988	22.4
1985	6.3	1989	33.5
1986	10.1	1990	49.3
1987	17.3	1991	52.1

a. Is the time series increasing at an increasing rate?

b. Using a computerized statistical package, find the multiple regression equation to predict the population such that a quadratic curvature is taken into account. Do you think this community can continue to grow at this rate?

16.9 Explain why a prediction equation with a quadratic trend may be dangerous to use in forecasting even though a quadratic trend fits the historic data very well.

16.10 The amount of money deposited into savings accounts at a local bank has grown steadily over the years, as the following data indicate (deposits are the amount of money in savings accounts at the end of the year, in units of $100,000):

YEAR	DEPOSITS	YEAR	DEPOSITS
1984	2.1	1988	10.3
1985	4.2	1989	13.3
1986	6.4	1990	14.9
1987	8.5	1991	16.7

a. Does it appear that a quadratic trend exists in the data?

b. Calculate the equation you would use to describe the trend.

16.11 An insurance company would like to find the trend line for the amount of insurance sold annually (in millions of dollars) across time. The variable time is represented by t and is equal to $1, 2, \ldots, 8$ for the past eight years. The following statistics were collected:

$$\Sigma t y_t = 394.5$$
$$\Sigma y_t = 29.4$$

Find the trend line for these time series data.

16.12 Due to rising competition from overseas, an electronics firm has been losing its share of the market. The following data show the percent of the market that the firm has captured for the past seven years.

YEAR	SHARE OF MARKET
1985	4.7
1986	4.3
1987	3.9
1988	3.8
1989	3.6
1990	3.0
1991	2.9

a. Does the trend appear to be linear?

b. Find the equation to estimate the trend for the time series data.

c. What would be your estimate of the electronics firm's share of the market in 1992?

16.13 Total disposable personal income (in billions of dollars) for the United States from 1980–1989 is given below:

YEAR	DISPOSABLE PERSONAL INCOME
1980	1,828.9
1981	2,041.7
1982	2,180.5
1983	2,428.1
1984	2,668.6
1985	2,838.7
1986	3,013.3
1987	3,205.9
1988	3,479.2
1989	3,725.2

Find the least squares prediction equation that best describes this trend. What is your estimate of disposable personal income in 1990?

(*Source: The World Almanac and Book of Facts* (1991): 111.)

16.14 The number of airline departures in the United States for the years 1976 to 1989 follows (units are millions of departures):

YEAR	DEPARTURES	YEAR	DEPARTURES
1976	4.8	1983	5.0
1977	4.9	1984	5.4
1978	5.0	1985	5.8
1979	5.4	1986	6.4
1980	5.4	1987	6.6
1981	5.2	1988	6.7
1982	5.0	1989	6.6

a. Estimate the trend line equation for the data assuming that a linear trend is present.

b. Estimate the trend line equation for the data, assuming that a quadratic trend is present. Use a computerized statistical package.

c. Estimate the number of departures for 1990 using the equations obtained in parts a and b.

(*Source: The World Almanac and Book of Facts* (1991): 179.)

16.15 The total number of employees in the U.S. retail trade service sector of the economy is presented below (in thousands) for the time period 1970–1989:

YEAR	EMPLOYEES ON THE PAYROLL OF THE RETAIL TRADE INDUSTRY
1970	11,047
1971	11,351
1972	11,836
1973	12,329
1974	12,554
1975	12,645
1976	13,209
1977	13,808
1978	14,573
1979	14,989
1980	15,035
1981	15,189
1982	15,179
1983	15,613
1984	16,545
1985	17,356
1986	17,930
1987	18,483
1988	19,110
1989	19,575

(*Source:* U.S. Department of Labor Statistics, *Employment and Earnings,* (July 1990): 81.)

a. Estimate the trend line equation for the data assuming that a linear trend is present.

b. Estimate the trend line equation for the data assuming that a quadratic trend is present. Use a computerized statistical package.

c. Using the equations obtained in parts a and b, estimate the number of employees on the payroll of the retail trade industry for the year 1990.

16.16   The value of the Canadian dollar in terms of U.S. dollars is given below from 1975 to 1988.

YEAR	U.S. DOLLARS PER CURRENCY UNIT
1975	.9863
1976	.9820
1977	.9113
1978	.8476
1979	.8547
1980	.8397
1981	.8438
1982	.8074
1983	.8030
1984	.7570
1985	.7151
1986	.7240
1987	.7692
1988	.8387

(*Source: Chase Global Investment Almanac* (1989): 407.)

a. Estimate the trend line equation for the data assuming that a linear trend is present.

b. Assuming that a quadratic trend is present, estimate the trend. Use a computerized statistical package.

c. Use the equations found in parts a and b to estimate the value of the Canadian dollar in terms of U.S. dollars for the year 1989. Comment on the differences.

◢ 16.3
Measuring Cyclical Activity: No Seasonality

Practically every time series in a business setting contains a certain amount of cyclical activity. Cyclical activity is a gradual movement about the trend. It is generally due to economic or other long-term conditions. The overall U.S. economy tends to fluctuate through "good times" and "bad times," producing (rather unpredictable) upward and downward variation about the long-term growth or decline in a time series.

One way of describing the cyclical activity component is to represent it as a fraction of the trend. This procedure provides accurate measures of the cyclical activity provided the time series contains *little irregular activity*. Assuming that each time series observation is the *product* of its components, then

$$y_t = TR_t \cdot C_t \cdot I_t$$

because we are dealing with data containing no seasonality.

If we represent a small irregular activity component as i_t (rather than I_t), then a time series containing little irregular variation (noise) can be written as

$$y_t = TR_t \cdot C_t \cdot i_t$$

The cyclical components are then obtained by dividing each observation, y_t, by its corresponding estimate using trend only, $\hat{y}_t$.

$$\text{ratio of data to trend} = \frac{y_t}{\hat{y}_t} = \frac{\cancel{TR_t} \cdot C_t \cdot i_t}{\cancel{TR_t}} = C_t \cdot i_t$$

■ FIGURE 16.12
A complete cycle within
a time series.

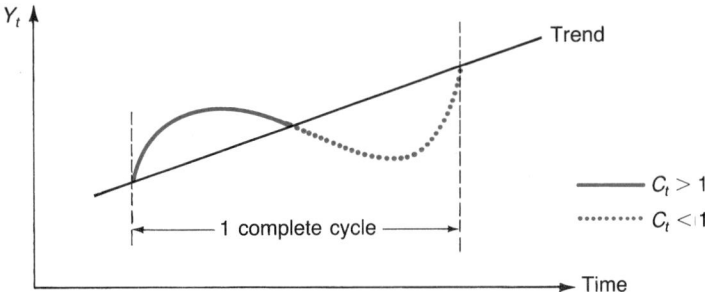

where y_t = actual time series observation at time period t and $\hat{y}_t = TR_t$ = the estimate of y_t using trend only.

Notice that the resulting ratios still contain some irregular activity. A method of reducing the irregular activity within these values is illustrated later.

An estimate of the cyclical components can be obtained by ignoring the irregular activity components in these ratios and defining

$$C_t \cong \frac{y_t}{\hat{y}_t} \qquad (16.6)$$

Assuming that we are dealing with data containing no seasonality (such as annual data), equation 16.6 provides a convenient method of determining the cycles present in the data. If $C_t > 1$, the actual y_t is larger than that predicted by trend alone. Consequently, this value is somewhere in a cycle *above* the trend line. A similar argument indicates a cycle below the trend line whenever $C_t < 1$ (Figure 16.12).

EXAMPLE 16.5

For the data in Figure 16.9, we determined a least squares trend line for the number of employees (y_t) over an eight-year period at Video-Comp. We observed a linear trend with the corresponding equation

$$\hat{y}_t = -.279 + 1.404t$$

where $t = 1$ represents 1983, $t = 2$ is for 1984, and so on. Determine and graph the cyclical activity over this period.

SOLUTION

We can obtain Table 16.1 by using the preceding trend line. Here $\hat{y}_1 = -.279 + 1.404(1) = 1.125$, $\hat{y}_2 = -.279 + 1.404(2) = 2.529$, and so on.

To examine the cyclical activity, we can describe each component as a percentage of the trend. For example, in Table 16.1, during the first time period, the

■ TABLE 16.1
Trend and cyclical
activity (Example 16.5).

t	y_t	$\hat{y}_t$	$C_t \cong y_t/\hat{y}_t$
1	1.1	1.125	.977
2	2.4	2.529	.949
3	4.6	3.933	1.169
4	5.4	5.337	1.012
5	5.9	6.741	.875
6	8.0	8.145	.982
7	9.7	9.549	1.016
8	11.2	10.953	1.022

The third column is the trend component, and the fourth column is the cyclical component as a fraction of the trend.

■ **FIGURE 16.13**
Cyclical activity about
trend line (Example
16.5).

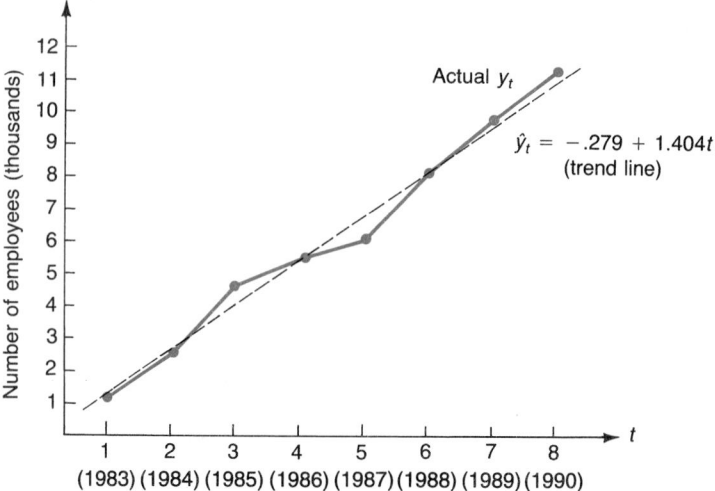

actual number of employees is 97.8% of the trend value: C_1 is .978. An illustration of the trend and cyclical activity is shown in Figure 16.13. The cycles fluctuate about the trend line. Between the years $t = 2$ (1984) and $t = 3$ (1985), $y_t = \hat{y}_t$ and a cycle begins. This cycle is completed somewhere between $t = 6$ (1988) and $t = 7$ (1989), where, once again, $y_t = \hat{y}_t$. As discussed earlier, you can also measure cycles from peak to peak or from trough to trough.

The summary of the cyclical variation (components) over the eight years is contained in Table 16.1 and Figure 16.14. The four-year cycle we described is more evident in this graph. The graph clearly indicates the beginning of the cycle, where $C_t = 1$. The cycle's peak occurs at $t = 3$ (1985), the trough is at $t = 5$ (1987), and the cycle is finally complete when C_t is again equal to 1, toward the end of 1988.

■

In summary, cyclical variation represents an upward or downward movement about the overall growth or decline (that is, the trend) in the time series data. Such cycles typically last more than one year. For annual data, these components can be

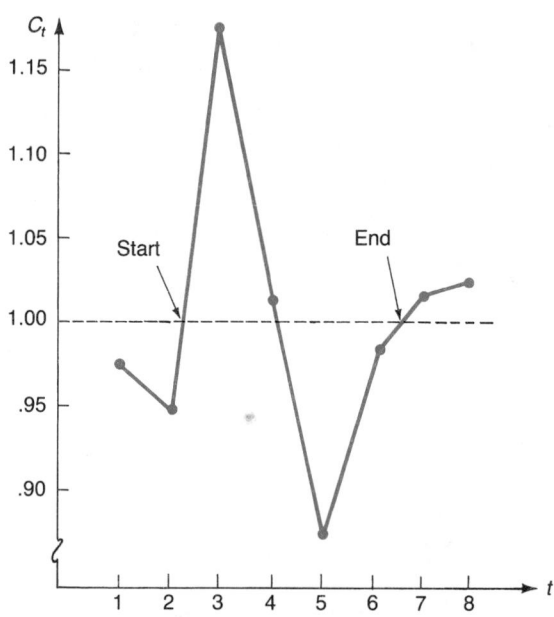

estimated by dividing each observation (y_t) by its corresponding estimate using the trend equation ($\hat{y}_t$).

Exercises

16.17 Using 15 years of data, a forecaster for an oil company found a trend line for the amount of the company's yearly contracts for oil service projects. For the past 5 years, the table gives the predicted value (in thousands of dollars) of the amount of annual contracts as well as the actual amount of annual contracts. The predicted value was arrived at by using the trend line based on 15 years of past data. Find the cyclical component for each of the past 5 years. Does the period of the cycle appear to be longer than 5 years or less than 5 years? Why?

t	y_t	$\hat{y}_t$
11	52	53
12	54	59
13	57	61
14	59	60
15	62	60

16.18 Explain the relationship between the actual value y_t and the value $\hat{y}_t$ from the trend line when $C_t > 1$ and when $C_t < 1$, where C_t is defined in equation 16.6.

16.19 A food-store chain has the following record for yearly sales volume (in hundreds of thousands of dollars) for the past 9 years:

YEAR	SALES VOLUME	YEAR	SALES VOLUME
1983	7	1988	17
1984	15	1989	12
1985	10	1990	8
1986	5	1991	17
1987	11		

a. Find the trend line.

b. Find the cyclical components.

c. Estimate the period of the cycle.

16.20 Residential Construction of America has been growing over the long term. Because the construction company is sensitive to cyclical variations in the economy, the level of employment for the company changes from year to year, as can be seen by the following data:

YEAR	FULL-TIME EMPLOYEES (IN HUNDREDS)	YEAR	FULL-TIME EMPLOYEES (IN HUNDREDS)
1979	2.4	1986	11.7
1980	9.2	1987	17.3
1981	11.1	1988	23.1
1982	8.5	1989	28.7
1983	10.5	1990	29.3
1984	6.8	1991	25.2
1985	5.4		

a. Find the trend line.

b. Find the cyclical components.

c. Estimate the period of the cycle.

16.21 For the data in Exercise 16.4, estimate the cyclical components.

16.22 Using the data from Exercise 16.15, determine and graph the cyclical activity for the years 1970 through 1988.

16.23 The president of Techronics, who is concerned about changes in the wholesale price of raw materials, gathers the following data on the wholesale price index for raw materials (WPI):

YEAR	WPI	YEAR	WPI
1982	105.0	1987	108.5
1983	106.0	1988	107.7
1984	105.0	1989	107.7
1985	104.9	1990	106.0
1986	105.6	1991	105.8

a. Estimate the trend line equation.

b. Determine the cyclical activity, C_t, where $C_t = y_t/\hat{y}_t$.

16.24 Using the data from Exercise 16.16, determine and graph the cyclical activity for the years 1975 through 1988.

16.25 Sales (in millions of dollars) of Konoco for the years 1982 through 1991 are as follows:

YEAR	SALES	YEAR	SALES
1982	151	1987	163
1983	194	1988	171
1984	177	1989	199
1985	157	1990	214
1986	188	1991	169

Determine and graph the cyclical activity, $C_t = y_t/\hat{y}_t$.

16.26 The total U.S. production of hydroelectric power (in quadrillion Btu) is presented below for the years 1978–1989:

YEAR	HYDROELECTRIC POWER
1978	2.94
1979	2.93
1980	2.90
1981	2.76
1982	3.26
1983	3.50
1984	3.31
1985	2.94
1986	3.03
1987	2.59
1988	2.31
1989	2.75

(Source: The World Almanac and Book of Facts
(1991): 183.)

Determine the cyclical component for each year.

◢ 16.4
Types of Seasonal Variation

Seasonality causes another type of variation about the trend in a time series. Seasonality generally is present when the data are quarterly or monthly. It can also occur for weekly or even daily data. For example, recurrent daily effects can be expected to occur in the check-processing volume in a bank. Seasonality is any recurrent, constant source of variation caused by events at the particular time of year rather than by any long-term influence (as in cyclical activity). For example, one would expect to sell more snowmobiles in January than in July. In a sense, the seasonal variation appears as a cycle within a year; we do not refer to this as cyclical variation, however, due to its recurrent nature.

We will discuss two types of seasonal variation: additive and multiplicative.

Additive Seasonal Variation

One encounters **additive seasonal variation** when the amount of the variation due to seasonality *does not depend on the level* y_t. This type of seasonal variation is illustrated in Figure 16.15, which shows the sales of snowmobiles over a three-year

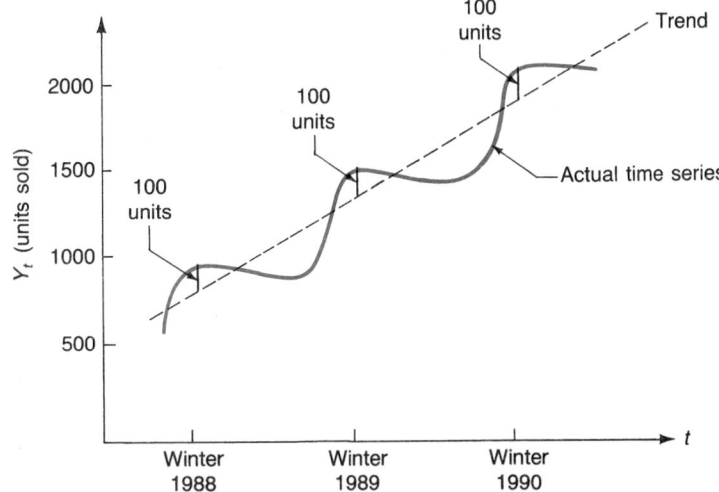

period at the Outdoor Shop. Notice that the amount of variation for each of the winter quarters remains the same (100 units), even as the unit sales increase over the three years. For an actual application, we assume an additive effect of seasonality if these increments are of nearly the same magnitude over the observed time series.

Assume that the sales data for Jetski snowmobiles from sales area 1 were recorded quarterly over a five-year period from 1986 through 1990. The following trend line was derived:

$$TR_t = \hat{y}_t = 100 + 20t$$

The seasonal indexes for a seasonal time series represent the incremental effect of the seasons alone, apart from any trend or cyclical activity. For the Jetski data in sales area 1, these indexes were found to be

$S_1 = +60$ (winter quarter) $S_3 = -40$ (summer quarter)

$S_2 = +30$ (spring quarter) $S_4 = -20$ (fall quarter)

In a time series decomposition (where we actually derive these seasonal indexes), additive seasonal variation assumes that the seasonal index for, say, the winter quarter is the *same* for each year. Using the additive model, this implies that the store will sell 60 more Jetski units in the winter quarter than would be predicted by trend alone during any year. This implies that $S_1 = S_5 = S_9 = \cdots = +60$.

To estimate y_t using only the trend and seasonality, we *add* the two corresponding components.

t(TIME)	$TR_t + S_t$ (SALES ESTIMATE)
1 (winter, 1986)	$[100 + 20(1)] + 60 = 180$
2 (spring, 1986)	$[100 + 20(2)] + 30 = 170$
3 (summer, 1986)	$[100 + 20(3)] - 40 = 120$
4 (autumn, 1986)	$[100 + 20(4)] - 20 = 160$
5 (winter, 1987)	$[100 + 20(5)] + 60 = 260$
6 (spring, 1987)	$[100 + 20(6)] + 30 = 250$
7 (summer, 1987)	$[100 + 20(7)] - 40 = 200$
8 (autumn, 1987)	$[100 + 20(8)] - 20 = 240$
⋮	⋮

A graph of the estimated sales is shown in Figure 16.16. Notice that as the overall level of sales increases, the deviation from the trend line (due to seasonality)

■ FIGURE 16.16
Jetski sales from sales
area 1 (additive seasonal
variation).

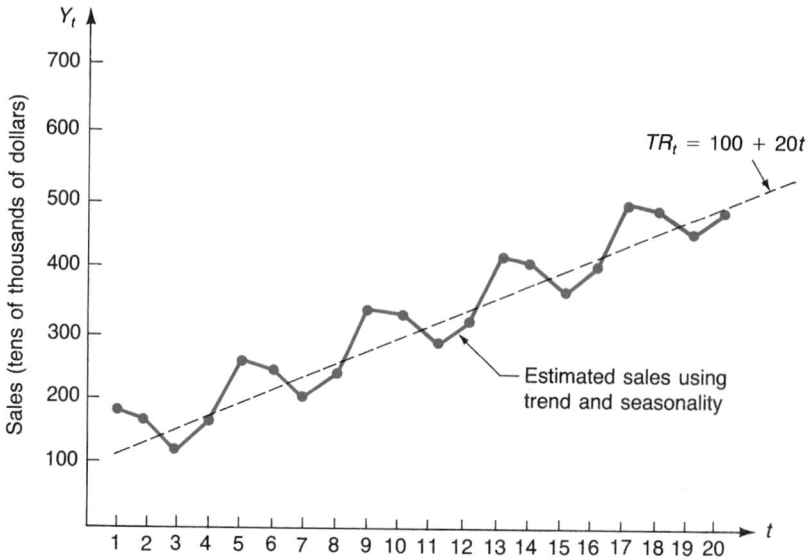

remains the same. If the past observations in the time series indicate that higher levels of sales produce wider seasonal fluctuations, this is an indication of multiplicative seasonal variation.

Multiplicative Seasonal Variation

Figure 16.6 shows **multiplicative seasonal variation** in the time series for the sale of Wildcat sailboats. Notice that in each successive year, the difference between the actual value and the trend value for July is larger. In multiplicative seasonal variation, the seasonal fluctuation is *proportional* to the trend level for each observation. Figure 16.17 is a general illustration of multiplicative seasonality; it shows the sales of heat pumps over a three-year period at Handy Home Center.

Considering only the effects of trend and seasonality, an estimate for a time series observation is given by

$$\text{estimate of } y_t = TR_t \cdot S_t$$

As in additive seasonal variation, the seasonal indexes, S_t, remain constant from one year to the next. When dealing with quarterly data, this means that $S_1 =$

■ FIGURE 16.17
Heat-pump sales
at Handy Home Center
(an illustration of
multiplicative seasonal
variation).

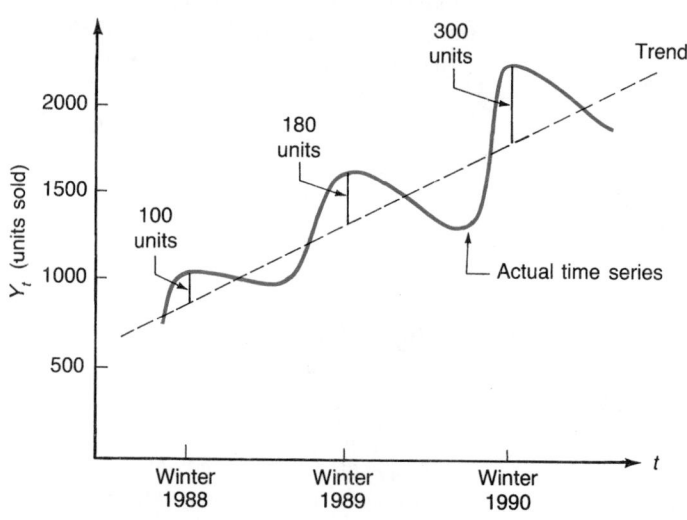

$S_5 = S_9 = \cdots$, $S_2 = S_6 = S_{10} = \cdots$, and so on. The next section discusses a method for determining these indexes for the case of multiplicative seasonality.

EXAMPLE 16.6   Suppose that the sales of Jetski snowmobiles from sales area 2 contain multiplicative seasonal effects with trend = $TR_t = 100 + 20t$ (as before) and seasonal indexes

$$S_1 = 1.4 \quad \text{(winter quarter)}$$

$$S_2 = 1.2 \quad \text{(spring quarter)}$$

$$S_3 = .6 \quad \text{(summer quarter)}$$

$$S_4 = .8 \quad \text{(autumn quarter)}$$

Determine the estimated sales using the trend and seasonal components.

SOLUTION   The calculations for the first two years are

t(TIME)	$TR_t \cdot S_t$(ESTIMATE)
1 (winter, 1986)	$[100 + 20(1)](1.4) = 168$
2 (spring, 1986)	$[100 + 20(2)](1.2) = 168$
3 (summer, 1986)	$[100 + 20(3)](.6) = 96$
4 (autumn, 1986)	$[100 + 20(4)](.8) = 144$
5 (winter, 1987)	$[100 + 20(5)](1.4) = 280$
6 (spring, 1987)	$[100 + 20(6)](1.2) = 264$
7 (summer, 1987)	$[100 + 20(7)](.6) = 144$
8 (autumn, 1987)	$[100 + 20(8)](.8) = 208$
$\vdots$	$\vdots$

A graph of the estimated sales over a five-year period is shown in Figure 16.18. Notice that seasonal patterns do exist, but (unlike additive variation) these fluctuations increase as the sales level rises. For a time series representing sales, this type of variation seems to make sense. If the volume of sales doubles, it is reasonable to expect a larger effect due to seasonality than occurred previously.

Remember that in practice, few time series exhibit exact additive or multiplicative seasonal effects. However, you can classify a great many time series as essentially belonging to one or the other of these two classes. ■

■ **FIGURE 16.18**
Jetski sales from sales area 2 (multiplicative seasonal variation).

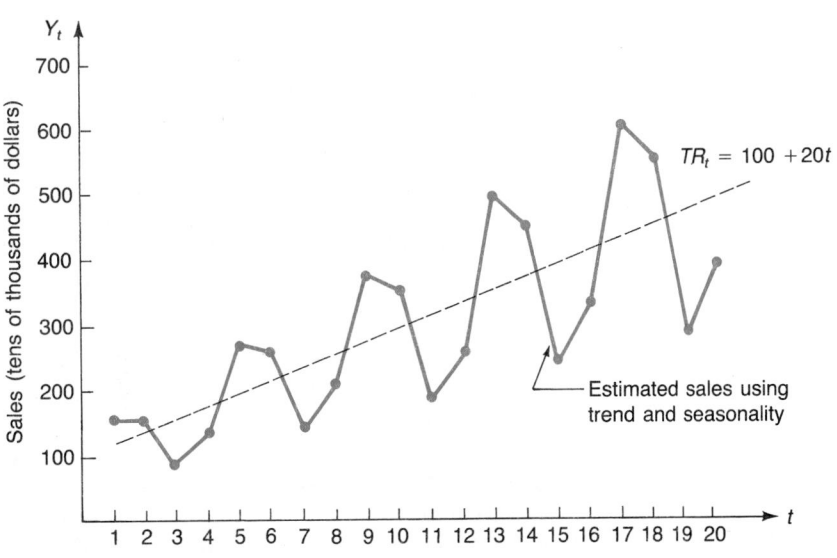

In the discussion to follow, we assume that any seasonality in the time series is *multiplicative*. Most analysts (including those in the U.S. Census Bureau) have had better success describing time series in this manner. The decomposition method to be discussed assumes that each observation is the *product* of its various components. So, the *component structure* is assumed to be

$$y_t = TR_t \cdot S_t \cdot C_t \cdot I_t \qquad \textbf{(16.7)}$$

where the components representing seasonality, trend, cyclical variation, and noise are multiplied by one another.*

Four-Step Procedure (Multiplicative Components)

Based on the multiplicative component structure in equation 16.7, the following four-step procedure can be used to decompose a time series containing the effects of all four components.

Step 1. *Determine a seasonal index, S_t, for each time period.* For quarterly data, this involves determining four such indexes, S_1, S_2, S_3, and S_4. When the time series contains monthly observations, 12 seasonal indexes (S_1 through S_{12}) must be calculated, one for each month.

Step 2. *Deseasonalize the data.* This step is often referred to as *adjusting for seasonality;* the seasonal component is eliminated. Because we are using a multiplicative structure, we divide each observation by its corresponding seasonal index. So

$$\text{deseasonalized observation} = d_t$$
$$= y_t/S_t$$

where

$$S_t = \begin{cases} S_1, S_2, S_3, \text{ or } S_4 & \text{(quarterly data)} \\ S_1, S_2, \ldots, \text{ or } S_{12} & \text{(monthly data)} \end{cases}$$

Because $y_t = TR_t \cdot S_t \cdot C_t \cdot I_t$,

$$d_t = \frac{y_t}{S_t} = \frac{TR_t \cdot \cancel{S_t} \cdot C_t \cdot I_t}{\cancel{S_t}} = TR_t \cdot C_t \cdot I_t$$

Step 3. *Determine the trend component, TR_t.* The trend is estimated by passing a least squares line through the *deseasonalized* data. The technique is identical to that discussed in Section 16.2 (which assumed no seasonality), except that we use the d_t values rather than the original time series. This process is illustrated in the next section.

Step 4. *Determine the cyclical component, C_t.* You obtain C_t by first dividing each deseasonalized observation, d_t, by the corresponding trend value from step 3. So the cyclical estimates are derived by first calculating (for each time period)

$$\frac{d_t}{\hat{d}_t} = \frac{d_t}{TR_t} = \frac{\cancel{TR_t} \cdot C_t \cdot I_t}{\cancel{TR_t}} = C_t \cdot I_t$$

---

*Similar methods for determining the components of a time series containing additive seasonality also exist. Chapter 17 contains a small discussion of this topic. For a complete discussion of such techniques, see B. L. Bowerman and R. T. O'Connell, *Forecasting and Time Series,* 2d ed., (North Scituate, Mass.: Duxbury, 1987).

Notice that the resulting series contains cycles and irregular activity (but no trend or seasonality). A method for reducing the irregular component in these ratios is demonstrated later. The resulting values are the cyclical components, C_t.

We do not use the cyclical components to attempt to forecast future values of the time series because their behavior (and period) generally cannot be predicted. The cyclical components can be used in forecasting if one is willing to assume a particular phase in the business cycle. If one assumes, for example, that the cycle is in the midst of an upturn, a value of C_t (such as $C_t = 1.2$) can be assigned to this particular time period. In the discussion to follow, the cyclical components are obtained strictly as a means of *describing* the cyclical activity within a recorded time series. ∎

Exercises

16.27 Explain the effect that seasonality has on the trend assuming additive and multiplicative seasonality. Would seasonality changes have a larger effect on an additive model or on a multiplicative model?

16.28 Riney, owner of Riney's Shoe Store, usually has a rush on shoe sales around September, when children are going back to school. Sales data (in thousands) for Riney's Shoe Store were recorded quarterly over a four-year period (1988 through 1991), and the following trend line was calculated from the data:

$$TR_t = 4 + 1.6t$$

where $t = 1, 2, 3, \ldots, 16$. Seasonal indexes were found to be the following:

$$\text{quarter 1 } S_1 = 1.3$$
$$\text{quarter 2 } S_2 = -1.4$$
$$\text{quarter 3 } S_3 = 4.2$$
$$\text{quarter 4 } S_4 = 2.6$$

Assuming an additive model, estimate the quarterly sales y_t using only trend and seasonality for the four quarters of 1989.

16.29 For a six-year period (1986 to 1991) quarterly sales data (in thousands) were used to arrive at the following trend line and seasonal indexes.

$$TR_t = 35 + 2.3t \qquad \text{for } t = 1, 2, \ldots, 24$$
$$S_1 = -8.7$$
$$S_2 = 2.5$$
$$S_3 = 8.4$$
$$S_4 = 3.1$$

Estimate the sales figures for the four quarters in 1990 using an additive equation containing only the trend and seasonality components.

16.30 Advanced Digital Components has experienced rapid growth during the past several years. The quarterly data for the past four years give the following trend line and seasonal indexes. Sales units are in tens of thousands.

$$TR_t = 0.85 + 0.8t \qquad \text{for } t = 1, 2, \ldots, 16$$
$$S_1 = 0.82$$
$$S_2 = 1.36$$
$$S_3 = 1.20$$
$$S_4 = 0.62$$

Estimate the sales figures for the four quarters in the most recent year using a multiplicative equation containing only the trend and seasonality components.

16.31 Monthly data from the years 1987 through 1991 were used to find the following trend line and seasonal indexes:

$$TR_t = 1.3 + 0.5t \quad \text{for } t = 1, 2, \ldots, 60$$

$S_1 = 0.5$	$S_7 = 2.4$
$S_2 = 0.8$	$S_8 = 3.1$
$S_3 = 0.6$	$S_9 = 0.3$
$S_4 = 1.3$	$S_{10} = 0.2$
$S_5 = 1.1$	$S_{11} = 0.2$
$S_6 = 1.4$	$S_{12} = 0.1$

Assuming a multiplicative model containing only the trend and seasonality components, estimate the data for the 12 months of 1990.

16.32 Refer to Exercise 16.65. Assuming that the sales of Luz Chemicals are subject to multiplicative seasonal variation, determine the deseasonalized sales for the 12 months of the year 1991. The seasonal indexes are as follows:

$S_1 = 0.8$	$S_7 = 0.9$
$S_2 = 0.8$	$S_8 = 0.9$
$S_3 = 1.2$	$S_9 = 0.7$
$S_4 = 1.2$	$S_{10} = 1.2$
$S_5 = 1.1$	$S_{11} = 0.9$
$S_6 = 1.0$	$S_{12} = 1.3$

16.33 Rework Exercise 16.32 assuming that the sales are subject to additive seasonal variation. The seasonal indexes are as follows:

$S_1 = -25$	$S_7 = 25$
$S_2 = -50$	$S_8 = 25$
$S_3 = -60$	$S_9 = -30$
$S_4 = 25$	$S_{10} = -40$
$S_5 = 25$	$S_{11} = -40$
$S_6 = 28$	$S_{12} = -30$

16.34 The nominal GNP quarterly estimates (in billions of dollars) for a certain third world nation for the years 1989 to 1991 are:

YEAR	QUARTER	NOMINAL GNP
1989	1	206.4
	2	226.3
	3	238.5
	4	254.2
1990	1	268.6
	2	274.8
	3	287.9
	4	297.9
1991	1	313.6
	2	312.8
	3	326.7
	4	346.8

Assuming multiplicative seasonality, calculate the deseasonalized data, given the following seasonal indexes:

$$S_1 = 1.30$$
$$S_2 = 0.80$$
$$S_3 = 1.20$$
$$S_4 = 0.70$$

■ **TABLE** 16.2
Time series with
quarterly observations.

TIME	QUARTER	t	y_t	MOVING TOTALS
1984	1	1	85	(1) 263
	2	2	41	(2) 268
	3	3	92	(3) 270
	4	4	45	and so on
1985	1	5	90	
	2	6	43	
	3	7	95	
	4	8	47	
1986	1	9	92	
	:	:	:	

▟ 16.5
Measuring Seasonality

Seasonality often is present in time series data collected over months or quarters. This effect is observed when, for example, some months are always higher than the average for the year. For example, the recorded values of the time series indicate that July sales are 25% higher than the average for the year, the July index should be 1.25 using the multiplicative structure.

We derive a seasonal index for each period during the year (4 for quarterly data, 12 for monthly data). We begin by developing a new series that contains *no seasonality*. This new series is obtained from the original time series and consists of the **centered moving averages.** This method provides an excellent way of isolating the seasonal components from the original time series. In addition to containing no seasonality, the centered moving averages are *smoother* (contain less irregular activity) than the original time series. Consequently, the moving averages give you a clearer picture of any existing trend within a time series containing significant seasonality and irregular activity. Other methods of smoothing a time series will be discussed in Chapter 17.

Centered Moving Averages

To illustrate the calculation of a moving average, consider a time series containing quarterly observations, as shown in Table 16.2.* Here,

$$(1) \doteq \text{sum of } y_1 \text{ through } y_4$$
$$= 85 + 41 + 92 + 45 = 263$$
$$(2) = \text{sum of } y_2 \text{ through } y_5$$
$$= 41 + 92 + 45 + 90 = 268$$
$$(3) = \text{sum of } y_3 \text{ through } y_6$$
$$= 92 + 45 + 90 + 43 = 270$$

and so on.

Because each total contains four observations (one from each quarter), any quarterly seasonal effects have been removed. Consequently, there is no seasonality in the moving totals 263, 268, 270, and so on, in Table 16.2.

The first moving total in Table 16.2 is equal to $(y_1 + y_2 + y_3 + y_4)$. If we were to position this total in the center of these values, it would lie between $t = 2$ and $t = 3$, at $t = 2.5$. The second moving total is equal to $(y_2 + y_3 + y_4 + y_5)$; again, we position this total in the center between $t = 3$ and $t = 4$, at $t = 3.5$.

We then add the first two moving totals. Notice that four values went into each

---

*An example using monthly data is contained in Section 16.6.

of these totals, so that a total of *eight* values makes up this sum. The sum of the first two moving totals is $263 + 268 = 531$. The average for the 8 months in the first two moving totals is $531/8 = 66.38$. This is a **centered moving average.** The position of this moving average is midway between $t = 2.5$ and $t = 3.5$, at $t = 3$. We therefore conclude that 66.38 is the centered moving average corresponding to $t = 3$.

EXAMPLE 16.7   Continue the procedure using Table 16.2 and determine the centered moving average for (1) $t = 4$ and (2) $t = 5$.

SOLUTION 1   Here we obtain

$$268 = y_2 + y_3 + y_4 + y_5$$

(positioned at $t = 3.5$) and

$$270 = y_3 + y_4 + y_5 + y_6$$

(positioned at $t = 4.5$). So the average of the eight numbers making up $268 + 270 = 538$ would be positioned midway between 3.5 and 4.5, at $t = 4$. Consequently, the centered moving average for $t = 4$ is

$$\frac{268 + 270}{8} = 67.25$$

SOLUTION 2   Proceeding as before,

$$270 = y_3 + y_4 + y_5 + y_6$$

(positioned at $t = 4.5$) and

$$273 = y_4 + y_5 + y_6 + y_7$$

(positioned at $t = 5.5$). Therefore, the centered moving average for $t = 5$ is

$$\frac{270 + 273}{8} = 67.88$$

∎

Assume quarterly sales data at Video-Comp were recorded over a four-year period. We now want to determine the centered moving averages for these data, shown in Table 16.3. There appears to be a definite seasonal effect within this time series; the highest sales occur in the fourth quarter of each year. Table 16.4 shows the centered moving averages for these data. The first *moving total* is

$$139 = y_1 + y_2 + y_3 + y_4$$
$$= 20 + 12 + 47 + 60$$

Its actual location is $t = 2.5$; it is positioned between $t = 2$ and $t = 3$. Similarly, the next moving total is centered at $t = 3.5$ and so appears between $t = 3$ and $t = 4$ in the table. This total is

$$159 = y_2 + y_3 + y_4 + y_5$$
$$= 12 + 47 + 60 + 40$$

■ **TABLE 16.3**
Sales data for Video-Comp (millions of dollars).

YEAR	QUARTER 1	QUARTER 2	QUARTER 3	QUARTER 4
1987	20	12	47	60
1988	40	32	65	76
1989	56	50	85	100
1990	75	70	101	123

Each moving total is centered midway between the values making up this total. For example, the last moving total, 369, is centered between $t = 14$ and $t = 15$ at $t = 14.5$. Here,

$$369 = y_{13} + y_{14} + y_{15} + y_{16}$$
$$= 75 + 70 + 101 + 123$$

The *centered moving average* at time t is the average of the moving total immediately preceding this time value and the total immediately following it. This means that, for $t = 3$,

$$37.25 = \frac{139 + 159}{8}$$

For $t = 4$,

$$42.25 = \frac{159 + 179}{8}$$

and so on. Consequently, for $t = 3$, $y_3 = 47$ and the centered moving average is 37.25.

This procedure produces 12 centered moving averages; we are unable to compute this value for $t = 1, 2, 15,$ or 16. Notice that the first two of these values of t are for quarters 1 and 2, whereas the remaining two correspond to quarters 3 and 4. In general, if our time series contains T observations, we can derive $T - 4$ centered moving averages using quarterly data or $T - 12$ averages for monthly data.

The moving totals and centered moving averages are formed by summing over the four quarters (seasons), so there is no seasonality present in these values. Furthermore, the irregular component has been reduced because averages always contain less random variation (noise) than do the individual values making up these averages. Representing this reduced irregular activity component as i_t (rather than I_t), we can represent a centered moving average at time t as

$$\text{centered moving average at time } t = TR_t \cdot C_t \cdot i_t$$

Because of this averaging procedure, the moving averages contain much less irregular activity and so are much "smoother" than the original time series. This procedure thus is referred to as **smoothing** the time series to get a clearer picture of any existing trend as well as of its shape (straight line or curve).

The centered moving averages in Table 16.4 show a steadily increasing trend.

■ TABLE 16.4
Moving averages for Video-Comp sales data.

YEAR	QUARTER	t	y_t	MOVING TOTAL	CENTERED MOVING AVERAGE	RATIO TO MOVING AVERAGE
1987	1	1	20	—	—	—
	2	2	12		—	—
				139		
	3	3	47		37.25	1.26
				159		
	4	4	60		42.25	1.42
				179		
1988	1	5	40		47.00	.85
				197		
	2	6	32		51.25	.62
				213		
	3	7	65		55.25	1.18
				229		
	4	8	76		59.50	1.28
				247		
1989	1	9	56		64.25	.87
				267		
	2	10	50		69.75	.72
				291		
	3	11	85		75.13	1.13
				310		
	4	12	100		80.00	1.25
				330		
1990	1	13	75		84.50	.89
				346		
	2	14	70		89.38	.78
				369		
	3	15	101		—	—
				—		
	4	16	123		—	—

■ FIGURE 16.19
Smoothing a time series
using moving averages
(Video-Comp
sales data).

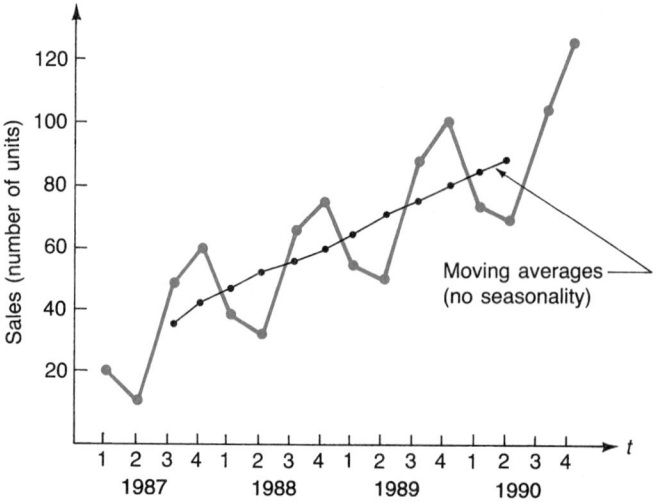

Because the differences between any two adjacent moving averages are nearly the same, this trend is very *linear*. The trend is more apparent in Figure 16.19, which contains the original data with the moving averages.

To determine the four quarterly seasonal indexes, the first step is to divide each observation, y_t, by its corresponding centered moving average; the result is shown in the last column in the table.

for $t = 3$:   ratio $= 47/37.25 = 1.26$ (belongs to quarter 3, 1987)

for $t = 4$:   ratio $= 60/42.25 = 1.42$ (belongs to quarter 4, 1987)

$\vdots$

for $t = 14$:   ratio $= 70/89.38 = .78$ (belongs to quarter 2, 1990)

When we divide y_t by its corresponding centered moving average, we obtain

$$\text{ratio} = \frac{y_t}{\text{centered moving average}} = \frac{\cancel{TR_t} \cdot S_t \cdot \cancel{C_t} \cdot I_t}{\cancel{TR_t} \cdot \cancel{C_t} \cdot i_t}$$
$$= S_t \cdot I_t$$

Consequently, these ratios contain the seasonal effects as well as the irregular activity (noise) components. The following discussion illustrates how you can reduce the effect of the irregular activity factor by combining these ratios into a set of four seasonal indexes, one for each quarter.

Computing a Seasonal Index

The purpose of a seasonal index is to indicate how the time series value for each quarter (or month) compares with the average for the year. The following discussion will assume that we are dealing with a time series containing quarterly data. In the next section, we illustrate this procedure using monthly data.

We begin by collecting the ratios to moving average, placing each of them in its respective quarter. In Table 16.4, we see that 1.26 belongs to quarter 3, 1.42 to quarter 4, .85 to quarter 1, and so on. Table 16.5 is the result. Notice that there are three ratios for each quarter. In general, you always will obtain (total number of years -1) ratios under each quarter (or month). The time series in this example

	QUARTER 1	QUARTER 2	QUARTER 3	QUARTER 4
	—	—	1.26	1.42
	.85	62	1.18	1.28
	.87	.72	1.13	1.25
	.89	.78	—	—
TOTAL	2.61	2.12	3.57	3.95
AVERAGE	.870	.707	1.190	1.317

■ TABLE 16.5
Ratios for each quarter.

contains four years; therefore, it has three ratios. To obtain a "typical" ratio for each quarter, you have several options, including

1. Determining an average of these ratios.
2. Finding the median of these values.
3. Eliminating the largest and smallest ratio within each quarter and computing a mean of the remaining ratios; this is called a **trimmed mean.**

We will follow the first procedure and calculate a mean ratio for each quarter, as illustrated in Table 16.5. When the time series contains five or more years of data, a trimmed mean offers you protection against an outlier ratio dominating the index for this quarter. Using the median ratios also helps guard against this type of situation.

A Final Adjustment

The last step in computing the seasonal indexes is to make sure that the four computed ratio averages **sum to 4** (or 12, for monthly indexes). This is accomplished by (1) adding the four averages computed in the table (call this SUM) and (2) multiplying each average by 4/SUM. The modified average obtained in this process is the seasonal index for that quarter.

EXAMPLE 16.8

Using Table 16.5, determine the four seasonal indexes.

SOLUTION   First,

$$SUM = .870 + .707 + 1.190 + 1.317$$
$$= 4.084$$

This means that we need to multiply each of the four averages in Table 16.5 by 4/4.084 = .9794.

QUARTER	SEASONAL INDEX
1	(.870)(.9794) = .852
2	(.707)(.9794) = .692
3	(1.190)(.9794) = 1.166
4	(1.317)(.9794) = 1.290
	4.0

The indexes for quarters 1 and 2 are below 1.0, so the sales during these quarters typically are below the yearly average. On the other hand, quarters 3 and 4 have seasonal indexes of 1.166 and 1.290, so the sales for these quarters are higher than the average for the year.   ■

This procedure for determining seasonal effects works well, provided the ratios in Table 16.5 are reasonably *stable*. In Example 16.8, all the ratios for quarter 2 are small (near .7) and all the ratios for quarter 4 are large (near 1.3). If strong seasonality is present, such will be the case.

Seasonal indexes can be updated as you obtain an additional year's observations on the variable of interest. You have the option of deleting the most distant year's observations prior to recalculating these values. This procedure leads to seasonal indexes that change slowly over the years.

In summary, to calculate the seasonal indexes:

1. Derive the *moving* totals by summing the observations for 4 (quarterly data) or 12 (monthly data) consecutive time periods.
2. Average and center the totals by finding the *centered moving averages.*
3. Divide each observation by its corresponding centered moving average.
4. Place the ratios from step 3 in a table headed by the 4 quarters or 12 months.
5. For each column in this table, determine the mean of these ratios; these are the unadjusted seasonal indexes.
6. Make a final adjustment to guarantee that the final seasonal indexes sum to 4 (quarterly data) or 12 (monthly data); these adjusted means are the seasonal indexes.

Deseasonalizing the Data

To remove the seasonality from the data, we deseasonalize the time series. The resulting series contains no seasonal effects and consists of the trend, cyclical activity, and, of course, irregular activity. We write deseasonalized data as d_t.

$$d_t = \frac{y_t}{\text{corresponding seasonal index}}$$
$$= \frac{TR_t \cdot S_t \cdot C_t \cdot I_t}{S_t} = TR_t \cdot C_t \cdot I_t$$

The deseasonalized sales values from Table 16.3 are contained in Table 16.6. These values contain trend, cyclical effects, and irregular activity. Notice how the trend is much more apparent in the deseasonalized values than in the original time series.

In Table 16.6, we obtained deseasonalized values for all 16 of the original observations, including the two quarters on each end. We can use the "new" deseasonalized series to determine the trend and cyclical components of the original time series. This will be illustrated in the next section, where we apply the four-step

	YEAR	t	y_t	SEASONAL INDEX (S_t)	DESEASONALIZED VALUES $d_t = y_t/S_t$
■ TABLE 16.6	1987	1	20	.852	23.47
Deseasonalized sales.		2	12	.692	17.34
		3	47	1.166	40.31
		4	60	1.290	46.51
	1988	5	40	.852	46.95
		6	32	.692	46.24
		7	65	1.166	55.75
		8	76	1.290	58.91
	1989	9	56	.852	65.73
		10	50	.692	72.25
		11	85	1.166	72.90
		12	100	1.290	77.52
	1990	13	75	.852	88.03
		14	70	.692	101.16
		15	101	1.166	86.62
		16	123	1.290	95.35

procedure: (1) computing seasonal indexes, (2) deseasonalizing the data, (3) computing the trend components from the deseasonalized time series (d_t), and finally (4) calculating the cyclical activity.

COMMENTS   Monthly values in the *Wall Street Journal* and other business publications are often stated as "seasonally adjusted." This simply means these values have been deseasonalized to reflect "real" changes in the variable. For example, the number of unemployed workers always increases in May because college students are looking for summer or long-term employment. To determine if there has actually been an increase in the unemployment rate, a deseasonalized (seasonally adjusted) value is generally quoted in the article or news release.

 *Exercises*   **16.35**   Mid-Cities Appliance store is interested in determining an approximate inventory level to control its overhead. Quarterly inventory levels (in units of ten thousand dollars) for five years are:

YEAR	QUARTER 1	QUARTER 2	QUARTER 3	QUARTER 4
1987	1	7	12	6
1988	4	11	17	9
1989	9	14	20	14
1990	12	16	24	18
1991	16	21	27	22

Find the four-quarter centered moving averages and ratios to moving average.

16.36   Several counties in Oregon and Washington depend heavily on the lumber industry. When there is little demand for lumber, the softness in the industry causes unemployment to increase. The following data represent the percentage of unemployment for certain counties in Oregon and Washington.

YEAR	JAN	FEB	MAR	APR	MAY	JUNE	JULY	AUG	SEPT	OCT	NOV	DEC
1988	10.8	9.6	8.7	7.5	6.4	5.4	6.1	7.3	8.5	8.9	10.1	10.9
1989	9.6	8.5	7.5	6.3	5.2	4.1	5.9	6.1	7.4	8.1	9.8	9.8
1990	6.9	7.4	6.3	7.5	4.9	3.7	4.2	5.4	6.5	7.7	8.0	9.1
1991	7.0	6.7	5.7	4.2	3.3	2.8	3.4	4.0	4.4	5.3	6.7	7.4

Find the 12-month centered moving averages and ratios to moving average.

16.37   Explain why a moving average is a smoothing technique.

16.38   The following table presents the ratio to moving average figures for sales at Zano Systems, a supplier of photocopy machines. Find the seasonal indexes.

YEAR	QUARTER 1	QUARTER 2	QUARTER 3	QUARTER 4
1986			.88	.87
1987	1.14	1.25	.83	.86
1988	1.19	1.22	.94	.88
1989	1.23	1.35	.90	.72
1990	1.16	1.32	.94	.81
1991	1.10	1.21		

16.39   The following table presents the ratio to moving average figures for the cost of a bushel of grapefruit in a certain county in Florida. Find the seasonal indexes.

YEAR	JAN	FEB	MAR	APR	MAY	JUNE	JULY	AUG	SEPT	OCT	NOV	DEC
1987							1.06	1.10	1.12	1.02	1.03	.99
1988	.90	.87	.95	.93	1.00	1.04	1.08	1.14	1.15	1.06	1.04	.97
1989	.87	.84	.81	.88	1.01	1.02	1.01	1.15	1.07	1.03	1.00	.90
1990	.81	.75	.82	.89	1.05	1.04	1.10	1.21	1.18	1.10	1.07	.97
1991	.87	.81	.77	.98	1.01	1.06						

16.40   The sale of grass sod is a seasonal business. Green Garden Supplies does most of its business in May, June, July, and August. The following table presents their monthly sales (in thousands of dollars). Find the seasonal indexes. For what month is the seasonal index the largest?

YEAR	JAN	FEB	MAR	APR	MAY	JUNE	JULY	AUG	SEPT	OCT	NOV	DEC
1987	.1	.1	1.2	2.2	4.1	4.5	5.5	5.3	3.5	1.1	.2	.1
1988	.1	.2	1.4	2.0	4.0	4.2	5.3	5.0	3.2	1.0	.1	.1
1989	.1	.2	1.3	2.2	4.3	4.4	5.6	5.3	3.5	1.1	.2	.1
1990	.1	.3	1.4	2.3	4.4	4.6	5.8	5.5	3.7	1.3	.3	.1
1991	.1	.3	1.5	2.3	4.6	4.8	6.0	5.6	3.7	1.4	.4	.1

16.41 The following table represents the ratio to moving average figures for the number of people below the poverty level in a certain county. Find the seasonal indexes.

YEAR	QUARTER 1	QUARTER 2	QUARTER 3	QUARTER 4
1987			.84	.83
1988	1.12	1.29	.91	.89
1989	1.17	1.24	.92	.90
1990	1.15	1.30	.92	.88
1991	1.13	1.26		

16.42 Seaside University has four quarterly semesters during the school year. Enrollment for each of these quarters is as follows. Find the seasonal indexes.

YEAR	QUARTER 1	QUARTER 2	QUARTER 3	QUARTER 4
1987	9,385	9,020	9,350	9,060
1988	9,970	9,671	9,928	9,701
1989	10,328	9,950	10,121	9,922
1990	10,411	9,995	10,250	9,998
1991	10,535	10,240	10,506	10,279

16.43 A major department store usually has a strong fourth quarter because of the holiday season. The company's earnings per share are as follows. Find the seasonal indexes.

YEAR	QUARTER 1	QUARTER 2	QUARTER 3	QUARTER 4
1987	.75	.60	.80	1.40
1988	.80	.55	.82	1.51
1989	.83	.59	.81	1.63
1990	.84	.62	.83	1.75
1991	.84	.61	.82	1.79

16.44 The average yields to maturity on AA industrial and utility bonds are presented below for the years 1986 to 1989. Find the seasonal indexes.

	AVERAGE YIELDS (IN PERCENT) ON AA BONDS			
	1986	**1987**	**1988**	**1989**
January	10.18	8.86	9.81	9.85
February	9.78	8.82	9.45	9.87
March	9.27	8.78	9.68	10.11
April	9.19	9.31	9.92	9.97
May	9.34	9.81	10.21	9.78
June	9.54	9.70	10.02	9.27
July	9.35	9.64	10.09	9.16
August	9.29	9.83	10.25	9.14
September	9.30	10.48	9.98	9.16
October	9.34	10.62	9.89	9.10
November	9.17	9.97	9.97	8.97
December	9.01	10.08	9.89	9.01

(*Source: Standard and Poor's Security Price Index Record, Statistical Service* (1990): 288.)

	1987	1988	1989	1990
J	106.39	113.64	122.47	132.56
F	105.80	115.10	118.90	127.34
M	120.44	131.59	139.76	148.33
A	125.37	131.00	137.92	144.96
M	129.07	137.56	148.17	154.14
J	128.98	139.05	147.06	153.47
J	128.95	135.37	142.63	148.93
A	131.02	140.20	150.86	157.43
S	123.77	133.00	142.11	145.57
O	127.21	135.90	140.22	150.68
N	125.38	140.25	146.27	154.64
D	154.75	170.81	174.82	181.37

■ **TABLE 16.7**
Total U.S. retail trade (sales and inventories) (billions of dollars).

Source: U.S. Department of Commerce, Bureau of the Census, *Current Business Reports* (Vols. BR-87-12, BR-88-12, BR-89-12, BR-90-12).

◢ 16.6

A Time Series Containing Seasonality, Trend, and Cycles

During the summer of 1991, the owner of an import/export company decided to investigate the past behavior of U.S. retail trade figures for the years 1987 through 1990. He collected the data in Table 16.7 using monthly figures released by the U.S. Department of Commerce. He suspected that these data would indicate high retail trade during December (due to holiday sales) with much lower activity during January and possibly February. For the remaining months, he had no idea whether seasonal effects would be present. He also suspected there would be a steadily increasing trend due to inflation and population growth.

We will perform a decomposition of the data in Table 16.7 and discuss the results. The four-step procedure for decomposing (a gruesome term, we'll admit) a time series into the seasonal, trend, and cyclical components was introduced in Section 16.4. We demonstrate this method of describing a time series using the monthly retail trade data.

Step 1. *Determine the seasonal indexes.* The first step is to determine the moving totals and centered moving averages for the 48 observations in Table 16.7. These calculations are summarized in Table 16.8. Notice that for the monthly data, there is no moving average for $t = 1$ through $t = 6$ (months 1 through 6, 1987) and for $t = 43$ through $t = 48$ (months 7 through 12, 1990). The first moving total is

$$1507.13 = y_1 + y_2 + \cdots + y_{12}$$
$$= 106.39 + 105.80 + \cdots + 154.75$$

This value is positioned midway between $t = 1$ and $t = 12$, at $t = 6.5$. The next moving total is

$$1514.38 = y_2 + y_3 + \cdots + y_{13}$$
$$= 105.80 + 120.44 + \cdots + 113.64$$

which is centered at $t = 7.5$. So the first centered moving *average* is positioned midway between $t = 6.5$ and $t = 7.5$, at $t = 7$. This is

$$125.90 = \frac{1507.13 + 1514.38}{24}$$

Notice that we divide by 24 because 24 observations went into the sum of these two moving totals.

The final moving average is

$$149.68 = \frac{1792.87 + 1799.42}{24}$$

and corresponds to $t = 42$.

■ TABLE 16.8
Moving averages for U.S. monthly retail trade.

YEAR	MONTH	t (1)	y_t (2)	MOVING TOTAL (3)	CENTERED MOVING AVERAGE (4)	RATIO TO MOVING AVERAGE (5)
1987	JAN	1	106.39			
	FEB	2	105.80			
	MAR	3	120.44			
	APR	4	125.37			
	MAY	5	129.07			
	JUNE	6	128.98	1507.13		
	JULY	7	128.95	1514.38	125.90	1.024
	AUG	8	131.02	1523.68	126.59	1.035
	SEPT	9	123.77	1534.83	127.44	0.971
	OCT	10	127.21	1540.46	128.14	0.993
	NOV	11	125.38	1548.95	128.73	0.974
	DEC	12	154.75	1559.02	129.50	1.195
1988	JAN	13	113.64	1565.44	130.19	0.873
	FEB	14	115.10	1574.62	130.84	0.880
	MAR	15	131.59	1583.85	131.60	1.000
	APR	16	131.00	1592.54	132.35	0.990
	MAY	17	137.56	1607.41	133.33	1.032
	JUNE	18	139.05	1623.47	134.62	1.033
	JULY	19	135.37	1632.30	135.66	0.998
	AUG	20	140.20	1636.10	136.18	1.029
	SEPT	21	133.00	1644.27	136.68	0.973
	OCT	22	135.90	1651.19	137.31	0.990
	NOV	23	140.25	1661.80	138.04	1.016
	DEC	24	170.81	1669.81	138.82	1.230
1989	JAN	25	122.47	1677.07	139.45	0.878
	FEB	26	118.90	1687.73	140.20	0.848
	MAR	27	139.76	1696.84	141.02	0.991
	APR	28	137.92	1701.16	141.58	0.974
	MAY	29	148.17	1707.18	142.01	1.043
	JUNE	30	147.06	1711.19	142.43	1.032
	JULY	31	142.63	1721.28	143.02	0.997
	AUG	32	150.86	1729.72	143.79	1.049
	SEPT	33	142.11	1738.29	144.50	0.983
	OCT	34	140.22	1745.33	145.15	0.966
	NOV	35	146.27	1751.30	145.69	1.004
	DEC	36	174.82	1757.71	146.21	1.196
1990	JAN	37	132.56	1764.01	146.74	0.903
	FEB	38	127.34	1770.58	147.27	0.865
	MAR	39	148.33	1774.04	147.69	1.004
	APR	40	144.96	1784.50	148.27	0.978
	MAY	41	154.14	1792.87	149.06	1.034
	JUNE	42	153.47	1799.42	149.68	1.025
	JULY	43	148.93			
	AUG	44	157.43			
	SEPT	45	145.57			
	OCT	46	150.68			
	NOV	47	154.64			
	DEC	48	181.37			

Table 16.8 also contains each ratio to moving average (column 2 divided by column 4). To illustrate,

$$1.024 = 128.95/125.90$$

$$1.035 = 131.02/126.59$$

and so on. These ratios are summarized in Table 16.9, which also shows the average of the three values for each time period.

The final step is to adjust each of the averages in Table 16.9 so that

■ **TABLE 16.9**   Summary of ratios.

Period	MONTH											
	1	**2**	**3**	**4**	**5**	**6**	**7**	**8**	**9**	**10**	**11**	**12**
Year												
1							1.024	1.035	0.971	0.993	0.974	1.195
2	0.873	0.880	1.000	0.990	1.032	1.033	0.998	1.029	0.973	0.990	1.016	1.230
3	0.878	0.848	0.991	0.974	1.043	1.032	0.997	1.049	0.983	0.966	1.004	1.196
4	0.903	0.865	1.004	0.978	1.034	1.025						
Average	0.885	0.864	0.998	0.981	1.036	1.030	1.006	1.038	0.976	0.983	0.998	1.207

they sum to 12 (because there are 12 time periods per year). Here,

$$\text{SUM} = .885 + .864 + \cdots + 1.207 = 12.002$$

and so

$$S_1 = \text{seasonal index for January}$$
$$= (.885) \cdot (12/12.002) = .88$$
$$S_2 = \text{seasonal index for February}$$
$$= (.864) \cdot (12/12.002) = .86$$
$$\vdots$$
$$S_{12} = \text{seasonal index for December}$$
$$= (1.207) \cdot (12/12.002) = 1.21$$

The final collection of seasonal indexes is:

MONTH	SEASONAL INDEX	MONTH	SEASONAL INDEX
Jan.	.88	July	1.01
Feb.	.86	Aug.	1.04
Mar.	1.00	Sept.	.98
Apr.	.98	Oct.	.98
May	1.04	Nov.	1.00
June	1.03	Dec.	1.21

The sum of the seasonal indexes ($S_1 + S_2 + \cdots + S_{12}$) is 12.01, which, due to the rounding of these values, is perfectly acceptable.

We observe (1) a large seasonal index for December ($S_{12} = 1.21$) indicating large retail trade for this month, (2) low indexes for January and February, and (3) very little seasonality for any of the remaining months.

Step 2. *Deseasonalize the data.* We obtain the deseasonalized values (which contain no seasonality) by dividing each observation by its corresponding seasonal index. These values are shown in Table 16.10. The trend is more apparent now because the deseasonalized values tend to increase over time.

Step 3. *Determine the trend components.* A common method (and the one we use) of estimating trend is to construct a least squares trend line (or curve) through the deseasonalized data. From the moving averages in Table 16.8, it appears that a straight line trend equation will be appropriate; these values tend to increase at a fairly steady rate.

The calculations for the trend line are identical to those discussed in Section 16.2, using the d_t values in place of the original observations, y_t. A summary of these calculations is given in Table 16.11. The least squares line through the deseasonalized data is given by

$$TR_t = \hat{d}_t = b_0 + b_1 t$$

■ TABLE 16.10
Deseasonalized monthly
retail trade values.

YEAR	MONTH	t	y_t	S_t	$d_t = y_t/S_t$
1987	JAN	1	106.39	0.88	120.26
	FEB	2	105.80	0.86	122.46
	MAR	3	120.44	1.00	120.66
	APR	4	125.37	0.98	127.89
	MAY	5	129.07	1.04	124.57
	JUNE	6	128.98	1.03	125.22
	JULY	7	128.95	1.01	128.15
	AUG	8	131.02	1.04	126.27
	SEPT	9	123.77	0.98	126.85
	OCT	10	127.21	0.98	129.46
	NOV	11	125.38	1.00	125.66
	DEC	12	154.75	1.21	128.23
1988	JAN	13	113.64	0.88	128.46
	FEB	14	115.10	0.86	133.22
	MAR	15	131.59	1.00	131.83
	APR	16	131.00	0.98	133.63
	MAY	17	137.56	1.04	132.76
	JUNE	18	139.05	1.03	135.00
	JULY	19	135.37	1.01	134.53
	AUG	20	140.20	1.04	135.11
	SEPT	21	133.00	0.98	136.31
	OCT	22	135.90	0.98	138.30
	NOV	23	140.25	1.00	140.56
	DEC	24	170.81	1.21	141.54
1989	JAN	25	122.47	0.88	138.44
	FEB	26	118.90	0.86	137.62
	MAR	27	139.76	1.00	140.01
	APR	28	137.92	0.98	140.69
	MAY	29	148.17	1.04	143.00
	JUNE	30	147.06	1.03	142.78
	JULY	31	142.63	1.01	141.74
	AUG	32	150.86	1.04	145.39
	SEPT	33	142.11	0.98	145.65
	OCT	34	140.22	0.98	142.70
	NOV	35	146.27	1.00	146.60
	DEC	36	174.82	1.21	144.87
1990	JAN	37	132.56	0.88	149.85
	FEB	38	127.34	0.86	147.39
	MAR	39	148.33	1.00	148.60
	APR	40	144.96	0.98	147.87
	MAY	41	154.14	1.04	148.76
	JUNE	42	153.47	1.03	149.00
	JULY	43	148.93	1.01	148.01
	AUG	44	157.43	1.04	151.72
	SEPT	45	145.57	0.98	149.20
	OCT	46	150.68	0.98	153.35
	NOV	47	154.64	1.00	154.99
	DEC	48	181.37	1.21	150.29

■ TABLE 16.11
Calculations for trend
line (U.S. monthly
retail trade data).

t	d_t	$t \cdot d_t$	t^2
1	120.26	120.26	1
2	122.46	244.92	4
3	120.66	361.98	9
4	127.89	511.56	16
⋮	⋮	⋮	⋮
45	149.20	6714.00	2025
46	153.35	7054.10	2116
47	154.99	7284.53	2209
48	150.29	7213.92	2304
1176	6635.45	168,767.49	38,024

where

$$b_1 = \frac{\Sigma td_t - (\Sigma t)(\Sigma d_t)/T}{\Sigma t^2 - (\Sigma t)^2/T}$$

$$= \frac{168,767.49 - (1176)(6635.45)/48}{38,024 - (1176)^2/48} = \frac{6198.96}{9212} = .6729$$

and

$$b_0 = \bar{d}_t - b_1 \bar{t}$$

$$= \frac{6635.45}{48} - (.6729)\left(\frac{1176}{48}\right)$$

$$= 138.24 - 16.49 = 121.75$$

Consequently, the trend equation is given by

$$TR_t = \hat{d}_t$$

$$= 121.75 + .6729t$$

This equation implies that, apart from seasonal fluctuations, the U.S. retail trade is increasing at an average rate of $672.9 million each month. A graph of the deseasonalized data and corresponding trend line is shown in Figure 16.20. A MINITAB solution for the trend line using the deseasonalized data as input is given in Figure 16.21.

Step 4. *Determine the cyclical activity.* We begin by following the procedure outlined in Section 16.3. We divide each deseasonalized observation by the corresponding trend value,

$$\frac{d_t}{TR_t} = \frac{d_t}{\hat{d}_t} = \frac{\cancel{TR_t} \cdot C_t \cdot I_t}{\cancel{TR_t}} = C_t \cdot I_t$$

The resulting values contain cyclical effects as well as an irregular activity component. One method of reducing the irregular activity effect is to compute a series of *three-period* moving averages on the $C_t \cdot I_t$ values. This procedure greatly reduces the irregular activity effect, and the moving

■ **FIGURE 16.20**
Deseasonalized data and trend line (monthly U.S. retail trade).

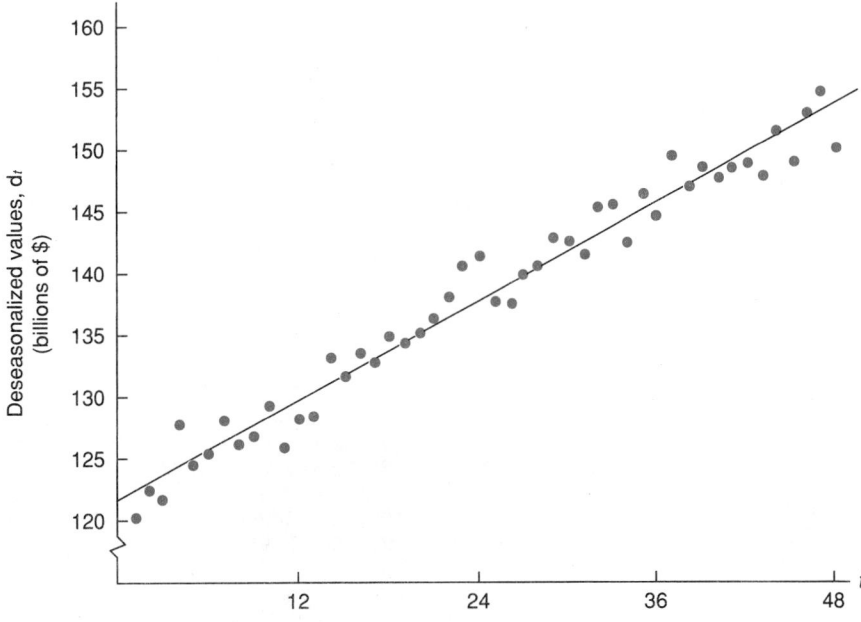

■ FIGURE 16.21
MINITAB solution
for trend line
(deseasonalized monthly
U.S. retail trade data).

```
MTB > SET INTO C1
DATA> 1:48 ─────────────────────────  This command generates
DATA> END                             integers 1 through 48
MTB > SET INTO C2
DATA> 120.26 122.46 120.66 127.89 124.57 125.22
DATA> 128.15 126.27 126.85 129.46 125.66 128.23
DATA> 128.46 133.22 131.83 133.63 132.76 135.00
DATA> 134.53 135.11 136.31 138.30 140.56 141.54
DATA> 138.44 137.62 140.01 140.69 143.00 142.78
DATA> 141.74 145.39 145.65 142.70 146.60 144.87
DATA> 149.85 147.39 148.60 147.87 148.76 149.00
DATA> 148.01 151.72 149.20 153.35 154.99 150.29
NAME C1 = 'TIME' C2 = 'DESEAS'
MTB > REGRESS Y IN C2 USING 1 PRED IN C1, RES IN C3, DHATS IN C4

The regression equation is
DESEAS = 122 + 0.673 TIME

Predictor        Coef        Stdev     t-ratio        p
Constant       121.752       0.534      228.17      0.000
TIME           0.67292      0.01896      35.49      0.000

s = 1.820       R-sq = 96.5%      R-sq(adj) = 96.4%

Analysis of Variance

SOURCE         DF          SS          MS          F          p
Regression      1        4171.4      4171.4     1259.79     0.000
Error          46         152.3         3.3
Total          47        4323.7
```

averages provide a much better estimate of the cyclical movement. The choice of a three-period moving average is somewhat arbitrary; however, when we use an odd number of terms, the moving averages need not be centered.

A partial solution is shown in Table 16.12. We see that the cyclical component for $t = 2$ is C_2, where

$$C_2 = \frac{.9823 + .9948 + .9748}{3} = .98$$

and the cyclical component for $t = 3$ is

$$C_3 = \frac{.9948 + .9748 + 1.0277}{3} = 1.00$$

Similarly,

$$C_4 = (.9748 + 1.0277 + .9956)/3 = 1.00$$

and

$$C_5 = (1.0277 + .9956 + .9955)/3 = 1.01$$

The complete set of cyclical components is contained in Table 16.13 and is plotted in Figure 16.22. Observe that for October, November, and December of 1987 and again midway through 1990, the retail trade is in a

■ TABLE 16.12
Calculating the cyclical components for the U.S. monthly retail trade data.

t	d_t	$\hat{d}_t$	$d_t/\hat{d}_t\,(=C_t \cdot I_t)$	THREE-MONTH MOVING AVERAGE (C_t)
1	120.26	$121.752 + .673(1) = 122.43$	.9823	—
2	122.46	$121.752 + .673(2) = 123.10$	.9948	.98
3	120.66	$121.752 + .673(3) = 123.77$	.9748	1.00
4	127.89	$121.752 + .673(4) = 124.44$	1.0277	1.00
5	124.57	$121.752 + .673(5) = 125.12$	.9956	1.01
6	125.22	$121.752 + .673(6) = 125.79$	.9955	1.00
⋮	⋮	⋮	⋮	⋮

■ **TABLE 16.13**
Cyclical components
(monthly U.S. retail
trade data).

YEAR	MONTH	d_t	$\hat{d}_t(TR_t)$	$d_t/\hat{d}_t(C_t \cdot I_t)$	THREE-MONTH MOVING AVERAGE (C_t)
1987	JAN	120.26	122.43	0.9823	—
	FEB	122.46	123.10	0.9948	0.98
	MAR	120.66	123.77	0.9748	1.00
	APR	127.89	124.44	1.0277	1.00
	MAY	124.57	125.12	0.9956	1.01
	JUNE	125.22	125.79	0.9955	1.00
	JULY	128.15	126.46	1.0133	1.00
	AUG	126.27	127.14	0.9931	1.00
	SEPT	126.85	127.81	0.9925	1.00
	OCT	129.46	128.48	1.0076	0.99
	NOV	125.66	129.16	0.9729	0.99
	DEC	128.23	129.83	0.9877	0.98
1988	JAN	128.46	130.50	0.9844	1.00
	FEB	133.22	131.17	1.0156	1.00
	MAR	131.83	131.85	0.9999	1.01
	APR	133.63	132.52	1.0084	1.00
	MAY	132.76	133.19	0.9968	1.00
	JUNE	135.00	133.87	1.0085	1.00
	JULY	134.53	134.54	0.9999	1.00
	AUG	135.11	135.21	0.9993	1.00
	SEPT	136.31	135.88	1.0032	1.01
	OCT	138.30	136.56	1.0128	1.01
	NOV	140.56	137.23	1.0243	1.02
	DEC	141.54	137.90	1.0264	1.02
1989	JAN	138.44	138.58	0.9990	1.00
	FEB	137.62	139.25	0.9883	1.00
	MAR	140.01	139.92	1.0007	1.00
	APR	140.69	140.59	1.0007	1.00
	MAY	143.00	141.27	1.0123	1.01
	JUNE	142.78	141.94	1.0059	1.00
	JULY	141.74	142.61	0.9939	1.00
	AUG	145.39	143.29	1.0147	1.01
	SEPT	145.65	143.96	1.0118	1.00
	OCT	142.70	144.63	0.9867	1.00
	NOV	146.60	145.30	1.0089	1.00
	DEC	144.87	145.98	0.9924	1.01
1990	JAN	149.85	146.65	1.0218	1.00
	FEB	147.39	147.32	1.0005	1.01
	MAR	148.60	148.00	1.0041	1.00
	APR	147.87	148.67	0.9946	1.00
	MAY	148.76	149.34	0.9961	0.99
	JUNE	149.00	150.01	0.9932	0.99
	JULY	148.01	150.69	0.9822	0.99
	AUG	151.72	151.36	1.0024	0.99
	SEPT	149.20	152.03	0.9813	1.00
	OCT	153.35	152.71	1.0042	1.00
	NOV	154.99	153.38	1.0105	1.00
	DEC	150.29	154.05	0.9756	—

■ **FIGURE 16.22**
Plot of cyclical activity
(monthly U.S. retail
trade data).

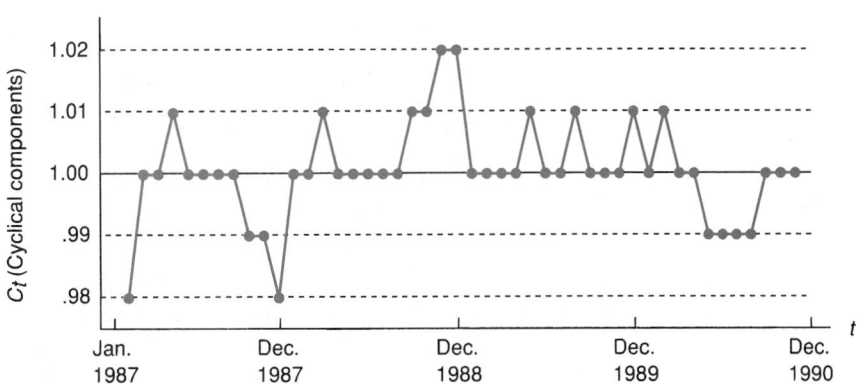

■ **FIGURE 16.23**
MINITAB solution for
the cyclical components
(monthly U.S. retail
trade data).

```
                                              ┌─────────────────────────────┐
                                              │ C2 contains the dᵢ values   │
                                      ┌────────│ C4 contains the d̂ᵢ values   │
                              ┌───────┘        │ See Figure 16.21            │
MTB > DIVIDE C2 BY C4, PUT INTO C6             └─────────────────────────────┘
MTB > LAG BY 1 VALUES IN C6, PUT INTO C7
MTB > LAG BY 2 VALUES IN C6, PUT INTO C8
MTB > LET C9 = (C6 + C7 + C8)/3
MTB > COPY C9 INTO C9;
SUBC> OMIT ROW 1.
MTB > MULTIPLY C9 BY 100, PUT INTO C9
MTB > ROUND C9, PUT INTO C9
MTB > DIVIDE C9 BY 100, PUT INTO C9
MTB > PRINT C9

C9
     *    0.98   1.00   1.00   1.01   1.00   1.00   1.00   1.00
   0.99   0.99   0.98   1.00   1.00   1.01   1.00   1.00   1.00
   1.00   1.00   1.01   1.01   1.02   1.02   1.00   1.00   1.00
   1.00   1.01   1.00   1.00   1.01   1.00   1.00   1.00   1.01
   1.00   1.01   1.00   1.00   0.99   0.99   0.99   0.99   1.00
   1.00   1.00
```

bit of a downturn, as evidenced by the below-normal cyclical components.*
A MINITAB solution for determining the cyclical components is
shown in Figure 16.23. MINITAB computes the $C_t \cdot I_t$ components (cycles
and irregular activity) and the three-month moving averages.

EXAMPLE 16.9   Once the steps in the previous sections have been completed, the various components can be combined for any specified value of t. Determine the four components
for (1) October 1987 and (2) August 1990.

SOLUTION 1   The value of t for October 1987 is $t = 10$. The seasonal index for October is
$S_{10} = .98$. The trend component (from Table 16.13) is $TR_{10} = 128.48$. The cyclical
component (from Table 16.13) is $C_{10} = .99$. The product of $S_{10} \cdot TR_{10} \cdot C_{10} =$
$(.98)(128.48)(.99) = 124.65$.

The actual observation during October 1987 is $y_{10} = 127.21$. Since $y_{10} =$
$S_{10} \cdot TR_{10} \cdot C_{10} \cdot I_{10}$,

$$I_{10} = \frac{y_{10}}{S_{10} \cdot TR_{10} \cdot C_{10}} = \frac{127.21}{124.65} = 1.0205$$

and the final decomposition is

$$y_{10} = 127.21 = S_{10} \cdot TR_{10} \cdot C_{10} \cdot I_{10} = (.98)(128.48)(.99)(1.0205)$$

SOLUTION 2   For $t = 44$ (August 1990), we have $S_{44} =$ seasonal index for August $= S_8 = 1.04$.
Also, $TR_{44} = 151.36$ and $C_{44} = .99$ from Table 16.13. Consequently,

$$I_{44} = \frac{y_{44}}{S_{44} \cdot TR_{44} \cdot C_{44}} = \frac{157.43}{(1.04)(151.36)(.99)} = 1.0102$$

The combined decomposition for this observation is

$$y_{44} = 157.43 = S_{44} \cdot TR_{44} \cdot C_{44} \cdot I_{44} = (1.04)(151.36)(.99)(1.0102)$$ ■

Summary of Time Series Decomposition

The time series decomposition procedure allows you to examine the presence of

Trend (a long-term growth or decline)

Seasonality (a within-year recurrent pattern)

Cyclical activity (upward and downward variation about the trend)

---

*In the first edition of this text, the U.S. retail trade data was examined from 1980 through
1983. For these data, a very clear cycle was observed between September 1980 and April
1983.

The remaining component (what is left after removing the effect of these three factors) is irregular activity. Having determined these components, you are able to describe a particular time series by carefully examining and plotting the calculated components.

A summary of the components for the U.S. retail trade time series is contained in Table 16.14. The irregular activity components (I_t) are determined by continuing the procedure in Example 16.9. Graphs of these components are shown in Figure 16.24. Notice that the graph of the irregular activity components contains no obvious pattern, as we would expect. By combining the various graphs of the time series components into a single set of graphs (Figure 16.24), we can tell at a glance

■ TABLE 16.14

Time series components for U.S. retail trade data.

YEAR	MONTH	y_t	TR_t	S_t	C_t	I_t
1987	JAN	106.39	122.43	0.88	—	—
	FEB	105.80	123.10	0.86	0.98	1.01
	MAR	120.44	123.77	1.00	1.00	0.98
	APR	125.37	124.44	0.98	1.00	1.03
	MAY	129.07	125.12	1.04	1.01	0.99
	JUNE	128.98	125.79	1.03	1.00	0.99
	JULY	128.95	126.46	1.01	1.00	1.01
	AUG	131.02	127.14	1.04	1.00	0.99
	SEPT	123.77	127.81	0.98	1.00	0.99
	OCT	127.21	128.48	0.98	0.99	1.02
	NOV	125.38	129.16	1.00	0.99	0.98
	DEC	154.75	129.83	1.21	0.98	1.01
1988	JAN	113.64	130.50	0.88	1.00	0.99
	FEB	115.10	131.17	0.86	1.00	1.02
	MAR	131.59	131.85	1.00	1.01	0.99
	APR	131.00	132.52	0.98	1.00	1.01
	MAY	137.56	133.19	1.04	1.00	0.99
	JUNE	139.05	133.87	1.03	1.00	1.01
	JULY	135.37	134.54	1.01	1.00	1.00
	AUG	140.20	135.21	1.04	1.00	1.00
	SEPT	133.00	135.88	0.98	1.01	1.00
	OCT	135.90	136.56	0.98	1.01	1.00
	NOV	140.25	137.23	1.00	1.02	1.00
	DEC	170.81	137.90	1.21	1.02	1.01
1989	JAN	122.47	138.58	0.88	1.00	0.99
	FEB	118.90	139.25	0.86	1.00	0.99
	MAR	139.76	139.92	1.00	1.00	1.00
	APR	137.92	140.59	0.98	1.00	1.00
	MAY	148.17	141.27	1.04	1.01	1.01
	JUNE	147.06	141.94	1.03	1.00	1.00
	JULY	142.63	142.61	1.01	1.00	0.99
	AUG	150.86	143.29	1.04	1.01	1.01
	SEPT	142.11	143.96	0.98	1.00	1.01
	OCT	140.22	144.63	0.98	1.00	0.98
	NOV	146.27	145.30	1.00	1.00	1.01
	DEC	174.82	145.98	1.21	1.01	0.98
1990	JAN	132.56	146.65	0.88	1.00	1.02
	FEB	127.34	147.32	0.86	1.01	0.99
	MAR	148.33	148.00	1.00	1.00	1.00
	APR	144.96	148.67	0.98	1.00	1.00
	MAY	154.14	149.34	1.04	0.99	1.00
	JUNE	153.47	150.01	1.03	0.99	1.00
	JULY	148.93	150.69	1.01	0.99	0.99
	AUG	157.43	151.36	1.04	0.99	1.01
	SEPT	145.57	152.03	0.98	1.00	0.99
	OCT	150.68	152.71	0.98	1.00	1.01
	NOV	154.64	153.38	1.00	1.00	1.01
	DEC	181.37	154.05	1.21	—	—

■ **FIGURE 16.24**   Illustration of time series components (monthly U.S. retail trade data).

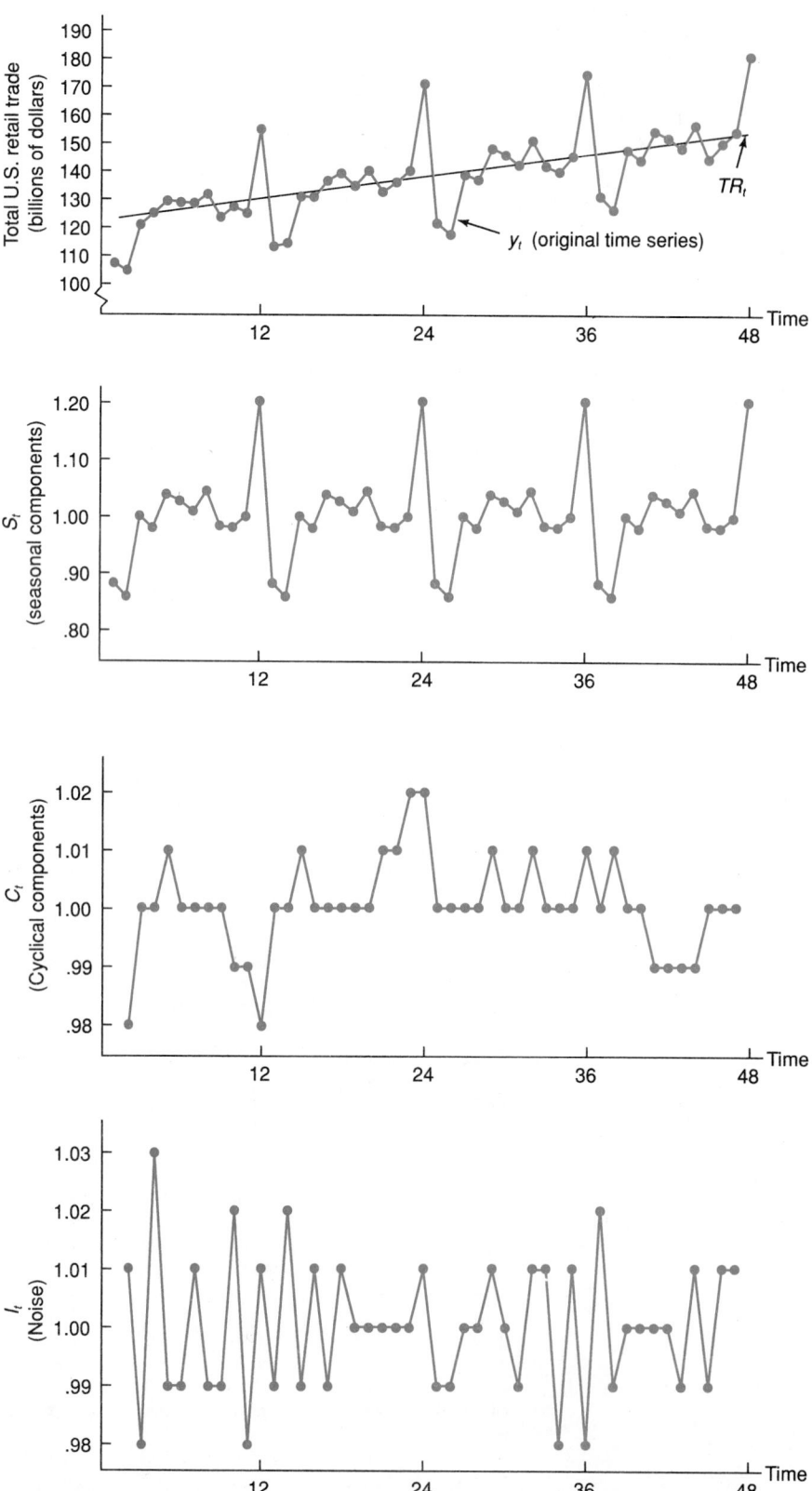

what is the nature of this series. The conclusions we can reach from this figure include

1. There is a strong linear trend that increases over the four-year period.
2. There is a strong retail trade peak each December, followed by weak trading in January and February.
3. There is a slight downturn in the retail trade during the latter months of 1987, a slight upturn during late 1988, and another very slight downturn midway through 1990.

As we discussed in Chapter 2, graphs offer an easy-to-comprehend method of summarizing data. It is time-consuming to construct this particular graph using pen and paper, but practically all microcomputer software packages have graph capabilities. Other methods of time series analysis are discussed in Chapter 17, where we examine time series forecasting.

Exercises

16.45 A real-estate broker, in order to understand the nature of the real-estate market in a growing suburb of New Orleans, collected data from the past four years on the price per square foot of the houses that had three bedrooms, two baths, and a two-car garage. The data are as follows:

YEAR	JAN	FEB	MAR	APR	MAY	JUNE	JULY	AUG	SEPT	OCT	NOV	DEC
1988	27.5	28.3	29.1	29.4	30.4	30.5	29.5	29.0	28.1	28.3	27.1	27.3
1989	27.1	28.6	29.5	30.4	31.4	31.8	31.3	30.7	30.4	29.5	29.6	29.0
1990	29.7	30.7	30.9	31.5	34.3	34.1	33.6	33.4	32.7	32.5	32.1	31.9
1991	31.8	32.7	35.6	36.1	36.8	36.7	35.4	36.5	35.0	34.8	34.7	34.7

a. Determine the seasonal indexes.

b. Determine the trend.

c. Determine the cyclical components for 1990 using a three-period moving average.

d. Determine the irregular components for the first three months of 1990.

16.46 For the month of July 1991, a researcher finds that the seasonal index is 1.14, the trend value is 65.4, and the cyclical component is .86. What is the irregular component if the actual observation for July 1991 is 80.42?

16.47 The following values represent deseasonalized observations. Find the corresponding trend values and cyclical components for each quarter.

YEAR	QUARTER 1	QUARTER 2	QUARTER 3	QUARTER 4
1987	1.79	1.77	1.75	1.68
1988	1.67	1.66	1.60	1.62
1989	1.63	1.71	1.72	1.70
1990	1.83	1.84	1.95	2.10
1991	2.13	2.11	2.15	2.16

16.48 Halston, a supplier of institutional food, sells to a restaurant in the metropolitan area of Memphis, Tennessee. Monthly data on sales (in units of ten thousand dollars) are gathered over the past five years to help describe the company's growth pattern.

YEAR	JAN	FEB	MAR	APR	MAY	JUNE	JULY	AUG	SEPT	OCT	NOV	DEC
1987	4.1	4.3	4.4	4.2	4.5	4.8	4.7	4.6	4.5	4.4	4.5	4.4
1988	4.3	3.6	4.7	4.9	5.3	5.5	5.9	5.7	5.5	5.3	5.2	5.0
1989	5.0	5.4	5.8	6.1	6.7	6.8	7.2	6.3	6.1	6.0	6.1	5.9
1990	5.8	6.3	6.7	6.9	7.4	7.9	8.5	7.9	7.8	7.7	7.4	7.3
1991	7.2	7.4	8.9	8.4	8.7	8.9	9.4	9.1	8.5	8.4	8.1	8.0

a. Compute the seasonal indexes.

b. Calculate the cyclical components for 1990 using a three-month moving average.

c. Calculate the irregular components for the last three months of 1990.

16.49 The manager of a large private golf course in southern California would like to examine the growth pattern of the number of golfers (given in hundreds) who would use the golf course. Monthly data were collected over a four-year period.

YEAR	JAN	FEB	MAR	APR	MAY	JUNE	JULY	AUG	SEPT	OCT	NOV	DEC
1988	4.3	4.4	4.7	4.5	4.8	5.1	5.4	5.6	5.7	5.0	4.7	4.7
1989	4.8	4.9	4.8	6.1	5.0	5.4	5.9	6.0	6.1	6.0	5.8	5.5
1990	5.0	5.1	5.2	5.3	5.5	5.7	6.1	6.3	6.1	5.9	5.7	5.3
1991	5.1	5.7	5.8	6.1	6.3	6.4	4.7	6.8	6.3	6.2	6.1	5.7

a. Determine the seasonal indexes.

b. Determine the trend.

c. Determine the cyclical components for 1990, using a five-period moving average.

d. Determine the irregular components for the first three months of 1990.

16.50 The quarterly averages of the yields (in percent) on 90-day treasury bills are presented below for the years 1983–1988.

YEAR	QUARTER 1	QUARTER 2	QUARTER 3	QUARTER 4
1983	8.11	8.40	9.14	8.80
1984	9.13	9.80	10.32	8.80
1985	8.18	7.46	7.10	7.17
1986	6.89	6.13	5.53	5.34
1987	5.53	5.73	6.03	6.00
1988	5.76	6.23	6.99	7.70

(*Source: Moody's Bank and Finance Manual* (1990): a11.)

a. Determine the seasonal indexes.

b. Determine the trend components.

c. Determine the cyclical components using a three-period moving average.

d. Determine the irregular components for the first two quarters of 1988.

16.51 Find the cyclical components (using a three-quarter moving average) and the irregular components of the data in Exercise 16.35 for the four quarters of 1988.

16.52 Deseasonalize the data in Exercise 16.36. Find the cyclical components using a three-month moving average for 1989. Find the irregular component for February of 1989.

◢ 16.7
Index Numbers

How many times have you heard a remark such as, "Fifteen years ago we could have bought that house for $20,000. Now it's worth $120,000." Or, "My weekly grocery bill used to be $25. Today, it's almost $100." Many people like to talk about the prices back in the "good old days," but were goods and services actually less expensive in those days?

Perhaps a particular item consumed a greater proportion of the typical consumer's consumable income (purchasing power) in years past. To compare effectively the change in the price or value of a certain item (or group of items) between any two time periods, we use an index number. An **index number** (or index) measures the change in a particular item (typically a product or service) or a collection of items between two time periods.

The average hourly wages for production employees at Kessler Toy Company in 1975, 1980, 1985, and 1990 are shown in Table 16.15. Suppose that we wish

■ TABLE 16.15
Average hourly wage of production employees at Kessler Toy Company.

	1975	1980	1985	1990
WAGE	6.40	7.05	8.50	10.90
INDEX (base = 1975)	100	110.2	132.8	170.3

to compare the average wages for 1980, 1985, and 1990 with those for 1975. By computing a ratio for each pair of wages (expressed as a *percentage* of the 1975 wage), we obtain the following set of index numbers:

$$\text{index number for 1980:} \quad \left(\frac{7.05}{6.40}\right) \cdot 100 = 110.2$$

$$\text{index number for 1985:} \quad \left(\frac{8.50}{6.40}\right) \cdot 100 = 132.8$$

$$\text{index number for 1990:} \quad \left(\frac{10.90}{6.40}\right) \cdot 100 = 170.3$$

When calculating an index number, we follow standard practice—round to the nearest tenth (as in Table 16.15) and omit the percent sign. For this application, all wages were compared to those in 1975, the **base year.** The index number for the base year is always 100.

When each index number uses the same base year, the resulting set of values is an **index time series.** An index time series is a set of index numbers determined from the same base year. The purpose of such a time series is to measure the yearly values in *constant* units (dollars, people, and so on). Because these values define a time series, they can be analyzed and decomposed by using the methods described previously. Our purpose in this section is simply to describe how to *construct* a time series of this type.

Price Indexes

Index numbers are derived for a variety of products (goods or services) as well as locations. For example, you may wish to compare the relative costs of consumer items in Los Angeles and Minneapolis if you are considering a move. Such information is readily available or can be determined from a number of business publications or government reports. The Department of Labor and the Bureau of Labor Statistics release reports (many of them monthly) on the price and quantity of many consumer items and agricultural commodities. Often these are recorded for specific U.S. cities, providing geographical comparisons.

We focus our attention on comparison of *prices* from one year to the next; such comparisons are **price indexes.** The most popular of these indexes is the Consumer Price Index (CPI), which combines a large number (over 400) of prices for consumer goods (such as food and housing) and family services (such as health care and recreation) into a single index. It is often called the cost-of-living index.

An index that includes more than one item is an **aggregate index.** We examine two methods of calculating an aggregate price index.

Say that we wish to measure the change in the prices of several items from 1980 to 1990, using a single price index. Table 16.16 shows four items; 1980 is the base year. Let P_0 denote the price for a particular item in the base year (1980) and P_1 represent this price during the reference year (1990). So

$$\Sigma P_0 = \text{sum of sampled prices for 1980}$$
$$= .75 + .95 + .89 + 31$$
$$= \$33.59$$

■ TABLE 16.16
Prices of four items in 1980 and 1990.

ITEM	1980	1990
Eggs	.75 (doz)	1.35 (doz)
Chicken	.95 (lb)	1.79 (lb)
Cheese	.89 (lb)	1.85 (lb)
Auto battery	$31 (each)	$55 (each)

and

$$\Sigma P_1 = \text{sum of sampled prices for 1990}$$
$$= 1.35 + 1.79 + 1.85 + 55$$
$$= \$59.99$$

The ratio of these sums represents the **simple aggregate price index** for this application.

$$\text{simple aggregate price index} = \left(\frac{\Sigma P_1}{\Sigma P_0}\right) \cdot 100 \qquad \textbf{(16.8)}$$

For our example,

$$\text{index} = \left(\frac{59.99}{33.59}\right) \cdot 100 = 178.6$$

It might be tempting to conclude that, based on the prices of these four items, all prices increased by 78.6% between 1980 and 1990. Two problems arise here. The first is whether these sampled items are *representative* of the population of all price changes over this ten-year period. This is not a new problem—the same concern arose when we first introduced statistical sampling.

The second problem is that this index does not take into account the amounts of these items that are typically purchased by consumers. A significant change in the price for any single item will have a dramatic effect on the simple aggregate index, regardless of the demand for this product. The increase of $24 in the price of an automobile battery dominated the computed value of the aggregate price index; however, a typical consumer will spend much more annually on chicken than on car batteries. *The simple aggregate price index assumes that equal amounts of each item are purchased.*

For this reason, the next step is to include a measure of the quantity (Q) of each item in the price index. (We discuss methods of selecting the item quantities later.) The resulting index is known as a **weighted aggregate price index.**

$$\text{weighted aggregate price index} = \left(\frac{\Sigma P_1 Q}{\Sigma P_0 Q}\right) \cdot 100 \qquad \textbf{(16.9)}$$

EXAMPLE 16.10   Assume that a representative family each year purchases 1 automobile battery and each month consumes 6 dozen eggs, 15 pounds of chicken, and 8 pounds of cheese. Using 1980 as the base year and equation 16.9, determine the weighted aggregate price index for 1990. Use the data in Table 16.16.

SOLUTION   The choice of time units on the quantities, Q, is arbitrary, but it is essential that you be consistent across all items. Converting the family purchases to annual units, we have $6 \cdot 12 = 72$ dozen eggs, $15 \cdot 12 = 180$ pounds of chicken, $8 \cdot 12 = 96$ pounds of cheese, and 1 car battery (Table 16.17).

The weighted aggregate price index for 1990 (using 1980 as the base year) is

$$\text{index} = \left(\frac{\Sigma P_1 Q}{\Sigma P_0 Q}\right) \cdot 100 = \left(\frac{652}{341.44}\right) \cdot 100 = 191.0$$

In this index, the increase of 91% between 1980 and 1990 is not as severely affected by the price change for the car battery as was the simple aggregate price

■ **TABLE 16.17**
Calculated aggregate
price index.

ITEM	1980			1990		
	P_0	Q	P_0Q	P_1	Q	P_1Q
Eggs	.75	72	$ 54.00	1.35	72	97.20
Chicken	.95	180	171.00	1.79	180	322.20
Cheese	.89	96	85.44	1.85	96	177.60
Auto battery	31	1	31.00	55	1	55.00
			$\Sigma P_0Q = 341.44$			$\Sigma P_1Q = 652.00$

index, which ignored annual demand for each item. All widely used business price indexes are based on some variation of the weighted aggregate price index in equation 16.9. ■

Selection of the Quantity, Q Because the weights in a weighted aggregate price index usually reflect the quantities consumed, a problem arises when these quantities cannot be assumed to remain constant over the time span of the index. In Example 16.10, the same quantities, Q, were applied to both time periods, so we are assuming an equal demand for the two years.

We have two options here: (1) use the quantities for the base year (1980, here) or (2) use the quantities for the reference year (1990, here). The first method is the *Laspeyres index;* the second is the *Paasche index.*

$$\text{Laspeyres index} = \left(\frac{\Sigma P_1Q_0}{\Sigma P_0Q_0}\right) \cdot 100 \qquad \textbf{(16.10)}$$

where Q_0 represents a base-year quantity.

$$\text{Paasche index} = \left(\frac{\Sigma P_1Q_1}{\Sigma P_0Q_1}\right) \cdot 100 \qquad \textbf{(16.11)}$$

where Q_1 represents a reference-year quantity.

Each of these indexes has strengths and weaknesses. The main advantage of the Laspeyres index is that the same base-year quantities apply to all future reference years. This greatly simplifies updating of this index, particularly given that most aggregate business indexes contain a large number of items. Its main disadvantage is that it tends to give more weight to those items that show a dramatic price increase. When a particular commodity's price increases sharply, it is typically accompanied by a decrease in the demand (measured by Q) for this item, or perhaps another item may be substituted by the consumer. The Laspeyres index fails to adjust for this situation. The advantages of this index outweigh its disadvantages, however, and it is more popular than the Paasche index.

The complexity of updating the reference-year quantities for the Paasche index make it difficult (and often impossible) to apply. Furthermore, because it reflects *both* price and quantity changes, we cannot use it to reflect price changes between two time periods. Its obvious advantage is that it uses current-year quantities, which provide a more realistic and up-to-date estimate of total expense.

We have seen that there is no completely reliable and accurate method of describing aggregate price changes. All such indexes include inaccuracies introduced by using a sample of items in the index as well as by the quantities to be used for weighting. Nevertheless, we treat such an index like any other sample estimate: We use the index as an estimate of relative price changes and realize that it is subject to a certain amount of error.

COMMENTS

1. The most widely used Laspeyres index is the Consumer Price Index (CPI), which is based on hundreds of items ranging from the price of housing to medical expenditures. The CPI is used as a measure of inflation and the cost of living in the United States. It is published monthly and is utilized by the Federal Government (and some private companies), which bases payment (or salary) adjustments on increases or decreases in the CPI. For example, social security payments and retirement benefits for federal civil service employees are tied into this index.

2. The CPI can be used to *deflate* a time series, providing a better comparison of dollar amounts across time. A value is deflated by *dividing* the actual dollar amount for this time period by the corresponding value of the CPI and then multiplying by 100. For example, if your current hourly wage is $12 and the CPI for this year is 150, then the deflated amount is (12/150)(100) = $8. Consequently, the $8 amount can be compared to the hourly wage for the base year (or any other deflated wage value) to determine if there has been a change in "real" wages. This technique can also be applied to the Gross National Product (GNP) to detect real changes in the total value of goods and services.

 *Exercises*

16.53 Lemer's Clothing Store has been selling the same style of men's slacks for six years. The average retail prices (in dollars) for the years 1986 to 1991 are as follows:

YEAR	PRICE
1986	12.75
1987	12.95
1988	13.95
1989	16.95
1990	19.95
1991	23.95

Compare the average prices for the years 1986, 1987, 1988, 1989, 1990, and 1991, using index numbers with 1986 as a base year.

16.54 The total annual profits (in millions of dollars) of car dealers in a large suburb of Chicago over a seven-year period are summarized as follows:

YEAR	TOTAL ANNUAL PROFITS
1985	2.13
1986	2.59
1987	3.60
1988	3.12
1989	3.33
1990	4.15
1991	4.54

Each total annual profit is an aggregate of profits. Find the simple aggregate price index for the years 1987, 1990, and 1991, using 1985 as a base year.

16.55 A typical family in Jackson, Mississippi, had the following weekly buying patterns in 1986 and 1991 (prices are in dollars). Use 1986 as the base year.

ITEM	1986 UNIT PRICE	1986 QUANTITY	1991 UNIT PRICE	1991 QUANTITY
Meat	1.03	2	1.25	2
Milk	.97	3	1.19	2
Fish	.98	2	1.05	3
Oranges	.65	3	.75	4
Bread	.40	1	.62	2

a. Find the simple aggregate price index.

b. Construct the Laspeyres index.

c. Construct the Paasche index.

16.56 Explain the meaning, including the advantages and disadvantages, of the Paasche and Laspeyres weighted indexes. Comment on whether the indexes can be used as a representation of buying pattern.

16.57 The following table reflects the typical family's buying habits per 6 months on repairs for the family car. Use 1985 as the base year.

ITEM	1985 PRICE	1985 QUANTITY	1991 PRICE	1991 QUANTITY
Lube job	3.50	2	5.00	1
Oil change	9.50	3	13.00	2
Tune up	29.95	1	39.95	1
New tires	35.95	2	49.00	2

a. Find the simple aggregate price index.

b. Construct the Laspeyres index.

c. Construct the Paasche index.

16.58 A conglomerate is considering buying one or more of three companies. The closing prices of the stocks of these three companies for the years 1983 to 1991 are:

YEAR	BETTER FOODS	FRIENDLY INSURANCE	CHOCK FULL OF COMPUTER CHIPS
1983	13.500	20.125	39.25
1984	13.750	20.250	35.50
1985	14.250	20.500	31.75
1986	15.125	21.750	34.25
1987	15.500	21.500	37.75
1988	16.000	21.750	39.75
1989	16.125	22.500	40.00
1990	16.250	23.750	39.50
1991	16.750	23.500	42.25

Find an appropriate index to measure the change in the price of these three stocks for the years 1986, 1987, 1989, and 1991 using 1983 as a base year.

16.59 Suppose that, for a certain basket of goods, the Paasche index for 1991 is 115 and the Laspeyres index is 97. Assuming that the base year is 1985, interpret the meaning of the value of the two indexes.

16.60 The number of housing starts for four counties for the years 1989, 1990, and 1991 is:

COUNTY	1989	1990	1991
Brooks	1304	1505	1580
Litton	1264	1759	1987
Riverbed	1135	1443	1565
Tannon	1401	1605	1615

a. Compare the housing starts for Litton county for the years 1990 and 1991 using 1989 as a base year.

b. Compare the aggregate of housing starts for the years 1990 and 1991 for the four counties using 1989 as a base year.

16.61 Data are listed below for Federal funds (in thousands of dollars) obligated for programs administered by the Department of Education during the years 1984 and 1990.

	1984	1990
Elementary and secondary education	4,294,269	7,171,545
Education for the handicapped	2,416,799	4,204,099
Vocational education and adult programs	954,320	1,202,736
Post-secondary student financial assistance	7,478,401	10,801,185

(*Source: The World Almanac and Book of Facts* (1991): 209.)

a. Use 1984 as a base year. Find the simple aggregate index for the total amount of Federal funds for the four categories listed above.

b. What can you conclude from the index calculated in part a?

16.62   A nursery purchases four different chemical ingredients to blend a certain popular fertilizer mixture. The data indicate the price per unit (PPU) paid for each ingredient and the quantity bought in 1989, 1990, and 1991.

INGREDIENTS	1989 PPU	1990 PPU	1991 PPU	1989 QUANTITY	1990 QUANTITY	1991 QUANTITY
A	.80	.81	.85	385	375	380
B	.51	.55	.60	345	360	379
C	.45	.50	.53	200	250	280
D	.37	.39	.40	150	180	195

a.   Calculate the Laspeyres index for 1990 and 1991 using 1989 as a base year.

b.   Calculate the Paasche index for 1990 and 1991 using 1989 as a base year.

c.   Compare the two indexes in parts a and b.

Summary

A variable recorded over time is a **time series.** You obtain a sample of values for this variable by recording its past observations. Because such a sample is not a random one, it is extremely difficult (if not impossible) to obtain any tests of hypothesis or confidence intervals. Consequently, we resort to describing the past observations by deriving the components of the time series. This process is called **time series decomposition.** The components of a time series are (1) **trend** (a long-term growth or decline in the observations), (2) **seasonality** (within-year recurrent fluctuations), (3) **cyclical activity** (upward and downward movements of various lengths about the trend), and (4) **irregular activity** (what remains after the other three components have been removed).

We described methods for estimating these components for a time series. We first specify how we believe the components interact with one another, thus describing the time series variable, y_t. The **additive** structure assumes that each observation is the *sum* of its components. In particular, this structure implies that seasonal fluctuations during a particular year are not affected by the base volume for that year. In the **multiplicative** structure, each value of y_t is the *product* of the four components. In this framework, the seasonal fluctuation for a specific month (or quarter) is more apt to be a constant *percentage* of the base volume for that year; for example, sales in December might be 35% higher than the average (base) sales for that particular year. The multiplicative structure was assumed for practically all of the illustrations in this chapter and is used more commonly in practice. The Bureau of the Census uses a variation of this procedure for their time series decomposition analyses.*

We described a four-step procedure for deriving these components for a particular time series, based on the multiplicative structure. The steps were: (1) determine a **seasonal index** for each month (monthly data) or quarter (quarterly data); (2) **deseasonalize** the data by dividing each observation by its corresponding seasonal index; (3) determine the **trend components** by deriving a least squares line or quadratic curve through the deseasonalized values; and (4) determine the **cyclical components** by, for each time period, dividing each deseasonalized value by its estimate using the trend equation and smoothing these values by computing three-period moving averages.

An **index time series,** often used by business analysts, is a time-related sequence of index numbers in which each value is a measure of the change in a

---

*The Bureau of the Census procedure is called the X11 program and is available on SAS/ETS. Consult the Econometric Time Series (ETS) user's guide (available from SAS) and the end-of-chapter appendix for a description of this procedure.

particular item (or group of items) from one year to the next. **Price indexes** are used to compare prices over time.

An **aggregate price index** is used to compare the relative price of a set of items for any year to the price during the base year. The index for the base year always is 100. The prices for the items can be averaged (**simple** aggregate price index) or weighted by the corresponding quantity of each item (**weighted** aggregate price index). Methods of selecting these quantities include using base-year quantities (the **Laspeyres index**) or using the reference-year quantities (the **Paasche index**). The most popular Laspeyres index in practice is the Consumer Price Index (CPI).

Further Reading

Bowerman, B. L., and R. T. O'Connell. *Forecasting and Time Series.* 2d ed. Boston: PWS-KENT, 1987.

Makridakis, S., S. C. Wheelwright, and V. E. McGee. *Forecasting: Methods and Applications.* 2d ed. New York: John Wiley, 1983.

Mendenhall, W., and J. E. Reinmuth. *Statistics for Management and Economics.* 6th ed. Boston: PWS-KENT, 1989.

Summary of Formulas

Linear trend line:

$$\hat{y}_t = b_0 + b_1 t$$

where

$$t = 1, 2, \cdots, T$$

$$b_1 = \frac{\Sigma t y_t - (\Sigma t)(\Sigma y_t)/T}{\Sigma t^2 - (\Sigma t)^2/T}$$

$$b_0 = \bar{y}_t - b_1 \bar{t}$$

Shortcut:

$$\Sigma t = 1 + 2 + \cdots + T$$
$$= \frac{T(T + 1)}{2}$$

$$\Sigma t^2 = 1 + 4 + \cdots + T^2$$
$$= \frac{T(T + 1)(2T + 1)}{6}$$

$$\bar{t} = \frac{\Sigma t}{T} = \frac{T + 1}{2}$$

Deseasonalized value of y_t:

$$d_t = \frac{y_t}{\text{corresponding seasonal index}}$$

Simple aggregate price index:

$$\left(\frac{\Sigma P_1}{\Sigma P_0}\right) \cdot 100$$

Weighted aggregate price index:

$$\left(\frac{\Sigma P_1 Q}{\Sigma P_0 Q}\right) \cdot 100$$

Laspeyres index:

$$\left(\frac{\Sigma P_1 Q_0}{\Sigma P_0 Q_0}\right) \cdot 100$$

where Q_0 represents a base-year quantity

Paasche index:

$$\left(\frac{\Sigma P_1 Q_1}{\Sigma P_0 Q_1}\right) \cdot 100$$

where Q_1 represents a reference-year quantity.

Review Exercises

16.63 Each of the following influences on the variation in profits of a national chain of department stores would contribute to which of the four components of a time series?

a. The long-term growth of the economy.

b. The resignation of top managers in the company.

c. Annual demand in spring and summer for garden equipment.

d. The closing of several other department stores.

16.64 A manufacturer of tractors has built a record number of tractors for every year for the past seven years. Given in thousands, the figures show the number of tractors built from 1985 to 1991.

YEAR	TRACTORS BUILT
1985	10.75
1986	11.78
1987	12.59
1988	13.4
1989	14.3
1990	15.7
1991	16.8

Find the least squares prediction equation that you would use to forecast the trend. What would you estimate the number of tractors built in 1991 to be?

16.65 Luz Chemicals, which manufactures a special-purpose baking soda, is interested in estimating the equation of the trend line for their monthly sales data (in tons) for the year 1991.

MONTH	BAKING SODA SALES	MONTH	BAKING SODA SALES
Jan.	28	July	34
Feb.	33	Aug.	34
Mar.	39	Sept.	35
Apr.	33	Oct.	36
May	38	Nov.	31
June	31	Dec.	37

a. Without considering the seasonality present in the monthly sales, estimate the trend line equation.

b. Using the equation obtained in part a, estimate the sales (in tons) for the month of February 1992.

16.66 Telemex, a supplier of telephone systems, has experienced moderate to rapid growth over a 12-year period. The data show the annual sales figures (in tens of thousands of dollars).

YEAR	SALES	YEAR	SALES
1980	3.1	1986	18.8
1981	6.3	1987	18.4
1982	10.5	1988	20.0
1983	10.2	1989	21.3
1984	11.5	1990	29.0
1985	14.7	1991	28.3

a. Find the trend line.

b. Find the cyclical components.

c. Graph the data and estimate the period of the cycle.

16.67 The number of trucks and buses produced in the United States from 1982 to 1988 are given below (in thousands).

YEAR	TRUCK AND BUS PRODUCTION
1982	1,912
1983	2,444
1984	3,151
1985	3,468
1986	3,506
1987	3,826
1988	4,080

(*Source: Statistical Abstract of the United States* (1990): 601.)

a. Graph the data. Does the trend appear linear?

b. Find the equation to estimate the trend for the time series data.

c. What would be your estimate of the number of trucks and buses produced in the United States in 1989?

16.68 Suppose that for the month of January 1991, the marketing department of a firm finds that the seasonal index is 1.20, the trend line value is $17,000 in sales, and the cyclical component is .79. What is the irregular component if the actual sales figure for January 1991 is $16,500?

16.69 Sales figures (in tens of thousands of dollars) for Dataphonics for a ten-year period follow. Find the corresponding trend values and cyclical components for each quarter of 1989 and 1990.

YEAR	QUARTER 1	QUARTER 2	QUARTER 3	QUARTER 4
1982	2.48	4.39	5.68	2.49
1983	2.76	4.86	5.69	2.73
1984	2.80	4.91	5.75	2.91
1985	2.90	5.10	5.85	2.95
1986	3.10	5.20	5.96	3.01
1987	3.15	5.21	6.04	3.10
1988	3.18	5.24	6.10	3.15
1989	3.20	5.30	6.14	3.19
1990	3.22	5.35	6.20	3.24
1991	3.25	5.36	6.23	3.25

16.70 The following table lists the number of building permits per month for nonresidential construction during the four-year period 1988 through 1991 in Parkins, Nebraska.

YEAR	JAN	FEB	MAR	APR	MAY	JUNE	JULY	AUG	SEPT	OCT	NOV	DEC
1988	21	22	23	24	25	28	29	30	27	26	20	20
1989	21	24	23	25	26	25	29	32	32	27	20	18
1990	17	18	21	24	22	28	29	30	27	26	22	20
1991	17	21	23	23	24	29	31	22	28	22	21	29

a. Determine the seasonal indexes.

b. Determine the cyclical components for 1990 using a three-month moving average.

c. Determine the irregular component for July of 1990.

16.71 The average monthly utility bill for residents of the small community of Ridgecrest for the years 1988 to 1991 is:

YEAR	JAN	FEB	MAR	APR	MAY	JUNE	JULY	AUG	SEPT	OCT	NOV	DEC
1988	190	180	179	130	135	145	148	153	145	153	170	185
1989	197	193	185	150	151	159	163	165	160	159	180	185
1990	215	205	193	175	171	179	185	184	180	180	173	190
1991	235	225	205	180	182	190	195	198	188	185	195	201

a. Determine the seasonal indexes.

b. Determine the trend.

c. Determine the cyclical components for 1989 using a three-period moving average.

d. Determine the irregular components for June and July 1989.

16.72 The weekly buying pattern of a typical family in a suburb of Atlanta, Georgia, for 1985 and 1991 follows.

ITEM	1985 UNIT PRICE	1985 QUANTITY	1991 UNIT PRICE	1991 QUANTITY
Chicken	2.40	1	2.75	2
Milk	1.02	3	1.19	2
Bread	.39	2	.45	2
Ground beef	1.59	3	1.89	2
Tomatoes	.39	2	.78	2

Using 1985 as a base year,

a. Find the simple aggregate price index.

b. Calculate the Laspeyres index.

c. Calculate the Paasche index.

d. Compare the indexes in questions b and c.

16.73 The president of R&B Home Builders uses a housing index to obtain information about the direction of the housing market. The index for the four quarters of 1988, 1989, 1990, and 1991 yields these data:

YEAR	QUARTER 1	QUARTER 2	QUARTER 3	QUARTER 4
1988	157	155	154	147
1989	142	145	140	142
1990	143	153	152	150
1991	163	165	162	160

a. Determine the seasonal indexes for each quarter.

b. Determine the trend line.

c. Determine the cyclical components for 1989 using a three-period moving average.

d. Determine the irregular components for the first and second quarter of 1989.

16.74 Ranton House, Inc., has been building a certain style of house for the past four years. This house has sold for various prices over the years 1988 to 1991, as shown (in thousands of dollars).

YEAR	JAN	FEB	MAR	APR	MAY	JUNE	JULY	AUG	SEPT	OCT	NOV	DEC
1988	49.5	51.3	51.3	51.5	52.0	57.3	57.4	58.3	57.2	56.3	55.4	58.6
1989	55.3	55.6	55.7	56.3	57.4	62.7	62.8	63.8	62.3	61.4	60.5	60.0
1990	60.5	61.3	62.4	62.7	63.0	68.6	68.9	70.1	68.6	68.4	67.4	67.0
1991	67.4	67.5	67.6	68.1	68.3	72.1	72.4	72.4	72.1	71.3	71.0	70.8

a. Determine the seasonal indexes.

b. Determine the trend.

c. Determine the cyclical components for 1990 using a three-period moving average.

d. Determine the irregular components for the months of September and October 1990.

16.75 The number of defaults per month on business loans at First State Bank are given over a five-year period. Find the seasonal indexes.

YEAR	JAN	FEB	MAR	APR	MAY	JUNE	JULY	AUG	SEPT	OCT	NOV	DEC
1987	54	53	52	50	48	46	48	50	52	56	58	60
1988	58	54	53	50	50	45	46	49	51	55	57	62
1989	53	52	48	47	47	44	45	48	49	52	55	60
1990	58	51	50	49	45	43	44	49	50	51	58	63
1991	59	58	56	52	54	49	50	51	54	58	60	64

16.76 The following data are the total amount of mortgage assets held by U.S. life insurance companies for 1979, 1982, 1985, and 1988 (in millions of dollars).

YEAR	TOTAL MORTGAGES
1979	111,421
1982	141,989
1985	171,797
1988	232,863

(Source: Moody's Bank and Finance Manual (1990): a12.)

The total amount of mortgages is an aggregate. Find the simple aggregate index for the years 1979, 1982, 1985, and 1988, using 1979 as the base year.

16.77 Monthly redemptions of mutual funds from 1984 through 1988 are given in units of millions of dollars.

	1984	1985	1986	1987	1988
January	1,338	2,158	4,555	6,878	6,966
February	1,680	2,433	3,511	6,595	7,085
March	1,518	2,454	4,770	8,763	8,763
April	1,570	2,763	5,571	13,569	8,084
May	1,727	2,624	5,252	9,666	7,988
June	1,532	2,674	5,567	8,113	7,922
July	1,350	3,229	5,561	8,332	6,898
August	1,604	3,008	5,156	9,162	8,309
September	1,662	2,469	6,423	12,617	7,264
October	2,039	2,702	5,901	15,826	6,975
November	1,828	2,712	5,772	7,543	7,534
December	2,369	4,425	8,923	9,172	8,716

(*Source: Standard and Poor's Statistical Service, Current Statistics* (January 1990): 5 and
Standard and Poor's Statistical Service, Current Statistics (April 1988): 5.)

a. Compute the seasonal indexes.

b. Determine the trend.

c. Calculate the cyclical components for 1987 using a three-month moving average.

d. Calculate the irregular components for the last three months of 1987.

16.78 The MINITAB computer printout shows two time series plots of the values in C1 and the values in C2. Would you say that these time series have multiplicative variation or additive variation? Draw a trend line for each time series. Estimate the length of the cycle.

```
MTB > PRINT C1-C3
ROW     C1       C2      C3

  1     4.45    17.22     1
  2     4.66    20.36     2
  3     3.00    18.75     3
  4     2.50    18.46     4
  5     7.00    27.00     5
  6    11.70    38.36     6
  7    12.30    38.77     7
  8    10.40    28.35     8
  9     9.20    25.36     9
 10    14.50    40.00    10

MTB > MPLOT C1 VS C3, AND C2 VS C3
```

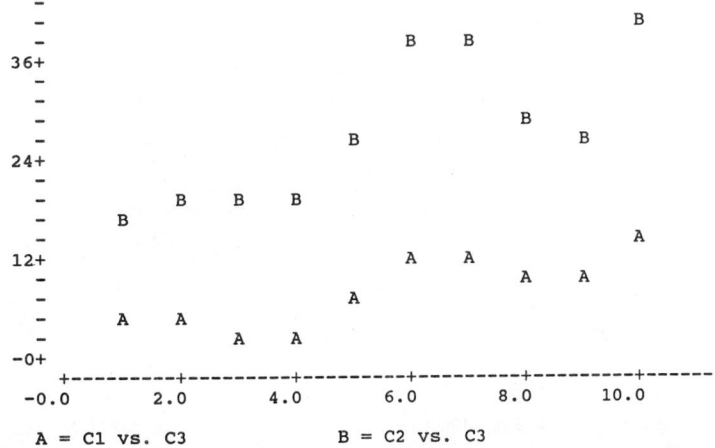

A = C1 vs. C3   B = C2 vs. C3

16.79 The following MINITAB computer printout shows a regression analysis of a time series with nine observations. From observing the plot of the residuals, comment on the validity of the statistical tests used in the regression analysis.

```
MTB > PRINT C1 C2
 ROW         C1    C2

   1     5.0062     1
   2     5.1638     2
   3     3.2737     3
   4     3.4650     4
   5     8.4375     5
   6    13.4437     6
   7    13.6125     7
   8    12.9825     8
   9    14.6250     9

MTB > REGRESS C1 ON 1 PREDICTOR IN C2, RESID IN C3

The regression equation is
C1 = 1.17 + 1.54 C2

Predictor       Coef        Stdev      t-ratio
Constant       1.174        1.736         0.68
C2             1.5431       0.3084        5.00

s = 2.389      R-sq = 78.1%    R-sq(adj) = 75.0%

Analysis of Variance

SOURCE      DF          SS            MS
Regression   1       142.87        142.87
Error        7        39.95          5.71
Total        8       182.82

MTB > PLOT C3 VS C2

C3      -
        -
        -            *                         *
     1.0+
        -
        -                                           *
        -                  *
        -
    -0.0+
        -                                *              *   *
        -
        -
        -
    -1.0+
        -               *
        -
        -
        -                    *
    -2.0+
        ------+---------+---------+---------+---------+---------+---------+C2
            1.5       3.0       4.5       6.0       7.5       9.0
```

◢ *Case Study*

Forecasting Water Usage of the Holly Water Treatment Complex, Fort Worth, Texas

The Water Department of the city of Fort Worth, Texas, has to purchase untreated lake water from several administrative entities around the city in order to fulfill the water requirements of its population. Most of the water is treated and pumped at one of two plants—the Holly Water Treatment Complex and the Rolling Hills Plant. The Rolling Hills Plant is located on the southeast side of Fort Worth and is on a higher plain than the west-side Holly Treatment Plant. Both plants are required to operate in order to meet the city's demand; however, the electrical cost

of pumping water out of the Holly Treatment Plant is approximately 15% higher than for the Rolling Hills Plant because of its lower elevation. Because of the cost differential, the water pumped from the two plants has been slowly approaching a 60–40 split in favor of the Rolling Hills Plant.

For budgetary reasons, the City Manager would like you to forecast the water processed and pumped at the Holly Treatment Plant. He will use your forecasts as a basis for estimating the associated electrical cost. You are informed that water consumption varies from fall to spring and summer to winter. There is also a gradual increase in the water demand at the Holly Treatment Plant because the total demand made by the population has slowly increased.

■ **Water Usage Data**
(In Tens of Millions of Gallons)

Month	YEAR				
	1985–86	1986–87	1987–88	1988–89	1989–90
OCT	365.5	361.3	429.7	354.6	312.9
NOV	319.9	307.7	331.2	292.2	368.3
DEC	309.6	312.8	291.3	279.2	316.3
JAN	327.0	294.0	296.4	268.5	316.3
FEB	285.6	258.2	267.4	251.8	275.2
MAR	358.9	300.2	319.4	287.8	283.3
APR	326.5	402.3	350.2	336.4	286.2
MAY	324.6	367.3	447.4	350.6	352.7
JUNE	352.5	349.5	476.0	371.5	512.6
JULY	616.4	469.5	540.3	479.2	529.2
AUG	531.2	628.7	617.5	462.5	488.6
SEPT	403.6	428.7	429.9	421.0	461.4

Note: The data have been adjusted for the case study.

Source: Fort Worth, TX Water Department

Case Study Questions

1. Using the data in the table, compute the centered moving averages. What is the benefit derived from obtaining these moving averages?

2. Compute the seasonal indexes for each month. When does the Water Department experience the largest seasonal effect? Does this seem reasonable? Does there seem to be a seasonal pattern?

3. Deseasonalize the data and use the deseasonalized time series to determine the linear trend equation. What is the average increase in water use per month? How far into the future do you think the trend could be projected?

4. Assuming a multiplicative model, obtain the cyclical components. Does there seem to be a cyclical pattern?

5. Prepare a forecast of water use in Fort Worth for the last three months of 1990 and the first month of 1991 using only the trend equation. Use the seasonal indexes to adjust the deseasonalized forecasts.

6. In January 1991, the actual water use turned out to be 325.6. Discuss the forecasting error encountered here.

SPSS

▮ **Solution**
Example in
Section 16.6

We used four-step decomposition on time series data to analyze the past behavior of U.S. trade from 1987 to 1990 (Table 16.7). The SPSS program listing in Figure 16.25 was used to perform the analysis.

The TITLE command names the SPSS run.

The DATA LIST command gives each variable a name and describes the data as being in free form.

The BEGIN DATA command indicates to SPSS that the input data immediately follow.

The next 48 lines contain the data values, which represent time series data over 48 time periods.

The END DATA statement indicates the end of the data.

DATE means the first data item entered will be the first month of the first year (January 1987).

The X11ARIMA command requests the X11 time series decomposition procedure be used on the dataset TRADE.

The MONTHLY statement means the data is monthly. TRADAYREG statement is used to adjust the time series.

Figure 16.26 shows a portion of the SPSS output obtained by executing the listing in Figure 16.25. The results differ slightly from the results in the text, since SPSS uses a more complicated time series decomposition procedure than the one outlined in this chapter.

■ **FIGURE 16.25**
Input for SPSS or SPSS/PC. Remove the periods for SPSS input.

```
TITLE X11 DECOMPOSITION.
DATA LIST FREE  /  YEAR MONTH MMMYY (A) TRADE.
BEGIN DATA.
1987   1 JAN87 106.39
1987   2 FEB87 105.80
1987   3 MAR87 120.44
1987   4 APR87 125.37
1987   5 MAY87 129.07
1987   6 JUN87 128.98
1987   7 JUL87 128.95
1987   8 AUG87 131.02
1987   9 SEP87 123.77
1987  10 OCT87 127.21
1987  11 NOV87 125.38
1987  12 DEC87 154.75
1988   1 JAN88 113.64
1988   2 FEB88 115.10
1988   3 MAR88 131.59
       .
       .
       .
1990   8 AUG90 157.43
1990   9 SEP90 145.57
1990  10 OCT90 150.68
1990  11 NOV90 154.64
1990  12 DEC90 181.37
END DATA.
DATE YEAR 1987 MONTH 1.
X11ARIMA TRADE
    /PERIOD=MONTHLY
    /TRADAYREG=SIGNIF.
```

■ **FIGURE 16.26** SPSS output.

B 1. Original series

Year	Jan	Feb	Mar	Apr	May	Jun	Jul	Aug	Sep
1987	106	106	120	125	129	129	129	131	124
1988	114	115	132	131	138	139	135	140	133
1989	122	119	140	138	148	147	143	151	142
1990	133	127	148	145	154	153	149	157	146
Avge	119	117	135	135	142	142	139	145	136

Year	Oct	Nov	Dec	Total
1987	127	125	155	1507
1988	136	140	171	1623
1989	140	146	175	1711
1990	151	155	181	1799
Avge	139	142	170	.

Original Time Series

Table total-   6641   Mean-   138   Std. deviation-   16

D10. Final seasonal factors
       Stable moving average selected.

Year	Jan	Feb	Mar	Apr	May	Jun	Jul	Aug	Sep
1987	88.1	86.5	99.2	98.5	103.8	102.8	100.8	104.0	97.2
1988	88.1	86.5	99.2	98.5	103.8	102.8	100.8	104.0	97.2
1989	88.1	86.5	99.2	98.5	103.8	102.8	100.8	104.0	97.2
1990	88.1	86.5	99.2	98.5	103.8	102.8	100.8	104.0	97.2
Avge	88.1	86.5	99.2	98.5	103.8	102.8	100.8	104.0	97.2

Year	Oct	Nov	Dec	Avge
1987	99.0	100.5	119.6	100.0
1988	99.0	100.5	119.6	100.0
1989	99.0	100.5	119.6	100.0
1990	99.0	100.5	119.6	100.0
Avge	99.0	100.5	119.6	

Seasonal Indexes

Table total-   4800.0   Mean-   100.0   Std. deviation-   8.0

■ FIGURE 16.26 (*continued*)

```
D11.  Final seasonally adjusted series

Year     Jan    Feb    Mar    Apr    May    Jun    Jul    Aug    Sep
1987     120    123    123    127    125    126    127    127    127
1988     129    130    132    132    134    135    135    135    136
1989     141    139    140    141    142    143    143    144    145
1990     150    149    148    149    148    148    149    150    151
Avge     135    135    136    137    137    138    138    139    140

Year     Oct    Nov    Dec    Total
1987     128    126    129    1506
1988     139    139    142    1617          ┌─────────────────────┐
1989     143    145    147    1712          │ Deseasonalized values │
1990     152    153    153    1801          └─────────────────────┘
Avge     140    141    143      .

    Table total-        6636    Mean-      138    Std. deviation-         9
```

SAS

▮ Solution
Example in
Section 16.6

We used four-step decomposition on time series data to analyze the past behavior of U.S. trade from 1987 to 1990 (Table 16.7). The SAS program listing in Figure 16.27 below was used to perform the analysis.

The TITLE command names the SAS run.

The DATA command gives the data a name.

The first DO statement sets a year loop from 1987 to 1990.

The second DO statement sets a month loop from 1 to 12.

SALES are to be entered as INPUT.

DATE means the first data item entered will be the first month of the first year (January 1987).

The OUTPUT statement requests that the input be printed.

The CARDS command indicates to SAS that the input data immediately follow.

■ FIGURE 16.27 Input for SAS (mainframe or micro version).

```
TITLE 'X11 DECOMPOSITION';
DATA RETAIL;
    DO YEAR=1984 TO 1987;
        DO MONTH=1 TO 12;
            INPUT SALES @@;
            DATE=MDY(MONTH,1,YEAR);
            N+1;
            OUTPUT;
        END;
    END;
    KEEP DATE SALES N;
    FORMAT DATE MONYY5.;
    LABEL SALES=' RETAIL TRADE';
    CARDS;
93.09 93.69 104.29 104.34 111.31 111.98 106.55 110.65 103.93 109.23
113.28 131.65 98.82 95.59 110.17 113.11 120.34 114.96 115.49 121.12
114.17 116.14 118.56 139.40 105.64 99.66 114.24 115.71 125.42 120.35
120.74 124.06 124.65 123.06 120.79 151.26 106.39 105.80 120.44 125.37
129.07 128.98 128.95 131.02 123.77 127.21 125.38 154.75
PROC X11 DATA=RETAIL;
    MONTHLY DATE=DATE CHARTS=STANDARD TDREGR=TEST PRINTOUT=STANDARD;
    VAR SALES;
    TITLE2 'MONTHLY EXAMPLE SHOWING STANDARD TABLES AND CHARTS';
```

The next five lines contain the data values, which represent time series data over 48 time periods.

The PROC X11 command requests the X11 time series decomposition procedure be used on the dataset RETAIL.

The MONTHLY statement means the data is monthly. DATE = DATE specifies when the time series starts and ends. CHART specifies charts to be produced by the X11 procedure. TDREGR = TEST statement is used to adjust the time series. PRINTOUT = STANDARD requests seasonal charts and trend cycle charts. VAR specifies that SALES is the input that will be analyzed by the procedure.

Figure 16.28 shows a portion of the SAS output obtained by executing the listing in Figure 16.27. The results differ slightly from the results in the text, since SAS uses a more complicated time series decomposition procedure than the one outlined in this chapter.

■ FIGURE 16.28   SAS output.

B 1. Original series

Year	Jan	Feb	Mar	Apr	May	Jun	Jul	Aug	Sep	Oct	Nov	Dec	Total
1987	106	106	120	125	129	129	129	131	124	127	125	155	1507
1988	114	115	132	131	138	139	135	140	133	136	140	171	1623
1989	122	119	140	138	148	147	143	151	142	140	146	175	1711
1990	133	127	148	145	154	153	149	157	146	151	155	181	1799
Avge	119	117	135	135	142	142	139	145	136	139	142	170	

Table total-   6641   Mean-   138   Std. deviation-   16

[Original Time Series]

D10. Final seasonal factors
       Stable moving average selected.

Year	Jan	Feb	Mar	Apr	May	Jun	Jul	Aug	Sep	Oct	Nov	Dec	Avge
1987	88.1	86.5	99.2	98.5	103.8	102.8	100.8	104.0	97.2	99.0	100.5	119.6	100.0
1988	88.1	86.5	99.2	98.5	103.8	102.8	100.8	104.0	97.2	99.0	100.5	119.6	100.0
1989	88.1	86.5	99.2	98.5	103.8	102.8	100.8	104.0	97.2	99.0	100.5	119.6	100.0
1990	88.1	86.5	99.2	98.5	103.8	102.8	100.8	104.0	97.2	99.0	100.5	119.6	100.0
Avge	88.1	86.5	99.2	98.5	103.8	102.8	100.8	104.0	97.2	99.0	100.5	119.6	

Table total-   4800.0   Mean-   100.0   Std. deviation-   8.0

[Seasonal Indexes]

D11. Final seasonally adjusted series

Year	Jan	Feb	Mar	Apr	May	Jun	Jul	Aug	Sep	Oct	Nov	Dec	Total
1987	120	123	123	127	125	126	127	127	127	128	126	129	1506
1988	129	130	132	132	134	135	135	135	136	139	139	142	1617
1989	141	139	140	141	142	143	143	144	145	143	145	147	1712
1990	150	149	148	149	148	148	149	150	151	152	153	153	1801
Avge	135	135	136	137	137	138	138	139	140	140	141	143	

Table total-   6636   Mean-   138   Std. deviation-   9

[Deseasonalized values]

C H A P T E R

17 *Quantitative Business Forecasting*

We have introduced you to several methods of capturing the behavior of a dependent variable, *Y*. The first procedure was linear regression, which used a set of predictor (independent) variables to explain the observed values of the dependent variable. In simple linear regression, a single predictor is used. When we had two or more predictor variables, we used a multiple linear regression model to attempt to account for the variation in the observed values of the dependent variable. These calculations were considerably more complex, and a computer solution was used to estimate the linear relationship between the dependent variable (*Y*) and the predictor variables (X_1, X_2, . . .)

The success or failure of this technique lies in your ability to arrive at a set of predictor variables that can accurately predict past (and future) values of the dependent variable. Suppose your model fails to fit adequately the observed values of *Y*, with a resulting large sum of squares for error (SSE) and a low value of R^2 (coefficient of determination). Do these results imply that multiple linear regression is not a reliable method of prediction for this situation? This could be the case, but it is just as likely that you omitted one or more key variables that would have significantly improved your prediction accuracy.

The time series decomposition technique, presented in the previous chapter, uses a different approach. This procedure attempts to explain each observed value by means of its various components. These components include trend (long-term growth or decline in the time series), seasonality (predictable variation within each year), and cyclical activity (generally due to unpredictable swings in the national or international economy).

The key distinction between these two procedures is that the time series approach does not search for explanatory (predictor) variables. Rather, it seeks to capture the past behavior of the time series by analyzing its various components. More complex time series techniques (which were not discussed) use past observations to predict the value for the future. You can use a time series approach to forecast future values by "extending" the pattern into the future. For example, if your company sales have been increasing approximately 150,000 units each year over the past six years, a reasonable forecast for next year would be a sales volume of 150,000 more than the present year's value.

Statistical forecasting is, in one sense, an extension of the prediction of a dependent variable. However, we now enter a more uncertain world—that of extrapolation. In previous chapters, we warned you of the dangers of this procedure, because outside the range of your data, the predicted values become less reliable. We can only hope that tomorrow's world will be similar to today's and that patterns observed over the past will continue. This uncertainty makes forecasting fascinating. We live in an uncertain world, and a reasonably accurate forecast can be extremely valuable for a marketing or production strategy.

This chapter introduces many (certainly not all) methods of using quantitative techniques for predicting future values of the variable of interest. We demonstrate how to forecast future values by using the past observations (the time series approach) as well as by using the multiple linear regression method. By applying the proper forecast method, you often can make the future considerably less uncertain.

◢ 17.1
Methods of Forecasting

Forecasting procedures come in a variety of shapes and colors. You can arrive at a sales forecast by simply assembling a panel of experts and arriving at a collective "guess" or by constructing a highly complex statistical model that attempts to predict the future using past data. In the broadest sense, forecasting methods can be classified as **qualitative** (the panel of experts procedure) or **quantitative** (the statistical forecasting procedure). Quantitative forecasting can be carried out using two

745

different approaches, namely, **regression** models (with several predictor variables) or **time series** models, which utilize past observations of the dependent variable to arrive at forecasted values.

Qualitative Forecasting

There are many instances when a qualitative approach to forecasting is appropriate. When no past data are available, it is impossible to construct a quantitative model to predict future values. This situation can occur when you intend to introduce a new product and no past sales data exist. Furthermore, when you introduce this product, it becomes a guessing game as to what the response will be from competitors in the field. Will they respond to your entry into the market? When will they respond? Will they lower their price to increase the demand for their product? Will they attempt to "copy" your product, and how soon can this be accomplished? Such questions do require expert opinion.

One popular method of qualitative forecasting is the **Delphi method.** With this procedure, you assemble individuals from the sales force and the market research staff and ask them to supply their predictions based on their knowledge of the area. This can be accomplished through a questionnaire or any other written set of specific questions. After this step, members of the team are informed as to the responses of the entire group and asked to reevaluate their opinions based on this new information. In this way, members of the team may be able to arrive at a "best educated" prediction of competitor response to their market entry. Of course, it is also entirely possible that no collective agreement will be reached after several rounds of this process.

We do not pursue qualitative forecasting methods in this chapter. The interested reader is referred to the text by Bowerman and O'Connell (see the Further Readings section at the end of the chapter) for additional qualitative procedures. The remainder of the chapter focuses on the use of quantitative forecasting techniques.

Quantitative Forecastings

With a quantitative forecasting procedure, you predict future behavior of a dependent variable using information from previous time periods. This can be accomplished in one of two ways: using a *regression* model or using a *time series* model.

Regression Models   The multiple regression approach consists of using regression models where variation of the dependent variable is explained using several independent (predictor) variables. One main advantage of this approach is that you can measure the effect of changes in one or more of the predictor variables. Furthermore, this type of model is generally easily understood by those individuals responsible for making the final forecast decision, since it is clear which variables are assumed to have an effect on the value of the dependent variable. A drawback of this type of model as a forecasting instrument is that to predict future values of the dependent variable, it is necessary to predict future values of the predictor variables, which, in many instances, may be as uncertain as future values of the dependent variable. This forecasting procedure is discussed in Section 17.9.

Time Series Models   A time series forecast is made by capturing the patterns that exist in the past observations and extending them into the future. Consider the annual data reflecting the sales of the Clayton Corporation between 1976 and 1990, shown in Figure 17.1. The data reflect a strong linear trend, as shown by the line passing through the points. To estimate the sales for 1991, one simple method

■ FIGURE 17.1
Sales for Clayton
Corporation.

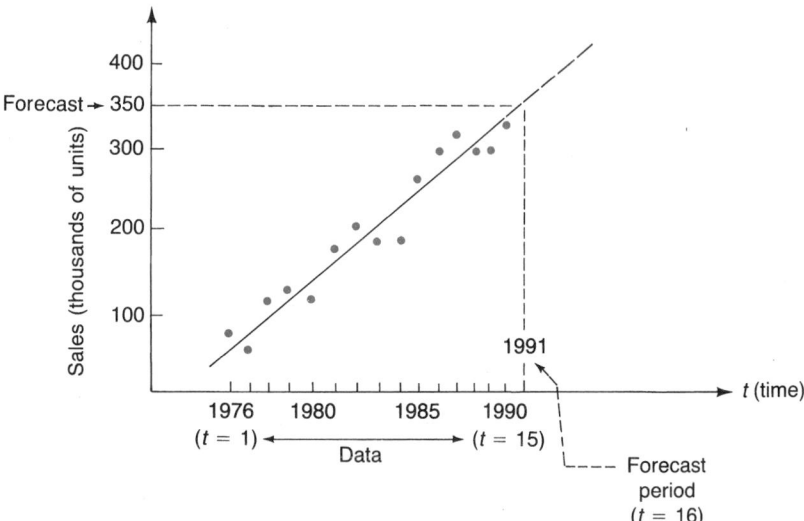

would be to extend this line to 1991, as illustrated in Figure 17.1. By graphically extending this line and observing the estimated value, we obtain

$$\hat{y}_{16} = \text{forecast for } 1991 \cong 350$$

that is, 350,000 units. This procedure, along with methods of dealing with trend and seasonality, is discussed later in the chapter.

At first glance, it might appear that time series forecasting is easier to apply than are multiple regression models. After all, there is no need to search for a reliable set of predictor variables. It is true that time series predictors can be simple and straightforward, as is the so-called naive forecast discussed in the next section. *Frequently, however, extracting the complex and interrelated structure of an observed time series requires sophisticated and complex prediction equations.*

As in Chapter 16, we do not put any statistical bounds (such as a 95% confidence interval) on the predicted values. Rather, we suggest alternative methods of forecasting and demonstrate a way of determining the "best" forecasting procedure for a particular set of data. Of course, all forecasts are subject to error and are based on the assumption that the past historical patterns (such as the straight line in Figure 17.1) continue into the future.

In Sections 17.2 through 17.8, we examine several time series models and methods of evaluating the predictive ability of each procedure when applied to a particular set of time series data.

The procedure for selecting a forecasting model is summarized in Figure 17.2. Steps 1 through 5 are the model selection and forecasting stage. Steps 6 and 7 are the model review phase, during which you reevaluate your forecasting procedure. This step allows you to update your model using the latest observations or to consider changing your forecasting model by returning to step 2. Any forecasting technique should be reviewed; you must reexamine the forecast errors (that is, the difference between the forecasted values and the previous observations).

Remember that any quantitative forecasting technique can never replace the forecast of an individual (or team of people) who uses his or her expertise and knowledge of unpredictable future events (such as strikes, wars, or market shifts) to make forecasts. Rather, the quantitative forecast is one tool the forecaster uses. A forecast is an excellent baseline that can be modified by informed judgment.

■ **FIGURE** 17.2
A step-by-step
procedure for
forecasting with time
series data.

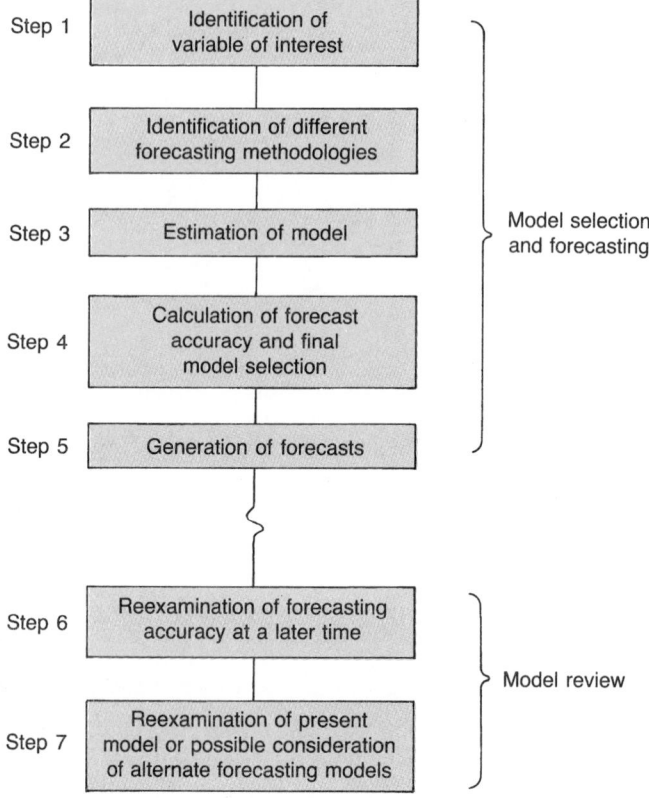

Step 1 — Identification of variable of interest

Step 2 — Identification of different forecasting methodologies

Step 3 — Estimation of model

Step 4 — Calculation of forecast accuracy and final model selection

Step 5 — Generation of forecasts

Model selection and forecasting

Step 6 — Reexamination of forecasting accuracy at a later time

Step 7 — Reexamination of present model or possible consideration of alternate forecasting models

Model review

17.2 The Naive Forecast

Simply put, the naive forecast procedure states that the estimate of Y for tomorrow is the actual value from today. In general,

$$\hat{y}_{t+1} = y_t \tag{17.1}$$

for any time period, t.

Once again, the "hat" notation is used to denote an estimate. Equation 17.1 reads: "y hat for time period $t + 1$ is y for time t." Here, $\hat{y}_{t+1}$ represents the *forecast* for time period $t + 1$.

This method of forecasting often works well for data that are recorded for smaller time intervals (such as daily or weekly) and contain no apparent upward or downward trend among the observed values. Data of this type are not apt to shift direction suddenly from one day to the next, and the naive forecast can provide a simple, yet fairly reliable, estimate of the next day's value. On more than one occasion, this predictor has outperformed much more complex forecasting equations—particularly when applied to a difficult-to-predict time series, such as an individual stock market price. It provides an inexpensive, easy method of forecasting.

EXAMPLE 17.1

The weekly closing price for a share of Keller Toy Company stock was recorded over a 12-week period. Using the following data, determine a forecast for week 13.

WEEK	PRICE	WEEK	PRICE	WEEK	PRICE
1	60	5	$64\frac{1}{2}$	9	$63\frac{1}{4}$
2	$62\frac{1}{4}$	6	62	10	$62\frac{1}{2}$
3	$61\frac{3}{4}$	7	$63\frac{1}{2}$	11	61
4	63	8	64	12	$61\frac{1}{2}$

SOLUTION The observed value for the last time period is $y_{12} = 61\frac{1}{2}$, so your forecast for the next time period is

$$\hat{y}_{13} = y_{12} = 61\frac{1}{2}$$

Notice that we are careful to distinguish between a *forecast*, such as $\hat{y}_{13}$, and an *observed* value, such as y_{12}.

One method of checking to see whether a particular forecasting technique is appropriate for your time series involves applying this procedure to each period of the observed data. For example, in Example 17.1, what would we have predicted for the fifth week using the naive forecasting equation 17.1? In other words, suppose we are at the end of the fourth week and need a forecast for $t = 5$. Using the naive predictor,

$$\hat{y}_5 = y_4 = 63$$

The actual value turned out to be $y_5 = 64\frac{1}{2}$, providing a **residual** of

$$\text{residual} = y_5 - \hat{y}_5 = 64\frac{1}{2} - 63 = 1\frac{1}{2} \qquad \blacksquare$$

EXAMPLE 17.2 Apply the naive forecasting procedure to the 12 time periods in Example 17.1 and determine the residual for each week.

SOLUTION The procedure cannot be applied during the first time period ($t = 1$) because $\hat{y}_1 = y_0$, where y_0 is the closing price for the week preceding the observations in the table. If this value is available, then the forecast value for $t = 1$ can be determined; it is equal to this value. Otherwise, the forecast for this time period is left blank. The forecasts and residuals for the remaining 11 time periods are given in Table 17.1. $\qquad \blacksquare$

When we first introduced the concept of a residual (or error) in the chapters dealing with linear regression, we stressed that small residuals were desirable. When the residuals were near zero for regression applications, this meant that the model did a good job of "fitting" the sample observations.

The same idea applies to evaluating the effectiveness of a forecasting procedure. Small residuals indicate that this particular forecast technique would have done a good job of predicting the past values of this time series. A method of

■ TABLE 17.1
Residuals for naive forecasts.

WEEK	y_t	$\hat{y}_t$	RESIDUAL $(y_t - \hat{y}_t)$
1	60	—	—
2	$62\frac{1}{4}$	60	$2.25 \ (= 62\frac{1}{4} - 60)$
3	$61\frac{3}{4}$	$62\frac{1}{4}$	$-.5 \ (= 61\frac{3}{4} - 62\frac{1}{4})$
4	63	$61\frac{3}{4}$	1.25 (and so on)
5	$64\frac{1}{2}$	63	1.5
6	62	$64\frac{1}{2}$	-2.5
7	$63\frac{1}{2}$	62	1.5
8	64	$63\frac{1}{2}$	.5
9	$63\frac{1}{4}$	64	$-.75$
10	$62\frac{1}{2}$	$63\frac{1}{4}$	$-.75$
11	61	$62\frac{1}{2}$	-1.5
12	$61\frac{1}{2}$	61	.5

combining these residuals into a single measure (much like the SSE in linear regression) will be introduced in a later section.

 *Exercises*   **17.1**   Explain the distinction between the technique of time series analysis and that of multiple regression analysis.

17.2   If a regression or time series model fits a set of data well, would the model necessarily provide small forecasting errors for future observations?

17.3   The price of the stock of Intersecond Bank has been cyclical over the years. From the following data, calculate the forecasted price of the stock using the naive model for the years 1981 to 1991. Also, calculate the residual for each forecast.

YEAR	PRICE OF STOCK	YEAR	PRICE OF STOCK
1981	28.50	1987	28.25
1982	29.25	1988	29.75
1983	31.75	1989	32.50
1984	29.50	1990	31.50
1985	28.00	1991	30.00
1986	27.50		

17.4   Sullivan's Mutual Fund invests primarily in technology stocks. The net asset value of the fund at the end of each month for the 12 months of 1991 is given. Find the forecasted value of the mutual fund for each month, starting with February, by using the naive model. Calculate the residuals.

MONTH	MUTUAL FUND PRICE	MONTH	MUTUAL FUND PRICE
Jan	8.43	July	8.35
Feb	8.10	Aug	9.45
Mar	7.15	Sept	9.01
Apr	6.95	Oct	10.31
May	7.25	Nov	10.25
June	7.95	Dec	11.04

17.5   What advantages and disadvantages can you think of in using the naive model to forecast?

▰ *17.3*
Projecting the Least Squares Trend Equation

For data containing a strong linear or curvilinear trend, one method of predicting future values of the time series is to extend the trend line (or curve) into the forecast periods. This method was illustrated in Figure 17.1, where the data from 1976 to 1990 demonstrated a very strong linear growth over these 15 years.

Suppose that a simple linear regression analysis is performed on these data, using the 15 sales values as the dependent variable and $t = 1, 2, \ldots, 15$ as the predictor variable (as discussed in Chapter 16). The resulting least squares line, shown in Figure 17.1, turns out to be

$$\hat{y}_t = 32 + 20t$$

The estimated forecast for 1991 in the earlier discussion was $\hat{y}_{16} = 350$. This value was determined simply by extending the least squares line into this time period and "eyeballing" the estimate for 1991. The actual forecast is

$$\hat{y}_{16} = 32 + 20(16) = 352$$

So, our estimate of sales for 1991 is 352,000 units, based on the linear trend equation.

A Time Series Containing Trend and Seasonality

The previous procedure can be adapted to situations in which the time series contains significant trend *and* seasonality. Such a situation can occur when the data are monthly or quarterly, with seasonal fluctuations about a linear or curvilinear trend.

The quarterly sales for Video-Comp over a four-year period (1987 to 1990) are contained in Table 16.3 on page 710 and are illustrated in Figure 17.3. The deseasonalized sales figures (often called *seasonally adjusted* sales) are summarized in Table 16.6 on page 714 and also are graphed in Figure 17.3. Notice that the extreme seasonal fluctuations of the original time series were removed when these values were divided by the appropriate seasonal index. The indexes for this application were derived in Example 16.8, and they indicated low sales for the first two quarters, above-average sales for the third quarter, and extremely high sales during the fourth (holiday) quarter. The corresponding indexes were

$$S_1 = .852$$
$$S_2 = .692$$
$$S_3 = 1.166$$
$$S_4 = 1.290$$

To forecast future values when using seasonal data, you once again determine the least squares line (or curve), except that now you use the *deseasonalized data* (say, d_t) as your dependent variable. Once you have calculated the trend forecast, you obtain your final forecast by multiplying this deseasonalized estimate by the corresponding seasonal index. So the procedure for extending trend and seasonal components is:

1. Calculate the deseasonalized (seasonally adjusted) data from the original time series ($y_1, y_2, \ldots, y_T$). Call these values $d_1, d_2, \ldots, d_T$.
2. Construct a least squares line through the deseasonalized data, where ($t = 1, 2, \ldots, T$)

$$\hat{d}_t = b_0 + b_1 t$$

3. Calculate the forecast for time period $T + 1$ using

$$\hat{y}_{T+1} = (\hat{d}_{T+1}) \cdot (\text{seasonal index for } t = T + 1)$$
$$= [b_0 + b_1(T + 1)] \cdot (\text{seasonal index for } t = T + 1)$$

■ FIGURE 17.3
Quarterly sales
at Video-Comp.

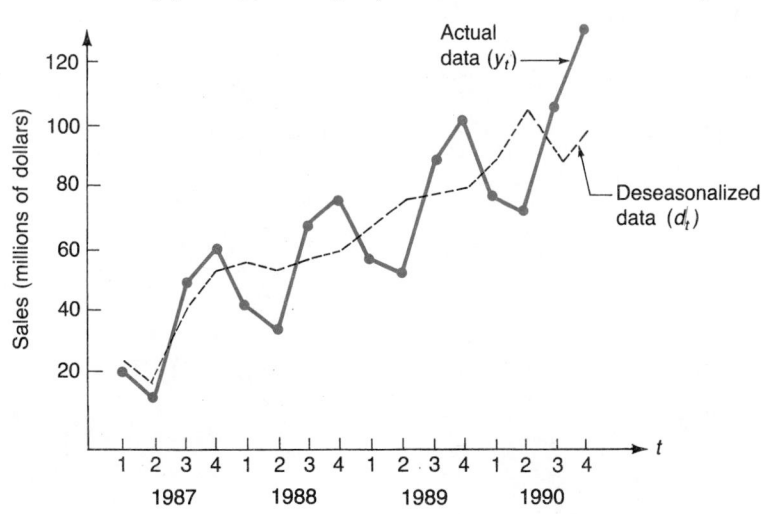

EXAMPLE 17.3   Using the Video-Comp data, what would be your forecast for the first-quarter sales for 1991? Second-quarter sales?

SOLUTION   The deseasonalized data and corresponding least squares line are shown in Figure 17.4. The MINITAB solution for the least squares line is shown in Figure 17.5, where

$$\hat{d}_t = 19.372 + 5.0375t$$

This equation tells us that, apart from seasonal variation, the sales at Video-Comp are increasing by approximately $5 million each quarter. Using this equation and Figure 17.4, your deseasonalized forecast for the first quarter of 1991 (time period 17) is

$$\hat{d}_{17} = 19.372 + 5.0375(17)$$
$$= 105.01$$

Now, the sales for the first quarter of each year are lower than the yearly average, as reflected in the seasonal index of $S_1 = .852$ (from Example 16.8). Consequently,

■ **FIGURE 17.4**
Trend line through deseasonalized data (quarterly sales, Video-Comp).

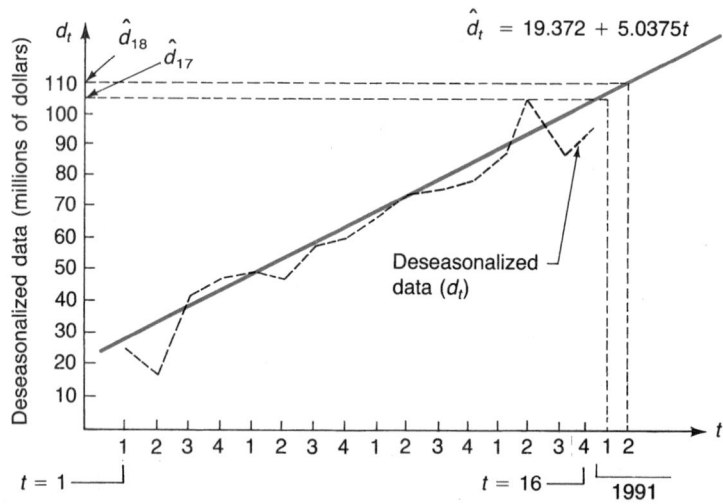

■ **FIGURE 17.5**
MINITAB solution for deseasonalized trend line.

```
MTB > SET INTO C1
DATA> 1:16
DATA> END
MTB > SET INTO C2
DATA> 23.47 17.34 40.31 46.51 46.95 46.24 55.75 58.91
DATA> 65.73 72.25 72.90 77.52 88.03 101.16 86.62 95.35
DATA> END
MTB > NAME C1 = 'TIME' C2 = 'DESEAS'
MTB > REGRESS Y IN C2 USING 1 PREDICTOR IN C1

The regression equation is
DESEAS = 19.4 + 5.04 TIME

Predictor      Coef      Stdev     t-ratio        p
Constant     19.372      3.111        6.23    0.000
TIME          5.0375     0.3217      15.66    0.000

s = 5.932       R-sq = 94.6%     R-sq(adj) = 94.2%

Analysis of Variance

SOURCE       DF         SS          MS        F        p
Regression    1       8627.9      8627.9   245.21    0.000
Error        14        492.6        35.2
Total        15       9120.5
```

your actual forecast for this time period is

$$\hat{y}_{17} = \hat{d}_{17} \cdot (\text{seasonal index for quarter 1})$$
$$= 105.01 \cdot .852$$
$$= 89.5 \quad (\text{million dollars})$$

This procedure can be used to forecast any future time period. For the second quarter of 1991, the estimated sales will be

$$\hat{y}_{18} = \hat{d}_{18} \cdot (\text{seasonal index for quarter 2})$$
$$= [19.372 + 5.0375(18)] \cdot 692$$
$$= (110.05)(.692) = 76.2 \quad (\text{million dollars})$$

Do these estimates seem reasonable? Look at the observed (1987–1990) and forecast (1991) values for the first and second quarters:

YEAR	FIRST-QUARTER SALES	SECOND-QUARTER SALES
1987	20	12
1988	40	32
1989	56	50
1990	75	70
1991	89.5	76.2

The forecast for the first quarter of 1991 seems to be about what we would expect, based on the past first-quarter sales. The predicted sales value for the second quarter of 1991 seems to be on the low side, with an increase of only 6.2 from the second quarter of 1990. Remember, however, that this forecasting technique contains the effect of *all* the quarters observed over the four years. By examining the past sales during the second quarter only, we are ignoring the remaining quarters, and perhaps an explanation for this seemingly low forecast lies in these values.   ∎

It is possible that this forecasting procedure is not a good one for the application in Example 17.3—there may be a better way to obtain a forecast for this situation. We show you several ways to forecast a time series and then determine which of these does the best job for a particular set of observed values. *No one procedure always performs well for all applications.*

 **Exercises**

17.6   A set of quarterly data has been gathered for 3 years. The seasonal indexes are found to be $S_1 = .81$, $S_2 = .93$, $S_3 = 1.19$, $S_4 = 1.07$. The least squares line through the deseasonalized data is found to be

$$\hat{d}_t = 10.1 + 1.3t$$

from 12 quarterly periods. Find the forecast for quarterly periods 13, 14, 15, and 16.

17.7   Sands Motel, which usually has a busy summer season on the beaches of Atlantic City, New Jersey, would like to obtain a forecast of future business. Business (in thousands of dollars) for each month from 1985 to 1991 is given:

YEAR	JAN	FEB	MAR	APR	MAY	JUNE	JULY	AUG	SEPT	OCT	NOV	DEC
1985	.4	.5	.7	.9	1.3	1.8	2.5	2.9	2.3	2.0	1.2	.7
1986	.5	.6	.8	1.1	1.2	1.9	2.8	3.1	2.7	2.1	.9	.8
1987	.7	.6	.9	1.1	1.4	2.1	2.8	3.3	2.9	2.4	1.5	.9
1988	.8	.7	.9	1.3	1.6	2.3	2.9	3.4	3.2	2.6	1.7	1.2
1989	1.0	.9	1.1	1.4	1.8	2.4	3.0	3.4	3.3	2.7	1.9	1.4
1990	1.1	.8	.9	1.3	1.9	2.5	3.2	3.6	3.2	2.6	2.0	1.5
1991	1.3	1.1	1.2	1.5	2.2	2.7	3.4	3.8	3.4	2.8	2.1	1.6

a.   Find the seasonal indexes.

b.   Find the least squares trend line for the deseasonalized data.

c.   Find the forecast for March 1992, July 1992, and December 1992.

17.8   Slater Industries would like to cut costs on the amount of inventory it holds. Quarterly data have been gathered for five years from 1987 through 1991. The following table lists the dollar amount of inventory in units of 10,000.

YEAR	QUARTER 1	QUARTER 2	QUARTER 3	QUARTER 4
1987	.3	.5	.4	.2
1988	.4	.7	.5	.3
1989	.5	.9	.7	.4
1990	.7	1.1	.9	.8
1991	.8	1.5	1.0	.9

a.   Find the seasonal indexes.

b.   Find the least squares trend line for the deseasonalized data.

c.   Find the forecast for each quarter of 1992.

17.9   Is the experience of the managers of a company necessary to use in aiding the forecasting process, if the model fits the past data very well?

17.10   Refer to Exercise 16.45. What is the deseasonalized forecast for January of 1990 for the price per square foot of the typical three-bedroom, two-bath, two-car-garage house? What is the actual forecast?

◢ 17.4
Simple Exponential Smoothing

In Chapter 16, we introduced the concept of smoothing a time series by computing a set of centered *moving averages*. The moving averages were used to derive the various seasonal indexes, but they also provided a "new" time series with considerably less random variation (irregular activity) and no seasonality. Because the moving average series was much smoother, it provided a clearer picture of any existing trend or cyclical activity.

Another method of smoothing a time series, which also serves as a forecasting procedure, is **exponential smoothing.** Unlike moving averages, this technique uses all the preceding observations to determine a smoothed value for a particular time period. The method described in this section is called **simple** (or single) **exponential smoothing** and works well for a time series containing *no trend* (Figure 17.6). A time series (such as the one in this figure) is said to be **stationary** if the data exhibit no trend and the variance about the mean ($\bar{y}_t$) remains constant over time. *Simple exponential smoothing generally will track the original time series well, provided this series is stationary.* We extend the simple exponential smoothing procedure to a series containing trend and seasonality in later sections.

The simplest way to determine a smoothed value for time period t using exponential smoothing is to find a weighted sum of the actual observation for this time period, y_t, and the previous smoothed value, S_{t-1}.

■ **FIGURE 17.6**
Illustration of a stationary time series.

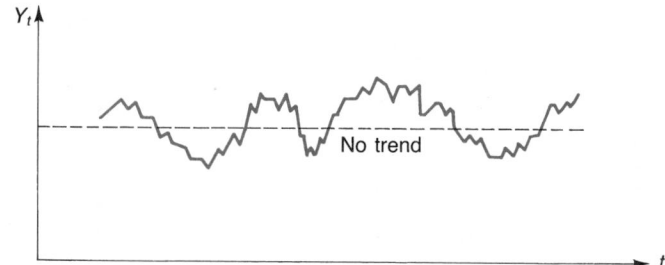

$$S_t = \text{smoothed value for time period, } t$$

$$= Ay_t + (1 - A)S_{t-1} \qquad (17.2)$$

where A is any number between 0 and 1.

The value of A is the **smoothing constant.** Small values of A produce smoothed values giving less weight to the corresponding observation, y_t. You should use such values (say, $A < .1$) for a volatile time series containing considerable irregular activity (noise). In this way, you give more weight to the previous smoothed value, S_{t-1}, rather than to the original observation, y_t. You can use larger values of A for a more stable time series.

The smoothing procedure used here begins by setting the first smoothed value, S_1, equal to the first observation, y_1. So,

$$S_1 = y_1$$

Then,

$$S_2 = Ay_2 + (1 - A)S_1$$

$$= Ay_2 + (1 - A)y_1$$

$$S_3 = Ay_3 + (1 - A)S_2$$

$$S_4 = Ay_4 + (1 - A)S_3$$

and so on.

The average attendance (y_t, in thousands) for major events held at the Jefferson County Civic Center for the past 13 years is contained in Table 17.2. We determine the exponentially smoothed values using three smoothing constants, $A = .1$, $A = .5$, and $A = .9$.

The actual time series and the three smoothed series are shown in Figure 17.7. For $A = .1$,

$$S_1 = y_1 = 5.0$$

$$S_2 = (.1)y_2 + (.9)S_1$$

$$= (.1)(8.0) + (.9)(5.0) = 5.3$$

$$S_3 = (.1)y_3 + (.9)S_2$$

$$= (.1)(2.1) + (.9)(5.3) = 4.98$$

and so on.

	YEAR	t	y_t	$S_t(A = .1)$	$S_t(A = .5)$	$S_t(A = .9)$
■ TABLE 17.2 Actual and smoothed values for attendance at Jefferson Civic Center.	1978	1	5.0	5.0	5.0	5.0
	1979	2	8.0	5.3	6.5	7.7
	1980	3	2.1	4.98	4.3	2.66
	1981	4	7.1	5.19	5.7	6.66
	1982	5	4.8	5.15	5.25	4.99
	1983	6	2.0	4.84	3.62	2.30
	1984	7	7.8	5.13	5.71	7.25
	1985	8	5.0	5.12	5.36	5.23
	1986	9	14.1	6.02	9.73	13.21
	1987	10	13.0	6.72	11.36	13.02
	1988	11	13.5	7.39	12.43	13.45
	1989	12	14.2	8.07	13.32	14.12
	1990	13	14.0	8.67	13.66	14.01

■ FIGURE 17.7
Smoothed values
for attendance data
(Table 17.2).

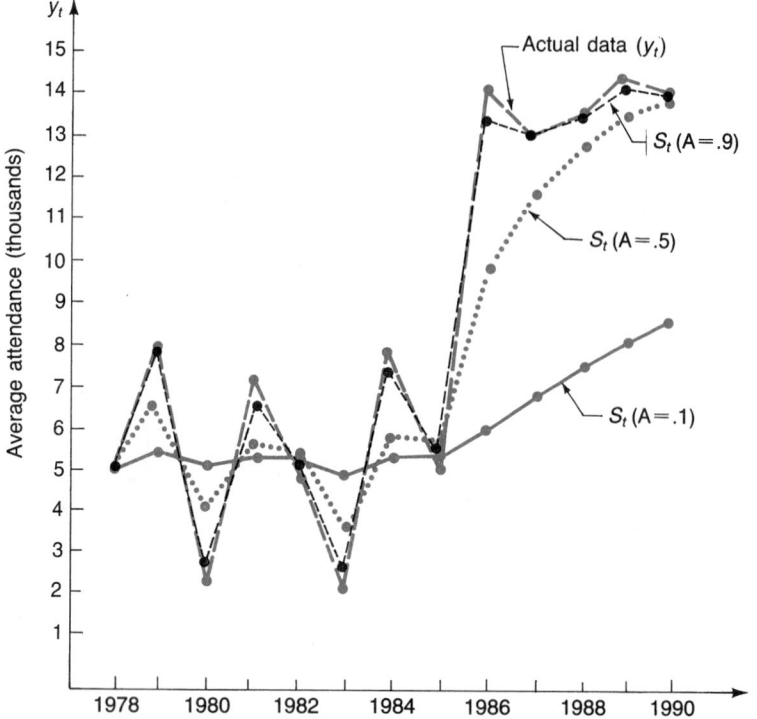

■ FIGURE 17.7
Smoothed values for attendance data (Table 17.2).

Notice that the average attendance, y_t, had a significant jump in 1986, when (it turns out) the facility was completely refurnished, providing better seating and more accessible snack booths. With the small value of $A = .1$, the smoothed values did not "track" the original series very well after this point. In general, when you use exponential smoothing with a small smoothing constant, the resulting series will be slow to detect any turning points or shifts in the observed values. However, such values of A provide considerable smoothing, as is evident from the values between the years 1978 and 1985.

The large value of $A = .9$ provides much better tracking (see Figure 17.7) but not much smoothing. Larger smoothing constants are more useful for a time series that does not contain a great deal of random fluctuation. Using $A = .5$ offers a compromise between these two extreme smoothing constants. Later we discuss methods of comparing the tracking ability of different values of A, in an effort to determine the best smoothing constant for a particular series.

To see why this procedure is called exponential smoothing, we look at how each smoothed value is obtained. First, $S_1 = y_1$. Then,

$$
\begin{aligned}
S_2 &= Ay_2 + (1 - A)S_1 \\
&= Ay_2 + (1 - A)y_1 \\
S_3 &= Ay_3 + (1 - A)S_2 \\
&= Ay_3 + (1 - A)[Ay_2 + (1 - A)y_1] \\
&= Ay_3 + A(1 - A)y_2 + (1 - A)^2 y_1 \\
S_4 &= Ay_4 + (1 - A)S_3 \\
&= Ay_4 + (1 - A)[Ay_3 + A(1 - A)y_2 + (1 - A)^2 y_1] \\
&= Ay_4 + A(1 - A)y_3 + A(1 - A)^2 y_2 + (1 - A)^3 y_1
\end{aligned}
$$

In general,

$$S_t = Ay_t + A(1 - A)y_{t-1} + A(1 - A)^2 y_{t-2} + \cdots + A(1 - A)^{t-2}y_2$$
$$+ (1 - A)^{t-1}y_1$$

For example, if $A = .5$, then

$$S_t = .5y_t + .25y_{t-1} + .125y_{t-2} + .062y_{t-3} + \cdots$$

Therefore, each smoothed value is actually a weighted sum of *all the previous observations*. Because the more recent observations have the largest weight, they have a larger effect on the smoothed value. Notice that the weights on the observations are decreasing exponentially. That is, the weight given to a particular observation is some constant (namely, $1 - A$) *times* the weight given to the preceding observation. That is why this procedure is called exponential smoothing.

Forecasting Using Simple Exponential Smoothing

The naive forecasting procedure introduced earlier predicts the time series value for tomorrow using the actual value for today. In other words, $\hat{y}_{t+1} = y_t$. The exponential smoothing process is similar, except now the forecast for tomorrow is the smoothed value from today. In general,

$$\hat{y}_{t+1} = S_t \tag{17.3}$$

For the special case where $A = 1$, we have

$$\hat{y}_{t+1} = S_t = 1y_t + (1 - 1)S_{t-1} = y_t$$

and the exponential smoothing forecast is the same as that provided by the naive predictor. Because A is considerably less than 1 in practice, the smoothed forecast makes use of all the past observations, rather than only the most recent measurement.

EXAMPLE 17.4

Using simple exponential smoothing with $A = .1$, what are the predicted values and residuals for the attendance data in Table 17.2?

SOLUTION

Suppose the year is 1978 ($t = 1$) and you want a forecast for 1979 ($t = 2$). You need the smoothed value for 1978: $\hat{y}_2 = S_1 = 5.0$. Next, the year is 1979, and you need a forecast for 1980. Here, $\hat{y}_3 = S_2 = 5.3$ (from Table 17.2). Continuing in this way, we obtain Table 17.3.

■ **TABLE 17.3**
Forecasts and residuals using simple exponential smoothing on attendance data ($A = .1$).

YEAR	t	y_t	$\hat{y}_t$	RESIDUAL ($y_t - \hat{y}_t$)
1978	1	5.0	—	—
1979	2	8.0	5.0	3.00
1980	3	2.1	5.3	−3.20
1981	4	7.1	4.98	2.12
1982	5	4.8	5.19	−.39
1983	6	2.0	5.15	−3.15
1984	7	7.8	4.84	2.96
1985	8	5.0	5.13	−.13
1986	9	14.1	5.12	8.98
1987	10	13.0	6.02	6.98
1988	11	13.5	6.72	6.78
1989	12	14.2	7.39	6.81
1990	13	14.0	8.07	5.93

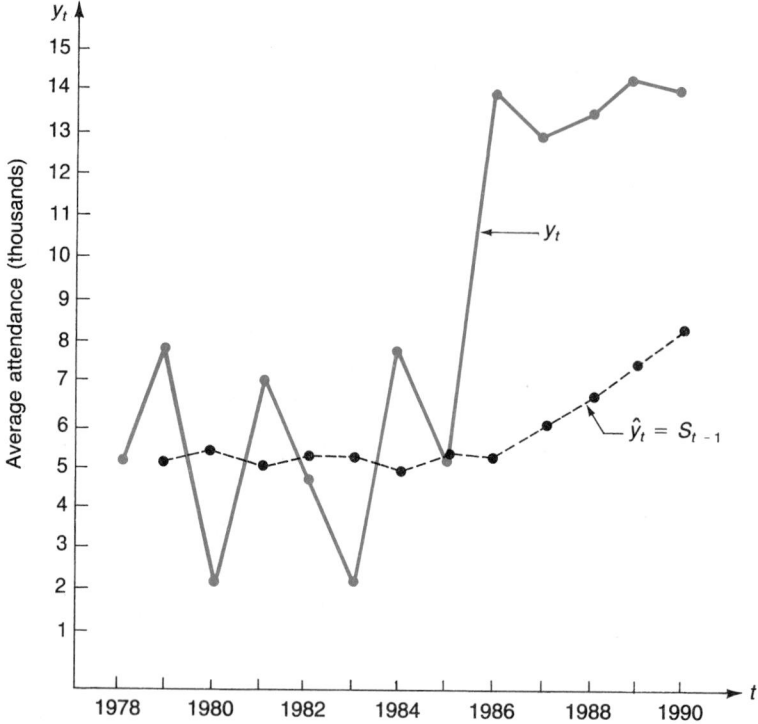

How well does this forecasting procedure perform here? We cannot use Figure 17.7 to compare the $\hat{y}$'s and the y's because, for each time period t, we have plotted y_t and S_t. The predicted value at t, however, is $\hat{y}_t = S_{t-1}$, not S_t. So we need to shift the smoothed values in Figure 17.7 one period to the right. This is shown in Figure 17.8, which contains a plot of the values in Table 17.3 for $A = .1$.

As we might expect from Figure 17.8, the residuals using this method are quite large from 1986 on because this value of A produces smoothed values that fail to adapt to the shift that occurred in 1986. This series was not a good one for simple exponential smoothing because of this sudden shift. However, between the years 1978 and 1985 (a relatively stationary set of observations), the smoothed time series contains much less noise using $A = .1$ and gives a clear indication of the lack of any trend.   ■

Simple exponential smoothing is a popular method of forecasting, particularly when there are hundreds or perhaps thousands of forecasts to be updated for each time period. Such is the case for many inventory-control systems, which are used to predict future demand levels for each item in inventory by means of a computerized forecasting procedure. Simple exponential smoothing often is used for such situations because each forecast, $\hat{y}_{t+1}$, requires only two values: the current observation, y_t, and the previous smoothed value, S_{t-1}. There is no need to store all the previous observations. *Computationally, this procedure is very simple and requires less computer time than do more sophisticated forecasting techniques.*

Exercises

17.11 If the smoothed value for time period $t = 5$ is $S_5 = 10$, find the forecast using simple exponential smoothing for time period $t = 7$ assuming that $y_6 = 14$ and $A = 0.2$.

17.12 The Fitness and Health Center has been increasing its membership over the years and is considering opening a new center. The following table lists the quarterly membership for the past four years.

YEAR	QUARTER 1	QUARTER 2	QUARTER 3	QUARTER 4
1988	105	117	120	115
1989	110	125	130	126
1990	120	135	140	132
1991	131	141	145	141

Using simple exponential smoothing with $A = .3$, find the forecasted membership for the first quarter of 1992.

17.13 The yield on a general obligation bond for Harrisville county fluctuates with the market. The following are the monthly quotations for the past year.

MONTH	YIELD	MONTH	YIELD
Jan	10.17	July	12.45
Feb	10.75	Aug	12.10
Mar	11.03	Sept	11.75
Apr	11.31	Oct	11.60
May	11.57	Nov	11.75
June	12.10	Dec	11.03

Using simple exponential smoothing with $A = .1$, find the forecasted yield for the months of October, November, and December using the data from the previous months. Calculate the residuals.

17.14 Refer to Exercise 17.3. Using simple exponential smoothing with $A = .3$, calculate the forecasted price of the stock of Intersecond Bank for the years 1981 to 1991. Calculate the residuals. Compare the residuals from simple exponential smoothing with those from the naive model.

17.15 The gold reserves of all countries that are members of the International Monetary Fund are given for the years 1973 to 1989.

YEAR	GOLD RESERVE	YEAR	GOLD RESERVE
1973	1,022.24	1982	949.16
1974	1,020.24	1983	947.84
1975	1,018.71	1984	946.79
1976	1,014.23	1985	949.39
1977	1,029.19	1986	949.11
1978	1,036.82	1987	944.49
1979	944.44	1988	944.92
1980	952.99	1989	938.95
1981	953.72		

(Source: The World Almanac and Book of Facts (1991): 119.)

Using simple exponential smoothing with $A = .1$, find the forecasted yield for 1988, 1989, and 1990.

◢ 17.5
Exponential Smoothing for a Time Series Containing Trend

The simple exponential smoothing technique discussed in the previous section always will lag behind a time series that contains a steadily increasing or decreasing trend. A procedure known as *Holt's two-parameter linear exponential smoothing* allows you to estimate separately the smoothed value of the time series as well as the average trend gain at each point in time. The resulting smoothed values track the past time series observations more accurately. We refer to this procedure as **linear exponential smoothing.** There are two equations for this method. The first, for smoothing the observations, is

$$S_t = Ay_t + (1 - A)(S_{t-1} + b_{t-1})$$ (17.4)

The second, for smoothing the trend, is

$$b_t = B(S_t - S_{t-1}) + (1 - B)b_{t-1} \qquad \textbf{(17.5)}$$

where (1) S_t is the smoothed value for time period t, (2) b_t is the smoothed *trend* estimate for this time period, and (3) A and B are smoothing constants between 0 and 1.

Smoothing the Observations (Equation 17.4)

Equation 17.4 is similar to the equation used for simple exponential smoothing, except that S_{t-1} is replaced by $(S_{t-1} + b_{t-1})$ to include the effect of the trend. The smoothing constant for this equation is $0 < A < 1$; typically, $A \le .3$.

Smoothing the Trend (Equation 17.5)

Equation 17.5 is a new addition to the smoothing process and represents the smoothed trend. It uses a separate smoothing constant, B, to smooth the trend values. This constant also is generally less than or equal to .3. This smoothed trend estimate is updated by using a weighted sum of (1) the difference between the last two smoothed values (an estimate of the current "trend") and (2) the previous smoothed trend estimate. Such a procedure significantly reduces any randomness (irregular activity) in the trend values across time.

Forecasting Using Linear Exponential Smoothing

Linear exponential forecasting uses both the smoothed observations and the smoothed trend estimates. The forecast for time period $t + 1$ is the current smoothed value plus the current smoothed trend value.

$$\hat{y}_{t+1} = S_t + b_t \qquad \textbf{(17.6)}$$

We also can use this procedure to forecast any number of time periods into the future, say, m periods. Here,

$$\hat{y}_{t+m} = S_t + mb_t \qquad \textbf{(17.7)}$$

The forecast using equation 17.6 is the **one-step ahead forecast** and the value from equation 17.7 is the **m-step ahead forecast**.

Summarizing the Results

To summarize the necessary calculations for linear exponential smoothing, you can use the format in Table 17.4. The initial year for this time series is 1978. As we did for simple exponential smoothing, we continue to set the first smoothed value, S_1, equal to the first observation, y_1. A new problem arises here: the initial estimate of the trend, b_1. We examine two procedures for estimating this value.

Procedure 1 Let $b_1 = 0$. Provided you have a large number of years in your observed time series, this procedure provides an adequate initial estimate for the

		ACTUAL OBSERVATION	SMOOTHED OBSERVATION	SMOOTHED TREND	FORECAST	RESIDUAL
YEAR	t	(y_t)	(S_t)	(b_t)	$(\hat{y}_t)$	$(y_t - \hat{y}_t)$
1978	1	y_1	$S_1 = y_1$	b_1	—	—
1979	2	y_2	S_2	b_2	$\hat{y}_2$	$y_2 - \hat{y}_2$
1980	3	y_3	S_3	b_3	$\hat{y}_3$	$y_3 - \hat{y}_3$
1981	4	y_4	S_4	b_4	$\hat{y}_4$	$y_4 - \hat{y}_4$
⋮						

■ **TABLE 17.4**
Summary for linear exponential smoothing.

trend. The smoothed trend value soon "catches up" with the actual trend of the series.

Procedure 2   You can obtain a more accurate estimate of b_1 by using the first five (or so) time periods to estimate the initial trend. A least squares line is constructed through these five observations (exactly as discussed in Chapter 16), with the resulting equation $\hat{y}_t = a + bt$. The value of b provides an initial trend estimate.

We demonstrate this technique in Example 17.5, which uses both procedures to obtain the initial trend estimate, b_1.

EXAMPLE 17.5

The time series in the following table contains the city taxes (in thousands of dollars) collected in Jackson City over the past 20 quarters. Using procedures 1 and 2 to calculate an initial trend estimate, obtain the smoothed values, S_t, for each time period. Also determine the predicted values, $\hat{y}_t$, using smoothing constants $A = .1$ and $B = .3$.

YEAR	QUARTER	TAXES COLLECTED	YEAR	QUARTER	TAXES COLLECTED
1986	1	76	1989	1	403
	2	93		2	282
	3	108		3	288
	4	128		4	387
1987	1	196	1990	1	484
	2	175		2	384
	3	141		3	330
	4	236		4	497
1988	1	256			
	2	190			
	3	227			
	4	299			

SOLUTION   A summary of the results using both initial trend estimates is shown in Table 17.5. For procedure 1, $b_1 = 0$. For procedure 2, the first 5 observations were used to obtain an initial trend estimate. The least squares line through these five observations is $\hat{y}_t = 37.7 + 27.5t$, and so we use $b_1 = 27.5$ for procedure 2.

To illustrate the necessary calculations here, consider $t = 10$, using procedure 1.

1. $y_{10} = 190$
2. $S_{10} = .1y_{10} + .9(S_9 + b_9)$
   $= .1(190) + .9(164.21 + 14.78) = 180.09$
3. $b_{10} = .3(S_{10} - S_9) + .7(b_9)$
   $= .3(180.09 - 164.21) + .7(14.78) = 15.11$
4. $\hat{y}_{10} = S_9 + b_9$ (from equation 17.6) $= 164.21 + 14.78 = 178.99$
5. Residual for $t = 10$ is
   $y_{10} - \hat{y}_{10} = 190 - 178.99 = 11.01.$

■ TABLE 17.5
Solution to Example
17.5 using linear
exponential smoothing
($A = .1, B = .3$).

		PROCEDURE 1					PROCEDURE 2		
t	y_t	S_t	b_t	$\hat{y}_t$	$y_t - \hat{y}_t$	S_t	b_t	$\hat{y}_t$	$y_t - \hat{y}_t$
1	76.0	76.00	0.0	—	—	76.00	27.50	—	—
2	93.0	77.70	0.51	76.00	17.00	102.45	27.18	103.50	−10.50
3	108.0	81.19	1.40	78.21	29.79	127.47	26.54	129.63	−21.63
4	128.0	87.13	2.77	82.59	45.41	151.41	25.76	154.01	−26.01
5	196.0	100.51	5.95	89.90	106.10	179.05	26.32	177.16	18.84
6	175.0	113.31	8.01	106.46	68.54	202.33	25.41	205.37	−30.37
7	141.0	123.29	8.60	121.32	19.68	219.07	22.81	227.74	−86.74
8	236.0	142.29	11.72	131.88	104.12	241.29	22.63	241.87	−5.87
9	256.0	164.21	14.78	154.01	101.99	263.13	22.39	263.92	−7.92
10	190.0	180.09	15.11	178.99	11.01	275.97	19.53	285.52	−95.52
11	227.0	198.38	16.06	195.20	31.80	288.65	17.47	295.49	−68.49
12	299.0	222.90	18.60	214.44	84.56	305.41	17.26	306.12	−7.12
13	403.0	257.65	23.44	241.50	161.50	330.70	19.67	322.67	80.33
14	282.0	281.18	23.47	281.09	0.91	343.53	17.62	350.37	−68.37
15	288.0	302.99	22.97	304.66	−16.66	353.83	15.42	361.15	−73.15
16	387.0	332.07	24.80	325.96	61.04	371.03	15.96	369.26	17.74
17	484.0	369.58	28.62	356.87	127.13	396.69	18.87	386.99	97.01
18	384.0	396.78	28.19	398.20	−14.20	412.40	17.92	415.55	−31.55
19	330.0	415.47	25.34	424.97	−94.97	420.29	14.91	430.32	−100.32
20	497.0	446.43	27.03	440.81	56.19	441.38	16.76	435.20	61.80

■ FIGURE 17.9
Predicted values using
linear exponential
smoothing ($A = .1$,
$B = .3$).

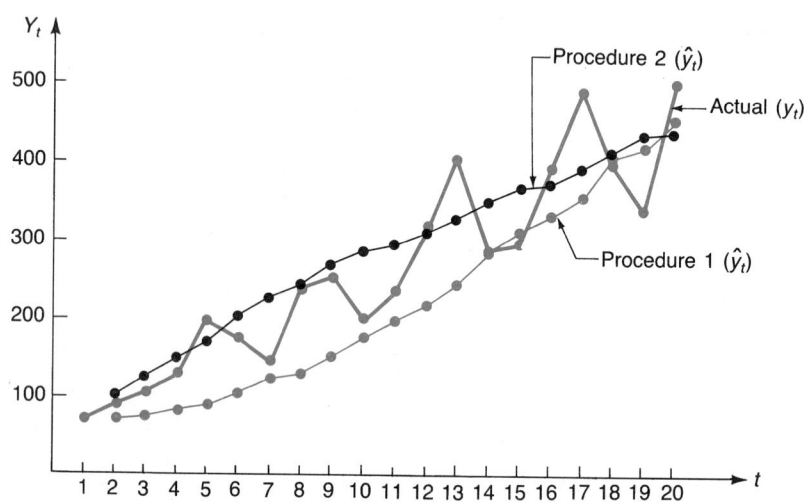

The values of y_t and $\hat{y}_t$ (the predicted value for that time period) are shown in Figure 17.9. For this particular example, procedure 2, which used the first 5 quarters to obtain the initial trend estimate, estimated (and smoothed) the past values more accurately.   ■

 *Exercises*

17.16 Using Holt's two-parameter linear exponential smoothing technique, find the smoothed value at time $t = 9$, using these values with $A = .1$ and $B = .2$.

$$Y_9 = 10.5$$
$$S_8 = 10.0$$
$$S_7 = 9.5$$
$$b_8 = 0.5$$

17.17 For U.S. life insurance companies, the total amounts of assets classified as government securities are listed below for the years 1970 to 1988. Units are in millions of dollars.

YEAR	GOVERNMENT SECURITIES
1970	11,068
1971	11,000
1972	11,372
1973	11,403
1974	11,965
1975	15,117
1976	20,260
1977	23,555
1978	26,552
1979	29,719
1980	33,015
1981	39,502
1982	55,516
1983	76,615
1984	99,769
1985	124,598
1986	144,616
1987	151,436
1988	159,781

(*Source: Moody's Bank and Finance Manual*
(1990): a12.)

a. Forecast the total amount of government securities for the year 1989. Use linear exponential smoothing with an initial estimate of zero for the slope with $A = .1$ and $B = .2$.

b. Rework part a using the least squares estimate of the slope from the first four years.

17.18 The amount of money in money market accounts at First Louisiana State Bank has fluctuated with the interest rate over the years. The bank would like to make a forecast using linear exponential smoothing.

YEAR	AMOUNT IN MONEY MARKET ACCOUNTS	YEAR	AMOUNT IN MONEY MARKET ACCOUNTS
1982	390,121	1987	435,495
1983	395,310	1988	440,370
1984	416,432	1989	444,184
1985	427,489	1990	458,543
1986	443,560	1991	451,967

a. Using zero as an initial estimate of the slope, find the predicted value for the amount of money in money market accounts at the end of 1992. Let $A = .1$ and $B = .2$.

b. Using the least squares estimate from the first three years for the slope, redo part a. What is the predicted value for the amount of money in the money market accounts at the end of 1992?

17.19 Using the data in Exercise 17.3, find the predicted value for the stock price of Intersecond Bank for 1992 using linear exponential smoothing. Let the initial value of the slope be zero, with $A = .09$ and $B = .15$.

17.20 Using the data in Exercise 17.12, find the forecasted membership values for the four quarters of 1992. Use the least squares estimate based on the first five observations for the initial value of the slope. Let $A = .3$ and $B = .2$. Compare these forecasted values with those obtained in Exercise 17.12.

▮ **17.6**

Exponential Smoothing Method for Trend and Seasonality

As we discussed earlier, seasonality is present in a time series whenever certain months or quarters are consistently higher or lower than the yearly average. In such cases, an extension of Holt's method, **Winter's linear and seasonal exponential smoothing,** offers additional flexibility. This three-parameter technique (that is, there are three smoothing constants) not only smooths the past observation and trend estimates (as does linear exponential smoothing) but also provides smoothed seasonality factors for each time period.

The smoothing equations for Winter's method are, for smoothing the observations,

$$S_t = A\left(\frac{y_t}{F_{t-L}}\right) + (1 - A)(S_{t-1} + b_{t-1}) \qquad \textbf{(17.8)}$$

for smoothing the seasonality factors,

$$F_t = B\left(\frac{y_t}{S_t}\right) + (1 - B)F_{t-L} \qquad \textbf{(17.9)}$$

and for smoothing the trend estimates,

$$b_t = C(S_t - S_{t-1}) + (1 - C)b_{t-1} \qquad \textbf{(17.10)}$$

Here, (1) S_t is the smoothed observation for time period t; (2) F_t is the smoothed seasonality factor for this time period; (3) b_t is the smoothed estimate of trend; (4) L is the number of periods per year ($L = 4$ for quarterly data and $L = 12$ for monthly data); and (5) A, B, and C are the three smoothing constants.

Equations 17.8 and 17.10 are similar to the corresponding equations from the linear exponential smoothing procedure, except that S_t now consists of deseasonalized smoothed values. These values are obtained by dividing each observation, y_t, by the smoothed seasonal factor from one year before that observation, F_{t-L}.

Forecasting Using Linear and Seasonal Exponential Smoothing

The procedure for forecasting using Winter's exponential smoothing method is similar to that used for Holt's. Here, the forecast for a particular quarter (month) includes the effect of all three smoothing equations. The forecast for m periods ahead is

$$\hat{y}_{t+m} = (S_t + mb_t) \cdot F_{t+m-L} \qquad \textbf{(17.11)}$$

The term $(S_t + mb_t)$ represents the smoothed *deseasonalized* estimate and includes the smoothed trend effect. The seasonality is included in the final estimate by multiplying by the smoothed seasonality factor for the quarter (or month) one year previous to the forecast time period, namely, F_{t+m-L}. This procedure is much like that used in Section 17.3, where the deseasonalized estimate was multiplied by the corresponding seasonal index to arrive at the final forecast.

When using this procedure on the past observations, you would, for example, determine $\hat{y}_{10}$ by assuming observations $y_1, y_2, \ldots, y_9$ are available. You would do a one-step ahead forecast ($m = 1$) using the smoothed seasonal value from the previous year; that is, $F_{10-4} = F_6$, assuming quarterly data. As a result,

$$\hat{y}_{10} = [S_9 + (1)b_9] \cdot F_6$$

Similarly,

$$\hat{y}_{11} = [S_{10} + (1)b_{10}] \cdot F_7$$
$$\hat{y}_{12} = [S_{11} + (1)b_{11}] \cdot F_8$$

and so on.

Forecasting *beyond* the range of your observations (extrapolating) is illustrated in the next section.

When dealing with quarterly data, your first set of predicted values will be

$$\hat{y}_5 = [S_4 + (1)b_4] \cdot F_1$$
$$\hat{y}_6 = [S_5 + (1)b_5] \cdot F_2$$
$$\hat{y}_7 = [S_6 + (1)b_6] \cdot F_3$$

and so on. If the time series consists of monthly observations ($L = 12$), then you begin your predicted values with

$$\hat{y}_{13} = [S_{12} + (1)b_{12}] \cdot F_1$$
$$\hat{y}_{14} = [S_{13} + (1)b_{13}] \cdot F_2$$
$$\hat{y}_{15} = [S_{14} + (1)b_{14}] \cdot F_3$$

and so on.

Summarizing the Results

A method of summarizing the necessary calculations is shown in Table 17.6. Suppose that the original year of the observed time series is 1986, with quarterly observations. Initial estimates must be supplied for (1) the seasonal factors for each quarter of 1985, (2) the trend estimate for quarter 4, 1985, and (3) the smoothed value corresponding to quarter 4, 1985.

Once again, we will examine two procedures for this situation—one is quick and easy, and the other is more accurate but requires additional calculations. Procedure 1 is used in Table 17.6. Both procedures are demonstrated in Example 17.6. These are not the only procedures—can you think of one or two others?

Procedure 1

1. Set the initial seasonal factors equal to 1.
2. Set the initial trend estimate (b_0) equal to 0.
3. Set the initial smoothed value for quarter 4, 1985 (S_0), equal to the actual value for quarter 4, 1986 (y_4). This value is also the *forecast value* ($\hat{y}_t$) for each of the four quarters in 1986.

■ **TABLE 17.6** Summary of linear and seasonal exponential smoothing (using procedure 1).

YEAR	QTR.	t	ACTUAL OBSERVATIONS (y_t)	SMOOTHED OBSERVATIONS (S_t)	SMOOTHED SEASONAL FACTORS (F_t)	SMOOTHED TREND (b_t)	FORECAST $(\hat{y}_t)$	RESIDUAL $(y_t - \hat{y}_t)$
1985	1				(1) 1.0			
(year 0)	2				1.0			
	3			(3)	1.0	(2)		
	4			$S_0 = y_4$	1.0	$b_0 = 0$		
1986	1	1	y_1	S_1	F_1	b_1	$\hat{y}_1 = S_0$	$y_1 - \hat{y}_1$
(year 1)	2	2	y_2	S_2	F_2	b_2	$\hat{y}_2 = S_0$	$y_2 - \hat{y}_2$
	3	3	y_3	S_3	F_3	b_3	$\hat{y}_3 = S_0$	$y_3 - \hat{y}_3$
	4	4	y_4	S_4	F_4	b_4	$\hat{y}_4 = S_0$	$y_4 - \hat{y}_4$
1987	1	5	y_5	S_5	F_5	b_5	$\hat{y}_5$	$y_5 - \hat{y}_5$
(year 2)	2	6	y_6	S_6	F_6	b_6	$\hat{y}_6$	$y_6 - \hat{y}_6$

Procedure 2

1. Use the first two years of data to determine the seasonal indexes. These are the four values for F_t in 1985. Actually, any number of years of data can be used here.
2. Deseasonalize the data for the first two years (or any number of years), and calculate the least squares line through these deseasonalized values, d_t. Call this line $\hat{d}_t = a + bt$. The initial trend estimate (b_0) is b.
3. The initial smoothed value for quarter 4, 1985, is $S_0 = [a + b(0)] \cdot$ (seasonal index for quarter 4 in step 1) $= a \cdot$ (seasonal index), where a is the intercept of the least squares line in step 2. Also, S_0 is the *forecast value* ($\hat{y}_t$) for each of the 4 quarters in 1986.

EXAMPLE 17.6

The quarterly taxes from Jackson City in Example 17.5 indicated significant seasonality. In particular, the first-quarter taxes appeared to be considerably larger than those for the yearly average. Using the linear and seasonal exponential smoothing procedures, determine the smoothed value, S_t, and predicted value, $\hat{y}_t$, for each time period. Use smoothing constants $A = .1$, $B = .3$, and $C = .2$.

SOLUTION

The computed results using procedure 1 are summarized in Table 17.7, where $b_0 = 0$, $S_0 = y_4$, and the initial seasonal factors are each 1.

    The procedure 2 solution is contained in Table 17.8. The first two years were used to obtain the initial seasonal factors by finding the four seasonal indexes as described in Chapter 16. These are

$$\text{quarter } 1 = 1.23 \quad\quad \text{quarter } 3 = .91$$
$$\text{quarter } 2 = .98 \quad\quad \text{quarter } 4 = .88$$

    Next, the data from the first two years were deseasonalized by dividing by the corresponding seasonal index to obtain the deseasonalized values, d_t. A least squares line through these eight values using the simple linear regression procedure from Chapter 16 produced:

$$\hat{d}_t = 44.03 + 23.02t$$

■ **TABLE 17.7**
Solution using
linear and seasonal
exponential smoothing,
procedure 1 ($A = .1$,
$B = .3$, $C = .2$).

	t	y_t	S_t	F_t	b_t	$\hat{y}_t$	$y_t - \hat{y}_t$
				1.0			
				1.0			
				1.0			
			128	1.0	0.0		
1986	1	76.0	122.80	0.89	−1.04	128.00	−52.00
	2	93.0	118.88	0.93	−1.62	128.00	−35.00
	3	108.0	116.34	0.98	−1.80	128.00	−20.00
	4	128.0	115.89	1.03	−1.53	128.00	0.0
1987	5	196.0	125.05	1.09	0.61	101.28	94.72
	6	175.0	131.81	1.05	1.84	117.45	57.55
	7	141.0	134.70	1.00	2.05	130.78	10.22
	8	236.0	145.95	1.21	3.89	141.03	94.97
1988	9	256.0	158.34	1.25	5.59	163.36	92.64
	10	190.0	165.59	1.08	5.92	172.55	17.45
	11	227.0	177.08	1.08	7.04	171.33	55.67
	12	299.0	190.48	1.32	8.31	222.23	76.77
1989	13	403.0	211.19	1.45	10.79	248.11	154.89
	14	282.0	225.87	1.13	11.57	239.97	42.03
	15	288.0	240.26	1.12	12.13	257.35	30.65
	16	387.0	256.57	1.37	12.97	332.12	54.88
1990	17	484.0	276.05	1.54	14.27	389.79	94.21
	18	384.0	295.23	1.18	15.25	328.43	55.57
	19	330.0	308.94	1.10	14.94	347.21	−17.21
	20	497.0	327.68	1.42	15.70	444.89	52.11

■ TABLE 17.8
Solution using linear and seasonal exponential smoothing, procedure 2 ($A = .1$, $B = .3$, $C = .2$).

	t	y_t	S_t	F_t	b_t	$\hat{y}_t$	$y_t - \hat{y}_t$
				1.23			
				.98			
				.91			
			39	.88	23		
1986	1	76.0	61.98	1.23	23.00	39.00	37.00
	2	93.0	85.97	1.01	23.19	39.00	54.00
	3	108.0	110.11	0.93	23.38	39.00	69.00
	4	128.0	134.69	0.90	23.62	39.00	89.00
1987	5	196.0	158.44	1.23	23.65	194.55	1.45
	6	175.0	181.19	1.00	23.47	184.00	−9.00
	7	141.0	199.34	0.86	22.40	190.59	−49.59
	8	236.0	225.76	0.94	23.21	199.81	36.19
1988	9	256.0	244.86	1.18	22.39	306.56	−50.56
	10	190.0	259.57	0.92	20.85	266.48	−76.48
	11	227.0	278.65	0.85	20.50	242.31	−15.31
	12	299.0	300.90	0.96	20.85	282.51	16.49
1989	13	403.0	323.85	1.20	21.27	378.24	24.76
	14	282.0	341.34	0.89	20.51	316.67	−34.67
	15	288.0	359.58	0.83	20.06	307.30	−19.30
	16	387.0	382.02	0.98	20.53	364.14	22.86
1990	17	484.0	402.76	1.20	20.58	481.55	2.45
	18	384.0	424.14	0.89	20.74	376.83	7.17
	19	330.0	439.92	0.81	19.75	371.36	−41.36
	20	497.0	464.65	1.00	20.74	448.32	48.68

The value of 23.02 (rounded to 23) became the initial slope estimate, b_0. Finally, the initial smoothed value for quarter 4, 1985, is

$$S_0 = (44.03)(\text{initial seasonal index for quarter 4})$$

$$= (44.03)(.88) = 38.7 \text{ (rounded to 39)}$$

Also, $S_0 = 39$ becomes the forecast value ($\hat{y}_t$) for each of the quarters in 1986.

The calculations required here can be illustrated using Table 17.7 and $t = 10$ for procedure 1.

1. $y_{10} = 190$

2. $S_{10} = .1\left(\dfrac{y_{10}}{F_{10-4}}\right) + .9(S_9 + b_9)$

$$= .1\left(\dfrac{y_{10}}{F_6}\right) + .9(S_9 + b_9)$$

$$= .1\left(\dfrac{190}{1.05}\right) + .9(158.34 + 5.59)$$

$$= 165.59$$

3. $F_{10} = .3\left(\dfrac{y_{10}}{S_{10}}\right) + .7F_6$

$$= .3\left(\dfrac{190}{165.59}\right) + .7(1.05)$$

$$= 1.08$$

4. $b_{10} = .2(S_{10} - S_9) + .8b_9$

$$= .2(165.59 - 158.34) + .8(5.59)$$

$$= 5.92$$

5. $\hat{y}_{10} = [(S_9 + (1)(b_9)]F_6$ (from equation 17.11)

$$= (158.34 + 5.59)1.05 \text{ (computer-stored value is 1.0526)}$$

$$= 172.55$$

■ **FIGURE 17.10**
Forecasted values using linear and seasonal exponential smoothing ($A = .1$, $B = .3$, $C = .2$).

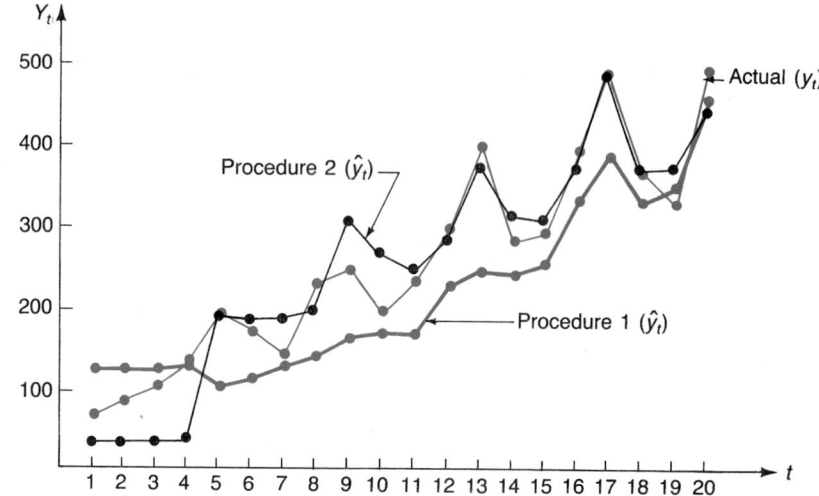

6. Residual for $t = 10$ is
$$y_{10} - \hat{y}_{10} = 190 - 172.55$$
$$= 17.45$$

A graphical illustration of the actual observations, y_t, and the predicted value for each time period, $\hat{y}_t$, is shown in Figure 17.10. Once again, the more complex procedure 2 performed better than did procedure 1; for the last ten quarters, procedure 2 tracked the actual time series extremely well.   ■

Exercises

17.21   Using Winter's linear and seasonal smoothing technique, answer the following questions using these values:

$$
\begin{array}{lll}
b_6 = 0.7 & F_8 = 0.9 & B = 0.1 \\
S_7 = 15.7 & F_4 = 0.75 & C = 0.1 \\
S_6 = 14.3 & Y_8 = 15.0 & L = 4 \\
S_5 = 14.0 & A = 0.2 &
\end{array}
$$

a.   Find the smoothed observation for time $t = 8$.

b.   Find the forecasted value for time period 8.

17.22   Ektronics manufactures electronic testing and measuring instruments. The company has managed to capture a large share of the market over the past four years. Sales of its equipment (in ten thousands) is recorded monthly for 1988 to 1991.

YEAR	JAN	FEB	MAR	APR	MAY	JUNE	JULY	AUG	SEPT	OCT	NOV	DEC
1988	1.0	1.1	1.2	1.7	1.9	2.3	2.7	3.1	2.5	2.3	2.0	1.9
1989	1.7	1.4	1.5	1.7	2.4	2.7	3.3	3.9	3.4	3.0	2.6	2.0
1990	1.9	2.0	2.1	2.3	3.1	3.5	4.1	4.7	4.3	3.4	2.9	2.8
1991	2.9	2.8	2.7	3.4	3.7	4.1	4.6	5.0	4.7	4.0	3.7	3.6

Use Winter's linear and seasonal smoothing technique to find the smoothed values for the first four months of 1991. Let $A = .2$, $B = .1$, and $C = .1$. Using procedure 1, set the initial estimates of the seasonal factors to 1.0 and let $b_0 = 0$.

17.23   The earnings per share of Mecta Mining, a large producer of silver, are:

YEAR	QUARTER 1	QUARTER 2	QUARTER 3	QUARTER 4
1988	.25	.20	.27	.30
1989	.26	.24	.34	.37
1990	.30	.27	.38	.45
1991	.36	.32	.47	.50

Using the linear and seasonal exponential smoothing procedure, determine the predicted value for each quarter of 1992. Use procedure 2 and the first two years of data to obtain b_0, S_0, and the four initial seasonal factors (F). Let $A = .3$, $B = .2$, and $C = .1$.

17.24   The average prime rates in the United States are given for each quarter from 1983 to 1988.

YEAR	QUARTER 1	QUARTER 2	QUARTER 3	QUARTER 4
1983	10.88	10.50	10.80	11.00
1984	11.07	12.30	12.99	11.80
1985	10.54	10.20	9.50	9.50
1986	9.37	8.61	7.85	7.50
1987	7.50	8.05	8.40	8.87
1988	8.59	8.78	9.71	10.18

(Source: Moody's Bank and Finance Manual (1990) 2: a11.)

Use Winter's linear and seasonal smoothing technique to find the smoothed values for the four quarters of 1988. Let $A = .2$, $B = .1$, and $C = .1$. Using procedure 1, set the initial estimates of the seasonal factors to 1.0 and let $b_0 = 0$.

17.25   Refer to Exercise 17.8. Find the predicted amount of inventory for the four quarters of 1992, using the linear and seasonal exponential smoothing procedure. Use procedure 2 and the first two years of data to obtain b_0, S_0, and the four initial seasonal factors (F). Let $A = .05$, $B = .1$, and $C = .1$.

17.26   Refer to Exercise 17.12. Find the forecasted memberships for the four quarters of 1992 using procedure 1 of the linear and seasonal exponential smoothing techniques. Compare these values to those obtained in Exercise 17.12. Let $A = .05$, $B = .1$, and $C = .1$.

◢ 17.7
Choosing the Appropriate Forecasting Procedure

Our purpose in showing you several different forecasting techniques is to point out that, unfortunately, no one procedure works well all the time. One method may work well on a particular steadily increasing time series that has little random fluctuation but perform poorly on a series that has considerable seasonality or random fluctuation.

As you gain more experience in time series applications, you will be better able to choose an appropriate forecasting technique. One factor to consider is the length of your forecast. We classify the forecast period as

Short-term forecast: one to three months

Medium-range forecast: greater than three months but less than two years

Long-range forecast: two years or more

The exponential smoothing procedures are excellent for *short-term* forecasts, whereas the component decomposition method (in Section 17.3) is useful in medium- and long-range forecasting. The latter also is a popular procedure for many short-term applications, including inventory control and production planning.

One method of deciding whether a certain forecast technique is appropriate in a particular situation is to determine how well the procedure "fits" the observed time series. You accomplish this by pretending that, in each time period, the next observation is unknown and letting the forecasting procedure "predict" the next value (the $\hat{y}_t$ values in the previous examples). Next, you compare the predicted $(\hat{y}_t)$ values with the observed values (y_t).

The three most popular methods of comparing the predicted and observed values use measures involving the residuals. These measures are the mean absolute deviation, the predictive mean squared error, and the mean absolute percentage error. The **mean absolute deviation (MAD)** is the average of the absolute values of each residual. Let

$$e_t = \text{residual at time } t$$

$$= y_t - \hat{y}_t$$

The mean absolute deviation is defined as

$$\text{MAD} = \frac{\sum |e_t|}{n} \qquad (17.12)$$

where n is the number of *predicted* values obtained from the past data. For example, when using linear exponential smoothing on 20 data values, you obtain 19 predicted values because $\hat{y}_1$ is unavailable, so n is 19 and not 20.

The **predictive mean squared error (MSE)** is similar to the MAD, except we find the average of the *squared* residuals.

$$(\text{predictive}) \text{ MSE} = \frac{\sum e_t^2}{n} \qquad (17.13)$$

where, again, n is the number of predicted values.*

The **mean absolute percentage error (MAPE)** considers the *relative* error of each forecast. The relative error at time period t is defined as e_t/y_t. The mean absolute percentage error is defined to be

$$\text{MAPE} = \frac{\sum \left| \frac{e_t}{y_t} \right|}{n} \qquad (17.14)$$

where n is the number of predicted values.

If, during a particular time period, the actual value is $y_t = 50$ and the forecast value is $\hat{y}_t = 60$, the absolute percentage error is

$$\left| \frac{50 - 60}{50} \right| = .2$$

So, the error at this time period is 20% of the actual value. Consequently, for a particular time series, the MAPE is the sum of the absolute percentage error for each predicted value divided by the number of predicted values.

The MSE severely penalizes large residuals because it *squares* each value. Consequently, you use the MSE for situations in which you prefer several small residuals to one large value and wish to be warned if there is one larger residual. The primary advantage of using the MAPE is that it can be used to compare the predictive ability of a certain forecasting technique on two different time series. By using relative error, rather than actual error, the effect of the magnitude of the time series observations has been removed from the predictive measure.

To illustrate these measures, consider Table 17.9. For forecasting method 1, there are no large residuals, whereas method 2 results in one large residual. So the MSE is smaller for method 1, but the MAD is smaller for method 2. When using any of these measures, the *smaller* the value, the *more accurate* your forecast procedure.

---

*The MSE that we compute as a measure of how well a forecasting procedure fits the observed data is not the same as the MSE computed in a normal ANOVA table. The ANOVA MSE is equal to SSE/(degrees of freedom for residual). In contrast, the predictive MSE is not used in any test of hypothesis and is merely the average of the squared deviations.

■ TABLE 17.9

Comparison of the mean absolute deviation (MAD), the mean squared error (MSE), and the mean absolute percentage error (MAPE).

FORECAST	y_t	$\hat{y}_t$	$e_t = y_t - \hat{y}_t$	$|e_t|$	e_t^2	$|e_t/y_t|$
Method 1	36	32	4	4	16	.111
	42	46	−4	4	16	.095
	45	49	−4	4	16	.089
				12	48	.295

MAD = 12/3 = 4.0
MSE = 48/3 = 16.0
MAPE = .295/3 = .098

FORECAST	y_t	$\hat{y}_t$	$e_t = y_t - \hat{y}_t$	$|e_t|$	e_t^2	$|e_t/y_t|$
Method 2	36	34	2	2	4	.056
	42	40	2	2	4	.048
	45	52	−7	7	49	.156
				11	57	.260

MAD = 11/3 = 3.67
MSE = 57/3 = 19.0
MAPE = .260/3 = .087

There is no consensus among statisticians as to which measure is preferable. Instead, it depends on the results of having large forecast residuals. If a large error is disastrous (such as in predicting the inventory level of an expensive product), then using the MSE is preferable. On the other hand, if you can afford to overlook a single severe miss provided the general tracking is close, then the MAD serves better. When comparing the predictive accuracy of two different time series, the MAPE is the appropriate measure.

EXAMPLE 17.7

We used two types of exponential smoothing to smooth (and predict) the city taxes collected in Jackson City over the past five years. Data from the past 20 quarters are contained in the table in Example 17.5, in which we used linear exponential smoothing (with smoothing constants $A = .1$ and $B = .3$) to reduce randomness within the observations and trend values. The results are summarized in Table 17.5, using the two procedures for providing initial estimates.

Example 17.6 examined the same data using linear and seasonal exponential smoothing, with smoothing constants $A = .1$, $B = .3$, and $C = .2$. A much better fit was obtained using the more sophisticated method of providing initial smoothed estimates (procedure 2). These results are summarized in Tables 17.7 and 17.8 and are presented graphically in Figure 17.10.

Determine the predictive MSE for each of these four methods. Using the appropriate procedure, determine the forecasted tax revenue for each quarter of 1991.

SOLUTION

1. *Linear exponential smoothing (procedure 1).* The residuals from this forecasting procedure are contained in Table 17.5. The computed predictive mean squared error is

$$\text{MSE} = \frac{(17.00)^2 + (29.79)^2 + \cdots + (56.19)^2}{19} = 5670.11$$

2. *Linear exponential smoothing (procedure 2).* Based on Figure 17.9, we would expect a much smaller predictive MSE here. There are no surprises, because

$$\text{MSE} = \frac{(-10.50)^2 + (-21.63)^2 + \cdots + (61.80)^2}{19} = 3426.46$$

3. *Linear and seasonal exponential smoothing (procedure 1).* These residuals are listed in Table 17.7, with a corresponding predictive mean squared error of

$$\text{MSE} = \frac{(-52.00)^2 + (-35.00)^2 + \cdots + (52.11)^2}{20} = 4414.95$$

(Note that we divide by 20 here because 20 predicted values are available using this procedure.)

A warning: It is not valid to conclude, based on the large MSE value, that this forecasting method is less appropriate than linear exponential smoothing. Remember that we are at the mercy of the particular values of the smoothing constants, A, B, and C. Perhaps a different set of constants would have resulted in a significantly smaller MSE. Finding the best set of constants for any one application involves finding the set of values for A, B, and C that *minimize* the resulting predictive MSE. This (not insignificant) computational burden is one of the drawbacks of using Holt's and Winter's exponential smoothing techniques.

4. *Linear and seasonal exponential smoothing (procedure 2).* In Figure 17.10, we observed excellent agreement between the actual time series, y_t, and the predicted series, $\hat{y}_t$, using the smoothed estimates. A very small predictive MSE value would be expected here, and such is the case:

$$MSE = \frac{(37.00)^2 + (54.00)^2 + \cdots + (48.68)^2}{20} = 1828.89$$

We conclude that the best choice of these four alternatives is the linear and seasonal exponential smoothing method using procedure 2 to derive the original estimates.

5. *Forecasted tax revenue.* Using equation 17.11 and the results in Table 17.8, the forecasts for 1991 would be as follows. For the first quarter (one step ahead): $t = 20$, $L = 4$, $m = 1$, and

$$\hat{y}_{21} = [S_{20} + (1)b_{20}] \cdot F_{17}$$
$$= [464.65 + (1)(20.74)](1.20) = 582$$

For the second quarter (two steps ahead): $t = 20$, $L = 4$, $m = 2$, and

$$\hat{y}_{22} = [S_{20} + (2)b_{20}] \cdot F_{18}$$
$$= [464.65 + (2)(20.74)] \cdot (.89) = 450$$

For the third quarter (three steps ahead): $t = 20$, $L = 4$, $m = 3$, and

$$\hat{y}_{23} = [S_{20} + (3)b_{20}] \cdot F_{19}$$
$$= [464.65 + (3)(20.74)](.81) = 427$$

For the fourth quarter (four steps ahead): $t = 20$, $L = 4$, $m = 4$, and

$$\hat{y}_{24} = [S_{20} + (4)b_{20}] \cdot F_{20}$$
$$= [464.65 + (4)(20.74)] \cdot (1.00) = 548 \qquad \blacksquare$$

Selecting the Smoothing Constants

As mentioned earlier, the computed MSE (or MAD) value for any exponential smoothing procedure is determined not only by the procedure itself but also by the value of the necessary smoothing constants. In Example 17.6, the smoothing constants were $A = .1$, $B = .3$, and $C = .2$, with a corresponding MSE value of 1828.89, using procedure 2. By changing these constants, you might improve the fit (lower the MSE), or you might obtain a less desirable solution (a larger MSE).

To illustrate this point, using Example 17.6 and procedure 2, for

$$A = .1, B = .4, C = .3: \quad MSE = 1748.10 \text{ (an improvement)}$$
$$A = .2, B = .2, C = .2: \quad MSE = 2127.67$$

To arrive at the smallest possible predictive MSE, you must examine a variety of values, compute the MSE for each combination, and select the set of values that provides the smallest MSE. For example, if you consider all nonzero values of A, B, and C between 0 and .4, in increments of .05, you will need $(.4/.05)^3 = 8^3 = 512$ different passes through the procedure to determine the corresponding 512 MSE values. The set of A, B, and C values that provides the smallest MSE is the one you should use in forecasting future values of the time series.

This procedure is not extremely difficult to perform with the help of a computer, but it takes away one main advantage of exponential smoothing—namely, the computational simplicity of this procedure in calculating and updating smoothed estimates. If you are using this method to perform a small number of forecasts, then this poses no problem. On the other hand, if the technique is being used to forecast future demand levels continuously for thousands of inventory items, then this added complexity is a cause for concern. You will have to consider complexity versus cost on an individual application basis.

As a final note here, you should know that computer packages exist for determining the optimal values of the smoothing constants for a given time series.* Such procedures fall under the heading of **automated forecasting procedures,** whereby the computer finds the best fit (using a specified measure of fit, such as MAPE, MSE, or MAD) for a particular forecasting model. By utilizing such procedures, you can drastically simplify the calculations necessary to fit a model to your time series, but, on the negative side, such a "black-box" procedure implies that you sacrifice some control and knowledge of the fitting process.

We can increase the computational burden (but also improve the accuracy) even more by using different values of the smoothing constant(s) at different times in the analysis of a time series. Such techniques are computer controlled. The constant(s) are changed automatically to adapt the process to shifts in the structure of the time series, using **adaptive control procedures.**

We showed you several forecasting procedures and methods for comparing the predictive accuracy of these techniques. Our purpose is to give you an arsenal of methodologies that will allow you to apply each procedure to a particular time series and then summarize and compare the resulting residuals. In this way, you can determine the most accurate procedure for a particular time series and use this method to arrive at a forecast.

Next we turn to another forecasting model, the autoregressive model. With this procedure, we again use the past observations to predict future values but in a slightly different way: we use the past values as variables in a regression equation.

Exercises

17.27 Consider the following forecasts for the yearly sales of Dentroff Wholesale Plumbing Supplies. Sales are given in units of 100,000.

YEAR	ACTUAL	FORECAST	YEAR	ACTUAL	FORECAST
1982	1.1	—	1987	2.1	2.0
1983	1.2	1.0	1988	2.0	2.2
1984	1.5	1.3	1989	2.5	2.3
1985	1.9	1.6	1990	2.7	2.5
1986	2.3	1.8	1991	3.4	2.9

---

*Two forecasting packages available for the microcomputer are FORECAST MASTER (Scientific Systems, Inc., Cambridge, Mass.) and SIBYL-RUNNER (Applied Decision Systems, Lexington, Mass.).

a. Compute the MAD, MAPE, and predictive MSE for the forecasts.

b. Compute the MAD, MAPE, and predictive MSE using the naive forecasts of sales.

c. Compare the forecasts in parts a and b.

17.28   The advertising expenditures for a local supermarket (in thousands of dollars) are as follows:

YEAR	JAN	FEB	MAR	APR	MAY	JUNE	JULY	AUG	SEPT	OCT	NOV	DEC
1988	.3	.4	.4	.5	.5	.6	.7	.7	.8	.7	.7	.6
1989	.5	.6	.6	.7	.6	.8	.9	1.0	1.2	1.1	1.1	1.0
1990	.7	.8	.9	.9	.9	.8	.9	1.0	1.1	1.4	1.3	1.2
1991	.9	1.0	1.0	1.2	1.1	1.3	1.4	1.6	1.6	1.4	1.5	1.3

a. Using the naive model, and omitting the first time period, obtain a forecast for the next 47 time periods. Find the MAD, MAPE, and the predictive MSE.

b. Obtain a forecast for these 47 time periods by using a least squares line that represents just trend. Find the MAD, MAPE, and the predictive MSE.

c. Obtain the forecast for these 47 time periods using only the trend and seasonal components to forecast. Find the MAD, MAPE, and the predictive MSE.

d. Compare the forecasts obtained in parts a, b, and c.

17.29   Two forecasting procedures produce the following sets of forecast errors:

YEAR	MONTH	PROCEDURE 1	PROCEDURE 2	YEAR	MONTH	PROCEDURE 1	PROCEDURE 2
1990	Jan	+5	+1		Nov	+6	+2
	Feb	+7	−2		Dec	+1	+1
	Mar	+6	+3	1991	Jan	−3	−3
	Apr	+2	−1		Feb	−5	+1
	May	−1	+2		Mar	−4	0
	June	−2	0		Apr	+3	−2
	July	−3	+1		May	+2	−19
	Aug	+2	+1		June	−1	−20
	Sept	+4	0				
	Oct	+7	−1				

Compute the MAD and predictive MSE for each forecasting procedure. Comment on the adequacy of the forecasting procedure.

17.30   The following data represent the number of single-family housing starts in a certain sector of the state of California. The units are in 10,000s.

YEAR	QUARTER 1	QUARTER 2	QUARTER 3	QUARTER 4
1988	.6	.8	1.4	.8
1989	.9	1.1	1.7	1.3
1990	1.2	1.4	2.1	1.6
1991	1.4	1.7	2.6	1.9

a. Using the simple exponential procedure with $A = .3$, find the predicted number of housing starts for each time period, omitting the first time period. Find the predictive MSE and MAD.

b. Use Holt's two-parameter linear exponential smoothing technique to obtain a forecast for each time period, omitting the first time period. Let $A = .3$ and $B = .2$. Use the least squares estimate of the slope from the first five periods for the initial value of the slope. Find the predictive MSE and MAD.

c. Compare the forecasts obtained in parts a and b.

17.31   An investor who invested $10,000 into the T. Krow long-term-growth mutual fund would have realized a gain of 93% after five years. The following table shows the value of the $10,000 investment over this period of time.

YEAR	QUARTER 1	QUARTER 2	QUARTER 3	QUARTER 4
1987	10,031	9,638	12,591	12,480
1988	12,691	11,745	13,721	13,980
1989	13,043	12,680	15,376	15,860
1990	14,932	14,280	17,035	17,210
1991	16,830	15,923	18,671	19,300

a. Use Winter's linear and seasonal smoothing technique to find the forecasted value of the original \$10,000 invested for each of the quarters of 1989, 1990, and 1991. Let $A = .2$, $B = .1$, and $C = .1$. Using procedure 1 and Winter's technique, set the initial estimates of the seasonal factors to 1.0 and $b_0 = 0$. Find the predictive MSE.

b. Redo question a with $A = .1$, $B = .2$, and $C = .2$. Find the predictive MSE.

c. Compare the forecasts from parts a and b using the predictive MSEs.

17.32 Explain how the MAD, MAPE, and the predictive MSE differ in what they measure. Why should the sum of the forecast errors divided by the number of forecasts not be used to compare two forecasting procedures?

17.33 Find the predictive MSE and MAPE for the forecasts of the amount of money in the money-market accounts at First Louisiana State Bank in Exercise 17.18b for the years 1983 to 1991 using the linear exponential smoothing technique specified in that exercise.

17.34 Find the MAPE and MAD for the forecast of Ektronics' monthly sales (Exercise 17.22) for the months of 1990 and 1991 using the linear and seasonal technique specified in that exercise.

◢ 17.8
Autoregressive Forecasting Techniques

So far, the forecasting procedures have used either a member of the exponential smoothing family or the method of time series decomposition. The exponential smoothing technique greatly reduces the randomness (irregular activity) within the observed time series, as well as smoothing any existing trend or seasonal effects.

For the case of simple exponential smoothing, the forecast for the next time period ($\hat{y}_{t+1}$) is the smoothed value for the current period (S_t). When you use the other exponential smoothing procedures, your forecast includes the effect of the smoothed seasonality or trend.

The time series decomposition method determines the various components in each observation, including seasonality, trend, cycles, and random activity. Forecasts are derived by extending the trend and seasonal components into the future. Unlike exponential smoothing, this method can provide reliable long-range forecasts. Naturally, the longer this forecast period is, the less reliable your forecasted value becomes.

This section examines yet another method of forecasting, a method that can be used when the time series variable is related to past values of itself. By regressing y_t on some combination of its past values, we are able to derive a forecasting equation. So we return to multiple linear regression, except now the dependent variable is y_t, and the predictor variables are the past values, y_{t-1}, y_{t-2}, ... This forecasting technique is **autoregression;** we are essentially regressing the time series variable on itself.

We can expect the autoregressive forecast technique to perform reasonably well for a time series that (1) is not extremely volatile and does not contain extreme amounts of random movement and (2) requires a short-term or medium-range forecast (that is, less than two years). The fact that the autoregressive procedure does not perform well on a time series containing a great deal of irregular activity is not a serious disadvantage; practically all forecasting techniques perform poorly in this situation.

Suppose we attempt to predict the values of y_t using the previous two observations. The prediction equation is

$$\hat{y}_t = b_0 + b_1 y_{t-1} + b_2 y_{t-2} \qquad (17.15)$$

The values of b_0, b_1, and b_2 are the least squares regression estimates, obtained from any multiple linear regression computer package. There are two predictor variables here: the **lagged variables,** y_{t-1} and y_{t-2}. Equation 17.15 is a **second-order** autoregressive equation because it uses the first two lagged terms. In general, a pth-order autoregressive equation is written

$$\hat{y}_t = b_0 + b_1 y_{t-1} + b_2 y_{t-2} + \cdots + b_p y_{t-p} \qquad (17.16)$$

We illustrate the computer-input procedure for the second-order equation with an example. Earlier, we used the naive forecasting procedure (forecast for tomorrow is the observed value for today) to predict the closing price of Keller Toy Company stock. The closing prices for a 12-week period are shown on page 749, and the predicted values are summarized in Table 17.1.

Suppose we use the second-order autoregressive equation (17.15) to predict these values. The input data required by the linear regression routine consist of the actual time series data and the two columns of lagged data, as illustrated in Table 17.10. The ten input rows are below the horizontal line. Notice that we lose the first two observations due to the missing values for the lagged variables. If these data are available from the two weeks prior to week 1, they can be used to fill in the missing values, providing 12 rows of data.

A computer solution to this problem using MINITAB is in Figure 17.11. The prediction equation is

$$\hat{y}_t = 45.5 + .278 y_{t-1} - .004 y_{t-2}$$

Also, $R^2 = .068$, indicating that the two lagged variables account for only 7% of the total variation in the ten time series values used as input (y_3 through y_{12}).

To determine whether this is the best way to forecast a particular time series, we can use the procedure discussed in the previous section. This involves calculating an MSE (or MAPE or MAD), using the autoregressive technique on the past observations and comparing this MSE with the MSE using other forecasting methods. For example, we obtain an improvement over the naive forecasting procedure here because, from Figure 17.11,

$$\text{(predictive) MSE} = \frac{10.926}{10} = 1.09 \text{ (for autoregressive forecaster)}$$

■ TABLE 17.10
Input for the second-order autoregressive predictor.

y_t	y_{t-1}	y_{t-2}
60	—	—
62.25	60	—
61.75	62.25	60
63	61.75	62.25
64.5	63	61.75
62	64.5	63
63.5	62	64.5
64	63.5	62
63.25	64	63.5
62.5	63.25	64
61	62.5	63.25
61.5	61	62.5

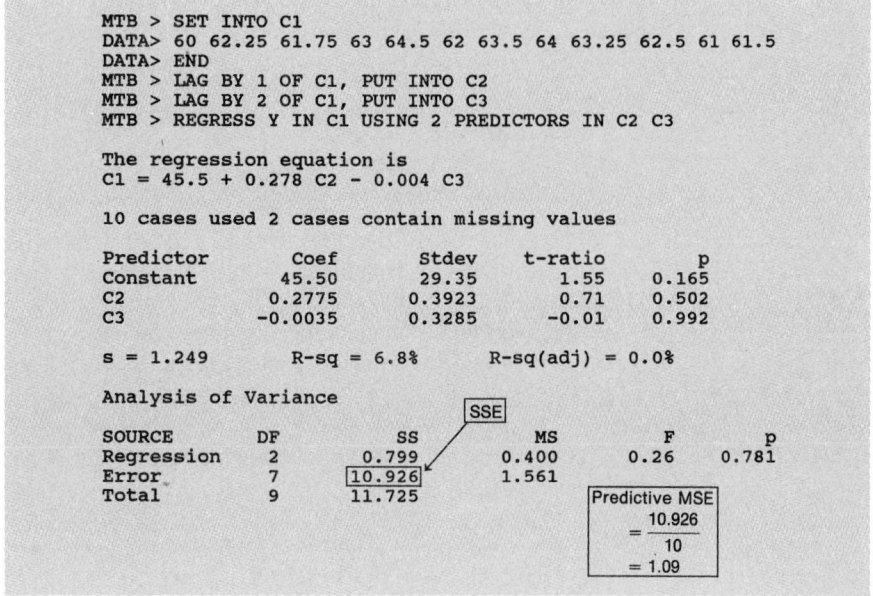

■ FIGURE 17.11
MINITAB procedure
for second-order
autoregression.

and from Table 17.1,

$$(\text{predictive}) \text{ MSE} = \frac{(2.25)^2 + (-.5)^2 + \cdots + (.5)^2}{11} = \frac{21.5}{11}$$

$$= 1.95 \text{ (for naive forecaster)}$$

So, despite the low value of R^2, we obtain a better fit to the observed data using the second-order autoregressive technique. This value of R^2, however, does indicate that the search for a more accurate forecasting procedure should continue.

Determining Autocorrelations

There are several methods of calculating the correlation between a time series, y_t, and its past values. For example, in Table 17.10,

y_t	y_{t-1}	y_{t-2}
61.75	62.25	60
63	61.75	62.25
⋮	⋮	⋮
61.5	61	62.5

To find the correlation between y_t and y_{t-1}, we could use the equation for the sample correlation coefficient, r, defined in Chapter 14. We could also find the correlation between y_t and y_{t-2} using the same procedure. However, the generally accepted method of computing such correlations is to use equation 17.17.

$$r_k = \frac{\sum_{t=1}^{T-k} (y_t - \bar{y})(y_{t+k} - \bar{y})}{\sum_{t=1}^{T} (y_t - \bar{y})^2} \qquad \textbf{(17.17)}$$

where (1) k is the lag under consideration, (2) r_k is the **sample autocorrelation** for lag k, (3) $\bar{y}$ is the average of the observed time series—that is,

$$\bar{y} = \frac{1}{T} \sum_{t=1}^{T} y_t$$

and (4) T is the number of observations in the time series.

There are three points to remember about the formula in equation 17.17:

1. The value of r_k using this formula will not agree with the value obtained using the correlation formula for r from Chapter 14, particularly for small sample sizes.
2. This formula is computationally more efficient.
3. This formula helps identify a time series that is not stationary.

EXAMPLE 17.8   Determine r_1 and r_2 using the following time series:

t	y_t
1	5
2	12
3	20
4	15
5	13

SOLUTION   Here $T = 5$ and

$$\bar{y} = \frac{1}{5}(5 + 12 + 20 + 15 + 13) = 13$$

Consequently,

r_1 (correlation between y_t and y_{t-1})

$$= \frac{(y_1 - \bar{y})(y_2 - \bar{y}) + (y_2 - \bar{y})(y_3 - \bar{y}) + (y_3 - \bar{y})(y_4 - \bar{y}) + (y_4 - \bar{y})(y_5 - \bar{y})}{(y_1 - \bar{y})^2 + (y_2 - \bar{y})^2 + \cdots + (y_5 - \bar{y})^2}$$

$$= \frac{(-8)(-1) + (-1)(7) + (7)(2) + (2)(0)}{(-8)^2 + (-1)^2 + (7)^2 + (2)^2 + (0)^2} = \frac{15}{118} = .13$$

Also,

$$r_2 = \frac{(y_1 - \bar{y})(y_3 - \bar{y}) + (y_2 - \bar{y})(y_4 - \bar{y}) + (y_3 - \bar{y})(y_5 - \bar{y})}{118}$$

$$= \frac{(-8)(7) + (-1)(2) + (7)(0)}{118} = \frac{-58}{118} = -.49$$

A graphical representation of these autocorrelations is a **correlogram.** The correlogram for Example 17.8 contains the values of r_1 and r_2, as illustrated in Figure 17.12. By inspecting a correlogram, you can determine which lagged variables appear to contribute to the prediction of your time series variable. *The autoregressive equation includes those lagged variables for which the corresponding autocorrelation is large in absolute value.* Statistical procedures for identifying significantly large autocorrelations are discussed in Bowerman and O'Connell, and also in Makridakis, Wheelright, and McGee (see Further Readings at the end of the chapter).

■ **FIGURE 17.12**
Correlogram for
Example 17.8.

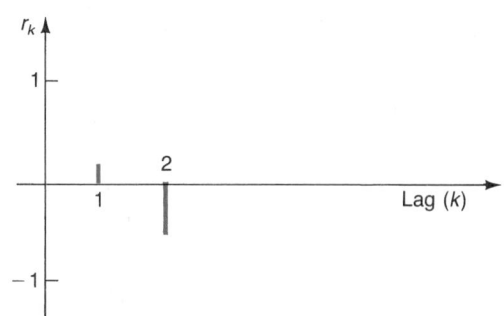

Detecting Seasonality

The autoregressive forecasting approach does allow you to detect seasonality in your time series data. If seasonality is present in quarterly data, we expect a significant positive correlation between y_t and y_{t-4}, that is, r_4 will be large, implying that the y value of one year ago is a good predictor of the y value today. Similarly, for monthly data, we can expect r_{12} to be large if there is significant seasonality.

EXAMPLE 17.9

Table 17.11 contains data for the quarterly profits from Ken's Auto Paint Shop. Which lagged variables appear to be correlated with y_t = profit during time period t?

SOLUTION

The MINITAB output for the autocorrelation equation (17.17) is shown in Figure 17.13. Here the *autocorrelation function* (called ACF by MINITAB) computes the first $10 + \sqrt{T}$ autocorrelations, where T (20, here) is the number of observations in the time series. If $10 + \sqrt{T}$ is not an integer, it is rounded *down* to the nearest integer, which is 14 in this case.

■ **TABLE 17.11**
Quarterly profits of
Ken's Auto Paint Shop
(in thousands of dollars).

QUARTER	1986	1987	1988	1989	1990
Spring	5.56	5.11	4.12	6.31	4.81
Summer	16.36	15.21	14.33	15.02	16.82
Fall	2.12	5.72	5.25	2.83	4.75
Winter	3.15	2.65	6.75	4.56	8.54

■ **FIGURE 17.13**
Autocorrelations
for Example 17.9
using MINITAB.

```
MTB > SET INTO C1
DATA> 5.56 16.36 2.12 3.15 5.11 15.21 5.72 2.65 4.12 14.33
DATA> 5.25 6.75 6.31 15.02 2.83 4.56 4.81 16.82 4.75 8.54
DATA> END
MTB > ACF OF C1 ←────────[ Generates the autocorrelations, r₁, r₂, . . . ]

ACF of C1

              -1.0 -0.8 -0.6 -0.4 -0.2  0.0  0.2  0.4  0.6  0.8  1.0
               +----+----+----+----+----+----+----+----+----+----+
    1  -0.315                    XXXXXXXX
    2  -0.260                    XXXXXXX
    3  -0.286                    XXXXXXX
    4   0.678                          XXXXXXXXXXXXXXXXXX ←
    5  -0.213                     XXXXX
    6  -0.168                     XXXXX
    7  -0.156                     XXXX
    8   0.512                          XXXXXXXXXXXXX ←        ┌───────┐
    9  -0.235                    XXXXXX                       │ Large │
   10  -0.134                      XXXX                       └───────┘
   11  -0.104                      XXXX
   12   0.388                         XXXXXXXXXXX ←
   13  -0.155                     XXXXX
   14  -0.060                       XX
```

■ **FIGURE 17.14**
MINITAB solution
for Example 17.9.

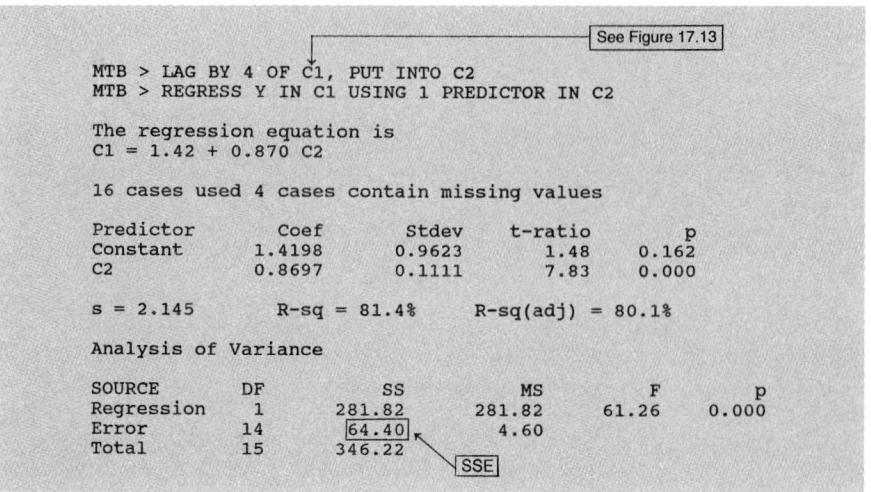

The seasonality pattern of four-quarter duration can be seen from the resulting autocorrelations in Figure 17.13. The large r_k values are $r_4 = .678$, $r_8 = .512$, and $r_{12} = .388$. Notice that $r_4 > r_8 > r_{12}$, which is typical of strong four-period seasonality. The large value of r_4 is your clue that such seasonality exists, and the large values of r_8 and r_{12} confirm this suspicion.

Using y_{t-4} as the predictor variable, we find from Figure 17.14 that the autoregression equation is

$$\hat{y}_t = 1.42 + .87y_{t-4}$$

The corresponding predictive MSE using this model is

$$(\text{predictive}) \text{ MSE} = \frac{\text{SSE}}{(\text{number of observations used})}$$

$$= \frac{64.40}{16} = 4.025$$

To decide whether this procedure performs well for this time series, we need to compare this MSE with the MSE obtained using other forecasting techniques.

■

Removing Nonstationarity

The autoregressive procedures discussed so far are effective if the time series is *stationary*—that is, if it contains no trend and has constant variance about the mean, $\bar{y}$. One method of detecting nonstationarity in a time series is to examine the correlogram. *If you notice that the autocorrelations in the correlogram do not die down rapidly (say after the second or third lag), then your time series is* not stationary.

With such a time series, an autoregressive procedure is *not appropriate*, unless you modify the time series to make it more stationary. That is, the data should be transformed to a stationary series before attempting to determine seasonality. This can be achieved by using the **differencing** method, which replaces y_t by the first difference, defined by

$$y_t' = y_t - y_{t-1}$$

To illustrate this technique, consider the series

$$2, 5, 8, 11, 14, \ldots$$

This series clearly contains a linear trend; each value is three more than the preceding value. The first differences are

$$y'_2 = 5 - 2 = 3$$
$$y'_3 = 8 - 5 = 3$$
$$y'_4 = 11 - 8 = 3$$
$$y'_5 = 14 - 11 = 3$$

and so on. These values contain no trend, and so the resulting series of first differences is stationary.

If this procedure has been successful in producing a stationary time series, the resulting correlogram (using the y'_t values) should die out quickly. If such is not the case, using the *second* differences of the original time series values ($y_1, y_2, \ldots, y_T$) often will produce a stationary series. Here, the second differences can be found by deriving first differences of the y'_t values, namely,

$$y''_t = y'_t - y'_{t-1} = y_t - 2y_{t-1} + y_{t-2}$$

You generally can achieve stationarity in your original time series by continuing to take differences until the autocorrelations of the "new" series drop to near zero after two or three time lags (except for possible large values, or spikes, due to seasonal effects). It usually is necessary to determine only first- or second-order differences when dealing with a nonstationary time series.

EXAMPLE 17.10

In Chapter 16, we examined the quarterly sales data of Video-Comp, shown in Table 16.3 on page 710. These data contained a strong linear trend and a definite seasonal pattern, with low sales in the first two quarters and high sales in the final two quarters. A graph of the data is contained in Figure 17.3. Is the seasonal effect more apparent using the original data or the first differences?

SOLUTION

The MINITAB autocorrelations using the original data are summarized and plotted in Figure 17.15. You can see the nonstationarity of this series—the autocorrelations fail to die out after two or three periods. Because of the strong trend component, the seasonal effect is not apparent.

■ **FIGURE 17.15**
MINITAB
autocorrelations using
sales data
(Example 17.10).

```
MTB > SET INTO C1
DATA> 20 12 47 60 40 32 65 76 56 50 85 100 75 70 101 123
DATA> END
MTB > ACF OF C1

ACF of C1

          -1.0 -0.8 -0.6 -0.4 -0.2  0.0  0.2  0.4  0.6  0.8  1.0
           +----+----+----+----+----+----+----+----+----+----+
    1    0.541                              XXXXXXXXXXXXXX
    2    0.105                              XXXX
    3    0.280                              XXXXXXX
    4    0.473                              XXXXXXXXXXXXX
    5    0.122                              XXXX
    6   -0.232                         XXXXXXX
    7   -0.087                          XXX
    8    0.063                              XXX
    9   -0.176                          XXXXX
   10   -0.422                    XXXXXXXXXXXX
   11   -0.279                        XXXXXXXX
   12   -0.121                          XXXX
   13   -0.234                         XXXXXX
   14   -0.343                      XXXXXXXXXX
```

■ FIGURE 17.16
MINITAB
autocorrelations using
first differences
(Example 17.10).

```
MTB > SET INTO C1
DATA> 20 12 47 60 40 32 65 76 56 50 85 100 75 70 101 123
DATA> END
MTB > DIFFERENCES OF LAG 1 FOR C1, PUT INTO C2
MTB > ACF OF C2

ACF of C2

               -1.0 -0.8 -0.6 -0.4 -0.2  0.0  0.2  0.4  0.6  0.8  1.0
                 +----+----+----+----+----+----+----+----+----+----+
     1    0.001                                     X
     2   -0.821            XXXXXXXXXXXXXXXXXXXXXXX
     3   -0.034                                    XX
     4    0.698                                     XXXXXXXXXXXXXXXXXX
     5    0.001                                     X
     6   -0.546                 XXXXXXXXXXXXXXX
     7   -0.024                                    XX
     8    0.431                                     XXXXXXXXXXXX
     9   -0.005                                     X
    10   -0.307                      XXXXXXXXX
    11   -0.019                                     X
    12    0.148                                     XXXXX
    13    0.010                                     X
```

The first differences are formed by subtracting adjacent y_t values:

FIRST DIFFERENCES

y_t	$y_t' = y_t - y_{t-1}$
20	*
12	−8
47	35
60	13
40	−20
⋮	⋮

The MINITAB autocorrelations using the first differences are computed in Figure 17.16. Now the seasonality pattern is much clearer, with large negative values for r_2, r_6, and r_{10} as well as large positive values for r_4, r_8, and r_{12}. The negative values are a result of a high (low) sales value in time period t followed by a low (high) sales figure two quarters later. Similarly, the large value of r_4 indicates a strong four-quarter seasonal effect; the large values of r_8 and r_{12} support this conclusion.

If you wished to use an autoregressive model for this application, you would use y_t' (not y_t) as your dependent variable because the y_t' values *are* stationary. An excellent set of predictor variables (using Figure 17.16) would be y_{t-2}' and y_{t-4}'. ■

Exercises   **17.35** Malcom Chemicals manufactures 12 different speciality chemicals. The company's net profit from these speciality chemicals has been stable over the past five years, as the data indicate. The figures in the tables are net profit per quarter in units of $10,000.

YEAR	QUARTER 1	QUARTER 2	QUARTER 3	QUARTER 4
1987	4.3	5.7	8.3	2.4
1988	4.7	5.9	8.0	2.7
1989	4.4	5.5	7.6	2.5
1990	4.9	5.9	8.4	2.8
1991	5.3	6.2	8.6	2.9

Fit the following two autoregressive processes to the data:

$$\hat{y}_t = b_0 + b_1 y_{t-1} + b_2 y_{t-2}$$

$$\hat{y}_t = b_0 + b_1 y_{t-1} + b_2 y_{t-2} + b_3 y_{t-3} + b_4 y_{t-4}$$

Find the predictive MSE for each autoregressive process and compare the values. Use a computerized statistical package to find the coefficients of the autoregressive equations.

17.36 The debt-to-equity capitalization ratio for Dooper Industries, a maker of machinery parts, has never been above 20% for the past six years, as a result of excellent management. These ratios (given as a percentage) for the past six years are as follows:

YEAR	QUARTER 1	QUARTER 2	QUARTER 3	QUARTER 4
1986	12	15	18	13
1987	11	14	16	12
1988	10	16	19	12
1989	11	17	20	14
1990	12	15	18	12
1991	11	14	17	13

a. Find the autocorrelations for lags of $k = 1, 2, 3,$ and 4.

b. Using a computer package, find the coefficients for the autoregressive process

$$\hat{y}_t = b_0 + b_1 y_{t-1} + b_2 y_{t-2} + b_3 y_{t-3} + b_4 y_{t-4}$$

17.37 The percent of Medical International's total revenue derived from freestanding centers for cardiac rehabilitation is given for the past 16 years:

YEAR	PERCENT OF REVENUE	YEAR	PERCENT OF REVENUE
1976	45	1984	42
1977	47	1985	43
1978	46	1986	40
1979	43	1987	37
1980	40	1988	35
1981	36	1989	37
1982	35	1990	40
1983	39	1991	43

a. Find the autocorrelations for lags $k = 1, 2, 3,$ and 4.

b. Fit a first-order autoregressive equation to the data.

17.38 How should the graph of a correlogram look if the time series is stationary?

17.39 Refer to Exercise 17.7. Find the autocorrelations for the seasonal data for Sands Motel, using lags of $k = 1, 2, 3, \ldots, 12$. Do the data appear to be stationary?

17.40 The manager of a children's clothing store has tabulated a sales index of children's clothes sold each quarter. The sales of children's clothes has been in an upward trend over the past five years, but the sales are also affected by seasonal variation.

YEAR	QUARTER 1	QUARTER 2	QUARTER 3	QUARTER 4
1987	108	256	201	190
1988	185	380	290	320
1989	280	421	360	331
1990	300	504	432	450
1991	400	862	510	480

a. Calculate the autocorrelations through lag 12.

b. Calculate the first differences for the time series.

c. Calculate the autocorrelations for the data obtained by differencing in part b.

d. Compare the autocorrelations in parts a and c. Comment on whether the autocorrelations describe a stationary time series.

17.41 Determine whether the following data are stationary. If the data are not stationary, take differences until it becomes clear that the resulting time series is stationary. The data represent quarterly interest (in thousands of dollars) paid by a local savings and loan association to depositors who have regular saving accounts.

YEAR	QUARTER 1	QUARTER 2	QUARTER 3	QUARTER 4
1987	25	38	15	21
1988	46	57	36	32
1989	65	80	54	49
1990	84	97	77	70
1991	104	109	96	90

17.42   Monthly averages of Standard and Poor's Utility Price Index are presented below from 1984 to 1989:

	1984	1985	1986	1987	1988	1989
January	68.50	75.83	92.06	120.1	106.1	114.4
February	66.25	78.14	97.51	119.9	110.7	116.9
March	65.25	78.89	102.0	117.7	107.2	116.7
April	64.34	81.25	103.8	110.0	104.1	119.9
May	64.94	83.60	102.4	108.1	103.1	127.7
June	64.00	86.90	106.7	112.6	109.9	133.5
July	64.66	87.22	112.1	110.9	108.5	137.2
August	68.11	83.21	118.5	117.7	107.9	140.5
September	69.71	81.46	113.1	115.0	109.7	141.0
October	72.02	81.49	114.0	111.7	113.0	142.7
November	73.58	86.80	114.1	106.5	111.7	143.4
December	74.43	90.83	115.5	102.4	113.0	152.2

(*Source: Standard and Poor's Security Price Index Record, Statistical Service* (1990): 108.)

a.   Calculate the autocorrelations through lag 15.

b.   Calculate the first differences for the time series.

c.   Calculate the autocorrelations for the difference data obtained in part b.

d.   Comment on the autocorrelations obtained in parts a and c.

17.43   The total number of stamps and stamped paper that the U.S. Postal Service sold in the years from 1970 to 1985 follows. Fit a second-order autoregressive process to the data. Sales are in units of millions of dollars.

YEAR	SALES	YEAR	SALES
1970	1936	1978	3943
1971	1999	1979	4382
1972	2371	1980	4287
1973	2399	1981	4625
1974	2504	1982	5559
1975	2819	1983	5709
1976	3155	1984	6023
1977	3658	1985	6520

(*Source: United States Statistical Abstracts,* (1987): 528.)

17.9
The Other Side of Forecasting: Linear Regression Using Time Series Data

We have already used linear regression procedures in many of the time series forecasting techniques we have discussed. For instance, simple linear regression was used to describe the *trend* present in time series data. Multiple linear regression was used in the previous section to predict a future time series value using the past one or more observations. This was the autoregressive forecasting method, where, for example,

$$\hat{y}_t = b_0 + b_1 y_{t-1} + b_2 y_{t-2}$$

Here the predictor variables are the *lagged* time series variables, y_{t-1} and y_{t-2}.

Often, you will wish to combine one or more autoregressive terms and several time-related predictor variables into the regression equation, such as

$$\hat{y}_t = b_0 + b_1 y_{t-1} + b_2 y_{t-2} + b_3 X_{1,t} + b_4 X_{2,t} + b_5 X_{3,t}$$

Or you can omit the autoregressive terms and use an equation such as

$$\hat{y}_t = b_0 + b_1 X_{1,t} + b_2 X_{2,t} + b_3 X_{3,t}$$

Data that are to be used in a multiple regression model and that are obtained at a single point in time are referred to as **cross-sectional data.** The resulting regression predictor variables are usually labeled as $X_1, X_2, \ldots, X_k$. Chapter 15 focused on regression applications dealing with cross-sectional data, such as data gathered from ten different households at one point in time. This chapter examines regression applications where the data are gathered across time, that is, **time series data.** The corresponding regression variables are written as $X_{1,t}, X_{2,t}, \ldots, X_{k,t}$ to denote the values of these variables during time period t. For example, $X_{2,5}$ represents the observed value of variable X_2 during the fifth time period. Whether using cross-sectional or time series data, the regression model (with corresponding formulas and assumptions) developed in Chapter 15 is applicable.

Linear regression techniques on time series data offer a variety of opportunities for better forecasting precision, including (1) the use of dummy variables to capture *additive* seasonality and (2) the use of lagged independent variables to allow for time-delay effects between the dependent and predictor variables. *On the other hand, when this technique is used on time series data, it becomes increasingly difficult to satisfy the linear regression assumptions discussed in Chapter 15.* In particular, the error term from one time period often is seriously affected by the previous errors, violating the assumption of independent errors. This means that we can expect to find that the *residuals* are *autocorrelated*. The degree of autocorrelation in the residuals can be measured and tested for significance by calculating the **Durbin-Watson statistic,** which is obtained from the residuals.

These matters are discussed in the remaining sections of this chapter. We look at how regression techniques expand the area of time series analysis but also present a new set of problems.

Use of Dummy Variables for Seasonality

We introduced the concept of additive seasonality previously. Essentially, this type of seasonal effect is present whenever the amount of the seasonal variation is unaffected by the underlying trend in the time series, as illustrated in Figure 16.15 on page 703. Notice that even as the sales grow over time, the seasonal effect remains the same.

Ignoring any cyclical activity, each observation, y_t, can be described by

$$y_t = TR_t + S_t + I_t$$

where (1) TR_t is the trend component described by a straight line ($TR_t = \beta_0 + \beta_1 t$) or a quadratic curve ($TR_t = \beta_0 + \beta_1 t + \beta_2 t^2$), (2) S_t is the seasonal effect, and (3) I_t is the irregular activity component.

Both the trend and seasonal components can be obtained by using multiple linear regression. The seasonal effects are captured by including a set of *dummy variables* in the regression equation. This type of variable was first introduced in Chapter 15, where we used a set of dummy variables to represent the categories of a *qualitative* variable—such as seasons of the year, in this application.

We use the same procedure for defining dummy variables here—we define one less dummy variable than the number of seasons (categories), L. Because $L = 4$

for quarterly data, we need $L - 1 = 3$ dummy variables. One possible scheme is to define

$$Q_1 = \begin{cases} 1 & \text{if quarter 1} \\ 0 & \text{otherwise} \end{cases} \quad Q_2 = \begin{cases} 1 & \text{if quarter 2} \\ 0 & \text{otherwise} \end{cases} \quad Q_3 = \begin{cases} 1 & \text{if quarter 3} \\ 0 & \text{otherwise} \end{cases}$$

$$(17.18)$$

With this procedure, no dummy variable is defined for the fourth quarter. The resulting coefficient of Q_1, Q_2, or Q_3 in the prediction equation will compare the effect of that quarter *against the fourth quarter*. For example, if the coefficient of Q_2 from your computer solution is -5, then, apart from any changes due to trend, quarter 2 produces a value of the dependent variable *five* less than during quarter 4. Quarter 4 is called the **base** quarter. Remember that the base period you select has absolutely *no effect* on the predicted values, $\hat{y}_t$.

For monthly data, you define $L - 1 = 12 - 1 = 11$ dummy variables. As before, you can omit the dummy variable for any one month, which then becomes the base month for all the computed dummy variable coefficients. If you omitted a variable for December, then December would be the base month, and your corresponding set of dummy variables would be

$$M_1 = \begin{cases} 1 \text{ if January} \\ 0 \text{ otherwise} \end{cases} \quad M_2 = \begin{cases} 1 \text{ if February} \\ 0 \text{ otherwise} \end{cases} \cdots M_{11} = \begin{cases} 1 \text{ if November} \\ 0 \text{ otherwise} \end{cases}$$

$$(17.19)$$

The quarterly sales at Video-Comp were examined in Examples 16.8 and 17.3, assuming *multiplicative* seasonality. The vice president of retail marketing, in reviewing this solution, thinks that the seasonal fluctuations do not appear to be increasing along with the trend and so believes that additive seasonality actually is present. We use the dummy variable approach to model the seasonality and calculate the forecasts for the first four quarters of 1991. We want to know whether this approach appears to provide a good fit to the four years of observed sales.

The prediction equation is

$$\hat{y}_t = TR_t + \hat{S}_t$$

where (1) $TR_t = b_0 + b_1 t$ (the trend appears to be nearly linear) and (2) $\hat{S}_t = b_2 Q_1 + b_3 Q_2 + b_4 Q_3$. Here, Q_1, Q_2, and Q_3 are the dummy variables defined in equation 17.18.

The input configuration used by the computer program consists of 16 rows and five columns:

y_t	t	Q_1	Q_2	Q_3	y_t	t	Q_1	Q_2	Q_3
20	1	1	0	0	56	9	1	0	0
12	2	0	1	0	50	10	0	1	0
47	3	0	0	1	85	11	0	0	1
60	4	0	0	0	100	12	0	0	0
40	5	1	0	0	75	13	1	0	0
32	6	0	1	0	70	14	0	1	0
65	7	0	0	1	101	15	0	0	1
76	8	0	0	0	123	16	0	0	0

■ FIGURE 17.17
MINITAB solution
using dummy variables
to represent quarterly
sales at Video-Comp.

```
MTB > READ INTO C1-C5
DATA> 20 1 1 0 0
DATA> 12 2 0 1 0
DATA> 47 3 0 0 1
DATA> 60 4 0 0 0
DATA> 40 5 1 0 0
DATA> 32 6 0 1 0
DATA> 65 7 0 0 1
DATA> 76 8 0 0 0
DATA> 56 9 1 0 0
DATA> 50 10 0 1 0
DATA> 85 11 0 0 1
DATA> 100 12 0 0 0
DATA> 75 13 1 0 0
DATA> 70 14 0 1 0
DATA> 101 15 0 0 1
DATA> 123 16 0 0 0
DATA> END
      16 ROWS READ
MTB > NAME C1 = 'SALES' C2 = 'TIME' C3 = 'QTR1'
MTB > NAME C4 = 'QTR2' C5 = 'QTR3'
MTB > REGRESS Y IN C1 USING 4 PREDICTORS IN C2-C5
```

> The regression equation is
> SALES = 41.7 + 4.80 TIME - 27.6 QTR1 - 39.2 QTR2 - 10.4 QTR3

```
Predictor      Coef      Stdev    t-ratio        p
Constant     41.750      1.688      24.74    0.000
TIME          4.8000     0.1258     38.16    0.000
QTR1        -27.600      1.635     -16.88    0.000
QTR2        -39.150      1.611     -24.30    0.000
QTR3        -10.450      1.596      -6.55    0.000

s = 2.250        R-sq = 99.6%     R-sq(adj) = 99.4%
                                          │
Analysis of Variance                     │R²

SOURCE       DF        SS         MS        F        p
Regression    4    13629.3     3407.3   672.90    0.000
Error        11       55.7        5.1
Total        15    13685.0    SSE
```

The MINITAB solution for $b_0, b_1, \ldots, b_4$ is provided in Figure 17.17. The resulting equation is

$$\hat{y}_t = 41.75 + 4.8t - 27.6Q_1 - 39.15Q_2 - 10.45Q_3 \qquad \textbf{(17.20)}$$

Notice that $R^2 = .996$, which indicates a strong fit to the 16 observations. How does this method of forecasting compare to the one used in Example 17.3, where we assumed *multiplicative* seasonality in the quarterly sales data? Comparing the two MSE's in Table 17.12, it appears that the seasonality effect is in fact additive, and the marketing vice president was correct.

From equation 17.20, we find that

1. The sales are increasing at an average rate of 4.8 (million dollars) per quarter.
2. Apart from trend effects, sales for the first quarter are 27.6 (million dollars) less than the sales for the fourth quarter.
3. Apart from trend effects, sales for the second quarter are 39.15 (million dollars) less than for the fourth quarter.
4. Sales for the third quarter are 10.45 (million dollars) lower than those during the fourth quarter, ignoring trend effects.

After you have decided to use the additive seasonality equation, to determine the 1991 forecasts you use the appropriate value for t with the 0 or 1 values for the dummy variables.

■ **TABLE 17.12**
Summary of
multiplicative versus
additive seasonal
forecasting (quarterly
sales of Video-Comp).

MULTIPLICATIVE SEASONALITY				ADDITIVE SEASONALITY			
t	y_t	$\hat{y}_t$	$y_t - \hat{y}_t$	t	y_t	$\hat{y}_t$	$y_t - \hat{y}_t$
1	20	20.80	−.80	1	20	18.95	1.05
2	12	20.38	−8.38	2	12	12.20	−.20
3	47	40.21	6.79	3	47	45.70	1.30
4	60	50.98	9.02	4	60	60.95	−.95
5	40	37.96	2.04	5	40	38.15	1.85
6	32	34.32	−2.32	6	32	31.40	.60
7	65	63.70	1.30	7	65	64.90	.10
8	76	76.98	−.98	8	76	80.15	−4.15
9	56	55.13	.87	9	56	57.35	−1.35
10	50	48.26	1.74	10	50	50.60	−.60
11	85	87.20	−2.20	11	85	84.10	.90
12	100	102.97	−2.97	12	100	99.35	.65
13	75	72.30	2.70	13	75	76.55	−1.55
14	70	62.21	7.79	14	70	69.80	.20
15	101	110.69	−9.69	15	101	103.30	−2.30
16	123	128.96	−5.96	16	123	118.55	4.45

Forecasting equation: $\hat{y}_t =$
$(19.372 + 5.0375t) \cdot S_t$

$S_1 = S_5 = S_9 = \cdots = .852$
$S_2 = S_6 = S_{10} = \cdots = .692$
$S_3 = S_7 = S_{11} = \cdots = 1.166$
$S_4 = S_8 = S_{12} = \cdots = 1.290$

predictive MSE $= \dfrac{\Sigma(y_t - \hat{y}_t)^2}{16}$

$= \dfrac{425.36}{16} = 26.58$

Forecasting equation: $\hat{y}_t =$
$41.75 + 4.8t - 27.6Q_1 -$
$39.15Q_2 - 10.45Q_3$

predictive MSE $= \dfrac{\Sigma(y_t - \hat{y}_t)^2}{16}$

$= \dfrac{55.7}{16} = 3.48$

Forecast for first quarter of 1991: here, $Q_1 = 1$, $Q_2 = 0$, $Q_3 = 0$, and

$$\hat{y}_{17} = 41.75 + 4.8(17) - 27.6(1) - 39.15(0) - 10.45(0)$$
$$= 95.75 \quad \text{(million dollars)}$$

Forecast for second quarter of 1991: here, $Q_2 = 1$ (all other Q's $= 0$) and

$$\hat{y}_{18} = 41.75 + 4.8(18) - 27.6(0) - 39.15(1) - 10.45(0)$$
$$= 89$$

Forecast for third quarter of 1991: now, $Q_3 = 1$ (all other Q's $= 0$) and

$$\hat{y}_{19} = 41.75 + 4.8(19) - 27.6(0) - 39.15(0) - 10.45(1)$$
$$= 122.5$$

Forecast for the fourth quarter of 1991: since $Q_1 = Q_2 = Q_3 = 0$,

$$\hat{y}_{20} = 41.75 + 4.8(20) - 27.6(0) - 39.15(0) - 10.45(0)$$
$$= 137.75$$

Exercises

17.44 The following multiple regression equation was used to fit quarterly sales data. The data are in units of 10,000.

$$\hat{y}_t = 0.5 + 1.8t + 3Q_1 - .6Q_2 + 1.1Q_3$$

where

$$Q_1 = \begin{cases} 1 & \text{if quarter 1} \\ 0 & \text{otherwise} \end{cases}$$

$$Q_2 = \begin{cases} 1 & \text{if quarter 2} \\ 0 & \text{otherwise} \end{cases}$$

$$Q_3 = \begin{cases} 1 & \text{if quarter 3} \\ 0 & \text{otherwise} \end{cases}$$

a. Apart from seasonality, how fast are sales increasing each quarter?

b. Apart from trend, how much are sales for the first quarter ahead of sales for the fourth quarter?

c. Apart from trend, how much lower are sales for the second quarter than for the fourth quarter?

d. What are the forecasted sales for time period 12, which is a fourth quarter?

17.45 Activity in the federal funds market consists of short-term (usually one day) loans of perhaps several million dollars by one commercial bank with surplus reserve funds to another bank that is short of reserves. The interest rate charged for these loans is the federal funds rate. The federal funds rate is given below from 1981 through 1986. Determine the multiple regression equation that takes into account trend and seasonality.

YEAR	QUARTER 1	QUARTER 2	QUARTER 3	QUARTER 4
1981	16.57	17.78	17.58	13.59
1982	14.23	14.51	11.01	9.29
1983	8.65	8.80	9.46	9.43
1984	9.69	10.56	11.39	9.27
1985	8.48	7.92	7.90	8.11
1986	7.83	6.92	6.21	6.27

(Source: Moody's Bank and Finance Manual (1987)1: a14.)

17.46 Quality Homes, Inc. builds single-family houses in several large cities. The number of carpenters that it hires fluctuates with the demand for housing. For the six years shown, find a multiple regression equation to predict the number of carpenters on the payroll at Quality Homes. The multiple regression equation should take into account the trend and the monthly seasonality. What percentage of the total variation has been explained using these variables?

YEAR	JAN	FEB	MAR	APR	MAY	JUNE	JULY	AUG	SEPT	OCT	NOV	DEC
1986	145	148	150	169	197	250	267	290	280	230	180	160
1987	155	150	166	178	220	290	320	325	300	270	200	190
1988	180	195	210	213	255	308	350	368	345	320	280	250
1989	230	245	258	290	330	342	394	405	380	350	310	290
1990	285	298	310	345	396	408	451	465	441	430	390	370
1991	350	361	372	395	420	439	480	495	483	450	420	410

17.47 Refer to Exercise 17.7 and the monthly data for Sands Motel. Using a computerized statistical package, find the coefficients of a multiple regression model that takes into account both the trend and the effect due to the particular month.

17.48 Explain the difference between multiplicative seasonality and additive seasonality. Which forecasting techniques are best suited for each of these situations?

17.49 Refer to Exercise 17.36. For the quarterly data from Dooper Industries, find the coefficients of the multiple regression model that takes into account both the trend and seasonality.

Use of Lagged Independent Variables

When using multiple linear regression on time series data, we can represent the model as

$$y_t = \beta_0 + \beta_1 X_{1,t} + \beta_2 X_{2,t} + \cdots + \beta_k X_{k,t} + e_t$$

where each $X_{i,t}$ represents the value of predictor variable X_i in time period t, and e_t is the error component for this period.

Look at the data in Table 17.13, which consist of semiannual figures from the past eight years. Our object is to predict the number of home loans financed by Liberty Savings and Loan. The predictor variables were chosen to be average interest rate (X_1), advertising expenditure (X_2), an election-year dummy variable ($X_3 = 1$ for a presidential election year and 0 otherwise), and a seasonal dummy variable, where $X_4 = 1$ for the first six months and $X_4 = 0$ for the final six months.

A financial analyst at Liberty Savings saw two problems with the data in Table 17.13. First, she thought that the number of home loans during a particular six-month period should be more affected by the *previous* six-month interest rate, due to the time delay between loan application and actual funding. The value of y_t increases by 1 each time a loan is funded, not when the application is turned in. This time delay generally runs between three and six months. For the same reason, she believed that the effect of any increased (or decreased) advertising would be reflected in the loan amounts of the next period.

The procedure to follow in this situation is to lag the predictor variable by the corresponding time lag. For this example, we can lag X_1 and X_2 by one time period, resulting in the following regression model:

$$y_t = \beta_0 + \beta_1 X_{1,t-1} + \beta_2 X_{2,t-1} + \beta_3 X_{3,t} + \beta_4 X_{4,t} + e_t$$

The other problem the financial analyst foresaw with using the regression variables in Table 17.13 is a common difficulty in applying regression techniques to a forecasting situation. If we had not lagged X_1 and X_2, then any forecast for 1991 would involve *specifying values for X_1 and X_2 for this future time period*. Because these values may be just as difficult to predict as the dependent variable, our model has little potential as a forecaster. By lagging these variables, we have removed this problem for a one-period-ahead forecast, because now the lagged values for tomorrow are the actual values for today.

■ TABLE 17.13
Housing data for
Liberty Savings and
Loan.

YEAR	t	NUMBER OF HOME LOANS Y_t	AVERAGE INTEREST RATE $(X_{1,t})$	ADVERTISING EXPENDITURE (THOUSANDS OF $) $(X_{2,t})$	ELECTION YEAR VARIABLE $(X_{3,t})$	SEASONAL VARIABLE $(X_{4,t})$
1983	1	122	11.8	10.4	0	1
	2	118	12.4	6.7	0	0
1984	3	106	11.0	7.5	1	1
	4	140	14.5	7.8	1	0
1985	5	86	11.0	5.1	0	1
	6	96	10.1	6.8	0	0
1986	7	110	12.9	6.8	0	1
	8	76	14.8	9.1	0	0
1987	9	62	10.5	7.5	0	1
	10	104	9.8	5.1	0	0
1988	11	135	10.1	8.8	1	1
	12	120	10.8	4.3	1	0
1989	13	105	10.0	7.1	0	1
	14	118	9.8	8.6	0	0
1990	15	121	9.5	8.8	0	1
	16	132	9.7	7.4	0	0

A portion of the input data using the lagged predictors is:

y_t	$X_{1,t-1}$	$X_{2,t-1}$	$X_{3,t}$	$X_{4,t}$
118	11.8	10.4	0	0
106	12.4	6.7	1	1
140	11.0	7.5	1	0
86	14.5	7.8	0	1
96	11.0	5.1	0	0
110	10.1	6.8	0	1
76	12.9	6.8	0	0
⋮				

The resulting prediction equation, shown in Figure 17.18, is

$$\hat{y}_t = 197.88 - 10.260 X_{1,t-1} + 2.839 X_{2,t-1}$$
$$+ 17.755 X_{3,t} + 1.471 X_{4,t} \quad \textbf{(17.21)}$$

■ **FIGURE 17.18**
MINITAB solution
using lagged interest
and advertising
variables (data from
Table 17.13).

```
MTB > READ INTO C1-C5
DATA> 122 1108 10.4 0 1
DATA> 118 12.4 6.7 0 0
DATA> 106 11.0 7.5 1 1
DATA> 140 14.5 7.8 1 0
DATA> 86 11.0 5.1 0 1
DATA> 96 10.1 6.8 0 0
DATA> 110 12.9 6.8 0 1
DATA> 76 14.8 9.1 0 0
DATA> 62 10.5 7.5 0 1
DATA> 104 9.8 5.1 0 0
DATA> 135 10.1 8.8 1 1
DATA> 120 10.8 4.3 1 0
DATA> 105 10.0 7.1 0 1
DATA> 118 9.8 8.6 0 0
DATA> 121 9.5 8.8 0 1
DATA> 132 9.7 7.4 0 0
DATA> END
      16 ROWS READ
MTB > LAG BY 1 OF C2, PUT INTO C12
MTB > LAG BY 1 OF C3, PUT INTO C13
NAME C1 = 'LOANS' C2 = 'INT' C3 = 'ADV' C12 = 'INTLAG1'
NAME C13 = 'ADVLAG1' C4 = 'ELECTION' C5 = 'SEASON'
MTB > REGRESS Y IN C1 USING 4 PREDICTORS IN C12, C13, C4, C5;
SUBC> DW.

The regression equation is
LOANS = 198 - 10.3 INTLAG1 + 2.84 ADVLAG1 + 17.8 ELECTION + 1.47 SEASON

15 cases used 1 cases contain missing values

Predictor      Coef      Stdev     t-ratio        p
Constant     197.88      23.09        8.57    0.000
INTLAG1      -10.260      1.945       -5.27    0.000
ADVLAG1        2.839      1.943        1.46    0.175
ELECTION      17.755      6.665        2.66    0.024
SEASON         1.471      6.390        0.23    0.823

s = 11.19       R-sq = 81.1%    R-sq(adj) = 73.6%
                                        R²

Analysis of Variance

SOURCE        DF        SS          MS        F        p
Regression     4      5386.0      1346.5    10.76    0.001
Error         10      1251.6       125.2
Total         14      6637.6
                          SSE

SOURCE        DF      SEQ SS
INTLAG1        1      4294.4
ADVLAG1        1       183.9
ELECTION       1       901.1
SEASON         1         6.6

Durbin-Watson statistic = 1.61
```

■ **FIGURE 17.19**
MINITAB solution
without lagging the
interest and advertising
variables (data from
Table 17.13).

```
MTB > REGRESS Y IN C1 USING 4 PREDICTORS IN C2-C5;
SUBC> DW.

The regression equation is
LOANS = 114 - 3.38 INT + 4.58 ADV + 24.6 ELECTION - 12.8 SEASON

Predictor        Coef       Stdev     t-ratio          p
Constant       113.65       39.44        2.88      0.015       ┌─────────┐
INT            -3.376        3.293      -1.03      0.327  ←─── │No longer│
ADV             4.583        3.446       1.33      0.210       │significant│
ELECTION       24.63        11.78        2.09      0.061       └─────────┘
SEASON        -12.83        10.68       -1.20      0.255

s = 19.94       ┌──────────────┐    R-sq(adj) = 12.3%
                │R-sq = 35.7%  │
                └──────────────┘
Analysis of Variance

SOURCE         DF           SS          MS          F         p
Regression      4        2430.9       607.7       1.53     0.261
Error          11        4375.0       397.7
Total          15        6805.9

SOURCE         DF        SEQ SS
INT             1          24.7
ADV             1         229.4
ELECTION        1        1602.5
SEASON          1         574.3

Durbin-Watson statistic = 1.89
```

We can draw several conclusions from Figure 17.18:

1. Based on the t values, the lagged interest rate variable ($X_{1,t-1}$) and the election year variable ($X_{3,t}$) are the only significant predictors of the number of home loans financed by Liberty.
2. Because $R^2 = .811$, 81.1% of the total variation of the y_t values has been explained using these four predictors.
3. The predictive mean squared error (for comparison purposes) is

$$\text{(predictive) MSE} = \frac{\text{SSE}}{15} = \frac{1251.6}{15} = 83.4$$

To illustrate the effect of lagging the independent variables, Figure 17.19 contains the solution to this example lagging neither X_1 nor X_2. Two things are striking. First, the R^2 value drops from .811 to .357. Second, the interest variable, X_1, is *no longer significant,* based on its small t-value.

In each application, try lagging the independent variables that could possibly have a delayed action on the dependent variable. You also should vary the lag period to account for predictor effects that show up several time periods later.

What would be the forecast for the first half of 1991 using prediction equation 17.21? Here, $X_{1,t-1} = X_{1,16} = $ interest rate for the last half of 1990 $= 9.7$, and $X_{2,t-1} = X_{2,16} = 7.4$. Also, $X_{3,17} = 0$ (1991 is not an election year) and $X_{4,17} = 1$ (for the first half of 1991). So,

$$\hat{y}_{17} = 197.88 - 10.260(9.7) + 2.839(7.4) + 17.755(0) + 1.471(1)$$

$$= 120.8$$

This result is approximately the same as the actual value for the first half of 1990, since the interest rate changed very little during 1989 and 1990 and neither 1990 nor 1991 was an election year.

Exercises

17.50 What is the importance of using lagged independent variables in a regression equation?

17.51 The following table lists the food price index (FPI) and the per-capita income (PCI) index for a certain third-world nation. The indexes are listed in six-month increments.

YEAR	MONTH	Y: FPI	X: PCI
1983	June	109	104
	Dec	101	116
1984	June	104	124
	Dec	107	120
1985	June	105	167
	Dec	114	133
1986	June	108	188
	Dec	126	148
1987	June	112	101
	Dec	100	158
1988	June	114	115
	Dec	104	127
1989	June	107	112
	Dec	102	141
1990	June	110	124
	Dec	105	162
1991	June	113	184
	Dec	120	140

a. Find the simple regression equation

$$\hat{y}_t = b_0 + b_1 X_t$$

b. Find the simple regression equation

$$\hat{y}_t = b_0 + b_1 X_{t-1}$$

c. Compare the R^2 for the two equations found in parts a and b.

17.52 Credit Corp finances small home-improvement projects for one year or less. Usually the company does not screen clients rigorously, because a mechanics lien is placed on the home. Credit Corp has found that a significant correlation exists between the interest rate on loans and the number of defaults. The following data give the average interest rate charged during that quarter and the number of times a loan holder was more than 30 days behind on a payment.

YEAR	QUARTER	AVERAGE INTEREST RATE (Y)	TIMES BEHIND ON A LOAN PAYMENT (X)
1988	1	8.5	53
	2	8.0	45
	3	9.0	44
	4	9.5	47
1989	1	9.0	49
	2	10.0	46
	3	10.5	49
	4	10.0	52
1990	1	10.5	50
	2	11.25	51
	3	11.50	55
	4	11.00	57
1991	1	12.25	52
	2	12.50	60
	3	12.00	63
	4	11.50	58

a. Find the simple regression equation

$$\hat{y}_t = b_0 + b_1 X_t$$

b. Find the simple regression equation

$$\hat{y}_t = b_0 + b_1 X_{t-1}$$

c. Compare the R^2 for the two equations.

17.53   Refer to Exercise 17.35. The following is the number of salespeople working for Malcom Chemical Company over the five-year period 1987 through 1991.

YEAR	QUARTER 1	QUARTER 2	QUARTER 3	QUARTER 4
1987	30	41	13	23
1988	31	39	15	22
1989	27	37	12	25
1990	30	42	14	26
1991	32	43	15	25

Find the multiple regression equation to predict the net profit per quarter for Malcom Chemical. Use dummy variables to represent seasonality and the variable, number of salespeople, lagged by one time period. Also find the predictive MSE.

17.54   Refer to Exercise 17.36. The following is the level of inventory for Dooper Industries in units of 10,000 for the years 1986 to 1991.

YEAR	QUARTER 1	QUARTER 2	QUARTER 3	QUARTER 4
1986	4.8	6.1	4.2	3.8
1987	4.2	5.2	4.1	3.3
1988	5.0	6.4	4.2	3.2
1989	5.5	6.8	4.2	3.1
1990	5.1	6.1	4.0	3.8
1991	4.2	5.9	4.1	3.7

Find the multiple regression equation to predict the debt-to-equity capitalization ratio for Dooper Industries using one variable to represent seasonality and a second variable, level of inventory lagged by one quarter. Find the R^2.

17.55   A local used-car dealer believes that advertising has greatly increased sales at the used car lot. The sales (in hundreds of dollars) of cars for each month of 1990 and 1991 along with the corresponding advertising expenditures follow:

YEAR	MONTH	SALES (Y)	ADVERTISING EXPENDITURE (X)	YEAR	MONTH	SALES (Y)	ADVERTISING EXPENDITURE (X)
1990	Jan	19	250	1991	Jan	16	300
	Feb	16	405		Feb	17	350
	Mar	20	308		Mar	19	401
	Apr	17	425		Apr	20	560
	May	21	550		May	23	630
	June	24	300		June	25	725
	July	16	450		July	28	630
	Aug	22	522		Aug	26	550
	Sept	23	630		Sept	23	430
	Oct	26	510		Oct	21	400
	Nov	23	320		Nov	21	350
	Dec	17	250		Dec	19	260

a.   Find the simple regression equation

$$\hat{y}_t = b_0 + b_1 X_t$$

b.   Find the simple regression equation

$$\hat{y}_t = b_0 + b_1 X_{t-1}$$

c.   Compare the R^2 for the regression equations in parts a and b.

◢ 17.10
The Problem of Autocorrelation: The Durbin-Watson Statistic

A problem you will encounter frequently when using multiple linear regression on time series data is that the residual terms (e_t) are not independent. We discussed autoregressive forecasting, in which an observation (y_t) is related to its past values. For this situation, we said that significant autocorrelation was present in the *observations*.

When we have an autocorrelated time series, we simply regress y_t on the past values. However, when a particular model dealing with least squares estimates of the unknown parameters (such as multiple linear regression) results in **autocorrelated residuals,** problems do arise. In particular, all tests of hypothesis, including the t-tests for individual predictors, become extremely suspect.

Detecting Autocorrelated Residuals

The Durbin-Watson statistic frequently is used to test for significant autocorrelation in the residuals. If its value is very small, significant *positive* autocorrelation exists. This means that each value of e_t is very close to its neighbors, e_{t-1} and e_{t+1}. A large value indicates high *negative* autocorrelation, where each e_t value is very different from the adjacent residuals.

The value of the Durbin-Watson statistic (DW) is determined using each residual value, e_t, and its previous value, e_{t-1}.

$$DW = \frac{\sum_{t=2}^{T} (e_t - e_{t-1})^2}{\sum_{t=1}^{T} e_t^2} \qquad (17.22)$$

where T is the number of observations in the time series.

The range of possible values for the Durbin-Watson statistic is from 0 to 4. The **ideal value of DW** is 2. When $DW = 2$, the errors are completely uncorrelated, and there is no violation of the independent errors assumption. As DW decreases from 2, positive autocorrelation of the errors increases. Values between 2 and 4 indicate various degrees of negative autocorrelation.*

The common problem of autocorrelated errors results from *positive* correlation between neighboring errors. When this situation occurs, the errors are not independent of one another; instead, each error is largely determined by its previous value. This implies that a similar behavior will exist for the estimated residuals in the regression model—that is, we can expect the estimated residuals to be positively correlated. The test for autocorrelation using the DW statistic is unique in that there is a certain range of DW values for which we can neither reject H_0: no autocorrelation exists, nor fail to reject it. The testing procedure uses Table A.9; the value of k in Table A.9 represents the number of predictor variables in the regression equation. The hypotheses are

H_0: no autocorrelation exists

H_a: positive autocorrelation exists

The testing procedure, using the values of d_L and d_U from Table A.9, is

Reject H_0 if $DW < d_L$.

Fail to reject H_0 if $DW > d_U$.

The test is inconclusive if $d_L \leq DW \leq d_U$.

The assumption behind the Durbin-Watson test is that the errors follow a normal distribution. An alternative to the Durbin-Watson procedure for detecting significant

---

*The Durbin-Watson statistic can be approximated using the autocorrelation of lag 1 (called r_1) discussed in Section 17.8. This approximation is $2(1 - r_1)$. Since r_1 ranges from -1 to 1, the value of DW ranges from 0 to 4 with positive autocorrelation occurring for values of r_1 close to 1, that is, for values of DW close to 0.

	t	y_t	$\hat{y}_t$	$e_t = y_t - \hat{y}_t$	$e_t - e_{t-1}$	$(e_t - e_{t-1})^2$	e_t^2
■ TABLE 17.14 Calculating the Durbin-Watson statistic for Example 17.11.	1	122	—	—	—	—	—
	2	118	106.338	11.6619	—	—	135.999
	3	106	108.903	−2.9028	−14.5647	212.129	8.426
	4	140	124.068	15.9325	18.8353	354.767	253.844
	5	86	72.724	13.2759	−2.6566	7.057	176.249
	6	96	99.498	−3.4979	−16.7738	281.359	12.235
	7	110	115.030	−5.0303	−1.5324	2.348	25.304
	8	76	84.830	−8.8301	−3.7998	14.439	77.971
	9	62	73.337	−11.3372	−2.5070	6.285	128.531
	10	104	111.442	−7.4425	3.8947	15.168	55.391
	11	135	131.037	3.9633	11.4058	130.091	15.708
	12	120	136.993	−16.9930	−20.9562	439.164	288.761
	13	105	100.750	4.2503	21.2433	451.276	18.065
	14	118	115.437	2.5631	−1.6872	2.847	6.569
	15	121	123.219	−2.2192	−4.7823	22.870	4.925
	16	132	125.394	6.6060	8.8252	77.885	43.639
						2017.7*	1251.6**

*Numerator for DW
**Denominator for DW (= SSE in Figure 17.18)

autocorrelations is the **runs test,** discussed in Chapter 19 (in Example 19.3). The runs test does *not* assume that the errors are normally distributed.

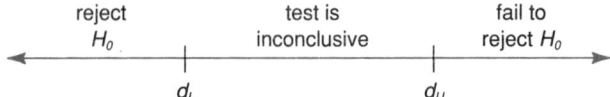

reject H_0 test is inconclusive   fail to reject H_0

d_L d_U

EXAMPLE 17.11   Determine the value of the Durbin-Watson statistic if equation 17.21 is used on the home-loan data in Table 17.13. Use $\alpha = .05$.

SOLUTION   Using Figure 17.18 to obtain the estimated values from equation 17.21, Table 17.14 shows the necessary calculations to compute the Durbin-Watson statistic. This statistic is a standard portion of the MINITAB output (see Figure 17.18).

Using Table 17.14, the Durbin-Watson statistic for this situation is

$$DW = \frac{2017.7}{1251.6} = 1.61$$

This value agrees with the boxed value for this statistic in Figure 17.18. Using $\alpha = .05$, $n = 15$, $k = 4$, and Table A.9, $d_L = .69$ and $d_U = 1.97$. Notice that the value of n is 15 because only 15 values of y_t were estimated. Also, equation 17.21 uses four variables to predict y_t, and so $k = 4$. The test of hypothesis will be to

Reject H_0 if $DW < .69$.

Fail to reject H_0 if $DW > 1.97$.

The test is inconclusive if $.69 \leq DW \leq 1.97$.

Because $DW = 1.61$ falls in the gray area between .69 and 1.97, positive autocorrelation *could* exist, but this test is inconclusive. We need additional information before we can draw any conclusion concerning possible error autocorrelation.   ■

Procedures for Correcting Autocorrelated Errors

All is not lost if the Durbin-Watson test concludes that significant autocorrelation is present in the residuals. We do not attempt to describe fully all the remedies for this situation. (For discussions of these methods, consult Bowerman and O'Connell,

Time Series and Forecasting, and Makridakis, Wheelright, and McGee, *Forecasting: Methods and Applications,* in the Further Reading at the end of this chapter.) However, the following procedures are often used to modify the model such that the "new" residuals are uncorrelated.

1. Replace y_t by the *first difference,* as discussed in Section 17.8. The new dependent variable is

$$y'_t = y_t - y_{t-1}$$

2. Replace y_t by the *percentage change* during year t,

$$z_t = \left(\frac{y_t - y_{t-1}}{y_{t-1}} \right) 100$$

3. Include the lagged dependent variables, $y_{t-1}, y_{t-2}, \ldots$ as predictors of y_t in the regression equation. This is a modification of the autoregressive technique in Section 17.8; now the lagged dependent variables are used with $X_{1,t}, X_{2,t}, \ldots$ (which might also be lagged) to predict the time series variable, y_t.

4. Improve the existing model by attempting to discover other significant predictor variables. Because the residuals include the effect of these missing variables, residual autocorrelation often can be improved by including these additional variables. This procedure offers the best solution to the autocorrelation problem but, unfortunately, is easier said than done.

5. Because the errors are autocorrelated, you can *model the error term* in much the same way we handled the situation of autocorrelated observations. This modeling involves describing each residual, e_t, by its previous values, such as

$$e_t = \phi_1 e_{t-1} + \phi_2 e_{t-2} + \cdots + \phi_j e_{t-j} + u_t$$

The value of j is arbitrary and represents the maximum period over which errors are correlated. Now the problem becomes estimating not only the coefficients of the predictor variables ($\beta_0, \beta_1, \beta_2, \ldots$) but also those of $\phi_1, \phi_2, \ldots$. The hope is that the new error term, u_t, will contain mostly noise with little autocorrelation.

COMMENTS Practically all computer packages automatically print out the Durbin-Watson statistic when performing multiple linear regression. When your data are *not* collected over time (but rather from different families, cities, companies, and so on), *this statistic is meaningless and should be ignored.*

 *Exercises*

17.56 If autocorrelation is present in a data set, what procedure used in multiple regression analysis would possibly become invalid?

17.57 What type of autocorrelation is indicated by a value of the Durbin-Watson statistic equal to zero? two? four?

17.58 Find the value of the Durbin-Watson statistic for the following yearly data.

YEAR	DATA	YEAR	DATA
1976	32	1984	42
1977	40	1985	48
1978	48	1986	54
1979	52	1987	64
1980	41	1988	42
1981	31	1989	35
1982	28	1990	32
1983	27	1991	27

a. Use the model $\hat{y}_t = b_0 + b_1 y_{t-1}$

b. Test for positive autocorrelation. Use a significance level of .05.

17.59   Test for positive autocorrelation in the residuals of the second-order process in Exercise 17.35. Use a significance level of .05.

17.60   Test for positive autocorrelation in the residuals of the fourth-order autoregressive process in Exercise 17.36. Use a significance level of .05.

17.61   Test for positive autocorrelation in the residuals of the multiple regression equation in Exercise 17.46. Use a significance level of .05.

17.62   If significant correlations are present in the residuals, what procedures can be used to modify the model so that the new residuals are uncorrelated?

Summary

In this chapter, we have looked briefly at several popular **forecasting** techniques. To cover all aspects of time series forecasting would fill an entire textbook. It is a fascinating side of statistics because anyone having a reliable "crystal ball" technique for predicting the future definitely is one step ahead of the game. We hope that this chapter has whetted your appetite to pursue further reading in this area.

Forecasting methods can be divided into two broad categories: qualitative procedures and quantitative techniques. When arriving at a **qualitative** forecast, expert opinion is used to arrive at a "best educated" estimate of future behavior. One such method is the **Delphi method,** which requires input from a team of experts. Each team member is then informed as to the responses from all other members and asked to reevaluate his or her opinion in light of this information. This process is continued for several rounds until each member of the team feels confident in his or her final decision.

Quantitative forecasting, the main emphasis of this chapter, dealt with two (sometimes overlapping) sets of procedures: time series techniques and multiple linear regression on time series data. **Time series procedures** attempt to capture the past behavior of the time series and use this information to predict future values. No external predictors are considered; only the past observations are used to describe and predict the future value of the time series variable. Time series methods include (1) the **decomposition** procedure, which extracts and extends the trend and seasonal components, (2) **exponential smoothing,** which reduces randomness and forecasts future values by using the smoothed values, and (3) **autoregressive** forecasting, which predicts future values by using a linear combination of past values.

There are various exponential smoothing procedures; the proper one to use depends on the nature of the time series. **Simple exponential smoothing** works best when the time series contains neither trend nor seasonality. **Linear exponential smoothing** is better for a time series that does contain trend but has no seasonality, and **linear and seasonal exponential smoothing** should be used for a time series that has both components.

■ **Exponential Smoothing**

	STRUCTURE OF TIME SERIES	
TYPE	Contains Trend	Contains Seasonality
Simple Exponential Smoothing	NO	NO
Linear Exponential Smoothing	YES	NO
Linear and Seasonal Exponential Smoothing	YES	YES

There are many factors that determine the strengths of any forecasting procedure, including the (1) time horizon of the forecast, (2) stationarity of the data, and (3) presence of trend, seasonality, or cyclical activity. To measure the forecast accuracy of a particular method, you can calculate the predictive mean squared error (MSE), the mean absolute deviation (MAD), or the mean absolute percentage error

(MAPE). The **MSE** is found by squaring each of the residuals obtained by applying this technique to the past observations and then deriving the average of these squared residuals. This measure is very sensitive to one or two very large residuals. The **MAD** is calculated by averaging the absolute values of the residuals and is less sensitive to a single large residual. The **MAPE** uses the *relative* error of each forecast value to arrive at a measure of prediction accuracy. It is very useful for comparing the accuracy of a particular forecasting technique on two different time series, since the effect of the magnitude of the observations has been removed.

The advantage of the time series methods is that there is no need to search for external predictors to explain the behavior of the dependent variable. One disadvantage is that the patterns within the observed values can be extremely complex and difficult to determine. Such methods often are hard to "sell" to managers, who may not be able to understand the technique.

Multiple linear regression forecasting requires additional input data; for each time period, data are recorded for each predictor (independent) variable as well as for the dependent variable, y_t. Such data are **time series data,** since the variables are observed across time. This situation differs from many of the earlier applications of linear regression, which used data gathered at one point in time, that is, **cross-sectional data.** In a time series regression model, the predictor variables can include **lagged** dependent or lagged predictor variables as well as **dummy** variables to represent seasonality or the occurrence (or nonoccurrence) of a particular event (such as an election year).

When you use multiple linear regression techniques on time series data, you often will violate the assumption of independent errors. The **Durbin-Watson statistic** can be used to test for significant autocorrelation in the regression residuals. If significant autocorrelation is present, the tests of hypothesis and confidence intervals contained in the regression output are unreliable.

The advantages of multiple linear regression on time series data include: (1) it is a very flexible approach, in that a wide variety of explanatory variables can be included in the model; (2) it allows for lagging the predictor variables, including lagged values of the dependent variable; and (3) it is generally easier to explain to managers, who can see easily which variables are predicting the behavior of the dependent variable. On the other hand, residual autocorrelation often is a problem. It may be caused by missing variables in the prediction equation, which typically are extremely difficult to determine. Also, a very complex pattern within the observed time series may be difficult to capture using a linear combination of predictor variables. Finally, forecasting with this technique becomes extremely difficult unless lagged variables are used. Dummy variables can be included, provided that they can be predicted with certainty. A dummy variable representing an election year would be acceptable, whereas one representing the occurrence (or nonoccurrence) of an earthquake would not be.

 *Further Reading*

Bowerman, B. L., and R. T. O'Connell. *Time Series and Forecasting.* 2d ed. North Scituate, Mass.: Duxbury, 1987.

Hanke, J., A. Reitsch, and J. P. Dickson. *Statistical Decision Models for Management.* Newton, Mass.: Allyn and Bacon, 1984.

Makridakis, S., S. C. Wheelright, and V. E. McGee, *Forecasting: Methods and Applications.* 2d. ed. New York: John Wiley, 1983.

Makridakis, S., and S. C. Wheelright, eds. *Handbook of Forecasting: A Manager's Guide.* 2d ed. New York: John Wiley, 1987.

Mendenhall, W. and T. Sincich. *A Second Course in Business Statistics: Regression Analysis*. 3d ed. San Francisco: Dellen, 1989.

Summary of Formulas

Forecasting Models

Naive:

$$\hat{y}_{t+1} = y_t$$

Using seasonality and trend:

$$\hat{y}_{t+1} = [b_0 + b_1(t + 1)]$$
$$\cdot \text{ (seasonal index for period } t + 1)$$

Simple exponential smoothing:

$$\hat{y}_{t+1} = S_t$$

where $S_t = Ay_t + (1 - A)S_{t-1}$ and
$0 < A < 1$

Using Holt's method (exponential smoothing for a time series containing trend):

$$\hat{y}_{t+1} = S_t + b_t$$

where

$$S_t = Ay_t + (1 - A)(S_{t-1} + b_{t-1})$$
$$b_t = B(S_t - S_{t-1}) + (1 - B)b_{t-1}$$

and $0 < A < 1, 0 < B < 1$

Using Winter's method (exponential smoothing for a time series containing trend and seasonality):

$$\hat{y}_{t+m} = (S_t + mb_t) \cdot F_{t+m-L}$$

where

m is the number of periods ahead to be forecast

L is 4 for quarterly data, 12 for monthly data

$$S_t = A\left(\frac{y_t}{F_{t-L}}\right) + (1 - A)(S_{t-1} + b_{t-1})$$

$$F_t = B\left(\frac{y_t}{S_t}\right) + (1 - B)F_{t-L}$$

$$b_t = C(S_t - S_{t-1}) + (1 - C)b_{t-1}$$

and $0 < A < 1, 0 < B < 1, 0 < C < 1$

Using a pth-order autoregressive model:

$$\hat{y}_{t+1} = b_0 + b_1 y_t + b_2 y_{t-1} + \cdots + b_p y_{t-p+1}$$

Additional Formulas

Sample autocorrelation of lag k:

$$r_k = \frac{\sum_{t=1}^{T-k} (y_t - \bar{y})(y_{t+k} - \bar{y})}{\sum_{t=1}^{T} (y_t - \bar{y})^2}$$

Durbin-Watson statistic:

$$DW = \frac{\sum_{t=2}^{T} (e_t - e_{t-1})^2}{\sum_{t=1}^{T} e_t^2}$$

Measurements of Forecast Error

$$MAD = \frac{\sum |e_t|}{n}$$

$$\text{(predictive) } MSE = \frac{\sum e_t^2}{n}$$

$$MAPE = \frac{\sum \left|\frac{e_t}{y_t}\right|}{n}$$

Review Exercises

17.63 A set of monthly data has been gathered over three years from January 1988 to December 1990. From these data, the seasonal indexes for the 12 months are found to be

$S_1 = 0.75$	$S_2 = 0.85$	$S_3 = 0.95$	$S_4 = 0.99$	$S_5 = 0.90$	$S_6 = 1.01$
$S_7 = 1.20$	$S_8 = 1.10$	$S_9 = 1.15$	$S_{10} = 1.05$	$S_{11} = 1.11$	$S_{12} = 0.94$

The least squares line through the deseasonalized data is found to be

$$\hat{d}_t = 2.73 + 0.62t$$

for the 36 monthly periods. Find the forecast for monthly periods 37, 38, 39, 40, and 41.

17.64   The total monthly trading volume for the stock of Xcon Corp is given for a four-year period. Units are in millions of shares.

YEAR	JAN	FEB	MAR	APR	MAY	JUNE	JULY	AUG	SEPT	OCT	NOV	DEC
1988	1.1	1.2	1.4	1.3	1.2	1.5	1.9	1.8	1.3	1.1	0.9	0.8
1989	1.3	1.4	1.5	1.4	1.4	1.7	2.1	1.9	1.6	1.4	1.1	1.1
1990	1.4	1.4	1.6	1.7	1.6	1.9	2.3	2.2	1.8	1.7	1.4	1.2
1991	1.3	1.5	1.5	1.6	1.7	2.0	2.2	2.3	1.9	1.7	1.5	1.4

Using simple exponential smoothing with the parameter $A = .3$, find the forecasted volume for the first month of 1992.

17.65   The number of employees at Computeron has fluctuated over the past nine years. The following table lists the number of employees on the payroll at Computeron at the end of each year for the years 1983 to 1991. The company would like you to forecast employment in 1992 by using the linear exponential smoothing technique.

YEAR	EMPLOYEES
1983	1030
1984	1020
1985	1041
1986	1050
1987	1062
1988	1075
1989	1130
1990	1135
1991	1175

a.   Using the initial estimate of the slope to be zero with $A = .2$ and $B = .2$, determine the predicted value for the employment at the end of 1992.

b.   Using the least squares estimate for the slope from the first three years of data, redo part a.

c.   Compare the predicted values in parts a and b.

17.66   Two forecasting procedures were used to forecast the 12 quarters from 1989 to 1991. The forecast errors for each quarter are given below. Compute the predictive MSE and MAD for each forecasting procedure. Interpret the results.

YEAR	QUARTER	PROCEDURE 1 FORECAST ERROR	PROCEDURE 2 FORECAST ERROR
1989	1	−1.0	.7
	2	.5	−.5
	3	−2.1	−.2
	4	2.5	.9
1990	1	.9	.1
	2	2.1	.2
	3	−1.3	−.3
	4	1.6	.9
1991	1	2.7	11.2
	2	−1.9	−.1
	3	2.4	.2
	4	−1.2	−.2

17.67   The amount of money spent on research and development by Energy Today in finding economical uses of alternative fuels is given over a four-year period. Units are in $10,000.

YEAR	QUARTER 1	QUARTER 2	QUARTER 3	QUARTER 4
1988	4.2	4.5	4.8	4.0
1989	4.3	4.7	5.6	4.4
1990	4.6	4.9	5.7	4.5
1991	4.7	5.0	5.8	4.7

Use Holt's two-parameter linear exponential smoothing technique to obtain a forecast for each of the quarters, omitting the first time period. Let $A = .3$ and $B = .2$. Use the least squares estimate of the trend from the first five periods for the initial value of the slope. Calculate the predictive MSE.

17.68   The following table represents the number of homes sold monthly in a growing community in California over the past five years:

MONTH	1987	1988	1989	1990	1991
Jan	48	83	117	156	192
Feb	51	85	121	158	195
Mar	65	98	133	170	208
Apr	67	102	134	173	209
May	69	105	139	176	214
June	78	114	149	184	221
July	81	118	153	187	224
Aug	85	123	155	191	227
Sept	67	105	132	173	210
Oct	71	109	142	175	213
Nov	72	113	145	179	217
Dec	75	115	148	184	220

a.   Calculate the autocorrelations through lag 15.

b.   Calculate the first differences for the time series.

c.   Calculate the autocorrelations for the difference data obtained in question b.

d.   Comment on the autocorrelations in parts a and c.

17.69   National Finance Company provides short-term loans to consumers to finance household goods. The amount of interest received quarterly is given below in units of $10,000.

YEAR	QUARTER 1	QUARTER 2	QUARTER 3	QUARTER 4
1987	20	31	39	42
1988	28	35	43	45
1989	31	38	45	49
1990	35	40	43	52
1991	38	44	48	56

Determine the multiple regression equation that takes into account trend and seasonality.

17.70   An independent gas station allows its customers to buy gasoline on credit. The amount of credit on the books for the 20 quarters of the years 1987 through 1991 follows. Find the multiple regression equation that takes into account trend and seasonality. The figures in the table are in units of $10,000.

YEAR	QUARTER 1	QUARTER 2	QUARTER 3	QUARTER 4
1987	2.3	2.7	3.4	3.0
1988	2.4	3.0	3.6	3.2
1989	2.6	3.1	3.8	3.4
1990	3.0	3.3	4.0	3.2
1991	3.2	3.4	4.4	3.5

17.71   Using Holt's two-parameter linear exponential technique, find the smoothed value at time $t = 12$, where the observed value at $t = 12$ is 13.6 and the observed value at $t = 11$ is 12.1. Also let $S_{11} = 7.4$, $S_{10} = 10.4$, $b_{11} = 0.6$, $A = .1$ and $B = .2$.

17.72   In Holt's two-parameter linear exponential smoothing technique, why do you think the values of A and B are typically less than or equal to .3?

17.73   Explain what one should look for in determining the appropriate forecasting procedure.

17.74   The total number of yearly work stoppages (strikes) in the United States is given below for the years 1967 through 1989. Find the first-order and second-order autoregressive models. Calculate the predictive MSE for each model.

YEAR	NUMBER OF STOPPAGES	YEAR	NUMBER OF STOPPAGES
1967	381	1978	219
1968	392	1979	235
1969	412	1980	187
1970	381	1981	145
1971	298	1982	96
1972	250	1983	81
1973	317	1984	62
1974	424	1985	54
1975	235	1986	69
1976	231	1987	46
1977	298	1988	40
		1989	51

(*Source: The World Almanac and Book of Facts* (1991): 142.)

17.75   The number of unemployed (in units of thousands) in the U.S. labor force is given below for the years 1970 to 1989. Using the linear exponential smoothing technique with an initial estimate of the slope to be zero with $A = .2$ and $B = .2$, determine the predicted value for the number of unemployed at the end of 1990.

YEAR	NUMBER OF UNEMPLOYED	YEAR	NUMBER OF UNEMPLOYED
1970	4,093	1980	7,637
1971	5,016	1981	8,273
1972	4,882	1982	10,678
1973	4,365	1983	10,717
1974	5,156	1984	8,539
1975	7,929	1985	8,312
1976	7,406	1986	8,237
1977	6,991	1987	7,425
1978	6,202	1988	6,701
1979	6,137	1989	6,528

(*Source:* U.S. Department of Labor, Bureau of Labor Statistics, *Employment and Earnings,* (July 1990): 11.)

17.76   The manager of an assembly plant is interested in predicting the time spent on maintenance of machines at the plant. The data in C1 of the MINITAB computer printout represents the number of hours spent on maintenance of machines at the assembly plant per quarter. The data in C2 represent the time, starting with period one and incrementing by one for each quarter. The dummy variables in C3 to C5 represent the quarter of the year. Test for positive autocorrelation in the residuals for the model in the MINITAB computer printout. Use a significance level of .05. Determine the MAD, MAPE, and predictive MSE for the forecasts.

```
MTB > NAME C1 = 'Y'
MTB > NAME C2 = 'TIME'
MTB > NAME C3 = 'DUMMY1'
MTB > NAME C4 = 'DUMMY2'
MTB > NAME C5 = 'DUMMY3'
MTB > NAME C6 = 'RESID'
MTB > PRINT C1-C5
  ROW       Y    TIME  DUMMY1  DUMMY2  DUMMY3

    1    77.10     1       1       0       0
    2    46.78     2       0       1       0
    3   179.43     3       0       0       1
    4   228.70     4       0       0       0
    5   152.90     5       1       0       0
    6   122.58     6       0       1       0
    7   247.65     7       0       0       1
    8   289.34     8       0       0       0
    9   213.54     9       1       0       0
   10   190.80    10       0       1       0
   11   323.45    11       0       0       1
   12   387.88    12       0       0       0
   13   285.55    13       1       0       0
   14   255.23    14       0       1       0
   15   384.09    15       0       0       1
   16   467.47    16       0       0       0
```

```
MTB > REGRESS Y IN C1 USING 4 PREDICTORS IN C2-C5;
SUBC> RESID IN C6;
SUBC> DW.

The regression equation is
Y = 163 + 18.0 TIME - 107 DUMMY1 - 153 DUMMY2 - 41.7 DUMMY3

Predictor         Coef        Stdev      t-ratio
Constant        163.086       6.975       23.38
TIME             18.0262      0.5199       34.67
DUMMY1         -106.996       6.759      -15.83
DUMMY2         -153.447       6.658      -23.05
DUMMY3          -41.666       6.597       -6.32

s = 9.300        R-sq = 99.5%     R-sq(adj) = 99.3%

Analysis of Variance

SOURCE         DF         SS           MS
Regression      4       197336        49334
Error          11          951           86
Total          15       198288

SOURCE         DF      SEQ SS
TIME            1      143889
DUMMY1          1        4227
DUMMY2          1       45770
DUMMY3          1        3451

Durbin-Watson statistic = 1.71

MTB > PRINT C6
RESID
    2.9846      1.0895      3.9322     -6.4902      6.6799      4.7849      0.0475
  -17.9551     -4.7847      0.9002      3.7427      8.4802     -4.8796     -6.7744
   -7.7222     15.9656
```

17.77   The data in Exercise 17.76 is analyzed in the following MINITAB computer printout with a model that contains the independent variables time and the *Y* variable lagged by four time periods. Test for positive autocorrelation in the residuals for the model in the MINITAB computer printout. Determine the MAD, MAPE, and predictive MSE for the forecasts and compare these values to those of the model in Exercise 17.76.

```
MTB > LAG BY 4 OF C1, PUT INTO C7
MTB > NAME C7 = 'LAG4'
MTB > PRINT C1 C2 C7
 ROW       Y      TIME      LAG4

   1     77.10       1        *
   2     46.78       2        *
   3    179.43       3        *
   4    228.70       4        *
   5    152.90       5      77.10
   6    122.58       6      46.78
   7    247.65       7     179.43
   8    289.34       8     228.70
   9    213.54       9     152.90
  10    190.80      10     122.58
  11    323.45      11     247.65
  12    387.88      12     289.34
  13    285.55      13     213.54
  14    255.23      14     190.80
  15    384.09      15     323.45
  16    467.47      16     387.88

MTB > REGRESS Y IN C1 USING 2 PREDICTORS IN C2, C7;
SUBC> RESIDUALS IN C6;
SUBC> DW.

The regression equation is
Y = 71.0 - 0.77 TIME + 1.04 LAG4

12 cases used 4 cases contain missing values

Predictor         Coef        Stdev      t-ratio
Constant         70.97       10.87        6.53
TIME             -0.770       1.679       -0.46
LAG4             1.04298     0.06086      17.14

s = 11.57        R-sq = 98.9%     R-sq(adj) = 98.7%
```

```
Analysis of Variance

SOURCE          DF              SS              MS
Regression      2           113258          56629
Error           9             1205            134
Total          11           114462

SOURCE          DF          SEQ SS
TIME            1           73946
LAG4            1           39311

Durbin-Watson statistic = 1.44

MTB > PRINT C6
RESID
        *           *           *           *       5.3684      7.4418     -5.0699
    -13.9978     -9.9693     -0.3159      2.6582    24.3765      1.8745     -3.9575
    -12.6794      4.2715
```

Case Study

Forecasting Business Cycles Using a Leading Economic Indicator

Many forecasters use barometric methods to forecast business and economic conditions. Barometric methods are based on the relatively consistent relationships that exist between the movement of different economic variables over time. That is, changes in some time series appear to consistently follow changes in one or more other time series. For example, lower interest rates "lead" an increase in housing starts.

Leading economic indicators are time series that anticipate business cycle turns. Other time series, called coincident indicators, roughly coincide with the business cycle. Lagging indicators lag behind the business cycle turns. Should the forecaster have the good happenstance to discover a time series that leads the one he or she is attempting to forecast, the leading series can be used as a barometer for forecasting a short-term change in the lagging time series.

The U.S. Department of Commerce publishes many time series of interest to analysts and forecasters in its monthly *Survey of Current Business*. Among these time series are a variety of common-stock indexes, including the NYSE composite in-

dex. This index covers all of the issues listed on the New York Stock Exchange and is based on a value of 50 for December 31, 1965. The NYSE composite index is considered a leading indicator for business cycles. The index reflects an aggregation of profit expectations by businesspeople and therefore a composite expectation of the level of business activity.

The following table of data gives a monthly NYSE composite index for January 1985 through December 1991, as reported by *Survey of Current Business*. Using the case study questions, analyze this time series to determine a good forecasting model that predicts future values for the NYSE index.

(*Source:* U.S. Department of Commerce, Bureau of Economic Analysis, *Survey of Current Business*, January 1986, 1987, 1988, 1989, 1990, 1991.)

Case Study Questions

1. Calculate 12-month moving averages for the time series and graph both the moving averages and the raw data. Is there a long-term trend for the series? Is there a seasonal component apparent in the series?
2. Based on your analysis in question 1, which of the three exponential smoothing techniques do you think would be most appropriate?
3. Use the three exponential smoothing techniques to predict the NYSE index for the months given in

■ NYSE Index Data.

MONTH/YEAR	1985	1986	1987	1988	1989	1990
Jan	99.11	120.16	151.17	140.55	160.40	187.96
Feb	104.73	126.43	160.23	145.13	165.08	182.55
Mar	103.92	133.97	166.43	149.88	164.60	186.26
Apr	104.66	137.27	163.88	148.46	169.38	185.61
May	107.00	137.37	163.00	144.94	175.30	191.35
June	109.52	140.82	169.58	152.72	180.76	196.68
July	111.64	138.32	174.28	152.12	185.15	196.61
Aug	109.09	140.91	184.18	149.25	192.94	181.46
Sept	106.62	137.06	178.39	151.47	193.02	173.24
Oct	107.57	136.74	157.13	156.36	192.49	168.05
Nov	113.93	140.84	137.21	152.67	188.50	172.21
Dec	119.23	142.12	134.88	155.35	192.67	179.57

the data. Use $A = .1$, $B = .3$, and $C = .1$. Use procedure 1 to set the initial estimates of the seasonal factors and trend.

4. Calculate the predictive MSE for each of the three exponential smoothing techniques. Which model best fits this time series data? Based on your answer, predict the NYSE index for the next two months.

5. Find the autocorrelations using the raw data and the first differences. Which of the two provides a stationary series?

6. Use a second-order autoregressive model with the stationary series found in question 5 to predict the NYSE index for the next two months. What is the predictive MSE for this model?

SPSS

⊿ Solution
Example 17.9

Example 17.9 is divided into two computer runs. In the first run we used the autocorrelation function to determine the significant autocorrelations among the given lag periods. In the second run, the significant lag period (lag 4) was included in the autoregression equation. The SPSS program listings in Figures 17.20 and 17.21 were used to perform the two procedures.

The TITLE command names the SPSS run.

The DATA LIST command gives each variable a name and describes the data as being in free form.

The BEGIN DATA command indicates to SPSS that the input data immediately follow.

The next 20 lines contain the data values, which represent time series data over 20 time periods. For example, 1 5.56 represents the observed value of 5.56 for time period 1.

The END DATA statement indicates the end of the data.

The VARIABLE LABELS statement assigns a descriptive label to the variables. This descriptive label is substituted for the variable name when output is printed.

The LIST command causes the printing of the variables T and YT.

■ **FIGURE 17.20**
Input for SPSS to determine significant autocorrelations.

```
TITLE    KENS AUTO PAINT SHOP - AUTOCORRELATION FUNCTION.
DATA LIST FREE / T YT.
BEGIN DATA.
1 5.56
2 16.36
3 2.12
4 3.15
5 5.11
6 15.21
7 5.72
8 2.65
9 4.12
10 14.33
11 5.25
12 6.75
13 6.31
14 15.02
15 2.83
16 4.56
17 4.81
18 16.82
19 4.75
20 8.54
END DATA.
VARIABLE LABELS T 'TIME'
                YT 'OBSERVED VALUES'.
LIST VAR=ALL.
ACF YT
   /MXAUTO=20.
```

■ **FIGURE 17.21**
Input for SPSS to determine the autocorrelation equation using one lagged variable.

```
TITLE   KENS AUTO PAINT SHOP - AUTOREGRESSIVE MODEL.
DATA LIST FREE / T YT.
BEGIN DATA.
1 5.56
2 16.36
3 2.12
4 3.15
5 5.11
6 15.21
7 5.72
8 2.65
9 4.12
10 14.33
11 5.25
12 6.75
13 6.31
14 15.02
15 2.83
16 4.56
17 4.81
18 16.82
19 4.75
20 8.54
END DATA.
VARIABLE LABELS T 'TIME'
              YT 'OBSERVED VALUES'.
COMPUTE YT4=LAG(YT,4).
PRINT / T YT YT4.
REGRESSION VARIABLES=YT YT4/DEP=YT/ENTER/SAVE PRED(PREDX).
PRINT / YT PREDX.
EXECUTE.
```

The ACF YT statement means that the YT time series is to be analyzed, and we wish to produce statistics for model identification.

The commands in Figure 17.21 are the same as those in Figure 17.20, with the following exceptions:

The COMPUTE YT4 = LAG(YT,4) command provides YT4 with the value of YT for the fourth case before the current one. For example, the value of YT in case 16 is 4.56. Therefore, the value of YT4 in case 20 is also 4.56.

■ **FIGURE 17.22**
SPSS output from Figure 17.20.

```
                 KENS AUTO PAINT SHOP - AUTOCORRELATION FUNCTION
Autocorrelations:   YT          OBSERVED VALUES

      Auto-  Stand.
Lag   Corr.    Err. -1   -.75   -.5 -.25   0   .25  .5   .75   1
                     |----+----+----+----+----+----+----+----|
  1   -.315   .208              . ******|         .
  2   -.260   .202              .  *****|         .
  3   -.286   .197              . ******|         .
 →4    .678   .191              .       |*******.******
  5   -.213   .185               .  ****|        .
  6   -.168   .178               .   ***|        .
  7   -.156   .172               .   ***|        .
  8    .512   .165               .      |******.***
  9   -.235   .158                .*****|       .
 10   -.134   .151                .  ***|       .
 11   -.104   .143                .   **|       .
 12    .388   .135                 .    |****.***
 13   -.155   .126                 . ***|      .
 14   -.060   .117                 .   *|      .
 15   -.123   .107                  . **|     .
 16    .206   .095                  .   |****
 17   -.101   .083                   .**|  .
 18    .031   .067                   .  |* .

Plot Symbols:       Autocorrelations *      Two Standard Error Limits .

Total cases:  20        Computable first lags:  19
```

r_k

Use this lagged variable in the autoregression equation

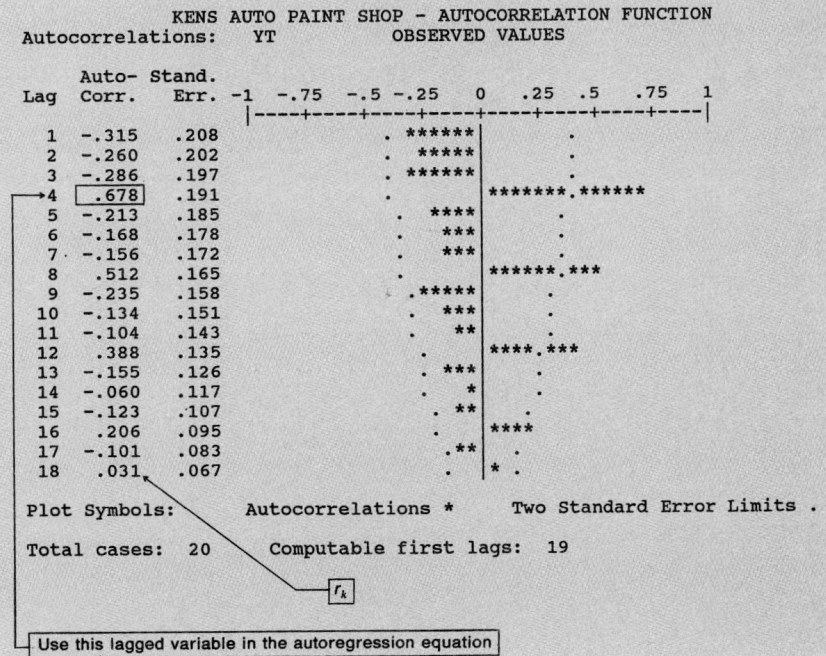

The REGRESSION VARIABLES statement specifies that variables YT and YT4 are to be used in the regression equation and that YT is the dependent variable.

The ENTER option specifies that all independent variables are to be applied in the equation simultaneously.

The SAVE option saves the predicted values under the name PREDX.

Figure 17.22 shows the SPSS output obtained by executing the listing in Figure 17.20, while Figure 17.23 shows the SPSS output obtained by executing the listing in Figure 17.21.

■ **FIGURE 17.23**   SPSS output from Figure 17.21.

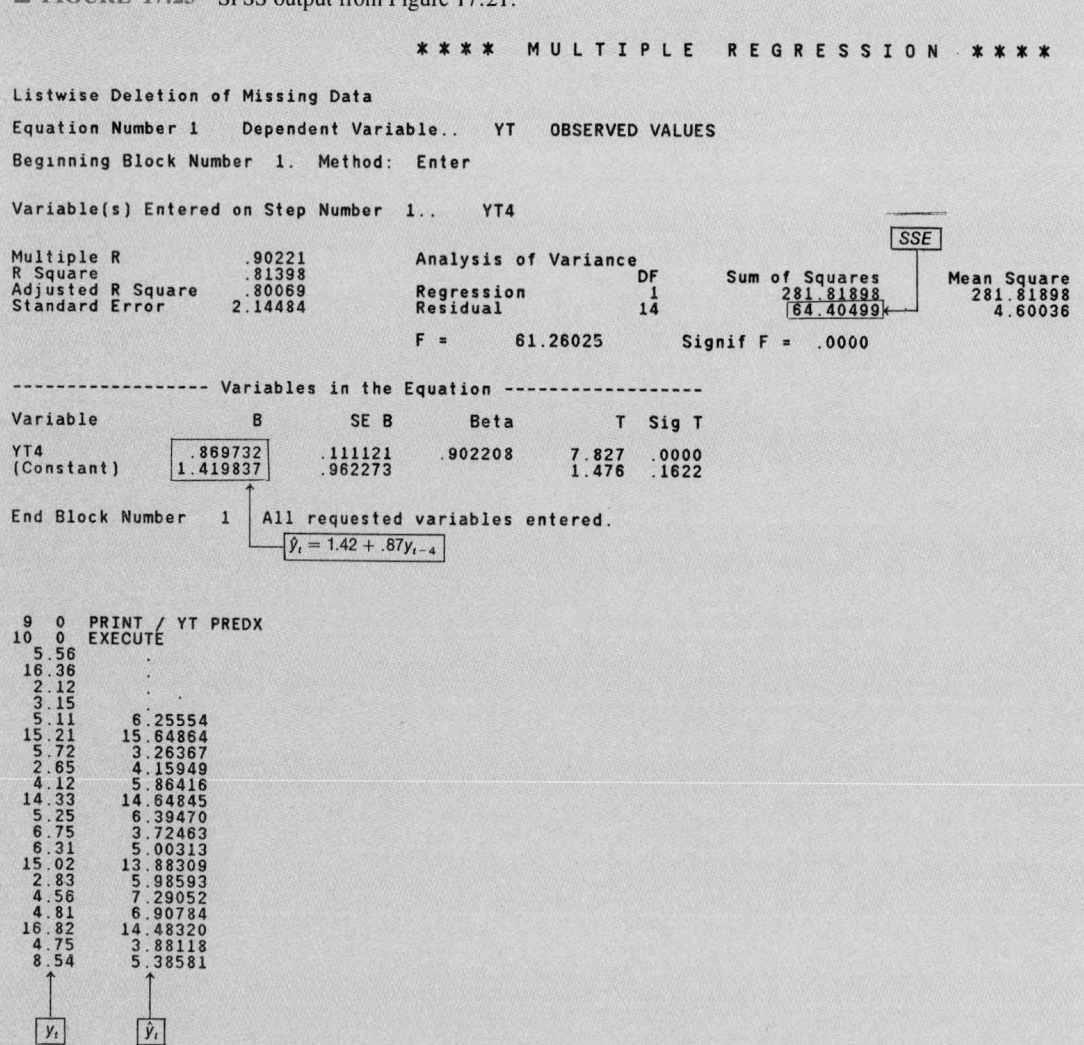

SPSS

Solution
Section 17.9
Example

We used multiple linear regression on time series data to predict the number of home loans financed by Liberty Savings and Loan. The predictor variables used were average interest rate, advertising expenditure, an election year dummy variable, and a seasonal dummy variable. The SPSS program listing in Figure 17.24 was used to compute the regression equation from the time series data.

The TITLE command names the SPSS run.

The DATA LIST command gives each variable a name and describes the data as being in free form.

The BEGIN DATA command indicates to SPSS that the input data immediately follow.

The next 16 lines contain the data values that represent time series data over 16 time periods. The first line, for example, represents the year 1983, the first time period, 122 home loans, an average interest rate of 11.8%, and so forth.

The END DATA statement indicates the end of the data.

The VARIABLE LABELS statement assigns a descriptive label to the variables. This descriptive label is substituted for the variable name when the output is printed.

The COMPUTE X1TLAG=LAG(X1T,1) command provides X1TLAG with the value of X1T for the case before the current one. For example, during the first half of 1983, the average interest rate (X1T) was 11.8%. Therefore, the value of X1TLAG for the last half of 1983 is also 11.8%.

The COMPUTE X2TLAG=LAG(X2T,1) command provides X2TLAG with the value of X2T for the case before the current one.

FIGURE 17.24
Input for SPSS or
SPSS/PC.

```
TITLE    LIBERTY SAVINGS AND LOAN.
DATA LIST FREE / YEAR TIME YT X1T X2T X3T X4T.
BEGIN DATA.
1983 1  122 11.8 10.4  0 1
1983 2  118 12.4  6.7  0 0
1984 3  106 11.0  7.5  1 1
1984 4  140 14.5  7.8  1 0
1985 5   86 11.0  5.1  0 1
1985 6   96 10.1  6.8  0 0
1986 7  110 12.9  6.8  0 1
1986 8   76 14.8  9.1  0 0
1987 9   62 10.5  7.5  0 1
1987 10 104 09.8  5.1  0 0
1988 11 135 10.1  8.8  1 1
1988 12 120 10.8  4.3  1 0
1989 13 105 10.0  7.1  0 1
1989 14 118 09.8  8.6  0 0
1990 15 121 09.5  8.8  0 1
1990 16 132 09.7  7.4  0 0
END DATA.
VARIABLE LABELS    YT 'NUMBER OF HOME LOANS'
                   X1T 'AVERAGE INTEREST RATE'
                   X2T 'ADVERTISING EXPENDITURE'
                   X3T 'ELECTION YEAR VARIABLE'
                   X4T 'SEASONAL VARIABLE'.
COMPUTE X1TLAG=LAG(X1T,1).
COMPUTE X2TLAG=LAG(X2T,1).
PRINT / YEAR TIME YT X1T X2T X3T X4T X1TLAG X2TLAG.
REGRESSION VARIABLES=YT X1TLAG X2TLAG X3T X4T/
           DEPENDENT=YT/ENTER/RESID=DURBIN.
```

The REGRESSION VARIABLES statement specifies that variables YT, X1TLAG, X2TLAG, X3T, and X4T are to be used in the regression equation and that YT is the dependent variable.

The ENTER option specifies that all independent variables are to be applied in the equation simultaneously.

The RESID option specifies that the DURBIN-WATSON statistic is to be computed.

Figure 17.25 shows the SPSS output obtained by executing the listing in Figure 17.24.

■ **FIGURE 17.25**   SPSS output.

```
* * * *   M U L T I P L E   R E G R E S S I O N   * * * *

Listwise Deletion of Missing Data

Equation Number 1    Dependent Variable..   YT   NUMBER OF HOME LOANS

Block Number  1.  Method: Enter

Variable(s) Entered on Step Number   1..    X4T        SEASONAL VARIABLE
                                     2..    X3T        ELECTION YEAR VARIABLE
                                     3..    X2TLAG
                                     4..    X1TLAG

Multiple R            .90080        Analysis of Variance
R Square              .81144                      DF    Sum of Squares     Mean Square
Adjusted R Square     .73601        Regression     4        5385.98216      1346.49554
Standard Error      11.18757        Residual      10        1251.61784       125.16178

                                    F =    10.75804    Signif F =  .0012

----------------- Variables in the Equation -----------------

Variable            B          SE B         Beta         T      Sig T

X4T          1.471218      6.390175      .034892       .230    .8226
X3T         17.755207      6.664778      .373251      2.664    .0237
X2TLAG       2.839342      1.942933      .217353      1.461    .1746
X1TLAG     -10.260348      1.945389     -.795239     -5.274    .0004
(Constant) 197.881074     23.091352                   8.569    .0000

End Block Number   1   All requested variables entered.
```

$$\hat{Y}_t = 197.88 - 10.26X_{1,t-1} + 2.84X_{2,t-1} + 17.76X_{3,t} + 1.47X_{4,t}$$

```
* * * *   M U L T I P L E   R E G R E S S I O N   * * * *
Equation Number 1    Dependent Variable..   YT   NUMBER OF HOME LOANS

Residuals Statistics:

             Min       Max      Mean   Std Dev    N

*PRED     72.7241  136.9930  108.6000   19.6141   15
*RESID   -16.9930   15.9325     .0000    9.4552   15
*ZPRED    -1.8291    1.4476     .0000    1.0000   15
*ZRESID   -1.5189    1.4241     .0000     .8452   15

Total Cases =      16

Durbin-Watson Test =   1.61206
```

SAS

Solution
Example 17.4

We used simple exponential smoothing on time series data to predict the average attendance at the Jefferson County Civic Center (Table 17.3). The SAS program listing in Figure 17.26 was used to compute the predicted values and residuals for the attendance data using a smoothing constant of $A = .1$.

The TITLE command names the SAS run.

The DATA command gives the data a name.

The INPUT command names and gives the correct order for the different fields on the data lines.

The LABEL statement assigns labels for variables TIME and A. These labels are substituted for the variable names when the output is printed.

The CARDS command indicates to SAS that the input data immediately follow.

The next 13 lines contain the data values that represent time series data over 13 time periods. The first line, for example, represents the year 1978 when the attendance was 5 (thousand).

The PROC FORECAST command requests the smoothed forecast.

OUT = B names an output data set to hold the forecast. OUTEST = C names an output data set to hold the parameter estimates. METHOD specifies the exponential smoothing method be used. TREND = 1 selects the constant trend model. WEIGHT = .1 specifies the smoothing weight. OUTALL requests all output. LEAD = 1 specifies the number of periods ahead for the forecast. ID and VAR mean identify by time and that A is the input data variable to be analyzed.

The PROC PRINT command requests that the generated data and estimated values be printed.

Figure 17.27 shows a portion of the SAS output obtained by executing the listing in Figure 17.26.

FIGURE 17.26
Input for SAS
(mainframe or
micro version).

```
TITLE 'SIMPLE EXPONENTIAL SMOOTHING';
DATA ATTEND;
  INPUT TIME A;
  LABEL TIME='YEAR'
             A='ATTENDANCE';
  CARDS;
1 5.0
2 8.0
3 2.1
4 7.1
5 4.8
6 2.0
7 7.8
8 5.0
9 14.1
10 13.0
11 13.5
12 14.2
13 14.0
;
PROC FORECAST DATA=ATTEND OUT=B OUTEST=C METHOD=EXPO
     TREND=1 WEIGHT=.1 OUTALL LEAD=1;
  ID TIME;
  VAR A;
PROC PRINT DATA=B;
  TITLE 'SMOOTHING FORECAST';
```

■ **FIGURE** 17.27
SAS output.

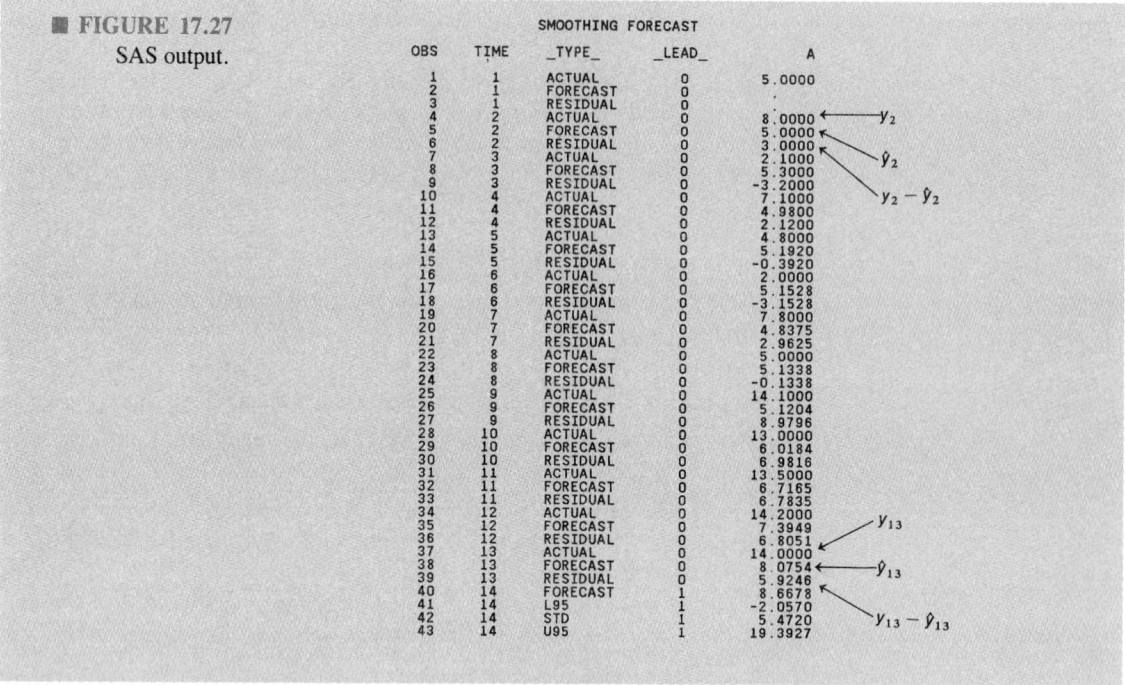

SMOOTHING FORECAST

OBS	TIME	_TYPE_	_LEAD_	A
1	1	ACTUAL	0	5.0000
2	1	FORECAST	0	.
3	1	RESIDUAL	0	.
4	2	ACTUAL	0	8.0000
5	2	FORECAST	0	5.0000
6	2	RESIDUAL	0	3.0000
7	3	ACTUAL	0	2.1000
8	3	FORECAST	0	5.3000
9	3	RESIDUAL	0	-3.2000
10	4	ACTUAL	0	7.1000
11	4	FORECAST	0	4.9800
12	4	RESIDUAL	0	2.1200
13	5	ACTUAL	0	4.8000
14	5	FORECAST	0	5.1920
15	5	RESIDUAL	0	-0.3920
16	6	ACTUAL	0	2.0000
17	6	FORECAST	0	5.1528
18	6	RESIDUAL	0	-3.1528
19	7	ACTUAL	0	7.8000
20	7	FORECAST	0	4.8375
21	7	RESIDUAL	0	2.9625
22	8	ACTUAL	0	5.0000
23	8	FORECAST	0	5.1338
24	8	RESIDUAL	0	-0.1338
25	9	ACTUAL	0	14.1000
26	9	FORECAST	0	5.1204
27	9	RESIDUAL	0	8.9796
28	10	ACTUAL	0	13.0000
29	10	FORECAST	0	6.0184
30	10	RESIDUAL	0	6.9816
31	11	ACTUAL	0	13.5000
32	11	FORECAST	0	6.7165
33	11	RESIDUAL	0	6.7835
34	12	ACTUAL	0	14.2000
35	12	FORECAST	0	7.3949
36	12	RESIDUAL	0	6.8051
37	13	ACTUAL	0	14.0000
38	13	FORECAST	0	8.0754
39	13	RESIDUAL	0	5.9246
40	14	FORECAST	1	8.6678
41	14	L95	1	-2.0570
42	14	STD	1	5.4720
43	14	U95	1	19.3927

y_2

$\hat{y}_2$

$y_2 - \hat{y}_2$

y_{13}

$\hat{y}_{13}$

$y_{13} - \hat{y}_{13}$

SAS

▨ **Solution**
Example 17.5

We used linear exponential smoothing on time series data to predict the city taxes collected in Jackson City over the past 20 quarters (Table 17.5). The SAS program listing in Figure 17.28 was used to compute an initial trend estimate and to obtain the smoothed values S_t, for each time period. It was also used to determine the predicted values, $\hat{y}_t$, using smoothing constants $A = .1$ and $B = .3$. This example will only run on the mainframe version of SAS.

The TITLE command names the SAS run.

The DATA command gives the data a name.

The INPUT command names and gives the correct order for the different fields on the data lines.

The LABEL statement assigns a label for the variable TAXES.

The CARDS command indicates to SAS that the input data immediately follow.

The next 20 lines contain the data values, which represent time series data over 20 time periods. The first line, for example, represents the year 1986 when the taxes collected were 76 (thousand).

The PROC FORECAST command requests the smoothed forecast.

OUT = B names an output data set to hold the forecast. OUTEST = C names an output dataset to hold the parameter estimates. METHOD specifies that the WINTERS smoothing method will be used. TREND = 2 selects the linear trend model. WEIGHT = (.1,.3,0.001) specifies the smoothing weights. OUTDATA requests that the observations used to fit the model be included in the data set. OUT1STEP requests that one step ahead forecasts be included in the dataset. OUTLIMIT requests that the forecast confidence limits be included in the data set. LEAD = 1 specifies the number of periods ahead for the forecast. ID and VAR mean identify by time and that T is the input data set to be analyzed.

The PROC PRINT command requests that the generated data and estimated values be printed.

Figure 17.29 shows a portion of the SAS output obtained by executing the listing in Figure 17.28.

■ FIGURE 17.28
Input for SAS
(mainframe or
micro version).

```
TITLE 'LINEAR EXPONENTIAL SMOOTHING';
DATA TAXES;
  INPUT TIME T;
  LABEL T='TAXES';
  CARDS;
1 76
2 93
3 108
4 128
5 196
6 175
7 141
8 236
9 256
10 190
11 227
12 299
13 403
14 282
15 288
16 387
17 484
18 384
19 330
20 497
;
PROC FORECAST DATA=TAXES OUT=B OUTEST=C METHOD=WINTERS TREND=2
  WEIGHT=(.1,.3,0.001) OUTDATA OUT1STEP OUTLIMIT LEAD=1;
  ID TIME;
    VAR T;
PROC PRINT DATA=B;
  TITLE 'SMOOTHING FORECAST';
```

■ FIGURE 17.29
SAS output.

SMOOTHING FORECAST

OBS	TIME	_TYPE_	_LEAD_	T
1	1	ACTUAL	0	76.000
2	1	FORECAST	0	.
3	2	ACTUAL	0	93.000 ← y_2
4	2	FORECAST	0	76.000
5	3	ACTUAL	0	108.000
6	3	FORECAST	0	78.210 $\hat{y}_2$
7	4	ACTUAL	0	128.000
8	4	FORECAST	0	82.593
9	5	ACTUAL	0	196.000
10	5	FORECAST	0	89.899
11	6	ACTUAL	0	175.000
12	6	FORECAST	0	106.458
13	7	ACTUAL	0	141.000
14	7	FORECAST	0	121.318
15	8	ACTUAL	0	236.000
16	8	FORECAST	0	131.882
17	9	ACTUAL	0	256.000
18	9	FORECAST	0	154.013
19	10	ACTUAL	0	190.000
20	10	FORECAST	0	178.990
21	11	ACTUAL	0	227.000
22	11	FORECAST	0	195.200
23	12	ACTUAL	0	299.000
24	12	FORECAST	0	214.443
25	13	ACTUAL	0	403.000
26	13	FORECAST	0	241.499
27	14	ACTUAL	0	282.000
28	14	FORECAST	0	281.094
29	15	ACTUAL	0	288.000
30	15	FORECAST	0	304.656
31	16	ACTUAL	0	387.000
32	16	FORECAST	0	325.963
33	17	ACTUAL	0	484.000
34	17	FORECAST	0	356.870
35	18	ACTUAL	0	384.000
36	18	FORECAST	0	398.201
37	19	ACTUAL	0	330.000
38	19	FORECAST	0	424.972
39	20	ACTUAL	0	497.000 ← y_{20}
40	20	FORECAST	0	440.817
41	21	FORECAST	1	473.463 $\hat{y}_{20}$
42	21	L95	1	312.851
43	21	U95	1	634.074

SAS

Solution
Example 17.6

We used Winter's linear and seasonal exponential smoothing on time series data to predict the city taxes collected in Jackson City over the past 20 quarters (Table 17.7). The SAS program listing in Figure 17.30 was used to compute an initial trend estimate and to obtain the smoothed values S, for each time period. It was also used to determine the predicted values, y_t, using smoothing constants $A = .1, B = .3$, and $C = .2$.

The TITLE command names the SAS run.

The DATA command gives the data a name.

The INPUT command names and gives the correct order for the different fields on the data lines.

The LABEL statement assigns a label for the variable TAXES.

The CARDS command indicates to SAS that the input data immediately follow.

The next 20 lines contain the data values, which represent time series data over 20 time periods. The first line, for example, represents the year 1986 when the taxes collected were 76 (thousand).

The PROC FORECAST command requests the smoothed forecast.

OUT = B names an output data set to hold the forecast. OUTEST = C names an output data set to hold the parameter estimates. METHOD specifies that the WINTERS smoothing method will be used. TREND = 2 selects the linear trend model. WEIGHT = (.1,.3,.2) specifies the smoothing weights. OUTDATA requests that the observations used to fit the model be included in the data set. OUT1STEP requests that one step ahead forecasts be in-

■ FIGURE 17.30
Input for SAS
(mainframe or
micro version).

```
TITLE 'WINTERS METHOD OF EXPONENTIAL SMOOTHING';
DATA A;
  INPUT TIME T;
  LABEL T='TAXES';
  CARDS;
1 76
2 93
3 108
4 128
5 196
6 175
7 141
8 236
9 256
10 190
11 227
12 299
13 403
14 282
15 288
16 387
17 484
18 384
19 330
20 497
;
PROC FORECAST DATA=A OUT=B OUTEST=C METHOD=WINTERS
     TREND=2 WEIGHT=(.1 .3 .2) OUTDATA OUT1STEP OUTLIMIT
     LEAD=1;
  ID TIME;
  VAR T;
PROC PRINT DATA=B;
  TITLE 'THE OUTPUT FROM PROC FORECAST: WINTERS METHOD';
```

cluded in the dataset. OUTLIMIT requests that the forecast confidence limits be included in the dataset. LEAD = 1 specifies the number of periods ahead for the forecast. ID and VAR mean identify by time and that T is the input data set to be analyzed.

The PROC PRINT command requests that the generated data and estimated values be printed.

Figure 17.31 shows a portion of the SAS output obtained by executing the listing in Figure 17.30. These results differ from those in Table 17.7 because a different initialization procedure was used.

■ FIGURE 17.31
SAS output.

```
            THE OUTPUT FROM PROC FORECAST: WINTERS METHOD

OBS     TIME     _TYPE_      _LEAD_       T

  1       1      ACTUAL          0       76.000
  2       1      FORECAST        0
  3       2      ACTUAL          0       93.000
  4       2      FORECAST        0       76.000
  5       3      ACTUAL          0      108.000
  6       3      FORECAST        0       78.210
  7       4      ACTUAL          0      128.000
  8       4      FORECAST        0       82.593
  9       5      ACTUAL          0      196.000
 10       5      FORECAST        0       89.899
 11       6      ACTUAL          0      175.000
 12       6      FORECAST        0      106.458
 13       7      ACTUAL          0      141.000
 14       7      FORECAST        0      121.318
 15       8      ACTUAL          0      236.000
 16       8      FORECAST        0      131.882
 17       9      ACTUAL          0      256.000
 18       9      FORECAST        0      154.013
 19      10      ACTUAL          0      190.000
 20      10      FORECAST        0      178.990
 21      11      ACTUAL          0      227.000
 22      11      FORECAST        0      195.200
 23      12      ACTUAL          0      299.000
 24      12      FORECAST        0      214.443
 25      13      ACTUAL          0      403.000
 26      13      FORECAST        0      241.499
 27      14      ACTUAL          0      282.000
 28      14      FORECAST        0      281.094
 29      15      ACTUAL          0      288.000
 30      15      FORECAST        0      304.656
 31      16      ACTUAL          0      387.000
 32      16      FORECAST        0      325.963
 33      17      ACTUAL          0      484.000
 34      17      FORECAST        0      356.870
 35      18      ACTUAL          0      384.000
 36      18      FORECAST        0      398.201
 37      19      ACTUAL          0      330.000
 38      19      FORECAST        0      424.972
 39      20      ACTUAL          0      497.000
 40      20      FORECAST        0      440.817
 41      21      FORECAST        1      473.463
 42      21      L95             1      312.851
 43      21      U95             1      634.074
```

SAS

⧄ Solution
Example 17.9

Example 17.9 is divided into two computer runs. In the first run we used the autocorrelation function to determine the significant autocorrelations among the given lag periods. In the second run, the significant lag period (lag 4) was included in the autoregression equation. The SAS program listings in Figures 17.32 and 17.33 were used to perform the two procedures.

The TITLE command names the SAS run.

The DATA command gives the data a name.

The INPUT command names and gives the correct order for the different fields on the data lines.

The LABEL statement assigns the labels TIME for variable T and OBSERVED VALUES for variable YT. These labels are substituted for the variable names when the output is printed.

■ FIGURE 17.32
Input for SAS to
determine
significant
autocorrelations.

```
TITLE  'KENS AUTO PAINT SHOP - AUTOCORRELATION FUNCTION';
DATA AUTO;
  INPUT T YT;
  LABEL T='TIME'
        YT='OBSERVED VALUES';
CARDS;
1 5.56
2 16.36
3 2.12
4 3.15
5 5.11
6 15.21
7 5.72
8 2.65
9 4.12
10 14.33
11 5.25
12 6.75
13 6.31
14 15.02
15 2.83
16 4.56
17 4.81
18 16.82
19 4.75
20 8.54
PROC ARIMA;
  IDENTIFY VAR=YT;
```

■ FIGURE 17.33
Input for SAS to
determine the
autocorrelation
equation using one
lagged variable.

```
TITLE  'KENS AUTO PAINT SHOP - AUTOREGRESSIVE MODEL';
DATA AUTO;
  INPUT T YT;
  LABEL T='TIME'
        YT='OBSERVED VALUES';
YT4=LAG4(YT);
CARDS;
1 5.56
2 16.36
3 2.12
4 3.15
5 5.11
6 15.21
7 5.72
8 2.65
9 4.12
10 14.33
11 5.25
12 6.75
13 6.31
14 15.02
15 2.83
16 4.56
17 4.81
18 16.82
19 4.75
20 8.54
PROC REG;
  MODEL YT=YT4;
  OUTPUT OUT=C PRED=P;
PROC PRINT;
  VAR YT P;
```

The CARDS command indicates to SAS that the input data immediately follow.

The next 20 lines contain the data values, which represent time series data over 20 time periods. For example, 1 5.56 represents the OBSERVED VALUE of 5.56 housing starts for the first time period.

The PROC ARIMA and IDENTIFY commands are used to analyze the time series YT and identify the significant autocorrelations.

The commands in Figure 17.33 are the same as those in Figure 17.32, with the following exceptions:

The YT4 = LAG4(YT) command provides YT4 with the value of YT for the fourth case before the current one. For example, the value of YT in case 16 is 4.56. Therefore, the value of YT4 in case 20 is also 4.56.

The PROC REG command and MODEL subcommand specify that variables YT and YT4 are to be used in the regression equation and that YT is the dependent variable.

The OUTPUT command specifies the output from PROC REG is to be stored in an output data set named C and that the predicting values in the data set are named PRED.

Figure 17.34 shows the SAS output obtained by executing the listing in Figure 17.32, while Figure 17.35 shows the SAS output obtained by executing the listing in Figure 17.33.

■ **FIGURE 17.34**
SAS output from
Figure 17.32.

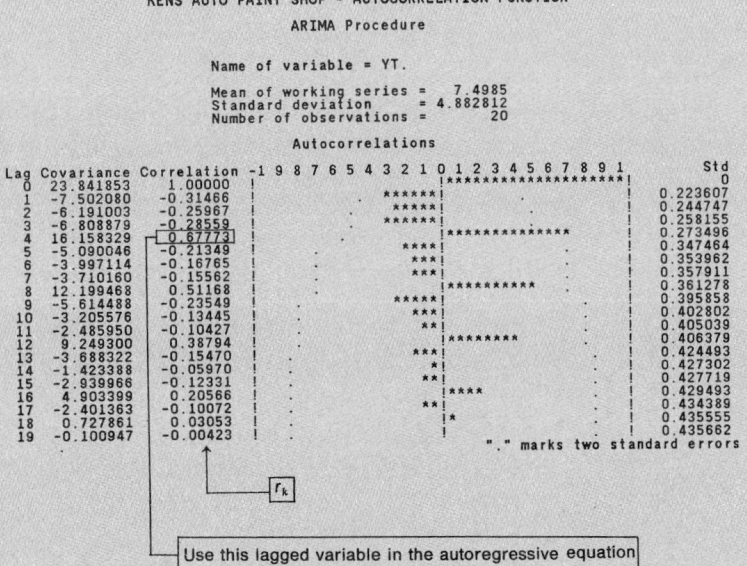

■ **FIGURE 17.35** SAS output from Figure 17.33.

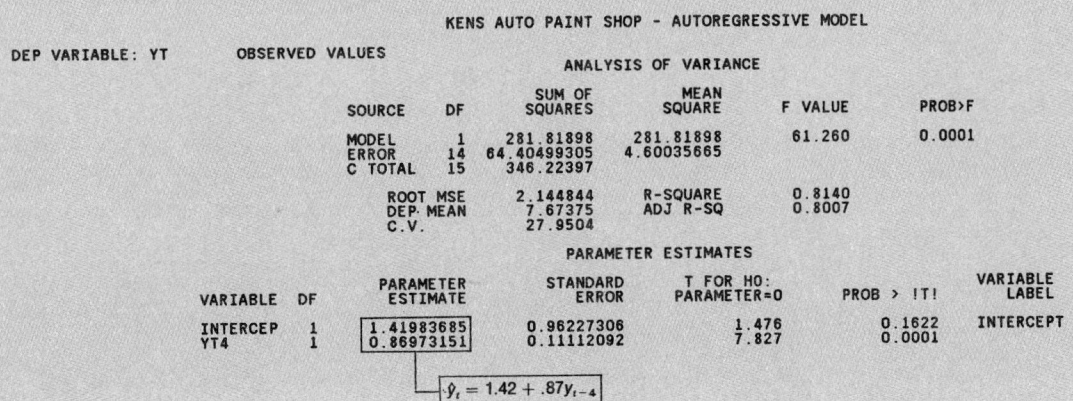

■ FIGURE 17.35 (*continued*)

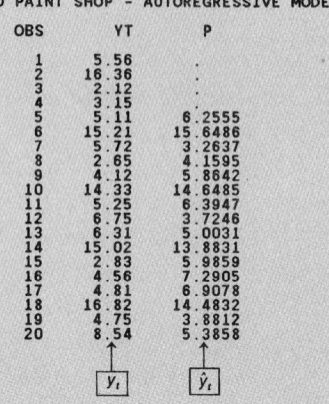

KENS AUTO PAINT SHOP - AUTOREGRESSIVE MODEL

OBS	YT	P
1	5.56	.
2	16.36	.
3	2.12	.
4	3.15	.
5	5.11	6.2555
6	15.21	15.6486
7	5.72	3.2637
8	2.65	4.1595
9	4.12	5.8642
10	14.33	14.6485
11	5.25	6.3947
12	6.75	3.7246
13	6.31	5.0031
14	15.02	13.8831
15	2.83	5.9859
16	4.56	7.2905
17	4.81	6.9078
18	16.82	14.4832
19	4.75	3.8812
20	8.54	5.3858

y_t $\hat{y}_t$

SAS

▨ Solution
Section 17.9
Example

We used multiple linear regression on time series data to predict the number of home loans financed by Liberty Savings and Loan. The predictor variables used were average interest rate, advertising expenditure, an election year dummy variable, and a seasonal dummy variable. The SAS program listing in Figure 17.36 was used to compute the regression equation from the time series data.

The TITLE command names the SAS run.

The DATA command gives the data a name.

The INPUT command names and gives the correct order for the different fields on the data lines.

The LABEL statement assigns labels for variables YT, X1T, X2T, X3T, and X4T. These labels are substituted for the variable names when output is printed.

The X1TLAG=LAG1(X1T) command provides X1TLAG with the value of X1T for the case before the current one. For example, the average interest rate (X1T) during the first half of 1983 was 11.8%. Therefore, the value of X1TLAG for the last half of 1983 is also 11.8%.

The X2TLAG=LAG1(X2T) command provides X2TLAG with the value of X2T for the case before the current one.

The CARDS command indicates to SAS that the input data immediately follow.

The next 16 lines contain the data values, which represent time series data over 16 time periods. The first line, for example, represents the year 1983, the first time period, 122 home loans, an average interest rate of 11.8%, and so on.

The PROC REG command and MODEL subcommand specify that variables YT, X1TLAG, X2TLAG, X3T, and X4T are to be used in the regression equation and that YT is the dependent variable.

The DW option requests that the Durbin-Watson statistic be computed.

Figure 17.37 shows the SAS output obtained by executing the listing in Figure 17.36.

■ **FIGURE 17.36**
Input for SAS
(mainframe or
micro version).

```
TITLE   'LIBERTY SAVINGS AND LOAN MODEL';
DATA AUTO;
 INPUT YEAR TIME YT X1T X2T X3T X4T;
 LABEL YT=NUMBER OF HOME LOANS
      X1T=AVERAGE INTEREST RATE
      X2T=ADVERTISING EXPENDITURE
      X3T=ELECTION YEAR VARIABLE
      X4T=SEASONAL VARIABLE;
 X1TLAG=LAG1(X1T);
 X2TLAG=LAG1(X2T);
 CARDS;
 1983 1   122 11.8 10.4 0 1
 1983 2   118 12.4  6.7 0 0
 1984 3   106 11.0  7.5 1 1
 1984 4   140 14.5  7.8 1 0
 1985 5    86 11.0  5.1 0 1
 1985 6    96 10.1  6.8 0 0
 1986 7   110 12.9  6.8 0 1
 1986 8    76 14.8  9.1 0 0
 1987 9    62 10.5  7.5 0 1
 1987 10  104 09.8  5.1 0 0
 1988 11  135 10.1  8.8 1 1
 1988 12  120 10.8  4.3 1 0
 1989 13  105 10.0  7.1 0 1
 1989 14  118 09.8  8.6 0 0
 1990 15  121 09.5  8.8 0 1
 1990 16  132 09.7  7.4 0 0
 PROC REG;
  MODEL YT=X1TLAG X2TLAG X3T X4T/ DW;
```

■ **FIGURE 17.37** SAS output.

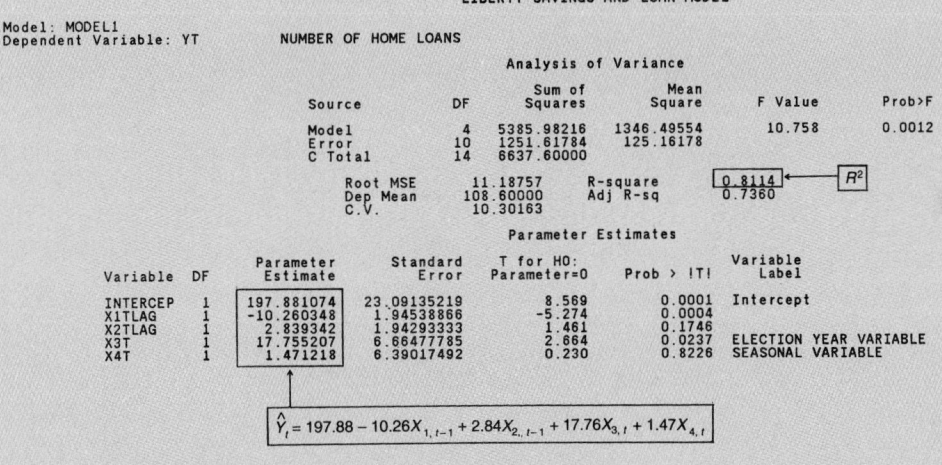

LIBERTY SAVINGS AND LOAN MODEL

Model: MODEL1
Dependent Variable: YT   NUMBER OF HOME LOANS

Analysis of Variance

Source	DF	Sum of Squares	Mean Square	F Value	Prob>F
Model	4	5385.98216	1346.49554	10.758	0.0012
Error	10	1251.61784	125.16178		
C Total	14	6637.60000			

Root MSE   11.18757   R-square   0.8114   ← R^2
Dep Mean   108.60000   Adj R-sq   0.7360
C.V.   10.30163

Parameter Estimates

Variable	DF	Parameter Estimate	Standard Error	T for H0: Parameter=0	Prob > |T|	Variable Label
INTERCEP	1	197.881074	23.09135219	8.569	0.0001	Intercept
X1TLAG	1	-10.260348	1.94538866	-5.274	0.0004	
X2TLAG	1	2.839342	1.94293333	1.461	0.1746	
X3T	1	17.755207	6.66477785	2.664	0.0237	ELECTION YEAR VARIABLE
X4T	1	1.471218	6.39017492	0.230	0.8226	SEASONAL VARIABLE

$$\hat{Y}_t = 197.88 - 10.26X_{1,\,t-1} + 2.84X_{2,\,t-1} + 17.76X_{3,\,t} + 1.47X_{4,\,t}$$

Durbin-Watson D ←   1.612
(For Number of Obs.)   15
1st Order Autocorrelation   0.122

CHAPTER

18

Decision Making Under Uncertainty

A LOOK BACK/INTRODUCTION

You have been exposed to decision making in which you use a statistical test of hypothesis. The final step for any test of hypothesis is to reject or fail to reject the null hypothesis, H_0. The previous chapters examined a number of such tests, ranging from a decision regarding two means (for example, is $\mu_1 > \mu_2$?) to selecting predictor variables for a multiple regression equation.

The previous tests of hypothesis concentrated on the *probabilistic* aspects of decision making. For example, if the probability of observing a t-value as large as the computed sample value was less than a predetermined significance level α, we rejected H_0 and decided to keep a particular predictor variable in the regression equation. Such procedures are intended to help you make statistical decisions regarding certain population **parameters,** such as the mean (μ) or standard deviation (σ) of a normal population, the population coefficient (β) for a certain independent variable in a regression equation, or the proportion of successes (p) for a binomial situation.

We now examine a different side of decision making, a side that is particularly useful when money is involved. This area of statistics allows the decision maker to consider the various benefits and losses associated with each possible alternative in an effort to find the best decision. For example, if your aging car one day rolls over and goes to that great garage in the sky, you may be faced with the decision of buying a new car or leasing one. If you are lucky enough to have the option of buying a new car that runs well and will incur few repair bills, perhaps your best bet would be to purchase it. Other factors to consider would be the length of time that you intend to keep the car and the tax advantages for each alternative. The main problem, however, is that the future reliability of the new car is uncertain.

The problem of decision making in such a situation would be greatly simplified if you had a crystal ball. A perfect predictor would tell you whether the new car you are thinking of buying is a lemon. Unfortunately, no such device exists, so you need to develop various decision strategies to deal with future uncertainties. Certain strategies are conservative; you stand neither to gain nor to lose a large amount. Other procedures attempt to measure the likelihood of future events by using probabilities. If you are a gambler at heart, strategies exist that allow you to defy odds in hopes of a big payoff.

If your life savings are at stake, you may elect to use a more conservative strategy in your investment decisions; gambling with these funds could be too dangerous. On the other hand, you may have certain reserve funds that you would be willing to invest in more speculative ventures, hoping for a large return at the risk of losing your investment. The purchase of a lottery ticket is a small-scale example of such an investment.

This chapter examines the various strategies available to the decision maker and illustrates these techniques for different business situations.

18.1
Defining the Decision Problem

When confronted with making a decision in the face of uncertainty, you essentially need to be concerned with two basic questions:

1. What are my possible **actions** (alternatives) for this problem?
2. What is it about the future that affects the desirability of each action?

For the buy versus lease example, you are faced with two possible actions,

ACTION	DESCRIPTION
A_1	Purchase the car
A_2	Lease the car

821

When you are trying the decide between these two options, what would you like to know? One possibility is to describe the future (as it applies to this decision) by means of three events, or **states of nature.**

S_1:   the new car will have less-than-average repair costs.

S_2:   the new car will have average repair costs.

S_3:   the new car will have above-average repair costs.

These future states of nature are **outcomes.** *The key distinction between an action and a state of nature is that the action taken is* under your control, *whereas the state of nature that occurs is strictly a matter of chance.* We will assume that *one and only one* of the states of nature will occur in the future; that is, they are *mutually exclusive.*

Other questions that require the decision maker to specify corresponding states of nature are ones such as:

What will be the future demand for a new computer software package?

How long will it be until the newly purchased electrical pump breaks down and needs to be replaced?

Will the stock market turn up or down in the next three months?

Will this year's winter be milder or colder than the average for the past 20 years?

Associated with each action (A) and state of nature (S) is a corresponding **payoff** or **profit.** We will assume that the payoff associated with each particular state of nature and action is known with certainty. These payoffs can be summarized in a **payoff table,** as shown in Table 18.1. The entry corresponding to action A_2 (row 2) and state of nature S_3 (column 3), for example, is denoted by π_{23} and is the payoff associated with action A_2 should state of nature S_3 occur.

Mr. Larson is owner of Sailtown, a store in southern Minnesota specializing in the sale of small sailboats. Each spring he is forced to place an order for his entire stock of Bluefin sailboats to be sold during the summer months because the Bluefin manufacturer is unable to supply any additional boats once the summer has begun.

Mr. Larson's main concern when ordering his summer inventory is the demand for his product during the next five months. He has discovered that this demand seems to be largely dependent on economic conditions, in particular on the prevailing interest rate. He has four possible actions (order quantities)

A_1:   purchase 50 boats

A_2:   purchase 75 boats

A_3:   purchase 100 boats

A_4:   purchase 150 boats

■ **TABLE 18.1**
Payoff table.

Action	STATES OF NATURE				
	S_1	S_2	S_3	$\cdots$	S_n
A_1	π_{11}	π_{12}	π_{13}		π_{1n}
A_2	π_{21}	π_{22}	π_{23}		π_{2n}
A_3	π_{31}	π_{32}	π_{33}		π_{3n}
$\vdots$					
A_k	π_{k1}	π_{k2}	π_{k3}		π_{kn}

The states of nature for this problem are

S_1: the interest rate increases significantly (more than 1.5%) from the current rate.

S_2: the interest rate holds steady.

S_3: the interest rate drops significantly (more than 1.5%).

Based on his expected sales under each condition, the payoff table in Table 18.2 was constructed. To demonstrate how the payoffs were determined, consider $\pi_{42} = 5$, which is the resulting profit if he orders 150 sailboats (action A_4) and if the interest rate remains basically unchanged (state of nature S_2). Mr. Larson believes that, for S_2, the resulting demand will be 100 sailboats. His profit per sale is $300, and his cost for holding and returning an unsold boat at the end of the fall season is $500. Consequently, if the demand is for 100 boats and he decides to stock 150, he ends up selling 100 boats and returning 50 of them to the manufacturer. The resulting dollar amounts are

$$\begin{aligned}
\text{profit for selling 100 boats} &= 100 \cdot \$300 = \$30,000 \\
\text{loss for returning 50 boats} &= 50 \cdot \$500 = \underline{\$25,000} \\
\text{net payoff} = \pi_{42} &= \$5,000
\end{aligned}$$

So Mr. Larson calculates that the interest rate holding steady is equivalent to a demand for 100 sailboats. Similarly, he determines that an increase in the interest rate will result in a demand for 50 boats, whereas a decrease in the interest rate will produce a demand for 150 boats (Table 18.3).

EXAMPLE 18.1

Using Table 18.3, determine the payoff for each of the four alternatives if the interest rate increases over the summer months. What would be the best action to take if you knew that this particular state of nature would occur?

SOLUTION

We are given that state of nature S_1 will occur. Mr. Larson thinks that under this state of nature, 50 people will walk in his front door wanting to purchase a Bluefin sailboat. Under this assumption, the payoffs in Table 18.4 can be derived.

If we know that the interest rate will increase during the summer months, Table 18.4 tells us that the ideal action is to purchase 50 boats (action A_1).

Is action A_1 the ideal action for each state of nature? Given state of nature S_2 and using Table 18.2, the payoffs are $15,000 (for A_1), $22,500 (for A_2), $30,000 (for A_3), and $5000 (for A_4). In this case, you would achieve maximum profit by purchasing 100 boats (action A_3). The third column of Table 18.2 shows that action

■ TABLE 18.2
Profit table for Sailtown
(thousands of dollars).

Amount Ordered	AVERAGE INTEREST RATE		
	Increases (S_1)	Steady (S_2)	Decreases (S_3)
50 (A_1)	15	15	15
75 (A_2)	2.5	22.5	22.5
100 (A_3)	−10	30	30
150 (A_4)	−35	5	45

■ TABLE 18.3
Demand for sailboats
for each state of nature.

STATE OF NATURE	INTEREST RATE	CORRESPONDING DEMAND
S_1	Increases	50
S_2	Holds steady	100
S_3	Decreases	150

■ TABLE 18.4
Payoff for Sailtown
under S_1
(Example 18.1).

ACTION	REVENUE FOR BOATS SOLD	LOSS DUE TO RETURNED BOATS	NET PAYOFF
A_1	$50 \cdot 300 = \$15,000$	$0 \cdot 500 = \$0$	$\$15,000$
A_2	$50 \cdot 300 = \$15,000$	$25 \cdot 500 = \$12,500$	$\$ 2,500$
A_3	$50 \cdot 300 = \$15,000$	$50 \cdot 500 = \$25,000$	$-\$10,000$
A_4	$50 \cdot 300 = \$15,000$	$100 \cdot 500 = \$50,000$	$-\$35,000$

A_4 (purchasing 150 boats) provides the maximum profit in the event that the interest rate declines.

■

This example illustrates **decision making under certainty.** Although the decision maker will never have this luxury, the technique at least enables him or her to determine the maximum profit under each state of nature. Also, if the *same action* provides the maximum payoff regardless of the state of nature, then this particular action is the obvious choice for this situation. Such was not the case in Table 18.2. There is no obviously superior action here, so we need to consider more elaborate decision strategies.

 *Exercises*

18.1   A stockbroker is trying to decide which of three possible actions to recommend to a client. One action (A_1) would be to invest 100% in a mutual fund that invests in stocks. Another action (A_2) would be to invest 50% in a fixed account yielding 10% per year and 50% in a stock market mutual fund. The third action (A_3) would be to invest 100% in the fixed account, yielding 10% per year. The stockbroker believes the following returns can be made in a particular mutual fund in one year with regard to the stock market's direction.

S_1: STOCK MARKET GOES UP	S_2: STOCK MARKET HAS NO DIRECTION	S_3: STOCK MARKET GOES DOWN
20%	5%	-10%

Using the three states of nature, S_1, S_2, and S_3 and the three actions A_1, A_2, and A_3, construct a payoff table if a client has $10,000 to invest for one year.

18.2   Would the following decisions be made under certainty or uncertainty?

a.   Whether to buy or lease a new car.

b.   Whether to accept a client who wants to pay you $40 per visit for five visits.

c.   Whether to borrow $10,000 at 12% interest per year for five years.

d.   Whether to switch jobs or stay at the same job.

18.3   A seminar on sales motivation is being given at a local hotel. The cost of handouts and materials per attendee is $10. The total cost of the hotel arrangements is $75. Each attendee pays $25 for the seminar. The coordinator of the seminar must plan the number of handouts and materials. The coordinator must plan for 7, 8, 9, 10, 11, 12, or 13 attendees. Construct the payoff table if the demand is 5, 6, 7, 8, 9, 10, 11, 12, or 13 attendees.

18.4   The Jones family has moved out of their house and now has it up for sale. They need to decide whether to price their house at the top of market ($88,000), at the market ($85,000), or below the market ($82,000). Each month that it takes to sell the house, the Jones family loses $700 from making the monthly payment. Construct a payoff table that shows the selling price of the home minus $700 for each month the house remains unsold. Use S_1 = sold in 1 month, S_2 = sold in 2 months, . . . , S_6 = sold in 6 months.

18.5   Mini-Super, a convenience food store, orders milk each week. The store pays $1.10 per gallon and sells the milk for $2.00 per gallon. Unsold milk at the end of the week is given to an orphanage. The manager must decide on ordering 100 gallons, 125 gallons, or 150 gallons per week. Construct the payoff table for this problem if demand is 100 gallons, 125 gallons, or 150 gallons a week.

18.2
Decision Strategies

When the action providing a maximum payoff depends on an uncertain state of nature, the decision maker is forced to consider all the values in the payoff table to choose the most attractive action. The various strategies discussed here allow you to choose a procedure that best fits your style of making decisions. We begin with a conservative strategy, the minimax procedure.

The Conservative (Minimax) Strategy

A conservative strategy is basically one that, when choosing between a savings account and an extremely risky venture, selects the savings account. It does this because, under the *worst* conditions, the loss is smaller with a savings account than with the high-risk venture.

We examined the ideal action for Mr. Larson to take in the event that he knew that the interest rate was going to increase (S_1 in Table 18.2). This action was to order 50 boats for a payoff of $\pi_{11} = 15$ (thousand). If he had taken action A_2 instead, his profit would be only \$2500. For this situation, we say that the opportunity loss is \$15,000 $-$ \$2500 $=$ \$12,500.

The **opportunity loss,** L_{ij}, is the difference between the payoff for action i and the payoff for the action that would have the largest payoff under state of nature j. *

The opportunity loss is not a loss in the accounting sense; rather, it describes how much more profit you *would have made* had you chosen the best action for this state of nature. In our example, the opportunity loss for action A_3 (assuming state of nature S_1) is

$$\text{opportunity loss} = L_{31} = 15 - (-10)$$
$$= 25 \quad \text{(thousand dollars)}$$

and this value for action A_4 (under S_1) is

$$\text{opportunity loss} = L_{41} = 15 - (-35)$$
$$= 50 \quad \text{(thousand dollars)}$$

EXAMPLE 18.2

Construct the remaining opportunity losses for Mr. Larson, and summarize them in an opportunity loss table.

SOLUTION

Keep in mind that opportunity losses are determined one *column* at a time in Table 18.2 by assuming that each individual state of nature occurs and then looking for the best action under this condition. This procedure is exactly the same as the one we used when discussing the unrealistic situation of decision making under certainty.

If the interest rate holds steady (S_2 occurs), the best action is to stock 100 boats (action A_3) with a payoff of $\pi_{32} = 30$. Table 18.5 shows the opportunity loss for this situation.

Similarly, action A_4 (stock 150 boats) is the ideal action in the event the interest rate decreases (S_3) and sales increase, as shown in Table 18.6.

We format these results in an **opportunity loss table,** as shown in Table 18.7. Notice that each column of an opportunity loss table contains a zero and that all values in this table are nonnegative (≥ 0).   ■

---

*Some textbooks refer to opportunity loss as *regret* and to the minimax strategy as the *minimax regret strategy.*

ACTION	PAYOFF	OPPORTUNITY LOSS
A_1	15	$L_{12} = 30 - 15 = 15$
A_2	22.5	$L_{22} = 30 - 22.5 = 7.5$
A_3	30	$L_{32} = 30 - 30 = 0$
A_4	5	$L_{42} = 30 - 5 = 25$

ACTION	PAYOFF	OPPORTUNITY LOSS
A_1	15	$L_{13} = 45 - 15 = 30$
A_2	22.5	$L_{23} = 45 - 22.5 = 22.5$
A_3	30	$L_{33} = 45 - 30 = 15$
A_4	45	$L_{43} = 45 - 45 = 0$

Action	STATE OF NATURE S_1	S_2	S_3
A_1	0	15	30
A_2	12.5	7.5	22.5
A_3	25	0	15
A_4	50	25	0

The Minimax Strategy   The minimax strategy is to:

1. Construct an opportunity loss table by using the maximum payoff for each state of nature.
2. Determine the maximum opportunity loss for each action.
3. Find the minimum value of the opportunity losses found in step 2; the corresponding action is the one selected by the minimax strategy.

The minimax strategy is a very conservative approach that does not search for large payoffs; rather, it selects the action that has the smallest "worst case" opportunity loss.

The minimax procedure begins by examining the worst possible situation for each action. So you examine Table 18.7 one *row* at a time and determine the largest opportunity loss for each action. Thus, we have

ACTION	MAXIMUM OPPORTUNITY LOSS
A_1	30
A_2	22.5
A_3	25
A_4	50

This is the *max* part of the minimax strategy. The *mini* side is finding the *minimum* of these four values. In this way, you attempt to offset the worst possible situation scenario. For this example, with values of 30, 22.5, 25, and 50, the minimum is 22.5, which belongs to action A_2, so the **minimax decision** is to order 75 sailboats.

In actual practice, a decision analysis rarely begins in the form of a payoff table. Instead, this table perhaps can be constructed using the information available and then the appropriate decision strategy can be applied to arrive at the corresponding action. This method is illustrated in the next example.

EXAMPLE 18.3

During 1990, GTE Southwest offered their residential and business customers a plan to help reduce long-distance costs. The plan was called *1 + Saver* and provided four options for each customer.

Option 1: a 10% discount at a cost of $1 per month (action A_1)

Option 2: a 15% discount at a cost of $6 per month (action A_2)

Option 3: a 20% discount at a cost of $15 per month (action A_3)

Option 4: a 30% discount at a cost of $50 per month (action A_4)

These discounts applied to any direct-dialed long-distance call placed with GTE Southwest that did not involve operator-handled or credit card calls. As an employee of Worldwide Import Distributors, you have been asked to select one of these options for your company's future long-distance calls. Company records show that previous long-distance charges can be summarized using four states of nature:

$$S_1: \quad \text{long-distance charges} = \$100$$

$$S_2: \quad \text{long-distance charges} = \$200$$

$$S_3: \quad \text{long-distance charges} = \$300$$

$$S_4: \quad \text{long-distance charges} = \$400$$

Construct a payoff table for this situation and select one of these four options using the minimax strategy.

SOLUTION   The payoff table is constructed by determining the discount amount and subtracting the corresponding monthly charge for this option. For example, if the monthly long-distance charges are approximately $300 ($S_3$) then using option 2 (action A_2), the payoff (savings) is ($300)(.15) − $6 = $39. The resulting payoff table is shown in Table 18.8.

If state of nature S_2 should occur, the best action is A_3, with a savings (payoff) of $25 and an opportunity loss of zero. Consequently, A_1 has an opportunity loss of $25 − $19 = $6. Continuing in this way, you can construct Table 18.9.

Next, we find the maximum opportunity loss for each action.

ACTION	MAXIMUM OPPORTUNITY LOSS
A_1	31
A_2	16
A_3	5
A_4	29

Finally we select the minimum of these values: $5, corresponding to action A_3. The minimax decision is to use option 3 with a 20% discount and a fixed charge of $15 per month. This action is fairly conservative here and is the one that minimizes the maximum difference between the savings received and the savings that could have been received if the state of nature (actual long-distance charges) had been known in advance.   ■

■ TABLE 18.8
Payoff table for
Worldwide Import
Distributors (dollars).

Action	STATE OF NATURE			
	$100 ($S_1$)	$200 ($S_2$)	$300 ($S_3$)	$400 ($S_4$)
Option 1 (A_1)	9	19	29	39
Option 2 (A_2)	9	24	39	54
Option 3 (A_3)	5	25	45	65
Option 4 (A_4)	−20	10	40	70

■ TABLE 18.9
Opportunity loss table
for Worldwide Import
Distributors (dollars).

Action	STATE OF NATURE				
	$100 ($S_1$)	$200 ($S_2$)	$300 ($S_3$)	$400 ($S_4$)	Maximum
Option 1 (A_1)	0	6	16	31	31
Option 2 (A_2)	0	1	6	16	16
Option 3 (A_3)	4	0	0	5	5
Option 4 (A_4)	29	15	5	0	29

The Gambler (Maximax) Strategy

The maximax strategy is the opposite of the minimax procedure and appeals to those who are gamblers at heart. The **maximax strategy** is to choose that action having the largest possible payoff. It is not a recommended procedure for most business decisions because, by choosing that action with the largest payoff, it fails to consider the possibility of large accounting losses or opportunity losses.

EXAMPLE 18.4   Using the information in Table 18.2, which action should Sailtown select using the maximax strategy?

SOLUTION   Of the 12 payoffs in Table 18.2, the largest is 45, which corresponds to action A_4. If Sailtown is desperate for a large payoff, the appropriate action using this strategy would be to order 150 sailboats for the summer months. Of course, the company also stands to lose the most using this action; the loss will be $35,000 if the interest rate increases.   ∎

EXAMPLE 18.5   Based on the payoff table in Table 18.8 and the maximax strategy, which of the four long-distance discount options would you select?

SOLUTION   The maximum savings is $70 per month, corresponding to option 4. So, if you want to gamble for a large savings, the corresponding action would be option 4, which has a 50% discount and a monthly charge of $50. Interestingly, the minimax procedure (typically, very conservative) selected option 3 with a maximum payoff of $65, only $5 less than that obtained using the maximax procedure.   ∎

Exercises

18.6   Why is the following table not an opportunity loss table? Give two reasons.

ACTION	S_1	S_2	S_3
A_1	4	6	80
A_2	29	0	150
A_3	57	−1	0

18.7   The owner of a newsstand orders the local daily newspapers for 10¢ per copy and charges 25¢ per copy. Each day, the owner orders either 70, 80, 90, or 100 copies. At the end of the day, the leftover newspapers are discarded. Construct the payoff table and opportunity loss table if the demand is 50, 60, 70, 80, 90, or 100 copies. What is the minimax decision? What is the maximax decision?

18.8   The owner of a small commercial building can either pay $1000 per year to insure the $200,000 building or not insure and save $1000 per year. If the states of nature are complete loss of the commercial building or no loss at all, what would the payoff table and opportunity loss table be for this situation? What is the minimax decision?

18.9   What would the opportunity loss table be using the following payoff table?

ACTION	S_1	S_2	S_3	S_4
A_1	150	2	89	5
A_2	−20	10	76	−10
A_3	0	15	94	−20

18.10   Refer to Exercise 18.1. Construct the opportunity loss table for the actions of the stockbroker and the states of nature of the stock market. What is the minimax decision? What is the maximax decision?

The Strategist (Maximizing Expected Payoff)

In many respects, a more sensible approach to any decision problem is to consider the likelihood that each state of nature will occur. In this way, you can use any information you have to help evaluate the possibilities of each of the states of nature.

If you believe strongly that the chance of the interest rate declining is small, a decision strategy that uses this information would be useful. The strategy discussed here differs from previous procedures, in that we begin by determining the *probability* associated with each state of nature.

Selecting the Probabilities   The probability for each state of nature measures to what degree you believe this state of nature will occur in the future. One way to obtain these probabilities is from past experience—referred to as **empirical evidence.** For example, if, under similar conditions in the past, the stock market declined 15% of the time, we would set

$$P(\text{stock market declines}) = .15$$

In this way, you can determine the probability for each state of nature. Because we are assuming that one (and only one) of these states *must* occur, these probabilities must sum to 1.

Another method of selecting these probabilities is the **subjective approach.** With this procedure, an individual or group of individuals will select each probability such that (1) each value represents their confidence that each state of nature will occur and (2) the probabilities sum to 1.

To someone unfamiliar with the concept of a probability, you can pose the question, given this set of circumstances 100 different times, how often do you think the stock market will decline? If the answer is about 15 times, then once again you have

$$P(\text{stock market declines}) = .15$$

If the resulting probabilities do not sum to 1 on the first pass, you can state that these probabilities are *all* a little too small (or large) and try again. By continuing in this manner, you eventually will arrive at a set of probabilities for this situation.

The strengths and weaknesses of using these probabilities in the decision process lie in the accuracy of their values. If they are inaccurate, you may well choose an action that incurs a small (or negative) payoff. As a result, this strategy can lead to poor decisions, particularly if the action chosen was based on unreliable subjective probabilities. Nevertheless, it continues to be a popular decision strategy because it allows the decision maker to place probabilities on the unknown future and consider the alternatives.

The Decision Strategy   When using probabilities for each of the states of nature, you determine the *average* payoff for each action in the long run—the average payoff if you repeatedly took this action. This is the **expected payoff** for each action. The strategy in this case is to choose that action having the *largest* expected payoff.

Consider Table 18.2. Suppose that the owner of Sailtown believes that there is a 30% chance that the interest rate will increase over the summer months. This can be written as

$$P(S_1) = .3$$

The chance of the interest rate holding steady is believed to be 20%, whereas the probability of a drop in the rate is 50% (Table 18.10).

■ **TABLE 18.10**
Probabilities for S_1, S_2, and S_3 from Table 18.2.

STATE OF NATURE		PROBABILITY
S_1	Interest rate increases	$P(S_1) = .3$
S_2	Interest rate remains unchanged	$P(S_2) = .2$
S_3	Interest rate decreases	$P(S_3) = .5$

One of the alternatives for this problem was to stock 150 sailboats (action A_4). Using Table 18.2, the respective payoffs are a loss of $35,000 should the interest rate increase (S_1), a profit of $5000 if it holds steady (S_2), and a profit of $45,000 if it decreases ($S_3$). So, if you repeatedly took this action (under the same conditions facing the owner of Sailtown), then you would

lose $35,000, 30% of the time (S_1 occurs)

make $5000, 20% of the time ($S_2$ occurs)

make $45,000, 50% of the time (S_3 occurs)

We discussed this situation in Chapter 5, where we examined *discrete random variables*. The random variable for this situation is

$$X = \text{payoff under action } A_4$$

Based on the preceding discussion, we have

$$X = \begin{cases} -35 \text{ with probability} & .3 \\ 5 \text{ with probability} & .2 \\ 45 \text{ with probability} & \underline{.5} \\ & 1.0 \end{cases} \tag{18.1}$$

The expected payoff for this action is simply the *mean of the random variable, X*. In Chapter 5, this was defined to be

$$\begin{aligned} \text{expected payoff for } A_4 &= \text{mean of } X \\ &= \Sigma(\text{each value of } X)(\text{its probability}) \\ &= (-35)(.3) + (5)(.2) + (45)(.5) \\ &= 13 \end{aligned}$$

This implies that, if the owner of Sailtown repeatedly ordered 150 sailboats (under similar conditions), he would make a profit of $13,000 on the average.

We next use these expected payoffs to form a decision strategy.

EXAMPLE 18.6   Determine the expected payoff for each of the actions in Table 18.2. Using this procedure, how many sailboats should Mr. Larson order?

SOLUTION   Based on the four expected payoffs, the appropriate action is to order 100 sailboats (A_3) with an expected (average) payoff of $18,000 (Table 18.11). If Mr. Larson chooses this alternative, his payoff for a one-time decision will be not a profit of $18,000 but, rather, a loss of $10,000 [with probability $P(S_1) = .3$] or a gain of $30,000 [with probability $P(S_2) + P(S_3) = .7$]. Mr. Larson will select this action if he believes that his long-term gain under this alternative has been maximized. In a sense, he has measured the uncertainty of the future in order to select the best action.   ■

■ TABLE 18.11
Expected payoffs for
Sailtown (thousands
of dollars).

ACTION		EXPECTED PAYOFF
A_1	Order 50 sailboats	$(15)(.3) + (15)(.2) + (15)(.5) = 15$
A_2	Order 75 sailboats	$(2.5)(.3) + (22.5)(.2) + (22.5)(.5) = 16.5$
A_3	Order 100 sailboats	$(-10)(.3) + (30)(.2) + (30)(.5) = 18$
A_4	Order 150 sailboats	$(-35)(.3) + (5)(.2) + (45)(.5) = 13$

EXAMPLE 18.7 In Example 18.3, the minimax strategy was used to select one of the four long-distance discounts offered by GTE Southwest. Suppose that you took a closer look at recent long-distance charges and arrived at the following set of probabilities for the four states of nature:

$$P(S_1) = .1 \qquad P(S_2) = .3 \qquad P(S_3) = .4 \qquad P(S_4) = .2$$

If you elect to use the expected payoff strategy, which of the long-distance options offers the largest expected savings?

SOLUTION The expected payoffs for the four options (actions) are:

$$A_1: \quad (.1)(9) + (.3)(19) + (.4)(29) + (.2)(39) = \$26.00$$
$$A_2: \quad (.1)(9) + (.3)(24) + (.4)(39) + (.2)(54) = \$34.50$$
$$A_3: \quad (.1)(5) + (.3)(25) + (.4)(45) + (.2)(65) = \$39.00$$
$$A_4: \quad (.1)(-20) + (.3)(10) + (.4)(40) + (.2)(70) = \$31.00$$

Consequently, A_3 has the largest expected payoff. Both the minimax and expected payoff strategies indicate that the best action is option 3, with the 20% discount and a fixed charge of $15 per month. ∎

EXAMPLE 18.8 Omega is about to introduce a new line of microcomputers. Their main concern is what selling price they should charge for their computers. The managers can estimate accurately the demand at each price; they are primarily concerned about the time it will take their competitors to catch up and introduce a similar product. They intend to determine a selling price and then not change it for the next two years. They decide to structure the decision problem using four possible alternatives (actions):

$$A_1: \quad \text{set selling price at } \$1500$$
$$A_2: \quad \text{set selling price at } \$1750$$
$$A_3: \quad \text{set selling price at } \$2000$$
$$A_4: \quad \text{set selling price at } \$2500$$

The states of nature specify the amount of time until a similar product is introduced by one of their competitors:

$$S_1 = \text{less than 6 months}$$
$$S_2 = \text{6 to 12 months}$$
$$S_3 = \text{12 to 18 months}$$
$$S_4 = \text{longer than 18 months}$$

The next step for this decision problem is to construct a payoff table. This is *not* an easy step because the managers must consider price-demand, cost-volume, and consumer-preference information in order to specify a payoff for each action under each state of nature. After many meetings between the production and marketing staffs, Table 18.12 was derived, showing projected profits over the next two years.

What is the appropriate action (selling price) if:

1. The minimax strategy is used?
2. Omega decides to maximize the expected payoff?

Use

$$P(S_1) = .1 \qquad P(S_2) = .5 \qquad P(S_3) = .3 \qquad P(S_4) = .1$$

■ **TABLE 18.12**
Profit table for Omega computer-price problem (millions of dollars).

SELLING PRICE	<6 MONTHS (S_1)	6–12 MONTHS (S_2)	12–18 MONTHS (S_3)	>18 MONTHS (S_4)
A_1 $1500	250	320	350	400
A_2 $1750	150	260	300	370
A_3 $2000	120	290	380	450
A_4 $2500	80	280	410	550

■ **TABLE 18.13**
Construction of opportunity loss table for Omega (Example 18.8).

STATE OF NATURE	ACTION WITH LARGEST PAYOFF	OPPORTUNITY LOSS
S_1	A_1	for A_1: 250 − 250 = 0 for A_2: 250 − 150 = 100 for A_3: 250 − 120 = 130 for A_4: 250 − 80 = 170
S_2	A_1	for A_1: 320 − 320 = 0 for A_2: 320 − 260 = 60 for A_3: 320 − 290 = 30 for A_4: 320 − 280 = 40
S_3	A_4	for A_1: 410 − 350 = 60 for A_2: 410 − 300 = 110 for A_3: 410 − 380 = 30 for A_4: 410 − 410 = 0
S_4	A_4	for A_1: 550 − 400 = 150 for A_2: 550 − 370 = 180 for A_3: 550 − 450 = 100 for A_4: 550 − 550 = 0

SOLUTION 1   Using the minimax strategy, we first construct an opportunity loss table for this situation (Table 18.13). We do this by considering each state of nature and finding the action with the largest payoff under each state. The opportunity loss for each action is the maximum payoff under this state of nature minus the payoff for this particular action.

Next, we find the maximum opportunity loss *for each action* (row in Table 18.14).

ACTION	MAXIMUM OPPORTUNITY LOSS
A_1	150
A_2	180
A_3	130
A_4	170

The minimum of these values is 130 for A_3. The minimax strategy would be to select a selling price of $2000 for the next two years.*

SOLUTION 2   The expected profit for each action is summarized in Table 18.15.

The maximum expected profit is 330, for action A_1. So the strategy here is to set the selling price at $1500, with an expected payoff of $330 million. Notice, however, that the three largest expected values are quite close to each other, imply-

---

*The minimax procedure is often confused with the *maximin* strategy, which examines the minimum payoff for each action and selects that action having the maximum of these minimum payoffs. For this application, the minimum payoffs for each action are A_1: 250, A_2: 150, A_3: 120, and A_4: 80. The maximum value here is 250 (belonging to A_1), and the maximin strategy is to select action A_1. The conclusions resulting from minimax and maximin are not the same here because the minimax strategy (which selects that action minimizing the maximum opportunity loss) is to use action A_3. Both strategies are typically very conservative.

■ **TABLE 18.14**
Opportunity loss table
for Omega computer-
price problem (millions
of dollars).

SELLING PRICE	<6 MONTHS (S_1)	6–12 MONTHS (S_2)	12–18 MONTHS (S_3)	>18 MONTHS (S_4)
A_1 $1500	0	0	60	150
A_2 $1750	100	60	110	180
A_3 $2000	130	30	30	100
A_4 $2500	170	40	0	0

■ **TABLE 18.15**
Expected profits for
Omega (Example 18.8).

ACTION	EXPECTED PROFIT
A_1	(.1)(250) + (.5)(320) + (.3)(350) + (.1)(400) = 330
A_2	(.1)(150) + (.5)(260) + (.3)(300) + (.1)(370) = 272
A_3	(.1)(120) + (.5)(290) + (.3)(380) + (.1)(450) = 316
A_4	(.1)(80) + (.5)(280) + (.3)(410) + (.1)(550) = 326

ing that one of the other alternatives might surpass A_1 if the state of nature probabilities are adjusted slightly. The preference for A_1 may be very sensitive to these probabilities. This situation should concern a decision maker, especially if the probabilities are determined subjectively. ■

Example 18.8 suggests another important element of the decision process—a sensitivity analysis.

Sensitivity Analysis

Typically, there is no way to determine a state of nature probability with certainty. You can consider past observations and derive an empirical estimate or merely make up a value that measures your belief that this event will occur (the subjective approach). The next step when using the maximum expected payoff strategy is to examine what happens to this solution under other sets of realistic probabilities. This examination is called a **sensitivity analysis.**

In Example 18.6, the expected payoff procedure selected action A_3. By ordering 100 sailboats, Sailtown achieved a maximum expected profit of $18,000. The state of nature probabilities used here were

$$P(S_1) = .3 \quad \text{(interest rate increases)}$$
$$P(S_2) = .2 \quad \text{(interest rate remains unchanged)}$$
$$P(S_3) = .5 \quad \text{(interest rate decreases)}$$

Although Mr. Larson and his financial advisor are uncertain as to the precise values of these probabilities, they believe that:

1. There is no more than a 50% chance that the interest rate will increase ($P(S_1) \le .5$).
2. There is no more than a 30% chance that the rate will remain unchanged ($P(S_2) \le .3$).
3. The probability that the rate will decrease is between .3 and .5.

They decide to examine the expected payoffs under the probability conditions listed in Table 18.16. The expected payoffs under each set of probabilities are determined as in Example 18.6. As an illustration, using $P(S_1) = .4$ and $P(S_2) = P(S_3) = .3$, the expected payoff for action A_4 is $(.4)(-35) + (.3)(5) + (.3)(45) = 1$ (that is, $1000).

The sensitivity summary in Table 18.16 indicates that action A_1 (ordering 50 sailboats) may be much more attractive than we thought. In fact, if there is more than a 30% chance that the interest rate will increase ($P(S_1) > .3$), this action produces the largest expected payoff. Under this decision, Mr. Larson can expect

■ TABLE 18.16
Summary of sensitivity analysis (values in color represent the action with the largest expected payoff).

$P(S_1)$	$P(S_2)$	$P(S_3)$	A_1	A_2	A_3	A_4
				EXPECTED PAYOFF		
.4	.2	.4	15	14.5	14	5
.4	.3	.3	15	14.5	14	1
.4	.1	.5	15	14.5	14	9
.5	.2	.3	15	12.5	10	−3
.5	.1	.4	15	12.5	10	1
.3	.3	.4	15	16.5	18	9
.3	.2	.5	15	16.5	18	13

to sell all his inventory, resulting in a profit of $15,000 *regardless of the state of nature*. Consequently, Sailtown would be seeking an expected gain of $3000 ($18,000 for A_3 minus $15,000 for A_1) by speculating on the uncertain future.

Without such a sensitivity analysis, Mr. Larson would not have noticed these results. In five of the six cases using probabilities other than those used in Example 18.6, action A_1 produced the maximum expected payoff. If Mr. Larson is uncertain in his original determination of these probabilities, this action is a better solution to his decision problem.

Evaluating Risk

Using the preceding sensitivity analysis and the payoffs in Table 18.2, we noticed that the payoff for action A_1 was 15 (thousand dollars), regardless of the state of nature. Action A_4, on the other hand, has possible payoffs of −35, 5, and 45, implying that you will encounter a higher risk using A_4 rather than A_1. In fact, action A_1 has *no risk* because its payoff is known with certainty.

Take a closer look at action A_4. In discussing equation 18.1, we remarked that the payoff for this action, X, is a random variable, where

$$X = \begin{cases} -35 & \text{with probability .3} \\ 5 & \text{with probability .2} \\ 45 & \text{with probability .5} \end{cases}$$

The expected payoff for this action is the *mean* of X.

A risky alternative (action) is one that has larger probabilities attached to extremely large or small payoffs. A good measure of this risk is simply the *variance* of X; the variance of the possible payoffs for each action is a measure of the risk associated with this alternative. The larger the variance is, the more risk will be incurred using this action. The variance of this *discrete* random variable is found in the same way as in Chapter 5 and is summarized here.

Let X_i be the payoff associated with action A_i. Then

$$X_i = \begin{cases} x_1 & \text{with probability } p_1 \\ x_2 & \text{with probability } p_2 \\ \vdots \\ x_n & \text{with probability } p_n \end{cases}$$

where n represents the number of states of nature.

The **expected payoff** for this action is the mean of X_i, where

$$\text{expected payoff} = \mu_i = \Sigma xp \qquad \textbf{(18.2)}$$

The **risk** associated with action A_i is the variance of X_i. So,

$$\text{risk} = \Sigma x^2 p - \mu_i^2 \qquad (18.3)$$

EXAMPLE 18.9   Compute the risk for each of the actions in Table 18.2, using the state of nature probabilities from Example 18.6. Based on these results and the sensitivity analysis in Table 18.16, which action appears to be the best one for this situation?

SOLUTION   Using equation 18.3, we can find the risk associated with each of the four alternatives, as shown in Table 18.17. The purpose of examining the risk for each action is that often the decision maker will prefer a less risky alternative over a riskier action with a larger expected payoff. In this example, action A_1 has no risk and also has a maximum expected payoff for most of the situations examined in the sensitivity analysis. On the other hand, action A_3 (the other suggested approach) carries the second-largest risk, as measured in Table 18.17. For these reasons, the soundest alternative appears to be action A_1, with a known payoff of $15,000. ∎

EXAMPLE 18.10   Which of the long-distance options facing Worldwide Import Distributors (Example 18.7) has the least amount of risk?

SOLUTION   The risk for each of the four options is computed in Table 18.18. The four options show an increasing amount of risk as the discount rate increases, with option 1 (a 10% discount and a monthly charge of $1) having the lowest risk but also the lowest expected monthly savings. ∎

EXAMPLE 18.11   Which of the alternatives in Example 18.8 has the least amount of risk?

SOLUTION   As before, the risk associated with each action is the variance of the corresponding random variable, shown in Table 18.19. The most desirable action for Omega is A_1 (selling price = $1500) because it wins on two counts: it not only has the largest

■ TABLE 18.17
Risk calculations for Sailtown decision problem.

ACTION	EXPECTED PAYOFF (μ_i)	RISK (USING EQUATION 18.3)
A_1	15	$[(15)^2(.3) + (15)^2(.2) + (15)^2(.5)] - 15^2 = 0$
A_2	16.5	$[(2.5)^2(.3) + (22.5)^2(.2) + (22.5)^2(.5)] - 16.5^2 = 84$
A_3	18	$[(-10)^2(.3) + (30)^2(.2) + (30)^2(.5)] - 18^2 = 336$
A_4	13	$[(-35)^2(.3) + (5)^2(.2) + (45)^2(.5)] - 13^2 = 1216$

■ TABLE 18.18
Risk calculations for Worldwide Import Distributors problem.

ACTION	EXPECTED PAYOFF (μ_i)	RISK (USING EQUATION 18.3)
A_1	26.00	$(9)^2(.1) + (19)^2(.3) + (29)^2(.4) + (39)^2(.2) - 26^2 = 81$
A_2	34.50	$(9)^2(.1) + (24)^2(.3) + (39)^2(.4) + (54)^2(.2) - 34.5^2 = 182.25$
A_3	39.00	$(5)^2(.1) + (25)^2(.3) + (45)^2(.4) + (65)^2(.2) - 39^2 = 324$
A_4	31.00	$(-20)^2(.1) + (10)^2(.3) + (40)^2(.4) + (70)^2(.2) - 31^2 = 729$

■ TABLE 18.19
Risk calculations for Omega decision problem.

ACTION	EXPECTED PAYOFF (μ_i)	RISK (USING EQUATION 18.3)
A_1	330	$[(250)^2(.1) + (320)^2(.5) + (350)^2(.3) + (400)^2(.1)] - 330^2 = 1,300$
A_2	272	$[(150)^2(.1) + (260)^2(.5) + (300)^2(.3) + (370)^2(.1)] - 272^2 = 2,756$
A_3	316	$[(120)^2(.1) + (290)^2(.5) + (380)^2(.3) + (450)^2(.1)] - 316^2 = 7,204$
A_4	326	$[(80)^2(.1) + (280)^2(.5) + (410)^2(.3) + (550)^2(.1)] - 326^2 = 14,244$

expected profit, it also has the smallest risk. For a great many decision problems, this will not be the case, and so the decision maker will have to decide how much risk he or she is willing to assume in an effort to gain a higher expected profit. *If a heavy loss would be devastating to a company, it may be forced into adopting strategies that select alternatives with reasonably attractive profits but considerably less risk.*   ■

Exercises

18.11   The following table gives the payoff for four different states of nature and three different actions. If each state of nature is equally likely, find the decision resulting from maximizing expected payoff. What is the risk for each action?

ACTION	S_1	S_2	S_3	S_4
A_1	180	150	10	50
A_2	55	55	55	55
A_3	80	160	100	40

18.12   The manager of a hardware store orders several cords of split logs to be sold to customers for firewood. More wood typically is sold when a winter is colder than usual. The manager figures that the chance of an extremely cold winter (S_1) is 0.25, the chance of a normal winter (S_2) is 0.50, and the chance of a relatively mild winter (S_3) is 0.25. The manager must decide on ordering either 50, 40, 30, or 20 cords of wood. The payoff table follows.

ACTION	S_1	S_2	S_3
A_1 (50)	5000	3000	1000
A_2 (40)	4400	3200	1200
A_3 (30)	3200	2800	1400
A_4 (20)	3000	2500	2000

a.   What is the minimax decision?

b.   What is the decision based on the maximum expected payoff?

c.   What is the risk of each action?

d.   What is the decision based on minimum risk?

18.13   Programs need to be printed for a theatrical performance. The programs are sold at the entrance of the theater before the performance starts. The director of the theater believes that there is a 35% chance that there will be a heavy turnout (S_1), a 50% chance for a normal turnout (S_2), and a 15% chance for a low turnout (S_3). The director must decide to have 200 copies (A_1), 300 copies (A_2), 400 copies (A_3), or 500 copies (A_4) of the program printed. The payoff table follows. Unsold programs would result in a loss.

ACTION	S_1	S_2	S_3
A_1	100	100	100
A_2	150	140	110
A_3	200	160	75
A_4	250	120	− 50

a.   Find the minimax decision.

b.   Find the decision based on the maximum expected payoff.

c.   Find the risk associated with each decision.

18.14   In Exercise 18.1, what is the decision based on the maximum expected payoff if the probability that the stock market goes up is 10%, the probability that the stock market has no direction is 50%, and the probability that the stock market goes down is 40%?

18.15   In Exercise 18.7, what is the maximum expected payoff if the demand for newspapers has the following probabilities?

DEMAND	PROBABILITY	DEMAND	PROBABILITY
50	0.10	80	0.25
60	0.10	90	0.20
70	0.25	100	0.10

18.16   In Exercise 18.8, what is the maximum expected payoff if the probability of a complete loss of the commercial building is .05? Assume that the probability of no loss is 0.95. Find the risk associated with each action.

Dominated Actions and the Value of a Crystal Ball

In a decision problem, we often can eliminate an action from consideration if another action in the problem has a larger payoff, regardless of the state of nature. Consider actions A_1 and A_2 from Table 18.12. Notice that the payoff for A_1 exceeds that for A_2 for all four states of nature. In this case, we say that A_1 **dominates** A_2. Action A_i dominates A_j if the payoff for A_i is greater than or equal to that for A_j under each state of nature. For at least one state of nature, the payoff for A_i must exceed that for A_j. In our example, there is no reason to consider A_2 for any decision strategy because A_1 produces a larger profit, regardless of what happens in the future. We say that action A_2 is inadmissible; it will not be included in the group of actions to be considered in the problem solution. Action A_i is **inadmissible** if it is dominated by any other action. Consequently, A_i is **admissible** if no other action under consideration dominates it.

EXAMPLE 18.12   Which of the actions in Table 18.2 are admissible?

SOLUTION   The procedure here is to determine whether any of these actions are dominated by any other action. For example, for actions A_2 and A_3, A_2 has a larger payoff, given state of nature S_1, but A_3 produces a bigger profit, given S_3. Therefore, neither of these two actions dominates the other. Comparing the remaining pairs of actions produces a similar argument; no one action produces a larger payoff for all states of nature. This implies that *all four* actions are admissible, and so they will all be considered in the search for the best action under the selected decision strategy.   ∎

Note that there is no serious harm in considering a dominated action; such an action never is selected by any of the decision strategies. By eliminating a dominated action from consideration, however, we simplify the decision process, since we have fewer actions to consider.

Expected Value of Perfect Information

When using the strategy of maximizing expected profit, one value of interest is how much you would be willing to pay for a predictor that could tell you the future state of nature correctly 100% of the time. For example, you might have the (very unrealistic) situation of a consulting firm that predicts the future correctly all the time or, just as farfetched, a crystal ball. Because the future is in fact never perfectly predictable, any information about the future will be imperfect. Consequently, you use the value of a perfect predictor to evaluate any cost that you might incur for such imperfect information.

In Example 18.6, what would Mr. Larson, the owner of Sailtown, expect to make if a perfect predictor existed? Referring to Table 18.2, because $P(S_1) = .3$, the crystal ball will predict state of nature S_1 30% of the time. In this case, Mr. Larson inspects this column of Table 18.2, realizes that action A_1 has the largest payoff, and so orders 50 sailboats. His profit for this decision will be 15 (thousand dollars).

Now suppose the crystal ball predicts that the interest rate will remain unchanged (S_2 occurs). In this event, Mr. Larson will order 100 sailboats (A_3), because this is the largest payoff in the column under S_2 in Table 18.2. His profit will be 30. Finally, if he is informed that S_3 will occur, he selects action A_4, with a payoff of 45, because this produces the largest profit given that the interest rate will decrease.

In the long run, with a perfect predictor, Mr. Larson would make

$15,000 30% of the time   (when S_1 occurs)

$30,000 20% of the time   (when S_2 occurs)

$45,000 50% of the time   (when S_3 occurs)

These results mean that his *expected payoff with a perfect predictor* is

$$(15,000)(.3) + (30,000)(.2) + (45,000)(.5) = \$33,000$$

Finally, recall that the action that maximized the expected payoff (from Example 18.6) was A_3, with a value of $18,000. So, Mr. Larson would make $33,000 on the average *with* a crystal ball, and he would earn $18,000 on the average *without* it by taking action A_3 each time. This means that the maximum price he should be willing to pay for a perfect predictor is

$$\$33,000 - 18,000 = \$15,000$$

This is the expected value of perfect information.

When you use expected payoffs in your decision strategy, you select that action (say, A') having the largest expected payoff. The **expected value of perfect information (EVPI)** is

$$EVPI = \frac{\text{(average payoff using a perfect predictor)} -}{\text{(average payoff for } A')} \qquad (18.4)$$

EXAMPLE 18.13   In Example 18.8, the managers of Omega attempted to choose a selling price for their new computer. The states of nature for this problem were concerned with the amount of time until a major competitor introduced a similar product. The payoffs for this situation are summarized in Table 18.12. Assuming that $P(S_1) = .1$, $P(S_2) = .5$, $P(S_3) = .3$, and $P(S_4) = .1$, determine the expected value of perfect information.

SOLUTION   The action having the largest expected profit (according to Example 18.8) is A_1, with an expected value of 330. Consequently, the payoff with a selling price of $1500 would be $330 million on the average.

Given a perfect predictor, the following payoffs are possible:

STATE OF NATURE (S_i)	MAXIMUM PAYOFF	PROBABILITY $P(S_i)$
S_1	250 (for A_1)	.1
S_2	320 (for A_1)	.5
S_3	410 (for A_4)	.3
S_4	550 (for A_4)	.1

Consequently, the expected payoff using a perfect predictor is

$$(.1)(250) + (.5)(320) + (.3)(410) + (.1)(550) = 363 \quad \text{(million dollars)}$$

From these results, the expected value of perfect information (from equation 18.4) is

$$EVPI = 363 - 330 = 33 \quad \text{(million dollars)}$$

So what is the maximum amount that Omega Corporation should be willing to pay an outside consulting firm for information regarding the time until a competitor introduces a similar model into the market? This is what the EVPI represents—an upper limit for the price of *any* information regarding the future. If Omega elects to pay an outside firm for information, they realize that the predicted state of nature could be wrong (that is, this information will be imperfect). For this reason, this information is worth *considerably less* than the EVPI of $33 million. Its value will depend in part on the reliability of the consulting firm, as measured by the latter's past performance in similar situations. This topic is pursued further in Section 18.4.　■

Exercises

18.17　Consider the following payoff table, in which each state of nature is equally likely. Find the EVPI. Are all actions admissible?

ACTION	S_1	S_2	S_3	S_4
A_1	40	10	4	15
A_2	30	15	0	35
A_3	20	10	-5	30
A_4	35	20	10	10

18.18　A builder usually builds 20-, 50-, or 100-unit apartments. The builder is concerned with three states of nature: low demand (S_1), medium demand (S_2), and high demand (S_3). The builder believes that the probability of S_1 is .40, the probability of S_2 is .30, and the probability of S_3 is .30. From the following payoff table, what is the maximum amount that the builder would be willing to pay a consultant for advice regarding the market demand for apartments? Are all the actions admissible?

ACTION	S_1	S_2	S_3
A_1 (20 units)	10,000	13,000	16,000
A_2 (50 units)	8,000	23,000	25,000
A_3 (100 units)	8,000	20,000	40,000

18.19　Greetings card shop must decide on whether to order 1500, 2000, or 2500 holiday cards before the holiday season. The card shop makes a profit of 50¢ on each card it sells and loses 30¢ on each card that remains unsold. If the demand for holiday cards is strong (S_1), 2500 cards should sell. If the demand is average (S_2), 2000 cards should sell. If the demand is weak (S_3), 1,500 cards should sell. The manager of Greetings believes that the following probabilities are representative of past sales: $P(S_1) = .40$, $P(S_2) = .40$, and $P(S_3) = .20$. What is the maximum amount that the manager of Greetings would be willing to pay for perfect information regarding the demand for holiday cards?

18.20　If the EVPI for a certain company is $40,000 and the average payoff based on the decision from the maximum expected payoff is $25,000, what is the maximum payoff using a perfect market predictor?

18.21　Refer to Exercise 18.1. What is the maximum that the stockbroker would be willing to pay a consultant regarding the direction of the stock market, assuming that all of the three states of nature are equally likely?

18.22　Refer to Exercise 18.12. What is the maximum that the manager of the hardware store would be willing to pay to obtain perfect information regarding the type of winter for the current year?

18.23　Refer to Exercise 18.13. What is the EVPI for the manager of the theater, regarding attendance turnout? Are any of the actions inadmissible?

18.3
The Concept
of Utility

We have concentrated on choosing the best action under various decision strategies by using the values contained in the payoff table. For example, one strategy determines the action having the largest expected payoff. Another (the minimax procedure) examines opportunity losses derived from the payoff table. Still another strategy examines expected payoffs. In other words, each action is evaluated by the corresponding *dollar amount* resulting from a particular strategy.

There are many instances in which it is more advantageous *not* to use expected payoffs, particularly when large amounts of money are at stake. Anyone who purchases an insurance policy or buys a lottery ticket generally is trying neither to minimize expected losses nor maximize expected gains. Rather, such a person *gambles* his or her money, trying to guard against a heavy loss (insurance) or hoping to strike it rich (lottery).

There is something else besides money involved in the decision to purchase an insurance policy or a lottery ticket. In the case of insurance, suppose you have a $100,000 home insured for the full amount. For most people, a gift of $100,000 would be nice (in fact, *very* nice) but a $100,000 loss would be totally devastating—this is the underlying concept behind the insurance philosophy. A gain of $100,000 does not have the same effect on the positive side as does a $100,000 loss on the negative side.

When you fail to purchase insurance, you are betting that your house will not go up in smoke. Your *risk* here is that the house may burn. When we look at expected payoffs only, we ignore risk. On the other hand, we also discussed a method of examining the risk of each action by finding the variance of the respective payoffs. What we need is a method that combines the decision maker's attitude toward the payoff with the corresponding risk of each alternative. An action with a possible higher payoff (or loss, in the case of insurance) often contains more risk. We measure the attractiveness of each outcome using utility values, which we will now develop.

> The **utility value** of a particular outcome is used to measure both the attractiveness and the risk associated with this dollar amount.

Constructing a Utility Value

Suppose you have $10,000 saved up for college expenses one year before you begin your freshman year. A friend of yours has offered you part interest in an oil-drilling venture for your $10,000. If the venture fails, you lose your entire investment, but if it succeeds, you stand to gain $40,000. According to the latest geological survey, the probability of hitting oil is .3. Also, if oil exists, the expected life of the venture is one year, with the payoff of $40,000.

Your other option is to invest the money for one year in a money market account at an expected interest rate of 12%. If you choose to maximize your expected payoff (dollars on hand at the end of the year), which action should you select?

The decision problem involves two actions,

A_1:   put the $10,000 into the money market   (interest = 12%)

A_2:   invest $10,000 in the oil venture

with two states of nature

S_1:   oil does not exist on the site

S_2:   oil does exist on the site

Your corresponding payoffs, should you select the oil investment, are

$0   if S_1 occurs (and you lose your investment)

$40,000   if S_2 occurs

The payoff table is shown next with the corresponding state of nature probabilities in parentheses.

The expected payoffs here are

	$S_1(.7)$	$S_2(.3)$
A_1	11,200	11,200
A_2	0	40,000

for action A_1: $11,200(.7) + 11,200(.3) = 11,200$

for action A_2: $(0)(.7) + (40,000)(.3) = 12,000$

The oil venture (A_2) has a larger expected payoff, so, using this decision strategy, you would elect to gamble your money in hopes of a large payoff. But is this a realistic strategy? Assume that the loss of your $10,000 would result in your not going to college. All things considered, this would be disastrous. Although the large payoff would be terrific, the high probability of a heavy loss might make you wonder if the gamble is a good idea.

The problem in this illustration is that a large payoff often is very attractive, but it is offset by a risk associated with it. We say that the **utility** associated with a gain of, say, $100,000 without risk is *higher* than the utility of this amount with a high risk. For each decision problem, we ask the decision maker to determine the utility value associated with the various payoffs in the problem. In this way, the person can build in his or her attitudes with regard to avoiding risk or preferring a gamble with a big payoff (a *risk taker*).

To illustrate the construction of a utility value, consider the payoff table we just constructed. There are many ways to proceed here, although all the various ways of assigning utility values produce the *same* decision when using these values to arrive at the best alternative. We use a two-step procedure:

Step 1. Assign a utility value of zero to the smallest payoff amount (π_{min}) and a value of 100 to the largest (π_{max}). For this example, the utility values would be written as $U(0) = 0$ and $U(40,000) = 100$, because $\pi_{min} = 0$ and $\pi_{max} = 40,000$. *All utility values range from 0 to 100.* Whether you assign utility values from 0 to 1, 0 to 100, 1 to 5, or any range does not matter. It is not the actual value of the utility that is important but rather its value *relative* to the range of all values.

There is one other payoff in the table to consider, namely, $11,200. What is the utility of this dollar amount to the decision maker (you, in this case)? We consider both the attractiveness and the risk involved with this payoff in the following situation.

Consider the largest payoff of $40,000 and the smallest of $0. You need to decide what the probability, P, would have to be before you would consider

$11,200 with certainty

to be as attractive as

$40,000 with probability P and $0 with probability $1 - P$

Suppose you decide that you would need at least a 50% chance of striking oil. So, $P = .5$. We next define the utility of the $11,200 payoff by using

$$\text{utility value} = P \cdot 100$$
$$= .5 \cdot 100$$
$$= 50$$

That is,

$$U(11,200) = 50$$

■ **FIGURE** 18.1
Illustration of utility
values for oil venture.

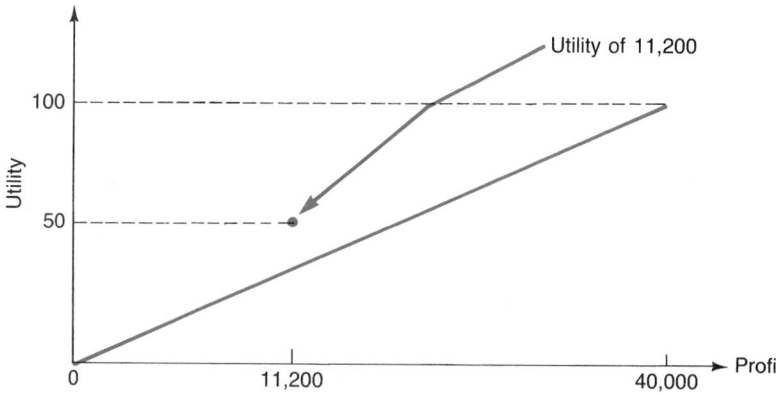

A graphical illustration of these utilities is shown in Figure 18.1. An easy way to measure your attitude toward risk is to connect the lower left (utility = 0) and upper right (utility = 100) corners. If the utility values you have assigned fall *above* this line, you tend to avoid risk. If they fall *below* the line, you are a risk taker. For this example, $U(11,200) = 50$ lies above the diagonal line, indicating that, for this situation, you are a risk avoider. A summary of this procedure is step 2 of the utility value assignment.

Step 2. The utility value for any payoff (say, π_{ij}) under consideration is found by using

$$U(\pi_{ij}) = P \cdot 100$$

where P is the probability such that

$$\pi_{ij} \text{ with certainty}$$

is equally as attractive as

$$\pi_{\max} \text{ with probability } P \text{ and } \pi_{\min} \text{ with probability } 1 - P$$

and $\pi_{\max}$ and $\pi_{\min}$ are determined in step 1.

The resulting table of utility values for this example is

ACTION	S_1: OIL VENTURE IS UNSUCCESSFUL (.7)	S_2: OIL VENTURE IS SUCCESSFUL (.3)
A_1 money-market account	50	50
A_2 oil venture	0	100

Notice that we return to the *original* probabilities for the states of nature (.7 and .3) when constructing this table. The values of .5 and .5 were used only to measure your willingness to take a risk. They do not change the fact that S_1 will occur 70% of the time (even though you may wish it would occur only 50% of the time).

To use the utility values in choosing one of the alternative actions, we proceed exactly as before, except that we select that action having the largest expected utility *rather than the largest expected payoff.*

expected utility for each action
$$= \Sigma(\text{each utility value}) \cdot (\text{its probability}) \qquad \textbf{(18.5)}$$

EXAMPLE 18.14 Using the preceding table, which action, based on expected utility, is the more attractive of the two?

SOLUTION

$$A_1: \text{ expected utility} = (50)(.7) + (50)(.3) = 50$$
$$A_2: \text{ expected utility} = (0)(.7) + (100)(.3) = 30$$

Because action A_1 (the money market account) has a larger expected utility, we choose this action over the riskier oil venture, A_2. This choice was also suggested by Figure 18.1, which indicated that, for this application, you are a risk avoider. ∎

Determining Utility Values for Large Decision Problems

Whenever your payoff table contains a large number of values, there is an alternative to the two-step procedure just described. The main problem here is the second step, which requires that the decision maker determine the utility of *each* payoff contained in the payoff table. This can be quite difficult, because there is a requirement for step 2: if

$$\text{payoff } \pi_{ij} < \text{payoff } \pi_{st}$$

then it is necessary that

$$U(\pi_{ij}) < U(\pi_{st})$$

This means that if one payoff is larger than another, the corresponding utility values must be in the *same order*.

When forced to determine the utility of many payoffs, some of which are nearly the same, the decision maker may rate one payoff lower than another, but the utility values may be in the opposite order. One way to avoid this situation is *not* to use step 2 on every payoff involved in the problem; rather, you use this step on *between five and ten payoff values over the range of payoffs* for this problem. Consequently, you would examine π_{min} and π_{max} from step 1 and select, say, six payoffs between these values. These *need not* be actual payoff values from the payoff table. You then use the step 2 procedure to determine the utility value (U) for each of these six payoffs.

Your next step is to plot these values in a graph and connect them to form a **utility curve.** The utility of each value in the payoff table can be obtained by approximating it from the resulting graph. Because of the requirement for step 2, you need to make sure that the utility curve always *increases as the payoff increases.* We will demonstrate this technique in Example 18.15.

EXAMPLE 18.15 Table 18.12 contains the various payoffs for the selling-price decision facing the Omega Corporation. The minimum payoff is $\pi_{min} = 80$ (million dollars), and the maximum is $\pi_{max} = 550$. So, for step 1, we have

$$U(80) = 0$$

and

$$U(550) = 100$$

Describe a procedure for determining the utility of the remaining 14 values.

SOLUTION One method, of course, is to have the decision maker choose 14 corresponding utility values using step 2 on each payoff. An easier procedure is to request this information for payoffs of, say, 100, 150, 200, 300, 400, and 500. Notice that these payoffs are not necessarily contained in Table 18.12, but they do cover the range

■ TABLE 18.20
Utilities for
Example 18.15.

	PAYOFF					
	100	150	200	300	400	500
Probability (P)	.20	.40	.55	.75	.90	.97
Utility [$P(100)$]	20	40	55	75	90	97

■ FIGURE 18.2
Utility curve for Omega
computer-price decision
(Example 18.15).

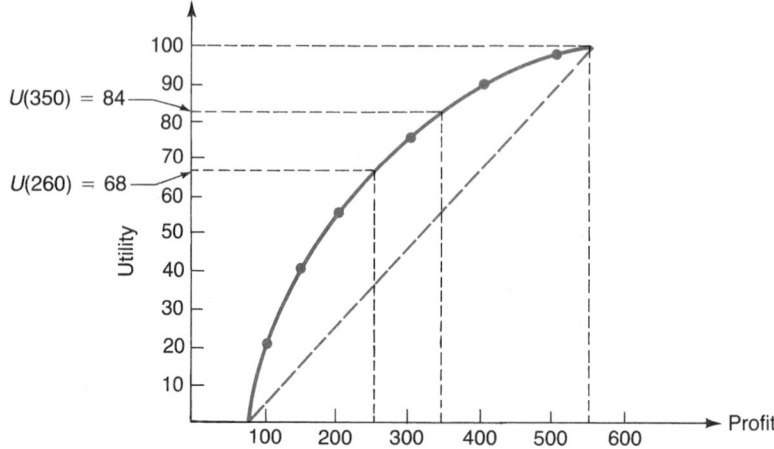

from 80 to 550. For a payoff of 200, we ask for that value of P such that a payoff of 200 (million dollars) with certainty is equally as attractive as a payoff of 500 with probability P and a payoff of 80 with probability $1 - P$. Suppose the decision maker's response is $P = .55$. Then the utility value of this payoff is

$$U(200) = .55 \cdot 100 = 55$$

Consider the set of probabilities (P) and corresponding utilities in Table 18.20. It will be much easier for the decision maker to supply these six values than to choose the 14 values remaining in Table 18.12. *A key ingredient to making any quantitative procedure usable is to keep it reasonably simple!*

The utilities for this problem are plotted in Figure 18.2; the curve through these points represents the decision maker's utility curve. Notice that the utility values *do* increase as the payoffs increase, so the requirement for step 2 is satisfied. As in Figure 18.1, the utility values lie *above* the line connecting the corners, indicating that this individual is a risk avoider.   ■

EXAMPLE 18.16

Using the utility curve in Figure 18.2, determine the utility for each payoff in Table 18.12. Which action (selling price) has the largest expected utility?

SOLUTION   From step 1, we have $U(80) = 0$ and $U(550) = 100$. The remaining utilities can be estimated from the utility curve constructed in Example 18.15. This process is illustrated for payoffs of 260 (action A_2, state of nature S_2) and 350 (action A_1, state of nature S_3) in Figure 18.2. Consequently,

$$U(260) = 68$$
$$U(350) = 84$$

Continuing this procedure results in Table 18.21. The expected utilities are, for example,

$$\text{expected utility for } A_2 = (.1)(40) + (.5)(68) + (.3)(75) + (.1)(87) = 69.2$$

	S_1 (.1)	S_2 (.5)	S_3 (.3)	S_4 (.1)	EXPECTED
ACTION					**UTILITY**
A_1: price = $1500	67	79	84	90	80.4
A_2: price = $1750	40	68	75	87	69.2
A_3: price = $2000	30	74	88	94	75.8
A_4: price = $2500	0	72	91	100	73.3

■ **TABLE 18.21**
Utility table for
Omega computer-price
decision problem
(Example 18.16).

If we choose that action with the largest expected utility, our decision is to select action A_1 (selling price $1500). For this application, A_1 maximizes both expected payoff (see Example 18.8) and expected utility. ■

Shape of Utility Curves

The shape of a decision maker's utility curve indicates his or her preference for or aversion to risk. There are essentially three categories of people in regard to risk: (1) the risk avoider, (2) the risk neutral, and (3) the risk taker.

The basic shapes of the utility curves for each of these classifications are shown in Figure 18.3. Notice that in all three situations, the utility curves increase as the

■ **FIGURE 18.3**
Three classes
of utility curves:
(a) the risk avoider,
(b) the risk neutral,
(c) the risk taker.

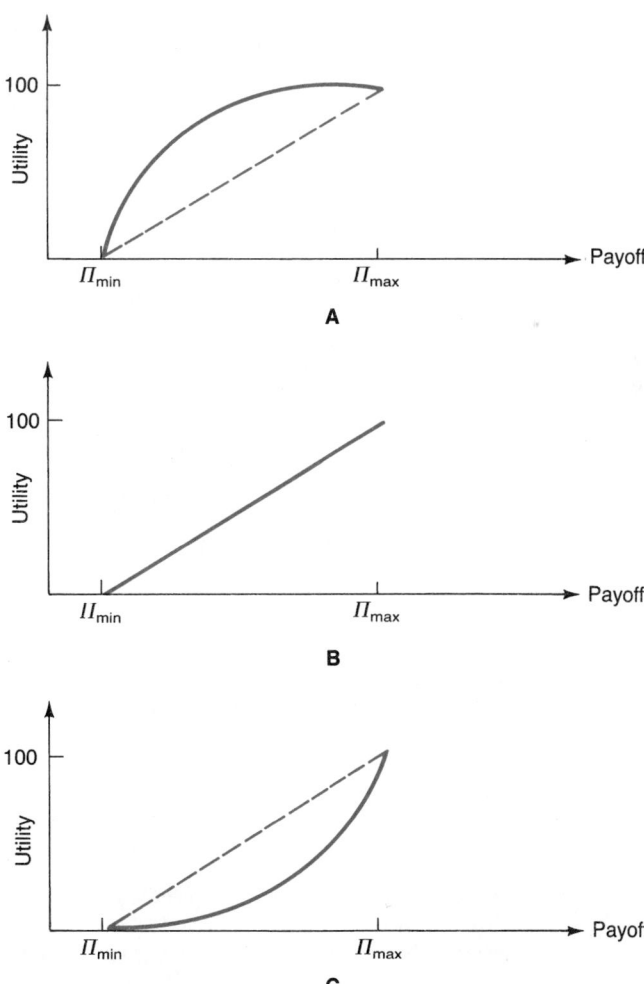

payoff increases. Variations of these curves also can occur; for example, a person may prefer a risk for small payoffs but then avoid a risk for large payoffs. A utility curve for such a person is S-shaped.

An individual who is **risk neutral** will have resulting utility values that lie close to the line connecting the corners. It makes no difference whether you maximize expected payoff or expected utility—the resulting best action for this person is the *same* in either case. Very wealthy people often demonstrate this behavior because, for them, the utility of each dollar remains nearly constant.

Most people are **risk avoiders,** particularly when large payoffs or losses are involved. For two actions with equal expected payoffs, the risk avoider prefers the one with the smaller risk. This person also prefers a smaller expected payoff with a small risk over a larger expected payoff with a large risk. The **risk taker,** on the other hand, is the gambler; he or she prefers an action with a possible large payoff, even if the risk is more severe.

In summary, utility values allow the decision maker to combine both payoff and risk into a single measure. However, the assignment of these values is subjective and special care must be taken in their determination.

Exercises

18.24 A utility function is assigned the value 100 for the largest payoff, which is $20,000, and the value 0 for the smallest payoff, which is $0. The following table gives the probability, P, that, if the amount in the left-hand column were certain, it would be equivalent to having a payoff of $20,000 with probability P and a payoff of $0 with probability $1 - P$. Graph the utility function and determine the attitude toward risk of a person with this utility curve. From the graph, determine the approximate utility value for a payoff of $17,000.

PAYOFF	PROBABILITY	PAYOFF	PROBABILITY
2,000	0.125	12,000	0.60
4,000	0.230	16,000	0.85
8,000	0.40	18,000	0.90

18.25 Suppose that a person has a utility function $U(x) = 2\sqrt{x}$, where x is assumed to be any value between 1 and 100. Consider the following payoff table, in which all four states of nature are equally likely.

ACTION	S_1	S_2	S_3	S_4
A_1	1	100	50	10
A_2	80	30	40	25
A_3	90	20	30	10

a.   Find the decision based on the maximum expected payoff.

b.   Find the decision based on the maximum expected utility of the payoff.

18.26 Suppose that a person is risk neutral and has the utility function $U(x) = 3x$. In the following payoff table, $P(S_1) = .20$, $P(S_2) = .40$, $P(S_3) = .30$, and $P(S_4) = .10$. Show that the decision based on the maximum expected payoff is equivalent to the decision based on the maximum expected utility of the payoff.

ACTION	S_1	S_2	S_3	S_4
A_1	50	10	30	10
A_2	20	20	30	60
A_3	10	50	10	20

18.27 If a person has a utility function $U(x) = x$, what probability, P, would the person have to assign to a maximum payoff of $10,000, with a $1 - P$ probability to a minimum payoff of $0, assuming that the person could receive $8100 with certainty?

18.28 The utilities of the payoffs given in the table in Exercise 18.11 are as follows:

ACTION	S_1	S_2	S_3	S_4
A_1	2.83	2.72	1.59	2.19
A_2	2.23	2.23	2.23	2.23
A_3	2.40	2.76	2.51	2.09

Find the decision based on the maximum expected utility of the payoff.

18.29 If the utility curve of the manager of the hardware store in Exercise 18.12 is $U(x) = \log_{10}(x)$, find the decision based on the maximum expected utility. Is the manager a risk neutral, a risk taker, or a risk avoider?

18.30 Suppose that a money manager is a risk avoider with the utility function $U(X) = 3 \cdot x\sqrt{x}$, where x can range from 0 to 200. In the following payoff table, $P(S_1) = .25$, $P(S_2) = .30$, $P(S_3) = .40$, $P(S_4) = .05$.

ACTION	S_1	S_2	S_3	S_4
A_1	151	33	95	40
A_2	75	75	97	180
A_3	29	162	30	50

a. Find the decision based on the maximum expected payoff.

b. Find the decision based on the maximum expected utility of the payoff.

18.4
Decision Trees and Bayes' Rule

This section describes the decision tree, a device useful for structuring and illustrating the uncertain outcomes associated with any decision problem. A decision tree graphically represents the entire decision problem, including a representation of:

1. The possible actions facing the decision maker.

2. The outcomes (states of nature) that can occur.

3. The relationships between these actions and outcomes.

The decision tree makes it easier for you to compute the expected values and to understand the process of making a decision. We will demonstrate how to construct a decision tree and discuss a procedure for using the tree to examine the alternatives and arrive at a decision.

The basic idea behind a decision tree was introduced in Chapter 4, which discussed **tree diagrams.** These diagrams can be very useful for determining probabilities in applications that lend themselves to a treelike structure. For example, if a defective part is discovered during final inspection, what is the probability that it was produced during the third shift? Because the Chapter 4 discussion was not concerned with making decisions, no mention was made then of "possible actions" or "states of nature."

Constructing Decision Trees

A convenient way of representing a set of alternatives and states of nature is to use a decision tree. A **decision tree** is a picture of the actions under consideration and the states of nature that affect the profitability of each action. It is a convenient way of illustrating the entire decision problem, because you can tell at a glance exactly which alternatives are being considered and what the payoff is under each state of nature.

In Example 18.8, Omega needed to make a decision about the selling price of their new computer. A decision tree for this situation is shown in Figure 18.4.

Decision tree for the
Omega computer-price
problem.

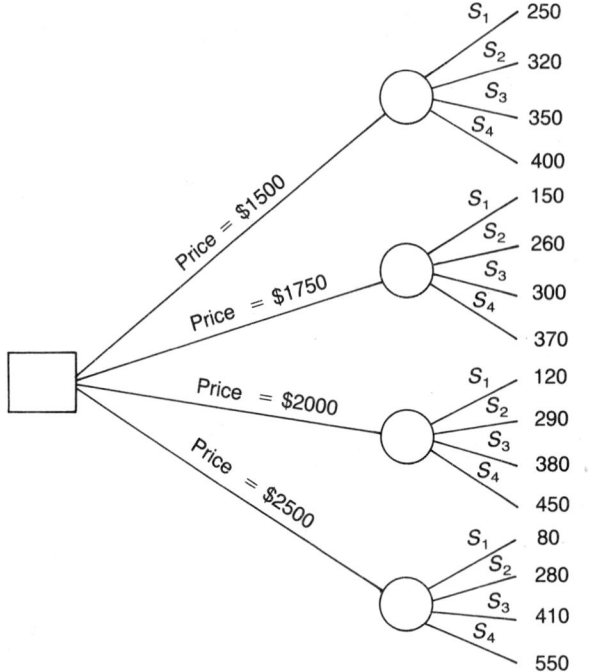

A decision tree represents a sequence of *decisions,* represented by boxes, and *outcomes* left strictly to chance, represented by circles. The boxes are decision nodes, and the circles are chance nodes.

When you reach a **decision node,** you need to make a decision at this point in the decision tree. The path you select reflects your choice of the best action to take at this point. This decision is under your control. The paths away from a **chance node** represent states of nature (S_1, S_2, . . .). There is no choice for you to make here; rather, each of these paths will occur with a certain probability, written as $P(S_1)$, $P(S_2)$, ...

The final step in completing a decision tree is to determine a dollar amount (or utility amount, if you are using utility values) within each chance node and decision node. The amount placed inside a chance node is the *expected payoff* at this point, using the probability for each state of nature. Consider the top chance node in Figure 18.4. Using the state-of-nature probabilities from Example 18.8, the completed tree (Figure 18.5) contains the expected payoff for each selling price. To illustrate, the expected payoff for a selling price of $1500 is

$$(.1)(250) + (.5)(320) + (.3)(350) + (.1)(400) = 330$$

In Figure 18.5, the amount in each *decision* node is not an expected value, because there are no probabilities associated with the paths leading away from this point. Instead, the dollar (or utility) amount, or the expected dollar (or utility) amount, associated with the *best* action at this point is contained within the box. Of the four paths leading away from the decision node in Figure 18.5, action A_1 (price = $1500) has the largest expected payoff, so this amount goes into the box. On the remaining three paths at this node, a double vertical bar across the path indicates that we have struck out these alternatives because they are not the ones to use at this point in the decision path.

Our conclusion from reading this tree would be to select a selling price of $1500 for an expected payoff of 330 (million dollars).

■ FIGURE 18.5
Completed decision
tree for the Omega
computer-price
problem.

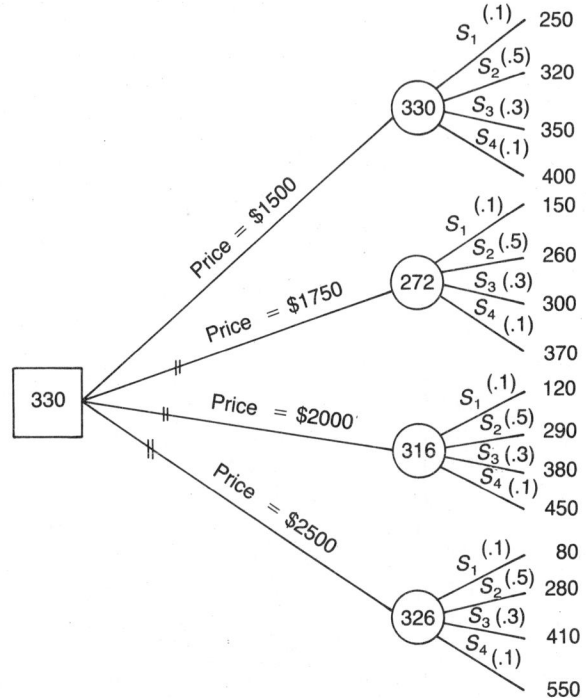

EXAMPLE 18.17   Structure the decision problem with the Sailtown data from Table 18.2 as a decision tree.

SOLUTION   The decision path begins with a decision node (how many sailboats to purchase), followed by a sequence of chance nodes reflecting the change in the interest rate. Figure 18.6 contains the completed tree for this problem. As in the previous analy-

■ FIGURE 18.6   Decision tree for Sailtown decision problem (Example 18.17).

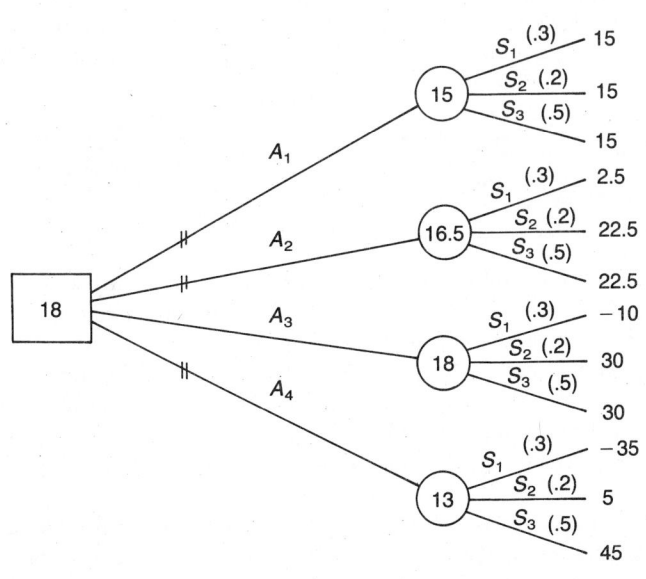

sis, when you are maximizing expected payoffs, your best alternative is to order 100 sailboats (action A_3) with an expected payoff of 18 (thousand dollars).   ∎

Once again you should perform a follow-up *sensitivity analysis* to determine how sensitive this solution is to the state-of-nature probabilities. When you summarize the results of a sensitivity analysis, you can construct a decision tree for each set of probabilities under consideration, and it will indicate the optimum path under this condition.

An Application of Decision Trees: Using Bayes' Rule to Maximize Profits

Bayes' Rule (attributed to the English clergyman, Thomas Bayes) was introduced in Chapter 4. This rule allows you to revise a probability in light of certain information that is provided. For example, suppose we know that 30% of the manufactured parts at your plant are produced during the third (late-night) shift. This probability is referred to as a **prior probability,** since it is a probability obtained *prior* to any known information, such as "I know this part is defective." The probability that the part was produced during the third shift *given that it is defective* is a **posterior probability.** It is a revised probability that a part was produced during the third shift, in light of the information that the part is defective.

To apply Bayes' Rule, a decision problem is structured into a decision tree (or tree diagram in Chapter 4) beginning with a set of paths, such as shift 1, shift 2, and shift 3. Determining a posterior probability is equivalent to finding the probability that the new information (part is defective) was arrived at along a certain path, such as path 3 belonging to shift 3. This type of situation was illustrated in Example 4.6, where, using Bayes' Rule,

$$P(\text{shift 3} \mid \text{part is defective}) = \frac{\text{third path}}{\text{sum of paths}}$$

$$= \frac{(.3)(.15)}{.091} = .495$$

That is, given that a defective part is discovered, there is nearly a 50–50 chance that it was produced during the third shift.

In general, Bayes' Rule states that given the final event (new information) B, the probability that the event was reached along the ith path corresponding to event E_i is

$$P(E_i \mid B) = \frac{P(E_i \text{ and } B)}{P(B)} = \frac{i\text{th path}}{\text{sum of paths}} \qquad (18.6)$$

In Table 18.2, the probability that the interest rate will increase (S_1) was $P(S_1) = .3$. What if a reliable consulting firm using various economic indicators predicted that the interest rate would increase? The value of $P(S_1) = .3$ was a subjective estimate, measuring Mr. Larson's belief that the rate would increase; it is a prior probability. In light of the *new information* obtained from the consulting firm, we would expect the probability of S_1 to increase (if, in fact, the firm is reliable). This probability, $P(S_1 \mid$ firm predicts a rate increase), is a posterior probability.

An excellent opportunity to use Bayes' Rule arises whenever you want to determine posterior probabilities based on recent information regarding the states of nature in your decision problem. This information can come from such sources as an outside consulting firm or a questionnaire developed by your company's marketing staff. Based on the new information, you can maximize your expected payoff (or utility) by replacing the prior probabilities with their corresponding posterior probabilities.

EXAMPLE 18.18

Now take another look at the Sailtown example. Mr. Larson, the owner, has decided to purchase the services of an outside consultant in an effort to determine more accurately the movement of the interest rate over the summer months. The information supplied by the consultant will be one of the following:

I_1:   consultant predicts an increase in the interest rate

I_2:   consultant predicts no change in this rate

I_3:   consultant predicts a drop in this rate

The information in Table 18.22 also was provided; it describes the past performance of this consultant when predicting interest rates. The values in the table contain conditional probabilities for the consultant's prediction under each state of nature. For example, $.7 = P(I_1 \mid S_1)$, $.4 = P(I_1 \mid S_2)$, and so forth. These probabilities mean that she predicted interest rate increases 70% of the times they actually occurred, and when there was no change in the interest rate, she predicted an increase 40% of the time. If the consultant is extremely reliable, the numbers from the upper left to the lower right (.7, .5, and .6) should be near 1, and the remaining values should be small.

The consultant predicts an increase (I_1) in the interest rate. What is the best action for Mr. Larson to take in light of this information? What is his expected profit?

SOLUTION

Figure 18.7 shows the new decision tree. Notice that Figures 18.6 and 18.7 are very similar, including the payoff amounts. The big difference is that the prior probabilities of $P(S_1) = .3$, $P(S_2) = .2$, and $P(S_3) = .5$ have been revised in light of the new information: $P(S_1)$ is replaced by $P(S_1 \mid I_1)$, $P(S_2)$ by $P(S_2 \mid I_1)$, and $P(S_3)$ by $P(S_3 \mid I_1)$. ∎

Deriving the Posterior Probabilities

To derive the posterior probabilities, we begin by constructing a tree diagram with the new information as the event on the *far right,* as shown in Figure 18.8. We can then obtain the probabilities along the various branches from the prior probabilities and the information in Table 18.22.

Using Bayes' rule,

$$P(I_1) = \text{sum of the paths}$$
$$= (.3)(.7) + (.2)(.4) + (.5)(.2)$$
$$= .39$$

■ **TABLE 18.22**
Conditional probabilities
for consultant
$[= P(I \mid S)]$
(Example 18.18).

Consultant Predicted	ACTUALLY OCCURRED		
	An Increase (S_1)	No Change (S_2)	A Decrease (S_3)
I_1 an increase	.7	.4	.2
I_2 no change	.2	.5	.2
I_3 a decrease	.1	.1	.6
	1.0	1.0	1.0

■ FIGURE 18.7   Decision tree, given information I_1.

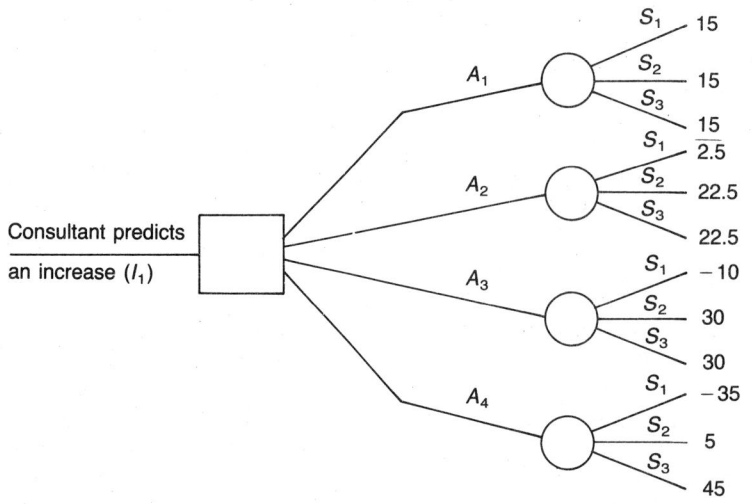

Actions
A_1: order 50 sailboats
A_2: order 75 sailboats
A_3: order 100 sailboats
A_4: order 150 sailboats
States of nature
S_1: interest rate increases
S_2: interest rate holds steady
S_3: interest rate decreases

■ FIGURE 18.8
Partial tree diagram
for deriving posterior
probabilities; new
information is the event
on the far right.

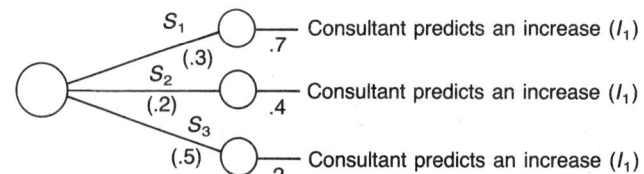

The posterior probabilities are given by

$$P(S_1 \mid I_1) = \frac{\text{first path}}{\text{sum of the paths}}$$

$$= \frac{.21}{.39} = .54$$

$$P(S_2 \mid I_1) = \frac{\text{second path}}{\text{sum of the paths}}$$

$$= \frac{.08}{.39} = .20$$

$$P(S_3 \mid I_1) = \frac{\text{third path}}{\text{sum of the paths}}$$

$$= \frac{.10}{.39} = .26$$

Placing these values in the decision tree results in Figure 18.9. The expected payoffs using the posterior probabilities are found in the usual manner. For action A_2,

$$\text{expected payoff} = (.54)(2.5) + (.20)(22.5) + (.26)(22.5) = 11.7$$

Our conclusion is that, given information that the consultant has predicted a rise in the interest rate, Mr. Larson's best alternative is to order 50 sailboats (action A_1) with an expected payoff of 15 (thousand dollars). Remember that, given no information at all, we use the *prior* probabilities to select that action having the largest expected payoff. These expected values were summarized in Table 18.11, where

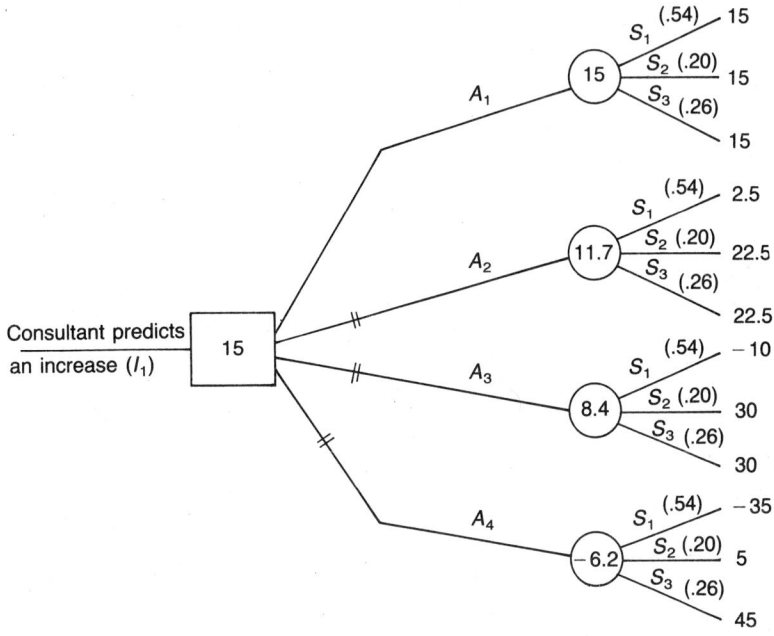

action A_3 (ordering 100 sailboats) provided the largest expected value. Notice also that in Table 18.11, the expected payoff for action A_2, given no information from the consultant, is 16.5. In other words, our revised expected payoff for this action, given the consultant's forecast, drops from 16.5 to 11.7.

Evaluating Sample Information

By combining a decision tree with Bayes' Rule for calculating posterior probabilities, the decision maker is able to determine whether purchasing new information is a good idea. We will refer to this new information as **sample information.** Such information may be collected from one of many sources, including a sample of questionnaires, a recently released government report, or, as in the previous example, an outside consultant. Typically, such information costs money; by using a decision-tree analysis, you will be able to decide between:

1. Not purchasing any additional information and using the prior probabilities to determine that action with the maximum expected payoff (or utility).
2. Purchasing this information because the expected payoff (or utility) for this decision is larger than that obtained using prior probabilities only.

In Example 18.18, the owner of Saitown used information provided by a consultant to revise his prior probabilities regarding a possible change in the interest rate. Based on the information provided (the interest rate will increase), Mr. Larson derived the posterior probabilities and decided to purchase 50 sailboats (action A_1).

Was it a good idea for Mr. Larson to purchase the consultant's services in the first place? The cost of this information was $2500. We previously found the expected value of perfect information (EVPI) for this situation to be $15,000. The consultant's fee is considerably less than this amount, so Mr. Larson was willing to evaluate the alternative of purchasing this information.

To construct a decision tree for the full problem, we begin exactly as we did in Figure 18.6. Our next step is to include an additional branch for puchasing information from the consultant. To complete this branch, we can use a two-step procedure (refer to Figure 18.10).

■ **FIGURE 18.10**   Completed decision tree for Sailtown decision problem.

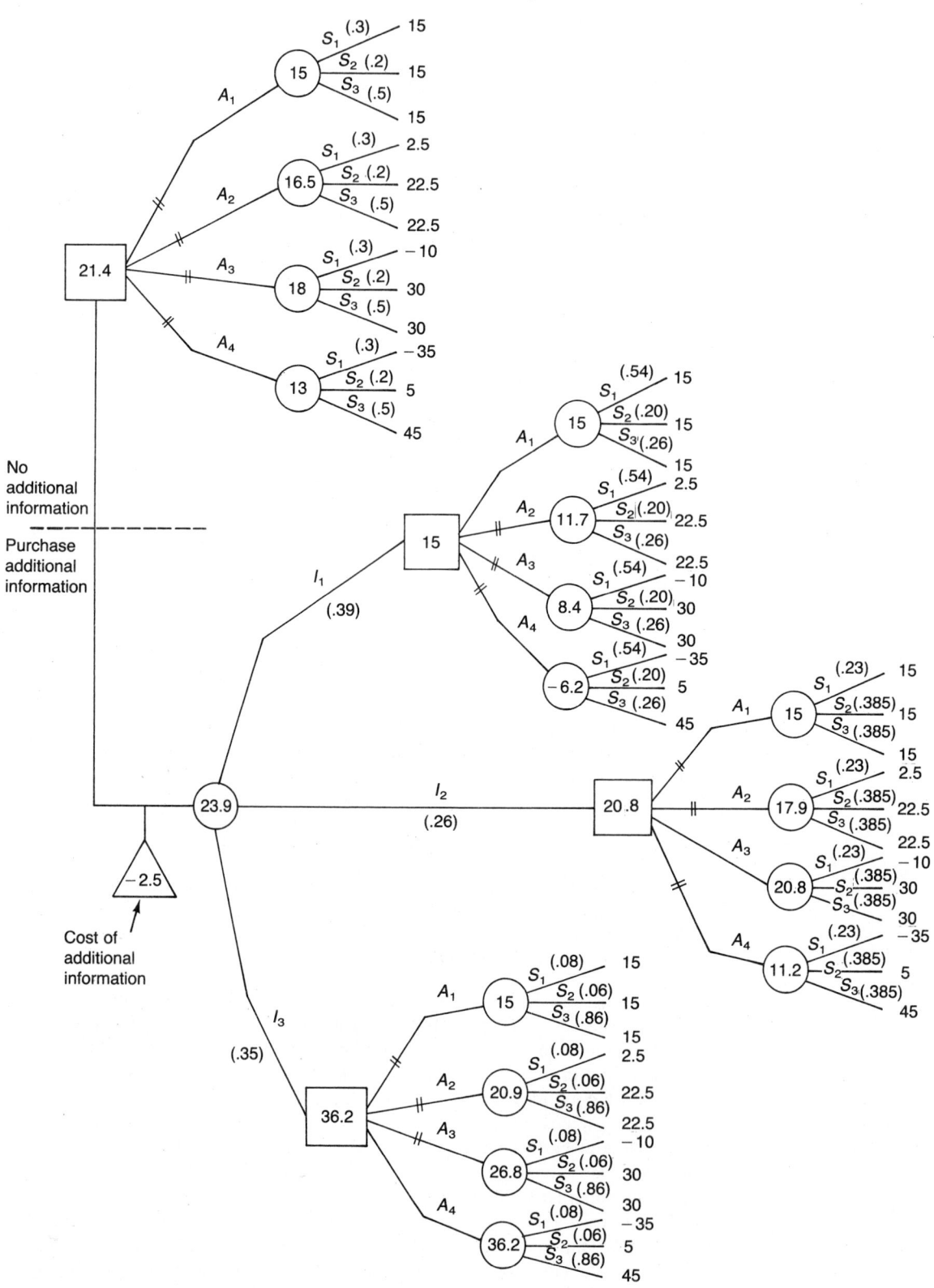

Step 1. The next node will be a *chance* node, representing the possible information to be provided. In Figure 18.10 this is

I_1: consultant predicts an increase in the interest rate

I_2: consultant predicts no change in the interest rate

I_3: consultant predicts a decrease in the interest rate

Step 2. For each branch representing I_1, I_2, . . . in step 1, we reconstruct the decision tree in Figure 18.6 because the possible actions and states of nature from this point on are the *same as before*. However, the probabilities for S_1, S_2, . . . will be the *posterior* probabilities rather than the prior probabilities in Figure 18.6.

Sum of branches $= P(I_1)$	Sum of branches $= P(I_2)$	Sum of branches $= P(I_3)$
$= .39$	$= .26$	$= .35$
$P(S_1\mid I_1) = .21/.39 = .54$	$P(S_1\mid I_2) = .06/.26 = .23$	$P(S_1\mid I_3) = .03/.35 = .08$
$P(S_2\mid I_1) = .08/.39 = .20$	$P(S_2\mid I_2) = .10/.26 = .385$	$P(S_2\mid I_3) = .02/.35 = .06$
$P(S_3\mid I_1) = .10/.39 = \underline{.26}$	$P(S_3\mid I_2) = .10/.26 = \underline{.385}$	$P(S_3\mid I_3) = .30/.35 = \underline{.86}$
1.0	1.0	1.0

Having constructed the decision tree, you next need to calculate the posterior probabilities. This calculation was illustrated in Example 18.18, where I_1 occurred (the consultant predicted an increase in the interest rate). Notice that the tree for this situation in Figure 18.9 becomes a portion of the large tree in Figure 18.10. A summary of the posterior probabilities is contained in Figure 18.11.

Also contained in Figure 18.11 are the probabilities for each of the possible predictions by the consultant. Here, $P(I_1) = .39$, $P(I_2) = .26$, and $P(I_3) = .35$. Because this prediction is *not* under your control, step 1 constructs a chance node including these three probabilities at this point.

We find the expected payoff given each of the consultant's predictions (15, 20.8, and 36.2) by using the posterior probabilities, as in Example 18.18 and Figure 18.11. We then calculate the expected payoff when using the consultant, where

expected payoff with consultant $= (.39)(15) + (.26)(20.8) + (.35)(36.2) = 23.9$

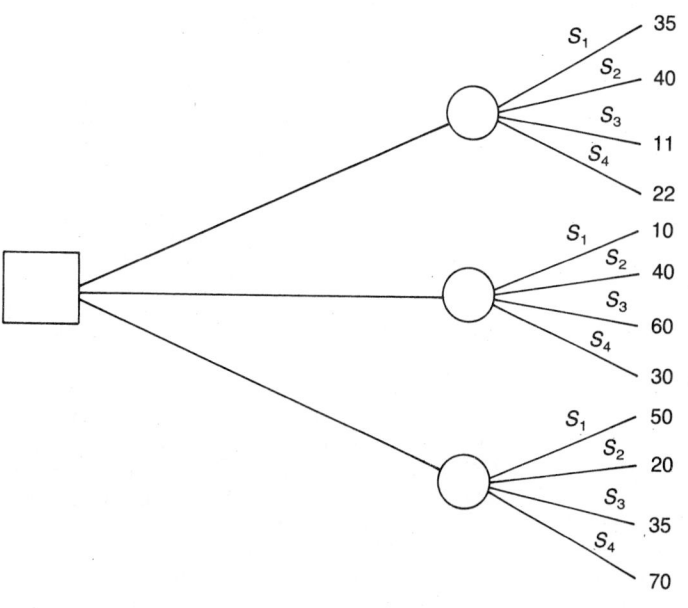

■ TABLE 18.23
Best actions for
Mr. Larson, given the
consultant's advice.

CONSULTANT PREDICTS	BEST ACTION
A rise in the interest rate (I_1)	Order 60 sailboats (A_1) Expected payoff: 15 Net profit: 12.5
No change in the interest rate (I_2)	Order 100 sailboats (A_3) Expected payoff: 20.8 Net profit: 18.3
A drop in the interest rate (I_3)	Order 150 sailboats (A_4) Expected payoff: 36.2 Net profit: 33.7

From this amount you need to subtract the cost of this information (2.5 thousand dollars), providing a net expected payoff of 21.4 (thousand dollars). Because this exceeds the four expected payoffs where no additional information is purchased, this action maximizes the expected payoff and provides the best alternative.

We thus conclude that the owner of Sailtown was right to purchase the services of the external consultant. The best action for him to take for each prediction is summarized in Table 18.23. The net profit is obtained for each case by subtracting the cost of information, 2.5 (thousand dollars).

Exercises

18.31 Complete the following decision tree and determine the decision based on the maximum expected payoff. Let $P(S_1) = .4$, $P(S_2) = .2$, $P(S_3) = .1$, and $P(S_4) = .3$.

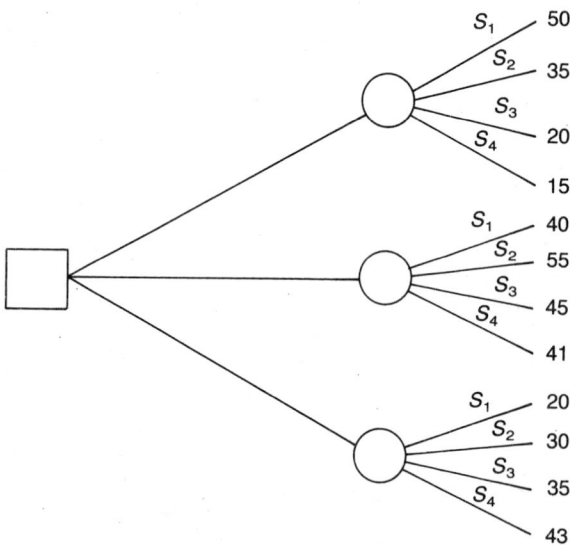

18.32 Security Designs sells alarm systems to businesses for protection against burglary and fire. If the number of small businesses continues to increase at a medium or fast rate, then Security Designs believes it should advertise more to increase its business. The manager at Security Designs believes that there is a 15% chance that the number of small businesses will decrease (S_1), a 25% chance that the number will stay the same (S_2), and a 25% chance and 35% chance tht the number of small businesses will increase moderately (S_3) and rapidly (S_4), respectively. Construct a decision tree using the following payoff table to find the decision based on the maximum expected payoff. The payoff is the amount of monthly profit, A_1: no advertising; A_2: keep advertising at current period; A_3: increase advertising 15%; A_4: increase advertising 30%.

ADVERTISING	S_1	S_2	S_3	S_4
A_1	1451	1840	2050	2300
A_2	−1091	1685	2430	2900
A_3	−2015	1100	3060	3561
A_4	−3460	−1350	3340	4300

18.33 Consider the following payoff table, which lists the utilities of the payoff to an investor who needs to decide on one of three different investment strategies: A_1, A_2, and A_3. The investor believes that there are four different states of nature that can affect the return on the investment. States S_1, S_2, S_3, and S_4 are believed to occur with probabilities .1, .3, .4, and .2, respectively.

ACTION	S_1	S_2	S_3	S_4
A_1	30	25	32	45
A_2	35	15	40	42
A_3	44	27	20	18

Construct a decision tree to find the decision based on the maximum expected utility of the payoff.

18.34 Sullivan and Orr, in an article in *Industrial Engineering* on using simulation analysis to determine optional decisions, describe a company that was contemplating making an investment of $150,000 in a new product line. Because of the uncertainty of the market, the following probabilities were assigned to the event that the product will be discontinued in either 8, 10, 12, or 14 years: .1, .2, .5. and .2, respectively. Suppose that a manager of the company could invest in one of three products: A_1, A_2, and A_3. Consider the following payoff table, which lists the utilities of the payoff to a manager who must decide on one of the three actions. States S_1, S_2, S_3, and S_4 refer to a market life of 8, 10, 12, or 14 years, respectively.

ACTION	S_1	S_2	S_3	S_4
A_1	480	750	2950	4900
A_2	−360	1240	3800	6210
A_3	−1180	1500	5300	7360

(*Source:* W. Sullivan and R. G. Orr, "Monte Carlo Simulation Analyzes Alternatives in Uncertain Economy," *Industrial Engineering* (November 1982): 43–49.)

Construct a decision tree to find the decision based on the maximum expected utility of the payoff.

18.35 Refer to Exercise 18.25. Construct the decision tree to find the decision based on the maximum expected utility of the payoff.

18.36 Refer to Exercise 18.26. Construct the decision tree to find the decision based on the maximum expected utility of the payoff.

18.37 Let A_1, A_2, A_3, A_4, and A_5 be a set of outcomes with $P(A_1) = .1$, $P(A_2) = .3$, $P(A_3) = .3$, $P(A_4) = .1$, and $P(A_5) = .2$. If it is given that $P(B \mid A_1) = .3$, $P(B \mid A_2) = .1$, $P(B \mid A_3) = .2$, $P(B \mid A_4) = .5$, and $P(B \mid A_5) = .5$, what are the probabilities $P(A_1 \mid B)$, $P(A_2 \mid B)$, $P(A_3 \mid B)$, $P(A_4 \mid B)$, and $P(A_5 \mid B)$?

18.38 A survey shows that 30% of the fashions that were found to be unprofitable were marketed by the major fashion clothes stores; 60% of the fashions found to be profitable were marketed by the major fashion clothes stores. If 70% of all fashions are profitable to market, find the probability that a fashion will be profitable if the major fashion clothes stores market it. What is the probability that the major fashion clothes stores market a particular fashion?

18.39 David, Harold, and Daniel are three salespeople at Southeast Insurance. David sells 40% of all insurance policies, Harold sells 33%, and Daniel sells 27%. The percent of policies sold by David that are whole-life insurance policies is 5%; for Harold this percentage is 8% and for Daniel, 10%. If a whole-life insurance policy at Southeast Insurance is selected at random, what is the probability that the insurance policy was sold by Harold?

18.40   Refer to Exercise 18.32. Assume that the manager at Security Designs has obtained some additional information from a consultant who predicts that the number of small businesses will stay the same; that is, state of nature S_2 is predicted to occur. The consultant has the following record, letting I be the event that the consultant predicts the number of small businesses will stay the same: $P(I \mid S_1) = .2$, $P(I \mid S_2) = .7$, $P(I \mid S_3) = .2$, and $P(I \mid S_4) = .1$. Find the decision based on the maximum expected payoff using revised probabilities with the additional information.

18.41   The investor in Exercise 18.33 subscribes to the stock market newsletter *Prudent Investor*. The newsletter forecasts state of nature S_1. Let I be the event that the newsletter forecasts S_1. The stock market newsletter has the following record: $P(I \mid S_1) = .6$, $P(I \mid S_2) = .2$, $P(I \mid S_3) = .3$, and $P(I \mid S_4) = .1$. Uisng revised probabilities based on the additional information, find the decision based on the maximum expected utility of the payoff.

18.42   Five legal secretaries type legal documents at a certain law firm. Secretary A types 10% of the work load, secretary B types 20%, secretary C types 20%, secretary D types 40%, and secretary E types 10%. The error rates for each of the secretaries are given below:

SECRETARY	PERFORMANCE
A	.04
B	.06
C	.08
D	.03
E	.09

If a typographical error is found on a legal document, which secretary is most likely responsible?

18.43   A decision maker must determine whether to conduct an experiment that will give one of three predictions, I_1, I_2, or I_3. If the experiment indicates I_1 and the decision maker uses this information, the expected profit is $15,000. If I_2 is indicated, the expected profit is $5000. But if I_3 is indicated, the expected profit is only $1000. If the experiment is not performed, the maximum expected profit would be $4000. Assuming that each of the predictions I_1, I_2, and I_3 is equally likely, is it worthwhile to conduct an experiment that costs $2000?

18.44   Nutritious Cereals would like to market a new multigrain cereal. The manager is trying to decide whether to produce the cereal in large quantities (A_1), moderate quantities (A_2), or small quantities (A_3). The manager believes that the probability of strong demand (S_1) is .4, of moderate demand (S_2) is .4, and of weak demand (S_3) is .2. A survey that can be conducted would predict strong demand (I_1), moderate demand (I_2), or weak demand (I_3). Historical data show the following conditional probabilities with regard to the predictions of the survey $[P(I \mid S)]$:

PREDICTION	S_1	S_2	S_3
I_1	0.8	0.3	0.3
I_2	0.1	0.5	0.1
I_3	0.1	0.2	0.6

The profit resulting from the different actions of Nutritious Cereal with regard to marketing the product is given in the following payoff table in thousands of dollars.

ACTION	S_1	S_2	S_3
A_1	88	53	20
A_2	75	66	32
A_3	57	50	39

If the survey costs $20,000 to conduct, should the management of Nutritious Cereals undertake it?

18.45   The following table lists conditional probabilities for certain predictions that are made by the consultant in Exercise 18.32, given the four states of nature of S_1, S_2, S_3, and S_4. Assume that I_1 represents the event that the consultant predicts the number of small businesses will decrease, I_2 that they will stay the same, I_3 that they will increase moderately, and I_4 that they will increase rapidly.

PREDICTION	S_1	S_2	S_3	S_4
I_1	0.80	0.10	0.20	0.10
I_2	0.10	0.70	0.20	0.20
I_3	0.05	0.10	0.50	0.30
I_4	0.05	0.10	0.10	0.40

Would the consultant's fee of $1200 be so high that Security Designs would not consider using the consultant's service?

18.46 Refer to Exercise 18.18. Let I_1, I_2, and I_3 be the event that an economist forecasts states of nature S_1, S_2, and S_3, respectively. The following table lists the conditional probabilities that the consultant makes one of these predictions given a particular state of nature. Would it be worth paying $2000 for the economist's services?

PREDICTION	S_1	S_2	S_3
I_1	.8	.4	.2
I_2	.1	.4	.2
I_3	.1	.2	.6

 *Summary*

This chapter presented a different approach to using probabilities—arriving at a **decision** when the future is uncertain. For example, should you lease a building or incur the extra expense of building one? Should your recently acquired inheritance be put in a money market account or should you take advantage of a reliable (in the past, at least) stockmarket report and invest in a newly formed corporation?

When facing such a problem, the decision maker must define the possible **actions** or **alternatives** (such as lease versus purchase or money market versus stocks) and **states of nature** that describe the uncertain future (such as company sales will be below expected, equal to expected, or greater than expected). For each action and state of nature, the decision maker must determine the corresponding **payoff** amount. These values can be summarized in a **payoff table.** This is certainly the most difficult and crucial step in the decision process, because each payoff value must reflect such factors as future costs to the company and responses of competitors. Any action whose payoff is less than that belonging to another action *regardless of the state of nature* is said to be **dominated** and can be removed from consideration.

Different strategies exist for any decision problem. If you elect to describe the uncertain future by assigning a probability to each state of nature, then a popular strategy is to select the action that **maximizes the expected payoff.** Typically, these probabilities are subjective, so any decision based on this method always should be followed up by a **sensitivity analysis** that repeats the decision procedure under various sets of probabilities. In other words, we use a "what-if" process that says, If the future is described by the following set of probabilities, then the best action using this strategy is . . . .

The minimax and maximax procedures do not require state-of-nature probabilities. The **minimax** strategy is very conservative. It begins by constructing an opportunity loss table that summarizes, for each state of nature, the loss the decision maker incurs by failing to take the most profitable action, given that this state of nature occurs. The action to take using this strategy is the one that minimizes the maximum opportunity loss for each of the actions under consideration. The **maximax** strategy is suited to the gambler; it selects that action having the largest possible payoff. Because it fails to take into consideration any heavy losses, it is not appropriate for most business decisions.

When using the expected payoff strategy, you should examine not only the payoffs that you can expect in the long run from each action but also the **risk** associated with each action. Here you measure the variation in the possible payoffs corresponding to each alternative. You often will select a less risky alternative and sacrifice a small amount of expected payoff. When you use the expected payoff strategy, a useful piece of information is the **expected value of perfect information** (EVPI), which is how much a decison maker should be willing to pay for a perfect prediction of tomorrow's state of nature, that is, for a crystal ball. Because any information about the future probably will be imperfect (for example, the consultant might be wrong), such information should cost considerably less than the EVPI.

It is not necessary to set up a decision problem by defining a payoff table in financial units. An alternative is to use **utility values,** which measure both the attractiveness and the risk associated with each dollar amount. For example, a $100,000 gain might be attractive, but a $100,000 loss may be disastrous to a struggling company. The utility value for each dollar amount can be summarized in a **utility curve.** You can use the shape of this curve to identify a decision maker as a **risk avoider,** a **risk neutral,** or a **risk taker.**

A complex decision problem can be summarized best using a **decision tree.** The tree identifies clearly the actions under consideration, the states of nature for the problem, and the expected payoffs for various segments of the decision analysis. **Bayes' Rule** puts such a tree to good use by allowing the decision maker to revise the subjective probability for each state of nature (the **prior probabilities**) in light of new information about the future. This new information could be a recent stock market analysis or predictions made by a consulting firm. The revised probabilities are **posterior probabilities.** Bayes' rule allows you to analyze a decision problem by determining the expected payoff of (1) not purchasing this information and using the prior probabilities or (2) purchasing this information and basing your decision on the results of this prediction.

Summary of Formulas

Expected payoff:

$$\mu_i = \Sigma x p$$

Risk:

$$\Sigma x^2 p - \mu_i^2$$

Expected value of perfect information:

EVPI = (expected payoff using
a perfect predictor)
− (largest expected payoff
of available actions)

Expected utility:

$$\Sigma (\text{utility}) \cdot (\text{corresponding probability})$$

Bayes' Rule:

$$P(E_i \mid B) = \frac{P(E_i \text{ and } B)}{P(B)}$$

$$= \frac{i\text{th path}}{\text{sum of paths}}$$

▧ Review Exercises

18.47 Pay-Lo drive-in grocery must decide how many loaves of bread to order each day. The demand per day is 29, 30, 31, 32, 33, 34, 35, or 36 loaves of bread. Given that 60¢ profit is made on each loaf of bread sold and a loss of 20¢ is incurred on each loaf not sold, construct the payoff table and the opportunity loss table for any number of loaves between 29 and 36 ordered for a particular day. What is the minimax decision? What is the maximax decision?

18.48 A computer company is considering marketing software that will take daily financial data, compute various statistics, and give a complete financial analysis as well as an up-to-date forecast of financial conditions. The introduction of the software will cost approximately $200,000 in fixed cost. A profit of $20 is expected from the sale of each financial software package. The vice president of the company believes that sales will amount to 5,000, 10,000, 15,000 20,000 or 25,000 packages of the software. Construct the opportunity loss table. What is the minimax decision?

18.49 Refer to Example 18.7. The managers at Omega have decided that a sensitivity analysis should be made before making a decision. Rework Example 18.7 using the following sets of probabilities. How sensitive is the decision based on the maximum expected payoff?

$P(S_1)$	$P(S_2)$	$P(S_3)$	$P(S_4)$
.1	.5	.2	.2
.1	.5	.1	.3
.1	.5	.3	.1
.1	.5	.2	.2
.2	.5	.2	.1

18.50 S & W Bookstore competes with the bookstore on a university campus for selling textbooks to students. *Introductory Statistics* is one of the textbooks that sells in large quantities. The manager of S & W Bookstore believes that there is a 30% chance that there will be a heavy enrollment (S_1) in this course. The probabilities for a normal enrollment (S_2) and a low enrollment (S_3) are .55 and .15, respectively. The manager must decide to order either 300, 400, or 500 copies of the textbooks. The payoff table follows.

ACTION	S_1	S_2	S_3
A_1(300)	830	750	710
A_2(400)	1230	1125	620
A_3(500)	1850	910	330

a. Are all the actions admissible?

b. What is the minimax decision?

c. What is the decision based on the maximum expected payoff?

d. What is the risk of each action?

e. What is the decision based on minimum risk?

f. What is the EVPI?

18.51 Suppose you have $5000 that you would like to invest in either a no-load mutual fund that invests completely in stocks (A_1) or a fixed money market account that yields 12% for one year (A_2). Assume that there are two states of nature: the stock market goes up (S_1), or the stock market goes down (S_2). An investment advisor gives you the following payoff table. Determine what value your personal utility function would have for the payoffs.

ACTION	S_1	S_2
A_1	1200	−575
A_2	600	600

18.52 You want to find several values of your utility for money from $0 to $2000. Find the value of your utility function at $0, $500, $1000, $1500, and $2000. Then, by graphing, approximate the value of your utility function at $700 and $1200.

18.53 Complete the decision tree in Figure 18.12 and determine the decision based on the maximum expected payoff. Assume $P(S_1) = .3$, $P(S_2) = .3$, $P(S_3) = .2$, and $P(S_4) = .2$. What is the EVPI?

18.54 Refer to Exercise 18.50. Assume that the manager at S & W Bookstore has obtained additional information from a consultant that the enrollment in the introductory statistics course will be heavy (I_1). Evidence from the consultant's previous performance indicates the following probabilities: $P(I_1 \mid S_1) = .6$, $P(I_1 \mid S_2) = .2$, and $P(I_1 \mid S_3) = .2$. Determine the decision based on the maximum expected payoff, using revised probabilities.

18.55 Refer to Exercise 18.54. Suppose that the consultant's predictions I_2 and I_3, which represent a normal enrollment and a low enrollment, respectively, have the following condi-

■ **FIGURE 18.12**
Decision tree for
Exercise 18.53.

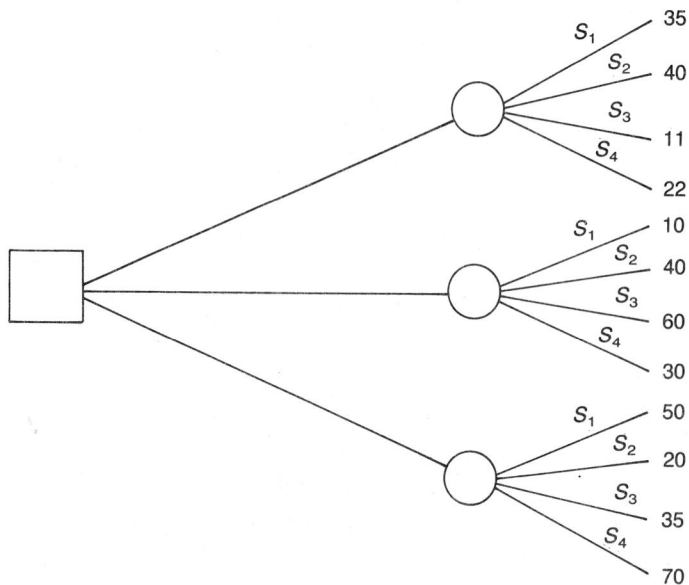

tional probabilities: $P(I_2 \mid S_1) = .1$, $P(I_2 \mid S_2) = .5$, $P(I_2 \mid S_3) = .3$, $P(I_3 \mid S_1) = .3$, $P(I_3 \mid S_2) = .3$, $P(I_3 \mid S_3) = .5$. Would the consultant's fee of \$350 make it worthwhile for S & W Bookstore to hire this consultant?

18.56 A manufacturing firm receives electronic components from three suppliers. Supplier A provides 48% of the components, supplier B provides 22%, and supplier C provides 30%. The proportion of defective products from suppliers A, B, and C are .0009, .005, and .002, respectively. If an electronic component is found to be defective, which of the three suppliers is most likely responsible?

18.57 Thirty percent of the clients of an investment broker invest in only long-term growth mutual funds. Of this group of clients, 60% are under the age of 40 years. Of those clients who do not invest solely in long-term mutual funds, 15% are under the age of 40 years. What is the probability that a randomly selected client under 40 years of age invests only in long-term-growth mutual funds?

18.58 A brokerage firm is conducting a seminar to explain the use of technical indicators in adjusting one's portfolio. The seminar director can decide to have either one (A_1), two (A_2), or three (A_3) guest speakers present at the seminar. Usually the more guest speakers, the more inclined the attendees will be to invest their money with the brokerage firm. The director believes that there is a 40% chance for a heavy turnout (S_1), a 35% chance for a normal turnout (S_2), and a 25% chance for a low turnout (S_3). The cost incurred to have the three guest speakers with a normal or low turnout would cause a loss. The payoff table is as follows.

ACTION	S_1	S_2	S_3
A_1	450	400	350
A_2	700	500	200
A_3	800	−100	−300

a.   Find the minimax decision.

b.   Find the decision based on the maximum expected payoff.

c.   Find the risk associated with each action.

18.59 Employees at a certain company use three methods of transportation to work. Fifty-five percent of the employees drive their own car, while 25% carpool and 20% use the city bus. Of the employees who drive their own car to work, 70% are male. Of the employees who carpool, 30% are male. Of the employees who use the city bus, 20% are male. What is the probability that a male employee drives his car to work?

Case Study

Whether to Use 100% Inspection: a Decision Theory Analysis

A great deal of the success of the Japanese in world markets has been attributed to the implementation of quality-control procedures developed by W. E. Deming. Deming has derived formulas for the total cost of inspection that include the cost of performing the initial inspection (say, k_1 dollars per unit) and the detrimental cost if faulty material is processed further (say, k_2 dollars per unit). The latter cost may include a cost due to damaged customer loyalty and reputation.

If we denote the proportion of nonconforming parts being produced by the process (while under control) as p, then a key element of the decision strategy is the value of $(k_2/k_1)p$. We will demonstrate this in the analysis that follows. Deming also advocates the use of control charts (a simple graph containing limits within which the process must fall) to detect the time when a process goes out of control, stopping to repair the process, and quarantining and sorting the parts made during the out-of-control period.

The total cost for incoming material is

$$TC = (Nk_1/q)\{1 + Qq[(k_2/k_1)p' - 1][1 - n/N]\}$$

where

$N =$ number of parts in the lot

$n =$ sample size drawn from each lot for inspection

$k_1 =$ cost to inspect one unit at the beginning of the process

$k_2 =$ cost to the firm when one nonconforming item goes farther into production

$p =$ average fraction of rejectable parts

$q = 1 - p$

$Q =$ fraction of lots accepted at initial inspection

$p' =$ the average fraction nonconforming in lots that are accepted and put straight into the production line.

For the sake of illustration, let us assume that p and p' are equal. In fact, p' can be smaller but only somewhat smaller than p. Consider a situation where $N = 1000$, $Q = .95$, $k_1 = \$2.00$, and $k_2 = \$100$. Consequently, the total cost per lot of incoming material is

$$TC = (2000/q)\{1 + .95q[50p - 1][1 - n/1000]\}$$

(*Source:* E. P. Papadakis, "The Deming Inspection Criterion for Choosing Zero or 100 Percent Inspection," *Journal of Quality Technology*, (July 1985): 17 no. 3.)

Case Study Questions

1. Suppose that the value of p is believed to be .01, .03, or .05; consequently we have three states of nature defined by these three proportions. The actions under consideration are to use sample sizes (n) of 0 (that is, inspection can be eliminated provided the process is assumed to be under control), 100, or 1000 (that is, 100% inspection). Construct a total cost table defined by these actions and states of nature. For example, the total cost (rounded to the nearest dollar) for a sample size of $n = 100$ and a proportion of $p = .03$ is

$$TC = (2000/.97)\{1 + (.95)(.97)[1.5 - 1][1 - .1]\}$$
$$= \$2917$$

2. Construct an opportunity loss table for this situation by defining the opportunity loss for each action and state of nature to be the TC for this situation minus the minimum TC given this state of nature. As a result, each opportunity loss is nonnegative.
3. Are any of the three actions inadmissible?
4. Using the minimax criterion, which action (sample size) would you select?
5. If it is believed that each of the values of p are equally likely, which action minimizes the expected total cost?
6. The main point of Deming's argument is that one should use 100% inspection whenever the value of $(k_2/k_1)p$ is larger than one and use 0% inspection otherwise. Is this consistent with the total cost table constructed in Question 1?

19 *Nonparametric Statistics*

A LOOK BACK/INTRODUCTION

Although last, this chapter is far from the least in importance. **Nonparametric** statistical techniques are used extensively for a variety of business-related applications.

In the previous chapters, we introduced a large assortment of tests of hypothesis. These tests generally were concerned with such quantities as the mean or variance of a population. A mean, variance, or proportion is referred to as a *parameter* in statistics, and so these tests are called *parametric* tests of hypothesis. The common underlying assumption in testing a parameter from a continuous population is that this population has a *normal* distribution. Any time you use a *t*-statistic, you assume a normal population distribution. When you test more than two means using the ANOVA procedure, you also assume that the shape of the populations is normal.

What can you do if you have reason to believe that the populations under study are not normally distributed? For example, suppose that data collected previously from these populations have been extremely skewed (not symmetric). One option when dealing with means or proportions is to collect large samples. In such a situation, the Central Limit Theorem assures us that the distribution of the sample estimators is approximately normal *regardless* of the population distribution. The other alternative, particularly for small or moderate sample sizes, is to use a nonparametric statistical procedure that deals with cases in which the assumptions of normality are not true.

Many of the nonparametric statistical tests try to answer the same sorts of questions as do those tests discussed previously. With these tests, however, the assumptions can be relaxed considerably. In earlier chapters, means and medians were referred to as *measures of central tendency*. A nonparametric test concerning such a measure does not assume an underlying normal population—unlike its parametric counterpart. Consequently, nonparametric methods are used for situations that violate the assumptions of the parametric procedures.

A common method for practically all nonparametric techniques is the use of **ranks.** Given a set of data, we obtain a set of ranks by replacing each data value by its relative *position*. To illustrate this idea, consider the following eight observations.

$$10.8, 6.4, 11.7, 5.3, 9.5, 2.5, 15.1, 10.4$$

Arranged in order, these are

Position	1	2	3	4	5	6	7	8
Value	2.5	5.3	6.4	9.5	10.4	10.8	11.7	15.1

We say that the rank of the value 2.5 is 1, the rank of the value 5.3 is 2, and so forth. Replacing each value by its rank and maintaining the original order produces

$$6, 3, 7, 2, 4, 1, 8, 5$$

Most nonparametric procedures use these eight *ranks* rather than the original data values. *By using ranks, we are able to relax the assumptions regarding the underlying populations and develop tests that apply to a wider variety of situations.*

You often will encounter an application in which a numeric measurement is extremely difficult to obtain, but a rank value is not. One example is a consumer taste test; each participant finds it much easier to rank several different brands of soft drinks than to assign a numeric value to each one. The data for analysis consist of the rank assigned to each brand.

Such data are said to be *ordinal* because only the relative position of each value has any meaning (see Chapter 1). This form of data is "weaker" than the *interval* form, for which not only the positions but also the *differences* between data values are meaningful. In this text, we have dealt mostly with interval or ratio data. Data consisting of temperatures, for example, are interval data; the difference between 60°F and 70°F is the same as the difference between 65°F and 75°F (10°F). When dealing with ranks, this is not the case; there is no reason to assume that the difference between ranks 1 and 3 is the same as that between ranks 3 and 5.

In Chapter 13 we introduced one nonparametric procedure that used the chi-square statistic to test for goodness-of-fit or independence between two classifications. In this chapter, we examine other popular nonparametric methods used in a business setting. These tests of hypothesis by no means constitute all the nonparametric techniques used in practice, but they should provide you with a basis for knowing how and when to apply such a method to a particular set of sample data.

19.1
A Test for Randomness: The Runs Test

The concept of *randomness* is a crucial assumption behind a great many statistical procedures. In the earlier chapters, all samples were assumed to be random. The reliability of any statistical test—even if run on a high-powered computer—is suspect if the sample was not obtained in a random manner. Similarly, the *t*- and *F*-tests in linear regression contain the assumption that the resulting sample residuals are *independent*, with no observable pattern. This assumption means that the *signs* of these errors should be random.

When you examine a sequence of observations or residuals, one method of detecting a lack of randomness is to observe the number of runs contained in the sequence. For a sequence containing two possible values (*A* and *B*, $+$ and $-$, and so on) a **run** consists of a string of identical values.

Suppose we flip a coin ten times, where each flip results in a head (H) or a tail (T). Consider the following three outcomes, each containing five heads and five tails.

Sequence 1	H	H	H	H	H	T	T	T	T	T
Sequence 2	H	T	H	T	H	T	H	T	H	T
Sequence 3	H	H	T	H	T	T	H	T	T	H

Only sequence 3 exhibits a random pattern. To see why, we will examine each sequence.

Sequence 1   These ten observations contain only two runs

$$\underbrace{\text{H H H H H}}_{\text{Run 1}} \quad \underbrace{\text{T T T T T}}_{\text{Run 2}}$$

The small number of runs is unlikely if this sequence was generated in a random manner.

Sequence 2   At first glance, this pattern may appear to be random, but there are an excessive number (ten) of runs

$$\underbrace{\text{H T H T H T H T H T}}$$
Run 1   Run 10

Once again, the process that generated this sequence is unlikely to be random because of the large number of runs.

Sequence 3   This sequence seems to be a compromise between the first two, exhibiting neither too few runs nor too many.

$$\underbrace{\text{H H T H T T H T T H}}$$
Run 1   Run 7

It appears that the sequence was generated in a random manner.

In this section, we use the runs test statistical procedure to test for randomness using the number of observed runs.

The Runs Test (Small Samples)

Consider a sequence of *n* observations, containing n_1 symbols of the first type (H, in our example) and n_2 symbols of the second type (T). So, $n = n_1 + n_2$. Let

R = number of runs within these *n* observations

■ **TABLE 19.1**
Arrangements when
$n_1 = 2$, $n_2 = 3$.

ARRANGEMENT	NUMBER OF RUNS (R)
H H T T T	2
H T H T T	4
H T T H T	4
H T T T H	3
T H H T T	3
T H T H T	5
T H T T H	4
T T H H T	3
T T H T H	4
T T T H H	2

The situation we consider here is for small samples, where $n_1 \leq 10$ and $n_2 \leq 10$, and we will demonstrate that this particular nonparametric technique is indeed distribution free, that is, it makes no assumptions about the population of H's and T's.

Consider the case where $n_1 = 2$ and $n_2 = 3$. In this section, we assume (without any loss of generality) that $n_1 \leq n_2$. So we have two H's and three T's. How many such arrangements (permutations) of these five symbols are there?* There are ten, provided in Table 19.1. This value in general can be found using

$$\text{number of arrangements} = A = \frac{n!}{n_1! n_2!} \qquad (19.1)$$

where $n = n_1 + n_2$. For this illustration,

$$A = \frac{5!}{2!3!} = \frac{(5)(\overset{2}{\cancel{4}})(\cancel{3})(\cancel{2})}{(\cancel{2})(\cancel{3})(\cancel{2})} = 10$$

Each of the ten arrangements in Table 19.1 is equally likely to occur *providing* the process generating this sequence is *random,* so each has probability .1. Some conclusions we can draw from Table 19.1 include:

1. For two of these sequences, there are two runs, and so $P(R = 2) = .2$.
2. For three of the sequences, there are three runs, and so $P(R = 3) = .3$.
3. For four of the sequences, there are four runs, and so $P(R = 4) = .4$.
4. For one of the sequences, there are five runs, and so $P(R = 5) = .1$.

Notice that these probabilities sum to 1, as they should.

Consequently, for this situation, we can make the following statements:

$$P(R \leq 2) = P(R = 2) = .2$$
$$P(R \leq 3) = P(R = 2) + P(R = 3) = .2 + .3 = .5$$
$$P(R \leq 4) = .2 + .3 + .4 = .9$$
$$P(R \leq 5) = .2 + .3 + .4 + .1 = 1.0$$

These probabilities mean, for example, that the probability of observing three or fewer runs if the sequence has been produced in a random manner, is .5. What we are seeing here is that these probabilities are obtained *without* assuming any probability distribution for the underlying population (process) that generated a sequence

---

*The formulas for the number of permutations given in Chapter 4 do not apply here because the n objects (symbols) are not all different (distinct).

of $n_1 = 2$ values of H and $n_2 = 3$ values of T. This is the beauty of nonparametric methods.

Probabilities such as those just discussed are summarized in Table A.15. The top portion of this table is reproduced in Table 19.2. The table entries contain the probability that $R \leq a$ for the possible values of a. Notice that the first row of this table is identical to the $\leq$ probabilities (called *cumulative* probabilities) that we just derived.

The hypotheses under investigation here are

$$H_0: \text{the sequence was generated in a random manner}$$

$$H_a: \text{the sequence was not generated in a random manner}$$

As we mentioned earlier, we reject H_0 whenever the number of runs is too *small* (say, whenever $R \leq k_1$) or too *large* (say, whenever $R \geq k_2$).

To illustrate the testing procedure here, consider our original sequences for the ten coins with $n_1 = 5$ and $n_2 = 5$. According to equation 19.1, there are $A = 10!/5!5! = 252$ possible arrangements here—sequences 1, 2, and 3 are three of these. We are looking for some "cutoff number" of runs, k_1, where we are fairly sure that fewer than k_1 are "too few" and more than k_2 are "too many." Using Table 19.2 with $n_1 = 5$ and $n_2 = 5$, we find, for example, that (assuming H_0 is true)

$$P(R \leq 3) = .04$$

$$P(R \leq 8) = .96$$

Consequently, $P(R > 8) = P(R \geq 9) = 1 - .96 = .04$. In other words, the event of observing three or fewer runs is very unlikely (with probability .04) if H_0 is true, so a value of $R \leq 3$ indicates that H_0 is not true and should be rejected. The same reasoning applies to $R \geq 9$.

These results mean that, with a significance level of $\alpha = .04 + .04 = .08$,

■ **TABLE 19.2**
A portion of Table A.15 for the runs test. Each entry is $P(R \leq a)$ where the values of a run across the table.

$(n_1 n_2)$	2	3	4	5	6	7	8	9	10
(2, 3)	.200	.500	.900	1.000					
(2, 4)	.133	.400	.800	1.000					
(2, 5)	.095	.333	.714	1.000					
(2, 6)	.071	.286	.643	1.000					
(2, 7)	.056	.250	.583	1.000					
(2, 8)	.044	.222	.533	1.000					
(2, 9)	.036	.200	.491	1.000					
(2, 10)	.030	.182	.455	1.000					
(3, 3)	.100	.300	.700	.900	1.000				
(3, 4)	.057	.200	.543	.800	.971	1.000			
(3, 5)	.036	.143	.429	.714	.929	1.000			
(3, 6)	.024	.107	.345	.643	.881	1.000			
(3, 7)	.017	.083	.283	.583	.833	1.000			
(3, 8)	.012	.067	.236	.533	.788	1.000			
(3, 9)	.009	.055	.200	.491	.745	1.000			
(3, 10)	.007	.045	.171	.455	.706	1.000			
(4, 4)	.029	.114	.371	.629	.886	.971	1.000		
(4, 5)	.016	.071	.262	.500	.786	.929	.992	1.000	
(4, 6)	.010	.048	.190	.405	.690	.881	.976	1.000	
(4, 7)	.006	.033	.142	.333	.606	.833	.954	1.000	
(4, 8)	.004	.024	.109	.279	.533	.788	.929	1.000	
(4, 9)	.003	.018	.085	.236	.471	.745	.902	1.000	
(4, 10)	.002	.014	.068	.203	.419	.706	.874	1.000	
(5, 5)	.008	.040	.167	.357	.643	.833	.960	.992	1.000
(5, 6)	.004	.024	.110	.262	.522	.738	.911	.976	.998

the values of k_1 and k_2 are $k_1 = 3$ and $k_2 = 9$. The corresponding testing procedure is to

$$\text{reject } H_0 \text{ if } R \leq 3 \text{ or } R \geq 9$$

We will formalize the procedure for using this information.

The overall significance level of this test is $.04 + .04 = .08$. One disadvantage of small-sample nonparametric procedures is that you cannot derive a test for any specified significance level (such as $\alpha = .05$ here). Rather, you are at the mercy of the available values in this table. We can summarize this testing procedure as follows:

Hypotheses

$$H_0: \text{pattern was generated in a random manner}$$

$$H_a: \text{pattern was not generated in a random manner}$$

Test Statistic (for small samples) R, where R denotes the number of runs in the sequence.

Procedure

$$\text{reject } H_0 \text{ if } R \leq k_1 \text{ or } R \geq k_2$$

where (1) k_1 is the value from Table A.15 such that $P(R \leq k_1) = \alpha/2$, and (2) k_2 is the value from Table A.15 such that $P(R \geq k_2) = \alpha/2$.

EXAMPLE 19.1 Using a significance level between .05 and .10 and as close to .05 as possible, determine which of the three sequences of five H's and five T's in the earlier discussion were generated in a random manner.

SOLUTION Using Table 19.2 (or Table A.15), the three smallest available significance levels for this test are

$$.008 + (1 - .992) = .008 + .008 = .016 \ (k_1 = 2, \ k_2 = 10)$$

$$.040 + (1 - .960) = .040 + .040 = .08 \ (k_1 = 3, \ k_2 = 9)$$

$$.167 + (1 - .833) = .167 + .167 = .334 \ (k_1 = 4, \ k_2 = 8)$$

Because $\alpha = .08$ comes closest to satisfying our desired significance level, we will

$$\text{reject } H_0 \text{ if } R \leq 3 \text{ or } R \geq 9$$

The results are

$$\text{for sequence 1: } R = 2, \text{ so reject } H_0$$

$$\text{for sequence 2: } R = 10, \text{ so reject } H_0$$

$$\text{for sequence 3: } R = 7, \text{ so fail to reject } H_0$$

For the first two sequences we conclude that these arrangements were not the result of a random process. For the third sequence, we have no reason to suspect the presence of a nonrandom process. ∎

We encounter another application of the runs test when we examine the residuals from a linear regression analysis. A key assumption when using linear regression is that the residuals are *independent*. Consequently, you should observe a random pattern in the sample residuals. If the observations in your data set are recorded across time (say, over 24 consecutive months), this often results in residuals that are *not* independent. In this case, we would say that the errors are correlated—more precisely, they are *autocorrelated:* they are correlated with each other. In Chapter 17, we computed the Durbin-Watson (DW) statistic to measure the degree of autocorrelation.

The DW statistic assumes that the errors follow a normal distribution, as do all the tests of hypothesis when using a linear regression equation. The nonparametric runs test also can be used to examine the residuals, by recording the *sign* ($+$ or $-$) of each residual and counting the number of runs. This test is valid regardless of the distribution of the residuals and can be used for any model that assumes the residuals are uncorrelated.

EXAMPLE 19.2

In Section 17.9, a linear regression model was used to predict the number of home loans financed by Liberty Savings and Loan over a six-month period. The predictor variables were the average interest rate lagged by one year, the advertising expenditure lagged by one year, a dummy variable to represent an election year, and a dummy variable to represent the first/last six-month period. The residuals from this analysis are shown in Table 17.14 and reproduced in Table 19.3. The Durbin-Watson test for *positive* autocorrelation performed on these residuals in section 17.10 was inconclusive using a significance level of .05. Based on the runs test, is there sufficient evidence to indicate that the residuals exhibit positive autocorrelation?

SOLUTION

The hypotheses under consideration are

$$H_0: \text{no autocorrelation exists}$$

$$H_a: \text{positive autocorrelation exists}$$

We begin by forming a sequence containing the sign of each residual.

$$+ \; - \; + \; + \; - \; - \; - \; - \; - \; + \; - \; + \; + \; - \; +$$

There are $n_1 = 7$ values of $+$, $n_2 = 8$ values of $-$, and $R = 9$ runs. If H_a is true then there should be too *few* runs since, on the average, each residual will be similar to its neighbors. Consequently, the test will be to

$$\text{reject } H_0 \text{ if } R \leq k_1 \text{ for some } k_1$$

A look at Table A.15 reveals that we can define a test with a significance level of .051 using $k_1 = 5$; that is, the test procedure is to

$$\text{reject } H_0 \text{ if } R \leq 5$$

Since $R = 9$, we fail to reject H_0 and conclude that there is insufficient evidence to indicate that the residuals are positively autocorrelated. ■

The Runs Test (Large Samples)

For large samples ($n_1 > 10$ and $n_2 > 10$), the approximate distribution for R if the generating process is random will be *normal* with mean

11.66	-2.90	15.93	13.28	-3.50	-5.03	-8.83	-11.34
-7.44	3.96	-16.99	4.25	2.56	-2.22	6.61	

■ **TABLE 19.3**
Residuals for multiple regression illustration in Section 17.9.

$$\mu_R = 1 + \frac{2n_1 n_2}{n_1 + n_2} \qquad \textbf{(19.2)}$$

and standard deviation

$$\sigma_R = \sqrt{\frac{2n_1 n_2 (2n_1 n_2 - n_1 - n_2)}{(n_1 + n_2)^2 (n_1 + n_2 - 1)}} \qquad \textbf{(19.3)}$$

By standardizing R in the usual way, we obtain the following summary.

Hypotheses

H_0: pattern was generated in a random manner

H_a: pattern was not generated in a random manner

Test Statistic (for large samples)

$$Z = \frac{R - \mu_R}{\sigma_R} \qquad \textbf{(19.4)}$$

where (1) R denotes the number of runs in the data sequence and (2) μ_R and σ_R are the mean and standard deviation of this random variable, defined in equations 19.2 and 19.3.

The testing procedure using the standard normal random variable is the same as in previous tests using Z. For the randomness test, a nonrandom pattern is indicated by a Z value in the right tail (too many runs) or in the left tail (too few runs).

EXAMPLE 19.3   The president of Northside National Bank requested the savings-account balances for 45 randomly selected accounts of nonmarried customers. When she examined the data, she began to question the randomness of the procedure used to select the accounts. Letting M denote the account of a male and F the account of a female, the following sequence was obtained, listed in the order in which they were selected for the supposedly random sample.

M M F F F F F M F F M M M M M M F F F F M M F

M M F F M F F F F F M M M M M F F F F M M M

Based upon this sequence, would you conclude that this sample consists of 45 randomly selected males and females? Use $\alpha = .05$.

SOLUTION   The preceding sequence contains $R = 15$ runs. Also,

$$n_1 = \text{number of males} = 22$$

$$n_2 = \text{number of females} = 23$$

For these values of n_1 and n_2, the mean number of runs if H_0 is true is

$$\mu_R = 1 + \frac{(2)(22)(23)}{45} = 23.49$$

This implies that, on the average, whenever $n_1 = 22$ and $n_2 = 23$, you will obtain 23.49 runs.

The sample contains only 15 runs, so it could be that this sequence exhibits a nonrandom pattern, due to insufficient runs. However, this depends heavily on the standard deviation of R; therefore, to complete the analysis, we next find

$$\sigma_R = \sqrt{\frac{(2)(22)(23)[(2)(22)(23) - 45]}{(45)^2(44)}} = \sqrt{10.9832} = 3.314$$

To determine whether $R = 15$ is sufficiently small to reject the random sequence hypothesis, we calculate the test statistic.

$$Z^* = \frac{15 - 23.49}{3.314} = -2.56$$

The test procedure here (using $\alpha = .05$) is to

$$\text{reject } H_0 \text{ if } |Z| > 1.96$$

The computed Z value does have an absolute value larger than 1.96, and so we reject H_0. There is evidence that the male-female sequence is nonrandom, indicating a lack of randomness in the sampling procedure used in selecting the individual accounts from the bank records.

A computer solution using MINITAB is contained in Figure 19.1. A value of -1 is used for F and $+1$ for M. This solution contains the number of runs as well as the mean of the runs statistic if H_0 is true. Also contained in Figure 19.1 is the p-value for this test, $p = .0106$. Because this value is smaller than the significance level of $\alpha = .05$, we reject H_0 and again conclude that there is sufficient evidence to indicate that the male-female sequence is nonrandom. ■

COMMENTS

1. When you perform a one-tailed runs test, you should divide the MINITAB p-value by two and then compare it to the level of significance (α) to make a decision.
2. For a one-tailed test, be sure that the sign of the Z value is compatible with your one-tailed alternative hypothesis, that is, it should be positive when testing H_a: too many runs and negative when testing H_a: too few runs.

■ **FIGURE 19.1**
MINITAB solution
for runs test.

```
MTB > SET INTO C1
DATA> 1 1 -1 -1 -1 -1 -1 1 1 -1 -1 1 1 1 1 1 1 -1 -1 -1 -1 1 1 -1
DATA> 1 1 -1 -1 1 -1 -1 -1 -1 -1 1 1 1 1 1 -1 -1 -1 -1 1 1 1
DATA> END
MTB > RUNS ABOVE AND BELOW 0, USING C1    (Abbreviated:  RUNS 0 C1)

    C1

    K =      0.0000
                                              R              μ_R
    THE OBSERVED NO. OF RUNS =    15
    THE EXPECTED NO. OF RUNS =    23.4889
    22 OBSERVATIONS ABOVE  K    23  BELOW
                 THE TEST IS SIGNIFICANT AT   0.0106

   Here, K is set equal to 0
```

Exercises

19.1   What assumptions need to be made about the data when using the runs test?

19.2   A jar contains two balls, a red one and a blue one. A person is asked to draw a ball at random. The ball is then replaced in the jar and the experiment is repeated. The results of repeating the experiment 17 times follow, where R represents the red ball and B represents the blue ball. At the .05 level of significance is there any evidence that the sequence is not randomly generated?

$$R\ B\ B\ R\ R\ B\ R\ R\ B\ B\ B\ R\ B\ R\ R\ B\ R$$

19.3   Are the negative and positive numbers randomly ordered in the following sequence? Use a .10 level of significance.

$$-1, 2, -5, 4, -10, 3, -1, 4, 6, 9, -7, 8, -3, 5$$

19.4   Ozark County Bank is taking applications for the position of loan officer. The following sequence lists the order in which either a male (M) or a female (F) applied for the position. Is there evidence to indicate that the sequence is not randomly generated? Use a .05 significance level.

$$M\ M\ F\ M\ M\ F\ M\ M\ F\ M\ F\ F\ M\ M\ F\ M\ M\ M\ F\ F\ F\ F\ M\ M\ F\ F\ F$$

19.5   After a television debate between two political candidates, a telephone line is open to viewers wishing to express their opinions on whether the Democratic (D) or the Republican (R) candidate won the debate. The following sequence represents 19 opinions of viewers in the order in which they telephoned. Using a runs test and a significance level of 5%, does the sequence indicate a nonrandom order?

$$R\ R\ D\ D\ R\ D\ D\ R\ R\ R\ R\ D\ D\ R\ D\ R\ D\ D\ D$$

19.6   Conduct a runs test on the following sequence of 3s and 4s to see if there is evidence that the sequence is not randomly generated. Use a 5% significance level.

$$3\ 3\ 3\ 4\ 4\ 3\ 4\ 3\ 4\ 3\ 3\ 3\ 4\ 4\ 3\ 4\ 4\ 3\ 3\ 3\ 3\ 4\ 3$$

19.7   A certain computer program generates a sequence of random digits. Test whether there is any evidence that the following sequence of numbers is nonrandom by considering the sequence of odd and even numbers. Use a 10% significance level.

$$4\ 8\ 7\ 9\ 3\ 2\ 1\ 6\ 7\ 9\ 4\ 1\ 8\ 3\ 2\ 5$$

19.8   The total U.S. energy consumption of coal in units of quadrillion Btu are given below for 1978–1989. A regression equation is fit through the data with time as the independent variable. The residuals for this regression are also given. Does the sequence of positive and negative residuals appear to be random with respect to time? Use a 5% significance level. Use the RUNS command in MINITAB.

YEAR	COAL CONSUMPTION	RESIDUAL
1978	13.77	−.435
1979	15.04	.404
1980	15.42	.352
1981	15.91	.410
1982	15.32	−.612
1983	15.90	−.463
1984	17.07	.275
1985	17.48	.253
1986	17.26	−.398
1987	18.01	−.080
1988	18.85	.328
1989	18.92	−.034

(*Source: The World Almanac and Book of Facts* (1991): 183.)

19.9   Thirty-five true-or-false questions were given on a history test. The following sequence contains the answers to the questions in the order in which they appeared. At the 5% level, is there evidence that the true and false answers are not randomly assigned?

F F F F T T F T T T F F F F T F T F F T T T T T F F T T F F T F T T T F

19.10   The composite index values of the Toronto Stock Exchange are listed for January 1988 through November 1990. Letting $t = 1$ for January 1988 and incrementing t by 1 for each month, the regression equation for predicting the value of the composite index is found to be $y_t = 2882.18 + 90.41t - 2.3343t^2$. Find the residuals for each time t. At the .05 significance level, is there evidence that the positive and negative residuals are not random with respect to time?

	1988	**1989**	**1990**
January	3057.2	3617.6	3704.4
February	3204.8	3562.0	3686.7
March	3313.8	3578.2	3639.5
April	3339.8	3627.9	3340.9
May	3249.2	3683.2	3582.0
June	3441.5	3760.9	3559.9
July	3376.7	3970.8	3561.1
August	3285.8	4010.2	3346.3
September	3283.7	3943.0	3159.4
October	3395.5	3918.6	3081.3
November	3294.7	3942.8	3151.0
December	3389.9	3969.8	—

(*Source: Standard and Poor's Statistical Service, Current Statistics* (December 1990): 27.)

▰ 19.2
Nonparametric Tests of Central Tendency

Chapter 3 introduced you to measures of central tendency. The more commonly used measures are the mean and median, which attempt to identify the middle of a set of sample data. In Chapter 9, we introduced two populations, where the question of interest was whether the two means were the same (a two-tailed test) or whether one mean exceeded the other (a one-tailed test). A two-population test is illustrated in Figure 19.2, where the variable of interest is height.

The main assumption in Figure 19.2 is that the two populations are normally distributed. When you sample from these populations, if both sample sizes (n_1 and n_2) are *large*, you can remove this assumption. However, there is a need for a nonparametric technique for this two-population situation when (1) you have small samples and you suspect that one or both populations do not follow a normal distribution or (2) your data are such that only the relative ranks are available within each sample, such as in a consumer taste test (in other words, you are dealing with *ordinal data*). *The t-tests from Chapter 9 assumed that the measurement scale of the data was at least interval, so these tests are inappropriate for data consisting of ranks.*

■ **FIGURE 19.2**
Two-population test of hypothesis for means.

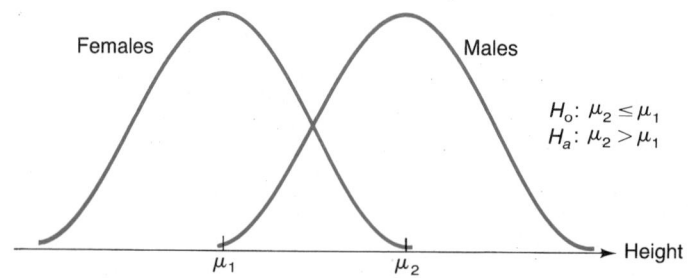

When dealing with samples from two populations in Chapter 9, we also looked at two situations:

1. The two samples are *independent*. In Figure 19.2, this would mean that a sample of n_1 female heights is obtained independently of the n_2 male heights. There is no reason to match up the first male height with the first female height, the second male height with the second female height, and so on in the two samples.
2. The two samples are *dependent* or *paired*. This might occur in Figure 19.3 if the question were, are husbands taller than wives? The data then consist of n_1 wives and $n_2 = n_1$ husbands.

This portion of the chapter discusses the nonparametric counterparts to these two parametric tests of hypothesis. The assumptions behind the application of these methods are considerably weaker than those for the *t*-tests in Chapter 9. These nonparametric techniques, named after the people responsible for their development, are the Mann-Whitney U test—a nonparametric procedure for situation 1 (two independent samples)—and the Wilcoxon signed rank test—a nonparametric procedure for situation 2 (two paired samples).

The parametric tests in Chapter 9 were concerned with population means. If two populations have different means, we say that these populations differ in **location.** This implies that population 1 is shifted to the left or right of population 2. When defining the corresponding nonparametric test, the hypotheses will be stated in terms of differing location rather than differing means.

The Mann-Whitney U Test for Independent Samples

The **Mann-Whitney U test** is named after H. B. Mann and R. Whitney, who developed this test in the 1940s. The purpose of this procedure is to provide a test for differing location that does not require the assumption of normal populations. The test is an alternative to the *t*-tests from Chapter 9, which do contain this assumption.

The two-tailed hypotheses for the Mann-Whitney test can be written

H_0: the two populations have identical probability distributions

H_a: the two populations differ in location

To use the Mann-Whitney nonparametric technique, we begin by combining (pooling) the two samples into one large sample and then determining the rank of each observation in the pooled sample. Next, let

T_1 = sum of the ranks of the observations from the first sample in this pooled sample

T_2 = sum of the ranks of the observations from the second sample

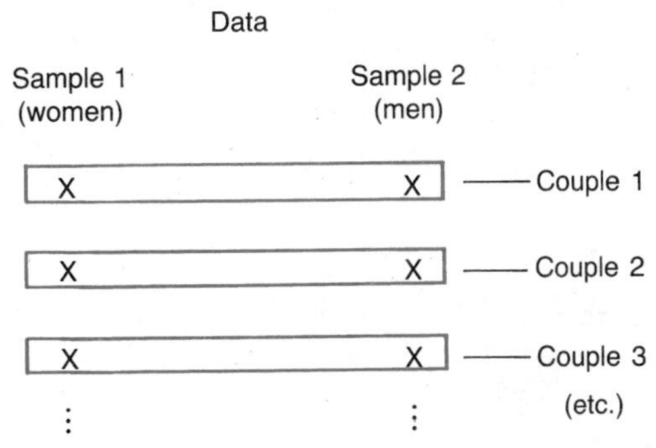

■ **FIGURE 19.3**

Illustration of dependent (paired) samples.

Data

Sample 1 (women)	Sample 2 (men)	
X	X	—— Couple 1
X	X	—— Couple 2
X	X	—— Couple 3
⋮	⋮	(etc.)

The procedure is (if $n_1 = n_2$)

reject H_0 if T_1 is "too much different" from T_2

To illustrate this technique, consider the following pooled sample, where the pooled observations have been arranged in order from smallest to largest. Here A represents a value from population A and B is a value from population B.

Value	A	A	A	A	A	B	B	B	B	B
Rank	1	2	3	4	5	6	7	8	9	10

For this pooled sample, we have

$n_1 = 5$

$T_1 = $ sum of ranks of the five A observations in the pooled sample

$= 1 + 2 + 3 + 4 + 5 = 15$

$n_2 = 5$

$T_2 = $ sum of the ranks of the B observations

$= 6 + 7 + 8 + 9 + 10 = 40$

Now consider another pooled sample:

Value	A	B	B	A	B	A	A	A	B	B
Rank	1	2	3	4	5	6	7	8	9	10

For this situation, the values are $T_1 = 1 + 4 + 6 + 7 + 8 = 26$ and $T_2 = 2 + 3 + 5 + 9 + 10 = 29$.

In the first pooled sample, there is clear evidence that the second population is shifted to the right of the first population, as indicated by the large difference between $T_1 = 15$ and $T_2 = 40$. This difference is also evident when you examine this pooled sample because the values from population A are all less than those from population B. From a parametric view, this implies that $\mu_B > \mu_A$. The Mann-Whitney procedure will result in rejecting H_0: the two populations have identical probability distributions in favor of H_a: the two populations differ in location (or H_a: population A is shifted to the left of population B, had we used a one-tailed test). For the second set of ten pooled observations, there is no indication of a difference in population location; the A and B values are fairly well mixed in the combined sample, as evidenced by the values of $T_1 = 26$ and $T_2 = 29$, which are nearly equal. The Mann-Whitney test will lead to a failure to reject the null hypothesis.

Mann-Whitney Test for Small Samples For this test of hypothesis, small samples are defined as both $n_1 \le 10$ and $n_2 \le 10$. *Regardless* of the sample sizes, the procedure begins by finding T_1 and T_2 as described previously and then letting

$$U_1 = n_1 n_2 + \frac{n_1(n_1 + 1)}{2} - T_1 \qquad (19.5)$$

and

$$U_2 = n_1 n_2 + \frac{n_2(n_2 + 1)}{2} - T_2 \qquad (19.6)$$

The Mann-Whitney test is summarized in the accompanying box.

The Mann-Whitney Test for Small Samples

Null hypothesis:

H_0: the two populations have identical probability distributions

Assumptions:

1. Random samples are obtained from each population.
2. The two samples are independent of one another—respective observations are not paired.
3. The sample data are at least ordinal.

Procedure:

1. Assume that $n_1 \leq n_2$ (if this is not the case, reverse your populations, so that n_1 is the smaller sample size).
2. Determine U_1 and U_2 from equations 19.5 and 19.6.
3. Use the value from Table A.10 to test H_0 versus H_a, where, once again, small p-values lead to rejecting H_0.

TWO-SIDED TEST	**ONE-SIDED TEST**	
H_a: the two populations differ in location	H_a: population 1 is shifted to the right of population 2	H_a: population 1 is shifted to the left of population 2
Reject H_0 if Table A.10 value for U is less than $\alpha/2$, where U = minimum of U_1 and U_2	Reject H_0 if Table A.10 value for U is less than α, where $U = U_1$.	Reject H_0 if Table A.10 value for U is less than α, where $U = U_2$.

EXAMPLE 19.4

A local auto dealer wants to know whether single male buyers purchase the same amount of "extras" (such as air conditioning, power steering, exterior trim) as do single females when ordering a new car. The researcher in charge of the analysis has no reason to believe the amounts are normally distributed and so elects to test the hypotheses using the Mann-Whitney procedure. A sample of eight males and nine females was obtained; the data consist of the dollar amounts of the ordered extras.

| Male purchases | 2450 | 1436 | 850 | 1240 | 3645 | 1766 | 1226 | 2840 | |
| Female purchases | 1742 | 3146 | 2740 | 2160 | 3436 | 2750 | 562 | 1290 | 2060 |

Use $\alpha = .05$ to test for a difference between the amounts purchased by the male and female buyers.

SOLUTION

The hypotheses are

H_0: the two populations have identical probability distributions

H_a: the two populations differ in location

The pooled sample here is

562, 850, 1226, 1240, 1290, 1436, 1742, 1766, 2060, 2160, 2450, 2740, 2750, 2840, 3146, 3436, 3645

Next, we indicate from which sample each value came in the pooled sample.

RANK	MALE SAMPLE	FEMALE SAMPLE	RANKS FOR MALE SAMPLE	RANKS FOR FEMALE SAMPLE
1		562		1
2	850		2	
3	1226		3	
4	1240		4	
5		1290		5
6	1436		6	
7		1742		7
8	1766		8	
9		2060		9
10		2160		10
11	2450		11	
12		2740		12
13		2750		13
14	2840		14	
15		3146		15
16		3436		16
17	3645		17	
			$T_1 = 65$	$T_2 = 88$

Using equations 19.5 and 19.6,

$$U_1 = (8)(9) + \frac{(8)(9)}{2} - 65 = 43$$

$$U_2 = (8)(9) + \frac{(9)(10)}{2} - 88 = 29$$

Because this is a two-sided alternative, we let $U =$ the minimum of 29 and 43, so $U = 29$.

For $n_1 = 8$, $n_2 = 9$, and $U = 29$, the value in Table A.10 is .2707. Because this value is greater than $\alpha/2 = .025$, we fail to reject H_0. Based on these data, there is insufficient evidence to indicate a difference between male and female purchase amounts.

The p-value for this test is $(2)(.2707) = .5414$, which is extremely large. For a one-sided test, the p-value would be obtained by finding the value from Table A.10 and *not* doubling it.   ■

Ties   When the pooled sample contains two or more identical observations, each is assigned a rank equal to the *average* of the ranks of the tied observations. For example, if there are two observations tied for sixth and seventh place, each is assigned a rank of 6.5. The rank of the next largest sample value is 8. We illustrate this procedure in the next section.

Mann-Whitney Test for Large Samples   Whenever n_1 or n_2 is greater than ten, a large-sample approximation can be used for the distribution of the Mann-Whitney U statistic. For this case, we can use either U_1 or U_2 in the test statistic for both one-sided *and* two-sided tests. The following discussion uses U_2.

In the event that the two populations have identical probability distributions (that is, H_0 is true), the U_2 statistic is approximately *normally* distributed with mean

$$\mu_{U_2} = \frac{n_1 n_2}{2} \tag{19.7}$$

and standard deviation

$$\sigma_{U_2} = \sqrt{\frac{n_1 n_2 (n_1 + n_2 + 1)}{12}} \qquad \textbf{(19.8)}$$

The rejection region for the various alternative hypotheses are defined in the accompanying box.

The corresponding test statistic here is

$$Z = \frac{U_2 - \mu_{U_2}}{\sigma_{U_2}} \qquad \textbf{(19.9)}$$

The Mann-Whitney Test for Large Samples

Null hypothesis:

H_0: the two populations have identical probability distributions

Assumptions: Same as for small samples
Procedure: Determine

$$U_2 = n_1 n_2 + \frac{n_2(n_2 + 1)}{2} - T_2$$

where T_2 = sum of the ranks for the second sample in the pooled sample.

TWO-SIDED TEST	ONE-SIDED TEST	
H_a: the two populations differ in location	H_a: population 1 is shifted to the right of population 2	H_a: population 1 is shifted to the left of population 2
Reject H_0 if $|Z| > Z_{\alpha/2}$ where (1) Z is defined in equation 19.9 and (2) $Z_{\alpha/2}$ is the value from Table A.4 having a right-tail area of $\alpha/2$.	Reject H_0 if $Z > Z_\alpha$.	Reject H_0 if $Z < -Z_\alpha$.

EXAMPLE 19.5

Food World operates two supermarkets in a large metropolitan area. One of their services to customers is to cash personal checks at no charge. The owner of Food World is concerned that one of the stores (store A), situated in a low-income neighborhood, may have a greater number of checks returned due to insufficient funds in the customers' checking accounts than does store B, which is located in a higher-income area. Data were collected for 12 randomly selected six-month periods from store A; the data consisted of the number of returned checks over this period. Similar data were collected for 15 randomly selected six-month periods for store B.

Store A	42	65	38	55	71	60	47	59	68	57	76	42			
Store B	22	17	35	19	8	24	42	14	28	17	10	15	20	45	50

The pooled sample and corresponding ranks are summarized in Table 19.4. Notice that there are two values of 17, which are tied for fifth and sixth place. Con-

RANK	STORE A SAMPLE	STORE B SAMPLE	RANKS FOR STORE A	RANKS FOR STORE B
1		8		1
2		10		2
3		14		3
4		15		4
5		17		5.5
6		17		5.5
7		19		7
8		20		8
9		22		9
10		24		10
11		28		11
12		35		12
13	38		13	
14	42		15	
15		42		15
16	42		15	
17		45		17
18	47		18	
19		50		19
20	55		20	
21	57		21	
22	59		22	
23	60		23	
24	65		24	
25	68		25	
26	71		26	
27	76		27	
			$T_1 = 249$	$T_2 = 129$

■ TABLE 19.4
Pooled sample for
Example 19.5.

sequently, each is given a rank of $(5 + 6)/2 = 5.5$. Similarly, there is a three-way tie for fourteenth, fifteenth, and sixteenth place, so a rank of $(14 + 15 + 16)/3 = 15$ is given to each.

Using $\alpha = .05$, is there sufficient evidence to indicate that store A has a larger number of returned checks than does store B?

SOLUTION   The hypotheses for this situation are

H_0: the two populations have identical probability distributions

H_a: population A is shifted to the right of population B

The test procedure is to

$$\text{reject } H_0 \text{ if } Z > 1.645$$

where $1.645 = Z_{.05}$ is obtained from Table A.4. From Table 19.4, we find that

$$T_2 = \text{sum of ranks for store B}$$

$$= 1 + 2 + 3 + 4 + 5.5 + 5.5 + \cdots + 15 + 17 + 19 = 129$$

and so

$$U_2 = (12)(15) + \frac{(15)(16)}{2} - 129 = 171$$

Also, the mean and standard deviation of the U_2 statistic are

$$\mu_{U_2} = \frac{(12)(15)}{2} = 90$$

and

$$\sigma_{U_2} = \sqrt{\frac{(12)(15)(28)}{12}} = 20.49$$

The value of the resulting test statistic is

$$Z = \frac{U_2 - \mu_{U_2}}{\sigma_{U_2}} = \frac{171 - 90}{20.49} = 3.95$$

This exceeds 1.645, and so we reject H_0 and conclude that store A does in fact have a larger volume of returned checks than store B. ∎

A MINITAB solution for Example 19.5 is contained in Figure 19.4. The Mann-Whitney statistic is denoted by W, which is actually the sum of the first sample ranks, T_1. T_2 can be obtained by using the identity

$$T_1 + T_2 = \frac{n(n + 1)}{2}$$

where n = pooled sample size = $n_1 + n_2$.

Wilcoxon Signed Rank Test for Paired Samples

When your sample data consist of *paired* observations from two populations, the Mann-Whitney procedure from the previous section does not apply because it assumes *independent* samples. By *paired observations,* we mean that respective observations from each sample are matched with one another. Examples of paired observations include husband-wife, brother-sister, and before-after combinations.

A method of testing population means under this type of sampling procedure was introduced in Chapter 9, where we used a *t*-test on the sample differences. However, as in all *t*-tests, a key assumption using this method of testing two means is that the differences are *normally distributed. When small samples from suspected nonnormal populations are used, a nonparametric technique is required.* The **Wilcoxon signed rank test** is used for such situations.

■ **FIGURE 19.6**   MINITAB commands for Wilcoxon signed rank test (Example 19.7).

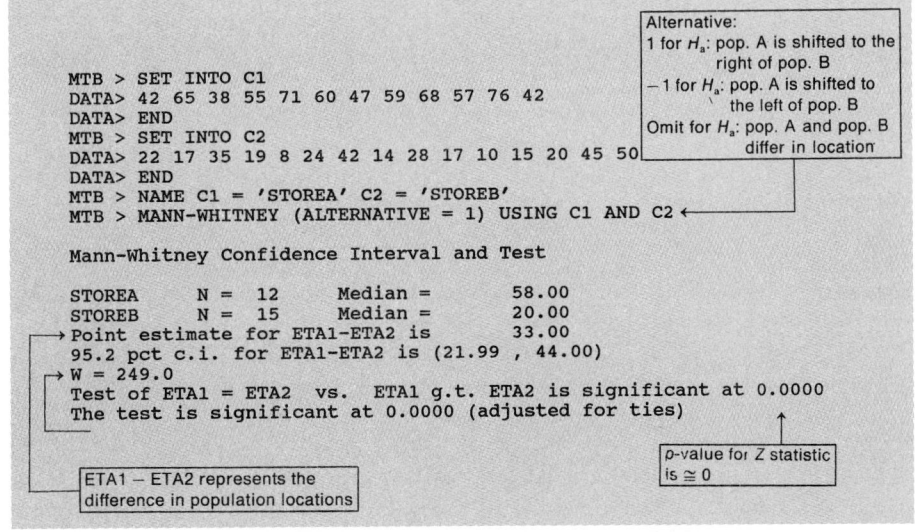

The Wilcoxon test begins like its parametric counterpart, the paired-sample *t*-test, by subtracting the data pairs and using the differences to perform the test. As in the paired-sample *t*-test, the hypotheses are written in terms of the location of the probability distribution for the population differences.

The steps involved in applying the Wilcoxon test are:

1. Determine the difference for each sample pair.
2. Arrange the *absolute value* of these differences in order, assigning a rank to each.
3. Let T_+ = sum of the ranks having a positive value and T_- = sum of the ranks for the negative values.
4. T_+, T_-, or T = the minimum of T_+ and T_- is used to define a test of H_0 versus H_a.

To demonstrate the test, suppose we are interested in determining the effects of a vigorous six-month advertising campaign. Sales figures are collected before and after the campaign from ten different cities. The results are shown in Table 19.5.

We determine the paired differences and rank the corresponding absolute values in order. Ties are handled as before by assigning a rank equal to the average of the tied positions. Also, if a pair of observations has a difference equal to zero, then this pair is *deleted* from the sample, and *n* is reduced by one. Other methods exist for handling zero differences, but this procedure is the simplest, and it works well provided there are not many zero differences.

According to Table 19.6, the negative differences are -6 and -8. Their corresponding ranks are 2 and 4. Therefore,

$$T_- = 2 + 4 = 6$$

A rule that can simplify the calculations here and serve as a check for arithmetic is that

$$T_+ + T_- = \frac{n(n + 1)}{2}$$

■ TABLE 19.5
Sales (thousands of dollars) for ten cities.

CITY	SALES BEFORE	SALES AFTER
Denver	61	63
Boston	50	57
Salt Lake City	18	34
Seattle	56	48
Miami	29	44
Dallas	25	38
Atlanta	34	28
Baltimore	48	68
Topeka	37	57
Minneapolis	14	26

■ TABLE 19.6
Illustration of Wilcoxon signed rank procedure (Example 19.6).

SALES BEFORE	SALES AFTER	DIFFERENCE (AFTER–BEFORE)	|DIFFERENCE|	RANK
61	63	2	2	1
50	57	7	7	3
18	34	16	16	8
56	48	-8	8	4($-$)
29	44	15	15	7
25	38	13	13	6
34	28	-6	6	2($-$)
48	68	20	20	9.5
37	57	20	20	9.5
14	26	12	12	5

where n = the number of sample pairs. In our example, $n = 10$, so

$$T_+ + T_- = \frac{(10)(11)}{2} = 55$$

which means that $T_+ = 55 - T_- = 55 - 6 = 49$

The Wilcoxon Signed Rank Test for Small Samples (Paired) Once T_+ and T_- have been obtained, you can use the Wilcoxon signed rank test for testing hypotheses about the location of the population differences.

The Wilcoxon Signed Rank Test for Small Samples (Paired)

Null Hypothesis:

H_0: the population differences are centered at 0

Assumptions:

1. Each data pair is randomly selected.
2. The absolute values of the differences can be ranked.

Procedure:

1. Determine the n differences using each sample pair, where each difference is defined to be sample 1 − sample 2.
2. Assign a rank to the absolute value of each difference; define T_+ = sum of the ranks of the positive values and T_- = sum of the ranks of the negative values.

Table A.11 is used to define the rejection region for the following tests.

TWO-SIDED TEST	ONE-SIDED TEST	
H_a: the population differences are not centered at 0	H_a: the population differences are centered at a value > 0	H_a: the population differences are centered at a value < 0
Using the two-sided value from Table A.11, reject H_0 if $T \le$ table value, where T = minimum of T_+ and T_-.	Using the one-sided value from Table A.11, reject H_0 if $T_- \le$ table value.	Using the one-sided value from Table A.11, reject H_0 if $T_+ \le$ table value.

EXAMPLE 19.6

Table 19.5 contains the sales results from ten cities before and after the six-month advertising campaign. Using $\alpha = .05$, are we able to conclude that there was a significant increase in sales after the advertising campaign?

SOLUTION

The hypotheses here can be stated as (A = after, B = before)

H_0: the population differences are centered at 0.

H_a: the population differences are centered at a value > 0.

We refer to the "after" population as population 1 and the "before" population as population 2 to correspond to the difference column in Table 19.6 (our procedure assumes that each difference is sample 1 (A) − sample 2 (B)).

The values of T_+ and T_- also are derived from Table 19.6, where

$$T_- = 6$$

and

$$T_+ = 49$$

The one-sided value in Table A.11 corresponding to $n = 10$ and $\alpha = .05$ is 11. Consequently, the test is to

$$\text{reject } H_0 \text{ if } T_- \leq 11$$

Because the value of T_- is smaller than 11, we reject H_0 and conclude that there is sufficient evidence of a sales increase after the advertising campaign. ∎

The Wilcoxon Signed Rank Test for Large Samples (Paired)   For samples consisting of $n > 15$ pairs, a large-sample approximation to the Wilcoxon test statistic can be used. An advantage to using this procedure is that p-values are much easier to determine. (A p-value is once again a measure of the strength of your conclusion.)

When using the large sample procedure, we can define a test using either T_+ or T_-. The following hypothesis tests use T_+, the sum of the ranks for the positive differences. If the population differences are centered at zero (that is, H_0 is true), then T_+ is approximately a normal random variable with mean

$$\mu_{T_+} = \frac{n(n + 1)}{4} \tag{19.10}$$

and standard deviation

$$\sigma_{T_+} = \sqrt{\frac{n(n + 1)(2n + 1)}{24}} \tag{19.11}$$

The corresponding test statistic is

$$Z = \frac{T_+ - \mu_{T_+}}{\sigma_{T_+}} \tag{19.12}$$

The one- and two-sided large sample procedures are summarized in the accompanying box.

The Wilcoxon Signed Rank Test for Large Samples (Paired)

Null Hypothesis:

H_0: the population differences are centered at 0.

Assumptions: Same as for small samples
Procedure: (1) and (2) are the same as for small samples. Each paired difference is defined to be sample 1 − sample 2.

TWO-SIDED TEST	ONE-SIDED TEST	
H_a: the population differences are not centered at 0	H_a: the population differences are centered at a value > 0	H_a: the population differences are centered at a value < 0
Reject H_0 if $|Z| > Z_{\alpha/2}$, where Z is defined in equation 19.12 and $Z_{\alpha/2}$ is the value from Table A.4 having a right-tail area of $\alpha/2$.	Reject H_0 if $Z > Z_\alpha$.	Reject H_0 if $Z < -Z_\alpha$.

EXAMPLE 19.7

The paper produced by Glendale Container Corporation has historically contained 2% hardwood in the paper pulp. A quality engineer at Glendale believed that a 10% hardwood concentration would improve the tensile strength of the paper. Management has decided to adopt the 10% concentration, despite the slightly higher cost, if in fact the tensile strength can be demonstrated to be larger at the 10% level. An experiment is performed in which the plant runs two batches of paper a day, one at 2% concentration and one at 10% concentration. A section of paper from each of the two batches is cut, dried, and tested. The following tensile strength data were obtained:

DAY	2%	10%	DAY	2%	10%
1	125	119	11	113	148
2	133	120	12	131	116
3	132	139	13	128	107
4	116	148	14	119	142
5	130	148	15	112	106
6	135	109	16	111	142
7	119	137	17	128	112
8	112	116	18	135	145
9	122	122	19	106	131
10	137	131	20	118	146

Use the Wilcoxon signed rank test to determine whether the quality engineer's belief—that the tensile strength is larger for the 10% hardwood concentration—is correct. Let $\alpha = .05$.

SOLUTION

If we let the 2% population be population 1, then the correct hypotheses are

H_0: the population differences are centered at 0

H_a: the population differences are centered at a value < 0

The alternative hypothesis agrees with Table 19.7, in which each difference is calculated using the 2% value (sample 1) minus the 10% value (sample 2). Because the values for day 9 are the same, the difference is zero, so this sample is removed from the analysis, leaving $n = 19$ pairs in the sample.

Based on a significance level of $\alpha = .05$, the proper test is to

$$\text{reject } H_0 \text{ if } Z < -1.645$$

For a value of $n = 19$, the mean of T_+ (assuming H_0 is true) is

$$\mu_{T_+} = \frac{(19)(20)}{4} = 95$$

with a standard deviation of

$$\sigma_{T_+} = \sqrt{\frac{(19)(20)(39)}{24}} = 24.85$$

■ **TABLE 19.7**
Paired samples for
Example 19.7.

HARDWOOD CONCENTRATION		DIFFERENCE (2% − 10%)	RANK OF ABSOLUTE VALUE	+ RANKS	− RANKS
2%	10%				
125	119	6	3	3*	
133	120	13	7	7	
132	139	−7	5(−)		5
116	148	−32	18(−)		18
130	148	−18	10.5(−)		10.5**
135	109	26	15	15	
119	137	−18	10.5(−)		10.5**
112	116	−4	1 (−)		1
122	122	0	—	removed, so use $n = 19$ pairs	
137	131	6	3	3*	
113	148	−35	19(−)		19
131	116	15	8	8	
128	107	21	12	12	
119	142	−23	13(−)		13
112	106	6	3	3*	
111	142	−31	17(−)		17
128	112	16	9	9	
135	145	−10	6(−)		6
106	131	−25	14(−)		14
118	146	−28	16(−)		16
				$T_+ = 60$	$T_- = 130$

*three-way tie; assigned rank = (2 + 3 + 4)/3 = 3.
**two-way tie; assigned rank = (10 + 11)/2 = 10.5.

Table 19.7 informs us that $T_+ = 60$, and so the value of the test statistic here is

$$Z^* = \frac{60 - 95}{24.85} = -1.41$$

This value is not less than -1.645, so there is insufficient evidence to conclude that the 10% concentration produces a larger tensile strength. Glendale Container should continue to use the less expensive 2% hardwood concentration.

The p-value for this test is obtained in the usual manner by finding, in this case, the area under a standard normal curve to the left of the calculated test statistic of $Z^* = -1.41$. According to Figure 19.5 and Table A.4, the p-value is .0793. Using our rule-of-thumb procedure from before, this p-value is neither large ($>.1$) nor small ($<.01$), but it *is* greater than $\alpha = .05$, which leads us to fail to reject H_0.

■

To use MINITAB for the Wilcoxon procedure, begin by subtracting the two samples and then using the WTEST command on the differences, as illustrated in Figure 19.6. Note that the p-value in this figure is slightly different from our previous result because the MINITAB procedure includes a continuity adjustment

■ **FIGURE 19.5**
p-value for
Example 19.7.

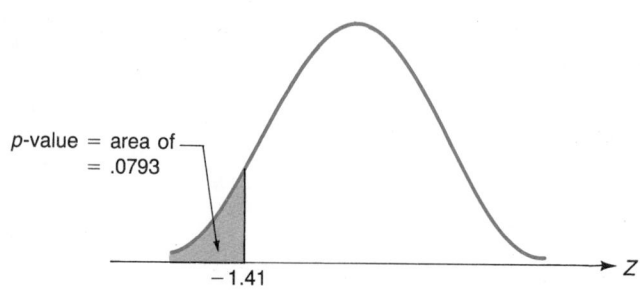

p-value = area of
= .0793

−1.41

■ **FIGURE 19.6** MINITAB commands for Wilcoxon signed rank test (Example 19.7).

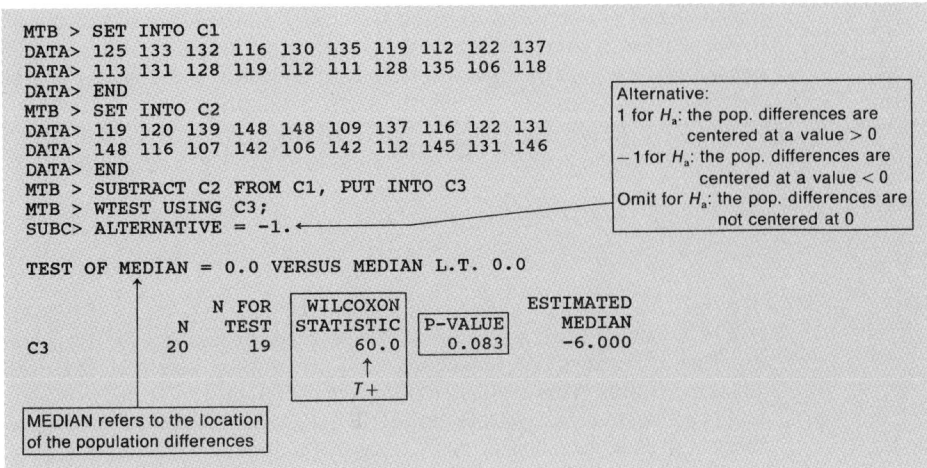

```
MTB > SET INTO C1
DATA> 125 133 132 116 130 135 119 112 122 137
DATA> 113 131 128 119 112 111 128 135 106 118
DATA> END
MTB > SET INTO C2
DATA> 119 120 139 148 148 109 137 116 122 131
DATA> 148 116 107 142 106 142 112 145 131 146
DATA> END
MTB > SUBTRACT C2 FROM C1, PUT INTO C3
MTB > WTEST USING C3;
SUBC> ALTERNATIVE = -1.◄
```

Alternative:
1 for H_a: the pop. differences are centered at a value > 0
-1 for H_a: the pop. differences are centered at a value < 0
Omit for H_a: the pop. differences are not centered at 0

```
TEST OF MEDIAN = 0.0 VERSUS MEDIAN L.T. 0.0
```

	N	N FOR TEST	WILCOXON STATISTIC	P-VALUE	ESTIMATED MEDIAN
C3	20	19	60.0	0.083	-6.000

$\uparrow$
$T+$

MEDIAN refers to the location of the population differences

(similar to that used in Chapter 6) and is slightly more accurate. If you have the MINITAB package available, use it to obtain more accurate *p*-values for this nonparametric test.

Exercises

19.11 What assumptions need to be made about the type and distribution of the data when the Mann-Whitney test is used?

19.12 If 15 observations are randomly selected from each of two populations and the sum of the ranks for the sample from population 1 (T_1) is 230, what is the sum of the ranks (T_2) for the sample from population 2?

19.13 Two groups of randomly selected students are given an aptitude test on understanding the financial markets. The first group has eight students, selected from second-semester freshmen, and the sum of the ranks (T_1) of these eight students is 65. If the second group has nine students selected from first-semester freshmen, is there sufficient evidence to conclude that the aptitude of the second-semester freshmen is higher than the aptitude of the first-semester freshmen on understanding the financial markets? Use a 5% significance level.

19.14 A real-estate agent claims that the homes in two neighborhoods have the same average value. A random sample from neighborhood A contains the following dollar values: 85,000, 70,000, 74,000, 69,000, 88,000, 89,000. A random sample from neighborhood B consists of the following values: 71,000, 64,000, 68,000, 73,000, 81,000, 69,000, 72,000. Test the alternative hypothesis that the value of homes in neighborhood B is less than the value of homes in neighborhood A. Use a 10% significance level.

19.15 An engineer is proposing a new manufacturing process to increase the tensile strength of a certain wire. Eleven samples of wire manufactured under the proposed process are collected, and 14 samples are collected from wire manufactured under the existing process. Using the following data and a 10% significance level, test the hypothesis that there is greater tensile strength in the proposed technique.

PROPOSED TECHNIQUE (PSI)	EXISTING TECHNIQUE (PSI)	PROPOSED TECHNIQUE (PSI)	EXISTING TECHNIQUE (PSI)
1.4	1.2	2.2	1.8
2.1	1.6	2.0	1.7
1.8	1.7	1.9	2.0
1.7	1.7	2.0	1.8
1.6	2.0		1.9
1.9	1.3		1.6
1.4	1.4		2.1

19.16   The head lawyer of the Brown and Smith firm would like to know whether there is a difference in the number of errors made by the two secretaries employed in the firm. Five randomly selected documents are given to secretary A to type, and five are given to secretary B. The number of errors per document is shown in the following table. Using a 5% level of significance, test the hypothesis that there is no difference in the number of errors made by each secretary.

SECRETARY A	SECRETARY B
3	2
5	0
4	4
2	3
0	1

19.17   A nursery is experimenting with two blends of fertilizer for fertilizing lawns in a certain area. Twenty-four randomly selected patches of grass are selected for experimenting with the fertilizer. Twelve patches are randomly assigned to fertilizer A and another 12 patches are randomly assigned to fertilizer B. The increase in the height of the grass after 2 weeks is given in the following table. Using a 10% level of significance, test the hypothesis that fertilizer B is more effective than fertilizer A.

FERTILIZER A	FERTILIZER B	FERTILIZER A	FERTILIZER B
1.2	1.0	0.8	1.2
0.9	1.1	1.4	1.3
1.3	1.0	0.7	0.8
0.5	0.9	0.9	1.0
0.3	0.7	0.8	0.8
0.9	0.8	0.7	1.1

19.18   Two samples of light bulbs are taken from the brands Everglo and Britelite. The following table gives the life of the bulbs for the collected sample. At the 1% level, is there sufficient evidence to indicate that Britelite bulbs last longer than do Everglo bulbs?

EVERGLO (HOURS)	BRITELITE (HOURS)	EVERGLO (HOURS)	BRITELITE (HOURS)
1134	1405	1107	1290
1255	1251	1095	1210
1313	1106	1401	1198
1012	1384	1109	1203
1265	1193	1150	1295
1375	1208		1102
1102	1110		1185

19.19   An economist wishes to compare the percent increase in personal income for two suburbs of Chicago. Using the data in the following table, test at the 5% significance level the hypothesis that there is no difference in the percent increase in personal income.

SUBURB A (%)	SUBURB B (%)	SUBURB A (%)	SUBURB B (%)
5.2	2.6	1.3	11.3
3.1	9.7	8.4	8.1
10.6	1.2	9.1	4.2
11.4	1.4	11.3	1.6
1.2	5.0	12.1	2.7
0.0	9.8	9.8	7.9

19.20   A supermarket manager was curious as to which of the two vending machines located at opposite ends of the store was used more during peak hours. During the peak hours, on 12 randomly selected days, the number of users were counted for machine A. During the peak hours on another randomly selected 12 days, the number of users were counted for machine B. From the following data, test at the 5% level of significance the hypothesis that there is no difference in the use of the two vending machines.

VENDING MACHINE A	VENDING MACHINE B	VENDING MACHINE A	VENDING MACHINE B
10	9	13	9
12	11	19	8
13	14	11	12
11	10	10	13
10	13	15	14
15	14	12	11

19.21 At the 5% significance level, test the hypothesis that there is no significant difference in the time for machine 1 and machine 2 in Exercise 9.34, to produce an item, using the Mann-Whitney test. Compare to the results obtained using the t-test.

19.22 The number of defective electronic components in lots of two different processes are given below.

PROCESS 1	PROCESS 2
10	13
15	19
13	14
25	18
18	20
16	19
15	5
12	19
11	20
17	15

a. Use a t-test, assuming equal population variances, to test that the two processes differ in location with respect to the number of defective items. Use a .05 significance level.

b. Use the Mann-Whitney test to perform part a.

c. Which test is preferable for this type of data?

19.23 What assumption needs to be made about the sample data used in the Wilcoxon signed rank test? What assumption needs to be made about the distributions of the population?

19.24 From 12 paired observations, it is found that by ranking the magnitude of the differences of the observations in each pair, T_+ (the sum of the ranks of the positive differences) is 27. Using a .05 significance level, can it be concluded that there is a difference in the location of the two populations?

19.25 A psychologist conducts a seminar to increase a person's self-esteem. A before-and-after test that measures each person's self-esteem is given to nine individuals. Using the following scores and a 5% significance level, is there evidence to conclude that the scores after the seminar are greater than the scores before the seminar?

BEFORE	AFTER	BEFORE	AFTER
70	74	55	58
72	88	43	41
75	71	51	63
61	62	84	80
82	89		

19.26 An insurance company believes that employees who have a college degree when hired progress faster in the company than those who do not. Pairs of employees are randomly selected; each pair consists of two people hired at the same time, one person with a college degree and the other without a college degree. The percent increase in pay for these employees after three years is recorded below. At the 10% level of significance, can you conclude that employees who have a college degree when hired progress faster than those who do not?

WITHOUT COLLEGE DEGREE (%)	WITH COLLEGE DEGREE (%)	WITHOUT COLLEGE DEGREE (%)	WITH COLLEGE DEGREE (%)
10	13	12	13
9	10	9	8
8	6	18	16
13	13	9	12
14	18	15	17
7	10	10	9
12	11	11	13
11	15	10	9
16	20		

19.27 Seven randomly selected faculty members were asked to evaluate two research project proposals on a scale from 0 to 10, with a higher score indicating a more acceptable proposal. The scores follow. Using a 5% significance level, can you conclude that the proposal for research project 2 is more acceptable than the proposal for research project 1?

RESEARCH PROJECT 1	RESEARCH PROJECT 2
5	7
3	5
6	9
7	6
8	9
4	6
7	10

19.28 Martin's Weight Control Center claims that if a woman maintains the same diet but attends aerobic classes three times per week, she will definitely lose weight. Seventeen women who attended the program for three months were randomly selected. From the table, which gives the participants' weights at the beginning and end of the three months, test the claim of Martin's Weight Control Center. Use a 5% significance level.

BEFORE	AFTER	BEFORE	AFTER	BEFORE	AFTER
119	117	148	140	114	108
131	130	152	138	122	114
135	125	180	171	125	111
125	121	110	114	120	118
140	143	130	132	112	113
119	114	118	110		

18.29 The manager of a calculator-assembly plant wanted to know whether machine operators with little experience produced more defective calculators than did the experienced machine operators. The number of defective calculators produced by 20 randomly selected experienced machine operators in one week was recorded. Then, these 20 experienced operators were replaced by inexperienced machine operators and the number of defective calculators produced at these positions was recorded for one week. If the operators at each position can be considered to be a pair, use the Wilcoxon test to test that the experienced operators produced fewer defective calculators. Use a 5% significance level.

EXPERIENCED EMPLOYEES	INEXPERIENCED EMPLOYEES	EXPERIENCED EMPLOYEES	INEXPERIENCED EMPLOYEES
10	14	13	19
13	14	18	21
15	12	19	18
18	25	10	13
14	13	19	26
10	15	25	26
30	21	15	17
14	18	21	20
22	23	20	28
15	13	12	19

19.30 The manager of an insurance company sent ten randomly selected salespeople to a sales-motivation lecture given by several top-selling insurance salespeople. The manager recorded the dollar amount (in hundreds of thousands) of insurance sold by the ten salespeople during the four months prior to attending the lecture and the four months after the lecture. At the 5% significance level, is there evidence to suggest that the sales-motivation lecture improved sales?

BEFORE LECTURE	AFTER LECTURE	BEFORE LECTURE	AFTER LECTURE
1.2	1.9	6.2	6.3
1.8	3.4	1.5	1.4
3.8	3.1	3.3	4.9
1.9	4.5	2.4	3.5
5.8	5.0	3.1	3.0

19.31 A paired-difference experiment yielded a value of 280 for the sum of the ranks of the positive differences from 30 observations. Using a 5% significance level, is there evidence to suggest that the population differences are not centered at zero?

19.32 A large discount department store chain decided to rearrange the layout of its merchandise at six of its stores to encourage customers to buy more on impulse. Sales from the six months prior to the rearrangement and sales from the six months after the rearrangement are given below. Using a 5% significance level, can it be concluded that there is an increase in sales after the rearrangement?

SALES BEFORE THE REARRANGEMENT ($\times \$100,000$)	SALES AFTER THE REARRANGEMENT ($\times \$100,000$)
3.4	3.9
2.8	2.9
4.1	4.0
3.6	3.8
4.8	5.6
5.1	5.4

19.33 The following values are the differences from 17 pairs of observations. At the 10% level, determine whether there is sufficient evidence to indicate that the population differences are not centered at zero.

-1.1, -2.3, 4.5, 1.6, 2.3, -4.3, 1.9, -2.6, 1.8, 1.6, -2.7, 1.8, 2.1, -3.8, -1.0, 1.4, 2.5

19.34 For Exercise 9.47, test at the 5% significance level that the blood pressure is less after the drug is administered. Use the Wilcoxon signed rank test. Compare the results using the paired t-test and a 5% significance level.

19.35 The productivity of ten employees is compared for two situations. In situation A, the noise level is high, and in situation B, the noise level is low. The data are given below in coded form, with large values indicating higher productivity.

EMPLOYEE	PRODUCTIVITY SITUATION A	PRODUCTIVITY SITUATION B
1	87	89
2	75	78
3	92	83
4	84	85
5	91	92
6	96	95
7	90	94
8	87	85
9	97	99
10	78	80

a. Use the Wilcoxon signed rank test to test the hypothesis that there is no difference in the productivity of an employee in the two situations. Use a .10 significance level.

b.   Use the paired t-test to test the hypothesis in part a.

c.   Compare the results in parts a and b and explain any difference.

▌ 19.3
Comparing More Than Two Populations: The Kruskal-Wallis Test and the Friedman Test

The Kruskal-Wallis Test

When comparing the means of more than two populations, a popular technique is the ANOVA procedure discussed in Chapter 11. One of the assumptions behind this technique is that you are dealing with normally distributed populations; the F test used in the ANOVA table is invalid unless all of the populations are nearly normally distributed with equal variances.

The nonparametric counterpart to the one-way ANOVA method is the **Kruskal-Wallis test.** It is named after W. H. Kruskal and W. A. Wallis, who published their results in 1952. This test, like many other nonparametric procedures, is relatively new, unlike most of the parametric hypothesis tests, which were developed much earlier. *The assumption of normal populations is* not necessary *for the Kruskal-Wallis test, making it an ideal technique for samples exhibiting a nonsymmetric (skewed) pattern*. It is also less sensitive than the ANOVA procedure to the assumption of equal variances. This test also is useful when the data consist of rankings (ordinal data) within each sample.

The assumption of normal populations becomes quite critical when dealing with small samples. As we've seen in many of the earlier tests of hypothesis, this assumption can be relaxed when larger samples are used, due to the Central Limit Theorem. However, many experiments of a business nature dealing with product comparisons result in the destruction of the product being tested. Consequently, small samples are often a necessity for such experiments, and nonparametric techniques are widely used to analyze the resulting data.

The Kruskal-Wallis test is actually an extension of the Mann-Whitney U test discussed earlier for *two* independent samples. Both procedures require that the sample values have a measurement scale that is at least ordinal (that is, each sample can be ranked from smallest to largest).

The hypotheses for this situation are similar to the Mann-Whitney hypotheses in that they are stated in terms of differing population locations. The Kruskal-Wallis hypotheses are

H_0: the k populations have identical probability distributions

H_a: at least two of the populations differ in location

Procedure   You first obtain random samples of size $n_1, n_2, \ldots, n_k$ from each of the k populations. The total sample size is $n = n_1 + n_2 + \cdots + n_k$. As with the Mann-Whitney procedure, you next pool the samples and arrange them in order, assigning a rank to each. For ties, you assign the average rank to the tied positions.

Let T_i = the total of the ranks from the ith sample. The Kruskal-Wallis test statistic (KW) is

$$KW = \frac{12}{n(n+1)} \sum_{i=1}^{k} \frac{T_i^2}{n_i} - 3(n+1) \qquad \textbf{(19.13)}$$

The distribution of the KW statistic approximately follows a chi-square distribution with $k - 1$ df. This approximation is good even if the sample sizes are small. To test H_0 versus H_a, the procedure is to

reject H_0 if KW is "large"

that is, if *KW* is in the right tail of the chi-square curve. This right-tail critical value is obtained from Table A.6, using a significance level = α and df = $k - 1$.

The Kruskal-Wallis Test

Hypotheses:

H_0: the *k* populations have identical probability distributions

H_a: at least two of the populations differ in location

Assumptions:

1. Random samples are obtained from each of the *k* populations.
2. The individual samples are obtained independently.
3. Values within each sample can be ranked.

Procedure: The individual samples are pooled and then ranked from smallest to largest. Letting T_i = the sum of the ranks of the *i*th sample, the *KW* statistic is determined using equation 19.13. The null hypothesis, H_0, is rejected if

$$KW > \chi^2_{\alpha, df}$$

where $\chi^2_{\alpha, df}$ is the value from Table A.6 corresponding to df = $k - 1$, with a right-tail area = α.

EXAMPLE 19.8

Drexton Industries has a number of different brands of copying machines at their main facility. A critical factor in the attractiveness of each brand is the amount of time in which a machine is not working and is waiting for repair (downtime). Management requested a study to be made on four different brands of machines to determine whether there is a difference in the amount of downtime for these brands. Data were collected by finding the total downtime per month for 20 randomly selected months. In this way, the downtimes for five randomly selected months were obtained for each of the four brands of machine. These results are shown in Table 19.8.

Do these data indicate a difference in the amount of downtime for the four brands? Use $\alpha = .05$.

SOLUTION

There are $k = 4$ populations here, so we need the $\chi^2_{.05, 3}$ value from Table A.6. Based on this value, the testing procedure is to

reject H_0 if $KW > \chi^2_{.05, 3} = 7.81$

From Table 19.8, we are able to compute the value of the *KW* statistic using the ranks of the observations in the pooled sample:

$$KW = \frac{12}{(20)(21)} \left[\frac{74^2}{5} + \frac{29^2}{5} + \frac{27^2}{5} + \frac{80^2}{5} \right] - 3(21)$$
$$= 13.83$$

■ TABLE 19.8	BRAND 1	RANK	BRAND 2	RANK	BRAND 3	RANK	BRAND 4	RANK
Amount of downtime for copying machines (Example 19.8).	28	12	5	1	10	3	45	18
	41	17	16	6	8	2	30	13
	34	15	20	8	18	7	49	19
	52	20	24	9	14	4	32	14
	25	10	15	5	26	11	36	16
		$T_1 = 74$		$T_2 = 29$		$T_3 = 27$		$T_4 = 80$

■ **FIGURE 19.7**
p-value for *KW* statistic;
χ² curve with 3 df
(Example 19.8).

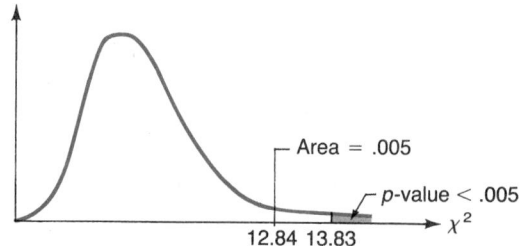

As a check of your calculations at this point, make sure the ranks sum to $n(n + 1)/2$. For this example, n = total number of observations = 20, and so the total of the ranks should be $(20)(21)/2 = 210$; here, $74 + 29 + 27 + 80 = 210$.

The calculated *KW* value exceeds 7.81, and so our conclusion is that there is a difference in downtime among the four brands. From the small values of T_2 and T_3, it appears that these two brands have much less downtime and are superior in this respect to brands 1 and 4.

Finally, the *p*-value here is <.005, indicating a very strong conclusion. In other words, these data indicate a clear difference in location for the four brands. This *p*-value is illustrated in Figure 19.7.   ■

The Kruskal-Wallis test statistic is computed using MINITAB in Figure 19.8. Notice that the downtime values are stored in column C1, and column C2 contains the sample number of each observation (1, 2, 3, or 4). The value of the Kruskal-Wallis statistic is called *H;* it agrees with the previous result.

The Friedman Test

The topic of comparing more than two population means using dependent samples was introduced in Chapter 11, where we examined the randomized block design. Such a design consists of a single factor of interest with *k* levels, with the sample data organized into blocks. For this situation, the samples are not independently obtained, but data within the same block may be gathered from the same city or person or at the same point in time.

The key assumption behind the use of the randomized block design is that the variable being measured (the *dependent* variable) is normally distributed within each factor level/block combination. There are two situations for a blocked design

```
MTB > SET INTO C1
DATA> 28 41 34 52 25 5 16 20 24 15
DATA> 10 8 18 14 26 45 30 49 32 36
DATA> END
MTB > SET INTO C2
DATA> 1 1 1 1 1 2 2 2 2 2 3 3 3 3 3 4 4 4 4 4
DATA> END
MTB > KRUSKAL-WALLIS USING DATA IN C1, LEVELS IN C2

LEVEL     NOBS     MEDIAN   AVE. RANK    Z VALUE
  1         5       34.00       14.8        1.88
  2         5       16.00        5.8       -2.05
  3         5       14.00        5.4       -2.23
  4         5       36.00       16.0        2.40
OVERALL    20                   10.5

H = 13.83   d.f. = 3   p = 0.003
     ↑
    KW
```

where the use of the parametric randomized block technique is inappropriate, requiring the use of a nonparametric procedure:

1. You have no evidence to support the assumption of normality.

or

2. The sample data are ordinal (that is, consist of rankings).

When confronted with either of these two situations, the nonparametric **Friedman** test provides a correct method of testing for differences in location for the k factor level populations. The corresponding hypotheses are

H_0: the k populations have identical probability distributions

H_a: at least two of the populations differ in location

Procedure   The k observations *within each block* are rank ordered, using the usual procedure of assigning the average of the tied positions in the event of ties within a block. Of course, this step is omitted if ordinal data are obtained initially, such as asking 50 people to rank order four particular products (or brands) according to a specified set of criteria. For this illustration, you would have $b = 50$ blocks and $k = 4$ populations defined by the four products. In addition, to remove potential bias introduced by fatigue, familiarity, and so on, the order in which the four brands are evaluated should be *randomly* determined for each person.

Define T_i = the total of the ranks for the ith population. The test statistic for the Friedman test is defined as

$$FR = \frac{12}{bk(k + 1)} \sum_{i=1}^{k} T_i^2 - 3b(k + 1) \qquad \textbf{(19.14)}$$

where b = number of blocks and k = number of factor levels (populations).

The distribution of the FR statistic approximately follows a chi-square distribution with $k - 1$ df. This approximation works well provided the number of blocks (b) or the number of factor levels (k) exceeds five. Like the Kruskal-Wallis test procedure, the Friedman test procedure is to reject H_0 if FR lies in the right-tail of the chi-square curve; that is,

$$\text{reject } H_0 \text{ if } FR > \chi^2_{\alpha, \text{df}}$$

where $\chi^2_{\alpha, \text{df}}$ is the value from Table A.6 corresponding to df $= k - 1$ and right-tail area $= \alpha$. This procedure is summarized in the accompanying box.

The Friedman Test

Hypotheses:

H_0: the k populations have identical probability distributions

H_a: at least two of the populations differ in location

Assumptions:

1. The factor levels are applied in a random manner within each block.
2. The number of blocks (b) or the number of factor levels (k) exceeds five.
3. Values within each block can be ranked.

Procedure: Values are ranked within each block. Letting T_i = sum of the ranks within the ith factor level, the FR statistic is calculated using equation 19.14. The null hypothesis is rejected if

$$FR > \chi^2_{\alpha, df}$$

where $\chi^2_{\alpha, df}$ is the chi-square value (Table A.6) with df = $k - 1$ and right-tail area = α.

EXAMPLE 19.9

In a study of the perceived attractiveness of new car warranties, a newspaper writer sent the warranties of three competing brands of automobiles (Henry, GA, and Roadster) to ten different editors of automotive magazines. At first, it was decided to have each editor assign a score from 0 (worst) to 100 (best) for each of the three warranties and use a randomized block analysis. Here the ten editors would represent the $b = 10$ blocks, and the factor of interest would be brand of automobile, consisting of $k = 3$ levels. However, after much discussion, it became clear that each editor had his or her own set of criteria for judging a warranty, not to mention different weights for each of these various criteria.

Consequently, it was decided simply to ask each editor to rank the three warranties, rather than determine a score for each one. The warranties were assigned in a random manner for each editor. The results are summarized in Table 19.9. Use the Friedman test and a significance level of .10 to determine if there is a difference in the perceived quality of the three warranties.

SOLUTION

The computed value of the Friedman statistic is

$$FR = \frac{12}{(10)(3)(4)} [22^2 + 14^2 + 24^2] - 3(10)(4)$$

$$= 125.6 - 120$$

$$= 5.6$$

Referring to the chi-square table (A.6), using $k - 1 = 2$ df and right-tail area = .10, the test procedure is to

reject H_0 if $FR > 4.60517$

Since $5.6 > 4.60517$, we reject H_0 and conclude that there *is* a difference in the perceived quality of the three warranties. The apparent reason for this result is the fact that the GA warranty was ranked first or second by all ten editors and far outranked the warranties for Henry and Roadster. Finally, the p-value here is between .05 and .10, indicating a fairly weak result, yet one that is statistically significant at a level of $\alpha = .10$. ∎

■ **TABLE 19.9**
Results of new-car warranty ranking (Example 19.9).

HENRY	GA	ROADSTER
3	2	1
2	1	3
3	1	2
2	1	3
3	2	1
2	1	3
1	2	3
2	1	3
3	1	2
1	2	3
$T_1 = 22$	$T_2 = 14$	$T_3 = 24$

Exercises

19.36 Samples of 9 observations were taken from each of three populations, making a total of 27 observations. When the observations were pooled and then ranked from smallest to largest, the sum of the ranks for the first, second, and third samples was 120, 148, and 110, respectively. At a significance level of .05, do the data indicate a difference in location for the three populations?

19.37 Ron, Ted, and Janet are three sales representatives covering three separate territories for a company that sells farm equipment at the wholesale level. To test the hypothesis that there is no difference in the monthly commission of the three salespeople, eight months were chosen at random for each and their commissions were recorded in thousand-dollar units. Use the Kruskal-Wallis test on the data to test this hypothesis. Use a 1% significance level.

MONTH	RON	TED	JANET	MONTH	RON	TED	JANET
1	2.3	2.8	2.0	5	4.6	1.7	4.4
2	2.1	2.9	1.9	6	3.8	3.4	2.2
3	1.5	3.1	2.6	7	3.1	2.3	3.8
4	2.3	1.8	3.9	8	2.6	2.1	2.5

19.38 The following table lists observations from a sample of four populations. Is there reason to believe that the four populations differ in location? Use a 1% significance level.

POPULATION 1	POPULATION 2	POPULATION 3	POPULATION 4
101	104	104	105
110	99	102	110
120	86	100	120
105	105	111	121
100	110	103	127
107	120	102	118
106	114		112
	110		

19.39 A survey was taken of the starting salaries of students who had completed a degree in business administration at either Oceanspray College, Stanton University, or Hillside College. Do the following data indicate, at the .05 level, a difference in the starting salaries of students with a degree in business administration from one of the three schools?

OCEANSPRAY COLLEGE	STANTON UNIVERSITY	HILLSIDE COLLEGE
27,100	24,500	22,500
25,300	27,250	23,000
22,450	26,700	26,000
26,800	27,000	25,750
25,100	26,500	21,630
21,350	22,300	22,500
22,500	27,600	23,650
25,000	28,150	24,180
		22,750

19.40 Four machines are used to package 16-ounce bags of puffed wheat. Each machine is designed to package the bags so that the average bag has 16 ounces of cereal in it. From the data, in which samples of eight bags were randomly selected from each machine, is there an indication that the amount of puffed wheat packaged is not the same for all four machines? Use a .05 significance level.

MACHINE 1	MACHINE 2	MACHINE 3	MACHINE 4
15.9	16.1	15.8	16.4
15.8	16.3	15.9	16.5
16.0	16.0	16.0	16.0
15.7	15.9	16.1	16.1
16.1	16.4	16.0	16.4
16.2	15.8	16.4	16.3
15.6	16.2	16.1	16.1
15.8	16.1	15.7	16.4

19.41 Thirty new employees were selected to test two training programs. Ten of them (group A) were randomly selected for a self-paced training program. Another ten (group B) were randomly selected for a classroom training program. The remaining ten employees (group C) were not given any training. After the completion of the experiment, the manager evaluated the 30 employees on their productivity over a two-week span. The following ranks were given by the manager, with the highest rankings being given to those who were not productive.

PROGRAM A	PROGRAM B	PROGRAM C	PROGRAM A	PROGRAM B	PROGRAM C
6	5	1	30	12	3
22	21	10	16	18	7
25	15	11	23	24	14
26	4	13	28	27	17
20	8	2	9	29	19

From the data, is there a significant difference in the productivity of the three groups? Use a .05 significance level.

19.42 Joe's Delicatessen sells cheese sandwiches, ham sandwiches, and roast-beef sandwiches. Joe would like to know if there is a significant difference in the number of sandwiches of each type sold. On 30 randomly selected days, the following number of sandwiches of each type was sold (ten days were selected for each type of sandwich). Is there a significant difference at the 10% level?

CHEESE	HAM	ROAST BEEF	CHEESE	HAM	ROAST BEEF
27	30	33	26	22	34
24	21	25	25	24	31
23	20	23	24	21	30
29	31	22	25	22	30
21	32	27	28	20	27

19.43 The management of a company that markets Soft and Fresh Detergent would like to increase sales of detergent by including a free drinking glass in the box, including a coupon worth 50 cents toward the next purchase, or using a colorful see-through plastic container for the detergent. Thirty stores in different cities were randomly selected to market the detergent in one of the three ways (10 stores for each way). The number of boxes sold in these stores over a one-month period follows. Do the data indicate a difference in the number of boxes sold for each of the marketing strategies? Use a 5% significance level.

FREE GLASS	COUPON	SEE-THROUGH PLASTIC CONTAINER	FREE GLASS	COUPON	SEE-THROUGH PLASTIC CONTAINER
350	320	374	270	311	349
310	315	371	340	318	331
250	300	332	310	330	322
380	315	361	290	340	368
290	390	356	375	314	351

19.44 The number of hours it takes three workers to complete a task is given in the following table. The task is assigned to each worker four times. Do the data indicate a significant difference in the time it takes each worker to complete the task? Use a significance level of .05.

WORKER 1	WORKER 2	WORKER 3
3.1	3.4	3.5
3.4	3.3	3.2
3.0	3.4	3.3
3.1	3.2	3.1

19.45 The number of cars passing each of three different intersections in Crossroads City between 5:00 P.M. and 5:30 P.M. is given in the following table for randomly selected days. Fifteen days were randomly selected, and then the amount of traffic was recorded for 5 of the 15 days at each intersection. Test the null hypothesis that there is no difference in the amount of traffic at each intersection between 5:00 P.M. and 5:30 P.M. Use a 10% significance level.

INTERSECTION 1	INTERSECTION 2	INTERSECTION 3
440	480	433
420	392	406
530	386	427
401	456	338
454	427	397

19.46 A manager believes that the higher-salaried employees in a certain company are more satisfied with their jobs than are the lower-salaried employees. A sample of ten employees from each of the salary levels indicated by the following table was taken. Is there a significant difference in the satisfaction level, measured on a scale of 1 to 10 (10 being a perfectly satisfied employee), for the three groups? Use a significance level of 5%.

$25,000 TO $40,000	$40,001 TO $60,000	OVER $60,000	$25,000 TO $40,000	$40,001 TO $60,000	OVER $60,000
4	7	8	9	3	9
3	8	7	1	4	10
7	6	6	8	9	3
6	7	7	7	6	8
5	9	5	6	7	7

19.47 For the data in Exercise 11.1, use the Kruskal-Wallis test statistic to test that there is no difference in the amount of wear on three different designs of rubber soles. Use a significance level of .05. Compare with the results obtained from the ANOVA procedure.

19.48 In Exercise 11.7, can you conclude that there is a significant difference in the monthly sales of the three salespeople using the Kruskal-Wallis test statistic? Use a 5% significance level. When would you prefer the Kruskal-Wallis test to the usual ANOVA procedure?

19.49 The vice president of quality assurance at an airline company is interested in whether its three quality engineers are usually in agreement on the ratings they give to different airplane seating designs. Seven different designs are chosen at random and the quality engineers are asked to rate the comfort to the passengers on a scale from one to ten, with ten representing the highest level of comfort possible. Do the given data indicate that there is a difference in the ratings of the three quality engineers? Use a .10 significance level.

DESIGN	QUALITY ENGINEER 1	QUALITY ENGINEER 2	QUALITY ENGINEER 3
1	5	7	8
2	4	3	5
3	6	5	4
4	9	7	8
5	5	7	4
6	8	7	6
7	9	6	8

19.50 The manager at a manufacturing plant is interested in whether there is a significant difference in the number of times four machines need to be readjusted after going out of control. Eight months are randomly selected and the number of times the machines are readjusted per month is recorded. Do the data support the conclusion that some machines need more adjusting than others? Use a .05 significance level.

MONTH	MACHINE 1	MACHINE 2	MACHINE 3	MACHINE 4
1	5	6	3	6
2	4	3	7	5
3	4	5	6	3
4	5	3	4	7
5	4	4	5	6
6	10	11	9	11
7	15	18	13	16
8	3	4	5	4

19.51   The manager of a management information systems department wishes to know if there is a preference in the level of color and graphics used in the computer-aided instruction courses that are available for the programmer. Eight programmers are asked to rank the three different levels of color and graphics, with the rank of 1 representing the most desirable level. Do the data support the conclusion that not all three levels of color and graphics are equally preferred? Use a .05 significance level.

PROGRAMMER	LEVEL 1	LEVEL 2	LEVEL 3
1	1	3	2
2	2	1	3
3	1	2	3
4	1	3	2
5	1	3	2
6	1	2	3
7	2	1	3
8	1	3	2

19.52   The vice president of a chain of convenience stores is interested in whether the sale of three brands of a certain product differ significantly. Convenience stores are chosen at random and a week is randomly selected to record the sales of each of the three brands of the products. Do the data indicate that not all brands are equally preferred? Use a .10 significance level.

STORE	BRAND A	BRAND B	BRAND C
1	25	38	41
2	20	31	35
3	15	19	13
4	10	8	12
5	49	51	60
6	26	23	29
7	37	33	35
8	7	10	15
9	12	10	14

19.53   A property manager wished to know if there was a significant difference in the bids of three independent contractors. The property manager recorded the bids by the contractors on seven randomly selected jobs. Do the data support the conclusion that the bids are significantly different for the contractors? Use a .05 significance level.

JOB	CONTRACTOR 1	CONTRACTOR 2	CONTRACTOR 3
1	6500	7310	7400
2	5130	4950	5800
3	2500	2300	2650
4	2900	2800	2675
5	7800	8000	8300
6	4650	4500	4725
7	1250	1050	1400

19.54   Seven financial advisors were randomly selected and asked to rank their choice of three investments for the next year: bonds, blue-chip stocks, and small company stocks. Do the data given below support the conclusion that not all three investments are expected to perform equally as well? Use a .05 significance level.

FINANCIAL ADVISOR	BONDS	BLUE-CHIP STOCKS	SMALL COMPANY STOCKS
1	2	1	3
2	3	1	2
3	1	2	3
4	2	3	1
5	2	1	3
6	3	1	2
7	2	1	3

◢ 19.4
A Measure of Association: Spearman's Rank Correlation

Whenever you encounter data describing two variables (say, X and Y), one measure of interest is the degree of **association** between X and Y. Are large values of X associated with large values of Y (a *positive* relationship)? Or do you observe smaller values of Y with larger values of X (a *negative* relationship)? Another possibility is that no relationship is observed between these two variables.

Consider the following data, in which a sample of ten families is used to determine the relationship (if any) that exists between X = market value of the family's home and Y = their total indebtedness (excluding the home mortgage; in thousands of dollars). Included in Y are any charge accounts, automobile loans, and other current liabilities.

FAMILY	X (MARKET VALUE OF HOME)	Y (TOTAL INDEBTEDNESS)
1	85	12
2	147	27
3	340	45
4	94	10
5	120	17
6	105	4
7	135	20
8	162	25
9	480	35
10	88	14

The president of Metro Savings and Loan believes that larger home values are associated with larger indebtedness, that is, a positive relationship exists between these two variables. His belief is confirmed by the scatter plot contained in Figure 19.9.

One method of measuring the association between two variables is the correlation coefficient, r, introduced in Chapter 14; r is also referred to as the Pearson product moment correlation. The equation used to determine this measure is

$$r = \frac{\Sigma xy - (\Sigma x)(\Sigma y)/n}{\sqrt{\Sigma x^2 - (\Sigma x)^2/n}\ \sqrt{\Sigma y^2 - (\Sigma y)^2/n}} \qquad (19.15)$$

where n represents the number of observations (pairs).

The value of r, often called the *sample correlation coefficient*, measures the amount of linearity that exists between the sample values of X and Y. It is used to measure ρ (rho), the *population* correlation coefficient. The value of ρ can be thought of as the correlation between *all* possible X, Y pairs, not just those contained in the sample.

In the previous discussions, a significant relationship between X and Y existed if we were able to reject H_0: $\rho = 0$. The test statistic here was a t-statistic, so this test assumes a normal distribution for the X, Y variables.

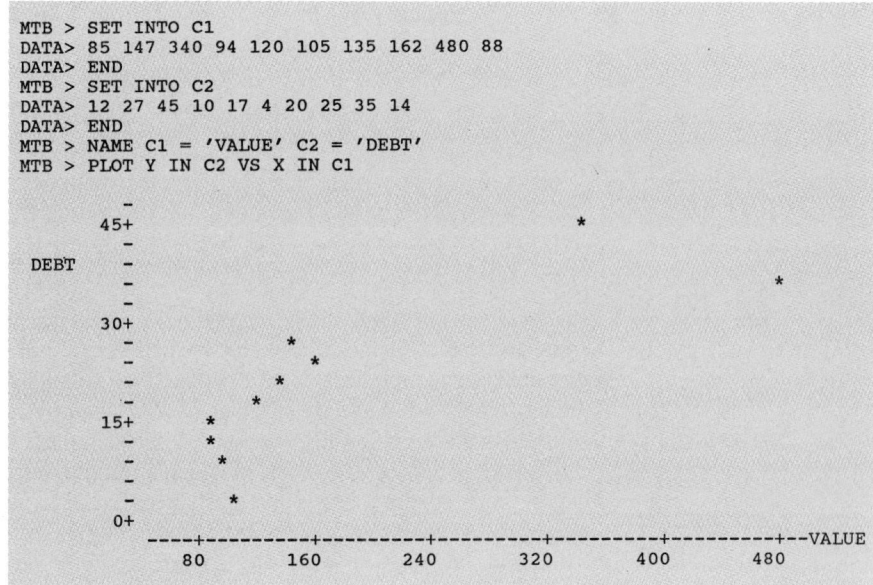

```
MTB > SET INTO C1
DATA> 85 147 340 94 120 105 135 162 480 88
DATA> END
MTB > SET INTO C2
DATA> 12 27 45 10 17 4 20 25 35 14
DATA> END
MTB > NAME C1 = 'VALUE' C2 = 'DEBT'
MTB > PLOT Y IN C2 VS X IN C1
```

An alternative to this procedure is a measure of association derived from the *ranks* of the X and Y variables. *This nonparametric measure does not assume a normal distribution; it assumes only that the values within the X and Y samples can be ranked.* For data such as the home price versus debt values, each of the X and Y values can also be replaced by their ranks. If we use these ranks in place of the actual data in equation 19.15, we obtain another measure of association, called the **Spearman rank correlation coefficient,** r_s:

$$r_s = \frac{\sum R(x)R(y) - [\sum R(x)][\sum R(y)]/n}{\sqrt{\sum R^2(x) - [\sum R(x)]^2/n}\ \sqrt{\sum R^2(y) - [\sum R(y)]^2/n}} \quad \textbf{(19.16)}$$

where $R(x)$ = rank of the X observation and $R(y)$ = rank of the Y observation.

If there are no ties, a second formula provides a much easier method of finding r_s. If there are a few ties, this formula still serves as a very good approximation to r_s. The shortcut method of finding r_s is

$$r_s = 1 - \frac{6\sum d^2}{n(n^2 - 1)} \quad \textbf{(19.17)}$$

where, for each observation, d is the difference between the X and Y ranks; that is, $d = R(x) - R(y)$.

The Pearson product moment correlation, r, measures the amount of *linear* relationship between X and Y and ranges from -1 to 1. Also, $r = 1$ or -1 only if all of the points fall exactly on a straight line. Similarly, the range for the Spearman rank correlation is

$$-1 \leq r_s \leq 1$$

One difference here is that r_s will equal 1 if Y increases in the sample observations every time X increases. This rate of increase need not be linear, as is illustrated in Figure 19.10 for a sample of five observations that do not lie on a straight line. Consequently, r is less than 1 but, as the following table shows, r_s does equal 1.

X	RANK R(X)	Y	RANK R(Y)	DIFFERENCE OF RANKS (d)	d^2
3	1	4	1	$1 - 1 = 0$	0
5	2	7	2	$2 - 2 = 0$	0
7	3	8	3	$3 - 3 = 0$	0
9	4	10	4	$4 - 4 = 0$	0
11	5	16	5	$5 - 5 = 0$	0
				$\Sigma d^2 = 0$	

You can also see that in Figure 19.10 the pairwise ranks *are* perfectly linear; this is why $r_s = 1$.

$$r_s = 1 - \frac{6\Sigma d^2}{n(n^2 - 1)}$$

$$= 1 - \frac{(6)(0)}{(5)(24)} = 1$$

When it is possible to calculate both, the values of r and r_s are generally not the same, although they are usually quite close and they will have the same sign. A

■ FIGURE 19.10
Measure of association.
Pearson product
moment correlation, r,
and Spearman rank
correlation, r_s.

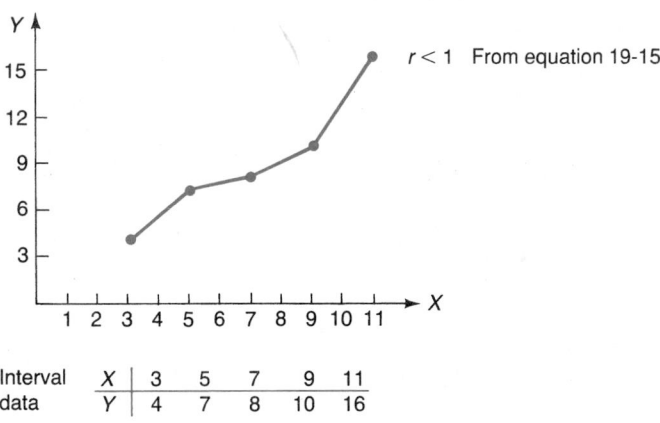

Interval data	X	3	5	7	9	11
	Y	4	7	8	10	16

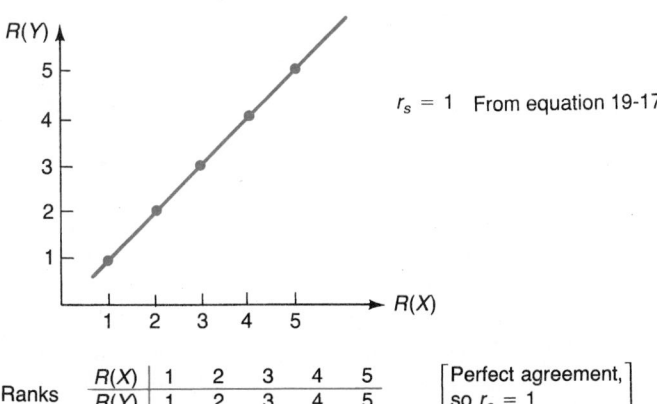

Ranks	R(X)	1	2	3	4	5
	R(Y)	1	2	3	4	5

[Perfect agreement, so $r_s = 1$]

positive value indicates a *positive* relationship between the two variables. Similarly, if r_s and r are negative, then a *negative* relationship exists (Y decreases as X increases). If the values of r and r_s are nearly the same, there is usually high linearity between the two variables.

As we discussed earlier, *nonparametric methods are well suited for situations in which (1) the data are* ordinal *or (2) the distribution of the population(s) from which the data are obtained is suspected to be nonnormal.* When using data of the ordinal type, it is no longer appropriate to use the Pearson coefficient of correlation (r) to perform a t-test to determine whether a significant linear relationship exists. This procedure requires interval or ratio data. Similarly, when dealing with nonnormal populations, the t-test is no longer valid because the normality assumption is a vital part of this procedure.

Suppose that Mr. Roberts and Mr. Clauson each evaluated eight brands of television sets by inspecting eight different sets. After inspecting the sets, rather than assigning each of them a score of some kind, they merely ranked them from 1 (best) to 8 (worst). The results were

BRAND	X (ROBERTS)	Y (CLAUSON)	DIFFERENCE
A	1	2	−1
B	4	3	1
C	2	1	1
D	6	6	0
E	8	7	1
F	3	5	−2
G	7	8	−1
H	5	4	1

Notice that each set of rankings consists of ordinal data because the only meaningful information contained within these values is the *order,* not the difference between them. For example, Mr. Robert's three favorite brands were brands A (first), C (second), and F (third). We have no way of knowing whether brand A was much better than brand C or only slightly better; the same is true for brand C versus brand F. Consequently, there is no way of knowing if the distances between ranks 1 and 2 and between ranks 2 and 3 are the same, and so these differences are meaningless.

The Spearman rank correlation, r_s, is also a measure of how well the two people agree. The larger this value is, the more agreement there is between the two sets of rankings. A value of $r_s = 1$ would indicate perfect agreement, whereas perfect disagreement would result in a value of $r_s = -1$. Here the rank correlation is

$$r_s = 1 - \frac{6\Sigma d^2}{n(n^2 - 1)}$$

$$= 1 - \frac{6[(-1)^2 + (1)^2 + \cdots + (-1)^2 + (1)^2]}{8(64 - 1)}$$

$$= 1 - \frac{(6)(10)}{(8)(63)} = .881$$

This value appears to be quite large, although we will need a formal testing procedure to determine whether there is significant agreement between the two testers (discussed next).

EXAMPLE 19.10   Refer to the table for the home values and debt data. The president of Metro Savings and Loan would like a measure of association between these two variables. Based on past experience, he is reluctant to assume that the home values are normally distributed; they usually are skewed right. There generally are enough homes with an extremely large market value (two, here) to produce a skewed distribution (see Figure 19.9). He asks you to use the Spearman measure of correlation and determine the value of r_s.

SOLUTION   The ranks are calculated as follows:

FAMILY	X	RANK $R(X)$	Y	RANK $R(Y)$	DIFFERENCE (d)	d^2
1	85	(1)	12	(3)	−2	4
2	147	(7)	27	(8)	−1	1
3	340	(9)	45	(10)	−1	1
4	94	(3)	10	(2)	1	1
5	120	(5)	17	(5)	0	0
6	105	(4)	4	(1)	3	9
7	135	(6)	20	(6)	0	0
8	162	(8)	25	(7)	1	1
9	480	(10)	35	(9)	1	1
10	88	(2)	14	(4)	−2	4

$$\Sigma d^2 = 22$$

Then,

$$r_s = 1 - \frac{(6)(22)}{(10)(99)} = .867$$

Based on this value, it appears that a significant positive relationship exists between the market value of a family's home and their total debts.

To determine whether a derived value of r_s is "large enough" to support a conclusion, as in Example 19.10, we develop a test of hypothesis that uses the rank correlation, r_s, as the test statistic.

To obtain a computer solution for Spearman's rank correlation, you ask MINITAB to rank the X and Y values and to compute r for the ranks. Recall that equation 19.17 was a shortcut for determining the rank correlation. This procedure is illustrated in Figure 19.11, where, as before, $r_s = .867$.   ■

A Test of Hypothesis Using the Rank Correlation

In Chapter 14, we used the Pearson correlation coefficient, r, to determine whether a linear relationship existed between two variables, X and Y. This relationship could be either positive (Y increases as X increases) or negative (Y decreases as X increases). When using Spearman's rank correlation, r_s, we drop the "linear" term in the hypotheses and test

H_0: no association exists between the X and Y variables

H_a: association does exist between the X and Y variables

■ **FIGURE 19.11**
MINITAB procedure for finding the Spearman rank correlation (Example 19.10).

```
MTB > SET INTO C1
DATA> 85 147 340 94 120 105 135 162 480 88
DATA> END
MTB > SET INTO C2
DATA> 12 27 45 10 17 4 20 25 35 14
MTB > RANK C1, PUT INTO C11
MTB > RANK C2, PUT INTO C12
MTB > CORRELATION OF C11 AND C12        rs

Correlation of C11 and C12 = 0.867
```

You can also perform one-sided tests, as summarized in the accompanying box. In this way, you can test for a significant positive or negative relationship between the two variables.

The Spearman Test for Rank Correlation

TWO-SIDED TEST	ONE-SIDED TEST	
H_0: no association exists between X and Y	H_0: no association exists between X and Y	H_0: no association exists between X and Y
H_a: association does exist between X and Y	H_a: a positive relationship exists between X and Y	H_a: a negative relationship exists between X and Y
Assumption: Sample values for each variable can be ranked.		
Procedure: Determine the value from Table A.12 using the sample size, n, and the column corresponding to $\alpha/2$.	Use the column in Table A.12 corresponding to α.	Use the column in Table A.12 corresponding to α.
Reject H_0 if $|r_s| >$ (table value).	Reject H_0 if $r_s >$ (table value).	Reject H_0 if $r_s <$ $-$(table value).

EXAMPLE 19.11   In the television-ranking example, is there sufficient evidence to indicate that there was general agreement between the rankings made by Mr. Roberts and those made by Mr. Clauson? Use $\alpha = .05$.

SOLUTION   The appropriate hypotheses here are

$$H_0: \text{no association exists between the two ranks}$$

$$H_a: \text{a positive association exists between the two ranks}$$

According to Table A.12, using $\alpha = .05$ and $n = 8$, the testing procedure is to

$$\text{reject } H_0 \text{ if } r_s > .643$$

Because the computed value of r_s is .881 (as we derived previously), we reject H_0 and conclude that there *was* significant agreement between the two sets of rankings.

The p-value for this result can be obtained by looking across the row in Table A.12 corresponding to $n = 8$. Here we find that .881 corresponds to $\alpha = .005$. Consequently, the p-value here is .005.   ■

EXAMPLE 19.12   The president of Metro Savings and Loan is attempting to demonstrate that a positive relationship exists between X = market value of a family's home and Y = their total indebtedness (excluding the home mortgage). Using the results of Example 19.10, is there sufficient evidence of a positive relationship between these two variables? Use $\alpha = .05$.

SOLUTION   In Example 19.10, the sample rank correlation was found to be $r_s = .867$. Using Table A.12 for $\alpha = .05$ and $n = 10$, we test

H_0: no association exists between the home market value and total indebtedness

H_a: a positive relationship exists

using

$$\text{reject } H_0 \text{ if } r_s > .564$$

The computed value of .867 exceeds the table value, so we reject H_0 and conclude that there is a tendency for larger values of X = home value and Y = family indebtedness to be paired together. This large value of r_s also is off the right side of Table A.12, indicating that for this test the p-value is $<.005$. This extremely small value is strong evidence of a positive relationship between these two variables. ■

Exercises

19.55 A market stand sells watermelons each week at different prices, depending on the supply of watermelons. Calculate the Pearson product moment correlation and the Spearman rank correlation of the quantity sold and the price.

PRICE	QUANTITY SOLD	PRICE	QUANTITY SOLD
1.80	53	2.50	46
2.00	45	2.10	46
1.50	60	1.50	70
1.25	75	1.75	65
1.75	50	1.90	47
2.25	48		

19.56 The rank correlation coefficient between 15 pairs of observations is .31. Using a significance level of .05, test the null hypothesis that there is no positive relationship between the two variables sampled.

19.57 The following data represent the high temperature of the day (°F) and the number of sno-cones sold at Dairy Freeze. Using a nonparametric procedure and a 5% significance level, test the hypothesis that there is a positive relationship between the two variables.

DAILY HIGH TEMPERATURE	NUMBER OF SNO-CONES SOLD	DAILY HIGH TEMPERATURE	NUMBER OF SNO-CONES SOLD
90	49	93	55
91	48	92	51
86	40	90	52
85	38	88	46
84	40		

19.58 A factory wants to know what the relationship is between the age of its machines and the number of breakdowns per year. Ten machines were selected at random, and the following table was constructed. Using the Spearman test for rank correlation, is there a relationship between the age of the machine and the number of breakdowns per year? Use a 10% significance level.

AGE (YEARS)	BREAKDOWNS PER YEAR	AGE (YEARS)	BREAKDOWNS PER YEAR
2	6	7	20
4	10	5	16
5	12	6	15
8	24	3	10
6	17	4	12

19.59 A physician would like to know the relationship between a person's diastolic blood pressure and the average number of hours spent exercising each week. Twenty people of age 30 years were selected randomly. Is there an indication from the data that there may be a negative relationship between exercise and blood pressure? Test with the Spearman test for rank correlation and use a 1% significance level.

DIASTOLIC BLOOD PRESSURE	HOURS EXERCISED	DIASTOLIC BLOOD PRESSURE	HOURS EXERCISED
74	9	80	8
70	10	64	18
62	16	56	20
58	15	68	10
82	6	72	11
84	3	78	8
90	0	84	7
84	4	88	4
72	12	70	12
70	9	66	15

19.60 Two taste-testers were asked to rank ten beers in order to taste preference, with the best tasting beers receiving the highest ranks.

BEER	TASTE TESTER 1	TASTE TESTER 2	BEER	TASTE TESTER 1	TASTE TESTER 2
1	5	7	6	7	5
2	3	3	7	2	4
3	6	6	8	10	9
4	1	2	9	4	1
5	8	10	10	9	8

Using a .05 significance level, is there a significant agreement between the first beer tester's rankings and the second beer tester's ranking?

19.61 The manager of Sales Unlimited wanted to know whether there was a significant positive relationship between a salesperson's travel expenses (in thousands of dollars) and his or her sales (in tens of thousands of dollars). Using the following data, test that there is a positive relationship at the .05 significance level.

TRAVEL EXPENSES	SALES	TRAVEL EXPENSES	SALES
1.5	3.4	2.5	4.1
2.0	3.9	2.2	3.6
3.5	6.1	3.8	5.7
1.6	2.5	4.1	4.3
1.8	3.4	2.6	4.6

19.62 A supervisor at a computer firm wanted to determine whether a relationship existed between the level of an employee's job satisfaction (measured from 1 to 10) and the number of years the employee has been employed with the firm. Test the null hypothesis that there is no relationship and use a 10% significance level.

LEVEL OF SATISFACTION	NUMBER OF YEARS EMPLOYED AT THE FIRM	LEVEL OF SATISFACTION	NUMBER OF YEARS EMPLOYED AT THE FIRM
6	1	3	2
3	5	9	10
7	10	7	7
5	12	8	12
8	6	7	16
4	5		

19.63 The management of a firm wishes to know whether there is a positive relationship between the length of time a certain product has been on the market and the percent of the market that the product has captured. Do the data indicate that a positive relationship exists? Use a 5% significance level.

TIME ON MARKET (YEARS)	PERCENT OF MARKET	TIME ON MARKET (YEARS)	PERCENT OF MARKET
1	1.2	6	3.9
2	2.6	7	3.2
3	1.8	8	4.1
4	2.7	9	3.8
5	2.9	10	4.6

19.64 The owner of a used-car lot believes that there is a negative relationship between the number of cars sold monthly and the average monthly interest rate for financing the cars. Twelve months were randomly selected to obtain the following data. Use a 5% significance level to determine whether the owner's belief is correct.

NUMBER OF CARS SOLD MONTHLY	AVERAGE MONTHLY INTEREST RATE (%)	NUMBER OF CARS SOLD MONTHLY	AVERAGE MONTHLY INTEREST RATE (%)
35	12	50	11
25	15	20	15
28	16	35	11.5
31	12	42	12.5
40	11.5	46	12
48	11	28	15

19.65 The number of cigarettes shipped (in units of billions) are given for the most popular cigarette brands of 1987 and 1989. The ranks for the brands are also given for each of the two years. At the .10 significance level, do the data indicate that a relationship exists between the rankings?

BRAND	NUMBER OF CIGARETTES SHIPPED 1987	RANK FOR 1987	NUMBER OF CIGARETTES SHIPPED 1989	RANK FOR 1989
Marlboro	134.57	1	138.02	1
Winston	63.28	2	47.40	2
Salem	43.80	3	32.34	3
Kool	34.33	4	31.18	4
Newport	24.23	6	24.66	5
Camel	24.18	7	20.37	6
Benson & Hedges	24.36	5	20.33	7
Merit	22.17	8	20.12	8
Doral	17.22	12	19.16	9
Virginia Slims	17.51	10	16.54	10
Pall Mall	17.42	11	13.95	11
Vantage	17.80	9	13.17	12

(*Source:* "If It's Legal, Cigarette Makers Are Trying It," *Business Week* (February 19, 1990): 52.)

◢ Summary

A key step in applying any statistical technique correctly is to make sure that it is appropriate for the type of data that is involved. For example, performing a *t* test using a small sample containing *ordinal* data (such as a set of consumer rankings) is never correct. For situations in which your data are ordinal or from populations that you suspect are nonnormally distributed, a **nonparametric technique** is often preferable. This chapter has introduced some (certainly not all) of the more popular nonparametric procedures.

The **runs test** examines a sequence containing an arrangement of two symbols (M or F, yes or no, + or −, and so on) to determine whether the sequence was generated in a random manner. We defined tests for both small samples (using Table A.15) and large samples (using a test statistic having an approximate normal distribution and Table A.4).

The **Mann-Whitney *U* test** is a nonparametric procedure for determining whether two populations differ in location using two independent samples. Unlike

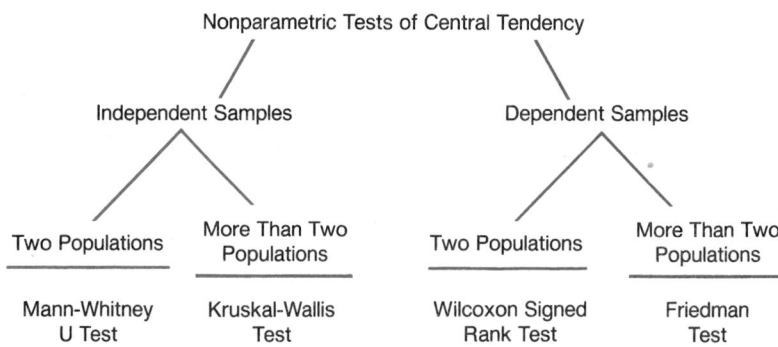

its counterpart, the *t* test, this test does not require that the populations be normally distributed. By combining (pooling) the samples and finding the ranks of the combined sample, you can calculate a value of the test statistic. This method can be applied both to small samples (using Table A.10) and to large samples (using an approximate normal distribution and Table A.4).

The **Wilcoxon signed rank test** is a nonparametric procedure used for determining whether the population of differences is centered at zero when dealing with two dependent (paired) samples. The Wilcoxon technique determines the differences of the paired observations and then calculates a value of the test statistic using the ranks of these differences. Both small samples (using Table A.11) and large samples (using an approximate normal distribution and Table A.4) can be tested.

The **Kruskal-Wallis test** is an extension of the Mann-Whitney test. It is used to test whether two or more populations differ in location when using independent samples. As in the Mann-Whitney procedure, the samples are pooled and then ranked from smallest to largest. The resulting ranks are then used to define a test statistic. This statistic has an approximate chi-square distribution (tabulated in Table A.6), even for fairly small sample sizes.

A nonparametric alternative to the randomized block technique (discussed in Chapter 11) is the **Friedman test.** The Friedman test does not require normal populations with equal variances, unlike the randomized block procedure. Values are ranked within each block and then summed for the various populations under consideration. The Friedman test statistic is calculated using these sums and has an approximate chi-square distribution.

These tests are summarized in Figure 19.12.

The final nonparametric technique, the **Spearman rank correlation,** allows you to measure the *association* between sample values on two variables that consist of ordinal data. If the data are of the interval or ratio type, they can be converted to ordinal data by using the ranks of these values. The Spearman rank correlation coefficient, denoted as r_s, is a nonparametric measure of association between the observations of these two variables. This statistic is computed using the ranks of the observations and can be used to test for a significant relationship between the two variables.

Summary of Formulas

1. Runs Test (large sample, R = number of runs):

$$\mu_R = 1 + \frac{2n_1 n_2}{n_1 + n_2}$$

$$\sigma_R = \sqrt{\frac{2n_1 n_2(2n_1 n_2 - n_1 - n_2)}{(n_1 + n_2)^2(n_1 + n_2 - 1)}}$$

Test statistic: $Z = \dfrac{R - \mu_R}{\sigma_R}$

2. Mann-Whitney U Test:

$$U_1 = n_1 n_2 + \frac{n_1(n_1 + 1)}{2} - T_1$$

$$U_2 = n_1 n_2 + \frac{n_2(n_2 + 1)}{2} - T_2$$

$$\mu_{U_2} = \frac{n_1 n_2}{2}$$

$$\sigma_{U_2} = \sqrt{\frac{n_1 n_2(n_1 + n_2 + 1)}{12}}$$

Test statistic (large samples): $Z = \dfrac{U_2 - \mu_{U_2}}{\sigma_{U_2}}$

3. Wilcoxon Signed Rank Test:

$$\mu_{T_+} = \frac{n(n + 1)}{4}$$

$$\sigma_{T_+} = \sqrt{\frac{n(n + 1)(2n + 1)}{24}}$$

Test statistic (large samples): $Z = \dfrac{T_+ - \mu_{T_+}}{\sigma_{T_+}}$

4. Kruskal-Wallis Test:
Test statistic:

$$KW = \frac{12}{n(n + 1)} \sum_{i=1}^{k} \frac{T_i^2}{n_i} - 3(n + 1)$$

5. Friedman Test:
Test statistic:

$$FR = \frac{12}{bk(k + 1)} \sum_{i=1}^{k} T_i^2 - 3b(k + 1)$$

6. Spearman rank correlation:

$$r_s =$$

$$\frac{\sum R(x)R(y) - [\sum R(x)][\sum R(y)]/n}{\sqrt{\sum R^2(x) - [\sum R(x)]^2/n} \, \sqrt{\sum R^2(y) - [\sum R(y)]^2/n}}$$

$$= 1 - \frac{6\sum d^2}{n(n^2 - 1)} \quad \text{(if no ties)}$$

Review Exercises

19.66 Using the runs test, is there evidence that the following sequence of A's and B's is not randomly generated? Use a 10% significance level.

A A B A B B A A B B B A A B A B A A B B A A B A B A A A A B

19.67 A radio station requests that people telephone the station to express their opinion on the new property tax that the city is levying. An F represents a person telephoning in who is for the tax, and an A represents one against it. Test whether the following sequence of people telephoning the radio station is nonrandomly generated. Use a 10% significance level.

F A F A A A F A A F A A A F F A F A F A F A A A F A A

19.68 The manager of a small-town savings and loan association is interested in finding out whether there is a relationship between the average monthly balance of a savings account and the age of the savings account. Fifteen accounts were selected at random. Do the data indicate a relationship at the .05 significance level?

AVERAGE MONTHLY BALANCE	AGE OF ACCOUNT (YEARS)	AVERAGE MONTHLY BALANCE	AGE OF ACCOUNT (YEARS)
2510	1.5	6148	3.8
3612	2.6	5134	4.7
5634	3.5	2614	1.1
3698	1.8	2581	1.9
3978	2.1	2501	0.5
6751	4.3	3986	4.2
5869	10.1	6645	4.1
		3582	2.3

19.69 Explain the differences in the assumptions necessary to perform parametric tests and nonparametric tests on two population means.

19.70 A statistician wants to determine whether two populations differ in location. After collecting a sample of size eight from each population, the statistician finds that the sum of the ranks of the observations in population 1 is 60. Is there evidence to indicate that there is a difference in location for the two populations? Use a 5% significance level.

19.71 An economist believes that the cost of a typical basket of goods bought by a family of four costs more in Atlanta, Georgia, than it does in Houston, Texas. Seven grocery stores were randomly selected in each of the two cities to collect the following data (cost of basket of goods, in dollars). Test that there is no difference in the cost of the basket of goods for the two cities. Use a 10% significance level.

ATLANTA	HOUSTON
30	37
33	36
41	39
43	40
37	32
39	36
41	32

19.72 Jeff's Auto Parts store special-orders parts from either warehouse A or warehouse B. Warehouse A claims that it delivers parts faster than any other warehouse. Jeff special-ordered 18 auto parts—9 from warehouse A and 9 from warehouse B—and noted the number of days it took to receive them. Using the following data, test warehouse A's claim. Use a 5% significance level.

WAREHOUSE A	WAREHOUSE B	WAREHOUSE A	WAREHOUSE B
2.50	3.00	2.25	2.00
2.25	3.25	3.00	2.25
3.50	2.50	2.50	2.75
2.00	2.75	2.50	2.25
2.75	3.00		

19.73 Vulcan Construction believes that a minicourse in safety to increase the employees' awareness of potential accidents would reduce the incidence of on-the-job accidents. The following data represent the number of accidents reported at 15 construction sites for the month before and the month after the workers attended the minicourse. Do the data indicate that the minicourse reduced accidents? Use a 5% significance level.

BEFORE	AFTER	BEFORE	AFTER
4	3	11	6
5	6	7	7
7	6	8	5
8	5	5	3
7	8	10	11
10	3	9	4
12	2	11	9
9	8		

19.74 A cooking contest was conducted; two top chefs baked chicken and then asked 12 tasters to judge the quality of the cooking on a scale from 1 to 10, 10 being the highest score. Test the null hypothesis that there was no difference between the taster's judgment of the two chefs' quality of cooking at a significance level of .05.

TASTER	CHEF A	CHEF B	TASTER	CHEF A	CHEF B
1	8	7	7	9	9
2	6	10	8	10	7
3	5	9	9	9	10
4	10	4	10	8	6
5	5	8	11	4	9
6	3	7	12	7	3

19.75 Paint A, paint B, and paint C were painted on metallic surfaces and then subjected to high temperatures. Nine replications of the experiment were made. A measure of the cohesiveness of the paint was then taken. The following coded data represent the cohesiveness of the individual paints. Test the hypothesis that there was no difference in the cohesiveness for the three paints. Use a 10% significance level.

PAINT A	PAINT B	PAINT C	PAINT A	PAINT B	PAINT C
1.3	2.1	3.4	1.6	1.9	2.5
1.6	2.7	2.8	2.6	2.3	2.4
3.1	1.6	1.9	2.7	1.8	1.9
2.6	1.9	2.0	1.9	1.6	1.7
4.3	1.6	2.8			

19.76 A chemist was interested in knowing whether three different drugs used for insomnia were equally effective. Three groups of mice, with ten mice to a group, were used. Each group of mice was given the adult-equivalent dosage of one of the three drugs. The time in minutes it took for the mice to fall asleep was recorded. Using the following data, test the null hypothesis that each drug is equally effective in reducing sleep latency. Use a significance level of .05.

DRUG 1	DRUG 2	DRUG 3	DRUG 1	DRUG 2	DRUG 3
32	38	31	41	35	30
35	37	33	28	39	28
40	42	29	34	40	33
30	44	34	39	41	37
33	37	31	28	42	35

19.77 An automobile worker would like to know whether there is any difference in the comfort of three different cars. Three groups of five drivers were selected to judge the riding comfort of the three cars; one of the three cars was assigned to each group. The drivers rated the comfort of each car on a scale from 1 to 5, 5 being the most comfortable. Test that there is no difference in the comfort of the three cars from these data. Use a .05 significance level.

CAR 1	CAR 2	CAR 3
2	5	4
4	3	1
4	4	2
3	3	5
5	2	4

19.78 A microbiologist wants to know whether there is any difference in the time in hours it takes to make yogurt from three different starters. Seven batches of yogurt were made with each of the starters. The following data give the times it took to make each batch. At the .05 significance level, do the data indicate a difference in the time it takes to make yogurt from the three starters?

LACTOBACILLUS ACIDOPHILUS	BULGARIUS	MIXTURE OF ACIDOPHILUS AND BULGARIUS
6.8	6.1	7.3
6.3	6.4	6.1
7.4	5.7	6.4
6.1	6.5	7.2
8.2	6.9	7.4
7.3	6.3	6.5
6.9	6.7	6.8

19.79 A company uses four advertising methods to increase sales: newspaper, mailers, television, and radio. Four 6-month periods were randomly selected, and sales (in thousands of dollars) were recorded monthly for one of the 6-month periods for each advertising method. Do the following data indicate a difference in sales promotion for the four advertising methods? Use a .05 significance level.

NEWSPAPER	MAILERS	TELEVISION	RADIO
2.1	4.1	5.1	4.2
5.6	2.6	4.2	4.3
6.2	2.1	3.7	4.0
3.4	3.3	2.0	3.1
3.1	3.0	3.6	3.8
4.1	3.1	3.1	3.7

19.80   The following data represent the years of job-related experience and last year's salary (in thousands of dollars) for ten randomly selected realtors. Do the data indicate a positive relationship? Use a 5% significance level.

YEARS OF EXPERIENCE	ANNUAL SALARY	YEARS OF EXPERIENCE	ANNUAL SALARY
13.2	42.5	7.8	38.4
10.1	36.8	5.4	29.6
4.6	15.9	3.8	21.6
5.7	27.6	9.6	33.7
6.7	34.3	8.4	35.3

19.81   The owner of a convenience food store is interested in whether there is a positive relationship between the number of cars that pass the store weekly (given in thousands) and the weekly sales of the store (in thousands of dollars). Test the hypothesis that there is a positive relationship using a 5% significance level.

WEEK	NUMBER OF CARS	SALES	WEEK	NUMBER OF CARS	SALES
1	24.6	1.3	6	24.6	1.2
2	29.7	2.6	7	32.7	2.9
3	22.6	1.4	8	35.1	3.4
4	30.4	2.8	9	29.9	3.9
5	20.1	1.1			

19.82   A cable television company is interested in whether there is a difference in the viewers' preference for three movie channels. Listed are rankings from 1 to 3 given to the three movie channels—channel 22, channel 29, and channel 32—by each household. A ranking of one indicates the most desirable channel. Do the data provide sufficient evidence to indicate that there is a difference in the preference for these channels? Use a .10 significance level.

HOUSEHOLD	CHANNEL 22	CHANNEL 29	CHANNEL 32	HOUSEHOLD	CHANNEL 22	CHANNEL 29	CHANNEL 32
1	1	3	2	6	1	3	2
2	3	2	1	7	3	1	2
3	1	2	3	8	2	3	1
4	2	3	1	9	1	3	2
5	3	2	1	10	3	2	1

19.83   Lifetime Exterior sells three types of siding: vinyl, aluminum, and steel. The manager is interested in whether one type of siding is sold more than another. Twelve salespersons are randomly selected and the number of sales of each type of siding is recorded over a 6-month period. From the data, can the manager conclude that not all types of siding sell equally well? Use a .05 significance level.

SALESPERSON	VINYL	ALUMINUM	STEEL	SALESPERSON	VINYL	ALUMINUM	STEEL
1	9	5	8	7	8	14	10
2	6	10	11	8	9	8	3
3	13	9	7	9	6	9	5
4	11	10	6	10	11	15	17
5	15	12	11	11	25	21	15
6	21	22	14	12	10	19	20

19.84   The number of employees on the payrolls of companies in the transportation and public utilities sector of the United States' economy is given for the years 1972 to 1989. Fit a regression equation to the data with the time variable t being the only independent variable ($t = 1, 2, \ldots, 18$). Is there sufficient evidence that the sequence of positive and negative residuals is not random with respect to time? Use a .05 significance level.

YEAR	NUMBER OF EMPLOYEES (UNITS ARE IN THOUSANDS)	YEAR	NUMBER OF EMPLOYEES (UNITS ARE IN THOUSANDS)
1972	4541	1981	5165
1973	4656	1982	5082
1974	4725	1983	4954
1975	4542	1984	5159
1976	4582	1985	5238
1977	4713	1986	5255
1978	4923	1987	5372
1979	5136	1988	5548
1980	5146	1989	5705

(*Source:* U.S. Department of Labor, Bureau of Labor Statistics, *Employment and Earnings* (July 1990): 81.)

19.85 The top 22 United States industrial exporters for 1988 and 1989 are listed below. Ranks are assigned to each company for these years, with rank 1 representing the company with the largest amount of export sales. Is there sufficient evidence of a positive relationship between the ranks of the companies for 1988 and 1989? Use a significance level of .05.

RANK 1989	RANK 1988	COMPANY	RANK 1989	RANK 1988	COMPANY
1	3	Boeing	12	13	Hewlett-Packard
2	1	General Motors	13	14	Unisys
3	2	Ford Motor	14	16	Motorola
4	4	General Electric	15	15	Philip Morris
5	5	International Business Machines	16	12	Digital Equipment
6	7	E. I. Dupont DeNemours	17	17	Occidental Petroleum
7	6	Chrysler	18	19	Allied-Signal
8	10	United Technologies	19	20	Weyehaeuser
9	9	Caterpillar	20	21	Union Carbide
10	8	McDonnell Douglas	21	18	General Dynamics
11	11	Eastman Kodak	22	22	Raytheon

(*Source: The World Almanac and Book of Facts* (1991): 164.)

19.86 The vice president of a credit union wishes to determine the relationship between the number of automobile loan applications per month (Y) and the loan interest rate (X). Twenty months were randomly selected. The data over these months is printed in the following MINITAB computer printout. Comment on the models used in the regression analysis given in the MINITAB printout. Which model appears to be more appropriate? Why? Comment on the significance of the rank correlation of the Y and X variables.

```
MTB > name c1 = 'Y'
MTB > NAME C2 = 'X'
MTB > PRINT C1 C2
ROW     Y      X
  1    143    7.0
  2    144    7.1
  3    136    8.0
  4    135    8.5
  5    137    8.8
  6    131    9.0
  7    126    9.5
  8    127   10.0
  9    125   10.6
 10    116   11.8
 11    117   12.0
 12    113   12.4
 13    110   13.0
 14    104   13.8
 15     96   14.8
 16     95   15.0
 17     91   16.0
 18     88   16.2
 19     79   16.7
 20     75   17.0
MTB > REGRESS Y IN C1 ON 1 PREDICTOR IN C2, RESIDUALS IN C5
```

```
The regression equation is
y = 190 - 6.37 X

Predictor        Coef         Stdev       t-ratio
Constant      189.919         2.470         76.88
X              -6.3676        0.2009       -31.69

s = 2.914        R-sq = 98.2%      R-sq(adj) = 98.1%

Analysis of Variance

SOURCE          DF           SS            MS
Regression       1        8528.0        8528.0
Error           18         152.8           8.5
Total           19        8680.8
```

```
MTB > NAME C5 = 'RESID'
MTB > PRINT C5
RESID
 -0.87983   -0.26538   -1.09030   -0.28820    1.12352   -0.57931   -1.22401
  0.26863    0.91094    0.42885    1.22942    0.71828    1.00997    0.69426
  0.11544    0.21458    1.09068    0.45702   -1.71673   -2.52136
```

```
MTB > RUNS ABOVE AND BELOW 0 USING C5

    RESID
    K =      0.0000

    THE OBSERVED NO. OF RUNS =    5
    THE EXPECTED NO. OF RUNS =   10.6000
    12 OBSERVATIONS ABOVE K     8 BELOW
         THE TEST IS SIGNIFICANT AT   0.0073
```

```
MTB > LET C3 = C2 * C2
MTB > NAME C3 = 'X*X'
MTB > REGRESS Y IN C1 USING 1 PREDICTOR IN C3, RESIDUALS IN C6

The regression equation is
y = 154 - 0.264 X*X

Predictor        Coef         Stdev       t-ratio
Constant      154.284         1.091        141.44
X*X            -0.263823      0.006402     -41.21

s = 2.249        R-sq = 99.0%      R-sq(adj) = 98.9%


Analysis of Variance

SOURCE          DF           SS            MS
Regression       1        8589.7        8589.7
Error           18          91.1           5.1
Total           19        8680.8
```

```
MTB > NAME C6 = 'RESID2'
MTB > PRINT C6
RESID2
  0.78555    1.43934   -0.65991   -0.10429    1.46990   -0.89202   -2.07377
 -0.41584    0.16503   -0.70701    0.32249   -0.32766    0.13812   -0.01895
 -0.23080    0.03571    2.03880    1.42479   -0.83880   -1.51430
```

```
MTB > RUNS ABOVE AND BELOW 0, USING C6

    RESID2
    K =      0.0000

    THE OBSERVED NO. OF RUNS =   12
    THE EXPECTED NO. OF RUNS =   10.9000
     9 OBSERVATIONS ABOVE K    11 BELOW
            THE TEST IS SIGNIFICANT AT   0.6096
            CANNOT REJECT AT ALPHA = 0.05
```

```
MTB > RANK C1, PUT INTO C10
MTB > RANK C2, PUT INTO C11
MTB > CORRELATION BETWEEN C10 AND C11

Correlation of C10 and C11 = -0.991
```

19.87   A real-estate agent was interested in whether there was a significant difference in the sizes of families in three neighborhoods. The data collected on family size are given in C1 below, and the neighborhoods are represented by 1, 2, or 3 in C2. Using the results from the Kruskal-Wallis test given in the following analysis, what can the real-estate agent conclude? Find the p-value.

```
MTB > PRINT C1 C2
  ROW   C1   C2

   1     2    1
   2     4    1
   3     5    1
   4     3    1
   5     2    1
   6     4    1
   7     2    2
   8     1    2
   9     2    2
  10     3    2
  11     2    2
  12     2    2
  13     3    2
  14     1    2
  15     4    3
  16     5    3
  17     3    3
  18     2    3
  19     3    3
  20     5    3
  21     7    3
  22     3    3
  23     4    3
  24     2    3

MTB > KRUSKAL-WALLIS USING DATA IN C1, LEVELS IN C2

  LEVEL    NOBS    MEDIAN   AVE. RANK   Z VALUE
    1        6      3.500     14.3       0.70
    2        8      2.000      7.0      -2.69
    3       10      3.500     15.9       1.96
  OVERALL   24                12.5

  H = 7.452
  H(ADJ. FOR TIES) = 7.913
```

Computer Exercises Using the Database

Exercise 1 -- Appendix I Choose ten observations at random from the database of households that own their home and another ten observations at random from households that rent their home (refer to variable OWNORENT). Using the Mann-Whitney test, is there sufficient evidence to conclude that the income of the principal wage earner is significantly different for the households that own their home and for the households that rent their home? Use a .05 significance level.

Exercise 2 -- Appendix I Choose 12 observations at random from the database of households that have a secondary income (INCOME2) above $20,000. Choose another 12 observations from households that have a secondary income (INCOME2) that is positive and less than $20,000. Also, choose a random sample of 12 observations

from households with no secondary income. Can one conclude that there is a difference in the house payment or apartment/house rent (HPAYRENT) for the three groups? Use the Kruskal-Wallis test with a .05 significance level.

Exercise 3 -- Appendix J Randomly select 12 observations from the database of companies with an A bond rating, 12 of companies with a B bond rating, and 12 of companies with a C bond rating. Can one conclude that there is a difference in the TOTAL (long-term assets) for the three groups? Use the Kruskal-Wallis test with a .05 significance level.

Exercise 4 -- Appendix J Choose 25 observations at random from the database. Is there evidence that a positive relationship exists between EMPLOYEE (the number of employees) and SALES (the amount of sales)? Use Spearman's rank correlation coefficient, and test at the .01 significance level.

Case Study

Identifying Markets Segments

Firms offering products or services need to know the nature and needs of their consumer market. The market may be composed of distinct groups that have marketing characteristics different from those of other groups. These different groups are called market segments. Market segmentation is the process of identifying and analyzing these different market groups. Marketing programs can then be designed to meet the needs of one or more particular segments.

There are various ways of identifying market segments. Common bases used for differentiation are geographic, demographic, and product usage. After analyses of these characteristics and of the expectations the consumer has for the product, firms are in a better position to modify their products. The intent is to match the demand of a market segment and to fashion an effective marketing program.

A market segmentation study was conducted in a large luxury hotel that is part of an international chain. The purpose was to investigate differences in the way the hotel's various market segments perceive the importance of various attributes of a hotel. The basis for the differentiation of market segments was the purpose of visit. One sample ($n = 73$) was drawn from a segment of the hotel's patrons whose purpose of travel was business related. A second sample ($n = 81$) was drawn from a segment of the hotel's patrons whose purpose of travel was pleasure. The questionnaire required the respondents to rate the importance of 12 attributes in their choice of a hotel on a five-point scale (5 = very important). The responses were averaged for each group. The following table summarizes the results.

Case Study Questions

1. Employing a nonparametric procedure avoids the necessity of assuming normally distributed populations. However, because of the nature of the data, even if the responses could be assumed to be normally distributed, the equivalent parametric procedure (paired difference t-test) still might not be appropriate. Can you explain why?

2. Rank the average responses for the attributes for each group. Measure the degree of agreement between the two groups in ranking the 12 attributes.

3. Explain why the Mann-Whitney test would not be appropriate to detect differences in the rankings made by the business patrons and those patrons traveling for pleasure.

4. At a .05 significance level, is there sufficient evidence to reject the null hypothesis that no significant difference exists between the two groups in their rankings of the attributes?

5. From your analysis in question 4, what are the attributes that the business patrons rate as more important than those traveling for pleasure? What attributes do both groups agree on? Do these observations seem reasonable to you?

PURPOSE OF TRAVEL AND ATTRIBUTE IMPORTANCE

Attribute	Business	Pleasure
1. Security in the hotel	4.38	4.80
2. Room furniture and decor	4.18	4.52
3. Overall furnishing decor	4.05	4.37
4. Telephone service	4.03	3.97
5. Bathroom decor/furnishing	3.96	4.41
6. Service in restaurants/bars/rooms	3.92	4.31
7. Reputation/image of hotel	3.73	4.29
8. Building (architecture/maintenance)	3.73	4.19
9. Business-center facilities	3.62	2.10
10. Class appeal of hotel	3.61	4.24
11. Low-price food and beverages	2.96	3.78
12. Variety of restaurants/bars	2.78	3.53

(*Source:* Adapted and modified from Subhash C. Mehta and Ariel Vera, "Segmentation in Singapore," *The Cornell H.R.A. Quarterly* (May 1990): 80. Used by permission. All rights reserved.)

SPSS

Solution
Example 19.3

Example 19.3 used the nonparametric runs test to determine whether a sample of savings accounts at the Northside National Bank was determined in a random manner. The runs test was used to see if the sample consisted of 45 randomly selected males and females. The SPSS program listing in Figure 19.13 was used to compute the number of runs and the Z value. In this problem the SPSS commands are the same for both the mainframe and PC versions (remember to end each command line with a period when using the PC version).

The TITLE command names the SPSS run.

The DATA LIST command gives each variable a name and describes the data as being in free form.

The BEGIN DATA command indicates to SPSS that the input data immediately follow.

FIGURE 19.13
Input for SPSS or SPSS/PC. Remove the periods for SPSS input.

```
TITLE    RUNS TEST FOR NORTHSIDE NATIONAL BANK.
DATA LIST FREE/SEX.
BEGIN DATA.
1
1
-1
-1
-1
-1
-1
1
-1
-1
1
1
1
1
1
1
-1
-1
-1
-1
1
1
-1
1
1
-1
-1
1
-1
-1
-1
-1
1
1
1
1
1
-1
-1
-1
-1
1
1
1
END DATA.
VALUE LABELS SEX -1 'FEMALE' 1 'MALE'/.
NPAR TESTS RUNS(1)=SEX/.
```

The next 45 lines contain the data values, with each line containing either a -1 to represent a female or a $+1$ for a male.

The END DATA statement indicates the end of the data.

The VALUE LABELS statement assigns labels to the values of the variable SEX. The value -1 is assigned the label FEMALE, while the value $+1$ is assigned the label MALE.

The NPAR TESTS command indicates that we wish to do a nonparametric test (in this case, it is the runs test). We are testing for runs that are greater than 0 and less than 0. In other words, runs of females (-1) or males $(+1)$.

Figure 19.14 shows the SPSS output obtained by executing the listing in Figure 19.13.

■ **FIGURE 19.14**
SPSS output.

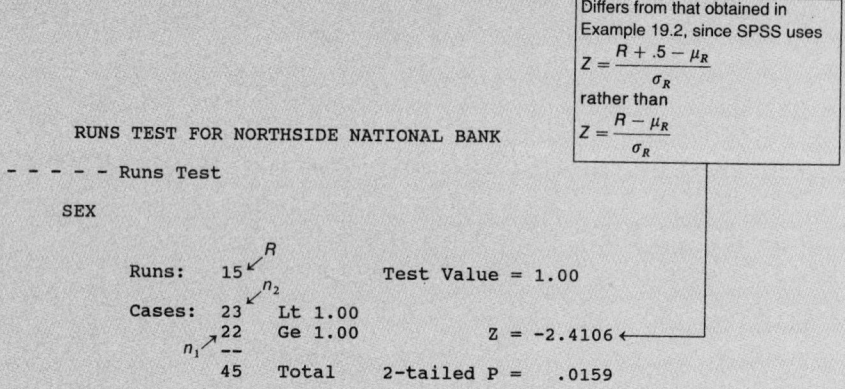

RUNS TEST FOR NORTHSIDE NATIONAL BANK

- - - - - Runs Test

        SEX

    Runs: 15   Test Value = 1.00

    Cases: 23   Lt 1.00
             22   Ge 1.00   Z = -2.4106
             --
             45   Total   2-tailed P =   .0159

Differs from that obtained in Example 19.2, since SPSS uses
$$Z = \frac{R + .5 - \mu_R}{\sigma_R}$$
rather than
$$Z = \frac{R - \mu_R}{\sigma_R}$$

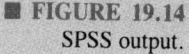

SPSS

✍ **Solution**
Example 19.5

Example 19.5 used the nonparametric Mann-Whitney U test to determine whether there was sufficient evidence to indicate that Food World's store A had a larger number of returned checks than did their store B. The SPSS program listing in Figure 19.15 was used to compute the Mann-Whitney test statistic. In this problem the SPSS commands are the same for the mainframe and PC versions (remember to end each command line with a period when using the PC version).

The TITLE command names the SPSS run.

The DATA LIST command gives each variable a name and describes the data as being in free form.

The BEGIN DATA command indicates to SPSS that the input data immediately follow.

The next 27 lines contain the data values, with each line representing the number of returned checks per month per store and the group to which the specific store belongs (0 = store A, 1 = store B).

The END DATA statement indicates the end of the data.

The NPAR TESTS command indicates that we wish to do a nonparametric Mann-Whitney test, comparing the number of returned checks (variable STORE) by the store category (variable GROUP).

Figure 19.16 shows the SPSS output obtained by executing the listing in Figure 19.15.

■ FIGURE 19.15
Input for SPSS or
SPSS/PC. Remove
the periods for
SPSS input.

```
TITLE    FOOD WORLD SUPERMARKETS.
DATA LIST FREE/STORE GROUP.
BEGIN DATA.
42 0
65 0
38 0
55 0
71 0
60 0
47 0
59 0
68 0
57 0
76 0
42 0
22 1
17 1
35 1
19 1
8  1
24 1
42 1
14 1
28 1
17 1
10 1
15 1
20 1
45 1
50 1
END DATA.
NPAR TESTS M-W = STORE BY GROUP (0,1).
```

■ FIGURE 19.16 SPSS output.

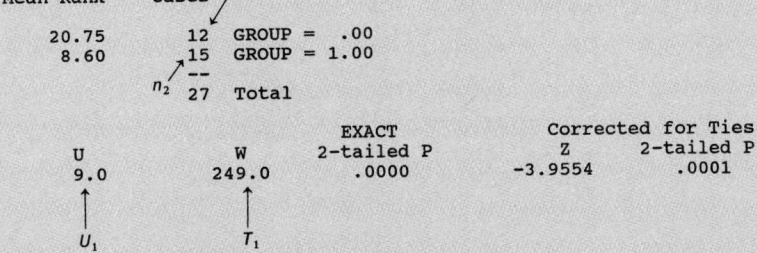

$$T_2 = \frac{n(n+1)}{2} - T_1$$

$$= \frac{(27)(28)}{2} - 249 = 129$$

and

$$U_2 = n_1 n_2 + \frac{n_2(n_2+1)}{2} - T_2$$

$$= (12)(15) + \frac{(15)(16)}{2} - 129 = 171$$

```
               FOOD WORLD SUPERMARKETS

- - - - - Mann-Whitney U - Wilcoxon Rank Sum W Test

     STORE
  by GROUP

                        n₁
    Mean Rank    Cases  ╱

       20.75        12   GROUP =  .00
        8.60        15   GROUP = 1.00
                 n₂  --
                    27   Total

                       EXACT        Corrected for Ties
       U          W    2-tailed P      Z     2-tailed P
      9.0       249.0    .0000     -3.9554    .0001

      ↑           ↑
      U₁          T₁
```

SPSS

⫻ Solution
Example 19.7

Example 19.7 used the nonparametric Wilcoxon signed ranks test to determine whether increasing from a 2% to a 10% concentration of hardwood would improve the tensile strength of the paper that the Glendale Container Corporation was manufacturing. The SPSS program listing in Figure 19.17 was used to compute the Wilcoxon Z statistic for the two concentrations. In this problem the SPSS commands are the same for the mainframe and PC versions (remember to end each command line with a period when using the PC version).

The TITLE command names the SPSS run.

The DATA LIST command gives each variable a name and describes the data as being in free form.

The BEGIN DATA command indicates to SPSS that the input data immediately follow.

The next 20 lines contain the data values, with each line representing the tensile strength at 2% and 10% concentrations of hardwood.

The END DATA statement indicates the end of the data.

The VARIABLE LABELS statement assigns a descriptive label to the variables. This descriptive label is substituted for the variable name when output is printed.

The NPAR TESTS command indicates that we wish to do the nonparametric Wilcoxon test as well as generate both the Z value and two-tailed p-value.

Figure 19.18 shows the SPSS output obtained by executing the listing in Figure 19.17.

■ FIGURE 19.17
Input for SPSS or
SPSS/PC. Remove
the periods for
SPSS input.

```
TITLE    GLENDALE CONTAINER CORPORATION.
DATA LIST FREE/TWOPC TENPC.
BEGIN DATA.
125 119
133 120
132 139
116 148
130 148
135 109
119 137
112 116
122 122
137 131
113 148
131 116
128 107
119 142
112 106
111 142
128 112
135 145
106 131
118 146
END DATA.
VARIABLE LABELS
          TWOPC    ' 2% HARDWOOD'
          TENPC    '10% HARDWOOD'.
NPAR TESTS  WILCOXON=ALL.
```

■ **FIGURE 19.18**
SPSS output.

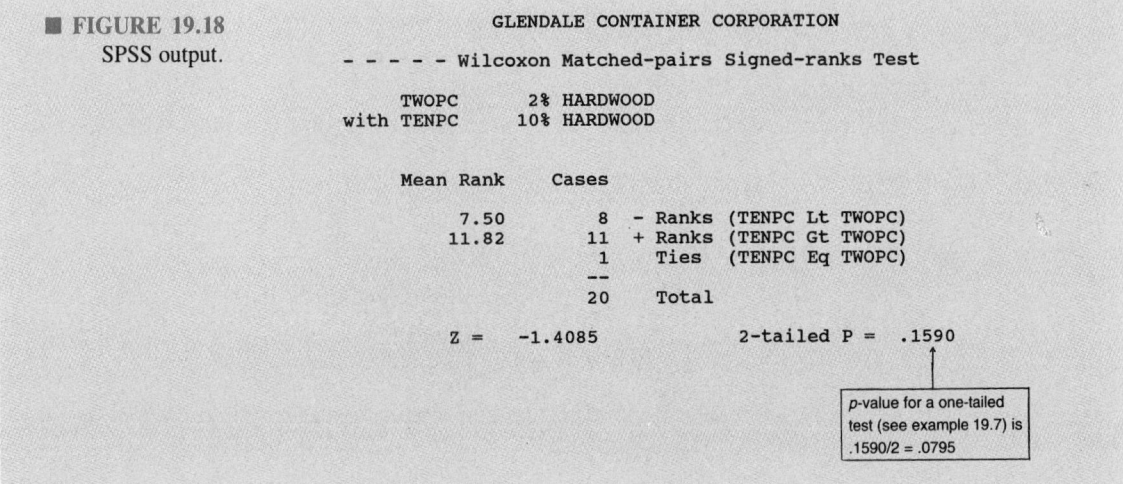

```
                        GLENDALE CONTAINER CORPORATION

            - - - - - Wilcoxon Matched-pairs Signed-ranks Test

        TWOPC        2% HARDWOOD
with TENPC       10% HARDWOOD


        Mean Rank     Cases

            7.50          8   - Ranks (TENPC Lt TWOPC)
           11.82         11   + Ranks (TENPC Gt TWOPC)
                          1     Ties  (TENPC Eq TWOPC)
                         --
                         20     Total

          Z =    -1.4085               2-tailed P =   .1590
```

> *p*-value for a one-tailed
> test (see example 19.7) is
> .1590/2 = .0795

SPSS

⊿ **Solution**
Example 19.8

Example 19.8 used the nonparametric Kruskal-Wallis one-way test to determine whether there was a difference in the amount of downtime among four different copying machines. Data were collected by observing the downtime for 20 randomly selected months. In this way, five randomly selected months were observed for each of the four brands of machines. The SPSS program listing in Figure 19.19 was used to compute the Kruskal-Wallis statistic and the *p*-value. In this problem the SPSS commands are the same for the mainframe and PC versions (remember to end each command line with a period when using the PC version).

■ **FIGURE 19.19**
Input for SPSS or
SPSS/PC. Remove
the periods for
SPSS input.

```
TITLE    DREXTON INDUSTRIES COPYING MACHINES.
DATA LIST FREE/DOWNTIME GROUP.
BEGIN DATA.
28 1
41 1
34 1
52 1
25 1
5  2
16 2
20 2
24 2
15 2
10 3
8  3
18 3
14 3
26 3
45 4
30 4
49 4
32 4
36 4
END DATA.
NPAR TESTS   K-W = DOWNTIME BY GROUP (1,4).
```

The TITLE command names the SPSS run.

The DATA LIST command gives each variable a name and describes the data as being in free form.

The BEGIN DATA command indicates to SPSS that the input data immediately follow.

The next 20 lines contain the data values, with each line representing the length of downtime (in hours) of a machine as well as the machine type (1, 2, 3, or 4).

The END DATA statement indicates the end of the data.

The NPAR TESTS command indicates that we wish to do a Kruskal-Wallis test, analyzing DOWNTIME by GROUP.

Figure 19.20 shows the SPSS output obtained by executing the listing in Figure 19.19.

■ **FIGURE 19.20**   SPSS output.

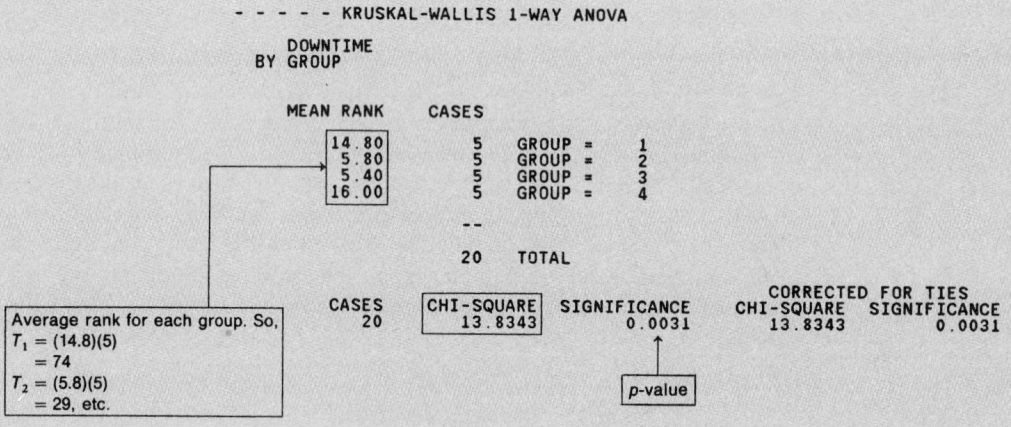

SPSS

☑ Solution
Example 19.9

Example 19.9 used the nonparametric FRIEDMAN test to determine whether there was a difference in the perceived quality of three automobile manufacturers' warranties. The SPSS program listing in Figure 19.21 was used to compute the Friedman test statistic. In this problem the SPSS commands are the same for the mainframe and PC versions (remember to end each command line with a period when using the PC version).

The TITLE command names the SPSS run.

The DATA LIST command gives each variable a name and describes the data as being in free form.

The BEGIN DATA command indicates to SPSS that the input data immediately follow.

The next ten lines contain the data values, with each line representing the rankings of one editor for the three automobile warranties. The first line indicates that editor 1 rated the Roadster first, the GA second, and the Henry third.

The END DATA statement indicates the end of the data.

The NPAR TESTS command indicates that we wish to do a nonparametric FRIEDMAN test, analyzing the rankings of the warranties of the three types of automobiles.

Figure 19.22 shows the SPSS output obtained by executing the listing in Figure 19.21.

■ FIGURE 19.21
Input for SPSS or
SPSS/PC. Remove
the periods for
SPSS input.

```
TITLE    WARRANTY ASSESSMENTS.
DATA LIST FREE/HENRY GA ROADSTER.
BEGIN DATA.
3  2  1
2  1  3
3  1  2
2  1  3
3  2  1
2  1  3
1  2  3
2  1  3
3  1  2
1  2  3
END DATA.
NPAR TESTS FRIEDMAN = HENRY GA ROADSTER.
```

■ FIGURE 19.22
SPSS output.

```
            WARRANTY ASSESSMENTS

- - - - - Friedman Two-way ANOVA


Mean Rank    Variable

    2.20     HENRY
    1.40     GA
    2.40     ROADSTER

Cases           Chi-Square        D.F.    Significance
  10             5.6000             2         .0608
```

SPSS

▨ Solution
Examples 19.10
and 19.12

Examples 19.10 and 19.12 used the nonparametric Spearman rank correlation coefficient to determine the relationship between the market value of homes and the personal indebtedness of the homeowners. The SPSS program listing in Figure 19.23 was used to compute the Spearman rank correlation coefficient r_s.

The TITLE command names the SPSS run.

The DATA LIST command gives each variable a name and describes the data as being in free form.

The BEGIN DATA command indicates to SPSS that the input data immediately follow.

The next ten lines contain the data values, with each line representing the total market value of a home and the total indebtedness of the homeowner.

The END DATA statement indicates the end of the data.

The VARIABLE LABEL statement assigns a descriptive label to the variables. This descriptive label is substituted for the variable name when the output is printed.

The NONPAR CORR statement requests that the rank correlation be computed for variables MKTVALUE and DEBT.

Figure 19.24 shows the SPSS output obtained by executing the listing in Figure 19.23.

■ **FIGURE 19.23**
Input for SPSS or
SPSS/PC. Remove
the periods for
SPSS input.

```
TITLE   HOME VALUE AND FAMILY INDEBTEDNESS ANALYSIS.
DATA LIST FREE/MKTVALUE DEBT.
BEGIN DATA.
85   12
147  27
340  45
94   10
120  17
105  4
135  20
162  25
480  35
88   14
END DATA.
VARIABLE LABELS
        MKTVALUE 'MARKET VALUE OF HOME'
        DEBT     'TOTAL INDEBTEDNESS'.
NONPAR CORR MKTVALUE WITH DEBT.
```

■ FIGURE 19.24   SPSS output.

```
- - - - - - - - - - - - - -  S P E A R M A N   C O R R E L A T I O N   C O E F F I C I E N T S  - - -


                DEBT

MKTVALUE        .8667  ←——— r_s
          N(   10)
          SIG .001  ←——— p-value for H_0: no association exists between X and Y

" . " IS PRINTED IF A COEFFICIENT CANNOT BE COMPUTED.
```

SAS

Solution
Example 19.5

Example 19.5 used the nonparametric Mann-Whitney U test to determine whether there was sufficient evidence to indicate that Food World's store A had a larger number of returned checks than did their store B. The SAS program listng in Figure 19.25 was used to compute the Mann-Whitney test statistic. In this problem the SAS commands are the same for the mainframe and PC versions.

The TITLE command names the SAS run.

The DATA command gives the data a name.

The INPUT command names and gives the correct order for the different fields on the data lines.

The CARDS command indicates to SAS that the input data immediately follow.

The next 27 lines contain the data values, with each line representing the number of returned checks per month per store and the group to which the store belongs (0 = store A, 1 = store B).

The Mann-Whitney test is sometimes called the Wilcoxon test. The following command set was used to obtain the statistics:

```
PROC NPAR1WAY WILCOXON
  VAR STORE
  CLASS GROUP
```

Using the above command set, we can compare the number of returned checks (variable STORE) by the store category (variable GROUP).

Figure 19.26 shows the SAS output obtained by executing the listing in Figure 19.25.

■ FIGURE 19.25
Input for SAS
(mainframe or
micro version).

```
TITLE  'FOOD WORLD SUPERMARKETS';
DATA STORES;
 INPUT STORE GROUP;
CARDS;
42 0
65 0
38 0
55 0
71 0
60 0
47 0
59 0
68 0
57 0
76 0
42 0
22 1
17 1
35 1
19 1
8 1
24 1
42 1
14 1
28 1
17 1
10 1
15 1
20 1
45 1
50 1
PROC NPAR1WAY WILCOXON;
 VAR STORE;
 CLASS GROUP;
```

■ FIGURE 19.26
SAS output.

This value is slightly different from the Z value in Example 19.5 since SAS adds .5 to the numerator of Z as well as adjusting this Z value for any ties obtained in the pooled sample.

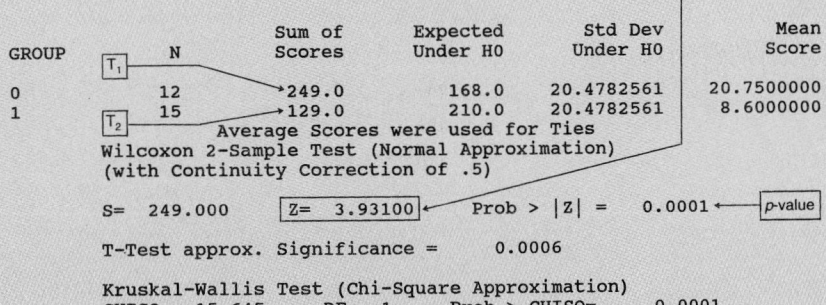

```
                     FOOD WORLD SUPERMARKETS


                 N P A R 1 W A Y   P R O C E D U R E

            Wilcoxon Scores (Rank Sums) for Variable STORE
                    Classified by Variable GROUP


                           Sum of      Expected      Std Dev        Mean
          GROUP     N       Scores     Under H0      Under H0       Score

            0      12       249.0        168.0      20.4782561    20.7500000
            1      15       129.0        210.0      20.4782561     8.6000000
                        Average Scores were used for Ties
            Wilcoxon 2-Sample Test (Normal Approximation)
            (with Continuity Correction of .5)

            S=  249.000      Z=  3.93100      Prob > |Z| =   0.0001      p-value

            T-Test approx. Significance =        0.0006

            Kruskal-Wallis Test (Chi-Square Approximation)
            CHISQ= 15.645        DF= 1      Prob > CHISQ=    0.0001
```

SAS

Solution
Example 19.8

Example 19.8 used the nonparametric Kruskal-Wallis one-way test to determine whether there was a difference in the amount of downtime among four different copying machines. Data were collected by observing the downtime for 20 randomly selected months. In this way, five randomly selected months were observed for each of the four brands of machines. The SAS program listing in Figure 19.27 was used to compute the Kruskal-Wallis statistic and the p-value. In this problem the SAS commands are the same for the mainframe and PC versions.

The TITLE command names the SAS run.

The DATA command gives the data a name.

The INPUT command names and gives the correct order for the different fields on the data lines.

The CARDS command indicates to SAS that the input data immediately follow.

The next 20 lines contain the data values, with each line representing the downtime, in hours, of a copier and the brand (1, 2, 3, or 4) of the copier.

The following command set was used to obtain the Kruskal-Wallis statistic:

PROC NPAR1WAY WILCOXON
   VAR DOWNTIME
   CLASS BRAND

Using the above command set, we are analyzing the difference in average downtime for each of the four brands of copiers.

Figure 19.28 shows the SAS output obtained by executing the listing in Figure 19.27.

■ **FIGURE 19.27**
Input for SAS (mainframe or micro version).

```
TITLE  'DREXTON INDUSTRIES COPYING MACHINES';
DATA MACHINES;
 INPUT DOWNTIME BRAND;
CARDS;
28 1
41 1
34 1
52 1
25 1
5  2
16 2
20 2
24 2
15 2
10 3
8  3
18 3
14 3
26 3
45 4
30 4
49 4
32 4
36 4
PROC NPAR1WAY WILCOXON;
 VAR DOWNTIME;
 CLASS BRAND;
```

■ **FIGURE 19.28**
SAS output.

```
                         DREXTON INDUSTRIES COPYING MACHINES


                              N P A R 1 W A Y   P R O C E D U R E

                       Wilcoxon Scores (Rank Sums) for Variable DOWNTIME
                                Classified by Variable BRAND


                                  Sum of      Expected        Std Dev          Mean
             BRAND       N         Scores      Under H0        Under H0        Score

               1         5         74.0      52.5000000      11.4564392    14.8000000
               2         5         29.0      52.5000000      11.4564392     5.8000000
               3         5         27.0      52.5000000      11.4564392     5.4000000
               4         5         80.0      52.5000000      11.4564392    16.0000000


              Kruskal-Wallis Test (Chi-Square Approximation)
              CHISQ=  13.834      DF=  3      Prob > CHISQ=      0.0031

                                                                        p-value
      Rank totals
```

SAS

■ **Solution**
Examples 19.10
and 19.12

Examples 19.10 and 19.12 used the nonparametric Spearman rank correlation coefficient to determine the relationship between the market value of homes and the personal indebtedness of the homeowners. The SAS program listing in Figure 19.29 was used to compute the Spearman rank correlation coefficient r_s. In this problem the SAS commands are the same for the mainframe and PC versions.

The TITLE command names the SAS run.

The DATA command gives the data a name.

The INPUT command names and gives the correct order for the different fields on the data lines.

The LABEL statements assign a descriptive label to the variables. This descriptive label is substituted for the variable name when output is printed.

The CARDS command indicates to SAS that the input data immediately follow.

The next ten lines contain the data values, with each line representing the market value of a home and the personal indebtedness of the homeowner.

■ **FIGURE 19.29**
Input for SAS
(mainframe or
micro version).

```
TITLE  'HOME VALUE AND FAMILY INDEBTEDNESS ANALYSIS';
DATA HOMES;
  INPUT MKTVALUE DEBT;
  LABEL MKTVALUE='MARKET VALUE OF HOMES'
        DEBT='TOTAL INDEBTEDNESS';
CARDS;
85   12
147  27
340  45
94   10
120  17
105  4
135  20
162  25
480  35
88   14
PROC CORR NOSIMPLE SPEARMAN;
  VAR MKTVALUE DEBT;
```

The following command set was used to calculate the Spearman rank correlation coefficient, r_s, between the variables MKTVALUE and DEBT:

PROC CORR NOSIMPLE SPEARMAN
    VAR MKTVALUE DEBT

Figure 19.30 shows the SAS output obtained by executing the listing in Figure 19.29.

■ FIGURE 19.30
SAS output.

```
                        HOME VALUE AND FAMILY INDEBTEDNESS ANALYSIS

                                 CORRELATION ANALYSIS

                          2 'VAR' Variables:  MKTVALUE DEBT


        Spearman Correlation Coefficients / Prob > |R| under Ho: Rho=0   N = 10

                                            MKTVALUE              DEBT        rs

                MKTVALUE                     1.00000            0.86667
                MARKET VALUE OF HOMES        0.0                0.0012

                DEBT                         0.86667            1.00000
                TOTAL INDEBTEDNESS           0.0012             0.0
```

p-value for H_0:
no association exists
between X and Y

Appendixes

TABLE A-1   Binomial Probabilities $[_nC_x p^x (1 - p)^{n-x}]$

n	x	0.01	0.05	0.10	0.20	0.30	0.40	0.50	0.60	0.70	0.80	0.90	0.95	0.99	x
2	0	980	902	810	640	490	360	250	160	090	040	010	002	0+	0
	1	020	095	180	320	420	480	500	480	420	320	180	095	020	1
	2	0+	002	010	040	090	160	250	360	490	640	810	902	980	2
3	0	970	857	729	512	343	216	125	064	027	008	001	0+	0+	0
	1	029	135	243	384	441	432	375	288	189	096	027	007	0+	1
	2	0+	007	027	096	189	288	375	432	441	384	243	135	029	2
	3	0+	0+	001	008	027	064	125	216	343	512	729	857	970	3
4	0	961	815	656	410	240	130	062	026	008	002	0+	0+	0+	0
	1	039	171	292	410	412	346	250	154	076	026	004	0+	0+	1
	2	001	014	049	154	265	346	375	346	265	154	049	014	001	2
	3	0+	0+	004	026	076	154	250	346	412	410	292	171	039	3
	4	0+	0+	0+	002	008	026	062	130	240	410	656	815	961	4
5	0	951	774	590	328	168	078	031	010	002	0+	0+	0+	0+	0
	1	048	204	328	410	360	259	156	077	028	006	0+	0+	0+	1
	2	001	021	073	205	309	346	312	230	132	051	008	001	0+	2
	3	0+	001	008	051	132	230	312	346	309	205	073	021	001	3
	4	0+	0+	0+	006	028	077	156	259	360	410	328	204	048	4
	5	0+	0+	0+	0+	002	010	031	078	168	328	590	774	951	5
6	0	941	735	531	262	118	047	016	004	001	0+	0+	0+	0+	0
	1	057	232	354	393	303	187	094	037	010	002	0+	0+	0+	1
	2	001	031	098	246	324	311	234	138	060	015	001	0+	0+	2
	3	0+	002	015	082	185	276	312	276	185	082	015	002	0+	3
	4	0+	0+	001	015	060	138	234	311	324	246	098	031	001	4
	5	0+	0+	0+	002	010	037	094	187	303	393	354	232	057	5
	6	0+	0+	0+	0+	001	004	016	047	118	262	531	735	941	6
7	0	932	698	478	210	082	028	008	002	0+	0+	0+	0+	0+	0
	1	066	257	372	367	247	131	055	017	004	0+	0+	0+	0+	1
	2	002	041	124	275	318	261	164	077	025	004	0+	0+	0+	2
	3	0+	004	023	115	227	290	273	194	097	029	003	0+	0+	3
	4	0+	0+	003	029	097	194	273	290	227	115	023	004	0+	4
	5	0+	0+	0+	004	025	077	164	261	318	275	124	041	002	5
	6	0+	0+	0+	0+	004	017	055	131	247	367	372	257	066	6
	7	0+	0+	0+	0+	0+	002	008	028	082	210	478	698	932	7
8	0	923	663	430	168	058	017	004	001	0+	0+	0+	0+	0+	0
	1	075	279	383	336	198	090	031	008	001	0+	0+	0+	0+	1
	2	003	051	149	294	296	209	109	041	010	001	0+	0+	0+	2
	3	0+	005	033	147	254	279	219	124	047	009	0+	0+	0+	3
	4	0+	0+	005	046	136	232	273	232	136	046	005	0+	0+	4
	5	0+	0+	0+	009	047	124	219	279	254	147	033	005	0+	5
	6	0+	0+	0+	001	010	041	109	209	296	294	149	051	003	6
	7	0+	0+	0+	0+	001	008	031	090	198	336	383	279	075	7
	8	0+	0+	0+	0+	0+	001	004	017	058	168	430	663	923	8
9	0	914	630	387	134	040	010	002	0+	0+	0+	0+	0+	0+	0
	1	083	299	387	302	156	060	018	004	0+	0+	0+	0+	0+	1
	2	003	063	172	302	267	161	070	021	004	0+	0+	0+	0+	2
	3	0+	008	045	176	267	251	164	074	021	003	0+	0+	0+	3
	4	0+	001	007	066	172	251	246	167	074	017	001	0+	0+	4
9	5	0+	0+	001	017	074	167	246	251	172	066	007	001	0+	5
	6	0+	0+	0+	003	021	074	164	251	267	176	045	008	0+	6
	7	0+	0+	0+	0+	004	021	070	161	267	302	172	063	003	7
	8	0+	0+	0+	0+	0+	004	018	060	156	302	387	299	083	8
	9	0+	0+	0+	0+	0+	0+	002	010	040	134	387	630	914	9
10	0	904	599	349	107	028	006	001	0+	0+	0+	0+	0+	0+	0
	1	091	315	387	268	121	040	010	002	0+	0+	0+	0+	0+	1
	2	004	075	194	302	233	121	044	011	001	0+	0+	0+	0+	2
	3	0+	010	057	201	267	215	117	042	009	001	0+	0+	0+	3
	4	0+	001	011	088	200	251	205	111	037	006	0+	0+	0+	4
	5	0+	0+	001	026	103	201	246	201	103	026	001	0+	0+	5
	6	0+	0+	0+	006	037	111	205	251	200	088	011	001	0+	6
	7	0+	0+	0+	001	009	042	117	215	267	201	057	010	0+	7
	8	0+	0+	0+	0+	001	011	044	121	233	302	194	075	004	8
	9	0+	0+	0+	0+	0+	002	010	040	121	268	387	315	091	9
	10	0+	0+	0+	0+	0+	0+	001	006	028	107	349	599	904	10

From Mosteller, Rourke, & Thomas, *Probability with Statistical Applications*, 2d ed. © 1970, Addison-Wesley, Reading, Mass. Table on pp. 475–477. Reprinted with permission.

■ TABLE A-1 (*continued*)

n	x	0.01	0.05	0.10	0.20	0.30	0.40	P 0.50	0.60	0.70	0.80	0.90	0.95	0.99	x
11	0	895	569	314	086	020	004	0+	0+	0+	0+	0+	0+	0+	0
	1	099	329	384	236	093	027	005	001	0+	0+	0+	0+	0+	1
	2	005	087	213	295	200	089	027	005	001	0+	0+	0+	0+	2
	3	0+	014	071	221	257	177	081	023	004	0+	0+	0+	0+	3
	4	0+	001	016	111	220	236	161	070	017	002	0+	0+	0+	4
	5	0+	0+	002	039	132	221	226	147	057	010	0+	0+	0+	5
	6	0+	0+	0+	010	057	147	226	221	132	039	002	0+	0+	6
	7	0+	0+	0+	002	017	070	161	236	220	111	016	001	0+	7
	8	0+	0+	0+	0+	004	023	081	177	257	221	071	014	0+	8
	9	0+	0+	0+	0+	001	005	027	089	200	295	213	087	005	9
	10	0+	0+	0+	0+	0+	001	005	027	093	236	384	329	099	10
	11	0+	0+	0+	0+	0+	0+	0+	004	020	086	314	569	895	11
12	0	886	540	282	069	014	002	0+	0+	0+	0+	0+	0+	0+	0
	1	107	341	377	206	071	017	003	0+	0+	0+	0+	0+	0+	1
	2	006	099	230	283	168	064	016	002	0+	0+	0+	0+	0+	2
	3	0+	017	085	236	240	142	054	012	001	0+	0+	0+	0+	3
	4	0+	002	021	133	231	213	121	042	008	001	0+	0+	0+	4
	5	0+	0+	004	053	158	227	193	101	029	003	0+	0+	0+	5
	6	0+	0+	0+	016	079	177	226	177	079	016	0+	0+	0+	6
	7	0+	0+	0+	003	029	101	193	227	158	053	004	0+	0+	7
	8	0+	0+	0+	001	008	042	121	213	231	133	021	002	0+	8
	9	0+	0+	0+	0+	001	012	054	142	240	236	085	017	0+	9
	10	0+	0+	0+	0+	0+	002	016	064	168	283	230	099	006	10
	11	0+	0+	0+	0+	0+	0+	003	017	071	206	377	341	107	11
	12	0+	0+	0+	0+	0+	0+	0+	002	014	069	282	540	886	12
13	0	878	513	254	055	010	001	0+	0+	0+	0+	0+	0+	0+	0
	1	115	351	367	179	054	011	002	0+	0+	0+	0+	0+	0+	1
	2	007	111	245	268	139	045	010	001	0+	0+	0+	0+	0+	2
	3	0+	021	100	246	218	111	035	006	001	0+	0+	0+	0+	3
	4	0+	003	028	154	234	184	087	024	003	0+	0+	0+	0+	4
	5	0+	0+	006	069	180	221	157	066	014	001	0+	0+	0+	5
	6	0+	0+	001	023	103	197	209	131	044	006	0+	0+	0+	6
	7	0+	0+	0+	006	044	131	209	197	103	023	001	0+	0+	7
	8	0+	0+	0+	001	014	066	157	221	180	069	006	0+	0+	8
	9	0+	0+	0+	0+	003	024	087	184	234	154	028	003	0+	9
	10	0+	0+	0+	0+	001	006	035	111	218	246	100	021	0+	10
	11	0+	0+	0+	0+	0+	001	010	045	139	268	245	111	007	11
	12	0+	0+	0+	0+	0+	0+	002	011	054	179	367	351	115	12
	13	0+	0+	0+	0+	0+	0+	0+	001	010	055	254	513	878	13
14	0	869	488	229	044	007	001	0+	0+	0+	0+	0+	0+	0+	0
	1	123	359	356	154	041	007	001	0+	0+	0+	0+	0+	0+	1
	2	008	123	257	250	113	032	006	001	0+	0+	0+	0+	0+	2
	3	0+	026	114	250	194	085	022	003	0+	0+	0+	0+	0+	3
	4	0+	004	035	172	229	155	061	014	001	0+	0+	0+	0+	4
	5	0+	0+	008	086	196	207	122	041	007	0+	0+	0+	0+	5
	6	0+	0+	001	032	126	207	183	092	023	002	0+	0+	0+	6
	7	0+	0+	0+	009	062	157	209	157	062	009	0+	0+	0+	7
	8	0+	0+	0+	002	023	092	183	207	126	032	001	0+	0+	8
	9	0+	0+	0+	0+	007	041	122	207	196	086	008	0+	0+	9
	10	0+	0+	0+	0+	001	014	061	155	229	172	035	004	0+	10
	11	0+	0+	0+	0+	0+	003	022	085	194	250	114	026	0+	11
	12	0+	0+	0+	0+	0+	001	006	032	113	250	257	123	008	12
	13	0+	0+	0+	0+	0+	0+	001	007	041	154	356	359	123	13
	14	0+	0+	0+	0+	0+	0+	0+	001	007	044	229	488	869	14
15	0	860	463	206	035	005	0+	0+	0+	0+	0+	0+	0+	0+	0
	1	130	366	343	132	031	005	0+	0+	0+	0+	0+	0+	0+	1
	2	009	135	267	231	092	022	003	0+	0+	0+	0+	0+	0+	2
	3	0+	031	129	250	170	063	014	002	0+	0+	0+	0+	0+	3
	4	0+	005	043	188	219	127	042	007	001	0+	0+	0+	0+	4
	5	0+	001	010	103	206	186	092	024	003	0+	0+	0+	0+	5
	6	0+	0+	002	043	147	207	153	061	012	001	0+	0+	0+	6
	7	0+	0+	0+	014	081	177	196	118	035	003	0+	0+	0+	7
	8	0+	0+	0+	003	035	118	196	177	081	014	0+	0+	0+	8
	9	0+	0+	0+	001	012	061	153	207	147	043	002	0+	0+	9
	10	0+	0+	0+	0+	003	024	092	186	206	103	010	001	0+	10
	11	0+	0+	0+	0+	001	007	042	127	219	188	043	005	0+	11
	12	0+	0+	0+	0+	0+	002	014	063	170	250	129	031	0+	12
	13	0+	0+	0+	0+	0+	0+	003	022	092	231	267	135	009	13

■ TABLE A-1 (*continued*)

n	x	0.01	0.05	0.10	0.20	0.30	0.40	P 0.50	0.60	0.70	0.80	0.90	0.95	0.99	x
	14	0+	0+	0+	0+	0+	0+	001	005	031	132	343	366	130	14
	15	0+	0+	0+	0+	0+	0+	0+	0+	005	035	206	463	860	15
16	0	852	440	185	028	003	0+	0+	0+	0+	0+	0+	0+	0+	0
	1	138	371	329	113	023	003	0+	0+	0+	0+	0+	0+	0+	1
	2	010	146	274	211	073	015	002	0+	0+	0+	0+	0+	0+	2
	3	0+	036	142	246	146	047	008	001	0+	0+	0+	0+	0+	3
	4	0+	006	051	200	204	101	028	004	0+	0+	0+	0+	0+	4
	5	0+	001	014	120	210	162	067	014	001	0+	0+	0+	0+	5
	6	0+	0+	003	055	165	198	122	039	006	0+	0+	0+	0+	6
	7	0+	0+	0+	020	101	189	175	084	018	001	0+	0+	0+	7
	8	0+	0+	0+	006	049	142	196	142	049	006	0+	0+	0+	8
	9	0+	0+	0+	001	018	084	175	189	101	020	0+	0+	0+	9
	10	0+	0+	0+	0+	006	039	122	198	165	055	003	0+	0+	10
	11	0+	0+	0+	0+	001	014	067	162	210	120	014	001	0+	11
	12	0+	0+	0+	0+	0+	004	028	101	204	200	051	006	0+	12
	13	0+	0+	0+	0+	0+	001	008	047	146	246	142	036	0+	13
	14	0+	0+	0+	0+	0+	0+	002	015	073	211	274	146	010	14
	15	0+	0+	0+	0+	0+	0+	0+	003	023	113	329	371	138	15
	16	0+	0+	0+	0+	0+	0+	0+	0+	003	028	185	440	852	16
17	0	843	418	167	022	002	0+	0+	0+	0+	0+	0+	0+	0+	0
	1	145	374	315	096	017	002	0+	0+	0+	0+	0+	0+	0+	1
	2	012	158	280	191	058	010	001	0+	0+	0+	0+	0+	0+	2
	3	001	042	156	239	124	034	005	0+	0+	0+	0+	0+	0+	3
	4	0+	008	060	209	187	080	018	002	0+	0+	0+	0+	0+	4
	5	0+	001	018	136	208	138	047	008	001	0+	0+	0+	0+	5
	6	0+	0+	004	068	178	184	094	024	003	0+	0+	0+	0+	6
	7	0+	0+	001	027	120	193	148	057	010	0+	0+	0+	0+	7
	8	0+	0+	0+	008	064	161	186	107	028	002	0+	0+	0+	8
	9	0+	0+	0+	002	028	107	186	161	064	008	0+	0+	0+	9
	10	0+	0+	0+	0+	010	057	148	193	120	027	001	0+	0+	10
	11	0+	0+	0+	0+	003	024	094	184	178	068	004	0+	0+	11
	12	0+	0+	0+	0+	001	008	047	138	208	136	018	001	0+	12
	13	0+	0+	0+	0+	0+	002	018	080	187	209	060	008	0+	13
	14	0+	0+	0+	0+	0+	0+	005	034	124	239	156	042	001	14
	15	0+	0+	0+	0+	0+	0+	001	010	058	191	280	158	012	15
	16	0+	0+	0+	0+	0+	0+	0+	002	017	096	315	374	145	16
	17	0+	0+	0+	0+	0+	0+	0+	0+	002	022	167	418	843	17
18	0	834	397	150	018	002	0+	0+	0+	0+	0+	0+	0+	0+	0
	1	152	376	300	081	013	001	0+	0+	0+	0+	0+	0+	0+	1
	2	013	168	284	172	046	007	001	0+	0+	0+	0+	0+	0+	2
	3	001	047	168	230	105	025	003	0+	0+	0+	0+	0+	0+	3
	4	0+	009	070	215	168	061	012	001	0+	0+	0+	0+	0+	4
	5	0+	001	022	151	202	115	033	004	0+	0+	0+	0+	0+	5
	6	0+	0+	005	082	187	166	071	014	001	0+	0+	0+	0+	6
	7	0+	0+	001	035	138	189	121	037	005	0+	0+	0+	0+	7
	8	0+	0+	0+	012	081	173	167	077	015	001	0+	0+	0+	8
	9	0+	0+	0+	003	039	128	186	128	039	003	0+	0+	0+	9
	10	0+	0+	0+	001	015	077	167	173	081	012	0+	0+	0+	10
	11	0+	0+	0+	0+	005	037	121	189	138	035	001	0+	0+	11
	12	0+	0+	0+	0+	001	014	071	166	187	082	005	0+	0+	1
	13	0+	0+	0+	0+	0+	004	033	115	202	151	022	001	0+	13
	14	0+	0+	0+	0+	0+	001	012	061	168	215	070	009	0+	14
	15	0+	0+	0+	0+	0+	0+	003	025	105	230	168	047	001	15
	16	0+	0+	0+	0+	0+	0+	001	007	046	172	284	168	013	16
	17	0+	0+	0+	0+	0+	0+	0+	001	013	081	300	376	152	17
	18	0+	0+	0+	0+	0+	0+	0+	0+	002	018	150	397	834	18
19	0	826	377	135	014	001	0+	0+	0+	0+	0+	0+	0+	0+	0
	1	159	377	285	068	009	001	0+	0+	0+	0+	0+	0+	0+	1
	2	014	179	285	154	036	005	0+	0+	0+	0+	0+	0+	0+	2
	3	001	053	180	218	087	018	002	0+	0+	0+	0+	0+	0+	3
	4	0+	011	080	218	149	047	007	0+	0+	0+	0+	0+	0+	4
	5	0+	002	027	164	192	093	022	002	0+	0+	0+	0+	0+	5
	6	0+	0+	007	096	192	145	052	008	0+	0+	0+	0+	0+	6
	7	0+	0+	001	044	152	180	096	024	002	0+	0+	0+	0+	7
	8	0+	0+	0+	017	098	180	144	053	008	0+	0+	0+	0+	8
	9	0+	0+	0+	005	051	146	176	098	022	001	0+	0+	0+	9
	10	0+	0+	0+	001	022	098	176	146	051	005	0+	0+	0+	10
	11	0+	0+	0+	0+	008	053	144	180	098	017	0+	0+	0+	11

■ **TABLE A-1** (*continued*)

n	x	0.01	0.05	0.10	0.20	0.30	0.40	P 0.50	0.60	0.70	0.80	0.90	0.95	0.99	x
	12	0+	0+	0+	0+	002	024	096	180	152	044	001	0+	0+	12
	13	0+	0+	0+	0+	0+	008	052	145	192	096	007	0+	0+	13
	14	0+	0+	0+	0+	0+	002	022	093	192	164	027	002	0+	14
	15	0+	0+	0+	0+	0+	0+	007	047	149	218	080	011	0+	15
	16	0+	0+	0+	0+	0+	0+	002	018	087	218	180	053	001	16
	17	0+	0+	0+	0+	0+	0+	0+	005	036	154	285	179	014	17
	18	0+	0+	0+	0+	0+	0+	0+	001	009	068	285	377	159	18
	19	0+	0+	0+	0+	0+	0+	0+	0+	001	014	135	377	826	19
20	0	818	358	122	012	001	0+	0+	0+	0+	0+	0+	0+	0+	0
	1	165	377	270	058	007	0+	0+	0+	0+	0+	0+	0+	0+	1
	2	016	189	285	137	028	003	0+	0+	0+	0+	0+	0+	0+	2
	3	001	060	190	205	072	012	001	0+	0+	0+	0+	0+	0+	3
	4	0+	013	090	218	130	035	005	0+	0+	0+	0+	0+	0+	4
	5	0+	002	032	175	179	075	015	001	·0+	0+	0+	0+	0+	5
	6	0+	0+	009	109	192	124	037	005	0+	0+	0+	0+	0+	6
	7	0+	0+	002	054	164	166	074	015	001	0+	0+	0+	0+	7
	8	0+	0+	0+	022	114	180	120	036	004	0+	0+	0+	0+	8
	9	0+	0+	0+	007	065	160	160	071	012	0+	0+	0+	0+	9
	10	0+	0+	0+	002	031	117	176	117	031	002	0+	0+	0+	10
	11	0+	0+	0+	0+	012	071	160	160	065	007	0+	0+	0+	11
	12	0+	0+	0+	0+	004	036	120	180	114	022	0+	0+	0+	12
	13	0+	0+	0+	0+	001	015	074	166	164	054	002	0+	0+	13
	14	0+	0+	0+	0+	0+	005	037	124	192	109	009	0+	0+	14
	15	0+	0+	0+	0+	0+	001	015	075	179	175	032	002	0+	15
	16	0+	0+	0+	0+	0+	0+	005	035	130	218	090	013	0+	16
	17	0+	0+	0+	0+	0+	0+	001	012	072	205	190	060	001	17
	18	0+	0+	0+	0+	0+	0+	0+	003	028	137	285	189	016	18
	19	0+	0+	0+	0+	0+	0+	0+	0+	007	058	270	377	165	19
	20	0+	0+	0+	0+	0+	0+	0+	0+	001	012	122	358	818	20

■ **TABLE A-2**   Values of e^{-a}

a	e^{-a}	a	e^{-a}	a	e^{-a}	a	e^{-a}
0.00	1.000000	2.60	.074274	5.10	.006097	7.60	.000501
0.10	.904837	2.70	.067206	5.20	.005517	7.70	.000453
0.20	.818731	2.80	.060810	5.30	.004992	7.80	.000410
0.30	.740818	2.90	.055023	5.40	.004517	7.90	.000371
0.40	.670320	3.00	.049787	5.50	.004087	8.00	.000336
0.50	.606531	3.10	.045049	5.60	.003698	8.10	.000304
0.60	.548812	3.20	.040762	5.70	.003346	8.20	.000275
0.70	.496585	3.30	.036883	5.80	.003028	8.30	.000249
0.80	.449329	3.40	0.33373	5.90	.002739	8.40	.000225
0.90	.406570	3.50	.030197	6.00	.002479	8.50	.000204
1.00	.367879	3.60	.027324	6.10	.002243	8.60	.000184
1.10	.332871	3.70	.024724	6.20	.002029	8.70	.000167
1.20	.301194	3.80	.022371	6.30	.001836	8.80	.000151
1.30	.272532	3.90	.020242	6.40	.001661	8.90	.000136
1.40	.246597	4.00	.018316	6.50	.001503	9.00	.000123
1.50	.223130	4.10	.016573	6.60	.001360	9.10	.000112
1.60	.201897	4.20	.014996	6.70	.001231	9.20	.000101
1.70	.182684	4.30	.013569	6.80	.001114	9.30	.000091
1.80	.165299	4.40	.012277	6.90	.001008	9.40	.000083
1.90	.149569	4.50	.011109	7.00	.000912	9.50	.000075
2.00	.135335	4.60	.010052	7.10	.000825	9.60	.000068
2.10	.122456	4.70	.009095	7.20	.000747	9.70	.000061
2.20	.110803	4.80	.008230	7.30	.000676	9.80	.000056
2.30	.100259	4.90	.007447	7.40	.000611	9.90	.000050
2.40	.090718	5.00	.006738	7.50	.000553	10.00	.000045
2.50	.082085						

■ TABLE A-3   Poisson Probabilities $\left[\dfrac{e^{-\mu}\mu^{x}}{x!}\right]$

					μ					
x	0.005	0.01	0.02	0.03	0.04	0.05	0.06	0.07	0.08	0.09
0	0.9950	0.9900	0.9802	0.9704	0.9608	0.9512	0.9418	0.9324	0.9231	0.9139
1	0.0050	0.0099	0.0192	0.0291	0.0384	0.0476	0.0565	0.0653	0.0738	0.0823
2	0.0000	0.0000	0.0002	0.0004	0.0008	0.0012	0.0017	0.0023	0.0030	0.0037
3	0.0000	0.0000	0.0000	0.0000	0.0000	0.0000	0.0000	0.0001	0.0001	0.0001

x	0.1	0.2	0.3	0.4	0.5	0.6	0.7	0.8	0.9	1.0
0	0.9048	0.8187	0.7408	0.6703	0.6065	0.5488	0.4966	0.4493	0.4066	0.3679
1	0.0905	0.1637	0.2222	0.2681	0.3033	0.3293	0.3476	0.3595	0.3659	0.3679
2	0.0045	0.0164	0.0333	0.0536	0.0758	0.0988	0.1217	0.1438	0.1647	0.1839
3	0.0002	0.0011	0.0033	0.0072	0.0126	0.0198	0.0284	0.0383	0.0494	0.0613
4	0.0000	0.0001	0.0002	0.0007	0.0016	0.0030	0.0050	0.0077	0.0111	0.0153
5	0.0000	0.0000	0.0000	0.0001	0.0002	0.0004	0.0007	0.0012	0.0020	0.0031
6	0.0000	0.0000	0.0000	0.0000	0.0000	0.0000	0.0001	0.0002	0.0003	0.0005
7	0.0000	0.0000	0.0000	0.0000	0.0000	0.0000	0.0000	0.0000	0.0000	0.0001

x	1.1	1.2	1.3	1.4	1.5	1.6	1.7	1.8	1.9	2.0
0	0.3329	0.3012	0.2725	0.2466	0.2231	0.2019	0.1827	0.1653	0.1496	0.1353
1	0.3662	0.3614	0.3543	0.3452	0.3347	0.3230	0.3106	0.2975	0.2842	0.2707
2	0.2014	0.2169	0.2303	0.2417	0.2510	0.2584	0.2640	0.2678	0.2700	0.2707
3	0.0738	0.0867	0.0998	0.1128	0.1255	0.1378	0.1496	0.1607	0.1710	0.1804
4	0.0203	0.0260	0.0324	0.0395	0.0471	0.0551	0.0636	0.0723	0.0812	0.0902
5	0.0045	0.0062	0.0084	0.0111	0.0141	0.0176	0.0216	0.0260	0.0309	0.0361
6	0.0008	0.0012	0.0018	0.0026	0.0035	0.0047	0.0061	0.0078	0.0098	0.0120
7	0.0001	0.0002	0.0003	0.0005	0.0008	0.0011	0.0015	0.0020	0.0027	0.0034
8	0.0000	0.0000	0.0001	0.0001	0.0001	0.0002	0.0003	0.0005	0.0006	0.0009
9	0.0000	0.0000	0.0000	0.0000	0.0000	0.0000	0.0001	0.0001	0.0001	0.0002

x	2.1	2.2	2.3	2.4	2.5	2.6	2.7	2.8	2.9	3.0
0	0.1225	0.1108	0.1003	0.0907	0.0821	0.0743	0.0672	0.0608	0.0050	0.0498
1	0.2572	0.2438	0.2306	0.2177	0.2052	0.1931	0.1815	0.1703	0.1596	0.1494
2	0.2700	0.2681	0.2652	0.2613	0.2565	0.2510	0.2450	0.2384	0.2314	0.2240
3	0.1890	0.1966	0.2033	0.2090	0.2138	0.2176	0.2205	0.2225	0.2237	0.2240
4	0.0992	0.1082	0.1169	0.1254	0.1336	0.1414	0.1488	0.1557	0.1622	0.1680
5	0.0417	0.0476	0.0538	0.0602	0.0668	0.0735	0.0804	0.0872	0.0940	0.1008
6	0.0146	0.0174	0.0206	0.0241	0.0278	0.0319	0.0362	0.0407	0.0455	0.0504
7	0.0044	0.0055	0.0068	0.0083	0.0099	0.0118	0.0139	0.0163	0.0188	0.0216
8	0.0011	0.0015	0.0019	0.0025	0.0031	0.0038	0.0047	0.0057	0.0068	0.0081
9	0.0003	0.0004	0.0005	0.0007	0.0009	0.0011	0.0014	0.0018	0.0022	0.0027
10	0.0001	0.0001	0.0001	0.0002	0.0002	0.0003	0.0004	0.0005	0.0006	0.0008
11	0.0000	0.0000	0.0000	0.0000	0.0000	0.0001	0.0001	0.0001	0.0002	0.0002
12	0.0000	0.0000	0.0000	0.0000	0.0000	0.0000	0.0000	0.0000	0.0000	0.0001

x	3.1	3.2	3.3	3.4	3.5	3.6	3.7	3.8	3.9	4.0
0	0.0450	0.0408	0.0369	0.0334	0.0302	0.0273	0.0247	0.0224	0.0202	0.0183
1	0.1397	0.1304	0.1217	0.1135	0.1057	0.0984	0.0915	0.0850	0.0789	0.0733
2	0.2165	0.2087	0.2008	0.1929	0.1850	0.1771	0.1692	0.1615	0.1539	0.1465
3	0.2237	0.2226	0.2209	0.2186	0.2158	0.2125	0.2087	0.2046	0.2001	0.1954
4	0.1734	0.1781	0.1823	0.1858	0.1888	0.1912	0.1931	0.1944	0.1951	0.1954
5	0.1075	0.1140	0.1203	0.1264	0.1322	0.1377	0.1429	0.1477	0.1522	0.1563
6	0.0555	0.0608	0.0662	0.0716	0.0771	0.0826	0.0881	0.0936	0.0989	0.1042
7	0.0246	0.0278	0.0312	0.0348	0.0385	0.0425	0.0466	0.0508	0.0551	0.0595
8	0.0095	0.0111	0.0129	0.0148	0.0169	0.0191	0.0215	0.0241	0.0269	0.0298
9	0.0033	0.0040	0.0047	0.0056	0.0066	0.0076	0.0089	0.0102	0.0116	0.0132
10	0.0010	0.0013	0.0016	0.0019	0.0023	0.0028	0.0033	0.0039	0.0045	0.0053
11	0.0003	0.0004	0.0005	0.0006	0.0007	0.0009	0.0011	0.0013	0.0016	0.0019
12	0.0001	0.0001	0.0001	0.0002	0.0002	0.0003	0.0003	0.0004	0.0005	0.0006
13	0.0000	0.0000	0.0000	0.0000	0.0001	0.0001	0.0001	0.0001	0.0002	0.0002
14	0.0000	0.0000	0.0000	0.0000	0.0000	0.0000	0.0000	0.0000	0.0000	0.0001

x	4.1	4.2	4.3	4.4	4.5	4.6	4.7	4.8	4.9	5.0
0	0.0166	0.0150	0.0136	0.0123	0.0111	0.0101	0.0091	0.0082	0.0074	0.0067
1	0.0679	0.0630	0.0583	0.0540	0.0500	0.0462	0.0427	0.0395	0.0365	0.0337
2	0.1393	0.1323	0.1254	0.1188	0.1125	0.1063	0.1005	0.0948	0.0894	0.0842
3	0.1904	0.1852	0.1798	0.1743	0.1687	0.1631	0.1574	0.1517	0.1460	0.1404
4	0.1951	0.1944	0.1933	0.1917	0.1898	0.1875	0.1849	0.1820	0.1789	0.1755

From Robert Parsons, *Statistical Analysis: A Decision Making Approach*, 2d ed. (New York: Harper & Row, 1978). Reprinted with permission of Harper & Row.

■ TABLE A-3 (*continued*)

					μ					
x	4.1	4.2	4.3	4.4	4.5	4.6	4.7	4.8	4.9	5.0
5	0.1600	0.1633	0.1662	0.1687	0.1708	0.1725	0.1738	0.1747	0.1753	0.1755
6	0.1093	0.1143	0.1191	0.1237	0.1281	0.1323	0.1362	0.1398	0.1432	0.1462
7	0.0640	0.0686	0.0732	0.0778	0.0824	0.0869	0.0914	0.0959	0.1002	0.1044
8	0.0328	0.0360	0.0393	0.0428	0.0463	0.0500	0.0537	0.0575	0.0614	0.0653
9	0.0150	0.0168	0.0188	0.0209	0.0232	0.0255	0.0280	0.0307	0.0334	0.0363
10	0.0061	0.0071	0.0081	0.0092	0.0104	0.0118	0.0132	0.0147	0.0164	0.0181
11	0.0023	0.0027	0.0032	0.0037	0.0043	0.0049	0.0056	0.0064	0.0073	0.0082
12	0.0008	0.0009	0.0011	0.0014	0.0016	0.0019	0.0022	0.0026	0.0030	0.0034
13	0.0002	0.0003	0.0004	0.0005	0.0006	0.0007	0.0008	0.0009	0.0011	0.0013
14	0.0001	0.0001	0.0001	0.0001	0.0002	0.0002	0.0003	0.0003	0.0004	0.0005
15	0.0000	0.0000	0.0000	0.0000	0.0001	0.0001	0.0001	0.0001	0.0001	0.0002

x	5.1	5.2	5.3	5.4	5.5	5.6	5.7	5.8	5.9	6.0
0	0.0061	0.0055	0.0050	0.0045	0.0041	0.0037	0.0033	0.0030	0.0027	0.0025
1	0.0311	0.0287	0.0265	0.0244	0.0225	0.0207	0.0191	0.0176	0.0162	0.0149
2	0.0793	0.0746	0.0701	0.0659	0.0618	0.0580	0.0544	0.0509	0.0477	0.0446
3	0.1348	0.1293	0.1239	0.1185	0.1133	0.1082	0.1033	0.0985	0.0938	0.0892
4	0.1719	0.1681	0.1641	0.1600	0.1558	0.1515	0.1472	0.1428	0.1383	0.1339
5	0.1753	0.1748	0.1740	0.1728	0.1714	0.1697	0.1678	0.1656	0.1632	0.1606
6	0.1490	0.1515	0.1537	0.1555	0.1571	0.1584	0.1594	0.1601	0.1605	0.1606
7	0.1086	0.1125	0.1163	0.1200	0.1234	0.1267	0.1298	0.1326	0.1353	0.1377
8	0.0692	0.0731	0.0771	0.0810	0.0849	0.0887	0.0925	0.0962	0.0998	0.1033
9	0.0392	0.0423	0.0454	0.0486	0.0519	0.0552	0.0586	0.0620	0.0654	0.0688
10	0.0200	0.0220	0.0241	0.0262	0.0285	0.0309	0.0334	0.0359	0.0386	0.0413
11	0.0093	0.0104	0.0116	0.0129	0.0143	0.0157	0.0173	0.0190	0.0207	0.0225
12	0.0039	0.0045	0.0051	0.0058	0.0065	0.0073	0.0082	0.0092	0.0102	0.0113
13	0.0015	0.0018	0.0021	0.0024	0.0028	0.0032	0.0036	0.0041	0.0046	0.0052
14	0.0006	0.0007	0.0008	0.0009	0.0011	0.0013	0.0015	0.0017	0.0019	0.0022
15	0.0002	0.0002	0.0003	0.0003	0.0004	0.0005	0.0006	0.0007	0.0008	0.0009
16	0.0001	0.0001	0.0001	0.0001	0.0001	0.0002	0.0002	0.0002	0.0003	0.0003
17	0.0000	0.0000	0.0000	0.0000	0.0000	0.0001	0.0001	0.0001	0.0001	0.0001

x	6.1	6.2	6.3	6.4	6.5	6.6	6.7	6.8	6.9	7.0
0	0.0022	0.0020	0.0018	0.0017	0.0015	0.0014	0.0012	0.0011	0.0010	0.0009
1	0.0137	0.0126	0.0116	0.0106	0.0098	0.0090	0.0082	0.0076	0.0070	0.0064
2	0.0417	0.0390	0.0364	0.0340	0.0318	0.0296	0.0276	0.0258	0.0240	0.0223
3	0.0848	0.0806	0.0765	0.0726	0.0688	0.0652	0.0617	0.0584	0.0552	0.0521
4	0.1294	0.1269	0.1205	0.1162	0.1118	0.1076	0.1034	0.0992	0.0952	0.0912
5	0.1579	0.1549	0.1519	0.1487	0.1454	0.1420	0.1385	0.1349	0.1314	0.1277
6	0.1605	0.1601	0.1595	0.1586	0.1575	0.1562	0.1546	0.1529	0.1511	0.1490
7	0.1399	0.1418	0.1435	0.1450	0.1462	0.1472	0.1480	0.1486	0.1489	0.1490
8	0.1066	0.1099	0.1130	0.1160	0.1188	0.1215	0.1240	0.1263	0.1284	0.1304
9	0.0723	0.0757	0.0791	0.0825	0.0858	0.0891	0.0923	0.0954	0.0985	0.1014
10	0.0441	0.0469	0.0498	0.0528	0.0558	0.0588	0.0618	0.0649	0.0679	0.0710
11	0.0245	0.0265	0.0285	0.0307	0.0330	0.0353	0.0377	0.0401	0.0426	0.0452
12	0.0124	0.0137	0.0150	0.0164	0.0179	0.0194	0.0210	0.0227	0.0245	0.0264
13	0.0058	0.0065	0.0073	0.0081	0.0089	0.0098	0.0108	0.0119	0.0130	0.0142
14	0.0025	0.0029	0.0033	0.0037	0.0041	0.0046	0.0052	0.0058	0.0064	0.0071
15	0.0010	0.0012	0.0014	0.0016	0.0018	0.0020	0.0023	0.0026	0.0029	0.0033
16	0.0004	0.0005	0.0005	0.0006	0.0007	0.0008	0.0010	0.0011	0.0013	0.0014
17	0.0001	0.0002	0.0002	0.0002	0.0003	0.0003	0.0004	0.0004	0.0005	0.0006
18	0.0000	0.0001	0.0001	0.0001	0.0001	0.0001	0.0001	0.0002	0.0002	0.0002
19	0.0000	0.0000	0.0000	0.0000	0.0000	0.0000	0.0000	0.0001	0.0001	0.0001

x	7.1	7.2	7.3	7.4	7.5	7.6	7.7	7.8	7.9	8.0
0	0.0008	0.0007	0.0007	0.0006	0.0006	0.0005	0.0005	0.0004	0.0004	0.0003
1	0.0059	0.0054	0.0049	0.0045	0.0041	0.0038	0.0035	0.0032	0.0029	0.0027
2	0.0208	0.0194	0.0180	0.0167	0.0156	0.0145	0.0134	0.0125	0.0116	0.0107
3	0.0492	0.0464	0.0438	0.0413	0.0389	0.0366	0.0345	0.0324	0.0305	0.0286
4	0.0874	0.0836	0.0799	0.0764	0.0729	0.0696	0.0663	0.0632	0.0602	0.0573
5	0.1241	0.1204	0.1167	0.1130	0.1094	0.1057	0.1021	0.0986	0.0951	0.0916
6	0.1468	0.1445	0.1420	0.1394	0.1367	0.1339	0.1311	0.1282	0.1252	0.1221
7	0.1489	0.1486	0.1481	0.1474	0.1465	0.1454	0.1442	0.1428	0.1413	0.1396
8	0.1321	0.1337	0.1351	0.1363	0.1373	0.1382	0.1388	0.1392	0.1395	0.1396
9	0.1042	0.1070	0.1096	0.1121	0.1144	0.1167	0.1187	0.1207	0.1224	0.1241
10	0.0740	0.0770	0.0800	0.0829	0.0858	0.0887	0.0914	0.0941	0.0967	0.0993
11	0.0478	0.0504	0.0531	0.0558	0.0585	0.0613	0.0640	0.0667	0.0695	0.0722

■ **TABLE A-3** (*continued*)

					μ					
x	7.1	7.2	7.3	7.4	7.5	7.6	7.7	7.8	7.9	8.0
12	0.0283	0.0303	0.0323	0.0344	0.0366	0.0388	0.0411	0.0434	0.0457	0.0481
13	0.0154	0.0168	0.0181	0.0196	0.0211	0.0227	0.0243	0.0260	0.0278	0.0296
14	0.0078	0.0086	0.0095	0.0104	0.0113	0.0123	0.0134	0.0145	0.0157	0.0169
15	0.0037	0.0041	0.0046	0.0051	0.0057	0.0062	0.0069	0.0075	0.0083	0.0090
16	0.0016	0.0019	0.0021	0.0024	0.0026	0.0030	0.0033	0.0037	0.0041	0.0045
17	0.0007	0.0008	0.0009	0.0010	0.0012	0.0013	0.0015	0.0017	0.0019	0.0021
18	0.0003	0.0003	0.0004	0.0004	0.0005	0.0006	0.0006	0.0007	0.0008	0.0009
19	0.0001	0.0001	0.0001	0.0002	0.0002	0.0002	0.0003	0.0003	0.0003	0.0004
20	0.0000	0.0000	0.0001	0.0001	0.0001	0.0001	0.0001	0.0001	0.0001	0.0002
21	0.0000	0.0000	0.0000	0.0000	0.0000	0.0000	0.0000	0.0000	0.0001	0.0001

x	8.1	8.2	8.3	8.4	8.5	8.6	8.7	8.8	8.9	9.0
0	0.0003	0.0003	0.0002	0.0002	0.0002	0.0002	0.0002	0.0002	0.0001	0.0001
1	0.0025	0.0023	0.0021	0.0019	0.0017	0.0016	0.0014	0.0013	0.0012	0.0011
2	0.0100	0.0092	0.0086	0.0079	0.0074	0.0068	0.0063	0.0058	0.0054	0.0050
3	0.0269	0.0252	0.0237	0.0222	0.0208	0.0195	0.0183	0.0171	0.0160	0.0150
4	0.0544	0.0517	0.0491	0.0466	0.0443	0.0420	0.0398	0.0377	0.0357	0.0337
5	0.0882	0.0849	0.0816	0.0784	0.0752	0.0722	0.0692	0.0663	0.0635	0.0607
6	0.1191	0.1160	0.1128	0.1097	0.1066	0.1034	0.1003	0.0972	0.0941	0.0911
7	0.1378	0.1358	0.1338	0.1317	0.1294	0.1271	0.1247	0.1222	0.1197	0.1171
8	0.1395	0.1392	0.1388	0.1382	0.1375	0.1366	0.1356	0.1344	0.1332	0.1318
9	0.1256	0.1269	0.1280	0.1290	0.1299	0.1306	0.1311	0.1315	0.1317	0.1318
10	0.1017	0.1040	0.1063	0.1084	0.1104	0.1123	0.1140	0.1157	0.1172	0.1186
11	0.0749	0.0776	0.0802	0.0828	0.0853	0.0878	0.0902	0.0925	0.0948	0.0970
12	0.0505	0.0530	0.0555	0.0579	0.0604	0.0629	0.0654	0.0679	0.0703	0.0728
13	0.0315	0.0334	0.0354	0.0374	0.0395	0.0416	0.0438	0.0459	0.0481	0.0504
14	0.0182	0.0196	0.0210	0.0225	0.0240	0.0256	0.0272	0.0289	0.0306	0.0324
15	0.0098	0.0107	0.0116	0.0126	0.0136	0.0147	0.0158	0.0169	0.0182	0.0194
16	0.0050	0.0055	0.0060	0.0066	0.0072	0.0079	0.0086	0.0093	0.0101	0.0109
17	0.0024	0.0026	0.0029	0.0033	0.0036	0.0040	0.0044	0.0048	0.0053	0.0058
18	0.0011	0.0012	0.0014	0.0015	0.0017	0.0019	0.0021	0.0024	0.0026	0.0029
19	0.0005	0.0005	0.0006	0.0007	0.0008	0.0009	0.0010	0.0011	0.0012	0.0014
20	0.0002	0.0002	0.0002	0.0003	0.0003	0.0004	0.0004	0.0005	0.0005	0.0006
21	0.0001	0.0001	0.0001	0.0001	0.0001	0.0002	0.0002	0.0002	0.0002	0.0003
22	0.0000	0.0000	0.0000	0.0000	0.0001	0.0001	0.0001	0.0001	0.0001	0.0001

x	9.1	9.2	9.3	9.4	9.5	9.6	9.7	9.8	9.9	10.0
0	0.0001	0.0001	0.0001	0.0001	0.0001	0.0001	0.0001	0.0001	0.0001	0.0000
1	0.0010	0.0009	0.0009	0.0008	0.0007	0.0007	0.0006	0.0005	0.0005	0.0005
2	0.0046	0.0043	0.0040	0.0037	0.0034	0.0031	0.0029	0.0027	0.0025	0.0023
3	0.0140	0.0131	0.0123	0.0115	0.0107	0.0100	0.0093	0.0087	0.0081	0.0076
4	0.0319	0.0302	0.0285	0.0269	0.0254	0.0240	0.0226	0.0213	0.0201	0.0189
5	0.0581	0.0555	0.0530	0.0506	0.0483	0.0460	0.0439	0.0418	0.0398	0.0378
6	0.0881	0.0851	0.0822	0.0793	0.0764	0.0736	0.0709	0.0682	0.0656	0.0631
7	0.1145	0.1118	0.1091	0.1064	0.1037	0.1010	0.0982	0.0955	0.0928	0.0901
8	0.1302	0.1286	0.1269	0.1251	0.1232	0.1212	0.1191	0.1170	0.1148	0.1126
9	0.1317	0.1315	0.1311	0.1306	0.1300	0.1293	0.1284	0.1274	0.1263	0.1251
10	0.1198	0.1210	0.1219	0.1228	0.1235	0.1241	0.1245	0.1249	0.1250	0.1251
11	0.0991	0.1012	0.1031	0.1049	0.1067	0.1083	0.1098	0.1112	0.1125	0.1137
12	0.0752	0.0776	0.0799	0.0822	0.0844	0.0866	0.0888	0.0908	0.0928	0.0948
13	0.0526	0.0549	0.0572	0.0594	0.0617	0.0640	0.0662	0.0685	0.0707	0.0729
14	0.0342	0.0361	0.0380	0.0399	0.0419	0.0439	0.0459	0.0479	0.0500	0.0521
15	0.0208	0.0221	0.0235	0.0250	0.0265	0.0281	0.0297	0.0313	0.0330	0.0347
16	0.0118	0.0127	0.0137	0.0147	0.0157	0.0168	0.0180	0.0192	0.0204	0.0217
17	0.0063	0.0069	0.0075	0.0081	0.0088	0.0095	0.0103	0.0111	0.0119	0.0128
18	0.0032	0.0035	0.0039	0.0042	0.0046	0.0051	0.0055	0.0060	0.0065	0.0071
19	0.0015	0.0017	0.0019	0.0021	0.0023	0.0026	0.0028	0.0031	0.0034	0.0037
20	0.0007	0.0008	0.0009	0.0010	0.0011	0.0012	0.0014	0.0015	0.0017	0.0019
21	0.0003	0.0003	0.0004	0.0004	0.0005	0.0006	0.0006	0.0007	0.0008	0.0009
22	0.0001	0.0001	0.0002	0.0002	0.0002	0.0002	0.0003	0.0003	0.0004	0.0004
23	0.0000	0.0001	0.0001	0.0001	0.0001	0.0001	0.0001	0.0001	0.0002	0.0002
24	0.0000	0.0000	0.0000	0.0000	0.0000	0.0000	0.0000	0.0001	0.0001	0.0001

■ TABLE A-4   Areas of the Standard Normal Distribution.

The entries in this table are the probabilities that a standard normal random variable is between 0 and z (the shaded area).

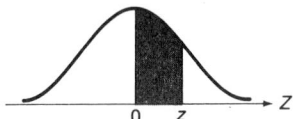

	Second Decimal Place in z									
z	0.00	0.01	0.02	0.03	0.04	0.05	0.06	0.07	0.08	0.09
0.0	0.0000	0.0040	0.0080	0.0120	0.0160	0.0199	0.0239	0.0279	0.0319	0.0359
0.1	0.0398	0.0438	0.0478	0.0517	0.0557	0.0596	0.0636	0.0675	0.0714	0.0753
0.2	0.0793	0.0832	0.0871	0.0910	0.0948	0.0987	0.1026	0.1064	0.1103	0.1141
0.3	0.1179	0.1217	0.1255	0.1293	0.1331	0.1368	0.1406	0.1443	0.1480	0.1517
0.4	0.1554	0.1591	0.1628	0.1664	0.1700	0.1736	0.1772	0.1808	0.1844	0.1879
0.5	0.1915	0.1950	0.1985	0.2019	0.2054	0.2088	0.2123	0.2157	0.2190	0.2224
0.6	0.2257	0.2291	0.2324	0.2357	0.2389	0.2422	0.2454	0.2486	0.2517	0.2549
0.7	0.2580	0.2611	0.2642	0.2673	0.2704	0.2734	0.2764	0.2794	0.2823	0.2852
0.8	0.2881	0.2910	0.2939	0.2967	0.2995	0.3023	0.3051	0.3078	0.3106	0.3133
0.9	0.3159	0.3186	0.3212	0.3238	0.3264	0.3289	0.3315	0.3340	0.3365	0.3389
1.0	0.3413	0.3438	0.3461	0.3485	0.3508	0.3531	0.3554	0.3577	0.3599	0.3621
1.1	0.3643	0.3665	0.3686	0.3708	0.3729	0.3749	0.3770	0.3790	0.3810	0.3830
1.2	0.3849	0.3869	0.3888	0.3907	0.3925	0.3944	0.3962	0.3980	0.3997	0.4015
1.3	0.4032	0.4049	0.4066	0.4082	0.4099	0.4115	0.4131	0.4147	0.4162	0.4177
1.4	0.4192	0.4207	0.4222	0.4236	0.4251	0.4265	0.4279	0.4292	0.4306	0.4319
1.5	0.4332	0.4345	0.4357	0.4370	0.4382	0.4394	0.4406	0.4418	0.4429	0.4441
1.6	0.4452	0.4463	0.4474	0.4484	0.4495	0.4505	0.4515	0.4525	0.4535	0.4545
1.7	0.4554	0.4564	0.4573	0.4582	0.4591	0.4599	0.4608	0.4616	0.4625	0.4633
1.8	0.4641	0.4649	0.4656	0.4664	0.4671	0.4678	0.4686	0.4693	0.4699	0.4706
1.9	0.4713	0.4719	0.4726	0.4732	0.4738	0.4744	0.4750	0.4756	0.4761	0.4767
2.0	0.4772	0.4778	0.4783	0.4788	0.4793	0.4798	0.4803	0.4808	0.4812	0.4817
2.1	0.4821	0.4826	0.4830	0.4834	0.4838	0.4842	0.4846	0.4850	0.4854	0.4857
2.2	0.4861	0.4864	0.4868	0.4871	0.4875	0.4878	0.4881	0.4884	0.4887	0.4890
2.3	0.4893	0.4896	0.4898	0.4901	0.4904	0.4906	0.4909	0.4911	0.4913	0.4916
2.4	0.4918	0.4920	0.4922	0.4925	0.4927	0.4929	0.4931	0.4932	0.4934	0.4936
2.5	0.4938	0.4940	0.4941	0.4943	0.4945	0.4946	0.4948	0.4949	0.4951	0.4952
2.6	0.4953	0.4955	0.4956	0.4957	0.4959	0.4960	0.4961	0.4962	0.4963	0.4974
2.7	0.4965	0.4966	0.4967	0.4968	0.4969	0.4970	0.4971	0.4972	0.4973	0.4974
2.8	0.4974	0.4975	0.4976	0.4977	0.4977	0.4978	0.4979	0.4979	0.4980	0.4981
2.9	0.4981	0.4982	0.4982	0.4983	0.4984	0.4984	0.4985	0.4985	0.4986	0.4986
3.0	0.4987	0.4987	0.4987	0.4988	0.4988	0.4989	0.4989	0.4989	0.4990	0.4990
3.1	0.4990	0.4991	0.4991	0.4991	0.4992	0.4992	0.4992	0.4992	0.4993	0.4993
3.2	0.4993	0.4993	0.4994	0.4994	0.4994	0.4994	0.4994	0.4995	0.4995	0.4995
3.3	0.4995	0.4995	0.4995	0.4996	0.4996	0.4996	0.4996	0.4996	0.4996	0.4997
3.4	0.4997	0.4997	0.4997	0.4997	0.4997	0.4997	0.4997	0.4997	0.4997	0.4998
3.5	0.4998									
4.0	0.49997									
4.5	0.499997									
5.0	0.4999997									

APPENDIX A

■ **TABLE A-5**   Critical Values of *t*.

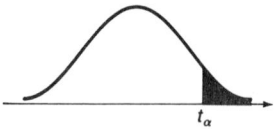

DEGREES OF FREEDOM	$t_{.100}$	$t_{.050}$	$t_{.025}$	$t_{.010}$	$t_{.005}$
1	3.078	6.314	12.706	31.821	63.657
2	1.886	2.920	4.303	6.965	9.925
3	1.638	2.353	3.182	4.541	5.841
4	1.533	2.132	2.776	3.747	4.604
5	1.476	2.015	2.571	3.365	4.032
6	1.440	1.943	2.447	3.143	3.707
7	1.415	1.895	2.365	2.998	3.499
8	1.397	1.860	2.306	2.896	3.355
9	1.383	1.833	2.262	2.821	3.250
10	1.372	1.812	2.228	2.764	3.169
11	1.363	1.796	2.201	2.718	3.106
12	1.356	1.782	2.179	2.681	3.055
13	1.350	1.771	2.160	2.650	3.012
14	1.345	1.761	2.145	2.624	2.977
15	1.341	1.753	2.131	2.602	2.947
16	1.337	1.746	2.120	2.583	2.921
17	1.333	1.740	2.110	2.567	2.898
18	1.330	1.734	2.101	2.552	2.878
19	1.328	1.729	2.093	2.539	2.861
20	1.325	1.725	2.086	2.528	2.845
21	1.323	1.721	2.080	2.518	2.831
22	1.321	1.717	2.074	2.508	2.819
23	1.319	1.714	2.069	2.500	2.808
24	1.318	1.711	2.064	2.492	2.797
25	1.316	1.708	2.060	2.485	2.787
26	1.315	1.706	2.056	2.479	2.779
27	1.314	1.703	2.052	2.473	2.771
28	1.313	1.701	2.048	2.467	2.763
29	1.311	1.699	2.045	2.462	2.756
30	1.310	1.697	2.042	2.457	2.750
40	1.303	1.684	2.021	2.423	2.704
60	1.296	1.671	2.000	2.390	2.660
120	1.289	1.658	1.980	2.358	2.617
∞	1.282	1.645	1.960	2.326	2.576

■ TABLE A-6   Critical Values of χ^2

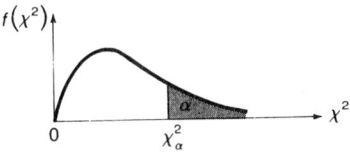

DEGREES OF FREEDOM	$\chi^2_{.995}$	$\chi^2_{.990}$	$\chi^2_{.975}$	$\chi^2_{.950}$	$\chi^2_{.900}$
1	0.0000393	0.0001571	0.0009821	0.0039321	0.0157908
2	0.0100251	0.0201007	0.0506356	0.102587	0.210720
3	0.0717212	0.114832	0.215795	0.351846	0.584375
4	0.206990	0.297110	0.484419	0.710721	1.063623
5	0.411740	0.554300	0.831211	1.145476	1.61031
6	0.675727	0.872085	1.237347	1.63539	2.20413
7	0.989265	1.239043	1.68987	2.16735	2.83311
8	1.344419	1.646482	2.17973	2.73264	3.48954
9	1.734926	2.087912	2.70039	3.32511	4.16816
10	2.15585	2.55821	3.24697	3.94030	4.86518
11	2.60321	3.05347	3.81575	4.57481	5.57779
12	3.07382	3.57056	4.40379	5.22603	6.30380
13	3.56503	4.10691	5.00874	5.89186	7.04150
14	4.07468	4.66043	5.62872	6.57063	7.78953
15	4.60094	5.22935	6.26214	7.26094	8.54675
16	5.14224	5.81221	6.90766	7.96164	9.31223
17	5.69724	6.40776	7.56418	8.67176	10.0852
18	6.26481	7.01491	8.23075	9.39046	10.8649
19	6.84398	7.63273	8.90655	10.1170	11.6509
20	7.43386	8.26040	9.59083	10.8508	12.4426
21	8.03366	8.89720	10.28293	11.5913	13.2396
22	8.64272	9.54249	10.9823	12.3380	14.0415
23	9.26042	10.19567	11.6885	13.0905	14.8479
24	9.88623	10.8564	12.4011	13.8484	15.6587
25	10.5197	11.5240	13.1197	14.6114	16.4734
26	11.1603	12.1981	13.8439	15.3791	17.2919
27	11.8076	12.8786	14.5733	16.1513	18.1138
28	12.4613	13.5648	15.3079	16.9279	18.9392
29	13.1211	14.2565	16.0471	17.7083	19.7677
30	13.7867	14.9535	16.7908	18.4926	20.5992
40	20.7065	22.1643	24.4331	26.5093	29.0505
50	27.9907	29.7067	32.3574	34.7642	37.6886
60	35.5346	37.4848	40.4817	43.1879	46.4589
70	43.2752	45.4418	48.7576	51.7393	55.3290
80	51.1720	53.5400	57.1532	60.3915	64.2778
90	59.1963	61.7541	65.6466	69.1260	73.2912
100	67.3276	70.0648	74.2219	77.9295	82.3581

■ TABLE A-6 (*continued*)

DEGREES OF FREEDOM	$\chi^2_{.100}$	$\chi^2_{.050}$	$\chi^2_{.025}$	$\chi^2_{.010}$	$\chi^2_{.005}$
1	2.70554	3.84146	5.02389	6.63490	7.87944
2	4.60517	5.99147	7.37776	9.21034	10.5966
3	6.25139	7.81473	9.34840	11.3449	12.8381
4	7.77944	9.48773	11.1433	13.2767	14.8602
5	9.23635	11.0705	12.8325	15.0863	16.7496
6	10.6446	12.5916	14.4494	16.8119	18.5476
7	12.0170	14.0671	16.0128	18.4753	20.2777
8	13.3616	15.5073	17.5346	20.0902	21.9550
9	14.6837	16.9190	19.0228	21.6660	23.5893
10	15.9871	18.3070	20.4831	23.2093	25.1882
11	17.2750	19.6751	21.9200	24.7250	26.7569
12	18.5494	21.0261	23.3367	26.2170	28.2995
13	19.8119	22.3621	24.7356	27.6883	29.8194
14	21.0642	23.6848	26.1190	29.1413	31.3193
15	22.3072	24.9958	27.4884	30.5779	32.8013
16	23.5418	26.2962	28.8454	31.9999	34.2672
17	24.7690	27.5871	30.1910	33.4087	35.7185
18	25.9894	28.8693	31.5264	34.8053	37.1564
19	27.2036	30.1435	32.8523	36.1908	38.5822
20	28.4120	31.4104	34.1696	37.5662	39.9968
21	29.6151	32.6705	35.4789	38.9321	41.4010
22	30.8133	33.9244	36.7807	40.2894	42.7956
23	32.0069	35.1725	38.0757	41.6384	44.1813
24	33.1963	36.4151	39.3641	42.9798	45.5585
25	34.3816	37.6525	40.6465	44.3141	46.9278
26	35.5631	38.8852	41.9232	45.6417	48.2899
27	36.7412	40.1133	43.1944	46.9630	49.6449
28	37.9159	41.3372	44.4607	48.2782	50.9933
29	39.0875	42.5569	45.7222	49.5879	52.3356
30	40.2560	43.7729	46.9792	50.8922	53.6720
40	51.8050	55.7585	59.3417	63.6907	66.7659
50	63.1671	67.5048	71.4202	76.1539	79.4900
60	74.3970	79.0819	83.2976	88.3794	91.9517
70	85.5271	90.5312	95.0231	100.425	104.215
80	96.5782	101.879	106.629	112.329	116.321
90	107.565	113.145	118.136	124.116	128.229
100	118.498	124.342	129.561	135.807	140.169

■ **TABLE A-7**   Percentage Points of the *F* Distribution.

(a) $\alpha = .10$

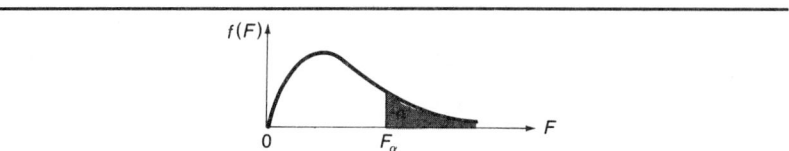

	v_1	**NUMERATOR DEGREES OF FREEDOM**								
v_2		**1**	**2**	**3**	**4**	**5**	**6**	**7**	**8**	**9**
	1	39.86	49.50	53.59	55.83	57.24	58.20	58.91	59.44	59.86
	2	8.53	9.00	9.16	9.24	9.29	9.33	9.35	9.37	9.38
	3	5.54	5.46	5.39	5.34	5.31	5.28	5.27	5.25	5.24
	4	4.54	4.32	4.19	4.11	4.05	4.01	3.98	3.95	3.94
	5	4.06	3.78	3.62	3.52	3.45	3.40	3.37	3.34	3.32
	6	3.78	3.46	3.29	3.18	3.11	3.05	3.01	2.98	2.96
	7	3.59	3.26	3.07	2.96	2.88	2.83	2.78	2.75	2.72
	8	3.46	3.11	2.92	2.81	2.73	2.67	2.62	2.59	2.56
	9	3.36	3.01	2.81	2.69	2.61	2.55	2.51	2.47	2.44
	10	3.29	2.92	2.73	2.61	2.52	2.46	2.41	2.38	2.35
	11	3.23	2.86	2.66	2.54	2.45	2.39	2.34	2.30	2.27
	12	3.18	2.81	2.61	2.48	2.39	2.33	2.28	2.24	2.21
	13	3.14	2.76	2.56	2.43	2.35	2.28	2.23	2.20	2.16
	14	3.10	2.73	2.52	2.39	2.31	2.24	2.19	2.15	2.12
	15	3.07	2.70	2.49	2.36	2.27	2.21	2.16	2.12	2.09
	16	3.05	2.67	2.46	2.33	2.24	2.18	2.13	2.09	2.06
	17	3.03	2.64	2.44	2.31	2.22	2.15	2.10	2.06	2.03
	18	3.01	2.62	2.42	2.29	2.20	2.13	2.08	2.04	2.00
	19	2.99	2.61	2.40	2.27	2.18	2.11	2.06	2.02	1.98
	20	2.97	2.59	2.38	2.25	2.16	2.09	2.04	2.00	1.96
	21	2.96	2.57	2.36	2.23	2.14	2.08	2.02	1.98	1.95
	22	2.95	2.56	2.35	2.22	2.13	2.06	2.01	1.97	1.93
	23	2.94	2.55	2.34	2.21	2.11	2.05	1.99	1.95	1.92
	24	2.93	2.54	2.33	2.19	2.10	2.04	1.98	1.94	1.91
	25	2.92	2.53	2.32	2.18	2.09	2.02	1.97	1.93	1.89
	26	2.91	2.52	2.31	2.17	2.08	2.01	1.96	1.92	1.88
	27	2.90	2.51	2.30	2.17	2.07	2.00	1.95	1.91	1.87
	28	2.89	2.50	2.29	2.16	2.06	2.00	1.94	1.90	1.87
	29	2.89	2.50	2.28	2.15	2.06	1.99	1.93	1.89	1.86
	30	2.88	2.49	2.28	2.14	2.05	1.98	1.93	1.88	1.85
	40	2.84	2.44	2.23	2.09	2.00	1.93	1.87	1.83	1.79
	60	2.79	2.39	2.18	2.04	1.95	1.87	1.82	1.77	1.74
	120	2.75	2.35	2.13	1.99	1.90	1.82	1.77	1.72	1.68
	∞	2.71	2.30	2.08	1.94	1.85	1.77	1.72	1.67	1.63

DENOMINATOR DEGREES OF FREEDOM

(a) $\alpha = .10$ (*continued*)

v_2 \ v_1	10	12	15	20	24	30	40	60	120	∞
1	60.19	60.71	61.22	61.74	62.00	62.26	62.53	62.79	63.06	63.33
2	9.39	9.41	9.42	9.44	9.45	9.46	9.47	9.47	9.48	9.49
3	5.23	5.22	5.20	5.18	5.18	5.17	5.16	5.15	5.14	5.13
4	3.92	3.90	3.87	3.84	3.83	3.82	3.80	3.79	3.78	3.76
5	3.30	3.27	3.24	3.21	3.19	3.17	3.16	3.14	3.12	3.10
6	2.94	2.90	2.87	2.84	2.82	2.80	2.78	2.76	2.74	2.72
7	2.70	2.67	2.63	2.59	2.58	2.56	2.54	2.51	2.49	2.47
8	2.54	2.50	2.46	2.42	2.40	2.38	2.36	2.34	2.32	2.29
9	2.42	2.38	2.34	2.30	2.28	2.25	2.23	2.21	2.18	2.16
10	2.32	2.28	2.24	2.20	2.18	2.16	2.13	2.11	2.08	2.06
11	2.25	2.21	2.17	2.12	2.10	2.08	2.05	2.03	2.00	1.97
12	2.19	2.15	2.10	2.06	2.04	2.01	1.99	1.96	1.93	1.90
13	2.14	2.10	2.05	2.01	1.98	1.96	1.93	1.90	1.88	1.85
14	2.10	2.05	2.01	1.96	1.94	1.91	1.89	1.86	1.83	1.80
15	2.06	2.02	1.97	1.92	1.90	1.87	1.85	1.82	1.79	1.76
16	2.03	1.99	1.94	1.89	1.87	1.84	1.81	1.78	1.75	1.72
17	2.00	1.96	1.91	1.86	1.84	1.81	1.78	1.75	1.72	1.69
18	1.98	1.93	1.89	1.84	1.81	1.78	1.75	1.72	1.69	1.66
19	1.96	1.91	1.86	1.81	1.79	1.76	1.73	1.70	1.67	1.63
20	1.94	1.89	1.84	1.79	1.77	1.74	1.71	1.68	1.64	1.61
21	1.92	1.87	1.83	1.78	1.75	1.72	1.69	1.66	1.62	1.59
22	1.90	1.86	1.81	1.76	1.73	1.70	1.67	1.64	1.60	1.57
23	1.89	1.84	1.80	1.74	1.72	1.69	1.66	1.62	1.59	1.55
24	1.88	1.83	1.78	1.73	1.70	1.67	1.64	1.61	1.57	1.53
25	1.87	1.82	1.77	1.72	1.69	1.66	1.63	1.59	1.56	1.52
26	1.86	1.81	1.76	1.71	1.68	1.65	1.61	1.58	1.54	1.50
27	1.85	1.80	1.75	1.70	1.67	1.64	1.60	1.57	1.53	1.49
28	1.84	1.79	1.74	1.69	1.66	1.63	1.59	1.56	1.52	1.48
29	1.83	1.78	1.73	1.68	1.65	1.62	1.58	1.55	1.51	1.47
30	1.82	1.77	1.72	1.67	1.64	1.61	1.57	1.54	1.50	1.46
40	1.76	1.71	1.66	1.61	1.57	1.54	1.51	1.47	1.42	1.38
60	1.71	1.66	1.60	1.54	1.51	1.48	1.44	1.40	1.35	1.29
120	1.65	1.60	1.55	1.48	1.45	1.41	1.37	1.32	1.26	1.19
∞	1.60	1.55	1.49	1.42	1.38	1.34	1.30	1.24	1.17	1.00

NUMERATOR DEGREES OF FREEDOM (column headers) / DENOMINATOR DEGREES OF FREEDOM (row labels v_2)

(b) $\alpha = .05$

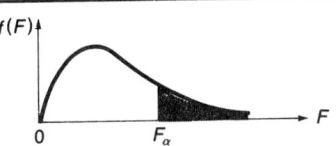

v_1 v_2	NUMERATOR DEGREES OF FREEDOM								
	1	2	3	4	5	6	7	8	9
1	161.4	199.5	215.7	224.6	230.2	234.0	236.8	238.9	240.5
2	18.51	19.00	19.16	19.25	19.30	19.33	19.35	19.37	19.38
3	10.13	9.55	9.28	9.12	9.01	8.94	8.89	8.85	8.81
4	7.71	6.94	6.59	6.39	6.26	6.16	6.09	6.04	6.00
5	6.61	5.79	5.41	5.19	5.05	4.95	4.88	4.82	4.77
6	5.99	5.14	4.76	4.53	4.39	4.28	4.21	4.15	4.10
7	5.59	4.74	4.35	4.12	3.97	3.87	3.79	3.73	3.68
8	5.32	4.46	4.07	3.84	3.69	3.58	3.50	3.44	3.39
9	5.12	4.26	3.86	3.63	3.48	3.37	3.29	3.23	3.18
10	4.96	4.10	3.71	3.48	3.33	3.22	3.14	3.07	3.02
11	4.84	3.98	3.59	3.36	3.20	3.09	3.01	2.95	2.90
12	4.75	3.89	3.49	3.26	3.11	3.00	2.91	2.85	2.80
13	4.67	3.81	3.41	3.18	3.03	2.92	2.83	2.77	2.71
14	4.60	3.74	3.34	3.11	2.96	2.85	2.76	2.70	2.65
15	4.54	3.68	3.29	3.06	2.90	2.79	2.71	2.64	2.59
16	4.49	3.63	3.24	3.01	2.85	2.74	2.66	2.59	2.54
17	4.45	3.59	3.20	2.96	2.81	2.70	2.61	2.55	2.49
18	4.41	3.55	3.16	2.93	2.77	2.66	2.58	2.51	2.46
19	4.38	3.52	3.13	2.90	2.74	2.63	2.54	2.48	2.42
20	4.35	3.49	3.10	2.87	2.71	2.60	2.51	2.45	2.39
21	4.32	3.47	3.07	2.84	2.68	2.57	2.49	2.42	2.37
22	4.30	3.44	3.05	2.82	2.66	2.55	2.46	2.40	2.34
23	4.28	3.42	3.03	2.80	2.64	2.53	2.44	2.37	2.32
24	4.26	3.40	3.01	2.78	2.62	2.51	2.42	2.36	2.30
25	4.24	3.39	2.99	2.76	2.60	2.49	2.40	2.34	2.28
26	4.23	3.37	2.98	2.74	2.59	2.47	2.39	2.32	2.27
27	4.21	3.35	2.96	2.73	2.57	2.46	2.37	2.31	2.25
28	4.20	3.34	2.95	2.71	2.56	2.45	2.36	2.29	2.24
29	4.18	3.33	2.93	2.70	2.55	2.43	2.35	2.28	2.22
30	4.17	3.32	2.92	2.69	2.53	2.42	2.33	2.27	2.21
40	4.08	3.23	2.84	2.61	2.45	2.34	2.25	2.18	2.12
60	4.00	3.15	2.76	2.53	2.37	2.25	2.17	2.10	2.04
120	3.92	3.07	2.68	2.45	2.29	2.17	2.09	2.02	1.96
∞	3.84	3.00	2.60	2.37	2.21	2.10	2.01	1.94	1.88

DENOMINATOR DEGREES OF FREEDOM

(b) $\alpha = .05$ (*continued*)

v_2 \ v_1	NUMERATOR DEGREES OF FREEDOM									
	10	12	15	20	24	30	40	60	120	∞
1	241.9	243.9	245.9	248.0	249.1	250.1	251.1	252.2	253.3	254.3
2	19.40	19.41	19.43	19.45	19.45	19.46	19.47	19.48	19.49	19.50
3	8.79	8.74	8.70	8.66	8.64	8.62	8.59	8.57	8.55	8.53
4	5.96	5.91	5.86	5.80	5.77	5.75	5.72	5.69	5.66	5.63
5	4.74	4.68	4.62	4.56	4.53	4.50	4.46	4.43	4.40	4.36
6	4.06	4.00	3.94	3.87	3.84	3.81	3.77	3.74	3.70	3.67
7	3.64	3.57	3.51	3.44	3.41	3.38	3.34	3.30	3.27	3.23
8	3.35	3.28	3.22	3.15	3.12	3.08	3.04	3.01	2.97	2.93
9	3.14	3.07	3.01	2.94	2.90	2.86	2.83	2.79	2.75	2.71
10	2.98	2.91	2.85	2.77	2.74	2.70	2.66	2.62	2.58	2.54
11	2.85	2.79	2.72	2.65	2.61	2.57	2.53	2.49	2.45	2.40
12	2.75	2.69	2.62	2.54	2.51	2.47	2.43	2.38	2.34	2.30
13	2.67	2.60	2.53	2.46	2.42	2.38	2.34	2.30	2.25	2.21
14	2.60	2.53	2.46	2.39	2.35	2.31	2.27	2.22	2.18	2.13
15	2.54	2.48	2.40	2.33	2.29	2.25	2.20	2.16	2.11	2.07
16	2.49	2.42	2.35	2.28	2.24	2.19	2.15	2.11	2.06	2.01
17	2.45	2.38	2.31	2.23	2.19	2.15	2.10	2.06	2.01	1.96
18	2.41	2.34	2.27	2.19	2.15	2.11	2.06	2.02	1.97	1.92
19	2.38	2.31	2.23	2.16	2.11	2.07	2.03	1.98	1.93	1.88
20	2.35	2.28	2.20	2.12	2.08	2.04	1.99	1.95	1.90	1.84
21	2.32	2.25	2.18	2.10	2.05	2.01	1.96	1.92	1.87	1.81
22	2.30	2.23	2.15	2.07	2.03	1.98	1.94	1.89	1.84	1.78
23	2.27	2.20	2.13	2.05	2.01	1.96	1.91	1.86	1.81	1.76
24	2.25	2.18	2.11	2.03	1.98	1.94	1.89	1.84	1.79	1.73
25	2.24	2.16	2.09	2.01	1.96	1.92	1.87	1.82	1.77	1.71
26	2.22	2.15	2.07	1.99	1.95	1.90	1.85	1.80	1.75	1.69
27	2.20	2.13	2.06	1.97	1.93	1.88	1.84	1.79	1.73	1.67
28	2.19	2.12	2.04	1.96	1.91	1.87	1.82	1.77	1.71	1.65
29	2.18	2.10	2.03	1.94	1.90	1.85	1.81	1.75	1.70	1.64
30	2.16	2.09	2.01	1.93	1.89	1.84	1.79	1.74	1.68	1.62
40	2.08	2.00	1.92	1.84	1.79	1.74	1.69	1.64	1.58	1.51
60	1.99	1.92	1.84	1.75	1.70	1.65	1.59	1.53	1.47	1.39
120	1.91	1.83	1.75	1.66	1.61	1.55	1.50	1.43	1.35	1.25
∞	1.83	1.75	1.67	1.57	1.52	1.46	1.39	1.32	1.22	1.00

DENOMINATOR DEGREES OF FREEDOM

(c) $\alpha = .025$

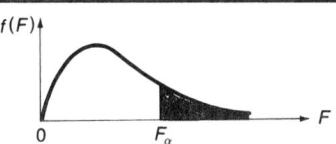

v_2 \ v_1	NUMERATOR DEGREES OF FREEDOM								
	1	2	3	4	5	6	7	8	9
1	647.8	799.5	864.2	899.6	921.8	937.1	948.2	956.7	963.3
2	38.51	39.00	39.17	39.25	39.30	39.33	39.36	39.37	39.39
3	17.44	16.04	15.44	15.10	14.88	14.73	14.62	14.54	14.47
4	12.22	10.65	9.98	9.60	9.36	9.20	9.07	8.98	8.90
5	10.01	8.43	7.76	7.39	7.15	6.98	6.85	6.76	6.68
6	8.81	7.26	6.60	6.23	5.99	5.82	5.70	5.60	5.52
7	8.07	6.54	5.89	5.52	5.29	5.12	4.99	4.90	4.82
8	7.57	6.06	5.42	5.05	4.82	4.65	4.53	4.43	4.36
9	7.21	5.71	5.08	4.72	4.48	4.32	4.20	4.10	4.03
10	6.94	5.46	4.83	4.47	4.24	4.07	3.95	3.85	3.78
11	6.72	5.26	4.63	4.28	4.04	3.88	3.76	3.66	3.59
12	6.55	5.10	4.47	4.12	3.89	3.73	3.61	3.51	3.44
13	6.41	4.97	4.35	4.00	3.77	3.60	3.48	3.39	3.31
14	6.30	4.86	4.24	3.89	3.66	3.50	3.38	3.29	3.21
15	6.20	4.77	4.15	3.80	3.58	3.41	3.29	3.20	3.12
16	6.12	4.69	4.08	3.73	3.50	3.34	3.22	3.12	3.05
17	6.04	4.62	4.01	3.66	3.44	3.28	3.16	3.06	2.98
18	5.98	4.56	3.95	3.61	3.38	3.22	3.10	3.01	2.93
19	5.92	4.51	3.90	3.56	3.33	3.17	3.05	2.96	2.88
20	5.87	4.46	3.86	3.51	3.29	3.13	3.01	2.91	2.84
21	5.83	4.42	3.82	3.48	3.25	3.09	2.97	2.87	2.80
22	5.79	4.38	3.78	3.44	3.22	3.05	2.93	2.84	2.76
23	5.75	4.35	3.75	3.41	3.18	3.02	2.90	2.81	2.73
24	5.72	4.32	3.72	3.38	3.15	2.99	2.87	2.78	2.70
25	5.69	4.29	3.69	3.35	3.13	2.97	2.85	2.75	2.68
26	5.66	4.27	3.67	3.33	3.10	2.94	2.82	2.73	2.65
27	5.63	4.24	3.65	3.31	3.08	2.92	2.80	2.71	2.63
28	5.61	4.22	3.63	3.29	3.06	2.90	2.78	2.69	2.61
29	5.59	4.20	3.61	3.27	3.04	2.88	2.76	2.67	2.59
30	5.57	4.18	3.59	3.25	3.03	2.87	2.75	2.65	2.57
40	5.42	4.05	3.46	3.13	2.90	2.74	2.62	2.53	2.45
60	5.29	3.93	3.34	3.01	2.79	2.63	2.51	2.41	2.33
120	5.15	3.80	3.23	2.89	2.67	2.52	2.39	2.30	2.22
∞	5.02	3.69	3.12	2.79	2.57	2.41	2.29	2.19	2.11

DENOMINATOR DEGREES OF FREEDOM

(c) $\alpha = .025$ *(continued)*

v_2 \ v_1	NUMERATOR DEGREES OF FREEDOM									
	10	12	15	20	24	30	40	60	120	∞
1	968.6	976.7	984.9	993.1	997.2	1001	1006	1010	1014	1018
2	39.40	39.41	39.43	39.45	39.46	39.46	39.47	39.48	39.49	39.50
3	14.42	14.34	14.25	14.17	14.12	14.08	14.04	13.99	13.95	13.90
4	8.84	8.75	8.66	8.56	8.51	8.46	8.41	8.36	8.31	8.26
5	6.62	6.52	6.43	6.33	6.28	6.23	6.18	6.12	6.07	6.02
6	5.46	5.37	5.27	5.17	5.12	5.07	5.01	4.96	4.90	4.85
7	4.76	4.67	4.57	4.47	4.42	4.36	4.31	4.25	4.20	4.14
8	4.30	4.20	4.10	4.00	3.95	3.89	3.84	3.78	3.73	3.67
9	3.96	3.87	3.77	3.67	3.61	3.56	3.51	3.45	3.39	3.33
10	3.72	3.62	3.52	3.42	3.37	3.31	3.26	3.20	3.14	3.08
11	3.53	3.43	3.33	3.23	3.17	3.12	3.06	3.00	2.94	2.88
12	3.37	3.28	3.18	3.07	3.02	2.96	2.91	2.85	2.79	2.72
13	3.25	3.15	3.05	2.95	2.89	2.84	2.78	2.72	2.66	2.60
14	3.15	3.05	2.95	2.84	2.79	2.73	2.67	2.61	2.55	2.49
15	3.06	2.96	2.86	2.76	2.70	2.64	2.59	2.52	2.46	2.40
16	2.99	2.89	2.79	2.68	2.63	2.57	2.51	2.45	2.38	2.32
17	2.92	2.82	2.72	2.62	2.56	2.50	2.44	2.38	2.32	2.25
18	2.87	2.77	2.67	2.56	2.50	2.44	2.38	2.32	2.26	2.19
19	2.82	2.72	2.62	2.51	2.45	2.39	2.33	2.27	2.20	2.13
20	2.77	2.68	2.57	2.46	2.41	2.35	2.29	2.22	2.16	2.09
21	2.73	2.64	2.53	2.42	2.37	2.31	2.25	2.18	2.11	2.04
22	2.70	2.60	2.50	2.39	2.33	2.27	2.21	2.14	2.08	2.00
23	2.67	2.57	2.47	2.36	2.30	2.24	2.18	2.11	2.04	1.97
24	2.64	2.54	2.44	2.33	2.27	2.21	2.15	2.08	2.01	1.94
25	2.61	2.51	2.41	2.30	2.24	2.18	2.12	2.05	1.98	1.91
26	2.59	2.49	2.39	2.28	2.22	2.16	2.09	2.03	1.95	1.88
27	2.57	2.47	2.36	2.25	2.19	2.13	2.07	2.00	1.93	1.85
28	2.55	2.45	2.34	2.23	2.17	2.11	2.05	1.98	1.91	1.83
29	2.53	2.43	2.32	2.21	2.15	2.09	2.03	1.96	1.89	1.81
30	2.51	2.41	2.31	2.20	2.14	2.07	2.01	1.94	1.87	1.79
40	2.39	2.29	2.18	2.07	2.01	1.94	1.88	1.80	1.72	1.64
60	2.27	2.17	2.06	1.94	1.88	1.82	1.74	1.67	1.58	1.48
120	2.16	2.05	1.94	1.82	1.76	1.69	1.61	1.53	1.43	1.31
∞	2.05	1.94	1.83	1.71	1.64	1.57	1.48	1.39	1.27	1.00

DENOMINATOR DEGREES OF FREEDOM

(d) $\alpha = .01$

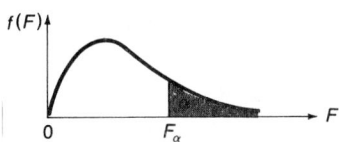

	v_1	NUMERATOR DEGREES OF FREEDOM								
	v_2	1	2	3	4	5	6	7	8	9
	1	4,052	4,999.5	5,403	5,625	5,764	5,859	5,928	5,982	6,022
	2	98.50	99.00	99.17	99.25	99.30	99.33	99.36	99.37	99.39
	3	34.12	30.82	29.46	28.71	28.24	27.91	27.67	27.49	27.35
	4	21.20	18.00	16.69	15.98	15.52	15.21	14.98	14.80	14.66
	5	16.26	13.27	12.06	11.39	10.97	10.67	10.46	10.29	10.16
	6	13.75	10.92	9.78	9.15	8.75	8.47	8.26	8.10	7.98
	7	12.25	9.55	8.45	7.85	7.46	7.19	6.99	6.84	6.72
	8	11.26	8.65	7.59	7.01	6.63	6.37	6.18	6.03	5.91
	9	10.56	8.02	6.99	6.42	6.06	5.80	5.61	5.47	5.35
DENOMINATOR DEGREES OF FREEDOM	10	10.04	7.56	6.55	5.99	5.64	5.39	5.20	5.06	4.94
	11	9.65	7.21	6.22	5.67	5.32	5.07	4.89	4.74	4.63
	12	9.33	6.93	5.95	5.41	5.06	4.82	4.64	4.50	4.39
	13	9.07	6.70	5.74	5.21	4.86	4.62	4.44	4.30	4.19
	14	8.86	6.51	5.56	5.04	4.69	4.46	4.28	4.14	4.03
	15	8.68	6.36	5.42	4.89	4.56	4.32	4.14	4.00	3.89
	16	8.53	6.23	5.29	4.77	4.44	4.20	4.03	3.89	3.78
	17	8.40	6.11	5.18	4.67	4.34	4.10	3.93	3.79	3.68
	18	8.29	6.01	5.09	4.58	4.25	4.01	3.84	3.71	3.60
	19	8.18	5.93	5.01	4.50	4.17	3.94	3.77	3.63	3.52
	20	8.10	5.85	4.94	4.43	4.10	3.87	3.70	3.56	3.46
	21	8.02	5.78	4.87	4.37	4.04	3.81	3.64	3.51	3.40
	22	7.95	5.72	4.82	4.31	3.99	3.76	3.59	3.45	3.35
	23	7.88	5.66	4.76	4.26	3.94	3.71	3.54	3.41	3.30
	24	7.82	5.61	4.72	4.22	3.90	3.67	3.50	3.36	3.26
	25	7.77	5.57	4.68	4.18	3.85	3.63	3.46	3.32	3.22
	26	7.72	5.53	4.64	4.14	3.82	3.59	3.42	3.29	3.18
	27	7.68	5.49	4.60	4.11	3.78	3.56	3.39	3.26	3.15
	28	7.64	5.45	4.57	4.07	3.75	3.53	3.36	3.23	3.12
	29	7.60	5.42	4.54	4.04	3.73	3.50	3.33	3.20	3.09
	30	7.56	5.39	4.51	4.02	3.70	3.47	3.30	3.17	3.07
	40	7.31	5.18	4.31	3.83	3.51	3.29	3.12	2.99	2.89
	60	7.08	4.98	4.13	3.65	3.34	3.12	2.95	2.82	2.72
	120	6.85	4.79	3.95	3.48	3.17	2.96	2.79	2.66	2.56
	∞	6.63	4.61	3.78	3.32	3.02	2.80	2.64	2.51	2.41

(d) $\alpha = .01$ (*continued*)

v_2	NUMERATOR DEGREES OF FREEDOM									
	10	12	15	20	24	30	40	60	120	∞
1	6,056	6,106	6,157	6,209	6,235	6,261	6,287	6,313	6,339	6,366
2	99.40	99.42	99.43	99.45	99.46	99.47	99.47	99.48	99.49	99.50
3	27.23	27.05	26.87	26.69	26.60	26.50	26.41	26.32	26.22	26.13
4	14.55	14.37	14.20	14.02	13.93	13.84	13.75	13.65	13.56	13.46
5	10.05	9.89	9.72	9.55	9.47	9.38	9.29	9.20	9.11	9.02
6	7.87	7.72	7.56	7.40	7.31	7.23	7.14	7.06	6.97	6.88
7	6.62	6.47	6.31	6.16	6.07	5.99	5.91	5.82	5.74	5.65
8	5.81	5.67	5.52	5.36	5.28	5.20	5.12	5.03	4.95	4.86
9	5.26	5.11	4.96	4.81	4.73	4.65	4.57	4.48	4.40	4.31
10	4.85	4.71	4.56	4.41	4.33	4.25	4.17	4.08	4.00	3.91
11	4.54	4.40	4.25	4.10	4.02	3.94	3.86	3.78	3.69	3.60
12	4.30	4.16	4.01	3.86	3.78	3.70	3.62	3.54	3.45	3.36
13	4.10	3.96	3.82	3.66	3.59	3.51	3.43	3.34	3.25	3.17
14	3.94	3.80	3.66	3.51	3.43	3.35	3.27	3.18	3.09	3.00
15	3.80	3.67	3.52	3.37	3.29	3.21	3.13	3.05	2.96	2.87
16	3.69	3.55	3.41	3.26	3.18	3.10	3.02	2.93	2.84	2.75
17	3.59	3.46	3.31	3.16	3.08	3.00	2.92	2.83	2.75	2.65
18	3.51	3.37	3.23	3.08	3.00	2.92	2.84	2.75	2.66	2.57
19	3.43	3.30	3.15	3.00	2.92	2.84	2.76	2.67	2.58	2.49
20	3.37	3.23	3.09	2.94	2.86	2.78	2.69	2.61	2.52	2.42
21	3.31	3.17	3.03	2.88	2.80	2.72	2.64	2.55	2.46	2.36
22	3.26	3.12	2.98	2.83	2.75	2.67	2.58	2.50	2.40	2.31
23	3.21	3.07	2.93	2.78	2.70	2.62	2.54	2.45	2.35	2.26
24	3.17	3.03	2.89	2.74	2.66	2.58	2.49	2.40	2.31	2.21
25	3.13	2.99	2.85	2.70	2.62	2.54	2.45	2.36	2.27	2.17
26	3.09	2.96	2.81	2.66	2.58	2.50	2.42	2.33	2.23	2.13
27	3.06	2.93	2.78	2.63	2.55	2.47	2.38	2.29	2.20	2.10
28	3.03	2.90	2.75	2.60	2.52	2.44	2.35	2.26	2.17	2.06
29	3.00	2.87	2.73	2.57	2.49	2.41	2.33	2.23	2.14	2.03
30	2.98	2.84	2.70	2.55	2.47	2.39	2.30	2.21	2.11	2.01
40	2.80	2.66	2.52	2.37	2.29	2.20	2.11	2.02	1.92	1.80
60	2.63	2.50	2.35	2.20	2.12	2.03	1.94	1.84	1.73	1.60
120	2.47	2.34	2.19	2.03	1.95	1.86	1.76	1.66	1.53	1.38
∞	2.32	2.18	2.04	1.88	1.79	1.70	1.59	1.47	1.32	1.00

DENOMINATOR DEGREES OF FREEDOM

■ TABLE A-8 Confidence Interval for a Population Proportion, Small Sample.

n = 5	α = .05		α = .10	
	P_L	P_U	P_L	P_U
x = 1	0.005	0.716	0.010	0.657
2	0.053	0.853	0.076	0.811
3	0.147	0.947	0.189	0.924
4	0.284	0.995	0.343	0.990

n = 6	α = .05		α = .10	
	P_L	P_U	P_L	P_U
x = 1	0.004	0.641	0.009	0.582
2	0.043	0.777	0.063	0.729
3	0.118	0.882	0.153	0.847
4	0.223	0.957	0.271	0.937
5	0.359	0.996	0.418	0.991

n = 7	α = .05		α = .10	
	P_L	P_U	P_L	P_U
x = 1	0.004	0.579	0.007	0.521
2	0.037	0.710	0.053	0.659
3	0.099	0.816	0.129	0.775
4	0.184	0.901	0.225	0.871
5	0.290	0.963	0.341	0.947
6	0.421	0.996	0.479	0.993

n = 8	α = .05		α = .10	
	P_L	P_U	P_L	P_U
x = 1	0.003	0.527	0.006	0.471
2	0.032	0.651	0.046	0.600
3	0.085	0.755	0.111	0.711
4	0.157	0.843	0.193	0.807
5	0.245	0.915	0.289	0.889
6	0.349	0.968	0.400	0.954
7	0.473	0.997	0.529	0.994

n = 9	α = .05		α = .10	
	P_L	P_U	P_L	P_U
x = 1	0.003	0.482	0.006	0.429
2	0.028	0.600	0.041	0.550
3	0.075	0.701	0.098	0.655
4	0.137	0.788	0.169	0.749
5	0.212	0.863	0.251	0.831
6	0.299	0.925	0.345	0.902
7	0.400	0.972	0.450	0.959
8	0.518	0.997	0.571	0.994

n = 10	α = .05		α = .10	
	P_L	P_U	P_L	P_U
x = 1	0.003	0.445	0.005	0.394
2	0.025	0.556	0.037	0.507
3	0.067	0.652	0.087	0.607
4	0.122	0.738	0.150	0.696
5	0.187	0.813	0.222	0.778
6	0.262	0.878	0.304	0.850
7	0.348	0.933	0.393	0.913
8	0.444	0.975	0.493	0.963
9	0.555	0.997	0.606	0.995

n = 11	α = .05		α = .10	
	P_L	P_U	P_L	P_U
x = 1	0.002	0.413	0.005	0.364
2	0.023	0.518	0.033	0.470
3	0.060	0.610	0.079	0.564
4	0.109	0.692	0.135	0.650
5	0.167	0.766	0.200	0.729
6	0.234	0.833	0.271	0.800
7	0.308	0.891	0.350	0.865

n = 11	α = .05		α = .10	
8	0.390	0.940	0.436	0.921
9	0.482	0.977	0.530	0.967
10	0.587	0.998	0.636	0.995

n = 12	α = .05		α = .10	
	P_L	P_U	P_L	P_U
x = 1	0.002	0.385	0.004	0.339
2	0.021	0.484	0.030	0.438
3	0.055	0.572	0.072	0.527
4	0.099	0.651	0.123	0.609
5	0.152	0.723	0.181	0.685
6	0.211	0.789	0.245	0.755
7	0.277	0.848	0.315	0.819
8	0.349	0.901	0.391	0.877
9	0.428	0.945	0.473	0.928
10	0.516	0.979	0.562	0.970
11	0.615	0.998	0.661	0.996

n = 13	α = .05		α = .10	
	P_L	P_U	P_L	P_U
x = 1	0.002	0.360	0.004	0.316
2	0.019	0.454	0.028	0.410
3	0.050	0.538	0.066	0.495
4	0.091	0.614	0.113	0.573
5	0.139	0.684	0.166	0.645
6	0.192	0.749	0.224	0.713
7	0.251	0.808	0.287	0.776
8	0.316	0.861	0.355	0.834
9	0.386	0.909	0.427	0.887
10	0.462	0.950	0.505	0.934
11	0.546	0.981	0.590	0.972
12	0.640	0.998	0.684	0.996

n = 14	α = .05		α = .10	
	P_L	P_U	P_L	P_U
x = 1	0.002	0.339	0.004	0.297
2	0.018	0.428	0.026	0.385
3	0.047	0.508	0.061	0.466
4	0.084	0.581	0.104	0.540
5	0.128	0.649	0.153	0.610
6	0.177	0.711	0.206	0.675
7	0.230	0.770	0.264	0.736
8	0.289	0.823	0.325	0.794
9	0.351	0.872	0.390	0.847
10	0.419	0.916	0.460	0.896
11	0.492	0.953	0.534	0.939
12	0.572	0.982	0.615	0.974
13	0.661	0.998	0.703	0.996

n = 15	α = .05		α = .10	
	P_L	P_U	P_L	P_U
x = 1	0.002	0.319	0.003	0.279
2	0.017	0.405	0.024	0.363
3	0.043	0.481	0.057	0.440
4	0.078	0.551	0.097	0.511
5	0.118	0.616	0.142	0.577
6	0.163	0.677	0.191	0.640
7	0.213	0.734	0.244	0.700
8	0.266	0.787	0.300	0.756
9	0.323	0.837	0.360	0.809
10	0.384	0.882	0.423	0.858
11	0.449	0.922	0.489	0.903
12	0.519	0.957	0.560	0.943
13	0.595	0.983	0.637	0.976
14	0.681	0.998	0.721	0.997

■ TABLE A-8 (*continued*)

n = 16

x	P_L (α=.05)	P_U (α=.05)	P_L (α=.10)	P_U (α=.10)
1	0.002	0.302	0.003	0.264
2	0.016	0.383	0.023	0.344
3	0.040	0.456	0.053	0.417
4	0.073	0.524	0.090	0.484
5	0.110	0.587	0.132	0.548
6	0.152	0.646	0.178	0.609
7	0.198	0.701	0.227	0.667
8	0.247	0.753	0.279	0.721
9	0.299	0.802	0.333	0.773
10	0.354	0.848	0.391	0.822
11	0.413	0.890	0.452	0.868
12	0.476	0.927	0.516	0.910
13	0.544	0.960	0.583	0.947
14	0.617	0.984	0.656	0.977
15	0.698	0.998	0.736	0.997

n = 17

x	P_L (α=.05)	P_U (α=.05)	P_L (α=.10)	P_U (α=.10)
1	0.001	0.287	0.003	0.250
2	0.015	0.364	0.021	0.326
3	0.038	0.434	0.050	0.396
4	0.068	0.499	0.085	0.461
5	0.103	0.560	0.124	0.522
6	0.142	0.617	0.166	0.580
7	0.184	0.671	0.212	0.636
8	0.230	0.722	0.260	0.689
9	0.278	0.770	0.311	0.740
10	0.329	0.816	0.364	0.788
11	0.383	0.858	0.420	0.834
12	0.440	0.897	0.478	0.876
13	0.501	0.932	0.539	0.915
14	0.566	0.962	0.604	0.950
15	0.636	0.985	0.674	0.979
16	0.713	0.999	0.750	0.997

n = 18

x	P_L (α=.05)	P_U (α=.05)	P_L (α=.10)	P_U (α=.10)
1	0.001	0.273	0.003	0.238
2	0.014	0.347	0.020	0.310
3	0.036	0.414	0.047	0.377
4	0.064	0.476	0.080	0.439
5	0.097	0.535	0.116	0.498
6	0.133	0.590	0.156	0.554
7	0.173	0.643	0.199	0.608
8	0.215	0.692	0.244	0.659
9	0.260	0.740	0.291	0.709
10	0.308	0.785	0.341	0.756
11	0.357	0.827	0.392	0.801
12	0.410	0.867	0.446	0.844
13	0.465	0.903	0.502	0.884
14	0.524	0.936	0.561	0.920
15	0.586	0.964	0.623	0.953
16	0.653	0.986	0.690	0.980
17	0.727	0.999	0.762	0.997

n = 19

x	P_L (α=.05)	P_U (α=.05)	P_L (α=.10)	P_U (α=.10)
1	0.001	0.260	0.003	0.226
2	0.013	0.331	0.019	0.296
3	0.034	0.396	0.044	0.359
4	0.061	0.456	0.075	0.419
5	0.091	0.512	0.110	0.476
6	0.126	0.565	0.147	0.530
7	0.163	0.616	0.188	0.582
8	0.203	0.665	0.230	0.632
9	0.244	0.711	0.274	0.680
10	0.289	0.756	0.320	0.726
11	0.335	0.797	0.368	0.770
12	0.384	0.837	0.418	0.812
13	0.435	0.874	0.470	0.853
14	0.488	0.909	0.524	0.890
15	0.544	0.939	0.581	0.925
16	0.604	0.966	0.641	0.956
17	0.669	0.987	0.704	0.981
18	0.740	0.999	0.774	0.997

n = 20

x	P_L (α=.05)	P_U (α=.05)	P_L (α=.10)	P_U (α=.10)
1	0.001	0.249	0.003	0.216
2	0.012	0.317	0.018	0.283
3	0.032	0.379	0.042	0.344
4	0.057	0.437	0.071	0.401
5	0.087	0.491	0.104	0.456
6	0.119	0.543	0.140	0.508
7	0.154	0.592	0.177	0.558
8	0.191	0.639	0.217	0.606
9	0.231	0.685	0.259	0.653
10	0.272	0.728	0.302	0.698
11	0.315	0.769	0.347	0.741
12	0.361	0.809	0.394	0.783
13	0.408	0.846	0.442	0.823
14	0.457	0.881	0.492	0.860
15	0.509	0.913	0.544	0.896
16	0.563	0.943	0.599	0.929
17	0.621	0.968	0.656	0.958
18	0.683	0.988	0.717	0.982
19	0.751	0.999	0.784	0.997

■ TABLE A-9 Critical Values for the Durbin-Watson DW Statistic

(a) $\alpha = .05$

n	k = 1		k = 2		k = 3		k = 4		k = 5	
	d_L	d_U	d_L	d_U	d_L	d_U	d_L	d_U	d_L	d_U
15	1.08	1.36	0.95	1.54	0.82	1.75	0.69	1.97	0.56	2.21
16	1.10	1.37	0.98	1.54	0.86	1.73	0.74	1.93	0.62	2.15
17	1.13	1.38	1.02	1.54	0.90	1.71	0.78	1.90	0.67	2.10
18	1.16	1.39	1.05	1.53	0.93	1.69	0.82	1.87	0.71	2.06
19	1.18	1.40	1.08	1.53	0.97	1.68	0.86	1.85	0.75	2.02
20	1.20	1.41	1.10	1.54	1.00	1.68	0.90	1.83	0.79	1.99
21	1.22	1.42	1.13	1.54	1.03	1.67	0.93	1.81	0.83	1.96
22	1.24	1.43	1.15	1.54	1.05	1.66	0.96	1.80	0.86	1.94
23	1.26	1.44	1.17	1.54	1.08	1.66	0.99	1.79	0.90	1.92
24	1.27	1.45	1.19	1.55	1.10	1.66	1.01	1.78	0.93	1.90
25	1.29	1.45	1.21	1.55	1.12	1.66	1.04	1.77	0.95	1.89
26	1.30	1.46	1.22	1.55	1.14	1.65	1.06	1.76	0.98	1.88
27	1.32	1.47	1.24	1.56	1.16	1.65	1.08	1.76	1.01	1.86
28	1.33	1.48	1.26	1.56	1.18	1.65	1.10	1.75	1.03	1.85
29	1.34	1.48	1.27	1.56	1.20	1.65	1.12	1.74	1.05	1.84
30	1.35	1.49	1.28	1.57	1.21	1.65	1.14	1.74	1.07	1.83
31	1.36	1.50	1.30	1.57	1.23	1.65	1.16	1.74	1.09	1.83
32	1.37	1.50	1.31	1.57	1.24	1.65	1.18	1.73	1.11	1.82
33	1.38	1.51	1.32	1.58	1.26	1.65	1.19	1.73	1.13	1.81
34	1.39	1.51	1.33	1.58	1.27	1.65	1.21	1.73	1.15	1.81
35	1.40	1.52	1.34	1.58	1.28	1.65	1.22	1.73	1.16	1.80
36	1.41	1.52	1.35	1.59	1.29	1.65	1.24	1.73	1.18	1.80
37	1.42	1.53	1.36	1.59	1.31	1.66	1.25	1.72	1.19	1.80
38	1.43	1.54	1.37	1.59	1.32	1.66	1.26	1.72	1.21	1.79
39	1.43	1.54	1.38	1.60	1.33	1.66	1.27	1.72	1.22	1.79
40	1.44	1.54	1.39	1.60	1.34	1.66	1.29	1.72	1.23	1.79
45	1.48	1.57	1.43	1.62	1.38	1.67	1.34	1.72	1.29	1.78
50	1.50	1.59	1.46	1.63	1.42	1.67	1.38	1.72	1.34	1.77
55	1.53	1.60	1.49	1.64	1.45	1.68	1.41	1.72	1.38	1.77
60	1.55	1.62	1.51	1.65	1.48	1.69	1.44	1.73	1.41	1.77
65	1.57	1.63	1.54	1.66	1.50	1.70	1.47	1.73	1.44	1.77
70	1.58	1.64	1.55	1.67	1.52	1.70	1.49	1.74	1.46	1.77
75	1.60	1.65	1.57	1.68	1.54	1.71	1.51	1.74	1.49	1.77
80	1.61	1.66	1.59	1.69	1.56	1.72	1.53	1.74	1.51	1.77
85	1.62	1.67	1.60	1.70	1.57	1.72	1.55	1.75	1.52	1.77
90	1.63	1.68	1.61	1.70	1.59	1.73	1.57	1.75	1.54	1.78
95	1.64	1.69	1.62	1.71	1.60	1.73	1.58	1.75	1.56	1.78
100	1.65	1.69	1.63	1.72	1.61	1.74	1.59	1.76	1.57	1.78

From J. Durbin and G. S. Watson, "Testing for Serial Correlation in Least Squares Regression, II," *Biometrika*, 1951, *30*, 159–178. Reproduced by permission of the *Biometrika* trustees.

■ **TABLE A-9**   (*continued*)

(b) $\alpha = .05$

n	k = 1 d_L	k = 1 d_U	k = 2 d_L	k = 2 d_U	k = 3 d_L	k = 3 d_U	k = 4 d_L	k = 4 d_U	k = 5 d_L	k = 5 d_U
15	0.81	1.07	0.70	1.25	0.59	1.46	0.49	1.70	0.39	1.96
16	0.84	1.09	0.74	1.25	0.63	1.44	0.53	1.66	0.44	1.90
17	0.87	1.10	0.77	1.25	0.67	1.43	0.57	1.63	0.48	1.85
18	0.90	1.12	0.80	1.26	0.71	1.42	0.61	1.60	0.52	1.80
19	0.93	1.13	0.83	1.26	0.74	1.41	0.65	1.58	0.56	1.77
20	0.95	1.15	0.86	1.27	0.77	1.41	0.68	1.57	0.60	1.74
21	0.97	1.16	0.89	1.27	0.80	1.41	0.72	1.55	0.63	1.71
22	1.00	1.17	0.91	1.28	0.83	1.40	0.75	1.54	0.66	1.69
23	1.02	1.19	0.94	1.29	0.86	1.40	0.77	1.53	0.70	1.67
24	1.04	1.20	0.96	1.30	0.88	1.41	0.80	1.53	0.72	1.66
25	1.05	1.21	0.98	1.30	0.90	1.41	0.83	1.52	0.75	1.65
26	1.07	1.22	1.00	1.31	0.93	1.41	0.85	1.52	0.78	1.64
27	1.09	1.23	1.02	1.32	0.95	1.41	0.88	1.51	0.81	1.63
28	1.10	1.24	1.04	1.32	0.97	1.41	0.90	1.51	0.83	1.62
29	1.12	1.25	1.05	1.33	0.99	1.42	0.92	1.51	0.85	1.61
30	1.13	1.26	1.07	1.34	1.01	1.42	0.94	1.51	0.88	1.61
31	1.15	1.27	1.08	1.34	1.02	1.42	0.96	1.51	0.90	1.60
32	1.16	1.28	1.10	1.35	1.04	1.43	0.98	1.51	0.92	1.60
33	1.17	1.29	1.11	1.36	1.05	1.43	1.00	1.51	0.94	1.59
34	1.18	1.30	1.13	1.36	1.07	1.43	1.01	1.51	0.95	1.59
35	1.19	1.31	1.14	1.37	1.08	1.44	1.03	1.51	0.97	1.59
36	1.21	1.32	1.15	1.38	1.10	1.44	1.04	1.51	0.99	1.59
37	1.22	1.32	1.16	1.38	1.11	1.45	1.06	1.51	1.00	1.59
38	1.23	1.33	1.18	1.39	1.12	1.45	1.07	1.52	1.02	1.58
39	1.24	1.34	1.19	1.39	1.14	1.45	1.09	1.52	1.03	1.58
40	1.25	1.34	1.20	1.40	1.15	1.46	1.10	1.52	1.05	1.58
45	1.29	1.38	1.24	1.42	1.20	1.48	1.16	1.53	1.11	1.58
50	1.32	1.40	1.28	1.45	1.24	1.49	1.20	1.54	1.16	1.59
55	1.36	1.43	1.32	1.47	1.28	1.51	1.25	1.55	1.21	1.59
60	1.38	1.45	1.35	1.48	1.32	1.52	1.28	1.56	1.25	1.60
65	1.41	1.47	1.38	1.50	1.35	1.53	1.31	1.57	1.28	1.61
70	1.43	1.49	1.40	1.52	1.37	1.55	1.34	1.58	1.31	1.61
75	1.45	1.50	1.42	1.53	1.39	1.56	1.37	1.59	1.34	1.62
80	1.47	1.52	1.44	1.54	1.42	1.57	1.39	1.60	1.36	1.62
85	1.48	1.53	1.46	1.55	1.43	1.58	1.41	1.60	1.39	1.63
90	1.50	1.54	1.47	1.56	1.45	1.59	1.43	1.61	1.41	1.64
95	1.51	1.55	1.49	1.57	1.47	1.60	1.45	1.62	1.42	1.64
100	1.52	1.56	1.50	1.58	1.48	1.60	1.46	1.63	1.44	1.65

■ **TABLE A-10** Distribution Function for the Mann-Whitney U Statistic*.

This table contains the value of $P(U \leq U_0)$ where $n_1 \leq n_2$.

$n_2 = 3$ U_0	1	n_1 2	3
0	.25	.10	.05
1	.50	.20	.10
2		.40	.20
3		.60	.35
4			.50

$n_2 = 4$ U_0	1	2	n_1 3	4
0	.2000	.0667	.0286	.0143
1	.4000	.1333	.0571	.0286
2	.6000	.2667	.1143	.0571
3		.4000	.2000	.1000
4		.6000	.3143	.1714
5			.4286	.2429
6			.5714	.3429
7				.4429
8				.5571

$n_2 = 5$ U_0	1	2	n_1 3	4	5
0	.1667	.0476	.0179	.0079	.0040
1	.3333	.0952	.0357	.0159	.0079
2	.5000	.1905	.0714	.0317	.0159
3		.2857	.1250	.0556	.0278
4		.4286	.1964	.0952	.0476
5		.5714	.2857	.1429	.0754
6			.3929	.2063	.1111
7			.5000	.2778	.1548
8				.3651	.2103
9				.4524	.2738
10				.5476	.3452
11					.4206
12					.5000

*Computed by M. Pagano, Dept. of Statistics, University of Florida. Reprinted by permission from *Statistics for Management and Economics*, 5th ed., by William Mendenhall and James E. Reinmuth. Copyright © 1986 by PWS-KENT Publishers, Boston.

■ **TABLE A-10**   (*continued*)

$n_2 = 6$ U_0	1	2	3	4	5	6
0	.1429	.0357	.0119	.0048	.0022	.0011
1	.2857	.0714	.0238	.0095	.0043	.0022
2	.4286	.1429	.0476	.0190	.0087	.0043
3	.5714	.2143	.0833	.0333	.0152	.0076
4		.3214	.1310	.0571	.0260	.0130
5		.4286	.1905	.0857	.0411	.0206
6		.5714	.2738	.1286	.0628	.0325
7			.3571	.1762	.0887	.0465
8			.4524	.2381	.1234	.0660
9			.5476	.3048	.1645	.0898
10				.3810	.2143	.1201
11				.4571	.2684	.1548
12				.5429	.3312	.1970
13					.3961	.2424
14					.4654	.2944
15					.5346	.3496
16						.4091
17						.4686
18						.5314

$n_2 = 7$ U_0	1	2	3	4	5	6	7
0	.1250	.0278	.0083	.0030	.0013	.0006	.0003
1	.2500	.0556	.0167	.0061	.0025	.0012	.0006
2	.3750	.1111	.0333	.0121	.0051	.0023	.0012
3	.5000	.1667	.0583	.0212	.0088	.0041	.0020
4		.2500	.0917	.0364	.0152	.0070	.0035
5		.3333	.1333	.0545	.0240	.0111	.0055
6		.4444	.1917	.0818	.0366	.0175	.0087
7		.5556	.2583	.1152	.0530	.0256	.0131
8			.3333	.1576	.0745	.0367	.0189
9			.4167	.2061	.1010	.0507	.0265
10			.5000	.2636	.1338	.0688	.0364
11				.3242	.1717	.0903	.0487
12				.3939	.2159	.1171	.0641
13				.4636	.2652	.1474	.0825
14				.5364	.3194	.1830	.1043
15					.3775	.2226	.1297
16					.4381	.2669	.1588
17					.5000	.3141	.1914
18						.3654	.2279
19						.4178	.2675
20						.4726	.3100
21						.5274	.3552
22							.4024
23							.4508
24							.5000

■ **TABLE A-10**   (*continued*)

$n_2 = 8$	U_0	1	2	3	4	5	6	7	8
					n_1				
	0	.1111	.0222	.0061	.0020	.0008	.0003	.0002	.0001
	1	.2222	.0444	.0121	.0040	.0016	.0007	.0003	.0002
	2	.3333	.0889	.0242	.0081	.0031	.0013	.0006	.0003
	3	.4444	.1333	.0424	.0141	.0054	.0023	.0011	.0005
	4	.5556	.2000	.0667	.0242	.0093	.0040	.0019	.0009
	5		.2667	.0970	.0364	.0148	.0063	.0030	.0015
	6		.3556	.1394	.0545	.0225	.0100	.0047	.0023
	7		.4444	.1879	.0768	.0326	.0147	.0070	.0035
	8		.5556	.2485	.1071	.0466	.0213	.0103	.0052
	9			.3152	.1414	.0637	.0296	.0145	.0074
	10			.3879	.1838	.0855	.0406	.0200	.0103
	11			.4606	.2303	.1111	.0539	.0270	.0141
	12			.5394	.2848	.1422	.0709	.0361	.0190
	13				.3414	.1772	.0906	.0469	.0249
	14				.4040	.2176	.1142	.0603	.0325
	15				.4667	.2618	.1412	.0760	.0415
	16				.5333	.3108	.1725	.0946	.0524
	17					.3621	.2068	.1159	.0652
	18					.4165	.2454	.1405	.0803
	19					.4716	.2864	.1678	.0974
	20					.5284	.3310	.1984	.1172
	21						.3773	.2317	.1393
	22						.4259	.2679	.1641
	23						.4749	.3063	.1911
	24						.5251	.3472	.2209
	25							.3894	.2527
	26							.4333	.2869
	27							.4775	.3227
	28							.5225	.3605
	29								.3992
	30								.4392
	31								.4796
	32								.5204

■ **TABLE A-10**   (*continued*)

$n_2 = 9$ U_0	1	2	3	4	n_1 5	6	7	8	9
0	.1000	.0182	.0045	.0014	.0005	.0002	.0001	.0000	.0000
1	.2000	.0364	.0091	.0028	.0010	.0004	.0002	.0001	.0000
2	.3000	.0727	.0182	.0056	.0020	.0008	.0003	.0002	.0001
3	.4000	.1091	.0318	.0098	.0035	.0014	.0006	.0003	.0001
4	.5000	.1636	.0500	.0168	.0060	.0024	.0010	.0005	.0002
5		.2182	.0727	.0252	.0095	.0038	.0017	.0008	.0004
6		.2909	.1045	.0378	.0145	.0060	.0026	.0012	.0006
7		.3636	.1409	.0531	.0210	.0088	.0039	.0019	.0009
8		.4545	.1864	.0741	.0300	.0128	.0058	.0028	.0014
9		.5455	.2409	.0993	.0415	.0180	.0082	.0039	.0020
10			.3000	.1301	.0559	.0248	.0115	.0056	.0028
11			.3636	.1650	.0734	.0332	.0156	.0076	.0039
12			.4318	.2070	.0949	.0440	.0209	.0103	.0053
13			.5000	.2517	.1199	.0567	.0274	.0137	.0071
14				.3021	.1489	.0723	.0356	.0180	.0094
15				.3552	.1818	.0905	.0454	.0232	.0122
16				.4126	.2188	.1119	.0571	.0296	.0157
17				.4699	.2592	.1361	.0708	.0372	.0200
18				.5301	.3032	.1638	.0869	.0464	.0252
19					.3497	.1942	.1052	.0570	.0313
20					.3986	.2280	.1261	.0694	.0385
21					.4491	.2643	.1496	.0836	.0470
22					.5000	.3035	.1755	.0998	.0567
23						.3445	.2039	.1179	.0680
24						.3878	.2349	.1383	.0807
25						.4320	.2680	.1606	.0951
26						.4773	.3032	.1852	.1112
27						.5227	.3403	.2117	.1290
28							.3788	.2404	.1487
29							.4185	.2707	.1701
30							.4591	.3029	.1933
31							.5000	.3365	.2181
32								.3715	.2447
33								.4074	.2729
34								.4442	.3024
35								.4813	.3332
36								.5187	.3652
37									.3981
38									.4317
39									.4657
40									.5000

■ TABLE A-10 (*continued*)

$n_2 = 10$ U_0	1	2	3	4	5	6	7	8	9	10
0	.0909	.0152	.0035	.0010	.0003	.0001	.0001	.0000	.0000	.0000
1	.1818	.0303	.0070	.0020	.0007	.0002	.0001	.0000	.0000	.0000
2	.2727	.0606	.0140	.0040	.0013	.0005	.0002	.0001	.0000	.0000
3	.3636	.0909	.0245	.0070	.0023	.0009	.0004	.0002	.0001	.0000
4	.4545	.1364	.0385	.0120	.0040	.0015	.0006	.0003	.0001	.0001
5	.5455	.1818	.0559	.0180	.0063	.0024	.0010	.0004	.0002	.0001
6		.2424	.0804	.0270	.0097	.0037	.0015	.0007	.0003	.0002
7		.3030	.1084	.0380	.0140	.0055	.0023	.0010	.0005	.0002
8		.3788	.1434	.0529	.0200	.0080	.0034	.0015	.0007	.0004
9		.4545	.1853	.0709	.0276	.0112	.0048	.0022	.0011	.0005
10		.5455	.2343	.0939	.0376	.0156	.0068	.0031	.0015	.0008
11			.2867	.1199	.0496	.0210	.0093	.0043	.0021	.0010
12			.3462	.1518	.0646	.0280	.0125	.0058	.0028	.0014
13			.4056	.1868	.0823	.0363	.0165	.0078	.0038	.0019
14			.4685	.2268	.1032	.0467	.0215	.0103	.0051	.0026
15			.5315	.2697	.1272	.0589	.0277	.0133	.0066	.0034
16				.3177	.1548	.0736	.0351	.0171	.0086	.0045
17				.3666	.1855	.0903	.0439	.0217	.0110	.0057
18				.4196	.2198	.1099	.0544	.0273	.0140	.0073
19				.4725	.2567	.1317	.0665	.0338	.0175	.0093
20				.5275	.2970	.1566	.0806	.0416	.0217	.0116
21					.3393	.1838	.0966	.0506	.0267	.0144
22					.3839	.2139	.1148	.0610	.0326	.0177
23					.4296	.2461	.1349	.0729	.0394	.0216
24					.4765	.2811	.1574	.0864	.0474	.0262
25					.5235	.3177	.1819	.1015	.0564	.0315
26						.3564	.2087	.1185	.0667	.0376
27						.3962	.2374	.1371	.0782	.0446
28						.4374	.2681	.1577	.0912	.0526
29						.4789	.3004	.1800	.1055	.0615
30						.5211	.3345	.2041	.1214	.0716
31							.3698	.2299	.1388	.0827
32							.4063	.2574	.1577	.0952
33							.4434	.2863	.1781	.1088
34							.4811	.3167	.2001	.1237
35							.5189	.3482	.2235	.1399
36								.3809	.2483	.1575
37								.4143	.2745	.1763
38								.4484	.3019	.1965
39								.4827	.3304	.2179
40								.5173	.3598	.2406
41									.3901	.2644
42									.4211	.2894
43									.4524	.3153
44									.4841	.3421
45									.5159	.3697
46										.3980
47										.4267
48										.4559
49										.4853
50										.5147

■ **TABLE A-11**   Critical Values of the Wilcoxon Signed Rank Test
($n = 5, \ldots , 50$)

1-sided	2-sided	$n = 5$	$n = 6$	$n = 7$	$n = 8$	$n = 9$	$n = 10$
$\alpha = .05$	$\alpha = .10$	1	2	4	6	8	11
$\alpha = .025$	$\alpha = .05$		1	2	4	6	8
$\alpha = .01$	$\alpha = .02$			0	2	3	5
$\alpha = .005$	$\alpha = .01$				0	2	3

1-sided	2-sided	$n = 11$	$n = 12$	$n = 13$	$n = 14$	$n = 15$	$n = 16$
$\alpha = .05$	$\alpha = .10$	14	17	21	26	30	36
$\alpha = .025$	$\alpha = .05$	11	14	17	21	25	30
$\alpha = .01$	$\alpha = .02$	7	10	13	16	20	24
$\alpha = .005$	$\alpha = .01$	5	7	10	13	16	19

1-sided	2-sided	$n = 17$	$n = 18$	$n = 19$	$n = 20$	$n = 21$	$n = 22$
$\alpha = .05$	$\alpha = .10$	41	47	54	60	68	75
$\alpha = .025$	$\alpha = .05$	35	40	46	52	59	66
$\alpha = .01$	$\alpha = .02$	28	33	38	43	49	56
$\alpha = .005$	$\alpha = .01$	23	28	32	37	43	49

1-sided	2-sided	$n = 23$	$n = 24$	$n = 25$	$n = 26$	$n = 27$	$n = 28$
$\alpha = .05$	$\alpha = .10$	83	92	101	110	120	130
$\alpha = .025$	$\alpha = .05$	73	81	90	98	107	117
$\alpha = .01$	$\alpha = .02$	62	69	77	85	93	102
$\alpha = .005$	$\alpha = .01$	55	61	68	76	84	92

1-sided	2-sided	$n = 29$	$n = 30$	$n = 31$	$n = 32$	$n = 33$	$n = 34$
$\alpha = .05$	$\alpha = .10$	141	152	163	175	188	201
$\alpha = .025$	$\alpha = .05$	127	137	148	159	171	183
$\alpha = .01$	$\alpha = .02$	111	120	130	141	151	162
$\alpha = .005$	$\alpha = .01$	100	109	118	128	138	149

1-sided	2-sided	$n = 35$	$n = 36$	$n = 37$	$n = 38$	$n = 39$	
$\alpha = .05$	$\alpha = .10$	214	228	242	256	271	
$\alpha = .025$	$\alpha = .05$	195	208	222	235	250	
$\alpha = .01$	$\alpha = .02$	174	186	198	211	224	
$\alpha = .005$	$\alpha = .01$	160	171	183	195	208	

1-sided	2-sided	$n = 40$	$n = 41$	$n = 42$	$n = 43$	$n = 44$	$n = 45$
$\alpha = .05$	$\alpha = .10$	287	303	319	336	353	371
$\alpha = .025$	$\alpha = .05$	264	279	295	311	327	344
$\alpha = .01$	$\alpha = .02$	238	252	267	281	297	313
$\alpha = .005$	$\alpha = .01$	221	234	248	262	277	292

1-sided	2-sided	$n = 46$	$n = 47$	$n = 48$	$n = 49$	$n = 50$	
$\alpha = .05$	$\alpha = .10$	389	408	427	446	466	
$\alpha = .025$	$\alpha = .05$	361	379	397	415	434	
$\alpha = .01$	$\alpha = .02$	329	345	362	380	398	
$\alpha = .005$	$\alpha = .01$	307	323	339	356	373	

From F. Wilcoxon and R. A. Wilcox, "Some Rapid Approximate Statistical Procedures," 1964.
Reprinted by permission of Lederle Labs, a division of the American Cyanamid Co.

▣ TABLE A-12 Critical Values of Spearman's Rank Correlation Coefficient

n	$\alpha = .05$	$\alpha = .025$	$\alpha = .01$	$\alpha = .005$
5	0.900	—	—	—
6	0.829	0.886	0.943	—
7	0.714	0.786	0.893	—
8	0.643	0.738	0.833	0.881
9	0.600	0.683	0.783	0.833
10	0.564	0.648	0.745	0.794
11	0.523	0.623	0.736	0.818
12	0.497	0.591	0.703	0.780
13	0.475	0.566	0.673	0.745
14	0.457	0.545	0.646	0.716
15	0.441	0.525	0.623	0.689
16	0.425	0.507	0.601	0.666
17	0.412	0.490	0.582	0.645
18	0.399	0.476	0.564	0.625
19	0.388	0.462	0.549	0.608
20	0.377	0.450	0.534	0.591
21	0.368	0.438	0.521	0.576
22	0.359	0.428	0.508	0.562
23	0.351	0.418	0.496	0.549
24	0.343	0.409	0.485	0.537
25	0.336	0.400	0.475	0.526
26	0.329	0.392	0.465	0.515
27	0.323	0.385	0.456	0.505
28	0.317	0.377	0.448	0.496
29	0.311	0.370	0.440	0.487
30	0.305	0.364	0.432	0.478

*From E. G. Olds, "Distribution of Sums of Squares of Rank Differences for Small Samples," *Annals of Mathematical Statistics*, Vol. 9 (1938). Reprinted with permission of the Institute of Mathematical Statistics.

■ TABLE A-13   Random Numbers

12651	61646	11769	75109	86996	97669	25757	32535	07122	76763
81769	74436	02630	72310	45049	18029	07469	42341	98173	79260
36737	98863	77240	76251	00654	64688	09343	70278	67331	98729
82861	54371	76610	94934	72748	44124	05610	53750	95938	01485
21325	15732	24127	37431	09723	63529	73977	95218	96074	42138
74146	47887	62463	23045	41490	07954	22597	60012	98866	90959
90759	64410	54179	66075	61051	75385	51378	08360	95946	95547
55683	98078	02238	91540	21219	17720	87817	41705	95785	12563
79686	17969	76061	83748	55920	83612	41540	86492	06447	60568
70333	00201	86201	69716	78185	62154	77930	67663	29529	75116
14042	53536	07779	04157	41172	36473	42123	43929	50533	33437
59911	08256	06596	48416	69770	68797	56080	14223	59199	30162
62368	62623	62742	14891	39247	52242	98832	69533	91174	57979
57529	97751	54976	48957	74599	08759	78494	52785	68526	64618
15469	90574	78033	66885	13936	42117	71831	22961	94225	31816
18625	23674	53850	32827	81647	80820	00420	63555	74489	80141
74626	68394	88562	70745	23701	45630	65891	58220	35442	60414
11119	16519	27384	90199	79210	76965	99546	30323	31664	22845
41101	17336	48951	53674	17880	45260	08575	49321	36191	17095
32123	91576	84221	78902	82010	30847	62329	63898	23268	74283
26091	68409	69704	82267	14751	13151	93115	01437	56945	89661
67680	79790	48462	59278	44185	29616	76531	19589	83139	28454
15184	19260	14073	07026	25264	08388	27182	22557	61501	67481
58010	45039	57181	10238	36874	28546	37444	80824	63981	39942
56425	53996	86245	32623	78858	08143	60377	42925	42815	11159
82630	84066	13592	60642	17904	99718	63432	88642	37858	25431
14927	40909	23900	48761	44860	92467	31742	87142	03607	32059
23740	22505	07489	85986	74420	21744	97711	36648	35620	97949
32990	97446	03711	63824	07953	85965	87089	11687	92414	67257
05310	24058	91946	78437	34365	82469	12430	84754	19354	72745
21839	39937	27534	88913	49055	19218	47712	67677	51889	70926
08833	42549	93981	94051	28382	83725	72643	64233	97252	17133
58336	11139	47479	00931	91560	95372	97642	33856	54825	55680
62032	91144	75478	47431	52726	30289	42411	91886	51818	78292
45171	30557	53116	04118	58301	24375	65609	85810	18620	49198
91611	62656	60128	35609	63698	78356	50682	22505	01692	36291
55472	63819	86314	49174	93582	73604	78614	78849	23096	72825
18573	09729	74091	53994	10970	86557	65661	41854	26037	53296
60866	02955	90288	82136	83644	94455	06560	78029	98768	71296
45043	55608	82767	60890	74646	79485	13619	98868	40857	19415
17831	09737	79473	75945	28394	79334	70577	38048	03607	06932
40137	03981	07585	18128	11178	32601	27994	05641	22600	86064
77776	31343	14576	97706	16039	47517	43300	59080	80392	63189
69605	44104	40103	95635	05635	81673	68657	09559	23510	95875
19916	52934	26499	09821	97331	80993	61299	36979	73599	35055
02606	58552	07678	56619	65325	30705	99582	53390	46357	13244
65183	73160	87131	35530	47946	09854	18080	02321	05809	04893
10740	98914	44916	11322	89717	88189	30143	52687	19420	60061
98642	89822	71691	51573	83666	61642	46683	33761	47542	23551
60139	25601	93663	25547	02654	94829	48672	28736	84994	13071

■ **TABLE A-14** Critical Values of Hartley's *H*-statistic, $\alpha = .05$

n = number of observations in each sample
k = number of samples

n	2	3	4	5	6	k 7	8	9	10	11	12
3	39.0	87.5	142	202	266	333	403	475	550	626	704
4	15.4	27.8	39.2	50.7	62.0	72.9	83.5	93.9	104	114	124
5	9.60	15.5	20.6	25.2	29.5	33.6	37.5	41.1	44.6	48.0	51.4
6	7.15	10.8	13.7	16.3	18.7	20.8	22.9	24.7	26.5	28.2	29.9
7	5.82	8.38	10.4	12.1	13.7	15.0	16.3	17.5	18.6	19.7	20.7
8	4.99	6.94	8.44	9.70	10.8	11.8	12.7	13.5	14.3	15.1	15.8
9	4.43	6.00	7.18	8.12	9.03	9.78	10.5	11.1	11.7	12.2	12.7
10	4.03	5.34	6.31	7.11	7.80	8.41	8.95	9.45	9.91	10.3	10.7
11	3.72	4.85	5.67	6.34	6.92	7.42	7.87	8.28	8.66	9.01	9.34
13	3.28	4.16	4.79	5.30	5.72	6.09	6.42	6.72	7.00	7.25	7.48
16	2.86	3.54	4.01	4.37	4.68	4.95	5.19	5.40	5.59	5.77	5.93
21	2.46	2.95	3.29	3.54	3.76	3.94	4.10	4.24	4.37	4.49	4.59
31	2.07	2.40	2.61	2.78	2.91	3.02	3.12	3.21	3.29	3.36	3.39
61	1.67	1.85	1.96	2.04	2.11	2.17	2.22	2.26	2.30	2.33	2.36
∞	1.00	1.00	1.00	1.00	1.00	1.00	1.00	1.00	1.00	1.00	1.00

■ **TABLE A-15**   Distribution Function for the Number of Runs R, in Samples of Size (n_1, n_2).

Each entry is $P(R \le a)$.

(n_1, n_2)	2	3	4	5	6	7	8	9	10
(2, 3)	.200	.500	.900	1.000					
(2, 4)	.133	.400	.800	1.000					
(2, 5)	.095	.333	.714	1.000					
(2, 6)	.071	.286	.643	1.000					
(2, 7)	.056	.250	.583	1.000					
(2, 8)	.044	.222	.533	1.000					
(2, 9)	.036	.200	.491	1.000					
(2, 10)	.030	.182	.455	1.000					
(3, 3)	.100	.300	.700	.900	1.000	1.000			
(3, 4)	.057	.200	.543	.800	.971	1.000			
(3, 5)	.036	.143	.429	.714	.929	1.000			
(3, 6)	.024	.107	.345	.643	.881	1.000			
(3, 7)	.017	.083	.283	.583	.833	1.000			
(3, 8)	.012	.067	.236	.533	.788	1.000			
(3, 9)	.009	.055	.200	.491	.745	1.000			
(3, 10)	.007	.045	.171	.455	.706	1.000			
(4, 4)	.029	.114	.371	.629	.886	.971	1.000		
(4, 5)	.016	.071	.262	.500	.786	.929	.992	1.000	
(4, 6)	.010	.048	.190	.405	.690	.881	.976	1.000	
(4, 7)	.006	.033	.142	.333	.606	.833	.954	1.000	
(4, 8)	.004	.024	.109	.279	.533	.788	.929	1.000	
(4, 9)	.003	.018	.085	.236	.471	.745	.902	1.000	
(4, 10)	.002	.014	.068	.203	.419	.706	.874	1.000	
(5, 5)	.008	.040	.167	.357	.643	.833	.960	.992	1.000
(5, 6)	.004	.024	.110	.262	.522	.738	.911	.976	.998
(5, 7)	.003	.015	.076	.197	.424	.652	.854	.955	.992
(5, 8)	.002	.010	.054	.152	.347	.576	.793	.929	.984
(5, 9)	.001	.007	.039	.119	.287	.510	.734	.902	.972
(5, 10)	.001	.005	.029	.095	.239	.455	.678	.874	.958
(6, 6)	.002	.013	.067	.175	.392	.608	.825	.933	.987
(6, 7)	.001	.008	.043	.121	.296	.500	.733	.879	.966
(6, 8)	.001	.005	.028	.086	.226	.413	.646	.821	.937
(6, 9)	.000	.003	.019	.063	.175	.343	.566	.762	.902
(6, 10)	.000	.002	.013	.047	.137	.288	.497	.706	.864
(7, 7)	.001	.004	.025	.078	.209	.383	.617	.791	.922
(7, 8)	.000	.002	.015	.051	.149	.296	.514	.704	.867
(7, 9)	.000	.001	.010	.035	.108	.231	.427	.622	.806
(7, 10)	.000	.001	.006	.024	.080	.182	.355	.549	.743
(8, 8)	.000	.001	.009	.032	.100	.214	.405	.595	.786
(8, 9)	.000	.001	.005	.020	.069	.157	.319	.500	.702
(8, 10)	.000	.000	.003	.013	.048	.117	.251	.419	.621
(9, 9)	.000	.000	.003	.012	.044	.109	.238	.399	.601
(9, 10)	.000	.000	.002	.008	.029	.077	.179	.319	.510
(10, 10)	.000	.000	.001	.004	.019	.051	.128	.242	.414

From F. Swed and C. Eisenhart, "Tables for Testing Randomness of Grouping in a Sequence of Alternatives," Annals of Mathematical Statistics, Vol. 14 (1943). Reproduced with permission of the Institute of Mathematical Statistics.

■ TABLE A-15   (*continued*)

(n_1, n_2)	11	12	13	14	15	16	17	18	19	20
(2, 3)										
(2, 4)										
(2, 5)										
(2, 6)										
(2, 7)										
(2, 8)										
(2, 9)										
(2, 10)										
(3, 3)										
(3, 4)										
(3, 5)										
(3, 6)										
(3, 7)										
(3, 8)										
(3, 9)										
(3, 10)										
(4, 4)										
(4, 5)										
(4, 6)										
(4, 7)										
(4, 8)										
(4, 9)										
(4, 10)										
(5, 5)										
(5, 6)	1.000									
(5, 7)	1.000									
(5, 8)	1.000									
(5, 9)	1.000									
(5, 10)	1.000									
(6, 6)	.998	1.000								
(6, 7)	.992	.999	1.000							
(6, 8)	.984	.998	1.000							
(6, 9)	.972	.994	1.000							
(6, 10)	.958	.990	1.000							
(7, 7)	.975	.996	.999	1.000						
(7, 8)	.949	.988	.998	1.000	1.000					
(7, 9)	.916	.975	.994	.999	1.000					
(7, 10)	.879	.957	.990	.998	1.000					
(8, 8)	.900	.968	.991	.999	1.000	1.000				
(8, 9)	.843	.939	.980	.996	.999	1.000	1.000			
(8, 10)	.782	.903	.964	.990	.998	1.000	1.000			
(9, 9)	.762	.891	.956	.988	.997	1.000	1.000	1.000		
(9, 10)	.681	.834	.923	.974	.992	.999	1.000	1.000	1.000	
(10, 10)	.586	.758	.872	.949	.981	.996	.999	1.000	1.000	1.000

◼ TABLE A-16   Critical Values of the Studentized Range (Q) Distribution.

The values listed in the table are the critical values of Q for $\alpha = .05$ and .01, as a function of degrees of freedom for MS(error) and k (the number of means)

| df for MS(ERROR) (v) | α | \multicolumn{10}{c|}{k (NUMBER OF MEANS)} |
		2	3	4	5	6	7	8	9	10	11
5	.05	3.64	4.60	5.22	5.67	6.03	6.33	6.58	6.80	6.99	7.17
	.01	5.70	6.98	7.80	8.42	8.91	9.32	9.67	9.97	10.24	10.48
6	.05	3.46	4.34	4.90	5.30	5.63	5.90	6.12	6.32	6.49	6.65
	.01	5.24	6.33	7.03	7.56	7.97	8.32	8.61	8.87	9.10	9.30
7	.05	3.34	4.16	4.68	5.06	5.36	5.61	5.82	6.00	6.16	6.30
	.01	4.95	5.92	6.54	7.01	7.37	7.68	7.94	8.17	8.37	8.55
8	.05	3.26	4.04	4.53	4.89	5.17	5.40	5.60	5.77	5.92	6.05
	.01	4.75	5.64	6.20	6.62	6.96	7.24	7.47	7.68	7.86	8.03
9	.05	3.20	3.95	4.41	4.76	5.02	5.24	5.43	5.59	5.74	5.87
	.01	4.60	5.43	5.96	6.35	6.66	6.91	7.13	7.33	7.49	7.65
10	.05	3.15	3.88	4.33	4.65	4.91	5.12	5.30	5.46	5.60	5.72
	.01	4.48	5.27	5.77	6.14	6.43	6.67	6.87	7.05	7.21	7.36
11	.05	3.11	3.82	4.26	4.57	4.82	5.03	5.20	5.35	5.49	5.61
	.01	4.39	5.15	5.62	5.97	6.25	6.48	6.67	6.84	6.99	7.13
12	.05	3.08	3.77	4.20	4.51	4.75	4.95	5.12	5.27	5.39	5.51
	.01	4.32	5.05	5.50	5.84	6.10	6.32	6.51	6.67	6.81	6.94
13	.05	3.06	3.73	4.15	4.45	4.69	4.88	5.05	5.19	5.32	5.43
	.01	4.26	4.96	5.40	5.73	5.98	6.19	6.37	6.53	6.67	6.79
14	.05	3.03	3.70	4.11	4.41	4.64	4.83	4.99	5.13	5.25	5.36
	.01	4.21	4.89	5.32	5.63	5.88	6.08	6.26	6.41	6.54	6.66
15	.05	3.01	3.67	4.08	4.37	4.59	4.78	4.94	5.08	5.20	5.31
	.01	4.17	4.84	5.25	5.56	5.80	5.99	6.16	6.31	6.44	6.55
16	.05	3.00	3.65	4.05	4.33	4.56	4.74	4.90	5.03	5.15	5.26
	.01	4.13	4.79	5.19	5.49	5.72	5.92	6.08	6.22	6.35	6.46
17	.05	2.98	3.63	4.02	4.30	4.52	4.70	4.86	4.99	5.11	5.21
	.01	4.10	4.74	5.14	5.43	5.66	5.85	6.01	6.15	6.27	6.38
18	.05	2.97	3.61	4.00	4.28	4.49	4.67	4.82	4.96	5.07	5.17
	.01	4.07	4.70	5.09	5.38	5.60	5.79	5.94	6.08	6.20	6.31
19	.05	2.96	3.59	3.98	4.25	4.47	4.65	4.79	4.92	5.04	5.14
	.01	4.05	4.67	5.05	5.33	5.55	5.73	5.89	6.02	6.14	6.25
20	.05	2.95	3.58	3.96	4.23	4.45	4.62	4.77	4.90	5.01	5.11
	.01	4.02	4.64	5.02	5.29	5.51	5.69	5.84	5.97	6.09	6.19
24	.05	2.92	3.53	3.90	4.17	4.37	4.54	4.68	4.81	4.92	5.01
	.01	3.96	4.55	4.91	5.17	5.37	5.54	5.69	5.81	5.92	6.02
30	.05	2.89	3.49	3.85	4.10	4.30	4.46	4.60	4.72	4.82	4.92
	.01	3.89	4.45	4.80	5.05	5.24	5.40	5.54	5.65	5.76	5.85
40	.05	2.86	3.44	3.79	4.04	4.23	4.39	4.52	4.63	4.73	4.82
	.01	3.82	4.37	4.70	4.93	5.11	5.26	5.39	5.50	5.60	5.69
60	.05	2.83	3.40	3.74	3.98	4.16	4.31	4.44	4.55	4.65	4.73
	.01	3.76	4.28	4.59	4.82	4.99	5.13	5.25	5.36	5.45	5.53
120	.05	2.80	3.36	3.68	3.92	4.10	4.24	4.36	4.47	4.56	4.64
	.01	3.70	4.20	4.50	4.71	4.87	5.01	5.12	5.21	5.30	5.37
∞	.05	2.77	3.31	3.63	3.86	4.03	4.17	4.29	4.39	4.47	4.55
	.01	3.64	4.12	4.40	4.60	4.76	4.88	4.99	5.08	5.16	5.23

From E. S. Pearson and H. O. Hartley (eds.), *Biometrika Tables for Statisticians*, 3rd ed., 1966. Reproduced by permission of *Cambridge University Press*.

◢ Appendix B   Derivation of Minimum Total Sample Size

Claim: When obtaining two independent samples, the maximum error for the difference of the two population means $\mu_1 - \mu_2$, is

$$E = Z_{\alpha/2} \sqrt{\frac{\sigma_1^2}{n_1} + \frac{\sigma_2^2}{n_2}} \qquad (\sigma_1, \sigma_2 \text{ known})$$

or estimated using

$$E = Z_{\alpha/2} \sqrt{\frac{s_1^2}{n_1} + \frac{s_2^2}{n_2}} \qquad (\sigma_1, \sigma_2 \text{ unknown})$$

For a specific value of E, the sample sizes, n_1 and n_2, that minimize the total sample size, $n = n_1 + n_2$, are given by

$$n_1 = \frac{Z_{\alpha/2}^2 s_1 (s_1 + s_2)}{E^2}$$

and

$$n_2 = \frac{Z_{\alpha/2}^2 s_2 (s_1 + s_2)}{E^2}$$

Proof: For ease of notation, define

$$Z = Z_{\alpha/2}$$
$$a = s_1$$
$$b = s_2$$
$$x = n_1$$
$$y = n_2$$

(For the case where the σ's are known, then $a = \sigma_1$ and $b = \sigma_2$.) Now,

$$E = Z \sqrt{\frac{a^2}{x} + \frac{b^2}{y}}$$

is fixed. Solving for y yields

$$y = \frac{Z^2 b^2 x}{E^2 x - Z^2 a^2}$$

The total sample size is $n = x + y$, and so

$$n = f(x) = x + \frac{Z^2 b^2 x}{E^2 x - Z^2 a^2}$$

To determine the value of x that minimizes $n = f(x)$, the procedure will be to find $f'(x)$, set it to zero, and solve for x.

$$f'(x) = 1 + \frac{(E^2 x - Z^2 a^2)(Z^2 b^2) - Z^2 b^2 x(E^2)}{(E^2 x - Z^2 a^2)^2} = 1 - \frac{Z^4 a^2 b^2}{(E^2 x - Z^2 a^2)^2}$$

Now,

$$f'(x) = 0$$

iff $x^2(E^4) + x(-2Z^2E^2a^2) + (Z^4a^4 - Z^4a^2b^2) = 0$

iff $x = \dfrac{2Z^2E^2a^2 \mp \sqrt{4Z^4E^4a^4 - 4Z^4E^4a^2(a^2 - b^2)}}{2E^4} = \dfrac{Z^2a^2 \mp Z^2ab}{E^2}$

Now,

$$f''(x) = \frac{2Z^4E^2a^2b^2(E^2x - Z^2a^2)}{(E^2x - Z^2a^2)^4}$$

Consequently,

$$f''(x) > 0 \qquad \text{iff } (E^2x - Z^2a^2) > 0$$

Letting

$$x = \frac{Z^2a^2 + Z^2ab}{E^2}$$

then

$$E^2x - Z^2a^2 = Z^2ab > 0$$

because a and b are > 0. Letting

$$x = \frac{Z^2a^2 - Z^2ab}{E^2}$$

then

$$E^2x - Z^2a^2 = -Z^2ab < 0$$

Conclusion:

1. $f(x)$ has a local minimum at

$$x = \frac{Z^2a(a + b)}{E^2}$$

2. $f(x)$ has a local maximum at

$$x = \frac{Z^2a(a - b)}{E^2}$$

Because we are restricted to values of x (that is, n_1) such that $f(x) = $ total sample size is positive, and because $f(x)$ approaches ∞ as x approaches ∞, for the admissible values of x, $f(x)$ has a global minimum at

$$x = n_1 = \frac{Z^2a(a + b)}{E^2}$$

$$= \frac{Z^2s_1(s_1 + s_2)}{E^2}$$

Solving for n_2, we previously stated that

$$y = \frac{Z^2b^2x}{E^2x - Z^2a^2}$$

Substituting

$$x = \frac{Z^2s_1(s_1 + s_2)}{E^2}$$

into this expression produces

$$y = n_2 = \frac{Z^2 s_2 (s_1 + s_2)}{E^2}$$

◢ Appendix C   Introduction to MINITAB— Mainframe or Microcomputer Versions

MINITAB is an easy-to-use, flexible statistical package. It was originally designed for students; over the years, it has been constantly improved. It is one of the more powerful statistical systems currently available. MINITAB will allow you to "speak" to the computer using commands that are similar to English sentences. The sequence of steps in a typical problem resembles the same steps you would take if solving the problem by hand.

MINITAB consists of a worksheet containing rows and columns. The data for each variable are stored in a particular *column*. The following discussion will provide a brief introduction of data entry and use of MINITAB commands. No distinction is made between mainframe and microcomputer MINITAB commands since they can be used on either system.

Entering the Data: Using MINITAB Commands   Data for each of the variables in your data set are stored in columns. For example, suppose you have four scores (80, 75, 43, and 91) that you want to enter as one variable. These can be entered as shown in Figure C.1 using the SET command. Here, the four test scores are stored in column C1. Notice that (1) the data values are separated by blanks (commas are OK) and (2) at the end of the data string, the word END is entered on the next line. If your data will not fit on a single line, type as many values as you wish, enter these, continuing typing on the next line, enter another line of data, and so forth until you have entered all of your data. Your last line always will be END.

■ **FIGURE C.1**
Commands to enter
four test scores.

```
MTB > SET INTO C1
DATA> 80 75 43 91
DATA> END
```

Suppose that you have test scores for three people (Joe, Mary, and Al), each of whom has four scores. The previous four test scores (80, 75, 43, and 91) belong to Joe. The other method of entering data is to read in your data one *row* at a time using the READ command. Each line contains a single row of data on each of the variables (three variables, here). This is shown in Figure C.2. These steps will input the 12 test scores one row (exam) at a time. Note that C1–C3 means C1 through C3.

■ **FIGURE C.2**
Commands to enter
four test scores for
each of three students.

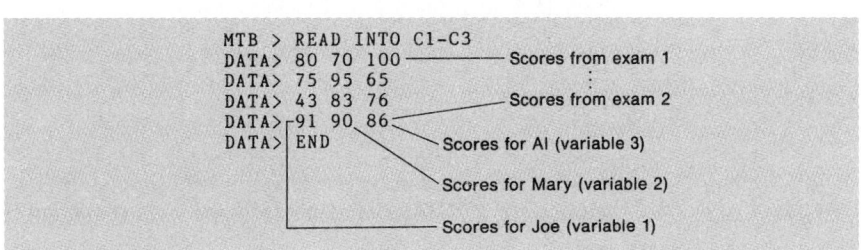

Entering the Data: Using the Full Screen Editor   A nice feature of MINITAB is the ability to use the MINITAB full screen editor when entering or editing your data. To obtain this screen (illustrated in Figure C.3), press the ESCAPE (Esc) key. By moving the cursor, enter each data value, pressing the RETURN or ENTER key after each value. To execute a MINITAB command, return to the normal (command) screen by pressing the ESCAPE key once again. At any point in the session, you can flip back and forth between the data editor and the command screen using the ESCAPE key.

■ **FIGURE C.3**   Input screen using data editor.

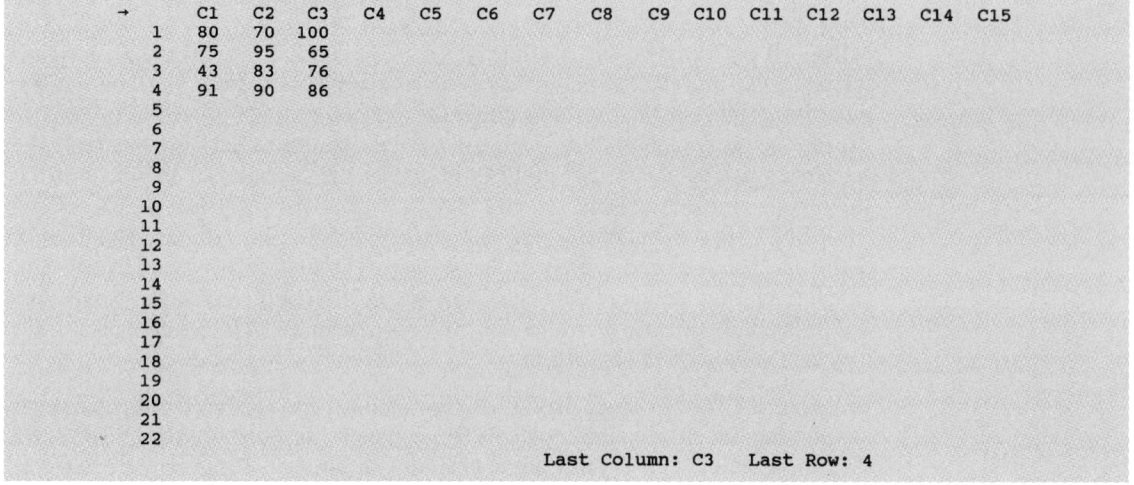

Output of the Data   To display the data, you should use the PRINT command, as shown in Figure C.4. When using the microcomputer version of MINITAB, to send your results to the printer and screen simultaneously, type

MTB> PAPER

To stop sending output to the printer at any time, type

MTB> NOPAPER

■ **FIGURE C.4**
Commands to display
(or print) data.

```
MTB > PRINT C1

C1
      80        75        43        91

MTB > PRINT C1-C3

ROW       C1        C2        C3

  1       80        70       100
  2       75        95        65
  3       43        83        76
  4       91        90        86
```

MINITAB Shorthand   One exceptionally nice feature of MINITAB is that most of the commands can be shortened to save time and effort. You are able to do this because the MINITAB system does not process the entire line that you enter—it reads only those pieces of information that it needs to know and ignores everything else! Consequently, you can misspell words or leave out unnecessary words and MINITAB will still execute your command. For example, in Figure C.1, you could have used the command

```
MTB> SET C1
```

That is, the word "into" was not necessary. We used it originally because it makes the statement easier to comprehend for someone who wants to know what we are doing at this step.

Many of the MINITAB commands are illustrated on the inside front cover of this book. Only those portions of each statement that are colored must be included. What you put between the colored portions is up to you—you can leave them blank (the shorthand version) or put in any words you wish to make the statement easier to understand. For example, in Figure C.1, we could have used

```
MTB> SET THE EXAM SCORES FOR JOE INTO C1
```

Here, all MINITAB needs to see is SET and C1. What you put in between is your decision.

In addition, for the commands on the inside front cover, any portion of the statement enclosed in brackets [ ] need not be included in the statement. If you want to omit this portion of the statement, you may do so. The information in the brackets generally allows you to be more specific in your input to MINITAB or informs MINITAB in which columns you would like certain information from the output to be stored.

Finally, MINITAB only looks at the first *four* characters of the first word in your command. For example, if you wanted to multiply columns one and two, placing the product into column three, the full command could be

```
MULTIPLY C1 BY C2, PLACE INTO C3
```

The abbreviated (shorthand) version would be

```
MULT C1 C2 C3
```

MINITAB Subcommands   Some of the more sophisticated MINITAB commands allow you to specify further information by using one or more *subcommands*. For those commands that allow subcommands (not all of them do), you should end the main command line with a *semicolon, ;*. This informs MINITAB that subcommands will follow. Each subcommand line should end with a semicolon unless it is the last subcommand—the last one ends with a *period*. If you forget to type the period at the end of the last subcommand, simply type a period on the next line.

An example of a command utilizing subcommands (called REGRESS, discussed in Chapters 14 and 15) is shown in Figure C.5. The MINITAB solutions contained throughout the text will clarify what commands allow the use of subcommands and what these possible subcommands are.

■ FIGURE C.5
The REGRESS command
with subcommands.

```
MTB > REGRESS Y IN C1 USING 2 PREDICTORS IN C2 C3;
SUBC> NOCONSTANT;
SUBC> COEF INTO C4;
SUBC> RESIDS INTO C5.
```

Informs MINITAB that this
is the end of the subcommands

Informs MINITAB that one or
more subcommands will follow

Appendix D   Introduction to SPSS—Mainframe Version

Computer packages often are used to perform various statistical analysis procedures. When used properly, these computer packages can save time and decrease the probability of human error. The purpose of this appendix is to provide a basic overview of one such mainframe package, the Statistical Package for the Social Sciences (SPSS).

To solve a statistical problem using SPSS, you must define the data and the format in which it is to be interpreted, and specify the statistical procedure to be performed. For example, the data in Figure D.1 are test grades for three students. If you wish to find the mean grade for each of the students, you define each individual test score and specify the SPSS procedure to obtain mean values (DESCRIPTIVE). Figure D.2 shows the statements and data required to perform this task under SPSS.

■ FIGURE D.1
Test grades for
three students.

80	70	100
75	95	65
43	83	76
91	90	86

■ FIGURE D.2
Statements and data
to find mean grades
for three students.

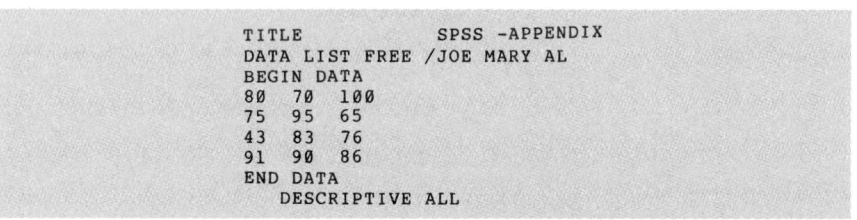

```
TITLE              SPSS -APPENDIX
DATA LIST FREE /JOE MARY AL
BEGIN DATA
80  70  100
75  95  65
43  83  76
91  90  86
END DATA
     DESCRIPTIVE ALL
```

Analysis of Statements   The following is an analysis of each of the statements used to obtain the mean grades for each student:

TITLE SPSS-APPENDIX	Defines the name of the SPSS run.
DATA LIST FREE / JOE MARY AL	Defines three variables (students) named Joe, Mary, and Al. FREE specifies the data is not in any format, other than it is to be read sequentially, starting at column one.
BEGIN DATA	Specifies that the following lines are input data lines.
80 70 100	Gives the test-score values for the first test (Joe scored 80, Mary scored 70, and Al scored 100).
75 95 65	Gives the test-score values for the second test.
43 83 76	Gives the test-score values for the third test.
91 90 86	Gives the test-score values for the fourth test.
	As shown, the values must be separated by one or more spaces, and there must be a value for all four tests for each of the three students.
END DATA	Indicates the end of the data card images.
DESCRIPTIVE ALL	Specifies the SPSS procedure to be performed. In this program, the mean, as well as other statistics, will be calculated for all variables.

Figure D.3 shows the printed results of running this procedure under SPSS.

■ **FIGURE D.3**
Results of running
listing in Figure D.2.

```
NUMBER OF VALID OBSERVATIONS (LISTWISE) =        4.00

VARIABLE        MEAN      STD DEV    MINIMUM    MAXIMUM VALID N    LABEL

JOE            72.250     20.614      43.00      91.00      4
MARY           84.500     10.847      70.00      95.00      4
AL             81.750     14.886      65.00     100.00      4
```

Most SPSS procedures require additional statements, either to describe the problem or to request different options for solving the problem in different ways. All procedures, statements, and options necessary for solving the problems in this text are discussed in the end-of-chapter appendixes. For further information on SPSS, refer to the *SPSS Introductory Statistics Guide* (Marija Norusis, New York: McGraw-Hill, 1990).

Basic SPSS Rules   As with all computer packages, you must observe certain rules when running SPSS programs. The following is a list of the basic rules for SPSS:

1. All commands must begin in column 1 and cannot exceed column 15.
2. Any additional information must appear in columns 16 through 80.
3. All command keywords must be separated by only one blank.
4. Multiple blanks are allowed beyond column 15.
5. Include the decimal point when entering decimal data (such as 38.95).

Additional rules concerning individual procedures are contained in the *SPSS Introductory Statistics Guide*.

JCL Statements   Job Control Language (JCL) statements must be included with SPSS statements in order for the SPSS program to execute properly. These statements identify the user and the procedure to be performed. Typically, you need only a job statement, an execute statement (EXEC SPSS), and a card image input statement (SYSIN DD*). The format and order of these statements may differ, so consult your computer center before attempting to run your first program.

▨ *Appendix E   Introduction to SPSS/PC—Microcomputer Version*

Computer packages often are used to perform various statistical analysis procedures. When used properly, these computer packages can save time and decrease the probability of human error. The purpose of this appendix is to provide a basic overview of one such microcomputer package, the Statistical Package for the Social Sciences (SPSS/PC).

To solve a statistical problem using SPSS/PC, you must define the data and the format in which it is to be interpreted, and specify the statistical procedure to be performed. For example, the data in Figure E.1 are test grades for three students. If you wish to find the mean grade for each of the students, you define each individual test score and specify the SPSS procedure to obtain mean values (DESCRIPTIVE). Figure E.2 shows the statements and data required to perform this task under SPSS/PC.

■ **FIGURE E.1**
Test grades for
three students.

80	70	100
75	95	65
43	83	76
91	90	86

■ **FIGURE E.2**
Statements and data
to find mean grades
for three students.

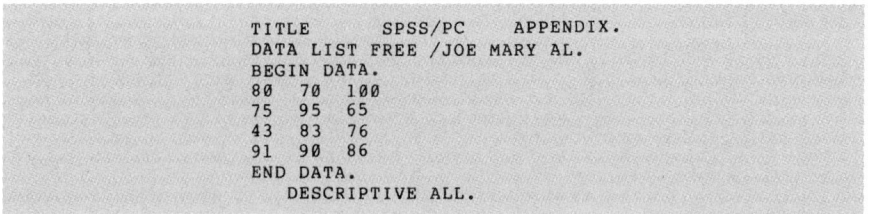

```
TITLE      SPSS/PC    APPENDIX.
DATA LIST FREE /JOE MARY AL.
BEGIN DATA.
80   70   100
75   95   65
43   83   76
91   90   86
END DATA.
    DESCRIPTIVE ALL.
```

Analysis of Statements   The following is an analysis of each of the statements used to obtain the mean grades for each student. Each statement *must* end with a period.

TITLE SPSS/PC APPENDIX.	Defines the name of the SPSS/PC run.
DATA LIST FREE / JOE MARY AL.	Defines three variables (students) named Joe, Mary, and Al. FREE specifies the data is not in any format, other than it is to be read sequentially, starting at column one.
BEGIN DATA.	Specifies that the following lines are input data lines.
80 70 100	Gives the test-score values for the first test (Joe scored 80, Mary scored 70, and Al scored 100).
75 95 65	Gives the test-score values for the second test.
43 83 76	Gives the test-score values for the third test.
91 90 86	Gives the test-score values for the fourth test.
	As shown, the values must be separated by one or more spaces, and there must be a value for all four tests for each of the three students.
END DATA.	Indicates the end of the data card images.
DESCRIPTIVE ALL.	Specifies the SPSS/PC procedure to be performed. In this program, the mean, as well as other statistics, will be calculated for all variables.

■ **FIGURE E.3**
Results of running
listing in Figure E.2.

```
-------------------------------------------------------------------------
            SPSS/PC     APPENDIX


Number of Valid Observations (Listwise) =        4.00

Variable      Mean     Std Dev    Minimum    Maximum     N   Label

JOE          72.25      20.61      43.00      91.00      4
MARY         84.50      10.85      70.00      95.00      4
AL           81.75      14.89      65.00     100.00      4
-------------------------------------------------------------------------
```

Figure E.3 shows the printed results of running this procedure under SPSS/PC.

Most SPSS/PC procedures require additional statements, either to describe the problem or to request different options for solving the problem in different ways. All procedures, statements, and options necessary for solving the problems in this text are discussed in the end-of-chapter appendixes. For further information on SPSS/PC refer to the *SPSS/PC for the IBM PC/XT/AT* (Marija Norusis, Chicago: SPSS Inc., 1990).

Basic SPSS/PC Rules   As with all computer packages, you must observe certain rules when running SPSS/PC programs. The following is a list of the basic rules for SPSS/PC.

1. All SPSS/PC statements must end with a period.
2. Lines containing data do not end with a period.
3. All command keywords must be separated by only one blank.

4. Multiple blanks are allowed beyond column 15.
5. Include the decimal point when entering decimal data (such as 38.95).
6. All data must be defined before requesting SPSS/PC to perform a statistical procedure.
7. To return to DOS, type FINISH at the end of your SPSS/PC session.

Additional rules concerning individual procedures are contained in the *SPSS/PC for the IBM PC/XT/AT.*

Obtaining Paper Output   At any point in the SPSS/PC session, you can obtain output on your printer by typing

SET PRINTER = ON,

All subsequent output will be sent to the screen and the printer. To terminate the printed output, type

SET PRINTER = OFF,

Appendix F   Introduction to SAS—Mainframe Version

Computer packages often are used to perform various statistical analysis procedures. When used properly, these computer packages can save time and decrease the probability of human error. The purpose of this appendix is to provide a basic overview of one such mainframe package, the Statistical Analysis System (SAS).

To solve a statistical problem using SAS, you must define the data to be used and specify the statistical procedure to be performed. For example, the data in Figure F.1 lists test grades for three students. If you wish to find the mean grade for each of the students, you define each individual test score and specify the SAS procedure to obtain mean values (PROC MEANS). Figure F.2 shows the statements and data required to perform this task under SAS.

■ **FIGURE F.1**
Test grades for
three students.

80	70	100
75	95	65
43	83	76
91	90	86

■ **FIGURE F.2**
Statements and data
to find mean grades
for three students.

```
TITLE     'SAS-APPENDIX';
DATA      GRADES;
INPUT     JOE MARY AL;
CARDS;
80   70   100
75   95   65
43   83   76
91   90   86
PROC  MEANS;
```

Analysis of Statements   The following is an analysis of each of the statements used to obtain the mean grades for each student:

TITLE 'SAS APPENDIX';	Defines the name of the SAS run. Enclose the title in single quotes.
DATA GRADES;	Defines a dataset named grades.
INPUT JOE MARY AL;	Defines three variables (students) named Joe, Mary, and Al.
CARDS;	Defines the input medium as card images.

80 70 100	Gives the text-score values for the first test (Joe scored 80, Mary scored 70, and Al scored 100).
75 95 65	Gives the test-score values for the second test.
43 83 76	Gives the test-score values for the third test.
91 90 86	Gives the test-score values for the fourth test.
	As shown, the values must be separated by one or more spaces, and there must be a value for all four tests for each of the three students.
PROC MEANS;	Specifies the SAS procedure to be performed.

Figure F.3 shows the printed results of running this procedure under SAS.

Most SAS procedures require additional statements, either to describe the problem or to request different options for solving the problem in different ways. All procedures, statements, and options necessary for solving the problems in this text are discussed in the end-of-chapter appendixes. For further information on SAS, refer to the *SAS User's Guide: Statistics Version,* 5th ed. (SAS Institute, Cary, NC, 1990).

■ **FIGURE F.3**   Results of running listing in Figure F.2.

SAS-APPENDIX									
VARIABLE	N	MEAN	STANDARD DEVIATION	MINIMUM VALUE	MAXIMUM VALUE	STD ERROR OF MEAN	SUM	VARIANCE	C.V.
JOE	4	72.25000000	20.61350690	43.00000000	91.00000000	10.30675345	289.00000000	424.91666667	28.531
MARY	4	84.50000000	10.84742673	70.00000000	95.00000000	5.42371337	338.00000000	117.66666667	12.837
AL	4	81.75000000	14.88567544	65.00000000	100.00000000	7.44283772	327.00000000	221.58333333	18.209

Basic SAS Rules   As with all computer packages, you must observe certain rules when running SAS programs. The following is a list of the basic rules for SAS:

1. SAS statements must begin in column 1 and cannot exceed column 72. Any SAS statement longer than 72 columns may be continued on the next line.
2. All SAS statements must end with a semicolon (;).
3. Lines containing data do not end with a semicolon.
4. All data must be defined before a procedure can be run.
5. More than one procedure can be performed by adding more procedure statements.
6. You must include the decimal point when entering decimal data (such as 38.95).

Additional rules concerning individual procedures are contained in the *User's Guide*.

JCL Statements   Job Control Language (JCL) statements must be included with SAS statements in order for the SAS program to execute properly. These statements identify the user and the procedure to be performed. Typically, you need only a job statement, an execute statement (EXEC SAS), and a data input statement (SYSIN DD*). The format and order of these statements may differ, so consult your computer center before attempting to run your first program.

▮ *Appendix G   Introduction to SAS—Microcomputer Version*

Computer packages often are used to perform various statistical analysis procedures. When used properly, these computer packages can save time and decrease the probability of human error. The purpose of this appendix is to provide a basic overview of one such microcomputer package, the Statistical Analysis System (SAS).

To solve a statistical problem using SAS, you must define the data to be used and specify the statistical procedure to be performed. For example, the data in Figure G.1 lists test grades for three students. If you wish to find the mean grade for each of the students, you define each individual test score and specify the SAS procedure to obtain mean values (PROC MEANS). Figure G.2 shows the statements and data required to perform this task under SAS.

■ **FIGURE G.1**
Test grades for
three students.

80	70	100
75	95	65
43	83	76
91	90	86

■ **FIGURE G.2**
Statements and data
to find mean grades
for three students.

```
TITLE     'SAS-APPENDIX';
DATA      GRADES;
INPUT     JOE MARY AL;
CARDS;
80  70  100
75  95  65
43  83  76
91  90  86
PROC  MEANS;
```

Analysis of Statements The following is an analysis of each of the statements used to obtain the mean grades for each student.

TITLE 'SAS APPENDIX';	Defines the name of the SAS run. Enclose the title in single quotes.
DATA GRADES;	Defines a dataset named grades.
INPUT JOE MARY AL;	Defines three variables (students) named Joe, Mary, and Al.
CARDS;	Defines the input medium as card images.
80 70 100	Gives the text-score values for the first test (Joe scored 80, Mary scored 70, and Al scored 100).
75 95 65	Gives the test-score values for the second test.
43 83 76	Gives the test-score values for the third test.
91 90 86	Gives the test-score values for the fourth test.
	As shown, the values must be separated by one or more spaces, and there must be a value for all four tests for each of the three students.
PROC MEANS;	Specifies the SAS procedure to be performed.

Figure G.3 shows the printed results of running this procedure under SAS.

Most SAS procedures require additional statements, either to describe the problem or to request different options for solving the problem in different ways. All procedures, statements, and options necessary for solving the problems in this text are discussed in the end-of-chapter appendixes. For further information on SAS, refer to the *SAS Introductory Guide for Personal Computers Version,* 6th ed. (SAS Institute, Inc., Cary, NC, 1990).

■ **FIGURE G.3** Results of running listing in Figure G.2.

N Obs	Variable	N	Minimum	Maximum	Mean	Std Dev
4	JOE	4	43.0000000	91.0000000	72.2500000	20.6135069
	MARY	4	70.0000000	95.0000000	84.5000000	10.8474267
	AL	4	65.0000000	100.0000000	81.7500000	14.8856754

Basic SAS Rules As with all computer packages, you must observe certain rules when running SAS programs. The following is a list of the basic rules for SAS:

1. All SAS statements must end with a semicolon (;).
2. Lines containing data do not end with a semicolon.
3. All data must be defined before a procedure can be run.
4. More than one procedure can be performed by adding more procedure statements.
5. You must include the decimal point when entering decimal data (such as 38.95).
6. To return to DOS, at the COMMAND = prompt, enter the single letter X.

Additional rules concerning individual procedures are contained in the *SAS Introductory Guide for Personal Computers*.

Obtaining Paper Output   To obtain printed output during interactive full screen sessions, in the OUTPUT window type FILE 'PRN:'. This should be typed after the output has appeared in the window.

If submitting the SAS job from a previously created ASCII file (say, SASRUN) containing the SAS statements and data, the SAS output will be contained in a newly created file (created by SAS), called SASRUN.LST. To submit the SAS job, at the hard disk drive prompt, type

```
SAS  B:SASRUN
```

This assumes you have saved SASRUN on a floppy diskette, contained in drive B. No output is routed to the screen; instead it will be routed to the B drive into the file SASRUN.LST. This will be an ASCII file and can be printed in the usual manner (such as the DOS command PRINT B:SASRUN.LST).

◢ *Appendix H   Introduction to MYSTAT*

Computer packages are often used to perform various statistical analysis procedures. When used properly, these computer packages can save time and decrease the probability of human error. The purpose of this appendix is to provide a basic overview of one such package, MYSTAT, a subset of SYSTAT.

In order to solve a statistical problem using MYSTAT, the data to be used must be defined, and the statistical procedure to be performed must be specified. If one wished to find the mean grade for each of the students listed below, one would define, to MYSTAT, each individual test score and specify the MYSTAT procedure to obtain mean values (i.e., STATS).

JOE	MARY	AL
80	70	100
75	95	65
43	83	76
91	90	86

sample data

Run Procedures   Following are the steps to be followed in order to run a MYSTAT program. The following program is designed to obtain the mean grade for each student:

At the C:\SYSTAT> prompt, type MYSTAT, return, return.

This will give the main command menu shown in Figure 1.

IF YOU ARE A NEW USER TYPE 'DEMO', PRESS THE ENTER KEY, AND FOLLOW THE INSTRUCTIONS. WHEN YOU RETURN TO THE MAIN COMMAND MENU, PROCEED WITH THE FOLLOWING EXAMPLE.

■ **FIGURE H.1**
Main Command Menu
for MYSTAT.

```
BUSINESS MYSTAT --- An Instructional Version of SYSTAT

  DEMO        EDIT        MENU        PLOT        STATS       LOG
  HELP                    NAMES       BOX         TABULATE    MEAN
  SYSTAT      USE         LIST        HISTOGRAM   TTEST       SQUARE
              SAVE        FORMAT      STEM        CORRELATE   TREND
              PUT         NOTE                                PCNTCHNG
              SUBMIT                                          DIFFRNCE
                                                              INDEX
  QUIT        OUTPUT      SORT        CHARSET     MODEL       SMOOTH
                          RANK                    CATEGORY    ADJSEAS
                          WEIGHT                  ANOVA       EXP
                                                 ESTIMATE    TPLOT
                                                             ACF,PACF
                                                             CLEAR

 >
```

Type EDIT, return.
This will give a data entry screen.
Enter your data as in Figure 2. (names such as 'JOE' must be enclosed in quotes when you enter them as labels).

■ **FIGURE H.2**
Data Entry Screen.

```
          MYSTAT Editor
          Case        JOE        MARY         AL
            1       80.000      70.000     100.000
            2       75.000      95.000      65.000
            3       43.000      83.000      76.000
            4       91.000      90.000      86.000
            5
            6
            7
            8
            9
           10
           11
           12
           13
           14
           15

          >SAVE GRADES
```

Press the ESC key, and at the prompt, type SAVE GRADES, return.
Type QUIT, return, to go back to the main menu.
At the main menu prompt type USE GRADES, return, return.
Then type STATS, return, to get the output in Figure 3.

■ **FIGURE H.3**
Output Using STATS
Command.

```
     TOTAL OBSERVATIONS:      4

                           JOE        MARY         AL

     N OF CASES              4           4           4
     MINIMUM            43.000      70.000      65.000
     MAXIMUM            91.000      95.000     100.000
     MEAN               72.250      84.500      81.750
     STANDARD DEV       20.614      10.847      14.886
```

Press return, to get back to the main menu.
Type QUIT, return, to leave the MYSTAT program.
If you wish to have your analysis directed to a printer as well as to the screen, you may type OUTPUT @, return, at the main menu prompt when you start MYSTAT. Type OUTPUT *, to turn the printer off, or type OUTPUT <filename> to save your analysis in a PC file.
For more information on MYSTAT, please see the user's manual.

Appendix I   Database Using Household Financial Variables

VARIABLE	DESCRIPTION
INCOME1	Income of principal wage earner
INCOME2	Income of secondary wage earner
FAMLSIZE	Family size
OWNORENT	Own or rent (1 = own, 0 = rent)
TOTLDEBT	Total indebtedness (excluding home mortgage)
HPAYRENT	House payment or apartment/house rent
UTILITY	Average monthly utility expenditure
LOCATION	Location of residence (1 = NE sector, 2 = NW sector, 3 = SW sector, 4 = SE sector)

OBS	INCOME1	INCOME2	FAMLSIZE	OWNORENT	TOTLDEBT	HPAYRENT	UTILITY	LOCATION
1	36741	20691	4	1	14634	1138	295	1
2	27242	25454	4	1	14445	1108	260	2
3	39076	0	6	1	10383	1012	274	1
4	41633	0	3	1	9569	867	211	2
5	32980	25396	3	1	17490	1026	245	1
6	44143	24302	4	0	20354	1201	306	2
7	32082	25715	5	1	11870	1100	261	3
8	43450	0	4	1	12226	1035	281	2
9	41534	22739	5	1	15023	926	284	1
10	40070	32128	2	1	15889	1494	384	1
11	31603	0	4	1	16222	975	206	1
12	45092	15136	5	1	16753	1312	346	4
13	29857	0	2	0	15479	362	95	1
14	36594	23431	2	0	13993	1246	291	3
15	26937	20418	2	1	15407	465	148	1
16	35066	0	3	0	10596	908	213	1
17	46632	0	3	1	16936	965	238	1
18	45212	23103	3	1	16336	1037	259	1
19	51952	2502	4	1	19307	935	228	1
20	55899	0	1	1	15751	1646	431	1
21	31989	0	7	1	12973	1515	164	1
22	36905	0	4	1	17537	728	390	3
23	45322	1772	4	1	11512	803	213	2
24	28022	1904	5	1	10293	906	179	1
25	35210	0	4	1	14694	555	163	2
26	35454	22311	5	0	7832	943	259	3
27	55392	0	4	1	12497	1018	250	1
28	34732	1651	2	1	8123	679	193	1
29	35484	0	5	1	12489	102	102	2
30	35864	0	4	0	9966	1067	277	3
31	50577	0	4	1	18138	1139	273	4
32	31876	13937	2	1	13442	775	183	2
33	29438	13865	3	1	10062	505	133	2
34	38364	1850	5	1	12599	579	159	2
35	30993	0	3	1	8924	1624	160	2
36	40365	15323	2	1	15534	552	182	4
37	42082	23054	2	1	14887	648	136	4
38	39816	0	3	0	6598	527	159	3
39	23878	0	2	1	6994	648	168	1
40	31882	0	2	1	15881	894	212	2
41	31307	2223	2	1	18138	1114	232	2
42	38612	0	3	1	13442	816	293	3
43	31603	0	3	1	10027	1332	247	2
44	29438	0	5	1	10062	1077	245	1
45	38364	24519	3	1	16615	1097	269	1
46	20141	0	4	1	17235	943	296	2
47	41609	0	4	1	10780	1176	272	1
48	40365	24519	5	0	2429	1362	353	2
49	40966	14458	2	1	10915	738	203	2
50	39816	19396	4	1	16726	918	134	1
51	38312	14486	2	0	10938	612	156	4
52	28864	0	3	1	10767	704	226	2
53	44604	15729	2	1	13903	892	307	1
54	37954	17879	6	1	12325	1206	319	1
55	26882	31060	4	1	17021	1409	269	2
56	35449	21741	5	1	16149	1104	221	1
57	37275	26655	2	1	14825	585	159	2
58	28205	22201	2	1	12233	558	140	1
59	40497	0	3	1	10421	967	273	1
60	26622	11982	3	0	10482	952	202	2
61	32008	24140	3	1	23384	520	142	1
62	29971	21459	5	1	34981	1357	315	4
63	31637	15961	2	1	9605	799	190	2
64	26583	21979	2	1	16079	894	272	1
65	28847	23477	2	1	20081	540	116	1
66	23384	24656	2	1	12568	1138	459	4
67	34981	0	7	1	20043	877	207	1
68	41180	21196	2	0	9796	657	211	2
69	41397	22843	2	1	16332	768	385	1
70	40479	0	6	1	14872	1657	118	2
71	31870	0	1	1	12311	1399	249	2
72	41779	16284	4	0	11314	1156	175	2
73	40405	0	4	1	9834	722		
74	40405	0	4	0				
75	39044	0	4	1				
76	32559	0	7	1				
77	31905	0	1	0				

OBS	INCOME1	INCOME2	FAMLSIZE	OWNORENT	TOTLDEBT	HPAYRENT	UTILITY	LOCATION
83	15802	0	5	0	2987	311	127	3
84	28990	0	5	1	4781	417	120	4
85	21374	21302	2	1	8739	429	76	1
86	28261	18313	5	1	11938	1192	315	1
87	34402	0	2	1	10350	372	87	2
88	26236	13286	2	0	8968	708	162	2
89	33935	19533	5	1	13164	1289	295	2
90	21208	0	1	1	6942	329	134	1
91	46367	24913	2	1	5596	1058	108	1
92	37574	19785	2	1	10338	763	235	1
93	32442	20439	3	1	13795	929	379	2
94	46144	12271	4	1	8786	1511	219	1
95	29707	0	1	1	18699	1283	311	1
96	31498	13275	2	1	6940	904	122	4
97	32080	26546	2	0	9352	606	136	1
98	31715	0	1	0	12068	713	123	1
99	33182	21576	2	0	13902	349	146	1
100	22264	17153	7	1	12292	477	406	2
101	30757	0	3	1	8613	1642	166	1
102	25922	0	3	1	2254	640	183	1
103	37858	12108	4	1	7154	859	163	3
104	36812	27812	4	1	7661	894	346	1
105	30782	14429	2	1	16996	1457	136	1
106	40463	0	5	0	8148	1251	278	2
107	45213	12267	5	1	9642	430	103	1
108	26590	0	4	1	11088	312	112	1
109	27727	9836	6	1	8605	1297	299	2
110	32400	0	2	1	8605	799	203	2
111	47121	0	3	1	11969	840	229	2
112	36050	18619	3	1	6127	872	225	2
113	29265	27822	9	1	10839	716	134	3
114	32398	19829	2	1	14430	885	247	1
115	35888	0	6	1	13734	1379	192	1
116	23968	0	2	4	15237	578	138	1
117	23938	26966	2	1	10261	542	143	1
118	44453	23240	5	0	8678	613	265	1
119	27949	15559	5	0	5465	890	166	1
120	29622	0	2	0	15582	1083	230	2
121	43379	0	3	0	12709	1244	258	2
122	41292	16231	4	1	10159	972	250	1
123	23425	16917	4	1	8974	869	211	2
124	32964	23540	3	0	10493	651	173	2
125	30210	0	3	1	12146	960	214	1
126	34325	20091	2	1	12340	689	273	2
127	33162	17543	5	1	15347	857	195	1
128	31418	0	2	1	6911	808	199	2
129	58868	15012	2	1	9732	1030	171	1
130	33239	22205	3	1	13194	515	268	2
131	31957	0	3	1	15612	998	219	3
132	67301	0	6	1	18605	1472	382	2
133	35258	25137	2	1	17655	1616	111	1
134	36992	1997	4	1	19200	743	255	1
135	31629	0	2	1	9689	972	467	1
136	35188	2072	2	1	10313	673	236	2
137	27814	0	1	1	14867	1798	142	4
138	27122	2008	3	1	11053	912	187	2
139	29519	14213	6	1	9983	604	199	1
140	33012	0	3	1	9689	531	96	2
141	29511	13960	2	1	12857	742	176	1
142	36083	25059	2	1	13395	518	106	4
143	44939	20009	2	1	9155	507	237	1
144	32587	0	2	1	8482	564	222	2
145	36003	16455	4	1	12802	967	181	4
146	35626	15952	3	1	21227	355	179	3
147	21167	13408	4	1	10724	917	202	1
148	31226	25857	2	1	13754	775	343	1
149	32822	18529	7	0	4665	860	207	1
150	29995	20112	3		10124	703		

OBS	INCOME1	INCOME2	FAMLSIZE	OWNORENT	TOTLDEBT	HPAYRENT	UTILITY	LOCATION
247	33467	22385	4	1	10755	1097	279	2
248	40159	30462	4	1	8537	805	202	1
249	32285	19681	3	1	16407	883	212	3
250	41630	23091	2	1	13909	600	224	4
251	32506	14967	4	1	15555	736	202	2
252	44612		3	1	16139	861	279	1
253	47479	19333	2	1	13168	1137	279	2
254	42788	23080	2	0	12731	336	93	3
255	24021		5	0	10871	549	104	1
256	32998		2	0	2407	1078	229	1
257	37490	1643	1	0	8143	714	209	3
258	37406		2	1	13552	958	241	1
259	25243	21605	4	0	7830	979	253	1
260	51213		3	1	13462	683	186	3
261	51727	22758	1	1	14097	849	220	1
262	43322	19323	2	0	14880	780	212	2
263	21716		3	1	17311	602	179	2
264	21952	16558	3	0	16093	940	179	1
265	41772	29001	3	1	12558	951	253	2
266	29236		2	1	6232	419	139	1
267	27413	8361	4	0	8409	579	304	2
268	23203	23209	4	1	16112	1328	235	3
269	30473		2	1	7990	1072	193	4
270	34701	22935	2	0	11756	662	241	3
271	32552	25880	3	0	22431	457	110	2
272	46698		2	1	9146	659	184	4
273	28347		2	1	13871	522	135	1
274	23747	22908	3	0	11901	584	134	2
275	34271	21371	2	1	13328	1020	198	2
276	38953	20626	2	0	14202	936	252	2
277	34284	24450	3	1	17210	746	267	3
278	43599	28752	4	1	16755	682	169	1
279	30264		3	0	11635	524	200	1
280	26033	18812	5	0	8771	1297	323	4
281	23237		5	1	9448	1286	329	1
282	33176	16174	2	0	13507	424	127	4
283	15610		3	0	6446	668	135	1
284	35215	20905	3	1	4817	1301	286	3
285	22931	16138	3	0	9823	1107	279	2
286	35061	18880	5	1	17082	1063	217	2
287	35428	16366	4	0	17002	601	152	2
288	40706	17934	3	0	14885	931	259	3
289	49969	13903	2	1	12485	488	144	1
290	35176		4	1	2429	1460	328	1
291	15514		4	0	6554	611	142	4
292	32162		7	1	6121	583	149	2
293	27953		3	1	11465	411	108	1
294	30132	19985	2	1	16464	752	204	4
295	30047		1	0	12887	668	163	2
296	22300		3	0	18853	1233	305	1
297	33633		2	0	10407	763	162	2
298	37544	15983	6	0	4739	939	200	3
299	27715	19936	2	1	6626	1364	235	1
300	26533		4	1	10449	317	94	2
301	33732		5	1	6117	902	206	4
302	40805	17125	3	0	16821	1212	375	2
303	35878	15882	2	0	9616	866	232	1
304	38749	19701	5	1	7682	933	259	3
305	28964		4	1	11883	1154	271	1
306	27356	22975	1	0	6972	1305	240	4
307	37356	2344	4	1	12834	702	112	3
308	48700		1	1	12024	1045	112	1
309	25758	14630	3	1	16779	1364	249	1
310	31172	17049	3	0	16821	1591	258	2
311	38943	15883	2	0	9349	394	186	1
312	45151		2	1	8555	996	207	2
313	28392	23814	3	0	13014	808	343	1
314	34088	27849	2	1	16098	777	112	1
315	30379	24441	2	1	4412	512		2
316	44281		6	0	11103			
325	26309	23149						
327	24612	23696	5		9669	1705	381	2

OBS	INCOME1	INCOME2	FAMLSIZE	OWNORENT	TOTLDEBT	HPAYRENT	UTILITY	LOCATION
165	36582	22969	4	0	14593	1250	283	1
166	54131		4	1	11101	1174	256	1
167	39248		4	1	21629	784	241	1
168	42107	32401	3	1	27582	785	161	2
169	27401		2	0	8888	760	187	4
170	27401	18159	3	0	7582	766	156	4
171	38947		3	0	15641	1040	233	4
172	47566		3	1	4001	821	192	1
173	36022	24830	4	1	18761	1387	363	1
174	47566	19247	6	1	13525	1635	152	1
175	36185		3	0	10114	643	152	4
176	25143		2	1	13773	430	110	1
177	26068	21342	4	0	14504	980	225	1
178	36548	22518	5	1	13773	813	178	1
179	29712	18405	2	0	11271	385	116	1
180	25912	22958	4	1	14342	1204	283	2
181	37104		3	1	7901	820	171	1
182	16401		2	0	7627	1536	369	2
183	16401	14081	3	0	2703	767	167	3
184	28707		6	1	10165	478	160	2
185	34495		3	1	9330	711	126	4
186	32418		1	0	7026	910	246	3
187	38171	20902	2	0	11574	382	113	1
188	35304	22870	2	1	12888	960	224	2
189	31222	15209	3	1	18346	553	261	1
190	46646	20772	3	0	10154	822	146	1
191	47880	17637	2	1	12620	777	188	4
192	27419	19462	3	0	12675	783	257	1
193	27047		2	1	10690	482	152	1
194	33875	21965	6	1	10232	1240	326	1
195	40282		7	0	10103	672	156	1
196	36018		2	1	12499	969	294	1
197	35259		3	1	11646	443	113	4
198	43155	15224	2	0	9383	980	218	2
199	21916	18874	2	1	8891	485	178	1
200	21937	13634	5	1	17929	574	165	1
201	49246		2	1	11568	320	204	1
202	31934	21440	2	0	13704	787	143	1
203	33652	24023	3	0	14201	651	164	2
204	29301	22873	2	1	7716	442	190	2
205	29301	22823	3	1	13788	991	302	1
206	34861	25263	4	1	14903	853	135	1
207	29379	23904	3	0	14883	856	184	1
208	22310	23368	5	1	10199	925	232	1
209	35247		2	0	14705	491	101	4
210	38106		2	1	12291	497	163	1
211	41441	21252	2	1	16664	633	317	4
212	38235	23504	3	1	7892	1251	229	2
213	43058		2	0	8510	1930	223	2
214	30669	9894	3	1	3300	935	312	4
215	37471		5	1	11503	1263	105	3
216	35162		2	0	11120	542	202	2
217	30081	11841	6	1	15469	418	295	1
218	28892	14419	2	1	6703	740	89	3
219	32009	26251	2	0	12231	1195	414	1
220	30974		7	1	15389	1043	219	4
221	33919		2	0	9370	1583	244	1
222	27041		4	1	9864	899	336	1
223	40138	24547	4	0	17265	1453	229	2
224	28564		6	1	18872	858	223	1
225	31630		3	0	16549	1263	312	2
226	26184		3	1	6703	193	105	3
227	26203	12231	2	1	4609	542	202	1
228	35892		4	1	15389	740	295	4
229	32513		3	0	9370	367	89	1
230	35010	24125	4	1	15525	1043	219	4
231	33197	19086	4	1	9864	1583	241	2
232	31009	21059	3	0	17265	899	182	1
233	24292	12123	7	1	16858	1113	236	1
234	19293		4	1	14436	1663	265	3
235	39379	21413	2	0	14207	1056	242	1
236	32810	20260	3	1	14207	812	297	2
237	35010	26843	4	1	5554	730	228	1
238	26309		2	1	13975	1004	228	1
239	36162		2	0	11556	375	125	1
245	38210	23149	4	1				
246	36162	12234	2	0	11556	375	125	1

OBS	INCOME1	INCOME2	FAMLSIZE	OWNORENT	TOTLDEBT	HPAYRENT	UTILITY	LOCATION
329	29969	0	3	0	6233	519	159	3
330	27484	26902	2	1	15794	872	208	1
331	29375	0	1	0	5958	458	209	3
332	38188	0	2	0	11031	636	203	2
333	37704	20769	4	1	15794	1880	239	2
334	49501	22058	3	0	19989	930	249	1
335	23468	0	3	1	8815	738	168	1
336	22260	26538	3	1	11062	595	191	4
337	33949	0	2	0	10029	718	155	1
338	34455	29146	2	0	10035	498	100	4
339	34698	0	2	1	9542	420	308	2
340	28763	28424	4	1	13713	1123	100	2
341	26734	0	1	1	10988	460	209	1
342	42234	17722	3	1	18255	585	162	2
343	35711	21099	3	0	18255	819	229	1
344	24837	19991	5	1	13182	1269	280	1
345	25544	17096	2	1	13114	1213	301	3
346	31257	0	4	0	8154	547	153	2
347	40570	0	2	1	12391	1218	263	1
348	41795	13727	3	1	13782	539	149	2
349	31941	0	4	1	4161	710	118	2
350	30859	21663	4	0	16252	810	208	4
351	29952	14933	4	1	11820	755	194	2
352	38867	19936	6	1	15862	1324	209	1
353	22975	14002	2	1	6336	449	321	4
354	32096	22422	3	1	15251	934	140	2
355	30227	22776	4	1	14465	612	200	1
356	40148	15120	4	1	5461	861	221	1
357	28649	0	2	1	10652	616	100	4
358	25720	21857	5	1	16245	344	101	2
359	31941	0	4	1	10716	750	207	4
360	57626	0	3	0	6982	1008	201	1
361	36064	27583	3	1	15271	592	146	2
362	35065	15828	4	1	10048	692	147	1
363	62146	23107	3	1	11152	1298	325	4
364	37415	16190	1	0	15795	886	214	2
365	35418	2227	4	0	14876	736	197	3
366	27191	17399	3	1	9589	802	300	2
367	34354	0	5	1	16418	1164	153	1
368	25461	13346	2	1	3507	643	116	4
369	31502	21732	5	0	9392	332	260	3
370	29519	0	2	1	10542	869	303	2
371	33502	22336	2	1	13288	537	98	1
372	29671	12017	5	1	11862	858	244	2
373	17339	22018	6	1	10971	780	209	4
374	24063	22366	3	1	11986	1006	258	2
375	35681	0	1	1	8540	581	237	2
376	41210	0	3	1	10933	619	123	1
377	45819	23533	5	1	11484	471	172	3
378	24319	21516	3	0	5602	570	149	4
379	26594	17507	5	1	18110	1122	281	2
380	37897	0	5	1	8170	1167	294	4
381	23146	13560	8	1	9281	853	164	2
382	26057	0	2	0	7700	1757	102	1
383	23878	0	2	0	2815	447	97	2
384	26195	21368	2	0	7700	1046	195	4
385	27973	17888	2	1	11418	1580	225	2
386	20089	19562	3	1	10440	800	382	3
387	17763	0	2	1	14200	561	137	1
388	37613	0	2	0	6100	510	94	2
389	28286	0	2	0	5747	698	205	2
390	25745	21433	5	1	12239	818	212	2
391	26195	0	2	1	19294	1352	117	1
392	29973	22743	2	1	12953	1306	107	3
393	21763	0	2	1	4323	440	327	1
394	37613	0	3	1	8627	888	217	2
395	28286	0	6	0	3341	320	105	1
396	35828	21433	4	1	8776	534	194	4
397	32528	23321	2	1	8564	708	224	2
398	36947	18010	3	1	16228	927	230	1
399	33186	21433	2	1	4323	818	105	3
400	44556	0	6	0	12200	561	97	2
401	31217	18010	2	1	10440	800	219	1
402	33043	23321	3	1	14200	510	137	1
403	35720	0	3	1	6100	698	94	2
404	32540	22743	6	1	5747	447	205	2
405	35828	0	1	1	12239	698	117	2
406	32605	18010	2	1	19294	818	327	1
407	35828	23321	3	1	8627	1352	107	1
408	36947	15888	3	1	16228	440	194	4
409	36947	15888	3	1	18194	708	224	1
410	47932	15888	3	1	18194	927	230	1
411	32882	7882	2	1	10538	521	101	1
412	53779	0	3	1	9486	817	236	3
413	31788	0	2	1	14959	546	114	1
414	31289	0	5	0	10953	1251	188	2
415	37409	20948	4	1	7226	791	291	3
416	24220	18987	4	1	7226	1762	297	1
417	26952	0	5	1	15447	1420	342	1
418	23468	27834	3	0	14082	1532	381	2
419	24770	11034	2	1	9524	977	230	2
420	42500	0	3	0	10449	1260	147	4
421	31648	0	5	1	6243	1555	309	2
422	46508	0	2	0	13098	1364	163	3
423	30035	0	3	0	6205	696	100	2
424	28596	20171	2	1	5681	561	253	3
425	29310	16062	3	1	11690	1543	142	2
426	29151	20962	5	1	18255	622	256	2
427	39333	0	4	0	13054	1036	182	1
428	16502	23429	3	1	10352	915	191	3
429	35574	0	3	1	18072	829	141	2
430	36347	22725	2	1	16593	1421	358	3
431	36433	0	6	1	16406	470	277	2
432	34737	27183	4	1	17406	1095	249	1
433	31611	15808	1	0	13447	1014	246	1
434	34815	25964	4	0	12655	824	314	1
435	26570	18000	6	1	12382	1347	209	3
436	36995	0	2	0	10352	903	227	1
437	34718	21483	2	1	18700	817	99	1
438	32216	0	6	1	13396	498	305	2
439	20165	18945	2	1	13065	1204	207	2
440	25805	13231	2	1	9250	1053	207	2
441	30095	13833	5	1	11665	829	233	2
442	27274	18833	3	1	10712	717	101	2
443	34604	25727	2	1	15578	718	359	3
444	34647	0	8	1	15412	454	101	3
445	38611	0	7	1	8154	1457	373	2
446	38549	0	4	1	9224	464	193	1
447	29236	0	4	1	5468	785	102	3
448	34579	0	4	1	11967	3100	264	1
449	22185	12357	4	1	11935	623	173	3
450	32031	21001	2	1	15935	1090	134	1
451	37405	22862	3	1	15615	506	184	3
452	30435	23011	5	1	13682	683	349	1
453	33718	14978	5	1	14886	636	86	2
454	46123	15228	2	1	17365	970	233	2
455	46390	12141	2	1	17348	776	259	2
456	41936	0	6	1	12657	1104	259	3
457	30550	15940	2	1	6857	513	323	1
458	18085	0	3	1	6317	1448	364	1
459	33327	0	5	1	15045	1726	265	1
460	31112	21647	3	1	15726	1260	184	2
461	42887	18430	4	1	12087	1777	233	1
462	26726	25125	5	1	16743	853	253	2
463	20720	12215	4	1	10802	967	230	1
464	33426	20711	2	1	8123	977	157	2
465	20358	10108	4	1	10529	852	101	4
466	36180	19915	4	1	13691	382	266	2
467	28585	0	2	1	9916	570	210	2
468	45816	0	3	1	9932	469	85	1
469	37662	21786	4	1	4990	879	271	4
470	34391	0	3	1	5539	632	154	1
471	23982	1287	2	1	15631	700	228	3
472	19647	0	1	1	14503	978	149	1
473	27088	0	3	1	7736	1060	259	3
474	20495	21154	2	1	12803	993	270	2
475	40013	17174	3	1	18798	983	93	1
476	32602	23359	2	1	12123	339	92	2
477	41316	0	4	1	16770	1341	327	1
478	33662	21380	7	1	12083	369	253	2
479	28286	17174	2	0	18193	983	253	4
480	44556	23359	2	1	7736	927	230	1
481	31217	0	4	1	18194	1405	92	1
482	33043	0	3	1	16770	339	253	2
483	35720	0	7	1	12083	1341	93	1
484	32540	21154	2	1	18193	369	92	2
485	35828	17174	2	1	7736	983	327	1
486	32605	23359	4	1	16770	339	253	2
487	35828	21380	7	1	12083	1060	93	1
488	36947	21329	2	1	18798	978	92	2
489	36947	14201	4	1	12123	1405	253	2
490	47932	21387	3	1	16770	983	92	1

OBS	INCOME1	INCOME2	FAMLSIZE	OWNORENT	TOTLDEBT	HPAYRENT	UTILITY	LOCATION
493	40364	0	6	1	11330	1024	291	2
494	29959	0	4	0	16507	737	145	3
495	30455	14775	3	0	14224	806	203	4
496	24476	0	1	1	5459	450	203	2
497	31819	0	2	0	6263	675	101	1
498	45305	0	1	1	12202	398	140	1
499	37149	22913	3	1	14226	1068	143	2
500	35317	0	2	1	5456	322	270	1
501	37913	0	2	0	7443	697	94	2
502	38893	13109	1	1	13482	1031	203	1
503	32250	0	4	0	10519	697	258	1
504	37293	29326	5	1	15830	535	98	3
505	22265	19706	5	1	10539	1371	337	1
506	26998	23537	7	1	15306	478	95	2
507	35680	16227	3	1	11433	960	245	2
508	35360	0	2	0	11703	1505	389	1
509	32271	0	4	0	4707	944	168	3
510	30690	0	3	1	6828	363	163	2
511	33987	19743	2	1	7295	554	166	1
512	43254	18563	5	0	14448	1091	290	4
513	36554	13212	4	0	12233	1315	198	2
514	42250	22472	2	1	17081	795	120	2
515	32082	0	3	0	13195	1035	257	1
516	22151	18309	3	0	13326	766	120	2
517	40771	19914	2	1	17313	784	186	1
518	34432	20107	2	1	14124	850	198	4
519	36456	26948	4	1	14877	306	122	1
520	20085	14452	2	1	9770	714	210	4
521	35167	0	4	0	7885	1534	382	3
522	41118	27421	5	0	14925	570	113	1
523	41102	20803	3	1	12461	1733	180	1
524	33228	0	6	1	12369	1483	340	1
525	61879	11717	1	0	13865	903	203	4
526	25786	0	5	0	1084	1025	275	2
527	25542	0	3	0	14810	939	217	1
528	25871	18477	4	1	11760	1130	217	3
529	24809	27082	2	0	15345	823	222	2
530	36457	23159	2	1	1407	475	122	1
531	22778	0	2	0	8003	513	103	1
532	33228	20657	3	1	15970	956	235	3
533	32778	23221	4	1	15298	513	253	2
534	32416	9643	7	0	16990	1329	254	1
535	61958	0	3	0	17246	1133	272	4
536	44523	21745	2	1	10684	1409	210	1
537	48249	0	1	1	10790	904	203	3
538	50138	0	4	1	12278	771	269	3
539	32628	21575	2	0	8392	967	135	2
540	22432	0	2	0	9408	673	261	1
541	21042	18895	3	1	15900	1548	353	4
542	42080	23336	2	1	10661	567	290	1
543	34516	13614	5	1	14516	469	130	2
544	32741	0	2	0	14088	452	111	2
545	26741	18337	1	1	8559	1016	290	3
546	61588	0	4	1	5931	886	179	1
547	36110	26312	2	1	12382	725	168	1
548	42478	0	2	0	5851	697	230	4
549	33276	16981	1	0	12223	1067	110	2
550	23548	21575	2	1	10205	1060	234	1
551	34441	0	5	1	9095	434	365	1
552	28102	18895	3	1	10657	826	292	2
553	37316	19195	3	1	15491	1283	257	2
554	43254	27414	4	1	8771	1133	266	2
555	28004	21873	4	1	17446	1045	184	1
556	23597	0	3	1	10737	1015	115	3
557	51722	0	2	1	12686	857	310	2
558	25770	18371	5	1	11140	1236	130	2
559	25913	14872	4	0	11140	601	115	4
560	39772	0	2	1	1817	458	392	2
561	19262	21384	2	1	18860	1098	258	3
562	21090	0	4	1	11669	1637	250	4
563	58884	0	3	1	14129	554	160	1
564	38582	0	4	0	13996	1095	163	2
565	48950	0	2	1	13585	872	186	3
566	30386	24517	3	0	12336	773	199	2
567	31062	14152	5	0	9252	841	293	3
568	35181	17886	4	1	15420	1181	199	2
569	35642	0	3	1	13585	773	263	2
570	12801	0	3	0	12336	872	186	1
571	35181	24517	4	0	9252	841	199	3
572	34282	14152	5	0	12336	1181	293	3
573	48554	0	4	1	15420	714	269	2
574	48554	17886	4	1	15420	714	269	2

OBS	INCOME1	INCOME2	FAMLSIZE	OWNORENT	TOTLDEBT	HPAYRENT	UTILITY	LOCATION
575	41156	13463	4	1	12552	1094	287	1
576	39066	0	4	0	13374	894	257	1
577	37341	18244	2	0	13539	995	233	1
578	25287	20735	6	0	16272	1115	288	2
579	30761	0	1	1	6224	321	104	2
580	39883	0	1	1	11774	1210	267	2
581	27792	26150	5	1	15046	1643	372	1
582	34170	21048	4	1	14826	343	105	2
583	28132	0	2	0	9305	1253	363	1
584	18729	17719	6	0	8443	792	212	3
585	43294	19521	2	0	16008	819	196	1
586	33266	19812	2	0	11425	972	264	2
587	46759	0	3	1	13701	685	156	1
588	35981	18257	2	1	14508	1014	162	1
589	38596	0	4	1	5797	909	246	3
590	38297	21831	2	1	10416	1243	197	2
591	40385	0	4	0	9881	648	163	4
592	33365	0	4	1	17149	767	366	3
593	36986	21380	3	1	7652	774	180	3
594	41690	0	2	1	4584	1368	171	1
595	24509	0	1	0	21519	1168	215	1
596	50672	0	7	1	10785	1772	405	3
597	20672	18632	5	1	15928	1259	317	1
598	37813	15923	2	1	7458	1504	143	2
599	35223	0	4	1	8445	1346	289	1
600	42489	0	4	1	15585	410	162	1
601	35996	19674	2	1	9250	1025	283	2
602	21566	0	2	1	9353	349	141	2
603	49294	19389	3	0	14381	797	256	1
604	30154	13984	6	0	19984	956	185	4
605	32341	25892	1	1	5204	569	187	3
606	27736	0	4	1	9236	602	112	2
607	24457	0	2	0	9245	901	271	4
608	38234	0	2	1	10513	242	121	2
609	48802	0	3	1	12676	319	162	4
610	38680	0	5	1	8632	380	242	2
611	32046	0	7	1	11250	286	286	2
612	44422	23250	4	1	9752	1218	126	1
613	30154	1657	2	0	15899	523	131	3
614	39109	0	1	1	15802	388	236	3
615	39035	20112	2	1	14809	873	131	3
616	24619	0	3	0	6399	577	134	2
617	22063	16662	2	1	9245	591	363	4
618	41238	26216	5	1	17242	1489	109	2
619	33064	13623	5	1	14890	609	136	1
620	35257	17467	2	0	9264	647	314	1
621	40004	22753	5	1	17229	1286	299	1
622	38860	25874	4	1	10768	1966	264	1
623	42574	21643	3	1	15573	1345	324	3
624	32942	0	2	1	10963	674	166	1
625	41107	30505	7	0	14680	1093	273	1
626	60047	0	3	1	10663	1291	296	2
627	43899	24689	2	1	14680	1162	224	1
628	44467	8487	3	1	12906	1079	217	2
629	36773	0	4	0	11753	667	313	2
630	24895	16880	3	1	13285	1296	371	2
631	33950	24354	2	1	12011	890	202	2
632	43882	19248	4	1	12089	957	247	1
633	21883	28249	3	0	20711	1032	261	1
634	46733	23325	3	0	12351	658	229	3
635	35642	0	5	1	12641	819	197	2
636	25980	27404	3	0	20163	1269	325	4
637	40703	26231	3	1	10788	1466	354	4
638	33701	15069	1	1	20165	947	240	4
639	42141	0	2	1	3631	1029	236	2
640	31539	23873	5	0	16223	981	132	3
641	15208	17385	3	1	16203	626	202	4
642	33380	0	2	1	6203	406	172	2
643	36148	13096	4	1	13914	459	143	1
644	29247	12295	3	0	8163	874	236	1
645	33920	8865	3	1	6898	915	198	1
646	32455		2	0	12352	760	161	1

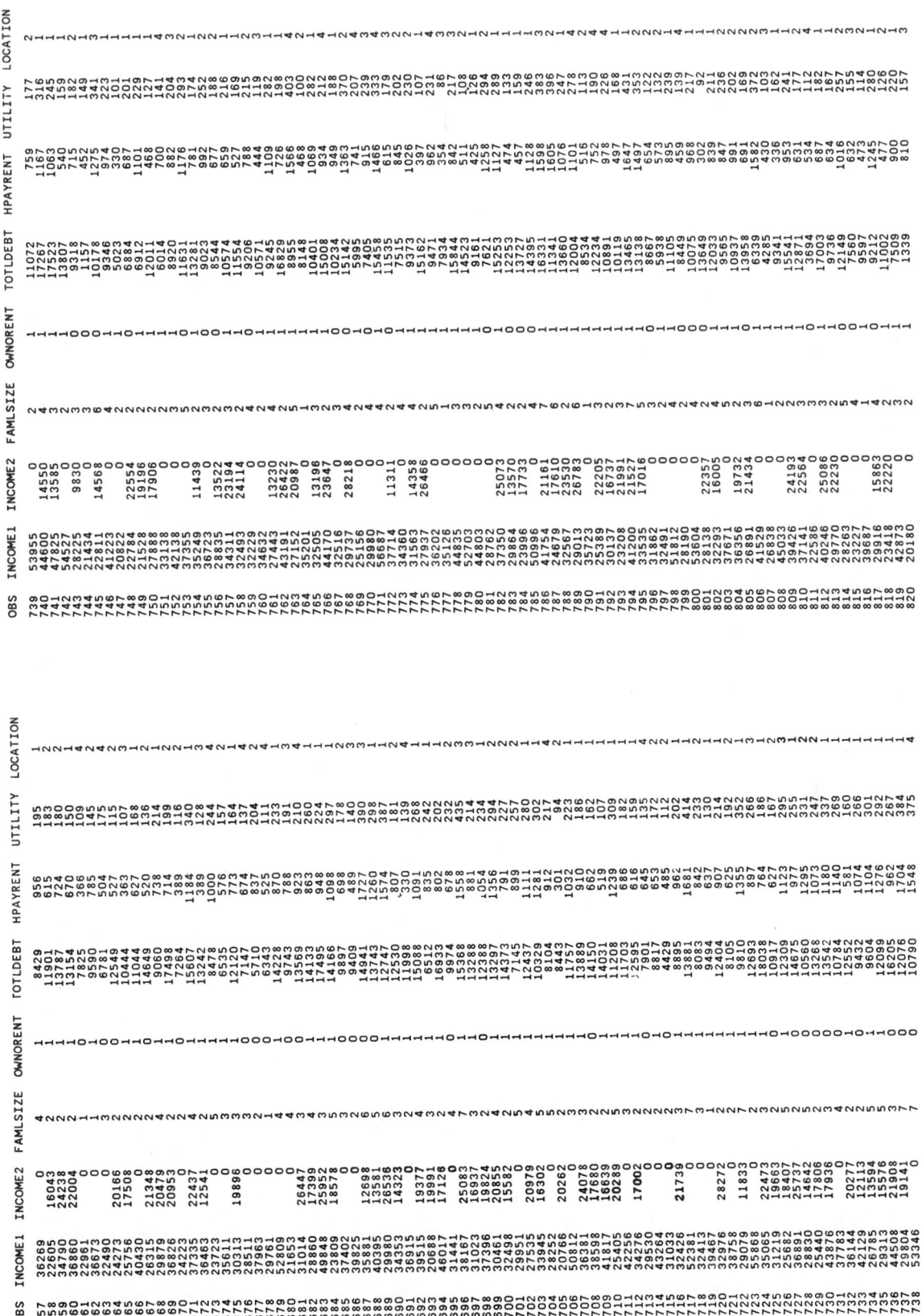

OBS	INCOME1	INCOME2	FAMLSIZE	OWNORENT	TOTLDEBT	HPAYRENT	UTILITY	LOCATION
739	53955	0	2	1	11072	759	177	2
740	44600	14550	4	1	17267	1167	316	1
741	47825	13585	3	1	17523	1063	245	1
742	24527	0	3	1	13807	540	159	2
743	28225	9830	3	0	9318	715	182	1
744	28434	0	6	0	1157	452	149	2
745	42213	14568	4	0	10178	1275	341	3
746	20822	0	2	0	9346	974	223	1
747	22784	22554	2	1	5023	330	101	1
748	21528	19196	2	1	6884	687	229	3
749	27888	17906	2	0	6912	1101	201	1
750	33138	0	2	1	12011	468	127	4
751	42138	0	3	1	16014	700	141	4
752	33135	0	3	1	8920	882	204	3
753	35249	11439	3	0	11631	1176	293	2
754	36723	0	2	1	9023	992	174	1
755	28835	13522	3	1	8544	781	252	2
756	34311	23194	2	0	10574	677	188	2
757	32493	24114	4	0	9206	659	216	1
758	32229	0	2	0	10571	527	169	3
759	34632	0	4	1	10571	788	182	1
760	43191	13230	2	1	6829	444	403	2
761	27152	26422	5	0	16829	1109	400	4
762	35201	20987	1	1	8955	1566	282	2
763	32505	0	3	0	8148	1069	212	1
764	44170	13196	2	0	10401	934	188	4
765	25737	23647	3	1	10234	949	370	2
766	25556	0	4	1	15142	1363	207	4
767	39807	0	2	1	5995	741	333	3
768	56584	28218	4	1	7405	915	239	4
769	34360	0	4	1	15458	1466	179	3
770	34463	0	4	0	11535	615	205	2
771	37937	11311	2	1	9373	845	202	2
772	36202	0	5	1	15162	1026	230	1
773	44835	14358	1	1	962	397	107	4
774	52703	26466	3	0	7934	354	231	3
775	43126	0	3	0	14523	842	86	3
776	27762	0	4	1	9161	511	217	2
777	37950	0	4	1	7621	425	108	1
778	22864	0	2	1	12253	1258	126	4
779	23996	0	4	1	14395	1127	289	3
780	41754	0	6	0	13727	474	133	2
781	32567	25073	2	1	8534	457	246	2
782	24679	13570	1	0	12004	1128	159	4
783	25389	13773	3	0	10891	1598	383	4
784	30137	0	3	0	10165	1076	396	2
785	23208	21661	5	1	13138	1101	247	4
786	41200	23530	3	1	8667	1447	110	2
787	32491	26783	5	0	5938	353	226	4
788	51842	0	5	1	8449	122	168	2
789	23008	0	2	1	10075	239	453	1
790	53048	22357	4	1	13649	968	132	2
791	32293	16005	4	0	12033	302	239	1
792	33671	0	2	1	9565	839	217	2
793	38356	19732	2	0	9937	847	92	2
794	41529	21434	3	1	11958	991	216	4
795	26891	0	4	1	6339	497	202	2
796	45033	0	2	1	4285	1582	168	2
797	37141	24193	3	1	9241	430	372	3
798	22536	25564	3	1	15547	353	431	1
799	40246	0	2	1	12871	351	122	1
800	29770	25086	5	1	17003	637	239	2
801	23227	2230	4	1	19149	634	217	1
802	38263	0	2	0	9560	1016	92	2
803	39887	0	3	1	8957	632	236	2
804	37671	0	3	1	9212	167	202	2
805	26891	15863	6	1	11002	1245	169	3
806	41529	22220	2	1	11509	477	372	2

OBS	INCOME1	INCOME2	FAMLSIZE	OWNORENT	TOTLDEBT	HPAYRENT	UTILITY	LOCATION
657	36269	0	4	1	8429	956	195	1
658	22605	16043	2	1	11901	615	183	2
659	34790	24238	2	1	13787	724	180	2
660	38860	22004	2	1	13154	670	150	1
661	22961	0	1	0	7825	366	109	4
662	36673	0	1	1	9590	504	145	4
663	22490	0	3	0	6781	175	175	2
664	24273	20166	2	0	12549	504	107	4
665	25756	17508	2	0	10444	363	168	3
666	40430	0	2	1	11044	627	136	1
667	26315	21348	4	0	14649	520	199	2
668	29979	20479	2	1	9060	738	214	1
669	36826	20953	4	1	17498	718	116	2
670	34423	0	2	1	17264	1389	110	2
671	37335	22437	2	1	15607	1189	244	3
672	36463	12541	5	0	13242	1300	157	2
673	23723	0	5	1	8478	776	164	2
674	30313	0	3	1	12147	773	137	1
675	28511	19896	2	0	5710	674	204	4
676	29761	0	1	1	6343	525	111	2
677	37963	0	4	1	11247	870	233	4
678	52809	0	4	1	14228	923	191	4
679	21653	26447	3	0	9743	788	210	1
680	31060	17399	4	1	13569	948	224	3
681	32688	25592	4	1	14133	1098	157	4
682	48848	18578	5	0	17495	698	178	1
683	23809	0	2	1	14166	1098	140	1
684	37402	0	5	1	9997	678	390	4
685	39925	0	3	1	9409	489	298	3
686	35881	12698	4	1	14941	1727	387	1
687	29830	13581	2	0	12743	1574	181	4
688	40395	26536	4	1	12536	807	139	1
689	29995	14323	3	1	11998	1091	268	2
690	34353	0	4	1	15088	835	242	4
691	36915	19377	3	0	6512	802	202	2
692	20688	19991	3	0	16933	678	435	4
693	46017	17126	2	0	31441	1558	234	3
694	38167	25083	4	0	15368	884	234	1
695	31023	16937	3	1	13288	1356	257	1
696	37396	19824	2	1	13297	1791	280	2
697	30461	20855	4	1	14115	899	217	1
698	32498	15582	2	1	12177	1211	94	2
699	29951	0	4	1	10239	1912	223	1
700	27535	20979	5	1	8134	301	217	4
701	33945	16302	2	0	8843	1032	186	2
702	20766	0	3	1	11757	910	162	1
703	28252	20262	2	1	13889	662	107	2
704	36381	0	2	1	14150	519	309	3
705	50812	24078	4	1	14031	1239	182	4
706	38598	17680	2	1	11208	686	159	1
707	41817	16639	2	0	11703	616	135	1
708	42405	20289	2	1	12595	645	172	1
709	29530	0	3	1	8817	653	112	2
710	31219	17002	2	0	7901	485	202	3
711	38890	0	2	1	14675	962	414	2
712	25440	0	2	1	10560	842	230	1
713	43876	14642	2	1	13542	637	214	2
714	36144	17806	4	1	10784	625	192	1
715	42125	17936	2	1	12552	1355	266	1
716	52381	0	3	1	13881	616	167	2
717	32413	21739	2	1	8083	627	295	2
718	39467	0	3	1	9494	977	241	2
719	32976	2827	2	1	12406	1295	337	3
720	38758	0	2	1	12603	1073	160	1
721	29079	11833	2	1	10917	1180	301	2
722	55868	22473	5	0	12679	581	266	2
723	31219	19663	3	1	10560	1074	186	4
724	35065	25707	2	1	13542	1104	295	1
725	38890	25440	4	1	10784	1276	241	4
726	43876	17806	2	1	13542	1180	337	3
727	37783	17936	3	0	10784	1140	269	1
728	36144	20277	3	1	12552	581	266	2
729	42125	22113	4	0	9609	1074	301	2
730	26735	13576	2	0	16205	962	292	1
731	29313	21908	5	0	12076	1704	384	4
732	44504	21941	7	0	10790	1548	375	4

OBS	INCOME1	INCOME2	FAMLSIZE	OWNORENT	TOTLDEBT	HPAYRENT	UTILITY	LOCATION
903	23555	13624	3	0	11138	614	160	2
904	47238	0	7	0	8943	1398	363	4
905	24237	16730	3	1	7078	616	173	2
906	28530	22096	2	1	8442	984	219	4
907	48290	14753	6	1	15116	1014	206	1
908	45208	0	1	1	10964	1128	277	1
909	30993	0	2	1	12832	492	101	2
910	33020	28851	5	1	14885	1487	316	1
911	26334	17516	4	0	17216	720	244	4
912	29628	20998	4	0	19624	883	124	2
913	41238	17225	7	1	11648	680	429	1
914	42382	24944	4	1	18500	1096	293	1
915	29491	0	3	1	9106	810	200	1
916	34298	19085	5	0	8919	859	209	4
917	25443	0	2	1	5534	478	251	4
918	31030	0	5	0	11317	871	254	3
919	45779	0	2	1	11739	689	221	2
920	34460	16868	2	1	9190	807	229	1
921	39587	0	3	0	8205	1191	299	4
922	43406	28236	2	1	12087	815	214	1
923	41536	24736	2	1	14606	899	257	1
924	33080	16849	3	1	11923	1077	257	1
925	36610	0	3	1	7944	908	228	1
926	39587	0	3	1	7089	832	191	1
927	31008	0	3	0	5033	703	145	2
928	36570	0	4	1	8695	867	222	1
929	31192	17592	5	1	6689	1021	272	1
930	33923	21506	4	1	14848	1082	274	2
931	33024	12441	7	1	14126	1214	301	1
932	41211	0	2	1	10036	1275	237	2
933	28731	0	3	0	7952	1114	96	2
934	28806	23615	2	1	3332	335	391	4
935	22349	11830	1	0	6755	620	325	2
936	28224	0	6	0	12734	1090	221	1
937	32181	18183	5	0	10734	817	137	4
938	37051	0	2	1	11649	462	224	2
939	22077	0	3	1	7636	946	130	3
940	26115	29927	4	0	9198	521	317	1
941	31840	0	5	1	6027	1258	261	3
942	45769	20089	3	1	11038	1040	238	1
943	34817	12881	5	1	11038	1095	260	1
944	35155	12866	4	0	16774	934	176	2
945	26001	0	2	1	5407	634	235	4
946	31388	16214	5	1	12823	1291	296	3
947	40778	11337	3	0	11751	1278	64	4
948	26050	0	2	0	8948	505	122	3
949	37089	0	2	0	8205	1111	300	1
950	41176	22881	2	1	14611	866	194	3
951	22064	16683	4	0	6158	655	182	2
952	32439	15574	4	1	6722	747	161	4
953	42486	7204	2	1	6526	806	177	2
954	43817	20102	4	1	15804	1354	314	3
955	41919	26322	2	0	24587	1091	261	1
956	26001	21053	3	1	9601	957	238	1
957	29191	2037	3	0	12978	594	120	2
958	24452	0	4	1	14569	1229	250	4
959	37898	0	2	0	8259	733	166	3
960	30958	11908	4	0	9455	709	217	2
961	30841	19500	7	1	10842	1076	266	1
962	33886	0	2	0	14281	1793	417	1
963	30786	17196	5	1	17359	1485	105	3
964	24602	19922	2	0	13379	816	351	2
965	33897	17954	7	1	12427	722	165	3
966	20632	18744	4	1	9638	1273	283	2
967	41919	22188	4	0	5969	742	201	1
968	32701	21870	5	1	8365	1101	140	1
969	37506	18847	2	0	19627	612	308	2
970	49582	0	2	1	13128	800	146	3
971	28040	20628	4	0	12611	442	209	3
972	53856	0	2	1	12006	1564	336	2
973	54034	15143	4	1	11931	1106	286	2
974	36678		2	1	12920	823	165	4
975			3		12108	794	189	1

OBS	INCOME1	INCOME2	FAMLSIZE	OWNORENT	TOTLDEBT	HPAYRENT	UTILITY	LOCATION
821	25752	14218	2	1	12492	844	205	3
822	38967	19992	1	0	10921	641	154	3
823	46653	15554	4	0	15936	785	173	3
824	32162	0	4	1	13735	619	293	2
825	54793	25698	3	0	16362	1042	267	1
826	25503	0	2	1	16787	1295	110	1
827	26077	14048	2	0	8402	596	161	4
828	26334	17930	2	0	7998	720	166	2
829	24451	20891	3	1	15434	951	234	4
830	29208	0	5	1	14962	460	162	2
831	42382	0	2	1	9481	1140	301	1
832	51130	0	5	1	11054	687	149	1
833	31243	30226	1	0	8873	567	93	3
834	31030	0	2	0	17756	817	215	1
835	50616	0	7	1	10991	1392	368	1
836	33490	28758	2	1	5244	442	127	1
837	31444	24568	2	0	12483	584	149	4
838	25471	0	2	0	16237	644	97	4
839	31862	23151	1	1	8421	452	164	3
840	30927	16884	6	1	17735	1508	244	3
841	29185	30055	3	0	10177	1894	217	1
842	35463	0	2	0	8049	839	235	2
843	32112	20987	5	1	12617	510	144	2
844	42389	22910	2	1	14826	395	102	1
845	46093	24734	5	1	15534	1073	266	5
846	30557	25528	2	1	14257	786	352	2
847	29580	0	1	0	9045	812	208	3
848	29273	18797	6	1	11581	1388	192	3
849	35663	14798	3	0	11581	709	110	1
850	38067	0	2	1	57143	713	370	2
851	33024	17055	2	1	17126	456	195	2
852	41313	24999	4	0	11623	819	181	4
853	28005	26243	2	1	16644	588	170	1
854	35361	25829	6	0	13059	1109	124	4
855	27970	18650	2	0	8452	1733	271	2
856	34740	20236	3	0	12308	1214	168	2
857	47463	20052	6	1	10575	1379	333	2
858	28751	24052	3	1	14382	808	282	1
859	25101	18783	2	1	13963	490	225	1
860	28976	0	2	1	8927	531	243	1
861	33357	13791	4	1	10748	959	242	1
862	30178	20688	5	0	13644	844	226	1
863	32067	14309	2	1	12544	766	321	2
864	28110	0	2	1	8549	451	183	1
865	30416	17290	2	1	7225	1196	285	3
866	38113	0	2	0	8130	642	138	3
867	45114	0	4	0	10224	1158	269	2
868	42022	25434	3	0	13018	1129	246	1
869	55414	0	6	0	5802	438	127	4
870	26405	17709	5	0	12999	559	148	1
871	27711	20240	4	1	9681	657	183	2
872	33260	19202	6	1	12308	1049	246	1
873	41416	0	3	0	10575	808	132	1
874	28165	21128	2	1	14382	531	257	3
875	20837	0	7	1	7991	1557	386	2
876	39868	0	2	0	20612	959	208	4
877	31878	0	5	1	6376	844	386	2
878	45795	12432	2	0	9163	559	119	3
879	25090	23650	4	1	9538	1416	136	2
880	22999	24011	4	0	10508	1026	265	2
881	30820	0	3	1	14734	632	198	2
882	30828	0	3	1	10327	742	272	3
883	47432	0	2	1	13520	985	362	1
884	40780	19654	7	0	12945	1296	325	1
885	37794	0	3	0	6111	800	207	2
886	22551	0	6	1	7614	444	111	2
887	51699	20621	3	0	14990	993	249	3
888	33215	0	4	0	28040	412	499	4
889	32190	0	4	1	8535	734	149	3
890	32706	0	1	1	8359	412	499	2
891	36201	17994	7	1	13222	1455	379	2
892	31968							

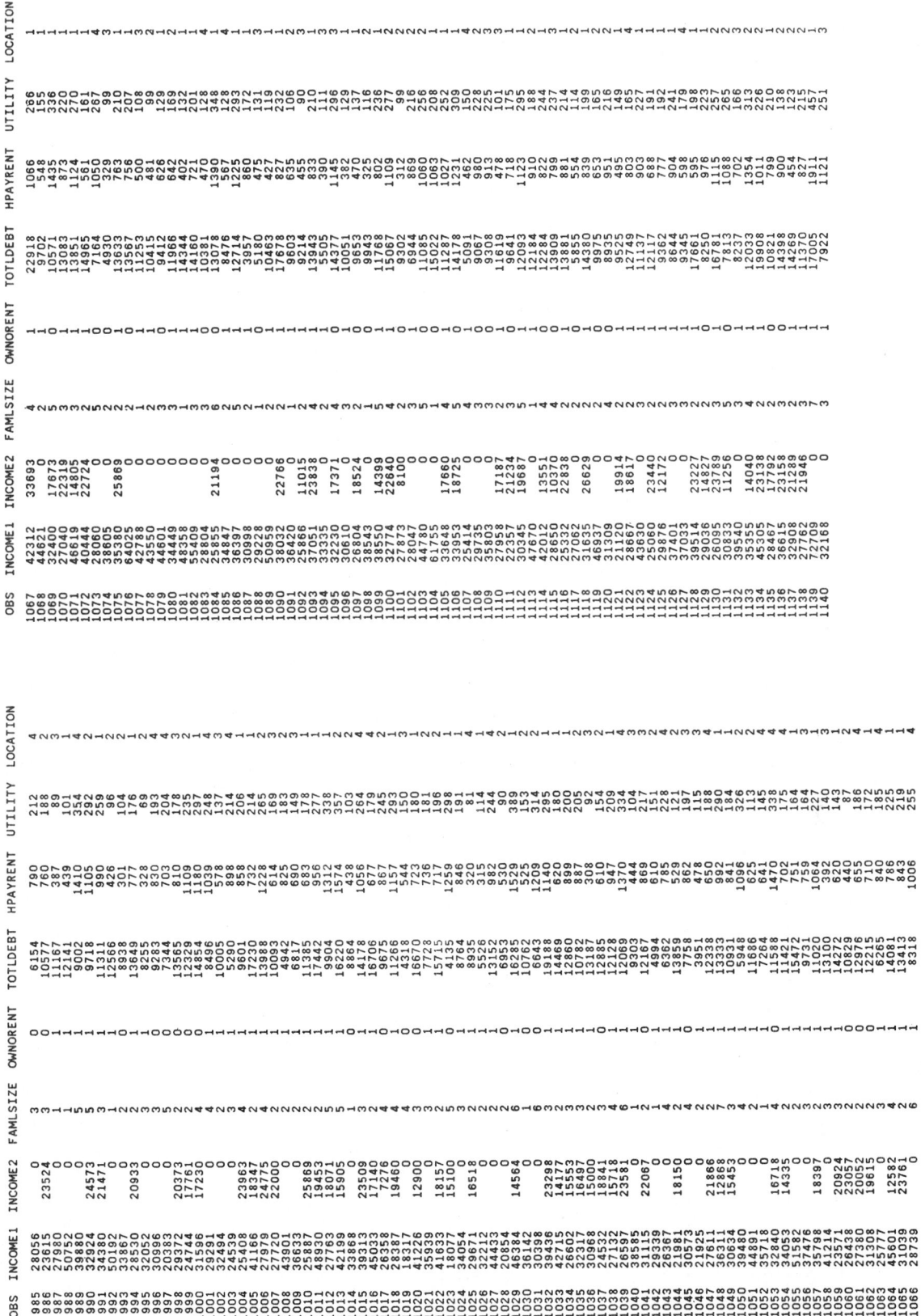

Appendix J   Database Using
Financial Variables of Companies

The observations that comprise the database are random selections of companies listed in the Moody's Investor Service Industrial Manual. Each observation includes the following variables.

NUMBER	NAME	DESCRIPTION
1	BONDRATE	Bond rating as given by Moody's where Aaa-A = 1, Baa-B = 2, Caa-C = 3
2	REGION	Region of United States where main office is located, where Northeast = 1, Southeast = 2, Southwest = 3, Northwest = 4
3	EMPLOYEE	Number of employees
4	SALES	Gross sales in thousands of dollars
5	COSTSALE	Cost of sales in thousands of dollars, i.e. the cost to the company to produce or manufacture the products sold
6	NETINC	Net income
7	ASSETS	Current assets
8	LIABIL	Current liabilities
9	TOTAL	Total assets or total liabilities

VARIABLES

OBS.	1	2	3	4	5	6	7	8	9
1	1	1	11600	1204236	932014	38378	355606	167371	650812
2	2	1	23000	2303731	1713703	-107331	947651	662653	1989945
3	1	1	4000	911002	692227	3116	389884	196724	511393
4	3	4	15592	9259100	6705200	310000	2617400	1424600	5280000
5	3	1	6100	1268580	991863	40723	221308	112590	407926
6	2	1	37481	3516289	3237018	-229627	20296124	1509417	4974267
7	1	1	8300	701059	460513	31215	279361	114844	469184
8	3	1	5320	413668	266110	34905	244063	76659	409064
9	1	3	51300	11113000	8332000	732000	4569000	2808000	12242000
10	1	1	141268	27148000	15129000	1538000	8960000	5636000	26733000
11	1	1	6200	1155711	857101	19088	196566	144692	401135257
12	3	1	13200	901890	613606	48844	413346	196566	685750
13	1	3	60700	4952900	3208100	408900	2432300	867200	3769000
14	3	3	34173	2094900	1407900	-10200	930800	581200	1856700
15	3	2	11082	635076	461671	25611	113650	67696	248067
16	3	1	7000	239352	119407	1798	84052	43298	114079
17	1	1	7830	725241	538399	24082	271643	131605	739445
18	2	1	1401	100094161	200072391	20000360	60359060	181676	449348
19	2	3	53000	3501000	2711000	85000	1296000	844000	2593000
20	3	2	9880	698394	368149	29531	123998	80256	377849
21	2	1	28064	3002700	2237000	152500	1101247	1138923	2685797
22	3	1	382274	62715800	51866000	3285100	18458000	15625600	37933000
23	2	3	16623	1549290	1081574	146391	455580	315248	2096345
24	3	2	750	543986	334305	9272	335606	133545	459164
25	2	2	5614	629700	722000	25700	282933	650487	235840
26	3	2	16800	692900	425400	-12100	273500	208000	1608900
27	3	2	9174	701194	628523	39575	320402	789164	1347069
28	2	2	4250	753774	496184	293161	659121	150600	691546
29	3	1	9400	648419	525101	4674	331613	762800	2086200
30	1	4	62056	4586600	2563900	-183500	804400	22848100	72593000
31	1	1	876000	102813700	88298000	2944700	26768400	68024	290059
32	2	3	7000	514589	325545	-20426	241193	432600	1819700
33	2	1	11914	2553800	1887300	-32700	611700	266800	1265100
34	2	1	18605	908800	497900	-101800	631900	1118700	4097200
35	3	1	41400	3725700	2517500	-472300	1496400	43615	155585
36	3	4	2110	225071	109915	-14296	110466	178399	819962
37	3	1	5578	1159595	773141	86112	582649	276100	2077000
38	2	1	14000	2039200	1637800	85300	470600	75305	80053
39	3	1	3650	366205	236911	7600	177766	246642	1709495
40	1	2	16000	968555	431093	70476	458596	341659	966924
41	1	4	10000	713685	566787	-1310	637237	30770	194958
42	1	2	7000	154536	98537	16701	131889	20676	181853
43	4	3	800	91570	70185	-21304	71073	4692	46474
44	4	3	265	104122	80944	839	44776	8216	67293
45	2	3	305	54263	45068	8498	18696	41474	92145
46	2	1	2200	198038	133083	-10873	66041	143208	728855
47	2	4	27477	93282	73104	-182761	176568	37358	686575
48	2	1	31024	189074	40353	4336	45085	2060	8860
49	1	2	2442	6902	1098	-3440	5460	1371	10270
50	2	3	11	3601	2398	20250	2846	20874	128118
51	3	1	1809	134189	80551	-3744	117317	5975	14093
52	3	2	167	11629	8603	-1116	7149	4147	11716
53	3	2	2839	7459	6775	-7927	1546	18696	156946
54	3	3	3100	229738	188830	-4245	76659	143351	541438
55	3	3	109	2404	6056	341	343142	614	2065
56	3	1	12	416	142	-7374	1836	3316	4815
57	3	4	867	1458	13277	-560	3001	5393	13211
58	3	4	240	23897	13277	-7389	9395	11574	79385
59	2	2	1086	27481	31798	-20	13865	738	1576
60	2	4	130	5046	3059	—	1040	—	—

VARIABLES

OBS.	1	2	3	4	5	6	7	8	9
61	2	3	27	2369	1459	307	1643	248	2226
62	3	1	198	784	383	-567	160	230	434
63	3	2	22000	1012451	416322	74425	561254	178721	873302
64	3	1	24	4873	2268	1509	6174	396	6353
65	1	4	298	16084	12418	1217	20635	1380	27607
66	3	1	350	52684	42682	1762	34316	27731	80770
67	3	1	11	216	183	-1367	785	304	1955
68	2	4	18	1345	376	-104	8357	3139	8357
69	1	3	575	28079	18922	1148	25677	13935	33712
70	1	4	261	25244	20085	1667	7179	5842	42725
71	3	1	1222	18874	11144	-1855	985	3057	12894
72	1	2	322	29575	22577	5470	27578	7659	38873
73	1	3	17	4935	4130	-635	3529	3642	4700
74	3	1	5	18404	232346	-218346	119139	173732	331874
75	1	1	19	311813	130158	-458900	1947351	1203353	2307941
76	3	1	311	52888	39948	1837	13548	2661	9667
77	3	4	520	11803	2118	-363	521	5749	35890
78	3	3	22	3153	21515	669	8698	668	597
79	3	2	550	25911	339	-71	93	4014	34817
80	2	3	92	339	9248	72	2126	50	186
81	2	2	292	17733	18648	793	3396	3197	22637
82	3	3	278	36308	36964	5801	7613	3378	14586
83	3	3	1226	50217	401	-1386	13471	5602	88957
84	3	3	36	734	13169	2601	70772	7296	9080
85	3	1	750	19266	7507	890	2385	7251	14160
86	1	1	239	23297	21440	710	35503	5714	17433
87	3	1	175	16705	49128	-21103	37718	9800	18709
88	2	1	12200	240314	226140	-680	34047	44819	177085
89	3	1	46	4703	2686	-5624	45569	1180	6502
90	3	4	900	16895	5624	4950	101	5322	37860
91	1	1	448	65412	21440	-29806	76174	12166	54499
92	2	4	460	56979	49128	-5394	2057	27093	46611
93	2	1	4100	37050	26020	28	3111	14330	68953
94	2	1	5	306	27	7998	72341	153	259
95	2	1	200	150118	60118	113	248242	44215	121398
96	2	4	333	3055	1796	-5980	1533	802	2648
97	1	1	133	6600	7085	-1575	1837	9891	12117
98	3	4	28	67	1642	-25478	72341	10542	577697
99	1	1	36	26048	26048	901	248242	10100	2520969
100	1	1	16	2609	1610	-1144	1533	1394	1984
101	1	1	22	2534	244	901	1837	403	2079
102	1	1	12	32672	26336	-119237	58478	67334	112820
103	1	1	606	64505	42653	987	18715	9605	400027514
104	1	1	115	13575	5165	1201	6125	12715	8836
105	3	2	164	28536	21493	-400	19095	12108	23173
106	2	1	213	9193	9193	107	7038	3210	8934
107	2	1	250	2406	1378	1061	4511	277	16667
108	3	3	14	1199	963	137	383	99636	2253
109	2	3	1470	161106	92321	-18175	25897	924882	219786
110	2	1	22	53794	20000	-187850	33961	8167	235433
111	2	4	550	25406	15329	1152	15852	5046	19135
112	2	4	600	13552	10303	1079	11224	4119	13329
113	1	2	38	221	388	-2403	407	1850	948
114	3	2	667	16453	13360	-2711	4333	1636	18989
115	2	1	470	33109	14368	3656	28010	261	34215
116	1	1	3900	1162	994	2463	1552	8625	17887
117	1	4	412	47040	31478	-7941	21768	274	26998
118	3	3	37	892	947	-309	271	547	405
119	3	1	2896	17	42	811	505	9519	809
120	2	4	170	12103	10744	-1416	9990	—	17152

VARIABLES

OBS.	1	2	3	4	5	6	7	8	9
181	2	1	37500	4332000	3913100	-152700	1183900	940300	4668900
182	2	1	21700	1791194	1160379	6304	1048104	692976	1580571
183	3	4	124196	16341000	15711000	665000	8478000	5659000	11068000
184	2	4	23333	3739970	2925870	101540	860785	547902	3533647
185	1	1	4800	919690	623817	49444	254380	115535	1600665
186	1	2	25500	1400196	808453	40138	493963	159392	658318
187	3	1	2953	216336	172221	8896	70971	35223	132180
188	2	2	46976	4378714	3173491	223225	1334792	626052	2762785
189	1	1	53731	7321000	5952000	76000	3363000	2180000	6288000
190	3	1	18300	4753700	4469100	375100	2063600	1299600	3370300
191	1	3	7074	1114242	872913	57274	436601	156671	812734
192	2	1	32200	4387623	3590044	200832	801502	730474	6025690
193	2	1	132422	22586500	18635200	1403600	5364000	730474	14463200
194	3	3	433	35635	18045	2330	31136	14012	44947
195	1	3	52	158	309	4637	1411	937	2986
196	3	1	8207	266800	300900	-41700	228700	98900	585200
197	3	3	5100	1089020	525076	95610	417237	177057	849225
198	1	2	28030	8669000	4645000	934000	3739412	2754814	8373438
199	3	1	18400	1616267	1199454	89270	533599	325157	1145018
200	3	3	2957	885411	805473	14226	200087	94085	293559
201	2	4	2342	466320	380831	58948	257731	106031	1214177
202	3	4	42176	5911046	5150120	105285	1283476	926158	1819696
203	1	1	25100	1856300	1220400	177100	860500	486000	2360800
204	3	1	38000	4548000	3168000	219200	1283100	1469500	3650600
205	3	1	3900	294297	215147	6487	95564	48215	152350
206	2	1	273	2535000	1834000	-292000	699000	559000	3090000
207	1	1	51300	8742200	5970600	4132000	4749300	1211300	2383000
208	1	1	29100	3720400	2860100	558200	2072200	1162500	4595800
209	2	4	17500	1058702	543761	-951	587517	252545	705561
210	2	3	22000	1975221	1938202	-215742	820900	1213631	2974542
211	2	1	20600	3217700	2609940	36300	975564	635100	2721200
212	3	2	7000	634627	609940	17107	143959	91500	740485
213	2	1	638	359638	219468	70605	404602	212473	2620608
214	2	4	6965	4541296	4210826	-44957	923301	634312	2098619
215	2	1	24000	3644410	2756150	15643	1247572	1310080	2194882
216	3	1	24300	2955900	2685400	154200	1701800	503100	2932500
217	1	1	40000	4476000	16370000	660000	2339900	1170600	5163700
218	1	1	111000	25409000	16370000	1478000	5914000	4482000	17642000
219	3	1	29166	1239496	643823	167924	791584	671288	2027577
220	2	4	2600	310228	297427	15815	76635	38782	207158
221	1	2	36500	4687100	2799600	316400	1615900	975700	4461400
222	1	2	124617	15978000	8383000	1064000	5934	4319	17019
223	1	1	121000	12295700	9913000	611200	3837500	3411400	7703400
224	3	1	3300	36356	167411	28254	79085	37097	212762
225	1	1	87000	7937722	5403149	223455	1779481	1151067	3503106
226	1	1	485400	44281500	4289710	135130	2104140	1425810	6599460
227	2	1	13706	1552931	916886	105952	753492	430290	1145647
228	3	2	67174	8577749	6603164	200445	746811	952541	3421088
229	1	1	39700	9218956	7592469	-344450	3454795	2445542	15955241
230	3	3	30350	9065819	3173368	89301	854565	742197	1797887
231	3	1	10700	3172260	2578671	58328	513939	252864	781291
232	1	1	72000	5023300	4011900	239200	2281200	1285200	5557100
233	1	1	21500	3762000	3454000	376000	1431000	1032000	4230000
234	1	2	6307	957796	717817	93217	315008	131912	834659
235	1	1	24485	2017775	1341293	370930	637626	473800	2795038
236	1	1	15110	1811937	1345108	108096	549808	197541	2060066
237	1	1	48000	5958000	3965200	17700	168400	88100	2850000
238	3	1	77100	7002900	2630100	-329500	3201600	2292700	5876700
239	3	3	754	113655	269234	-184224	85803	28029	453542
240	2	1	17383	3340700	1744600	-318900	729600	686400	2084200

VARIABLES

OBS.	1	2	3	4	5	6	7	8	9
121	1	4	444	19821	11676	-7423	5225	10709	17116
122	1	1	116	2720	1236	-376	1645	1579	2934
123	3	3	2050	68028	77415	4795	74395	32028	125990
124	3	1	1649	125790	86436	2402	53453	20154	72129
125	2	4	800	54662	23634	5138	28333	12705	75935
126	1	1	5065	396403	150208	54149	279364	91014	417016
127	1	1	722	110719	55927	-2228	31937	12003	106879
128	2	2	4300	616463	341375	38275	168173	88248	429618
129	1	1	68143	3923220	2342926	176260	933803	441291	1492533
130	3	3	694	61376	31528	2910	34821	5720	45944
131	3	4	19000	1625958	1095989	83222	450855	211377	928426
132	2	1	36600	3246139	1263369	430855	4214698	997955	4214698
133	2	4	2844	447755	231038	33289	131829	70662	415657
134	2	3	241	9048	529	8519	4405	2624	4405
135	1	1	83	16100	10019	1043	7792	1636	14411
136	1	1	109	14874	10818	1136	13818	3139	17877
137	2	2	23	16419	13045	900	5492	2265	19308
138	2	2	4640	532754	237817	102178	251591	237027	1310572
139	1	1	92	5429	3314	658	1663	834	6418
140	2	2	3800	192032	139229	-328	125691	68463	223898
141	1	1	1350	94570	81336	6347	40461	8456	58348
142	1	1	1200	91245	57542	2601	76845	52684	102512
143	1	3	810	83923	57456	1454	33198	12525	43078
144	2	1	183	188207	175456	-379	44878	24714	107854
145	2	4	118	40530	18645	10309	17342	13617	130776
146	2	4	150	1115406	561883	35455	501915	237036	985774
147	3	1	1227	222375	129290	16471	71770	17608	114270
148	2	2	4006	336963	183587	-18674	226914	17608	300617
149	2	3	803	62375	34255	3394	36882	12477	58566
150	2	1	31	6179	3515	594	1810	3896	2536
151	1	1	85700	7039000	4163000	667000	3119000	1208000	5760000
152	2	2	492	64076	59623	-1555	21165	13344	55256
153	1	1	3400	930708	559949	72778	241162	230860	1981396
154	1	1	1012	56997	34530	2677	40891	7365	63105
155	2	3	1627	123325	68926	6725	73249	26255	117438
156	1	1	40	419	2120	-2742	865	354	1651
157	2	3	2464	63064	50913	667	24790	24815	44005
158	1	1	4252	169168	129118	-3763	68369	19793	126061
159	2	1	1073	74947	54525	-23804	59407	23810	74767
160	1	2	11750	643831	488472	13096	228206	88742	369265
161	2	1	5400	216985	94206	12045	15801	31828	342772
162	2	3	375	39755	26113	-6156	7827	47365	60415
163	1	1	2700	423444	328995	11533	190262	78461	967444
164	1	1	215	11905	6840	867	6095	2388	16660
165	2	1	16600	1982134	1146089	4735	704536	471908	2705730
166	3	1	5400	423220	290583	-8553	209471	105311	334065
167	1	1	35700	4667200	3509600	254100	1551800	944300	6766700
168	1	1	34462	3816000	2086000	203000	1891000	1126000	3667000
169	1	1	38900	2997692	2227551	188467	1305174	712121	2208436
170	2	4	129000	14021484	10578985	144528	1638472	1312524	3590174
171	2	1	7505	800136	605518	-37608	279131	143352	521482
172	1	1	17823	995620	821493	41207	418692	198523	803310
173	1	2	2200	195010	144619	1559	98483	56593	185193
174	1	1	20000	1204246	543875	105960	769465	273284	1313414
175	1	3	3133	1132120	400075678	40048624	398629	243313	2296300
176	1	1	35200	1885400	1034000	158700	1046700	672900	732700
177	3	1	6000	1010300	872500	58900	331200	165300	277828
178	3	1	4697	439727	269089	16603	157847	103188	404158
179	3	3	2708	370882	292054	-13053	257724	117991	277828
180	2	4	7947	864670	674397	25631	242637	138025	495472

VARIABLES

OBS.	1	2	3	4	5	6	7	8	9
301	3	1	900	901875	345587	78808	787621	603680	1699314
302	2	1	1070	199246	154581	-6258	74946	30810	261306
303	2	1	15565	868629	738231	28980	823096	319922	1257428
304	3	1	20600	2233511	1776514	158268	663759	367130	1741251
305	3	1	58000	3113506	2359278	339990	1509141	982785	3117664
306	2	1	7900	648337	585573	38808	274425	108928	477078
307	1	1	43428	3811000	2875000	137600	1676500	646300	3025100
308	3	1	23000	1145122	951664	44897	264191	161937	445173
309	2	1	1010	205624	154826	8836	63095	43234	161588
310	2	1	26100	2683961	2270866	-60900	936188	699894	2002894
311	2	1	1900	335000	136600	28700	1386200	2023400	2234400
312	1	1	12500	997837	678598	125820	144028	161744	1042389
313	1	1	482000	28139000	20757000	249000	14288000	11461000	34591000
314	1	1	39000	7223000	5783000	296000	1420000	837000	5114000
315	2	1	32100	2818300	1183800	15800	1489700	900700	2535000
316	2	1	7200	350587	103802	12515	248606	127784	336884
317	2	1	33400	3440125	3159806	78660	1305388	624229	1963670
318	2	1	600	103044	76499	-3676	222543	81461	4243400
319	1	1	47000	4366177	2622239	301734	1614694	910407	2837364
320	1	1	35000	740366	442746	97839	301188	127551	1302273
321	3	1	38162	1865700	225200	-136700	1394100	937500	4712900
322	1	1	29857	2799481	2105587	93974	1441196	711009	2355273
323	1	1	403508	34276000	16197000	4879000	27749000	12743000	57814000
324	1	1	13900	1177700	1047400	-217000	768100	337100	2427100
325	1	1	4400	5500000	3991000	305000	1628000	1332000	7848000
326	3	1	4004	401109	315456	6092	112571	59048	318045
327	3	1	1392	138158	89442	-873	111120	50195	140670
328	2	1	4000	422600	308100	-50700	271700	91900	317100
329	3	1	7200	321769	1-56183	22623	216282	93048	274956
330	3	3	70	2477	1038	-33224	233808	55975	240518
331	3	2	123	1911664	4219291	-1809951	1921227	3040119	12958397
332	3	3	1000	243542	175685	6236	54507	26882	86033
333	3	1	32000	1091675	788550	-59259	140683	273444	568493
334	3	1	5500	314429	209241	1610	86029	88019	234095
335	2	1	16300	919818	703576	42426	305882	120168	587945
336	3	1	4500	528483	357597	18922	158816	83884	291180
337	2	1	6333	587820	431961	20256	208205	98854	315642
338	3	1	525000	1033560	772020	36436	184877	65540	213423
339	3	1	1000	772241	603988	6695	66463	50268	172134
340	3	1	7350	975727	637382	29739	324689	234328	628924
341	3	1	4997	884726	474275	73753	466805	266556	849814
342	3	1	6371	561858	329341	32666	262230	69973	192257
343	3	1	2100	553068	479267	13045	124750	69244	401892
344	3	1	110	532	589	-8793	3614	1150	4813
345	2	1	1729	143677	73908	11968	54938	18428	85079
346	2	1	1187	221	257	-4101	4889	566	8560
347	1	1	1864	108382	82877	3321	52080	25880	90447
348	1	1	2200	1285811	1172318	15600	332149	151798	385194
349	1	1	265	34722	31232	1137	65080	5750	96809
350	2	2	5500	587492	501072	22137	125724	120154	278656
351	1	3	156	23035	5822	3328	11648	2219	16052
352	2	2	967	30463	9047	-6490	12046	14182	31476
353	3	1	4	646	600	-691	352	764	1061
354	3	3	75	2158	5859	-5859	16526	1411	26414
355	3	3	190	8131	4394	-2915	8539	4552	15110
356	2	2	23	5975	5675	182	3013	2674	10427
357	3	1	5000	50811	39297	1125	18468	13379	39821
358	2	2	3917	200602	171578	6670000	92354	30300	186680
359	3	1	457	36605	26327	1530	15542	4188	21081
360	1	4	154	20813	12837	1967	17255	1239	22469

VARIABLES

OBS.	1	2	3	4	5	6	7	8	9
241	3	1	4800	355377	216322	571	156951	55509	300024
242	1	3	36490	4303100	2712200	269400	1032900	804800	3676000
243	2	1	10944	1396401	1029008	77480	455659	278743	1016624
244	2	4	55500	4521002	3592465	71137	2717684	1747786	4569321
245	2	1	22950	8629988	7941301	545503	19024309	16107685	19024309
246	1	1	847	826500	881000	-20600	173600	210970	1680200
247	3	4	12714	1506200	1197110	187910	309720	210970	1886970
248	3	3	44000	6440871	4917211	225940	564050	757685	1551671
249	2	1	8600	817797	388933	59526	393187	277000	870130
250	3	3	19400	1920262	1451786	81228	913677	304393	2513343
251	2	3	5105	1430036	335530	78282	386802	234967	1404615
252	2	3	2567	399267	90210	55049	156000	80835	300351
253	1	1	68500	4752537	4421794	202344	1005376	829641	2470745
254	1	1	21000	1452010	929650	203420	786850	209765	2226150
255	1	3	151700	10376000	7533000	381000	3525000	1604000	6209000
256	2	1	9200	741586	502326	19461	423097	208670	691685
257	2	1	124400	49865000	48458000	1407000	10869000	10432000	39412000
258	1	2	51703	6879000	4341000	433000	2808000	1716000	8269000
259	3	1	51660	2110348	1985022	39411	585915	369101	969393
260	2	1	12895	1729600	1413000	75600	1206800	1124000	2904400
261	2	3	16000	972819	716833	5778	340000	172494	1656127
262	3	3	8465	374563	181210	17103	195609	74359	272127
263	3	3	4940	402357	283221	24009	164384	76143	335282
264	2	1	6000	549314	531246	-324195	477970	415135	1149950
265	2	2	307	129285	242993	-64258	457764	63136	590179
266	2	4	51000	15344143	12640306	181064	3711130	3313121	17466777
267	3	1	16300	17072610	13715600	752260	6500980	390577	1544588
268	3	4	875	811292	765098	27783	238785	105881	478464
269	3	3	220	164991	158898	5004	29482	24020	64589
270	2	4	9600	1107564	717718	51947	466912	219168	996685
271	1	1	6257	1921223	995746	45443	851472	287827	3369278
272	1	1	214000	9290800	3731800	457800	2503800	2223100	8028600
273	3	3	14581	1290558	726975	72572	1032886	496958	1399613
274	2	1	21800	9786000	7914000	228000	2800000	2231000	12399000
275	2	1	6345	899065	644636	50228	200728	108165	665042
276	3	1	2044	217296	146359	-3304	156219	120288	222451
277	3	1	1484	158829	132669	-2070	56616	15924	86126
278	2	1	27100	1611281	1413389	-59562	2089903	940717	3619807
279	2	1	2445	243255	198246	11036	76552	40658	160699
280	2	2	25600	3638900	3028500	191800	1407100	814900	3708800
281	2	3	12060	2067000	1346000	138000	941000	434000	1842000
282	2	4	8000	634162	553910	47318	361758	154393	502370
283	1	3	32641	18220000	12792000	883000	4081000	3483000	26214000
284	2	4	1134	104192	39833	5769	13004	9246	185983
285	3	2	1100	50152	61776	7225	43096	9562	87166
286	2	1	28000	1725200	1464600	82600	681100	334800	1287900
287	1	1	35724	3807634	1868402	278238	1888872	1303035	3865609
288	1	1	13722	11794000	9111000	605000	4760600	365500	1126800
289	1	1	47000	2696993	2290943	-97279	604825	541357	63468
290	2	1	20900	2645500	2367800	-472000	949500	685300	264200
291	2	1	7000	1034953	1056538	9140	412333	265576	1840314
292	1	1	26600	14993000	9495000	615000	4743000	3750000	21604000
293	3	2	46500	221654	175615	23701	422872	413592	2313968
294	2	1	61000	5543000	3526000	444000	2344000	1881000	7068000
295	3	3	80185	9319800	7456900	1167100	2755600	3593800	21090900
296	1	1	34900	4835900	1515200	589500	2758600	1015200	4183000
297	3	1	2266	241428	162961	22675	157228	56782	335910
298	3	3	6600	1309915	967662	71051	565527	287037	1509744
299	1	1	51095	26245000	5272000	715000	9050000	6485000	34583000
300	1	1	24149	2551469	2163448	50886	1175044	1240412	2161408

VARIABLES

OBS.	1	2	3	4	5	6	7	8	9
421	3	1	8	17254	11630	2830	34163	34163	34163
422	3	1	127	16868	8781	2107	124992	2000	225622
423	2	3	7432	263952	213431	40952	185829	30749	567383
424	3	1	599	66549	45814	1696	20951	11396	31451
425	2	1	2100	96200	67497	3645	42137	8646	54848
426	2	1	3200	315357	227653	14320	194088	75744	308890
427	3	1	31	671	1504	-794	441	651	2294
428	3	2	308	15608	9186	1626	22632	1690	25975
429	2	1	1250	53829	39069	1466	24090	11798	35934
430	2	1	485	41446	30466	1308	22685	4738	25950
431	2	4	1522	207589	175901	850	45723	37833	76129
432	3	1	7500	523146	366301	9252	135982	45413	208291
433	3	3	570	50203	33446	1510	28503	16960	32337
434	3	1	1700	251815	192014	3055	112658	83482	147866
435	3	1	3273	228599	131243	2430	80396	58937	108751
436	1	1	431	29811	13243	1983	14915	8584	23517
437	3	1	1800	115078	91191	9648	56585	20528	83110
438	3	2	1000	101944	88035	-968	31321	21094	51699
439	1	1	2794	130854	101562	1004	275867	89871	1006511
440	3	2	860	32934	23365	2151	26547	4365	42301
441	1	1	1002	50303	34223	2211	16667	4370	24353
442	2	1	4560	174224	94237	-61670	329653	222512	629256
443	1	3	690	73825	47188	-5200	52693	18913	99205
444	3	2	556	20476	13920	-399	16269	2143	25074
445	3	2	1018	327272	313430	-650314	752206	157444	996267
446	3	1	2880	21710	12204	2866	30803	12257	64906
447	2	1	350	114057	69803	-4664	98908	56933	202084
448	3	1	450	42177	14428	2794	61490	33243	120691
449	1	3	30	40126	27733	2785	12128	7482	21203
450	1	2	117	117	106	594	480	920	1028
451	3	3	3307	294818	232110	19092	59755	35321	146714
452	3	2	583	47747	35752	629	16244	9556	20109
453	2	3	150	10163	7130	-1183	6475	3751	8538
454	3	1	159	2684	1076	-400	3143	904	3871
455	2	3	375	23566	16422	4740	49767	36723	241560
456	3	2	1786	11512	8653	682	9347	4853	16069
457	3	2	272	23439	17029	-3261	24301	19470	40562
458	3	3	6516	108767	110417	-800	28275	10089	622
459	3	1	898	58707	46379	90000174	84	319	28741
460	3	2	858	73	75	1745	28021	10493	205267
461	3	3	2600	192599	133555	16980	36932	22800	165861
462	3	2	2000	271947	165143	15728	127477	36953	75924
463	1	3	1177	142568	94404	8515	44162	12364	30498
464	3	1	362	149519	229685	5419	20819	5092	260908
465	3	1	6823	174644	129685	1743	29153	36943	9164
466	3	3	2641	120601	93567	18741	152852	96792	40014
467	3	3	128	10241	2318	861	6895	2330	79420
468	3	4	85	13309	11266	-2710	17431	4358	53528
469	3	3	27	747	549	2155	38279	11637	37126
470	3	3	14	1366381	744609	159	17	6	59385
471	3	1	400	32892	13668	107657	279206	1241383	502
472	1	4	1600	95993	60796	4248	19667	4173	16223
473	3	2	5	314	190	5865	35578	11607	311837
474	2	4	205	21020	11474	-273	10265	6730	128265
475	2	1	4303	531640	323196	-722	42652	62964	45601
476	3	1	1419	195453	170739	37818	185089	25602	79420
477	1	1	1300	150028	167113	4678	42652	25898	128265
478	2	4	208	31072	6514	-2421	5389	3096	11682
479	2	3	2	7703	2790	173	1043	1542	25749

VARIABLES

OBS.	1	2	3	4	5	6	7	8	9
361	1	2	425	24300	15172	1908	6556	5166	29941
362	1	1	75	1084	638	-1426	2510	7744	4193
363	3	3	2400	108280	2157	390610	91781	6609	120956
364	3	1	810	124399	100770	52610	51766	8679	71806
365	2	1	265	6685	2128	-6882	9117	4981	21039
366	3	3	2485	163005	107508	7196	131685	26313	161602
367	2	1	52	4148	3727	-52270	3813	3500	7077
368	2	2	3000	45098	34644	17671	133533	63272	640576
369	3	1	742	94205	68254	2495	37188	9980	60853
370	2	1	532	35824	21794	711	23628	6503	30518
371	2	2	10336	223918	190749	4656	32504	13743	149302
372	2	3	2365	303951	183230	16951	261860	69658	406338
373	2	2	691	146800	105906	6083	95213	42935	109090
374	3	1	1545	125668	98133	5070	36267	12798	89472
375	3	1	310	36949	10999	2859	27096	6140	33580
376	1	3	910	104949	53992	7489	60545	16544	79828
377	1	1	42	5397	5749	-212	6955	1692	33930
378	3	3	320	15016	10022	700	12868	4229	20622
379	3	2	4	221258	158669	-290324	2197508	1659957	9868389
380	3	2	550	56260	48928	1257	21212	12335	29732
381	3	2	215	20758	12544	720	6796	3868	11509
382	3	3	500	11217	8750	740	5006	5345	23976
383	2	3	1434	514090	453000	-16942	303344	533582	1317613
384	1	3	976	11739	5039	1521	6137	2232	6854
385	3	3	71	1734	125	-12084	1858	2372	14980
386	3	3	33	1541	820	746	2766	7224	6030
387	3	2	3420	80471	58394	2812	45762	15956	130543
388	2	2	480	19581	19581	1205	11096	4106	21197
389	2	1	383	19612	7808	188	25497	20804	34490
390	1	2	1700	119979	53942	6838	21295	6812	61474
391	1	3	6	101577	45975	45975	201938	2719	1160881
392	2	3	270	43809	35275	1694	11251	6608	20997
393	1	1	34	1235	361	-2849	2704	2719	5041
394	3	3	845	42004	19300	58	14630	8408	35739
395	1	2	94	4822	3737	-430	3974	1590	16873
396	3	2	104	52166	25587	4280	26725	14264	31709
397	2	1	22400	3172000	2700000	-426000	1578000	931000	4698000
398	1	4	148	181	179	-439	29	308	1641
399	1	1	1000	61805	48341	1393	14432	3750	16960
400	1	4	330	29655	22730	208	10538	7169	12902
401	2	1	261	46441	23126	6930	24176	6444	27412
402	3	2	29	335	395	-2305	2538	338	2892
403	2	1	25	156	105	-1949	2537	9472	3197
404	2	1	67	6197	3130	8661	638	2671	9472
405	1	4	111	6603	7874	-1415	63553	399	72661
406	3	1	972	1525	1227	-452	2624	27895	4378
407	3	3	886	155327	127156	1025	42095	652187	53790
408	1	1	6	5350638	4230629	88974	1188435	1879122	1879122
409	1	3	270	27908	20104	-1206	7591	7765	38707
410	3	3	325	410345	328248	-96405	112349	80355	405322
411	3	1	176	266788	111163	-23173	170198	72829	244062
412	2	4	2664	766217	656581	-58601	609464	323545	1170548
413	3	1	1800	106096	86213	2033	59861	44749	81378
414	3	2	931	23279	18036	1185	7033	3561	11741
415	1	2	460	64483	53636	1130	62627	16788	83804
416	3	1	1017	288664	210275	10193	180494	78229	292196
417	3	1	3645	4007	61787	-8190	7912	3699	14451
418	2	4	87	93285	61787	6520	41115	11170	107448
419	3	1	1100	39357	20223	4642	34653	10452	53478
420	2	1	520	87180	62975	-24594	189635	34427	453681

| | | | | | | VARIABLES | | | |
OBS.	1	2	3	4	5	6	7	8	9
480	1	1	724	82097	69112	1278	36433	23863	39489
481	3	4	200	1726	574	-214	425	306	665
482	3	4	1554	95637	52441	8346	73347	15104	113539
483	3	4	300	41527	40138	-156	42263	5030	61087
484	3	1	7	663875	412882	-2035663	1189863	1068744	4737485
485	3	2	2000	68274	46906	14923	43570	20578	233376
486	3	2	77	12236	8068	213	6300	5361	28639
487	3	1	792	9654	7338	351	7163	2963	12546
488	1	1	26700	2216636	1488646	59609	1041910	549747	1793160
489	1	1	25120	2615110	1998741	226733	1079821	419683	2914319
490	1	2	15980	2169614	1453440	132764	393420	272747	1356303
491	1	2	43600	1647788	342617	103370	2278189	477067	2278189
492	3	1	6000	1960237	1893884	39079	320723	124524	205614
493	1	1	89000	4930652	4109528	174644	1239606	715694	6793372
494	1	1	13700	961077	623328	79583	597837	304262	1276230
495	2	1	22459	3173242	3073622	-19264	894956	466954	2526557
496	2	1	7800	745049	505448	33345	279903	178944	718598
497	3	3	83	34830	8395	6142	23122	5197	30465
498	3	3	45	3	1	-983	474	137	2678
499	3	3	4959	334437	151442	23406	161769	59915	369031
500	3	1	500	46391	41375	-2810	43596	9504	95791
501	3	1	5100	195359	157650	-3089	28676	32489	251433
502	3	1	390	66475	48135	85590.	22435	9469	60517
503	3	4	280	31399	25941	405	22706	3777	52974
504	3	3	12400	433560	359396	71164	81007	51436	115065
505	3	1	14000	666186	502736	43655	211444	145569	924533
506	3	2	3500	87205	44056	-5323	13389	14219	40057
507	3	2	190	23950	16623	167	8741	6054	12479
508	2	3	5470	460279	203891	9000	46432	251415	851426
509	3	1	1797	184296	72009	15346	132060	63273	216741
510	3	1	1601	131755	59866	7750	105599	15763	130799
511	2	1	498	53259	47236	3434	55159	32369	69364
512	1	2	1700	68274	22637	3948	36435	15763	48261
513	2	1	1486	113184	89184	6328	124629	4632	173771
514	3	2	2530	190718	171782	10118	89694	51982	122575
515	3	3	80	114319	110855	17538	15001	111358	74903
516	2	2	988	473561	372511	17104	36598	35424	566769
517	2	1	1700	931934	854117	-27402	105670	84356	260126
518	3	3	250	22369	50953	-3639	6385	6385	82053
519	3	3	343	16372	11749	162	4789	1933	10646
520	3	2	46	9249	7569	5529	3768	1451	6844
521	3	2	14	121670	75259	21360	104797	51776	119456
522	3	3	23	212886	117964	1016571	959381	41371	175473
523	3	1	21800	1933055	1267918	-46294	174763	483918	1786870
524	2	1	4269	381046	305238	4073	30918	95580	340577
525	1	1	1600	180920	91568	6562	260558	28245	96853
526	1	1	9900	757820	603568	23665	40709	121721	508098
527	1	1	1619	63225	51524	-3170	15479	22813	84382
528	1	4	2150	23008	18002	-149032	442930	7212	26502
529	3	3	3300	294502	120687	521100	1892800	150405	674354
530	3	1	33747	3745400	1315200	-304300	1116568	1355300	4222000
531	2	2	4730	1792632	2308025	-13100	162864	1335022	3288484
532	2	3	1600	339190	241206	14335	481225	62846	565746
533	3	1	6277	341382	253913	98928	1505506	232787	274461
534	1	1	20500	1403472	934393	396291	530775	750631	1131988
535	1	1	16915	1784629	566049	26148	110259	401302	2408831
536	3	1	9411	2988396	2494357	25648	97055	90402	1770232
537	3	1	4000	380501	148029	-896	604797	49633	202200
538	2	2	2300	208025	152407	53687	97055	49633	131237
539	2	1	22800	1671871	1396682	53687	604797	220378	961058

| | | | | | | VARIABLES | | | |
OBS.	1	2	3	4	5	6	7	8	9
540	3	1	4700	267978	215548	-20861	147115	35358	188817
541	2	1	15500	2032325	1564619	-35415	530459	203410	1523600
542	1	1	23645	10047000	7109000	385000	3289000	2386000	11684000
543	1	1	16000	1433940	944297	45400	669418	255211	1399176
544	1	2	27941	2920310	1623351	408085	659082	458869	2929081
545	1	1	16565	1058055	875006	2736	406206	305490	1403529
546	2	1	2325	157232	103788	5664	94212	37008	130531
547	3	1	898	780278	743890	17580	114883	6976	491889
548	2	3	857	56530	38684	1352	40947	7061	56808
549	2	1	23000	2667912	2326185	12928	1304547	84835	3027518
550	2	1	78556	6035900	4678200	217700	1748900	1351700	3909300
551	1	1	10000	796351	627078	37500	387535	166797	591908
552	2	3	7500	961411	734983	11945	279713	105526	501707
553	2	3	17355	2045125	1340174	129934	568849	262070	2751482
554	1	1	50292	6343000	4343000	496000	2414000	1881000	7571000
555	2	1	98300	5115900	3681600	-43400	4536400	3342100	9408800
556	1	1	8982	766525	633564	-80071	67521	236991	11091787
557	2	3	18005	8091000	12373271	72727	6940346	4626477	10133000
558	1	1	20700	2291448	795614	176000	1201840	1175000	2664956
559	1	1	21700	2723664	1884596	252646	686547	617053	2006068
560	2	3	2080	1880083	1652081	225508	312085	323055	1697322
561	2	3	42100	3660553	2550072	-99888	720164	388834	1197082
562	3	3	53200	3102918	1052781	103137	1510068	969806	2515923
563	1	2	6400	1215064	671199	309484	219422	185113	1145227
564	1	3	2733	235635	160217	100173	78477	42847	237405
565	3	1	117267	10731000	7771200	17020	4635300	4196400	8481800
566	1	1	30956	4008700	3595500	670800	1088200	537300	2202300
567	2	3	5500	1858600	1480700	199700	1036600	876800	3907900
568	2	3	100367	4822000	2412000	-239700	3973000	2206000	10608000
569	1	3	30	2673	153000	465000	1175	986	2594
570	3	4	462	77636	46442	294	13615	5424	64836
571	3	4	160	22285	20973	16052	8216	3472	73871
572	3	4	39	1606	1556	1920	833	849	1399
573	4	4	8	1358	1106	46	760	2004	3443
574	3	2	135	182344	123542	-1242	65917	45053	80788
575	3	4	744	113	331	-2528	168	90	540
576	2	1	273	94031	71620	3210	37561	13321	38892
577	2	4	957	23321	14666	2633	14438	2777	24390
578	3	1	130	9484	6592	833	3266	3443	9458
579	2	1	1200	143356	118253	-2161	44006	17049	63271
580	3	1	221	17957	17966	-1000	8414	2851	19651
581	2	3	28	6387	4478	-1052	4319	3809	17797
582	3	3	5	28	208	-420	298	7	332
583	1	4	1400	126502	96362	4921	35509	18463	69321
584	3	3	24	1454	1454	-909	179	1524	5790
585	3	3	1509	86052	68440	1748	26074	10488	36386
586	3	2	16	410	296	-406	120	561	473
587	3	1	238	5679	4540	408	21652	5734	23937
588	3	3	26	1157	844	-852	1299	302	2019
589	3	1	2300	209787	147536	4469	72903	26021	94204
590	3	1	103	861630	632040	13220	117592	82287	205579
591	1	1	10	149	69	-46	859	36	1641
592	1	4	227	10	40	-30	181	12	232
593	4	4	976	31290	16445	617	10286	8663	35150
594	3	1	152	626900	613919	2906	158765	67538	12621
595	3	3	45	201	11853	879	5326	4757	164697
596	2	1	42	1257	889	-1341	1333	875	2008
597	2	1	75	5008	1095	486	3945	554	6204
598	2	1	45	1257	889	-1341	1333	875	2008
599	2	1	42	5008	1095	486	3945	554	6204
600	1	1	75	4887	2264	300	3171	413	4700

OBS.	1	2	3	4	5	6	7	8	9
					VARIABLES				
601	2	1	77	2563	2467	-3929	9808	1283	14582
602	2	1	29	1092	797	-1105	2005	608	3583
603	2	4	37	32	7	-6	10	5	195
604	2	1	526	44102	17914	5222	28373	5325	33469
605	3	2	6	380	196	-533	481	239	1076
606	1	1	600	17813	11255	-1098	6653	2603	12072
607	1	1	16	767	480	-534	583	356	780
608	2	1	270	7132	5891	309	2306	1092	3882
609	3	1	123	6245	2912	-119	7596	839	10302
610	3	1	103	7027	4562	-4015	24757	1713	33118
611	3	1	30	1423	603	-822	2006	541	2563
612	3	4	8	83	32	-3	47	25	147
613	2	2	23	1811	928	-931	1201	1594	2956
614	1	2	1700	151828	118070	7867	49505	23838	143602
615	1	4	94	7464	4473	-72	6933	4522	7379
616	2	1	486	45974	34068	-6613	21579	18289	32689
617	3	1	2600	274198	87677	28743	132373	50890	290083
618	3	1	6438	35914	25095	5556	19060	9166	22714
619	2	4	1860	201901	187380	806	44014	24961	219695
620	2	1	149	11088	8944	-12002	13340	2646	33738
621	2	4	15	1328	508	-319	1008	125	1303
622	1	4	85	193	197	541	452	976	4294
623	2	3	68	6361	5550	-179	2480	430	27772
624	1	4	298	34520	33226	1264	15104	14662	20295
625	1	1	984	9910	4651	282	6156	969	9693
626	3	1	115	14529	7953	1515	6993	892	7737
627	2	1	684	13558	11167	-614	13734	1409	17743
628	3	1	400	6162	4028	-132	2628	1563	8173
629	1	1	350	232487	224625	588	19616	14601	25794
630	1	4	1352	109899	49385	9447	55962	15261	94477
631	1	4	432	72209	19793	-19412	89212	36717	275663
632	1	1	334	119152	98577	3899	35980	20738	38889
633	3	1	88	17800	10731	1164	6166	3834	6568
634	3	2	302	25080	22258	667	18434	1798	40334
635	2	4	473	46180	18334	5065	51680	10864	59886
636	1	4	700	54111	41395	628	8813	5801	13295
637	2	1	4400	310668	234483	12710	129085	38506	190230
638	2	4	48	4177	3566	-351	2630	509	3924
639	3	2	3347	1018431	809854	30251	133841	85753	291459
640	3	4	14	80	131	-1350	119	88	2565
641	1	1	807	44291	32986	929	11817	5448	14140
642	3	1	186	32779	20027	3234	23329	3391	33399
643	3	1	1400	43379	22976	-1224	24671	7939	36663
644	1	1	169	22536	13537	2166	14791	4253	16830
645	2	1	2331	197659	134541	9518	144852	26544	193278
646	2	4	98	12040	6698	1120	6823	1098	10619
647	2	4	10	259	81	-8906	1761	23294	92886
648	2	1	336	16483	10225	-16920	4910	3876	8870
649	1	1	1080	125308	71221	13534	56242	16187	68824
650	3	1	2737	249570	207381	8233	70894	29324	101651
651	3	3	30	2268	2179	-644	1018	1045	1678
652	1	1	630	25872	17193	1655	11679	10782	29984
653	2	4	465	51748	22089	10367	80499	8547	90564
654	2	2	43	1161	515	-50	293	390	60
655	3	1	5	510	633	-425	2126	495	2442
656	3	3	175	263	460	-99	1118	522	1973
657	3	4	1415	149261	88902	1242	104166	32384	154138
658	3	1	20	50	774	-1034	53	1568	141
659	2	1	22	678	588	406	1483	887	1813
660	2	3	32	2882	1435	-4359	2773	3014	19232

OBS.	1	2	3	4	5	6	7	8	9
					VARIABLES				
661	1	4	600	78824	58592	2024	35519	10381	45688
662	1	1	124	9214	6917	93	1512	1458	3777
663	2	4	13	783	581	-375	530	239	1755
664	3	4	850	79533	58348	5243	56898	21963	72896
665	3	2	1250	4223	4135	-2398	1528	7770	5318
666	3	1	897	4253	2728	2577	5716	735	6633
667	3	4	129	10218	8154	-1980	5833	4043	7726
668	3	1	12	480	199	-1981	325	737	869
669	3	1	63	18034	15680	-199	10495	7478	11911
670	2	3	136	212	344	-141	898	82	968
671	2	4	1570	210799	38625	30100	129729	44740	174731
672	3	4	455	76019	36623	5889	88795	22969	96689
673	3	3	11	1011	463	190	693	333	3295
674	3	3	280	10383	7143	535	3946	1090	7599
675	3	1	79	1898	1179	-1021	3250	4064	29600
676	1	4	3300	81506	27445	-2913	16705	11234	54321
677	1	4	532	172299	76492	27177	95822	19436	108801
678	2	1	955	12134	7435	198	4373	1143	6987
679	2	1	200	17146	10457	969	17561	2169	30257
680	2	1	217	12212	7642	2828	42604	3574	42995
681	2	1	2224	458377	353315	-3688	187749	74452	241939
682	2	2	400	36901	26776	2542	19553	4070	23440
683	1	1	15	14167	230520	-244356	64393	243314	1430578
684	3	3	3787	263286	198047	14788	120620	55002	194341
685	2	3	550	35332	31575	-10489	46446	6407	124620
686	1	1	275	10021	8472	-2297	4663	1121	6147
687	2	4	112	318	19606	-19348	119853	100100	900891
688	3	2	111	994121	687580	129019	257449	100454	323540
689	1	3	1800	7935	4959	-228	5766	937	8845
690	1	1	326	10609	6332	115	2492	1066	2754
691	2	3	631	39051	34797	-7038	20808	13184	45162
692	3	4	58	4357	942	488	5818	1381	8743
693	2	2	836	22145	9557	-1044	2224	4764	6549
694	2	1	150	17545	11613	-7851	11489	3992	12859
695	2	1	181	16309	8190	-5154	14327	3212	18685
696	1	4	32	4686	1730	584	3673	2307	4422
697	2	3	68	3400	2504	-232	1661	1030	2575
698	1	3	192	10456	9424	-3935	5934	1924	17698
699	2	4	68	3818	2307	-596	1255	1610	3298
700	2	1	266	24575	15333	2771	12370	5507	22826
701	1	1	227	89528	79921	1388	26760	17033	27890
702	1	3	943	64038	60930	3815	50284	22732	62025
703	3	2	330	34685	22018	1375	9777	3641	12263
704	2	4	1100	20697	10133	1347	22572	3224	44390
705	3	1	66	6060	4426	-261	3061	1112	3640
706	3	4	900	31024	18122	7555	45022	8237	46966
707	2	3	14	549	471	9	1173	356	1316
708	1	3	118	7817	7930	-1947	6555	4699	14212
709	1	1	600	37255	23591	799	21347	9079	32545
710	2	3	8300	421612	343610	19478	194267	28031	277006
711	3	4	559	3000	1719	-1948	1593	1706	1964
712	2	2	159	10078	6671	332	3016	1939	7464
713	2	3	1501	112788	57338	3736	66001	24838	113771
714	1	2	30	720	126	-1076	1729	172	2176
715	1	1	13	976	536	-95	231	342	375
716	1	2	58	5415	3185	-294	3718	739	5165
717	1	4	660	57240	31026	715	11991	8304	15443
718	1	4	123	30965	25538	-943	14168	3911	15213
719	1	2	162	16169	13023	-4269	2990	1866	5324
720	1	2	2058	92743	67301	3778	24338	14161	57490

OBS	1	2	3	VARIABLES 4	5	6	7	8	9
781	3	2	1219	507670	227848	-12980	88534	64604	499292
782	3	1	78	6063	3698	-2469	6973	1935	7971
783	3	1	416	17173	20040	-201190	1058532	87312	1095518
784	2	4	200	21447	3546	1800	15449	6002059	32205
785	1	3	299	60914	46075	1201	21032	8142	23353
786	3	1	1027	11083	7433	228	4613	1615	7022
787	3	1	3300	274510	181191	1616	118673	55572	27449
788	1	3	302	42611	37324	-2308	22882	31624	24552
789	1	1	210	14813	12454	-230	8465	1667	13846
790	3	1	8	420	148	-523	550	416	843
791	1	2	311	1652	709	75	757	757	991
792	2	4	241	20412	11987	-3851	10160	6442	14775
793	2	4	127	10861	5199	1402	5620	1665	7276
794	1	2	60	320	337	-336	97	200	169
795	2	3	1298	93392	65291	215	27846	12722	70439
796	1	4	60	116201	98883	-18050	38181	8190	38181
797	3	1	321	9918	7250	439	3887	3176	9287
798	3	1	1130	9146	2209	251	273	50000180	100005028
799	2	1	231	17387	13466	-1657	3837	5632	8038
800	1	1	935	68872	23684	1342	19517	16313	43880
801	2	1	2700	354927	326544	-7338	123167	67891	155527
802	2	1	31	1824	142	-557	911	363	7761
803	2	1	47	3189	1442	182	5736	176	5915
804	3	2	350	30132	1467	2000261	21156	3988	28668
805	3	2	36	3041	1941	-638	2889	735	3812
806	3	1	65	4265	3561	-2292	5879	2883	7594
807	2	4	50	15938	12226	-688	8032	4846	22106
808	3	1	800	112361	98770	-9428	22376	26593	38048
809	2	2	312	4874	3971	318	764	1865	6745
810	2	1	4338	653040	461652	87542	184984	67879	723181
811	3	1	3600	84822	27773	878	4538	11364	47904
812	2	1	165	77587	22267	-5255	10231	13743	45352
813	2	1	420	28292	23712	524	10233	5051	22811
814	3	1	31	22394	16484	809	11862	10803	17543
815	2	1	1736	122517	102216	3711	32010	23001	78355
816	2	1	400	84622	47231	-6484	20004	16692	36811
817	3	1	79	4287	2791	-693	3125	1124	4952
818	1	2	4910	257547	55593	18371	59902	36108	139277
819	2	1	1000	116027	58300	-21500	130407	31078	176314
820	3	1	260	12768	8267	781	10639	5761	13246
821	3	3	1556	22873	20979	-3333	9938	8103	12862
822	2	4	56	263268	194671	13244	104609	74799	126950
823	2	4	2319	4538	6432	-698	5049	1603	5339
824	2	1	143	11710	51944	23382	4819	2913	7900
825	1	1	3600	82047	322660	189	68968	25754	142260
826	3	1	658	417628	21594	170	291587	155563	396325
827	2	1	1320	36769	109531	6680	9121	6623	17240
828	3	3	477	137245	78472	1626	102647	23898	113228
829	2	1	1600	78472	794759	18094	17829	11678	34021
830	1	1	1226	150747	119681	4339	55583	57343	240447
831	3	1	82	437	4972	-642	3302	21316	83348
832	3	4	14	23979	805	1481	515	1599	3926
833	3	4	280	810822	10834	1150	12877	1201	890
834	3	3	86088	399006	586649	21125	334533	6714	27700
835	3	3	5642	107149	284120	-3516	46464	99468	1106263
836	2	1	883	173476	34524	-1081	131946	45088	146542
837	3	1	631	128984	128984	-529	34114	18340	174447
838	3	1	65	493	24458	9	961	13917	57138
839	3	3	615	66464		4528	59389	170	1093
840	3	3						9991	81353

OBS	1	2	3	VARIABLES 4	5	6	7	8	9
721	2	3	236	32585	6822	5057	31402	4694	37598
722	3	4	360	14604	17563	-5689	4522	8049	10103
723	2	4	946	104421	75734	3463	46869	18470	67138
724	2	4	200	67828	17547	2741	13821	8288	81605
725	3	1	532	43306	36219	1457	15910	10058	34458
726	2	1	898	16593	17404	-2427	3115	4724	9693
727	3	1	146	148240	117507	-52723	162944	299776	274224
728	3	1	50	2949	1606	-227	1536	505	3282
729	1	1	696	1671	1604	-1554	1232	863	2540
730	1	1	487	197	1045	-1260	1054	1784	3046
731	2	2	281	7769	3960	-3677	2753	2492	8639
732	1	1	15	295	111	5	168	41	197
733	3	3	28	795	654	2383	907	1159	1168
734	1	1	1000	130427	94548	1150	111370	61862	132835
735	1	1	601	5878	3341	-1031	1334	1669	2129
736	1	4	895	98682	63840	5645	48793	19616	78608
737	3	3	1725	280274	237735	6249	15254	13893	72091
738	3	3	238	17938	15800	488	4365	2466	6595
739	3	1	83	9592	5870	-1102	4529	1322	6216
740	3	1	3500	1377202	1268137	1797	252644	140783	427078
741	1	1	300	45955	29465	2630	29900	7459	40771
742	1	3	665	122424	93816	507	21903	8004	29263
743	1	1	85	8716	5229	-2936	6966	1905	8169
744	2	3	433	253292	237142	-3447	34337	40007	223622
745	2	4	229	19134	16869	29	2550	15122	54067
746	1	4	105	5077	4735	-412	13643	1904	27115
747	1	4	730	7365	4485	106	60574	7524	77971
748	3	1	2443	110012	79851	613	7002	25224	15266
749	2	1	256	33642	27765	-242	135397	7002	237346
750	2	2	4061	521234	338904	40469	26733	67829	64178
751	2	2	356	55004	20153	4442	80	17137	2364
752	2	1	181	4128	3495	-61	25385	1093	36225
753	1	3	454	35644	15414	6208	11994	6131	28790
754	3	4	688	23044	19385	-5057	43317	12019	50079
755	3	3	813	104001	65123	6792	55506	19802	126685
756	3	2	2834	160555	117299	2732	11522	39862	229435
757	3	4	300	32788	25554	1006	96275	8241	47214
758	2	2	16400	269269	220083	9055	14015	26011	63110
759	2	4	1924	27525	9107	6957	14170	6148	229688
760	3	3	4500	106990	29504	4799	105654	9795	14620
761	2	3	460	275424	183459	12797	5966	39955	75527
762	3	4	1607	22983	17231	723	40210	4237	15456
763	3	2	177	102093	39789	11488	13907	8852	6218
764	3	1	262	18018	7305	2081	10982	2948	44033
765	3	3	107	15360	4919	560	2154	4977	4103
766	3	1	1660	21417	11790	89	75457	16025	122779
767	1	1	200	4709	4898	1093	14525	1358	16460
768	3	1	1500	150648	9924	14368	2781	13153	3588
769	2	1	27	35543	26736	-927	2312	6676	2467
770	2	3	4169	9098	6179	344	204831	1205	335307
771	1	1	50	3492	1329	360	468	1837	1908
772	1	4	1460	299524	227026	-11377	1858	50430	2175
773	1	4	579	345	230	-113	29301	711	53869
774	3	3	2700	1618	676	-1734	65844	790	131377
775	3	4	78	160334	87441	11908	31252	14124	41607
776	3	1	22	232282	126836	6838	28772	34843	34556
777	3	1	1210	102691	86763	327	1391	9003	9859
778	3	1	1780	53557	39610	3510	82950	12858	116994
779	3	3		4132	1837	261		940	
780	3	3		174012	104187	9484		18831	

VARIABLES

OBS.	1	2	3	4	5	6	7	8	9
841	2	1	20	591	452	-805	1553	96	1888
842	2	1	154	14207	7839	959	6169	1586	7329
843	2	1	790	67400	30175	3028	42958	18107	56259
844	2	2	170	19212	12411	858	15511	1293	18814
845	3	1	61	10211	6987	269	5208	1015	7515
846	3	2	367	18381	12683	1449	10677	6710	24200
847	3	2	470	101354	87491	1625	29422	14381	35658
848	3	2	32	1625	695	-2885	895	1779	3305
849	3	3	1888	221634	207625	833	217708	157612	293582
850	2	1	169	13428	10777	-164	11702	4382	32642
851	3	3	675	40515	12272	3629	10226	2645	21091
852	3	2	281	21410	16277	-1775	7804	3654	19848
853	2	3	22	676	137	-854	131	195	507
854	3	1	2200	40031	18180	-1384	15188	4667	24428
855	3	3	390	20781	10250	-2487	14568	6854	20739
856	3	1	429	17452	7527	1135	23121	9169	31697
857	3	4	911	108792	33699	18077	167615	16593	193830
858	3	2	100	2160	1622	203	1087	770	1223
859	2	4	126	6343	5601	-335	3908	1099	6436
860	3	4	444	96023	25149	8126	35078	20338	47466
861	3	1	1172	95400	58295	-12662	72201	16169	114874
862	3	4	94	3466	3157	-926	2705	2379	4499
863	2	1	16	446	210	-315	210	444	1037
864	3	2	1306	129298	64412	15730	81481	18163	154741
865	2	1	33	2877	992	-13600	992	3660	1644
866	2	1	218	169767	161097	-463	16731	23650	37355
867	1	1	324	9010	6648	-210	3129	2265	4395
868	3	1	55	6275	4648	96	2354	1390	4061
869	2	2	4	59433	64830	-63060	78925	87100	135226
870	1	1	175	22034	14387	-1987	7361	7022	10444
871	1	4	231	27025	18117	2412	28276	4271	34163
872	3	2	208	24867	10944	9487	27419	5829	32112
873	2	4	1050	75095	50935	8271	53988	8790	59294
874	2	1	2100	67009	59648	2222	18384	13117	33063
875	3	1	4300	431591	326402	6784	27875	29228	86249
876	2	1	1300	76608	57634	-1079	13085	13526	34060
877	3	1	497	95730	71424	2363	45379	19700	55121
878	2	1	12	171	121	-940	135	454	179
879	3	1	741	66257	47076	2614	27049	10154	38328
880	3	2	210	11493	6515	7173	16984	6662	26743
881	3	1	1368	130718	95708	91	62354	23223	79294
882	1	1	120	17334	12718	1365	15172	7384	23703
883	3	1	22	3814	1040	7110	3556	1037	4442
884	2	2	900	136063	100237	1785	168295	53773	258052
885	2	1	225	19689	9750	-340	11321	1900	18291
886	1	1	570	14770	12390	7485	6346	4627	7488
887	3	3	41	31147	1825	994	2084	267	2584
888	3	3	1082	16981	3766	2652	16162	2067	25201
889	3	1	52	5348	3705	-2458	1287	1160	3009
890	3	2	353	49377	30455	2957	18208	9824	34566
891	2	1	370	16277	10833	553	9495	2259	14553
892	3	3	3850	207493	142302	7485	30769	25696	72531
893	3	3	29	860	457	-708	1240	83	1333
894	2	4	373	46594	33696	1314	19772	10658	29531
895	2	1	173	1313	363	-621	7598	7598	7598
896	2	1	48	5489	3145	321	3083	682	5129
897	3	1	400	86498	34013	10943	35427	21397	63296
898	3	1	130	15731	13021	14	8667	3943	11114
899	1	4	2032	189538	93109	10789	109272	20638	186781
900	1	4	248	221418	200300	3012	65569	39152	68554

OBS.	1	2	3	4	5	6	7	8	9
901	2	1	21	769	654	411	440	925	485
902	3	3	331	21428	20457	-4477	6974	12560	27001
903	3	2	465	24734	20979	-6029	21873	2297	30733
904	2	2	722	48865	64808	-33941	32978	9658	132343
905	2	4	1050	38234	38027	18250	106297	21681	132839
906	3	4	22	1875	1090	-299	948	631	1265
907	2	3	118	107	87	-184	126	34	200
908	3	1	311	5301	2613	235	1870	688	3355
909	3	2	91	24609	16686	-1145	8033	15642	22091
910	1	1	300	3505	1539	275	1138	471	1162
911	3	1	84	62004	46810	176	26716	13996	29167
912	2	1	660	4408	3485	-683	2517	659	3454
913	2	1	300	443092	196362	39064	255495	90074	398065
914	2	1	2917	13301	45137	4013	42896	12514	51831
915	3	3	346	184861	9077	855	3634	2942	4879
916	3	3	3579	15652	137936	2430	102137	46686	136877
917	1	1	36	349027	13592	127	6384	1590	9440
918	1	1	30	1211	245360	19434	128644	62588	236630
919	1	4	4403	1100	12	539	1772	241	2060
920	2	1	1149	291913	533	-2078	2168	394	5287
921	2	1	1601	79769	158400	32039	185824	56318	328240
922	2	1	16000	383735	35257	4710	42594	6687	57169
923	2	2	100	524358	312202	10076	99849	27219	131412
924	3	1	451	2303	188923	10947	61373	39702	212460
925	2	1	841	2534	2335	-912	6250	936	8268
926	2	1	4	524	1566	98	906	670	2877
927	3	2	46	51869	1150	581	4079	759	4592
928	3	2	794	1901	40876	-1420	38864	20535	51374
929	2	1	218	7189	447	-306	603	99	614
930	3	3	12	30237	1648	-735	701	894	809
931	3	1	18	15993	4801	-809	4410	2029	6949
932	3	4	67	741	26058	-1008	50618	27056	95002
933	1	4	1500	11252	12195	439	9008	7819	22541
934	3	4	2	11530	42	-22	798	0	1800
935	3	1	1214	4036	587	-125	404	92	1226
936	2	3	7	125906	3573	1616	9403	3091	11643
937	1	1	1841	54	77139	3367	49243	22175	63970
938	3	2	671	166086	2867	307	1424	890	2161
939	3	1	78	13794	83917	6373	64236	29998	94970
940	2	4	22	6293	19	-423	1754	67	2195
941	3	1	24000	4705	87792	13855	112561	35252	154523
942	3	2	115	944356	7582	962	11585	4384	41176
943	2	1	618	10211	4499	11	5256	3059	6308
944	3	3	450	59323	2673	647	4992	321	5770
945	2	1	1200	27724	783649	28003	205730	161708	653369
946	3	2	370	236629	6993	405	5724	5383	8697
947	3	1	465	58717	42976	1063	27259	7459	42478
948	3	3	4000	42756	24381	-723	14610	3439	22069
949	3	1	4214	270325	140785	6762	54248	28541	111511
950	3	4	479	77925	39992	7461	12688	6910	36390
951	3	1	3000	47681	19046	5033	27121	4023	40421
952	2	2	85	8310	191747	39	135995	48588	224608
953	3	3	130	21363	59209	2909	37198	14635	45698
954	2	1	6300	394092	20050	-2121	25331	9298	33065
955	3	2	26	16736	817	646	510	3479	7925
956	3	1	1561	82526	5313	472	5924	1190	7571
957	3	1			11723	26056	11997	1907	15465
958	3	1			256490	-4807	140320	50767	255898
959	2	4			21523	9854	7955	821	30971
960	1	4			44008		73337	20477	90177

| | | | | | VARIABLES | | | | |
OBS.	1	2	3	4	5	6	7	8	9
961	1	3	139	14330	5495	1101	8142	2490	12095
962	1	1	870	74444	54034	4825	26051	5991	48720
963	1	1	3	614	348	-21	166	232	277
964	2	3	2205	9031	7722	-790	4695	1450	5468
965	1	4	376	891	581	-1362	777	2373	1944
966	3	2	3000	288160	228052	17971	69631	22634	168816
967	2	1	1607	90784	75611	1972	45421	27489	50354
968	4	4	13	482	254	-389	331	445	2061
969	2	3	1138	64771	33404	-7445	65664	13412	90318
970	3	1	5672	419991	289779	24664	186911	71576	269887
971	2	2	6720	298238	216796	22190	113435	35671	173281
972	2	1	3600	423413	317117	2607	152391	77618	250427
973	1	1	207	9281	5359	-366	4627	989	6256
974	1	1	733	16016	7783	1276	11113	4265	12773
975	2	1	46	1728	904	-759	391	830	1968
976	2	1	57	4363	2930	-75	3425	1117	4214
977	1	1	1438	8086	5755	1942	4809	2362	8160
978	1	3	467	27246	13935	1425	10417	8176	15112
979	1	1	3406	237686	174743	1970	76243	44776	158461
980	2	3	500	8510	2298	4559	2529	2324	9367
981	1	1	1043	95588	52030	5581	86205	17162	128819
982	1	1	45	3901	1767	-709	2131	589	3914
983	2	4	220	53615	34352	2247	23462	9288	26704
984	3	1	180	110890	50418	5818	55283	20356	67384
985	3	1	443	84604	74095	1877	40503	5205	58203
986	3	1	1343	263164	204696	5038	86799	31182	106804
987	2	1	234	16904	6682	436	9747	2466	14303
988	3	3	313	46006	30789	4246	29793	6491	49586
989	2	4	9	306	990	-1530	184	2900	1903
990	2	1	802	70667	49443	6042	14731	6747	36304
991	2	1	4119	543041	263681	31753	135165	110798	633558
992	2	1	7531	813497	502247	86194	294387	81216	355502
993	3	3	255	24834	17293	506	13524	4590	17277
994	3	1	75	7743	5545	795	6250	2251	7122
995	2	3	4534	391700	417100	-117100	152000	88700	480200
996	1	4	3300	487685	361921	33484	99710	75445	484975
997	1	1	16	362	44	-1155	726	59	1017
998	3	3	156	1826	1028	-324	1023	266	1559
999	2	1	135	27908	15965	2410	11519	3016	18394
1000	2	4	1540	194335	129150	3855	280546	32150	451404

◢ *Appendix K   Answers to Odd-Numbered Exercises*

Chapter 1

1.1 The population of interest is all small businesses in Los Angeles. The sample is the 50 randomly selected small businesses.

1.3 a. The manager used descriptive statistics by constructing a graph to get an idea of the distribution of the incomes.   **b.** The manager used inferential statistics by forming a conclusion about the population of interest. The manager concluded that mostly people in the middle income bracket fly Easy Fly Airlines.

1.5 a. continuous   **b.** discrete   **c.** continuous   **d.** discrete   **e.** continuous

1.7 nominal

1.9 discrete

1.11 a. ordinal   **b.** ratio   **c.** interval   **d.** nominal   **e.** nominal   **f.** ratio

Chapter 2

2.1 b. There is no "correct" number of classes. Consider using $K = 8$ classes.   **c.** $CW = (100 - 18)/8 = 10.25$ (round to 10)   **d.** Relative frequency is equal to the frequency divided by the total number of values in the data set.

2.3 a. Relative frequencies are .38, .31, .21, .06, .03, .01.   **b.** Lower class limits are 0, 2, 4, 6, 8, and 10, and upper class limits are 2, 4, 6, 8, 10, and infinity.   **c.** Class midpoints are 1, 3, 5, 7, 9, and undefined.   **d.** no

2.5 a. discrete data

2.7 a. $CW = (17.1 - 2.2)/5 = 2.98$ (round to 3)   **b.** Determine the basic shape of the data.   **c.** Consider $K = 6$ classes, 2 and under 5, 5 and under 8, and so forth.

2.13 b. The shape indicates that only one class has a low frequency and that most of the data fall in the larger class intervals.   **c.** No, the shape would not change.

2.15 a. Classes are 5 and under 10, 10 and under 15, . . . , 35 and under 40; frequencies are 1, 0, 7, 6, 2, 0, and 1.

2.17 a. The histogram has a U-shaped pattern.

2.19 a. Classes are 0–3, 4–7, 8–11, 12–15, and 16–19; frequencies are 12, 10, 7, 1, and 1; cumulative relative frequencies are .39, .71, .94, .97, and 1.00.   **b.** 29%

Note: Solutions are not given for all odd-numbered exercises.

2.23 c. The distribution peaks between 70 and 80, and 90% of the distribution is between 40 and 100.

2.25 b. from graph in part a, approximately 12

2.27 c. A "typical" automotive loan rate is between 11.00% and 11.50%.

2.29 Proprietorships are 76.97% (277°), partnerships are 7.82% (28°), and corporations are 15.21% (55°).

2.31 The class 10,000–14,999 is 17.5% (63°), the class 15,000–19,999 is 10.0% (36°), the class 20,000–24,999 is 32.5% (117°), the class 25,000–29,999 is 15.0% (54°), and the class 30,000–34,999 is 25.0% (90°).

2.33 A histogram cannot be used to illustrate categorical data.

2.35 c. The manager should take immediate action on B and D complaints.

2.39 a. The data can be considered to be continuous.

2.49 c. Thirty-two percent of the loan officers are between 45 and 54 years of age, with another 36% between 35 and 44.

2.51 c. Seventy-five percent of the boxes contain less than 15 defective fuses; 50% of the boxes contain less than 10 defective fuses.

2.53 b. For 15 days out of 30 there are at least 11 workers absent.

2.55 c. Most of the complaints result from either noise from other rooms, room not clean enough, or room not ready on time.

2.57 b. The six largest dollar investment sectors represent 69.44% of the total value of the portfolio.

2.61 Most of the observations fall close to the numbers 34.8, 34.9, 35.0, 35.10, 35.20, and 35.20. A shift would not help to reduce the number of nonconforming ball bearings.

Chapter 3

3.1 Mean is 13.8333, median is 13.5, mode is 12, and midrange is 14.

3.3 a. Mean is 101.9, median is 99, mode is 99, and midrange is 107.5.   **b.** Mean is 99.3333, median is 99, mode is 99, and midrange is 102.5.

3.5 The median would fall in the interval $15,000 to $19,999. An approximate estimate of the median could be taken to be the midpoint of the interval, which is $17,500.

3.7 a. Mean is 50; median is 50.   **b.** yes

3.9 Mean is 74.35, median is 57.5, and mode is 130.

3.11 a. Mean is 39.1538; median is 22.   **b.** Mean is

32.41667; median is 22.   **c.** Mean is 41.5, and median is 26 when Ecuador's debt is omitted. The mean is affected more than the median when Brazil's debt is omitted. The median is affected slightly more than the mean when Ecuador's debt is omitted.

3.13 a. Range is 87.   **b.** MAD is 24.09.   **c.** Variance is 794.08.   **d.** s is 28.18   **e.** CV is 56.27.

3.15 CV is 53.41% for project A. CV is 44.64% for project B.

3.19 a. Play A is better (on the average).   **b.** Play B is more consistent.

3.21 Variance is 1742.13.

3.23 approximately the 39th percentile

3.25 a. 32   **b.** 19   **c.** 17.5

3.27 a. $Z = .78$ **b.** $X = 63$ is .78 standard deviation to the right of the sample mean.

3.29 s is 2.5.

3.31 $(x - \bar{x}) = 14.24$

3.33 a. Mean is 3896.51, median is 3893.55, variance is 669322.9499, and $s = 818.1216$.   **b.** Sk = .01214

3.35 between 40 and 80

3.37 near zero

3.39 a. CVs are 49.425, 58.065, 62.857, 27.851, 35.127, 19.706, 58.333, and 203.7044.   **b.** Use $\bar{x} - 2s$ to $\bar{x} + 2s$.

3.41 15.8 to 40.6

3.43 Data do not appear to conform to empirical rule.

3.45 $200 to $360

3.47 Median is approximately $26,666.67.

3.49 Median is approximately 44.26 years.

3.51 a. Mean is 75.96; variance is 162.8049.   **b.** s is 12.76.

3.55 b. Twelve defectives are most frequently found.   **c.** Data are skewed right.

3.57 a. The median appears to be the appropriate measure of central tendency.   **b.** $s = 2.6089$; MAD = 1.94

3.61 Mean is 412.125; s is 7.53.

3.63 Mean is 1760; s is 2932.

3.65 Mean is .00133; s is .0001555.

3.67 a. Mr = 6.2   **b.** Mean is 5.677.   **c.** Median is 5.6.   **d.** Mode is 5.6.   **e.** $s = 1.6466$ **f.** CV is 29.005%.   **g.** $Q_1 = 4.4$ **h.** Ninetieth percentile is 7.7.   **i.** $IQR = 3.5$

3.69 b. .32   **c.** Between -1.155 and 1.655

3.71 $45 to $105

3.73 a. -4.0

3.75 Mean is 53.756; s is 2.78.

3.77 a. 167.758   **b.** 164.9   **c.** no mode

d. 3263.08   **e.** 57.123   **f.** 37.29   **g.** 280.8   **h.** 34   **i.** .15   **j.** 139.8   **k.** 221.5   **l.** $-.381$ **m.** 53.512 to 282.004   **n.** approximately 68%

3.79 a. .5736   **b.** Range is 17.8; midrange is 10.4; interquartile range is 5.25.   **c.** between 0 and 21.036

Chapter 4

4.1 5/4 cannot be a probability.

4.3 .286; yes, this is the relative frequency approach.

4.7 a. .16   **b.** .45   **c.** .676   **d.** .84

4.9 a. .541   **b.** .015   **c.** .5   **d.** .949   **e.** yes

4.11 a. .776   **b.** .403   **c.** .263   **d.** They are mutually exclusive but not independent.

4.13 a. .703   **b.** .447   **c.** .638   **d.** .768

4.17 a. .3   **b.** .8   **c.** .286   **d.** .5   **e.** .714   **f.** .2   **g.** .8

4.19 .75

4.21 no

4.23 .25

4.25 P(both nondefective) = .64; P(both defective) = .04

4.27 P(M and E) = .1

4.29 a. .54   **b.** .86   **c.** .675

4.31 .90, assuming they are independent

4.33 .676

4.35 .2278

4.37 .034

4.39 .48

4.41 .048

4.43 a. .0395   **b.** .405

4.47 a. yes   **b.** .849   **c.** 0

4.49 .00148

4.51 .86

4.53 24

4.55 120

4.57 210

4.59 90

4.61 20,000

4.63 60

4.65 45

4.67 1/595

4.69 Probability is 1/495. This does constitute a random sample.

4.71 1/45

4.73 .667

4.75 a. .091   **b.** .125   **c.** .273

4.77 a. .143   **b.** .143   **c.** .2857   **d.** 1.0
4.79 a. .195   **b.** .755   **c.** .354   **d.** no
4.81 847, 660, 528
4.83 a. .8   **b.** .2
4.85 a. The classifications are mutually exclusive.
b. .3705   **c.** .1316   **d.** .5107
4.87 a. .98   **b.** .10   **c.** .625
4.89 a. .216   **b.** .134   **c.** .8844   **d.** .0093
4.91 a. .75   **b.** .45   **c.** .25

Chapter 5

5.1 The number of customers or daily account size of its customers.

5.3 a. discrete   **b.** continuous   **c.** continuous   **d.** discrete   **e.** discrete   **f.** continuous

5.5 12 outcomes, each with probability $1/12$

5.7 .25

5.9 a. $P(X = x) = 1/6$ for $x = 1, 2, 3, 4, 5, 6$

5.11 $P(X = 0) = 9/25$; $P(X = 1) = 12/25$; $P(X = 2) = 4/25$

5.13 yes

5.15 rolling a die

5.17 $P(X = x) = 1/3$ for $x = 2, 4, 6$

5.19 $P(X = 1) = 1/6$, $P(X = 2) = 2/6$, $P(X = 3) = 3/6$; sum of probabilities is 1.

5.21 Mean is .2, variance is .18.

5.23 Mean is 3; variance is 3.

5.25 Mean is 17.4; standard deviation is 4.98.

5.27 Mean is 2.18; standard deviation is .7547.

5.29 .2963

5.31 $P(x \geq 5) = .991$; $P(4 \leq x \leq 10) = .913$; $P(x \leq 7) = .275$

5.33 .506

5.35 .404

5.37 a. .001   **b.** 3   **c.** 1.55   **d.** .982

5.39 .7636

5.41 a. .3185   **b.** Mean is 7.8; standard deviation is 1.31.

5.43 .8

5.45 .6353

5.47 .9286

5.49 a. .76766   **c.** .76102

5.51 Mean is 3; probability is .5768.

5.53 a. .9004   **b.** 2.83

5.55 a. .3711   **b.** .1954

5.57 yes

5.59 Mean is 3.5; variance is 18.85.

5.61 a. .029   **b.** .237   **c.** .117   **d.** .085

5.63 .623

5.65 Mean is 1.01; standard deviation is 1.3304.

5.67 .3658 (approximately)

5.69 Probability is .224; standard deviation is 1.73.

5.71 .1804

5.73 a. .818   **b.** .001

5.75 .2753

5.77 b. $\bar{x}$ is 8.508, μ is 8.5.   **c.** Sample variance is 4.595; population variance is 4.25.

Chapter 6

6.1 The mean and variance indicate where the curve is centered and how wide the curve is, respectively.

6.3 a. .3413   **b.** .0919   **c.** .6826   **d.** .0606

6.5 a. .9439   **b.** .0179   **c.** .1986   **d.** .9010

6.7 a. $z = .35$ **b.** $z = .67$ **c.** $z = 1.57$ **d.** $z = -1.62$

6.9 $z = .65$ and $z = 1.57$

6.11 a. .3632   **b.** .2119   **c.** .3842   **d.** .1791

6.13 8.08%

6.15 $82.50

6.17 .0574

6.19 a. .0329   **b.** $P(X \leq 60) = .0329$

6.21 a. .2514   **b.** .1056

6.23 Mean is 9.45.

6.25 Mean is 1.29; standard deviation is .556.

6.27 a. .688   **b.** .692

6.29 .9612

6.31 .8564

6.33 .0441 (Poisson approximation)

6.35 a. .5   **b.** Value is 1.8.

6.37 a. .524   **b.** 23.15   **c.** 1.819

6.39 .51

6.41 .2997

6.43 .3836

6.45 Answer varies for different data sets.

6.47 a. .0133   **b.** .8664   **c.** .2734

6.49 .6700

6.51 The values 97.92 and 102.08 bound the middle 40% of the distribution of X.

6.53 Value is 87.12.

6.55 Probability is .344; standard deviation is 30.

6.57 58.89%

6.59 .0668

6.61 .4168

6.63 a. Mean is 1.0; standard deviation is .289.   **b.** .3

6.65 .5

6.67 a. .6826   **b.** .3811   **c.** .3446   **d.** .1587

6.69 Variance is 43.03.

6.71 .1635

6.73 .0031

Chapter 7

7.1 a. .017   **b.** .1446   **c.** .2974   **d.** .7108

7.3 c.

$\bar{X}$	26.7	43.3	53.3	56.7	176.7
P	.1	.1	.1	.1	.1
$\bar{X}$	186.7	190.0	203.3	206.7	216.7
P	.1	.1	.1	.1	.1

7.5 .0022

7.7 .0222

7.9 without replacement, .0020; with replacement, .0031

7.11 .7960, assuming the population of electric bill amounts follows a normal distribution

7.13 .0823

7.15 .0268

7.17 For $n = 40$, probability is .8790; for $n = 20$, probability is .7852.

7.19 19.86 to 21.34

7.21 92.41 to 107.59

7.25 22.24 to 23.36

7.27 12.72 to 13.68

7.29 7.86 to 8.74

7.31 a. 1.311   **b.** 2.160   **c.** 1.734   **d.** −1.325
e. −1.708   **f.** 1.303

7.33 27.78 to 32.22

7.35 a. 62.24 to 73.76   **b.** 63.21 to 72.79

7.37 302.37 to 317.63

7.39 9.79 to 12.11

7.41 91.95 to 99.25

7.43 89

7.45 16

7.47 6.1033

7.49 24

7.51 57

7.53 601

7.55 Estimate is .80; confidence interval is .716 to .884 (in thousands).

7.57 Estimate is 5.0422; confidence interval is 4.47 to 5.61.

7.59 Estimate is 3.3923; confidence interval is 2.54 to 4.25.

7.61 a. .2266   **b.** .0329

7.63 .1994 to .2226

7.65 59

7.67 a. .0344   **b.** .0455

7.69 Estimate is 3.94; confidence interval is 3.46 to 4.42.

7.71 .0021

7.73 Estimate is 10.174; confidence interval is 9.56 to 10.79.

Chapter 8

8.1 a. When failing to reject the null hypothesis, the possible outcomes are: (1) the loan is made and paid back and (2) the loan is made and not paid back. When rejecting the null hypothesis, the possible outcomes are: (1) the loan is not made and would not have been paid back had it been granted and (2) the loan is not made and would have been paid back had it been granted.
b. Type II   **c.** No, a Type I error may have been made.

8.3 a. false   **b.** false   **c.** true   **d.** false

8.5 $Z^* = 4.65$; reject H_0.

8.7 $Z^* = -3.486$; reject H_0.

8.9 Reject H_0 since the 95% confidence interval does not contain 2.0.

8.11 a. 12.7274 to 12.7326   **b.** Reject H_0.

8.13 a. .9732   **b.** approximately 1   **c.** .9131

8.15 $P(Z < -15.29) + P(Z > -11.37) = 1$

8.17 a. $Z > 1.645$ **b.** $Z < -1.645$ or $Z > 1.645$
c. $Z < -2.33$

8.19 $Z^* = -2.93$; reject H_0.

8.21 $Z^* = -2.83$; reject H_0.

8.23 $Z^* = 3.64$; reject H_0.

8.25 $Z^* = -1.5$; fail to reject H_0.

8.27 a. Fail to reject H_0.   **b.** Reject H_0.   **c.** Reject H_0.   **d.** Fail to reject H_0.

8.31 $Z^* = 1.03$; p-value is .1515; fail to reject H_0.

8.33 $Z^* = -1.29$; p-value is .0985, fail to reject H_0 for a significance level of .05 and .01.

8.35 $Z^* = 2.80$, p-value is .0026.

8.37 a. $t < -1.729$ **b.** $t > 1.318$ **c.** $t > 2.145$ or $t < -2.145$

8.39 $t^* = .59$; fail to reject H_0.

8.41 46.39 to 90.41

8.43 $t^* = 3.0$; p-value $< .005$; reject H_0.

8.45 $t^* = 1.92$; reject H_0.

8.47 a. 15.987   **b.** 46.979   **c.** 7.261   **d.** 45.642

8.49 a. 7.612 to 27.439　　**b.** 2.76 to 5.24　　**c.** Fail to reject H_0.

8.51 Chi-square $= 19.81$; fail to reject H_0.

8.53 Chi-square $= 21.39$; p-value $> .10$; fail to reject H_0.

8.55 Chi-square $= 45.24$; fail to reject H_0.

8.57 a. Increasing (decreasing) the significance level will increase (decrease) the rejection region.
b. Increasing (decreasing) the significance level will decrease (increase) the probability of a Type II error.

8.59 $Z^* = -2.99$; p-value $= .0028$; reject H_0.

8.61 a. $P(Z < -.75) + P(Z > 3.17) = .2274$
b. $P(Z < -2.77) + P(Z > 1.15) = .1279$
c. $P(Z < -6.04) + P(Z > -2.12) = .9830$

8.63 $Z^* = -3.5$; p-value $= .0004$; reject H_0.

8.65 $t^* = -9.22$, reject H_0.

8.67 a. 14.193 to 14.807　　**b.** $Z^* = 3.19$; reject H_0.

8.69 27.726 to 38.274

8.71 a. 36.85 to 40.35　　**b.** $Z^* = -1.57$; fail to reject H_0.　　**c.** .0582

8.73 Chi-square $= 14.37$; fail to reject H_0.

8.75 Chi-square $= 27.22$; fail to reject H_0.

8.77 a. $Z^* = 4.615$; reject H_0.　　**b.** Assume n/N is small and that $\bar{x}$ is approximately normally distributed.

8.79 a. Mean is not equal to 1 hour.　　**b.** yes
c. large sample size

Chapter 9

9.1 a. paired　　**b.** independent　　**c.** independent

9.3 dependent samples

9.5 No, the samples are dependent.

9.7 The samples are independent.

9.9 -4.16 to -1.039

9.11 7.485 to 8.171

9.13 -4.849 to -2.951

9.15 $Z^* = .697$; fail to reject H_0.

9.17 a. $Z^* = 1.765$; p-value $= .0384$
b. $Z^* = -.19$; p-value $= .8494$　　**c.** $Z^* = -.063$; p-value $= .4761$

9.19 $Z^* = -3.20$; ; reject H_0; p-value $= .0007$.

9.21 $Z^* = -2.66$; reject H_0.

9.23 $Z^* = 2.19$; reject H_0.

9.25 $t^* = 1.53$; fail to reject H_0.

9.27 df $= 13$; $t'^* = .4595$; fail to reject H_0.

9.29 $t^* = .4595$; fail to reject H_0.

9.31 -48.364 to -1.636

9.33 -26.565 to 12.065

9.35 df $= 14$; confidence interval is -4.026 to $-.934$.

9.37 $F^* = 1.746$; fail to reject H_0.

9.39 .511 to 12.72

9.41 a. .23 to 1.91　　**b.** $F^* = 2.01$; fail to reject H_0.

9.43 .70 to 3.91

9.45 $F^* = 2.0$; fail to reject H_0.

9.47 a. $t^* = 3.5355$; reject H_0.　　**b.** Use dependent samples t statistic.

9.49 $t^* = .707$; fail to reject H_0.

9.51 $t^* = .71$; p-value $= .20$. Based on p-value, fail to reject H_0.

9.53 a. $t^* = .638$; fail to reject H_0.　　**b.** The samples are dependent.

9.55 a. $t^* = 2.67$; fail to reject H_0.　　**b.** Differences are independent and normally distributed.

9.57 The samples are dependent since both brands of tires are placed on the same car.

9.59 $t^* = 3.8443$; reject H_0.

9.61 $t^* = 1.038$; fail to reject H_0.

9.63 $t^* = .6286$; fail to reject H_0.

9.65 $F^* = 2.56$; fail to reject H_0.

9.67 $t^* = 3.90$; reject H_0.

9.69 a. $F^* = 2.1904$; reject H_0.　　**b.** $t^* = 2.329$; reject H_0.

9.71 a. As the confidence level increases, the widths of the confidence intervals increase.　　**b.** Data are from a normal population.

Chapter 10

10.1 .394 to .783

10.3 601

10.5 95% confidence interval is $(.465, .903)$; 90% confidence interval is $(.502, .884)$.

10.7 655

10.9 8141

10.11 2018

10.13 .001 to .069

10.15 .0444 to .196

10.17 p-value $= .2296$; fail to reject H_0.

10.19 Since $.10 < .123$, reject H_0.

10.21 Sample size must be at least 167.

10.23 $Z^* = 1.124$; fail to reject H_0.

10.25 p-value $= .3015$

10.27 p-value $= .121$; fail to reject H_0.

10.29 $Z^* = 1.59$; fail to reject H_0.

10.31 $Z^* = 2.38$; reject H_0.

10.33 The sign of Z^* is reversed.

10.35 $n_1 = 26$; $n_2 = 29$

10.37 $Z^* = -.91$; p-value $= .3628$; fail to reject H_0.

10.39 $n_1 = 35$; $n_2 = 42$

10.41 a. Since $.10 < .226$, fail to reject H_0 **b.** .003 to .226

10.43 .213 to .734

10.45 593

10.47 $Z^* = 1.3488$; reject H_0.

10.49 a. .23   **b.** .13   **c.** 41

10.53 p-value $= .3557$; fail to reject H_0.

10.55 $Z^* = -.512$; fail to reject H_0.

10.57 a. .0223 to .1277   **b.** $n_1 = 479$; $n_2 = 405$

10.59 .694 to .851

10.61 .4727 to .5673

10.63 $-.0398$ to .1864

Chapter 11

11.1 a. H_0: $\mu_1 = \mu_2 = \mu_3$; H_a: not all three means are equal   **b.** The samples are taken independently from normal populations with a common variance.   **c.** For μ_1, 4.533; for μ_2, 5.333; for μ_3, 4.633   **d.** 1.29

e.

Source	df	SS	MS	F
Factor	2	2.28	1.14	.88
Error	15	19.38	1.29	
Total	17	21.66		

$F^* = .88$; fail to reject H_0.

f. Not appropriate

11.3

Source	df	SS	MS	F
Factor	1	6.5333	6.5333	2.2603
Error	28	80.9334	2.8905	
Total	29	87.4667		

$F^* = 2.2603$; fail to reject H_0.

11.5 a.

Source	df	SS	MS	F
Factor	2	54.111	27.056	12.82
Error	15	31.667	2.111	
Total	17	85.778		

b. p-value $< .01$ **c.** The type of word processing software does significantly affect the performance.
d. The within-sample variation increases.   **e.** Means 1 and 3 differ; means 2 and 3 differ.

11.7 a.

Source	df	SS	MS	F
Factor	2	74026.1	37013.05	8.7670
Error	21	88659.2	4221.8667	
Total	23	162685.3		

$F^* = 8.7670$; reject H_0.
b. p-value $< .01$ **c.** Means for John and Randy differ; means for John and Ted differ.

11.9

Source	df	SS	MS	F
Factor	2	11.6667	5.8335	1.232
Error	27	127.8	4.7333	
Total	29	139.4667		

$F^* = 1.232$; fail to reject H_0; p-value $> .10$.

11.11 Means 1 and 3 differ; means 2 and 3 differ.

11.13 $F = 12.8164$; p-value $< .01$

11.15 $H^* = 4.591$; reject H_0.

11.17

Source	df	SS	MS	F
Factor	2	76.5	38.2	1.94
Error	27	532.5	19.7	
Total	29	609.0		

$F^* = 1.94$; fail to reject H_0.

11.19 b. randomized block design   **d.** one dependent variable   **e.** 4 treatments with 24 observations   **f.** 24 treatments with a minimum of 48 observations

11.21 a. completely randomized design
b. randomized block design

11.23 a. $t^* = -.8355$; fail to reject H_0.

b.

Source	df	SS	MS	F
Factor	1	517,750	517,750	.70
Blocks	10	1,730,841,990	173,084,199	
Error	10	7,349,590	734,959	
Total	21	1,738,709,330		

$F^* = .70$; fail to reject H_0.

c. The F value is the square of the t value.

11.25

Source	df	SS	MS	F
Factor	3	1998	666	57.61
Blocks	3	734	244.67	21.17
Error	9	104	11.56	
Total	15	2836		

11.27

Source	df	SS	MS	F
Factor	2	20.33	10.17	17.93
Blocks	5	113.5	22.5	39.68
Error	10	5.667	.567	
Total	17	138.5		

11.29 No change in sum of squares.

11.31

Source	df	SS	MS	F
Factor	3	130	43.33	4.73
Blocks	4	280	70.0	
Error	12	110	9.167	
Total	19	520		

11.33

Source	df	SS	MS	F
Factor	2	.8184	.4092	13.46
Blocks	12	5.5225	.4602	
Error	24	.7283	.0304	
Total	38	7.0692		

$F^* = 13.46$; reject H_0.

11.35

Source	df	SS	MS	F
Factor	3	293.1	97.7	15.4988
Blocks	9	5160.2	573.3556	90.955
Error	27	170.2	6.3037	
Total	39	5623.5		

Both F-values are significant.

11.37

Source	df	SS	MS	F
Factor	2	77.0556	38.5278	16.8582
Blocks	11	42.2223	3.8384	1.6795
Error	22	50.2777	2.2854	
Total	35	169.5556		

The factor F-value is significant. The block F-value is not significant.

11.39 a.

Source	df	SS	MS	F
A	2	242.67	121.335	68.242
B	2	8.22	4.11	2.312
Interaction	4	11.11	2.778	1.562
Error	18	32	1.778	
Total	26	294		

e. p-value for $F_1 < .01$; p-value for $F_2 > .10$; p-value for $F_3 > .10$

11.41 Power base is not significant. Dependency is significant. Interaction is not significant.

11.43 a. five replicates in each treatment combination

b.

Source	df	SS	MS	F
A	1	3.2	3.2	.27
B	1	105.8	105.8	8.82
Interaction	1	16.2	16.2	1.35
Error	16	192.0	12.0	
Total	19	317.2		

d. The mean of group instruction and computer-assisted training is significantly different from the mean of computer-assisted training and self-paced programs.

11.45 a. $H_0: \mu_1 = \mu_2 = \mu_3$; H_a: The means are not equal.   **b.** SS(factor) = 14.0952   **c.** 53.1429

d.

Source	df	SS	MS	F
Factor	2	14.0952	7.0477	2.3871
Error	18	53.1429	2.9524	
Total	20	67.2381		

$F^* = 2.3871$; fail to reject H_0; p-value > 0.1.

11.47 a.

Source	df	SS	MS	F
Factor	2	1591	795	4.55
Error	12	2098	175	
Total	14	3689		

b. $F^* = 4.55$; reject H_0.   **c.** 24.71 to 50.439
d. −13.75 to 36.95   **e.** Means 1 and 3 differ.

11.49 a.

Source	df	SS	MS	F
Manager level	2	780.8	390.4	5.93
Sex	1	122.7	122.7	1.86
Interaction	2	84.8	42.4	.64
Error	12	790.0	65.8	
Total	17	1778.3		

b. Reject H_0.   **c.** Fail to reject H_0.   **d.** Fail to reject H_0.   **e.** For part a, p-value is between .01 and .025; for part b, p-value $> .10$; for part c, p-value $> .10$.

11.51

Source	df	SS	MS	F
Factor	3	217.9	72.6333	706.549
Error	36	3.7	0.1028	
Total	39	221.6		

$F^* = 706.549$; reject H_0; p-value $< .01$.

11.53

Source	df	SS	MS	F
Factor	2	1.9233	.9617	14.68
Blocks	7	25.5383	3.6483	55.706
Error	14	.9167	.0655	
Total	23	28.3783		

Both the factor and the blocks are significant.

11.55

Source	df	SS	MS	F
Factor A	1	3088	3088	6.58
Factor B	3	3400	1133.3	2.41
Interaction	3	49000	16333.3	34.79
Error	16	7512	469.5	
Total	23	63000		

11.57 $H^* = 2.67$; fail to reject H_0.

11.59 a. 9   **b.** 15   **c.** Increase the sample size.

11.61

Source	df	SS	MS	F
Factor	2	294.9	147.5	10.61
Blocks	9	170.3	18.9	1.36
Error	18	249.6	23.9	
Total	29	714.8		

Factor is significant.

11.63 Since $F^* = 1.98$, fail to reject H_0.

11.65 $F_1 = 2.66$, $F_2 = 2.0$, and $F_3 = 3.14$. Only the interaction term is significant at the .10 level.

Chapter 12

12.3 a. UCL = 28.85; CL = 18.61; LCL = 8.37
b. no discernable pattern   **c.** UCL = 32.06;
CL = 14.05; LCL = 0

12.5 a. UCL = 1585.9; CL = 1529.8; LCL = 1473.7
b. Pattern 5 is detected.

12.7 a. UCL = 2.37; CL = 1.27; LCL = .17
b. Pattern 3 is detected.   **c.** likely to exceed the UCL
d. equipment drifting out of adjustment, tool wear, fatigue

12.9 a. UCL = 44.30; CL = 28.44; LCL = 12.58
b. UCL = 58.14; CL = 27.5; LCL = 0   **c.** Pattern 1 is detected in the $\bar{X}$ chart.

12.11 a. for the $\bar{X}$ chart: UCL = 82.66, CL = 68.20, LCL = 53.75   for the R chart: UCL = 59.93, CL = 29.90, LCL = 0   **b.** no patterns detected

12.13 a. p chart   **b.** c chart   **c.** c chart
d. p chart   **e.** c chart

12.15 UCL = 20.44; CL = 10.65; LCL = .86

12.17 a. UCL = .053; CL = .03; LCL = .007
b. UCL = .043; CL = .03; LCL = .017 **c.** Control limits become narrower as the sample size increases.

12.19 a. p chart **b.** UCL = .176; CL = .11; LCL = .044 **c.** out of control

12.21 a. .00873 **b.** UCL = .03369; CL = .00873; LCL = 0 **c.** out of control

12.23 process A: C_{pk}; process B: C_p; process C: C_{pk}; process D: C_p

12.25 C_{pk} = .46. The process is centered nearer the USL.

12.27 process A: C_p = .78 (inadequate); process B: C_p = 1.13 (adequate); process C: C_p = 2.16 (good); process D: C_p = 1.11 (adequate)

12.29 a. C_{pk} = .50 **b.** 133,614 **c.** Z_U = 1.49, Z_L = −2.34, number nonconforming ppm = 77,700

12.31 a. process A: spread = 2.36; process B: spread = 7.73; process C: spread = 2.50

12.33 MIL-STD-105D used for attribute data; MIL-STD-414 used for variable data

12.35 a. .7350 **b.** .7358

12.39 a. Accept the lot. **b.** Accept the lot. **c.** Accept the lot. **d.** Reject the lot. **e.** Reject the lot.

12.41 a. .9992 **b.** .7787 **c.** .4335 **d.** .1906 **e.** .0423

12.43 p = .005, β = .9963; p = .02, β = .8795; p = .04, β = .5697; p = .06, β = .3028; p = .09, β = .0948

12.45 p = .005, β = .9856; p = .01, β = .9197; p = .03, β = .4232; p = .05, β = .1246; p = .07, β = .0296

12.47 plan A: β = .4232; plan B: β = .2650; lower consumer risk using plan B.

12.51 a. variable **b.** attribute **c.** attribute **d.** variable **e.** variable

12.53 a. attribute data **b.** p chart **c.** UCL = .166; CL = .083; LCL = 0

12.55 a. UCL = 87.5; CL = 84.5; LCL = 81.5 **b.** UCL = 16.18; CL = 8.91; LCL = 1.64 **c.** Pattern 1 is detected. **d.** Search for an assignable cause.

12.57 Plan A: β = .0498; plan B: β = .2103; plan C: β = .3084; plan A has lower consumer risk.

12.59 a. p chart **b.** $\bar{X}$ chart **c.** R chart **d.** c chart **e.** $\bar{X}$ chart **f.** p chart

12.61 Take a second sample. Accept the lot if between 0 and 2 nonconforming units are found in the second sample. Reject the lot if 3 or more nonconforming units are found in the second sample.

12.63 a. Yes **b.** no pattern detected

12.65 a. No; one point (sample 7) is outside the UCL. **b.** Seek an assignable cause for sample 7.

Chapter 13

13.1 a. Z^* = −.8386; fail to reject H_0. **b.** Chi-square = .7033; fail to reject H_0. **c.** The chi-square value is the square of the Z value.

13.5 Chi-square = .8125; fail to reject H_0.

13.7 Chi-square = 6.0; fail to reject H_0.

13.9 Using four classes, chi-square = 5.5104; fail to reject H_0.

13.11 Estimate of proportion $\cong$ .20. Using the classes $\leq 1, 2, 3,$ and ≥ 4, chi-square = .4184; fail to reject H_0.

13.13 Chi-square = 2.24; fail to reject H_0.

13.15 Chi-square = 8.9839; reject H_0.

13.17 Pool "satisfied" and "not satisfied." Chi-square = .5641, fail to reject H_0.

13.19 Pool trained group and experienced group. Chi-square = 7.139; reject H_0.

13.21 a. Chi-square = 6.004; fail to reject H_0. **b.** p-value > .10.

13.23 Chi-square = 6.06514; fail to reject H_0.

13.25 Chi-square = 1.36; fail to reject H_0.

13.27 Using E_1 = 50, E_2 = 27, E_3 = 12, E_4 = 11, chi-square = 2.279; fail to reject H_0.

13.29 Chi-square = 2.133; fail to reject H_0.

13.31 Estimate of mean = 3. Using the classes $\leq 1, 2, 3, 4, \geq 5$, chi-square = .351; fail to reject H_0.

13.33 Using five classes, chi-square = 2.0749; fail to reject H_0.

13.35 Chi-square = 4.69; fail to reject H_0.

13.37 Chi-square = 15.32; fail to reject H_0, p-value slightly above .05.

13.39 Chi-square = .4144; fail to reject H_0.

13.41 Chi-square = 15.9; reject H_0.

13.43 Chi-square = 3.2511; reject H_0.

13.45 a. $\bar{x}$ = 28.35, s = 19.96 **b.** Chi-square = 36.881; reject H_0. **c.** p-value < .005 **d.** no change

13.47 a. Chi-square = 2.99; fail to reject H_0. **b.** p-value > .10 **c.** insufficient evidence to conclude that these qualities are not independent.

13.49 Chi-square = .220; fail to reject H_0.

13.51 Since the p-value is greater than .10, fail to reject the null hypothesis.

Chapter 14

14.1 b. r = .796

14.3 b. $\hat{Y}$ = 544.008 + 3.084X

14.7 EAMIS is the most influential with UIS.

14.9 a. .8024   **b.** $\hat{Y} = 9.47 + .299X$ **c.** 1.39

14.11 $\hat{Y} = 1.5823 + .7659X$; for $X = 9.67$, $\hat{Y} = 8.9886$

14.13 a. $\hat{Y} = 6.662 + 2.984X$

b.

Y: 10.125	10.000	10.250	10.750	10.500
$Y - \hat{Y}$: .210	.055	.246	-1.283	-1.981

Y: 14.000	14.250	14.370	15.000	14.550
$Y - \hat{Y}$: 1.370	1.321	.248	$-.167$	$-.020$

c. The error terms do not appear to be correlated.

14.15 a. $\hat{Y} = -3.889 + 5.820X$
c. SSE = 129.531, $s^2 = 25.906$

14.17 SSE = 155.94; $s = 2.6624$; all the sample residuals are within two standard deviations.

14.19 Yes, the variance of the error component is not constant.

14.21 no

14.23 a. SSE = 15.7775; $t^* = 9.9296$; there is a positive relationship.   **b.** p-value $< .005$; there is a positive relationship.

14.25 Using log Y as the dependent variable, SSE = .0285, $t^* = 27.1601$; there is a positive relationship between X and log Y.

14.27 $\hat{Y} = 215.683 + 2.046X$; $t^* = 17.30$; reject H_0.

14.29 .8665 to 1.8001

14.31 .2080 to .3898

14.33 $r = .8371$, $t^* = 5.5172$; there is a linear relationship.

14.35 $r = .9742$, $t^* = 15.5525$; there is a linear relationship.

14.37 a. no   **b.** $r = -.1225$; $r^2 = .015$; $t^* = -.4450$; fail to reject H_0.

14.39 $r = -.93167$; $t^* = -10.8799$; reject H_0 (the same result as before).

14.41 $t^* = 2.9344$; reject H_0.

14.43 a. $\hat{Y} = -201.7469 + 25.9865X$; for $X = 30$, $\hat{Y} = 577.8481$; confidence interval is 520.94 to 634.76.
b. Prediction interval is 317.92 to 837.78.

14.45 $\hat{Y} = 9.1036$; prediction interval is 7.426 to 10.781.

14.47 The prediction interval is wider. Both intervals are the narrowest for $X = \bar{x}$.

14.49 a. $\hat{Y} = .29255 + .0909X$. For $X = 75$, $\hat{Y} = 7.1101$, confidence interval is 6.3529 to 7.8673.
b. Prediction interval is 4.334 to 9.886.   **c.** 69.07

14.51 $\hat{Y} = -.0267 + 1.2233(.30) = .3403$; prediction interval is .2098 to .4708.

14.53 a. $\hat{Y} = 39.516 - .1557X$ **b.** $r = -.932$; coefficient of determination = .87   **c.** $-.204$ to $-.107$
d. There is a significant linear relationship between X and Y.   **e.** 19.226 to 22.438   **f.** 16.681 to 24.983

14.55 a. $\hat{Y} = 568.969 + .9664X$ **b.** .5613
c. SSE = 149.8681   **d.** 8.3260   **e.** $t^* = 4.7990$; reject H_0.   **f.** $\hat{Y} = 684.664$; prediction interval is 677.13 to 692.16.

14.57 $r = .9445$; $t^* = 7.04$; reject H_0.

14.59 a. $\hat{Y} = 10.8 + 5.38X$ **b.** $\hat{Y} = 1.05 + .99X^2$
c. Correlation between X and Y is .8278. Coefficient of determination is .685. Correlation between X^2 and Y is .9994. Coefficient of determination is .9988.

14.61 $\hat{Y} = .322 + 1.3602X$

14.63 SSE = 80.75; $s = 2.06155$; one of the sample residuals lies outside two standard deviations; the histogram appears to be nearly symmetric.

14.65 a. $\hat{Y} = .203 + .1484X$ **b.** $t^* = 26.44$; reject H_0.   **c.** 417.322 to 473.484   **d.** 438.020 to 452.786

14.67 a. $\hat{Y} = 110.428$ **b.** $t^* = 19.54$, reject H_0.
c. .8783 to 1.0949   **d.** $r^2 = .9928$ if observation 7 is removed.

Chapter 15

15.1 a. $\hat{Y} = 13.85$ **b.** 2.8

15.3 SSE = 191.16; $s = 3.4565$; all the residuals lie within two standard deviations; empirical rule approximately holds.

15.5 a. $\hat{Y} = .0090 + 1.1102X_1 + .13855X_2$
b. SSE = 4.7289

15.7 6.6646

15.9 a. $\hat{Y} = -.50 + 2.40X_1 + 2.95X_2$
b. $R^2 = .9381$ **c.** $n = 20$ **d. i.** 314.80   **ii.** 19.50
iii. 295.30   **iv.** 128.39   **v.** $F > 3.59$ **vi.** 14.048
vii. $t^* > 2.110$ **viii.** MSE = 1.15

15.11 a. $t^* = 3.1509$, reject H_0.   **b.** $t^* = -2.205$, reject H_0.

15.13 a. $t^* = 5.593$, X_2 contributes.   **b.** 3.4056 to 7.7944

15.15

Source	df	SS	MS	F
Regression	2	1165.66	582.83	30.17
Residual	7	135.24	19.32	
Total	9	1300.90		

15.17

Source	df	SS	MS	F
Regression	2	97.12	48.56	1.9994
Residual	17	412.88	24.287	
Total	19	510		

$F^* = 1.9994$; fail to reject H_0.

15.19 $F^* = 196.5$, reject H_0.

15.21 one

15.23 $F^* = 14.056$; X_2 and X_3 contribute.

15.25 $F^* = 2.718$; X_2 and X_3 do not contribute.

15.27 $F^* = 2.6367$; the three variables do not contribute.

15.29

Source	df	SS	MS	F
Regression	2	2.0834	1.0417	9.23
Residual	12	1.3539	.112825	
Total	14	3.4373		

p-value $< .01$; reject H_0.
R-square $= .606$; adjusted R-square $= .54$

15.31 presence of multicollinearity

15.33 Correlation $= .8625$; multicollinearity may be present.

15.35 a. Correlation matrix is

	Y	X_1	X_2
Y	1.0	.975	.9965
X_1		1.0	.955
X_2			1.0

b. Using X_1, .951; using X_2, .993   **c.** no, due to the high correlation between X_1 and X_2

15.37 no

15.39 a. $\hat{Y} = 7630$ **b.** $F^* = 10.0$; the dummy variables contribute.

15.41 $Y = \beta_0 + \beta_1 X_1 + \beta_2 X_2 + \beta_3 X_3 + \beta_4 X_4 + \beta_5 X_5 + \beta_6 X_6 + \beta_7 X_7 + \beta_8 X_8 + e$
Y = total amount of compensation paid for a claim
X_1 = age of employee (years)

$X_2 = \begin{cases} 1 \text{ if male} \\ 0 \text{ if female} \end{cases}$

$X_3 = \begin{cases} 1 \text{ if employee is single} \\ 0 \text{ if not} \end{cases}$

X_4 = length of employment (years)

$X_5 = \begin{cases} 1 \text{ if injury is to head} \\ 0 \text{ if not} \end{cases}$

$X_6 = \begin{cases} 1 \text{ if injury is to a limb} \\ 0 \text{ if not} \end{cases}$

$X_7 = \begin{cases} 1 \text{ if employee works for} \\ \quad \text{manufacturer \#1} \\ 0 \text{ if not} \end{cases}$

$X_8 = \begin{cases} 1 \text{ if employee works for} \\ \quad \text{manufacturer \#2} \\ 0 \text{ if not} \end{cases}$

15.43 a. X_3 **b.** X_3

15.45 Stepwise regression can remove variables previously included.

15.47 equal variance for error terms

15.49 a. $\hat{Y} = 14.9$; confidence interval is 8.984 to 20.816.   **b.** 4.895 to 24.905

15.51 The prediction interval is always wider than the corresponding confidence interval.

15.53 a. $\hat{Y} = 25.1434$; confidence interval is 22.9139 to 27.3729.   **b.** 18.7494 to 31.5374

15.55 a. $\hat{Y} = .1026 + .688X_1 + 3.98X_2$

b. $R^2 = .504$ **c.** $n = 20$ **d.** SST $= 170.7$; SSE $= 84.609$; MSE $= 4.977$ **e.** Model is significant.   **f.** X_1 contributes.   **g.** X_2 does not contribute.

15.57 a. yes, since the predictor variables will quite likely be highly correlated   **b.** 4.14 billion dollars; this is not a valid conclusion if multicollinearity is present.

15.59 a. $\hat{Y} = 17,357 - 1132X_1 - 33.2X_2 - 2556X_3 - 3275X_4 + 776X_5$ **b.** $F^* = 43.29$; reject H_0.   **c.** 5192 to 8437   **d.** Correlation matrix is

	Y	X_1	X_2	X_3	X_4	X_5
Y	1.0	$-.872$	$-.857$	$-.077$	$-.527$	.659
X_1		1.0	.845	.125	.309	$-.462$
X_2			1.0	$-.032$	.480	$-.599$
X_3				1.0	$-.619$	.177
X_4					1.0	$-.692$
X_5						1.0

e. Using the forward procedure at the .10 significance level, the model is $\hat{Y} = 14,510 - 1581X_1 + 2841X_5$.

15.61 a. $\hat{Y} = 15.24 + 4.8676X_1 - 5.802X_2 + 2.248X_3$ **b.** .987   **c.** $F^* = 18.1538$; X_2 and X_3 contribute.   **d.** $t^* = 22.778$; X_1 contributes.   **e.** The residuals appear to be random.

15.63 a. $F^* = 37.16$; reject H_0.   **b.** $F^* = 1.48$; X_2 does not contribute.   **c.** .0192 to .1465   **d.** 1.4757 to 2.3519   **e.** $\hat{Y} = 11.5657$

15.65

Source	df	SS	MS	F
Regression	2	49528576	24764288	483.46
Residual	7	358563	51223	
Total	9	49887139		

Since $F^* = 482.17 > F_{.05,2,7} = 4.74$, reject H_0.
b. R-square $= .993$; adjusted R-square $= .991$.
c. no   **d.** 16041.7 to 17174.3

Chapter 16

16.3 Due to seasonal effects, there may be a decline in demand.

16.5 a. cyclical   **b.** seasonal   **c.** trend

16.7 a. $\hat{y}_t = -45,585.691 + 23.0303t$
b. $\hat{y}_t = 37.333 + 23.0303t$ **c.** The predicted values are the same.

16.9 The nature of the quadratic curve is unknown outside the range of the time series data.

16.11 $\hat{y}_t = -24.418 + 6.2429t$

16.13 $\hat{y} = 1601.61 + 207.163t$

16.15 a. $\hat{y} = 10339.2 + 436.05t$ **b.** $\hat{y} = 10920.7 + 277.47t + 7.552t^2$ **c.** from part a, $\hat{Y} = 19496.4$; from part b, $\hat{Y} = 20077.8$

16.17 $C_{11} = .9811$; $C_{12} = .9152$; $C_{13} = .9344$; $C_{14} = .9833$; $C_{15} = 1.033$. The period of the cycle appears to be longer than five years.

16.19 a. $\hat{Y}_t = 8.417 + .5833t$, where $t = 1$ corresponds to 1983   **b.** $C_t = y_t/\hat{y}_t$. These components are .778, 1.565, .984, .464, .971, 1.427, .96, .612, 1.244.   **c.** approximately four years

16.21 The trend equation is $\hat{y}_t = 74.654 + .5220t$. The cyclical components are 1.062, .991, .932, . . . , .958, .976, 1.031.

16.23 a. The trend equation is $\hat{y}_t = 105.153 + .1940t$.
b. The cyclical components are .977, 1.004, .993, .990, .995, 1.021, 1.011, 1.009, .992, .988.

16.25 The trend equation is $\hat{y}_t = 164.0 + 2.6t$. The cyclical components are .906, 1.147, 1.030, .900, 1.062, .908, .939, 1.077, 1.142, .890.

16.29 65.4, 78.9, 87.1, 84.1

16.31 9.9, 16.24, 12.48, 27.69, 23.98, 31.22, 54.72, 72.23, 7.14, 4.86, 4.96, 2.53

16.33 53, 83, 99, 8, 13, 3, 9, 9, 65, 76, 71, 67

16.35 Centered moving averages are 6.88, 7.75, 8.88, . . . , 19.13, 20.13, 21.00.

16.39 Seasonal indexes are .8710, .8256, .8458, .9291, 1.0276, 1.0503, 1.0730, 1.1614, 1.1411, 1.0629, 1.0452, .9670.

16.41 Seasonal indexes are 1.0913, 1.2155, .8573, .8358.

16.43 Seasonal indexes are .87, .61, .86, 1.65.

16.45 a. Seaonsal indexes are .94, .97, 1.01, 1.02, 1.07, 1.06, 1.04, 1.02, .99, .98, .96, .94.   **b.** The de-seasonalized trend line is $\hat{d}_t = 27.1648 + .1773t$.
c. The cyclical components are .9888, .9825, .9694, .9711, .9817, .9932, .9937, .9959, .9992, 1.0019, 1.0053, 1.0048.   **d.** The irregular components are 1.0113, 1.0138, .9878.

16.47 The trend equation is $\hat{y}_t = 1.552 + .0263t$. The cyclical components for $t = 2, 3, 4, . . . , 17, 18, 19$ are 1.103, 1.063, 1.026, . . . , 1.059, 1.053, 1.043.

16.49 a. Seasonal indexes are .89, .94, .94, 1.04, .98, 1.02, 1.08, 1.10, 1.10, 1.02, .97, .93.
b. $\hat{d}_t = 3.7214 + .0836t$ **c.** Cyclical components are 1.0358, 1.0005, .9829, .9771, .9800, .9812, .9930, .9941, .9987, .9958, .9913, 1.0047.   **d.** Irregular components are .9850, .9800, 1.0122.

16.51 Cyclical components are .9465, .9879, 1.0627, 1.0356. Irregular components are .7508, 1.1694, 1.0340, .9012.

16.53 Index numbers are 100.0, 101.6, 109.4, 132.9, 156.5, 187.8.

16.55 a. 120.6   **b.** 119.0   **c.** 118.7

16.57 a. 135.6   **b.** 136.11   **c.** 135.9

16.61 a. 154.38   **b.** Total amount of federal funds for the four categories has increased by approximately 54%.

16.63 a. trend   **b.** irregular   **c.** seasonal
d. irregular

16.65 a. $\hat{y}_t = 32.697 + .213t$ **b.** $\hat{y}_{14} = 35.68$

16.67 b. $\hat{y}_t = 1823.4 + 343.68t$ **c.** $\hat{y} = 4573$

16.69 The deseasonalized trend equation is $\hat{d}_t = 3.8255 + .0215t$. The cyclical components are 1.0082, .9928, .9953, .9979, .9985, .9823, .9880, .9907.

16.71 a. Seasonal indexes are 1.20, 1.15, 1.07, .92, .91, .95, .96, .96, .92, .93, .99, 1.05.
b. $\hat{d}_t = 151.806 + 1.094t$ **d.** Irregular components are 1.00, 1.00.

16.73 a. Seasonal indexes are .99, 1.02, 1.00, .98.
b. $\hat{d}_t = 144.7426 + .8401t$ **c.** .9719, .9444, .9435, .9426   **d.** Irregular components for the first two quarters are .9909 and 1.0050.

16.75 Seasonal indexes are 1.10, 1.03, .99, .95, .94, .87, .89, .95, .98, 1.03, 1.10, 1.18.

16.77 a. Seasonal indexes are .944, .873, 1.033, 1.195, 1.031, .979, .943, .934, 1.010, 1.108, .824, 1.126.
b. $\hat{d}_t = 709.535 + 156.663t$ **c.** The cyclical components are 1.17, 1.17, 1.33, 1.40, 1.36, 1.21, 1.20, 1.36, 1.57, 1.52, 1.31, 1.00.   **d.** Irregular components are 1.187, .866, and .990.

16.79 Residuals follow a nonrandom pattern, violating the assumptions of the regression statistical tests.

Chapter 17

17.3 Estimates are 28.50, 29.25, 31.75, 29.50, 28.00, 27.50, 28.25, 29.75, 32.50, 31.50.

17.7 a. .48, .42, .52, .68, .88, 1.21, 1.62, 1.84, 1.63, 1.33, .83, .58   **b.** $\hat{d}_t = 1.295 + .0137t$
c. $\hat{y}_{87} = 1.275$; $\hat{y}_{91} = 4.060$; $\hat{y}_{96} = 1.493$

17.11 10.8

17.13 Estimated values are 11.06, 11.11, 11.18. Residuals are .54, .64, −.15.

17.15 977.39, 974.14, and 970.62

17.17 a. 111308.3   **b.** 111521.2

17.19 29.79

17.21 a. 17.176   **b.** 12.35

17.23 Initial seasonal factors are .95, .83, 1.07, and 1.16. Least squares line is $.2287 + .0109t$. Forecasts for each quarter are .404, .361, .491, .538.

17.25 Initial seasonal factors are .91, 1.50, 1.09, and .50. Least squares line is $.2709 + .0342t$. Forecasts for each quarter are 1.05, 1.78, 1.29, .72.

17.27 a. MAD = .267, MAPE = .125, MSE = .089
b. MAD = .32, MAPE = .144, MSE = .14

17.29 Procedure 1: MAD = 3.556, MSE = 16.556; procedure 2: MAD = 3.333, MSE = 44.556. Procedure 1 is superior overall, and procedure 2 is superior if the last two observations are ignored.

17.31 a. Predicted values are 12500.97, 12564.44, 13096.19, . . . , 15940.29, 17151.46, 17656.35; MSE = 2,049,098.2.   **b.** Predicted values are 12067.43, 11996.72, 13013.65, . . . , 15160.85, 17268.70, 17738.42; MSE = 2,694,686.4.   **c.** The procedure in part a is better.

17.33 MAD = 16,931.80; MSE = 472,762,368; MAPE = .0321

17.35 Second-order equation is $\hat{y}_t = 10.919 − .4813y_{t-1} − .5393y_{t-2}$. Fourth-order equation is

$\hat{y}_t = -.483 + .0171y_{t-1} + .0515y_{t-2} + .0376y_{t-3} + 1.0115y_{t-4}$. MSE for second-order equation is 2.9641; MSE for fourth-order equation is .1155.

17.37 a. $r_1 = .693$; $r_2 = .108$; $r_3 = -.355$; $r_4 = -.428$ **b.** One possible model is $\hat{y}_t = b_0 + b_1y_{t-1} = 11.542 + .7105y_{t-1}$.

17.39 $r_1 = .848$; $r_2 = .500$; $r_3 = .066$; $r_4 = -.334$; $r_5 = -.599$; $r_6 = -.688$; $r_7 = -.590$; $r_8 = -.320$; $r_9 = .046$; $r_{10} = .435$; $r_{11} = .726$; $r_{12} = .832$ The data do not appear to be stationary.

17.41 The series of second differences appears to be stationary with two-period seasonal spikes.

17.43 $\hat{y}_t = 184.1 + .7921y_{t-1} + .2645y_{t-2}$; $R^2 = .972$

17.45 $\hat{y}_t = 15.25 - .422t + .313Q_1 + .909Q_2 + .842Q_3$

17.47 Let $M_1 = 1$ for Jan., $M_2 = 1$ for Feb., . . . , $M_{11} = 1$ for Nov. The regression equation is $\hat{y}_t = .6143 - .2042M_1 - .3012M_2 - .1268M_3 + .1619M_4 + .5506M_5 + 1.5356M_6 + 1.8423M_7 + 2.2452M_8 + 1.8768M_9 + 1.3226M_{10} + .4685M_{11} + .01131t$

17.49 $\hat{y}_t = 12.617 - 1.489Q_1 + 2.507Q_2 + 5.337Q_3 + .0036t$

17.51 a. $\hat{y}_t = 108.52 + .0031X_t$ **b.** $\hat{y}_t = 74.454 + .252X_{t-1}$ **c.** R^2 for part a is approximately .0, R^2 for part b is .957.

17.53 $\hat{y}_t = .5134 + .1555X_{t-1} + .5783Q_1 + .6600Q_2 + 1.3822Q_3$; MSE $= .0091$

17.55 a. $\hat{y}_t = 14.532 + .01459X_t$ **b.** $\hat{y}_t = 9.5541 + .02569X_{t-1}$ **c.** R^2 for part a is .334; R^2 for part b is .972.

17.57 positive autocorrelation, no autocorrelation, negative autocorrelation

17.59 $DW = 2.66$; fail to reject H_0.

17.61 $DW = .56$; possible positive autocorrelation

17.63 $\hat{y}_{37} = 19.25$; $\hat{y}_{38} = 22.35$; $\hat{y}_{39} = 25.56$; $\hat{y}_{40} = 27.25$; $\hat{y}_{41} = 25.34$

17.65 a. 1123.2 **b.** 1135.78

17.67 MSE $= .32$

17.69 Let $Q_1 = 1$ for quarter 1, $Q_2 = 1$ for quarter 2, $Q_3 = 1$ for quarter 3. The regression equation is $\hat{y}_t = 39.275 - 16.019Q_1 - 9.613Q_2 - 4.406Q_3 + .79375t$

17.71 8.56

17.75 8660.75

17.77 MAD $= 7.665$; MSE $= 100.40$; MAPE $= .029$

Chapter 18

18.1

	States of Nature		
Action	S_1	S_2	S_3
A_1	2,000	500	−1,000
A_2	1,500	750	0
A_3	1,000	1,000	1,000

18.3

	States of Nature								
Action	5	6	7	8	9	10	11	12	13
7	−20	5	30	30	30	30	30	30	30
8	−30	−5	20	45	45	45	45	45	45
9	−40	−15	10	35	60	60	60	60	60
10	−50	−25	0	25	50	75	75	75	75
11	−60	−35	−10	15	40	65	90	90	90
12	−70	−45	−20	5	30	55	80	105	105
13	−80	−55	−30	−5	20	45	70	95	120

18.5

	States of Nature		
Action	$S_1(100)$	$S_2(125)$	$S_3(150)$
$A_1(100)$	90	90	90
$A_2(125)$	62.5	112.5	112.5
$A_3(150)$	35	85	135

18.7 a.

	States of Nature					
Action	50	60	70	80	90	100
70	5.5	8	10.5	10.5	10.5	10.5
80	4.5	7	9.5	12	12	12
90	3.5	6	8.5	11	13.5	13.5
100	2.5	5	7.5	10	12.5	12.5

b.

	States of Nature					
Action	50	60	70	80	90	100
70	0	0	0	1.5	3.5	4.5
80	1	1	1	0	1.5	3
90	2	2	2	1	0	1.5
100	3	3	3	2	1	0

Minimax decision is to order 90 copies. Maximax decision is to order 100 copies.

18.9

	States of Nature			
Action	S_1	S_2	S_3	S_4
A_1	0	13	5	0
A_2	170	5	18	15
A_3	150	0	0	25

18.11 a. $E(A_1) = 97.5$, $E(A_2) = 55$, $E(A_3) = 95$; expected payoff is maximized using A_1.
b. risk $(A_1) = 4868.75$; risk $(A_2) = 0$; risk $(A_3) = 1875$

18.13 a. A_3 **b.** A_3 **c.** risk $(A_1) = 0$; risk $(A_2) = 169$; risk $(A_3) = 1642.1875$; risk $(A_4) = 9850$

18.15 Maximum expected payoff is 10.125.

18.17 EVPI $= 6.25$; action A_3 is inadmissible.

18.19 Expected payoff by ordering 2000 cards is $840; expected payoff with a perfect predictor is $950; maximum amount is $110.

18.21 Expected payoff using A_3 is 1000; expected payoff with a perfect predictor is 1333.33; maximum amount is 333.33.

18.23 EVPI $= 22.75$; A_1 is inadmissible.

18.25 a. A_2

b. Table of utility values:

States of Nature

Action	S_1	S_2	S_3	S_4
A_1	2	20	14.14	6.32
A_2	17.89	10.95	12.65	10
A_3	18.97	8.94	10.95	6.32

$E(A_1) = 10.615$; $E(A_2) = 12.873$; $E(A_3) = 11.295$; the decision is A_2.

18.27 $p = .81$

18.29 $E(A_1) = 3.413$; $E(A_2) = 3.433$; $E(A_3) = 3.386$; $E(A_4) = 3.394$. Because the decision is A_2, the manager is a risk avoider.

18.31

18.33

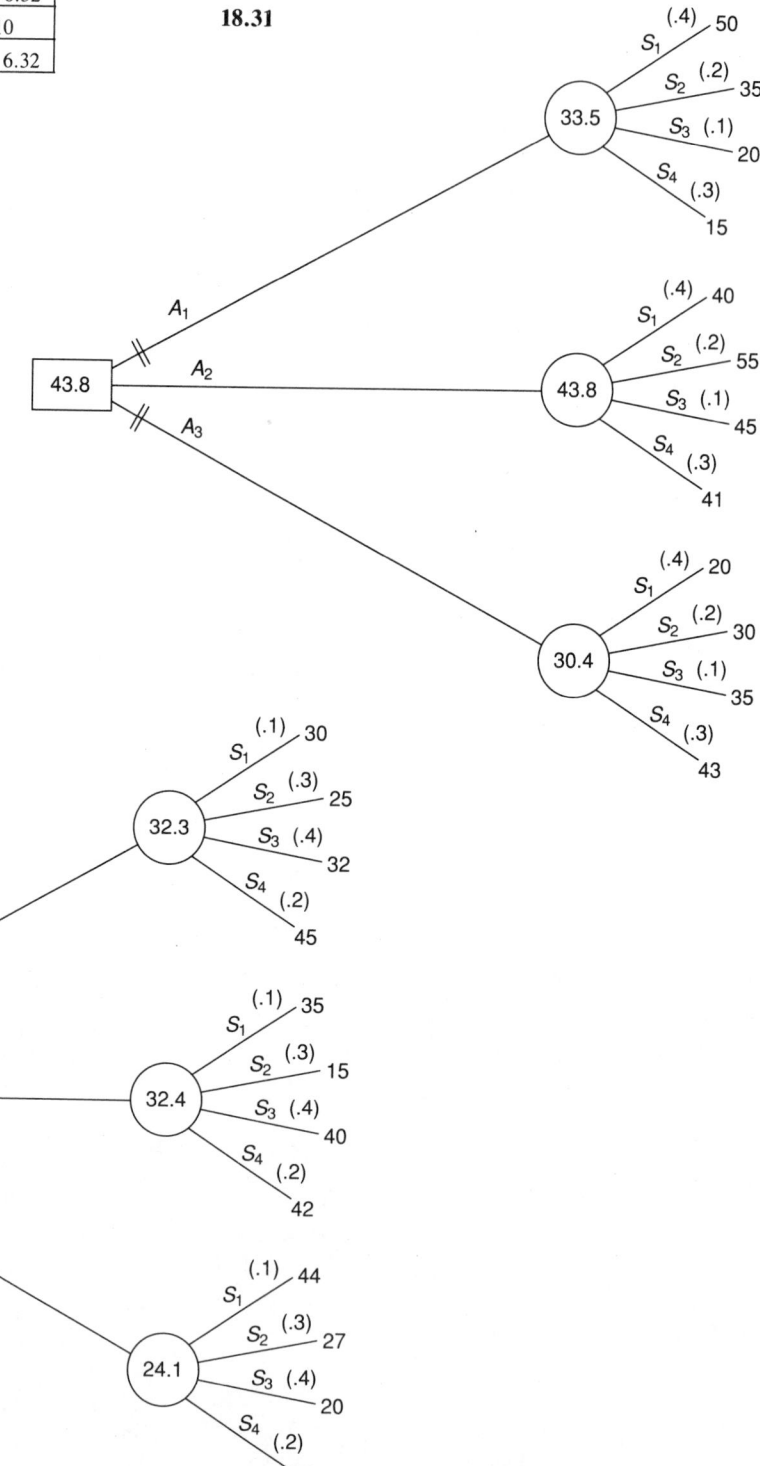

18.35

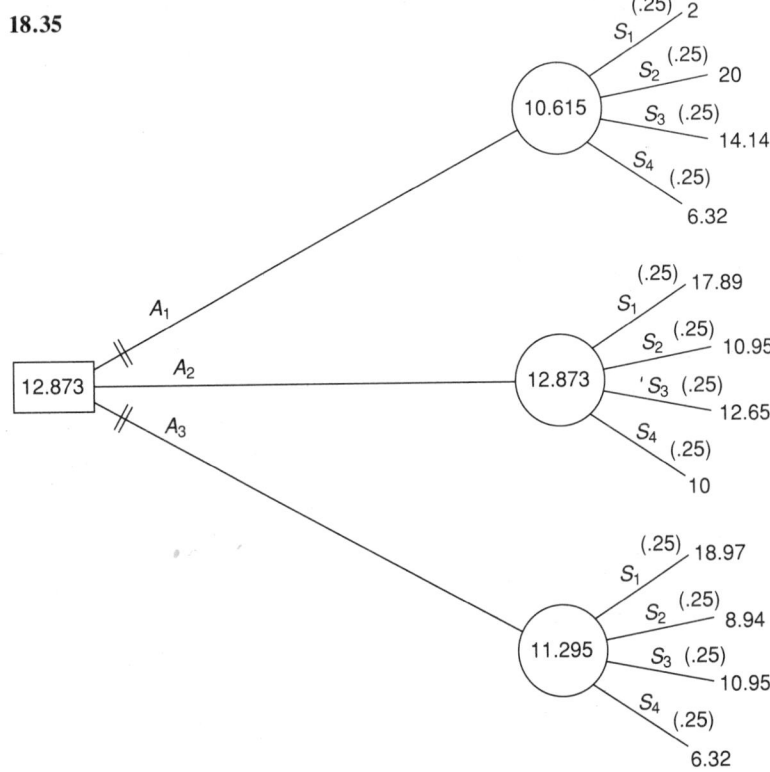

18.37 $P(A_1 \mid B) = .111; P(A_2 \mid B) = .111;$
$P(A_3 \mid B) = .222; P(A_4 \mid B) = .185; P(A_5 \mid B) = .370$

18.39 $.0264/.0734 = .3597$

18.41 $P(B) = .26; P(S_1 \mid B) = .2308;$
$P(S_2 \mid B) = .2308; P(S_3 \mid B) = .4615;$
$P(S_4 \mid B) = .0769;$ maximum expected utility of 33.23
occurs for A_2.

18.43 EVPI $= 3000$; yes

18.45 EVPI $= 2592.98 - 1995.15 = 597.83$. It is not
worthwhile to hire the consultant.

18.47 Minimax decision is to order 34 or 35 loaves;
maximax decision is to order 36 loaves.

18.49 The change of probabilities will not affect the
minimax decision. Using the expected payoffs, action A_4
is the optimal action for sets 1, 2, and 4, and action A_1 is
optimal for sets 3 and 5.

18.51 One example would be to let $U(x) = 0$ for $x < 0$,
$U(x) = 2x$ for $0 \le x \le 1000$, and $U(x) = 2000 +$
$(x - 1000)^2$ for $x > 1000$.

18.53

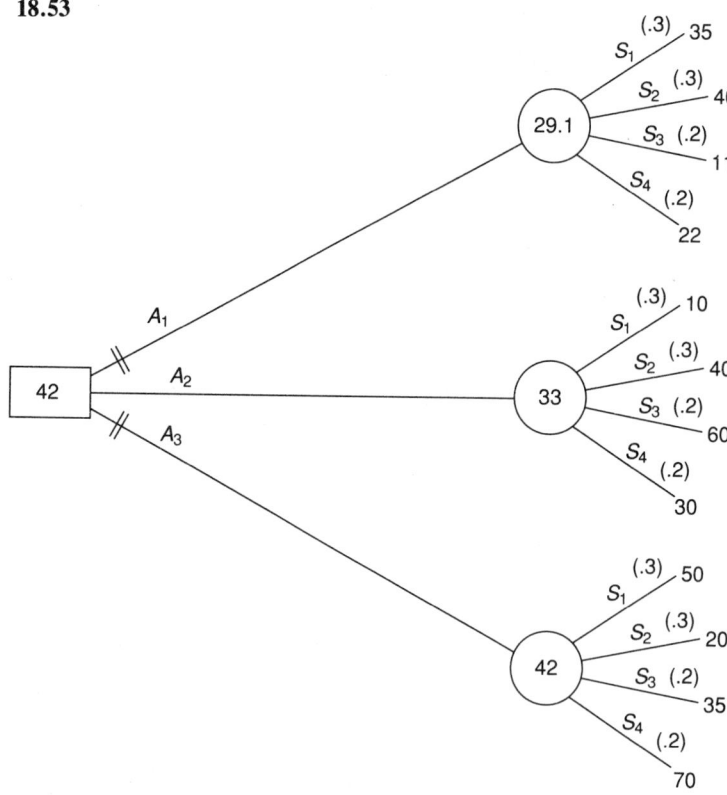

18.55 The maximum payoff with the consultant is 1157.08. The maximum payoff without the consultant is 1105 using A_3. Since $1157.08 - 1105 = 52.08$, which is less than 350, it is not worthwhile to use the consultant's service.

18.57 $P(\text{long}|\text{under }40) = .632$

18.59 .77

Chapter 19

19.1 The only assumption needed is that you have a sequence of n observations containing n_1 symbols of the first type and n_2 symbols of the second type.

19.3 $R = 12$; reject H_0.

19.5 $R = 10$; fail to reject H_0.

19.7 $R = 12$; fail to reject H_0.

19.9 $R = 17$; $Z^* = (17 - 18.486)/2.91 = -.51$; fail to reject H_0.

19.13 $U = U_1 = 43$; fail to reject H_0.

19.15 $Z^* = (97 - 77)/18.267 = 1.09$; fail to reject H_0.

19.17 $Z^* = (53.5 - 72)/17.32 = -1.068$; fail to reject H_0.

19.19 $Z^* = (88.5 - 72)/17.32 = .953$; fail to reject H_0.

19.21 $U = 14$; reject H_0.

19.25 $T_+ = 12$; fail to reject H_0.

19.27 $T_- = 26.5$; reject H_0.

19.29 $Z^* = (48 - 105)/26.786 = -2.13$; reject H_0.

19.31 $Z^* = (280 - 232.5)/48.618 = .9769$; fail to reject H_0.

19.33 $Z^* = (81.5 - 76.5)/21.12 = .24$; fail to reject H_0.

19.35 $T = 17.5$; fail to reject H_0.

19.37 $KW = .485$; fail to reject H_0.

19.39 $KW = 6.167$; reject H_0.

19.41 $KW = 7.649$; reject H_0.

19.43 $KW = 6.83$; reject H_0.

19.45 $KW = 2.105$; fail to reject H_0.

19.47 $KW = 1.51$; fail to reject H_0.

19.49 $FR = 2$; fail to reject H_0.

19.51 $FR = 7$; reject H_0.

19.53 $FR = 6$; reject H_0.

19.55 $r = -.83$; rank correlation $= -.865$

19.57 Rank correlation $= .874$; reject H_0.

19.59 Rank correlation $= -.944$; reject H_0.

19.61 Rank correlation $= .88$; reject H_0.

19.63 Rank correlation $= .93$; reject H_0.

19.65 Rank correlation $= .916$; reject H_0.

19.67 $R = 16$; $Z^* = (16 - 12.52)/2.25 = 1.55$; fail to reject H_0.

19.71 $U = 16$; fail to reject H_0.

19.73 $T = 10.5$; reject H_0 (use $n = 14$).

19.75 $KW = 2.93$; fail to reject H_0.

19.77 $KW = .14$; fail to reject H_0.

19.79 $KW = 4.10$; fail to reject H_0.

19.81 Rank correlation $= .845$; reject H_0.

19.83 $FR = 31.2$; reject H_0.

19.85 Rank correlation $= .97$; reject H_0.

19.87 $KW = 7.452$; reject H_0. The p-value is between .01 and .025.

Index

Areas of the Standard Normal Distribution

The entries in this table are the probabilities that a standard normal random variable is between 0 and z (the shaded area).

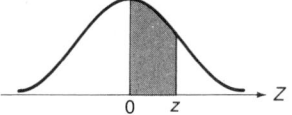

z	Second Decimal Place in z									
	0.00	0.01	0.02	0.03	0.04	0.05	0.06	0.07	0.08	0.09
0.0	0.0000	0.0040	0.0080	0.0120	0.0160	0.0199	0.0239	0.0279	0.0319	0.0359
0.1	0.0398	0.0438	0.0478	0.0517	0.0557	0.0596	0.0636	0.0675	0.0714	0.0753
0.2	0.0793	0.0832	0.0871	0.0910	0.0948	0.0987	0.1026	0.1064	0.1103	0.1141
0.3	0.1179	0.1217	0.1255	0.1293	0.1331	0.1368	0.1406	0.1443	0.1480	0.1517
0.4	0.1554	0.1591	0.1628	0.1664	0.1700	0.1736	0.1772	0.1808	0.1844	0.1879
0.5	0.1915	0.1950	0.1985	0.2019	0.2054	0.2088	0.2123	0.2157	0.2190	0.2224
0.6	0.2257	0.2291	0.2324	0.2357	0.2389	0.2422	0.2454	0.2486	0.2517	0.2549
0.7	0.2580	0.2611	0.2642	0.2673	0.2704	0.2734	0.2764	0.2794	0.2823	0.2852
0.8	0.2881	0.2910	0.2939	0.2967	0.2995	0.3023	0.3051	0.3078	0.3106	0.3133
0.9	0.3159	0.3186	0.3212	0.3238	0.3264	0.3289	0.3315	0.3340	0.3365	0.3389
1.0	0.3413	0.3438	0.3461	0.3485	0.3508	0.3531	0.3554	0.3577	0.3599	0.3621
1.1	0.3643	0.3665	0.3686	0.3708	0.3729	0.3749	0.3770	0.3790	0.3810	0.3830
1.2	0.3849	0.3869	0.3888	0.3907	0.3925	0.3944	0.3962	0.3980	0.3997	0.4015
1.3	0.4032	0.4049	0.4066	0.4082	0.4099	0.4115	0.4131	0.4147	0.4162	0.4177
1.4	0.4192	0.4207	0.4222	0.4236	0.4251	0.4265	0.4279	0.4292	0.4306	0.4319
1.5	0.4332	0.4345	0.4357	0.4370	0.4382	0.4394	0.4406	0.4418	0.4429	0.4441
1.6	0.4452	0.4463	0.4474	0.4484	0.4495	0.4505	0.4515	0.4525	0.4535	0.4545
1.7	0.4554	0.4564	0.4573	0.4582	0.4591	0.4599	0.4608	0.4616	0.4625	0.4633
1.8	0.4641	0.4649	0.4656	0.4664	0.4671	0.4678	0.4686	0.4693	0.4699	0.4706
1.9	0.4713	0.4719	0.4726	0.4732	0.4738	0.4744	0.4750	0.4756	0.4761	0.4767
2.0	0.4772	0.4778	0.4783	0.4788	0.4793	0.4796	0.4803	0.4808	0.4812	0.4817
2.1	0.4821	0.4826	0.4830	0.4834	0.4838	0.4842	0.4846	0.4850	0.4854	0.4857
2.2	0.4861	0.4864	0.4868	0.4871	0.4875	0.4878	0.4881	0.4884	0.4887	0.4890
2.3	0.4893	0.4896	0.4898	0.4901	0.4904	0.4906	0.4909	0.4911	0.4913	0.4916
2.4	0.4918	0.4920	0.4922	0.4925	0.4927	0.4929	0.4931	0.4932	0.4934	0.4936
2.5	0.4938	0.4940	0.4941	0.4943	0.4945	0.4946	0.4948	0.4949	0.4951	0.4952
2.6	0.4953	0.4955	0.4956	0.4957	0.4959	0.4960	0.4961	0.4962	0.4963	0.4974
2.7	0.4965	0.4966	0.4967	0.4968	0.4969	0.4970	0.4971	0.4972	0.4973	0.4974
2.8	0.4974	0.4975	0.4976	0.4977	0.4977	0.4978	0.4979	0.4979	0.4980	0.4981
2.9	0.4981	0.4982	0.4982	0.4983	0.4984	0.4984	0.4985	0.4985	0.4986	0.4986
3.0	0.4987	0.4987	0.4987	0.4988	0.4988	0.4989	0.4989	0.4989	0.4990	0.4990
3.1	0.4990	0.4991	0.4991	0.4991	0.4992	0.4992	0.4992	0.4992	0.4993	0.4993
3.2	0.4993	0.4993	0.4994	0.4994	0.4994	0.4994	0.4994	0.4995	0.4995	0.4995
3.3	0.4995	0.4995	0.4995	0.4996	0.4996	0.4996	0.4996	0.4996	0.4996	0.4997
3.4	0.4997	0.4997	0.4997	0.4997	0.4997	0.4997	0.4997	0.4997	0.4997	0.4998
3.5	0.4998									
4.0	0.49997									
4.5	0.499997									
5.0	0.4999997									